IIIb	IVa	Va	VIa	VIIa	
						2 4.00260 He $1s^2$ 24.5 0.93
	5 10.81 B $(He)2s^2 2p^1$ 14.0 8.3 0.82 2.01	6 12.011 C $(He)2s^2 2p^2$ 19.5 10.7 0.77 2.50 2.60(4−)	7 14.0067 N $(He)2s^2 2p^3$ 25.5 13.1 0.75 3.07 1.71(3−)	8 15.9994 O $(He)2s^2 2p^4$ 32.3 15.9 0.73 3.50 1.40(2−)	9 18.9984 F $(He)2s^2 2p^5$ 46.4 18.7 0.72 4.10 1.36(−1)	10 20.179 Ne $(He)2s^2 2p^6$ 48.4 21.6 1.31
Ib	13 26.9815 Al $(Ne)3s^2 3p^1$ 11.3 6.0 1.18 1.47 0.50(3+)	14 28.086 Si $(Ne)3s^2 3p^2$ 15.0 7.8 1.74	15 30.9738 P $(Ne)3s^2 3p^3$ 18.7 10.2 1.17 2.06	16 32.06 S $(Ne)3s^2 3p^4$ 20.7 11.7 1.06 2.44 1.84(2−)	17 35.453 Cl $(Ne)3s^2 3p^5$ 25.3 13.8 1.02 2.83 1.81(−1)	18 39.948 Ar $(Ne)3s^2 3p^6$ 29.2 15.9 0.99 1.74

Ni	29 63.546 Cu $(Ar)3d^{10}4s^1$ 10.7 7.7 (4p 4.0) 1.20 1.75 0.69(2+)	30 65.37 Zn $(Ar)3d^{10}4s^2$ 9.4 1.38 1.66 0.74(2+)	31 69.72 Ga $(Ar)3d^{10}4s^2 4p^1$ 12.7 6.0 1.31 1.82	32 72.59 Ge $(Ar)3d^{10}4s^2 4p^2$ 15.6 7.6 1.26 2.02	33 74.9216 As $(Ar)3d^{10}4s^2 4p^3$ 17.6 9.1 1.22 2.20	34 78.96 Se $(Ar)3d^{10}4s^2 4p^4$ 20.8 10.8 1.20 2.48	35 79.904 Br $(Ar)3d^{10}4s^2 4p^5$ 24.0 12.5 1.16 2.74 1.95(−1)	36 83.80 Kr $(Ar)3d^{10}4s^2 4p^6$ 27.5 14.3 1.14 1.89
Pd 1.31 0.86(2+)	47 107.868 Ag $(Kr)4d^{10}5s^1$ 1.42 1.26(+1)	48 112.40 Cd $(Kr)4d^{10}5s^2$ 1.53 1.46 0.97(2+)	49 114.82 In $(Kr)4d^{10}5s^2 5p^1$ 1.48 1.49	50 118.69 Sn $(Kr)4d^{10}5s^2 5p^2$ 1.44 1.72 1.12(2+)	51 121.75 Sb $(Kr)4d^{10}5s^2 5p^3$ 1.41 1.82	52 127.60 Te $(Kr)4d^{10}5s^2 5p^4$ 1.40 2.01	53 126.9045 I $(Kr)4d^{10}5s^2 5p^5$ 1.36 2.21 2.16(−1)	54 131.30 Xe $(Kr)4d^{10}5s^2 5p^6$ 1.33 2.09 (1.31 cov.)
Pt $5d^{10}$ 1.28	79 196.9665 Au $(Pt)6s^1$ 1.42	80 200.59 Hg $(Pt)6s^2$ 1.43 1.44	81 204.37 Tl $(Pt)6s^2 6p$ 1.51 1.44	82 207.2 Pb $(Pt)6s^2 6p^2$ 1.52 1.55	83 208.9806 Bi $(Pt)6s^2 6p^3$ 1.47 1.67	84 (210) Po $(Pt)6s^2 6p^4$ 1.46 1.76	85 (210) At $(Pt)6s^2 6p^5$ 1.46 1.96	86 (222) Rn $(Pt)6s^2 6p^6$ 1.45 2.14

| Eu
1.85
2+)
1.11 | 64
157.25 Gd
$(Xe)6s^2 5d^1 4f^7$
1.61
1.02(3+) | 65
158.9254 Tb
$(Xe)6s^2 4f^9$
1.59
1.10 1.00(3+) | 66
162.50 Dy
$(Xe)6s^2 4f^{10}$
1.59
1.10 0.99(3+) | 67
164.9303 Ho
$(Xe)6s^2 4f^{11}$
1.58
1.10 0.97(3+) | 68
167.26 Er
$(Xe)6s^2 4f^{12}$
1.57
1.11 0.96(3+) | 69
168.9342 Tm
$(Xe)6s^2 4f^{13}$
1.56
1.11 0.95(3+) | 70
173.04 Yb
$(Xe)6s^2 4f^{14}$
1.72
1.06 1.13(2+) | 71
174.97 Lu
$(Xe)6s^2 4f^{14} 5d^1$
1.56
1.14 0.93(3+) |
| m
3+)
1.2 | 96
(247) Cm
$(Rn)7s^2 5f^7 6d^1$
1.2 | 97
(247) Bk
$(Rn)7s^2 5f^8 6d^2$
1.2 | 98
(251) Cf
$(Rn)7s^2 5f^9 6d^1$
1.2 | 99
(254) Es
$(Rn)7s^2 5f^{11}$
1.2 | 100
(253) Fm
$(Rn)7s^2 5f^{12}$
1.2 | 101
(256) Md
$(Rn)7s^2 5f^{13}$ | 102
(254) No | 103
(257) Lr |

Concepts of Chemistry

WILLIAM W. PORTERFIELD
HAMPDEN-SYDNEY COLLEGE

Chiu Po-Yuen
Pensacola, 1972

Concepts of Chemistry

W · W · NORTON & COMPANY · INC ·
NEW YORK

Copyright © 1972 by W. W. Norton & Company, Inc.

FIRST EDITION

Library of Congress Catalog Card No. 78-177441

SBN   393   09385   9

Printed in the United States of America

1 2 3 4 5 6 7 8 9 0

Contents

Preface ... xi

PART I: Chemical Measurement ... 1

1 Physical Measurement and Chemical Composition ... 3
- 1-1: The Operations of Science – Scientific Measurement and Explanation ... 3
- 1-2: Measurement, Physical Quantities and Relationships, and Dimensional Analysis ... 6
- 1-3: Nonmechanical Physical Relationships ... 10
- 1-4: Homogeneity, Elements, and Compounds ... 13
- 1-5: The Laws of Stoichiometry ... 17
- 1-6: Formulas and Atomic Weights ... 20
- 1-7: Gram-Atomic Weights, Gram-Molecular Weights, and the Mole ... 25
- 1-8: The Stoichiometry of Reactions ... 30
- 1-9: Aqueous Ions and Nomenclature ... 36
- 1-10: Stoichiometry, Measurement, and Interactions ... 41
- Study Problems ... 42
- Some Further Reading ... 44

2 Mathematical Functions in Chemistry ... 45
- 2-1: The Representation of Functions ... 46
- 2-2: Linear and Polynomial Functions ... 50
- 2-3: The Concept of Slope ... 53

2-4:	Sinusoidal Functions	55
2-5:	Exponential and Logarithmic Functions	59
2-6:	Slopes and Derivatives of Polynomials	61
2-7:	Derivatives of Other Functions	64
2-8:	Differentials	70
2-9:	Summation and Integration	72
2-10:	Integrals and Areas	74
2-11:	Statistics and Probability	79
2-12:	The Reliability of Measurements	79
2-13:	Significant Figures	83
2-14:	The Normal Distribution	84
Appendix:	The Random Walk and the Gaussian Distribution	91
	Study Problems	95
	Some Further Reading	96

PART II: Chemical Structure — 97

3 The Behavior of Gases — 99

3-1:	The Bulk Properties of Gases	100
3-2:	The Molecular Model and the Kinetic Theory of Gases	107
3-3:	Transport Properties, Gas Mixtures, and the Kinetic Theory	118
3-4:	Nonideal Gases — Experiment and Theory	131
3-5:	Summary	144
Appendix:	Gas Law Calculations	146
	Study Problems	149
	Some Further Reading	151

4 The Behavior of Solids — 152

4-1:	Experimental Properties of Crystals	153
4-2:	Differences Between Crystals	162
4-3:	Order, Wave Diffraction, and X Rays	168
4-4:	Close Packing and Crystal Lattices	176
4-5:	The Absorption of Heat by Crystals	185
4-6:	Crystal Defects	190
4-7:	Summary	193
	Study Problems	195
	Some Further Reading	196

5 Atomic Structure — 197

5-1:	The Nuclear, Electrostatic Atom	198
5-2:	Light Quanta and the Black-Body Experiment	204
5-3:	The Bohr Atom and Hydrogen Spectra	207

5-4:	Defects of the Bohr Model – Wave Properties of Particles and Uncertainty of Measurement	214
5-5:	The Classical Wave Equation and the Schrödinger Equation	217
5-6:	One-Electron Solutions: Algebraic Wave Functions and Graphical Orbitals	221
5-7:	The Physical Meaning of the Wave Function – Normalization and Orthogonality	233
5-8:	The Many-Body Problem, Screening, and Slater Orbitals	237
5-9:	Spin, the Exclusion Principle, and the Periodic Table	240
5-10:	Ionization and Ionization Energies	246
5-11:	Summary	249
	Study Problems	250
	Some Further Reading	251

6 Molecular Structure 252

6-1:	Prediction of Molecular Geometry	252
6-2:	Diatomic Molecules – Potential Energy of Atom Pairs	260
6-3:	Diatomic Molecules – The LCAO Principle	262
6-4:	Diatomic Molecules – Bonding and Antibonding Orbitals	265
6-5:	Diatomic Molecules – The Algebra of Electron Energies	269
6-6:	Diatomic Molecules – Symmetry of Bonds	275
6-7:	Diatomic Molecules – Heteronuclear Systems	288
6-8:	Diatomic Molecules – Ionicity, Polarization, and Hydrogen Bonding	293
6-9:	Triatomic Molecules	298
6-10:	Tetraatomic Molecules	315
6-11:	Tetrahedral Molecules, Donor–Acceptor Interactions, and Hybridization	321
6-12:	Many-Atom Systems and Metals	327
6-13:	Summary	331
	Study Problems	333
	Some Further Reading	335

PART III: Chemical Dynamics 337

7 The Behavior of Liquids 339

7-1:	Experimental Properties of Liquids	340
7-2:	A Theory of Liquid Structure	350
7-3:	Liquids in Chemical Mixtures – Solutions	355
7-4:	Liquids in Physical Mixtures – Liquid–Vapor Equilibrium	362
7-5:	Liquids in Physical Mixtures – Liquid–Solid Equilibrium	390

viii | Contents

	7-6:	Statistics and Spontaneity — A Summary	401
		Study Problems	402
		Some Further Reading	405

8 Spontaneity and Equilibrium — 406

	8-1:	Thermodynamic Definitions	407
	8-2:	The Potential of Spontaneous Transition	410
	8-3:	Entropy and Randomness	427
	8-4:	Entropy and Temperature	432
	8-5:	Summary	436
		Study Problems	438
		Some Further Reading	440

9 Energy in Chemical Processes — 441

	9-1:	Heat, Work, and Internal Energy	441
	9-2:	The Enthalpy Function and Heat Capacities	446
	9-3:	Heat Flow in Chemical Reactions	450
	9-4:	Enthalpies and Molecular Properties	459
	9-5:	Low Energy, High Probability, and Spontaneity	462
	9-6:	Free Energy, Equilibrium, and the Equilibrium Constant	469
	9-7:	Summary — The Laws of Thermodynamics	474
		Study Problems	478
		Some Further Reading	480

10 Equilibrium Constants and Electrochemistry — 481

	10-1:	Equilibrium Constants for Gaseous Reactions	482
	10-2:	Ionic Equilibrium in Solution — Stability Constants	492
	10-3:	Proton Transfer in Solution	499
	10-4:	Acid–Base Equilibria	506
	10-5:	Solubility Equilibria in Aqueous Solution	527
	10-6:	Electron-Transfer Reactions in Solution	538
	10-7:	Electrochemistry and Thermodynamics	547
	10-8:	Summary	563
		Study Problems	563
		Some Further Reading	566

11 The Mechanism of Chemical Change — 567

	11-1:	Rates and Rate Constants	568
	11-2:	Molecular Mechanisms and Macroscopic Rates	573
	11-3:	Gaseous Reaction Rates — Collision Theory	577
	11-4:	Gaseous Reaction Rates — Transition-State Theory	582
	11-5:	Reaction Rates in Solution	595
	11-6:	Heterogeneous Reactions and Catalysis	605

11-7:	Summary	611
	Study Problems	613
	Some Further Reading	617

PART IV: Chemical Synthesis 619

12 Periodicity and Electronegativity 621

12-1:	The Periodic Table	621
12-2:	Ionization Potentials	622
12-3:	Electron Affinities	631
12-4:	Differential Ionization Energies	633
12-5:	Electronegativity	639
12-6:	Electronegativity Scales	640
12-7:	Electronegativity and Bond Energies	642
12-8:	Electronegativity and Bond Lengths	644
12-9:	Periodicity of Density and Atomic Volume	648
12-10:	Periodicity of Other Physical Properties	651
12-11:	Periodicity of Acid–Base Properties	654
12-12:	Electronegativity and Periodicity of Oxidation States	658
12-13:	Summary	661
	Study Problems	661
	Some Further Reading	662

13 Ionic Compounds 664

13-1:	Charge Separation in Molecular Orbitals and Molecular Energies	664
13-2:	Systems of Charged Particles and Madelung Energies	668
13-3:	Born–Haber Cycle Calculations	676
13-4:	Solvation Energies and Solubility	683
13-5:	Chemistry of Ionic Compounds—Electronegativity Differences Greater than 2.0 Units	690

Ionic Hydrides 692 · Ionic Carbides 693 · Ionic Nitrides 695
Ionic Oxides 698 · Ionic Sulfides 703 · Ionic Halides 706
Ionic Fluorides 708 · Polyatomic Positive Ions 709

13-6:	Summary	711
	Study Problems	712
	Some Further Reading	713

14 Semicovalent Compounds 715

| 14-1: | Charge Separation in Partly Covalent Systems | 716 |
| 14-2: | Lattice Geometry, Directional Overlap, and Lattice Energies | 719 |

x | Contents

14-3: Polymeric Structures in Solids and Solutions — 725
14-4: Chemistry of Partly Ionic Compounds—Electronegativity Differences Between 1.0 and 2.0 Units — 732

    Partly Ionic Hydrides 733 · Partly Ionic Borides, Carbides, Nitrides, Silicides, Phosphides 734 · Partly Ionic Oxides 738 · Partly Ionic Sulfides 752 · Partly Ionic Halides 755

14-5: Summary — 759
Study Problems — 760
Some Further Reading — 761

15 Covalent Compounds — 762

15-1: Characteristics of Covalent Bonding — 762
15-2: Criteria for Catenation — 766
15-3: Molecular Orbitals for Catenated Systems — 769
15-4: Organic Compounds and Their Reactions — 781
15-5: The Formation of Polymers — 795
15-6: Nonorganic Covalent Systems — 810
15-7: Summary — 822
Study Problems — 823
Some Further Reading — 826

16 d-Electron Compounds — 828

16-1: The Transition Metals as Metals — 828
16-2: The Nature of d Orbitals and d Electrons — 831
16-3: Molecular Orbitals for d-Electron Compounds — 841
16-4: Coordination Numbers and Molecular Geometries — 851
16-5: Multiple Oxidation States and Nonstoichiometric Compounds — 855
16-6: Ionic and Partly Ionic Transition-Metal Compounds — 859
16-7: Covalent Transition-Metal Compounds and Coordination Compounds — 863
16-8: Summary — 879
Study Problems — 881
Some Further Reading — 884

Tables — 885
Answers to Problems — 893
Index — 937

Preface

In writing this book, I took the basic premise that an introductory text ought to be as complete and as fully integrated as possible. This requires several features of the book that perhaps deserve comment. First, completeness means that the mathematical and physical models—indeed, the very process of modeling—need to be fully explained. Concepts that are familiar and self-evident to the experienced chemist, so that they require no derivation or justification, are unfamiliar and arbitrary to the beginning student. The phrase "It can be shown that . . ." is anathema to students and to me, and at every possible point I have developed concepts from first principles.

Second, completeness means that the concepts presented need to be those chemists really use, as opposed to subterfuges or anachronisms. Modern chemists work and think in terms of Boltzmann energy distributions, molecular-orbital structures and energies for molecules, thermodynamic interpretations of equilibrium and electrochemistry, and reactions governed in many cases by available reaction mechanisms. All of these concepts can be presented at an elementary level without discouraging students if a strong physical flavor is maintained throughout the mathematical discussions. However, these discussions do require the use of some calculus, which most beginning students have not had. Therefore an introductory section on calculus is provided in order to bring the student to an appreciation of the chemist's mathematical arguments. With this background it has been possible, I believe, to produce a sophisticated yet understandable account of the basic concepts of chemistry.

Third, and most important, integration means that the descriptive chemistry and the principles of chemical synthesis that are developed must deal not only with exciting areas of academic chemical research, but also with chemistry as it appears in the world and the society around the student. Such discussions are available, of course, but rarely are they based on a unifying framework of theoretical concepts. Conversely, most theoretically elegant texts develop the subject in abstract terms, discussing silanes but not silicate minerals, and amides but not nylon. There is no reason for a sound introduction to theory to be unrelated to chemistry in our society. In fact, it is important that the correlation between the two, which is genuine, be demonstrated to a society increasingly unimpressed with science for its own sake.

The text follows my own preference for the order of introduction of topics. Even though the chapters are rather closely linked together, however, some instructors may find an alternative sequence more attractive: Chapters 7–10 may be taken up immediately after Chapter 4. In effect this takes the student through gases, solids, liquids, phase changes, spontaneity, thermodynamics, and equilibrium calculations before discussing atomic structure, molecular structure, and molecular reactivity in the chapter on kinetics and mechanisms. By either means the student is given a coherent introduction to the theory that underlies the ensuing descriptive chemistry. The book is intended to be completed in two semesters or three quarters; its length suggests this, but is perhaps misleading in that there is an unusually large number of illustrations. Often the more ominous mathematical aspects of theoretical chemistry can be defused by graphical or schematic presentation, without sacrificing accuracy if the illustrations are well prepared. Considerable care has gone into the preparation of the book's numerous two-color illustrations, and I hope the student will find them helpful.

Most of the problems at the end of each chapter require the student to produce at least a small synthesis of ideas, which seems particularly important in a conceptually oriented course. Indeed, in several problems the student will develop concepts that are necessary in succeeding chapters. Since student participation in the study problems is so desirable, the number has been kept at about two thoughtful problems per class meeting, with full answers, and the student can thus be expected to work all the study problems in each chapter. If this number seems inadequate, several excellent paperback problem books are available. In addition, the tabular material on general-purpose data such as bond energies, free energies of formation, and such is much more extensive than the chapter usage requires; this allows the instructor to assign problems for which the data are available to the student but which are not part of the text discussion.

Finally, I have the particularly pleasant duty of acknowledging incisive but kind and helpful criticisms from many of my colleagues, both at this college and at other institutions: Professors John R. Barker, Jerry A. Bell, Seth Boorstein, John R. Butcher, Derek A. Davenport, Thomas Mayo, Richard A. Palmer, Herbert J. Sipe, Jr., and Homer A. Smith. They have influenced the writing in many beneficial ways—but the errors and misleading statements that will undoubtedly be found must be laid at the door of the author, who has been known to resist even good advice. I owe a debt of gratitude to Professor Harris Burns, with whom I originally conceived the book and who had a strong influence on its sequence of ideas (besides writing part of Chapter 2), and to Norris Jeffrey and Professor S. Y. Tyree for a chemical education extending over many years. My thanks also go to Mary Pell and the staff of W. W. Norton for their untiring efforts. And last and most important, there is my wife, who took the children into another room and didn't complain about my disappearance.

William W. Porterfield

Hampden-Sydney, Virginia
October 1971

I CHEMICAL MEASUREMENT

1 Physical Measurement and Chemical Composition

Man's curiosity is universal; in all times and in all civilizations he has wondered about the world he lives in. It is three thousand years since Solomon wrote "... the glory of kings is to search things out." The discipline of chemistry is the structure of ideas resulting from one kind of inquiry, one area of understanding. Chemistry deals with the reasons why matter has its particular forms and properties, and with the ways in which it changes those forms and properties. Although much remains to be discovered, 200 years of investigation have developed an intellectual structure astonishing in its diversity and in its capacity to accommodate and relate experimental facts. It is this structure that we shall look at in this book. These—as we shall presently understand them—are the concepts of chemistry.

1-1 The Operations of Science—Scientific Measurement and Explanation

As we investigate modern chemistry, our pattern will be first to discuss the fundamental laws and theories, then to explore their effects in descriptive or synthetic chemistry. This is a reasonably efficient procedure; it is rather like learning a language by first mastering the grammar. But just as languages are not formed by people deciding upon a grammar and then constructing the language, so the structure of chemistry is not the result of some agreement on fundamental causes followed by experiments to show the natural results of these causes. Rather, the fundamental causes—our basic concepts—have been developed and understood only as a result of centuries of observation and experimentation.

Theory and Law

The process of reasoning from a set of experimental facts to a generalization or **natural law** that describes the facts, and then to a hypothetical principle or **theory** that provides a unified model or explanation for the facts, is the inductive process. Scientific explanation is entirely inductive, which places narrow bounds on the acceptable hypotheses. If a hypothesis disagrees with even a single experimental fact, the hypothesis must be modified or restricted. We rely entirely on experimental observations, and we formulate hypotheses and theories so as to correlate all our observations, or at least so as not to contradict any of them.

What is a theory, and how does it differ from a natural law? A natural law summarizes our experimental experience without attempting to demonstrate *why* the experiments should come out the way they do. For instance, the second law of thermodynamics says that if a hot object and a cold one are placed together the hot one always cools off and the cold one always warms up. The law does not attempt to rationalize this fact; it simply says that we have always found it to be true. A theory, by contrast, exists only for the purpose of correlating and rationalizing our experimental observations—*why* should the hot object always cool? A law may have exceptions but it does not change. The experimental facts are there, and if they are properly verified they will not change. On the other hand, a theory is always subject to change or refinement as new experimental evidence accumulates that the theory must accommodate.

Measurement is the most basic process of natural science. It is only after we have performed a series of experiments that we can take the resulting data as raw material for the formulation of a law or for the construction of a theory. A theory will in general be a sort of mathematical model for an experimental process or set of properties. For the mathematical model to be meaningful we must have some numbers with which to work. Thus our experiments will generally consist of measurements of different kinds, which will provide numerical data to feed into our model. All this amounts simply to saying that quantitative thinking is central to any scientific experiment.

The Limits of Measurement

If experimental data are the fundamental building blocks of science, and if they are characteristically quantitative statements involving measurement, then there are two basic limitations on the theories that we can construct from them. The first is a practical limitation: measurements are subject to error. In this sense error does not mean the possibility of a mistake; rather, it means that the accuracy of any measurement is limited by the delicacy of the measuring instrument. If we measure the length of a line of type on this page with an ordinary yardstick, we are limited by the fact that the finest subdivisions on a yardstick are 1/8 in. apart. Our best measurement, then, still involves an uncertainty of approximately half the finest division, or 1/16 in. We can say that there is an error on the order of 1/16 in. in the measurement. On the other hand, if we use a machinist's ruler with graduations of 1/100 in., we can do a much better job—the error is on the order of half that division, or 5/1000 in. Depending on what measurement we want to make, we may find that our instruments are very precise or very crude. Clearly this will have a pronounced effect on the use of these data in the construction of a theory. We shall return to this problem of error and averaging techniques in Chapter 2.

The second basic limitation on the use of experimental data in constructing theories is a more philosophical one, but nevertheless one with great practical effect. If observation and measurement are the fundamental operations on which our theoretical structure is based, then our theories must be constructed so as to deal only with the measured quantities and to predict the measured results. We must exclude theories that rely on quantities that by their very nature cannot be measured either directly or indirectly. We say that we rely on **operational definitions**; all our quantities are defined so as to refer to the process of measurement. This seems fairly obvious, but it has some subtle effects. For instance, when we deal experimentally with a microscopic particle such as the electron, it is necessary to realize that there is nothing smaller than the electron for us to use as a probe. This means that the very act of measurement is going to strongly influence or perturb the measured quantity itself or some related quantity. Since our measurements always influence the measured system, the theory we construct from these data must allow for this unknown degree of influence. The operational definitions we insist on in our theory will be just those that *do* predict the results of our actual measurements.

1-2 Measurement, Physical Quantities and Relationships, and Dimensional Analysis

The concepts of chemistry are the unifying theories developed over the last century or so that have gone far toward forming a logical structure and rationale for the experimental facts of chemistry. But we have seen that these theories exist at the sufferance of experimental results; they rely, of necessity, on operational definitions. We need, then, to look closely at the experimental process of measurement. Since measurement is a quantitative process, each measured point or **datum** is a number. If we were compiling highway distances in the United States, one datum would be the distance from Washington to Baltimore — 39 mi. The number 39 is the central quantity in the measurement, but if the United States were on the metric system, the number would be quite different; we would say that the distance was 63 km. So the numbers we use in our measurements are dictated not only by the nature of the measured quantity but also by its relationship to the units in which we choose to express it. It is generally true that measured physical quantities have dimensions and units. For the numerical measure — the number — to be meaningful, it must have the appropriate units specified. I could say accurately that the distance from Washington to Baltimore is 206,000, but this would be meaningless to you unless you knew that the units were feet. It is appropriate to state at the very beginning, then, the units we shall use in discussing the measurements and the resulting theories throughout the book.

Fundamental Physical Quantities and Dimensional Analysis

There are only five fundamental physical quantities whose units we need in order to express all the measurable physical properties of matter. They are mass, distance, time, electric charge, and temperature. All the others, such as energy, density, frequency, viscosity, and so on, can be expressed in terms of these fundamental quantities; Table 1-1 gives some of the useful relationships, particularly those involving motion or mechanical quantities. Using the units in Tables 1-1 and 1-2 we can express any physical property we choose; the way in which we express a particular physical property (by assigning it dimensions and units) has a profound effect on the mathematical manipulations we can perform on the measurement. For instance, consider the physical property density, shown in the center of Table 1-1 with dimensions of mass per unit volume. Its units are commonly grams per cubic centimeter (g/cm^3), as indi-

Table 1-1
Fundamental Physical Quantities and Relationships

Each of the derived physical quantities (in boxes) is related to one or more of the five basic quantities (shown in red) as indicated by the arrows. The basic dimensions of any derived quantity may be found by reducing each of its component quantities to a combination of basic quantities. For example, substituting simple defining equations:

$$\begin{aligned}
\text{work} &= Fd \\
&= (ma)(d) \\
&= (m)(v/t)(d) \\
&= (m)(d/t^2)(d) \\
&= md^2/t^2 = \text{mass-distance}^2/\text{time}^2
\end{aligned}$$

Where different physical relationships lead to quantities with the same dimensions, the equivalence is indicated by a double red arrow.

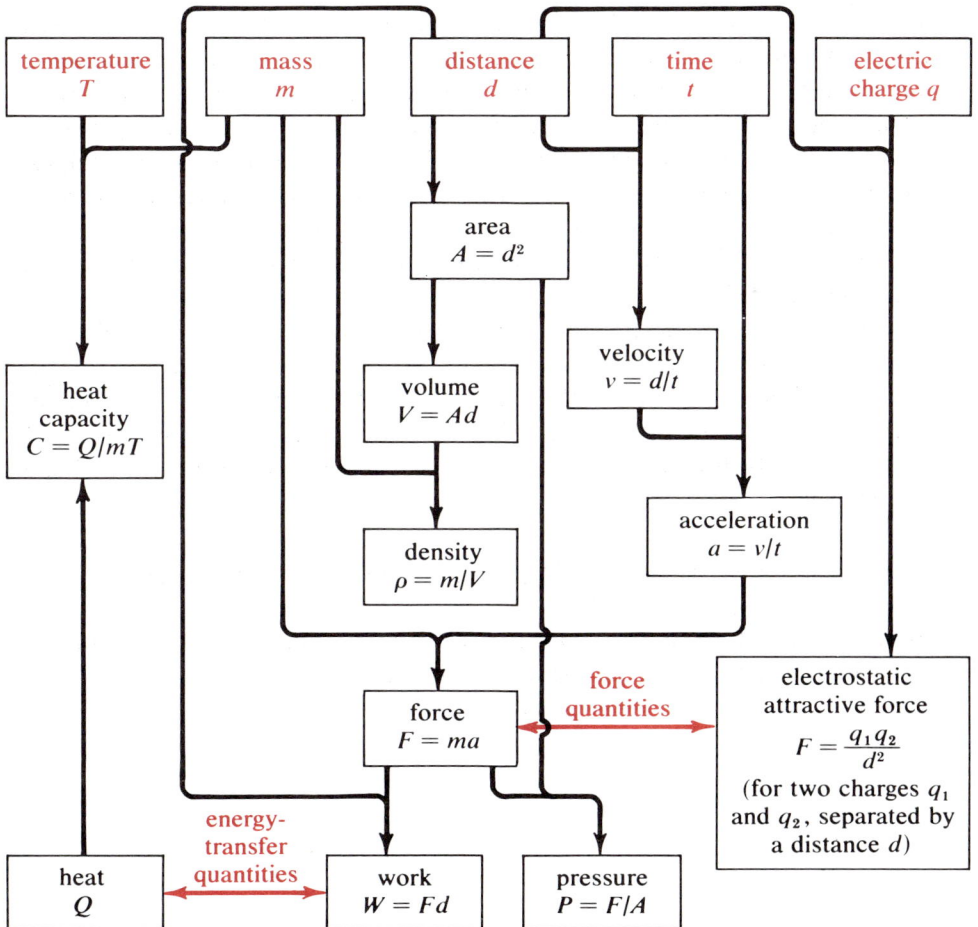

cated in Table 1-2. Now suppose we know that the density of liquid mercury metal is 13.59 g/cm³, and we wish to know the volume of 500 g of mercury. We can avoid making the wrong mathematical operation on these two numbers by noting that the answer we want is volume, which has units of cubic centimeters. Just as we multiply and divide the numbers in our data, we also multiply and divide the units. Since our answer must not contain grams, we must either divide 500 g by 13.59 g/cm³,

$$\frac{500 \text{ g}}{13.59 \text{ g/cm}^3} = \frac{500 \text{ g-cm}^3}{13.59 \text{ g}} = 36.9 \text{ cm}^3$$

so that the gram units cancel out, or else divide 13.59 g/cm³ by 500 g to get the same cancellation:

$$\frac{13.59 \text{ g/cm}^3}{500 \text{ g}} = \frac{13.59 \text{ g}}{500 \text{ g-cm}^3} = 0.0271 \frac{1}{\text{cm}^3}$$

Looking at these two possible operations, we see that only the first one gives the units of volume in the answer; therefore it is correct and the other operation is meaningless. This process of establishing the correct mathematical treatment of data by observing the effect of possible operations on the units or dimensions of the data is called **dimensional analysis**. Dimensional analysis is very helpful in guiding us through problems in which the appropriate mathematical treatment is not clear from the physical nature of the problem. It amounts simply to recognizing that the units of a datum are as much a part of its nature as its numerical value.

We can also use dimensional analysis in a simpler way to convert equivalent units for the same physical quantity. Suppose (for a frivolous example) we want the distance from Baltimore to Washington in centimeters, and we know only that the distance is 39 mi and that 1 in. = 2.54 cm. We can set up a series of unit-conversion factors in which all the units cancel except those that should be in our answer:

$$\text{distance} = 39 \text{ mi} \times 5280 \frac{\text{ft}}{\text{mi}} \times 12 \frac{\text{in.}}{\text{ft}} \times 2.54 \frac{\text{cm}}{\text{in.}} = 6.39 \times 10^6 \text{ cm}$$

A routine use of this sort of analysis will be helpful, in fact indispensable, in solving numerical problems throughout the book.

Measurement, Physical Quantities and Relationships, and Dimensional Analysis

Table 1-2
Dimensions and Units

Quantity	Dimensions	CGS Units	SI Units
distance	distance	centimeter (cm)	meter (m)
mass	mass	gram (g)	kilogram (kg)
time	time	second (sec)	second (s)
electric charge	electric charge	statcoulomb or electrostatic unit (esu)	coulomb (C)
temperature	temperature	degree (deg)	kelvin (K)
area	distance^2	cm^2	m^2
volume	distance^3	cm^3	m^3
velocity	distance/time	cm/sec	m/s
acceleration	distance/time^2	cm/sec^2	m/s^2
force	mass-distance/time^2	dyne	newton (N)
work	mass-distance^2/time^2 (energy)	erg	joule (J)
heat	mass-distance^2/time^2 (energy)	calorie (cal)	joule (J)
pressure	mass/distance time^2	dyne/cm^2	N/m^2
density	mass/distance^3	g/cm^3	kg/m^3
heat capacity	distance^2/time^2-temperature (energy/mass-temperature)	cal/g-deg	J/kg-K

Subsidiary CGS Units

Quantity	Unit	Fundamental Unit Equivalence
distance	angstrom unit (Å)	1 Å = 10^{-8} cm
volume	liter (l)	1 l = 10^3 cm^3
pressure	torr or millimeter of mercury, atmosphere (atm)	1 torr = 1333.21 dynes/cm^2 1 atm = 1.01325 × 10^6 dynes/cm^2
energy	electron volt (eV)	1 eV = 1.60210 × 10^{-12} erg 1 eV/atom = 23.061 kcal/mole

Unit Prefixes for Decimal Fractions and Multiples

Multiple	Prefix	Fraction	Prefix
10^3	kilo- (k)	1/10 ≡ 10^{-1}	deci- (d)
10^6	mega- (M)	10^{-2}	centi- (c)
10^9	giga- (G)	10^{-3}	milli- (m)
10^{12}	tera- (T)	10^{-6}	micro- (μ)
		10^{-9}	nano- (n)
		10^{-12}	pico- (p)

Table 1-2 (continued)
Numerical Unit Conversion Factors

---

distance
    2.54 cm/in.; 30.48 cm/ft; 5280 ft/mi; 39.37 in./m
mass
    453.59 g/lb; 28.35 g/oz; 907.2 kg/ton
electric charge
    2.9979×10^9 esu/C
temperature
    1.8 (or 9/5) °F/°C; 1 °C/K
area
    6.452 cm^2/in.^{2}; 929.0 cm^2/ft^2; 43,560 ft^2/acre
volume
    1.0567 quarts/l; 16.387 cm^3/in.^{3}; 28.316 l/ft^3; 42 gal/barrel (petroleum)
velocity
    44.70 (cm/sec)/(mi/hr); 1.46667 (ft/sec)/(mi/hr)
force
    10^5 dynes/N
work (energy)
    10^7 ergs/J; 4.184×10^7 ergs/cal; 3.600×10^{13} ergs/kW-hr
pressure
    760 torr/atm; 14.70 (lb/in.^{2})/atm

1-3 Nonmechanical Physical Relationships

Most of the relationships in Table 1-1 deal with mechanical quantities and simple motion, which we shall need repeatedly. There are some other physical quantities and relationships, however, that will also be important to us and that we shall examine briefly here.

Wave Motion

We shall need to be familiar with the properties of waves, particularly light waves, since the interaction of electromagnetic radiation (such as light waves) with matter is one of the most powerful techniques for investigation of the structure of matter. Figure 1-1 shows a typical wave and defines the quantity **wavelength**. This wave, which we can envision as a ray of light, is traveling along the direction in which the wavelength is measured. It has a velocity that we can define as the distance an individual "hump" on the wave moves in a unit time; for a light wave moving in vacuum, this velocity is a constant 2.99793×10^{10} cm/sec. Since the humps are all alike, we might be more interested in the time required for the passage of one wavelength — called the **period** of the wave.

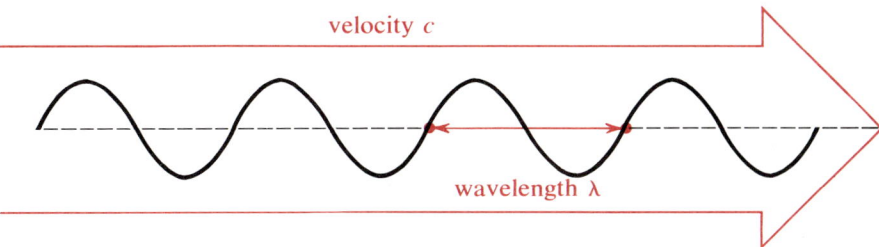

Figure 1-1 Wave motion.

The period and the wavelength are related to the velocity just as their dimensions would suggest:

$$c = \frac{\lambda}{t} \tag{1-1}$$

where c is the velocity of light in centimeters per second, λ is the wavelength in centimeters, and t is the period in seconds. The quantity $1/t$ is defined as the **frequency** of the light wave and is denoted by ν; it is the number of wavelengths that pass a given point in one second. Using the frequency we can rewrite the above equation as

$$\nu(\text{sec}^{-1}) = \frac{c \ (\text{cm/sec})}{\lambda \ (\text{cm})} \tag{1-2}$$

Since c is a constant, it is clear that wavelength and frequency are inversely related to each other.

For visible light, the frequency is an inconveniently large number—in the vicinity of 10^{15} oscillations/sec. To make it more manageable, spectroscopists often define a new quantity, the **wave number**, which is the frequency divided by the velocity of light:

$$\text{wave number} \equiv \nu'(\text{cm}^{-1}) = \frac{1}{\lambda(\text{cm})} \tag{1-3}$$

Note that this equation is the same as the previous one, but divided by c; to be consistent, wave numbers must have the units of reciprocal centimeters. The unit cm^{-1} is used for wave number, but frequency is officially measured in hertz (Hz) after one of the pioneers in the study of electromagnetic radiation. The unit is defined by $1 \text{ Hz} = 1 \text{ sec}^{-1}$, so that all the numbers remain the same. For visible light, the frequency range is about 4.5–7.5×10^{14} Hz, but the equivalent wave-number range is about $15{,}000$–$25{,}000 \text{ cm}^{-1}$, a more convenient set of numbers.

Electrostatic Attraction and Repulsion

We also need to concern ourselves a little further with the relationships between measurements of electrical quantities. Table 1-1 indicated that the force between two electric charges is found to obey the **Coulomb Law**:

$$\text{force} = \frac{q_1 q_2}{r^2} \tag{1-4}$$

Here q_1 and q_2 are the two charges involved and r is the distance separating them; the units are dynes, electrostatic units (esu), and centimeters. If the two charges are alike in their polarity (both positive or both negative), the force is a repulsion. This means that an energy change is involved in moving one charge while holding the other still, simply because work equals force times distance. We must do work on the charges to move them closer together, but work is done for us if we allow them to move farther apart. If we had opposite electric charges, positive and negative, they would attract each other instead of repelling, but Coulomb's law would still hold. How do we describe the work which must be done in moving a single charge? It is clearly proportional to the magnitude of the charge, since the bigger the charge, the stronger will be the force. Thus we can write the following relationship:

$$E = qV \tag{1-5}$$

where E is the work done (an energy quantity), q is the magnitude of the charge, and V is the proportionality constant. The constant V is called the **electric potential** and is measured in **volts** (V). A volt is that difference in electric potential which enables a charge of 1 coulomb (C) to do 1 joule (J) of work on moving through it. Later we shall find it convenient to use a different energy unit, the **electron volt** (eV), which is the energy effect produced by moving the charge on 1 electron (1.6021×10^{-19} C) through a potential difference of 1 V. Division shows immediately that 1 eV = 1.6021×10^{-12} erg, a very small unit of energy. But (if we may anticipate the later discussion a little) there are many electrons in an ordinary sample of matter; in a convenient unit called the mole there are 6.023×10^{23} atoms or molecules. If we are interested in 1 electron per atom, then, the total energy effect for a mole is much larger:

$$1 \frac{\text{eV}}{\text{atom}} \times 6.023 \times 10^{23} \frac{\text{atoms}}{\text{mole}} = 6.023 \times 10^{23} \frac{\text{eV}}{\text{mole}}$$

$$6.023 \times 10^{23} \frac{\text{eV}}{\text{mole}} \times 1.602 \times 10^{-12} \frac{\text{erg}}{\text{eV}} \times 10^{-7} \frac{\text{J}}{\text{erg}} = 96{,}487 \frac{\text{J}}{\text{mole}}$$

Thermal Energy

Finally, since we have said that a unit for temperature is necessary to describe the effect of heat on matter, let us try to relate heat units to our other units. Heat is the natural form of energy to which other forms are converted when they are dissipated in some way; for instance, when a car stops, the energy of motion is converted into heat in the brake linings. The fundamental unit of heat is the **calorie** (cal), which is defined as the amount of heat necessary to raise the temperature of 1 g of water by 1 degree centigrade (from 14.5 °C to 15.5 °C, to be precise). Comparing the calorie with the electron volt, we have

$$1\frac{\text{eV}}{\text{atom}} = 96{,}487\frac{\text{J}}{\text{mole}} \times \frac{1}{4.184}\frac{\text{cal}}{\text{J}} = 23{,}061\frac{\text{cal}}{\text{mole}}$$

Since convenient measurements often involve large quantities of heat, it is common practice to use the **kilocalorie** (kcal), which is just 1000 cal. Using this unit we can say that 1 eV = 23.061 kcal/mole.

1-4 Homogeneity, Elements, and Compounds

All of our discussion so far has dealt with the scientist's approach to the process of measurement. We can proceed now to look at the basic measurements of chemistry and see how the resulting calculations are performed.

The matter we see around us can be described in terms of its chemical composition. The ground we stand on, the air we breathe, the paper in this book—all have their own unique chemical composition, which is basically responsible for the properties they possess. We shall see eventually that we account for the fact that the ground is firm and that the paper is flexible in terms of a theoretical model that we shall construct; but to construct the model, we need first of all a measurement of the composition of these substances. Expressing these measurements requires some new definitions. How do we specify the particular bit of matter in which we are interested? What do we mean by chemical composition?

Almost anything we look at shows diversity. This paper is white, but it has black marks on it, and we suspect instinctively that the chemical composition of the paper is different from that of the ink. To explain the properties of the paper or the properties of the ink, then, we should consider them separately. In other words, we should restrict ourselves to the examination of systems that are uniform throughout. When we are interested in a system showing diversity we shall break it down into smaller systems, each of which is uniform throughout.

States of Matter and Phases

The most obvious differences we see in nature are those between gases, liquids, and solids. These are the three physical **states** of matter. Matter in the solid state has a definite volume and a definite shape. In the liquid state, it has a definite volume but will take on the shape of its container to the extent that the liquid is large enough to fill the container. In the gaseous state, it has neither a definite volume nor a definite shape and will not only take on the shape of its container but will expand to fill it. In restricting ourselves to uniform systems, we can thus agree to consider only systems consisting of a single physical state.

But even a system having only one physical state can show diversity. A mixture of oil and water is entirely liquid but not chemically uniform, because oil and water are chemically different. If we expand our definition to say that we shall consider only a system that is both physically and chemically uniform throughout, we have defined a **phase**. The mixture of oil and water is a two-phase system; one phase is liquid water, the other is liquid oil. The distinction between these two phases is that, although they are physically uniform (both are liquids), they are chemically different. On the other hand, a mixture of crushed ice and water is also a two-phase system; one phase is solid water (ice), the other is liquid water. Here the distinction between phases is that although they are chemically uniform (both are water) they are physically different.

Now for another definition: a sample of matter consisting of only one phase is said to be **homogeneous**. If more than one phase is present, the system is **heterogeneous**.

The question of homogeneity is sensitive to the scale of the experiment in which we are interested. For instance, a solution of salt, NaCl, in water is homogeneous on the ordinary scale of observation; we cannot detect any chemical or physical differences between adjacent regions of the solution. But on a microscopic scale — the scale of atoms and molecules — the solution is a mixture of water molecules and electrically charged atoms or **ions** of sodium and chlorine. The distinction between physical uniformity and chemical uniformity that we have made is also influenced by scale, and chemical differences often prove to be physical differences, or vice versa, when examined on a microscopic scale. There are "gray areas" in which any distinction would be arbitrary. So the concept of homogeneity is relative strictly to our level of observation, but the concept is useful nonetheless.

Throughout Parts I and II of this book, dealing with chemical structure, we shall be interested mostly in homogeneous systems. Later, in dealing with chemical reactivity and synthesis, we shall frequently be interested in heterogeneous systems.

Elements and Compounds

It is clear that a heterogeneous substance is a mixture; what is perhaps not so clear is that a homogeneous substance can also be a mixture. Speaking chemically, a pure homogeneous substance is either a chemical **element** or a chemical **compound**. A homogeneous mixture (on the laboratory scale) is a **solution**. Thus rock salt is a compound, sodium chloride; an iron bar is an element, iron; a steel bar is a solution of an element, carbon, dissolved in another element, iron. All three of these substances are homogeneous, but we have distinguished between them. What definitions allow us to make the distinction?

A **chemical element** is a substance that cannot be produced by combination of any other substances. We can avoid this rather negative approach by using another very old operational definition: an element always gains weight in undergoing any chemical change. Even this is somewhat negative and incomplete, since not all chemical changes have been studied, but the definition is useful. In microscopic terms, an element is a substance composed entirely of the same kind of atoms. There are presently 105 known elements, of which 90 occur naturally. The others have been synthesized on a relatively small scale by atomic physicists. These elements are shown inside the front cover of the book arranged in a *periodic table*, about which we shall have a great deal more to say later. Table 1-3 may also be useful for purposes of nomenclature.

A **chemical compound** is a pure homogeneous substance with an unvarying composition, made up of elements. We shall examine this definition carefully, along with some laws about the composition of compounds, in the next section (1-5). We admit at once, however, that some compounds do have a partially variable composition, depending on their past history. It is exactly true within the limits of our best experiments, however, that most compounds do have an unvarying composition, and we adopt this definition to emphasize the distinction between compounds and solutions.

A **solution** is a homogeneous substance with a variable composition. It can be made up of elements, as in the case of the steel bar mentioned earlier, or compounds, as in the case of salt water (the compound sodium chloride dissolved in the compound water), or both, as in the case of oxygenated water (the element oxygen dissolved in the compound water). Solutions can exist in all three physical states. Air is a gaseous solution of oxygen (plus other gases) in nitrogen, while salt water and the steel bar represent liquid and solid solutions, respectively. A great deal of synthetic chemistry is performed using solutions, and we shall study them more thoroughly in Chapter 8.

Table 1-3
Nomenclature of the Chemical Elements

Element	Symbol	Different Usage in Compound Names	Element	Symbol	Different Usage in Compound Names
actinium	Ac		indium	In	
aluminum	Al		iodine	I	
americium	Am		iridium	Ir	
antimony	Sb	stib-	iron	Fe	ferr-
argon	Ar		krypton	Kr	
arsenic	As		lanthanum	La	
astatine	At		lawrencium	Lw	
barium	Ba		lead	Pb	plumb-
berkelium	Bk		lithium	Li	
beryllium	Be		lutetium	Lu	
bismuth	Bi		magnesium	Mg	
boron	B		manganese	Mn	
bromine	Br		mendelevium	Md	
cadmium	Cd		mercury	Hg	
calcium	Ca		molybdenum	Mo	
californium	Cf		neodymium	Nd	
carbon	C		neon	Ne	
cerium	Ce		neptunium	Np	
cesium	Cs		nickel	Ni	
chlorine	Cl		niobium	Nb	
chromium	Cr		nitrogen	N	az- or am-
cobalt	Co		nobelium	No	
copper	Cu	cupr-	osmium	Os	
curium	Cm		oxygen	O	
dysprosium	Dy		palladium	Pd	
einsteinium	Es		phosphorus	P	
erbium	Er		platinum	Pt	
europium	Eu		plutonium	Pu	
fermium	Fm		polonium	Po	
fluorine	F		potassium	K	
francium	Fr		praseodymium	Pr	
gadolinium	Gd		promethium	Pm	
gallium	Ga		protactinium	Pa	
germanium	Ge		radium	Ra	
gold	Au	aur-	radon	Rn	
hafnium	Hf		rhenium	Re	
helium	He		rhodium	Rh	
holmium	Ho		rubidium	Rb	
hydrogen	H	isotope with atomic weight = 2 called deuterium, symbol D	ruthenium	Ru	
			samarium	Sm	
			scandium	Sc	

Table 1-3 (continued)

Element	Symbol	Different Usage in Compound Names	Element	Symbol	Different Usage in Compound Names
selenium	Se		thulium	Tm	
silicon	Si		tin	Sn	stann-
silver	Ag	argent-	titanium	Ti	
sodium	Na		tungsten	W	
strontium	Sr		uranium	U	
sulfur	S	thio-	vanadium	V	
tantalum	Ta		xenon	Xe	
technetium	Tc		ytterbium	Yb	
tellurium	Te		yttrium	Y	
terbium	Tb		zinc	Zn	
thallium	Tl		zirconium	Zr	
thorium	Th				

1-5 The Laws of Stoichiometry

The Composition of Compounds

Since most of the pure substances we encounter are compounds, the definition of a compound is worth examining further. We have indicated the importance of measurement and have specified that compounds have an unvarying composition, but we have not said anything about the measurement of this composition. The statement of the composition of compounds or solutions in terms of the proportion of chemical elements present is called **stoichiometry**. In general it is possible, by taking advantage of the characteristic chemical reactions of each element, to obtain a quantitative analysis for the amount of each element in a given compound. The results are usually expressed in weight units or in weight percent: a 1.000-g sample of a copper compound may prove to contain 0.434 g of copper, which corresponds to 43.4% Cu by weight.

When we say that compounds have an unvarying composition, we are summarizing the results of analysis of a very large group of compounds — so large that we can formulate the **law of definite proportions**:

> When two or more elements combine into a compound, they normally do so in a very specific proportion by weight.

Thus when sodium and chlorine combine to form a compound, sodium chloride, it always has the composition 39.3% Na, 60.7% Cl. We have never observed a compound of sodium and chlorine with any other composition.

18 | Physical Measurement and Chemical Composition

In some cases, however, more than one compound may form between the same two (or more) elements. For instance, when we prepare a compound of copper and chlorine, it may have either of two possible compositions, depending on how we prepared it. It may analyze as 64.2% Cu and 35.8% Cl, or it may analyze as 47.3% Cu and 52.7% Cl; there are no other possibilities. Now we can ask a pertinent question about the compositions of these two copper chlorides: What weight of chlorine, in each case, combines with each gram of copper?

In the first compound there are 35.8 g of Cl for 64.2 g of Cu; dividing, we have

$$\frac{35.8 \text{ g Cl}}{64.2 \text{ g Cu}} = 0.554 \frac{\text{g Cl}}{\text{g Cu}}$$

In the second compound there are 52.7 g of Cl for 47.3 g of Cu and, dividing again,

$$\frac{52.7 \text{ g Cl}}{47.3 \text{ g Cu}} = 1.113 \frac{\text{g Cl}}{\text{g Cu}}$$

Looking at these two ratios of combining weights, we see that one is just twice the other:

$$\frac{1.113}{0.554} = 2.01$$

In examining many sets of compounds between the same elements, we find that it is quite generally true that the ratios of their combining weights are in proportion to two small integers; in this case, $1.113:0.554 = 2:1$. As another example, consider the two carbon–hydrogen compounds methane and ethane. Methane analyzes as 74.8% C and 25.2% H; ethane analyzes as 79.9% C and 20.1% H. We can compare the two compounds:

<table>
<tr><td align="center">methane</td><td align="center">ethane</td></tr>
<tr><td>$\dfrac{25.2 \text{ g H}}{74.8 \text{ g C}} = 0.337 \dfrac{\text{g H}}{\text{g C}}$</td><td>$\dfrac{20.1 \text{ g H}}{79.9 \text{ g C}} = 0.252 \dfrac{\text{g H}}{\text{g C}}$</td></tr>
</table>

$$\frac{0.337}{0.252} = 1.34 = \tfrac{4}{3}$$

From these and many other analyses, we can formulate the **law of multiple proportions**:

> When two or more compounds that contain two given elements are analyzed, the ratios of the combining weights of the elements in any two of the compounds are in proportion to two small integers.

It is important to distinguish between the law of definite proportions and the law of multiple proportions. The first serves simply to define a compound and to distinguish it from a solution, which presumably can have a more or less continuously variable composition by weight. The second law has a further significance; it suggests that somewhere in these ratios of combining weights there is a fundamental unit of each element, and that compounds of these elements form by combining proper numbers of these units. If we inspect the ratios that have this integral relationship, we see that they are basically dimensionless (weight divided by weight leaves no dimensions). This means that the basic combining unit of each element is not a standard weight but is instead a number unit that has a different weight for each element.

There is another law that emphasizes even more strongly the combining of elements in compounds in essentially integral units. This is the **law of equivalent proportions**:

> If three elements A, B, and C form three compounds A—B, B—C, and A—C, the combining weights of the elements in each of the three compounds follow the relationship
>
> $$\frac{\text{ratio of combining weights in A—C}}{\text{ratio of combining weights in A—B}} = \text{ratio of two small integers}$$
>
> (with ratio of combining weights in B—C in the numerator)

Consider the compound methane from the previous example, which is a compound of carbon and hydrogen. Both carbon and hydrogen can form compounds with oxygen; an example of a C—O compound is carbon dioxide, and an example of a H—O compound is water. For carbon dioxide, analysis indicates a composition of 27.2% C, 72.8% O, which gives a ratio of combining weights of $27.2/72.8 = 0.374$ g C/g O. For water, analysis gives 11.2% H, 88.8% O, or a ratio of combining weights of $11.2/88.8 = 0.125$ g H/g O. Placing these numbers into the law of equivalent proportions,

$$\frac{(0.125 \text{ g H/g O})/(0.374 \text{ g C/g O})}{\text{ratio of combining weights in methane}} = \text{ratio of two small integers}$$

But in our previous example we showed that the ratio of combining weights for methane, the C—H compound, is 0.337. If we insert this value, we see that a ratio of small integers does indeed result:

$$\frac{0.125}{0.374} = 0.334 \frac{\text{g H}}{\text{g C}} \qquad \frac{0.334 \text{ g H/g C}}{0.337 \text{ g H/g C}} = \frac{1}{1}$$

The existence and general validity of the law of equivalent proportions show that it is not just a peculiarity of two compounds of the same two elements that the elements are combined in an intrinsically integral fashion. The crux of a measurement of stoichiometry, then, is the weight of the fundamental combining unit of each element present, since we anticipate from these laws that any compound, no matter how small a sample we take, consists of a combination of an integral number of such units of each element present.

Atoms as a Model for Matter

This is precisely where we begin to use atoms as a model for the ultimate chemical structure of elements, and molecules as a model for the ultimate chemical structure of most compounds. If a molecule is composed of several atoms of element A bonded to some other atoms of element B, then the chemical and physical properties of the compound will grow out of the precise combination of A and B atoms involved. If m atoms of A and n atoms of B are involved, we can write a **chemical formula** for the compound: $A_m B_n$. It is clear that the hypothetical existence of atoms of each element gives a model that accounts for the integral relationships between combining weights of elements in compounds. However, it is a rather daring hypothesis, in spite of the fact that our society bandies about the word "atom" rather indiscriminately. Since our whole understanding of chemistry rests on understanding the nature of atoms, we shall spend a considerable portion of the next part (Chapters 3, 4, 5, and 6) in developing the experimental background for the model using atoms and molecules as the ultimate chemical structure of matter. At this point we must operate on faith, since the whole process of examining the stoichiometry of chemical compounds and chemical reactions rests on the counting of atoms and molecules.

1-6 Formulas and Atomic Weights

The laws of stoichiometry suggest that molecules are formed by combining integral numbers of atoms, if we provisionally accept the atomic model, and our initial instinct is to ask what these numbers are for a given compound. How do we decide what m and n are in the formula $A_m B_n$? Unfortunately, analysis alone will not tell us. If we look at the laws of stoichiometry, which *are* based solely on analytical results, we see that they apply only to the ratio of weights in two compounds related because they contain similar atoms. They have nothing at all to say about the numbers of combining units — the numbers of atoms — in a single compound. The reason analysis does not give us molecular formulas directly is that the weight percentages of the elements present are influenced both by the formula (the ratios of the relative numbers of atoms in the molecule) and by the weight of each individual atom. We can establish a formula only if we have **atomic weights** available to us. Of course, if we know a formula we can extract relative atomic weights from the analysis, but that's just where the problem lies. To establish formulas we must look to other means of establishing atomic weights.

Relative Atomic Weights

The first thing we must realize in setting up a scale of atomic weights is that atoms are too small to establish an absolute atomic weight based on the mass of a single atom. Through modern instrumentation such weights or masses are available, but they are far too small to use in the practical manipulation of chemicals. Instead, we resort to the establishment of relative atomic weights, with the element having the smallest observed combining weight being arbitrarily assigned a **gram-atomic weight** of 1 and other gram-atomic weights being scaled up from that value. Hydrogen has the smallest characteristic combining weight; no compound of hydrogen with any other element has more than 25% H by weight. Accordingly, hydrogen was originally assigned a gram-atomic weight of 1; we shall see shortly that this has been modified slightly by modern definitions. This still does not solve the problem completely, since in order to get other gram-atomic weights in compounds of other elements with hydrogen a formula is necessary to get the ratio of numbers of atoms. So even agreeing to accept relative atomic weights does not solve the problem.

Atomic Weights for Gaseous Compounds

The earliest progress toward experimental establishment of gram-atomic weights was made by the measurement of gas densities—the weight of gas present in 1 l volume at standard atmospheric pressure. Joseph Louis Gay-Lussac in 1808 and Amadeo Avogadro in 1811 showed that when gases react to give chemically different products, the ratios of volumes that react and are formed (at a constant pressure and temperature) bear the same integral relationship as the ratios of combining weights. This strongly suggests what is called Avogadro's hypothesis: equal volumes of gas at the same pressure and temperature contain the same number of molecules. However, even this did not remove the possibility that molecules of gaseous elements might be polyatomic and thereby cause uncertainty in ascertaining atomic weights. If 1 l of oxygen gas weighs 16 times as much as 1 l of hydrogen gas, is it because the oxygen atom is 16 times as heavy as the hydrogen atom or because, say, the oxygen atom is 8 times as heavy and there are two oxygen atoms per oxygen molecule but only one hydrogen atom per hydrogen molecule? Stanislao Cannizzaro, in 1860, supplied the convincing argument in a statistical manner by pointing out that, given an assumed **gram-molecular weight** of 2 for hydrogen gas, other gram-molecular weights for gases follow immediately from the gas densities and Avogadro's hypothesis. If analytical results are obtained for a series of compounds of the same element, the number of grams of the element present in 1 gram-molecular weight of each compound must be equal either to the gram-atomic weight or to a multiple of it. Since all gaseous compounds of oxygen contain a multiple of 16 g of oxygen per gram-molecular weight of gas, he assumed the gram-atomic weight of oxygen to be 16 g/g-at. wt. No larger number would fit the data, and a smaller number, such as 8 g/g-at. wt., would require the additional postulate that an even number of oxygen atoms are combined in every oxygen compound. Following this pattern of reasoning, he was able to establish gram-atomic weights consistent with his assumption for all the elements that form series of volatile compounds—generally, the nonmetallic elements.

Atomic Weights for Solid Elements

For the metallic elements and some of the other solid elements, Pierre Louis Dulong and Alexis Petit in 1819 were able to remove uncertainty as to whether a multiple or submultiple might be involved in a given atomic weight. They observed that the heat capacity—the amount of heat energy necessary to raise the temperature of the solid by one degree—of solid elements was the same for 1 gram-atomic weight of each of many elements studied, about 6 cal/g-at. wt.-deg (within about 10%). As a result of early analyses, many elements had been assigned rather speculative atomic weights based on supposed formulas of their compounds. The principal criterion for selection of a formula was simplicity, but similarities with other elements were also considered. Not surprisingly, many of the resulting atomic weights were in error, because simplicity can be carried too far, but many were correct. For the resulting large number of elements whose atomic weights were correct by modern standards, the figure of 6 cal/g-at. wt.-deg held quite generally, with many metals in particular having this value. Elements for which this was not true seemed to fall into line if their supposed atomic weights were multiplied or divided by an integer. This observation did not really determine atomic weights, since there is a substantial variation around the rough value of 6 cal/g-at. wt.-deg, but it removed uncertainty in interpreting analytical results since both the formula and the accurate atomic weight could be established.

For instance, suppose we know from Cannizzaro's work that the atomic weight of oxygen is 16, and we find that the oxide of lead known as litharge has the composition 7.15% O, 92.85% Pb. What is the atomic weight of lead and the formula of the compound? If the formula is Pb_2O, the two lead atoms present must weigh 92.85/7.15 times as much as the one oxygen atom:

$$2 \text{ g-at. wt. Pb} = \frac{92.85}{7.15} \times 16 = 207.2 \text{ g}$$

$$\text{g-at. wt. Pb} = \tfrac{1}{2}(207.2) = 103.6 \text{ g/g-at. wt.}$$

On the other hand, if the formula is PbO, then only one lead atom is present for one oxygen:

$$1 \text{ g-at. wt. Pb} = \frac{92.85}{7.15} \times 16 = 207.2 \text{ g/g-at. wt.}$$

If the formula is Pb_2O_3 or PbO_2,

$$2 \text{ g-at. wt. Pb} = \frac{92.85}{7.15} \times 3 \times 16 = 621.6$$

$$\text{g-at. wt. Pb} = \tfrac{1}{2} \times 621.6 = 310.8 \text{ g/g-at. wt.}$$

$$1 \text{ g-at. wt. Pb} = \frac{92.85}{7.15} \times 2 \times 16 = 414.4 \text{ g/g-at. wt.}$$

Dulong and Petit's observation permits us to make a rough estimate of the gram-atomic weight of lead from the heat capacity of lead metal: 0.0306 cal/deg/g. Their rule indicates that

$$\text{g/g-at. wt. Pb} \cong \frac{6 \text{ cal/deg/g-at. wt.}}{0.0306 \text{ cal/deg/g}}$$

$$\cong 196 \text{ g/g-at. wt.}$$

This allows us to say from the analytical data that the gram-atomic weight of lead is 207.2 g/g-at. wt., and in an obvious manner allows us to specify the formula of the oxide as PbO.

It can be seen that the establishment of atomic weights rests on assumptions we make to have our atomic model agree with experimental results and still be as simple as possible. Cannizzaro assumed that the gram-atomic weight of an element was the largest common submultiple of the combining weights of the element in compounds. Dulong and Petit assumed that all solid elements should have the same heat capacity per gram-atomic weight. These are assumptions, features of a model that we construct—they are not laws. But they help us to understand the broad sweep of chemical combination by providing unifying concepts in a reasonably simple model. We therefore think they are useful, important, and fundamental concepts of chemistry.

Very accurate relative gram-atomic weights can be established for gaseous elements by working at low pressures, for reasons we shall discuss in Chapter 3. With these, other atomic weights can be established very accurately by careful analytical determination of combining weights, as we did for the example of PbO. T. W. Richards, the first American to receive the Nobel prize in chemistry (in 1914), was cited for a series of beautifully precise determinations of this sort.

Modern Determinations of Atomic Weight

Since Richards' time, atomic weights have been established using a **mass spectrometer**, which places a specific electric charge on each atom and measures its mass by measuring the strength of its interaction, while in motion, with an electric or magnetic field. Such an instrument is shown schematically in Fig. 5-20. The most refined of these instruments can measure relative masses with an accuracy of one part per million, which corresponds to an error of about one in the fourth decimal place of most gram-atomic weights. It is curious but true that this accuracy is in most cases wasted, in that most gram-atomic weights are quoted only to two or three decimal places. The reason for this is that most elements exist as mixtures of **isotopes**, meaning chemically identical atoms with slightly different masses. For instance, there is a naturally occurring isotope of hydrogen, **deuterium**, with a mass of 2.00 units, in addition to the one with a mass of 1.00 unit. The mixture of these two that occurs in nature is much richer in the ^{1}H isotope than in the ^{2}H (the superscripts indicate the masses), and the resulting mixture has an average atomic weight of 1.008 unit. Since the isotopic abundances of an element can differ slightly depending on the origin of the natural deposit of the element (from different radioactive transformations, for instance), a completely uniform gram-atomic weight cannot be quoted, even though a very accurate "local" atomic weight can be measured. For instance, Sicilian sulfur has a different atomic weight than Texas sulfur. For most elements, the quoted atomic weight (as given inside the front cover) is limited by natural irreproducibility, not measurement error.

The Modern Basis for Atomic Weights

We are now in a position to discuss the atomic weight standard. As all the atomic weights are established only with reference to a standard, its definition is a matter of some importance. Since both ^{1}H and ^{2}H exist, the natural mixture of hydrogen is not really ideal. For many years chemists used the natural mixture of oxygen isotopes in the atmosphere as a standard, defined exactly at 16 g/g-at. wt. On the other hand, physicists used the ^{16}O isotope as a standard, defined exactly at 16 g/g-at. wt. Since there are other naturally occurring isotopes of oxygen, the two scales were not the same. Agreement was reached only by moving to a different element; the ^{12}C isotope of carbon is now defined as the standard, at exactly 12 g/g-at. wt. This standard was chosen because of the ease with which mass spectrometry can be performed on carbon compounds. Taking the other carbon isotopes into account, this makes the gram-atomic weight of carbon 12.0111 g. The other atomic weights inside the front cover are scaled from this value.

From the establishment of accurate atomic weights we can obtain correct molecular formulas, as the PbO example has shown. We need to look at some examples of how this particular calculation is done, but before we do let us consider the physical significance of the gram-atomic weight.

1-7 Gram-Atomic Weights, Gram-Molecular Weights, and the Mole

Following Cannizzaro in a "nonatomic" definition, we say that the atomic weight of an element is the smallest combining weight of that element observed in any of its gaseous compounds that are equivalent in pressure, volume, and temperature to 2 g of hydrogen gas. Of course, this quantity represents many atoms (or perhaps molecules).

Avogadro's Number

What is the significance of the gram-atomic weight in terms of the atomic model? Compounds of hydrogen that contain only one hydrogen atom per molecule must give the smallest combining weight for hydrogen, since they involve the smallest total mass of hydrogen. Similarly, compounds of any other element that contain only one atom of the element per molecule give the smallest observed combining weights. In other words, the gram-atomic weight for any element is the weight of a specified number of atoms of that element, and *the number is the same from one element to another.* In terms of our modern definition, a gram-atomic weight of any element contains exactly the same number of atoms of that element as there are carbon atoms in exactly 12 g of ^{12}C. This number is called **Avogadro's number**; it is difficult to determine experimentally because it is so large, but in succeeding chapters we shall discuss ways in which it is established. The modern value for Avogadro's number, N, is $N = 6.02217 \times 10^{23}$ atoms/g-at. wt.

The Mole

The idea of a constant number of atoms in a gram-atomic weight is obviously useful in establishing the stoichiometry of compounds. If a specified number of atoms combine to form a molecule, then the weights of the atoms add to give the weight of the molecule, and the gram-atomic weights add in the same way (since they are related to the weights of the atoms themselves by a constant ratio, Avogadro's number) to give a gram-molecular weight, which contains Avogadro's number of molecules. In fact, the number is useful even for some particles which are not atoms or molecules; if 1 gram-atomic weight of silver nitrate ($AgNO_3$) in solution is plated out as silver metal by passage of an electric current, the silver has reacted with Avogadro's number of electrons. The quantity of anything—atoms, molecules, electrons, or any other microscopic particle of interest to us—that contains Avogadro's number of particles is called a **mole**. One gram-atomic weight of an element is thus a mole; so is one gram-molecular weight of a compound, and we can even speak of a mole of electrons in an electrochemical reaction. What the mole does is transform the number relationships of reacting atoms or molecules to macroscopic (large-scale) quantities that are convenient to handle. For instance, if we write the equation for a chemical reaction,

$$Xe + 3F_2 \rightarrow XeF_6$$

we could mean that 1 xenon atom is reacting with 3 fluorine molecules to give 1 molecule of xenon hexafluoride. But this is not what the chemist normally means. Since the scale-up factor of Avogadro's number applies to all the species in the reaction, whether atom or molecule, what the chemist means when he writes this reaction is that 1 mole of xenon atoms will react with 3 moles of fluorine molecules to produce 1 mole of xenon hexafluoride molecules. The reaction can be scaled up or down, but the reacting ratios remain the same: 3 molecules to 1 atom, 3 moles to 1 mole. This convenience is the reason for defining the mole as we do. Students sometimes complain that if the mole is a standard unit, it ought to have a standard weight; but the mole is *not* a weight unit. Rather, it is a number unit—just as we expect 6 cannonballs to weigh more than 6 corks, we should not be surprised to find that a mole of platinum weighs more than a mole of ethyl alcohol. In this comparison, the word "mole" carries the same sort of numerical connotation that "6" does.

Stoichiometric Calculations Using the Mole

Using the concept of the mole we can approach the problem of calculating molecular formulas from analytical data and atomic weights. We shall use

Gram-Atomic Weights, Gram-Molecular Weights, and the Mole | 27

atomic weights to convert the weight of each element present to the number of moles of each element present; then, since the ratio of the number of moles of each element is the same as the ratio of the number of atoms (because the mole is a number multiple) the mole ratio will give us the atom ratio in the molecular formula.

Consider the two copper chlorides we used in an earlier example. One was analyzed as 64.2% Cu, 35.8% Cl. How can we convert these figures into a stoichiometric formula (or **empirical formula**)?

If we had a 100-g sample of the compound, 64.2 g of it would be Cu. Since the atomic weight of copper is 63.54 g/mole, this amount of the compound would contain

$$\frac{64.2 \text{ g Cu}/100\text{-g sample}}{63.54 \text{ g Cu/mole Cu}} = 1.01 \text{ moles Cu}/100\text{-g sample}$$

Similarly, the 100-g sample would contain 35.8 g of Cl, and since the atomic weight of Cl is 35.46 g/mole the sample would contain

$$\frac{35.8 \text{ g Cl}/100\text{-g sample}}{35.46 \text{ g Cl/mole Cl}} = 1.01 \text{ moles Cl}/100\text{-g sample}$$

Now, since the ratio of these combining numbers of moles is clearly $1:1$, the ratio of atoms is $1:1$, and we write the empirical formula as CuCl.

Going on to the second chloride, we found that it analyzed as 47.3% Cu and 52.7% Cl; in a 100-g sample, 47.3 g would be Cu, and converting to numbers of moles we have

$$\frac{47.3 \text{ g Cu}/100\text{-g sample}}{63.54 \text{ g Cu/mole Cu}} = 0.744 \text{ mole Cu}/100\text{-g sample}$$

The same sample would contain 52.7 g of Cl, which can be converted similarly:

$$\frac{52.7 \text{ g Cl}/100\text{-g sample}}{35.46 \text{ g Cl/mole Cl}} = 1.487 \text{ moles Cl}/100\text{-g sample}$$

The ratio of combining numbers of moles in the arbitrarily chosen 100-g sample is

$$\frac{1.487 \text{ mole Cl}/100\text{-g sample}}{0.744 \text{ mole Cu}/100\text{-g sample}} = 2.00 \frac{\text{moles Cl}}{\text{mole Cu}}$$

so the ratio of combining numbers of atoms must also be 2.00, and we conclude that the empirical formula of this chloride is $CuCl_2$.

Compounds with Several Elements

We need not restrict ourselves to binary compounds in calculating the empirical formula of a compound from its analyses. In 1968, Crouch, Fowles, and Walton reported the synthesis of a compound which had the analysis 41.4% C,

4.1% H, 31.5% Cl, 8.1% N, and 14.5% V (vanadium). What was its empirical formula?

Imagine that we have a 100-g sample and can convert the appropriate weights of each element to the number of moles of each element:

$$\text{C:} \quad \frac{41.4 \text{ g C}}{12.01 \text{ g C/mole C}} = 3.45 \text{ moles C}$$

$$\text{H:} \quad \frac{4.1 \text{ g H}}{1.008 \text{ g H/mole H}} = 4.07 \text{ moles H}$$

$$\text{Cl:} \quad \frac{31.5 \text{ g Cl}}{35.45 \text{ g Cl/mole Cl}} = 0.885 \text{ mole Cl}$$

$$\text{N:} \quad \frac{8.1 \text{ g N}}{14.01 \text{ g N/mole N}} = 0.578 \text{ mole N}$$

$$\text{V:} \quad \frac{14.5 \text{ g V}}{50.94 \text{ g V/mole V}} = 0.284 \text{ mole V}$$

By convention we write empirical formulas with integral numbers of units of each element, since the numbers of atoms combined in any molecule are integers. Accordingly, in this case we divide each of these numbers representing moles of element per 100-g sample by the smallest number (0.284 in this case) to get the number of moles of each element per mole of the least abundant:

$$\text{C:} \quad \frac{3.45 \text{ moles C}}{0.284 \text{ mole V}} = 12.13 \frac{\text{moles C}}{\text{mole V}}$$

$$\text{H:} \quad \frac{4.07 \text{ moles H}}{0.284 \text{ mole V}} = 14.31 \frac{\text{moles H}}{\text{mole V}}$$

$$\text{Cl:} \quad \frac{0.885 \text{ mole Cl}}{0.284 \text{ mole V}} = 3.12 \frac{\text{moles Cl}}{\text{mole V}}$$

$$\text{N:} \quad \frac{0.578 \text{ mole N}}{0.284 \text{ mole V}} = 2.03 \frac{\text{moles N}}{\text{mole V}}$$

$$\text{V:} \quad \frac{0.284 \text{ mole V}}{0.284 \text{ mole V}} = 1.000 \frac{\text{mole V}}{\text{mole V}}$$

Crouch, Fowles, and Walton concluded that this corresponded to the formula $C_{12}H_{14}Cl_3N_2V$, which they had reason to suspect from their preparation of other similar compounds.

Experimental Accuracy and Rounding Error

There is a significant amount of "rounding-off error" in treating the experimental results above as integers, particularly for hydrogen. Because integral numbers of atoms combine in any molecule, we expect the numbers arrived at from analysis to be integers (unless the empirical formula contains more than one unit of each element). For the analysis above the rounding off is justified because the amount of rounding off is less than the experimental error in the analyses. That is, the hydrogen analysis is known only to an accuracy of about 1 part in 41. This corresponds to about 35 parts in 1431:

$$\frac{0.1}{4.1} = \frac{1}{41} = \frac{35}{1431} = \frac{0.35}{14.31}$$

Since the number of units of hydrogen is calculated to be 14.31, a possible error of 0.35 means the ideal value of 14.00 is within the accuracy of the analysis.

Beware of rounding off these ratios too freely, however. For example, aluminum carbide analyzes as 75.0% Al and 25.0% C; let us find its formula.

If we convert to moles of each element in a given sample,

$$\text{Al:} \quad \frac{75.0 \text{ g Al}}{26.98 \text{ g Al/mole Al}} = 2.78 \text{ mole Al}$$

$$\text{C:} \quad \frac{25.0 \text{ g C}}{12.01 \text{ g C/mole C}} = 2.08 \text{ mole C}$$

and then take the ratio of moles of each element per mole of the least abundant,

$$\frac{2.08 \text{ mole C}}{2.08 \text{ mole C}} = 1.000 \frac{\text{mole C}}{\text{mole C}} \qquad \frac{2.78 \text{ mole Al}}{2.08 \text{ mole C}} = 1.335 \frac{\text{mole Al}}{\text{mole C}}$$

we have a case in which the rounding off would seem to be similar to that in the previous case. Here the rounding off is *not* justified, however. To round the number 1.33 down to 1.00 implies an experimental error of 33 parts in 133; but the worst analysis (carbon) is given to 1 part in 250. So the experimental error of analysis is nowhere near large enough to justify such a rounding off (which would suggest a formula AlC). Instead, we must multiply *both* numbers of moles by an integer selected so as to make both become integers themselves. In this case, the integer 3 is needed:

$$1.000 \frac{\text{mole C}}{\text{mole C}} \times 3 \frac{\text{mole C}}{\text{mole compound}} = 3.000 \frac{\text{mole C}}{\text{mole compound}}$$

$$1.335 \frac{\text{mole Al}}{\text{mole C}} \times 3 \frac{\text{mole C}}{\text{mole compound}} = 4.004 \frac{\text{mole Al}}{\text{mole compound}}$$

Accordingly, we write the formula of aluminum carbide as Al_4C_3, not AlC. A little thought will reveal that for complicated compounds, the accuracy of the analysis is often not good enough to distinguish between two closely related formulas.

Empirical Formulas and Molecular Formulas

There is one more aspect of establishing molecular formulas through use of the mole property that deserves comment. In a number of cases the empirical formula obtained by the method above proves to be only a submultiple of the true molecular formula. For instance, methane has the empirical formula CH_4 and the true molecular formula CH_4, but the similar compound ethane, with an empirical formula CH_3, has a true molecular formula C_2H_6—two units of the empirical formula. How is this recognized when it occurs?

We must realize first that the analysis cannot distinguish between CH_3 and C_2H_6, or between multiples of any empirical formula. Our calculations thus far give only the *ratio* of atoms in a given molecule, and of course this is unaffected by multiplying both numbers by a constant. Rather, we must follow the lead of Gay-Lussac, Avogadro, Dulong, and Petit in seeking regularities in the *molar* properties of compounds. A number of physical properties, such as gas density, the vapor pressure over a solution, and the melting point of a mixed solid, depend not on the *mass* of material present but on the number of *moles* of material present. If we know the number of moles present and the weight of compound present, we can immediately get the molecular weight. This will either match the empirical formula's total of atomic weights or will be some multiple of it, in which case we obtain the true molecular formula by multiplying accordingly. For example, benzene has the empirical formula CH, which would have a molecular weight of $12.01 + 1.01 = 13.02$. Yet its experimental molecular weight is about 78 g/mole. Since $78/13 = 6.0$, we write the true molecular formula of benzene as C_6H_6, and its true molecular weight as 78.11 g/mole.

In succeeding chapters, particularly Chapter 7, we shall look at the various methods for establishing molecular weights from molar properties. Here we need only note that the number aspect of the mole is basic to all these methods and is necessary to establish the true proportions of the molecule.

1-8 The Stoichiometry of Reactions

Stoichiometry is the measurement of the proportions of elements or compounds in a system in which we are interested. It is divided naturally into the stoichiometry of homogeneous systems (the proportions of compounds and solutions) and that of heterogeneous systems. We have been describing the stoichiometry of compounds, and in Chapter 7 we shall discuss the stoichiometry of solutions. Here we look at the stoichiometry of heterogeneous systems, in particular as it applies to chemical reactions.

Our interest in heterogeneous systems will be limited to those that have chemically different phases (as opposed to those that have physically different phases). We shall also limit ourselves to those systems that are capable of changing their chemical composition; a system that is changing its chemical composition is said to be undergoing a chemical reaction. Thus we are not interested in the heterogeneous system ice–water, since the phases do not differ chemically, nor in the heterogeneous system oil–water, since it cannot react chemically to change its composition.

Writing Reaction Equations

Let us take a heterogeneous system that can undergo chemical reaction and look at its stoichiometry. If, for instance, we take elemental boron, which is a very high-melting solid, and pass elemental chlorine gas over it we observe no change. But if we heat the system to a temperature of several hundred degrees centigrade, we observe the formation of a new gaseous substance, and the solid boron is gradually used up. Elemental analysis indicates that the new material is a compound with the empirical formula BCl_3, boron trichloride. We represent the transformation by a chemical equation:

$$B + Cl_2 = BCl_3$$

Here boron and chlorine are said to be **reactants**, and boron trichloride is a **product**. The equality symbol is often replaced by an arrow to represent the predominant direction of transformation, since many reactions will go only in one direction:

$$B + Cl_2 \rightarrow BCl_3$$

At this point we have still not said anything about the actual stoichiometry, the numerical proportions, of the chemical reaction (as represented by the chemical equation). Since we have fortified ourselves with the concept of the mole as a number-of-atoms or number-of-molecules unit, however, the stoichiometry of the reaction is fairly obvious. If the product molecule contains 3 chlorine atoms for each boron atom, then the gaseous product as a whole will contain 3 moles of chlorine atoms for each mole of boron atoms. A mole of product will contain 3 moles of chlorine atoms and 1 mole of boron atoms if BCl_3 is the true molecular formula. So we must have provided 1 mole of boron atoms and 3 moles of chlorine atoms as reactants; since there are 2 chlorine atoms in a chlorine molecule (as we could determine, say, from its gas density), we need only provide $\frac{3}{2}$ moles of chlorine molecules to give 3 moles of chlorine atoms:

$$B + \tfrac{3}{2}Cl_2 \rightarrow BCl_3$$

This is an acceptable form for the equation, but in most cases we prefer to have integral units of each reactant and product; we can achieve this by multiplying the whole equation through by 2:

$$2B + 3Cl_2 \rightarrow 2BCl_3$$

For either of these cases we say that we have **balanced** the equation, by choosing coefficients such that an equal number of mole units of each element appears on each side of the arrow or equality sign. The basic principle we are operating under is the **law of the conservation of mass**:

> In a chemical reaction (as opposed to a nuclear reaction), the total mass of each element present does not change, so the total number of atoms of each element does not change.

We find this to be very uniformly true for a wide variety of chemical reactions, and we use this law as a basis for expressing the stoichiometry of the reaction by coefficients in the equation representing the reaction. So whenever we are interested in the stoichiometry of a reaction, we first identify the formulas of each of the reactants and products by chemical analysis and some sort of physical determination of the molecular weight. Then we write a chemical equation for the reaction and balance it by providing an equal number of moles of each element on each side of the equation.

Some chemical reactions can be made to proceed in either direction, depending on the conditions under which the reaction is carried out. For instance, at about 300°C metallic mercury reacts with atmospheric oxygen to form mercury(II)oxide, HgO. We represent this observed reaction by an (unbalanced) equation:

$$Hg + O_2 \rightarrow HgO$$

However, if we raise the temperature to 400°C or above we can see the HgO decompose to give mercury metal and oxygen. Changing the conditions has caused the reactants in the equation to become more favorable — more stable — than the product. Under the new conditions we could represent the observed reaction by the equation

$$HgO \rightarrow Hg + O_2$$

Since the reaction can be made to go in either direction, we use the equal sign or a pair of arrows ($\rightleftharpoons$) to show the change instead of a single arrow:

$$Hg + O_2 = HgO \quad \text{or} \quad Hg + O_2 \rightleftharpoons HgO$$

We have not balanced this equation, since the left side contains 2 moles of oxygen while the right side has only 1 mole. So we must take 2 moles of the compound HgO:

$$Hg + O_2 = 2HgO$$

To balance this completely we need only provide 2 moles of Hg on the left side to balance the 2 moles on the right:

$$2Hg + O_2 = 2HgO$$

As before, the equal sign indicates that the reaction can be made to go in either direction. We shall have occasion in Chapters 9 and 10 to examine the idea that such a reaction, viewed on a microscopic scale, actually does go in both directions at the same time. The result of these competing reactions, forward and backward, is a **chemical equilibrium** with the more stable species in the system predominating (as, for instance, HgO at 300°C) but with the actual large-scale composition of the system being an average of the possibilities. The law of conservation of mass and the counting of moles to balance the equation in accordance with this law are, of course, quite independent of whether the reaction is represented as an equilibrium or proceeds in only one direction.

Most reactions are not as simple as the ones described above, but the equations representing them can be balanced by the same methods. If we take C_8H_{18}, the formula for octane, as an average formula for gasoline, then we can represent the burning of gasoline in an engine as

$$C_8H_{18} + O_2 \rightarrow CO_2 + H_2O$$

To balance this equation we first get the same numbers of moles of carbon and hydrogen on each side of the equation:

$$C_8H_{18} + O_2 \rightarrow 8CO_2 + 9H_2O$$

But this leaves an odd number of moles of oxygen atoms on the right while O_2 can give only an even number, so we must multiply the other coefficients by 2, including the coefficient of C_8H_{18}:

$$2C_8H_{18} + O_2 \rightarrow 16CO_2 + 18H_2O$$

Now we balance the moles of oxygen atoms; there are $(16 \times 2) + 18 = 50$ on the right, so we must provide 50 on the left to yield that number on the right:

$$2C_8H_{18} + 25O_2 \rightarrow 16CO_2 + 18H_2O$$

Sometimes we might be interested in, for instance, the energy yield for each mole of C_8H_{18} which burns, and in this case it would be appropriate to write the equation for that amount using fractional coefficients:

$$C_8H_{18} + \frac{25}{2}O_2 \rightarrow 8CO_2 + 9H_2O$$

Ionic Equations

Many reactions are carried out in aqueous solution. For reasons we shall discuss in Chapter 13, many common compounds separate into positively charged and negatively charged fragments when they dissolve in water:

$$NaCl_{(s)} \xrightarrow{H_2O} Na^+_{(aq)} + Cl^-_{(aq)}$$

Here (s) represents solid NaCl, while (aq) indicates that the separate charged fragments or *ions* are in aqueous solution. Many of the reactions that can occur in such solutions of ions involve only one of the ionic species. For instance, we can **precipitate** the chloride ion from the above solution as solid silver chloride (AgCl) by adding another solution of silver nitrate ($AgNO_3$):

$$AgNO_{3(s)} \longrightarrow Ag^+_{(aq)} + NO^-_{3(aq)}$$

$$Ag^+_{(aq)} + Cl^-_{(aq)} \rightarrow AgCl_{(s)}$$

The second reaction equation is the best representation of what is happening in the system; it is an **ionic equation**. It is also possible to write a **molecular equation** for this reaction by ignoring the fact that separation of ions has occurred in solution, but this leads to misinterpretation of the real chemical process and should be avoided:

$$NaCl + AgNO_3 \xrightarrow{H_2O} AgCl + NaNO_3$$

It is important to be able to recognize when a given reaction should be written in ionic form. To this end, we need to know which ionic species are stable in water solution, and Tables 1-4 and 1-5 in Section 1-9 provide this information. If salts involving any of the ions in Tables 1-4 and 1-5 are involved in a reaction in aqueous solution, the reaction should normally be represented by an ionic equation. For instance, if aqueous solutions of hydrochloric acid and sodium carbonate are mixed, carbon dioxide gas is evolved. The reaction does not involve the sodium ion or the chloride ion; we write

$$2H_3O^+_{(aq)} + CO^{2-}_{3(aq)} = CO_{2(g)} + 3H_2O_{(l)}$$

Note that reactants and products of the reaction need not all be ions for the ionic form of the equation to be appropriate. To take another example, if we mix aqueous solutions of lead(II) nitrate and potassium chromate, solid lead(II) chromate is formed:

$$Pb^{2+}_{(aq)} + CrO^{2-}_{4(aq)} = PbCrO_{4(s)}$$

The nitrate ion and potassium ion do not appear in the equation at all, since they are essentially unaffected by the reaction.

Reaction Stoichiometry Calculations

An appropriate way to close this section is to work out some problems that are typical of the way in which a chemist uses the stoichiometry of chemical reactions. In each of the following cases, we shall see that the fundamental process is the counting of moles, for comparison of products with reactants.

Suppose n-propyl iodide, C_3H_7I, is allowed to react to produce propene, C_3H_6, according to the following reaction equation:

$$C_3H_7I + OH^- \rightarrow C_3H_6 + H_2O + I^-$$

If 17.0 g of n-propyl iodide are used, how much propene should result if the reaction is complete? If 1.56 g of propene are actually formed, what is the **percent yield** relative to the expected amount?

We attack the problem by converting everything to mole amounts, since the balanced equation deals with combinations of moles, not combinations of grams. In order to calculate the number of moles of *n*-propyl iodide present, we need the molecular weight (MW) of that compound, calculated from the atomic weights (AW) of the elements:

$$MW(C_3H_7I) = [3 \times AW(C)] + [7 \times AW(H)] + [1 \times AW(I)]$$
$$= (3 \times 12.01) + (7 \times 1.008) + (126.90)$$
$$= 169.99 \text{ g/mole}$$

This means that the number of moles of *n*-propyl iodide reactant initially present was

$$\text{number of moles} = \frac{17.0 \text{ g}}{169.99 \text{ g/mole}} = 0.100 \text{ mole}$$

Now, the stoichiometry of the reaction equation tells us that for every mole of *n*-propyl iodide reactant 1 mole of propene product will form; so 0.100 mole of *n*-propyl iodide should lead to 0.100 mole of propene. This amount should weigh 0.100 as much as the molecular weight of propene:

$$\text{wt. propene formed} = 0.100 \times MW(C_3H_6)$$
$$= 0.100\{[3 \times AW(C)] + [6 \times AW(H)]\}$$
$$= 0.100[(3 \times 12.01) + (6 \times 1.008)]$$
$$= 0.100(42.08)$$
$$= 4.208 \text{ g}$$

This is the answer to the first part of the problem; the answer to the second part simply involves taking the percent ratio of the experimental yield to the theoretical yield:

$$\text{percent yield} = \frac{\text{experimental yield}}{\text{theoretical yield}} \times 100\%$$
$$= \frac{1.56 \text{ g}}{4.208 \text{ g}} \times 100\%$$
$$= 37.1\%$$

Notice that the whole basis of the problem is the conversion of all the quantities to a mole basis; the reaction equation, when properly balanced, counts moles for us, and this is precisely how we use it in stoichiometric problems.

A problem that looks rather different, but actually amounts to doing exactly the same thing, is as follows. A mixture of Mn_2O_3 and MnO [manganese(III) oxide and manganese(II) oxide, respectively] is heated in a stream of hydrogen gas, which reacts selectively with the Mn_2O_3:

$$Mn_2O_{3(s)} + H_{2(g)} \rightarrow 2MnO_{(s)} + H_2O_{(g)}$$

If 14.85 g of the mixture is used, the water that results weighs 1.043 g. What is the weight percent Mn_2O_3 in the original mixture?

The balanced equation tells us that for every mole of water that results, 1 mole of Mn_2O_3 must have reacted. If we know the number of moles of water, that will also be the number of moles of Mn_2O_3 present, and from the number of moles we can find the weight and the weight percent:

$$\text{number of moles } Mn_2O_3 = \text{number of moles } H_2O = \frac{1.043 \text{ g}}{MW(H_2O)}$$

$$= \frac{1.043 \text{ g}}{18.02 \text{ g/mole}} = 0.0578 \text{ mole}$$

$$0.0578 \text{ mole} \times MW(Mn_2O_3) = \text{wt. } Mn_2O_3$$

$$MW = [2 \times AW(Mn)] + [3 \times AW(O)] = 157.88 \text{ g/mole}$$

$$\text{wt. } Mn_2O_3 = 0.0578 \text{ mole} \times 157.88 \text{ g/mole}$$

$$= 9.12 \text{ g } Mn_2O_3$$

$$\text{weight \% } Mn_2O_3 = \frac{\text{wt. } Mn_2O_3 \times 100\%}{\text{total sample wt.}} = \frac{9.12}{14.85} \times 100\% = 61.5 \text{ wt. \%}$$

Again we have seen that the number concept of the mole translates numbers of moles of product directly into the number of moles of reactant that must have been present. Stoichiometry problems involving chemical reactions necessarily rest on a balanced reaction equation, whose only function is to count moles of reactants and products.

1-9 Aqueous Ions and Nomenclature

In Section 1-8 we noted that many reactions that occur in water solution (and even in some other solvents) involve electrically charged ions. For reasons we shall explore in Chapter 13, many inorganic compounds dissociate into ions when they dissolve in water. Since a reaction equation should be written to include only the species that are actually reacting, it is important to be able to predict from the formula of a compound which ions it may yield in aqueous solution. Tables 1-4 and 1-5 indicate the more common aqueous ions with their names and the formulas of the corresponding acids.

Table 1-4
Positively Charged Species[a]

Ion	Name	Ion	Name
Li^+	lithium ion	Ni^{2+}	nickel(II)
Na^+	sodium	Cu^{2+}	copper(II)
K^+	potassium	Zn^{2+}	zinc
Rb^+	rubidium	Cd^{2+}	cadmium
Cs^+	cesium	Hg^{2+}	mercury(II)
Ag^+	silver	Hg_2^{2+}	mercury(I)
NH_4^+	ammonium	Sn^{2+}	tin(II)
H_3O^+	hydronium (or oxonium)	Pb^{2+}	lead(II)
		Al^{3+}	aluminum
Be^{2+}	beryllium	Sc^{3+}	scandium
Mg^{2+}	magnesium	(rare earth)$^{3+}$	(elements 57–71)
Ca^{2+}	calcium	Ga^{3+}	gallium
Sr^{2+}	strontium	Cr^{3+}	chromium(III)
Ba^{2+}	barium	Fe^{3+}	iron(III)
Mn^{2+}	manganese(II)	Ce^{4+}	cerium(IV)
Fe^{2+}	iron(II)	Sn^{4+}	tin(IV)
Co^{2+}	cobalt(II)	Th^{4+}	thorium

[a] In aqueous solution all ions include hydrated water molecules.

Table 1-5
Negatively Charged Species[a]

Symbol	Ion Name	Symbol	Corresponding Acid Name
F^-	fluoride	HF	hydrofluoric acid
Cl^-	chloride	HCl	hydrochloric acid
Br^-	bromide	HBr	hydrobromic acid
I^-	iodide	HI	hydriodic acid
I_3^-	tri-iodide	HI_3	
ClO^-	hypochlorite	HClO	hypochlorous acid
ClO_2^-	chlorite	$HClO_2$	chlorous acid
ClO_3^-	chlorate	$HClO_3$	chloric acid
ClO_4^-	perchlorate	$HClO_4$	perchloric acid
BrO^-	hypobromite	HBrO	hypobromous acid
BrO_3^-	bromate	$HBrO_3$	bromic acid
IO_3^-	iodate	HIO_3	iodic acid
IO_4^-	periodate	HIO_4	periodic acid

[a] In aqueous solution all ions include hydrated water molecules.

Table 1-5 (continued)

Symbol	Ion Name	Symbol	Corresponding Acid Name
MnO_4^-	permanganate	$HMnO_4$	permanganic acid
OH^-	hydroxide	H_2O	water
HO_2^-	hydroperoxide	H_2O_2	hydrogen peroxide
OCN^-	cyanate	$HOCN$	cyanic acid
SH^-	hydrosulfide	H_2S	hydrosulfuric acid
HSO_3^-	hydrogen sulfite	H_2SO_3	sulfurous acid
SCN^-	thiocyanate	$HSCN$	thiocyanic acid
N_3^-	azide	HN_3	hydrazoic acid
NO_2^-	nitrite	HNO_2	nitrous acid
NO_3^-	nitrate	HNO_2	nitric acid
$H_2PO_4^-$	dihydrogen phosphate	H_3PO_4	phosphoric acid
$H_2AsO_4^-$	dihydrogen arsenate	H_3AsO_4	arsenic acid
HSO_3^-	hydrogen sulfite	H_2SO_3	sulfurous acid
HSO_4^-	hydrogen sulfate	H_2SO_4	sulfuric acid
CN^-	cyanide	HCN	hydrocyanic acid
HCO_3	hydrogen carbonate	H_2CO_3	carbonic acid
$C_2H_3O_2^-$	acetate	$HC_2H_3O_2$	acetic acid
BH_4^-	borohydride		(unstable)
$B(OH)_4^-$	borate	H_3BO_3	boric acid
BF_4^-	fluoborate	HBF_4	fluoboric acid
SiF_6^{2-}	fluosilicate	H_2SiF_6	fluosilicic acid
S^{2-}	sulfide	HS^-	hydrogen sulfide or hydrosulfide
SO_3^{2-}	sulfite	HSO_3^-	hydrogen sulfite
SO_4^{2-}	sulfate	HSO_4^-	hydrogen sulfate
$S_2O_3^{2-}$	thiosulfate	$HS_2O_3^-$	hydrogen thiosulfate
O_2^{2-}	peroxide	HO_2^-	hydroperoxide
SeO_4^{2-}	selenate	$HSeO_4^-$	hydrogen selenate
HPO_4^{2-}	hydrogen phosphate	$H_2PO_4^-$	dihydrogen phosphate
$HAsO_4^{2-}$	hydrogen arsenate	$H_2AsO_4^-$	dihydrogen arsenate
CO_3^{2-}	carbonate	HCO_3^-	hydrogen carbonate
$C_2O_4^{2-}$	oxalate	$H_2C_2O_4$	oxalic acid
CrO_4^{2-}	chromate		(unstable)
MoO_4^{2-}	molybdate		(unstable)
WO_4^{2-}	tungstate		(unstable)
PO_4^{3-}	phosphate	HPO_4^-	hydrogen phosphate
AsO_4^{3-}	arsenate	$HAsO_4^{2-}$	hydrogen arsenate
VO_4^{3-}	vanadate		(unstable)
SiO_4^{4-}	silicate		(unstable)

Aqueous Ions and Names of Compounds

Many compounds are named as if they formed ions even though they may be very unstable in aqueous solution so the charged fragments do not actually separate as aqueous ions. For instance, barium sulfate is quite insoluble in water but is still named as if it dissociated into the barium ion and the sulfate ion. This is not unreasonable since we could still think of the charged ions as existing in the solid compound. But compounds such as S_4N_4, tetrasulfur tetranitride, are also named in this way even though there is no experimental reason to believe that any significant charge exists on any of the atoms. In naming these compounds, it is generally true that if the compound is composed of a metal and a nonmetal, so that we can reasonably think of it as being composed of ions, the positive charge on the metal will be indicated by Roman numerals in parentheses, as in Table 1-3. If there is only one charge that an element adopts, the Roman numerals are omitted and the charge on the ion is understood to be the usual one; thus zinc always has a charge of 2+ and $ZnCl_2$ is called zinc chloride, not zinc(II) chloride. The positive ions in Table 1-4 are all shown with their normal charge if no Roman numerals are shown by the name. Using Tables 1-4 and 1-5 we can write

copper(II) chloride for $CuCl_2$
tin(IV) oxide for SnO_2
chromium(III) fluoride for CrF_3

but

potassium cyanide for KCN
calcium sulfate for $CaSO_4$
cadmium iodide for CdI_2

If the compound is composed of two nonmetals, prefixes are added to the names of each element to indicate the number of moles of that element in a mole of the compound, without suggesting a fictitious charge. The appropriate prefixes are

di- = 2
tri- = 3
tetra- = 4
penta- = 5
hexa- = 6
hepta- = 7
octa- = 8

For 1 mole of an element per mole of compound, no prefix is used. Note that, for nonmetal oxides involving 2 moles of nonmetal per mole of compound (in the empirical formula), the di- prefix is often omitted on the assumption that it may be inferred from the normally divalent nature of oxygen (see Table 1-5). Naming nonmetal compounds in this way, we have

>boron tribromide for BBr_3
>oxygen difluoride for OF_2
>phosphorus pentoxide for P_2O_5 (empirical formula)
>tetraphosphorus decoxide for P_4O_{10} (true molecular formula)

Here we have extended the Greek roots of the prefixes to provide "deca-" for the subscript 10; note also that we have dropped the terminal "a" on the prefixes "penta-" and "deca-" since the "o" in "oxide" would otherwise give a difficult diphthong.

Nonaqueous Ions

In this discussion we have mentioned the oxide ion several times although it does not appear in Table 1-5. The omission is deliberate, because the oxide ion is not stable in water solution; it reacts with water to give the hydroxide ion:

$$O^{2-} + H_2O = 2OH^-$$

There are several species that are usually regarded as ions in at least some of their compounds, but that are not stable in water for much the same reasons that the oxide ion is not. Table 1-6 gives the names of some of these ions. The reasons that some ions are stable in water and others are not will be explored in

Table 1-6
Ions Unstable in Aqueous Solution

Symbol	Ion Name	
O^{2-}	oxide	
N^{3-}	nitride	
NH_2^-	amide	
ClF_4^-	tetrafluorochlorate(III)	(and many similar ions)
OCH_3^-	methoxide	
PF_6^-	hexafluorophosphate(V)	
$NiCl_4^{2-}$	tetrachloronickelate(II)	(and many similar ions)
AlH_4^-	tetrahydridoaluminate	
XeO_6^{4-}	xenate(VIII) or perxenate	
XeF_8^{2-}	octafluoroxenate(VI)	
NO^+	nitrosonium	
NO_2^+	nitronium	
PCl_4^+	tetrachlorophosphorus(V)	

Chapters 10, 13, and 14; they are included here for the sake of completeness even though they do not appear as reactants or ultimate products in ionic reactions in water solution.

1-10 Stoichiometry, Measurement, and Interactions

All of the discussion thus far has emphasized the central nature of quantitative measurement in any science, and chemistry in particular. The stoichiometric relationships of compounds and reactions are only the first examples of this sort of measurement. In the discussion of gases and solids, in Chapters 3 and 4, many other quantitatively measured properties will be important. We shall discover that for a given system there are usually a number of properties that can be characterized quantitatively, and that these usually have an influence on each other. In other words, the physical properties of a system interact with each other. The ways in which they interact are our clearest guide to the nature of the system.

For instance, we shall see in Chapters 3 and 4 that increasing the pressure on a sample of gas decreases its volume sharply, while increasing the pressure on a sample of solid decreases its volume only very slightly. If we can understand the interaction between the two properties pressure and volume, in a quantitative sense, and build a mathematical model for it, we shall have gone a long way toward understanding the nature of a gas and the nature of a solid.

Although the physical properties seem simple enough to measure, so that the experimental facts should not be too hard to establish, the interactions between them seem difficult to grasp or describe. Here mathematics comes to our rescue. The interactions in which we are interested are meaningful associations between two (or more) numerical quantities. The mathematical concept of **function** has precisely this meaning, and throughout this book we shall have frequent recourse to functions to establish the interactions that characterize chemical systems. For this reason, even before we attack the problem of the behavior of gases and solids, it will be profitable to look at the simple mathematics of functions and the ways functions change.

Also, because we are dealing with very large numbers of atoms, it will be important for us to appreciate the mathematical tools for handling the statistics of large numbers of individual objects, and the techniques for predicting the most probable average of all these objects' behavior. Accordingly, Chapter 2 deals with the rudiments of two types of mathematics: functions and calculus on the one hand, and statistics on the other. Fortunately, only a very modest amount of mathematics is necessary for our purposes, so it should not prove too burdensome. But when we need it, we need it; and so it seems appropriate to develop the mathematics chemists use before trying to pursue the chemistry itself.

Study Problems

1. Oil wells in Texas are limited by law to a production of 50 barrels/day. Given that crude oil has a density of 0.705 g/cm³, how many wells' production would be required to fill a 300,000-ton tanker in 48 hr?
2. What is the wavelength range of visible light? What are the appropriate units for this distance measurement?
3. The element gallium has an unusually long temperature range in which it is a liquid; its melting point is 30°C and its boiling point is 2237°C. Suppose we wish to define a new temperature scale in which the melting point of gallium is defined as 0° Chaud (a properly Gallic expression) and the boiling point of gallium is defined as 1000 °Chaud. What would be the boiling point of water in degrees Chaud?
4. If a gallon of paint covers 385 ft², how thick is the coat of paint?
5. It has been established that the triviance of the standard laboratory nard is directly proportional to the nard's feebity in poods, but inversely proportional to the square of its frumiance in arbols. If the usual units of triviance are nils, what must be the units of the proportionality constant?
6. An international committee of distinguished scientists met to establish the standard snail's pace (P_{sn}). They observed that the laboratory snail covered 1.000 ft in 26 min 18 sec. What is P_{sn} in miles per hour and in cm/sec?
7. There are four chemical reactions commonly used to prepare arsenic trichloride. Equations for these are written below; balance each:

$$As_4 + Cl_2 \rightarrow AsCl_3$$

$$As_4O_6 + Cl_2 \rightarrow AsCl_3 + As_4O_{10}$$

$$As_4O_6 + NaCl + H_2SO_4 \rightarrow AsCl_3 + H_2O + NaHSO_4$$

$$As_4O_6 + S_2Cl_2 + Cl_2 \rightarrow AsCl_3 + SO_2$$

8. Molybdenum(IV) chloride formed a compound with dimethylsulfoxide [$(CH_3)_2SO$, molecular weight 78.0 g/mole]. Analysis of the compound showed that it contained 27.0% Mo and 20.0% Cl. Was all of the original chlorine still present in the new compound? Assuming that only one Mo is present per molecule of the new compound, what is its molecular weight?
9. A high-resolution mass spectrometer is used to establish precise atomic weights. Although ^{12}C is defined as having a mass of exactly 12 amu (atomic mass units) the measurement of the relative mass of ^{16}O does not involve measuring the ratio of masses of ^{12}C and ^{16}O. Instead, mass "doublets" are used, in which two species have very nearly the same mass but different proportions of elements. For instance, the

mass of ^{1}H can be obtained from the ratio of masses of nonane (C_9H_{20}) and naphthalene ($C_{10}H_8$), both of which are very near 128 amu; these two differ in having 12 H for 1 C. For a mass doublet in which the two species have masses M and $M+m$, the mass spectrometer should yield

$$\frac{M}{M+m} = \frac{V_{M+m}}{V_M}$$

where V is the electric voltage deflecting the flight of the species' ions in the instrument. Solve this expression for m, the mass difference. If this treatment gives a mass of 1.007825 amu for ^{1}H, and the ratio of voltages for the two species C_3H_8 (propane) and CO_2 (carbon dioxide) is

$$\frac{V_{C_3H_8}}{V_{CO_2}} = 1.001655$$

what is the mass difference between C_3H_8 and CO_2? What is the mass of ^{16}O?

10. Write chemical reaction equations and balance them for the following reactions:
    (a) Sulfur dichloride is allowed to react with sodium fluoride to give sulfur tetrafluoride, disulfur dichloride, and sodium chloride.
    (b) Potassium thiocyanate and potassium hydrogen sulfate yield thiocyanic acid and potassium sulfate.
    (c) Mercury(I) chromate decomposes on heating to give mercury metal, chromium trioxide, and oxygen gas (O_2).
    (d) Cobalt(II) chloride plus hydrogen fluoride gas yields cobalt(II) fluoride and hydrogen chloride.
11. Write a balanced equation for the reaction between FeS_2 and O_2 to give Fe_2O_3 and SO_2. If all the iron in the Fe_2O_3 is later converted to iron metal, how many tons of SO_2 gas will have been discharged into the atmosphere for each ton of iron produced? If this were converted quantitatively to sulfuric acid instead, how many tons of sulfuric acid would it yield?
12. In 1970 Roesky et al. [*Inorg. Chem.* **9**, 831 (1970)] reported that the reaction of cesium fluoride with thiophosphoryl fluoride, SPF_3, gave $CsPF_6$ and another compound which analyzed as follows: Cs, 49.8%; F, 14.3%; P, 12.0%; S, 24.0%. What was the empirical formula of the new compound?
13. The compound triethylphosphine, $(C_2H_5)_3P$, adds to some metal chlorides to give new compounds containing the triethylphosphine molecule as an electron donor to the metal atom. One such compound recently reported, made with chromium(III) chloride, analyzed as Cr, 13.0%; Cl, 26.6%; C, 36.7%; H, 7.7%; P by difference from 100% [Bennett et al., *J. Chem. Soc.*, 541 (1970)]. What is the empirical formula of the new compound? How many triethylphosphine donor molecules are there for each chromium atom? If the molecular weight of the compound in benzene solution is 800 ± 10 amu, what is the true molecular formula?
14. If a mixture of the minerals orpiment (As_4S_6) and realgar (As_4S_4) weighing 1.000 g is heated in a current of air it oxidizes:

$$As_4S_4 + O_2 \rightarrow As_4O_6 + SO_2$$

$$As_4S_6 + O_2 \rightarrow As_4O_6 + SO_2$$

Balance these equations. If the resulting solid As_4O_6 weighs 0.905 g, what was the original composition by weight of the mixture?

15. The mole is a number quantity, like a six-pack, but a very large one. We use it only for microscopic particles because otherwise it is inconveniently large. If a standard beer can is $2 \times 2 \times 5$ in, how deep would a neatly stacked mole of beer cans cover the continental United States? Assume a square array and vertical sides to the stack; the area of the United States is 3.6×10^6 mi^2.

Some Further Reading

Kieffer, W. F., *The Mole Concept in Chemistry*, New York: Van Nostrand Reinhold, 1962. This is a very easygoing, systematic treatment of a topic that looks deceptively simple but is actually difficult to place on a sound foundation.

Cahn, R. S., *An Introduction to Chemical Nomenclature*, London: Butterworths, 1968 (3rd ed.). A lot more than you need to know at this point about nomenclature, but well done and worth coming back to later on for organic nomenclature and co-ordination-compound nomenclature.

Sienko, M. J., *Stoichiometry and Structure*, New York: Benjamin, 1964. Strictly a problems book, mostly on topics introduced in this chapter. Lots of worked examples. You don't understand this material until you can put numbers to it.

Nash, L. K., *Stoichiometry*, Reading, Mass.: Addison-Wesley, 1966. Much like Kieffer's book; very lucid and readable.

Beynon, J. H., *Mass Spectroscopy and Applications to Organic Chemistry*, Amsterdam: Elsevier, 1960. This is very difficult, but Chapter 2 on the measurement of mass will give you some insight into how atomic weights are presently determined by instrumental methods.

Kemeny, J. G., *A Philosopher Looks at Science*, New York: Van Nostrand Reinhold, 1959. Very much off the subject of chemistry, but a good book to familiarize yourself with the way we shall try to ask questions.

2 | Mathematical Functions in Chemistry

Many kinds of measurements are possible in the chemical laboratory, but no matter what measurement is being made a single question presents itself: How can we interpret the significance of this measurement? To put it another way: What does this measurement tell us about the nature of the thing being measured? If we measure some property of a physical object, it does us no good unless we can relate it to the nature of the object. We expect to be able to reason logically from the measurement itself and the conditions under which the measurement was made to an understanding of the nature of the object whose properties were being measured; this is the inductive reasoning process we have mentioned. So the most important thing we do is to understand why our measurements have the values they do.

To begin, we ask: What could cause this measurement to have a *different* value? If we changed one particular condition of the experiment, would that change the value of the measurement? If so, then that particular condition and the property being measured are linked by the nature of the object whose property we are measuring. Since this nature is the very thing we hope to learn about, we attach considerable importance to the relationship or association between measurable properties. How do we represent the interaction between them for a given system?

2-1 The Representation of Functions

What we are going to need, as we have already said, is the mathematical concept of **function**. For chemical purposes we can define a function as an association between two quantities, each of which has a numerical measure. Let us look at a very simple example of a function as a means of describing the results of an experimental measurement.

Tabular Representations of Functions

If we put ice into a warm drink, the drink will cool. The cooling begins immediately when we add the ice, of course, but if we want the drink really cold we must wait. This suggests that we can measure the temperature of the drink and relate it to the time that has passed since the ice was added. The result will be a function relating temperature to time. Suppose we have a glass of water at room temperature, an ice cube, a thermometer to measure the temperature of the water, and a stopwatch to measure the time. The thermometer reads 72°F initially. We drop in the ice cube and simultaneously start the watch; thus at zero time the temperature is 72°F. After 10 sec we look at the thermometer, which now reads 64°F. After another 10 sec we look at the thermometer again; it now reads 57°F. We can continue this as long as we like, and the results can be indicated conveniently by Table 2-1. This table of pairs of numbers repre-

Table 2-1
The Relation Between Time and Temperature in a Cooling Liquid

Total Time (sec)	Water Temperature (°F)
0	72
10	64
20	57
30	52
40	48
50	44
60	42
70	40
80	38
90	37
100	36

sents a function relating time and temperature. The two quantities time and temperature are called **variables**. Since time passed without reference to anything in our experiment, it is clearly an **independent variable**; since the temperature we observed depended on the time at which we chose to observe it, we regard temperature as a **dependent variable** in this experiment. Sometimes it is not so easy to identify which is the independent variable and which is the dependent variable, and sometimes we may even be free to choose either quantity in an experiment as the independent variable. We record the independent variable in the left column of such a table and the dependent variable on the right as we make each measurement. When the series of measurements is complete, the table of pairs of numbers gives us the form of the function.

Graphical Representations of Functions

We can also represent a function by drawing a graph. We can draw a horizontal axis with a uniform scale of units and a vertical axis with a uniform scale, although they need not have the *same* scale. By convention we plot the independent variable along the horizontal axis and the dependent variable along the vertical axis. Thus we can indicate the function describing this experiment by the graph shown in Fig. 2-1. If we make the usually reliable assumption that

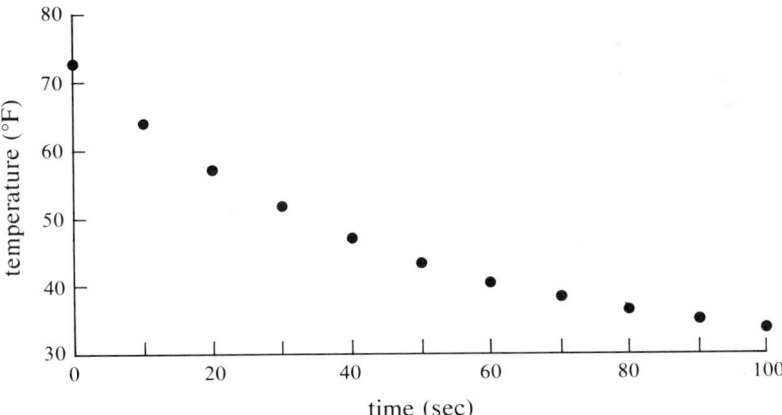

Figure 2-1 Individual points belonging to the time–temperature function.

physical properties change smoothly, we can fill in the spaces between the points representing the actual measurements (Fig. 2-2). This is a graphical representation of a function. It is equivalent to the table of pairs of numbers; the function is the same, and only the representation has changed. The function is the relationship itself. We speak of the temperature being a function of the time, but this usage expresses only one direction of a two-way relationship.

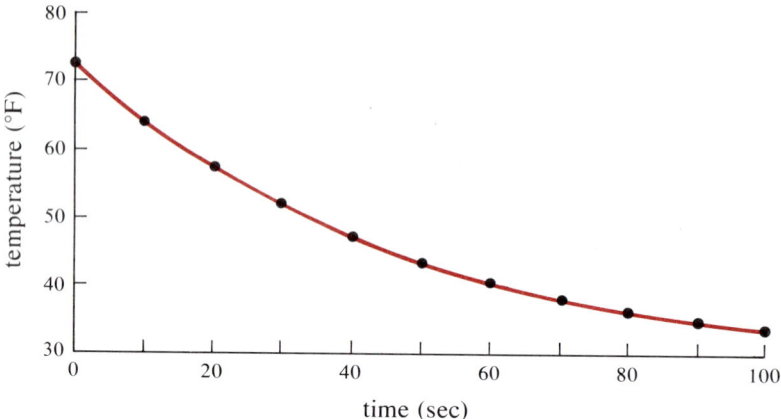

Figure 2-2 Continuity or smoothness assumption for the time–temperature function.

Units and Dimensions in Functions

Another point is that when a function involves physical quantities, the variables have units or dimensions. In this case, the units of time are seconds and the units of temperature are degrees Fahrenheit. The units are important because if they are changed the representation of the function is different even though the function itself—in this case the relationship between the quantity temperature and the quantity time—does not change. If, for instance, we use minutes as the units of time, the tabular representation of the function changes to that of Table 2-2. The graphical representation changes correspondingly to that of Fig. 2-3. But even though these representations have a different appearance from the original ones, we are certain that the water and the ice cube are unaffected by the particular gears we choose to put in the stopwatch, so that the function itself—the interaction between physical properties—does not change.

Table 2-2
The Relation Between Time and Temperature in a Cooling Liquid

Total Time (min)	Water Temperature (°F)
0	72
1	42
2	33
3	32
4	32
5	32
6	32

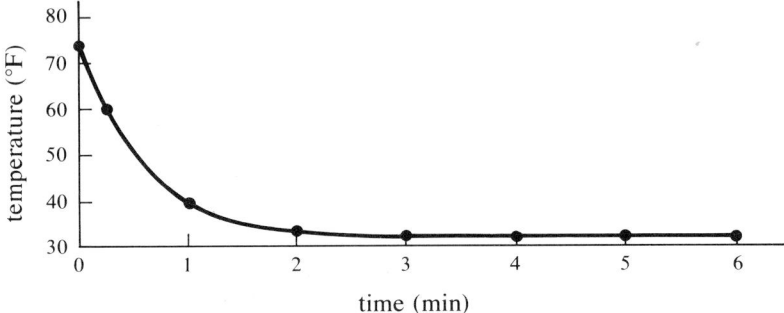

Figure 2-3 Effect of changing units of measurement on the time–temperature function.

Algebraic Representations

We can also represent a function algebraically. Again, the algebraic representation is completely equivalent to the graphical representation, which is completely equivalent to the tabular representation. Thus if we write the algebraic equation $y = ax + b$, we are representing a function that can also be represented by a graph (Fig. 2-4) or a table of pairs of numbers:

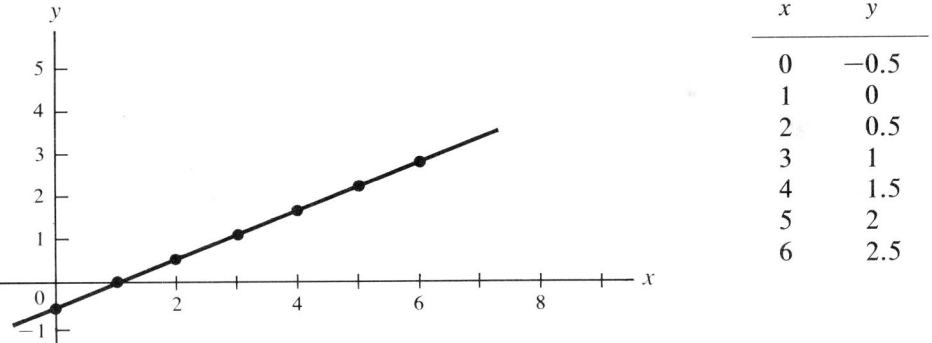

Figure 2-4 Graphical representation of $y = ax + b$.

When we use functions to represent relationships between experimentally determined physical quantities, we usually go in the opposite direction; that is, we usually obtain the tabular representation first, then convert this to a graphical representation, and finally, if we choose, convert the graph to an algebraic representation. Each of these representations has certain advantages. The tabular representation is usually the most convenient to use in the laboratory. The graphical representation is most pictorial and often provides a feel for the overall form of a function that is otherwise difficult to achieve. The algebraic representation of a function is the most compact and the most mathematically flexible, so if we wish to rearrange this relationship to indicate some other relationship that is of interest to us, we shall usually choose the algebraic representation of the function and the manipulations of algebra to achieve this end. In this book we shall find occasion to use all three forms, but most frequently we shall use graphical and algebraic functions and hope to find their properties complementing each other.

2-2 Linear and Polynomial Functions

The Linear Polynomial Function

We should examine some kinds of functions that the chemist finds useful. The function we have just represented algebraically by the equation $y = ax + b$ should be familiar to you as the **linear** function. If it is not familiar, a glance at the graph should justify the name. In this algebraic representation of the linear function y is the dependent variable and x the independent variable. The quantity b is a constant called the **intercept**; the name derives from the fact that in the graphical representation b is the point at which the straight line representing the function intercepts the y axis. In other words, b is the value y assumes when $x = 0$. The other quantity in the algebraic equation representing our function is a, which is a constant called the **slope**. Again the name is taken from the graphical representation and can be readily recognized as the amount by which y increases for a given increase in x; the larger a is, the more y increases. It is possible for a to be a negative number. For instance, we could have $y = -4x + 12$, in which the quantity a, the slope, is equal to -4. This is perfectly acceptable algebraically; if we try to interpret it on the graph of the same function (Fig. 2-5), we see that if the increase in y for a given increase in x is negative, it is, of course, a decrease. If x increases going to the right on the graph (which is conventional), then a positive slope means that the function or curve is going uphill to the right. A negative slope means that the curve is going downhill to the right. (The graphical representation of a function is generally called a curve even if it is a straight line.)

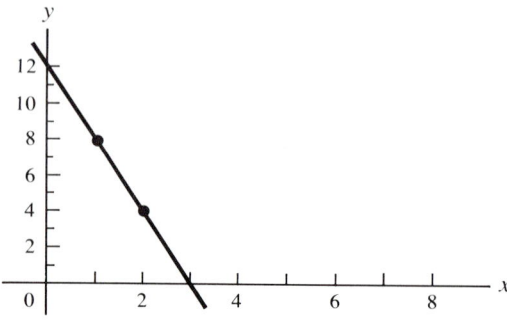

Figure 2-5 Linear function with negative slope.

The Quadratic Polynomial Function

Another function that is sometimes useful to chemists is the **quadratic** function, $y = ax^2 + bx + c$. The graphical representation is shown in Fig. 2-6a for the special case in which $b = c = 0$ and should be familiar to you as a parabola. The more general case, also a parabola, in which neither a, b, nor $c = 0$ is shown in Fig. 2-6b. In producing the general curve for $y = ax^2 + bx + c$ we have in effect added three very simple functions together. We can think of y as having

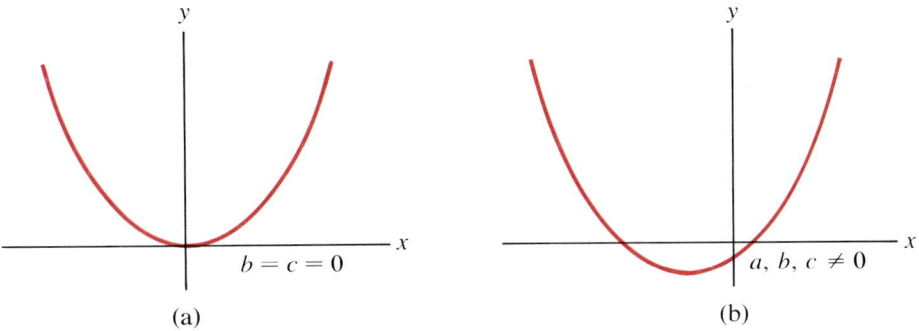

Figure 2-6 The quadratic function $y = ax^2 + bx + c$ for a special case (a) and the general form (b).

three components: a constant contribution c, a contribution proportional to the independent variable x (through the **proportionality constant** b), and a contribution proportional to the square of the independent variable x, through the proportionality constant a. We can produce graphical representations of each of these three contributions (Fig. 2-7), and can combine them to give the

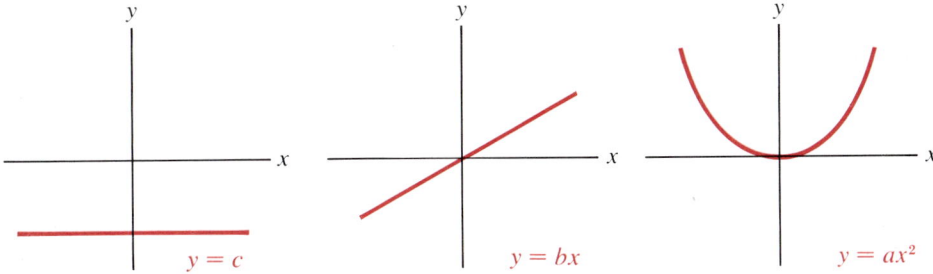

Figure 2-7 Components of the general quadratic function.

general curve (or function) by adding the contributions together at each value of the independent variable x. For example, in the combination shown in Fig. 2-8 the value of y at $x = 2$ is $4a + 2b + c$, which we could get directly from the algebraic representation of the function. We can also get it by adding together

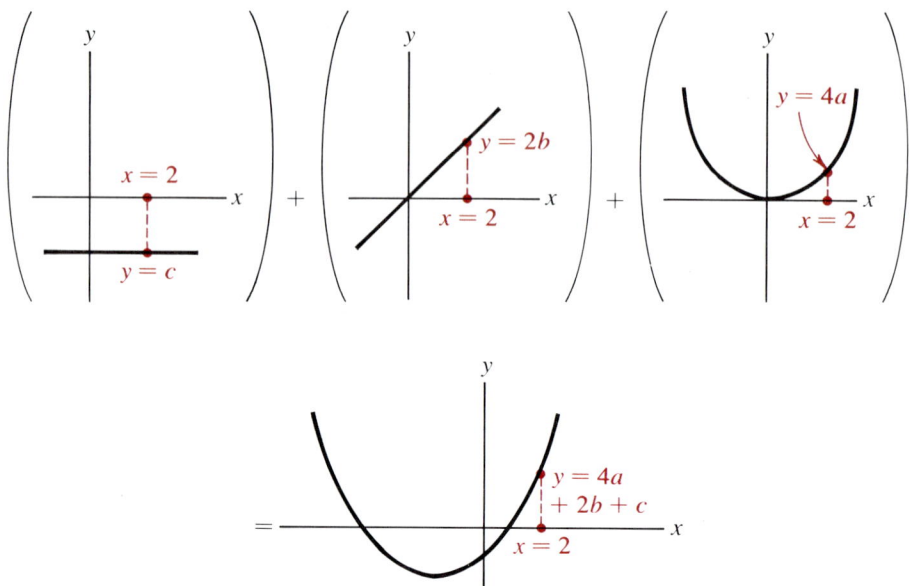

Figure 2-8 Linear combination of functions to give a new function.

the parabolic contribution, which at $x = 2$ is equal to a times 2^2 or $4a$, the linear contribution $2b$, and the constant contribution c. We shall later have frequent occasion, in considering some of the functions of chemistry, to use the technique of adding functions together to develop a mathematical model of experimental chemical reality.

The General Polynomial Function

It should now be apparent that we can extend these ideas as far as we like to polynomial functions such as $y = a + bx + cx^2 + dx^3 + ex^4 + \cdots$. There are numerous occasions for the chemist to use polynomial functions, particularly the two simplest ones, the linear and quadratic. In working with these functions there will be occasions when we need to find **roots** of polynomials. A root of a polynomial is a value of the independent variable that causes the dependent variable to equal zero. If $y = 6x - 3$, clearly $y = 0$ when $x = \frac{1}{2}$; then $\frac{1}{2}$ is a root (the only root) of the equation for the linear function $y = 6x - 3$. You should also be familiar with the **quadratic formula**, which is an algebraic expression for the roots of a general quadratic function $y = ax^2 + bx + c$:

$$x_{\text{root}} = \frac{-b \pm \sqrt{b^2 - 4ac}}{2a} \tag{2-1}$$

This expression indicates the values of x for which $y = 0$ in a general quadratic function, which will be useful in several applications.

2-3 The Concept of Slope

In dealing with the linear function we developed the concept of slope as the amount by which the dependent variable increases for a given increase in the independent variable. The slope is thus the **rate of change** of the dependent variable with respect to the independent variable. If, in using functions to correlate physical properties of objects or systems, we have established an independent and a dependent variable, then we can deal immediately with the effect on one property of changing the other property. In the case involving the glass of water and the ice cube, we could define the rate of change of temperature with respect to time. This is the crux of the problem we set out to discuss, namely, the relationship between measurable quantities. A large number of physical properties are best described with respect to other properties in terms of their rates of change. This means that we shall be very much interested in examining rates of change of functions, particularly those that are useful to the chemist.

The Slope of a Curving Line

For the linear function this rate-of-change quantity, the slope, is the constant a in the algebraic representation of the function $y = ax + b$. It turns out that we can define the slope for functions whose graphs are not straight lines, but the slope is no longer a constant if the curve is no longer going uphill or downhill at the same rate everywhere. We can produce a definition of the slope of a curve that is independent of whether the curve is a straight line, however:

$$\text{slope} \equiv \frac{y_2 - y_1}{x_2 - x_1} \qquad (2\text{-}2)$$

Here y_2 and x_2 are the graphical coordinates of point 2, and y_1 and x_1 are the coordinates of point 1; in the simple functions we shall be dealing with, point 2 is always to the right of point 1. From the graphs in Fig. 2-9 we can see that this definition matches our intuitive notion of slope. Function (b) has a higher slope than function (a), as we expect it to, and function (c) has a negative slope since y_2 is actually a smaller number than y_1.

What about functions whose graphical representations are not linear? The same definition will still work, but now it produces an average slope over the range of the independent variable between x_1 and x_2. For a nonstraight line this average slope will depend on the range we select—both on its size and on

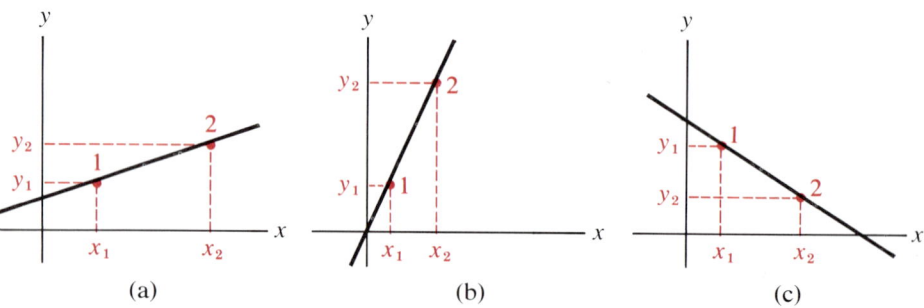

Figure 2-9 Definition of slope for the linear function.

where we choose to put it (Fig. 2-10). Thus the graphs of (d), (e), and (f) are all of the same function, but the slopes are different. In graph (d), a relatively wide range of the independent variable x has been selected, and the resulting average slope is that of the straight line through points 1 and 2. In graph (e), the points 1 and 2 have been selected much closer together, but point 1 has not been moved. From the straight line with a slope equal to our average slope, however, we can see that this average slope is less than that in (d). In graph (f), the range between points 1 and 2 on the x axis (i.e., the size of the range) has not been changed from graph (e), but the range has been moved to the right; we see that the average slope (as represented by the straight line through 1 and 2) is now greater than in (e). As a matter of fact, the coordinate x_2 on graphs (d) and (f) is the same, so that both (e) and (f) represent a shrinking of the range over which the average slope is determined in graph (d); but neither of these average slopes is equal to the original average slope, and they are not equal to each other. Clearly the notion of slope of a curved line is a slippery one.

It is apparent, however, that the smaller the range between points 1 and 2, the better off we are. The appearance of graphs (d), (e), and (f) suggests, correctly,

Figure 2-10 Average-slope definition applied to a nonlinear function.

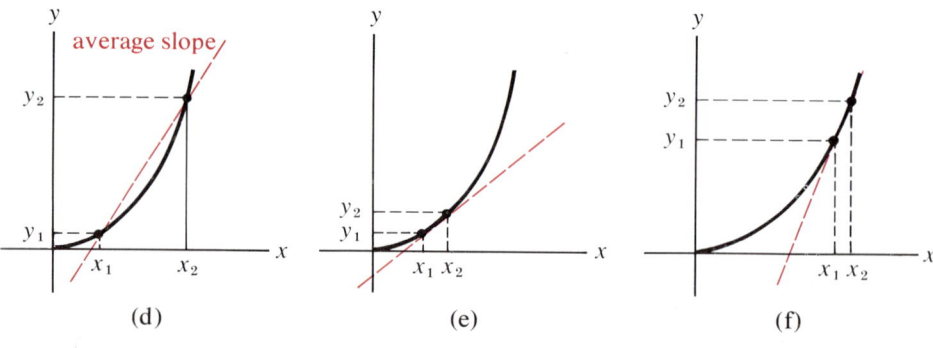

that the shorter the range between points 1 and 2, the more closely the straight line representing the average slope comes to the line representing the function itself. Of course, if the straight line were to lie exactly on top of the curve, we would have an exact value for the slope of the curve itself, since the slope of the straight line would then be equal to the slope of the curve (at the point where they match). We notice also, in connection with shrinking the range over which the average slope is taken, that when we go to a very small range for the independent variable x we also obtain, in all but the most exceptional cases, a very small range for the dependent variable y. That is, if x_2 is very close to x_1 so that $x_2 - x_1$ is very small, y_2 is very close to y_1 and $y_2 - y_1$ is also small. Their ratio, which is what we have defined as the slope in which we are interested, may still be a substantial and well-defined number even if the two quantities $y_2 - y_1$ and $x_2 - x_1$ become infinitesimally small. The criterion of smoothness we used earlier suggests that if we shrink the range of the average slope, the new value of the average slope will be a better approximation to the actual slope of the curve; and furthermore, that the smaller we shrink the range, the better the approximation becomes.

Shrinking the Range of an Average Slope to a Limit

This shrinking process is the basis for the concept of **limits**. Since the approximation improves as the range shrinks, we might say, speculatively, that the slope of the curve (the function) itself would be exactly equal to the limit of the ratio of $y_2 - y_1$ to $x_2 - x_1$ as x_2 approached x_1. This is in fact true, and we can write the following statement, which is in effect a definition of slope:

$$\lim_{x_2 \to x_1} \frac{y_2 - y_1}{x_2 - x_1} = \text{slope at point 1} \qquad (2\text{-}3)$$

where $\lim_{x_2 \to x_1}$ means the limiting value, as x_2 approaches x_1, of the expression after it. Equation 2-3 means that we can define the slope at a given point even for a very complicated curve and obtain a numerical value for it. We shall come back to this in section 2-6 after examining a few of the other types of functions chemists use.

2-4 Sinusoidal Functions

In general, polynomial functions change more or less uniformly as the independent variable increases in value. A polynomial function containing x^n as the highest power of x will have n roots, but after x has increased beyond the last of these the function will either increase uniformly toward plus infinity or decrease toward minus infinity. Since many physical properties have a repetitive nature—for instance, the height of waves on a pond as a function of distance across the water's surface—we shall need a new kind of function that repeats the same values indefinitely, and that never shows any signs of aspiring to infinity.

The Repeating Properties of a Circle

One of the best ways to deal with repetitive properties as dependent variables is to consider that the independent variable has the nature of an angle, and that the repetitive property passes through one cycle while this "angle" is passing through 360°. As the "angle" continues to increase, the value of the dependent variable takes on the same values all over again. For instance, when the "angle" has reached 410°, the physical property represented by the dependent variable will have the same value it did when the "angle" was 50°, as shown in Fig. 2-11.

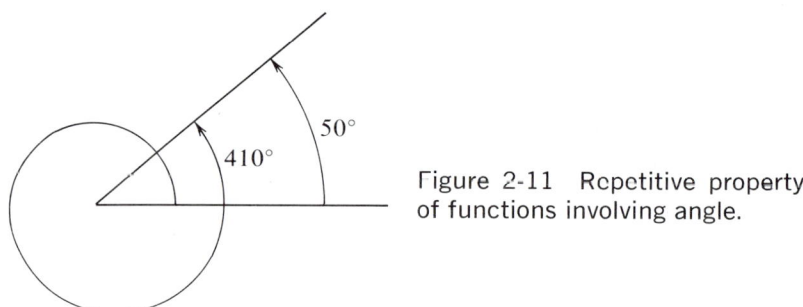

Figure 2-11 Repetitive property of functions involving angle.

Let us plot a graph of a circle and follow the coordinates of a point going around the circle (i.e., as we increase the angle from some arbitrary starting point). Suppose the circle has a radius of 1 unit and we allow the angle to increase in a counterclockwise fashion from a starting position where the circle crosses the x axis. The y coordinate of points on the circle will have a repetitive nature, since on the second time around we pass through the original points as the angle θ continues to increase (Fig. 2-12). The y coordinate will clearly

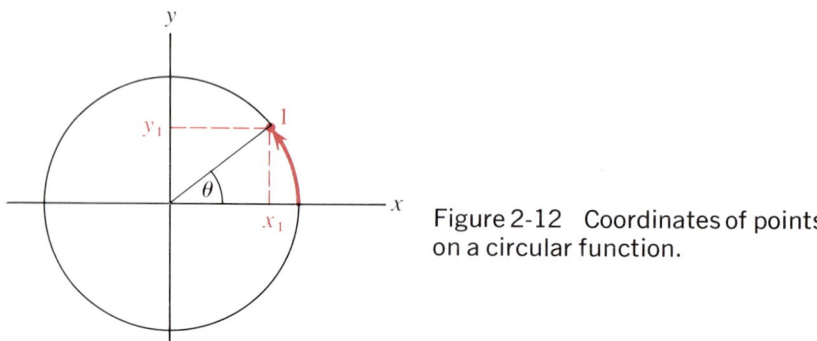

Figure 2-12 Coordinates of points on a circular function.

be zero at the beginning, since we are starting on the x axis. When θ has reached 90°, the y coordinate will be equal to the radius of 1 unit. When θ has reached 180°, the y coordinate will have decreased to a value of 0 again. When θ has reached 270°, the y coordinate will have decreased further to a value of −1 unit, and so on. Suppose we plot a graph (Fig. 2-13) of the y coordinate of the points on this circle as it relates to this angle; it increases quite rapidly at first, then more slowly as θ approaches 90°. This pattern is repeated as we go around the circle, even if we go around an indefinite or an infinite number of times.

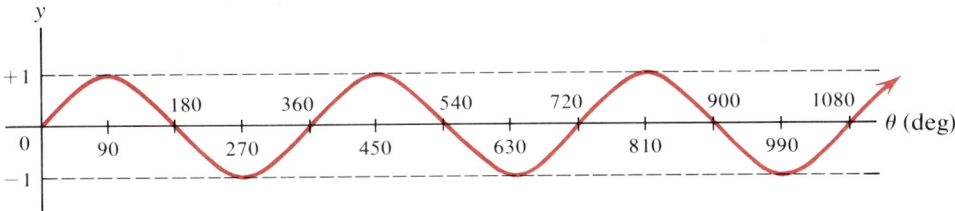

Figure 2-13 Dependence of the y coordinate of a circular function on angle.

This repetitive or periodic curve clearly represents the sort of repetitive function we set out to look for. It is called the **sine** function, written in abbreviated fashion as $y = \sin \theta$. Physical properties that relate to other properties in this fashion are said to be **sinusoidal**.

There is a quite similar function that we can obtain (Fig. 2-14) simply by looking at the x coordinate of points on the circle instead of the y coordinate.

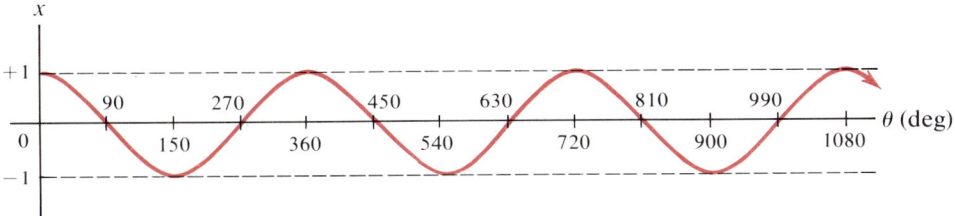

Figure 2-14 Dependence of the x coordinate of a circular function on angle.

This function starts at 1, decreases to 0 when $\theta = 90°$, decreases further to −1 when $\theta = 180°$, increases to 0 when $\theta = 270°$, and so on. This is the **cosine** function, abbreviated $y = \cos \theta$. We notice that there is an obvious relationship between these two functions, sine and cosine, in that one of them is just the other moved sideways by 90° on the graph.

The wave character of these two functions can be represented by two properties we shall later have occasion to use. The **wavelength** λ is the range of the independent variable over which one cycle of the dependent variable occurs; in this case, in which we are dealing with angle, the wavelength is 360°, but in the case of waves on a pond, the wavelength might be 8 in. It is helpful to incorporate the wavelength into the function by writing (if y is the dependent variable and x the independent variable)

$$y = \sin\frac{x}{\lambda/360°} \quad \text{or} \quad y = \sin\frac{360x}{\lambda}$$

Written this way the periodic property goes through 1 cycle in 1 unit of the independent variable. The other important property is **amplitude**, which is simply how high the wave is. As we have written these functions, they reach a maximum of 1 unit, but since the units are arbitrary we can modify the function by multiplying the whole thing by an amplitude factor to get it into whatever units we please: $y = A \sin(360x/\lambda)$. In this way we can represent the waves on the pond as being 3 in. high, or $\frac{1}{4}$ ft high, or in whatever unit of length we choose.

Radian Units

Finally, we need to recognize the usefulness of another set of units for the independent variable. We have been using degrees of angle, and at the same time assuming that the radius of the circle was 1 unit, but we can generalize both of these quantities if we let the radius of the circle be r and measure angle in units of r around the perimeter of the circle. This unit of angle is the **radian**; since there are 2π radii of a circle in its perimeter, there will be 2π radians in 360°. Thus 1 radian is equivalent to about 57°. Using these more general units we can write the conventional wave equation as

$$y = A \sin\frac{2\pi x}{\lambda}$$

The quantities A and λ will generally be physical properties. If the wave in which we are interested is a sound wave, for instance, the pitch is related to $1/\lambda$ and the loudness or intensity is proportional to A^2. For a light wave, color is related to $1/\lambda$ and brilliance (or intensity, again) to A^2. We shall have occasion to use the wave equation later, in discussing the wave properties of matter (in Chapter 5).

2-5 Exponential and Logarithmic Functions

There are two other types of function we need to examine for chemical purposes: the exponential function and the logarithmic function.

Exponential Functions

The **exponential function** can be written $y = n^x$, where n is some constant number and x is the independent variable. It is important to realize the difference between x^n and n^x; if $n = 2$ and we let the variable take a value of 1000, then x^n (which is the polynomial form) $= 1000^2$ or 1,000,000. But n^x will be 2^{1000}, which is a great deal bigger than 1,000,000.

The significant properties of the exponential function can be seen in its graph (Fig. 2-15). Since we are not certain about n, let us try two values of n separate-

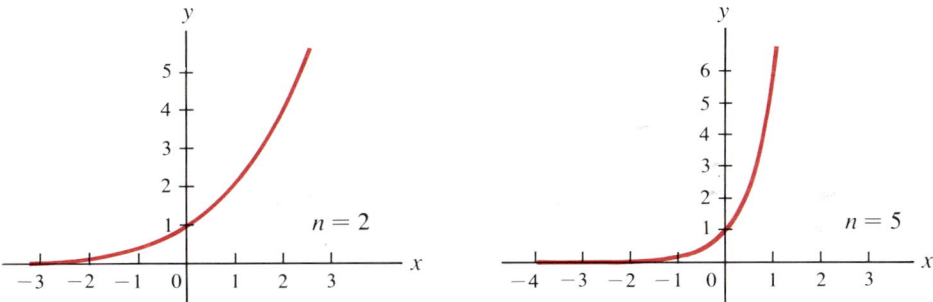

Figure 2-15 The exponential function for two different bases.

ly, $n = 2$ and $n = 5$. Obviously the curves have basically the same shape and what varies is the slope; whatever n may be, n^{+x} represents a function that increases faster and faster as x increases and n^{-x} represents a function that gradually decreases from its initial value toward zero but never quite gets there. Since n determines the slope, we can make the slope anything we please by choosing n appropriately (Fig. 2-16).

An interesting possibility is to choose a value of n (call it e) such that, for the function $y = e^x$, the value of the slope of the function at any point is always equal to the value of the function itself at that point. That is, a number e ought to exist such that

$$\lim_{x_2 \to x_1} \left(\frac{e^{x_2} - e^{x_1}}{x_2 - x_1} \right) = e^{x_1}$$

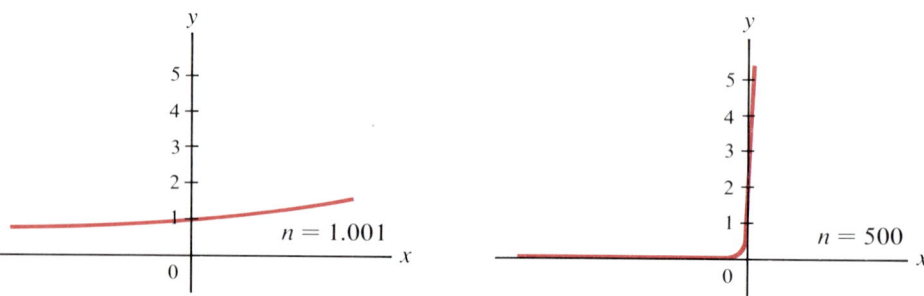

Figure 2-16 The exponential function with extreme bases.

using the definition of slope we developed earlier. There is such a number: $e = 2.71828\ldots$. Like π, e is irrational and continues indefinitely. The function $y = e^x$, where e is defined this way, is useful simply because there are many physical properties whose rates of change with respect to some other property are proportional to their own magnitudes — exactly the property this function has.

Logarithmic Functions

The **logarithmic function** is related to the exponential function. **Common logarithms** are defined by $x = 10^{\log x}$, and **natural logarithms** are defined by $x = e^{\log_e x}$ or $e^{\ln x}$, where e is the number we just discussed, $2.71828\ldots$. Although common logarithms (to the base 10) are much more convenient for numerical calculations, the natural log function is more appropriate in describing the relationship of physical properties to each other, for essentially the same reason that the number e is the appropriate base for exponential relationships. Fortunately, the two types of logarithms can be simply interconverted:

$$10^{\log x} = e^{\ln x}$$

$$\ln 10^{\log x} = \ln e^{\ln x}$$

$$(\ln 10)(\log x) = \ln x$$

$$2.303 \log x = \ln x$$

If we construct a graph (Fig. 2-17) of the logarithmic function $y = \ln x$, we see that it rises more and more slowly as x becomes larger. We have not so far dealt with a function that had this general shape, but there are several relationships between physical properties that require this pattern. Some examples of these functions — polynomial, sine, cosine, exponential, and logarithmic — will appear as we move on through the book.

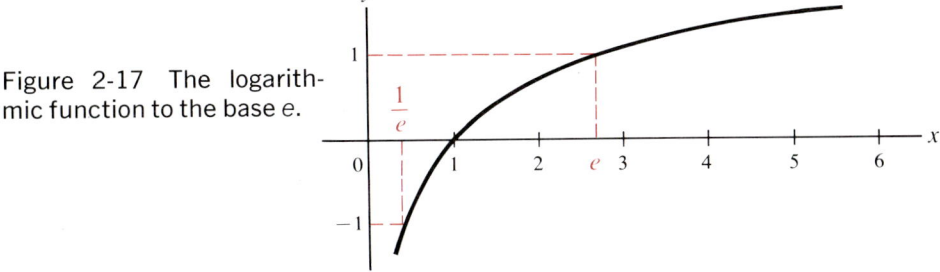

Figure 2-17 The logarithmic function to the base e.

2-6 Slopes and Derivatives of Polynomials

Let us return to the question of dealing with slopes or rates of change. For the straight line of the linear function we saw that the slope can be defined as $(y_2-y_1)/(x_2-x_1)$. This definition was valid because the slope of a straight line is constant. For a nonlinear function, such as the quadratic, we proposed to retain the same definition of slope, interpreting it as an average slope over a given range of the independent variable. The exact slope, which varies from point to point on the curve, can be obtained by shrinking the range over which the average slope is calculated and taking the limiting value of the average slope as the range shrinks toward zero:

$$\lim_{x_2 \to x_1} \frac{y_2-y_1}{x_2-x_1} = \text{slope at } x_1$$

We can actually obtain some slopes this way by simple algebra, if we make a change in notation. Call the range, x_2-x_1, over which the average slope is taken Δx. Then if y is some function of x, which we denote by $f(x)$, we can write

$$\text{slope} = \lim_{\Delta x \to 0} \frac{f(x+\Delta x) - f(x)}{\Delta x} \qquad (2\text{-}5)$$

Let us try this out on a few simple examples. If $y = 3x+4$, what will be the slope of the graph of this function — or, in terms of the function itself, what will be the rate of change of y with respect to x?

$$y = f(x) = 3x+4$$

$$\text{slope} = \lim_{\Delta x \to 0} \frac{3(x+\Delta x)+4-(3x+4)}{\Delta x}$$

$$= \lim_{\Delta x \to 0} \frac{3x+3\Delta x+4-3x-4}{\Delta x}$$

$$= \lim_{\Delta x \to 0} \frac{3\Delta x}{\Delta x}$$

$$= \lim_{\Delta x \to 0} 3$$

$$= 3 \quad \text{(constant slope)}$$

This checks with what we already know about the slope of this linear function. Now find the slope of the quadratic function $y = f(x) = x^2$.

$$\text{slope} = \lim_{\Delta x \to 0} \frac{(x+\Delta x)^2 - x^2}{\Delta x}$$

$$= \lim_{\Delta x \to 0} \frac{x^2 + 2x(\Delta x) + (\Delta x)^2 - x^2}{\Delta x}$$

$$= \lim_{\Delta x \to 0} \frac{2x(\Delta x) + (\Delta x)^2}{\Delta x}$$

$$= \lim_{\Delta x \to 0} (2x + \Delta x)$$

$$= 2x$$

We see that in the quadratic case the slope is not constant, but depends on the particular value of the independent variable we choose. Notice that the slope or rate of change of a given function is itself a function, so that we can represent a function graphically and also represent its slope graphically, as in Fig. 2-18.

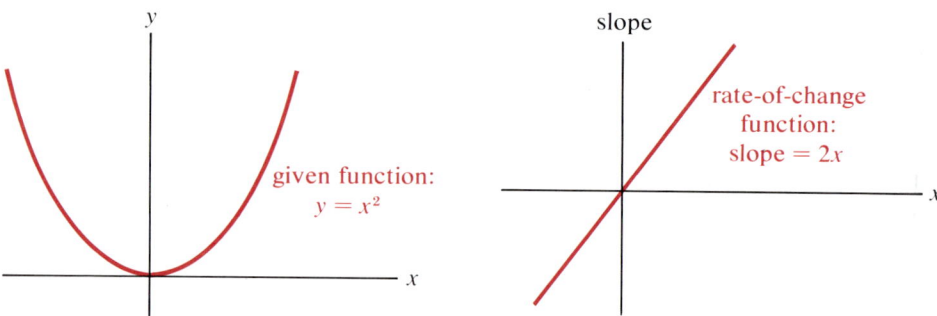

Figure 2-18 Relation between a function and its slope function.

Let us find the slope of the cubic polynomial function $y = f(x) = x^3$.

$$\text{slope} = \lim_{\Delta x \to 0} \frac{(x+\Delta x)^3 - x^3}{\Delta x}$$

$$= \lim_{\Delta x \to 0} \frac{x^3 + 3x^2(\Delta x) + 3x(\Delta x)^2 + (\Delta x)^3 - x^3}{\Delta x}$$

Slopes and Derivatives of Polynomials | 63

$$= \lim_{\Delta x \to 0} \frac{3x^2(\Delta x) + 3x(\Delta x)^2 + (\Delta x)^3}{\Delta x}$$

$$= \lim_{\Delta x \to 0} 3x^2 + 3x(\Delta x) + (\Delta x)^2$$

$$= 3x^2$$

Looking at this result and the previous one, we see that a general rule is developing: The slope of a function $y = x^n$ is nx^{n-1}.

The Derivative

Our notation is still somewhat cumbersome. Let us introduce a new symbol, dy/dx, which we shall define as

$$\frac{dy}{dx} = \lim_{\Delta x \to 0} \frac{\Delta y}{\Delta x} \qquad (2\text{-}6)$$

This does not change any of the algebra we just performed, of course, but it makes it possible to write a very compact expression for the rate at which the dependent variable in a function changes as the independent variable changes: dy/dx = rate of change of y with respect to x. The quantity dy/dx is called the **derivative of y with respect to x**. Another way to write a derivative with exactly the same meaning is

$$\frac{dy}{dx} = \frac{d}{dx}(y) = \frac{d}{dx}[f(x)]$$

The process of obtaining the derivative of a function is called **differentiation**. We already know how to differentiate the function $y = x^n$:

$$\frac{d}{dx}(x^n) = nx^{n-1} \qquad (2\text{-}7)$$

Some other rules apply to differentiating polynomials (or, indeed, any functions), which you can verify by going through the limiting process with $f(x)$ and $f(x + \Delta x)$:

$\frac{d}{dx}(\text{constant}) = 0$ \qquad (try $y = f(x) = 4$)

$\frac{d}{dx}(ay) = a\frac{dy}{dx}$, where a is a constant \qquad (try $ay = a[f(x)] = 3x^2$)

$\frac{d}{dx}[f(x) + g(x) + h(x)] = \frac{df}{dx} + \frac{dg}{dx} + \frac{dh}{dx}$, where f, g, and h are all functions of x
\qquad (try $y = x^2 + x + 1$)

2-7 Derivatives of Other Functions

Can we differentiate the other types of functions we have developed? The easiest way to do so will be graphically, since the limiting process with Δx is more involved for these other functions. We need only remember that the derivative, in graphical terms, is the slope of a curve representing a given function, and will itself be another function. Just as we picked two points on the curve and drew a straight line representing the average slope, we can take the limit of this average slope by bringing the two points closer and closer together so that eventually we are drawing a **tangent** to the curve. A tangent is a straight line drawn so that it touches a curve at only one point and has the same slope as the curve at that point.

Derivatives of Sinusoidal Functions

Let us differentiate the function $y = \sin x$ this way. Again—in graphical terms this means we are trying to produce, from a graph of the function $y = \sin x$, another graph of a function whose height at any point x_0 is the slope of the original function at x_0. First we draw the original function (Fig. 2-19) and draw tangents to it at several locations. Then we plot another curve with the slopes of each of these tangents, as in Fig. 2-20. Finally, when enough slopes have

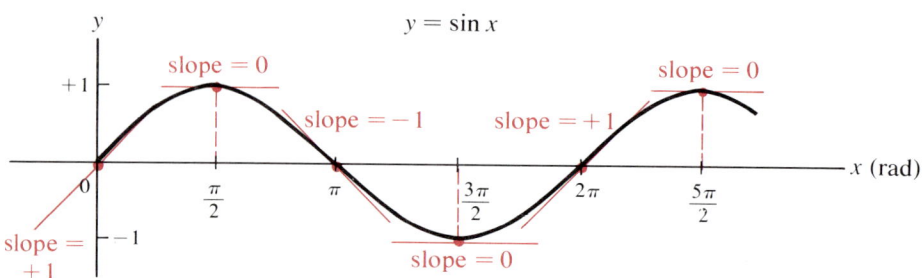

Figure 2-19 Approximating a slope function or derivative for the sine function by constructing tangents.

Figure 2-20 Points belonging to the slope function (derivative) of the sine function.

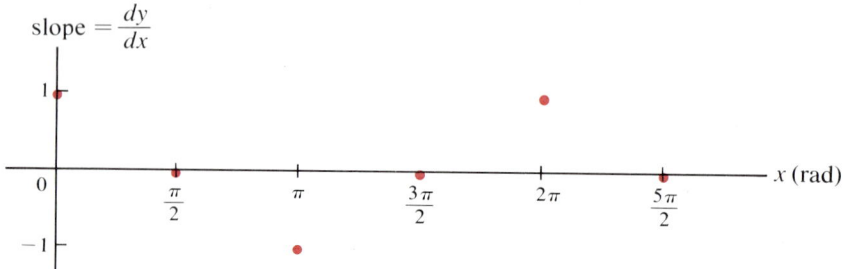

been obtained, we fill in the curve smoothly as in Fig. 2-21.

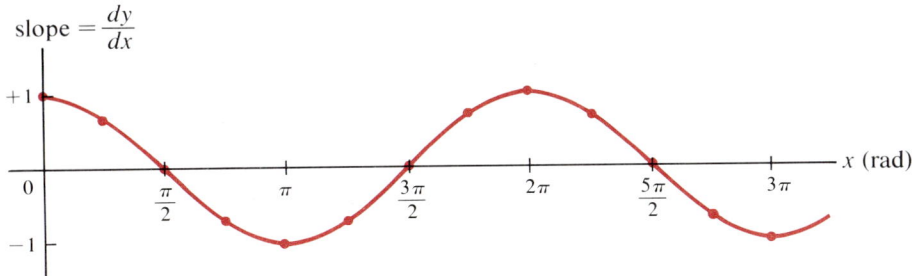

Figure 2-21 The slope function (derivative) for the sine function.

This graph of the slope (the derivative) of the function $y = \sin x$ looks very much like the cosine function. As a matter of fact, if we construct the graph very carefully, we can verify that it is. In other words,

$$\frac{d}{dx}(\sin x) = \cos x \qquad (2\text{-}8)$$

We can go through the same process for the function $y = \cos x$. First we draw the curve (Fig. 2-22) representing the original function $y = \cos x$, then we

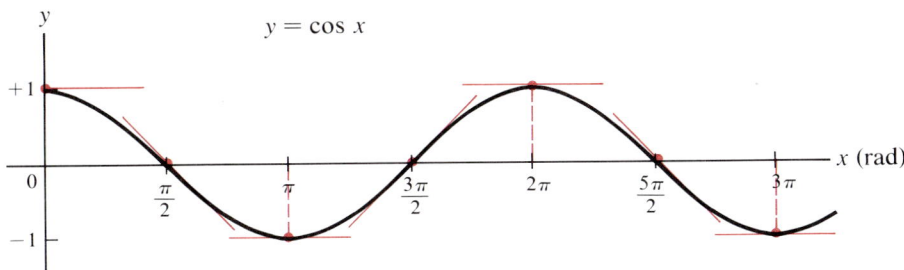

Figure 2-22 Constructing the slope function (derivative) for the cosine function, using tangents.

draw in several tangents to it. We plot the slopes of each of these tangents on a new graph, with each tangent plotted at the same value of x as in the original graph, and fill in the curve smoothly as in Fig. 2-23. This new function does not

Figure 2-23 The slope function (derivative) for the cosine function.

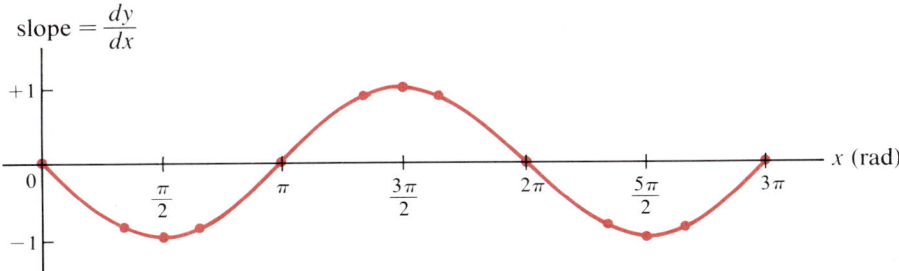

look exactly like any we have developed before, but it does oscillate, just as the sine and cosine functions themselves do, so it should be possible to express it in terms of either the sine or cosine function, or perhaps in terms of both. The new function (the slope or derivative function of the cosine function) has a value of 0 when $x = 0$, just as the sine function does, but it becomes more and more negative as x increases (initially), while the sine function becomes more positive. Then perhaps this slope function is $slope = -\sin x$; again, by very careful graphing we can verify this. So we can write for the derivative of the cosine function

$$\frac{d}{dx}(\cos x) = -\sin x \tag{2-9}$$

The derivatives or rate-of-change functions for the sine and cosine functions have a rather ingrown appearance. Since the derivative of a function (a curve) is another function (another curve), it occurs to us that it should be possible to take a **second derivative** or a derivative of a derivative. We can symbolize this as d^2y/dx^2 or $d^2/dx^2(\ )$, and we see that for the sine and cosine functions the second derivative has a remarkable property:

$$\frac{d}{dx}(\sin x) = \cos x$$

$$\frac{d^2}{dx^2}(\sin x) = \frac{d}{dx}\left[\frac{d}{dx}(\sin x)\right] = \frac{d}{dx}(\cos x) = -\sin x$$

$$\frac{d^2}{dx^2}(\sin x) = -\sin x$$

and

$$\frac{d}{dx}(\cos x) = -\sin x$$

$$\frac{d^2}{dx^2}(\cos x) = \frac{d}{dx}\left[\frac{d}{dx}(\cos x)\right] = \frac{d}{dx}(-\sin x) = -\frac{d}{dx}(\sin x) = -\cos x$$

$$\frac{d^2}{dx^2}(\cos x) = -\cos x$$

That is, the second derivative of the sine or cosine function is equal to minus the original function. No other function has this property, and this is often taken as the fundamental wave property. In Chapter 5 we shall find that the electron has wave properties, so that we apply this mathematical treatment. If we have any physical property that oscillates in a wavelike fashion, we can immediately write its description (call it y) as

$$\frac{d^2y}{dx^2} = -y$$

The Wave Equation

Let us pursue this matter of wave behavior a little further. When we developed the sine and cosine functions we produced the conventional wave function, $y = A \sin 2\pi x/\lambda$. This is the function for which we really want to take a second derivative, but it is complicated by the presence of some extra factors; what will happen to them when we differentiate? Here we need the assistance of an expression known as the **chain rule**, which is very helpful in differentiating complicated expressions. The chain rule simply says that if u is a function of x and y is a function of u, then

$$\frac{dy}{dx} = \frac{dy}{du}\frac{du}{dx} \tag{2-11}$$

This seems obvious if we regard the individual derivatives as fractions, but beware of overconfidence. A derivative is a single symbol for a limit, and it is not at all obvious that the quantities du in the two derivatives refer to the same thing. Nevertheless, the chain rule is valid for all the functions in which we shall be interested, and we use it here in the following way. Define a new quantity u such that $u = 2\pi x/\lambda$. Now, since

$$y = A \sin \frac{2\pi x}{\lambda}$$

it is also true that

$$y = A \sin u$$

and

$$\frac{dy}{du} = A \cos u$$

We can also get du/dx:

$$\frac{du}{dx} = \frac{d}{dx}\left(\frac{2\pi}{\lambda}x\right) = \frac{2\pi}{\lambda}$$

so that

$$\frac{dy}{dx} = \left(\frac{2\pi}{\lambda}\right)\left(A \cos \frac{2\pi x}{\lambda}\right)$$

Here we have used the chain rule in the last line in multiplying dy/du by du/dx to get dy/dx. To get the second derivative of $y = A \sin 2\pi x/\lambda$, we need only differentiate dy/dx again, using the chain rule in exactly the same way:

$$\frac{dy}{dx} = \frac{2\pi A}{\lambda}\left(\cos \frac{2\pi x}{\lambda}\right) = \frac{2\pi A}{\lambda} \cos u$$

$$\frac{d}{dx}\left(\frac{dy}{dx}\right) = \frac{d}{du}\left(\frac{dy}{dx}\right)\frac{du}{dx} \quad \text{(chain rule)}$$

$$\frac{d}{du}\left(\frac{dy}{dx}\right) = \frac{d}{du}\left(\frac{2\pi A}{\lambda}\cos u\right) = \left(\frac{2\pi A}{\lambda}\right)\frac{d}{du}(\cos u) = -\frac{2\pi A}{\lambda}\sin u$$

$$\frac{du}{dx} = \frac{d}{dx}\left(\frac{2\pi}{\lambda}x\right) = \frac{2\pi}{\lambda}$$

$$\frac{d^2y}{dx^2} = \frac{d}{dx}\left(\frac{dy}{dx}\right) = \left(-\frac{2\pi A}{\lambda}\sin\frac{2\pi x}{\lambda}\right)\left(\frac{2\pi}{\lambda}\right)$$

$$= -\frac{4\pi^2}{\lambda^2}\left(A\sin\frac{2\pi x}{\lambda}\right) = -\frac{4\pi^2}{\lambda^2}y$$

This last expression is the **classical wave equation**, and we shall use it in discussing chemical properties in Chapter 5.

Differentiating Exponential Functions

Now let us return to the problem of differentiating the other two functions we developed earlier, the exponential function and the logarithmic function. In dealing with the exponential function $y = n^x$, we noticed that the effect of changing the base, n, is to change the slope of the curve without changing its general shape. In differentiating, of course, the slope or rate of change of a function is exactly what we are interested in. We observed that there was a number, e, which when used as the base of an exponential function gave a function having a slope at any point equal to the function's own value at that point. The function $y = e^x$ is graphed in Fig. 2-24; let us examine the tangents A, B, and C. Line A is tangent to the curve at $x = -1$; this is the point at which the function has the value $y = e^{-1} = 1/e$. The slope, if we measure it with a ruler, is such as to rise 1 unit in a horizontal range of 2.7 units. Using the

Figure 2-24 Constructing the slope function (derivative) for the exponential function, using tangents.

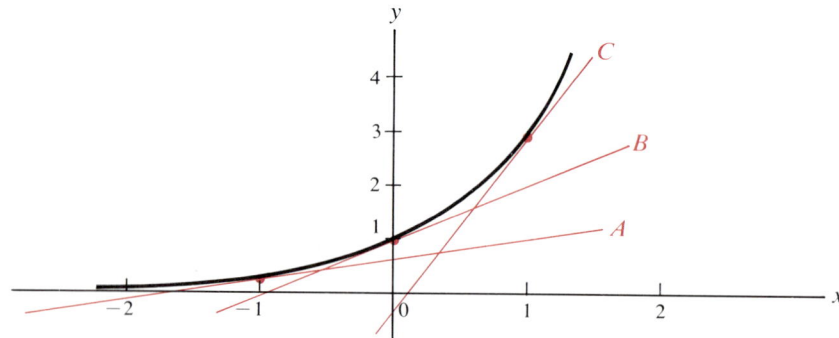

definition of slope as $(y_2-y_1)/(x_2-x_1)$, this tangent has a slope of 1/2.7 or 1/e. Tangent B touches the curve at $x=0$, where $y=e^0=1$. A glance at the graph shows B rising 1 unit in a horizontal range of 1 unit, which corresponds to a slope of 1/1 or 1. Similarly, tangent C, which touches the curve at $x=1$ (where $y=e^1=2.718\ldots$), has a slope of 2.7/1 or 2.7. We see that it is indeed possible to produce an exponential function with a rate of change at any point equal to its own value at that point. In other words, the derivative of $y=e^x$ is

$$\frac{d}{dx}(e^x) = e^x \qquad (2\text{-}12)$$

The properties of exponentials will be useful in describing the properties of gases; indeed, the appendix to this chapter deals with the statistical interpretation of random changes in just this way.

Differentiating Logarithmic Functions

Taking the same approach to the logarithmic function, $y=\log x$, we have already observed that using this same number e as a base for the natural logarithm function $y=\ln x$ gives a function with unique properties. Let us examine the tangents to the natural logarithmic function plotted in Fig. 2-25.

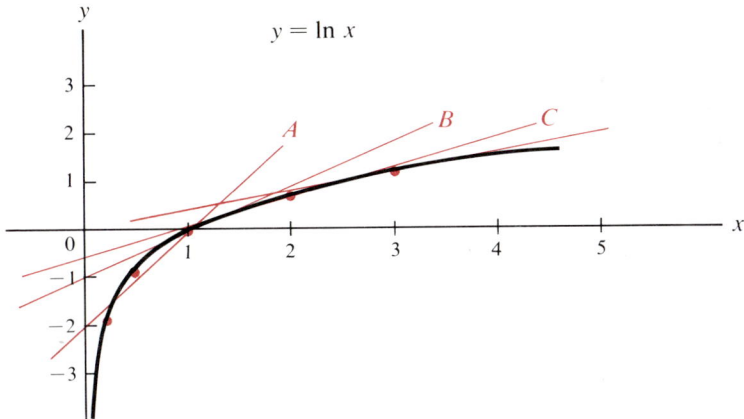

Figure 2-25 Constructing the derivative for the logarithmic function.

Measuring the slopes of the tangents as before, we see that the slope of the tangent A at $x=\frac{1}{2}$ is 2. The slope of tangent B, touching the curve at $x=1$, is 1, and the slope of tangent C, touching the curve at $x=2$ is $\frac{1}{2}$. These three values, together with others we could construct at some of the other points shown on the curve, suggest that the slope of the $y=\ln x$ curve is $1/x$. As in previous cases, we can establish this as exactly as we please by constructing a very careful graph with a large number of tangents. The net result is an expression for $(d/dx)(\ln x)$:

$$\frac{d}{dx}(\ln x) = \frac{1}{x} \tag{2-13}$$

Only the logarithmic function to the base e has this property; as we showed earlier, $\log_{10} x = \ln x/2.303$, so $(d/dx)(\log_{10} x) = 1/2.303x$, a somewhat less convenient quantity.

At this point we have established rules for differentiating all the functions we expect to use; it may be appropriate to summarize them here.

$$\frac{d}{dx}(x^n) = nx^{n-1}$$

$$\frac{d}{dx}(\sin x) = \cos x$$

$$\frac{d}{dx}(\cos x) = -\sin x$$

$$\frac{d}{dx}(e^x) = e^x$$

$$\frac{d}{dx}(\ln x) = \frac{1}{x}$$

$$\frac{d}{dx}(\text{constant}) = 0$$

$$\frac{d}{dx}[af(x)] = a\frac{df(x)}{dx} \quad (a = \text{constant})$$

$$\frac{d}{dx}[f(x) + g(x) + h(x)] = \frac{df(x)}{dx} + \frac{dg(x)}{dx} + \frac{dh(x)}{dx}$$

$$\frac{d^2f(x)}{dx^2} = \frac{d}{dx}\left[\frac{df(x)}{dx}\right]$$

$$\frac{dy}{dx} = \frac{dy}{du}\frac{du}{dx} \quad (\text{chain rule})$$

2-8 Differentials

The form of the chain rule strongly suggests that the two quantities dy and dx in the derivative dy/dx can *under proper circumstances* be regarded as separate entities. If we can treat the quantity dx as being a small change in the independent variable x, then the quantity dy takes on a specific meaning defined by the process of taking limits. These two quantities dx and dy are **differentials**, and their meaning may be made clearer by the graph of an arbitrary curve as in Fig. 2-26. We see that the small range of the independent variable can be called either dx or Δx, but that dy is not the same as Δy. Δy is just the quantity $y_2 - y_1$, which we originally regarded as contributing to the average slope of a curve.

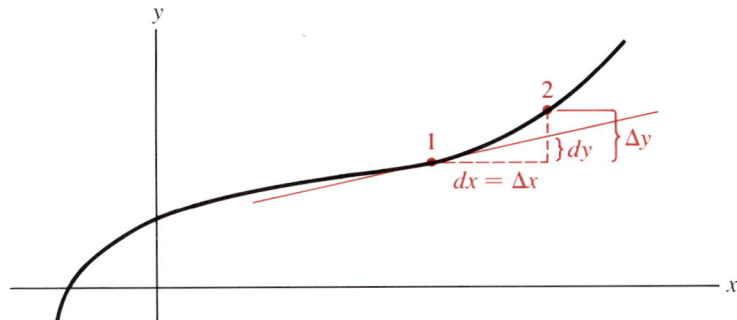

Figure 2-26 The distinction between increments and differentials (Δy and dy).

No limiting process is involved. The quantity dy, on the other hand, has a ratio to dx that is equal to the slope of the curve at point 1, *after the limiting process has taken place* (as represented by the tangent, which touches the curve at only one point). If we wanted to deal with dy separately, we could take this ratio into account and write $dy = $ (slope of tangent)dx or

$$dy = \frac{dy}{dx} dx$$

Note that since the two differentials dy and dx are meaningful only in that their ratio is the slope of the curve, both must appear in any expression involving differentials. Thus if we are dealing with the function $y = \sin x$, we can write $dy = \cos x \, dx$.

How can we use differentials? In physical problems we often have experimental evidence of the rate of change of some physical quantity, but we would like to have not the rate of change but a function relating the property itself to its environment. That is, we often have the derivative of a function and want the function itself. There ought to be—and is—a method of undoing what we have just done. We have developed rules for going from a function to its derivative; can we start with the derivative and obtain the function? The procedure for this is called **integration** and involves the differential notation.

2-9 Summation and Integration

If $dy/dx = 3$, what is the function $y = f(x)$? Remembering the rules for differentiation, we can say at once that $y = 3x + C$, where C is a **constant of integration**. There is an easy check for this. Differentiate the function $y = 3x + C$ to see if it gives the same derivative: $(d/dx)(3x) = 3$, $(d/dx)(C) = 0$, and $3 + 0 = 3$. Don't forget the constant C; since the derivative of a constant is 0, we can never know what the added constant is if the derivative of the function is all we have. In consequence, we must add the constant of integration C and hope to evaluate it from some of the other conditions of the experiment. The constant may be 0, but we cannot omit it until we have evaluated it.

Summation

In the limiting process by which we obtained the derivative, we conceptually cut the curve up into smaller and smaller segments. Over any given range Δx along the independent variable, then, the increments Δy and Δx indicate the direction in which the curve is heading. If we could add up all the little Δy's and Δx's for all the intervals along the independent variable x (assuming we knew them) we could generate an approximation to the whole original curve from the many straight-line segments $\Delta y / \Delta x$. This process of summation is represented by the Greek letter Σ: If we had cut the curve into 20 pieces we could write

$$\sum_{i=1}^{20} \Delta y_i = \sum_{i=1}^{20} \frac{\Delta y}{\Delta x_i} \Delta x_i$$

Here i is just a numerical label or index for each of the 20 pieces and $\Sigma_{i=1}^{20}$ means there are 20 pieces to be added together.

Integration

For differentials the process is conceptually the same, but now we are adding up an infinite number of infinitesimally small pieces, and we use a different symbol, a flattened S (for sum): $\int$. Now we can write

$$\int dy = \int \frac{df(x)}{dx} dx$$

and since this generates the original curve we can say that this expression is exactly equivalent to $y = f(x)$. Using this notation we can immediately produce integration rules for undoing the five differentiation rules we obtained earlier:

$$\int x^n \, dx = \frac{1}{n+1} x^{n+1} + C \qquad (\text{except for } n = -1) \qquad (2\text{-}14)$$

$$\int \cos x \, dx = \sin x + C \qquad (2\text{-}15)$$

$$\int \sin x \, dx = -\cos x + C \qquad (2\text{-}16)$$

$$\int e^x \, dx = e^x + C \qquad (2\text{-}17)$$

$$\int \frac{1}{x} \, dx = \ln x + C \qquad (2\text{-}18)$$

The three graphs in Fig. 2-27 illustrate the desirability of increasing the number of segments in the summation and suggest that the summation of an infinite number of infinitesimal differentials (the integration process) ought to duplicate the curve exactly.

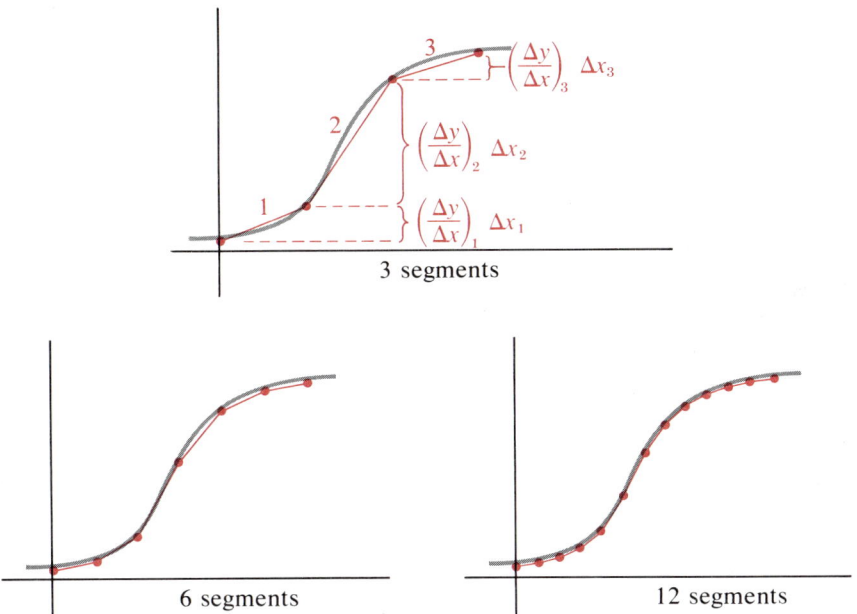

Figure 2-27 Successive approximations to the reconstruction of a function, using its average slope over smaller and smaller increments.

What do these integration rules do for us in physical problems? If, by experiment or postulate, we have come up with an expression that is a function of x and that represents the rate of change of some quantity in which we are interested with respect to x, we can call the quantity y and we should recognize that the function of x with which we are dealing is just dy/dx. Here, of course, y is any independent variable with which we might be concerned, such as concentration of a solution or pressure of a gas, and x is any independent variable that affects y, such as temperature, or time since the beginning of the experiment, or distance from an electrode surface. When we have established the function of x that represents dy/dx, we can immediately obtain another function of x that represents y, the dependent variable itself, simply by plugging the derivative function into the appropriate integration rule:

$$y = \int dy = \int \frac{dy}{dx} dx = f(x) + C$$

If, for instance, we say that the rate of change of the work done by an expanding gas on its surroundings, as a function of the volume of the gas at any given time, is inversely proportional to the volume itself, we can represent this by the algebraic statement $dW/dV = 1/V$, where W represents work done and V the volume of the gas. Then if we want an expression representing the work done by the gas (not the rate of change of the work) as a function of its volume, we can immediately write

$$W = \int dW = \int \frac{dW}{dV} dV = \int \frac{1}{V} dV = \ln V + C$$

A more accurate statement of the proportionality relationship is $dW/dv = K(1/V)$, where K is the proportionality constant. To integrate this we need only remember that $(d/dx)(ay) = a(dy/dx)$, so that

$$W = \int dW = \int \frac{dW}{dV} dV = \int K \frac{1}{V} dV = K \int \frac{1}{V} dV = K(\ln V + C)$$

That is, the integral of a constant times a function is equal to the constant times the integral of the function. Since our expressions will usually include proportionality constants, this property of integrals will be quite useful. It is true in general that we can derive from each property of the derivative a corresponding property of the integral. This should not be too surprising, since both differentiation and integration deal with the relationship between two functions, one of which represents the rate of change of the other.

2-10 Integrals and Areas

Integrals have another property that will be useful. Suppose we have a function $f(x)$ and its derivative, which is another function we shall call $f'(x)$. The function $f(x)$ is thus the integral of the function $f'(x)$. Let us examine the derivative function $f'(x)$ graphically (Fig. 2-28). At any point x_0 the quantity $f'(x_0)$ is the

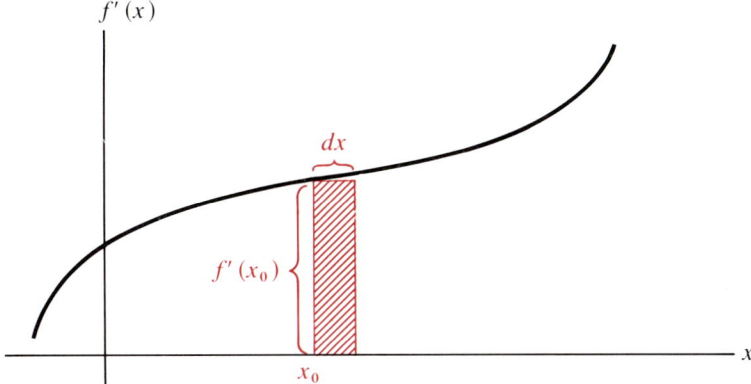

Figure 2-28 The product of $f'(x_0)$ and dx.

height of the curve above the x axis, by the nature of the graphical representation of functions. The differential quantity dx, to the extent that it has independent meaning, is simply a small segment of the x axis—a small horizontal distance. Then the product of $f'(x)$ and dx is the *area* of a tall, thin rectangle located at or near x_0. If we construct a number of such rectangles as in Fig. 2-29 by dividing the x axis between x_0 and x_f into a number of segments of width dx and producing the appropriate heights for each, the sum of the areas

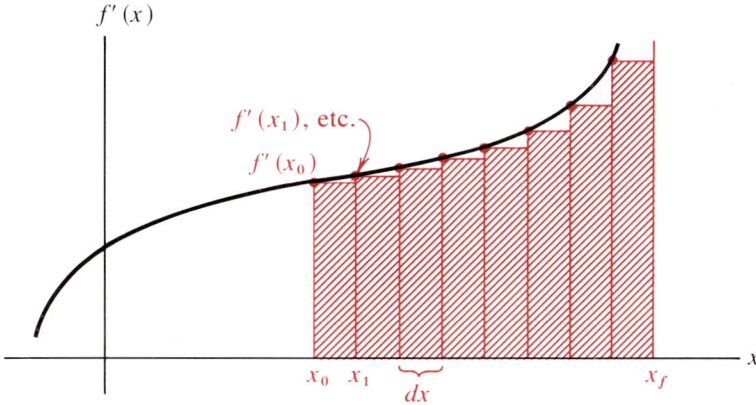

Figure 2-29 Approximating the area under a curve by summing products of its derivative and dx.

of the rectangles is an approximation to the total area under the curve between x_0 and x_f. The reason this sum is only an approximation is that it does not include the area of the small triangles at the top of each rectangle. But if we return to the concept of limits, we see that as the number of rectangles increases—as dx decreases—the area of the triangles becomes smaller and the sum of the areas of the rectangles becomes a better approximation, so that in the limit in which each dx is infinitesimal and the summation is over an infinite number of rectangles, the sum will be equal to the area under the curve; see Fig. 2-30. The

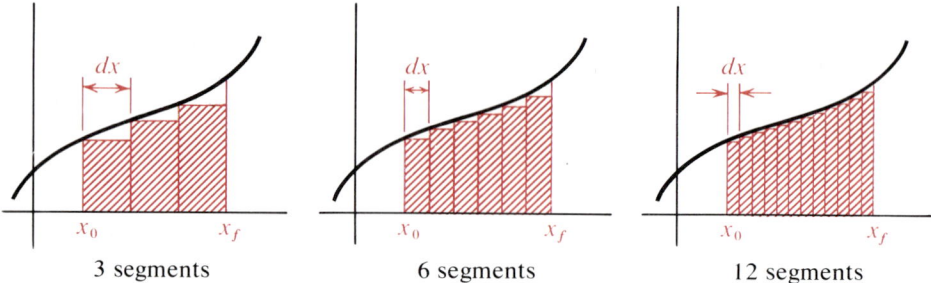

Figure 2-30 Successive approximations to the area under a curve, using smaller and smaller increments dx.

summation of an infinite number of products of $f'(x)$ and dx is exactly what is represented by the integral

$$f(x) = \int f'(x)\, dx \qquad (2\text{-}19)$$

This result is important: The integral of a function is equal to the area under the curve that is the graphical representation of that function.

Boundaries of Integration

There is an obvious objection to this. An integral includes an undetermined constant, and if this constant is to appear in the expression for the area, the area will also be undetermined by the amount of the constant. For this reason, the kind of integral we have been discussing is called an **indefinite integral**. Fortunately, if we consider the properties of the area under a curve, we can do something about making this indefinite area more definite, at least under certain circumstances. Consider the graph in Fig. 2-31. The area under the

Figure 2-31 The area under a curve between boundary values of the independent variable.

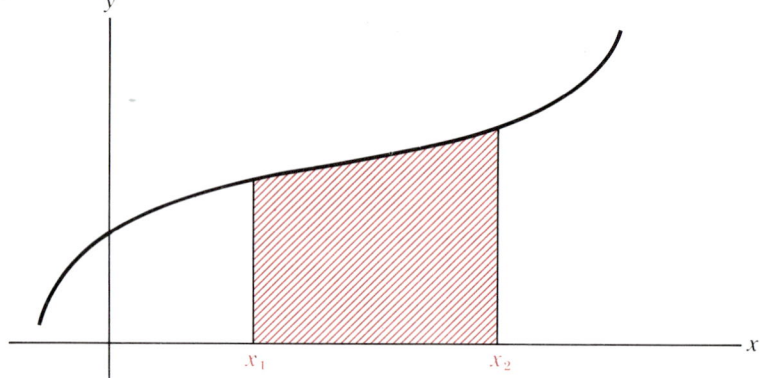

curve *between* x_1 *and* x_2 is perfectly well defined; there is no indeterminacy about it. How can we express this well-defined area in terms of the integral of the function the curve represents? If we took the algebraic representation of the function and integrated it, we would get a perfectly well-defined algebraic expression except for the constant of integration C. For any given value of the independent variable x, we could plug into the algebraic expression for the integral and get a definite answer for the area except for the constant C. But here we are defining only *one* value of x, and the reason the area is indeterminate is just because there is no other defined value of x that limits the area to be described; the situation is rather like the graph in Fig. 2-32. But the curve—

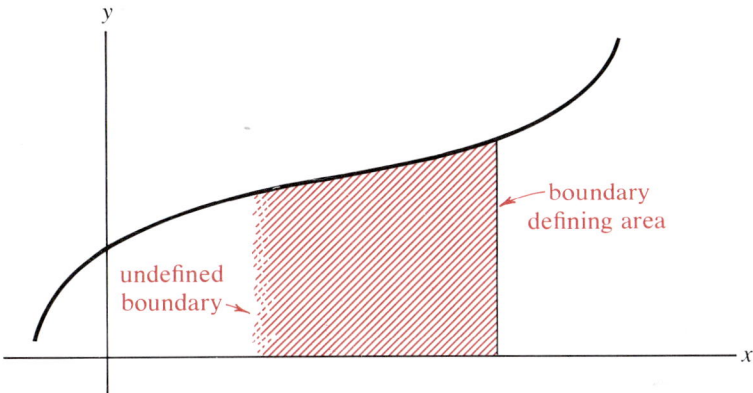

Figure 2-32 The failure of an indefinite integral to specify boundaries for the area it represents.

the function to be integrated—remains the same curve no matter where we take the value of x that limits the area at the right end, and the area under it from wherever the beginning is to the defined limit must always be the same for a given curve. So we can expect to get an exact area by defining *two* boundaries in such a way that we subtract the integral at the left boundary (with its constant C) from the same integral evaluated at the right boundary (with the same constant C):

exact area between x_1 and $x_2 =$

$$= \left[\int f(x)\, dx, \text{evaluated at } x = x_2 \right] - \left[\int f(x)\, dx, \text{evaluated at } x = x_1 \right] \quad (2\text{-}20)$$

The Definite Integral

The sort of integral in which two boundaries on the range of possible values of the independent variable are defined so as to get a definite value of the integral without the undefined constant of integration is called a **definite integral**.

If the two limits are x_1 and x_2, the definite integral of the function $f(x)$ is written $\int_{x_1}^{x_2} f(x)\,dx$. Another way to indicate the process of evaluating the integral function at two points is as follows. Suppose $f(x) = \int f'(x)\,dx$. Then

$$\int_{x_1}^{x_2} f'(x)\,dx = f(x_2) - f(x_1)$$

and

$$\int_{x_1}^{x_2} f'(x)\,dx = \left|f(x)\right|_{x_1}^{x_2}$$

Suppose we look at an example—the same one we used on p. 74. Remember that we had the function $dW/dV = K(1/V)$, which expressed the fact that the rate of change of the work done by an expanding gas on its surroundings, as a function of the volume of the gas, is inversely proportional to the volume. In integrating this, we saw that

$$W = \int \frac{dW}{dV}\,dV = \int K\frac{1}{V}\,dV = K(\ln V + C)$$

Suppose we state the problem a little more exactly: How much work is done when the gas in question expands from volume V_1 to volume V_2? Clearly the two limits of integration are being defined here, and we write

$$W = \int_{V_1}^{V_2} \frac{dW}{dV}\,dV = K(\ln V_2 + C) - K(\ln V_1 + C) = K(\ln V_2 - \ln V_1) = K \ln \frac{V_2}{V_1}$$

It should be apparent that the algebraic representation of the functions and integrals involved allows us to construct a very compact yet flexible statement of the relationship in which we are interested.

Notice also that interpreting integrals as areas in this way will occasionally require us to use negative areas. This should not be too disturbing if we note that in Fig. 2-33 the rectangular unit of area at x_1 is positive since it represents the product of y_1 (positive) and dx_1 (positive), but the rectangular unit of area at x_2 is negative since it represents the product of y_2 (negative) and dx_2 (positive). Clearly, any area below the x axis will be negative, while any area above the x axis will be positive.

Figure 2-33 The significance of positive and negative areas in integration.

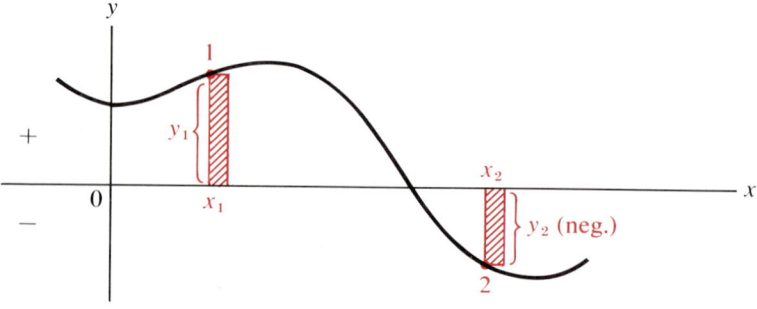

2-11 Statistics and Probability

Before the mathematical techniques just described can be applied to chemical problems, it is necessary to assign numbers to the properties of chemical systems. This is the function of measurement. We have already discussed measurement and the errors associated with it. The treatment of measurements and errors falls within the province of statistics and probability. These studies serve two distinct functions in physical science. First, if several measurements are made, all giving slightly different values for a quantity, statistics provides a basis for choosing the best value for the quantity based on the set of measurements, and a basis for judging the **reliability** of the measurements. Second, when the quantity whose value is to be determined cannot be measured exactly, or when the number of measured quantities required is so large as to make measurement impracticable, probability theory may provide a means for indirect estimation of the values without direct measurement. We shall discuss both of these applications, the first in some detail in this chapter, and the second briefly here and at length in later sections of the book.

2-12 The Reliability of Measurements

Let us consider the matter of errors of measurement in some detail. We shall exclude **systematic** errors since these properly fall within the province of experimental design rather than probability or statistics. **Random errors** may arise from a multitude of sources. The nature of these sources is unimportant for our purposes here, although it may be of supreme importance to the experimentalist who is trying to improve his experimental accuracy. Errors that are truly random are equally likely to lead to high or low results. We say that the **distribution** of measurements is **symmetrical** about some "true" value. Let it be clear immediately that this "true" value cannot be determined except by analysis of measurements, so no absolute value can ever be assigned to it without error. We might define it provisionally as the limit of the best value as the number of measurements approaches infinity, although such a definition may not be entirely acceptable from a strictly operational point of view. From any finite set of measurements, we can only find the "best value" based on that set.

The Distribution of Error

To obtain the distribution of a set of measurements, we divide the range of values encompassed by our set of measurements into a number of equal intervals. Then we count the number of measured values that fall within each such interval. By plotting the number of measurements in each interval as a function of the position of the interval, we obtain a kind of bar graph called a **histogram**, which is one way of specifying a distribution.

In Table 2-3 there are some measured values for the volume of a certain piece of apparatus. Below the list of measured values there is a table of **frequencies**, or numbers of values that occur in a set of intervals covering the range of values.

Table 2-3
Volume Measurements (ml)

46.03   45.95   46.01   46.04   46.10   46.07   46.06   46.00   45.99   46.05

Frequencies

Range	Number of Measurements
45.91–45.94	0
45.95–45.98	1
45.99–46.02	3
46.03–46.06	4
46.07–46.10	2
46.11–46.14	0

A histogram of these measurements is shown in Fig. 2-34. The dashed curve superimposed on this figure represents the **normal distribution** for these measurements, assuming that the measurements obtained accurately represent the parent distribution. We assume that a distribution obtained from a very large number of measurements would have this form. Because we have only a small number of measurements, our experimental distribution differs considerably from the normal distribution.

Let us digress for a moment to point out that no single experimental result can have its validity analyzed statistically. However carefully the work may have been done, it is not possible to be sure that such a measurement does not contain serious errors. For any measurement to be valid in scientific work, it must be made more than once — preferably several times. It is not always necessary, however, to repeat the measurement under the identical conditions.

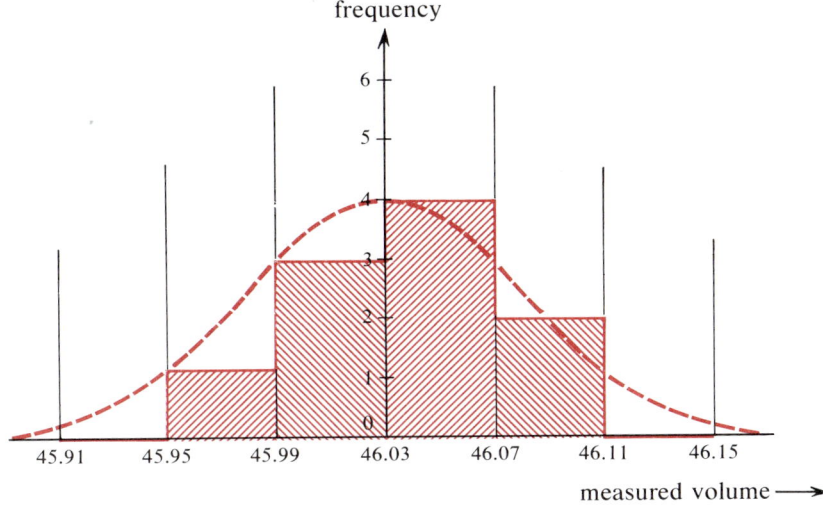

Figure 2-34   A histogram and the normal distribution for a set of measured values.

As we have seen, it is often desirable to permit one experimental variable to change and to obtain the measured quantity as a function of this changing quantity. The methods of statistics can be applied to this process.

If the experiment is well designed and the investigator is skillful, we expect that most of the measured values will be near the best value, and that the random error will be large in only a small fraction of the measurements. (What we mean by "near" and "large" will depend on the experiment. We shall define these terms more formally a little later.) Thus we intuitively anticipate that a large number of measurements will give the "bell-shaped" distribution exemplified by the normal distribution curve in our example.

Averaging

Since the normal distribution is symmetrical and the observed distribution approximately so, we might expect that the best value for our measurement would be the **average** or **arithmetic mean** of all the measured values. In this, intuition and theory are in agreement. For any set of measurements of a single quantity that has a normal distribution, the best value is obtained by dividing the sum of all the measured values by the number of measurements taken.

Even when we have determined the best value for our set of measurements, we still have not given enough information to satisfy the demands of scientific observation. We must specify the reliability of the measurements, which means we must provide some information concerning the average error in the measurements. Our first inclination might be to take the **average deviation** from the mean, that is, the algebraic sum of the differences between the measured values and the mean, divided by the number of measurements. Unfortunately, this value is always zero, simply because of the definition of the arithmetic mean. An obvious modification of this procedure is to take the average of the absolute values of the deviations. Since none of the numbers in this average is less than zero, its magnitude will be zero only when the measurements are all identical; and the average deviation computed in this manner is sometimes used as a measure of reliability.

Establishing the Likely Error and the Range of Reliability

A measure of reliability that is of greater theoretical significance is the **root-mean-square deviation** or **standard deviation**. The standard deviation is given the symbol s, and is defined by the relation

$$s = \sqrt{\frac{\sum_i (\bar{x} - x_i)^2}{n - 1}} \qquad (2\text{-}21)$$

where $\bar{x}$ is the arithmetic mean of all measurements, x_i is an individual measurement, n is the number of measurements taken, and the symbol Σ_i indicates that the sum of all terms of the form $(\bar{x} - x_i)^2$ is to be taken.

The actual significance that can be attached to either of these measures of reliability depends upon the number of measurements taken. Clearly, if one measurement has no statistical validity, two or three do not lead to perfection. We expect that the reliability of our best value will increase as the number of measurements increases. Accordingly, statisticians have developed the concept of **confidence levels** and **confidence intervals**. The confidence level is simply an arbitrarily chosen percentage that gives the odds that the best value based on a given set of measurements is close to the true value, or the value that would be obtained from a much larger set of measurements. The confidence interval that corresponds to a given confidence level is simply the range on either side of the best value within which the true value is expected to be found.

If we are to have a high degree of confidence — say 99% — we must allow a larger interval about the mean value than we would if we are satisfied with a lower confidence level. For a given confidence level and a given number of measurements, the confidence interval is proportional to the standard deviation; that is, if we say

$$\text{true value } (z) = \text{average measurement } (\bar{x}) \pm \text{confidence interval } (p)$$

with a certain degree of confidence, then

$$p = ts \qquad (2\text{-}22)$$

where t is a function only of the confidence level and the number of measurements taken. An abbreviated list of t values is shown in Table 2-4. A more complete table of t values is found at the end of the book.

Table 2-4
t Values

Number of Measurements	Confidence Level (%)			
	50	80	90	99
3	0.816	1.886	2.920	9.925
4	0.765	1.638	2.353	5.841
10	0.703	1.383	1.833	3.250

For the set of measurements discussed in Table 2-3, the mean value may be shown to be 46.03 and the standard deviation 0.042. If we set our confidence level at 90%, we can say that for that confidence level, the probability that a single measurement in the set is within 1.833×0.042 (i.e., $t \times s$) of the true value is 90%. Our confidence interval for 90% confidence is 0.077.

Observe that the t values given in the table are for single measurements. We can place considerably more confidence in the reliability of the mean. Indeed, at a given confidence level, the confidence interval for the mean is reduced by a factor of $1/\sqrt{n}$. Thus, for 90% confidence, the confidence interval for the mean in our set of measurements is $0.077/\sqrt{10}$ or 0.024.

2-13 Significant Figures

The reason for considering confidence intervals was to provide a formal mechanism for evaluating the reliability of a measured datum point. There is a less formal way to indicate this, using the concept of **significant figures**. A number is not quoted to any more digits than are reliable according to the above treatment; usually one digit is included that changes when the full range of the confidence interval is covered.

Although it is often convenient to discuss reliability in terms of significant figures, it should be remembered that the really important criterion of reliability is the percentage of error involved in a measurement. Students occasionally get the impression that it is worthwhile to weigh out 1.0006 g of sample instead of 0.9992 because in doing so they gain an extra significant digit. Such a distinction is futile since no real improvement in the relative reliability of the measurement is obtained in this way.

The real importance of the concept of significant figures is in deciding how many digits to report in a calculated result. When calculations are performed with measured quantities, it is probable that extra digits will be acquired in the course of computation. *These extra digits have no physical significance.* At the end of the calculation (but only at the end) all digits beyond the significant digits should be dropped. Several extra digits should be carried throughout the calculation to minimize calculation errors. The errors in calculated results may be obtained by application of the following rules: 1. in addition or subtraction, the absolute value of the errors should be added; 2. in multiplication or division, the percentage errors should be added.

At the end of the calculation, it should be a simple matter to compute the magnitude of the error in the calculated result, decide which figure in the calculated result represents the first uncertain figure, and drop the rest. In computing the standard deviation and confidence interval, it is common to retain one digit past the last significant figure in the measured or calculated value. This procedure should mean that the standard deviation normally has two significant figures, although more than two are frequently reported.

2-14 The Normal Distribution

We have discussed the normal distribution at some length without having defined it except in the most intuitive manner. As a matter of fact, the normal distribution can be described formally by a well-known mathematical function, known as the **error function** or the **Gaussian distribution**. A derivation of this equation is given in the appendix to this chapter. For our purpose here, it is sufficient to state its form:

$$\frac{dN}{N} = \frac{1}{\sqrt{2\pi}\sigma} \exp\frac{-(\bar{x}-x)^2}{2\sigma^2} dx \qquad (2\text{-}23)$$

The left-hand side of Eq. 2-23 is the number of values of the measured quantity that lie between x and $x + dx$, divided by the total number of measurements, N. On the right-hand side, σ is a constant whose significance will become apparent presently, and $\bar{x} - x$ is the difference between x and the mean value of the distribution, $\bar{x}$. The use of the differential notation implies the limiting process and thus a continuous distribution. Obviously we cannot reproduce a continuous distribution with a finite number of measurements, so any distribution obtained from a set of measurements can be at best an approximation to the normal distribution. Note that the use of the number e as a base for exponentials appears here in a particularly significant form; in what follows we shall sometimes use "exp" to mean that the subsequent expression is the exponent of the number e.

The Standard Deviation and the Error Function

Let us consider the properties of the function $f(x)$ defined by the relation

$$f(x) = \frac{1}{\sqrt{2\pi}\,\sigma} \exp \frac{-(\bar{x}-x)^2}{2\sigma^2}$$

If we differentiate f with respect to x using the chain rule, we obtain

$$\frac{df}{dx} = -\frac{1}{\sqrt{2\pi}} \frac{(\bar{x}-x)}{\sigma^2} \exp \frac{-(\bar{x}-x)^2}{2\sigma^2}$$

The function $f(x)$ is a maximum when df/dx is zero (when the slope tangent is horizontal), but df/dx can be zero only when $x = \bar{x}$. This implies that the frequency of measured values is greatest in the vicinity of the mean value of the distribution.

Suppose we compute the second derivative of $f(x)$:

$$\frac{d^2f}{dx^2} = \frac{1}{\sqrt{2\pi}\,\sigma} \left(\frac{(\bar{x}-x)^2}{\sigma^4} - \frac{1}{\sigma^2} \right) \exp \frac{-(\bar{x}-x)^2}{2\sigma^2}$$

When the second derivative of a function is zero, the slope of the function has a maximum or minimum value, which means that the curvature of the function changes from concave upward to concave downward, or vice versa (see Fig. 2-35). The points for which d^2f/dx^2 is zero are the points for which

$$\frac{(\bar{x}-x)^2}{\sigma^4} = \frac{1}{\sigma^2}$$

Figure 2-35 Standard deviation and the shape of the error function.

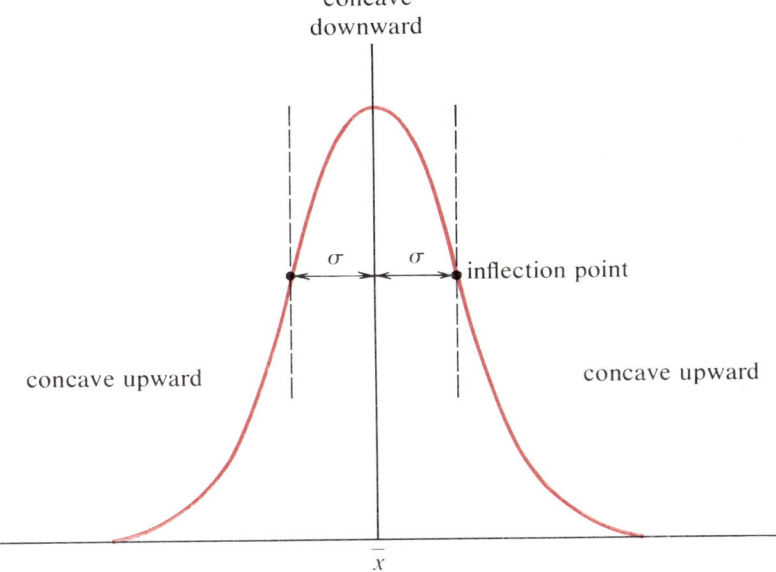

or
$$(\bar{x}-x)^2 = \sigma^2$$

But these are the points at which $x = \bar{x} \pm \sigma$. The points where the curvature of any function changes in this manner are called **points of inflection**. Here the significance of σ is that if we fit the Gaussian function to a large number of measurements, $\sigma = s$ (the standard deviation).

Probability

Our discussion so far has been limited to statistics obtained experimentally, after the fact. It is useful, and frequently possible, to predict in advance the sort of distribution that will be obtained from a measurement. Such a priori calculations are the province of the study of **probability**. The probability that a measurement will yield a certain result may be defined in two ways. The experimental, or a posteriori probability is defined as the fraction of a large number of measurements that give the specified result. The theoretical, or a priori probability is the number of different ways in which the desired result may be obtained, divided by the total number of ways in which all results may be obtained. The first kind of probability is determined by statistical analysis, the second by analysis by the mathematical discipline known as probability.

The simplest applications of probability are to games of chance. Indeed, the origin of the study was an effort by some French gamblers to improve their chances of winning such games through scientific analysis. Fortunately, many problems in chemistry and physics have much in common with games of chance, and a brief consideration of one or two games is instructive.

Probabilities of Small Numbers of Events

We shall limit ourselves here to two very simple games: coin tossing and dice throwing. The behavior of a shuffled deck of cards is another obvious application.

If we toss a "fair" coin, that is, a coin that is equally likely to land on either side and not at all likely to balance on its edge, the probability of throwing heads on a single toss is 0.5. Suppose we toss this coin many times. Certainly we do not expect to see it turn up heads every time. In fact, we expect to get heads half of the time. Now, if we toss a coin six times and get five heads, are we justified in concluding that the coin is not a fair one? Of course not. It certainly is possible to toss five heads in six tosses with a fair coin—it is even possible to toss six out of six heads.

What is the a priori probability of tossing five heads in six tries? The probability of tossing one head in one try is 0.5. If we toss the coin twice, the probability that both tosses will give heads is the probability that the first toss will give heads, multiplied by the probability that the second will give heads. This should be evident since there is a 0.5 probability that the first toss will be heads. If the first toss *is* heads, the probability of tossing a second head is again 0.5, since the two tosses do not influence one another. There are four possible outcomes of two tosses:

First Toss	Second Toss
heads	heads
heads	tails
tails	heads
tails	tails

We see, then, that there is one chance in four (i.e., a probability of 0.25) of tossing two heads, and that this is just the product of the two independent probabilities. Therefore, before the toss of the first coin, the probability of tossing two successive heads was 0.5×0.5 or 0.25.

The probability of any particular sequence of tosses is just the product of the probabilities of each outcome. Thus the probability of tossing heads first, then tails, is also 0.25. However, the probability of tossing one head and one tail is 0.50, since this may be done by tossing heads first and then tails *or vice versa*. The probability that one of two mutually exclusive results will occur is the sum of the separate probabilities.

From the table above, we can make the following statements about the various probabilities:

1. The probability of any particular sequence is 0.25.
2. The probability of tossing two heads is 0.25.
3. The probability of tossing one head and one tail is 0.50.
4. The probability that both tosses will yield the same result is 0.50.

Let us now return to the more complicated problem of tossing the coin six times. Since there are two possible outcomes at each toss, the probability of any given sequence must be $\frac{1}{2} \times \frac{1}{2} \times \frac{1}{2} \times \frac{1}{2} \times \frac{1}{2} \times \frac{1}{2}$, or $(\frac{1}{2})^6$. Since every particular sequence is equally probable we may find the probability of tossing five heads in six tosses simply by counting the number of possible sequences. Now there are 2^6 or 64 possible sequences of six tosses. Only one of these sequences leads to six heads, so the probability of tossing six heads is $\frac{1}{64}$. If we wish to know the probability of tossing five heads, we must count the number of sequences that will give five heads. All this amounts to is counting the number of different positions the single tail can have in the sequence. This is obviously six, since it can occur at any toss. Accordingly, there are six ways to obtain five heads, or the probability of five heads is $\frac{6}{64}$ or 0.094. Thus approximately one sequence of six tosses out of every 10 such sequences will yield five heads.

Counting the number of sequences that will yield four heads is slightly more complicated. Again, it amounts to determining the number of ways of arranging two tails in six tosses. If the first tail appears on the first toss, there are five places remaining for the second. If the first tail appears on the second toss, there are four places remaining. If the first tail appears on the fifth toss, there is only one place for the second. Thus the number of possible sequences that will give four heads is $5+4+3+2+1$ or 15; and the probability of tossing four heads in six tosses is $\frac{15}{64}$ or 0.234.

It is left to the student to calculate the probability of tossing three heads and three tails in a sequence of six.

A more complicated game involves throwing dice. A die has six sides, all of which are equally likely to turn up on a toss (assuming the dice are not loaded). In the usual game, two dice are thrown. The probability that any particular combination of numbers will turn up—assuming distinguishable dice—is just $\frac{1}{6} \times \frac{1}{6}$ or $\frac{1}{36}$. In a game involving several dice, the probability of tossing a particular set of numbers is $(\frac{1}{6})^N$, where N is the number of dice thrown.

Since there is only one way to throw a 2 with a pair of dice, the probability of throwing a 2 is $\frac{1}{36}$. There are six ways to throw a 7: 1-6, 2-5, 3-4, 4-3, 5-2, and 6-1, so the probability of throwing a 7 is $\frac{6}{36}$ or $\frac{1}{6}$.

Probabilities of Large Numbers of Events

In problems in chemistry which involve considerations of probability, the number of possible results is usually extremely large, and the differences between different results extremely small. For instance, the many molecules in a gas sample can have different velocities, but the difference between any two may be very small. Accordingly, it is convenient to use continuous functions to calculate probability. Such functions are called **probability distributions** or **probability densities**. We have already met one such distribution function, the Gaussian distribution function. The probability distribution specifies the fraction, dN/N, of measurements that will fall between x and $x+dx$, and may be represented as $f(x)\,dx$. The function $f(x)$ is the probability density. Note that the differentials come in pairs, as we have previously indicated.

Statistical Averaging

We obtained the arithmetic mean of a set of measurements by adding all the measured values and dividing by the number of measurements. Suppose we have a set of measurements in which each value occurs several times. Then an equivalent method of finding the arithmetic mean is to multiply each value by the number of times it occurs, add the products, and divide by the total number of measurements. This may be expressed mathematically as

$$\bar{x} = \frac{\sum_i n_i x_i}{N} \tag{2-24}$$

where x_i is any particular value measured and n_i is the number of times that value was obtained. Now we have the relationship

$$N = \sum_i n_i$$

If we wished we could have written the equation for the average as

$$\bar{x} = \frac{\sum_i n_i x_i}{\sum_i n_i}$$

If we wished to obtain the average value of the *squares* of all the measurements, we could write

$$(\bar{x}^2) = \frac{\sum_i n_i x_i^2}{\sum_i n_i}$$

Another kind of average value of x can be obtained from the equation

$$\bar{x}' = \frac{\sum_i n_i x_i^2}{\sum_i n_i x_i}$$

This value differs from the arithmetic mean in that it is always greater than the arithmetic mean. (It is equal to the arithmetic mean if all values of x are the same.) This is because squaring the value of x tends to weight the larger values of x more heavily than the smaller values.

Averages from Distribution Functions

The numbers n_i in the foregoing discussion correspond to frequencies or probability distribution functions in cases in which the distribution is regarded as continuous. Since integration of a continuous function corresponds to summation of a set of individual numbers (a **discrete set**), we can find the mean value of a continuous distribution from the formula

$$\bar{x} = \frac{\int_a^b x f(x)\, dx}{\int_a^b f(x)\, dx} \tag{2-25}$$

The constants a and b in this equation are the limiting values on the range of x. Since the value so obtained represents our prediction of the best value from a set of measurements, it is often referred to as the **expected value** or the **expectation value** of x.

Normal Probability Distributions

It is convenient to arrange distribution functions in such a way as to make the integral in the denominator equal to one, that is,

$$\int_a^b f(x)\,dx = 1$$

A distribution function for which this equality holds is said to be **normalized to unity**, or, more commonly, simply **normalized**. If the form of a distribution function is known, it may be readily normalized by dividing the function by its definite integral. Thus, if

$$\int_a^b f(x)\,dx = N \neq 1$$

then a normalized distribution function $F(x)$ may be obtained by dividing $f(x)$ by N:

$$F(x) = \frac{f(x)}{N}$$

The number N is called the **normalization constant**.

The determination of averages in this manner plays an important role in the study of the kinetic-molecular theory as well as in the study of chemical bonding. We shall frequently have occasion to calculate averages of various kinds by the methods described in this section. The mathematical notation used in such calculations often has the appearance of great complexity and difficulty, and the computations are often very tedious. However, from the point of view of understanding what is involved in such work, it is helpful to keep in mind that it amounts to no more than finding an average.

Appendix: The Random Walk and the Gaussian Distribution

The Gaussian distribution has a curious status in scientific work. The operational connection between its mathematical form and the problems to which it is applied is tenuous at best, yet its rigorous applicability is seldom called into question. It is said that "everyone believes that the Gauss distribution describes the distribution of random errors, mathematicians because they think physicists have verified it experimentally, and physicists because they think mathematicians have verified it theoretically."[1]

In fact, the best justification for the use of the Gaussian distribution is that it seems to work better than anything else that has yet been thought of. However, it is possible to derive the equation based on a set of assumptions that may be considered applicable to experimental measurements, although the applicability is open to some question.

Basically, the assumptions are these: random errors are equally likely to make a measurement larger or smaller than the "true" value; random errors may be divided into a number of small individual contributions, all of the same size; these individual contributions are very small and very numerous.

The problem, then, may be described in terms of a "random walk." An excellent description of the random-walk problem is found in Moore's *Physical Chemistry*,[2] and this description will be summarized below.

The problem is often stated somewhat colorfully as follows. A sailor, considerably the worse for drink, wishes to walk a straight line from the Waterfront Bar back to his ship. Unfortunately, he is not able to control his direction, and every step he takes forward also takes him a step to the right or left. At any given step, he is equally likely to go right or left. The question is, how far from the straight and narrow is he likely to stray in a given number of steps?

His distance from his chosen path after he has taken several steps will obviously depend upon the actual sequence of right and left steps, but it may be that many different sequences could get him to the same place. The probability that he will be a certain number of steps n from the path after N total steps is equal to the product of the probability of his following *any* particular

---

1. Young, Hugh D., *Statistical Treatment of Experimental Data*, New York: McGraw-Hill, 1962, p. 65.
2. Moore, Walter J., *Physical Chemistry*, Englewood Cliffs, N.J.: Prentice-Hall, 1962, pp. 232 ff.

sequence times the number of sequences that would lead him n steps off of the path.

The probability of a particular sequence is just the same as the probability of a given sequence of coin tosses, $(\frac{1}{2})^N$. The number of possible sequences is readily obtained, but the method of obtaining it is somewhat subtle and should be examined carefully.

If he is n steps to the right of the path, he must have taken n more steps right than left. If he took L steps left and $L+n$ steps right, and the total number of steps was N we have

$$L+L+n = N$$

or

$$2L+n = N$$

Then

$$L = \frac{N-n}{2}$$

and

$$L+n = \frac{N+n}{2}$$

Thus our friend has taken $(N+n)/2$ steps right and $(N-n)/2$ steps left. How many ways are there in which to do this?

Let us digress. How many ways are there in which we can arrange a deck of cards? We have our choice of 52 cards to go at the bottom of the stack. The next card must be one of only 51, so that there are 52×51 ways to arrange the first two cards. There are only 50 to choose from for the third card, and so there are $52 \times 51 \times 50 \times 49 \times \cdots \times 3 \times 2 \times 1$ ways, or 52! (52 factorial) ways in which the deck can be arranged. Brief consideration will convince you that this is a very large number.

Suppose now we are interested only in the sequences of red and black cards and will ignore the denominations and actual suits. In other words, we consider all red cards indistinguishable and all black cards indistinguishable. Now many previously distinguishable sequences of cards become indistinguishable. In fact, the number of sequences of black cards is 26! and the number of sequences of red cards is 26! The number of distinguishable sequences in which color is the only distinction between cards is

$$\text{number of sequences} = \frac{52!}{26! \times 26!}$$

Although it may not be obvious from inspection, due consideration should show that this is still a large number.

In general, the number of ways in which N objects can be distributed among a set of groups, having $n_1, n_2, n_3, \ldots$ objects, such that $\Sigma_i n_i = N$ is given by

$$\frac{N!}{n_1! \times n_2! \times n_3! \times \cdots}$$

Now let us return to our wandering sailor: he has taken N steps in two groups. One group contains $(N+n)/2$ steps and the other $(N-n)/2$. (You should convince yourself, by the way, that these two numbers are both integers.) The number of possible sequences of steps, then, is just the number of ways in which N steps can be divided into $(N+n)/2$ right steps and $(N-n)/2$ left steps, or

$$\frac{N!}{[(N+n)/2]![(N-n)/2]!}$$

The probability that our sailor is n steps right of the path after N steps is defined as $p(n, N)$. [The probability that he is n steps left of the path would be $p(-n, N)$.] We have the equation

$$P(n, N) = (\tfrac{1}{2})^N \frac{N!}{[(N+n)/2]![(N-n)/2]!}$$

In order to make further progress, it is convenient to take the natural logarithm of both sides of this equation. The natural logarithm of the factorial of a large number may be obtained from Stirling's approximation:

$$\ln N! \cong (N + \tfrac{1}{2}) \ln N - N + \tfrac{1}{2} \ln 2\pi$$

This is an approximate formula valid only for large values of N. We expect that in most cases, our sailor will take about as many steps left as right, so he will stay reasonably close to the path. We can assume then, that if N is very large, n will be very much smaller than N. The logarithm of $p(n, N)$ is

$$\ln p(n, N) = (N + \tfrac{1}{2}) \ln N - \tfrac{1}{2}(N + n + 1) \ln \left(\frac{N+n}{2}\right)$$

$$- \tfrac{1}{2}(N - n + 1) \ln \left(\frac{N-n}{2}\right) - \tfrac{1}{2} \ln 2\pi - N \ln 2$$

Now $(N \pm n)/2$ can be written as $\tfrac{1}{2} N(1 \pm n/N)$, and the logarithm of this expression is just $\ln (N/2) + \ln (1 \pm n/N)$. Because n is much smaller than N, n/N is much smaller than 1. The logarithm of $(1 \pm n/N)$ can be expanded as an infinite series; and for small values of n/N, the series may be truncated after the second term without introducing an unacceptable error:

$$\ln \left(1 \pm \frac{n}{N}\right) = \pm n/N - n^2/2N^2 \pm \cdots$$

Making this substitution, the equation for $p(n, N)$ becomes

$$\ln p(n, N) = (N+\tfrac{1}{2}) \ln N - \tfrac{1}{2} \ln 2\pi - N \ln 2$$

$$-\tfrac{1}{2}(N+n+1)\left(\ln N - \ln 2 + \frac{n}{N} - \frac{n^2}{2N^2}\right)$$

$$-\tfrac{1}{2}(N-n+1)\left(\ln N - \ln 2 - \frac{n}{N} - \frac{n^2}{2N^2}\right)$$

This equation may be simplified immediately to

$$\ln p(n, N) = -\tfrac{1}{2} \ln N + \ln 2 - \tfrac{1}{2} \ln 2\pi - \frac{n^2}{2N}$$

Taking antilogarithms of both sides gives us an expression for the probability in question:

$$p(n, N) = \left(\frac{2}{\pi N}\right)^{1/2} \exp \frac{-n^2}{2N}$$

This expression begins to resemble the Gaussian expression. If we now make our sailor's erratic path a nearly continuous rather than a stepwise progress, we can approach the final result. If the (lateral) distance traversed at each step is s, his distance from the path is ns. If we call this distance x, we can calculate the probability that after N steps he is between x and $x + \Delta x$ from the path. Here we assume that Δx is a small interval but considerably larger than s.

N and n are integers, and n must be even when N is even and odd when N is odd. Accordingly, for a fixed value of N we must consider increments in n to be in even numbers of steps, since one step more to the right means one step fewer to the left. Therefore, the number of positions accessible within a range of Δn steps is $\Delta n/2$, that is, the number of even or odd numbers in the range Δn. If $\Delta x = s \Delta n$, the number of positions accessible within a range Δx is $\Delta x/2s$. Since N is large and Δx is small, the probability of finding the sailor at any position within the range Δx is about the same as that of any other. Accordingly, this probability is

$$p(x, N) = \tfrac{1}{2} p(n, N) \Delta n$$

or

$$p(x, N) = \tfrac{1}{2} p(n, N) \frac{\Delta x}{s}$$

But we may substitute the expression for $p(n, N)$ and rearrange this equation to read

$$p(x, N) = (\tfrac{1}{2} \pi N s^2)^{1/2} \exp\left(\frac{-x^2}{2Ns^2}\right) \Delta x$$

If we identify σ with $(Ns^2)^{1/2}$, this equation is identical to the equation for the Gaussian distribution.

Study Problems

1. Suppose you are given the equation

$$\frac{dP}{dT} = \frac{\Delta H_{vap}}{RT^2/P}$$

   Rearrange the equation to get only P and dP on the left side. ΔH_{vap} and R are constants and should be on the right side with terms in T and the differential dT. Evaluate the definite integral of the left side between limits P_1 and P_2; evaluate the definite integral of the right side between limits T_1 and T_2. The resulting equation is very important in describing the evaporation of a liquid.
2. If the rate of change of the concentration C (of a substance in solution) with respect to time is proportional to the concentration C itself, write an equation describing this situation. Use k_1 for the proportionality constant. Assume C is decreasing so that the rate of change is negative. Separate the variables by getting only terms in the dependent variable C on the left side, only k_1 and terms in t (time) on the right side. Take the definite integral of each side between limits C_0 and C, t_0 and t. Express the answer as an exponential. This expression is useful in describing the changing concentration of a reacting chemical species in solution.
3. Show that the function $\psi = A \sin[(2\pi/h)\sqrt{2mE}]x$ is a solution of the equation $(d^2\psi/dx^2) + (8\pi^2 mE/h^2)\psi = 0$, by differentiating the function and substituting appropriately in the equation. All symbols are constants except ψ and x. This function describes the behavior of a microscopic particle in a box.
4. The enthalpy change in calories, ΔH, of a substance when it is heated is given by $\Delta H = \int_{T_1}^{T_2} C_p \, dT$, where T represents temperature. If CO_2 is to be heated from 300°K until its enthalpy has changed by 1000 cal, what temperature will it reach? $C_p(CO_2) = 6.214 + 10.396 \times 10^{-3} T$.
5. Suppose you have a histogram that is a good approximation to the Gaussian function. The center (highest) bar is 100 units high. How high are the two bars that are one standard deviation away from the mean? How high are the two bars that are 10 standard deviations away from the mean? What does this say about the chances of a random error causing a deviation of 10σ?
6. What are the chances that the top five cards in a well-shuffled deck are hearts?
7. If your two opponents at bridge hold six spades between them, is a 4-2 or a 3-3 split more likely, and why?
8. On p. 85 we stated that the frequency of measured values in a normal distribution of measurements is greatest near the mean value of the distribution. Suppose n careful measurements have been made of a single unknown quantity whose true magnitude is x. The measured values are $a_1, a_2, a_3, \ldots, a_n$, with individual measurement errors of $(x - a_1), (x - a_2), \ldots, (x - a_n)$. A criterion for the most probable value of x is that

it minimizes the sum of the squares of the errors:

$$S = (x-a_1)^2 + (x-a_2)^2 + \cdots + (x-a_n)^2$$

where S is obviously a function of x. We take x as a variable since it is unknown and must be fitted to our measurements. For the function S to be at a minimum its rate of change with respect to x must be zero (otherwise, by changing x a lower value of S could be achieved). In other words, $dS/dx = 0$ at the desired minimum. Show that for the "least-squares" criterion above, the most probable value of x is the algebraic mean.

9. A student measures the molecular weight of a compound by the vapor-density method and obtains the following values: 80.0, 80.0, 81.1, and 81.7 g/mole. Calculate the mean, the average deviation, and the standard deviation for these data. If the student wishes to have a confidence interval for the mean which is no greater than 1.0 g/mole, what is the greatest confidence level he may select?

10. Evaluate $Q = \int_0^\infty e^{-ax} \, dx$. (Think about what happens to the function at the limits.)

Some Further Reading

Kleppner, D., and Ramsey, N., *Quick Calculus*, New York: Wiley, 1965. This is a programmed text, so there is not really as much there as the book appears to have. Very good introductory treatment of the same sort of material treated in the first part of this chapter, done with just a little more rigor.

Butler, J. N., and Bobrow, D. G., *The Calculus of Chemistry*, New York: Benjamin, 1965. Introductory calculus, with some computer programming thrown in. Not quite as easy to follow as Kleppner and Ramsey.

Bauer, E. L., *A Statistics Manual for Chemists*, New York: Academic, 1960. Higher level than the very brief account in this chapter, but not too forbidding. A worthwhile reference.

Hodgman, C. D., ed., *CRC Standard Mathematical Tables*, Cleveland: Chemical Rubber Publishing Co., various dates. Numerous editions of this are around; it has tables and formulas for almost any application. If an integral can be done, it already has been done and the answer is here.

II | CHEMICAL STRUCTURE

3 The Behavior of Gases

In Part II we shall try to develop the basic ideas of our present understanding of the form and structure of matter. Chemistry itself, of course, is concerned predominantly with the ways matter *changes* its form, and we shall move toward an understanding of these changes as we explore the structure of matter. But we cannot hope to understand why the changes occur, or how they can be made to proceed in a desirable fashion, unless we understand the basic structure of matter—what it is that is changing. So for the time being we shall restrict ourselves to inquiring into the basic structure of two of the states of matter, gases in Chapter 3 and solids in Chapter 4. In examining the stoichiometry of compounds and reactions in Chapter 1 we have already encountered the idea that atoms, as fundamental chemical units of matter, provide a simple model for the laws of chemical stoichiometry. One of our concerns will be to explore the possibility that atoms, as a model for the structure of matter, provide a satisfactory explanation for the physical (not chemical) properties of gases and solids. We shall see that the simple model has a great deal of diversity, which will improve our confidence in the adequacy of the model. Having done this, we can begin to look into the internal structure of the atom itself, and then into the ways in which atoms combine into chemically different compounds—this latter is obviously a step toward understanding the ways matter changes its form, or **chemical dynamics**, which is the subject of Part III.

To begin at the beginning, however, we must first inquire into the basic structure of gases. If this is the case, why is the chapter titled "The Behavior of Gases"? The word "behavior" implies change, and we usually think of structure as being relatively static. The answer lies in some of the remarks at the beginning of Chapter 2: the way we find out about the correlations between physical quantities is to change one of them and see how the others change. This is precisely why mathematical functions are useful—as representations of the correlations between these physical properties. So our first task is to express the relationships between the physical properties of gases in the form of mathematical functions or equations. From these relationships we can begin to propose an atomic model for the relationships, and also begin to stipulate the properties the atoms must have.

3-1 The Bulk Properties of Gases

If we isolate a sample of a gas for study we can take four of its physical properties as fundamental: mass, volume, temperature, and pressure. Man has always had crude, qualitative ideas about these properties, but the measurement of them is a relatively recent concept.

Pressure and Boyle's Law

In 1643 Evangelista Torricelli, a pupil of Galileo, first demonstrated that the atmosphere itself has a pressure. He filled a long glass tube with mercury, inverted the tube with the open end (see Fig. 3-1) inside a pool of mercury, and

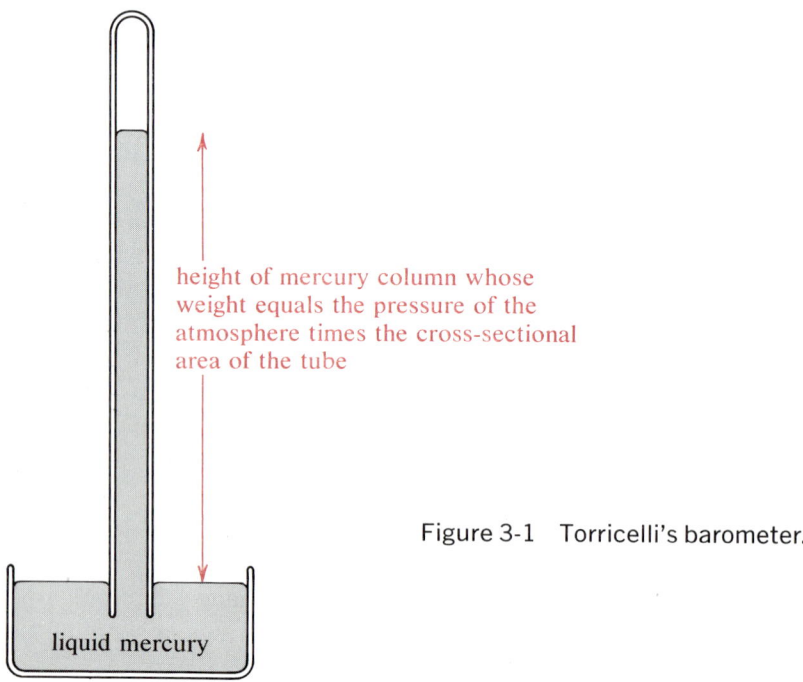

Figure 3-1 Torricelli's barometer.

observed that the mercury inside the tube did not sink to the level of the mercury in the pool. Instead, there was a characteristic height of the mercury surface (or **meniscus**) in the tube above the pool surface. This device, of course, is a **barometer**, and he was measuring the **barometric pressure**. He was able to establish that he was dealing with a pressure (force per unit area), rather than simply with the force necessary to hold up the column of mercury, because the characteristic height, about 76 cm in modern units, remained constant no matter what diameter of tube he used; if, for instance, the cross-sectional area of the tube were doubled, then the volume of mercury inside the tube—and its weight—doubled, and the force necessary to hold it up doubled. But since both the force and the area are doubled, there is a constant quantity that we define as force per unit area or pressure. We can say, then, that the atmospheric pressure (which remained constant through the series of experiments) was equal and opposite to the pressure due to the weight of the column of mercury.

All this suggests that we can measure the pressure of a gas sample by measuring the height of the mercury column it will support; this is the principle of the **manometer**, which is the basic gas-pressure instrument. A **closed-end manometer** is shown in Fig. 3-2; it is essentially the same as Torricelli's barometer. Torricelli's name has survived in the **torr**, defined as a pressure sufficient to support a column of mercury exactly 1 mm high.

In 1660 Robert Boyle, using a U-shaped tube like the manometer in Fig. 3-2, observed that if he added mercury to an empty tube, thereby trapping air in the

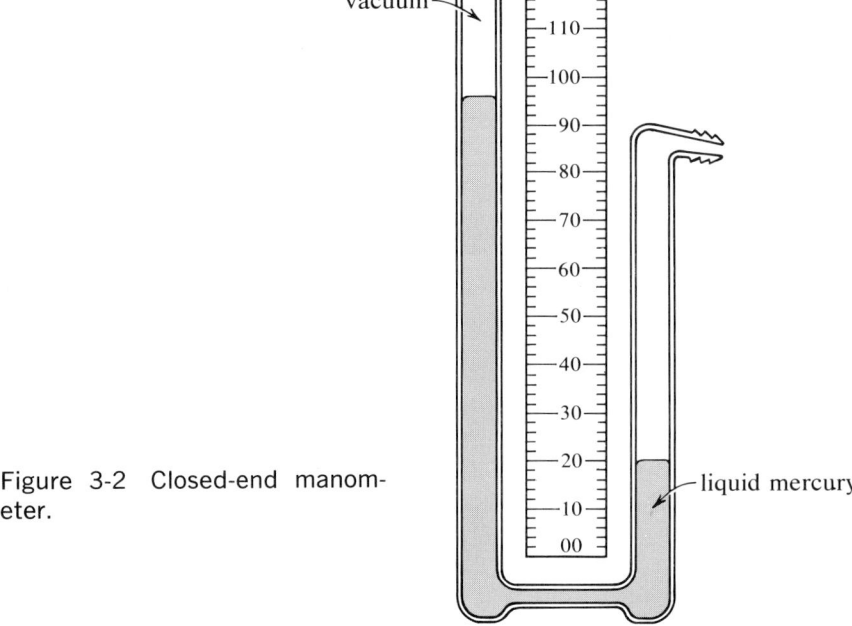

Figure 3-2 Closed-end manometer.

closed arm of the tube, he could halve the volume of the trapped air by doubling the height of mercury it was forced to support in the open arm, as in Fig. 3-3.

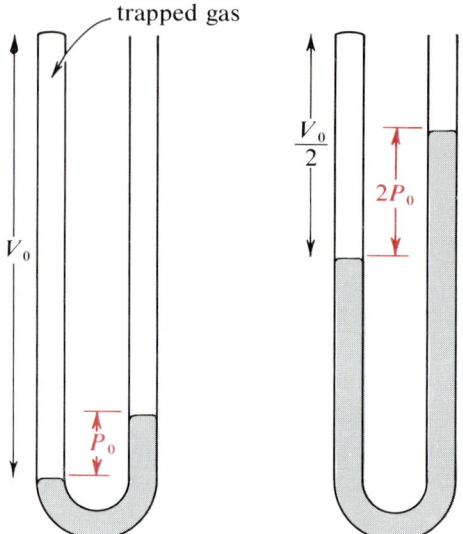

Figure 3-3   Boyle's experiment.

As the pressure on the gas increased, its volume decreased. This relationship is summarized in the equation expressing **Boyle's law**:

$$PV = \text{constant} \tag{3-1}$$

Here the constant is not really constant, but instead depends on the temperature, the mass of the gas, and its identity; however, if we express the quantity of gas by the number of moles of gas present rather than by its mass, the identity of the gas no longer matters—all gases follow Boyle's law reasonably well. We note in passing that it is significant that a physical property should depend on the mole quantity we developed to express chemical combination properties; this suggests that atoms, which serve as a model for the mole, are also involved in the pressure–volume relationship.

Gas Temperature and Charles' Law

Interest in the properties of gases continued, but a considerable time passed before further progress was made in correlating their properties by mathematical functions. In 1802 Joseph Louis Gay-Lussac reported and extended some unpublished work of Jacques Alexandre Cesar Charles (1787). Charles had investigated the effect of changing the temperature of a gas while holding its pressure and mass constant. Using a graduated bulb immersed in a constant-temperature bath, as in Fig. 3-4, he was able to show that the ratio of the volume of a sample of gas at the temperature of boiling water to its volume at the

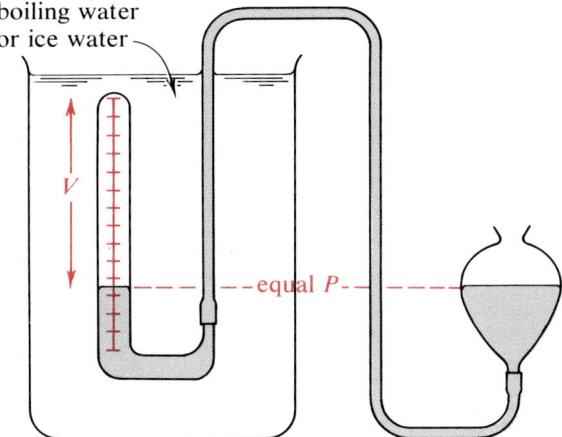

Figure 3-4 Charles' experiment.

temperature of freezing water was constant for a number of gases,

$$\frac{V_b}{V_f} = K \tag{3-2}$$

where V_b is the volume in boiling water and V_f is the volume in freezing water. Gay-Lussac's work indicated a value of 1.375 for K, but more modern measurements give 1.36609 for all gases at sufficiently low pressure. Since this does seem to be a reliable property of gases, we might propose using a gas as a thermometer; once we define a temperature zero, we can call the gas sample's volume at that temperature V_0:

$$V_b = K V_f$$
$$V = K(T) V_0$$

where $K(T)$ is a function of the temperature. Unless we know something about the temperature scale, we have no idea what the function $K(T)$ is like. However, we are entitled to set up a linear **ideal-gas temperature scale** in which we define the temperature scale itself by assuming that $K(T) = 1 + cT$, where c is a constant.

We could set up any kind of temperature scale we liked: cubic, logarithmic, and so on. The linear scale, however, matches a large number of physical properties (such as expansion of mercury in a conventional thermometer). Thus we have a temperature scale in which

$$V = V_0 (1 + cT)$$

Absolute Temperature

Using Gay-Lussac's law and the centigrade temperature scale (T_c), we can evaluate c:

$$\frac{V_{100°}}{V_{0°}} = \frac{V_0(1+100c)}{V_0(1+0c)} = \frac{1+100c}{1} = 1.36609$$

$$c = \frac{1.36609-1}{100} = \frac{0.36609}{100} = \frac{1}{273.16}$$

Gases, then, decrease in volume by an amount $V_0/273.16$ for each centigrade degree of temperature change below the freezing point of water, at least in an idealized sense. It is useful to imagine a gas with these idealized properties, which would reach zero volume at a centigrade temperature of $-273.16°$. Since negative volume would result from further cooling, and we don't understand what that would be, we stipulate that there is an absolute minimum that temperatures can reach, $-273.16°$, and we can set up an absolute temperature scale that has degrees of the same size as centigrade degrees but that has its zero point at this absolute minimum. This is the Kelvin scale (T_K), which we discussed in Chapter 1. Now, we have already written

$$V = V_0(1+cT_c)$$

where V_0 is the volume of the gas sample at 0 °C, not 0 °K. We can rewrite this as $V = V_0 + V_0 cT_c$, and we see that the temperature dependence of the gas volume is given by the term $V_0 cT_c$. Since the degrees are the same on each scale, we can also write

$$V = V(0 °K) + V_0 cT_K$$

but if the gas sample's volume at this absolute temperature zero is zero, then $V(0 °K) = 0$, and

$$V = V_0 cT_K$$

Since V_0 and c are constants, we can see that

$$\frac{V}{T} = \text{constant} \tag{3-3}$$

if we use the Kelvin temperature scale.

Combining Boyle's and Charles' Laws

This new constant in Eq. 3-3, like that in Boyle's law, is not really a constant but depends on the pressure exerted on the gas and the mass of the gas. But, of course, Boyle's law involves the same physical quantities or variables; can we combine them into a single expression? This new expression would involve, explicitly, the pressure P, the volume V, and the Kelvin temperature T, and would be true for the idealized gas that follows Gay-Lussac's law at all temperatures. Let us rewrite Boyle's and Gay-Lussac's laws (Eqs. 3-1 and 3-3) to show that the respective constants really vary with temperature and mass, and pressure and mass:

$$PV = F(T, m)$$

$$\frac{V}{T} = G(P, m)$$

Here $F(T, m)$ is a function of the temperature and mass, but not pressure or volume, and $G(P, m)$ is a function of the pressure and mass, but not volume or temperature. We can solve each expression for V:

$$V = \frac{F(T, m)}{P}$$

$$V = G(P, m)T$$

and set them equal to each other:

$$\frac{F(T, m)}{P} = G(P, m)T$$

We can rearrange this to an expression in which the left side depends only on mass and temperature, and the right side depends only on mass and pressure:

$$\frac{F(T, m)}{T} = G(P, m)P \tag{3-4}$$

Suppose the temperature varied but the pressure remained constant in our gas sample; then the right side of Eq. 3-4 would be constant (or at least would depend only on mass), a fact that we can express by replacing it by a new quantity that is a function of mass only, $H(m)$:

$$\frac{F(T, m)}{T} = H(m)$$

Multiplying this out, we have

$$F(T, m) = H(m)T$$

and if we insert this in Boyle's law we have

$$PV = H(m)T \tag{3-5}$$

which represents the combination of Boyle's and Gay-Lussac's laws we sought.

The Ideal Gas Law

In 1811 Amadeo Avogadro proposed a hypothesis which has since been fully accepted and is an integral part of our understanding of the behavior of gases, even though it is not quite in the nature of a law. He suggested that at equal temperature and pressure, any two gases contain the same number of molecules in the same volume. Algebraically,

$$V = V_0(P, T)n \qquad (3\text{-}6)$$

where $V_0(P, T)$ is a standard volume for any gas and n is a measure of the number of molecules present. This is where the mole reenters our discussion. If n is stipulated to be the number of moles of gas present, then $V_0(P, T)$ is simply the volume of 1 mole of gas; it varies with P and T, but should be the same for all gases at a standard P and T. As a matter of fact, careful measurements indicate a volume of 22.414 l as V_0 for 1 mole at 0 °C and 1 atm pressure. We may combine the algebraic statement of Avogadro's hypothesis (see Study Problem 1) with the combined statement of Boyle's and Gay-Lussac's laws (Eqs. 3-5 and 3-6) to produce a single algebraic equation that contains no undetermined functions at all:

$$PV = nRT \qquad (3\text{-}7)$$

in which R is now a universal constant for all gases and all conditions. If pressure is expressed in atmospheres, volume in liters, n in gram moles, and temperature in degrees Kelvin, $R = 0.08205$ l atm/mole °K. This Eq. 3-7, which relates all the characteristic physical properties of gases to each other, is an **equation of state** for gases, known as the **ideal gas law**. Remember that since all gases liquefy at temperatures above 0 °K, they are not really ideal, and it should not be surprising that deviations from ideal-gas behavior can reach several percent even under rather ordinary conditions of temperature and pressure. Other equations of state have been proposed to account for these deviations, and we shall look at some of them later in this chapter. It is true, however, that for most purposes the ideal gas law is quite adequate to describe the behavior of real gases, and we shall later have occasion to use it in several contexts, particularly in connection with thermodynamics (Chapters 8 and 9). Because of its general use, the student should be sure that he understands how to apply it and manipulate it for any desired purpose; the appendix to this chapter and some study problems are provided for this sort of practice. It is probably the most fundamental example of a mathematical function as a description of the interactions of physical properties, and so has considerable importance.

3-2 The Molecular Model and the Kinetic Theory of Gases

All of Section 3-1 has been simply a description of experimental fact; we have, in effect, been developing a certain portion of natural law. But at the end of Chapter 1 we indicated that our purpose in Chapters 3 and 4 would be not only to describe the behavior of gases and solids, but to show that the diverse relationships that appear experimentally can be explained—or rationalized—by the atomic or molecular model for the structure of matter. At this point we have certainly not exhausted the experimental properties of bulk gases (and we shall return to discuss more properties), but we have produced a very fundamental result in the ideal gas law. What does the molecular model say about it? Perhaps a better way of asking the question is to say, what must be true of our hypothetical molecules if they are to explain the experimental facts? We shall see that forcing the molecular model to fit the observed properties of gases gives us some very interesting information about what molecules must be like.

The Mechanical Properties of Molecules and Boyle's Law

A gas has a very low density compared to that of the same compound condensed to a liquid or solid; also, all gases are completely miscible, unlike some liquids (e.g., oil and water). Then if a gas is made up of molecules that contain all the mass of the gas, they must be far apart—meaning many molecular diameters. Only if there is a lot of open space between the molecules can oil vapor and water vapor mix and the density be so low. Since what seems to be important is the open space, we can simplify our model by assuming provisionally that each molecule is an infinitesimal point mass. There will be very many molecules in any reasonable sample, so many that we can assume that statistical averages are very accurate. We can number the molecules, in principle at least, from 1 to N; then the ith molecule will have mass m_i. Gases exert pressure on their container walls, which the walls balance by an inward force; so there must be a force acting on the molecules that shows up on a large scale as pressure. One way to look at force is to regard it as the time rate of change of momentum (see Table 1-1), so if we assume that each molecule is moving with velocity v_i, then changing the momentum $m_i v_i$ provides the force we observe. Since this seems to provide us with the force we need, let us keep the model simple by assuming that there are no other forces—the molecules neither attract nor repel each other. Therefore, we have widely separated, noninteracting particles, each with position x_i, y_i, z_i, mass m_i, and velocity v_i.

In this large collection of molecules there will be motion in all three directions. For simplicity, however, consider only those moving in the x direction inside the cubical container shown in Fig. 3-5, with length L and area of a single face $A = L^2$. If they are moving only in the x direction, we need consider only the x coordinate, x_i, of each molecule, and the x velocity, V_{x_i}. We can expand to three dimensions later.

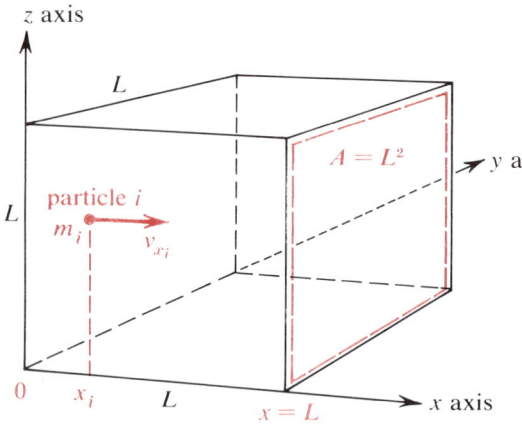

Figure 3-5 Gas molecule in motion inside a cubical container with dimensions $L \times L \times L$.

Define a quantity M as follows:

$$M = \tfrac{1}{2} \sum_i m_i x_i^2$$

M doesn't represent any physical quantity, but let's see what we can get from it. Use the chain rule (with $u = x$) to differentiate M with respect to time:

$$\frac{dM}{dt} = \frac{dM}{dx}\frac{dx}{dt} = \left[\tfrac{1}{2}\sum_i m_i(2x_i)\right]\left(\frac{dx_i}{dt}\right) = \sum_i m_i x_i \frac{dx_i}{dt}$$

Now differentiate this again, using the fact that the derivative of the product of two functions u and v with respect to t is $u(dv/dt) + v(du/dt)$:

$$u = x_i \qquad v = \frac{dx_i}{dt}$$

$$\frac{d^2M}{dt^2} = \sum_i m_i x_i \frac{d^2 x_i}{dt^2} + \sum_i m_i\left(\frac{dx_i}{dt}\right)^2$$

What is the average of this second derivative of M over a long period of time?

We can use the averaging procedure we defined in Chapter 2:

$$\left\langle \frac{d^2M}{dt^2} \right\rangle_{time} = \frac{\int_0^\tau (d^2M/dt^2)\,dt}{\int_0^\tau dt} = \frac{\int_0^\tau (d^2M/dt^2)\,dt}{\tau - 0} = \frac{1}{\tau}\left[\left(\frac{dM}{dt}\right)_\tau - \left(\frac{dM}{dt}\right)_0\right]$$

where the angular brackets indicate average. Looking at the earlier expression for dM/dt, we see that neither m_i, x_i, nor dx_i/dt (the x velocity) can be infinite; so when the time becomes very long (τ becomes large), the time average of the second derivative approaches zero, since τ is in the denominator. Rewriting d^2M/dt^2 as a time average, then,

$$\left\langle \frac{d^2M}{dt^2} \right\rangle_{time} = \left\langle \sum_i m_i \left(\frac{dx_i}{dt}\right)^2 \right\rangle_{time} + \left\langle \sum_i m_i x_i \frac{d^2x_i}{dt^2} \right\rangle_{time} = 0$$

In the first of the time-average terms on the right we have the product of the particle's mass and the square of its velocity in the x direction. Although the quantity M does not have physical significance, this new quantity does; it is very closely related to the energy of motion or **kinetic energy** of the particle in the x direction. From elementary physics we define kinetic energy as

$$\text{kinetic energy (KE)} \equiv \tfrac{1}{2}mv^2$$

Clearly the first time-average summation is equal to twice the total kinetic energy of the gas molecules in the x direction, so we can write

$$2\langle KE_x \rangle_{time} + \left\langle \sum_i m_i x_i \frac{d^2x_i}{dt^2} \right\rangle_{time} = 0$$

$$\langle KE_x \rangle_{time} = -\frac{1}{2}\left\langle \sum_i m_i x_i \frac{d^2x_i}{dt^2} \right\rangle$$

Now $m_i\,d^2x_i/dt^2$ is the mass times the acceleration of particle i, which is the force acting on particle i. Let us insert F_{x_i} for $m_i\,d^2x_i/dt^2$, and at the same time expand the whole thing to three dimensions (since y and z motions are the same):

$$\langle KE_x \rangle_{time} + \langle KE_y \rangle_{time} + \langle KE_z \rangle_{time} = \langle KE_{total} \rangle_{time}$$

$$= -\tfrac{1}{2}\left[\left\langle \sum_i x_i F_{x_i}\right\rangle_{time} + \left\langle \sum_i y_i F_{y_i}\right\rangle_{time} + \left\langle \sum_i z_i F_{z_i}\right\rangle_{time}\right]$$

The quantity on the right side of the equation is called the **virial**; it is used in several kinds of mathematical treatments of physical problems. Interpreting it here, we observe that no force acts on the molecules in our gas except at $x_i = L$, $y_i = L$, and $z_i = L$, where the molecules hit the wall of the container and are turned around, transferring momentum to the walls of the container. Force is also exerted on the faces of the box where x_i, y_i, and z_i are zero, but these terms in the virial are zero. The force, of course, is just the pressure *in* on the gas $(-P)$ times the area of the face of the container that the molecules are hitting, $\Sigma_i A_i = A$ or L^2:

$$\langle KE_{total} \rangle_{time} = -\tfrac{1}{2}\left[\left\langle \sum_i L(-PA_i)\right\rangle_{time} + \left\langle \sum_i L(-PA_i)\right\rangle_{time} + \left\langle \sum_i L(-PA_i)\right\rangle_{time}\right]$$

$$= \tfrac{1}{2}PL^3 + \tfrac{1}{2}PL^3 + \tfrac{1}{2}PL^3$$

$$= \tfrac{3}{2}PV \tag{3-8}$$

So the average kinetic energy of our system of molecules is proportional to the product of the pressure and the volume of our system. If we consider the kinetic energy of the system as a function of the temperature, we have an expression exactly like Boyle's law (Eq. 3-1). Our model, then, will predict Boyle's law. Of course, we have taken a very special condition (cubical box, no diagonal motion, etc.), but if the problem is generalized to remove these special aspects, the result is exactly the same.

Energy and Degrees of Freedom

We shall want to examine the significance of the average kinetic energy, but this is a good place to make an observation about a general property of systems containing a very large number of molecules. Each of the three directions in which a molecule can move in the above example is an independent possibility for the development of kinetic energy; a given molecule may or may not express its kinetic energy by moving in, say, the y direction. This motion through space is known as **translation**, and we say that there are three **translational degrees of freedom** open to each molecule. For the system as a whole, each of these degrees of freedom contributes $\tfrac{1}{2}PV$ (or, if we make the jump to the ideal gas law, $\tfrac{1}{2}nRT$) to the total kinetic energy of the system. In other words, the kinetic energy is partitioned equally among all the degrees of freedom open to each molecule at $\tfrac{1}{2}nRT$ per degree of freedom; this is known as the **equipartition principle**. If we anticipate our discussion of molecular structure by noting that two atoms that are bonded together may vibrate in and out, or rotate, we may expect to find vibrational and rotational degrees of freedom; under appropriate circumstances each rotational degree of freedom will also contribute $\tfrac{1}{2}nRT$ to the energy of the system and each vibrational degree, nRT. The vibrational degrees of freedom are worth twice as much because they can express two types of energy: kinetic energy and potential energy. This will be important to us when we consider the addition of heat energy to the gas.

The Velocity of Gas Molecules

Returning to the theoretical expression we have developed to reproduce Boyle's law,

$$PV = \tfrac{2}{3} \langle KE_{total} \rangle_{time} \tag{3-9}$$

what is meant by the total kinetic energy, averaged over time? The total kinetic energy is just the sum of the individual molecular kinetic energies, with the N individual molecules' velocities being time-averaged:

$$\langle KE_{total} \rangle_{time} = \sum_{i=1}^{N} \tfrac{1}{2} m_i \langle v_i^2 \rangle_{time}$$

From the form of this summation we see that if we define a new kind of average velocity, the **root-mean-square velocity** v_{rms},

$$v_{rms} \equiv \sqrt{\langle v_i^2 \rangle_{time}}$$

then Boyle's law can be interpreted very simply in terms of molecular motion:

$$PV = \tfrac{2}{3} \left[\sum_{i=1}^{N} \tfrac{1}{2} m_i v_{rms}^2 \right] = \tfrac{1}{3} N m_i v_{rms}^2$$

It is interesting to calculate an approximate magnitude for the root-mean-square velocity from this equation. First, note that the total number of molecules times the mass of one molecule is equal to the total mass of the system, which we can also express as the number of moles times the gram-molecular weight:

$$N m_i = \text{total mass} = n(\text{MW})$$

Now, using the ideal gas law, substitute nRT for PV and observe that the numbers of moles cancel (meaning that the velocity does not depend on how large the system is):

$$v_{rms}^2 = \frac{3PV}{Nm_i} = \frac{3PV}{n(\text{MW})} = \frac{3nRT}{n(\text{MW})} = \frac{3RT}{(\text{MW})}$$

So

$$v_{rms} = \sqrt{\frac{3RT}{(\text{MW})}} \tag{3-10}$$

From this expression we could immediately obtain a value for the magnitude of the rms velocity if we had appropriate values to plug in for the quantities involved. Unfortunately, the only value we have established for R has units of liter-atmospheres per mole-degree Kelvin. However, if we express atmospheric pressure as a force in dynes per square centimeter (which we can do by establishing the weight of 760 mm of mercury), we find that we can convert liter-atmospheres directly to ergs (see Study Problem 2): $R = 0.08205$ l-atm/mole-°K $= 8.314 \times 10^7$ ergs/mole-°K. Using this value and taking nitrogen gas (N_2) at room temperature (300 °K),

$$v_{\text{rms}} = \sqrt{\frac{3RT}{(\text{MW})}} = \sqrt{\frac{3 \times 8.314 \times 10^7 \text{ ergs/mole-°K} \times 300 \text{ °K}}{28.0 \text{ g/mole N}_2}}$$

$$= \sqrt{2.68 \times 10^9 \frac{\text{g-cm}^2/\text{sec}^2}{\text{g}}} = 5.17 \times 10^4 \text{ cm/sec}$$

This corresponds to approximately 1150 mph (an interesting unit-conversion problem in itself), so it is clear that our molecular model means that each molecule must be moving very rapidly. Incidentally, since liter-atmospheres convert directly to ergs, the product of pressure and volume must have the dimensions of energy; this is also apparent from the form of Eq. 3-9.

The Distribution of Gas-Molecule Velocities

Up to this point, our derivation would have worked equally well if we had assumed that each molecule had an identical velocity equal to the rms velocity. However, we have not *had* to assume this in order to carry out the derivation, and if we think about it this must not be the case in a real sample of gas. For if we set up an idealized gas sample with billions of molecules, all moving with the same speed but in random directions, they will be very likely to collide (they wouldn't if they really were infinitesimal mass points, but that is obviously an oversimplification). The collisions tend to slow a molecule down if it is hit from the front, but speed it up if it is hit obliquely or from behind—so the initial uniform velocity will shortly turn into a range of velocities both greater and less than the initial velocity. Now, since it is equally likely that a given molecule will be hit from the front or rear, the spreading out of molecular velocities (for a very large number of collisions) will be governed by the same considerations as the random walk, in which a step to the right or left was equally likely. So we can apply the Gaussian distribution function we quoted in Chapter 2 and derived in its appendix.

What we are going to do, then, is develop a **velocity distribution function** that describes the relative probability that a molecule will have a given velocity; in other words, the relative number of molecules as a function of velocity. A word of caution before we begin: the derivation is not intended as something to be memorized for repetition. Try only to follow the train of thought so that when we arrive at a velocity distribution function its form will seem logical.

To begin, we quote the Gaussian function from Chapter 2:

$$\frac{dN}{N} = \left(\frac{1}{2\pi\sigma^2}\right)^{1/2} e^{-(\bar{x}-x)^2/2\sigma^2} \, dx$$

This is a one-dimensional function, so we begin by considering motion in only one dimension, say the x direction. We replace x in the Gaussian function by v_x, the velocity in the x direction, dx by dv_x, and $\bar{x}$ by $\langle v_x \rangle$, the long-time average of v_x after many collisions. The quantity dN is replaced by dN_x, the probability of finding a molecule with a certain x-direction velocity. All this gives us

$$\frac{dN_x}{N} = \left(\frac{1}{2\pi\sigma^2}\right)^{1/2} e^{-(\langle v_x \rangle - v_x)^2/2\sigma^2} \, dv_x$$

The quantity σ^2 is a measure of the broadening of the range of velocities (see Fig. 2-35). Now $\langle v_x \rangle$, after a very large number of collisions, will be zero because there will be as many molecules moving in the $+x$ direction as in the $-x$ direction. But this algebraic-sign effect disappears if we square the velocity, so $\langle v_x^2 \rangle$ will *not* equal zero; it will instead be a measure of the range of observed velocities, since the larger $\langle v_x^2 \rangle$ is, the larger the range (from plus to minus) is. So we can identify σ^2, which must have dimensions of (velocity)^2, with $\langle v_x^2 \rangle$. Thus we need an algebraic expression for $\langle v_x^2 \rangle$ in terms of observable physical quantities. From Eq. 3-8 we have

$$\langle KE_x \rangle = \tfrac{1}{2} PV$$

But if we have N molecules in the gas sample and $\tfrac{1}{2} m \langle v_x^2 \rangle$ is the kinetic energy of a single average molecule,

$$\langle KE_x \rangle = \tfrac{1}{2} Nm \langle v_x^2 \rangle$$

and from the ideal gas law, $\tfrac{1}{2} PV = \tfrac{1}{2} nRT$. Substituting these two quantities in Eq. 3-8,

$$\tfrac{1}{2} Nm \langle v_x^2 \rangle = \tfrac{1}{2} nRT$$

Now let us consider a 1-mole sample; in this case, $n = 1$, and $N = N_0$, Avogadro's number (6.0222×10^{23} molecules/mole). This gives us

$$N_0 m \langle v_x^2 \rangle = RT$$

or

$$\langle v_x^2 \rangle = \frac{RT}{N_0 m}$$

Suppose we now define a new constant $k = R/N_0$; in effect this is the gas constant for a single molecule instead of a mole. This is called **Boltzmann's constant** after its original user. If we divide R in erg/mole-°K by N_0 in molecules/mole we get $k = 1.3806 \times 10^{-16}$ erg/molecule-°K. Using Boltzmann's constant and equating σ^2 with $\langle v_x^2 \rangle$, we have

$$\sigma^2 = \langle v_x^2 \rangle = \frac{kT}{m}$$

and inserting this in the Gaussian function for our velocity distribution gives

$$\frac{dN_x}{N} = \left(\frac{m}{2\pi kT}\right)^{1/2} e^{-m(\langle v_x \rangle - v_x)^2/2kT}\, dv_x$$

But we have already observed that $\langle v_x \rangle = 0$, so

$$\frac{dN_x}{N} = \left(\frac{m}{2\pi kT}\right)^{1/2} e^{-mv_x^2/2kT}\, dv_x \qquad (3\text{-}11)$$

Third, we wish to expand our description to three dimensions. The three degrees of freedom x, y, and z can each be described by a distribution function exactly like Eq. 3-11 (with appropriate replacement of x by y or z). Equation 3-11 represents the probability that a molecule will have an x velocity near v_x. Suppose we look at all the molecules with v_x; they will have a complete distribution of velocities in the y direction, and we can define a new distribution function that represents the probability that a molecule will simultaneously have an x velocity near v_x and a y velocity near v_y. As we saw in Chapter 2, probabilities of independent events multiply to give the probability of a multiple event, so the new distribution function is the product of the independent x and y distribution functions:

$$\frac{dN_{x,y}}{N} = \frac{dN_x}{N}\frac{dN_{x,y}}{dN_x} = \left(\frac{m}{2\pi kT}\right)^{1/2} e^{-mv_x^2/2kT}\, dv_x \left(\frac{m}{2\pi kT}\right)^{1/2} e^{-mv_y^2/2kT}\, dv_y$$

Here dN_x is the number of molecules with v_x, and $dN_{x,y}$ is the smaller number with both v_x and v_y. Clearly we can go through the whole process again for the distribution of z velocities to get a still smaller number $dN_{x,y,z}$, which is the number of molecules simultaneously having v_x and v_y and v_z. Again the simultaneous probability is a product, and

$$\frac{dN_{x,y,z}}{N} = \frac{dN_x}{N}\frac{dN_{x,y}}{dN_x}\frac{dN_{x,y,z}}{dN_{x,y}}$$

$$= \left[\left(\frac{m}{2\pi kT}\right)^{1/2}\right]^3 e^{-mv_x^2/2kT}\, e^{-mv_y^2/2kT}\, e^{-mv_z^2/2kT}\, dv_x\, dv_y\, dv_z$$

But multiplying means adding exponents:

$$\frac{dN_{x,y,z}}{N} = \left(\frac{m}{2\pi kT}\right)^{3/2} e^{-m(v_x^2+v_y^2+v_z^2)/2kT}\, dv_x\, dv_y\, dv_z \qquad (3\text{-}12)$$

This equation, then, represents the probability that a molecule will simultaneously have an x velocity near v_x, a y velocity near v_y, and a z velocity near v_z.

The distribution function we want does not involve the various components of the velocities, only the overall magnitudes of the velocities themselves—v as opposed to v_x, v_y, and v_z. Suppose we draw the graph shown in Fig. 3-6, in

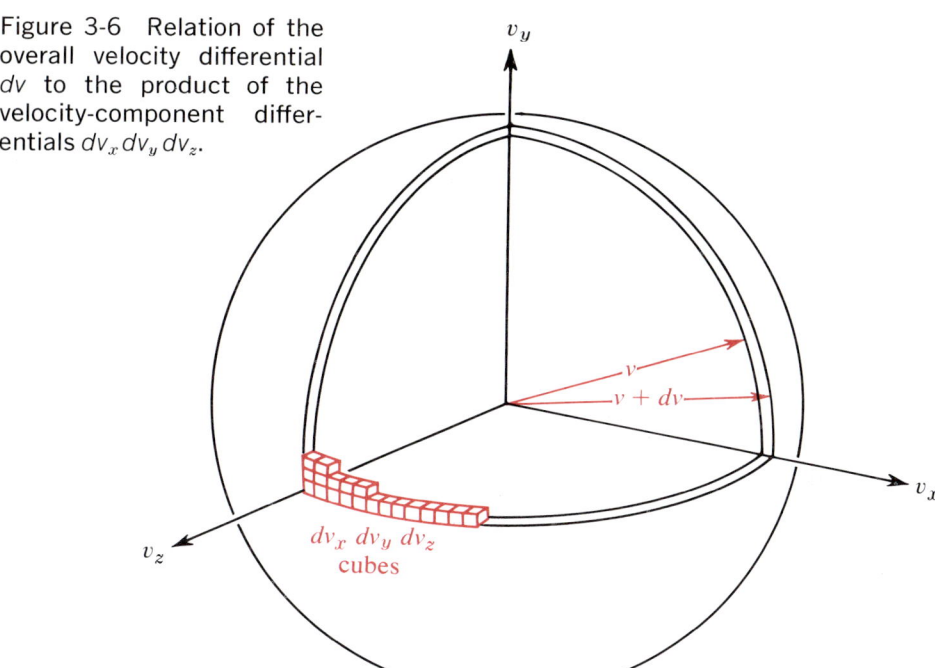

Figure 3-6 Relation of the overall velocity differential dv to the product of the velocity-component differentials $dv_x\, dv_y\, dv_z$.

which the three independent velocities, not the coordinates themselves, are plotted along the three axes. The sphere is simply a sketch of the Pythagorean theorem at work, since $v^2 = v_x^2 + v_y^2 + v_z^2$. The velocity distribution function we want is one that gives the relative probability of finding a molecule with a velocity between v and $v + dv$, a velocity range that is different from that indicated above by $dv_x\, dv_y\, dv_z$. Looking at Fig. 3-6, however, we can see that this new velocity range is just the volume of the thin shell surrounding the sphere, which is equal in turn to the surface area of the sphere times the thickness dv of the shell:

$$(\text{velocity range corresponding to } dN) = 4\pi v^2\, dv$$

In Fig. 3-6 we see that the spherical shell representing the velocity range $4\pi v^2\, dv$ can in a sense be built up from the little blocks $dv_x\, dv_y\, dv_z$ just as an igloo is built up from individual blocks. This corresponds to saying that there are all kinds of ways to get different v_x's, v_y's, and v_z's to add up to a given v (using the Pythagorean theorem). Accordingly, for this velocity range we expect a different and larger number of molecules than for the range $dv_x\, dv_y\, dv_z$; we have called this number dN. In Eq. 3-12, we can replace $dN_{x,y,z}$ by the larger and more general number dN (which does not depend on the components v_x, v_y, v_z) if we replace the velocity range $dv_x\, dv_y\, dv_z$ by the larger and more general velocity range $4\pi v^2\, dv$:

$$\frac{dN}{N} = \left(\frac{m}{2\pi kT}\right)^{3/2} e^{-mv^2/2kT} 4\pi v^2 \, dv$$

or (finally)

$$\frac{dN}{N} = 4\pi \left(\frac{m}{2\pi kT}\right)^{3/2} e^{-mv^2/2kT} v^2 \, dv \qquad (3\text{-}13)$$

The Maxwell–Boltzmann Function

Equation 3–13 is the **Maxwell–Boltzmann velocity distribution function**, named after the two men who developed the statistical treatment of molecular motion in gases, James Clerk Maxwell and Ludwig Boltzmann. It is particularly important and widely applicable since it gives us a means of understanding how some molecules can have a velocity much greater than the average velocity of the whole group of molecules, and also describes the effect of temperature on this distribution. Notice that the function is the product of two factors, a negative exponential in v^2 and also a v^2 parabola, as shown in Fig. 3-7. When these two are multiplied together, a humped curve results, since the parabola pre-

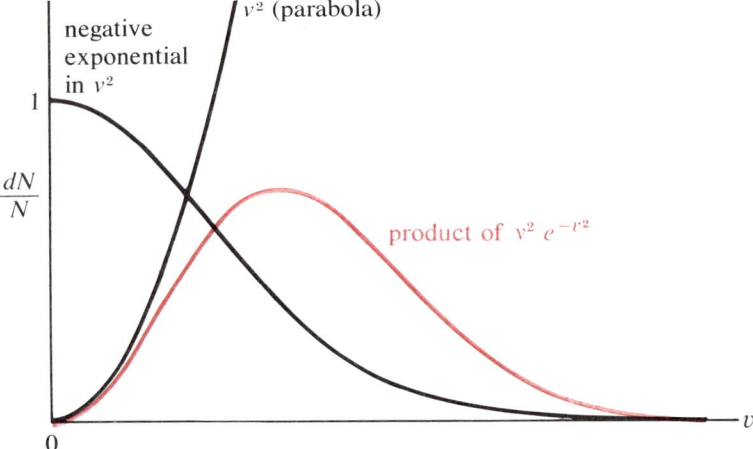

Figure 3-7 Factors influencing Maxwell–Boltzmann velocity-distribution curve shape.

dominates at small v and the negative exponential predominates at large v; Fig. 3-8 shows the result at three different temperatures for nitrogen. Notice first that the peak of the curve, which we can call the **most probable velocity**, is very near the root-mean-square velocity we calculated earlier; it is not exactly the same, however, since these are basically two different ways of defining an average velocity. Second, note that there is a significant probability that a molecule will have twice the most-probable velocity, but only a very small

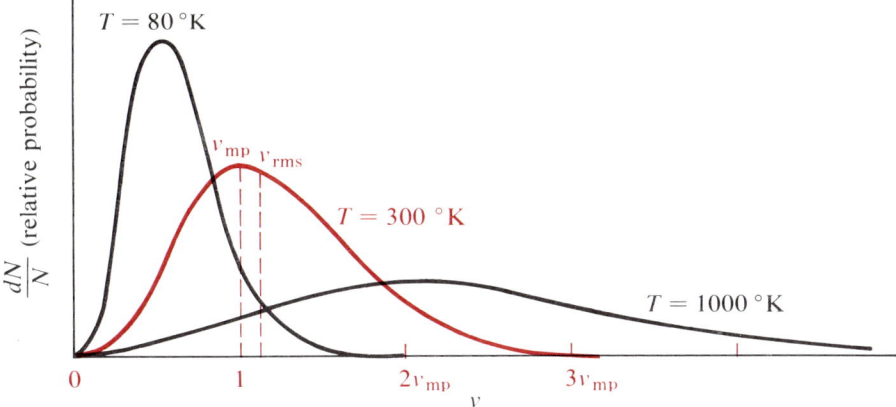

Figure 3-8 Temperature dependence of the Maxwell–Boltzmann velocity distribution.

chance that it will have three or four times that velocity. Third, observe that the effect of raising the temperature of the gas sample is to broaden the curve and shift the maximum to a higher velocity. The area under the curve is constant for all three curves, since it represents the total probability of the molecules' having any velocity between zero and infinity. This must be 1.00, since every molecule must have some sort of velocity, even if it is zero. So the area under all three curves is equal to 1.00 unit. As we raise the temperature, then, the number of molecules having a velocity less than the most-probable velocity at the lower temperature shrinks, but the number having the higher velocities increases sharply. This is a very important concept; it bears directly on the evaporation of liquids and the effect of heating on chemical reaction rates, among many other things.

As a matter of fact, the Maxwell–Boltzmann velocity distribution function is only a special case of a more general concept. The v^2 term in Eq. 3-13 results from the resolution of velocity into x, y, and z components, not from any basic statistical law. In effect, the statistics of the problem is represented by the factor $e^{-mv^2/2kT}$. Looking at the exponent, we see that the quantity $mv^2/2$ is the kinetic energy of a molecule, so the exponent is just the ratio of the kinetic energy to kT, which is the available thermal energy. It is quite generally true that for any system that has different arrangements leading to different energies, the number of molecules having any particular arrangement is proportional to $e^{-\Delta E/kT}$, where ΔE is the difference in energy between the particular arrangement in which we are interested and the most stable possible arrangement. In this case, the most stable (lowest-energy) possible arrangement is zero kinetic energy, so $\Delta E = \frac{1}{2}mv^2$ for any particular v. Later we shall see that many different energy relationships are governed by $e^{-\Delta E/kT}$, the **Boltzmann energy-distribution factor**.

In this section we have developed a lot of theory about the behavior of a gas, all supported by a single experimental relationship, the ideal gas law. If the theory is successful, it must deal with a diverse assortment of experimental results. Let us see what other aspects of gas behavior we can relate to this model.

3-3 Transport Properties, Gas Mixtures, and the Kinetic Theory

A Molecular Model for Gas Viscosity

Gases do a lot of things besides respond to pressure and temperature changes, and we have several possibilities to inspect. One property of gases which, superficially at least, seems least related to the pressure–volume relationship is viscosity. We are familiar with the fact that liquids have viscosities — molasses is very viscous but water is not — but it is also true that gases have viscosities, although much smaller ones than liquids. Viscosity is the drag which opposes a shearing (sideways) force applied to a fluid; it results in smaller and smaller fluid velocities the greater the distance through the fluid from the point at which the force is applied, as Fig. 3-9 indicates. If a gas is flowing past a stationary solid

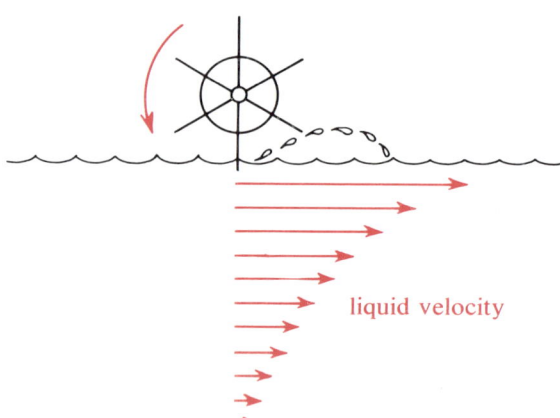

Figure 3-9   Effect of viscosity on the velocity of moving liquid.

surface (for instance, through a tube) the layer of gas next to the stationary surface will also be stationary, and the velocity will increase as the distance from the surface increases. We can regard this as a series of layers of gas, with each layer moving progressively faster the farther it is from the stationary surface, as shown in Fig. 3-10. Experimentally the viscous drag force has been found to

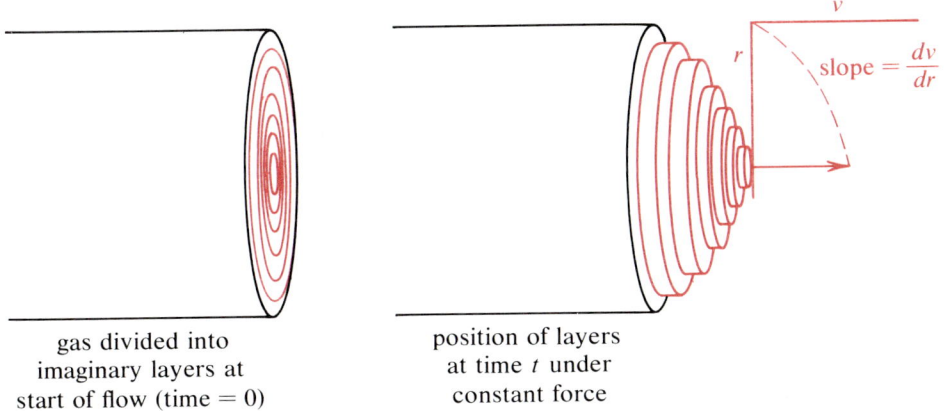

gas divided into imaginary layers at start of flow (time = 0)

position of layers at time t under constant force

Figure 3-10 Velocity profile of moving gas in a tube, with viscous drag exerted by the walls.

be proportional to the area of the interface between layers, and also proportional to the velocity of the fastest-moving layer, but inversely proportional to the distance of the fast-moving layers from the stationary surface (it takes more force to make the gas that is very near the surface move). We can combine the last two factors by defining the **velocity gradient**, which is the rate at which velocity in a direction parallel to the stationary surface changes with distance from the stationary surface: dv/dr, as in Fig. 3-10. We can represent this relationship algebraically:

$$f = \eta S \frac{dv}{dr}$$

where f is the viscous drag force, S is the layer interface area, and dv/dr is the velocity gradient in the flowing gas. The proportionality constant η (eta) is called the **coefficient of viscosity**; the larger it is, the greater the viscous drag force. The value of η is characteristic of the fluid, and our theory should predict it and also its relationship to temperature and pressure changes.

Collision Number and Mean Free Path

To produce a theoretical interpretation of the coefficient of viscosity, it is necessary to look more closely at a feature of the model we have already mentioned, the collisions between molecules. The existence of collisions requires that the molecules have a finite diameter, and this diameter is one of the features of the model that viscosity experiments will dictate to us. It is intuitively reasonable (although it can also be proved) that a spherical molecule with diameter D moving through a random assortment of other similar molecules undergoes the same number of collisions per second as a spherical molecule with diameter $2D$ moving through other geometric-point molecules (see Fig. 3-11). This being the case, we need only calculate the volume swept out by

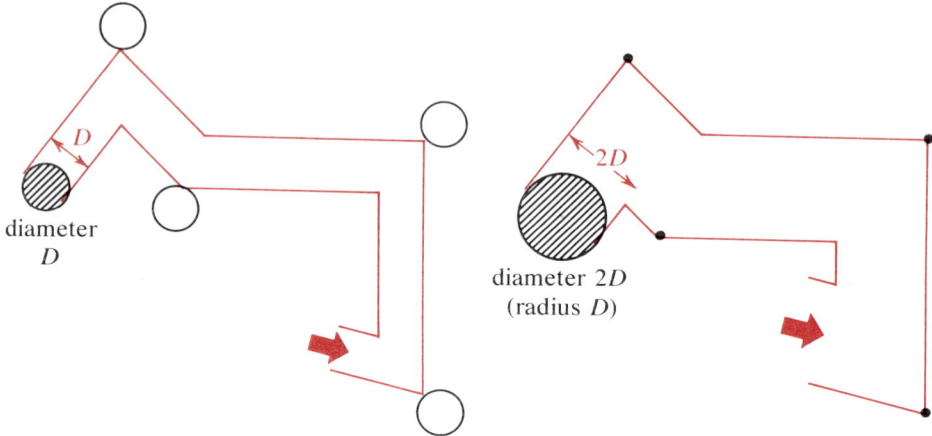

Figure 3-11 Equivalence of a sphere with diameter $2D$ in point masses to uniform spheres with diameter D.

such a molecule in 1 sec and multiply it by the number of molecules per unit volume to get the number of collisions per second, a quantity we shall call the **collision number**, Z_c. The volume of the spherical molecule itself (with *radius* $= D$) is $\frac{4}{3}\pi D^3$, and the effective volume it sweeps out in 1 sec is simply this volume times the number of radii it moves through in 1 sec, which is $v_{\rm rms}/D$. Here we assume that the molecule is moving with the root-mean-square velocity. So the effective volume swept out per second is

$$\text{volume swept} = \frac{4}{3}\pi D^3 \frac{v_{\rm rms}}{D} = \frac{4}{3}\pi D^2 v_{\rm rms}$$

and, since the number of molecules per unit volume is N/V,

$$Z_c = \frac{4\pi D^2 v_{\rm rms} N}{3V}$$

With an expression for Z_c, we can define another quantity that will be useful in discussing viscosity. The **mean free path** is the average distance a molecule travels between collisions. If we let λ be the mean free path, then we can see immediately that

$$\lambda = \frac{\rm cm}{\rm collision} = \frac{\rm cm/sec}{\rm collisions/sec} = \frac{v_{\rm rms}}{Z_c} = \frac{3V}{4\pi D^2 N} \tag{3-14}$$

Now let us look at the viscosity of a moving gas. Consider, as in Fig. 3-12, two adjacent layers of gas molecules separated by a distance λ, so that on the average molecules from the bottom layer that are moving up collide with molecules in the top layer and transfer momentum to them, and vice versa for the

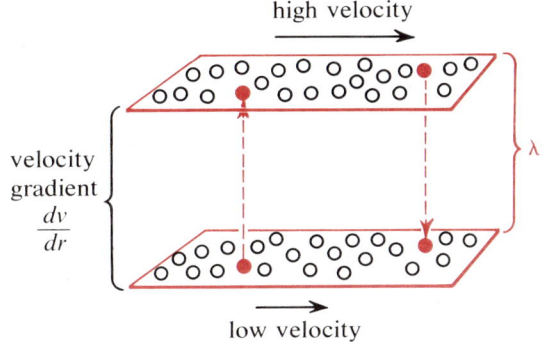

Figure 3-12 Transfer of molecules through the mean free path to a layer having a different velocity.

molecules in the top layer. As a simple approximation we say that at any given time one-third of the molecules in any sample are moving up and down, so the number going back and forth between layers in each second is one-third the number of molecules per unit volume, times the area of the layers, times the velocity of the molecules:

$$\text{number of molecules transferred per second} = \frac{1}{3}\frac{N}{V}Av_{\text{rms}}$$

The force one layer exerts on the other is equal to the time rate of change of momentum, which is equal to the number of molecules transferred per second times the change in momentum of each molecule. Since momentum is mv, each molecule suffers a momentum change of its mass, m, times the change in velocity between the two layers; but the velocity difference is just the velocity gradient, dv/dr, times the distance between layers, λ. So the force one layer exerts on the next layer is

force = number of molecules transferred per second × molecular mass × velocity difference

$$= \frac{NAv_{\text{rms}}}{3V} m\lambda \frac{dv}{dr}$$

$$= \frac{Nm}{V} \frac{\lambda v_{\text{rms}}}{3} A \frac{dv}{dr}$$

Looking at this equation we can see two things. First, Nm/V is just the total mass of the system divided by its volume, or the density of the gas, which we call ρ. Second, this equation looks exactly like the experimental one if we let

$$\eta = \tfrac{1}{3}\rho v_{\text{rms}}\lambda$$

Here, then, is our theoretical prediction of the coefficient of viscosity. If we use Eq. 3-14 to substitute for λ, and also the previously implied relationship

$$v_{\text{rms}} = \sqrt{\frac{3kT}{m}}$$

we can rearrange the coefficient-of-viscosity expression:

$$\eta = \tfrac{1}{3}\rho \frac{3V}{4\pi D^2 N}\sqrt{\frac{3kT}{m}}$$

$$= \frac{\rho}{4\pi D^2}\frac{V}{N}\sqrt{\frac{3mkT}{m^2}}$$

$$= \frac{V}{Nm}\frac{\rho}{4\pi D^2}\sqrt{3mkT}$$

$$= \frac{1}{\rho}\frac{\rho}{4\pi D^2}\sqrt{3mkT}$$

$$= \frac{\sqrt{3mkT}}{4\pi D^2} \tag{3-15}$$

Although this result is straightforward, it has some remarkable features. First, the density has cancelled out, so the viscosity of a gas should not depend on the density of the gas. This theoretical prediction was made by Maxwell before any experimental studies of the effect of pressure on viscosity had been made; he carried them out over the pressure range 0.001–100 atm, and found that to a very good approximation it was indeed true that the viscosity was independent of density. This intuitively surprising result represented a triumph for the theoretical model. Second, since the temperature appears in the numerator of the expression, it appears that gases become more viscous, not less, as they warm up. Of course, this is exactly opposite to our experience with liquids, and is again surprising; but again, this prediction has been verified experimentally, although the $T^{1/2}$ dependence is slightly different.

Molecular Diameters

If we use the experimental value for η and assume that we know Avogadro's number, we can use Eq. 3-15 to solve for the diameter of a molecule, D:

$$\eta = \frac{\sqrt{3mkT}}{4\pi D^2}$$

$$84.2 \times 10^{-6} = \frac{\sqrt{3(2.016/(6.023 \times 10^{23}))(1.38 \times 10^{-16})273}}{4(3.142)D^2} \quad \text{(for H}_2\text{ at 0°C)}$$

$$D^2 = \frac{\sqrt{3.75 \times 10^{-37}}}{1.058 \times 10^{-3}} = \frac{6.12 \times 10^{-19}}{1.058 \times 10^{-3}} = 5.80 \times 10^{-16} \text{ cm}^2$$

$$D = 2.41 \times 10^{-8} \text{ cm}$$

So, in addition to the high velocities that experiment requires of our model, it appears that the molecules are very small. They don't travel very far, either: inserting the appropriate numbers in Eq. 3-14 gives $\lambda = 1.7 \times 10^{-5}$ cm. Our

A Molecular Model for Gas Diffusion

There are two other **transport properties** of gases besides viscosity which we shall examine briefly. One is diffusion of one gas into another; the other is heat conductivity. In dealing experimentally with diffusion we start with a sharp boundary between two different gases and allow them to penetrate each other; this is a process in which two layers of gas are exchanging mass, just as in viscosity the two layers exchange momentum. Accordingly, it is not surprising to find that diffusion is governed by an equation very similar to that governing viscosity:

viscosity: viscous force $\left[\dfrac{d(mv)}{dt}\right] = \eta A \dfrac{dv}{dr}$

diffusion: mass flow rate $\left(\dfrac{dm}{dt}\right) = DA \dfrac{d(\text{concentration})}{dr}$

In this equation, m is usually measured in moles, concentration in moles per liter, and D is the **diffusion coefficient**—entirely analogous to the coefficient of viscosity. The kinetic-theory derivation is very similar to the derivation for viscosity, and the theoretical expressions that result are very similar:

viscosity: $\eta = \tfrac{1}{3}\rho v_{\text{rms}} \lambda$

diffusion: $D = \tfrac{1}{3} v_{\text{rms}} \lambda$

A Molecular Model for Heat Conduction in Gases

Heat conductivity is another process similar to viscous flow. If we start with two layers of gas having different temperatures, (in terms of our molecular model) the two layers of gas exchange kinetic energy, just as in viscosity they exchange momentum and in diffusion they exchange mass. (Remember that our model requires that kinetic energy be proportional to temperature.) We may compare the three experimental equations:

viscosity: viscous force $= \eta A \dfrac{dv}{dr}$

diffusion: mass flow rate $= DA \dfrac{dc}{dr}$

heat conductivity: heat flow $= \kappa A \dfrac{dT}{dr}$

In the heat-conductivity equation, heat flow is usually measured in calories per second and κ (kappa) is the **thermal conductivity**. The equations have an obvious similarity; the three factors dv/dr, dc/dr, and dT/dr are called, respectively, the velocity gradient, the concentration gradient, and the temperature gradient. Figure 3-13 also indicates the experimental similarities. Again the

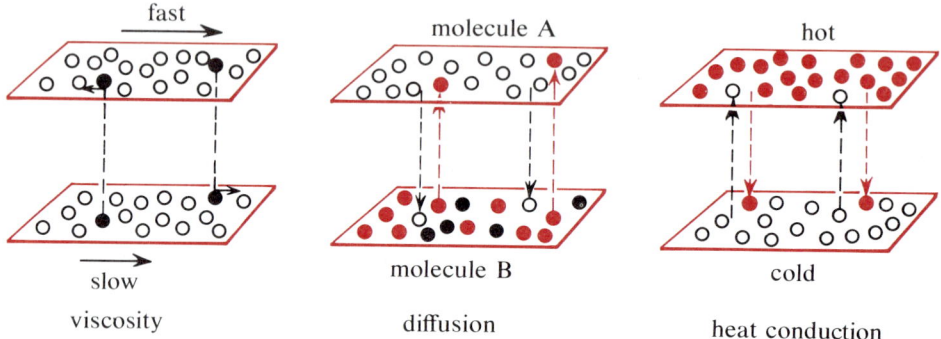

Figure 3-13   Equivalence (in molecular terms) of transport properties.

theoretical derivation is quite similar, and we can compare the theoretical results:

$$\text{viscosity:} \qquad \eta = \tfrac{1}{3}\rho v_{\text{rms}}\lambda$$

$$\text{diffusion:} \qquad D = \tfrac{1}{3}v_{\text{rms}}\lambda$$

$$\text{heat conductivity:} \qquad \kappa = \tfrac{1}{3}\rho v_{\text{rms}}\lambda C_v$$

The Heat Capacity for Gases

There is one very important difference in the result for κ, the thermal conductivity. In deriving the viscosity expression we pointed out that the momentum change for each molecule moving from one layer to another is equal to its mass times the velocity difference between layers; in dealing with heat conduction, at the corresponding point in the derivation we must multiply the temperature difference between layers by a quantity called C_v. This new quantity, which we call the **heat capacity**, must correspond to the rate at which the energy of the molecule (or a mole of the molecules) changes with temperature. Now we describe rates of change as derivatives, so

$$C_v = \frac{d}{dT}(\text{molecular energy}) \equiv \frac{dE}{dT}$$

For a simple, monatomic gas molecule, which cannot vibrate or rotate, we can easily interpret C_v. From Eq. 3-8, we have

$$\langle KE \rangle \equiv E = \tfrac{3}{2}PV$$

or, substituting from the ideal gas law for 1 mole of gas:

$$E = \tfrac{3}{2}PV = \tfrac{3}{2}RT$$

The derivative of this is very simple:

$$C_v = \frac{dE}{dT} = \frac{d}{dT}(\tfrac{3}{2}RT) = \tfrac{3}{2}R$$

However, suppose we have a diatomic molecule, which can vibrate in and out and also rotate in two different ways, as shown in Fig. 3-14. Then the equiparti-

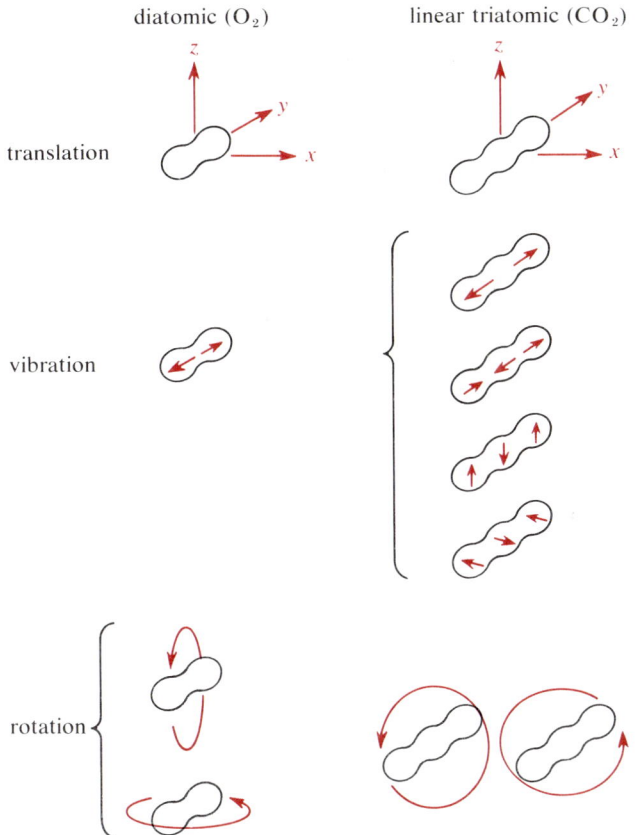

Figure 3-14 Possible modes of motion (degrees of freedom) of diatomic and linear triatomic molecules.

tion principle says that each of these three new degrees of freedom should contribute to the energy of a mole of the gas. The total kinetic energy is

$$E = (\tfrac{3}{2}RT)_{\text{transl}} + (\tfrac{2}{2}RT)_{\text{rot}} + (RT)_{\text{vib}}$$

or

$$E = (\tfrac{3}{2} + \tfrac{2}{2} + 1)RT = \tfrac{7}{2}RT$$

so

$$C_v = \frac{dE}{dT} = \tfrac{7}{2}R$$

Because we usually express heat flow in calories, it is preferable to express R in calories instead of liter-atmospheres or ergs. If we make this conversion, $R = 1.987$ cal/mole-°K. In these units, then, we expect the heat capacity of a monatomic gas to be $(\tfrac{3}{2})1.987$, or 2.98 cal/mole-°K; for the diatomic gas, $C_v = (\tfrac{7}{2})1.987 = 6.95$ cal/mole-°K. For a linear triatomic molecule, such as CO_2 in Fig. 3-14, there are again three translational degrees of freedom and two rotational degrees of freedom, but now there are four vibrational possibilities (degrees of freedom); accordingly, we expect $C_v = \tfrac{3}{2}R + \tfrac{2}{2}R + 4R = \tfrac{13}{2}R$ = 12.90 cal/mole-°K. Table 3-1 shows theoretical and experimental heat capacities for a number of molecules, and we can see that the monatomic gases do fit the theory. But the theory clearly breaks down for the diatomic molecules, and is almost useless for triatomic molecules.

Table 3-1

Heat Capacities of Gases: Experimental and Calculated From the Equipartition Principle

	C_v Calculated (25 °C, cal/mole-°C)	C_v Observed (25 °C, cal/mole-°C)
Monatomic gases		
helium	2.98	2.99
argon	2.98	2.99
mercury	2.98	2.98
Diatomic gases		
H_2	6.95	4.86
O_2	6.95	4.97
Cl_2	6.95	6.15
HCl	6.95	5.07
Triatomic gases		
CO_2	12.90	6.74
SO_2	11.90	7.70
H_2O	11.90	7.67 (100 °C)

Although the theory behind the equipartition principle had been well established, no refinement of the kinetic theory could improve the discrepancy in the heat capacities. One refinement we could obviously make would be to treat molecules as having the Maxwell–Boltzmann distribution of velocities instead of all having the rms velocity. This was done during the nineteenth century, and the numerical factors in front of some quantities, such as the number of collisions per second, changed slightly, but not the predicted heat capacities. At the beginning of the twentieth century it was generally felt that the kinetic theory accounted for the properties of nearly ideal gases in a satisfactory way except for their heat capacities. Inspecting the numbers in Table 3-1, it seems to be true that rotation is occurring but vibration is not, and there was no explanation for this that classical physics or chemistry could provide. A modification of the model was clearly necessary, but one that would not damage the good agreement the kinetic theory provided for Boyle's law, viscosity, heat transfer, and so on. When the answer was provided, it involved **quantization** of the vibrational energy of the molecule—requiring it to vibrate only at certain frequencies with widely separate energies. If this were the case, the thermal energy kT near room temperature would not be sufficient to allow many molecules to vibrate: in terms of the Boltzmann distribution,

$$\frac{\text{number of molecules in high-energy state}}{\text{number of molecules in lowest-energy state}} = e^{-\Delta E/kT}$$

the energy separation ΔE between states of vibration would be so great that few molecules could vibrate in any but the lowest-energy fashion, and all the apparent degrees of freedom could not be exercised. And in fact, gas heat capacities do increase with temperature, and appear to reach limiting values (at temperatures high enough to allow all the degrees of vibrational freedom) near those predicted by the kinetic theory, as shown in Fig. 3-15. We shall discuss quantization in some depth in Chapter 5.

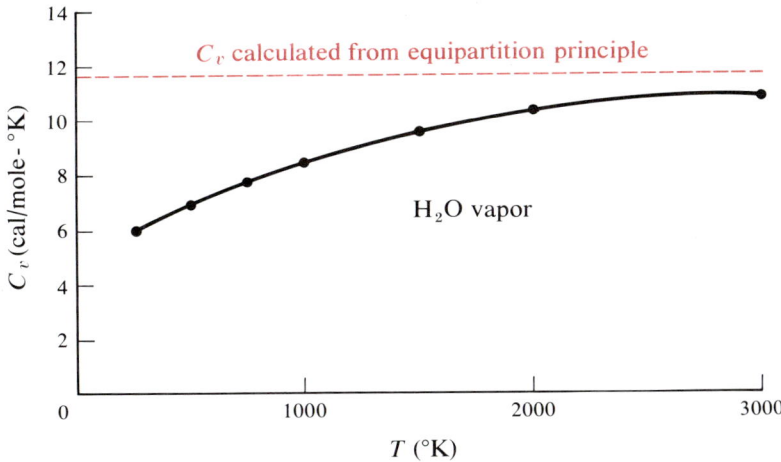

Figure 3-15 Temperature dependence of gas heat capacity.

Effusion and Graham's Law

The property of diffusion is closely related to another property, which is somewhat simpler to describe, called **effusion**. If, as in Fig. 3-16a, we provide a very small hole in the side of a gas container, the gas will gradually escape or effuse through the hole, and the rate of effusion—the volume of gas that escapes per second—can be measured. If the container is then refilled with a different gas, but kept at the same temperature and pressure, the effusion rate of the second gas is, in general, different from that of the first. If the effusion is occurring through a hole in the wall separating two different gases, as in Fig. 3-16b, the composition of the gas mixture on each side will not change at the same rate.

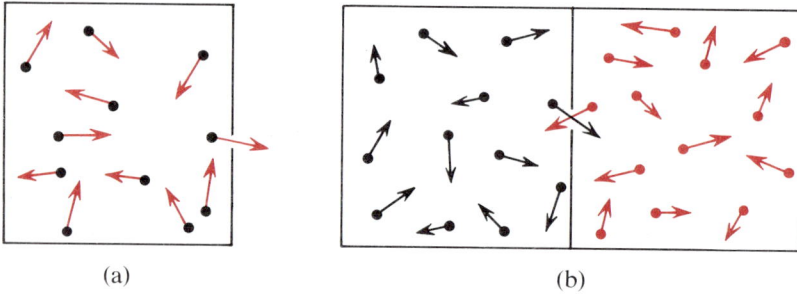

(a) (b)

Figure 3-16 Effusion of gases and gas mixtures.

Specifically, the rate of effusion is inversely proportional to the square root of the molecular weight of the gas; if the gases are labeled 1 and 2, then

$$\frac{\text{rate}_1}{\text{rate}_2} = \sqrt{\frac{\text{MW}_2}{\text{MW}_1}} \qquad (3\text{-}16)$$

This relationship is called **Graham's law**, after Thomas Graham, a Scottish chemist who discovered it in 1848. Graham's law is very easy to reconcile with the kinetic theory. In the first place, it is found experimentally that the law will not accurately describe the rates unless the hole is smaller than the mean free path one could calculate for the gases involved at that temperature and pressure. This suggests that the law holds only for an experimental geometry in which no collisions occur. If no collisions occur, the rate of effusion must be strictly proportional to the velocity of the molecules. Now, we have already derived an expression for the rms velocity of gas molecules (Eq. 3-10):

$$v_{\text{rms}} = \sqrt{\frac{3RT}{(\text{MW})}}$$

From this it is immediately obvious that we can set up a ratio for the two gases 1 and 2 that will give us Graham's law:

$$\frac{\text{rate}_1}{\text{rate}_2} = \frac{v_{\text{rms}}(1)}{v_{\text{rms}}(2)} = \frac{\sqrt{3RT/\text{MW}_1}}{\sqrt{3RT/\text{MW}_2}} = \sqrt{\frac{\text{MW}_2}{\text{MW}_1}}$$

This explanation was one of the earliest successes of the kinetic theory; here we see it as another example of successful prediction and correlation. Because gases of differing molecular weights can be differentiated this way even though they may have no chemical differences, the technique was used to separate ^{238}U from ^{235}U for the production of the first atomic bomb. Gaseous uranium compounds are hard to find, but uranium hexafluoride, UF_6, is a very volatile solid that has a vapor pressure of 1 atm at about 56°C. Using this compound, what would be the ratio of the effusion rates for the two isotopes?

$$\frac{\text{rate }^{235}U}{\text{rate }^{238}U} = \sqrt{\frac{\text{MW}(^{238}UF_6)}{\text{MW}(^{235}UF_6)}} = \sqrt{\frac{238 + 6(19.00)}{235 + 6(19.00)}}$$

$$= \sqrt{\frac{352.0}{349.0}} = \sqrt{1.0086} = 1.0043$$

Clearly, if a separation is to be achieved by this modest difference, many stages will be required.

Gas Mixtures and Partial Pressure

The discussions of gas diffusion and effusion bring us to the topic of mixtures of gases. So far in this chapter we have, without saying so, been careful to deal only with samples of gas in which all the molecules were identical. How can we describe mixtures of gases? The most important property of a mixture to specify is its composition. Since the physical properties of gases in general depend not on the mass of the gas but on the number of moles, the most convenient way of specifying the composition of a gas mixture is to give the number of moles of each component of the mixture. Then, since it is a good approximation to assume that the molecules do not interact, and since one gas introduced into a container full of another always diffuses to occupy the whole volume, we can apply the ideal gas law to each component of the mixture individually:

for component 1: $\quad P_1 V_{\text{total}} = n_1 RT$

for component 2: $\quad P_2 V_{\text{total}} = n_2 RT$

In these expressions, P_1 and P_2 are new quantities called **partial pressures**, which we shall examine more closely; note first, however, that we need not specify R, which is a universal constant for all gases, nor T, since with collisions frequently occurring between unlike molecules they will very quickly achieve the same average kinetic energy and thus the same temperature. Returning to the partial pressures, since Avogadro's hypothesis indicates that all gases should show identical pressure–volume relationships per mole, regardless of their identity, we can define a new quantity n_t, which is the total number of moles of all gases in the sample, and which will also fit into the ideal gas law:

$$P_{\text{total}} V_{\text{total}} = n_t RT$$

Now, since $n_t = n_1 + n_2$ if there are only two components,

$$\begin{aligned}
P_{\text{total}} V_{\text{total}} &= (n_1 + n_2) RT \\
&= n_1 RT + n_2 RT \\
&= P_1 V_{\text{total}} + P_2 V_{\text{total}} \\
P_{\text{total}} &= P_1 + P_2
\end{aligned} \qquad (3\text{-}17)$$

The reason for the name "partial pressure" is obvious from this relationship, which is called **Dalton's law** after its discoverer, John Dalton. We can express it in another useful way if, for the ith component of a mixture, we write

$$P_i V = n_i RT$$

and divide this equation by

$$P_{\text{total}} V = (n_1 + n_2 + \cdots + n_i + \cdots) RT$$

$$\frac{P_i \cancel{V}}{P_{\text{total}} \cancel{V}} = \frac{n_i \cancel{RT}}{(n_1 + n_2 + \cdots + n_i + \cdots) \cancel{RT}}$$

$$\frac{P_i}{P_{\text{total}}} = \frac{n_i}{n_1 + n_2 + \cdots n_i + \cdots}$$

The quantity on the right is a measure of the composition of the system that we call the **mole fraction** and symbolize by x_i. Then

$$P_i = x_i P_{\text{total}}$$

The mole fraction is a very useful measure of concentration, defined as the number of moles of a particular component of a mixture divided by the total number of moles of all the components in the mixture. For instance, consider the mixture of gases that results from the reaction

$$2 H_2 O_2 \rightarrow 2 H_2 O + O_2$$

If the mixture is taken at 1-atm pressure and constant T and V and dried, what is the pressure of the remaining oxygen?

First we notice that the chemical reaction produces 2 moles of water for every mole of oxygen, so, no matter what the scale of the reaction is, the mole fraction of oxygen is $1/(1+2)$ or $\tfrac{1}{3}$. Immediately we write

$$P_{O_2} = x_{O_2} P_{total}$$
$$= \tfrac{1}{3}(1 \text{ atm})$$
$$= 0.33 \text{ atm}$$

In other applications, Dalton's law is more conveniently applied as the sum of the partial pressures: At 24 °C, the vapor pressure of water is 22.4 torr. If the relative humidity of the atmosphere at 24°C is 100%, it means the vapor pressure of water has reached that level in the atmosphere. If a constant-volume container, which has been open to the atmosphere under these conditions, is sealed, with the barometric pressure equal to 747.8 torr, and the air in the container is dried by circulation over a desiccant, what will be the pressure in the container? The total pressure of the initial humid air is just the barometric pressure, 747.8 torr, and the partial pressure of the water in the air is 22.4 torr; so the sum of the remaining partial pressures (of nitrogen, oxygen, CO_2, etc.) is

$$P_{total} = P_{H_2O} + P_{dry\ air}$$
$$P_{dry\ air} = P_{total} - P_{H_2O}$$
$$= 747.8 - 22.4$$
$$= 725.4 \text{ torr}$$

Throughout this discussion we have assumed that all the gases could be adequately described in terms of the properties of an ideal gas, one that follows Gay-Lussac's law at all temperatures. But this is not strictly true for any gas, since all gases condense to a liquid at some temperature above 0 °K. Hydrogen and helium, two of the most nearly ideal gases, boil at 20.4°K and 4.2°K, respectively. How severe is this problem of nonideality, how do we describe it experimentally, and what features of our model need to be changed to account for it?

3-4 Nonideal Gases — Experiment and Theory

To begin, we need a criterion of nonideality so that we can measure departures from ideal behavior. What is most commonly done is to take a 1-mole sample, so that presumably $PV = RT$ or $PV/RT = 1.000$, and measure PV/RT at different pressures. To the extent that PV/RT as a function of pressure departs from 1.000, the gas is nonideal; some gases are worse than others, as Fig. 3-17 shows.

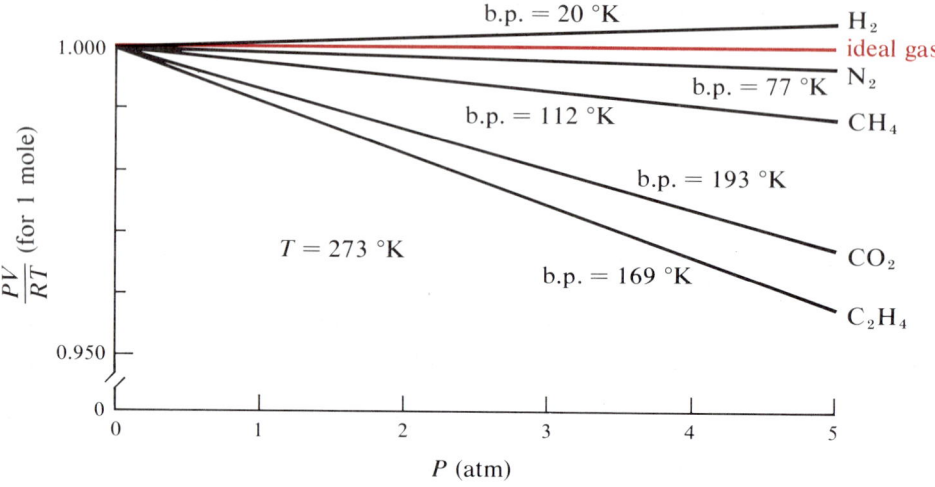

Figure 3-17 Nonideality of various gases at low pressures. (Redrawn from Kauzmann, W., *Kinetic Theory of Gases*, New York: W. A. Benjamin, 1966.)

The quantity PV/RT, because of its relationship to pressure applied to the gas, is called the **compressibility factor**. Fortunately for the ideal gas law, these deviations are relatively small at pressures up to about 5 atm (i.e., about 75 psi), as a glance at the vertical scale of the graph will show; at 5 atm none of the deviations exceeds 4%. However, all these gases have very low boiling points, as the graph indicates. If we lower the temperature at which the measurements are made, or if we go to much higher pressures, the deviations become very serious indeed, as Figs. 3-18 and 3-19 show. As we lower the temperature, we

Figure 3-18 Temperature dependence of nonideality for NH_3. (Redrawn from Kauzmann, W., *Kinetic Theory of Gases*, New York: W. A. Benjamin, 1966.)

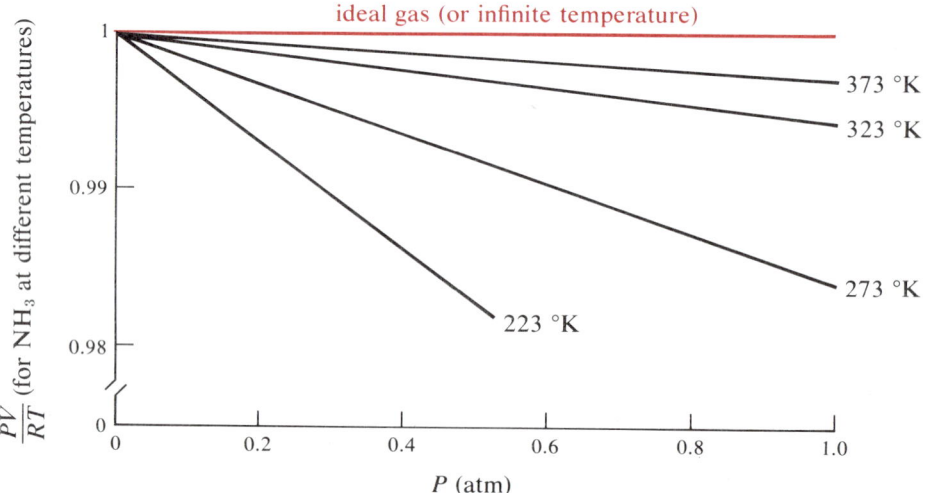

eventually reach a point at which sufficient pressure will actually liquefy the gas, which causes a sharp drop in the volume and hence in PV/RT; this is seen for CO_2 in Fig. 3-19. The highest temperature at which this will occur is called the **critical temperature**, about which we shall have more to say later and in

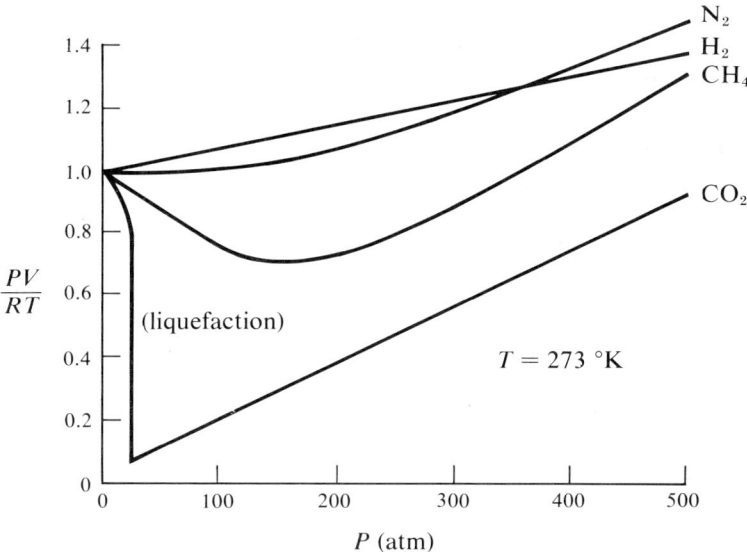

Figure 3-19 Pressure dependence of nonideality at high pressures. (Redrawn from Kauzmann, W., *Kinetic Theory of Gases*, New York: W. A. Benjamin, 1966.)

Chapter 7. Even for gases that do not liquefy, however, the deviations at high pressures are sometimes as great as 75%. For these extreme conditions, the ideal gas law is a completely inadequate equation of state, and we need some replacement. How can we reduce the relationships indicated graphically in these figures to algebraic form?

The van der Waals Equation for Nonideal Gases

There are many ways to approach the problem of representing graphical data as an algebraic equation. The earliest and best known of the equations of state for real gases is the **van der Waals equation**:

$$\left[P + a\left(\frac{n}{v}\right)^2\right](V - bn) = nRT \tag{3-18}$$

It is not obvious on casual inspection why van der Waals, a Dutch physicist of the late nineteenth and early twentieth century, chose this form. He was guided by the basic structure of the kinetic theory, and indeed the form of the equation can be reproduced by adding intermolecular forces to the virial expression that we derived earlier. However, the constants a and b, which are characteristic of individual gases, are usually not calculated but determined by adjusting them to provide the best experimental fit; Table 3-2 gives some values for these constants for different gases. The van der Waals equation provides a much improved fit for the observed compressibility factors of gases, but still leaves something to be desired, as Fig. 3-20 shows for the particular case of C_2H_6. We shall return shortly to the reasons for the form of this equation.

Table 3-2
van der Waals Constants

Gas	a (l²-atm/mole²)	b (l/mole)
acetone, $(CH_3)_2CO$	13.91	0.0994
ammonia, NH_3	4.170	0.03707
argon, Ar	1.345	0.03219
n-butane, C_4H_{10}	14.47	0.1226
carbon dioxide, CO_2	3.592	0.04267
chlorine, Cl_2	6.493	0.05622
ethane, C_2H_6	5.489	0.06380
helium, He	0.03412	0.02370
hydrogen, H_2	0.2444	0.02661
hydrogen chloride, HCl	3.667	0.04081
krypton, Kr	2.318	0.03978
mercury, Hg	8.093	0.01696
methane, CH_4	2.253	0.04287
neon, Ne	0.2107	0.01709
nitric oxide, NO	1.340	0.02789
nitrogen, N_2	1.390	0.03913
nitrogen dioxide, NO_2	5.284	0.04424
oxygen, O_2	1.360	0.03183
phosphine, PH_3	4.631	0.05156
propane, C_3H_8	8.664	0.08445
silane, SiH_4	4.320	0.05786
sulfur dioxide, SO_2	6.714	0.05636
water, H_2O	5.464	0.03049
xenon, Xe	4.194	0.05105

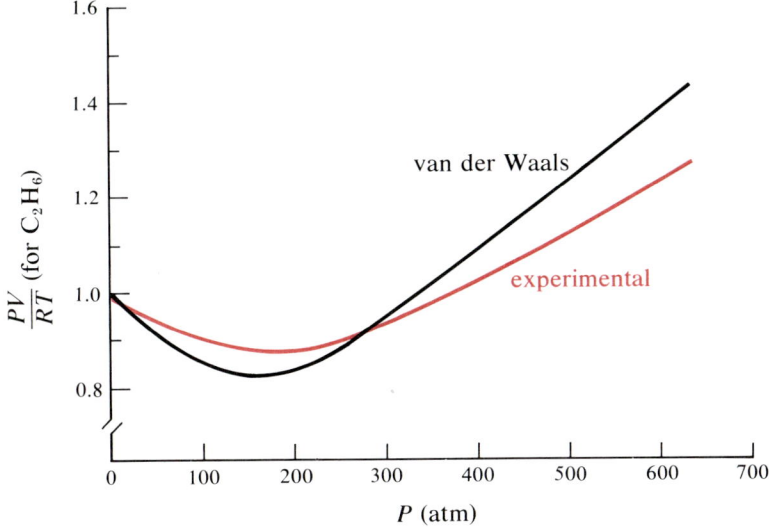

Figure 3-20 Fit of van der Waals equation of state to experimental data for ethane, C₂H₆. (Redrawn from Kauzmann, W., *Kinetic Theory of Gases*, New York: W. A. Benjamin, 1966.)

Other Equations for Nonideal Gases

A more mathematically straightforward but less physically meaningful equation is obtained if one simply represents the graph of the compressibility factor for a particular gas as a polynomial with an indefinite number of coefficients:

$$\frac{PV}{nRT} = 1 + B\left(\frac{n}{V}\right) + C\left(\frac{n}{V}\right)^2 + D\left(\frac{n}{V}\right)^3 + \cdots$$

or

$$\frac{PV}{nRT} = 1 + B'P + C'P^2 + D'P^3 + \cdots$$

Both of these equations are called **virial equations**, because their forms can be reproduced by adding attraction force terms to the virial derivation of the ideal gas law; however, it would probably be fairer to say that they represent an efficient curve-fitting method and let it go at that. By using enough coefficients, the compressibility-factor curve for any gas can be fitted to any desired degree of accuracy, but this does not seem very elegant. It should be noted that the coefficients in both equations are not constants but are instead temperature dependent, so that presumably each of them must be represented somehow as a function of temperature, thereby increasing the number of arbitrary constants still further.

Many other equations of state have been proposed over the years. Table 3-3 indicates some of the more successful ones, including the best all-purpose equation, that of Beattie and Bridgman. This latter is capable of representing PV data with excellent accuracy (less than 1% error) up to about 300 atm, even near critical temperatures for those gases which do liquefy. PV data near critical temperatures and pressures are very difficult to fit to a model because the liquid state is not well understood, so this is a genuine triumph, if an empirical one. However, the large number of terms in the more accurate equations of state makes them awkward to use, and frequently the most satisfactory procedure is a graphical one, which we shall discuss shortly.

Table 3-3
Equations of State for Real Gases

Van der Waals equation	$\left(P + \dfrac{a}{V^2}\right)(V - b) = RT$
Dieterici equation	$P = \left(\dfrac{RT}{V - b}\right) e^{-a/VRT}$
Virial equation	$\dfrac{PV}{RT} = 1 + B\left(\dfrac{1}{V}\right) + C\left(\dfrac{1}{V}\right)^2 + \cdots$
Redlich–Kwong equation	$P = \dfrac{RT}{V - b} - \dfrac{a}{V(V - b)\sqrt{T}}$
Beattie–Bridgman equation	$P = \dfrac{RT}{V} - \dfrac{A_0 - B_0 RT + cR/T^2}{V^2}$ $+ \dfrac{aA_0 - bB_0 RT - cB_0 R/T^2}{V^3} + \dfrac{bcB_0 RT}{V^4}$

Reduced Variables and the Dimensionless van der Waals Equation

At the critical temperature there is a pressure that is just sufficient to liquefy the gas involved, called the **critical pressure**; and there is a characteristic volume of 1 mole of gas at that temperature and pressure called the **critical volume**; these three quantities are often represented by the symbols T_c, P_c, and V_c, respectively. Values for these critical parameters can be obtained experimentally for any gas. In the van der Waals equation of state there are also three constants, a, b, and R; by considering what the equation of state must predict at the critical point (if it is to be valid at all) it is possible to express T_c, P_c, and V_c in terms of a, b, and R. Without going through the derivation, the results for 1 mole are

$$T_c = \frac{8a}{27Rb} \qquad P_c = \frac{a}{27b^2} \qquad V_c = 3b$$

But an alternative to this representation is to express the constants a, b, and R themselves in terms of the three critical parameters. If we rearrange the above expressions, we get

$$a = 3P_c V_c^2 \qquad b = \frac{V_c}{3} \qquad R = \frac{8 P_c V_c}{3 T_c}$$

Now suppose we substitute these expressions into the van der Waals equation for 1 mole:

$$\left(P + \frac{3 P_c V_c^2}{V^2}\right)\left(V - \frac{V_c}{3}\right) = \frac{8 P_c V_c}{3 T_c} T$$

Divide both sides of this by P_c (using only the first factor on the left):

$$\left[\frac{P}{P_c} + 3\left(\frac{V_c}{V}\right)^2\right]\left(V - \frac{V_c}{3}\right) = \frac{8 V_c}{3 T_c} T$$

Now divide both sides by V_c, using only the second factor on the left:

$$\left[\frac{P}{P_c} + 3\left(\frac{V_c}{V}\right)^2\right]\left(\frac{V}{V_c} - \frac{1}{3}\right) = \frac{8}{3}\frac{T}{T_c}$$

If we define three new dimensionless quantities called the **reduced temperature**, **reduced pressure**, and **reduced volume** as,

$$T_r \equiv \frac{T}{T_c} \qquad P_r \equiv \frac{P}{P_c} \qquad V_r \equiv \frac{V}{V_c}$$

so each is the ratio of a physical property to the value that physical property has at the critical point, the van der Waals equation takes on a very interesting form:

$$\left(P_r + \frac{3}{V_r^2}\right)(V_r - \tfrac{1}{3}) = \tfrac{8}{3} T_r \qquad (3\text{-}19)$$

Because we have absorbed the three constants into the "reduced" properties, the equation no longer contains any arbitrary constants. Accordingly, if we express temperature, pressure, and volume as the reduced variables, we should see identical behavior for all gases just as we do (to a poorer approximation) in the ideal gas law.

Graphical Treatment of Nonideal Gases

Now, this form of the van der Waals equation is no better and no worse than the original form; we have not changed the physical approximation at all. But if we were to replot the compressibility-factor curve from, say, Fig. 3-19, against the *reduced* pressure for several different *reduced* temperatures, all gases should follow the same curves. Figure 3-21 shows such a set of curves (on a logarithmic P_r scale); this is called a **Hougen–Watson chart** and is one of the most convenient means of quantitatively estimating real-gas behavior. For instance, consider a gas cylinder with an internal volume of 10 l and a pressure rating of 2000 psi (133 atm) at 40°C. What weight of ethylene, C_2H_4, can safely be contained in the cylinder at that temperature?

From Table 3-2 we can get the T_c and P_c values for C_2H_4: $T_c = 283°K$, $P_c = 50.9$ atm. From these we calculate the reduced temperature and pressure so that we can use the Hougen–Watson chart: $T_r = (40+273)/283 = 1.10$; $P_r = 133/50.9 = 2.62$. Selecting the $T_r = 1.10$ line on the chart and reading it at $P_r = 2.62$, the value of PV/RT is approximately 0.45 (remember this is for 1 mole). If we solve this for the volume of 1 mole, we can divide that volume into the total volume of the cylinder to get the total number of moles allowed:

$$\frac{PV}{RT} = 0.45$$

$$V = 0.45\frac{RT}{P} = 0.45\frac{0.08205(313)}{133} = 0.0866 \text{ l/mole}$$

$$\frac{10 \text{ l}}{0.0866 \text{ l/mole}} = 115.5 \text{ moles allowed}$$

From this we can immediately get the allowed weight of C_2H_4:

$$(115.5 \text{ moles})(28.0 \text{ g/mole}) = 3230 \text{ g } C_2H_4 \text{ allowed}$$

By comparison, if we work the problem using the ideal gas law the prediction is quite different:

$$PV = nRT$$

$$n = \frac{PV}{RT} = \frac{133(10)}{0.08205(313)} = 52.0 \text{ moles}$$

$$52.0(28.0) = 1456 \text{ g } C_2H_4 \text{ allowed}$$

Hougen–Watson charts are not flawless, principally because an equation of state with more than three constants cannot be transformed into dimensionless form by introducing reduced variables, but they are an excellent approximation for many purposes.

Forces Between Gas Molecules

Having developed all this elaborate procedure for describing nonideality of gases—why should it be true? Looking at the Hougen–Watson chart, we can see two areas of concern. At relatively low temperatures and moderate pressures, it appears that the volume of a given gas sample is distinctly smaller

Nonideal Gases – Experiment and Theory | 139

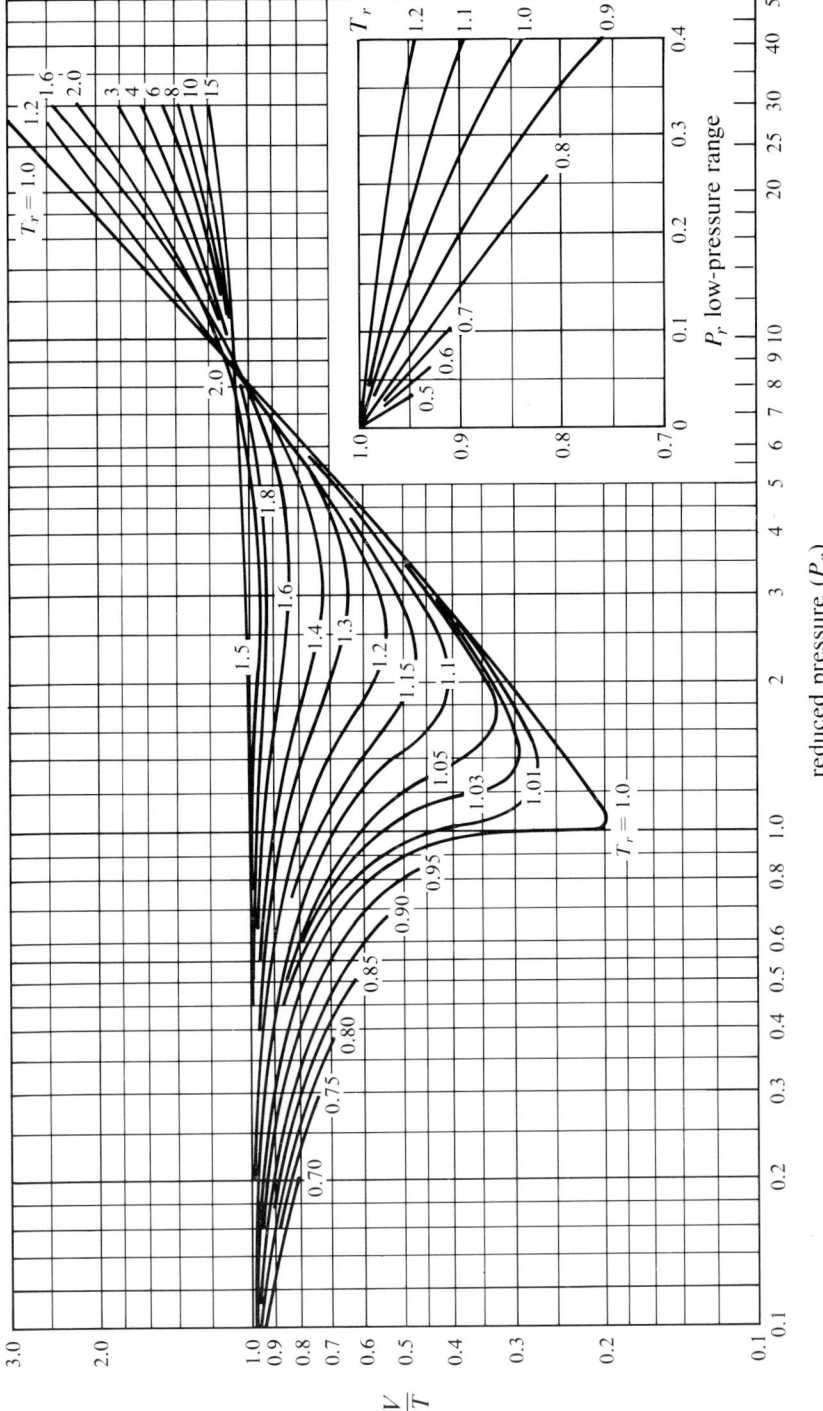

Figure 3-21 Hougen–Watson chart for nonideal gases (from Hougen, O. A. and Watson, K. M., *Chemical Process Principles*, New York: Wiley, 1947).

than the ideal gas law suggests; apparently there is an attractive force tending to hold the molecules together. This would be an alteration in our model, because we stipulated initially that the molecules would be noninteracting. If there is such a force, why does it not show up at low pressures? We must surmise that the attractive force is not effective at long range, and becomes important only when the molecules are compressed fairly close together; such a force might be proportional to $1/r_{ij}^n$, where r_{ij} is the distance between molecule i and molecule j, and n is some number greater than 1. The other area of concern on the Hougen–Watson chart is the very high pressure region, where the volume is much larger than the ideal gas law suggests. When the molecules become packed extremely close together, it apparently becomes much more difficult to squeeze them still further. We have noted that for collisions to occur molecules must have finite size, and we have even calculated an approximate value for that size. Naturally, when the pressure becomes so high that the molecules are essentially packed tight, further compression will be very difficult, since it will be the molecules themselves and not free space that are being compressed. But to keep this consistent with the attractive force we have already noted, we can express this finite size as a very strong repulsive force which operates over an even shorter range than the attractive force. If the attractive force has the form $1/r_{ij}^n$, then such a repulsive force would have to be something like $1/r_{ij}^m$, where m is a larger number than n. As a matter of fact, this is more satisfactory than a sort of solid–billiard-ball model, because we never reach a pressure so high that the gas is *completely* incompressible, as a billiard-ball model would suggest.

The Potential Energy of Pairs of Gas Molecules

It appears, qualitatively at least, that we can account for the departures from ideality in gases by proposing forces that exist between molecules. To make any further progress toward a quantitative description, we shall need to take a small detour. So far, when we have discussed energy, we have spoken exclusively of kinetic energy — energy of motion. But if we hold a ball high in the air, and hold it still, it has no energy of motion; yet when we drop it, it acquires more energy of motion the farther it drops and the faster it goes. Where is this energy coming from? We can keep our thinking orderly if we assume that while we were holding it, it had energy of position or **potential energy**. When we drop it, its height decreases, and thus its potential energy of position decreases, but this loss of potential energy shows up as a gain in kinetic energy, so we may consider that energy as a whole is neither created nor destroyed. Now, if we try to extend the idea of potential energy to our molecular model, we can define a zero of energy as being two molecules an infinite distance from each other, each sitting still. As we move the molecules closer together, the medium-range attractive force we have proposed begins to attract them, and it begins to be true that the molecules have energy of position. But this must be a *negative* potential energy, because we would have to do work to overcome the attractive force in order to pull the molecules apart and get back to the zero-energy condition; if we put energy in and wind up at zero, we must have been in a negative-energy condition initially. If the attractive force is really inversely proportional to some power of the distance between the molecules, then the closer together they get the more negative the potential energy becomes (as the denominator in the force expression gets smaller, the force gets bigger and becomes harder to overcome). But at short range the repulsive force begins to take effect, reducing the overall net attraction, and as the range becomes very short it will even be possible for the repulsive force to predominate over the attractive force (a small fractional number in the denominator of the repulsive force expression, raised to a large power, is a very small number indeed). At very close range the potential energy of position can even become positive, so that if we released two molecules in that position they would fly apart spontaneously, developing kinetic energy as they went. Figure 3-22 shows the potential energy of a pair of molecules as a function of their separation. Notice that there are two characteristic distances shown. One, σ, is the closest the two molecules could approach if they started with zero energy and coasted together; the other is the distance at which the tradeoff between the attractive and repulsive forces is most desirable, the potential energy minimum. This is the origin of the potential term $\frac{1}{2}RT$, which appears in the vibrational part of the equipartition principle.

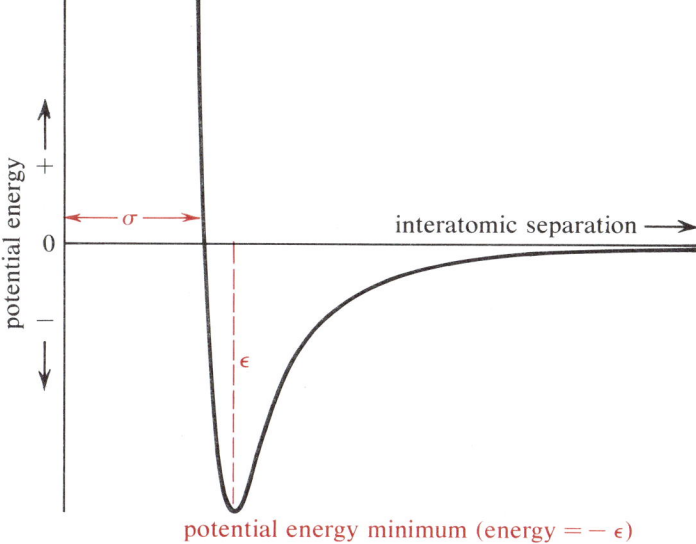

Figure 3-22 Lennard-Jones potential for a pair of neighbor molecules.

It is appropriate to think of this potential energy curve as a curved trough along which a ball is rolling while being attracted by the earth's gravity; the ball will spontaneously roll down from large r_{ij} toward the minimum, developing kinetic energy as it goes, then will coast up the sharp incline using up its kinetic energy and slowing down, unable to get farther up the trough than the height at which it started. Another crude analogy is to think of the two molecules as similar to two tennis balls with powerful magnets inside them — at moderate distances the magnets will attract the balls together, but when the balls get so close together as to begin to compress each other, this compression overcomes the attraction of the magnets, and there is an effective distance of closest approach of the centers of the balls.

If we symbolize potential energy (a function of r_{ij}) by $V(r)$, we can write an algebraic expression for this combination of attractive and repulsive potential energy:

$$V(r) = \frac{p}{r^{12}} - \frac{q}{r^6}$$

Here the term with r^{12} is the repulsive energy contribution and that with r^6 is the attractive energy contribution; p and q are the respective proportionality constants. The powers 12 and 6 have been chosen to provide the best fit with experimental PV data for most molecules, but with more information about the structure of molecules than we have developed so far it is possible to predict on theoretical grounds that the attractive potential energy should be $1/r^6$. This expression is known as the **Lennard-Jones potential**. It can be rewritten in a slightly different form in which the Greek letters σ (sigma) and ϵ (epsilon) now refer to the quantities shown in Fig. 3-22; they are, respectively, the distance at which the attractive and repulsive contributions are equal, and the depth of the potential energy well:

$$V(r) = 4\epsilon \left[\left(\frac{\sigma}{r}\right)^{12} - \left(\frac{\sigma}{r}\right)^6 \right] \tag{3-20}$$

The Lennard-Jones potential is extremely useful to us in accounting for the tendency of molecules to attract each other, yet remain at finite separations even while the attraction continues. We shall use it in accounting for the structures of crystalline solids in Chapter 4 and in dealing with liquids and the transition between liquid and vapor in Chapter 7.

Intermolecular Forces and the van der Waals Equation

Let us take another look at the van der Waals equation of state in light of all this development. In the equation

$$\left[P + a\left(\frac{n}{V}\right)^2 \right](V - bn) = nRT$$

there are two parameters added beyond the form of the ideal gas law. One is a pressure correction proportional to $(n/V)^2$. This basically reflects the attraction of the molecules. While we have described only two molecules in the potential energy expression, the attraction exists between every pair of molecules — between one molecule and all its neighbors. For a molecule in the interior of the gas sample, the overall attraction of all its neighbors averages out to zero, since they are in all directions (as in Fig. 3-23); but a molecule on the surface of the gas sample experiences a net force tending to pull it back into the gas. Taking

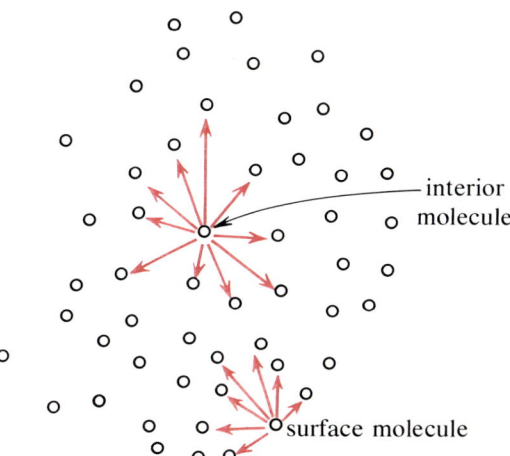

Figure 3-23 Imbalance of forces on a surface molecule in a gas.

the force per unit area, which the pressure represents, the net attraction of all molecules per square centimeter of gas surface is proportional to the density of the gas (the number of molecules — or moles — per unit volume) because as the density increases the number of surface molecules per unit area increases. But at the same time the distance to the neighbors is shortening in proportion to the density, which also increases the force of attraction. So the net pressure correction due to the molecular attraction must be proportional to the square of the density, which of course is represented by n/V, the number of moles per unit volume. The correction is added instead of subtracted because the quantity in square brackets is what the ideal pressure ought to be, and if it were not for this attraction the pressure would be larger than we really observe — so the corrected pressure must be larger than the observed pressure.

The other change from the ideal gas equation to the van der Waals equation is a volume correction, which is subtracted. The volume available for molecules is smaller than the volume of the container would suggest by a factor that is proportional to the number of moles — or molecules — present. Clearly this is related to the repulsive force, and is simply the volume excluded from the motion of other molecules by the presence of a particular molecule with its characteristic Lennard-Jones radius σ. The total volume excluded, of course, is proportional to the total number of molecules. So the form of the van der Waals equation has a physical rationale that takes into account precisely the forces between molecules we have been proposing. We have been able to produce a consistent model even for the departures from ideal behavior.

3-5 Summary

We have taken a fairly thorough look at the properties of gases in this chapter. Experimentally, they display a number of physical characteristics that are quite diverse in nature but that we can describe by appropriate mathematical functions. In this fashion, we have in varying degrees of detail accounted experimentally for Boyle's law, Gay-Lussac's (or Charles') law, and the combination of them and Avogadro's hypothesis that is generally familiar as the ideal gas law. Seeking diversity, we have treated viscosity, diffusion, effusion, heat transfer, heat capacity, and even the pattern of deviation from the ideal gas law that gases display. For all of these properties save one — the heat capacity of polyatomic molecules — we have been able to produce a satisfactory model by attributing certain properties to the atoms or molecules that we propose as the internal structure of gases. If we suppose that a gas consists of many molecules moving through free space as infinitesimal mass points without interacting with each other, we can predict Boyle's law from the simple mechanical properties of such a collection. If we further suppose that the kinetic energy of the collection is proportional to the absolute temperature of the system, we can match the ideal gas law. We can characterize the kinetic energy by using an average velocity for the collection, v_{rms}; this turns out on calculation from experiment to be on the order of 10^4 cm/sec near room temperature, so that the molecules are moving very rapidly. However, we have seen that it is not reasonable to suppose that all molecules have identical velocities and have developed a function, the Maxwell–Boltzmann distribution function, to describe the relative probability of a certain velocity among all those possible for a gas sample. This distribution function makes it clear how some molecules in a large system can have kinetic energies substantially above the average, which is a useful property to keep in mind.

The distribution function, however, and also the transport properties of gases, depend on the molecules of the gas having frequent collisions. This can occur only if the molecules have a finite size, which necessitates changing the original model; from viscosities, we can calculate that molecules must have a diameter on the order of 10^{-8} cm or 1 Å. The collisions can be characterized by a collision number (per second) and a mean free path between collisions, both of which we can calculate; the collision number is on the order of 10^{10} collisions/sec, and the mean free path λ is on the order of 10^{-5} cm, or about a thousand molecular diameters, when the gas is at ordinary atmospheric temperature and pressure.

Another modification in the original model becomes necessary when we try to describe theoretically the pattern of deviations from the ideal gas law. The pattern is perhaps most easily recognized from a Hougen–Watson chart of generalized gas behavior; from this we can see that there must be an attractive force between molecules that operates only at relatively short range, and another very strong repulsive force between molecules that operates at very short range. The distance "limit" of the repulsive force may be thought of as characterizing the radius of the molecule. These forces together, and thus the nonideality of gases as observed experimentally, are described by the Lennard-Jones potential energy function, which we shall use again in dealing with condensed phases in which molecules are touching several neighbors at all times.

The model has progressed from noninteracting point masses, all moving at the same speed, to a statistical collection of molecules that show an attractive force toward surrounding molecules but repel them at very short distances, and have a statistically reproducible but somewhat complicated range of velocities. With this model, however, we have shown that we can account for a remarkable variety of gas behavior. The one failure we have encountered so far is the failure of the model to predict the variety of heat capacities that polyatomic molecules display; the problem will recur in Chapter 4, with crystalline solids. There we shall see, as we have hinted here, that the model will have to be altered further by proposing that energy must exist, for microscopic systems, in small integral units, which cannot be further subdivided, called quanta. By making this change, the molecular theory can be shown to account satisfactorily for all the experimental phenomena with which we have dealt here. The question now is, can the properties of molecules account for the behavior of a quite different state of matter, the crystalline solid, and what additional features must the molecules display in so doing?

Appendix: Gas Law Calculations

Using the ideal gas law, we can obtain any of the quantities it describes if we have information on the other variables:

$$PV = nRT \begin{cases} P = \dfrac{nRT}{V} \\ V = \dfrac{nRT}{P} \\ n = \dfrac{PV}{RT} \\ T = \dfrac{PV}{nR} \end{cases}$$

For all these calculations, we normally use R in liter-atmosphere units ($R = 0.08205$ l-atm/mole-°K) because of the convenience of expressing the measured pressures and volumes this way. The calculations are very straightforward.

Example 1 What is the pressure inside a cylinder that contains 100 g of nitric oxide, NO, at 25 °C, if the volume of the cylinder is 2.00 l?

The volume is in the correct units, and we can calculate the pressure in any units we please; however, the temperature must be converted to degrees Kelvin and the weight of NO must be converted to the number of moles of NO.

$$T = 25°C + 273 = 298°K$$

$$n = \frac{\text{wt NO}}{\text{wt/mole NO}} = \frac{100 \text{ g NO}}{30.0 \text{ g NO/mole}} = 3.33 \text{ moles NO}$$

Now we can substitute directly into the ideal gas law:

$$P = \frac{nRT}{V} = \frac{3.33 \times 0.08205 \times 298}{2.00} = 40.8 \text{ atm}$$

Example 2 A gas chromatograph is an instrument that separates volatile substances in a stream of an inert gas. If a cylinder containing 3.00 lb of helium provides the inert gas for 4000 hr of operation, and the temperature is 200°C inside the chromatograph, what volume of helium is flowing out of the chromatograph into the atmosphere per second?

We can solve for the volume of helium that 3.00 lb would occupy under these conditions and then divide by the time of operation to get the flow rate. First, however, we need to get all the quantities in the proper units:

$$n = 3.00 \text{ lb He} \times 454 \frac{g}{\text{lb}} \times \frac{1}{4.003} \frac{\text{mole He}}{g}$$

$$= 340.2 \text{ moles He}$$

$$T = 200°C + 273 = 473°K$$

P is in effect defined as being 1 atm at the point where the hot helium leaves the gas chromatograph. Now solve for the volume occupied by the helium:

$$V = \frac{nRT}{P} = \frac{340.2 \times 0.08205 \times 473}{1}$$

$$= 13{,}200 \text{ l} \quad \text{or} \quad 1.320 \times 10^4 \text{ l}$$

Finally, express the time of operation in seconds and divide to get the flow rate:

$$\text{time} = 4000 \text{ hr} \times 3600 \text{ sec/hr} = 1.440 \times 10^7 \text{ sec}$$

$$\text{flow rate} = \frac{\text{volume}}{\text{time}} = \frac{1.320 \times 10^4 \text{ l}}{1.440 \times 10^7 \text{ sec}}$$

$$= 0.916 \times 10^{-3} \text{ l/sec} = 0.916 \text{ ml/sec}$$

Example 3 A sample of a nitrogen-containing compound weighing 40.0 g is carefully burned to yield 5.12 ml of N_2 at 747 torr and 22°C. What weight percent nitrogen is in the compound?

We are trying to determine the number of moles of gas we have, since that number can be converted into a weight by using the formula weight, and the weight can be readily converted into a weight percent value. Again we need to express all the data in the proper units. Since 1 atm is defined as 760.00 torr, the pressure in atmospheres will be 747/760 or 0.984 atm; the temperature in degrees Kelvin will be $22 + 273 = 295°K$; and the volume will be $5.12 \text{ ml} \times 1 \text{ l}/1000 \text{ ml} = 5.12 \times 10^{-3}$ l. Now we can substitute directly into the ideal gas law again:

$$\text{number of moles } N_2 = n = \frac{PV}{RT} = \frac{0.984 \times 5.12 \times 10^{-3}}{0.08205 \times 295}$$

$$n = 2.08 \times 10^{-4} \text{ mole } N_2$$

Translating the number of moles into the weight of nitrogen present, we have

$$2.08 \times 10^{-4} \text{ mole } N_2 \times \frac{28.0 \text{ g N}}{\text{mole } N_2} = 5.82 \times 10^{-3} \text{ g N}$$

or 5.82 mg N present. Using the original sample weight, we can convert this into the weight percent N:

$$\text{percent N} = \frac{\text{wt N present}}{\text{total wt present}} \times 100 = \frac{5.82 \text{ mg} \times 100}{40.0 \text{ mg}}$$
$$= 14.57\% \text{ N by weight}$$

Example 4 At what temperature would gaseous helium have a molar volume of 1.00 l/mole under 760 torr?

We may choose the size of our sample; if we take exactly 1 mole, the volume will be 1 l, and we can immediately substitute in the ideal gas law:

$$T = \frac{PV}{nR} = \frac{1.00 \times 1.00}{1.00 \times 0.08205} = 12.19°K$$

Since the molar volume of ideal gases at 0°C is 22.4 l, we can see that there is a substantial change in the density of gases at very low temperatures. We should check the calculated temperature against the boiling point of liquid helium, however, since the result is obviously nonsense if the gas condenses to a liquid before it can be chilled to this temperature; the standard boiling point of helium is 4.2°K, so this is still a reasonable calculation. Any other gas but helium would have condensed above 12°K, however.

There is one rearrangement of the ideal gas law that we have not looked at, and it has some convenient consequences. Since $PV = nRT$, $R = PV/nT$, and since R is constant for all gases and all conditions, we can sometimes save ourselves labor in converting a gas from one set of conditions to another by observing that PV/nT must be constant:

$$R = \text{constant} = \frac{PV}{nT}$$

so

$$\frac{P_1 V_1}{n_1 T_1} = \frac{P_2 V_2}{n_2 T_2}$$

In any particular problem we can usually simplify this expression by dividing out any quantities which remain unchanged, leaving a simple proportion.

Example 5 A gas cylinder is filled to a pressure of 2500 psi (lb/in.²) at 27°C; its rated pressure limit is 3000 psi. At what temperature is it likely to burst?

Here the number of moles of gas remains unchanged, as does the volume of the cylinder (at least until it bursts), so we can divide n and V out of the proportion:

$$\frac{P_1 V_1}{n_1 T_1} = \frac{P_2 V_2}{n_2 T_2} \qquad \frac{P_1}{T_1} = \frac{P_2}{T_2}$$

$$T_2 = \frac{P_2 T_1}{P_1} = \frac{3000 \times 300}{2500} = 360°K \quad \text{or} \quad 87°C$$

Note that in dealing with this proportion we need not convert pressure units, only express them the same way throughout the problem. Temperatures must still be in absolute degrees, however. Incidentally, the high pressure involved here makes the ideal gas law less accurate than other methods such as the Hougen–Watson chart, but it is still suggestive.

Example 6 A mixture of NO_2 and Cl_2 that is exactly stoichiometric for the following reaction is placed in a glass bulb at 1 atm total pressure and 27°C:

$$Cl_2 + 2NO_2 \rightarrow 2NO_2Cl \quad \text{(all are gases)}$$

If the bulb is heated to 300°C and the reaction proceeds completely to the right, what will be the pressure in the bulb?

The stoichiometry of the reaction tells us that a mixture originally containing three moles of gas will react to produce only two moles of gas, so n is changing along with the pressure and temperature. The volume is constant, however, and may be divided out of the proportion:

$$\frac{P_1 V_1}{n_1 T_1} = \frac{P_2 V_2}{n_2 T_2}$$

$$\frac{P_1}{n_1 T_1} = \frac{P_2}{n_2 T_2}$$

$$P_2 = \frac{P_1 n_2 T_2}{n_1 T_1} = \frac{1 \text{ atm} \times 2 \times 573}{3 \times 300} = 1.272 \text{ atm}$$

Practice with these manipulations as provided by the study problems is worthwhile; we shall also use the algebraic expressions for P, V, and T in later derivations.

Study Problems

1. Combine Eqs. 3-5 and 3-6 by eliminating V between them. Rearrange the result until only terms in the variable m are on the left side. Reason from this expression to the ideal gas law.
2. Express the atmosphere in dyne/cm² pressure units by calculating the weight of 760 mm height of mercury metal in a tube with a cross-sectional area of 1 cm². Remember that weight is a force equal to the mass of mercury times the acceleration of gravity (980 cm/sec²). Working from the universal gas constant expressed in liter-atmospheres ($R = 0.08205$ l-atm/mole-°K), convert R to erg units: 8.314×10^7 ergs/mole-°K. The density of mercury is 13.59 g/cm³.

3. In order for any object to escape from the earth's gravitational attraction it must have a velocity greater than 11 km/sec. What fraction of the nitrogen molecules (N_2, molecular weight 28.013 g/mole) in the earth's atmosphere have velocities in the range 11.00–11.05 km/sec? Let the fraction be dN/N and the velocity range dv in the Maxwell–Boltzmann velocity distribution function. Compare this fraction with the fraction having velocities in the range 450–500 m/sec, which is the same size range. Assume a temperature of 300°K for both ranges.

4. What is the molecular weight of a gas that (at 300°K) has v_{rms} = escape velocity?

5. Neon and mercury vapor have very nearly equal van der Waals b parameters, which implies nearly equal atomic volumes and radii. Would you expect any difference in the viscosities of these gases (at the same temperature)?

6. If two gas-containing vessels share a common wall in which a hole small enough to allow effusion has been punched, and one vessel initially contains helium at 25 °C and 1 atm while the other initially contains xenon at 25 °C and 1 atm, what, if anything, will happen to the pressure in each vessel as effusion occurs?

7. A good vacuum pump will bring a tight system down to 10^{-10} torr. How many gas molecules remain per milliliter at this pressure and 300°K?

8. Some laboratory gas-chromatograph instruments detect the presence of the vapor of a sample compound by comparing the heat conductivity of a small volume of the vapor (mixed with a flowing carrier gas) with the heat conductivity of the pure carrier gas. Since the largest possible thermal-conductivity *difference* is desirable in order to obtain maximum sensitivity in the instrument, the carrier gas should either be an exceptionally good heat conductor or an exceptionally bad one, relative to most samples. Usually helium is used; should it be a very good or a very poor conductor? Remember that its atomic weight is 4.0 g/mole.

9. The vapor pressure of water at room temperature (25°C) is 23.8 torr. What total weight of water is present in the atmosphere of a room 12 × 12 × 8 ft at 25 °C and maximum humidity (100% relative humidity)?

10. In a mass spectrometer a vapor sample at 10^{-1} torr effuses into a high-vacuum chamber at 10^{-6} torr, held at that pressure by a vacuum pump with a capacity of 100 l/sec. How long will a liter of the vapor sample last if its effusion rate requires the pump's full capacity?

11. A liquid organic compound analyzes 85.5% C, 14.5% H. Its boiling point is lower than that of water — a sample of it in a 100.0-ml flask held at 100°C and 740 torr thus evaporates completely. If it sweeps all air out of the flask in the process, the volume is completely filled at that temperature and pressure by the pure compound. On cooling, the compound in the flask condenses to a liquid weighing 0.2652 g. What is its molecular formula?

12. A McLeod gauge is a vacuum gauge that works by compressing a very rarefied gas sample from a standard volume into a smaller volume, until its pressure is multiplied to a value which can be measured with an ordinary manometer; the smaller volume is a closed-end capillary tube, as shown in Fig. 3-24. If the bulb has a volume of 100 ml exclusive of the capillary, and the capillary is 1.00 mm in diameter and 10 cm long, to what initial pressure in the bulb would a mark 1.00 mm from the top of the capillary correspond?

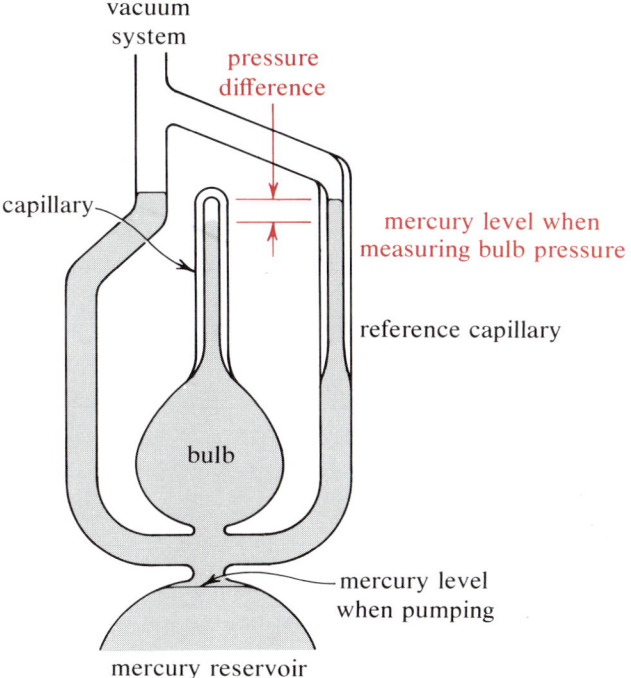

Figure 3-24 McLeod gauge.

Some Further Reading

Hildebrand, J. H., *An Introduction to Molecular Kinetic Theory*, New York: Van Nostrand Reinhold, 1963. Covers much the same materials as this chapter in a very clear style. Hildebrand is one of the country's most distinguished chemistry teachers.

Kauzmann, W., *Kinetic Theory of Gases*, New York: Benjamin, 1966. The same material again, but done with much more mathematical sophistication. Beautifully clear but only if you like mathematical arguments.

Hougen, O. A., Watson, K. M., and Ragatz, R. A., *Chemical Process Principles Charts*, New York: Wiley, 1960. Hougen–Watson charts, but also lots of charts for all kinds of chemical material properties. No explanation with the charts, but an engineering text by the same authors gives explanations.

Moore, W. J., *Physical Chemistry*, Englewood Cliffs, N.J.: Prentice-Hall, 1962 (3rd ed.). Chapters 1 and 7 of this standard text deal with the material from this chapter very nicely; in fact, it reads so well that you may think you understand something when you don't, so watch out.

4 | The Behavior of Solids

The study of the behavior of gases has given us some interesting ideas about what a molecule must be like if it is to prove a reliable model of the structure of matter. Gases, however, are only a part, and not even the most obvious part, of the matter in our experience. The other states of matter are the liquid state and the solid state. Both of these offer additional insights into the nature of molecules and atoms, but for the present we shall concentrate on the behavior of solids to provide maximum contrast to our previous discussion. Later, in Chapter 7, we shall see that the behavior of liquids offers a sort of bridge between the experimental properties of solids and gases and between our ideas of chemical structure and our ideas of chemical change or dynamics.

Solids, of course, are all around us. The vast majority of the material objects we encounter in our everyday lives are solids, and to a very substantial extent our technological culture has been developed by taking advantage of the differing properties of solids. Solids are defined as being in one of two categories: **crystalline** or **amorphous**. Crystalline solids display a very orderly structure; they include such obvious examples as quartz, diamond, rock salt, and snowflakes, and less obvious examples such as metals, minerals, and common dirt. All pure chemical elements form crystals in their solid state. Amorphous (having no form) solids have structures that do not display evidence of long-range order; they include glass, soot, wood, and plastics. It is possibly inaccurate and certainly unfair to describe biochemical systems such as wood as having no long-range order, since their cellular structure obviously evidences a high degree of order, but we shall see later how the division is made. Most of the solids that are not crystalline are polymers, either biopolymers (wood) or synthetic polymers (nylon), and we shall offer some comments on their nature both here and later (Chapter 15). We shall concentrate on crystals, however, because they have a number of important properties that shed light on our molecular model.

4-1 Experimental Properties of Crystals

A waggish physicist has written that solids are those portions of the physical universe that support when sat upon and hurt when kicked. This is probably the most basic property of crystals or other solids: they retain their shape even against fairly substantial forces. There are three ways we can apply a force to a crystal. We can try to pull the crystal apart; this is **tension**. We can try to crush the crystal; this is **compression**. We can apply a force sideways to make one part of the crystal slide over another part; this is a **shear** force. These three forces are shown in Fig. 4-1.

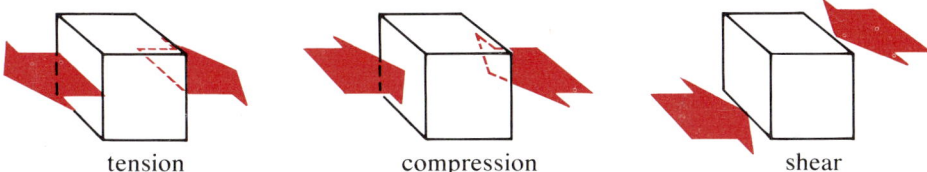

Figure 4-1 Forces applied to crystals.

Compressibility

Of these three forces, compression is the closest analogy to the pressure exerted on a gas that we discussed in Chapter 3. Intuitively, we know that solids respond far less readily to compression than gases, but a numerical comparison may be helpful. Suppose we take a nearly ideal gas, such as air, and compare it with a crystal, such as sodium chloride. If we take a sample of air at 1 atm pressure and apply an additional atmosphere of pressure, the volume will be halved; the fractional volume change of an ideal gas at 1 atm is 0.5 atm^{-1}. Repeating the experiment with sodium chloride gives a fractional volume change (or **compressibility**) of 4.25×10^{-6} atm^{-1}. In other words, starting from 1 atm total pressure, an additional atmosphere of pressure has about 100,000 times as much effect on a gas as on a solid. This suggests that the atoms or molecules in any model of a solid must be very close to each other, essentially touching (in contrast to the widely separated atoms in a gas). Only at this very short range could the large Lennard-Jones repulsive force—proportional to $1/r^{12}$—resist compression so strongly.

Density

This view of the model is substantiated by another easily measured property of solids, density. It is no surprise that solids are more dense than gases, since nearly everyone has perceived that a box of books being carried up the stairs is a lot heavier than the same box full of air being carried down; we may even have been so impressed by the difference in densities as to call the box full of air "empty." We can be a good deal more quantitative, however. Figure 4-2 shows two convenient devices for measuring the densities of gases and solids.

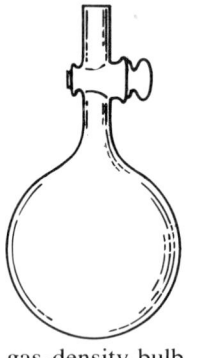

gas-density bulb

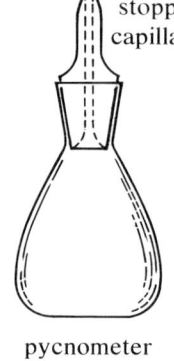

stopper with capillary neck

pycnometer

Figure 4-2 Density-measuring devices.

The bulb on the left is used for gases; to use it we weigh it full of water (which has an accurately known density), full of our gas sample, and evacuated to a reasonably good vacuum. From the weight of water we can establish the bulb's volume, and from the weight of the gas and the bulb's volume we can calculate the gas density. The device on the right, called a **pycnometer**, is used in much the same way to measure the density of a solid. Since it is difficult to fill the pycnometer with a solid, we use a small sample of the solid and fill out the rest of the volume of the pycnometer with a liquid of known density; by subtraction (see Study Problem 1) we can obtain the solid's density. Making such measurements, we observe that gases usually have densities in the range 1–5 g/l, while solids usually have densities in the range 1–5 g/ml; that is, solids are about 1000 times as dense as gases. This is entirely compatible with the idea that the atoms in a crystalline solid must be packed close together, of course, and as a matter of fact is in fairly good agreement with the van der Waals parameter b, which is the volume occupied by the molecules themselves and thus denied to the motion of other molecules. For instance, the van der Waals parameter b for benzene is 121 ml/mole, and from the density of solid benzene and its molecular weight we calculate a volume of about 89 ml/mole; for water, the comparison is 30.4 ml/mole (b) vs. 19.6 ml/mole (ice volume); for carbon dioxide, b is 42.7 ml/mole and the dry-ice volume is about 40 ml/mole. This reasonable agreement offers some assurance that closely packed molecules represent a valid extension of the molecular model from gases to crystalline solids.

Elasticity

Another property common to solids is that they retain their shapes under substantial forces. This is not the same as saying that they have high resistance to compression (low compressibilities), because liquids also have very low compressibilities—liquid mercury will split a glass thermometer if it is overheated. The difference between solids and liquids here is that solids resist shear forces as well as compression forces, while liquids do not. Toward small shear forces solids show an **elastic** response—that is, they deform but return immediately to their previous shape when the force is removed. By contrast, liquids show a **plastic** response toward even the smallest shear force; they yield to it and do not return to their previous shape when the force is removed. With only a few exceptions, crystals can be made to deform in a plastic fashion if sufficient shear force is applied, but the force is usually rather large. Note, however, that the reason we cannot make a lead bell is that the lead deforms plastically when the clapper hits it, which produces a "thunk" instead of a "bong." The resistance of solids to shear forces suggests, in terms of our model, one of two things. Either the atoms in the solid have specific directional chemical bonds between them (preventing motion sideways as well as in and out), or else—perhaps in addition—the atoms are very closely packed, with a high degree of order and few holes. Consider the alternative structure, in which there are no directional bonds, but only a general attraction that does not depend on angle, and that also has a large number of holes in it; these three possible structures are shown in Fig. 4-3. The unbounded, disorderly structure has essentially no forces tending

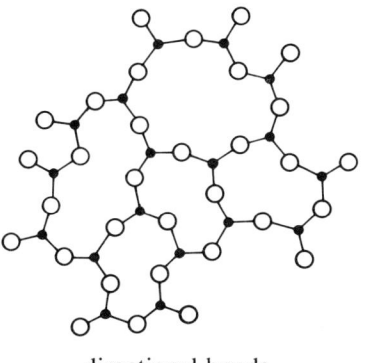

directional bonds
disorderly

Figure 4-3 Possible microscopic structures for condensed phases (liquids and solids).

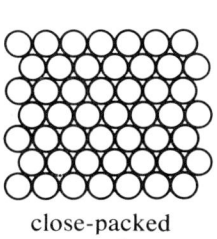

close-packed
orderly

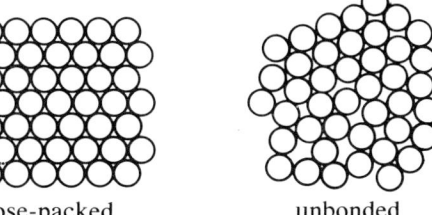

unbonded
disorderly

to prevent the lateral movement of atoms at the top of the figure relative to those at the bottom, since the atoms can move into holes in the structure, thereby creating more holes behind them for other atoms to move into; and there is no change in potential energy (with respect to the Lennard-Jones potential-energy curve) if the moving atom remains at the same distance from other atoms. By contrast, if the directional bonds in the structure possessing them can really exist only at certain angles, we can see that moving the atoms at the top of that structure to the right will require breaking numerous chemical bonds, requiring a large amount of energy. And if the solid were to have the very close-packed, orderly structure shown in Fig. 4-3, movement of the atoms at the top of the structure to the right would necessitate breaking down the orderly structure and pulling neighbor atoms away from each other. This would also require putting energy into the crystal, since the neighbor atoms attract each other through the Lennard-Jones potential. So we can see that either directional bonds or a high degree of order would tend to make a crystal retain its shape; we shall see in Chapter 7 that the unbonded, disorderly structure, which would not resist a shear force, is a reasonable model for a liquid.

Anisotropic Crystal Properties: Crystal Shapes

Crystalline solids have another property closely related to their retention of shape. A crystal's response to external influences such as the further growth of the crystal by deposition from solution, or an external applied force, or sometimes even the passage of light through the crystal, will depend on the angular orientation of the crystal. A sample of matter whose properties do *not* depend on direction or angular orientation is said to be **isotropic**; crystals, by this definition, are **anisotropic**. For instance, all crystals are easier to split (**cleave**) in some directions than others. Mica, which consists of thin flakes, is an extreme example of this, but diamond cutters also take advantage of it. When crystals are grown from solution, they do not grow into spheres, which we might expect an isotropic solid to do (growing equally fast in all directions), but rather into characteristic polyhedral shapes, as shown in Fig. 4-4 for some attractive examples. A crystal of given chemical composition, growing under identical solution conditions, will always grow in the same shape, regardless of the shape of the seed crystals used. Figure 4-5 shows the transition in growth of an alum crystal [$NaAl(SO_4)_2 \cdot 12H_2O$] from a cubic seed crystal to an octahedron. Light passing through some crystals will become plane polarized, as in "Polaroid" brand sunglasses, in a direction characteristic of the crystal's faces. Some crystals, in fact, will appear to change color if they are rotated while being viewed by polarized light. Many other properties of crystals show similar anisotropy, and, in fact, it is precisely this quality that allows us to distinguish

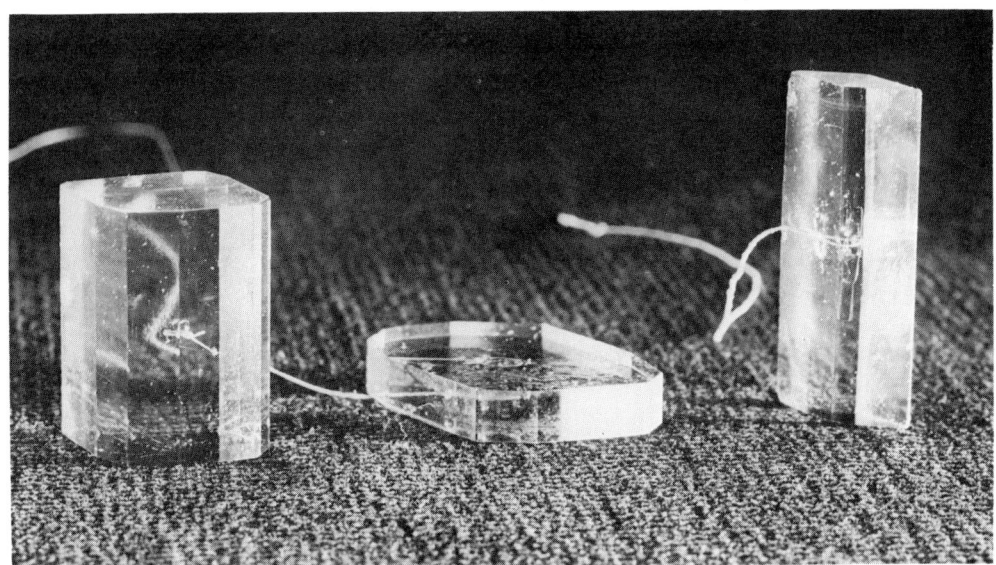

Figure 4-4 (a) Crystals of potassium sodium tartrate tetrahydrate or "Rochelle salt" (colorless). (b) Crystals of sodium chlorate (colorless). (c) Crystal of potassium aluminum sulfate dodecahydrate, "alum," or "potash alum" (colorless). (From Holden, A., and Singer, P., *Crystals and Crystal Growing*, New York: Doubleday, 1961.)

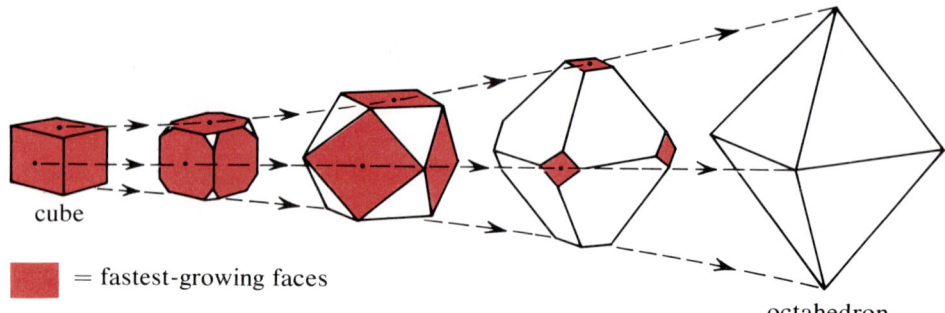

cube

■ = fastest-growing faces

octahedron

Figure 4-5 Change of shape for a growing crystal due to different rates of growth by different faces.

crystalline solids from noncrystalline solids, which are generally isotropic or much more nearly so.

Molecular Stacking in Crystals

What does this require from our molecular model? If we consider the three structures in Fig. 4-3 again, we can draw some inferences. The bonded but disorderly structure, if it were extended to include billions of atoms, would have bonds distributed uniformly in every direction. Insofar as the interatomic attractions or bonds influence the physical properties of that structure, then, the structure should be isotropic. Similarly, the unbonded disorderly structure, since it has no systematic pattern of rows or layers of atoms, will show no preferential directions and will be isotropic. Only the orderly, close-packed structure has its atoms arranged in a sufficiently long-range pattern to make the anisotropy of the rows of atoms show up in the macroscopic properties of the structure. For instance, Fig. 4-6 shows the three favorable directions for cleavage of the crystal. This is a key result. Only an orderly structure can result in a solid with anisotropic properties (regardless of whether or not directional

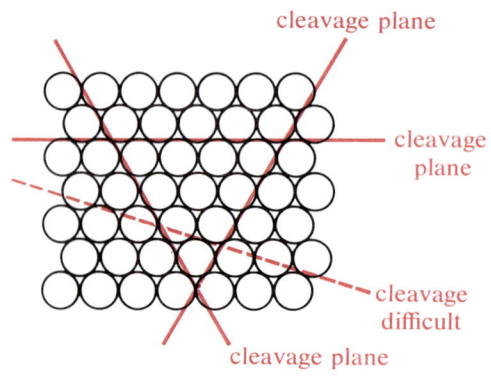

Figure 4-6 Cleavage possibilities for a close-packed orderly structure.

bonds exist within the orderly structure). So we say that a crystalline solid, regardless of its chemical composition, must consist of an orderly array of touching atoms or molecules. We have already noted that the disorderly, unbonded structure is that of a liquid; it is worth adding that the bonded disorderly structure is characteristic of glass, which does have isotropic properties.

The shapes of the polyhedra that crystals form shed further light on the property of close packing of atoms or molecules within a crystal. How can we characterize these shapes? This question is particularly vexing when we realize that crystals are frequently damaged or imperfect so that the beautiful sharp edges or corners are missing. Fortunately there is a property that is quite independent of the physical condition of the crystal, provided only that it is recognizable as a crystal: the angles between faces of a crystal do not change, even though the edges or corners may be damaged, as in Fig. 4-7. So we can characterize the shapes and symmetries of crystals by measuring these angles, which would be the angles between the plane surfaces of an ideal crystal if one were available.

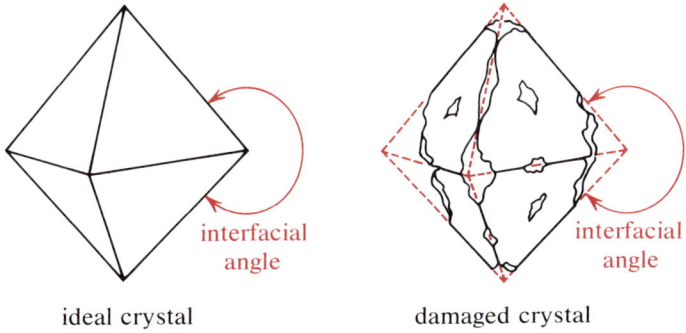

Figure 4-7 Constancy of interfacial angles of crystals.

Crystal Symmetries

It might seem that there would be a limitless variety of shapes and combinations of facial angles that crystals could adopt. As a result of careful examinations of thousands of crystals, however, the surprising fact has emerged that there are only seven possible symmetries displayed by crystals; these are called the seven **crystal systems**. We can define these crystal systems in terms of the angles between faces if we set up the algebraic definition of a plane surface in three-dimensional space. If we have three coordinate axes x, y, and z defined as in Fig. 4-8, then the function that is represented graphically by a plane surface may be written algebraically as

$$h_1 x + h_2 y + h_3 z = k$$

160 | The Behavior of Solids

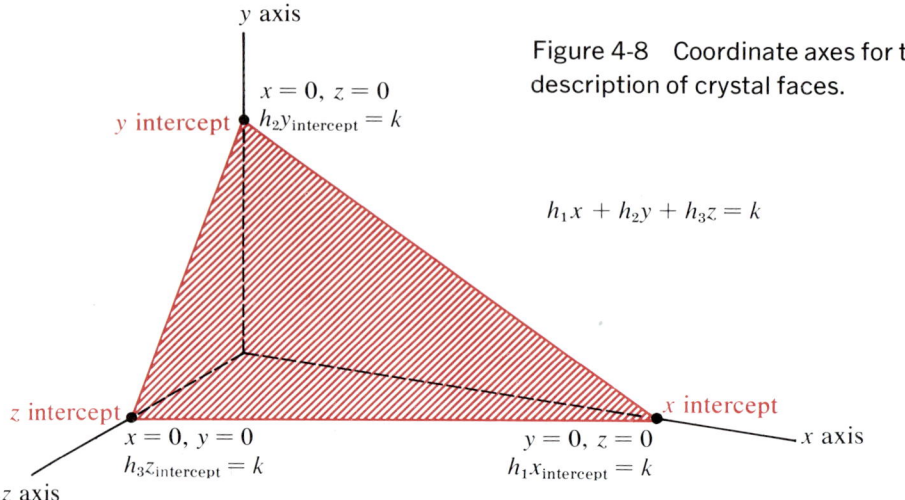

Figure 4-8 Coordinate axes for the description of crystal faces.

where h_1, h_2, h_3, and k are constants. The constant k has nothing to do with the orientation of the plane to the axes; it only moves the plane surface in and out from the origin. Since we are interested only in the orientation of the plane crystal face relative to other faces, we can afford to ignore k, concentrating on representing the plane of the crystal face by the three constants h_1, h_2, and h_3. If we choose the axes x, y, and z to coincide in the best possible fashion with the symmetry of the crystal (or the symmetry of its idealized shape), which may involve having the angles between them differ from 90°, and if we also choose the units along each axis in the simplest possible fashion for one observed face (which may make them different lengths), then the constants h_1, h_2, and h_3 for *all* faces are found experimentally to be three small integers. If the unit lengths along the x, y, and z axes are called a, b, and c respectively, then $h_1 a$ is the intercept plane with the x axis, and so on. Figure 4-9 illustrates this discussion for a particular crystal.

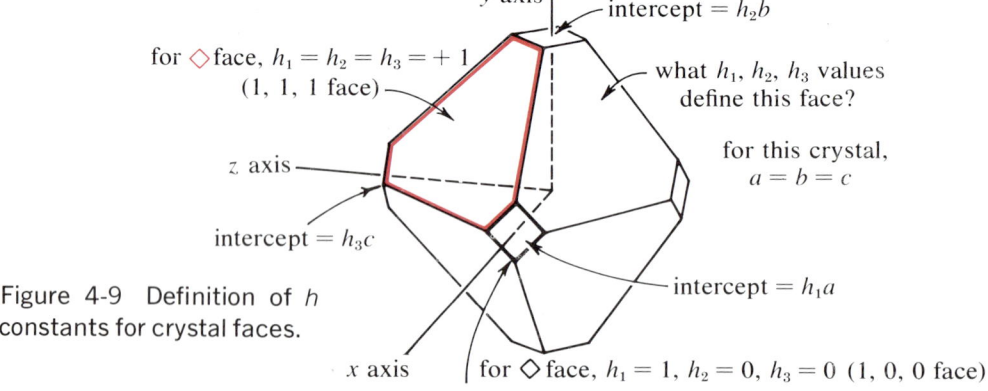

Figure 4-9 Definition of h constants for crystal faces.

Experimental Properties of Crystals | 161

Now that we have defined a coordinate system for each crystal, we can move to the crystal systems. Each crystal system consists of a particular set of coordinate axes with their unit lengths a, b, and c, and with angles between them that we can call α, β, and γ. Table 4-1 gives the seven crystal systems with the relationships between these unit lengths and angles, and also illustrates some possible crystal shapes for each system. We can see, for instance, that a tetrahedron is a truncated cube. Many of the other crystal shapes can also exist in truncated forms, but no other basic symmetries exist than those shown.

Table 4-1
The Possible Crystal Symmetries

Crystal System	Unit Length Relationships	Angle Relationships	Crystal Shape Examples
cubic	$a = b = c$	$\alpha = \beta = \gamma = 90°$	
tetragonal	$a = b \neq c$	$\alpha = \beta = \gamma = 90°$	
hexagonal	$a = b \neq c$	$\alpha = 120°, \beta = \gamma = 90°$	
trigonal (rhombohedral)	$a = b = c$	$\alpha = \beta = \gamma \neq 90°$	
orthorhombic	$a \neq b \neq c$	$\alpha = \beta = \gamma = 90°$	
monoclinic	$a \neq b \neq c$	$\alpha = \gamma = 90°, \beta \neq 90°$	
triclinic	$a \neq b \neq c$	$\alpha \neq \beta \neq \gamma \neq 90°$	

4-2 Differences Between Crystals

All of what has been said so far has described properties that all crystals display—in other words, has suggested their uniformity. But we have already noted that there is a very great variety of physical properties displayed by crystals. How can we classify the differences?

A crystal may be hard or soft. In practice, of course, this means that real crystals display a wide range of hardness, which we can define as resistance to elastic deformation. A closely related property, which also shows considerable variation, is brittleness. Brittleness is resistance to plastic deformation; if a force applied to a crystal permanently deforms it without cracking it, the crystal is said to be **malleable**, which is the opposite of brittle. In general we expect these properties to go together, so that a hard crystal will also be brittle and a soft crystal will also be malleable, but this is far from uniformly true. A common property of all crystals is that each has a sharply defined melting point, or temperature of transition to the liquid state, and this can also vary greatly. The element hydrogen, for instance, melts at 14°K, but the element tungsten melts at about 3650°K. Another property is electrical conductivity—the ease with which an electrical current passes through the crystal—measured as the reciprocal of the electrical resistance of a 1-cm³ cube of the solid. Electrical conductivity varies more, perhaps, than any other physical property; silver conducts electricity about 10^{24} times as well as quartz.

There are many other physical properties that differ substantially from one crystal to another, but these four can be used to define some categories into which all observed crystals seem to fall (with remarkably little confusion). Table 4-2 indicates the combinations of these properties that are usually observed, and introduces four new names for these categories: crystals are said to be **ionic**, **partly ionic**, **covalent-molecular**, or **metallic**. These names anticipate our discussion of bonding and chemical synthesis, and a full amplification of their significance must wait until Chapters 6, 13, 14, and 15. We can, however, apply our atomic model to each of these categories and obtain some additional information about what atoms must be like.

Ionic Crystals

In Chapter 1 we introduced Coulomb's law governing the force of attraction or repulsion between two electrically charged particles. This is quite a strong force, and if we propose a model for a crystal in which the Coulomb attraction of oppositely charged atoms (**ions**) is the force holding the crystal together, we

Table 4-2
Physical Properties of Crystal Types

Crystal Type	Physical Property			
	Hardness	Brittleness	Melting Point	Electrical Conductivity
ionic	high	very high	high	low (high when molten)
partly ionic (giant molecule)	very high	high	very high	very low
covalent molecular	low	medium	low	very low
metallic	varies	very low	varies	very high

may expect the attractive force to resist strongly any effort to deform the crystal. Figure 4-10 shows a two-dimensional version of an ionic crystal model; note the sandwich effect, in which each positively charged ion is surrounded by negatively charged ions, and vice versa. This particular geometry leads to the ions being strongly bound together, since the ions with similar charges are separated by a greater distance than the oppositely charged ions (remember that the force is inversely proportional to the square of the distance: $-q_1 q_2 / r^2$).

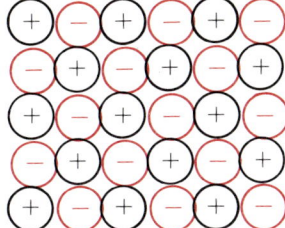

Figure 4-10 Ionic crystal model showing alternating charges on ions.

This energetically favorable situation is only possible, however, in a crystal whose structure is very orderly and symmetrical. If we try to deform the crystal, we see in Fig. 4-11 that we need move part of the crystal only one ionic diameter with respect to the other part in order to create a strong *repulsive* force that will break the crystal apart. So the strength of the Coulomb attractive force will make an ionic crystal quite hard, but the fact that a very small displacement creates a strong repulsive force makes the crystal quite brittle.

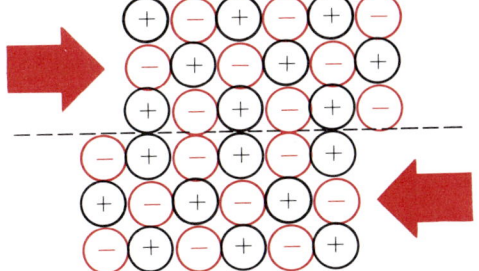

Figure 4-11 Shattering of an ionic crystal under shear force. Repulsions between similar charges shatter the crystal along the dashed line.

Going on to the melting point, the strength of the attractive force in the orderly crystal array of atoms will make it difficult to create the disorderly, open structure of a liquid. An ionic crystal will have to absorb a considerable amount of energy in order to melt. Since the energy of the ions in the crystal is proportional to kT, a high temperature will be required to provide the ions with enough energy to overcome the attractive force. So we can expect the melting point of a crystal roughly to parallel its hardness.

With respect to electrical conductivity, the carrying of an electric current means transporting electrical charge. Although we think of the atoms as being charged ions, they cannot move freely within the orderly lattice and so cannot—within the crystal—carry a current. The electrical conductivity of an ionic crystal should be low, which is experimentally true. However, if the crystal is melted, the ions can move much more freely through the disorderly, open structure of the liquid and carrying a current becomes possible; again this is exactly what is observed. As shown in Fig. 4-12, an electrical circuit can be completed by a molten ionic salt such as NaCl—the bulb will light. But if the molten salt is allowed to cool and crystallize, the bulb will dim and go out completely when crystallization is complete.

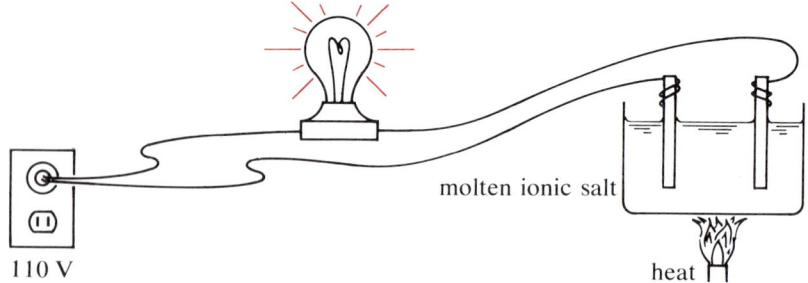

Figure 4-12 Conduction of an electrical current by a molten ionic salt.

All this suggests that we can account quite nicely for the properties of ionic crystals if we make a further assumption about the nature of atoms. Atoms must have an electrical nature; it must be possible to place either a positive or a negative electrical charge on an atom, as we have already suggested in Chapter 1, for this model to be legitimate. We have not previously had any intimation that atoms were electrical in nature, but we shall see in Chapter 5 that this is the key to understanding their internal structure.

Partly Ionic (Giant Molecule) Crystals

If we have a crystal with the structure of an ionic crystal, but which has in addition some directional chemical bonds tending to make the structure rigid, as in Fig. 4-13, we expect properties rather like those of ionic crystals. How-

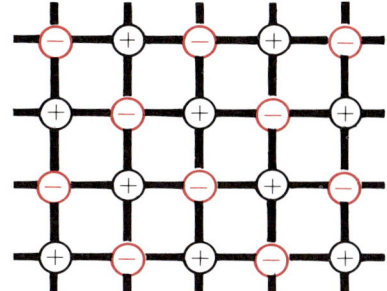

Figure 4-13 Partly ionic crystal structure with alternating charges and directional bonds.

ever, since the atoms are now held together not only by the Coulomb forces but by directional bonds, we should expect this crystal to have even greater resistance to deformation; the partly ionic crystal should be very hard, even more so than the ionic crystal. The ionic crystal was very brittle because its binding forces were converted into repulsive forces by moving a row of atoms only one diameter. Although we have as yet said nothing about the nature of the directional bonds in the partly ionic crystal, there is no reason to suppose that they become repulsions if a row of atoms is displaced; so it is not surprising that partly ionic crystals are somewhat less brittle than ionic crystals.

Melting these partly ionic crystals means breaking a large number of directional chemical bonds, since the disorderly liquid structure cannot contain the orderly pattern of bonds. This will tend to make the melting points of partly ionic crystals very high, since a large amount of thermal energy must be put into the crystal to overcome the bond energies. That is, the disorderly liquid structure not only gives each atom fewer neighbors, on the average, but also scatters or distorts the neighbor atoms out of the favorable (low-energy) bond directions. By way of comparison, NaCl (a typical ionic crystal) melts at 801 °C, but AlN (aluminum nitride, a typical partly ionic crystal) melts above 2200 °C.

To the extent that the electrically charged nature of the atoms in the crystal is diminished by the formation of the directional bonds, we expect the partly ionic crystal to be an even poorer electrical conductor than the ionic crystal. This is borne out, for instance, by the fact that quartz (SiO_2), which is a fairly typical partly ionic crystal, is one of the best insulators—poorest conductors—known. Most partly ionic crystals do not even become good conductors when melted, although the difficulty of working at very high temperatures has prevented the accumulation of much data. As would be expected, the conductivity when molten depends on the amount of charge remaining on individual atoms.

Covalent-Molecule Crystals

Suppose now that we consider a crystal made up of individual molecules, with directional bonding within them but no significant electrical charges on any atoms or on the whole molecule. There can thus be no Coulomb attraction between the molecules, and if we stipulate in addition that no directional bonding can exist between molecules we are left with no attraction between the molecules except for the relatively weak van der Waals forces (as represented by the attractive part of the Lennard-Jones potential). This is the situation shown in Fig. 4-14. If the attractive forces are weak, the molecules may be easily

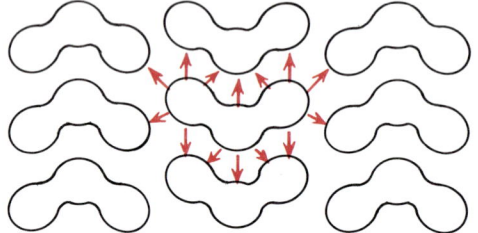

Figure 4-14 Covalent-molecule crystal structure with no charge attraction or directional bonds between molecules. Arrows indicate the directions of weak van der Waals forces.

displaced, and we expect such a crystal to be relatively soft. It should also be easily deformable or malleable, and easy to break apart, since no very great attraction opposes either operation. We may also expect that the melting point of such a crystal will be low, since not much thermal energy is necessary to break down the orderly structure if the binding energies are small. And finally, if there are no electrical charges on any of the atoms in the crystal, it will be virtually impossible to transport electrical charge through the crystal and conductivity will be very low. All these expectations are fully verified experimentally for crystals which we have reason to believe are covalent-molecular.

Metallic Crystals

To adapt the atomic model to the properties of the metallic crystal requires some adjustment of our thinking. The most remarkable feature of metallic crystals is their ability to conduct electrical currents readily in the solid state. It is inconceivable that this should be due to the motion of charged ions through the solid lattice, since metals characteristically conduct even better than molten ionic salts—perhaps a hundred or a thousand times as well. And metals conduct even better when they are chilled, which is the opposite from the tendency we would expect if thermal energy were responsible for removing charged ions from their sites in the crystal array and transporting them. As a matter of fact, many metals at very low temperatures (on the order of $1-10°K$) become superconducting, offering virtually no resistance to the passage of electrical current. Accordingly, we must propose that in a metallic crystal there is a sort of "electrical fluid" or gas of charged particles that is essentially free to move within the crystal and is not bound to individual atomic sites. In the light of other experiments, which we shall describe in Chapter 5, these particles can be identified as electrons, with a negative charge. This means that the metal atoms must all be positively charged ions, surrounded by the free-electron "gas" which, by virtue of its Coulomb attraction for the several positive ions, binds the crystal together. This situation is shown in Fig. 4-15.

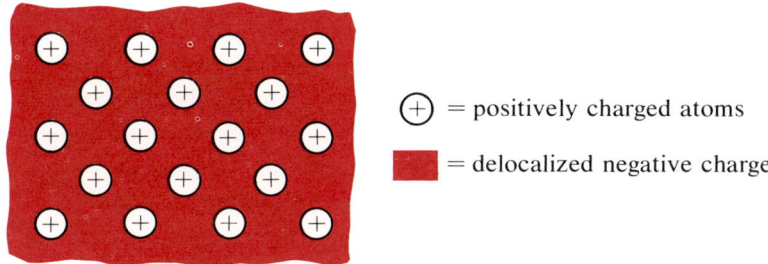

Figure 4-15 Metallic crystal with electron "gas" binding positively charged atoms.

This remarkable structure leads us directly to the unusual properties of metals. Since adjacent ions have identical charges, they have no Coulomb attraction for each other; thus the hardness of metals is not due to the same forces that make an ionic crystal hard. Rather, we must propose that the hardness of metals is due to the existence of directional chemical bonds between metal atoms, which add to the attraction of the "electron gas" in binding the crystal together. Since these bonds may be either strong or weak, the hardness of metals varies considerably. All metals are at least fairly malleable, however, and we may simply interpret this as being due to the fact that the "electron gas" binds the positive ions together equally well no matter what the shape of the metal crystal may become. The melting point, like the hardness, varies depending on the strength of the directional chemical bonds that determine the crystal geometry. In most cases these bonds are reasonably strong, which leads to fairly hard, high-melting metals; but gallium metal melts at 30°C, and mercury metal at −40°C, suggesting quite weak bonds in these cases. Finally there is the electrical-conductivity property we mentioned initially. Clearly if the electrons are really free or **delocalized** through the crystal the conductivity should be essentially infinite, which corresponds to the superconductivity we mentioned previously. Apparently the finite conductivity at higher temperatures is due to imperfections or defects in the crystal structure that tend to **localize** the free electrons and thus impede the flow of electric current. We shall have more to say about crystal defects presently.

4-3 Order, Wave Diffraction, and X Rays

We have stressed that atoms must exist in an orderly array in crystals to explain the physical properties the crystals display. This high degree of order and regular spacing between atoms lead to another experimentally accessible property of crystals: they will diffract X rays at very precisely determined angles. This is an extremely useful property, since from the diffraction pattern we can reason back to the necessary geometry of the atoms within the crystal. X-ray diffraction is the most important single source of experimental information on crystal and molecular structure. To appreciate the way the method works, we need first to look into the way waves interact with particles.

The Interaction of Waves with Ordered Systems of Particles

Figure 4-16 shows a wave traveling toward the right of the page. The regular nature of the wave is shown by the vertical lines, each pair of which is separated by a constant distance, λ, the wavelength. Let us agree to represent a traveling wave by a series of lines drawn perpendicular to the direction in which the wave is moving, as the lines passing through each crest of the wave do here.

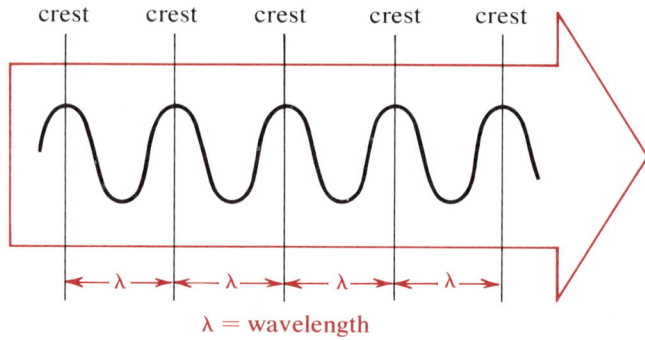

Figure 4-16 Representation of wave motion and associated quantities.

Each line thus represents an individual crest or wavelet of the wave. Now what happens if we allow this wave to interact with a particle? As a simple model, what pattern of ripples will we observe if a small wave crest in water meets a single post of a pier? Figure 4-17 shows the familiar result, which is an expression of **Huygens' principle**: each point in a wave front serves as a source of new waves. The post produces a circular pattern of ripples around it, with the same wavelength as the incoming wave and in the same *phase*. By phase we mean the relative position of a wave along its wavelength, which implies comparison with another wave. If two waves are moving so that their crests and troughs occur

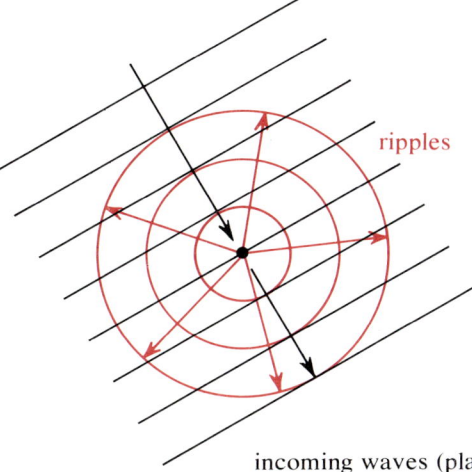

Figure 4-17 Interaction of plane waves with a single stationary particle.

exactly together, they are said to be *in phase*, a condition analogous to being in step when marching. If this is not true, the waves are *out of phase* and might even be opposed to each other so that the trough of one occurred with the crest of the other. If the two waves are moving in the same medium, their relative phase will have a substantial effect on the observed degree of wave motion. Figure 4-18 shows the effect of adding together two waves with varying phases;

Figure 4-18 Phase coherence and interference.

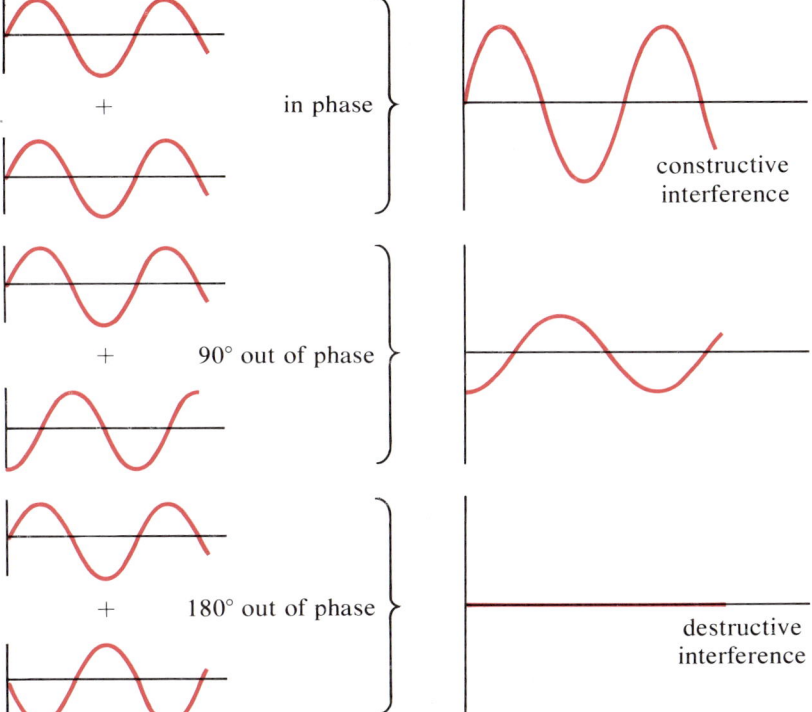

it is clear that the waves **interfere** with each other. When the two waves add to produce a larger wave than either alone (i.e., when the ripple and the original wave are nearly in phase with each other) the result is said to be **constructive interference**; when the waves tend to cancel each other (out of phase and nearly opposed) the result is **destructive interference**.

Looking back at Fig. 4-17, we can see that constructive interference or reinforcement occurs wherever the initial wave crests and the ripple wave crests coincide (where black lines coincide with red lines). Notice also that the fact that the ripples (the diffracted waves) are in phase with the incoming wave when they begin at the post (the diffracting particle) means that as successive wave crests come past, successive concentric diffracted waves are formed; the longer the time since the initial wave passed, the larger the circle representing the diffracted wave.

Suppose now that we put several more particles in the path of the incoming wave, arranged in a uniformly spaced straight line as shown in Fig. 4-19. This rather formidable figure simply shows that each of the five particles is forming a similar set of concentric circular diffracted waves. What is new is that the overlapping diffracted-wave crests tend to form wave fronts, as shown by the

Figure 4-19 Interaction of plane waves with multiple evenly spaced particles (diffraction).

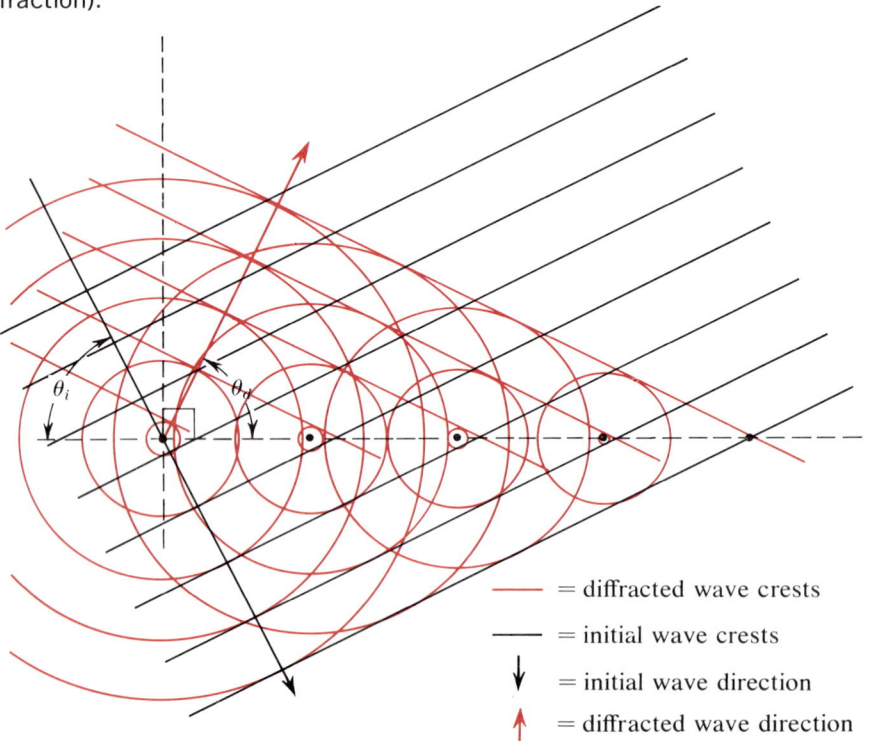

diagonal straight red lines. If the number of particles is very large, these diffracted wave fronts become very well defined, but *only* if the particles are uniformly spaced. For irregularly spaced particles, the same concentric diffracted wave patterns form around each particle, but the radii of the diffracted waves vary and no linear wave fronts are formed. For the regularly spaced case, however, the diffracted wave fronts move away from the row of particles at an angle indicated by θ_d in Fig. 4-19. Consideration of the geometry of the circles will show that θ_d, the **angle of diffraction**, must always equal θ_i, the **angle of incidence**, no matter what θ_i is. In other words, an incoming wave is "reflected" in much the same way that a mirror reflects a ray of light.

The next step in simulating a three-dimensional regular array of atoms in a crystal is to place another row of particles next to the first row with a uniform spacing d, as shown in Fig. 4-20. Here we have simplified the drawing by omitting the circles and drawing only the lines that represent the incoming wave and the diffracted wave fronts. What conditions must govern the spacing, d, in order that the diffracted wave fronts from the new row of particles will reinforce those from the first row of particles? This is the condition shown in the figure,

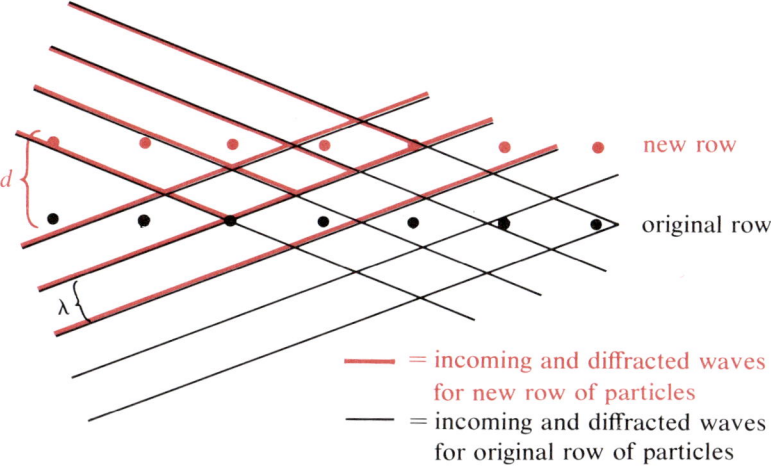

Figure 4-20 Reinforcement of diffracted wave fronts by additional rows of particles.

where the red lines coincide with the black lines. Clearly there must be a relationship between the spacing and the wavelength. Figure 4-21 indicates what the relation must be. In this figure the wave crests are omitted and the lines indicate the path of incoming and diffracted rays. For reinforcement to occur at points C and F, the waves that were originally in phase at A and D must be in phase again. Since DEF is a longer path than ABC, the only way the two rays can be in phase at C and F is for the extra distance traveled by DEF to be an

172 | The Behavior of Solids

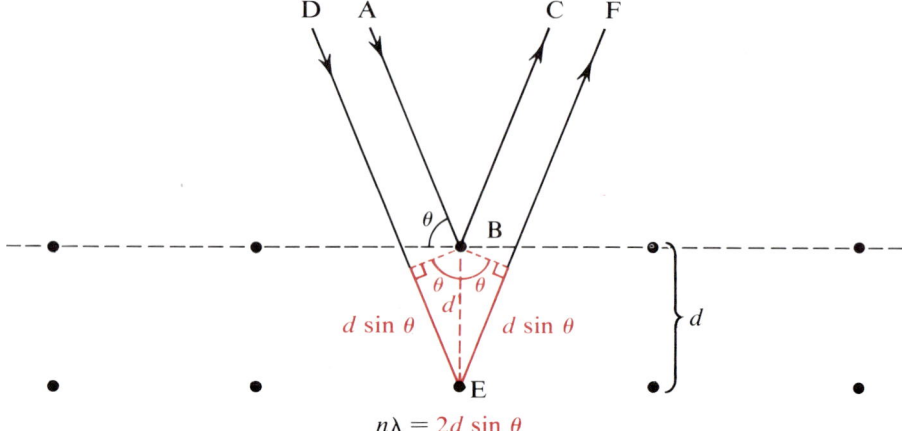

Figure 4-21 Reinforcement condition on angle of incidence and wavelength for a given spacing d.

integral number of wavelengths, $n\lambda$, where n is an integer. This is the distance shown in red on the figure. A basic trigonometric relation tells us that if the hypotenuse of the two triangles shown in red in the figure is d, then the extra distance must be $d \sin \theta + d \sin \theta$, or $2d \sin \theta$. Since this must be an integral number of wavelengths for reinforcement to occur, we have the relation

$$n\lambda = 2d \sin \theta \qquad (4\text{-}1)$$

This relationship, which is basic to X-ray crystallography, is known as the **Bragg equation**. Notice that the spacing between particles sideways in each row does not matter—the angle of diffraction will equal the angle of incidence as long as the spacing is regular. All that is important is that the distance *between* rows have the proper relationship to the wavelength of the incoming wave.

X-Ray Diffraction in Crystallography

We can use the Bragg equation to measure two features of the internal atomic structure of a crystal: first, the orientation of rows (in the three-dimensional real crystal, planes) of atoms within the crystal—this usually bears a straightforward relationship to the external shape of the crystal, which can be established by comparing the orientation of the crystal face with that of the diffracted beam of X rays; second, the observed angle together with the known wavelength give the interplanar spacing d. From this information we can reconstruct the entire internal geometry of the crystal.

Where do the X rays come in? The quantity $\sin \theta$ in the right-hand side of the Bragg equation can never exceed 1.00, so the right-hand side can never exceed $2d$. Therefore λ on the left-hand side must be small enough to be less than $2d$. If the centers of the atoms in the crystal are separated by something like the diameter of an atom, then $2d$ will be in the vicinity of 2–5 Å, or about 2–5×10^{-8} cm. Thus the wavelength of whatever radiation we use must be no greater than about 1–2 Å. Only X rays have such a small wavelength; visible light, for instance, has a wavelength on the order of 5000 Å, which is clearly much too long. Now, although λ must be smaller than $2d$ in order to observe diffraction in the way we have described, it does not need to be close to $2d$; it can be much smaller, in which case $\sin \theta$ will need to be quite small. This gives us a way to determine the wavelength of the X rays in the first place. We can rule a grating mechanically that has a spacing of 10^{-4} cm, and if we then direct a beam of X rays at the grating at a very small angle θ, the diffraction pattern can still be observed and the wavelength determined from the known spacing of the grating. This wavelength then becomes one of the constants for use in the Bragg equation when the same beam of X rays is used on a crystal.

On several previous occasions we have referred to Avogadro's number, $N = 6.022 \times 10^{23}$ atoms/mole, without demonstrating our experimental access to this number. Study Problem 2 indicates a way in which we can use X-ray diffraction to establish Avogadro's number; it should be considered carefully.

We have indicated that atoms in a crystal diffract X rays, without saying anything about how this occurs or what features of the structure of the atom itself are revealed. X rays are electromagnetic radiation, and they interact with atoms by encountering the electrical charge within each atom. Anticipating the discussion in Chapter 5, we can say that the electrically charged electrons in each atom occupy a much larger volume than does the nucleus of the atom, even though the nucleus contains most of the mass. Accordingly, it is the electrons that interact with—diffract—the X rays, and a diffraction pattern reveals the overall distribution of electrons within the atoms or molecules in the crystal. Figure 4-22 shows the interpreted result of a very careful modern study of NaCl by X-ray diffraction, an electron density map. It can be seen that the atoms are nearly spherical and that they are in some sense "touching" each other. The interatomic spacing d is not the distance between the outer fringes of the atoms, which is presumably zero, but rather the distance between the centers of the spheres. (What is the distance between Los Angeles and Pasadena?) The topographical-map appearance of the figure is the conventional way of indicating that electrons are distributed much more thickly near the center of the atom, more thinly near the outside. Similar diagrams result from the study of much more complicated molecules, in recent years including even large biochemical molecules such as enzymes; achieving such precision as this figure indicates, however, is extremely difficult or even impossible for very large molecules.

174 | The Behavior of Solids

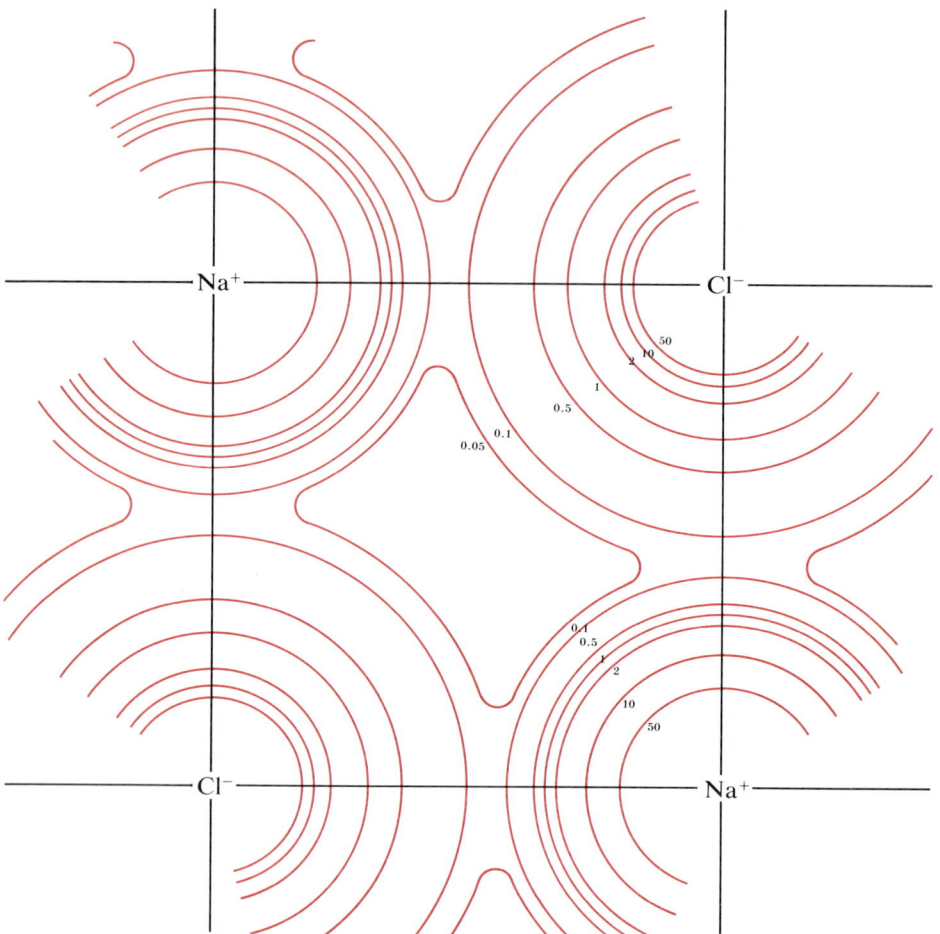

Figure 4-22 Electron map resulting from an X-ray diffraction study of NaCl. The numbers indicate relative electron densities. [From Schoknecht, V., *Z. Naturforsch.* **12a**, 983 (1957).]

How are X-ray diffraction experiments done? In the early **Laue method**, a single crystal of the desired substance was placed in a beam of X rays as in Fig. 4-23a. The crystal planes of atoms selected the angles at which diffraction would occur, and a large photographic plate recorded the locations of the diffracted beams. The fourfold symmetry of the atoms in the crystal is clear. Figure 4-23b shows the experimental arrangement for the **rotating-crystal method**, which is much more widely used. The crystal is mounted in a certain orientation and slowly rotated, so that all sets of planes are at the proper angle to "reflect" the incoming beam at some time. This allows the use of **monochromatic** X rays (those having only a single specified wavelength), which

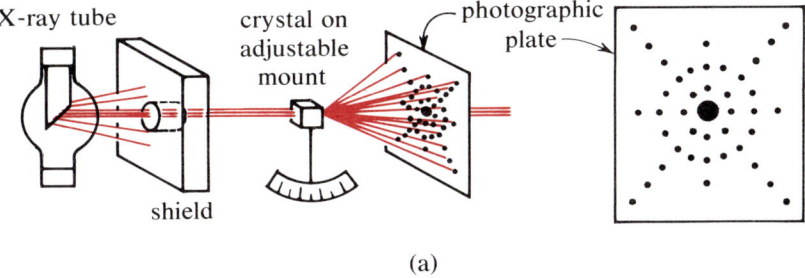

(a)

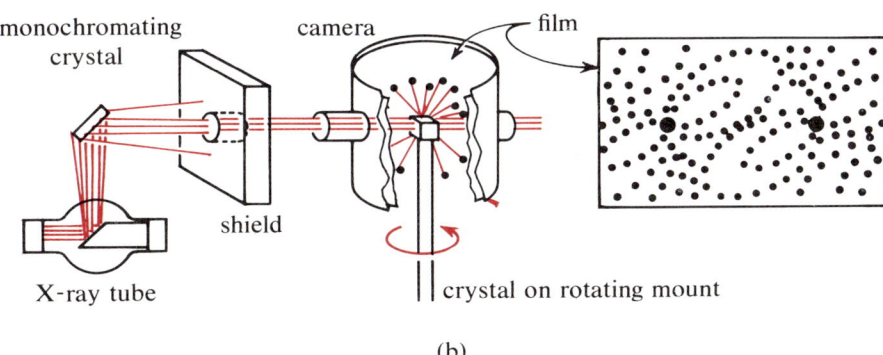

(b)

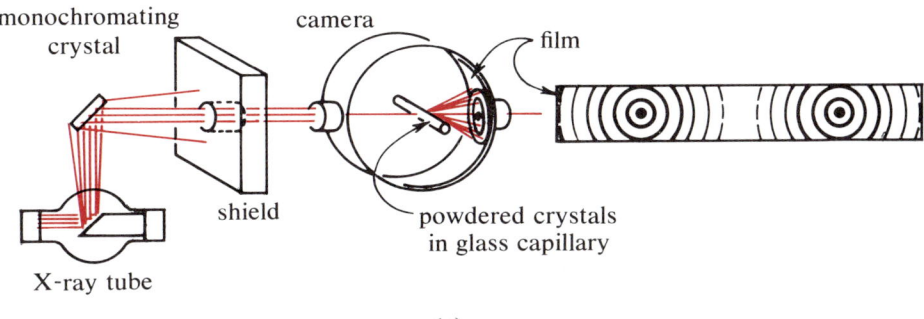

(c)

Figure 4-23 Experimental X-ray diffraction techniques. (a) Laue method. (b) Rotating-crystal method. (c) Powder method.

increases the amount of information available from the film. Finally, in Fig. 4-23c we see the basis of the **powder method**. Here a powder consisting of many randomly oriented crystals is placed in the beam; no matter what planes of atoms we wish to consider, there will always be some crystals having those planes exposed to the X-ray beam. The result will be a cone of diffracted X rays that intersects a strip of photographic film surrounding the powder sample. This method is somewhat more difficult to interpret than the others, but is very convenient because it does not require the preparation of a large single crystal—sometimes a very difficult task. By any method, the result is a determination of the angles at which X rays are diffracted, which can be used with the Bragg equation to establish spacings.

4-4 Close Packing and Crystal Lattices

Information from X-ray diffraction or neutron diffraction (a closely related method) has given us detailed information on many different crystal structures. If these were all unrelated, we would have a staggering task in trying to understand the ways atoms pack together in crystals. Fortunately, a great number of the observed crystal structures correspond (either exactly or with minor variations) to what are called **close-packed** structures. These are structures that give the maximum possible density for the crystal; in the process they usually give each atom the largest possible number of nearest neighbors. The detailed arrangement of atoms in a crystal is called the crystal **lattice**, the name arising from the network or lattice of atoms or molecules bound to their nearest neighbors. Let us look at the formation of a close-packed lattice.

Square Packing and Triangular Packing

In two dimensions, we could arrange circular "atoms" in a square array as shown in Fig. 4-24a. This is not the most efficient use of space, however, as we can see by comparing Fig. 4-24a with Fig. 4-24b; in the latter figure the same number of atoms are arranged in a smaller space, thereby achieving greater density. It follows, then, that a close-packed lattice must involve triangular

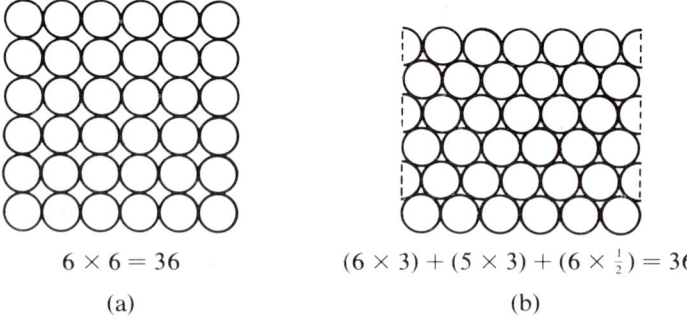

$6 \times 6 = 36$ $\qquad$ $(6 \times 3) + (5 \times 3) + (6 \times \frac{1}{2}) = 36$

(a) $\qquad\qquad\qquad\qquad$ (b)

Figure 4-24 Comparison of packing efficiency for (a) square arrays and (b) triangular arrays.

packing, not square packing. Given a layer of close-packed spherical atoms, how can we construct a three-dimensional lattice that preserves close packing? We can place another similar layer on top of the first, but we must not superimpose the atoms directly; viewed from the side this would be equivalent to the square packing in Fig. 4-24a. We must allow the atoms in the second layer to nestle in the recesses between the atoms in the first layer, as in Fig. 4-25. The recesses are all equivalent, and it does not matter how we do this.

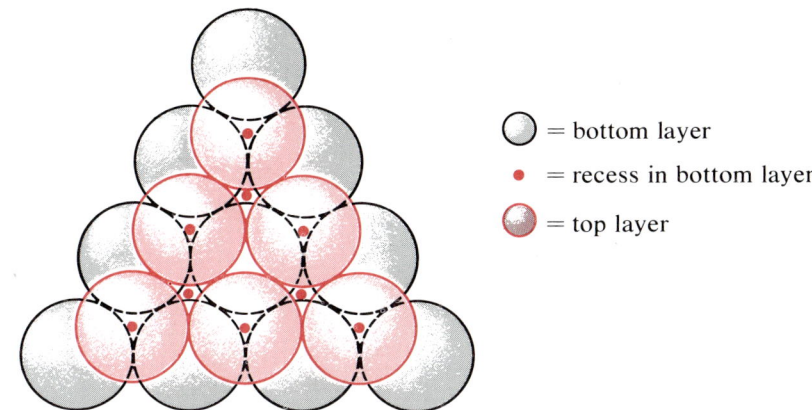

Figure 4-25 Layer close packing of triangular arrays.

Close-Packed Lattices

When we come to place a third layer on top of the first two, however, we have a choice to make even if we follow the practice of placing the atoms in the third layer over recesses in the second layer. Because of the size of the atoms, we cannot place an atom in every recess, only in every other recess. In the second layer there are two kinds of recesses or holes; one is over atoms in the first layer, the other is over holes in the first layer. Accordingly, there are two kinds of close-packed lattices, depending on whether we place the third layer over atoms in the first layer or over holes in the first layer. If we choose to place it over atoms, we have the **hexagonal close-packed** lattice shown in Fig. 4-26. If we call the positions of the atoms in the first layer a and the second layer b, the hexagonal close-packed (hcp) lattice consists of close-packed layers stacked *abababab*.

If, on the other hand, we choose to place the atoms in the third layer over the holes in the first layer, we have a fundamentally different lattice. Calling these previously unoccupied positions the c positions, we can have a lattice

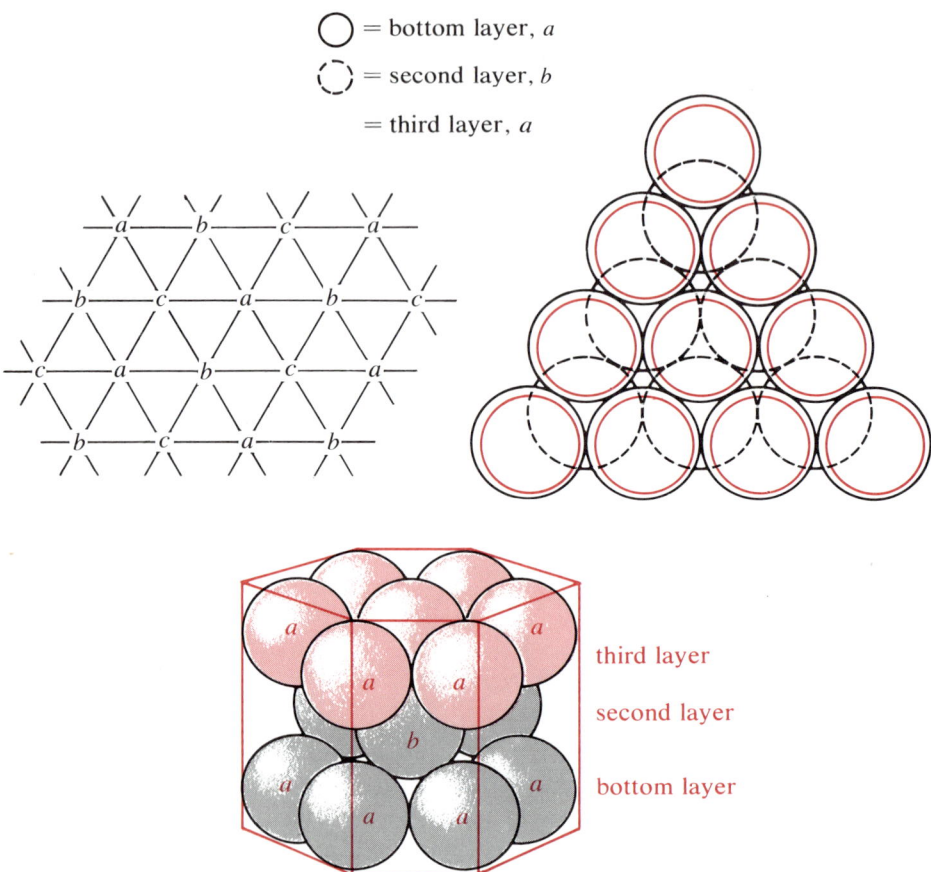

Figure 4-26 Placement of a third layer on two close-packed layers to give a hexagonal close-packed lattice.

stacked *abcabcabc*. This is the **cubic close-packed** (ccp) lattice shown in Fig. 4-27. It is not obvious that triangular packing leads to a lattice with the symmetry of a cube, but we should remember that looking down the body diagonal of a cube we see threefold symmetry about the corner nearest us; this is exactly the way the atoms are stacked. If your spatial visualization failed in the last page or two, try gluing together some plastic balls into layers or using gumdrops and toothpicks for the same purpose; stacking these is much more helpful than any two-dimensional picture.

Either of these two close-packed structures has 74.0% of the available volume filled by spherical atoms, and these are the only structures that achieve such a high efficiency in packing. Study Problem 3 derives this number for the ccp case; only simple geometry is involved. There is a related structure that is not close-packed but that uses 68% of the volume and is found fairly often,

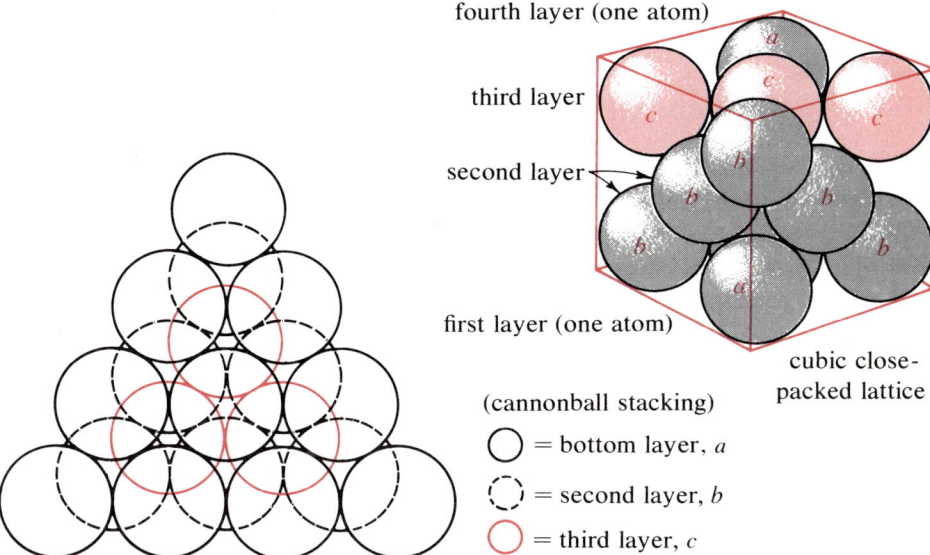

Figure 4-27 Alternate placement of a third layer on two close-packed layers to give a cubic close-packed lattice.

particularly among metals. This is the **body-centered cubic** lattice (bcc) shown in Fig. 4-28. Evidence that the bcc lattice is not close-packed comes from the lower percentage of volume used, of course, but also from the number of nearest neighbors each atom has, called its **coordination number**. In the bcc lattice, the coordination number is 8, while consideration of the hcp and ccp (sometimes called the face-centered cubic lattice for reasons that should be apparent from Fig. 4-27) shows that each of these gives every atom a coordination number of 12. The greater coordination number for the close-packed lattices correctly suggests that the packing of the bcc lattice is not the closest that could be achieved.

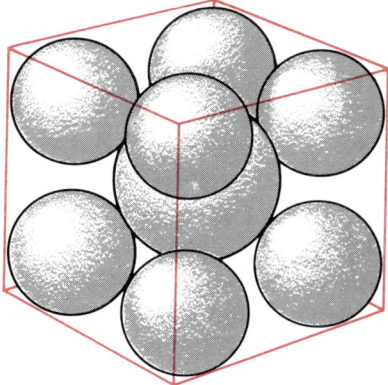

Figure 4-28 Body-centered cubic lattice.

Free Space in a Close-Packed Lattice

By subtraction, we can see that 26% of the volume inside a close-packed lattice is not filled by atoms. How is this space distributed? There are two kinds of holes in the lattice. The more easily visualized is the hole immediately underneath an atom sitting in a recess between three other atoms, as in Fig. 4-29.

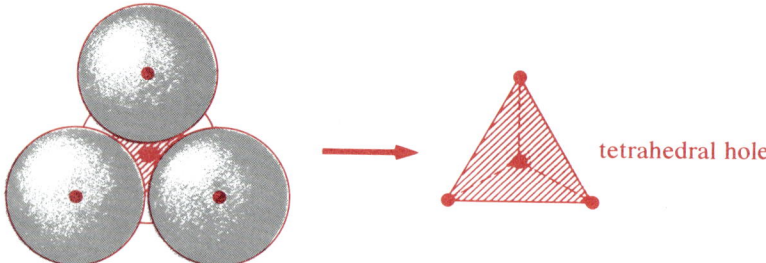

Figure 4-29 Interstitial space in a close-packed lattice; the tetrahedral site geometry.

We normally define its geometry in terms of the positions of the centers of the atoms surrounding the hole; these four atoms are arranged tetrahedrally, and the hole is said to be a **tetrahedral site**. The term "site" is used because this is a possible location for a smaller atom.

The other kind of hole is the one that exists in a recess that does *not* have an atom sitting in it, as in Fig. 4-30. Just as three atoms surround it on the bottom,

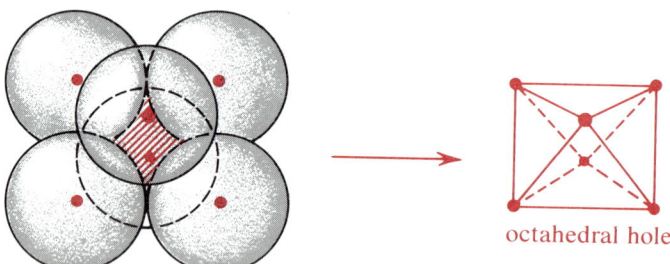

Figure 4-30 Octahedral site geometry.

three atoms surround it on the top, and the hole is said to represent an **octahedral site**. If we have two layers arranged *ab*, then tetrahedral sites exist both over the *a* atoms and under the *b* atoms, and octahedral sites exist in the *c* positions between the layers. (Fig. 4-25 may be helpful here.) So there are two tetrahedral sites for every octahedral site. Furthermore, there are two tetrahedral sites for each atom in the lattice, since there will be a tetrahedral site both over it and under it; there is, then, one octahedral site per atom in a close-packed lattice.

With this geometrical background, we can observe that if we have a close-packed lattice of atoms of element B, for example, and if we also have a supply of atoms of element A that are small enough to fit into the holes or sites, we can make a compound AB by fitting an A atom into each octahedral site in the close-packed B lattice. Alternatively, we can make the same compound by putting an A atom into half the tetrahedral sites. We can make A_2B by putting an A atom into half the tetrahedral and all the octahedral sites, or simply by filling all the tetrahedral sites. We can make AB_2 by filling half the octahedral sites, and so on. This actually occurs in many minerals, in which the close-packed lattice is formed by oxygen atoms, and the smaller metal atoms fit into the various sites. For instance, it is a fairly good approximation to regard quartz (SiO_2) as having Si atoms in one-fourth of the tetrahedral holes of a lattice of close-packed oxygen atoms.

We must remember that there are size limitations on the metal atoms, however, since they must fit into relatively small holes. Figure 4-31 shows a cross section of an octahedral site, with a metal atom in it. It is easy to show

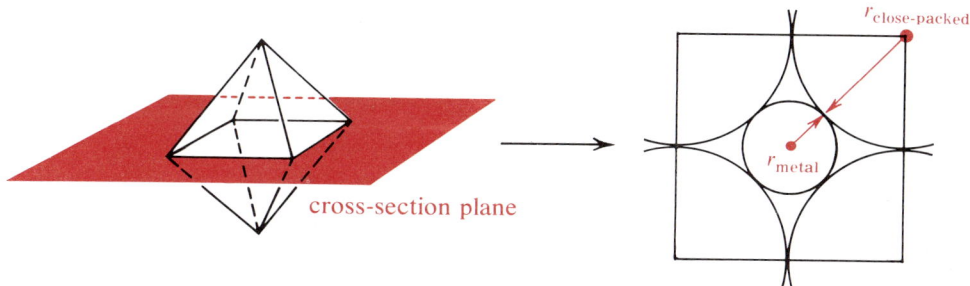

Figure 4-31 Radius relationships of a metal atom in an octahedral site of its host lattice.

(see Study Problem 4) that if the radius of the metal atom is greater than 0.414 of the radius of the close-packed atom, the close-packed atoms will be forced apart. Similarly, idealized tetrahedral sites limit the radius of the metal atom to 0.225 of the radius of the close-packed atom. One of the results of this limitation is that many of the lattices with tetrahedral sites filled are distorted due to the metal atom's radius being too large.

Frequently Observed Crystal Geometries

Some of the more common crystal structures are worth examining in light of their relationship to the close-packed models. Figure 4-32 shows the **NaCl structure**, which is also adopted by most of the alkali halides (Li, Na, K, and

182 | The Behavior of Solids

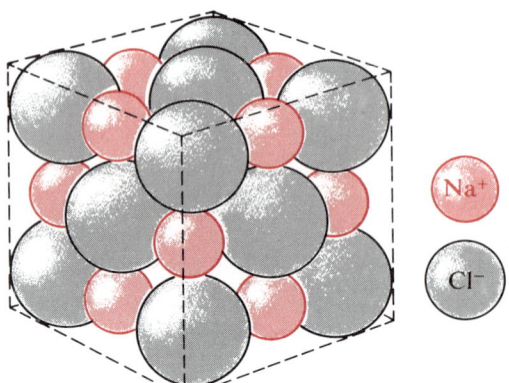

Figure 4-32   NaCl structure (compare with Fig. 4-27).

Rb compounds with F, Cl, Br, and I), as well as many other compounds with 1:1 stoichiometry. Notice that the Cl atoms are approximately cubic close-packed, but that they have been spread somewhat by the Na atoms, which are slightly too large. Each Na is surrounded by six Cl atoms, and each Cl by six Na atoms. We can thus say that all of the octahedral sites in the "close-packed" Cl lattice are filled by Na atoms.

Figure 4-33 shows the **CsCl (cesium chloride) structure**, in which the atoms are body-centered cubic; alternatively, we may say that the chlorides are in a simple cubic—not close-packed—lattice, spread apart by the relatively large cesium atoms. A number of other compounds involving large metal atoms, such as TlCl, adopt this lattice.

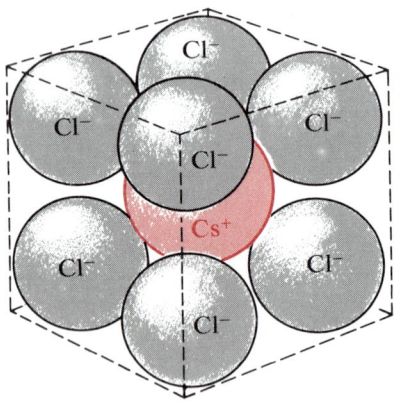

Figure 4-33   CsCl structure.

Figure 4-34 shows a structure related to that of CsCl, the CaF_2 or **fluorite structure**. The fluorides are again in a simple cubic lattice, but because of the 1:2 stoichiometry of the compound only half the metal-atom positions are filled. The metal atoms alternate in such a way that, although each metal has a coordination number of 8, each F has a coordination number of 4 and is sur-

Close Packing and Crystal Lattices | 183

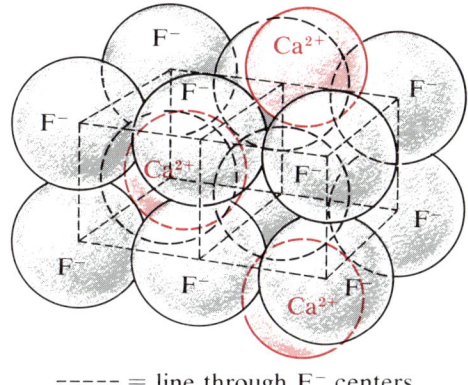

----- = line through F⁻ centers

Figure 4-34 CaF$_2$ structure (fluorite).

rounded tetrahedrally. There is also an antifluorite structure, adopted by compounds such as Na$_2$O, which is like the CsCl structure except that all of the metal atoms are present and half the other atoms are missing. These two structures are quite commonly adopted by compounds with AB$_2$ and A$_2$B stoichiometry, respectively.

Figure 4-35 shows the **zincblende (ZnS) structure**, which has an interesting relationship to both the fluorite and ccp structures. We can regard it either as

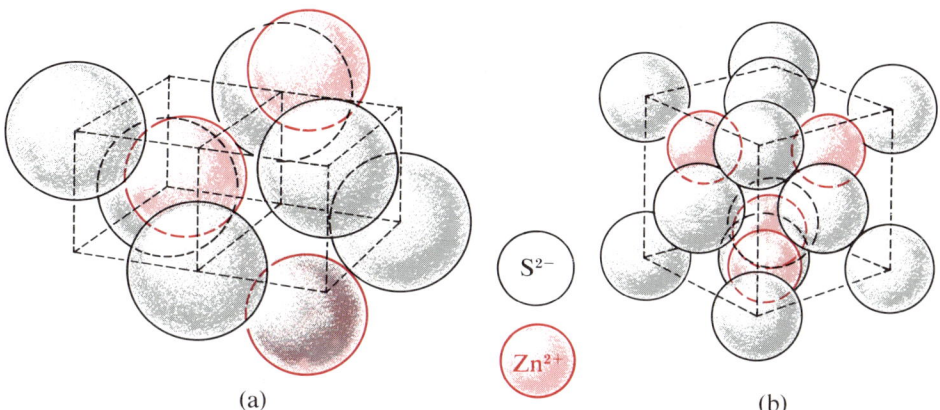

Figure 4-35 Zincblende structure of ZnS (a) as derived from the fluorite structure, and (b) as derived from ccp sulfide ions.

the fluorite structure with half the negative ions missing in order to satisfy the 1:1 stoichiometry, or as ccp sulfur atoms with metal atoms in half the tetrahedral holes. Another interesting comparison is that between the zincblende structure and the **wurtzite structure** (Fig. 4-36), both of which are forms of the same compound, zinc sulfide (ZnS). The difference is simply that the sulfur

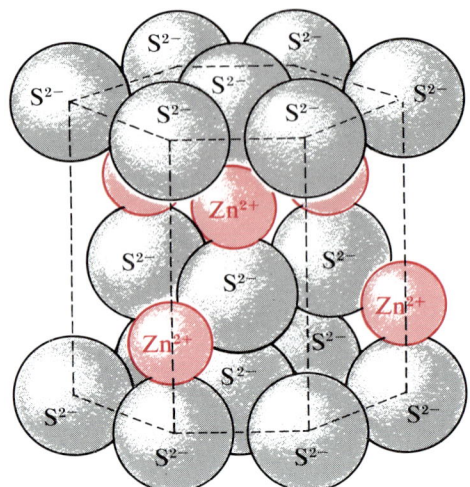

Figure 4-36   Wurtzite structure of ZnS (compare with the zincblende structure of Fig. 4-35).

atoms are cubic close-packed in zincblende, but hexagonal close-packed in wurtzite; this serves to emphasize the relationship between the two close-packed lattices.

As a final example, look at the **rutile (TiO_2) structure** in Fig. 4-37, which consists of distorted hexagonal close-packed oxygens with metal atoms in half the octahedral holes. This has approximately the same relationship to the fluorite structure that the wurtzite had to the zincblende; like the fluorite structure, it is exhibited by many AB_2 compounds.

Figure 4-37   Rutile structure of TiO_2. (a) Distorted hcp. (b) Ti atoms in bcc packing.

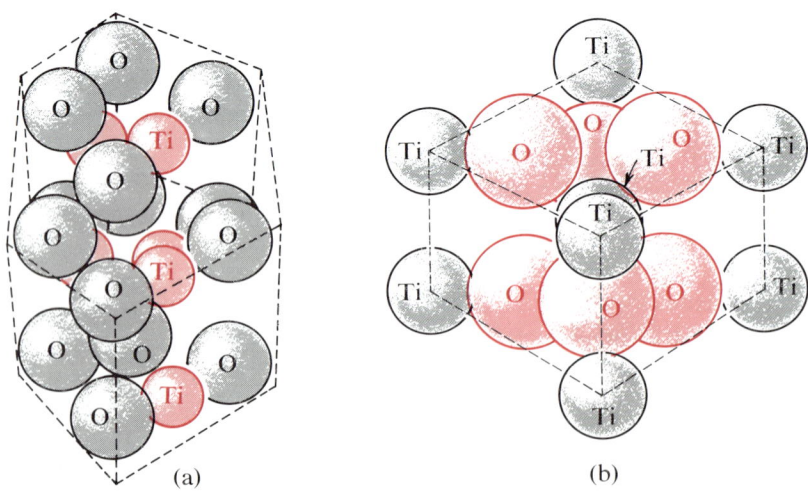

The Potential Energy of Close-Packed Systems

It is clear that the phenomenon of close-packing is of great importance in the formation of a crystal. Why should this be? The answer lies in the attractive forces we described previously: the force represented by the Lennard-Jones potential for uncharged atoms or molecules, and the Coulomb force for ions. Every pair of atoms that approach to the distance corresponding to lowest energy lower their potential energy and increase their stability in so doing. The more neighbors an atom has, the more its potential energy is lowered and its stability increased. So close packing, with its creation of maximum coordination numbers, corresponds to the situation in which each atom or molecule has decreased its potential energy as much as possible. It is not surprising that close-packed lattices or variations of them are a very common unifying feature of crystalline elements and compounds.

4-5 The Absorption of Heat by Crystals

One of the experimental properties that crystals have in common with all other material bodies is a finite heat capacity. We can set up an experiment in which we add heat to a crystal and see how rapidly its temperature changes. In Chapter 3 we defined the heat capacity for a system remaining at constant volume as $C_v = dE/dT$, where E is the energy of the molecules or atoms in the system—the crystal, in this case. We can measure the heat capacity of any sample by adding a known amount of heat energy to it and observing the magnitude of the temperature rise; usually the most convenient way to add a known amount of energy is by using a known electrical current through an electrical heater. If we do this for most crystals, or even noncrystalline solids, at about room temperature we find that the observed heat capacity is nearly always close to a value of 6.3 cal/deg *per mole of atoms*; for instance, near room temperature metallic silver has a heat capacity of 6.02 cal/mole, while silver chloride (AgCl) has a heat capacity of 13.00 cal/mole or 6.50 cal/mole-atoms (since there are two atoms in the formula). This is known as the **law of Dulong and Petit**; its uses have already appeared in Chapter 1. However, if we measure the heat capacity of solids at lower temperatures we always find that the heat capacity is smaller the lower the temperature, approaching zero heat capacity at absolute zero temperature; the colder the solid, the easier it is to warm it up. Figure 4-38 shows the heat capacity of metallic silver as a function of temperature as an example of this behavior.

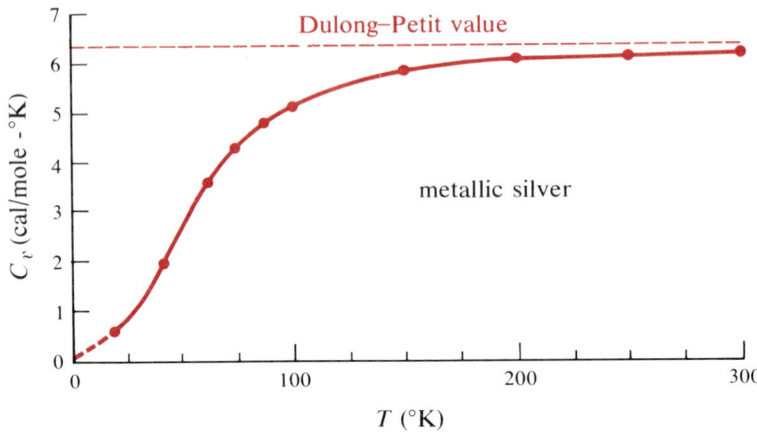

Figure 4-38 Temperature dependence of crystal heat capacity.

Just as was true for the heat capacities of gases, the classical theory of the nature of atoms was unable to account for this sort of temperature dependence. In 1900, faced with a similar discrepancy in an experiment we shall describe in Chapter 5, Max Planck made an imaginative proposal. He suggested that the energy of any electromagnetic radiation was proportional to its frequency, and further, that the energy of the oscillation that produced the radiation could only be some integral multiple of that frequency. In other words,

$$E_n = nh\nu$$

where E_n is the energy of the oscillator, ν is the frequency of the radiation, h is a proportionality constant (called **Planck's constant**), and n is an integer called a **quantum number**. This was a radical proposal and was hotly attacked for some years, but it worked as no other proposal did. Albert Einstein seized on the idea and modified it to apply to the problem of the heat capacity of crystals.

The Einstein Theory of Heat Capacities

Einstein proposed to treat a mole of crystal as a collection of N (Avogadro's number) atoms, each of which can oscillate in three directions; the crystal would then be equivalent to $3N$ one-dimensional oscillators. The energy of the crystal would then be equal to the total energy of all the oscillators, and the heat capacity would be the derivative of this energy with respect to temperature.

In following Einstein's treatment, we suppose first that there is some fundamental frequency ν that applies to all of the $3N$ oscillators, but that they can adopt different values of n, the quantum number. Then the total energy of the crystal is

$$E_{\text{total}} = 3N\langle n\rangle h\nu$$

where $\langle n\rangle$ is the average quantum number displayed by all the oscillators. If we want an expression for the heat capacity as a function of temperature, then, we must find $\langle n\rangle$ as a function of temperature. In Chapter 2 we developed (see Eq. 2-25) an expression for taking this sort of average. The distribution function will just be the Boltzmann function (see p. 117), and inserting this in the averaging expression, we have

$$\langle n\rangle = \frac{\sum_{n=0}^{\infty} n e^{-nh\nu/kT}}{\sum_{n=0}^{\infty} e^{-nh\nu/kT}}$$

To simplify the appearance of this expression, let $e^{-h\nu/kT} = u$:

$$\langle n\rangle = \frac{\sum_{n=0}^{\infty} n u^n}{\sum_{n=0}^{\infty} u^n}$$

We need to be able to evaluate these summations explicitly to get the sort of function we want in a closed form; let us look at the numerator and denominator separately, and take the denominator first because it is a little simpler:

$$\sum_{n=0}^{\infty} u^n = 1 + u + u^2 + u^3 \cdots$$

$$= \frac{1}{1-u}$$

This result may be verified by algebraic long division of the resulting fraction. Now, looking at the numerator, we see that we can express it in terms of a derivative of the denominator:

$$\sum_{n=0}^{\infty} n u^n = u \sum_{n=0}^{\infty} n u^{n-1} = u \frac{d}{du} \sum_{n=0}^{\infty} u^n = u \frac{d}{du}\left(\frac{1}{1-u}\right)$$

Using the chain rule to differentiate $(1-u)^{-1}$, we have

$$\sum_{n=0}^{\infty} n u^n = u\{-1[-(1-u)^{-2}]\} = \frac{u}{(1-u)^2}$$

Now, combining these expressions for the numerator and denominator of the n expression:

$$\langle n \rangle = \frac{u/(1-u)^2}{1/(1-u)} = \frac{u}{1-u} = \frac{1}{(1/u)-1}$$

(dividing by u), and

$$\langle n \rangle = \frac{1}{e^{h\nu/kT}-1}$$

since

$$\frac{1}{e^{-x}} = e^x$$

This is a simple expression for the average quantum number, which we can plug into the total energy expression with which we began:

$$E_{\text{total}} = 3N\langle n \rangle h\nu = \frac{3Nh\nu}{e^{h\nu/kT}-1}$$

With this expression for E_{total} as a function of T, we need only differentiate it with respect to T to get another expression for C_v, the heat capacity at constant volume:

$$C_v = \frac{dE_{\text{total}}}{dT} = 3Nh\nu \frac{d}{dT}[(e^{h\nu/kT}-1)^{-1}]$$

Using the chain rule again for the differentiation of the expression in square brackets:

$$C_v = 3Nh\nu[-(e^{h\nu/kT}-1)^{-2}]\frac{d}{dT}(e^{h\nu/kT}-1)$$

$$= \frac{-3Nh\nu}{(e^{h\nu/kT}-1)^2} e^{h\nu/kT} \frac{d}{dT}\left(\frac{h\nu}{kT}\right)$$

$$= \frac{-3Nh\nu}{(e^{h\nu/kT}-1)^2} e^{(h\nu/kT)} h\nu/-kT^2$$

$$= \frac{(3Nh^2\nu^2/kT^2) e^{h\nu/kT}}{(e^{h\nu/kT}-1)^2}$$

We have no reason to doubt the validity of this expression for C_v as a function of T, but it has T in too many places for us to be able to see what its behavior will be like as T changes. We can look at it in two extreme cases, however.

First, consider the behavior of the function when T is large, so that $(h\nu/kT) \ll 1$. Now, we can always express an exponential function as a power series:

$$e^u = 1 + \frac{1}{1!}u + \frac{1}{2!}u^2 + \frac{1}{3!}u^3 + \cdots$$

A common approximation that is made to this series when u is small is to say that u^2 and all further terms are negligible: $e^u \cong 1 + u$. In this case,

$$e^{h\nu/kT} \cong 1 + \frac{h\nu}{kT}$$

and substituting this approximation in the C_v expression gives

$$\begin{aligned}
C_v &\cong \frac{3N(h^2\nu^2/kT^2)(1+h\nu/kT)}{[1+(h\nu/kT)-1]^2} \\
&\cong \frac{3N(h^2\nu^2/kT^2)}{h^2\nu^2/k^2T^2}\left(1+\frac{h\nu}{kT}\right) \\
&\cong 3Nk\left(1+\frac{h\nu}{kT}\right)
\end{aligned}$$

or for very large T,

$$C_v = 3Nk$$

Since $Nk = R$, the gas constant, which in calorie units is 1.987 cal/mole-°K,

$$C_v \cong 3R \cong 6 \text{ cal/mole-°K}$$

This is entirely in accord with the law of Dulong and Petit. But what about the dropoff of C_v as the temperature falls? This is the other interesting extreme, where $(h\nu/kT) \gg 1$.

In this very-low-temperature case, $e^{h\nu/kT}$ will be a very large number, and $(e^{h\nu/kT} - 1)$ will be very nearly equal to $e^{h\nu/kT}$. Then, substituting the C_v expression again we have

$$\begin{aligned}
C_v &= \frac{3N(h^2\nu^2/kT^2)\,e^{h\nu/kT}}{(e^{h\nu/kT}-1)^2} \\
&\cong \frac{3N(h^2\nu^2/kT^2)\,e^{h\nu/kT}}{(e^{h\nu/kT})^2} \\
&\cong \frac{3N(h^2\nu^2/kT^2)}{e^{h\nu/kT}}
\end{aligned}$$

Now when T becomes small, the exponential in the denominator becomes very large and dominates this expression, so that for very small T the heat capacity approaches zero, which is precisely the sort of behavior we hoped to explain. Figure 4-39 shows a graph of the heat capacity of diamond, both theoretical and experimental, taken from Einstein's own calculation in *Annalen der Physik*; it can be seen that the agreement is quite good when ν is properly chosen.

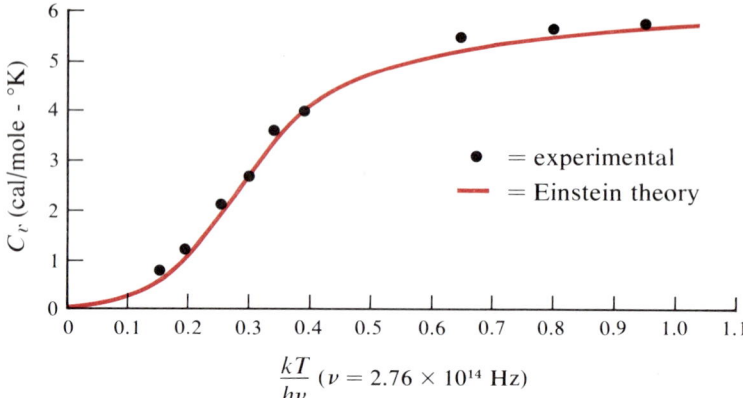

Figure 4-39 Comparison of experimental and theoretical values of heat capacities of diamond.

The Mathematical Necessity of Quantization

What is important about this theoretical triumph is not the derivation itself, because more refined models, particularly that of Debye, give even better agreement with observed data. Rather, the key point we should notice is that it was necessary to stipulate at the very beginning that the energy of the atoms in the crystal was quantized. Only under this unfamiliar assumption can we prevent $h\nu$ from being vanishingly small for *all* values of T, which would lead to a heat capacity of 6 cal/mole-°K for all temperatures. The observation that the quantization of the energy of atoms is a necessary part of our description of the nature of atoms was one of the most significant steps toward our present understanding of the nature of matter.

4-6 Crystal Defects

All of the discussion so far has assumed tacitly that the crystals under discussion are geometrically perfect. This is analogous to the way we explored the properties of ideal gases at length, finally dealing with their departures from ideality or imperfections. Accordingly, we need to look briefly at the nature of crystal imperfections. For gases, the imperfections took the form of departures from the model of randomly moving, noninteracting atoms. For crystals, the imperfections we should look for are departures from the perfect geometry of the ideal crystal. As for gases, the departures will not be great under ordinary conditions.

Chemical Defects

We can envision three kinds of crystal defects. First, there is the possibility that some A atoms in the crystal of element A or in the crystal of compound AB may be replaced by an equal number of atoms of another element C; this is a chemical substitution. The result is usually a solid solution. If a metallic element has a substantial number of atoms replaced by another metal, so that the percentage composition is noticeably affected, the resulting crystal is usually said to be an alloy. In more detailed treatments of this subject it is possible to predict limits of solubility fairly well by considering the energy relationships between the host lattice and the atoms of the entering element. This and other similar calculations are of great interest to theoretical metallurgists, and are very useful in our technological society. We shall have occasion to come back to this idea of chemical substitution in a crystal lattice in Chapters 14 and 16.

Vacancies

The second kind of crystal defect is a lattice vacancy, which is called a **Schottky defect**. This corresponds to an atom missing from where it ought to be inside the crystal; we can create a Schottky defect in a perfect crystal by taking a single atom out of the lattice and placing it on the surface of the crystal. Inside a close-packed or approximately close-packed crystal, each atom has its potential energy lowered by the presence of a large number of nearest neighbors. On the surface of a crystal the number of nearest neighbors is only about half what it is inside, and the energy of the atom is not lowered by as many attractive interactions. The net result is that it requires an energy input to create a Schottky defect. In the absence of any source of energy we do not expect to see any such defects in a crystal. At any temperature above absolute zero, though, there is the possibility that an atom may acquire enough thermal energy, kT, to escape. If the amount of energy necessary to bring one atom out to the surface (creating a Schottky defect) is E_S, then at any temperature we expect to see the relative number of defects in a crystal governed by the Boltzmann distribution:

$$\frac{n_S}{N} = e^{-E_S/kT}$$

where n_S is the number of Schottky defects and N is the total number of atoms in the lattice. We thus expect to see the number of Schottky defects increase as the temperature increases.

Interstitial Atoms

The third kind of crystal defect is also a lattice imperfection, rather than a chemical imperfection. If an atom is removed from its position in the lattice, but is only taken as far as a lattice site—an **interstitial** position—elsewhere in the crystal, the result is called a **Frenkel defect**. Figure 4-40 shows this and the other types of defects. It also requires energy to create Frenkel defects, since although the number of nearest neighbors may not change for the atom being moved, the atoms surrounding it in its interstitial position are compressed so that the repulsion between atoms at very short distances raises its energy.

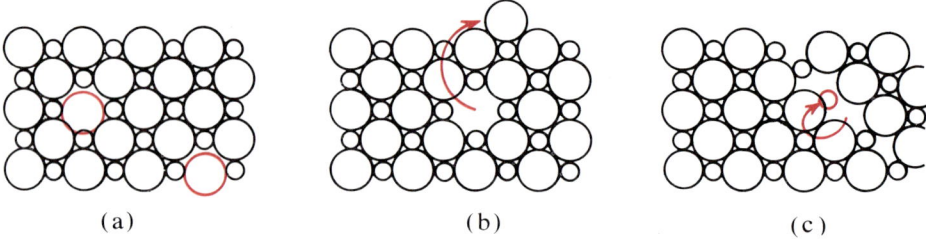

Figure 4-40 Types of crystal defects. (a) Chemical substitution defect. (b) Schottky defect. (c) Frenkel defect.

Frenkel defects occur in crystals because of the thermal energy of some atoms, just as for Schottky defects, and their occurrence is governed by the Boltzmann distribution in a similar way.

What evidence do we have that these defects actually occur? Perhaps the most striking evidence lies in a quantity we have already examined, the dependence of the crystal's heat capacity on temperature. The law of Dulong and Petit and the Einstein theory of heat capacities both suggest that the heat capacities of crystals at relatively high temperatures ought not to increase past about 6.3 cal/mole-°K. The experimental reality is sometimes quite different, however. Figure 4-41 shows the heat capacity of silver bromide (AgBr) at high temperatures, in calories per degree per mole *of atoms*. We attribute the sharp rise to the formation of lattice defects at the higher temperatures; since the formation of these consumes heat energy, the apparent heat capacity of the crystal rises as more defects are formed. Since a defect is obviously a step on the road to the disorderly structure that we have suggested is characteristic of liquids, in Chapter 7 we shall look more closely at the structure of liquids in terms of these defects or holes.

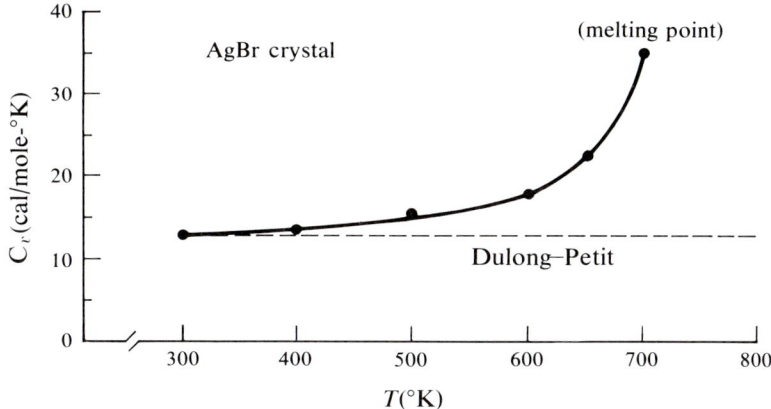

Figure 4-41 Energy of defect formation in crystals at high temperatures. (Modified from Christy, R. W., and Lawson, A. W., *J. Chem. Phys.* **19** (517), 1951.)

4-7 Summary

In this chapter we have still been engaged in a search for experimental clues to the conditions under which the atomic model is an adequate representation of the basic structure of the varied forms of matter. We have deliberately chosen crystalline solids for study because they are, in our intuitive common experience, completely different from gases. Yet it has been easy to show at each stage of our development of the properties of crystals that the atomic model as we developed it for gases is either already adequate or is readily adaptable to the property being considered. From the adaptations, as they became necessary, we have gathered some more information about what the atoms that comprise our model must be like.

We were able to see quickly that the incompressibility and high density of solids are compatible, both conceptually and in numerical magnitude, with the features of atoms we developed in Chapter 3. The rigidity of crystals toward shear forces, and their general anisotropy toward external influences, however, required the introduction of the concept of a highly symmetrical, ordered structure or array of atoms.

In searching for a pattern in the differences between crystals, we looked at the interrelationships between the four physical properties hardness, brittleness, melting point, and electrical conductivity. We were able to characterize crystals as belonging to one of four classes: ionic, partly ionic, covalent-molecular, and metallic. Again, in applying the atomic model to these classes we found it to be adequate only if we stipulated some additional properties of the atoms. Atoms have an electrical nature; they can be given a permanent electrical charge in a crystal, but need not have such a charge to form a crystal. Perhaps more surprising, it proved desirable to postulate a completely delocalized "gas" of electrical charge to explain the high electrical conductivity of solid metals; this means that there is some very small, electrically charged part of an atom that can be removed from the atom.

We saw that there is an experimental method for observing the order that we knew must exist in crystals. Only if the order is quite high can we get sharply defined diffraction of X-ray beams, but this is observed. Indeed, we can get enormous amounts of information from the X-ray diffraction data about the exact geometrical arrangement of atoms in crystals, by observing diffraction angles and working with these through the Bragg equation. In the results of X-ray studies, we detect a strong tendency of crystalline systems toward structures that approach close packing, and we looked at some of the features of such systems. The crystal structures of many compounds can be explained by assuming a close-packed lattice of the larger atom, with the smaller atom systematically (or sometimes unsystematically) arranged in the octahedral or tetrahedral sites that are characteristic of close-packed lattices.

Finally, in examining the absorption of heat by crystals, we found two more clues for further study and synthesis. To explain the fact that the heat capacity of all crystals approaches zero at the absolute zero of temperature, we postulated that the vibrational energy of atoms bound in crystals is quantized—that it can exist only as multiples of a fundamental quantity, which cannot be subdivided. And in the unexpected increase of the heat capacity of crystals at high temperatures we found evidence for the formation of crystal defects or lattice vacancies, which will be a useful starting point for the formulation of a theory of the structure of liquids.

At this point it seems that we have adequately documented the case for an atomic model for the structure of matter, over a range of properties from a gas under high pressure to the passage of electricity through a metal crystal. Now, if atoms are an appropriate model for the structure of matter, what can we propose as an appropriate model for the structure of the atom itself? Atoms are about 1 Å in diameter and are generally spherical. They are electrical in nature, and an electrically charged part of them, which is very small, can be removed under appropriate experimental circumstances. Their energy is quantized. They attract each other at close range, but repel each other strongly at very close range. In Chapters 5 and 6 we shall give a brief picture of present views as to the most reasonable model for the structure of an atom having all these properties, as well as some others with which we shall introduce the discussion. We shall also expand the discussion of atomic structure to explain the present view of the nature of chemical bonding, keeping it consistent with the experimental facts of molecular structure and also with the atomic structure we propose. But there are difficulties in achieving a sound intuitive grasp of phenomena that are many orders of magnitude smaller than those we can sense directly. We shall find that the model explains a remarkable diversity of chemical behavior, but we shall be approaching the frontiers of what we presently know, and we can only be sure that the ultimate model for the chemical bond will be different from the one we present. Perhaps not too much different, however—it is worth a look.

Study Problems

1. Describe the series of weighings necessary to measure the density of NaCl, using *m*-xylene, which has a density of 0.868 g/cm³, as the pycnometer fluid. Remember that a weight for the NaCl sample is necessary, and that its volume can be determined if the weight of *m*-xylene necessary to fill the pycnometer is known. How is the volume of the pycnometer to be established?
2. NaCl has the structure shown in Fig. 4-32; if X rays with $\lambda = 1.537$ Å are diffracted by the top two planes of atoms in the figure, with a smallest angle of diffraction of 15°50′, what is d (the Na–Cl spacing)? If this spacing is the same in all directions, what volume is occupied by a "formula unit," two adjacent atoms Na and Cl? If the density of solid NaCl is 2.165 g/cm³, what is the volume of 1 mole of NaCl? What is the value of Avogadro's number as determined from this experiment?
3. In the cubic close-packed structure shown in Fig. 4-42a, one-eighth of each of the corner spheres is inside the indicated cube; half of each facial sphere is inside the cube. Using this and the drawing of a single face of the cube (Fig. 4-42b), prove geometrically that the cubic close-packed spheres occupy 74% of the cube's volume.

Figure 4-42 Face centered cubic lattice geometry.

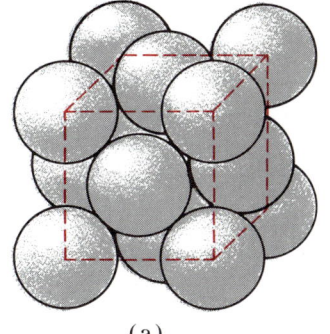
(a)

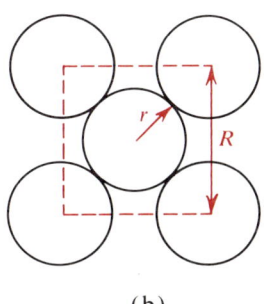
(b)

4. Working from Fig. 4-31, show that $r_{\text{metal}} = 0.414 r_{\text{close-packed}}$ for an octahedral hole in a close-packed structure.
5. Draw a 10 × 10 array of dots as a two-dimensional model of a crystal (as is done in Fig. 4-21). What are the most appropriate choices of coordinate axes for this crystal?

If the spacing between the dots is taken as a and b for the respective axes, draw in at different places on the array the "planes" (lines in this case) 10, 01, 11, 21, and 43. Draw in the nearest parallel "planes" to each. Comment on the spacing of the "planes" as it relates to the magnitude of h_1 and h_2, and similarly on the number of atoms in a given "area of each plane" (length of each line). What effect should this last observation have on the intensities of diffracted beams from different planes $h_1 h_2$?

6. What relationship would you expect to exist between cleavage planes in a crystal and the principal X-ray diffraction planes 100, 110, 111, and so on?

7. Although ionic crystals are generally quite brittle, for reasons shown in Fig. 4-11, it is possible to produce plastic deformation under certain conditions. The formation of "salt domes" in coastal areas involves this sort of flow in NaCl deposits. Inspect the NaCl structure in Fig. 4-32 and sketch the planes of atoms that can move relative to each other without producing strong electrostatic repulsion.

8. Why is "white" X radiation (a broad range of wavelengths) necessary for the Laue method, while monochromatic X radiation (a single wavelength or very narrow range) is necessary for the powder and rotating-crystal methods?

9. Consider the heat capacity of AgBr as shown in Fig. 4-41. Remembering that C_v is the rate of change of internal energy with temperature, what mathematical treatment would give the total internal energy change due to the creation of defects over the range from 300°K to 705°K (the melting point)? Assuming that the *excess* heat capacity (above the Dulong–Petit value) shown in the figure is given algebraically by $C_v(\text{excess}) = 23 e^{-10.8 + 0.0153 T}$ cal/mole-°K, what is the total internal energy change due to the creation of defects? If the energy used in creating a single defect were equal to the Ag—Br bond energy of 69 kcal/mole bond, what fraction of the solid lattice positions would show defects at the melting point?

Some Further Reading

Holden, A., and Singer, P., *Crystals and Crystal Growing*, Garden City, New York: Doubleday, 1960. Not much theory here, but an interesting discussion of how to grow crystals and what the basic forms are like.

Brey, W. S., *Physical Methods for Determining Molecular Geometry*, New York: van Nostrand Reinhold, 1965. Chapters 2 and 3 deal with the basic theory of X-ray diffraction in a very clear manner.

Sands, D. E., *Introduction to Crystallography*, New York: Benjamin, 1969. The same material as in Brey, but more of it and with a more mathematical treatment. This is the straight word, but will be slower reading.

Wells, A. F., *Structural Inorganic Chemistry*, Oxford: Clarendon Press, 1962 (3rd ed.). A staggering amount of information on observed crystal structures. There is a well-written discussion, but most people use the book to look up structures. Later chapters of this book will refer to Wells and will use some of the excellent pictures.

Daniels, F., and Alberty, R. A., *Physical Chemistry*, New York: Wiley, 1966 (3rd ed.). Chapter 18 has another nice treatment of crystal structure, brief and moderately mathematical. This is another of the better standard physical chemistry texts.

5 | Atomic Structure

We have seen that atoms or molecules provide a workable model for the basic structure of matter. We have examined the general properties of gases and solids and have been able to relate the experimental results in each case to this simple model. As we have examined the various experimental techniques in turn, our model has become more sophisticated. Let's review the features of the model that have been revealed by the experiments we have discussed so far.

We can account for the form of Boyle's and Charles' laws by assuming that quantities of gas large enough to work with in the laboratory consist of very large numbers of atoms or molecules, each having a well-defined mass and a velocity that varies with time because of collisions but with which we can deal as a statistical average. From the deviations from these laws (i.e., from the ideal gas law) we have established that the atoms or molecules are not infinitesimally small but have a reasonably well-defined size, and further, that when they are very close to each other they exert an attractive force on each other. Since they attract each other when they are not actually in contact, we surmise that they must have an electrical nature; only electrical or magnetic forces (which are related to each other) act over a distance and are as large as the force encountered here. The atom's approximate size is confirmed by the similarity between the van der Waals constant b and the volume of a mole of solid in a crystal. X-ray diffraction measurements provide a value for Avogadro's number and for the exact separation of the centers of the atoms and molecules in the measured crystals. The electrical nature of the atom is confirmed by the ability of certain solids to conduct an electrical current when melted. Finally, we have observed that the heat capacity of crystals (as a function of temperature) requires a very curious property of the atom: its vibrational energy cannot vary continuously but can be only one of a finite number of exactly defined energies. The atom's energy is said to be quantized.

We now need to look more closely at the nature of the atom itself. What experiments can we do that will shed light on the internal electrical structure of the atom? What sort of model can we devise for the structure of the atom that will rationalize and unify the results of these experiments? Let us begin by looking at the experiments that led to our present ideas of the structure of the atom.

5-1 The Nuclear, Electrostatic Atom

Four men are associated with the key experiments behind the concepts of a nuclear, electrostatic atom: Faraday, Thomson, Millikan, and Rutherford. Their contributions spanned a century—from 1813 to 1911—and during this time, of course, many other workers added to our understanding. It is important to recognize in advance, however, that in the model we develop the atom is held together by electrical forces, so understanding the structure of an atom really implies understanding electrical particles and electrical forces.

Michael Faraday

In 1813 Michael Faraday went to the British Royal Institution to work as a laboratory assistant to Humphry Davy, one of the leading investigators of the chemical effects of electricity. Davy had already, in 1807, demonstrated one of the most remarkable chemical effects of electricity by using the recently invented Voltaic piles or batteries to produce metallic sodium and potassium from their hydroxides; this was the first isolation of these elements, since no chemical reaction is capable of producing them.

The Electrical Nature of Matter Faraday extended Davy's work by studying the quantitative relationship between chemical change and the transport of electricity. After years of investigation of the "electrolysis" of solutions, he was able to show that the transport of a given amount of electrical charge will always deposit the same weight of an element on an electrode. He also developed the term **electrochemical equivalent** to refer to the ratios of weights deposited for different elements, and today we use the term **faraday** to refer to that quantity of electrical charge (96,487 ampere-seconds or coulombs) that will deposit 1 equivalent weight (in grams) of an element. He also suggested that the attraction between unlike charges was the source of the energy binding molecules together, and the very broad range of compounds he studied made it plain that an electrical nature was common to all elements.

Working from the idea of electrochemical equivalents, however, it is easy to see that there is some definite quantity of electrical charge that is added to or taken from a neutral atom to make an ion; we can readily calculate its magnitude by dividing the faraday by Avogadro's number:

$$\text{change in charge per atom} = \frac{\text{charge per equivalent}}{\text{atoms per mole}}$$

$$= \frac{96487 \text{ C/equivalent}}{6.0222 \times 10^{23} \text{ atoms/mole}} = 1.6021 \times 10^{-19} \text{ C/atom}$$

if 1 electrochemical equivalent equals 1 mole of the element. Frequently the change in charge per atom is a multiple of this value (times 2 or 3 or 4) but it is never less than this. So this quantity of electricity seems to be a basic unit or "atom" of electricity, and in 1874 a British physicist, G. J. Stoney, proposed that it was a charged particle, which he called an **electron**.

J. J. Thomson

Stoney's proposal was not accepted immediately, and was indeed quite controversial. It was generally accepted that the passage of electricity through an aqueous solution must involve some such mechanism as Stoney proposed, but physicists had for some years been aware that electricity could also be conducted through gases at very low pressure, and this conduction did not seem to represent the action of particles at all. If two electrodes were placed in an evacuated tube and one was given a negative charge at very high voltage relative to the other, the negative electrode or **cathode** emitted a sort of radiation. This radiation was called **cathode rays**, and most experiments done on these suggested that the cathode rays were some sort of electromagnetic radiation like visible light, but of a different frequency. J. J. Thomson established that the cathode rays were particles—electrons—by constructing the cathode-ray tube shown in Fig. 5-1. The negatively charged electrode or cathode is labeled C; the cathode rays pass through the two slits S_1 and S_2, which produce a narrow beam that then passes down the tube to strike the phosphor P at the far end of the tube. The path of the cathode rays can be accurately tracked since the phosphor emits visible light when struck by cathode rays (a television picture tube is a cathode-ray tube of very similar design). Using this device, Thomson was able to demonstrate the particulate nature of cathode rays by measuring e/m, the ratio of the charge on the electron to the mass of the electron.

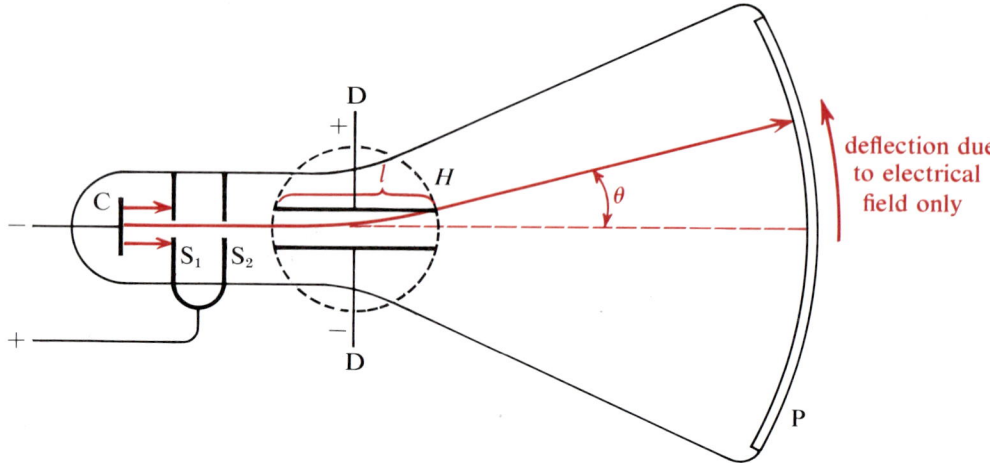

Figure 5-1 Experimental arrangement of Thomson's cathode-ray tube. The magnet pole faces are indicated by the dashed line; the magnetic field H is perpendicular to the plane of the paper.

The Measurement of e/m for the Electron   The two deflection plates D are separate electrodes that can be positively or negatively charged; Thomson found that the cathode rays were deflected toward the positively charged plate. Since opposite charges attract each other, this is the behavior to be expected from a negatively charged particle. The deflection of the beam could be measured by the movement of the lighted spot on the phosphorescent screen. If the magnetic field H were then turned on, the beam could be brought back to its original position by balancing the deflection due to the interaction of the cathode rays with the magnetic field against that due to their interaction with the electrical field. The derivation proceeds as follows [*Phil. Mag.* **44**, 293 (1897)].

The force on a particle with electrical charge e in an electrical field E is just Ee; correspondingly, there is a magnetic force on a moving charged particle that is passing through a magnetic field. The magnetic force is Hev, where H is the strength of the magnetic field, e is the charge, and v is the velocity of the particle. Now suppose that the cathode rays do consist of particles, each with a charge e and a mass m and having a horizontal velocity (parallel to the axis of the tube in Fig. 5-1) equal to v. In the electrical field the force Ee attracts the particle so that it acquires a vertical velocity v_E. How large is the velocity v_E? It is equal to the acceleration produced by the force Ee times the time the particle spends in the region between the deflection plates D:

$$v_E = \text{acceleration} \times \text{time}$$

But the definition of force (force = mass × acceleration) indicates that we can replace the acceleration in this expression by force/mass, or Ee/m:

$$v_E = \frac{Ee}{m} \times \text{time}$$

The angle of deflection in the electrical field, which we can call θ, is equal to the ratio of the vertical velocity v_E to the horizontal velocity v:

$$\theta = \frac{v_E}{v} = \frac{Ee}{mv} \times \text{time}$$

(This quantity is really the tangent of θ, which very nearly equals θ in radians for small angles.) The time the particle spends in the electrical field is equal to the length of the deflection plates, l, divided by the velocity of the particle, v. [Remember that distance/(distance/time) equals time.] Thus we have

$$\theta = \frac{Ee}{mv}\frac{l}{v} = \frac{Eel}{mv^2}$$

In exactly the same way we can show that the deflection angle φ of the particle in the magnetic field is

$$\varphi = \frac{Hev}{mv} \times \frac{l}{v} = \frac{Hel}{mv}$$

To use the same value l here we must make the magnetic field the same length as the electrical field, a detail of construction shown in Fig. 5-1.

Now let us divide the θ expression by the φ expression and solve the resulting expression for v:

$$\frac{\theta}{\varphi} = \frac{Eel/mv^2}{Hel/mv} = \frac{E}{Hv} \quad \text{or} \quad v = \frac{E\varphi}{H\theta}$$

We can rearrange the φ expression to give e/m and substitute this value of v into it:

$$\varphi = \frac{Hel}{mv} = \frac{Hl}{v} \times \frac{e}{m}$$

$$\frac{e}{m} = \frac{\varphi v}{Hl} = \frac{\varphi}{Hl}\frac{E\varphi}{H\theta} = \frac{E\varphi^2}{H^2 l\theta}$$

When the magnetic field is used to bring the beam back to its original position, $\theta = \varphi$, so

$$\frac{e}{m} = \frac{E\theta}{H^2 l}$$

an expression in which all of the quantities on the right side can be measured in the laboratory; θ, of course, must be measured with the magnetic field turned off. We have thus measured the ratio of the charge on this "atom" of electricity to its mass.

The measurement constituted a convincing demonstration that electrons were real particles, with measurable physical properties. In electrolysis measurements, following Faraday's work, a value for the same ratio e/m for the lightest ion, H^+, had been established as 95722 C/g. A refined value of e/m for the electron, following Thomson, is 1.759×10^8 C/g, which is 1836 times as large as the value for H^+. This presents us with the possibility that the charge on the electron is 1836 times as great as that on H^+, with the same mass, or that both the mass and the charge differ with only the ratio established, or that the charges on the two particles are of the same magnitude but e/m is greater for the electron because it is much lighter than H^+. Since the electron appears to be the elementary unit of charge, we surmise that the mass of the H^+ ion or **proton** must be 1836 times as large as the mass of the electron. From this assumption, then, we have values of the individual properties charge and mass for the electron.

Robert A. Millikan

As long as only a ratio could be measured, however, the individual properties and the particulate model for electricity remained in some doubt. In 1909 Robert Millikan produced the final demonstration of the appropriateness of the particle model. He measured the charge itself in an experiment that is elegant in its simplicity. Figure 5-2 shows the apparatus he used. A spray of very finely divided oil droplets is produced in the upper part of the chamber. The droplets fall under the influence of gravity into the lower part of the chamber, where some of them acquire a charge by collision with gaseous ions produced by X rays or radioactivity in the chamber. Those droplets that have no electrical charge continue to fall, but those having an electrical charge can be halted by

Figure 5-2 Experimental arrangement of Millikan's oil-drop charge measurement.

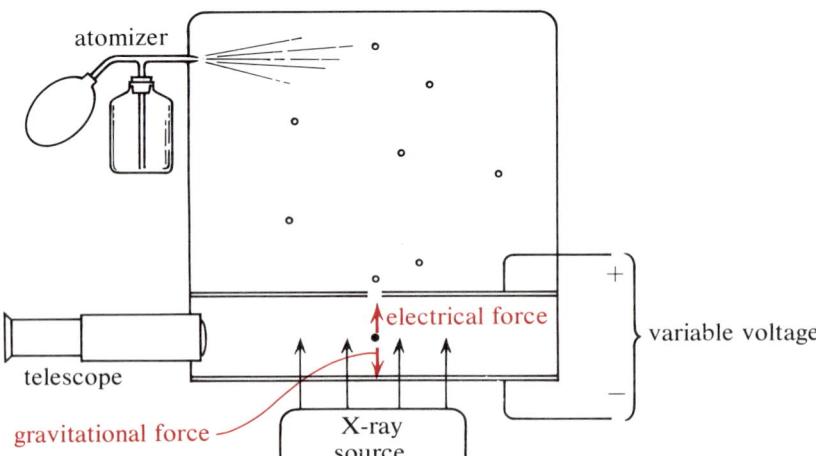

applying a voltage that attracts them to the upper electrode. Since the electrical field strength can be determined by measuring the voltage and the distance between the electrodes, and since the upward force is equal to the charge on the droplet times the electrical field strength, the charge on the droplet can be determined by setting the upward electrical force equal to the downward gravitational force. Millikan found that the charge on the droplets was always equal to a multiple of a basic unit that he took to be the charge of the electron. This is the same as the charge per atom equivalent obtained previously from electrolysis, and showed conclusively that the electron is the fundamental unit of charge existing in all atoms.

E. R. Rutherford

How many of these fundamental units are there in each atom, and how are they arranged? Some other work of Thomson's, notably on the scattering of beams of X rays, had suggested that the number of electrons in each atom was of roughly the same order as the atomic weight. This would require, for electrical neutrality, that there be the same number of positive charges, and Thomson proposed that the atom might consist of a diffuse sphere of positive charge with electrons embedded in it. This model seemed quite reasonable until Rutherford undertook a series of experiments that he expected to fail.

Hans Geiger and Ernest Marsden, of Rutherford's laboratory, began a study of the scattering of radioactive α particles by thin metal foils. α particles are helium nuclei, with an atomic weight of 4; they are therefore almost 10,000 times as heavy as an electron, and they also have high velocities. An electron, then, would not be able to deflect them significantly. Furthermore, if the Thomson model of the atom were correct, even the relatively massive atoms of the metal would not be able to deflect them significantly through the positive charge because that charge was thought to be spread out over a relatively large volume. In that case, the full repulsion of the positive charge of the atom for the positively charged α particles could not be experienced by the α particle at any given time; some of the positive charge would be on the other side of the atom. In the experimental arrangement, a beam of α particles was allowed to pass through a metal foil to strike a phosphor screen that would give off light in the same manner as Thomson's screen had detected cathode rays. Large deflections could, of course, be detected by a scintillation of the screen at some distance from the center of the undeflected beam. But it was not expected that any α particles would be scattered through large angles. Geiger and Marsden soon reported not only large deflections but in a few cases actual reversals of direction of the α particles, so that they came back in nearly the direction from which they had entered the foil. In Rutherford's words, "It was quite the most incredible event that has ever happened to me in my life. It was almost as incredible as if you fired a 15-inch shell at a piece of tissue paper and it came back and hit you." Calculations convinced Rutherford that it was statistically impossible for a long series of very small deflections to add up to these large deflections; only if all of the positive charge in the atom were concentrated on a single, very small, very massive nucleus could the α particle experience the full repulsion of the positive charge and be occasionally deflected so strongly. Furthermore, the scattering angles indicated that the total number of positive charges was only about half the atomic weight; thus magnesium, with an atomic weight of about 24, would have only 12 positive charges on its nucleus and, to maintain electrical neutrality, 12 electrons outside the nucleus. Rutherford proposed that the electrons orbited the massive nucleus, held in place by its positive charge, much as planets orbit the sun. At this point the stage was set for the modern concepts of atomic structure.

5-2 Light Quanta and the Black-Body Experiment

Our discussion so far has been limited to the experimental proof that the basic structure of the atom is that of a system of electrically charged particles. We have, however, overlooked another feature of experiments on atomic systems that must be accounted for in any model of the atom. As we have seen in describing the temperature dependence of the heat capacity of solids, it is necessary to assume that the atom's vibrational energy is quantized. This seems somehow arbitrary and unreasonable, and in fact it was an extremely controversial proposal when first put forward. But it is only a matter of recognizing that certain things come in discrete units and cannot be described in any other way. If you were asked to hold up the fourth finger on your left hand, the response would be obvious; if you were asked to hold up the 3.73 finger on your left hand, you would correctly say that the request was meaningless, since fingers come only in integral units. We might say that it is all right to talk about a continuity of fingers, with fractional numbers or decimals, but in terms of operational usefulness it is necessary to recognize the integral nature of fingers. Otherwise we cannot explain the fact that there is no way to hold up the 3.73 finger. Experimentally, it is the same way with the energy of microscopic systems. The reason the integral nature, the quantized nature, of energy had not been observed previously was that the experiments done previously had observed energy on too large a scale to detect the very small quanta. If 10,000 people put their hands down side by side, so that there is a very long row of fingers, it is a quite reasonable approximation to speak of the finger 1 mile from the beginning of the row, although we recognize that this is only an approximation; we can also speak of the finger 0.742 mile from the beginning, and we tend to think of the row as being a continuum of fingers. But of course this does not change the fact that when the experiment gets down to a sufficiently small scale the idea of a continuum of fingers becomes meaningless. The question of the scale of the experiment is all-important; energy quantization is observable only for very small masses and very low energies.

Black-Body Radiation

At the turn of the century physicists were bothered by just this sort of problem, although it was not apparent. Excellent experimental data were available on the frequency distribution of radiation from a black body, but they were inconsistent with the macroscopic concepts of energy of the time. A black body is one that does not reflect any radiation, so that all the radiation coming from it is emitted by the body itself. For instance, the sun is an excellent black body in the sense that virtually all of the sunlight we receive comes from the radiation of the sun itself rather than from any reflected starlight. On a laboratory scale we can construct a good black body by building a box of material that is a good heat insulator, drilling a hole in the side of the box, maintaining the box at any temperature we please, and looking at the hole in the side of the box. The hole will be reflecting essentially no radiation from outside, so that the radiation will be characteristic of the interior of the box alone. Figure 5-3 shows the frequency distribution of this sort of radiation for a very hot black body; the graph is consistent with our intuitive ideas of what "red hot" means. The visible range of the spectrum is indicated in the figure. Note that for a red-hot body a good deal of radiation is emitted at the red end of the spectrum but very little at the blue end; thus the red color. What is the classical description of this situation?

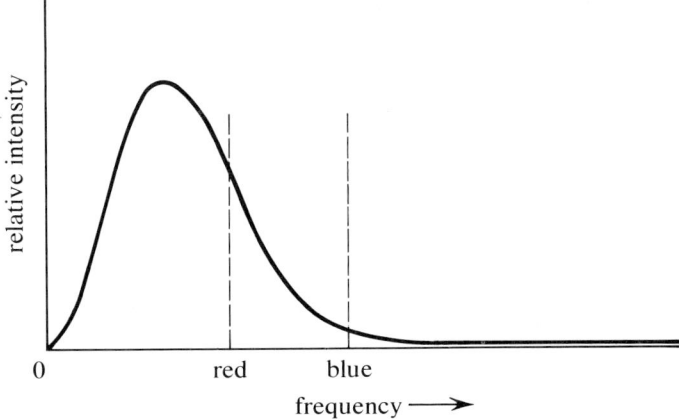

Figure 5-3 Frequency distribution for a "red-hot" black body.

The radiation from the body is an oscillation that can be described in terms of its wavelength or its frequency. Figure 5-3 shows that a broad spectrum of different frequencies is emitted. For a large collection of oscillators such as the many atoms in the black body, classical statistics tells us that each frequency should carry the same amount of energy; this is the equipartition principle referred to in our discussion of the kinetic theory of gases. But the number of possible frequencies of oscillations inside a box increases as the square of the frequency, so that the blue or ultraviolet end of the spectrum should have *more* intense radiation than the red or infrared end due to its higher frequency, rather than *less*. Figure 5-4 shows the classical prediction of intensity of radiation; the fact that it refuses to come back down to low intensity in the ultraviolet was referred to by nineteenth-century physicists as the "ultraviolet catastrophe."

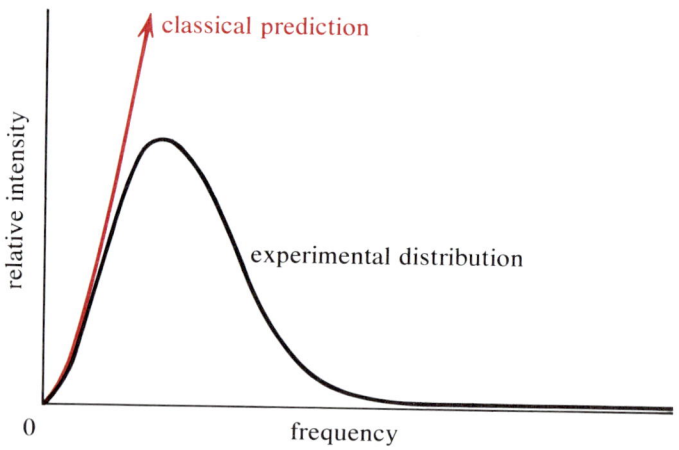

Figure 5-4 Comparison of experimental results with the classical nonquantized prediction of frequency distribution.

The Equivalence of the Energy and Frequency of Radiation

In 1900 Max Planck made a very imaginative and penetrating proposal. He suggested, first, that each frequency did not carry the same amount of energy but rather that the energy of an oscillation was proportional to its frequency: $E = h\nu$, where E is the energy carried by the oscillation, ν is the frequency, and h is the proportionality constant now known as Planck's constant, with a value of 6.625×10^{-27} erg-sec. Second, he proposed that the energy of these oscillations was quantized so that an oscillator could acquire or lose energy only in units of $h\nu$. We can see that this would cause low intensity of high-frequency radiation, since the oscillator could acquire the energy to radiate, $h\nu$, only from the thermal energy of the black body, kT. If kT is much smaller than $h\nu$, there is a very small chance that the oscillator will have enough thermal energy to emit the high-frequency radiation. As a matter of fact, the distribution of frequencies should be described by the Boltzmann function we have already encountered for the energy distribution of gas molecules. This is the case; the agreement between the experimental data and Planck's mathematical model is excellent. This result led Einstein to his treatment of the heat capacity of solids that we have described. It also led Niels Bohr to synthesize all these ideas into the first coherent model of the structure of an atom.

5-3 The Bohr Atom and Hydrogen Spectra

Rutherford's picture of the atom as a nucleus with electrons orbiting about it is by this time familiar to everyone. Much too familiar, in fact, since the model contains the seeds of its own destruction, and even with the additions and corrections of Bohr and, later, of Arnold Sommerfeld it has been shown to be quite generally inadequate and misleading. It is worth examining, however, not only for its historical value but because it provides us with insight toward the goal of constructing a more generally applicable and more rigorous model. Be warned, however, that the Bohr model is no longer accepted and that the whole idea of electron orbits must be abandoned eventually. But let us see how this next step toward the modern concept of the atom was made.

Since the advent of earth-orbiting capsules people have developed a better intuitive feel for the mechanics of one body orbiting another. In the case of the electron orbiting the nucleus, it is permissible to think of the electron's motion as a combination of a free fall toward the nucleus under the attraction of the opposite electrical charge and a rapid motion sideways that perpetually throws the electron out away from the nucleus. In free fall, a body accelerates; this is a severe problem since classical electromagnetic-radiation theory insists that an accelerating electrical charge must emit radiation, which uses up some of its energy. In other words, the orbiting electrons of Rutherford's model would continuously emit radiation, thereby using up energy so that they dropped further and further into the potential-energy well of the nucleus and would soon spiral in toward the nucleus, destroying the atom. But the hydrogen atom, the simplest one to examine since it has only one electron, does *not* emit radiation continuously and is obviously quite stable in that it does not collapse. Bohr realized that Rutherford's model was based on sound experimental fact, but that no account had been taken of the quantization of the electron's energy as the work of Planck and Einstein demanded. He therefore developed the following model.

The Bohr Model of the Atom

If the electron were not to fall into the nucleus, Bohr proposed that the radius at which it orbits must be constant. If the radius does not change, the forces on the electron in the radial direction must exactly cancel each other. The attractive force between two unlike charges, the **Coulomb attraction**, is proportional to the product of the charges and inversely proportional to the square of the distance between the charges: attractive force $= -q_1 q_2/r^2$, where q_1 is the charge on particle 1, q_2 is the charge on particle 2, r is the distance between the two particles or charges, and the minus sign accounts for the fact that if one charge is positive and the other negative their product will be negative but the attraction will be positive (a negative attraction is a repulsion). In this case q_1 is Ze, where Z is the number of positive charges on the nucleus and e is the magnitude of each charge unit; q_2 is $-e$, the charge on the electron. The attractive force, then, is Ze^2/r^2. This must be balanced by a force tending to pull the electron and nucleus apart, which we can identify as the centrifugal force on the rapidly circling electron. From elementary physics we can establish that the centrifugal force on a circling particle is proportional to its mass and to its velocity (actually to the square of its velocity), and inversely proportional to its distance from the center about which it is circling. Thus we write:

$$\text{centrifugal force} = \frac{mv^2}{r}$$

and since

$$\text{attractive (centripetal) force} = \text{centrifugal force}$$

$$\frac{Ze^2}{r^2} = \frac{mv^2}{r} \tag{5-1}$$

where m is the electron's mass, v is its velocity, r is its distance from the nucleus, $-e$ is its charge, and Ze is the nuclear charge, as shown in Fig. 5-5.

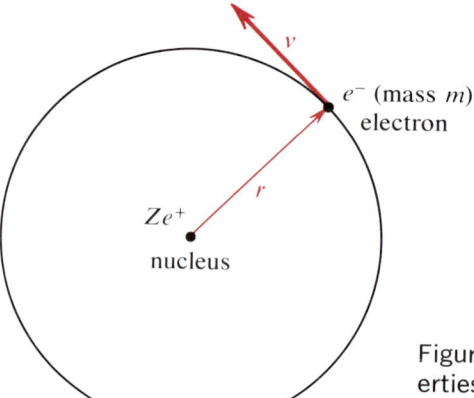

Figure 5-5 Geometry and mechanical properties of the Bohr model of the atom.

But this is only the basic classical expression for Rutherford's model, in which the electron would lose energy by radiation and spiral into the nucleus. At this point Bohr introduced the idea of quantization of energy. He proposed that the **angular momentum** of the electron was quantized in multiples of $h/2\pi$. The angular momentum of a circling particle is given by mvr, using our present symbols, and it can be seen that the units of these quantities are consistent:

$$mvr = \frac{nh}{2\pi}$$

where n is the number of quanta of angular momentum.

$$\text{mass}\left(\frac{\text{distance}}{\text{time}}\right)\text{distance} = \text{energy (time)}$$

$$= \text{mass}\left(\frac{\text{distance}}{\text{time}}\right)^2 \text{time}$$

$$\frac{\text{mass (distance}^2)}{\text{time}} = \frac{\text{mass (distance}^2)}{\text{time}}$$

If we combine Bohr's postulate with the balance-of-forces expression (Eq. 5-1), we see that r itself is quantized:

$$mvr = \frac{nh}{2\pi}$$

so

$$v = \frac{nh}{2\pi mr}$$

Plugging this into Eq. 5-1,

$$\frac{Ze^2}{r^2} = \frac{m}{r}\left(\frac{nh}{2\pi mr}\right)^2$$

$$= \frac{n^2 h^2}{4\pi^2 mr^3}$$

Cancelling r^2,

$$Ze^2 = \frac{n^2 h^2}{4\pi^2 mr}$$

or

$$r = \frac{n^2 h^2}{4\pi^2 mZe^2} = \frac{n^2}{Z}a_0 \qquad (5\text{-}2)$$

where

$$a_0 \equiv \frac{h^2}{4\pi^2 me^2}$$

Since all the quantities on the right side of Eq. 5-2 are constants, it follows that only certain radii are possible for the electron—those for which the **quantum number** n takes on the successive integral values 1, 2, 3,.... The smallest Bohr radius for a hydrogen atom is the quantity a_0 in Eq. 5-2; combining accepted values for the constants in it, $a_0 = 0.529$ Å.

Is the energy really quantized? The total energy of the electron is the sum of the energy due to its motion, or kinetic energy, and the energy it has by virtue of its position near the nucleus, or potential energy. As with gas molecules, we define the potential energy zero at infinite separation of the charges from each other. We have already seen in dealing with the kinetic theory of gases that the kinetic energy of a moving particle is $\frac{1}{2}mv^2$; the potential energy of the Coulomb attraction is $-Ze^2/r$. Notice that the Coulomb attractive energy is different from the Coulomb attractive force; as a matter of fact, the potential energy is just the integral of the force over the distance from the potential energy zero at infinite separation in to the Bohr radius r:

$$\int_{\infty}^{r} \frac{Ze^2}{r^2} dr = -\frac{Ze^2}{r}\bigg|_{\infty}^{r} = -\frac{Ze^2}{r} + \frac{Ze^2}{\infty} = -\frac{Ze^2}{r}$$

So the total energy of the electron, E, is given by

$$E = \frac{1}{2}mv^2 - \frac{Ze^2}{r}$$

But Eq. 5-1 tells us that mv^2 is just equal to Ze^2/r, so

$$E = \frac{1}{2}\frac{Ze^2}{r} - \frac{Ze^2}{r} = -\frac{1}{2}\frac{Ze^2}{r} \tag{5-3}$$

If we now substitute the quantized expression for r into this expression (from Eq. 5-2), we get

$$E = -\frac{1}{2}\frac{Ze^2}{(n^2h^2/4\pi^2mZe^2)}$$

which simplifies to

$$E = -\frac{2\pi^2mZ^2e^4}{n^2h^2} \tag{5-4}$$

Equation 5-4 shows that the total energy of the electron is indeed quantized, with the quantum number, which must be an integer, appearing squared in the denominator.

What experimental observations can be brought forward to confirm or deny the validity of this model? First we notice the curious fact that the energy is described as being negative, and the idea of negative energy seems unfamiliar. But we see from our definitions of kinetic energy and potential energy that zero energy would have to correspond to r being infinite and the charged particles at rest, so the negative energy when r is not infinite simply means that the electron is bound to the nucleus. Thus we would have to put energy *into* the atom in order to get the electron out to an infinite distance; and if we have to put energy in to get up to zero the initial energy must have been negative. Throughout our discussion of atomic and molecular structure we shall see that a negative energy corresponds to a system being bound together. In this case the energy that must be put in is the **ionization energy** needed to bring about the process $H^0 \rightarrow H^+ + e^-$.

Hydrogen Spectra

There is a more crucial test than the ionization energy. In 1885 J. J. Balmer discovered a mathematical relationship governing the wavelengths of the emission spectrum of hydrogen gas. At very high temperatures, gaseous atoms or molecules emit light; if this light is passed through a prism to separate it into the spectrum of its wavelengths, as shown in Fig. 5-6, it is found to consist, for

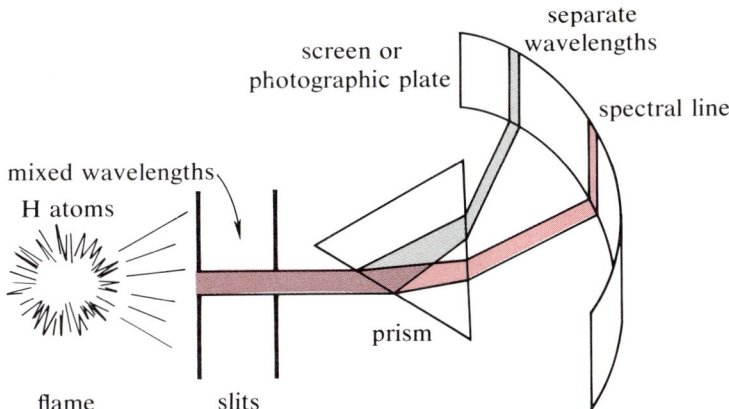

Figure 5-6 Experimental arrangement for observation of hydrogen spectra.

individual atoms, of lines (if the source is a narrow slit) at very precise wavelengths. For hydrogen atoms, the most readily observable lines had been established as having wavelengths of 6562.1 Å, 4860.7 Å, 4340.0 Å, and 4101.3 Å. Balmer showed that these wavelengths fit the formula

$$\lambda = 3645.6 \left(\frac{n_2^2}{n_2^2 - n_1^2} \right)$$

to better than five significant figures: the error was smaller than the accuracy with which the apparatus of that day could measure wavelength. In this expression n_1 and n_2 are integers; to fit the observed wavelengths, n_1 must be taken as 2 and n_2 may assume any integral value greater than 2. Thus we have

$$\lambda_1 = 6562.1 = 3645.6\left(\frac{3^2}{3^2-2^2}\right) = 3645.6\left(\frac{9}{5}\right) = 6562.1$$

$$\lambda_2 = 4860.7 = 3645.6\left(\frac{4^2}{4^2-2^2}\right) = 3645.6\left(\frac{16}{12}\right) = 4860.7$$

and so on. If Bohr's model of the atom were correct it should be able to produce this formula or an entirely equivalent one, and because of the precision of the numerical values of wavelengths the test is an exacting one.

It is convenient to convert the wavelengths of the emission lines to frequencies. This is easily done because $\nu = c/\lambda$, where ν is the frequency, λ is the wavelength, and c is the speed of light, 2.99792×10^{10} cm/sec. Inserting λ from Balmer's relationship, we have

$$\nu = \frac{c}{\lambda} = \frac{c}{3645.6[(n_2^2/(n_2^2-n_1^2))]}$$

$$= \frac{c}{3645.6}\left(\frac{n_2^2-n_1^2}{n_2^2}\right) = \frac{c}{3645.6}\left(1-\frac{n_1^2}{n_2^2}\right)$$

$$= \frac{c}{3645.6}(n_1^2)\left(\frac{1}{n_1^2}-\frac{1}{n_2^2}\right)$$

Since $n_1 = 2$ for all the spectral lines in the Balmer series, we write

$$\nu = \frac{4c}{3645.6}\left(\frac{1}{n_1^2}-\frac{1}{n_2^2}\right)$$

or

$$\nu = R\left(\frac{1}{n_1^2}-\frac{1}{n_2^2}\right)$$

where R is the **Rydberg constant** for hydrogen, 3.28805×10^{15} Hz. It is usually more convenient to use smaller numbers, and frequencies are often given as **wave numbers**, which are equal to the frequency divided by the speed of light; in these units, cm^{-1}, the Rydberg constant is 109,677.581 cm^{-1}. How does the Bohr model predict frequencies or wave numbers?

Bohr assumed that radiation could be emitted only as the electron "jumped" from an orbit at an allowed radius r to an inner orbit. Otherwise, the atom was stable for any fixed n and did not emit the classically expected radiation. Then, representing the energy difference for the electron between these two allowed radii by ΔE, and following Planck's assumption that $\Delta E = h\nu$, we have the following:

$$\nu = \frac{\Delta E}{h} = \frac{E_2 - E_1}{h} = \frac{1}{h}\left(-\frac{2\pi^2 m Z^2 e^4}{n_2^2 h^2} + \frac{2\pi^2 m Z^2 e^4}{n_1^2 h^2}\right)$$

$$= \frac{2\pi^2 m Z^2 e^4}{h^3}\left(\frac{1}{n_1^2} - \frac{1}{n_2^2}\right)$$

Since $Z = 1$ for H,

$$R = \frac{2\pi^2 m e^4}{h^3} \quad \text{or} \quad R = \frac{2\pi^2 m e^4}{ch^3} \quad (\text{in cm}^{-1})$$

So the Bohr model predicts the correct algebraic form of the Rydberg formula and has a collection of fundamental constants that should equal the Rydberg constant. If the constants are evaluated using their best current values, the predicted Bohr value for the Rydberg constant is 109,678 cm^{-1} after correcting for the slight motion of the hydrogen nucleus as the electron circles it. In other words, the value is exact within experimental error, although in fact the precision of measurement of the Rydberg constant is so great as to make it the preferred standard for establishing values of constants such as the mass of the electron. The internal consistency of the Bohr theory for the hydrogen atom was so striking as to make it one of the greatest theoretical triumphs of physics.

Atomic Numbers and Nuclear Screening

Another immediate by-product of the Bohr model of the atom was the firm establishment of atomic numbers in the periodic table of elements through Moseley's work on X-ray emission spectra. When an element is struck by an intense beam of high-energy electrons it emits X rays at frequencies that are characteristic of that element. Figure 5-7 shows the frequencies of X rays emitted by the first row of transition metals; it is clear, for instance, that nickel

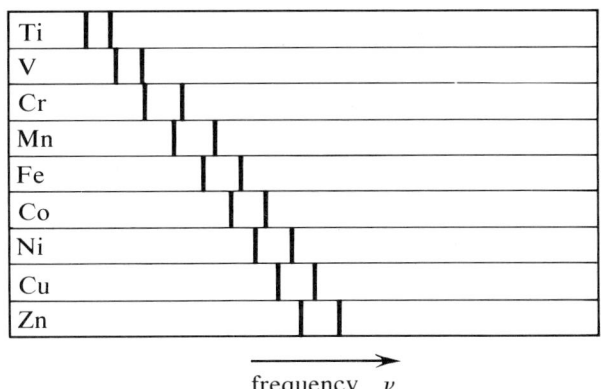

Figure 5-7 X-ray emission frequencies of first-row transition metals. (Redrawn from White, H. E., *Introduction to Atomic Spectra*, New York: McGraw-Hill, 1934.)

belongs after cobalt, not before it, even though nickel's atomic weight is slightly less than that of cobalt. By analogy with Rydberg's formula, Moseley wrote $\nu = R(Z-\sigma)^2[(1/1^2) - (1/2^2)]$ for the wave number of this particular series of X-ray lines, and the extension of the Bohr model is obvious. The beam of electrons at high energy knocks an electron with $n=1$ out of the atom, and the emission of X rays is caused by an electron with $n=2$ jumping into the $n=1$ vacancy. In Moseley's formula Z is the number of positive charges on the nucleus, which we can now think of as the atomic number of the element; σ is a "shielding constant" that expresses the fact that the electron in transition from $n=2$ to $n=1$ does not experience the full attraction of the charge on the nucleus but is partially shielded from it by the negative charges on the other electrons in the atom. The quantity $(Z-\sigma)$, then, is a sort of **effective nuclear charge**, a concept we shall have occasion to use frequently in discussing atoms with many electrons (Sections 5-8 and 5-9, for instance).

5-4 Defects of the Bohr Model — Wave Properties of Particles and Uncertainty of Measurement

With the brilliant success just described for the Bohr model of the atom, why was it necessary to make the earlier cautionary statement that the Bohr atom is no longer accepted and that the whole concept of electron orbits must be abandoned? For one thing, it proved impossible to describe spectra of atoms other than hydrogen according to the Bohr model except for the rather special case in which all but one electron had already been ionized away from the atom, so that in having only one electron it was formally similar to the hydrogen atom. With the failure of repeated, sophisticated attempts to refine the Bohr model in this regard, and with the failure of the model to describe chemical bonding, the realization grew that the Bohr atom was an ingenious oversimplification and was not really a fundamental description.

The Relation Between Momentum and Wavelength

In addition, some 10 years after Bohr's initial article, in 1923, Louis de Broglie proposed a novel theoretical concept that was subsequently verified by experiment. The concept was that of the **wave properties of matter**. De Broglie, following the insights of Planck and Einstein, noticed that energy was associated both with radiation (Planck's $E = h\nu$) and with the very existence of matter (Einstein's $E = mc^2$). He suggested that these two characteristic energies should be equivalent to each other, and that therefore a mass could be assigned to a small packet of radiation or a wavelength could be assigned to a particle: $E = h\nu$ and $E = mc^2$, so $h\nu = mc^2$. But $\nu\lambda = c$, so $hc/\lambda = mc^2$ and

$h/\lambda = mc$, or $\lambda = h/mc$. Now c is the speed of light, so mc for a "light particle" is equivalent to mv for an ordinary particle, which is its momentum:

$$\lambda = \frac{h}{mv}$$

If we symbolize momentum by p, we can write the de Broglie expression as $p\lambda = h$. How is this to be examined experimentally? Much earlier, Einstein had shown that light (radiation) must be considered to have particle properties in order to account for the photoelectric emission of electrons by irradiated metals; the converse demonstration of the wave properties of particles was demonstrated in 1927 by Clinton Davisson and Lester Germer in an elegant experiment in which they demonstrated that electrons can be diffracted—a property of waves—by a regularly spaced "grating" of atoms in a metal crystal. This property of electrons has been put to extensive use since in the electron microscope, which has become an important tool to biologists and medical scientists. The fantastic power of the electron microscope is demonstrated in Fig. 5-8, an electron micrograph recently obtained by workers at the University of Chicago, showing individual thorium atoms bonded to a molecule of a polymeric organic compound.

Figure 5-8 The strings of dots are chains of thorium atoms as revealed by a scanning electron microscope at the University of Chicago. (University of Chicago photo.)

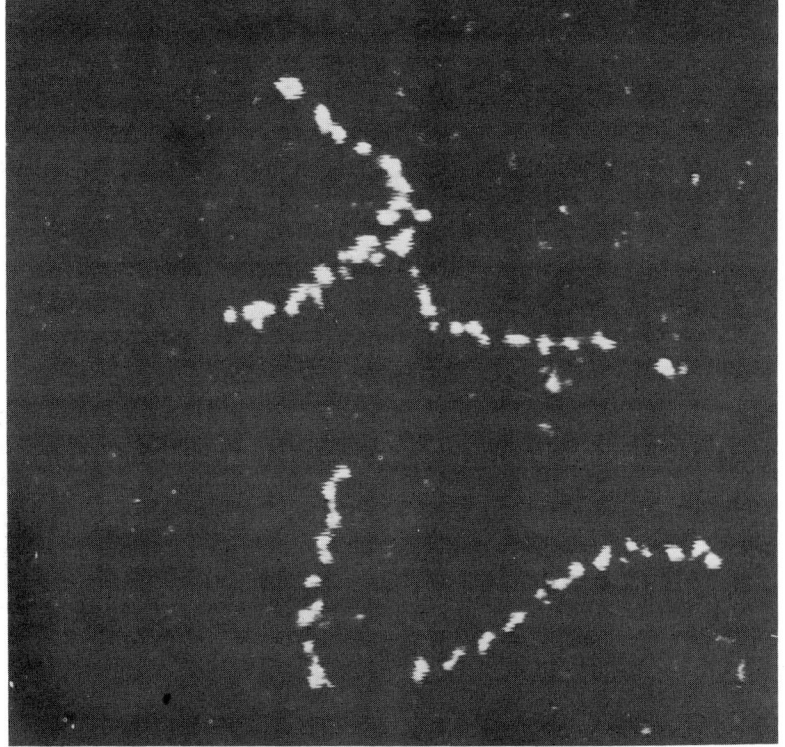

A word of caution is appropriate here. De Broglie's relationship suggests that *every* particle including baseballs, Polaris missiles, and planets, ought to have an associated wavelength and wave properties. This is presumably true in an abstract sense, but we can only *detect* wave properties by diffraction or interference, both properties that require physical bodies of the same approximate size as the wavelength of the "wave." For any macroscopic particle, even for a dust mote, the momentum is so great as to indicate a wavelength much shorter than the dimensions of any physical object; so these wave properties can never be observed. Only for the fundamental particles such as electrons and neutrons is this an observable phenomenon. For these particles, however, wave character is so important as to require us to take a whole new approach to the structure of the atom, starting from the description of a wave.

The Uncertainty Principle

There is, however, another objection to the Bohr model of the atom; perhaps it is the same objection approached differently. It is a very fundamental objection, however, since it strikes at the very foundation of operational definitions and experimental science. The Bohr model is overspecified. It assumes that we know quantities relating to the atom that we can never determine, not even with completely idealized apparatus. This objection can be stated as the **Heisenberg uncertainty principle**: it is not possible to determine simultaneously the position and the momentum of a particle with complete exactness. What we have not yet considered is the nature of the experiments we must do on a microscopic particle to determine its position or its momentum.

Suppose we are standing at the edge of a darkened basketball court. On the court is a basketball, somewhere. Since it is dark we cannot see where the ball is, although we can be sure its momentum is zero since the ball is not moving. We cannot turn on the lights, but we have a supply of basketballs. If we really want to know where the ball is on the court, we can roll the other balls across the court until one does not come out where we expect it to, then we can look at the angle at which it does come out, knowing that it must have bounced off the missing ball. We can very easily calculate where the missing ball must have been; but now there is another problem. By hitting the missing ball, we have imparted an unknown amount of momentum to it. Of course, if we had Ping-Pong balls to roll across the court, we could get the same information about the basketball's position without imparting much momentum to it at all; and if we could turn on the lights, thereby hitting it with "light particles" that we can call **photons**, we could impart still less. But that is just the problem; there is nothing smaller than an electron to throw at the electron. We are left forever in the position of having only basketballs to throw at the basketball. It may seem that a light wave—a photon—would be less massive than an electron, but remember de Broglie's relation, $\lambda p = h$. To impart a very small amount of momentum to the electron the momentum, p, of the photon we choose must be small; but then its wavelength will be large, and we cannot determine the position with any more accuracy than about 1 wavelength. To get the position exactly, we must use a small wavelength, which then means imparting a relatively large momentum to the electron. So even in this case we are stuck with a limit that we can express as $\Delta p \, \Delta x \cong h$, where Δp is the uncertainty in the momentum, Δx is the uncertainty in the position, and h is Planck's constant. As a matter of fact, this is the conventional expression of the uncertainty principle. The Bohr model is in complete violation of this, since it assumes the existence of well-defined orbits in which the position and the momentum of the electron are known exactly. We shall have to take this as well as the wave character of the electron into account in reconstructing our model. Fortunately, the wave approach has this uncertainty built into it, since the position of a wave is uncertain to the extent of its wavelength and its momentum—de Broglie again—is inversely proportional to its wavelength. Then let us see how we approach the behavior of an electron in an atom by treating it as a wave.

5-5 The Classical Wave Equation and the Schrödinger Equation

Let us review the classical wave equation, which we first examined in Chapter 2.

The Classical Wave Equation

The function we use to describe a periodic, repetitive, wave-like phenomenon is the sine (or the cosine) function; adding an amplitude factor A and expressing the wavelength explicitly in radian units, we have, for a wave function ψ, $\psi = A \sin 2\pi x/\lambda$. Here we are speaking of a one-dimensional wave running along the x direction. If we differentiate this wave function with respect to x, using the chain rule, we have

$$\frac{d\psi}{dx} = \frac{d\psi}{du}\frac{du}{dx} = A \cos \frac{2\pi x}{\lambda} \frac{2\pi}{\lambda} = \frac{2\pi A}{\lambda} \cos \frac{2\pi x}{\lambda}$$

Differentiating again,

$$\frac{d^2\psi}{dx^2} = \frac{2\pi A}{\lambda}\left(-\sin \frac{2\pi x}{\lambda}\right)\frac{2\pi}{\lambda}$$

$$\frac{d^2\psi}{dx^2} = -\frac{4\pi^2 A}{\lambda^2} \sin \frac{2\pi x}{\lambda} = -\frac{4\pi^2}{\lambda^2}\left(A \sin \frac{2\pi x}{\lambda}\right) = -\frac{4\pi^2}{\lambda^2}(\psi)$$

The classical differential equation describing a wave, then, is

$$\frac{d^2\psi}{dx^2} = -\frac{4\pi^2}{\lambda^2}\psi \quad \text{or} \quad \frac{d^2\psi}{dx^2} + \frac{4\pi^2}{\lambda^2}\psi = 0 \tag{5-5}$$

Since the electron in an atom must be thought of as having wave properties we shall assume that the function ψ in Eq. 5-5 describes the electron's distribution in some way. Of course we shall want a much more precise definition of ψ, but first let us build a few more desirable features into the differential equation.

Producing the Schrödinger Equation

The property the Bohr model described best was the energy of the hydrogen atom; it will be necessary for our wave-like description to treat the energy explicitly, so as to duplicate the Bohr predictions. We may also presume that the electron will be distributed differently in different systems, so that the differential equation governing ψ, which in turn describes the electron distribution, ought to have in it an explicit recognition of the forces or energy factors peculiar to the particular atom being described. Fortunately, this is a place at which we can directly adapt part of the Bohr development to serve these two purposes. From that previous discussion we know that

$$E = \frac{1}{2}mv^2 - \frac{Ze^2}{r} \tag{5-6}$$

But

$$\frac{1}{2}mv^2 = \frac{1}{2}\frac{m^2v^2}{m} = \frac{(mv)^2}{2m} \quad \text{and} \quad \lambda = \frac{h}{mv}$$

That is,

$$\frac{1}{2}mv^2 = \frac{(h/\lambda)^2}{2m} = \frac{h^2}{2m\lambda^2} \tag{5-7}$$

Rearranging Eq. 5-6,

$$\frac{1}{2}mv^2 = E + \frac{Ze^2}{r}$$

and substituting from Eq. 5-7,

$$\frac{1}{2}mv^2 = \frac{h^2}{2m\lambda^2} = E + \frac{Ze^2}{r} \quad \text{or} \quad \frac{1}{\lambda^2} = \frac{2m}{h^2}\left(E + \frac{Ze^2}{r}\right)$$

We can put this in

$$\frac{d^2\psi}{dx^2} + \frac{4\pi^2}{\lambda^2}\psi = 0$$

to get

$$\frac{d^2\psi}{dx^2} + \frac{8\pi^2 m}{h^2}\left(E + \frac{Ze^2}{r}\right)\psi = 0$$

but in doing this we have neglected the fact that the wave equation we started out with was for a one-dimensional system, a wave running only in the x

direction. The hydrogen atom is three-dimensional and we must allow for this in the wave equation. It is mathematically demonstrable but also intuitively reasonable that the three-dimensional analogy to the equation above should be

$$\frac{\partial^2 \psi}{\partial x^2} + \frac{\partial^2 \psi}{\partial y^2} + \frac{\partial^2 \psi}{\partial z^2} + \frac{8\pi^2 m}{h^2}\left(E + \frac{Ze^2}{r}\right)\psi = 0 \qquad (5\text{-}8)$$

This is the **Schrödinger equation** describing the behavior of the electron in the hydrogen atom or any other one-electron atom. The curly d's in the Schrödinger equation are the notation for partial derivatives — ψ is now a three-dimensional function in x, y, and z, and $\partial^2\psi/\partial x^2$ refers to the second derivative of ψ with respect to x only, treating the other coordinates as if they were constants. If we can solve the Schrödinger equation to establish the nature of the function (or functions), ψ, the result will be a description of the distribution of the electron in a hydrogen atom. This statement is the basic postulate of quantum mechanics.

Solutions of the Schrödinger Equation for a One-Electron Atom

The Schrödinger equation can be solved exactly, at least for the hydrogen atom, and it is found that there is not just a single function that fits the Schrödinger equation, but many — actually an infinite number. In solving for these acceptable wave functions, it turns out to be easier to change the coordinates x, y, and z into the polar coordinates r, θ, and φ. The definition of these coordinates in comparison with the familiar rectangular coordinates is shown in Fig. 5-9, and the algebraic representation of these wave functions is given in Table 5-1 for the ones most useful to chemists. Each of these functions satisfies the Schrödinger equation and thus may be considered to acceptably describe a possible distribution of the electron in a one-electron atom.

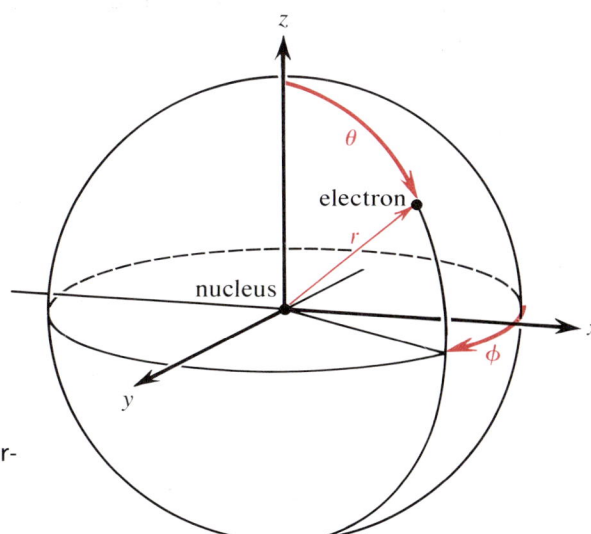

Figure 5-9 Definition of spherical polar coordinates.

Table 5-1
Normalized Wave Functions for a One-Electron Atom[a]

Quantum Numbers			Wave Function
n	l	m	
1	0	0	$\psi_{1s} = \dfrac{1}{\sqrt{\pi}} \left(\dfrac{4\pi^2 mZe}{h^2} \right)^{3/2} e^{-\rho}$
2	0	0	$\psi_{2s} = \dfrac{1}{4\sqrt{2\pi}} \left(\dfrac{4\pi^2 mZe^2}{h^2} \right)^{3/2} (2-\rho) e^{-\rho/2}$
2	1	0	$\psi_{2p_z} = \dfrac{1}{4\sqrt{2\pi}} \left(\dfrac{4\pi^2 mZe^2}{h^2} \right)^{3/2} \rho e^{-\rho/2} \cos\theta$
2	1	± 1	$\psi_{2p_x} = \dfrac{1}{4\sqrt{2\pi}} \left(\dfrac{4\pi^2 mZe^2}{h^2} \right)^{3/2} \rho e^{-\rho/2} \sin\theta \cos\varphi$
			$\psi_{2p_y} = \dfrac{1}{4\sqrt{2\pi}} \left(\dfrac{4\pi^2 mZe^2}{h^2} \right)^{3/2} \rho e^{-\rho/2} \sin\theta \sin\varphi$
3	0	0	$\psi_{3s} = \dfrac{2}{81\sqrt{3\pi}} \left(\dfrac{4\pi^2 mZe^2}{h^2} \right)^{3/2} (27 - 18\rho + 2\rho^2) e^{-\rho/3}$
3	1	0	$\psi_{3p_z} = \dfrac{2}{81\sqrt{\pi}} \left(\dfrac{4\pi^2 mZe^2}{h^2} \right)^{3/2} (6\rho - \rho^2) e^{-\rho/3} \cos\theta$
3	1	± 1	$\psi_{3p_x} = \dfrac{2}{81\sqrt{\pi}} \left(\dfrac{4\pi^2 mZe^2}{h^2} \right)^{3/2} (6\rho - \rho^2) e^{-\rho/3} \sin\theta \cos\varphi$
			$\psi_{3p_y} = \dfrac{2}{81\sqrt{\pi}} \left(\dfrac{4\pi^2 mZe^2}{h^2} \right)^{3/2} (6\rho - \rho^2) e^{-\rho/3} \sin\theta \sin\varphi$
3	2	0	$\psi_{3d_{z^2}} = \dfrac{1}{81\sqrt{6\pi}} \left(\dfrac{4\pi^2 mZe^2}{h^2} \right)^{3/2} \rho^2 e^{-\rho/3} (3\cos^2\theta - 1)$
3	2	± 1	$\psi_{3d_{xz}} = \dfrac{\sqrt{2}}{81\sqrt{\pi}} \left(\dfrac{4\pi^2 mZe^2}{h^2} \right)^{3/2} \rho^2 e^{-\rho/3} \sin\theta \cos\theta \cos\varphi$
			$\psi_{3d_{yz}} = \dfrac{\sqrt{2}}{81\sqrt{\pi}} \left(\dfrac{4\pi^2 mZe^2}{h^2} \right)^{3/2} \rho^2 e^{-\rho/3} \sin\theta \cos\theta \sin\varphi$
3	2	± 2	$\psi_{3d_{x^2-y^2}} = \dfrac{1}{81\sqrt{2\pi}} \left(\dfrac{4\pi^2 mZe^2}{h^2} \right)^{3/2} \rho^2 e^{-\rho/3} \sin^2\theta \cos 2\varphi$
			$\psi_{3d_{xy}} = \dfrac{1}{81\sqrt{2\pi}} \left(\dfrac{4\pi^2 mZe^2}{h^2} \right)^{3/2} \rho^2 e^{-\rho/3} \sin^2\theta \sin 2\varphi$

[a] For simplicity, $(4\pi^2 mZe^2/h^2) r$ has been indicated by ρ.

5-6 One-Electron Solutions: Algebraic Wave Functions and Graphical Orbitals

The Physical Meaning of ψ

What is ψ? How does it describe the distribution of the electron in an atom? When we described waves such as sound waves and light waves in Chapter 2, we saw that the intensity of the wave property is proportional to the square of the amplitude of the wave function. By analogy with these more familiar waves, we suspect that the "intensity" of the electron in the atom is proportional to the square of the wave function. For this to be meaningful, we must rely on a statistical interpretation of the "intensity" of the electron; by this we mean that if ψ^2 is larger near one point in the vicinity of the atom's nucleus than it is at another point, then if many experiments are done to locate the electron, it will be found more often near the point with large ψ^2 than near the point with small ψ^2. In other words, ψ^2 is proportional to the probability of finding the electron near any given point. ψ itself has no physical meaning, and all we can extract from this approach to the internal structure of the atom is, through ψ^2, the relative probability of finding the electron near the various points in the vicinity of the atom. It should be pointed out that this substantial uncertainty about the electron's position is entirely in keeping with the uncertainty principle. Since we expect to establish a very precise value for the electron's energy, which involves its momentum, it is apparent that we shall have to settle for considerable uncertainty as to the exact position of the electron.

Nodes and Quantum Numbers

Notice that we have not introduced any quantum postulate that would be analogous to Bohr's postulate for the angular momentum. But the model must take quantization into account, because of the work of Planck and Einstein. If we examine these algebraic wave functions, from Table 5-1 we see that all but the first have a factor somewhere in them that for certain values of the coordinates causes the wave function to equal zero. Thus the second function vanishes when the factor $(2-\rho)$ equals zero; that is, when

$$\frac{4\pi^2 mZe^2}{h^2} r = 2$$

Similarly, the third function vanishes when either $\sin\theta$ or $\cos\varphi$ equals zero at certain angles with respect to the coordinate axes. These coordinate values at which the wave function vanishes are called **nodes**; it is significant that we cannot be sure of where the electron is, only of where it is not. These nodes are the naturally arising expression of quantization in the atom. Each of these nodes will in general be a two-dimensional surface, as we shall see in looking at the graphical representations of these functions. If we take these wave functions and put them in the Schrödinger equation to get E, the energy will depend on the number of these nodal surfaces, which is obviously an integer. So solution of the postulated wave equation automatically results in a quantum number for the energy.

There are three quantum numbers applicable to the hydrogen atom, corresponding to the three coordinates r, θ, and φ. The quantum number that counts the number of nodes in the range of the φ angle is called m; it can be zero if there are no φ nodes, or $m = 1$ for one node, $m = 2$ for two nodes, and so on. But it turns out that there are two distinct ways to have one node, two ways to have two nodes, and so on. Mathematically this is expressed by allowing m to take the values $0, \pm 1, \pm 2, \ldots$. The quantum number that counts the number of nodes in the range of the θ angle is called l; because these nodes must be consistent with the geometry of the φ nodes, l must always be at least as large as m, but can be larger. Going on to the third coordinate, r, in the mathematical treatment it is seen that radial nodes do occur, but that their number is linked to the number of angular nodes; there is a third quantum number, n, the **principal quantum number**, that counts the total number of nodes, radial and angular together, and this total number is always $n - 1$. Usually we relate these quantum numbers to each other in reverse order:

$$n = 1, 2, 3, \ldots$$

$$l = 0, 1, 2, \ldots, n-1$$

$$m = 0, \pm 1, \pm 2, \ldots, \pm l$$

The Shapes of Orbitals

Let us look at the graphical representations of these wave functions, with their nodes and quantum numbers. Graphical representations of wave functions are usually called **orbitals**; it is important to distinguish between orbitals, which are statistical generalizations about where the electron usually is, and orbits, which imply exact knowledge of the electron's position. Orbitals are compatible with our experiments; orbits are not. If we look at Table 5-1, the first orbital will be that for which $n = 1$, so that l and m must both equal zero. There are, then, no nodes, and as Fig. 5-10 shows, the orbital is spherically symmetrical.

One-Electron Solutions: Algebraic Wave Functions and Graphical Orbitals | 223

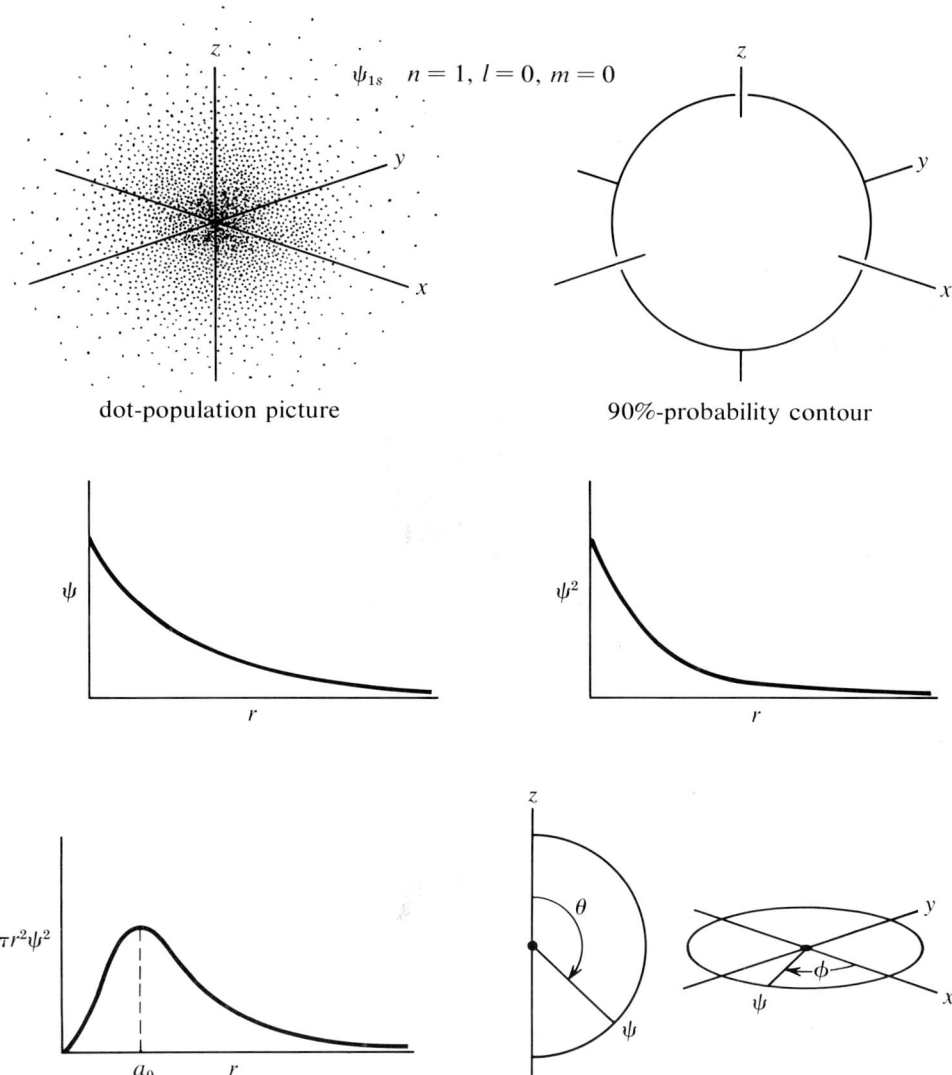

Figure 5-10   Graphical representations of the 1s orbital.

1s

There are several different aspects of the electron distribution explored in Fig. 5-10, and it might be well to discuss them individually. The first picture is a dot-population picture that attempts to represent the relative probability of finding an electron at a given location by the density of dots near that location; this is the most realistic depiction of the electron's time-average distribution, but it is difficult to draw for schematic purposes. The second picture, much easier to draw, introduces the idea of **equal-probability contours**. If we go far enough out from the nucleus to assure ourselves that 90% of the time the electron, when it is in that direction, is between our location and the nucleus, then we are at what is called a 90%-probability point. The surface that represents all of these points for all directions is called an equal-probability contour; in other words, 90% of the time the electron will be inside that surface, no matter what direction we picked. This form of graphical representation is very useful, but it is important to remember that the hard surface it suggests is not real—an orbital is not a little box containing an electron, and the distribution function really trails off toward infinity. This aspect of the function is shown by the plot of ψ vs. r, which is not physically meaningful, and also by the plot of ψ^2 vs. r, which is the probability-distribution function. The graph of $4\pi r^2 \psi^2$ vs. r is perhaps the most intuitively useful since it represents the probability of finding the electron in some thin shell at radius r from the nucleus (remember that the area of a spherical surface is $4\pi r^2$). Notice that this graph has a maximum that falls, for the hydrogen atom, at exactly a_0, the radius of the first Bohr orbit; so the most likely radius for the electron is exactly that which Bohr suggested as the only possible radius. Finally there are two graphs that, in this case, indicate only that there is no angular dependence of the wave function.

2s

Proceeding to the second function in Table 5-1, we see in Fig. 5-11 that it is also spherically symmetrical. There is still no angular dependence; this is true whenever $l = 0$ because l measures the number of angular nodes. We designate any orbital with $l = 0$ as an s orbital, and prefix the s with the value of the principal quantum number n, so the first orbital was the $1s$, and this is the $2s$. Since $n = 2$ and there are always $n - 1$ nodes, however, there will be a radial node, which as we have already seen comes at the radius at which $r = 2(h^2/4\pi^2 mZe^2)$. To display this, the dot-population picture in Fig. 5-11 is a cross section.

2p

Having a radial node is not the only way to have one node in a wave function, however, and the next three functions in Table 5-1 are shown in Fig. 5-12. They each have one node, so n still equals 2; but now $l = 1$, so the node is not radial. Orbitals with $l = 1$ are called p orbitals; these are the $2p$ orbitals. There is no way to decide experimentally which is the x and which the y axis, but the two wave functions with $m = \pm 1$ lie at 90° to each other in the xy plane, so either is legitimate. The subscript after the $2p$ designation indicates the direction in which that particular orbital sticks out. There are thus a total of four orbitals with $n = 2$.

One-Electron Solutions: Algebraic Wave Functions and Graphical Orbitals | 225

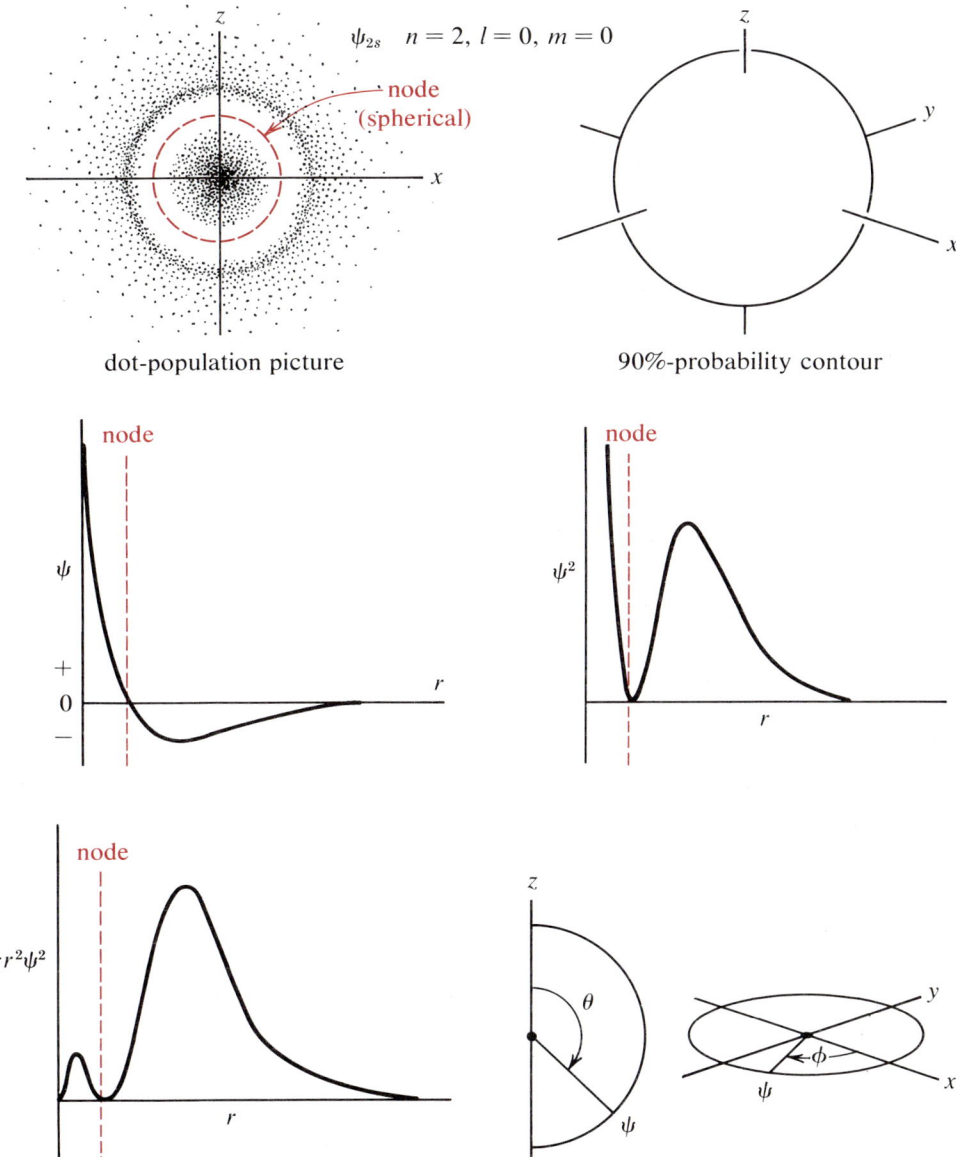

Figure 5-11 Graphical representations of the 2s orbital.

226 | Atomic Structure

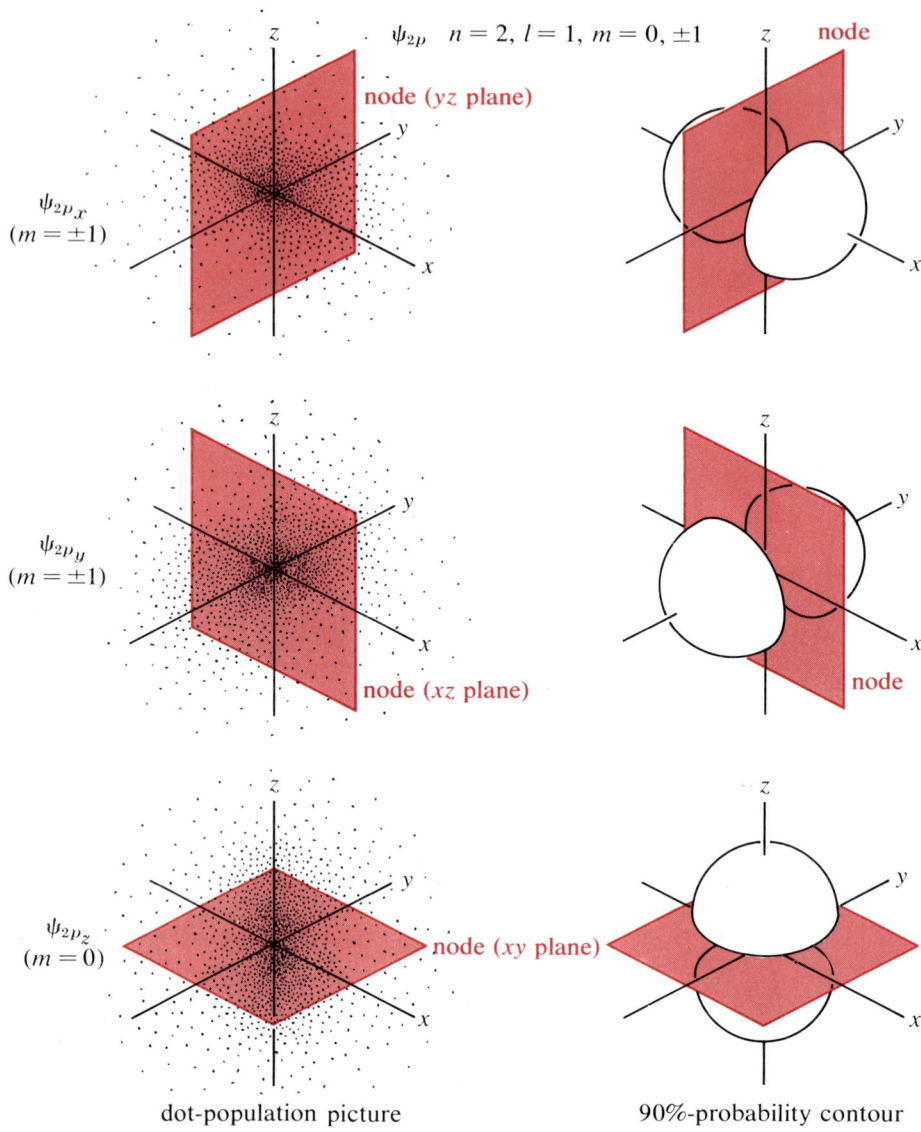

Figure 5-12 Graphical representations of the 2p orbitals.

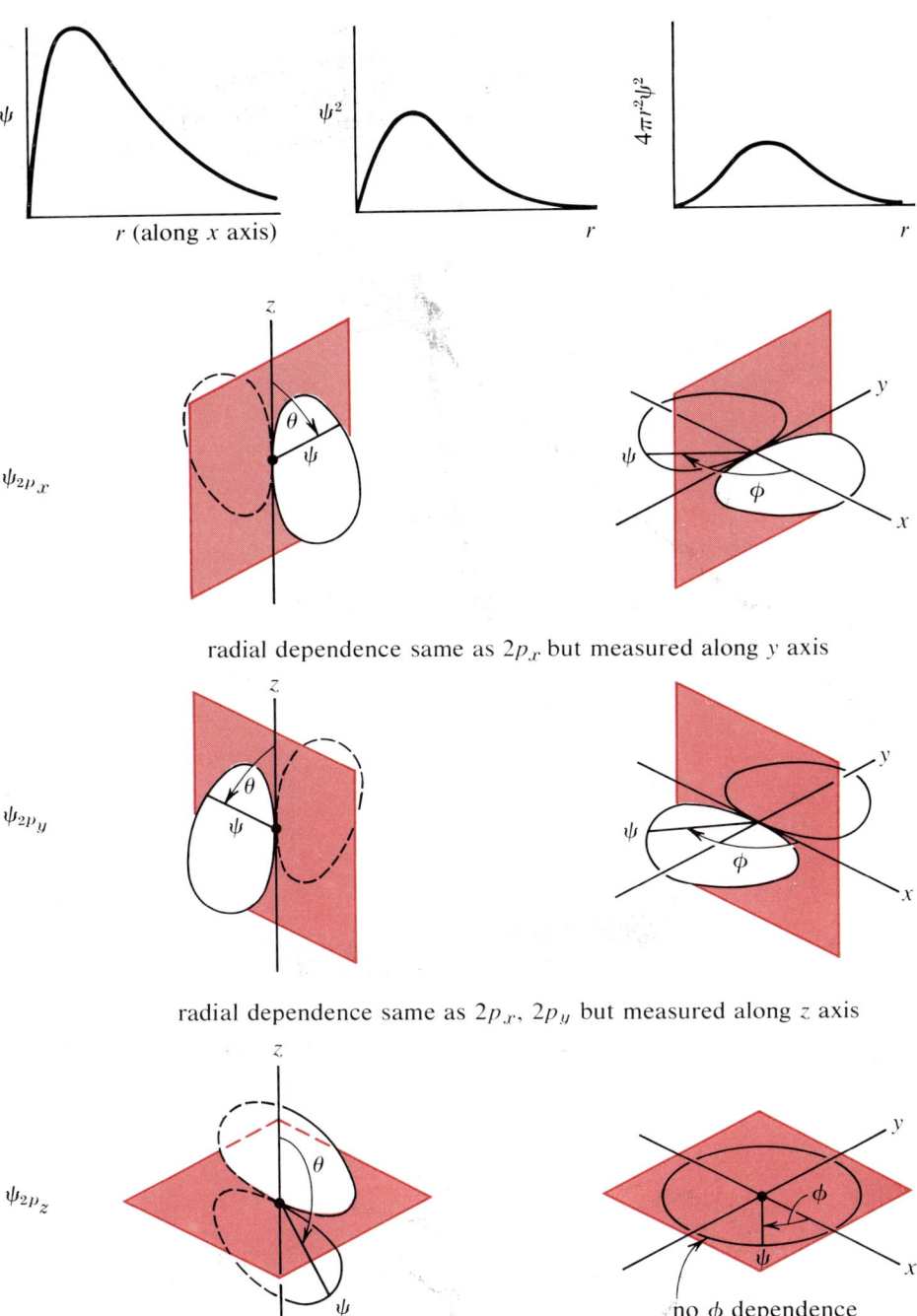

3s, 3p, and 3d

The last nine orbitals in Table 5-1 all have $n = 3$. The first has no angular dependence and thus must be the 3s orbital; since there are no angular nodes there must be two radial nodes, as shown in Fig. 5-13. The next three have one

Figure 5-13 Graphical representations of the 3s orbital.

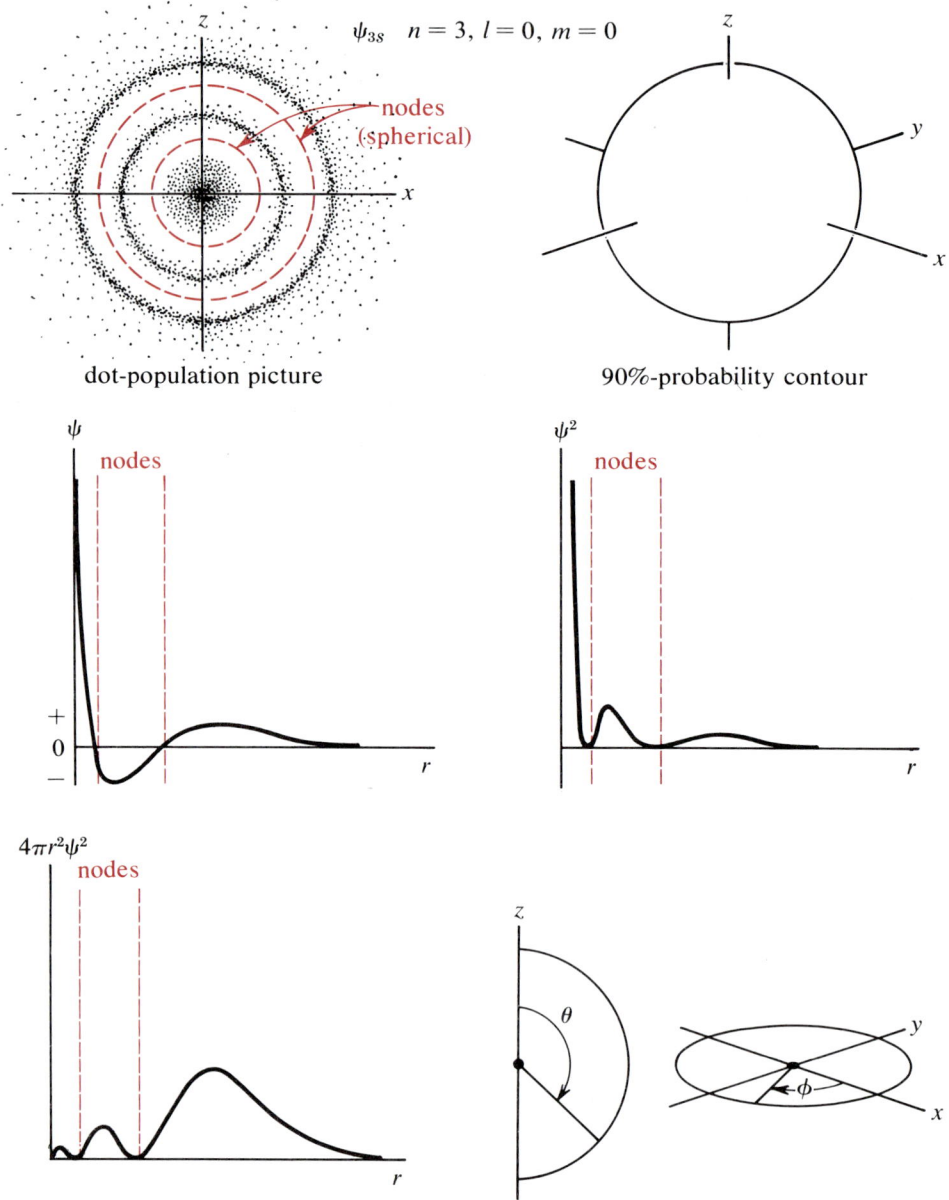

angular node, as in Fig. 5-14, and must have one additional radial node; they have $l = 1$, and are the 3p orbitals. Their angular dependence is the same as for the 2p orbitals and they are similarly called the $3p_x$, $3p_y$, and $3p_z$ orbitals. The last five have two angular nodes, so $l = 2$, and are known as the 3d orbitals. There are no radial nodes. For the case in which $m = 0$ there are no nodes that cut the xy plane (the range of the angle φ); looking down the z axis, then, we see circular symmetry. The two nodal surfaces are cones with their axes on the

Figure 5-14 Graphical representations of the 3p orbitals.

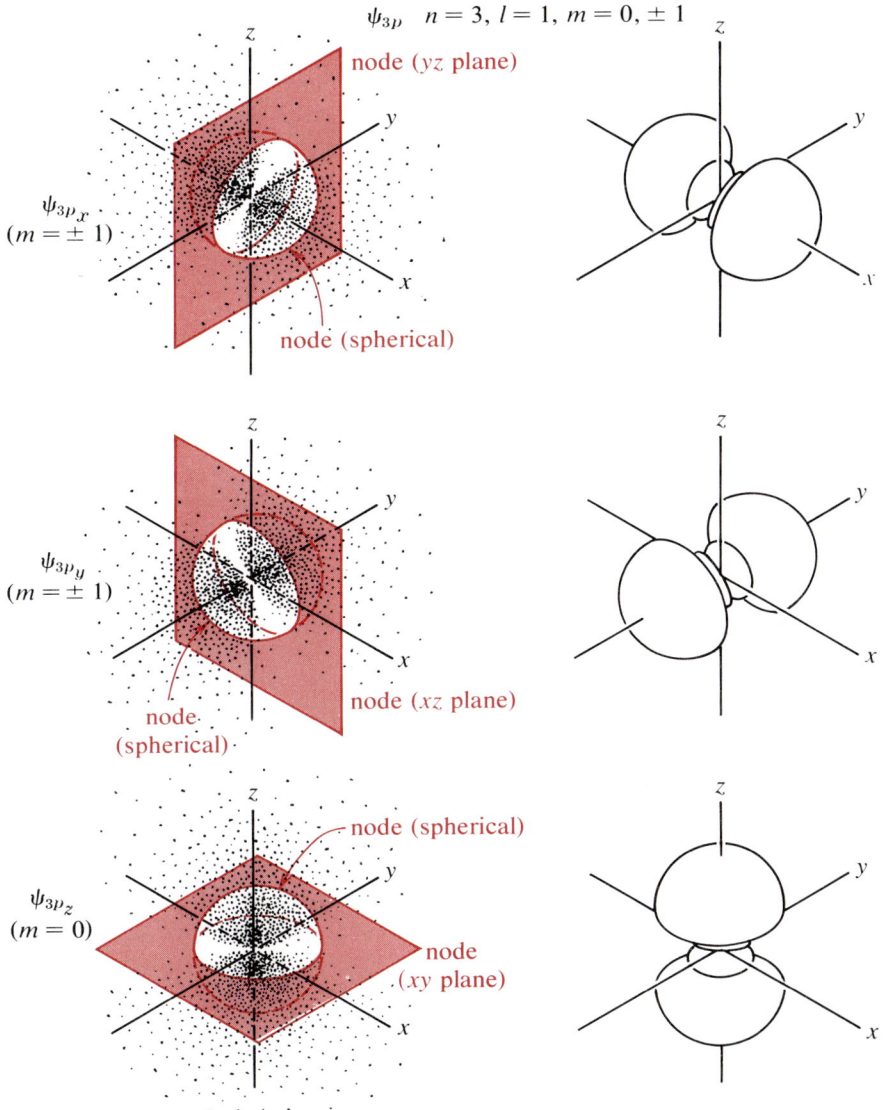

230 | Atomic Structure

z axis, as seen in Fig. 5-15, so there is a sort of belt around the middle of the orbital (the whole thing, belt and all, is the orbital). This is the $3d_{z^2}$ orbital. In

Figure 5-15 Graphical representations of the 3d orbitals.

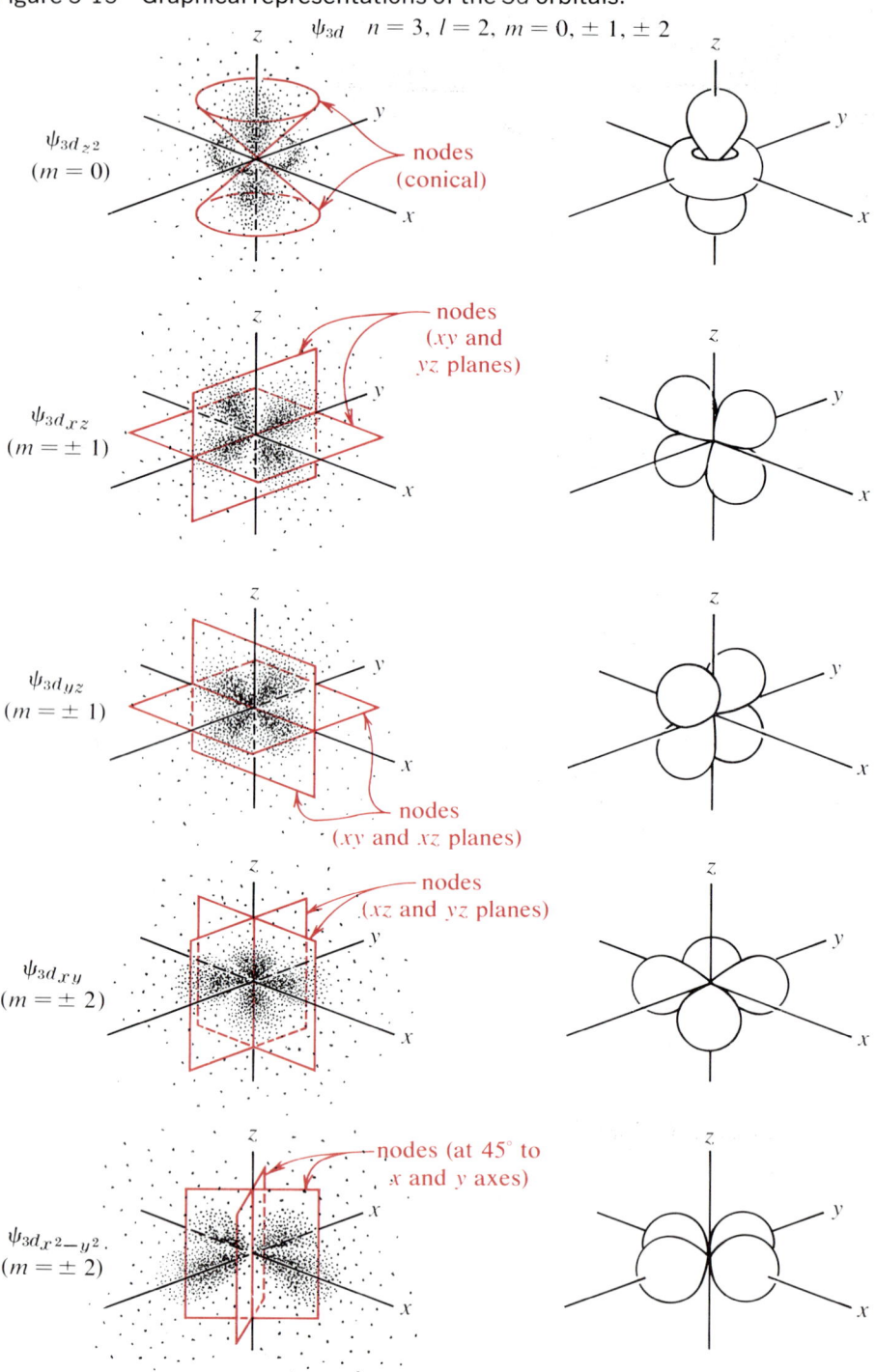

the same figure are shown the two 3d orbitals with $m = \pm 1$, which have one node cutting the xy plane and another lying in the xy plane. These are the $3d_{xz}$ and $3d_{yz}$ orbitals. Finally, there are the two 3d orbitals with $m = \pm 2$, having both nodes cutting the xy plane. Two independent ways to do this are shown; the orbitals are called $3d_{xy}$ and $3d_{x^2-y^2}$. It is true for d orbitals as for p orbitals that the subscript denotes the direction or directions in which the orbital sticks out most prominently.

Let us review and extend our nomenclature. An orbital is named according to the values of its quantum numbers. First the value of the principal quantum number n is given, then a letter indicating the value of the l quantum number. The letters are as follows (derived from spectroscopic notation even before the Bohr atom was devised):

$$l = 0 \quad 1 \quad 2 \quad 3 \quad 4 \quad 5 \quad 6$$
$$\text{letter designation} = s \quad p \quad d \quad f \quad g \quad h \quad i$$

Finally, the m quantum number is designated by a subscript indicating the directions in which that particular orbital sticks out. Thus we have $2p_x$ and $3d_{xy}$, but for 1s or 4s no direction is given since the orbital is spherically symmetrical. Given the restrictions on possible values of the quantum numbers, we can see that the following kinds of orbitals are possible: 1s; 2s, 2p; 3s, 3p, 3d; 4s, 4p, 4d, 4f; 5s, 5p, 5d, 5f, 5g; 6s, 6p, and so on. Note that it is not possible to have a 1p orbital since there is no way to have one angular node but no total nodes; similar arguments apply to the other possibilities.

Mathematical and Physical Properties of Orbitals

In Fig. 5-16 the sizes of the orbitals with the same l values are compared, and also the sizes of the orbitals with the same n but differing l. The principal quantum number n has a very striking effect on the size of the electron distribution, but l has a relatively small effect. We expect, then, that the major effect of the quantum numbers on the energy of the electron will come when n changes, since the distance of the electron from the nucleus is what determines the electron's energy through the Coulomb attraction. We shall see later that under certain circumstances orbitals of quite different sizes on two different atoms will not fit together well into a chemical bond, so that these sizes have some chemical significance. Notice also that all the s orbitals, for instance, have at least a small "hump" of probability density at very nearly the same distance from the nucleus, so even the outer ones have some tendency to penetrate to very near the nucleus; this "penetration" is discussed in Chapters 12 and 16.

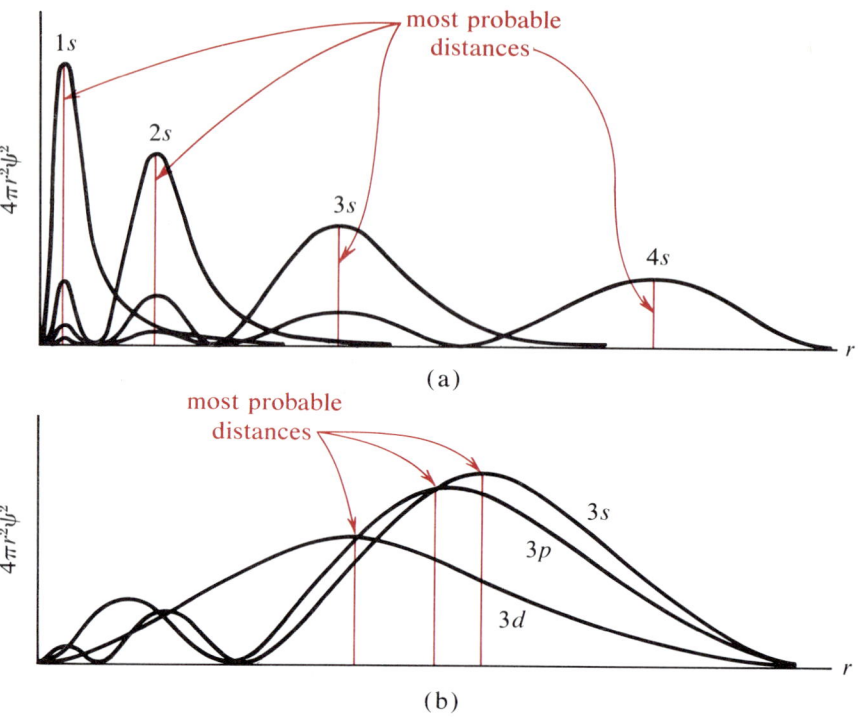

Figure 5-16 Radial distributions of orbitals with (a) changing n and (b) changing l.

With respect to the shapes of orbitals and their nodal surfaces, Fig. 5-17 may be helpful. Remember that the nodes arise from the places at which some factor in the algebraic wave function equals zero; at these places the wave function is changing sign. That is, the value of ψ is changing from positive to negative or vice versa. So every time we go across a nodal surface following any coordinate the algebraic sign of the wave function changes. It is common to put a plus or a minus sign inside each lobe of an orbital to indicate its sign. *This plus and minus refers only to the wave function's sign and never to the electrical charge, which for an electron is always negative.* Of course, these signs have no effect on the physically observable quantity that is the electron's probability distribution, because the wave function has to be squared to provide that quantity and squaring eliminates any effect of the algebraic sign. Drawings sometimes appear in which the wave function is shown squared so as to properly indicate probability, but that also include this sign notation. These drawings serve two purposes, and it is necessary to distinguish between them.

Figures 5-10 through 5-17 are worthy of close study. A substantial amount of the conceptual material in this book depends on the combining of atomic orbitals with symmetries such as those shown here for the hydrogen atom. We cannot overemphasize the need to develop an intuitive feel for the sizes and shapes of these orbitals.

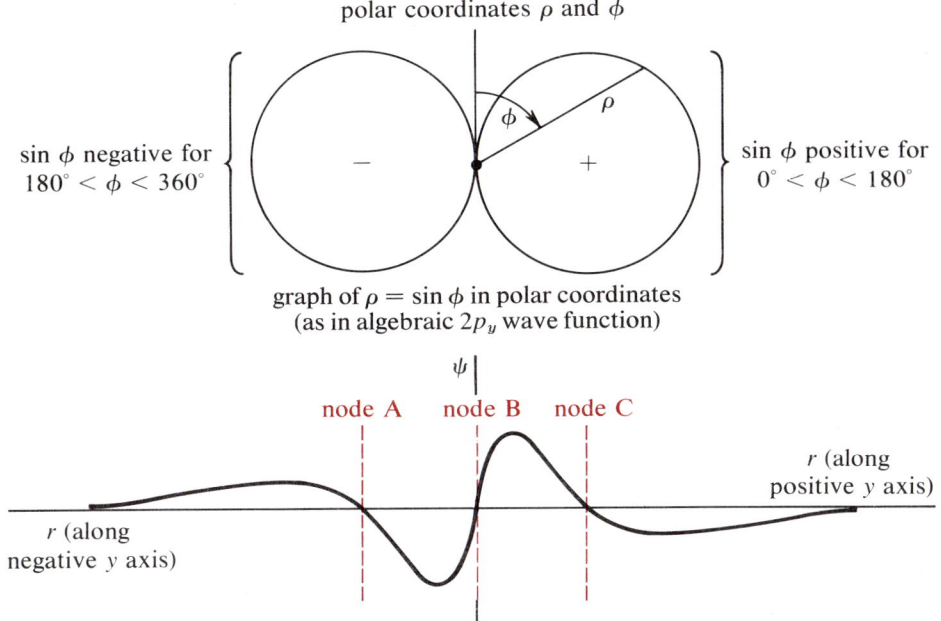

Figure 5-17 Angular dependence of orbitals with angular nodes only or with angular and radial nodes.

5-7 The Physical Meaning of the Wave Function — Normalization and Orthogonality

We need now to return to our discussion of the physical meaning of wave functions or orbitals. There are two additional properties of wave functions that we should explore; they relate not only to the physical nature of wave functions (i.e., to the probability represented by the square of the new function) but also to the ways we can combine wave functions when we begin to discuss bonding. Let us look at the probability density function, $4\pi r^2 \psi^2$, to try to extract a little more information from it.

Normal Probability for the Electron

We have defined ψ^2 as a function that gives us the relative probability that an electron will be found near a given point in space. Since there are an infinite number of geometric points in any finite volume, the probability of finding the electron exactly *at* a given point is infinitesimally small. What we really mean by the relative probability is the probability of finding the electron in an arbitrarily small volume—a box that is dx long, dy wide, and dz high. Alternatively, using polar coordinates we can speak of the probability of finding an electron in some arbitrarily thin spherical layer that is dr thick and lies at some distance r from the nucleus; this layer presumably has the geometry of the skin on an onion. This layer is what we mean by the radial probability density function $4\pi r^2 \psi^2 \, dr$. Just as adding up all the spherical layers of an onion from the center to the outside gives us the whole onion, it is also true that adding up all the probability layers from the nucleus ($r = 0$) as far out as we can go ($r = \infty$) gives us the total probability of finding the electron in all of space. This total probability must be 1; if we look everywhere, we're bound to find it somewhere. So we have a way to scale our relative probability function to an absolute probability. A moment's thought will show that the process of adding up little chunks of r, each dr long, from $r = 0$ to $r = \infty$ is really the process of taking a definite integral:

$$\text{total probability} = 1 = \int_0^\infty 4\pi r^2 \psi^2 \, dr$$

Since the factor $4\pi r^2$ is there for any orbital, we usually absorb it along with dr and the differential angle increments into a volume element $d\tau$. So we can write

$$\int_0^\infty \psi^2 \, d\tau = 1 \tag{5-9}$$

Equation 5-9 is called the **normalization condition**; it expresses the fact that the wave functions have been set to the proper "intensity" to give a meaningful overall probability for finding the electron. In other words, such ψ functions yield a normal probability.

But since we have not told the Schrödinger equation about this condition, the acceptable wave functions it describes may not meet the condition in Eq. 5-9. Fortunately, we can scale the wave functions up or down to give us any probability we please by simply multiplying each one by a **normalization constant** N:

$\psi(\text{correct probability}) = N\psi'$ (from the Schrödinger equation)

All of the hydrogen wave functions in Table 5-1 have already been normalized; that is the origin of the curious numerical factors such as $\sqrt{2}/81\sqrt{\pi}$ at the beginning of each wave function.

Orthogonality

We might also want to consider what relationship these wave functions bear to each other, since they have all been generated from the same differential equation (the Schrödinger equation) simply by allowing one constant, E, the elec-

The Physical Meaning of Wave Function – Normalization and Orthogonality | 235

tron's total energy, to assume different values. Are these wave functions really all necessary, all independent, or is one of them just three times another one, say, so that if we know one we know them all? This is like asking how many dimensions an object has. A straight line has an infinite number of points on it, but any of them can be defined in terms of one dimension since a line is one dimensional; thus one point could have a coordinate value three times that of another point, for instance. A flat surface is two dimensional and we can describe any point on it by using two coordinates, such as x and y. These two coordinates are completely independent of each other—relationships that hold for points along the x direction (or dimension) have nothing to do with relationships along the y direction. Thus we can resolve a line drawn on a two-dimensional surface into an x component and a y component, but these components are completely independent, and both are necessary to specify the line. This is because the two dimensions with which we deal are **orthogonal** (Greek "at right angles") to each other; the y axis has no x component, and vice versa. So the question we are really asking about these wave functions (are they all independent of each other?) could also be phrased: are they orthogonal to each other?

If the $2s$ wave function is orthogonal to the $1s$ wave function, it will have no $1s$ component; we couldn't get even part of the $2s$ function by multiplying any constant times the $1s$ function. Graphically, in terms of the orbitals we have developed, this means the two orbitals must not have any net overlap. But they do overlap, since all the orbitals are centered on the same nucleus. If they are really orthogonal, then the extent to which they overlap each other constructively is just cancelled by the extent to which they overlap destructively. We can express this by looking at the algebraic signs of the orbitals. If two orbitals overlap and both have the same algebraic sign (either plus or minus) the sign of the product of the two functions will be plus; if in a different region of space they have opposite signs, the sign of the product will be minus. If they are orthogonal, all the plus contributions to the overlap will be just balanced by all the minus contributions. So we find ourselves again wanting to add up a very great number of small contributions, just as we did for the normalization condition, and again this is the process of integration. We can say, then, that there is an **orthogonality condition**

$$\int_0^\infty \psi_a \psi_b \, d\tau = 0 \tag{5-10}$$

Figure 5-18 shows the graphical approach to this condition for the $1s$ and $2p_x$ orbitals; the orbitals are shown as their 90%-equal-probability contours and the area of positive overlap is shaded black, while the area of negative overlap is shaded red. Clearly the two areas just cancel each other, and we can say that the $1s$ and $2p_x$ orbitals are orthogonal. They are entirely independent, and neither can be expressed in terms of even a partial contribution from the other.

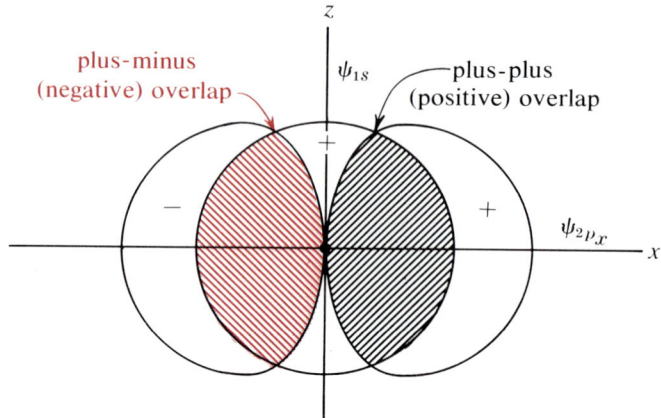

Figure 5-18   Orthogonality of the 1s and $2p_x$ orbitals.

By drawing similar graphs, we could establish that *all* the orbitals we have discussed are orthogonal to each other. In algebraic terms, the integral of one wave function times the other over all of space is just zero, as in Eq. 5-10, no matter which two wave functions we chose for ψ_a and ψ_b. If there were any non-independent (nonorthogonal) orbitals, we would want to eliminate them so as to deal with the smallest possible set, but the orbitals we have already found are indeed the most efficient set. The orthogonality relationship, however, will be important to us again in Chapter 6 when we consider bonding in molecules.

Energy Quantization in the Wave Model

Finally, there is the question of the energy of an electron in one of these orbitals. In beginning to discuss the wave picture of the hydrogen atom, we noted that it would be necessary to construct the wave model so as to preserve the superb agreement of the energies predicted by the Bohr model with the experimental energies obtained through spectroscopy. If we start with the Schrödinger equation (Eq. 5-8) and set out to find functions ψ that fit it, we obtain an infinite number of satisfactory functions. Not all of these, however, are physically meaningful; we insist, for instance, that it must be possible to integrate the square of the function to satisfy the probability interpretation represented by the normalization condition. If we restrict ourselves to ψ functions that are both mathematically and physically acceptable, we find that for negative electron energies E the quantum number n (which measures the total number of nodes) must satisfy the condition

$$n = \sqrt{\frac{2\pi^2 Z^2 m e^4}{h^2(-E_n)}}$$

But we can rearrange this to

$$E_n = -\frac{2\pi^2 Z^2 m e^4}{n^2 h^2}$$

and this is identical to the energy expression resulting from the Bohr model. Since the algebraic expression is the same, the splendid agreement with experimental spectra provided by the Bohr model is also provided by the wave model.

5-8 The Many-Body Problem, Screening, and Slater Orbitals

At this point we have established a modern and adequate description of the internal structure of hydrogen or any other one-electron atom. We have not shown it explicitly, but the wave model gives exactly the same energies as the Bohr model while allowing for the wave properties of microscopic particles and remaining within the confines of the uncertainty principle. It appears, then, that we potentially have a tool for dealing with any atom, specifically including all the ones of chemical interest that have many electrons. Unfortunately, as soon as we leave the one-electron system the mathematical problem becomes impossible to solve exactly and we have to resort to approximations. We shall look at the approximations shortly, but let us first see why the mathematical problem breaks down.

The Many-Body Problem

In developing the Schrödinger equation for the hydrogen atom we used an expression (Eq. 5-6) for the total energy of the electron in the hydrogen atom. Suppose we try to write the corresponding expression for the helium atom, which has two electrons. Equation 5-6 simply expressed the total energy as the sum of the kinetic energy of motion and the potential energy of position; we can write an equivalent expression for electron 1 in the helium atom, $\frac{1}{2}mv_1^2 - Ze^2/r_1$, where v_1 is the velocity of electron 1 and r_1 is its distance from the nucleus. Of course, there is an equivalent expression for electron 2, $\frac{1}{2}mv_2^2 - Ze^2/r_2$, and the total energy of the system is the sum of these two expressions. We could solve that exactly—but we have left out a factor affecting the energy in an important way. Because of the opposite charges on the electron and the nucleus, they attract each other, which is the potential energy we have already discussed; but the two electrons, with the same charge, also repel each other, and we have not dealt with this effect. This repulsion is governed by the Coulomb law, and we can write it as $+e^2/r_{12}$, where r_{12} is now neither r_1 nor r_2 but the distance between the two electrons. So for the total energy we have to write

$$E_{\text{electronic}} = \tfrac{1}{2}mv_1^2 + \tfrac{1}{2}mv_2^2 - \frac{Ze^2}{r_1} - \frac{Ze^2}{r_2} + \frac{e^2}{r_{12}}$$

The presence of this last energy term, the repulsion energy of the two electrons, is what makes the mathematics insoluble, because while we can separate the 1's and 2's from each other, r_{12} depends on both r_1 and r_2 so that we cannot conveniently separate it from either.

Obviously, as we proceed through the periodic table to elements with more electrons, the mathematics becomes messier and messier because of the very large number of repulsion energy terms: for copper, with 29 electrons, there would be 406 repulsion terms. This sort of problem is not uncommon in physics, and it turns out to be generally impossible to obtain an exact solution for problems involving more than two particles or bodies, such as the gravitational effects involving the earth, the moon, and a lunar vehicle rocket. While approximate solutions are possible, they are not very tidy, and the whole thing is referred to as the **many-body problem**.

Approximate Many-Electron-Atom Orbitals

What sort of approximations can we make to give us a reasonable approach to the electron distribution in polyelectronic atoms? One possibility is to ignore the repulsion energy completely; in certain specialized cases this is sometimes done with reasonable success, but it is not generally acceptable. Another possibility, and the one most often used, is to treat the electron–electron repulsion as a partial reduction of the attractive energy of the nucleus for the electrons. In other words, we consider that the negative charge on one electron partially screens or shields another electron from the full positive charge on the nucleus. There is some experimental reason to hope that this might be a practical approach, because we have already seen in Moseley's work on X-ray spectra of the elements (see p. 214) that his data agreed with an algebraic expression involving an effective nuclear charge $Z - \sigma$, where σ was a screening constant. So if we replace Z, the atomic number, in the hydrogen orbitals by a new quantity Z_{eff}, an effective atomic number that, following Moseley, is just the real atomic number reduced by a screening constant, we might have a workable approximation.

The best-known approximate orbitals were proposed by J. C. Slater in 1930, only four years after the introduction of the Schrödinger equation. **Slater orbitals** have angular distributions identical to the exact hydrogen orbitals, and only the radial distribution is changed. The same pattern of orbitals exists: $1s, 2s, 2p,$

3s, 3p, 3d, 4s, 4p, 4d, 4f, and so on. The radial part of each orbital has the following form, which can be seen in comparison with Table 5-1 to be very similar to, but a little simpler than, the hydrogen orbitals:

$$\text{radial part of wave function} = r^{(n^*-1)} \, e^{-(Z-\sigma)\rho/n^*Z}$$

Here σ is the screening constant we have already discussed and n^* is an effective quantum number (like n) that expresses the idea of screening and penetration (as in Fig. 5-16). There are simple rules for determining values of both n^* and σ:

1. n^* is assigned according to the real quantum number n for the orbital being considered:

   $n = 1 \quad 2 \quad 3 \quad 4 \quad 5 \quad 6$
   $n^* = 1 \quad 2 \quad 3 \quad 3.7 \quad 4.0 \quad 4.2$

2. To determine σ, the electrons are divided into the following groups from the inside of the atom to the outside: $1s$; $2s$–$2p$; $3s$–$3p$; $3d$; $4s$–$4p$; $4d$–$4f$; $5s$–$5p$; $5d$–$5f$; $6s$–$6p$.

3. For any given electron, the groups above contribute to σ as follows:
   (a) Nothing from any group outside the one considered.
   (b) 0.35 from each additional electron in the same group (except 0.30 in the $1s$ group).
   (c) For an s or p electron, an amount 0.85 from each electron with n less by 1, and 1.00 from each electron still farther in. For a d or f electron, 1.00 from each electron inside it.

An example may be worthwhile, although we must use the electron-orbital distributions we shall be describing in Section 5-9. Suppose we want to know the effective nuclear charge attracting a $2p$ electron in fluorine, which has atomic number 9 and nine electrons. In the $2s$–$2p$ group, fluorine has six electrons besides the one we are considering, so these six electrons contribute 6×0.35 or 2.10 toward the screening constant. The only inner electrons are two $1s$ electrons, with n less by 1 than it is for the electron we are considering, so they each contribute 0.85—a total of 1.70. The total screening constant, then, is $2.10 + 1.70 = 3.80$. So the $2p$ electron in fluorine feels an effective nuclear charge of $9 - 3.80$ or 5.20 units of positive charge. To show that the screening is quite effective, look at cesium, atomic number 55. Its outermost electron is a $6s$ electron, and there are no other electrons in the same group. There are eight electrons in a group with n less by 1 (the $5s$–$5p$ group), and their contribution to σ will be $8 \times 0.85 = 6.80$. There are 46 electrons still farther inside the atom, so they will contribute $46 \times 1.00 = 46.00$ to σ. The screening constant, then, is $6.80 + 46.00 = 52.80$, and the $6s$ electron in cesium feels an effective nuclear charge of $55 - 52.80 = 2.20$ units of positive charge, which is very small considering that the nucleus really has 55 units of positive charge.

This repulsion energy is a very large factor in the energy of the electrons in an atom, and we must not neglect it.

It must not be thought that Slater orbitals are the last word in dealing with polyelectronic atoms, however. Another mathematical approach has been devised that consists of repetitive calculations of the electron distribution until the assumed distribution gives the same distribution as an answer; this is known as the self-consistent field method and generally gives better numerical results than Slater orbitals, even though it is very difficult to use. Some other orbitals have been devised recently that are combinations of Slater orbitals adjusted to give the best possible fit with self-consistent field results; these seem to be superior to Slater orbitals. But for our purposes the ideas involved in Slater orbitals will be entirely adequate.

5-9 Spin, the Exclusion Principle, and the Periodic Table

Electron Spin

The picture we have developed so far is an adequate representation of current ideas about the distribution of electrons in the atom. It does not account, however, for a property of the electron that exists quite independent of whether the electron is in an atom or not. The electron's behavior, particularly in the presence of magnetic fields, makes it appear that it possesses a constant intrinsic angular momentum, quite apart from the angular momentum it displays in moving about the nucleus of an atom. A constant angular momentum can be achieved for a body of constant mass and speed of rotation in either of two ways; it can rotate clockwise or counterclockwise. Accordingly, given a reference direction such as a magnetic field (which amounts to defining a place from which to look at the rotation) this **electron spin** can take either of two values corresponding to the two possible directions of rotation. This is reminiscent of the space quantization of the orbitals, and so we say that there is an electron **spin quantum number** m_s that can assume only two values, $+\frac{1}{2}$ and $-\frac{1}{2}$. In this way the magnitude of the spin is always constant and there is a change of one unit between one spin state and the other. The spin property, as we have said, does not affect the orbitals describing the electron's distribution in space around the atom's nucleus, so we can afford to consider it separately. It is very important to the distribution of electrons in a polyelectronic atom or molecule, however, as we shall see.

The Pauli Exclusion Principle

In all of what has gone before we have tacitly assumed that we are dealing with a single electron. Even the Slater orbitals that we use to deal with polyelectronic atoms treat the electron–electron repulsion as if it were only a modification of the nuclear charge. We need to ask exactly how the electrons are distributed in a polyelectronic atom. The experimental data that are available on the energies of electrons in atoms suggest the validity of the following two statements:

1. The electrons in an atom will always seek to distribute themselves according to the wave function that gives them the lowest energy (greatest stability).
2. No two electrons in a given system (atom or molecule) can have all four quantum numbers n, l, m, and m_s the same.

The first of these two statements is not at all surprising—it is rather like the statement that water always runs downhill (i.e., to the position of lowest energy or greatest stability). The second statement is a little more surprising but fundamental nonetheless; it is known as the **Pauli exclusion principle** after its discoverer, Wolfgang Pauli. Its truth must be accepted empirically—it works. When we consider that electrons are identical particles that we cannot label, however, it seems reasonable that unless they differ in some aspect of their behavior (which the quantum numbers govern) we could not tell whether more than one was there or not. In any event, no exception to this rule has been found, and we therefore need to examine its operation in atoms.

Electrons in Polyelectronic Atoms

If we take a nucleus with Z positive charges and start to put in Z electrons (which would make it a neutral atom) one at a time, the first one will go into a distribution in space described by the $1s$ orbital; that is, with $n = 1, l = 0, m = 0, m_s = \frac{1}{2}$ (although the spin quantum number is arbitrary). This is because the $1s$ electron is closest, on the average, to the nuclear charge and thus is most strongly attracted. The second electron will also go into a $1s$ distribution. Again this is the lowest-energy distribution, and the exclusion principle is not violated since the spin quantum number can be different: $n = 1, l = 0, m = 0, m_s = -\frac{1}{2}$. The third electron cannot adopt a $1s$ distribution because it would have to have quantum numbers identical to one of the electrons already there. It does the next best thing, adopting the $2s$ distribution. The fourth electron will do the same, taking opposite spin to the third. Thus one will be (for the quantum numbers n, l, m, and m_s) 2 0 0 $\frac{1}{2}$, the other will be 2 0 0 $-\frac{1}{2}$. Because of the exclusion principle, the fifth electron will adopt the lowest possible energy distribution, the $2p$. Which of the three $2p$ distributions does not matter for a free atom; they have the same energy. The sixth will also adopt a $2p$ distribution, and we have a new question: will it pair spins with the existing $2p$ electron, or will it go into one of the other $2p$ distributions? The answer is one of what are known as **Hund's rules**: if an electron has the choice of pairing with an existing electron or adopting an alternative distribution with the same energy (which no other electron has adopted), it will always avoid pairing and adopt the alternative distribution, with its spin parallel to the existing spin. So in our case the sixth electron will adopt one of the other unused $2p$ distributions, and both of the $2p$ electrons will have parallel spins, with both m_s quantum numbers $+\frac{1}{2}$. Similarly, the seventh electron will adopt the third $2p$ distribution with parallel spin. The eighth electron will find it preferable to adopt a $2p$ distribution by pairing spins rather than adopt the higher-energy $3s$ distribution, and so on. In general, as we saw in Fig. 5-16, the lowest n quantum number will provide the lowest energy, but this is not always true as n increases because the possibility of penetration (also shown there) makes, for instance, $4s$ lower energy than $3d$ for the potassium atom. Figure 5-19 shows the order in which the energies of the atomic orbitals fall for most atoms. This is not absolutely uniform because as the atomic number increases the effective nuclear charge increases, particularly for orbitals in which the electron is likely to penetrate near the nucleus. A little thought will show that this means that the higher the atomic number, the more likely it is that the lower n will provide the lower energy. So Fig. 5-19 should be thought of as a sort of average, not as an absolute description.

The Wave Model and the Periodic Table

Using the relative energies of the orbitals, we can very quickly see why the periodic table has the structure it does. Table 5-2 lists the electron distributions of the naturally occurring elements, built up in exactly the way we have just described. Because the angular distribution functions repeat themselves exactly for each value of n (so that $1s$, $2s$, $3s$, $4s$, and $5s$ are all alike and $2p_x$, $3p_x$, and $4p_x$ are all alike, for instance), and because there is a tendency for the effective nuclear charge to repeat itself for similar outer-electron distributions, we might expect that an outer-electron configuration of $6s^2$ (the notation means two electrons are distributed according to the $6s$ wave function) would be very similar to a configuration of $5s^2$. This similarity means, in physical terms, that the two s electrons in each case will have the same geometry, although the $6s$ will be, on the average, a little farther from the nucleus than the $5s$, and also that the force

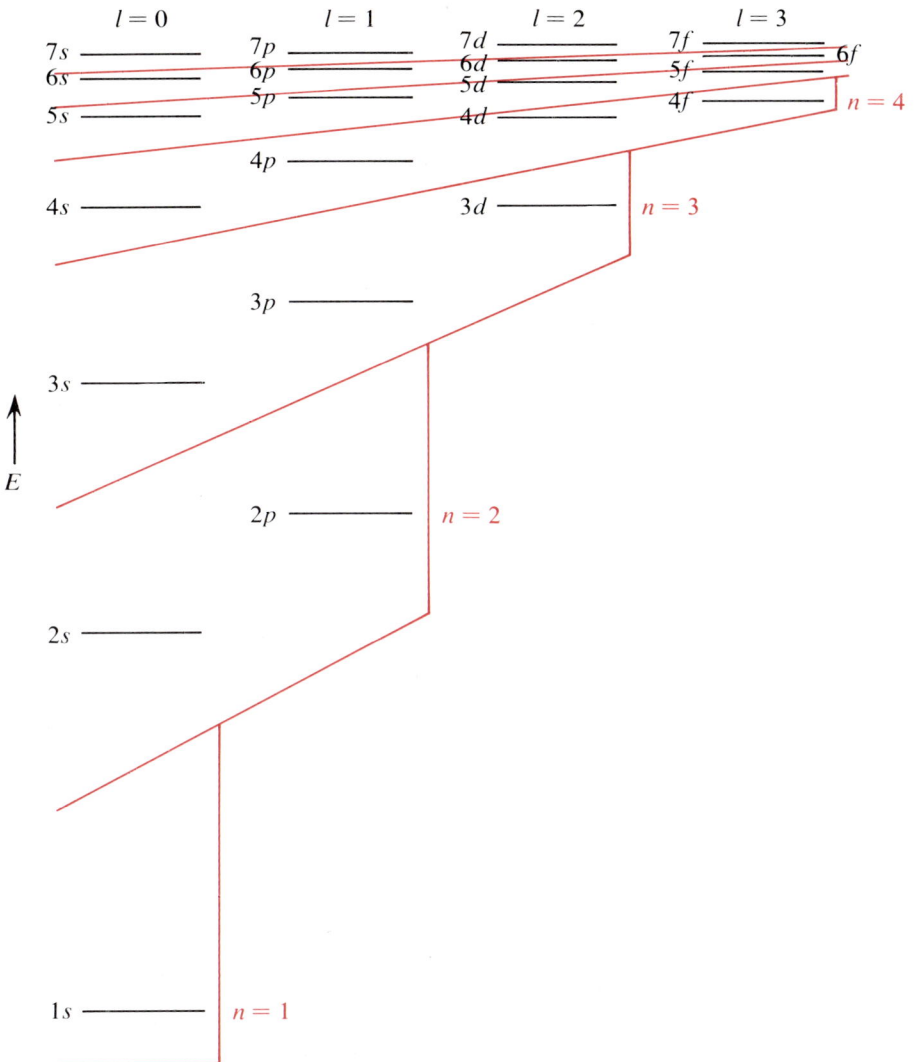

Figure 5-19 Relative energies of different orbitals in an average atom.

attracting them to the nucleus will be nearly the same. We shall see in coming chapters that chemical reactivity is largely a matter of the relative availability of electrons to be shared with or contributed to another atom, so that the similarity of electronic distributions indicates that the element with an outer-electron configuration $6s^2$ (barium) should have chemical properties very much like the element with configuration $5s^2$ (strontium). This is exactly the case; the

Table 5-2
Electron Distributions of Elements

Atomic number	Element	Electronic configuration	Atomic number	Element	Electronic configuration
1	H	$1s$	27	Co	$-3d^74s^2$
2	He	$1s^2$	28	Ni	$-3d^84s^2$
3	Li	[He]$2s$	29	Cu	$-3d^{10}4s$
4	Be	$-2s^2$	30	Zn	$-3d^{10}4s^2$
5	B	$-2s^22p$	31	Ga	$-3d^{10}4s^24p$
6	C	$-2s^22p^2$	32	Ge	$-3d^{10}4s^24p^2$
7	N	$-2s^22p^3$	33	As	$-3d^{10}4s^24p^3$
8	O	$-2s^22p^4$	34	Se	$-3d^{10}4s^24p^4$
9	F	$-2s^22p^5$	35	Br	$-3d^{10}4s^24p^5$
10	Ne	$-2s^22p^6$	36	Kr	$-3d^{10}4s^24p^6$
11	Na	[Ne]$3s$	37	Rb	[Kr]$5s$
12	Mg	$-3s^2$	38	Sr	$-5s^2$
13	Al	$-3s^23p$	39	Y	$-4d5s^2$
14	Si	$-3s^23p^2$	40	Zr	$-4d^25s^2$
15	P	$-3s^23p^3$	41	Nb	$-4d^45s$
16	S	$-3s^23p^4$	42	Mo	$-4d^55s$
17	Cl	$-3s^23p^5$	43	Tc	$-4d^55s^2$
18	Ar	$-3s^23p^6$	44	Ru	$-4d^75s$
19	K	[Ar]$4s$	45	Rh	$-4d^85s$
20	Ca	$-4s^2$	46	Pd	$-4d^{10}$
21	Sc	$-3d4s^2$	47	Ag	$-4d^{10}5s$
22	Ti	$-3d^24s^2$	48	Cd	$-4d^{10}5s^2$
23	V	$-3d^34s^2$	49	In	$-4d^{10}5s^25p$
24	Cr	$-3d^54s$	50	Sn	$-4d^{10}5s^25p^2$
25	Mn	$-3d^54s^2$	51	Sb	$-4d^{10}5s^25p^3$
26	Fe	$-3d^64s^2$	52	Te	$-4d^{10}5s^25p^4$

differences between barium chemistry and strontium chemistry are quite subtle (and they can, as we shall see later on, be rationalized by considering the radial differences). Furthermore, since the shielding by other electrons in the same group of Slater orbitals for a given atom is not very good (0.35 electron worth of shielding for each electron added) while the nuclear charge goes up by one full unit for each electron added, we expect to see the effective nuclear charge change substantially as we go across any row of the periodic table, increasing as we go to the right. So the $4s^2$ element might have quite different chemical properties from the $4s^23d^3$ element, for instance (calcium vs. vanadium), and they would certainly be quite different from the $4s^23d^{10}4p^5$ element

Table 5-2 (continued)

Atomic number	Element	Electronic configuration	Atomic number	Element	Electronic configuration
53	I	$-4d^{10}5s^25p^5$	78	Pt	$-4f^{14}5d^96s$
54	Xe	$-4d^{10}5s^25p^6$	79	Au	$[\text{Xe } 4f^{14}5d^{10}]\, 6s$
55	Cs	$[\text{Xe}]\, 6s$	80	Hg	$-6s^2$
56	Ba	$-6s^2$	81	Tl	$-6s^26p$
57	La	$-5d6s^2$	82	Pb	$-6s^26p^2$
58	Ce	$-4f^26s^2$	83	Bi	$-6s^26p^3$
59	Pr	$-4f^26s^2$	84	Po	$-6s^26p^4$
60	Nd	$-4f^46s^2$	85	At	$-6s^26p^5$
61	Pm	$-4f^56s^2$	86	Rn	$-6s^26p^6$
62	Sm	$-4f^66s^2$	87	Fr	$[\text{Rn}]\, 7s$
63	Eu	$-4f^76s^2$	88	Ra	$-7s^2$
64	Gd	$-4f^75d6s^2$	89	Ac	$-6d7s^2$
65	Tb	$-4f^96s^2$	90	Th	$-6d^27s^2$
66	Dy	$-4f^{10}6s^2$	91	Pa	$-5f^26d7s^2$
67	Ho	$-4f^{11}6s^2$	92	U	$-5f^36d7s^2$
68	Er	$-4f^{12}6s^2$	93	Np	$-5f^46d7s^2$
69	Tm	$-4f^{13}6s^2$	94	Pu	$-3f^67s^2$
70	Yb	$-4f^{14}6s^2$	95	Am	$-5f^77s^2$
71	Lu	$-4f^{14}5d6s^2$	96	Cm	$-5f^76d7s^2$
72	Hf	$-4f^{14}5d^26s^2$	97	Bk	$-5f^97s^2$
73	Ta	$-4f^{14}5d^36s^2$	98	Cf	$-5f^{10}7s^2$
74	W	$-4f^{14}5d^46s^2$	99	Es	$-5f^{11}7s^2$
75	Re	$-4f^{14}5d^56s^2$	100	Fm	$-5f^{12}7s^2$
76	Os	$-4f^{14}5d^66s^2$	101	Md	$-5f^{13}7s^2$
77	Ir	$-4f^{14}5d^76s^2$	102	No	$-5f^{14}7s^2$
			103	Lw	$-5f^{14}6d7s^2$

(bromine). On the other hand, bromine should be very much like iodine ($5s^24d^{10}5p^5$). And since the electron's average separation from the nucleus changes sharply when n increases, so that its energy does too, there should be a particularly marked change in chemical properties when an element's last electron has to enter a distribution with a new value of n that is higher than that of any of the electrons in the previous element. This is exactly why the periodic table begins on the left with lithium, sodium, potassium, rubidium, cesium, and francium; there is a profound difference between their chemistry and that of the element just before each one, which is one of the rare gases.

5-10 Ionization and Ionization Energies

There is a process on the atomic scale, which can be defined for each atom without having to worry about what other atom is involved in the same process, that defines the electron's availability. The process is ionization; we have written a chemical equation on p. 211 defining the process for the hydrogen atom. In the more general case of an atom M, we write

$$M^0 + \text{(ionization energy)} \rightarrow M^+ + e^-$$

All electrons that are bound to atoms have a negative energy, so energy must be put into the system to separate the electron from the rest of the atom. The energy necessary to remove an electron from a neutral atom is called the **first ionization energy**, that necessary to remove an electron from a 1+ ion to give a 2+ ion is called the **second ionization energy**, and so on.

Mass Spectrometry and Ionization Energies

Experimentally, the technique is easy to understand although it is expensive to perform. First a vacuum chamber is set up so gaseous ions can travel its length without colliding with any other atom. A gaseous sample of the atoms under investigation is introduced at one end of the vacuum chamber, as in Fig. 5-20. It is struck by a beam of electrons whose energy is closely controlled

Figure 5-20 Experimental arrangement for ionization potential or mass measurement.

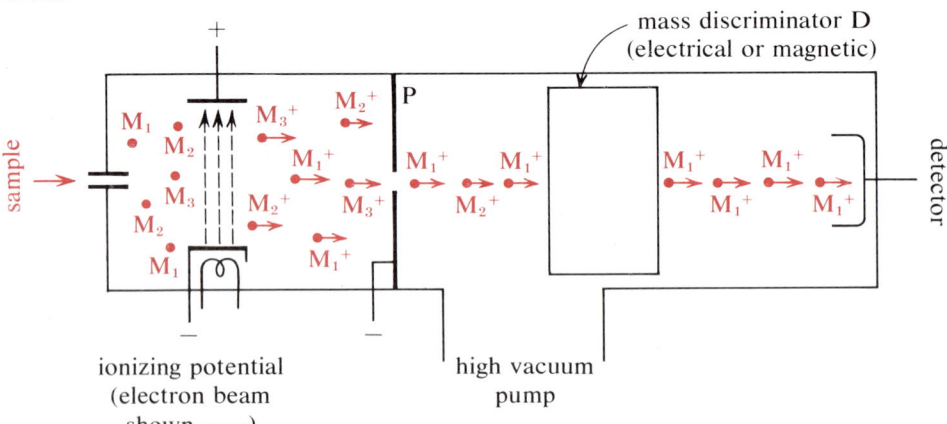

electrically, and if their energy exceeds the ionization energy some of the atoms will be ionized and develop a net positive charge. These positively charged ions, when they form, are strongly attracted by the negatively charged plate P, so that a beam of ions is formed moving to the right through the hole in the plate. The ions are sorted according to their mass by a discriminator D, which may be a magnetic field deflecting the ions of different mass through different angles (as Thomson's cathode-ray tube deflected electrons with characteristic e/m) or various electrical means, and finally arrive at a detector. Because of the mass discrimination, the detector can look at only one kind of atom — if the ionizing electron beam had energy greater than the ionization energy of that element, the detector will develop an electrical signal. The threshold of energy at which a signal first appears is the ionization energy of the atom. The ionizing electron beam's energy, in turn, is controlled by adjusting the voltage or electrical potential that accelerates the electrons in the beam, and ionization energies are usually expressed in an energy unit called the **electron volt** (eV) — the energy an electron acquires when accelerated by a potential of 1 V. Alternatively, ionization energies are sometimes called **ionization potentials** and expressed directly in volts. It should be noted that energy and potential are not the same thing and do not have the same units, but since one electron is being removed from the atom its ionization energy in electron volts and its ionization potential in volts will be numerically the same. The instrument we have just described is a **mass spectrometer**; since it can establish not only ionization energies but also masses of unknown atoms, it has obvious analytical applications in dealing with unknown compounds.

Valence Orbital Ionization Potentials

The tendency for an electron to adopt the orbital distribution that gives it the lowest energy is obviously related to the ionization energy. We have defined ionization energies or potentials so as to refer only to an atom with its overall electron configuration and the ion (with one less electron) and its overall electron configuration. These ionization energies are not necessarily related to any particular orbital, since we look only at the total energy of the atom before and after ionization. By various means, however, it is possible to establish particular energies that are characteristic of the orbital electron distributions themselves. That is, we can produce a figure for the energy of the $2p$ electrons in a carbon atom, for instance. This is rather more specific than the difference between the energies of a neutral carbon atom and a carbon +1 ion. Such specific orbital energies, called **valence orbital ionization potentials** and abbreviated VOIP, have been tabulated and are very useful in comparing electron energies to judge the relative availability of various electrons for chemical purposes. In Chapter 6 we shall explore some uses of VOIP's in predicting the nature of compounds between different atoms; presently our only purpose is to observe the way these numerical values reflect the periodic trends.

VOIP's and Periodicity

The periodic table inside the front cover gives a number of VOIP's; Fig. 5-21 reproduces some of these to make the periodic trends clearer. The lithium atom has a 2s electron that is some distance from the nucleus (making the denominator larger in the Coulomb energy expression $Z_{eff}e^2/r$) and is fairly well shielded by the two inner 1s electrons (tending to reduce Z_{eff}). It is not surprising, then, that it is relatively easy to remove the lithium 2s electrons; lithium has a small VOIP for 2s electrons, since not much energy is necessary to remove them.

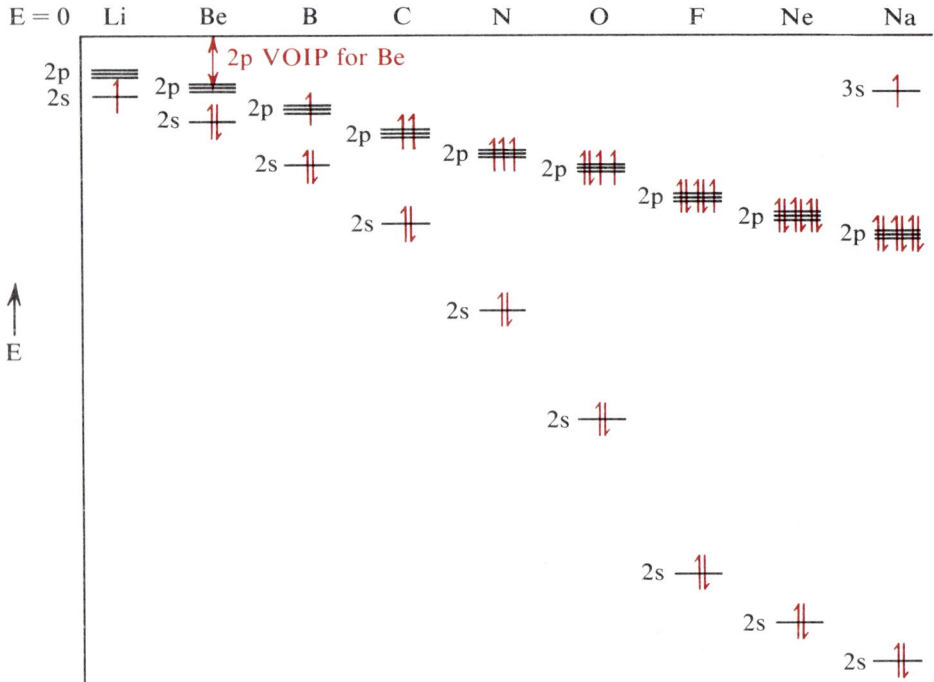

Figure 5-21 VOIP's and electron population for first-row elements.

(The VOIP is just the energy necessary to go from the orbital energy up to zero.) As we add electrons (and protons to the nucleus) in forming other elements in this row of the periodic table, two effects show up. The effective nuclear charge increases, since more electrons in the same Slater orbital group do not shield very well, and at the same time the 2s and 2p orbital distributions of electrons actually get closer, on the average, to the nucleus. (Remember that a Slater orbital has Z_{eff} in the exponential radial factor—the larger Z_{eff} is, the faster the wave function will drop off toward zero.) Both of these factors tend to make the Coulomb attractive energy and thus the ionization energy larger, and indeed we see a uniform increase going to the right in the row. But when we get to sodium, its outermost electron is required by the exclusion principle to adopt a 3s distribution, and all the remarks that applied to the lithium 2s electron now apply to the sodium 3s electron. So we see that the periodic trends we mentioned are well represented by the trends in numerical VOIP values. We shall examine all of this in more detail in Chapter 12.

5-11 Summary

The earlier chapters described the behavior of gases and solids, and discussed this behavior in terms of the atomic-molecular model currently used by chemists. In this chapter we have extended our appreciation of the atomic model, first by discussing some of the major experiments that shed light on necessary features of the atomic model, then by looking at the first modern approach to a model of the atom's internal structure, the Bohr atom. Perceiving some very basic faults in the Bohr atom, we developed a wave model of the atom to allow for the demonstrated wave nature of microscopic particles and at the same time for the uncertainty of measurement that is inherent to experiments on these particles.

This wave model, in which the distribution of the electron in an atom is described by a square of a function that is a solution to the Schrödinger equation, has been described at some length. We have taken some pains to describe the sizes and shapes of the various electron-orbital distributions, because these quantities will be of considerable importance to us in our discussion of bonding in molecules. We have also examined, both algebraically and graphically, the normalization and orthogonality conditions that apply to these atomic orbitals, again because these will have a considerable bearing on the results of our bonding discussion.

We expanded our discussion to polyelectronic atoms by discussing a simple way in which we can deal with the problem of describing the effects of electron–electron repulsion by modifying the hydrogen orbitals into Slater orbitals. Using these orbitals we examined the specific distribution of electrons in large atoms, first by establishing the electron spin property, then by using the exclusion principle to govern the configuration, that is, the assignment of electrons to orbitals.

Finally, we established the basic agreement between this model and the structure of the periodic table by a qualitative discussion of the availability of electrons for bonding from the various electron configurations. In developing a quantitative measurement, we saw that the valence orbital ionization potential gives us a numerical tool for dealing with electron energies and ultimately the tendency of electrons to change their distribution. This is the basis of chemical reaction.

Study Problems

1. In Rutherford's α-particle scattering experiment, suppose the large deflections were due to a series of small deflections, always by chance in the same direction. If the average individual deflection were 1° and we simplify the problem by assuming that there are only two choices for deflection at each collision (right or left, say), what is the probability that an α particle would be deflected in the same direction (to the right) in 180 consecutive collisions to give a 180° scattering? How many moles of α particles would have to be used in order, statistically, to observe one 180° scattering?
2. Lyman reported a series of spectral lines analogous to the Balmer series, but with wavelengths of 1215.7, 1025.7, 972.5, and 949.7 Å, plus others of shorter wavelength. To what values of n_1 and n_2 in the Rydberg expression do these lines correspond?
3. Derive an expression from the Bohr model for the velocity v of the electron. For the H $1s$ electron, what is the numerical result in centimeters per second; in orbit diameters per second? In a spectroscopic electronic transition that involves movement of the electron through a distance approximately equal to an orbit diameter, what characteristic time would be required for the transition? How does this time compare with the time required for the H atom to move the same distance at its rms velocity (see Eq. 3-10)?
4. Where will radial nodes occur in the $3s$ wave function? Use Table 5-1. How well does the innermost coincide with the $2s$ radial node?
5. How many radial nodes should there be in a $4f$ wave function? How many angular nodes? What should be the angle between the angular nodes?
6. Sketch the $2s$ and $3d_{xy}$ orbitals, centered on the same nucleus. (Use equal-probability contours.) Show graphically that the orbitals are orthogonal.
7. O^{2-}, F^-, Ne, Na^+, and Mg^{2+} are *isoelectronic* (have the same total number of electrons). Consider the mathematical form of the Slater orbitals and predict the trend in the sizes of these atoms or ions.
8. In the text it has been noted that the electron's average separation from the nucleus increases sharply when n increases. Why is the size difference between strontium ($5s^2$) and barium ($6s^2$) only about 3%?

9. Consider the algebraic form of the Slater orbitals and explain why the energy necessary to remove a valence electron from Be ($1s^2 2s^2$) is not the same as the VOIP of the $2s$ atomic orbital on Be.
10. Figure 5-21 shows a smooth trend in the VOIP's of the elements both for the $2s$ and $2p$ orbitals. Why, then, should the first ionization energy for Be be greater than that for either Li or B?

Some Further Reading

Daniels, F., and Alberty, R. A., *Physical Chemistry*, New York: Wiley, 1966 (3rd ed.). Chapter 12 is probably the best place to look for another discussion of this material. The coverage is nearly the same, but the treatment is more mathematical. Slow reading but worthwhile.

Day, M. C., and Selbin, J., *Theoretical Inorganic Chemistry*, New York: Van Nostrand Reinhold, 1969 (2nd ed.). Chapters 1 and 2 provide an alternative to Daniels and Alberty, but the mathematical transitions are a bit more abrupt.

Heitler, W., *Elementary Wave Mechanics*, Oxford: Clarendon Press, 1956 (2nd ed.). The first part of this little book is an excellent intuitive, nonmathematical introduction to the wave model of the atom. There are some equations, but the discussion is physical. This will improve your feel for what is happening even though it is not directly applicable to our discussion here.

Linnett, J. W., *Wave Mechanics and Valency*, New York: Wiley, 1960. A more mathematical version of Heitler. The early part of the book has a good discussion of atomic structure from the point of view of the current mathematical model, but some of the math will be heavy going.

6 | Molecular Structure

In Chapter 5 we developed the modern theory of atomic structure in a fairly extensive and quantitative way. This is the essential basis for all of chemistry. It is important to realize, however, that it is quite rare to find individual atoms in the chemical makeup of our environment. Only the rare gases — the group from helium through radon — are customarily found in a monatomic state, and at least three of them can be encouraged to form compounds. Everything else we encounter involves the bonding of atoms into molecules or other cohesive structures: air, water, chrome bumpers, porcelain steins, textbook pages, and professors. To understand the makeup of matter, then, we must develop a comprehensive theory of the combination of atoms into chemically bonded systems. The picture should be consistent with the ideas of atomic structure we have already developed, and it must give a unified and consistent description of the chemical properties of these bonded systems. However, it is difficult to develop all of the features of the present model of bonding from key experiments as we did for the atomic model. Therefore we shall construct the model and show that the various features of it are compatible with the chemical and physical properties of a variety of compounds.

6-1 Prediction of Molecular Geometry

The first thing we need is a reliable way to describe the geometry of atoms bonded together into molecules. There are other kinds of polyatomic systems than molecules, and we shall consider them in Section 6-12, but the number of covalent molecules is so great as to make them the most important part of the broad array of chemical compounds. The theoretical system we shall set up can predict the geometry of molecules, but the process is very complicated and time consuming. In general, the more common applications of bonding theory assume that the geometry of the molecules is known (i.e., that the **bond lengths** between atomic nuclei and the **bond angles** between the lines representing each bond axis are known—see Fig. 6-1). From the known geometry the atomic structure model of Chapter 5 allows us to predict the chemical and physical properties of a given compound to a substantial extent.

Figure 6-1 Definitions of bond length and bond angle.

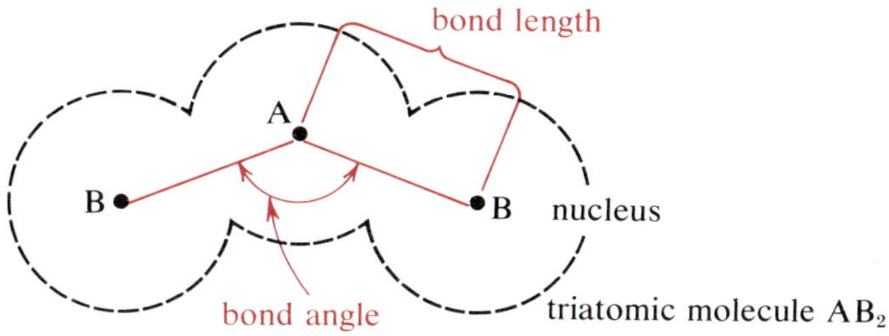

But how can we predict the geometry of a particular molecule in a reasonably simple way so we can describe the bonding? We shall see in Chapter 12 that bond lengths can be predicted quite accurately from the covalent radii tabulated in the periodic table. For the time being we need only assume that the bond lengths can be obtained, either in this way or directly from experiments such as X-ray diffraction or other diffraction or spectroscopic methods. We shall confine ourselves to the problem of predicting (or rationalizing) the bond angles of a given molecule. It will be easy to check the validity of our technique, because very accurate data are available for the bond lengths and angles of a large number of compounds. A convenient compilation of many of these is *Structural Inorganic Chemistry* by A. F. Wells (Clarendon Press, Oxford, 1962, 3rd ed.).

Obviously, the only possible geometry for a **diatomic** molecule (composed of two atoms) is a straight line between the two atoms; we cannot define any bond angles. What about polyatomic molecules? First we have to decide which is the central atom of the molecule, the one about which the others are arranged. Not all molecules can be described as having a central atom; in particular, complicated organic molecules contain chains of atoms. It would be necessary in a case such as this to define the geometry around one particular atom in the chain, then do this for the other atoms one at a time. For most inorganic compounds, however, we can define a central atom unequivocally. The empirical rules we are about to give do an excellent job of rationalizing the observed geometry of compounds in the liquid or vapor phase, except for the transition metal compounds (the metal elements from titanium across to copper, and the elements immediately below them). The rules are less reliable in the solid state, and as we progress through the chapter we shall see why. We shall defer any discussion of transition-metal compounds to Chapter 16; the internal structure of the transition metals places certain restrictions on the geometry of their compounds that do not seem to apply to the nontransition elements.

In many cases the compound's stoichiometry makes the identity of the central atom obvious, such as in NH_3 where the N is the central atom, or SF_6 where the S is the central atom. Sometimes there are three or more kinds of atoms present, as in phosphorus oxytrichloride, $POCl_3$; in these cases one should assume that F is never at the center of the molecule. Other halogens (Cl, Br, I) are usually not at the center except when combined with other halogens or oxygen, and oxygen is not usually at the center unless it is combined to two other atoms as in H_2O or OF_2. Usually these assumptions will correctly indicate the central atom; thus for $POCl_3$ we expect P to be the central atom, and it is.

Having established the central atom of the molecule, count up the number of valence electrons surrounding it in the molecule. Begin with the atom's own valence electrons; these will be all those with the highest value of the principal quantum number n (the others with lower n will be too tightly bound to the atom to participate in bonding). Add to this number the electrons contributed to the central atom by the outer, or **ligand**, atoms; assume that each ligand atom forms enough two-electron bonds (one electron of its own and one of the central atom's) to surround itself with eight valence electrons. For each of these two-electron bonds one ligand electron is shared by the central atom. Finally, add the number of electrons represented by the net charge on the molecule; if it has a 2− charge overall, two more electrons should be considered to be around the central atom, while if it has a + charge one electron less should be considered to be around the central atom.

Consider as an example OF_2, which we mentioned above as an example of oxygen as a central atom. Oxygen has six valence electrons (those with $n = 2$). Each of the F atoms has seven valence electrons (those with $n = 2$); thus each F can surround itself with eight valence electrons by forming one two-electron bond with O out of one of its own electrons and one of the O's. Then there will be a total of two F electrons contributed to the O. There is no net charge on the molecule OF_2, so no adjustment needs to be made to the number of electrons surrounding the O. If we represent the number of the central atom's own valence electrons by VE, the number of ligand-atom electrons contributed to it by LE, and the charge adjustment by CA, we can write

$$\text{central atom valence electrons} = VE + LE + CA$$

In the case of OF_2 the number of O valence electrons is $6 + 2 + 0 = 8$.

To take a slightly more difficult case, look at $POCl_3$. Phosphorus is the central atom, and it has five valence electrons ($n = 3$). The O has six valence electrons and to surround itself with eight it must form two two-electron bonds with the P atom, each with one of its electrons and one P electron; so the O contributes two electrons to the number around the P atom. Each Cl has seven valence electrons ($n = 3$) and needs to form one two-electron bond with the P atom. Thus each Cl contributes one electron to the P and all three contribute a total of three electrons in addition to the two from the O. There is no net electric charge on $POCl_3$, so we write P valence electrons $= 5 + 5 + 0 = 10$.

With the number of valence electrons about the central atom established, assume that they are paired whether or not they form a two-electron bond. Thus the eight electrons around O in OF_2 form four pairs; the 10 electrons around P in $POCl_3$ form five pairs. However, if one ligand atom forms two (or three) bonds to the central atom we consider all four (or six) electrons together. $POCl_3$, then, has five pairs but only four "bunches" or groups of electrons around it, since the O forms two bonds to the P. If the compound has an odd number of electrons, take the last electron as a "one-electron bunch."

Now distribute these pairs or groups of electrons around the central atom so that repulsion of like charges is minimized. For instance, for a triatomic molecule with two electron groups it is easy to see that the two groups should be on opposite sides of the central atom, with a bond angle of 180°. The geometries that minimize the repulsion have been mathematically established as shown in Fig. 6-2. From this figure we can see that the electron groups in both OF_2 and $POCl_3$ will be distributed tetrahedrally. The geometry of the molecule is conventionally considered to be *not* the geometry of the electrons but that of the atoms, so we have to establish this from the electrons' geometry. In the case of OF_2, since the bonding to each of the atoms is identical and the four positions on a tetrahedron are equivalent, it does not matter where we put the F atoms. The bonds, if we draw them as lines, will look like Fig. 6-3a, and the molecule will be bent. Figure 6-3b shows the geometry of $POCl_3$. Since there are no **nonbonding** pairs of electrons in this case it still does not matter how we distribute the three Cl atoms and the O atom on the tetrahedron because all positions are equivalent. The $POCl_3$ molecule is said to be tetrahedral since the positions of the atoms correspond exactly to the positions of all the electron groups.

There are some more subtle adjustments that we can make to the predicted geometries. First, nonbonding pairs are larger than bonding pairs, because they are held in place by the attraction of only one nucleus, not two as in a bonding pair. This means that the angle between a bonding and a nonbonding pair will be spread slightly, and the angle between two nonbonding pairs will be spread even more. Thus in OF_2 the angle F—O—F will be squeezed slightly, since all the other electron-pair angles are larger than 109°27′, and indeed the F—O—F angle is found experimentally to be only 103.2°. Second, the size of an electron group depends on how many electrons are in it: a four-electron group is larger

Number of Electron Groups	Distribution about Central Atom	Appearance	Bond Angles	Equivalent Positions
2	linear		180°	both
3	trigonal planar	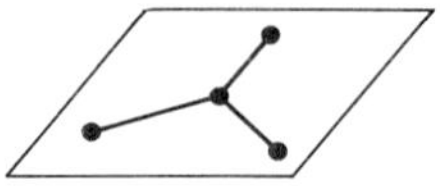	120°	all
4	tetrahedral	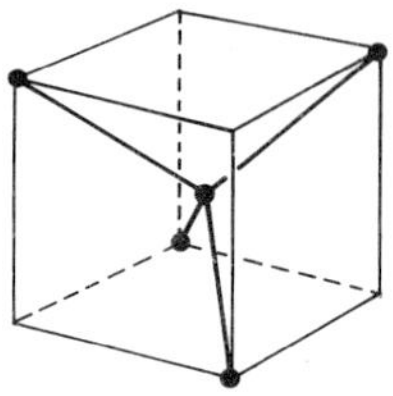	109° 27′	all
5	trigonal bipyramid	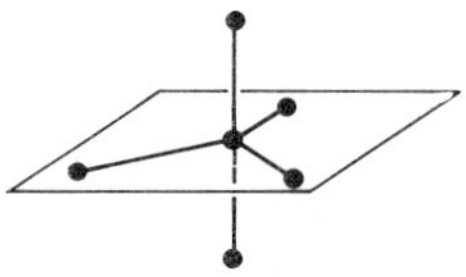	90° for polar groups, 120° in equatorial plane	polar ≠ equatorial
6	octahedral	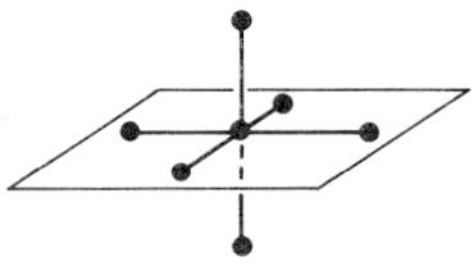	90°	all
7	pentagonal bipyramid	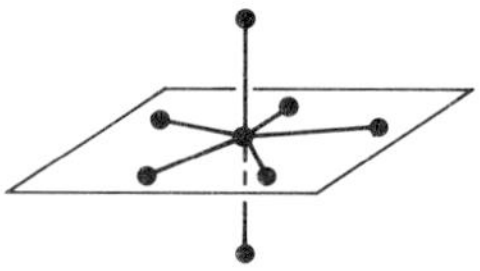	90° for polar groups, 72° in equatorial plane	polar ≠ equatorial
8	square antiprism	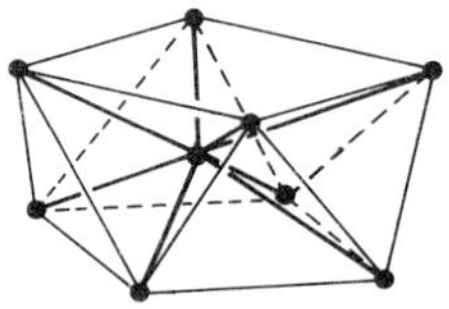	74° 8′ for nearest neighbors	all

Figure 6-2   Geometries of least repulsion.

than an electron pair, which is larger than a single electron. The angle between a four-electron group and a pair will be greater than the corresponding angle between two pairs, then, and the O—P—Cl angle in $POCl_3$ is larger than the Cl—P—Cl angle: experimentally the Cl—P—Cl angle is 103.5°, which is indeed smaller than the tetrahedral angle, while the O—P—Cl angle is correspondingly larger. Third, the greater the electronegativity of the ligand atom,

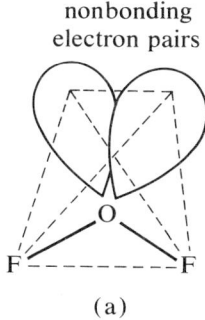 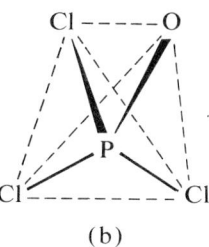

Figure 6-3 Molecular geometries of (a) OF$_2$ and (b) POCl$_3$.

the smaller the bond pair to that atom. We shall define electronegativity more precisely in Chapter 12, but here we can say that it is a measure of an atom's attraction for electrons near it, so that an atom with relatively high electronegativity tends to shrink its bonding electrons in toward it. The periodic table inside the front cover gives numerical values for the electronegativities of the elements; for instance, F is more electronegative than Cl, which is more electronegative than Br. This means that if we consider the series of compounds POF$_3$, POCl$_3$, and POBr$_3$, the F—P—F angle should be smaller than the Cl—P—Cl angle, which should be smaller than the Br—P—Br angle. Experimentally: F—P—F$\angle$ = 102.5°, Cl—P—Cl$\angle$ = 103.5°, Br—P—Br$\angle$ = 108°. Finally, when there are five or seven electron groups such that not every atom has the same number of neighbors, the ligand atoms with the largest number of neighbors are farthest from the central atom.

A few more examples will illustrate the use of this procedure for predicting geometry, known as the **valence-shell electron-pair repulsion** technique (VSEPR). Consider ammonia, NH$_3$. Nitrogen is the central atom. From the periodic table we see that N has five valence electrons of its own ($n = 2$). Hydrogen, alone among ligand atoms, has a stable electronic configuration with two valence electrons (rather than eight), so it will form one two-electron bond and thus contribute one electron to the N atom; the three H atoms will contribute three electrons. There is no charge on the NH$_3$ molecule. So there are $5 + 3 + 0 = 8$ electrons or four pairs around the nitrogen, and no double or triple bonds. The electrons' geometry, then, will be tetrahedral. Since there is one nonbonding pair, the molecule's geometry will be **trigonal pyramidal**, as shown in Fig. 6-4. Since the nonbonding pair is larger than the bonding pairs, the H—N—H angles will be squeezed noticeably from the tetrahedral angle of 109°27'. The experimental angle H—N—H is 106.6°.

As another example, take the three species NO$_2^+$, NO$_2$, and NO$_2^-$. Nitrogen is the central atom in each, and has five valence electrons ($n = 2$). Each O has

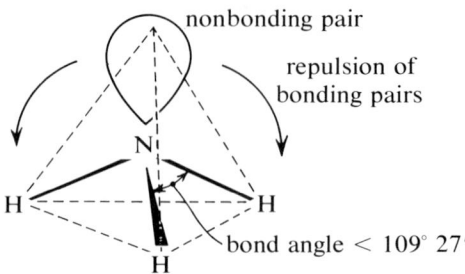

Figure 6-4   Molecular geometry of NH$_3$.

six valence electrons and can achieve a share in eight by forming two two-electron bonds with the N, thereby contributing two electrons to the N. So in all three of these compounds the two O atoms contribute four electrons to the N. The difference between the three species comes in the charge: for NO$_2^+$ we have $5+4-1 = 8$ electrons around the N atom; for NO$_2$ we have $5+4+0 = 9$ electrons in the valence shell; for NO$_2^-$ we have $5+4+1 = 10$ electrons. (There are valid reasons for worrying about putting more than eight electrons on the N atom, but they do not affect the accuracy of this scheme for predicting the geometry of the three species.) The eight electrons around the central atom in NO$_2^+$ are in two four-electron groups, since the oxygens are each double bonded to the N. There are thus two electron groups, which will be most favorably distributed at 180° to each other, and the bond angle in NO$_2^+$ should be 180°. In NO$_2$ there is an extra electron, which forms a small "bunch" by itself; the three-electron groups should then be at 120° to each other, but the bond angle (between the two four-electron groups) should be substantially expanded because the repulsion of the single electron for each of the four-electron groups will be relatively weak. In NO$_2^-$ the 10 electrons are arranged in two four-electron bonding groups and one nonbonding pair. The three-electron groups should be trigonal planar, but now the relatively large nonbonding pair can probably hold its own against the four-electron groups and the bond angle should not be too greatly distorted from 120°. So for these three species we expect bond angles of 180°, about 130°, and about 120°. The experimental bond angles are 180°, 134°, and 115°; see Fig. 6-5.

A good final example is bromine trifluoride, BrF$_3$. Bromine, the central atom, has seven valence electrons ($n = 4$). Each F, having seven valence electrons, adds an eighth by forming one two-electron bond with the Br atom, contributing one electron to the Br in the process. So the F atoms contribute three electrons to the Br, and the total number of valence shell electrons on the Br atom is $7+3+0 = 10$. These should be arranged in five pairs, since no ligand atom is multiply bonded. The electrons will thus be arranged in a trigonal bipyramid,

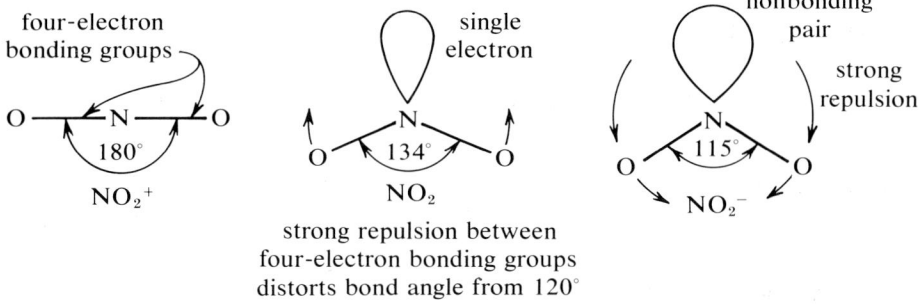

Figure 6-5 Molecular geometry of NO_2^-, NO_2, and NO_2^+.

and there will be three bonding pairs and two nonbonding pairs. Now, the five positions on a trigonal bipyramid are not equivalent; the two polar positions have three neighbors at 90° while the three equatorial positions have two neighbors at 90° and two more at 120°. We must therefore decide how to locate the F atoms to provide the least repulsion among the five electron pairs. The repulsion becomes most important at small bond angles (or electron-pair angles), and for the equatorial positions and two neighbors at 120° are not as important as a third neighbor at 90° would be (as in the polar positions). So the positions with the least overall repulsion will be the equatorial positions, and the larger nonbonding pairs will be most favorably located there. The three F atoms, then, will be in the two polar positions and in one equatorial position; the molecule will be T-shaped. Figure 6-6 shows the experimental result; note that the two polar F atoms, with more effective neighbor electron pairs, are farther from the Br atom, as the last generalization above suggests. The ideal bond angle of 90° has been squeezed somewhat by the two nonbonding pairs.

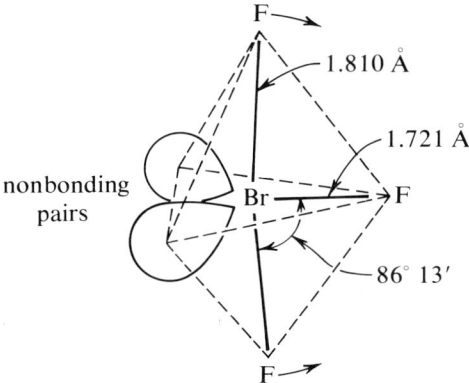

Figure 6-6 Molecular geometry of BrF_3.

The VSEPR Method

We can summarize the VSEPR method for predicting molecular geometry briefly as follows:

1. Establish the central atom in the molecule.
2. Construct two-electron bonds to give all ligand atoms the proper number of electrons (usually eight).
3. Add up the valence shell electrons about the central atom, counting its valence electrons and electrons contributed by ligand atoms, and adjusting for net charge.
4. Assume perfect pairing of electrons and consider double bonds to ligand atoms as single groups of four electrons.
5. Distribute electron pairs or groups about the central atom according to Fig. 6-2.
6. Modify this predicted electron-group geometry for any of the following possible effects:
   (a) Nonbonding pairs are larger than bonding pairs.
   (b) Double bond (four-electron) groups are larger than single bond pairs, which are larger than single electrons.
   (c) Bond pairs to very electronegative ligand atoms are smaller than those to less electronegative ligand atoms.
   (d) For geometries with nonequivalent positions (polar vs. equatorial) ligand atoms with greater numbers of neighbors are farther from central atom.
7. Describe the molecular geometry in terms of electron-group geometry.

6-2 Diatomic Molecules — Potential Energy of Atom Pairs

Now that we have a method for predicting the geometry of polyatomic molecules we can begin to consider the problem of describing bonding when we know the geometry the molecule assumes. The easiest way to begin is to consider the approach of two atoms to each other to form a diatomic molecule. The most obvious feature of such an approach is that the two atoms attract each other strongly when they are at short range, because the two atoms in the molecule stay together in most cases even when they are given considerable thermal energy (kT) by heating. For instance, the H_2 molecule can be heated to $2000°K$ with only about one of every thousand molecules dissociating. There is a considerable **bond energy**, then, that can be estimated roughly from the temperature necessary to dissociate the bond.

What do we know about the forces attracting atoms to each other? For gases, we have seen that a potential-energy function can be constructed that gives a reasonable picture of the departure of the gases from ideality; this is the Lennard-Jones potential (see p. 141) in which the attractive part of the potential is proportional to $1/r^6$. But this cannot be the chemical bonding energy—if it were, the hydrogen molecule would dissociate at the same temperature at which it boils, 20°K, since the boiling process corresponds to removing H_2 molecules from the Lennard-Jones potential well provided by the neighbor molecules in the liquid. From the temperature comparison we see that the chemical bond energy is at least 100 times as great as this dipole-induced dipole attractive energy. Since the distances, σ, at which the Lennard-Jones potential operates are at least of the same approximate size as the bonding distances, the fact that its attractive energy is much too small must be because the radius in the denominator is raised to too great a power. (All of the attractive forces must be essentially electrical in origin, ultimately expressions of Coulomb's law.) The temperature of H_2 bond dissociation is approximately the cube of the boiling point (Lennard-Jones dissociation), so we can tentatively guess that the potential at work in chemical bonding has as its denominator the cube root of the Lennard-Jones denominator, $1/r^2$.

If this is indeed the appropriate form of the attractive potential energy for chemical bonding, it is much like the potential that attracts an electron to a nucleus in an individual atom. It is true that it varies as $1/r^2$ while the potential energy in the atom varies as $1/r$, but the effective nuclear charge in a polyelectronic atom also varies approximately as $1/r$ (see Fig. 6-7) so $Z_{\text{eff}}e^2/r$ is

Figure 6-7 Approximate $1/r$ dependence of z_{eff}. (Redrawn from Slater, J. C., *Quantum Theory of Matter*, New York: McGraw-Hill, 1968.)

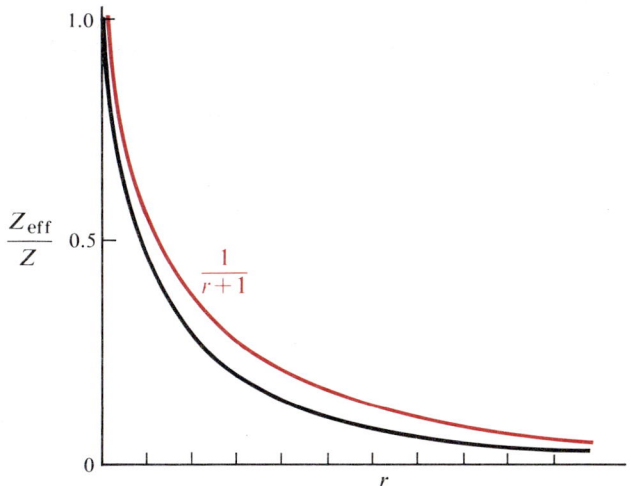

nearly the same as $(Ze^2/r)(1/r)$ or Ze^2/r^2. This makes it appear that the bonding must be due to the electrons (or at least some of them) on one atom being attracted to the other nucleus just as are the other nucleus' own electrons. For the diatomic molecule AB, then, a reasonable speculation as to the mechanism of bonding is that one or more of B's electrons are distributed in space around the two nuclei so as to be described not only by B's atomic orbitals but also by A's atomic orbitals. In this way A and B are held together by the force that attracted the electron to each one individually. Of course, the same argument applies whether the bonding electrons came originally from B or A. In fact, electrons originally belonging to each can participate in formation of a bond, although this is not necessary. We propose, then, that the overall attractive force that leads to the bond energy arises from the *mutual* attraction of an electron or electrons for two nuclei rather than only one.

6-3 Diatomic Molecules — The LCAO Principle

The attraction we have just described must be fitted into a mathematical model so as to be compatible with the model we have already constructed for the electronic structure of the atom. We could approach the problem by setting up an appropriate form of the Schrödinger equation for the two nuclei and the electrons other than the one being described by the wave function under consideration. If we could do this and work out the functions that were acceptable solutions to the Schrödinger equation, they would presumably describe the distribution of electrons in the molecule; by analogy we could call these **molecular orbitals**. Unfortunately, because of the many-body problem, it is impossible to solve the Schrödinger equation exactly for a molecular system, although some excellent approximations have been made for very small molecules. It is true for all these systems, however, no matter how complicated, that if we can choose a reasonable estimate of what the molecular orbital ought to look like and vary it over a small range, the estimated wave function that gives the most stable — lowest-energy — electronic distribution is the best approximation to what the real wave function is like. This means that we have only to make a reasonable guess for the molecular orbitals to be able to refine it into a good description, even though we cannot solve the Schrödinger equation exactly.

The LCAO Approximation

The key to the approximation we shall use (which is not the only possible one) is the fact that the bonding energy is due to one or more electrons being distributed so as to be described by the orbitals of *both* atoms. We shall assume that for the diatomic molecule AB,

$$\psi_{AB}^{\text{molecular orbital}} = c_1 \psi_A^{\text{atomic orbital}} + c_2 \psi_B^{\text{atomic orbital}}$$

In other words, we are taking a **linear combination of atomic orbitals** as an approximation to the molecular orbital. Making the obvious abbreviation, this is called the **LCAO principle**. The abbreviations AO and MO are often used for "atomic orbital" and "molecular orbital." In the linear combination above, the constants c_1 and c_2 are adjusted to give the resulting MO the lowest energy (greatest stability). This assures us the best approximation to the true electron distribution that an LCAO can give. We shall see repeatedly that, in terms of prediction of actual experimental results, this can be a very good approximation indeed. The mathematical procedure for determining c_1 and c_2 can become very tedious, and computers are usually used for this purpose; fortunately, a qualitative understanding of the MO model can be achieved without having to go through these calculations. In a few cases, the symmetry of the molecule's stoichiometry and geometry can give us the c_i values directly without having to go through the calculations, and we shall look at the results in those cases.

The simplest stable common molecule we can consider is the hydrogen molecule, H_2. Since the two atoms in the molecule are identical, both will contribute equally to the chemical bonding in the molecule; there is no reason for one to be favored over the other. Each hydrogen atom has one valence electron in a distribution described by the 1s atomic orbital. In forming the LCAO molecular orbital we shall want to combine the 1s or valence orbitals:

$$\psi^{MO}(H_2) = \psi_{1s}^{AO}(H_a) + \psi_{1s}^{AO}(H_b) \tag{6-1}$$

In general the important AO's to combine will be the valence orbitals. Usually these are the occupied AO's with the highest value of n, although if other AO's have very nearly the same energy we may need to include them. For the H_2 case the nearest AO in energy to the 1s is the 2s, and that is not very close, as another look at Fig. 5-19 will show. So we can afford to ignore all but the 1s AO's in forming H_2 MO's. Let us look at this LCAO molecular orbital more closely.

The Origin of Bonding in the LCAO Model

Figure 6-8 shows the plot of ψ^{AO} vs. r for the hydrogen 1s AO exactly as we drew it for Fig. 5-10. It also shows this same radial dependence drawn four times as it would appear along the H_2 bond axis: to the left and to the right from hydrogen a, and to the left and to the right from hydrogen b. Equation 6-1 tells us that the MO we are looking for is simply the sum of these two functions, as shown by the red line; the height of the red line, representing the molecular

264 | Molecular Structure

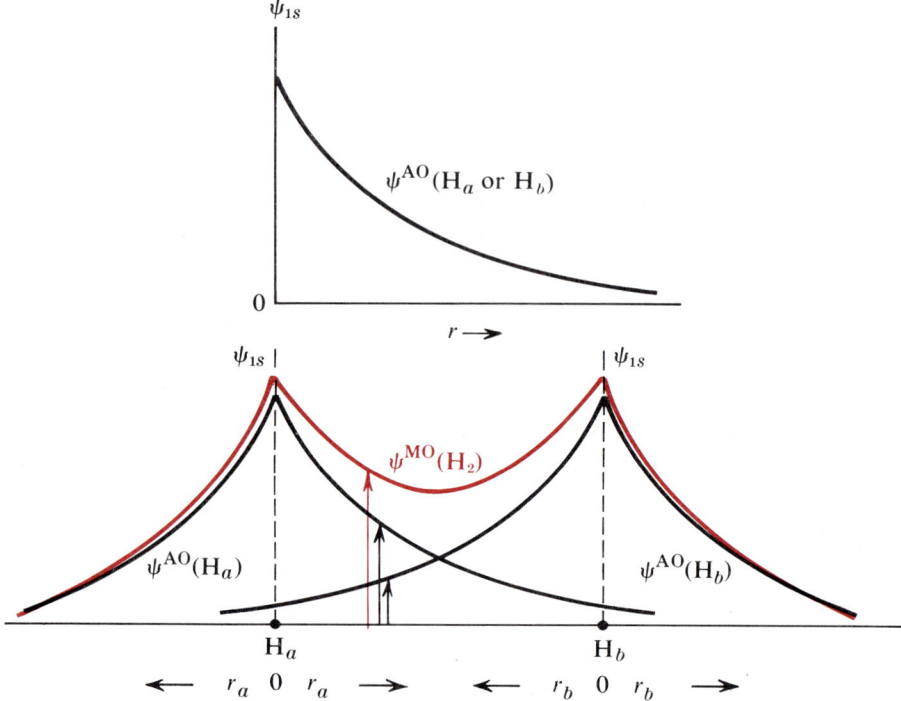

Figure 6-8 Radial dependence of overlapping 1s orbitals in H_2, and of the resulting MO.

wave function or orbital, above the r axis is the sum of the heights of the two atomic wave functions. Figure 6-9 shows the square of this MO, which is the probability-distribution function. Notice that there is a *greater* probability of

Figure 6-9 Probability comparison for an electron in the strongly bonding region between nuclei showing MO probability vs. AO probabilities.

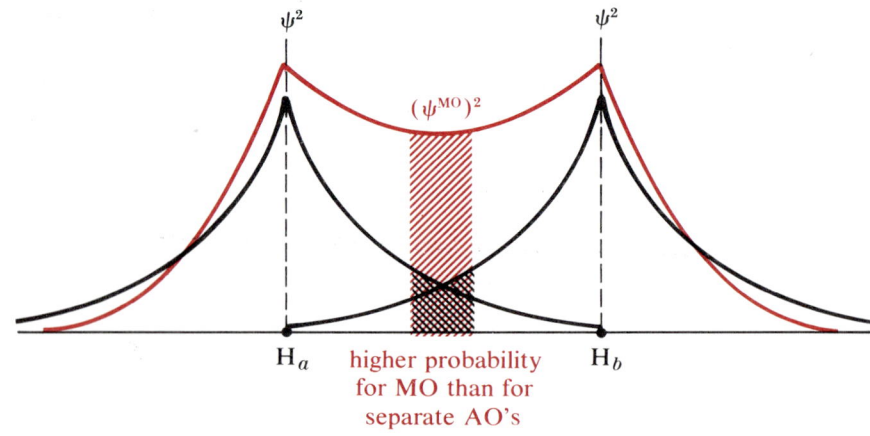

finding the electron between the two nuclei than there would be if the two AO's did not interact to form the MO. This is the origin of the bond energy. The electron is more often found in a position in which it experiences the favorable attraction of two nuclei, namely directly between the two nuclei on or near the bond axis. So the MO does correctly predict the existence of a bonding energy for the two atoms involved.

6-4 Diatomic Molecules — Bonding and Antibonding Orbitals

Independent LCAO's

We have constructed what seems to be a qualitatively satisfactory function describing a possible distribution of an electron around the two nuclei in a hydrogen molecule. There is a problem to consider before we go further, however. We have taken two atomic orbitals — possible distributions for electrons — and combined them to produce only one molecular orbital or possible electron distribution. But for our bonding model to be compatible with the model of atomic structure, the number of possible orbitals must remain constant; there is no way to use 12 bricks to build a wall that contains only six bricks. In other words, we need to find another combination of atomic orbitals that will give us a different, distinguishable molecular orbital. The number of molecular orbitals we form must exactly equal the number of atomic orbitals with which we start. In this case there is only one other independent (meaning orthogonal) possibility for combining the two orbitals linearly: we can subtract one atomic orbital from the other:

$$\psi^{MO}(H_2) = \psi^{AO}_{1s}(H_a) - \psi^{AO}_{1s}(H_b)$$

This is a quite different electron distribution from the first MO, as can be seen in Fig. 6-10, which is drawn in exactly the same way as Fig. 6-8. Notice that the two AO's cancel in the center of the bond axis, so that the square of the MO — the probability-distribution function — has a node halfway between the two nuclei. This is the very place where the electron should be to have maximum effectiveness in bonding the two nuclei together, and yet we see that according to this distribution function it will *never* be there. To see that the middle of the bond axis is the most effective place for the electron to be, we need only remember that the Coulomb law governs the attraction of the electron for each nucleus: $Z_{eff}e^2/r$. Since r is in the denominator, the potential

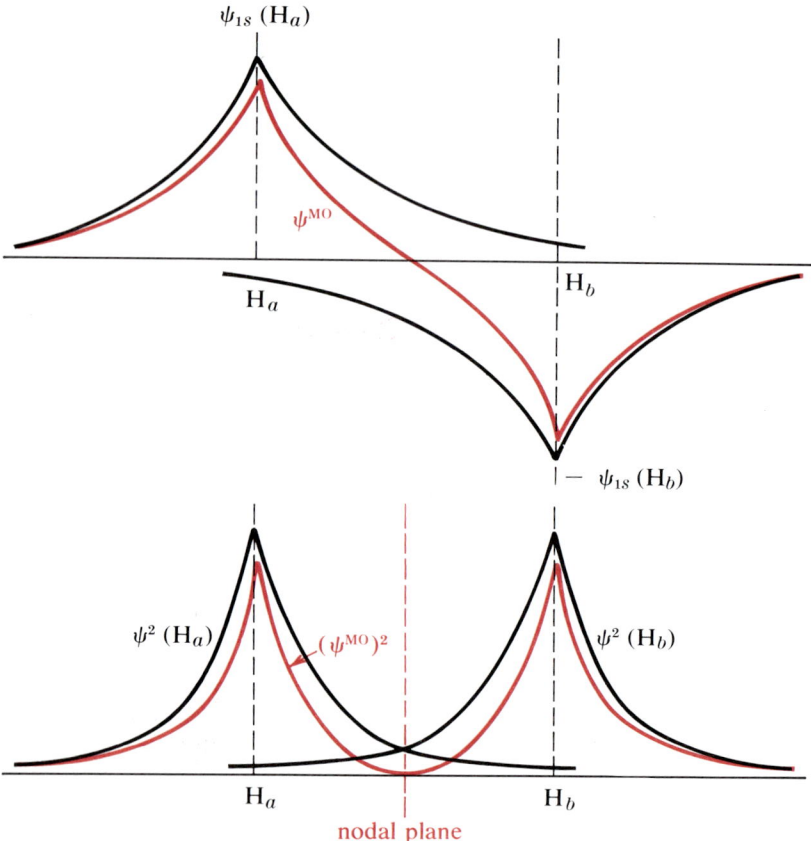

Figure 6-10 Orbital combination and the probability distribution for the H_2 antibonding MO.

energy will be most favorable when the electron is simultaneously closest to the two nuclei. This occurs, of course, when the electron is at the midpoint of the bond axis.

Bonding vs. Antibonding Combinations

This distribution function, then, represents an even less energetically stable arrangement of the electron than two AO's that do not interact. Figure 6-11 shows the three possible conditions, drawn in terms of their equal-probability contours. (This is a top view of what the previous graphs are a profile of, so to speak.) Again it is obvious that the first distribution, with both AO's having the same sign, concentrates the electron between the two nuclei in a favorable bonding position. For this reason it is called a **bonding orbital**. There is a possibility of the two AO's not combining; this situation is described as **nonbonding** although in this case it does not represent a separate MO. The last

bonding MO
(AO's combined
with same sign)

nonbonding
condition (AO's
do not combine)

antibonding MO
(AO's combined
with opposite signs)

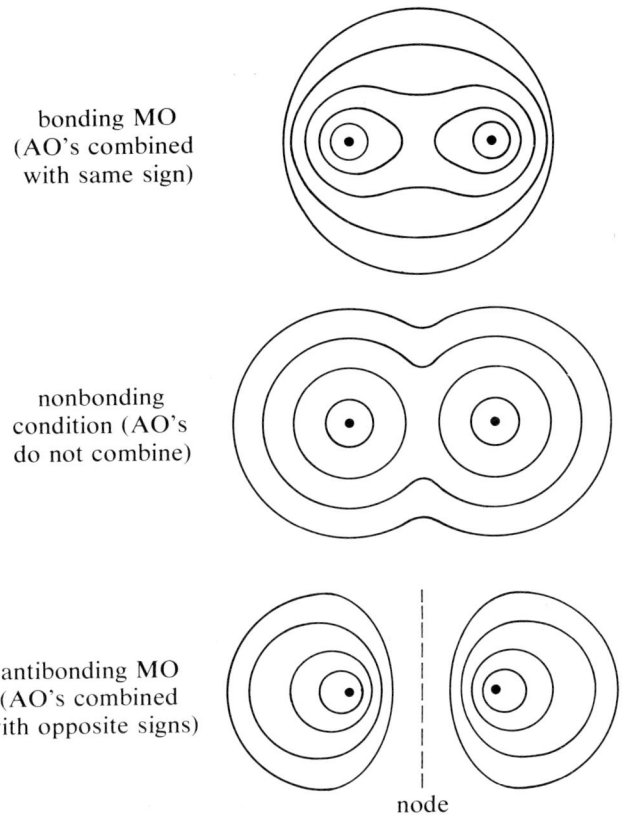

node

Figure 6-11 Electron-density maps for bonding and antibonding MO's and for the nonbonding condition.

possibility actually has the electron missing from the favorable location between the two nuclei, so that the repulsion of the two nuclei is enhanced. The situation is even worse than that in which the two atoms do not interact, and this distribution is an **antibonding orbital**.

Energy-Level Diagrams

We have taken the two possible atomic distributions of electrons and produced two possible molecular distributions, of different energies. We ought to set up an energy-level diagram for the molecule that will take these different energies into account. Figure 6-12 shows such a diagram for the H_2 molecule. As for atoms, the energies are sharply defined, and the agreement with the uncertainty principle comes in the statistical averaging of the electrons' position. The bonding in the molecule comes from the fact that, just as in atoms, the electrons will occupy the most stable distributions allowed by the

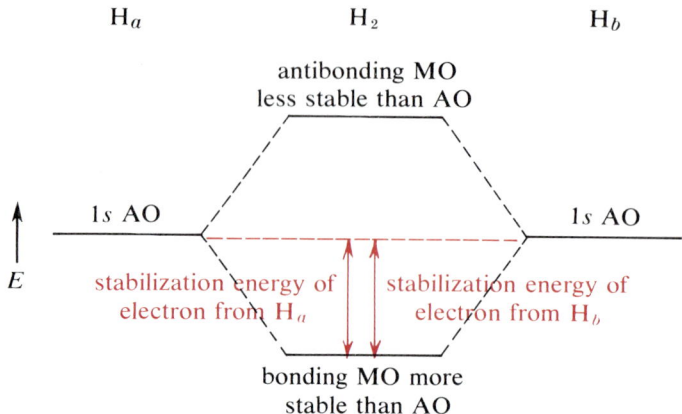

Figure 6-12 Energy-level diagram for the combination of H 1s AO's into bonding and antibonding MO's.

exclusion principle. H_2 has two electrons, and both can occupy the same spatial distribution provided their spin quantum numbers are different, that is, that their spins are **paired**. So both electrons will distribute themselves according to the bonding MO, and the molecule actually will be held together by a bond energy that is equal, to a first approximation, to the sum of the energies by which each is individually stabilized in going from the AO to the bonding MO, as shown in Fig. 6-12. If there were another valence electron to be accounted for, it would have to adopt the antibonding distribution (according to the exclusion principle), and this would be unfavorable. If there were two more electrons, both the bonding and antibonding orbitals would be completely filled, and there would be no net bonding effect; the atoms would bounce off each other. This is why bonding is so commonly described in terms of electron pairs — two electrons can both occupy a bonding distribution.

We are also in a position now to appreciate why not every pair of atoms that comes together bonds chemically. Helium atoms also have only 1s valence electrons. If we bring two He atoms together with the idea that they may form a molecule, the MO's that will be formed are presumably exactly like the H_2 MO's except for some difference in size and energy due to the larger Z_{eff} for the He nucleus. We would expect, then, to get a bonding and an antibonding orbital, just as in the H_2 case. But because He has two valence electrons, the hypothetical He_2 molecule would have four electrons to be distributed in the two orbitals, and the two in the antibonding MO would just cancel any favorable bonding effect due to the two electrons in the bonding MO. Thus it should not be possible to form a He_2 molecule (and this is entirely in agreement with our experimental experience). When two He atoms collide in the gas phase, then, they will just bounce apart, which is the process we described in Chapter 3.

They cannot combine into a molecule because the exclusion principle prevents their electrons from interpenetrating—from A's electrons being described also by B's atomic orbitals. Figure 6-13 shows the energy-level diagram and demonstrates the absence of any net bonding energy. Although we have not yet demonstrated it, this is exactly why gaseous molecules in general bounce off each other in collisions rather than reacting with each other to produce some new species. And conversely, it at least begins to suggest the way a chemical reaction goes on between two atoms or molecules in the gas phase. When the exclusion principle does *not* prohibit interpenetration of the electron-distribution functions, the approach of an electron to a positive nucleus becomes progressively more favorable as the atoms close in on each other.

Figure 6-13 Energy-level diagram for the hypothetical He₂ molecule, showing the absence of bonding.

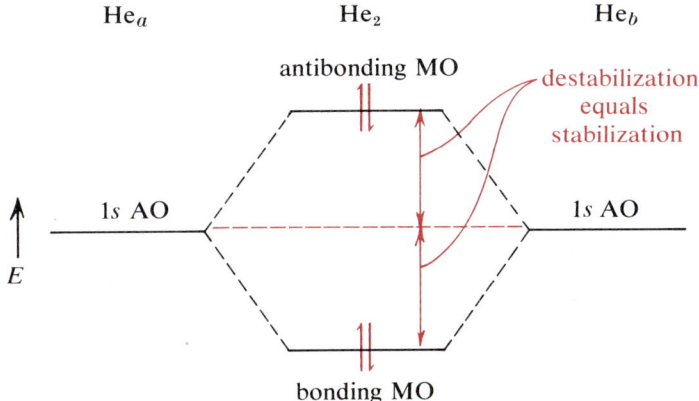

6-5 Diatomic Molecules—The Algebra of Electron Energies

Normalization and the Overlap Integral

Let us take a closer look at these MO's. We can write them algebraically as

$$\psi_{MO}^b = N^b[\psi_{1s}(H_a) + \psi_{1s}(H_b)]$$

$$\psi_{MO}^* = N^*[\psi_{1s}(H_a) - \psi_{1s}(H_b)]$$

where the superscript b refers to the bonding orbital and the asterisk identifies the antibonding orbital. The N constants are normalization constants, since it is true for MO's as for AO's that the total probability of finding the electron distributed according to the MO somewhere in space must be exactly 1.00. In algebraic terms, the normalization condition is

$$\int_{\text{all space}} \psi^2 \, d\tau = 1$$

where $d\tau$ is the volume element of space; since space is three-dimensional, we would have to integrate separately over all three coordinate differentials. Let us substitute the algebraic expression for ψ^b into the normalization condition and find N^b.

$$\int_{\text{all space}} (\psi^b)^2 \, d\tau = 1$$

$$\int_{\text{all space}} \{N^b[\psi_{1s}(H_a) + \psi_{1s}(H_b)]\}^2 \, d\tau = 1$$

$$(N^b)^2 \left[\int_{\text{all space}} \psi_{1s}^2(H_a) \, d\tau + 2 \int_{\text{all space}} \psi_{1s}(H_a)\psi_{1s}(H_b) \, d\tau + \int_{\text{all space}} \psi_{1s}^2(H_b) \, d\tau \right] = 1$$

But we know that we are working with normalized AO's, so the first and third integrals, which are just the normalization integrals for the individual atomic orbitals, each equal 1.00. This gives us

$$(N^b)^2 \left[1 + 2 \int_{\text{all space}} \psi_{1s}(H_a)\psi_{1s}(H_b) \, d\tau + 1 \right] = 1$$

The remaining integral expresses the extent to which the product of $\psi_{1s}(H_a)$ and $\psi_{1s}(H_b)$ is not zero over all of space. In other words, it expresses the extent to which the two AO's overlap and is called the **overlap integral**. If we take the Slater orbitals, multiply two together, separated by the bond distance, and integrate the product function on a computer, we can get a numerical value for the overlap integral. The value will always be between -1 and $+1$, since a perfect superposition of two orbitals would give $+1$, the same situation with opposite signs would give -1, and complete separation would give 0. Usually, for an ordinary bond such as the one in H_2, the overlap is about 0.2–0.3 (although in H_2, because there are no inner electrons to repel the approach of the bonding electron, the overlap of $1s$ AO's at the experimental bond distance is about 0.6). Knowing about what the degree of overlap is likely to be, we can symbolize the overlap integral by S and use it algebraically in our MO normalization expression:

$$(N^b)^2(1 + 2S + 1) = 1$$

$$(N^b)^2 = \frac{1}{2 + 2S}$$

$$N^b = \sqrt{\frac{1}{2(1+S)}}$$

We see that N^b can be determined if the original AO's are normalized. It is usually a decent approximation to ignore the relatively small value of S in the normalization constant and simply call the constant $\sqrt{1/2}$. If we go through the same process for the antibonding MO, N^* turns out to be $\sqrt{1/2(1-S)}$, or if we ignore S, $\sqrt{1/2}$. So the normalized MO's are

$$\psi_{MO}^b = \frac{1}{\sqrt{2}}[\psi_{1s}(H_a) + \psi_{1s}(H_b)]$$

and

$$\psi_{MO}^* = \frac{1}{\sqrt{2}}[\psi_{1s}(H_a) - \psi_{1s}(H_b)]$$

Molecular Electron Energies

How can we treat the energies of the MO's algebraically? We have assumed LCAO approximate MO's would be acceptable solutions to the Schrödinger equation for the molecule. The Schrödinger equation contains the energy, E, explicitly, as another look at Eq. 5-8 will show. If we rewrite the Schrödinger equation for the molecule and rearrange it so the energy is on the right side of the equation, we can summarize the various terms on the left as a single symbol $\mathscr{H}\psi^b$, in which ψ^b is a possible one-electron distribution corresponding to the bonding MO. We can write a shorthand form of the Schrödinger equation as

$$\mathscr{H}\psi^b = E\psi^b$$

We cannot just divide out ψ^b from this equation to get the energy, because $\mathscr{H}\psi^b$ is not a product but a single quantity in which ψ^b is being manipulated, but we can do something just as effective. If we multiply both sides of the above form of the Schrödinger equation by ψ^b and integrate over all of space we get

$$\int_{\text{all space}} \psi^b \mathscr{H} \psi^b \, d\tau = E \int_{\text{all space}} (\psi^b)^2 \, d\tau$$

ψ^b is normalized, so

$$\int_{\text{all space}} \psi^{b2} \, d\tau = 1$$

and

$$\int_{\text{all space}} \psi^b \mathscr{H} \psi^b \, d\tau = E(1) = E^b \quad (6\text{-}2)$$

If we substitute the algebraic form of ψ^b in this expression, we have

$$E^b = \int_{\text{all space}} \left\{ \frac{1}{\sqrt{2}} [\psi_{1s}(H_a) + \psi_{1s}(H_b)] \right\} \mathscr{H} \left\{ \frac{1}{\sqrt{2}} [\psi_{1s}(H_a) + \psi_{1s}(H_b)] \right\} d\tau$$

or

$$E^b = \tfrac{1}{2} \int_{\text{all space}} \psi_{1s}(H_a) \mathscr{H} \psi_{1s}(H_a) \, d\tau + \tfrac{1}{2} \int_{\text{all space}} \psi_{1s}(H_a) \mathscr{H} \psi_{1s}(H_b) \, d\tau$$

$$+ \tfrac{1}{2} \int_{\text{all space}} \psi_{1s}(H_b) \mathscr{H} \psi_{1s}(H_a) \, d\tau + \tfrac{1}{2} \int_{\text{all space}} \psi_{1s}(H_b) \mathscr{H} \psi_{1s}(H_b) \, d\tau$$

$$(6\text{-}3)$$

Electron Energy Integrals

The first and fourth of the integrals in Eq. 6-3, which contain only a single AO, represent simply the energy of that AO, by analogy with Eq. 6-2. This may be taken as the ionization energy of an electron distributed according to that AO in a single atom, the VOIP. This energy arises from the simple Coulomb attraction of the nucleus for the electron, so the integral is referred to as the **coulomb integral**. A numerical value for it is available to us from experimental measurements of ionization energies, so we can symbolize the whole integral by α in our algebra, remembering that α will be different for different atoms or different orbitals on the same atom.

The second and third of the integrals in Eq. 6-3 are a little more unusual. Each has *both* AO's in it, and if we were to try to draw another analogy to Eq. 6-2, we would have to say that this represented the energy of an electron experiencing the *mutual* attraction of two nuclei, since it is in two AO's at once. This, of course, is the bonding energy we have previously described graphically. We usually symbolize the integral for this mutual attraction energy as β and refer to it as a **resonance integral**. Using all this shorthand, then, Eq. 6-3 reduces to

$$E^b = \tfrac{1}{2}\alpha + \tfrac{1}{2}\beta + \tfrac{1}{2}\beta + \tfrac{1}{2}\alpha = \alpha + \beta$$

So the stabilization (negative) energy of an electron distributed according to the bonding MO in H_2, relative to the energy of separated nuclei and a free electron, is equal to the energy the electron would have on a free H atom *plus* some additional energy due to its mutual attraction to both nuclei. Figure 6-14 shows

Figure 6-14 Algebraic energy quantities for the H_2 energy-level diagram.

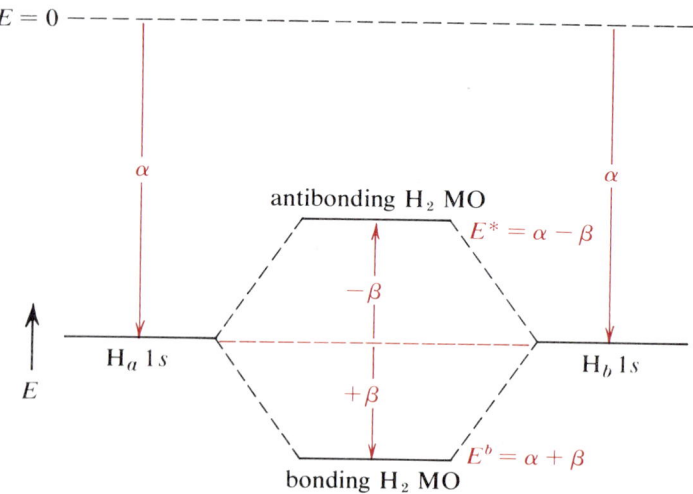

the energy-level diagram for H_2 again, but with these algebraic quantities identified.

Figure 6-14 also shows the energy of the antibonding MO. If we substitute the algebraic form of ψ^* in Eq. 6-2, the minus sign in the LCAO for the antibonding MO gives us

$$E^* = \alpha - \beta$$

It might be good practice to verify this before going any further. The energy quantities α and β both tend to bind the electron to the atom or molecule, so by the same convention we adopted previously we say that both energies are negative and $\alpha + \beta$ is a more negative number than $\alpha - \beta$.

The resonance integral that we have symbolized by β is worth investigating, since it is obviously the guts of the whole treatment of bond energies. It is extraordinarily difficult to work out in any reasonable mathematical approximation what $\int_{\text{all space}} \psi_a \mathcal{H} \psi_b \, d\tau$ should be, even if fairly simple approximate AO's are chosen (for ψ_a and ψ_b). Usually what is called a semiempirical approach is adopted, in which the valence-orbital ionization energy is taken as the proper value for that AO's Coulomb integral (as we have already indicated), and the resonance integral is estimated either from spectroscopic information or by an extension of the ionization-energy treatment. Let us look at the spectroscopic method first.

Estimation of Energy Integrals

Figure 6-15 shows the energy-level diagram for the H_2 molecule and indicates the amount of energy, 2β, that would have to be added to a molecule to **excite**

Figure 6-15 Spectroscopic method for establishing a value of β.

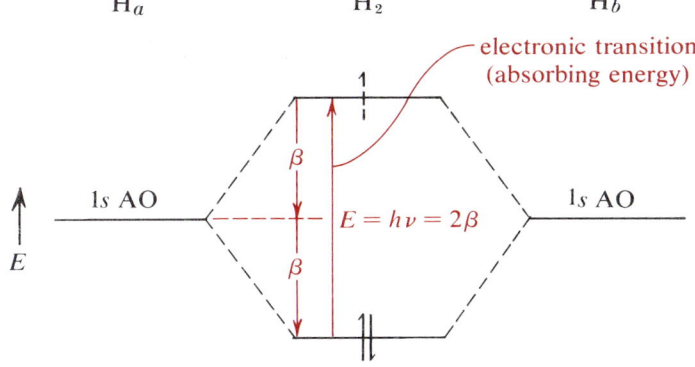

frequency is sharply defined since energy levels are sharply defined

an electron into the antibonding MO. If we shine light on the molecule we can supply it with this energy, since for a light wave $E = h\nu$. By varying ν, the frequency of the light, we can see what frequency is absorbed by the molecule. This frequency will be governed by $h\nu = 2\beta$, since the energy levels are well defined, so we can measure ν and substitute to get β. This is the general picture we have of the color of some compounds—their molecules absorb light from the visible spectrum at certain frequencies or colors, reflecting all other colors. In doing this an electron is excited from one MO to another MO of higher energy, and the energy difference between MO's is just equal to $h\nu$. Careful interpretation of molecular spectra requires us to make allowance for the repulsion energy of the electrons in the molecule in their initial and final states (which have different electron distributions, so that the net repulsion energy changes), but the general principle is the same.

The argument for the ionization-energy approach is as follows. The resonance integral β represents the mutual energy of attraction of an electron to two nuclei, and we consider that it is distributed around those two nuclei in the two orbitals ψ_a and ψ_b. Then the electron's energy of mutual attraction must be the average of the energies it would have in either AO alone; in other words, β must be proportional to the two valence-orbital ionization energies:

$$\beta \propto \frac{IE_a + IE_b}{2} \quad \text{or} \quad \beta = K\frac{IE_a + IE_b}{2}$$

But there is another factor at work. The average of the orbital ionization energies ought to apply only if the electron, no matter where it is found, is always in *both* AO's. This would be true if the two AO's were completely superimposed on each other. But they are not superimposed; they are on nuclei some distance apart and only overlap in part. In fact, if the two atoms were completely removed from each other, so that the orbitals did not overlap at all, the electron could never be found in a position described by AO's of both nuclei, and thus could never be attracted by both nuclei. In other words, the resonance integral β_{ab} must also be proportional to the overlap integral S_{ab} and we write the following expression as an approximation to the value of β:

$$\beta_{ab} \cong S_{ab}\left(\frac{IE_a + IE_b}{2}\right)$$

This is a rather crude approximate expression, but it can provide useful results, as we shall see in Study Problem 8.

6-6 Diatomic Molecules — Symmetry of Bonds

In the last section we discussed the MO model for the bonding in H_2, and have seen that the model correctly predicts that there will be no He_2 since no net bonding energy can exist. Let us move on through the periodic table to see what kinds of bonding we can predict for the lighter elements. The next eight elements after helium will have valence orbitals with $n = 2$, the $2s$ and $2p$ orbitals. We need not worry about the $n = 1$ orbital on each atom, because it is drawn in very close to the nucleus (on the average) by the large effective nuclear charge; thus it does not extend far enough from the nucleus to overlap other orbitals effectively and will not participate significantly in the bonding. We can also ignore the $n = 3$ orbitals because they represent distributions of the electron having much higher energy, so that it is very unlikely that the electron would be found in such a distribution. However, we do need to consider the $2s$ and $2p$ orbitals even for the atoms that do not have any $2p$ electrons in their most stable configuration (the **ground state**). This is because the relatively modest energy separation between the $2s$ AO and the $2p$ AO means both will be involved in establishing the molecular energy levels for the MO's. Taking the next element after He, which is lithium (Li), let us examine the LCAO procedure that describes the molecule Li_2.

Because the $2p$ AO's have directional properties relative to each nucleus, we shall first need a convenient coordinate system for the two nuclei. It is perfectly legitimate to choose the axes of the coordinate system entirely arbitrarily; physical reality cannot possibly depend on how we choose the location of these axes, so we are entitled to choose them in the most convenient way. This is only a matter of using AO's as building blocks, so to speak, and we shall "build" a description of the molecule out of these "blocks" in a convenient way, then see if it matches the experimental facts. The coordinate system we will need for the Li_2 molecule is shown in Fig. 6-16. It is usually

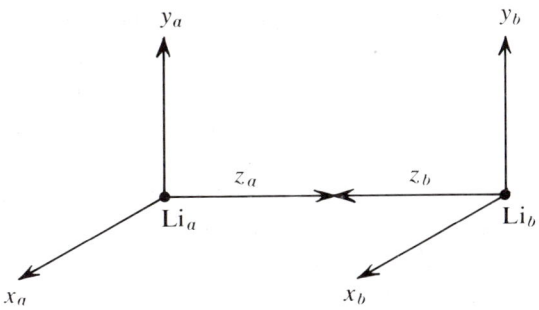

Figure 6-16 Coordinate system for orbital overlap in Li_2 (or any other diatomic molecule).

most convenient for us to have the bond axis be the z axis for each atom, and all of the coordinate systems we set up will be arranged this way insofar as possible.

Sigma Overlap of Orbitals

First let us consider the two atoms' 2s orbitals overlapping, as shown in Fig. 6-17. This situation is exactly like the overlapping of the 1s orbitals for the H_2 molecule (except for actual sizes and distances, of course), and we can

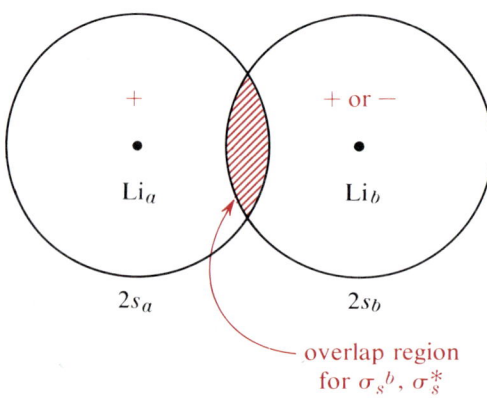

Figure 6-17 Sigma overlap of 2s AO's in Li_2.

expect to get a bonding and an antibonding MO from the combination of these. Algebraically, we write

$$\psi(\sigma_s^b) = \frac{1}{\sqrt{2}}(2s_a + 2s_b)$$

$$\psi(\sigma_s^*) = \frac{1}{\sqrt{2}}(2s_a - 2s_b)$$

In these two MO's the factor $1/\sqrt{2}$ is the normalization constant, and we have introduced some new nomenclature to simplify the notation; $2s_a$ refers to the 2s AO on lithium atom a, and $2s_b$ to the 2s AO on lithium atom b. The σ_s^b refers to the designation of the MO: the superscript b refers to the fact that the MO is a bonding orbital, the subscript s indicates that the MO is formed from s AO's, and the σ describes the symmetry of the MO to the extent that it indicates there are no nodal planes *containing the bond axis* (even though the σ_s^* has one that *cuts* the bond axis).

There are a number of other possibilities for AO overlap. One is the combination of the two atoms' $2p_z$ orbitals. As in the case of the s orbitals, this can be done in two ways, with the orbitals having the same algebraic sign where they overlap, or with the signs opposed. We write these two possible molecular wave functions as

$$\psi(\sigma_z^b) = \frac{1}{\sqrt{2}}(2p_{z_a} + 2p_{z_b})$$

$$\psi(\sigma_z^*) = \frac{1}{\sqrt{2}}(2p_{z_a} - 2p_{z_b})$$

and depict them graphically as in Fig. 6-18. Again we expect the MO with the opposed signs in the overlap area to be antibonding since it prevents the electron from ever being found in the favorable (bonding) location directly between the nuclei. In general, we expect these MO's formed from $2p$ AO's to be higher energy than the MO's formed from $2s$ AO's, because the $2p$ AO's themselves represent a higher-energy distribution than the $2s$ AO's. Notice again that the symmetry of these MO's is σ even though there are several nodal planes, because none of these nodes lie in the plane of the bond axis.

Figure 6-18 Sigma overlap of $2p_z$ AO's in Li_2.

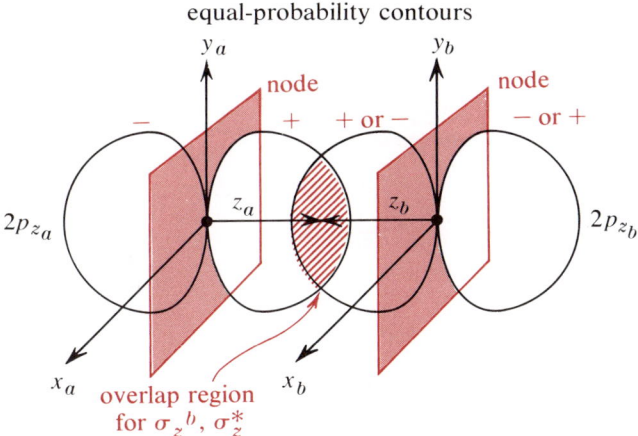

Pi Overlap

We can also allow the two atoms' $2p_x$ orbitals to overlap, or their $2p_y$ orbitals. We can do this with the overlap signs the same or opposed, and the result is a bonding and an antibonding orbital, for the same reasons as before. The algebraic description of these is the same as for the others:

$$\psi(\pi_x^b) = \frac{1}{\sqrt{2}}(2p_{x_a} + 2p_{x_b})$$

$$\psi(\pi_x^*) = \frac{1}{\sqrt{2}}(2p_{x_a} - 2p_{x_b})$$

$$\psi(\pi_y{}^b) = \frac{1}{\sqrt{2}}(2p_{y_a} + 2p_{y_b})$$

$$\psi(\pi_y^*) = \frac{1}{\sqrt{2}}(2p_{y_a} - 2p_{y_b})$$

The graphical representation is given in Fig. 6-19. Notice that the symmetry of the MO's (basically due to the symmetry of the AO's) in these cases is such as to have a nodal plane that contains the two nuclei and thus contains the bond axis. This is the reason for the π designation in the algebraic notation above. There is a very basic reason for distinguishing between σ and π symmetry, which we shall explore a little later in this section.

Figure 6-19 Pi overlapping of $2p_x$ and $2p_y$ AO's in Li_2. (a) Equal-probability contours in the xz plane. (b) Equal-probability contours in the yz plane.

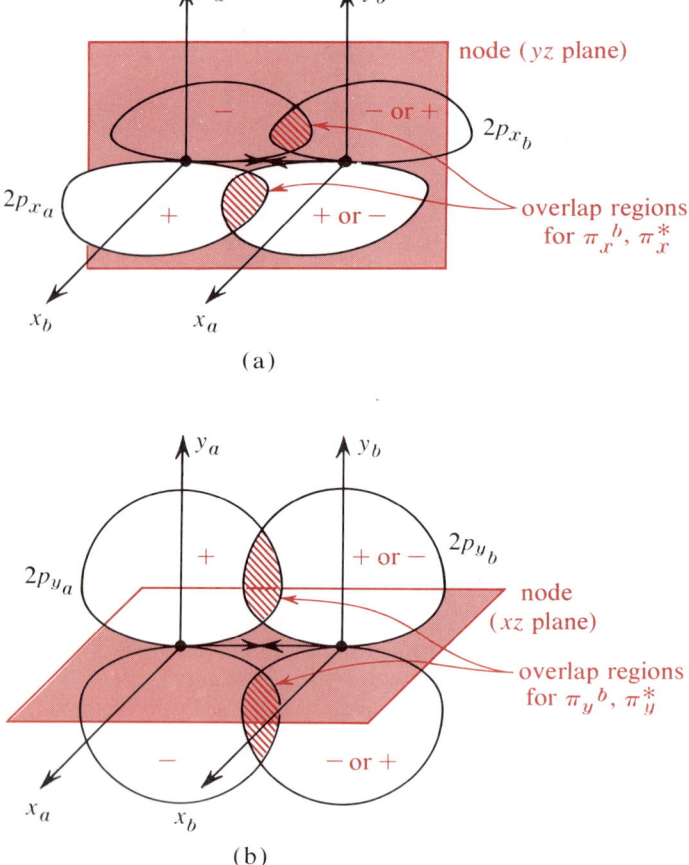

One of the characteristics of the π overlap occurring in these MO's is that it is not very good. That is, although the $2p_x$ orbitals in this LCAO are the same size as the $2p_z$ orbitals in the σ_x combination, they are overlapping sideways instead of head on. For the same internuclear separation (bond length), then, the overlap, as represented by the overlap integral S, is perhaps only half as great as for σ-type overlap. The result is that β_{ab} for π bonding or antibonding orbitals is substantially smaller than the corresponding β_{ab} for σ MO's because β_{ab} is proportional to S_{ab}. The overlap comparison is shown in Fig. 6-20. The net effect of the smaller β_{ab}, of course, is to give a smaller energy separation between the bonding and antibonding π MO's than between the bonding and antibonding σ MO's.

Figure 6-20 Comparison of (a) σ and (b) π overlap between 2p orbitals of equivalent size and internuclear separation.

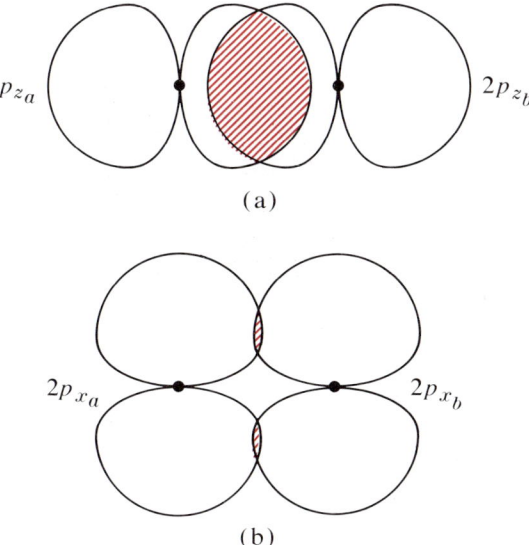

Another important feature of this π overlap is that the $2p_x$ and $2p_y$ orbitals overlap in exactly the same way—they are the same size, they are both pointed at right angles to the bond axis, and they are the same distance apart. Then we should expect that the bonding MO $\pi_x{}^b$ would have exactly the same energy as the bonding MO $\pi_y{}^b$. This is exactly analogous to the fact that the $2p_x$ and $2p_y$ AO's have exactly the same energy. When more than one orbital (whether atomic or molecular) have the same energy, the energy levels are said to be **degenerate**; in this case the energy level is doubly degenerate, since there are two orbitals with that energy. The antibonding MO's will show exactly the same condition—they are also degenerate.

The Set of *s*- and *p*-Based Diatomic Molecular Orbitals

We have said that the total number of MO's we generate must exactly equal the number of AO's with which we started. Originally each atom had four valence orbitals (one 2s and three 2p) and there are two atoms in the molecule, so we started with eight AO's. Since we have produced eight MO's this condition is satisfied. These eight are in general independent of each other—that is, orthogonal—as we can see graphically from Fig. 6-21. In this diagram, the black shading indicates an area in which the MO has a positive algebraic sign, and the red a negative sign. Since the black–black overlap (same sign = positive overlap) is equal to the black–red overlap (different sign = negative overlap) the total overlap is zero, which is the condition for orthogonality.

Figure 6-21   Orthogonality of MO's formed for Li_2. σ_s^b is orthogonal to π_y^b.

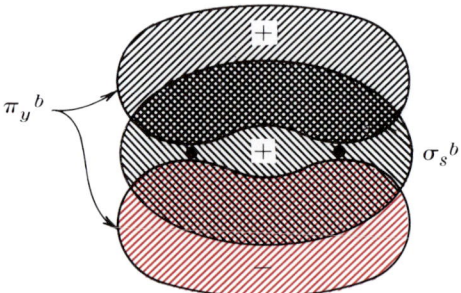

There are some further possibilities for combination. We could combine the 2s AO on one atom with the $2p_z$ on the other. This would be equivalent to combining the two MO's σ_s^b and σ_z^b, or to combining σ_s^* and σ_z^*, since in either case the first comprises the 2s AO's, the second the $2p_z$ AO's and they are not too far apart in energy. These two pairs of MO's are not orthogonal. So far, when we have combined two AO's we have always gotten a MO of lower energy than either of the AO's, and a MO of higher energy than either of the AO's; the same will be true here. That is, when this interaction is allowed for, both the σ_s^b and the σ_s^* that were originally a little lower in energy will be still lower, and the σ_z^b and σ_z^* that were originally a little higher will be still higher, as in Fig. 6-22.

Another possibility is that the 2s AO on one atom might combine with the $2p_x$ on the other atom. Figure 6-23 shows that in this case there would be no overlap at all (for the same reasons as in Fig. 6-21). Thus there is no energy effect: since $S_{ab} = 0$, $\beta_{ab} = 0$. This is the real reason for designating σ and π bonds separately; an AO with σ symmetry relative to the bond axis *cannot*

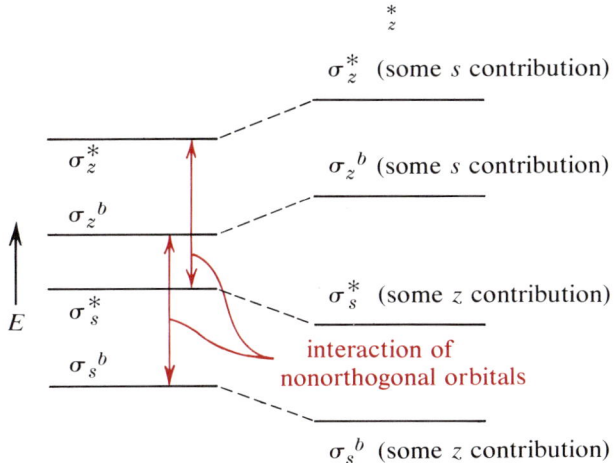

Figure 6-22 Further interaction of orbitals having the same overall symmetry.

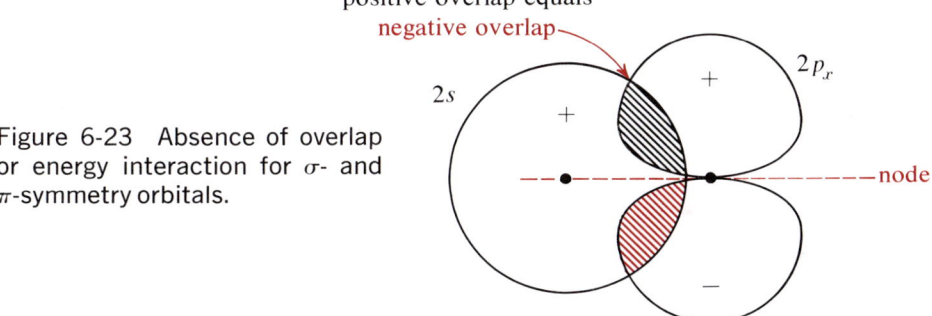

Figure 6-23 Absence of overlap or energy interaction for σ- and π-symmetry orbitals.

combine with another that has π symmetry relative to the same axis, so they cannot contribute to chemical bonding in a molecule. Following this same reasoning, the interaction shown in Fig. 6-22 for the σ_s^* and σ_z^b orbitals cannot occur for the $\pi_{x,y}^b$ orbitals, since there are no other MO's with π symmetry (except the $\pi_{x,y}^*$ already considered).

Diatomic Molecular-Orbital Energies

We are now in a position to draw an energy-level diagram for Li_2, although up to this point it should be realized that we have no explicit account of any of the features of the atomic structure of Li. Figure 6-24a is the appropriate diagram, with all of the MO energies displayed that we have just discussed. It will be worthwhile to compare the energy-level positions here with the previous discussion.

282 | Molecular Structure

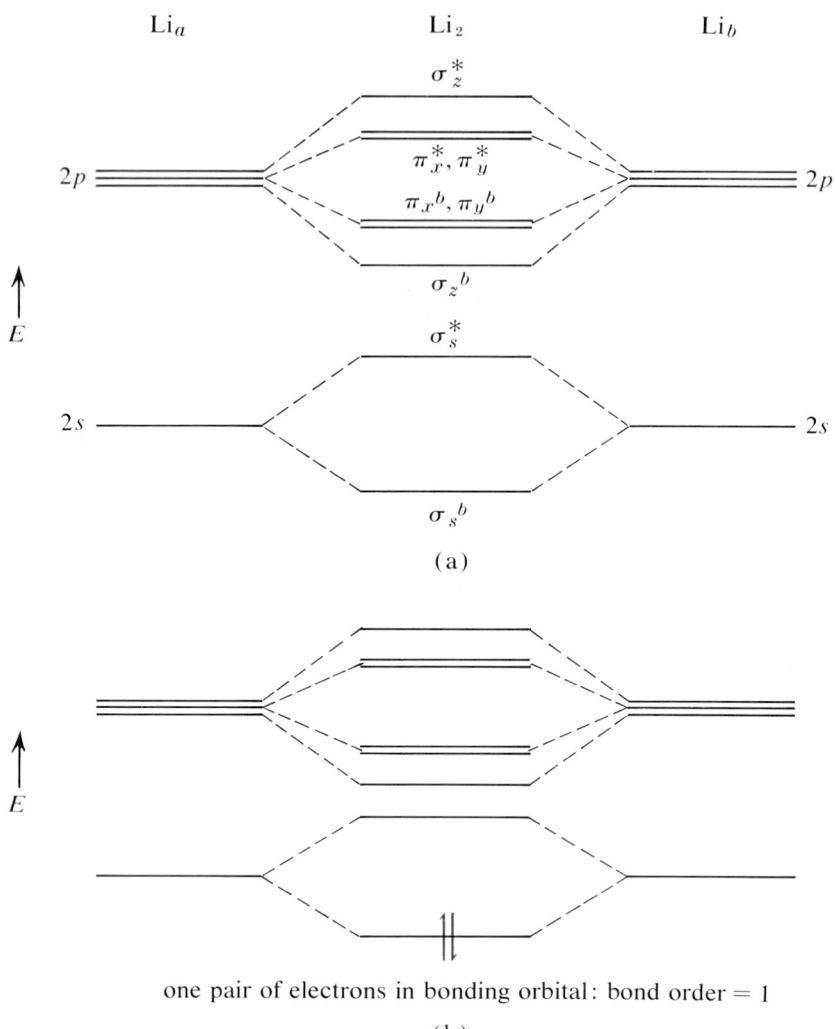

Figure 6-24 Energy-level diagrams for Li_2; (a) individual MO identification; (b) populated with Li_2 valence electrons.

Li_2 For the molecule Li_2 the diagram can be made meaningful by populating the orbitals with the valence electrons of the system, following the exclusion principle; this is done in Fig. 6-24b. Each Li atom has one valence electron (2s), so the system has two electrons that can both occupy the σ_s^b MO if their spins are opposed. Since there is a pair of electrons in a bonding orbital and no electrons in any antibonding orbital, and since we are accustomed to regarding an electron pair as a bond, we say that there is one bond in Li_2. Alternatively, we may say that in Li_2 the **bond order** is 1. The upshot of all this is that we can predict that Li$_2$ should be a stable molecule. This is in fact true; if metallic lithium is boiled, the vapor contains some atoms but also some diatomic Li$_2$ molecules, as we can learn from vapor-density measurements.

Be$_2$ Suppose we consider Be$_2$, the diatomic molecule that we could presumably form from the next element in the periodic table. Since it also has $2s$ and $2p$ valence orbitals, the whole previous discussion applies to this molecule as well; nothing is changed except the AO energies (which can be obtained from the periodic table inside the front cover) and the number of electrons in the diatomic system. Since Be has two valence electrons ($2s$), the system has four electrons. Putting these into the energy levels of Fig. 6-24a, we have the electron configuration shown in Fig. 6-25. There are two bonding electrons and two antibonding electrons. There are thus no net bonds, and we do not expect the molecule to stay together. Again, this is true. The molecule Be$_2$ has never been observed. So our model is holding up well; of the four molecules we have examined, two exist and two do not, and we have predicted this.

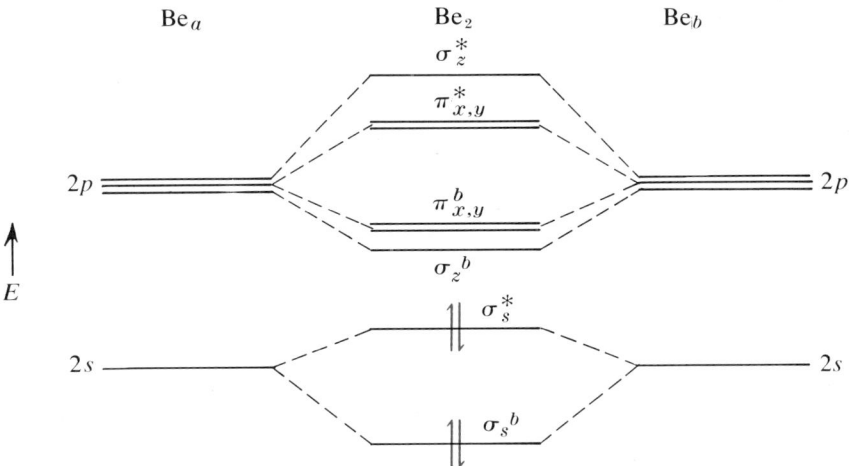

Figure 6-25 Energy-level diagram for the hypothetical Be$_2$ molecule.

B$_2$, C$_2$, N$_2$, O$_2$, and F$_2$ Proceeding across the first row of the periodic table we can inquire into the possibility of forming B$_2$, C$_2$, N$_2$, O$_2$, F$_2$, and Ne$_2$. All of these molecules would consist of atoms whose valence orbitals are $2s$ and $2p$; as for the case of Be$_2$, we expect that the same MO's will be formed as for Li$_2$ and that the energy-level diagram will differ only in the exact energies of the AO's and the resulting MO's. Figure 6-26 shows the energy-level diagrams for

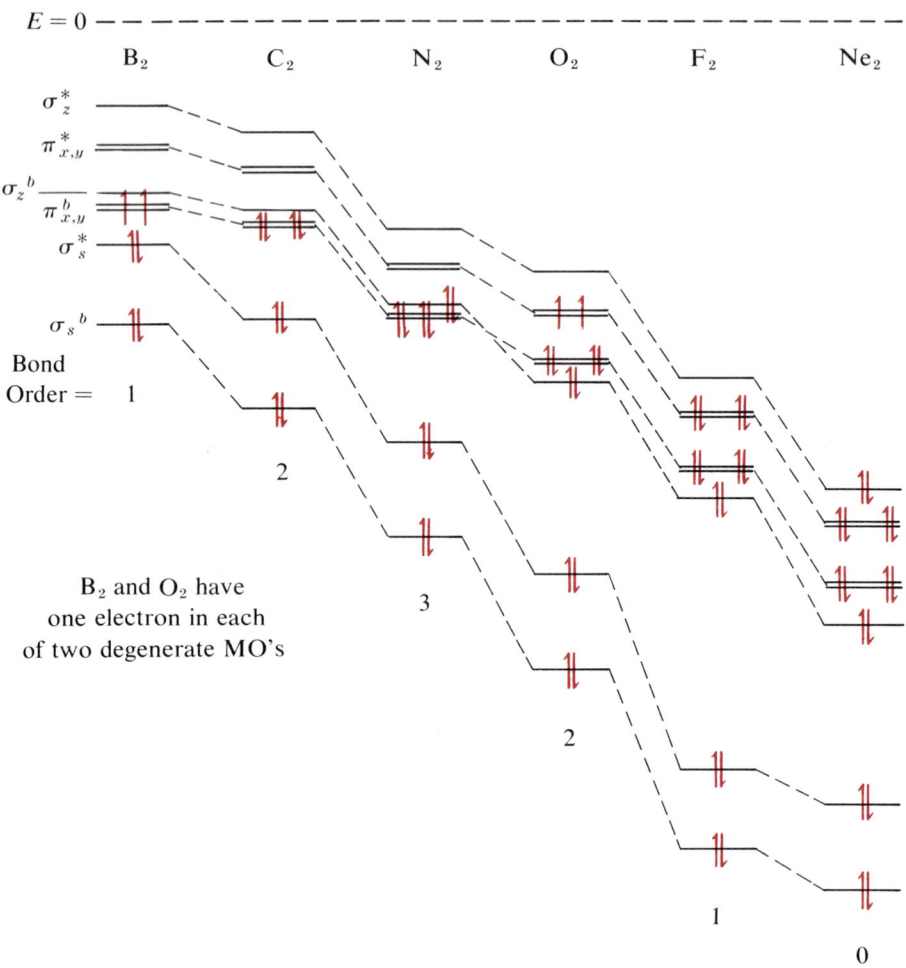

Figure 6-26 Energy-level diagrams for first-row diatomic molecules (and the hypothetical Ne_2).

these cases, all drawn to the same energy scale (although the values of β_{ab} are only approximate). The valence-orbital ionization energies for the AO's are taken from the values tabulated inside the front cover. Notice that, as we mentioned in Chapter 5, these ionization energies increase sharply (the orbitals become more stable) in moving to the right across any row of the periodic table. The interaction between the σ_s and σ_z MO's decreases going to higher atomic number because the energy gap between $2s$ and $2p$ is increasing, and the larger the energy difference the less the effect on the energy due to the interaction. It is also worth noticing that, although the F_2 electrons are more stable than the Li_2 electrons, the difference in molecular-electron stabilities is not nearly as great as the difference in the stabilities of the AO's themselves. In other words, forming these diatomic molecules has reduced the atoms' chemical reactivity; the potential electron donors have more stable electrons than they used to and the potential acceptors do not have such stable vacant orbitals as they used to.

Molecular Magnetism for Diatomic Compounds

All of the molecules whose energy levels are formulated in Fig. 6-26 are known except Ne_2, which is just the one the bond orders indicate as being nonbonded and presumably unstable. We can get a further verification of the experimental validity of our model by looking at the two molecules B_2 and O_2, which by virtue of their electronic structures have degenerate MO's only half occupied. Here Hund's rule takes over again (see p. 242) so that one electron goes into each of the degenerate orbitals. This does not affect the overall bond order of the molecule, but it does leave these two molecules, unlike all the others, with a net electron spin. A spinning or rotating electrical charge generates a magnetic field; remember that an electromagnet is made by passing a current of electrical charge through a coil, which in effect causes the charge to rotate. So the two molecules B_2 and O_2 should be attracted by an external magnetic field—behavior that is characteristic of unpaired electrons. Such compounds are said to be **paramagnetic**. On the other hand, all other compounds have all the electrons paired and should not be attracted by an external magnetic field at all. In fact, an external magnetic field will cause the electron pairs to circulate so as to produce a small opposed field; thus these compounds are actually slightly repelled by a magnetic field. This is the behavior of **diamagnetic** compounds. The distinction between these two types of magnetic behavior is a rather critical test of bonding theories, because the tendency of electrons to pair spins is so pronounced that it is difficult to explain exceptions unless by some argument very like this. The experimental fact is that of all the compounds we have dealt with so far, only B_2 and O_2 are paramagnetic.

Molecular-Orbital Contours

As a final step in considering the similarities and differences in the bonding of the molecules we have looked at so far, a close study of Fig. 6-27 may be helpful. This figure is a series of computer-generated equal-probability contours for this series of molecules, adapted from some work by A. C. Wahl of Argonne Laboratories [*Science* **151**, 961 (1966)]. All of the diatomic molecules we have thus far considered are shown except He_2, Be_2, and Ne_2, which show no net bonding and so do not form. For each of the molecules the contours of the electron-distribution function are shown for all of the MO's that are occupied in the ground state. The outermost contour in each case represents an electron

Figure 6-27 Computer-generated MO contours for H_2, Li_2, B_2, C_2, N_2, O_2, and F_2. (Reprinted from Wahl, A. C., *Science*, Vol. 151, pp. 961–967, February 25, 1966. Copyright 1966 by the American Association for the Advancement of Science.)

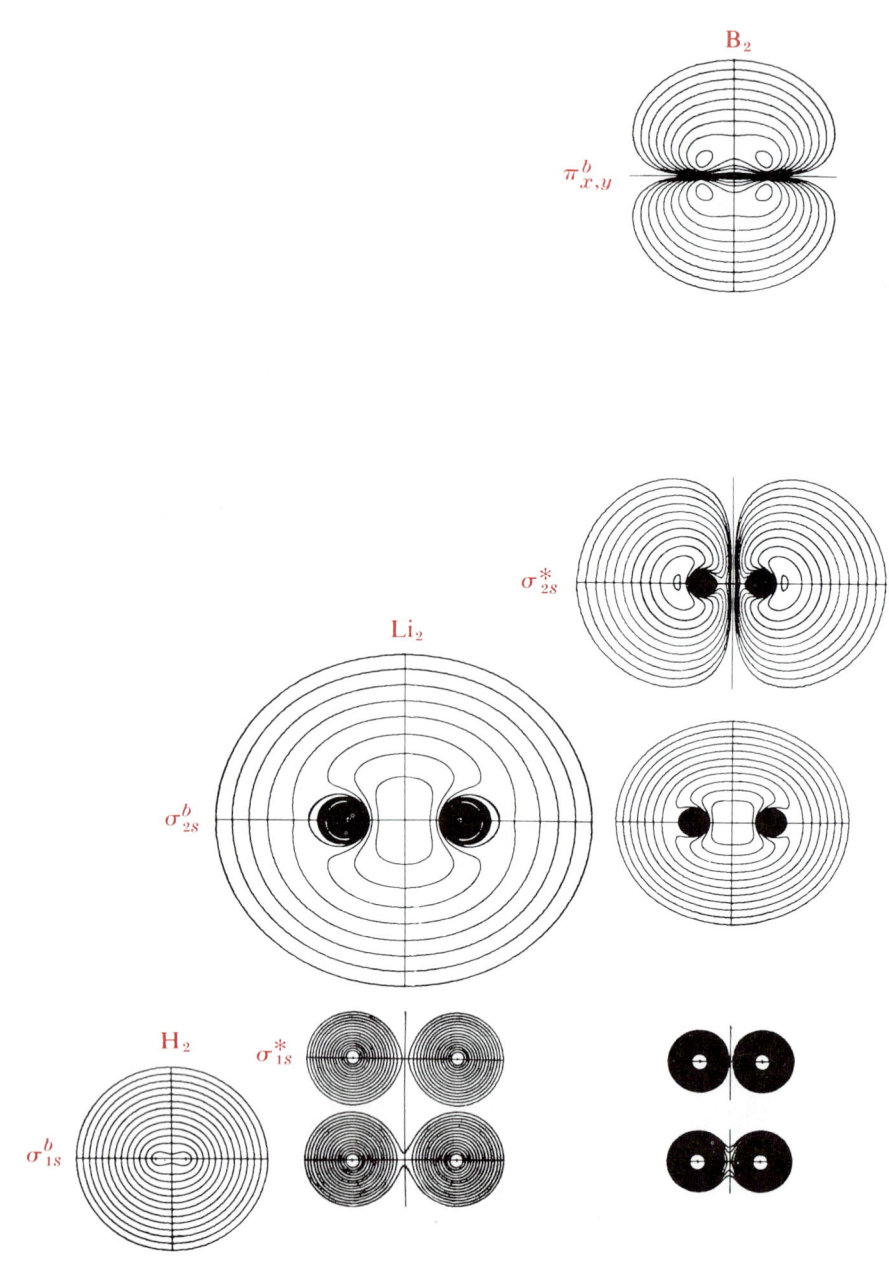

Diatomic Molecules — Symmetry of Bonds | 287

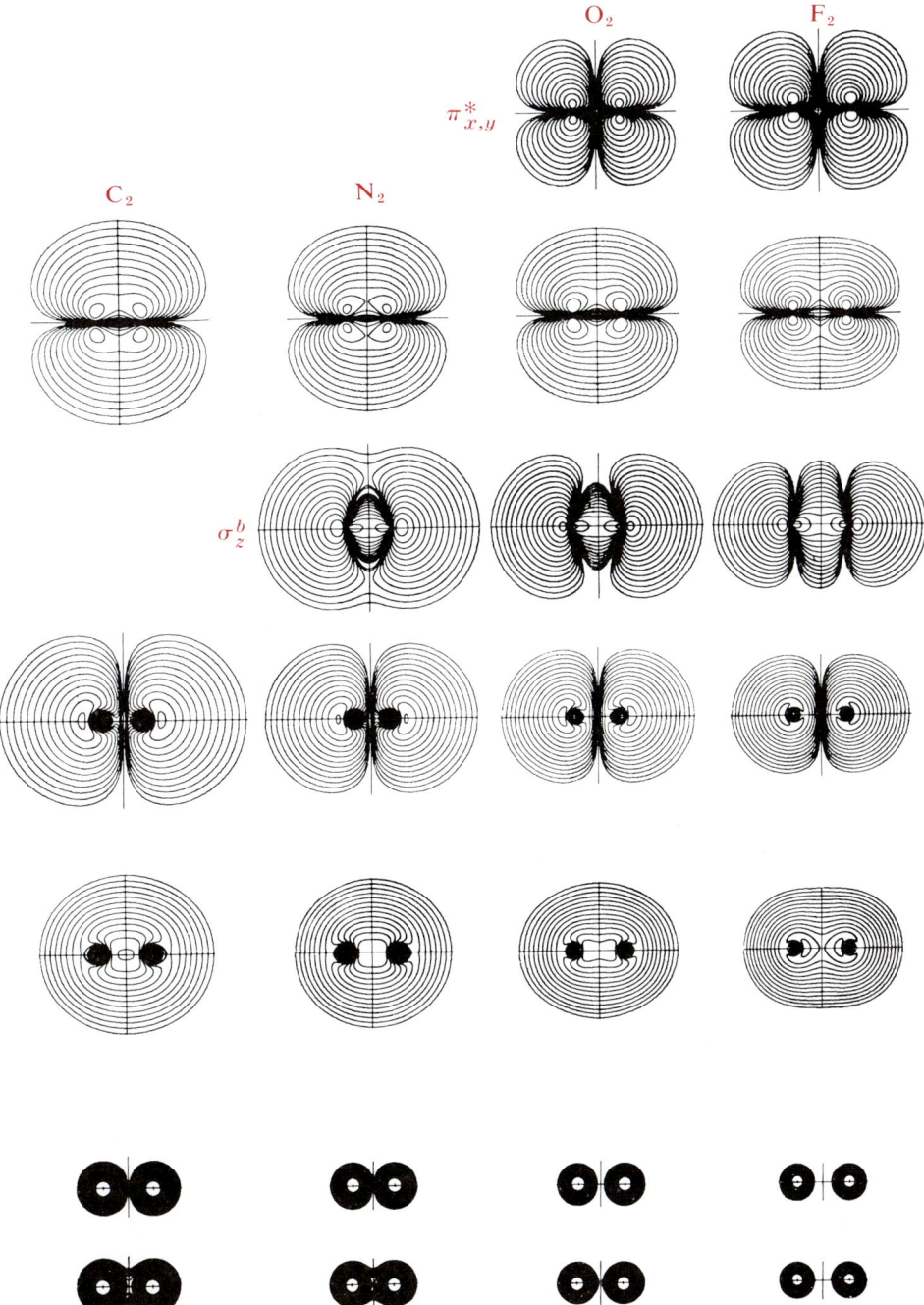

"probability density" of approximately 4.2×10^{-4} electron/Å^3, and each inner one represents a successive doubling of the "density" represented by the contour outside it. The overall effect, then, is like that of a topographic map. All the contours are plotted to the same scale, and one of the interesting observations that is immediately apparent is that the size of the same MO, say σ_s^b, shrinks rapidly as Z_{eff} increases. The consistency of this effect with the energy levels of Fig. 6-26 should be considered. Another point to notice is that the first two MO's in each case, which are the LCAO's for the 1s (i.e., nonvalence) electrons, are hardly affected at all by the combination process—they still look very much like the spherical 1s AO's. This is a clear illustration of the statement that the energy effect of bonding is very small if the overlap is poor. We seem to be justified in ignoring the 1s contribution to bonding (except, of course, for H_2).

6-7 Diatomic Molecules—Heteronuclear Systems

Having acquired some feel for the symmetry of the various σ and π MO's that can be formed in a diatomic system, we need to look at the more general case of molecules formed from unlike nuclei—**heteronuclear** diatomics. Although we have not covered more than a fraction of the periodic table, it should be clear that we have covered the general principles of bonding for any system using only s and p AO's for the linear combinations; bonding involving d AO's is exactly analogous but slightly more complicated geometrically, and we shall come to it somewhat later (Chapter 16).

The basic difference between homonuclear and heteronuclear molecules as far as our bonding theory is concerned is that when the atoms are identical we know they must contribute equally to the LCAO's, but when they are different we cannot be sure. That is, in the general equation we developed previously,

$$\psi^{\text{MO}} = c_1 \psi_1^{\text{AO}} + c_2 \psi_2^{\text{AO}}$$

the constants c_1 and c_2 must be the same for equivalent orbitals of two identical atoms, but for two different atoms they will, in general, *not* be the same. This is a result of the electron's search for the lowest possible energy. If two AO's of different energy combine to form a bonding MO, the electron distribution it describes will be more closely related to the lower-energy AO than to the higher-energy one; in an antibonding orbital, the reverse will be true. Suppose we look at a case that represents only a slight difference in valence-orbital ionization energies between the two atoms, and follow it up with another case involving a large difference.

Diatomic Molecular Orbitals Formed from Atomic Orbitals with Similar Energies

Carbon monoxide, CO, is a good case involving little difference in these energies. Figure 6-28 shows the energy-level diagram for CO, plotted in just the same way as those in Figs. 6-24, 6-25, and 6-26, but with the valence-orbital energies taken from the inside front cover. It is apparent that the same orbitals are overlapping with the same kinds of symmetry and that, in fact, the MO energy-level diagram is very much like that for homonuclear diatomics. The

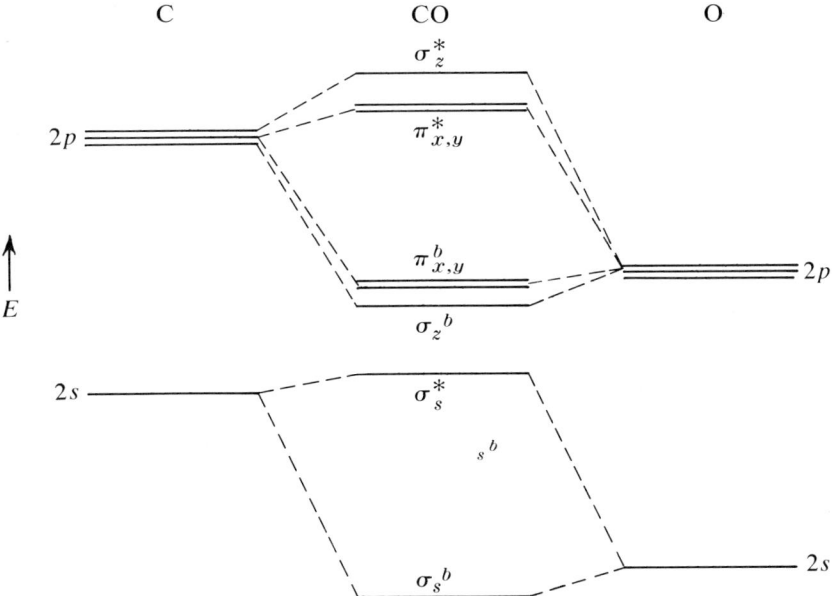

Figure 6-28 Energy-level diagram for CO (diatomic molecule with relatively close VOIP's on the two atoms).

difference lies in the fact that, for instance, the σ_s^b MO is now somewhat closer in energy to the O $2s$ energy than to the C $2s$ energy, so we can correctly infer that the O AO is more important in determining the electron distribution in the MO. Put another way, if the two electrons in the σ_s^b MO originally came one from each atom, then in this particular MO there will have been some charge transfer from the C to the O, since both electrons spend more than half their time near the O. On the other hand, the electrons in the σ_s^* (antibonding) orbital have an energy closer to that of the C $2s$ than to the O $2s$, so they will be distributed somewhat more according to the C $2s$ than according to the O $2s$.

Figure 6-29 shows the comparison between these two heteronuclear MO's and the corresponding homonuclear MO's; the differences are fairly subtle but worth noticing.

Figure 6-29 Comparison of homonuclear and heteronuclear electron-probability distributions.

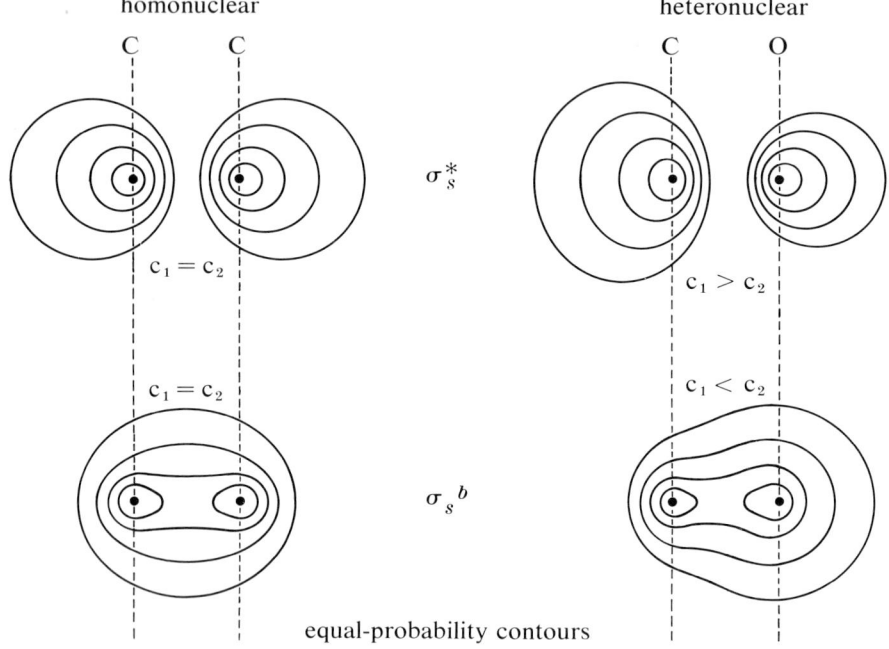

Diatomic Molecular Orbitals Formed from Atomic Orbitals with Quite Different Energies

Gaseous lithium fluoride (LiF) is a good example of a heteronuclear diatomic molecule having a large difference between the AO energies. Figure 6-30 shows the energy-level diagram for this case, with the valence-orbital energies taken again from the inside front cover. Note that the same orbitals are overlapping but that the resulting MO's now fall in a different order of energies. The reason is that when two AO's combine to form two MO's, one is always lower in energy than either AO, and the other is always higher in energy than either AO. So when the Li $2s$ turns out to be much higher in energy than any of the F valence orbitals, the antibonding MO formed from the $2s$ AO's is higher than any of the bonding MO's formed from the $2p$ AO's (since they must all be lower in energy than the F $2p$ AO's). The eight valence electrons in LiF (one from Li, seven from F) exactly fill all the bonding orbitals, which are very

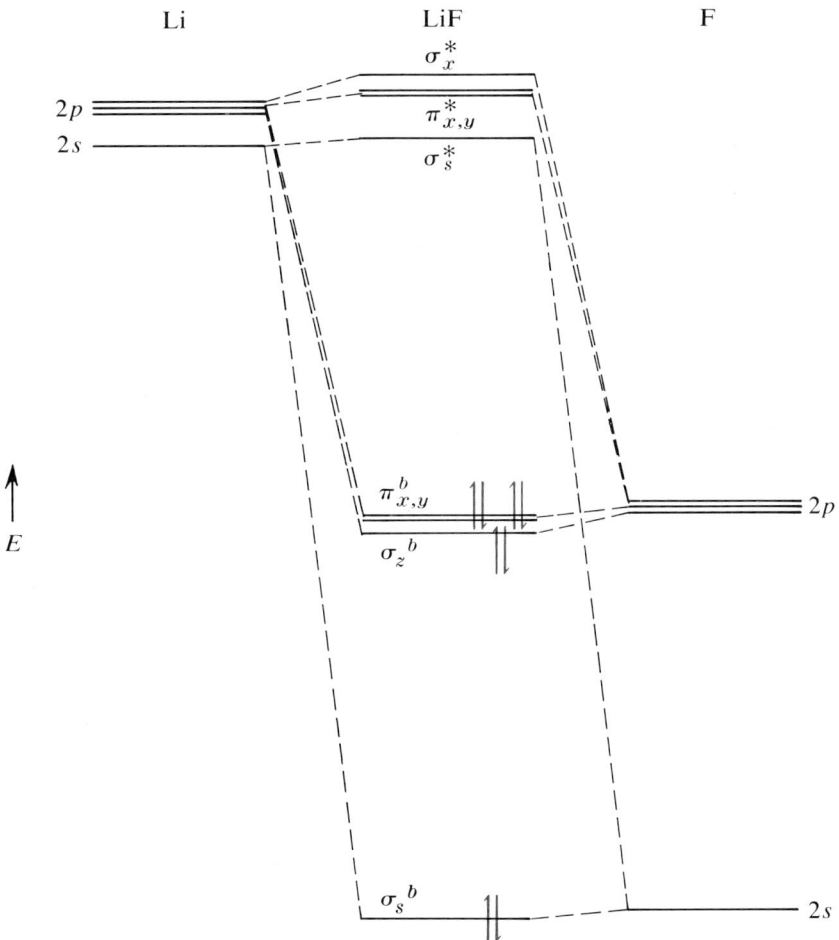

Figure 6-30 Energy-level diagram for LiF gaseous molecule (or any other diatomic molecule with widely separated VOIP's).

stable (low energy) and there is a large energy gap up to the lowest vacant orbital. So it will not be energetically favorable for LiF vapor either to donate or to accept electrons. In comparing this case with the previous one, notice that the energy differences between the bonding MO's and the corresponding F AO's (Fig. 6-30) are much smaller than the corresponding differences for CO (Fig. 6-28). This is another example of a comment made earlier: the larger the energy difference between two combining orbitals, the smaller the bonding or antibonding energy effect will be. The large energy effect observed in forming LiF comes not so much from the covalency of the bonding electrons (as represented by the resonance integral β), but from the transfer of Li's one valence electron from the high-energy Li AO's to the very low-energy bonding MO's. Since the relative energies are what they are, these are very largely based on F, and there is a substantial degree of electron transfer from the Li to the F. The Li atom acquires a partial positive charge, the F atom a partial negative charge, and we begin to think of the bonding as being significantly **ionic**. The variable size of the bonding-energy effect points up one additional important fact: by our formal definition of bond order, there are four bonds in LiF, but in fact the covalent-energy effect is rather small. The bonding effect of a pair of electrons is influenced by many factors and cannot be simply and completely defined.

Figure 6-31 illustrates another aspect of the definition of bond energy. Bond energy is necessarily defined as a difference; it is the energy difference between two bonded atoms and two separate atoms. But what are the accepted states of the separate atoms? The energy difference between LiF and $Li^0 + F^0$ is quite different from the energy difference between LiF and $Li^+ + F^-$. In general, bond energies are tabulated in terms of the electrically neutral atoms, and all

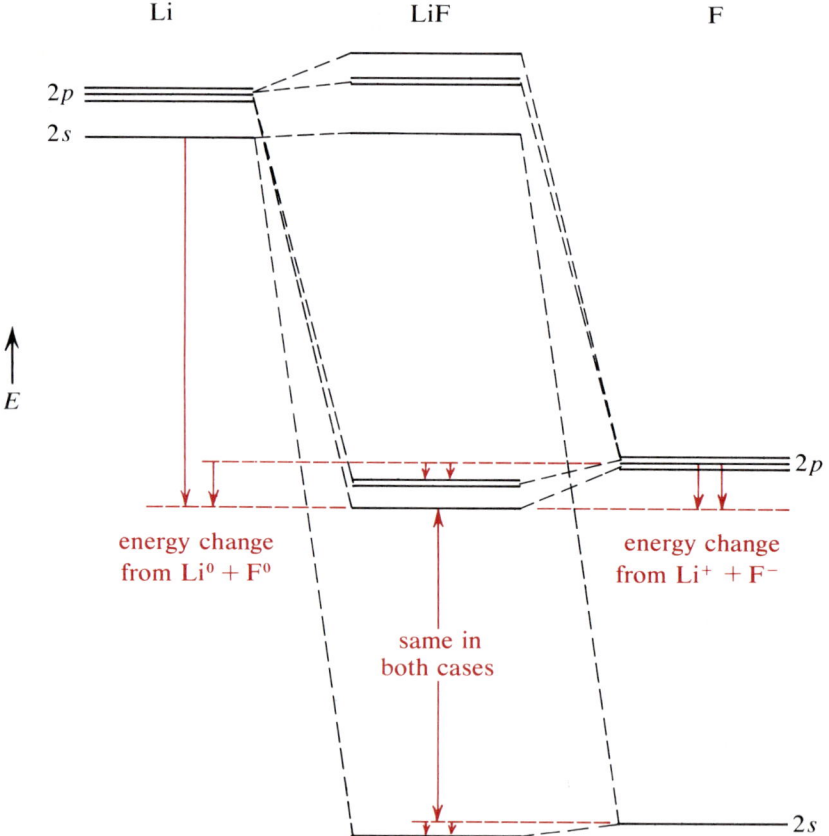

Figure 6-31 Comparison of bond energies for two different states of separated atoms.

the bond energies used in this book will be of this sort unless they are otherwise specified. However, it is important to realize that an energy expressed as a separation into ions may be more important for some particular case, since it may be a more faithful model of what really happens. We can also run into trouble of another sort in dealing with the breaking of successive bonds in polyatomic molecules; we shall return to this in Section 6-9.

6-8 Diatomic Molecules — Ionicity, Polarization, and Hydrogen Bonding

The Extent of Electron Transfer

We have just introduced without definition the very important concept of ionicity. All of our previous discussion has been based on the idea that the bonding between two atoms in a molecule is entirely **covalent**, that is, that the electrons doing the bonding are equally shared by the two atoms. As we have seen for LiF, however, the electrons doing the bonding in heteronuclear systems can be substantially transferred to one atom; this will be particularly important if the energies of the valence atomic orbitals are quite different for the two atoms. When this transfer from A to B has occurred to a substantial extent, it has obviously converted A into a species with a very significant positive charge, perhaps as large as 1.5+, and at the same time has converted B into a species with a negative charge, perhaps as large as 1.0−. For a diatomic system the positive charge is always equal to the negative charge, since both arise from the same electron-transfer process. The oppositely charged ions attract each other, following Coulomb's law, and the attraction contributes to the bond energy holding the atoms together. To the extent that the neutral atoms have been converted into charged ions in this way, we speak of the bonding as being ionic.

With the sole exception of the ionic crystals we mentioned in Chapter 4, which we shall discuss in Chapter 13, the charges on atoms (partial ions) in stable compounds are always small, simply because the range of chemical energies is not great enough to make substantial charge separation the most energetically favorable way to distribute electrons. Looked at another way, a large electrical charge on an atom causes such a strong attraction (or repulsion) on surrounding electrons that they will redistribute themselves to minimize the charge, causing changes in the chemical bonding in the process. Any hypothetical molecule with large charges on the atoms, then, can be expected to be very chemically reactive as these charges seek to neutralize themselves. This general idea is known as Pauling's **electroneutrality principle** after its formulator, Linus

Pauling. Following this principle, we can say that covalent bonding (as indicated by a significant value of the resonance integral β) is important for every molecular system, while ionic bonding is significant in only a relatively small number of compounds, and even in those is not exclusively responsible for chemical bonding. Within the framework of stable molecular species, then, there is a fairly wide range of electron transfer within stable bonds.

Polarization and Polarizability

The conventional way this process of partial charge transfer is described is in terms of **polarization**. A charged ion is called a **monopole**. A molecule with one end more electrically positive than the other (such as LiF) is called a **dipole**; it can be thought of as two point charges or monopoles separated by the distance between the centers of charge. A chemical bond that results in the formation of a dipole along the bond axis is described as **polar**; the larger the charge separation, the more polar the bond. For polyatomic molecules the geometry of the molecule can sometimes make a compound appear nonpolar even though the bonds have some polarity (e.g., octahedral SF_6, in which the positive "end" of each bond is inside the molecule and all the negative "ends" are spread out over the surface of the nearly spherical molecule). For diatomic molecules, however, if the bond is polar the compound's properties will show it.

For any atom, and for that matter for molecules, we can describe two quantities related to the process of polarization. The first is **polarizability**, which is the degree to which an atom's electron distribution deforms in response to an external electric field. The dimensions of polarizability are volume per molecule, which may be thought of as the change in volume swept out by the deforming electrons in response to a unit electrical field. We may say as an approximation that the larger the polarizability of an atom or molecule is, the "softer" it is. Table 6-1 gives a number of polarizabilities of atoms, ions, and simple molecules; they are all of the same rough order of magnitude, but there are some significant differences that are not difficult to account for. The larger the effective nuclear charge attracting an atom's electrons, the more resistant they are to deformation—we can see the polarizability decreasing in the series Cs^+, Ba^{2+}, La^{3+} as the charge increases, even though the number of electrons remains the same. By the same token, in negative ions where there is an overall repulsion of the ion's net charge for the electronic charge, the polarizability is larger the larger the negative charge is—as we can see by comparing Ar^0, Cl^-, and S^{2-}, which have the same number of electrons. Another factor affecting polarizabilities is the distance of the electrons from their attracting nucleus. The larger this distance is, the smaller the Coulomb force holding them in place and the more easily they are deformed. Thus Cs^+ is much more polarizable than Li^+ even though their effective nuclear charges are about the same.

Table 6-1
Polarizabilities of Simple Systems (in Å or cm³ × 10⁻²⁴)

			H₂				He
			0.79				0.20
Li⁺	Be²⁺			N₂	O₂	F⁻	Ne
0.03	0.01			1.76	1.60	0.81	0.39
					O²⁻		
					3.90		
						Cl₂	
Na⁺	Mg²⁺				S²⁻	4.61	Ar
0.24	0.10				10.2		1.62
						Cl⁻	
						2.98	
K⁺	Ca²⁺			Zn²⁺	Se²⁻	Br⁻	Kr
1.00	0.60			0.5	10.5	4.24	2.46
Rb⁺	Sr²⁺		Ag⁺		Te²⁻	I⁻	Xe
1.50	0.90		1.9		14.0	6.45	3.99
Cs⁺	Ba²⁺	La³⁺					
2.40	1.69	1.3					

The second quantity related to the polarization process is **polarizing power**, which in a sense is the inverse of polarizability. Polarizing power refers to the ability of an atom or ion to induce polarization in another atom. The smaller an ion, and the greater its positive charge, the stronger effect it has on the electrons of another atom it is placed next to. These are the factors that lead to a *small* polarizability, so we can see that the two properties are closely related.

What is the relationship of these quantities to the MO description of diatomic molecules? The degree of charge transfer from A to B in the molecule AB is related to the polarizabilities of the two atoms, but it must be remembered that the values in Table 6-1 are for the *ions* A⁺ and B⁻. That is, in considering the gaseous molecule LiF, for instance, we would have to start by assuming that the two atoms had *completely* transferred an electron to form Li⁺ and F⁻, then use the polarizabilities of these two ions to estimate the extent to which they had polarized each other back toward a covalent bond. The total degree of polarization of this ion pair will be proportional to the product of the polarizing power of the positive ion and the polarizability of the negative ion (since the degree of polarization increases if either of these quantities increases). We have already suggested that the polarizing power of an ion is essentially the reciprocal of its polarizability, so if we allow α_+ to represent the polarizability of the positive ion and α_- that of the negative ion, we can write

$$\text{degree of polarization of ions} = \frac{1}{\alpha_+} \alpha_- = \frac{\alpha_-}{\alpha_+}$$

This product (actually a ratio) represents the relative extent to which the ions are polarized from "complete charge transfer" back toward covalent bonding. It is an indication of the relative importance of covalent bonding in the molecule, and as such may be taken as a rough measure of the size of β, the resonance integral that (in our MO model) represents the source of covalent bonding energy. For the molecules LiI, LiF, CsI, and CsF, this product is as follows:

LiI	217
LiF	27
CsI	2.7
CsF	0.34

Compare this series of relative polarizations of ions with the MO energy levels for the same gaseous molecules in Fig. 6-32. The larger this polarization product is, the closer together the valence AO energies usually are, and the larger β is. So the result is basically the same whether we treat a molecule in which some electron transfer occurs by the MO method or by taking it as an ion pair and allowing these to polarize each other. In effect, we have the result that "soft" atoms tend to form stable compounds with each other because the covalent-bond energy effect is large; "hard" atoms tend to form stable compounds with each other because the electron-transfer energy is large in an essentially ionic system. We shall see in Chapter 10 that these ideas are important in developing the concept of acidity.

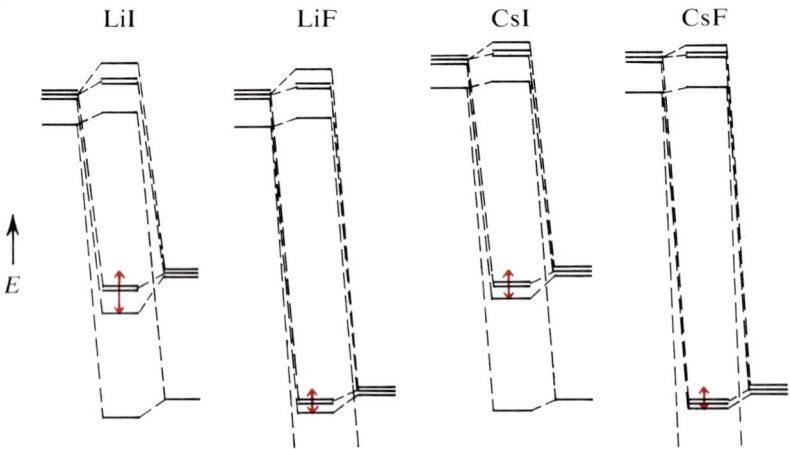

Figure 6-32 Comparison of covalent bonding energies for gaseous fluorides and iodides of lithium and cesium. The arrows indicate the size of the covalent bonding energy.

Hydrogen Bonding

The hydrogen atom is unique in that it has only one electron, so that if it were completely ionized the ion would be a bare nucleus, having only about 1/10,000 the radius of any other positive ion. When a hydrogen atom is bonded to another atom with much more stable valence orbitals (so that some electron transfer occurs from the hydrogen to the other atom), it develops a very great polarizing power because of its tiny radius. In these circumstances the partially positive hydrogen atom will attract (polarize) an unshared pair of electrons on an atom in another molecule nearby so strongly as to form a partial chemical bond with that atom. This bond, which is only about 1/10 as strong as most ordinary covalent bonds (about 5 kcal/mole vs. 50 kcal/mole), is called a **hydrogen bond**. Hydrogen bonding is important for only a few kinds of compounds, because only a few elements have valence orbitals of lower energy than hydrogen. By inspection of the valence-orbital ionization energies inside the front cover we can see that only F, O, N, and Cl qualify (remember to include the *p* AO's) and of these only F and O show a pronounced effect. As a matter of fact, in the compound HF this tendency to hydrogen bond is so strong that there is a very stable ion HF_2^- that has the structure F—H—F, with the H halfway between the two F's. In this case the bonding would have to be regarded as equal between the H and each F, but this is not usually true. The H is usually substantially closer to the atom to which it is "permanently" bonded.

We may deal with this phenomenon using the MO model. Since there is no inner core of nonvalence electrons in hydrogen, its $1s$ orbital can overlap much more favorably with another atom's valence orbitals. It is possible to have three atoms' orbitals overlap as in Fig. 6-33, and we see there that the total overlap will in general be better if the hydrogen atom is off-center, since the other two atoms' orbitals cannot overlap very much because of the exclusion principle. The overall result is again the same as if viewed from the polarization approach.

Hydrogen bonding, by producing a substantial attractive force between separate molecules in which it can occur, has a profound effect on the physical

Figure 6-33 Influence of overlap on the asymmetry of hydrogen bonds.

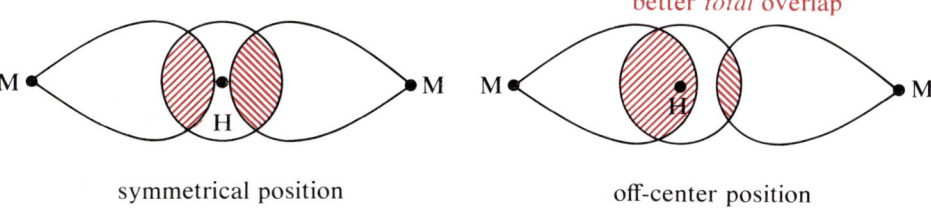

symmetrical position   off-center position

and chemical properties of these systems. One example is the crystal structure of ice. If we analyze the molecular geometry of H_2O according to the VSEPR technique, we expect it to be bent, with an H—O—H angle near 109°. Then if we fit these bent molecules together into a solid lattice, we might expect (from van der Waals or Lennard-Jones forces alone) to get close packing to maximize the number of neighbors, perhaps in a sort of chevron arrangement (Fig. 6-34a). But in fact the crystal structure of ice is such that each O is tetrahedrally surrounded by four H's, two of which are covalently bonded to it and the other two of which are hydrogen bonded to it (Fig. 6-34b). This is a very open structure and far from being close packed, but the energy of the hydrogen bonds makes it preferable over the close-packed structure.

Figure 6-34 Influence of hydrogen bonding on the open structure of ice. (a) Hypothetical close-packed structure. (b) Experimental (tetrahedral) structure.

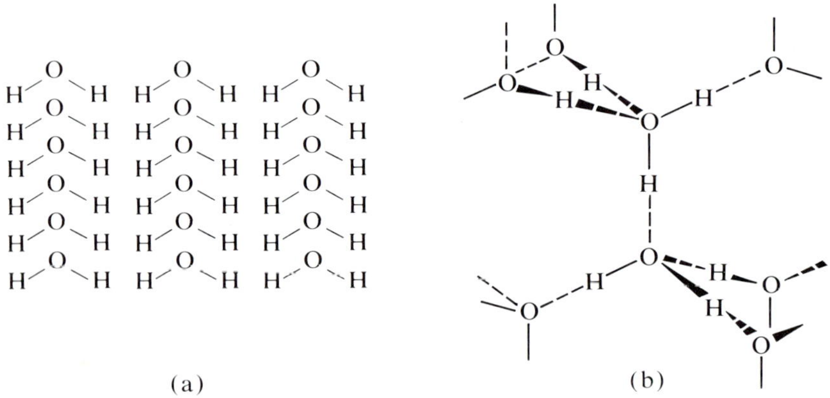

6-9 Triatomic Molecules

Having explored in some detail the MO model for diatomic molecules, let us go on to more complicated molecular systems. Next in the scale of complexity are triatomic molecules, such as $BeCl_2$, CO_2, H_2O, and OF_2. We can set up MO's for these molecules in exactly the same way as for diatomic molecules, provided we know the geometry of the system so we can accurately predict the nature of the overlap of the AO's. A little thought will show that if three atoms are to be bonded together into a molecule, there are only three possible arrangements: (1) linear—AB_2 with both B atoms bonded to A and angle B—A—B = 180°; (2) bent—AB_2 with angle B—A—B less than 180°; and (3) triangular—A_3 with each atom bonded to both of the other two. The triangular case is unknown (unless we count cyclopropane, C_3H_6, which really has nine atoms) so we shall concentrate on the other two cases.

Linear Triatomic Molecular Orbitals from s and p Atomic Orbitals

Consider as a first example beryllium chloride, $BeCl_2$. The molecule's geometry may be established as follows. The Be atom has two valence electrons (2s) and each Cl seven (3s and 3p). Each Cl can achieve a share in eight electrons by forming one two-electron bond with the Be, so the overall effect on the Be is to give it a share in two Cl electrons, for a total of four electrons in its valence shell. This corresponds to two pairs, both of which are bonding, so the equilibrium geometry will be linear, as indicated by Fig. 6-2. This is *not* true for the solid compound, for reasons we shall examine in Section 6-11, but there is reason to expect that it may be true for a single molecule in the gas phase.

Given a linear geometry, how can the various AO's combine? Let us take the valence orbitals on the central atom, Be, first, and see how various combinations of ligand-atom orbitals can overlap them. Taking the combinations with σ symmetry first (no nodal planes containing the bond axes), we see in Fig. 6-35 that, given the atom coordinates at the top of the figure, both the Be 2s and $2p_z$

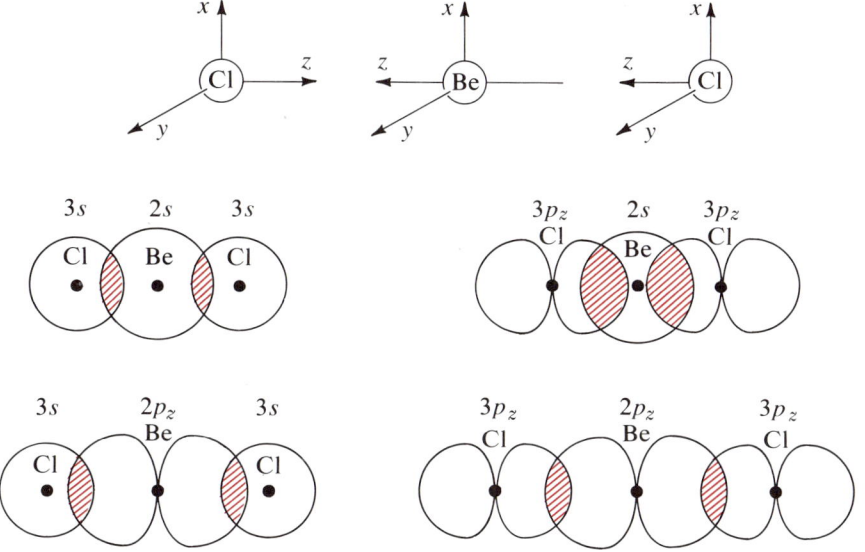

Figure 6-35 Coordinate axes and AO overlaps for $BeCl_2$ or any other linear triatomic molecule.

AO's can combine with both the Cl 3s and $3p_z$ orbitals. There are a total of six AO's involved, so we must produce six σ MO's. The lowest-energy, most stable AO's involved are the Cl 3s orbitals; since they overlap two different Be orbitals there will be two bonding MO's of even lower energy than the Cl 3s, with the bonding effect (β) being somewhat smaller for the one involving the

Be $2p_z$ AO because the energy separation between combining AO's is greater — see Fig. 6-36a. The highest-energy MO's will be antibonding combinations involving the Be $2p_z$, since those antibonding combinations must be even higher energy than that AO. Again there will be two of these since the Be $2p_z$ overlaps two different Cl orbitals; these antibonding orbitals are shown in Fig. 6-36b. Finally, there must be two more MO's, since we must produce six. The other two can be thought of as being mixtures in which the Cl $3p_z$ overlaps the Be AO's in a bonding fashion but the Cl $3s$ overlaps in an antibonding fashion; these two will both have energies higher than the Cl $3s$ but lower than the Cl $3p_z$, as in Fig. 6-36c.

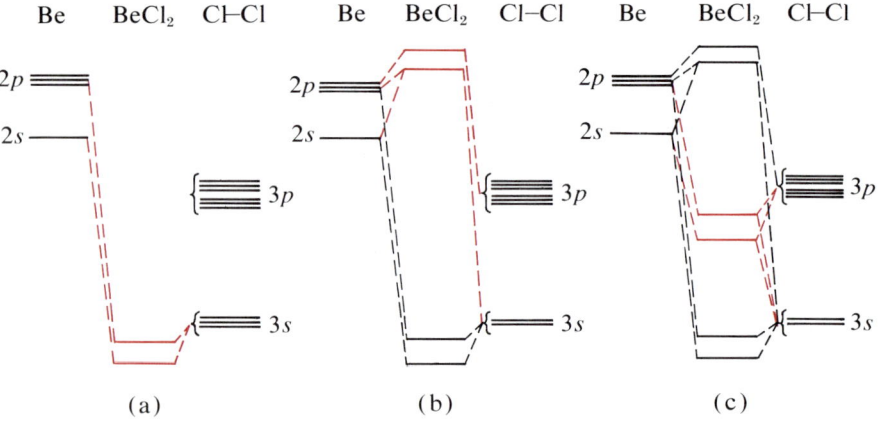

Figure 6-36 Development of the σ-overlap part of the $BeCl_2$ energy-level diagram. (a) Bonding overlap of Be $2s$ and $2p_z$ with Cl $3s$. (b) Antibonding overlap of Be $2p_z$ with Cl $3s$ and $3p_z$. (c) Mixed (nonbonding) overlap of Be $2s$ and $2p_z$ with Cl $3s$ and $3p_z$.

Next, taking combinations of AO's with π symmetry, we see in Fig. 6-37 what the possible π overlaps are. Notice that, as we have said before, only AO's that themselves have π symmetry relative to the bond axis can combine to produce π MO's. This immediately eliminates the s and p_z AO's on each atom, leaving only the p_x and p_y orbitals on each atom. Thus there are six AO's under consideration again, and we must produce six MO's with π symmetry. Because the set of p_x AO's is completely separate from (orthogonal to) the set of p_y AO's, we can consider each group separately; we shall need three MO's from each group. For the p_x group, then, there will be a bonding MO in which the overlap is uniformly favorable (tending to locate the electrons between the atoms), and there will be an antibonding orbital in which the overlap is uniformly unfavorable. There will also be a MO in which the favorable overlap of one Cl $3p_x$ with the Be $2p_x$ is just cancelled by the unfavorable overlap of the other

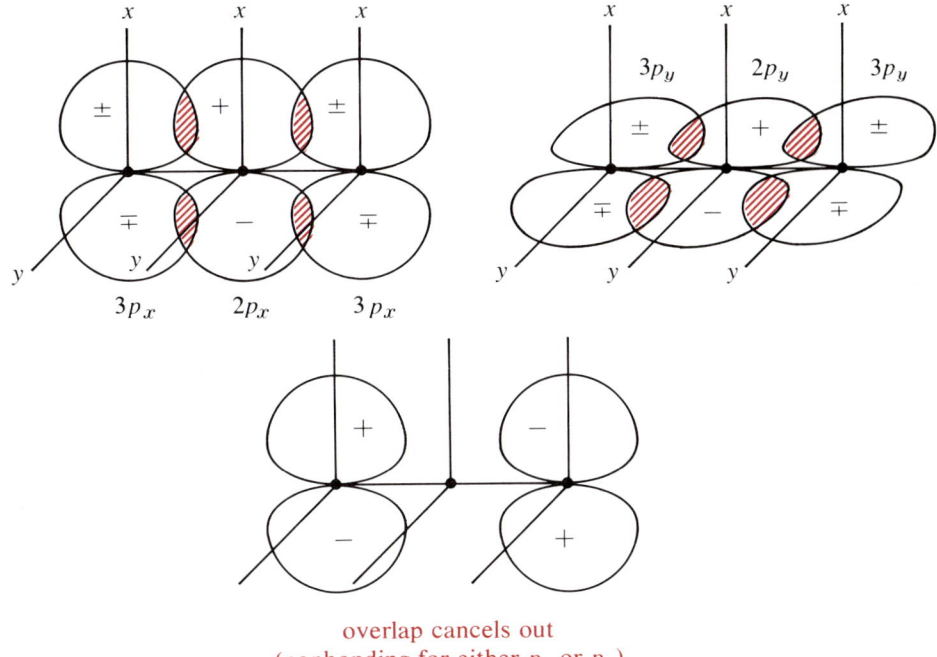

Figure 6-37 Pi overlap, bonding, antibonding, and nonbonding for BeCl$_2$.

Cl $3p_x$, so no net bonding occurs, but there is also no net antibonding effect. This is known as a **nonbonding** orbital, and is symbolized by the Greek letter denoting its symmetry, with no superscript. This nonbonding MO has exactly the energy of the Cl $3p_x$ orbitals, since the existence of the Be $2p_x$ has effectively been cancelled out. These MO energies are shown in Fig. 6-38a.

Looking at the combinations of the p_y AO's we see that they have exactly the same energy as the p_x combinations, because the AO p_x and p_y energies are the same and the geometry of the overlaps is identical. Thus the same bonding, antibonding, and nonbonding MO's will exist for the p_y orbitals as for the p_x's, with degenerate energies as shown in Fig. 6-38b.

Combining the energies of the σ MO's with those of the π MO's, we have the energy-level diagram for the linear molecule in Fig. 6-39, where the subscripts on the MO designations show the principal contributing AO's. There are 16 valence electrons in the BeCl$_2$ molecule, two from the Be and seven from each Cl, so the exclusion principle requires that the MO's be filled up through the π nonbonding orbitals, as the red arrows indicate. From the relative energies of the electrons and vacant orbitals, we can see that as long as the linear geometry is maintained, BeCl$_2$ will not be a very good electron donor, because all the electrons are in fairly low-energy MO's; it will also not be a good acceptor,

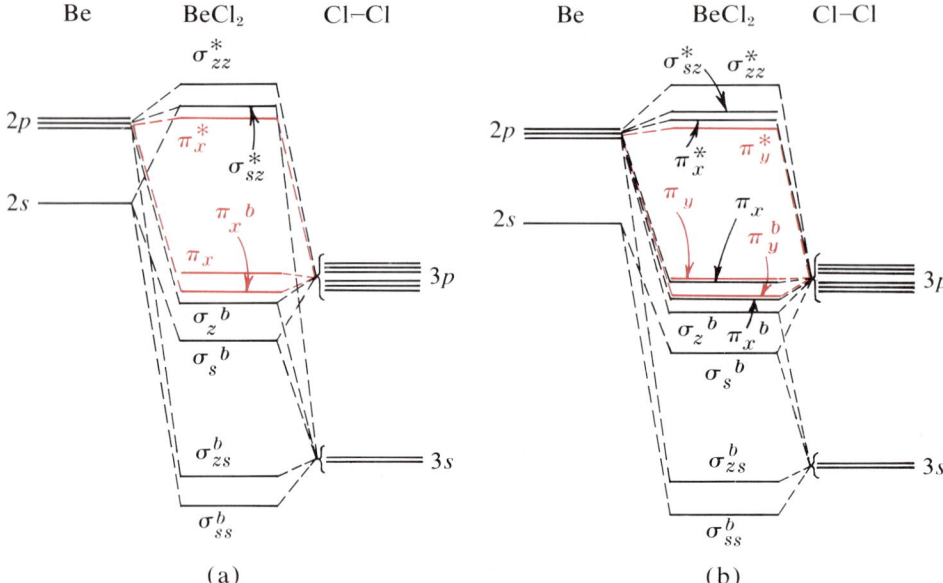

Figure 6-38 Development of the complete energy-level diagram for BeCl$_2$. (a) Pi overlap of Be $2p_x$ with Cl $2p_x$'s in bonding, nonbonding, and antibonding MO's. (b) Similar overlap of p_y AO's.

since the lowest-energy vacant MO is a quite unstable antibonding orbital. However, if the geometry changes, the situation might be different.

We can write out algebraic expressions for these LCAO's. In the following list it should be noticed that when an AO is taken with a negative sign the nature of its overlap with a neighbor orbital is reversed, and also that all the possible combinations have been included. That is, for every MO having two AO's combined with the same sign, there is one with those two AO's combined with opposite signs. This assures us that we have covered all the possibilities that are truly independent (orthogonal).

$\sigma_{ss}^b = c_1[2s(\text{Be})] + c_2[3s(\text{Cl}_1) + 3s(\text{Cl}_2)]$

$\sigma_{zs}^b = c_3[2p_z(\text{Be})] + c_4[3s(\text{Cl}_1) - 3s(\text{Cl}_2)]$

$\sigma_s^b = c_5[2s(\text{Be})] + c_6[3p_z(\text{Cl}_1) - 3p_z(\text{Cl}_2)] - c_7[3s(\text{Cl}_1) + 3s(\text{Cl}_2)]$

$\sigma_z^b = c_8[2p_z(\text{Be})] + c_9[3p_z(\text{Cl}_1) + 3p_z(\text{Cl}_2)] - c_{10}[3s(\text{Cl}_1) - 3s(\text{Cl}_2)]$

$\left. \begin{array}{l} \pi_x^b = c_{11}[2p_x(\text{Be})] + c_{12}[3p_x(\text{Cl}_1) + 3p_x(\text{Cl}_2)] \\ \pi_y^b = c_{11}[2p_y(\text{Be})] + c_{12}[3p_y(\text{Cl}_1) + 3p_y(\text{Cl}_2)] \end{array} \right\}$ same coefficients by symmetry

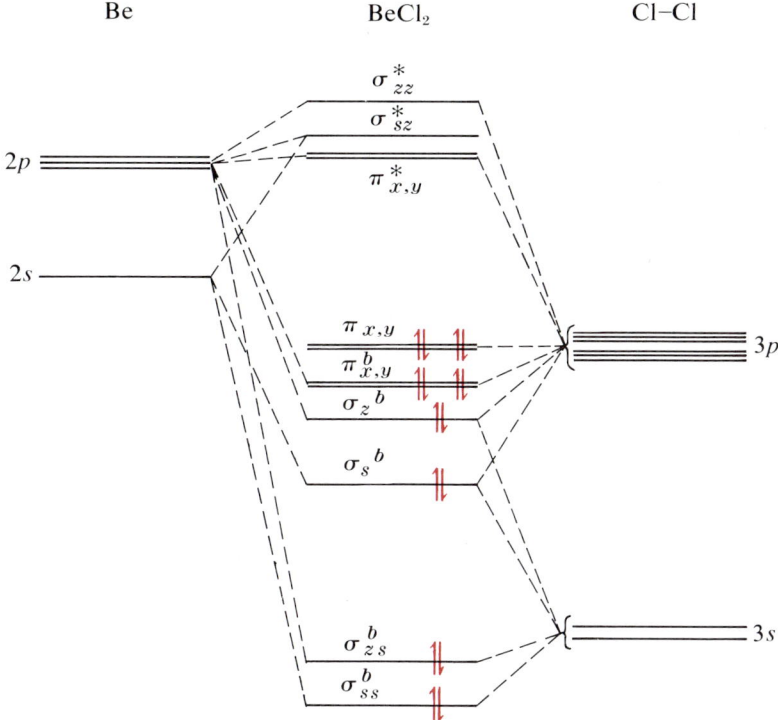

Figure 6-39 Population of linear triatomic energy levels by BeCl$_2$ valence electrons.

$$\left.\begin{array}{l}\pi_x = \dfrac{1}{\sqrt{2}}\,[3p_x(\text{Cl}_1) - 3p_x(\text{Cl}_2)] \\[1em] \pi_y = \dfrac{1}{\sqrt{2}}\,[3p_y(\text{Cl}_1) - 3p_y(\text{Cl}_2)]\end{array}\right\} \text{same coefficients by symmetry}$$

$$\left.\begin{array}{l}\pi_x^* = c_{13}[2p_x(\text{Be})] - c_{14}[3p_x(\text{Cl}_1) + 3p_x(\text{Cl}_2)] \\ \pi_y^* = c_{13}[2p_y(\text{Be})] - c_{14}[3p_y(\text{Cl}_1) + 3p_y(\text{Cl}_2)]\end{array}\right\} \text{same coefficients by symmetry}$$

$$\sigma_{sz}^* = c_{15}[2s(\text{Be})] - c_{16}[3p_z(\text{Cl}_1) - 3p_z(\text{Cl}_2)]$$

$$\sigma_{zz}^* = c_{17}[2p_z(\text{Be})] - c_{18}[3p_z(\text{Cl}_1) + 3p_z(\text{Cl}_2)]$$

At this point you should check to see that these really represent bonding, nonbonding, or antibonding situations by sketching a few orbital-overlap pictures like Fig. 6-23, crosshatching favorable-overlap areas (orbitals having same sign) in one direction and unfavorable-overlap areas (orbitals having opposite signs) in another direction to see whether favorable (bonding) overlap or unfavorable (antibonding) overlap predominates. The complete symmetry by which all linear combinations are taken with opposed signs as well as with the same sign may prove reassuring to the mathematically minded student, but is not crucial to a qualitative understanding of the MO model.

We should notice at this point that in producing these MO's we have made no assumptions about the identity of the molecule except that both the central atom and the ligand atoms have s and p valence AO's, and that the molecule's geometry is linear. Consequently, if we pick out another molecule meeting these conditions, the same MO's will apply, with the only change being the exact position of the AO energies and some resulting changes in the MO energies. Thus for the molecule CO_2 we can establish that the geometry is linear. The C atom has four valence electrons ($2s$ and $2p$) and the O atoms six valence electrons (each also $2s$ and $2p$). The O atoms can achieve a share in eight electrons each by forming two two-electron bonds each with the C. This will give the C eight electrons in its valence shell; they will not be arranged in four pairs, however, but in two double-bond groups of four electrons. Since there are only two groups of electrons, the geometry of the molecule will be linear. With this and the identity of the valence orbitals established, we can move directly to a MO energy-level diagram that differs from that for $BeCl_2$ only in detail, as seen in Fig. 6-40, which is constructed from the atomic valence-orbital ionization energies inside the front cover. We see that the CO_2 electrons are, on the whole, somewhat more stable than the $BeCl_2$ electrons. In addition, it should be evident from the relative AO energies that there will be a good deal more electron transfer (and thus ionic character) in $BeCl_2$ than in CO_2. We expect both the Be and the C to have a partial positive charge, but the Be much more so.

Bond-Energy Comparisons

This is a good point to take another look at the definition of bond energies. What is the energy of a C—O bond? If we try to define it by the energy change of the reaction

$$CO_2 \rightarrow CO + O$$

the experimental answer is 127 kcal/mole. But we could presumably equally well define it as half the energy change of the reaction

$$CO_2 \rightarrow C + O + O$$

in which case the experimental result is $\frac{1}{2}(383) = 192$ kcal/mole. Why should there be such a profound difference? In the first case, which is the **bond-**

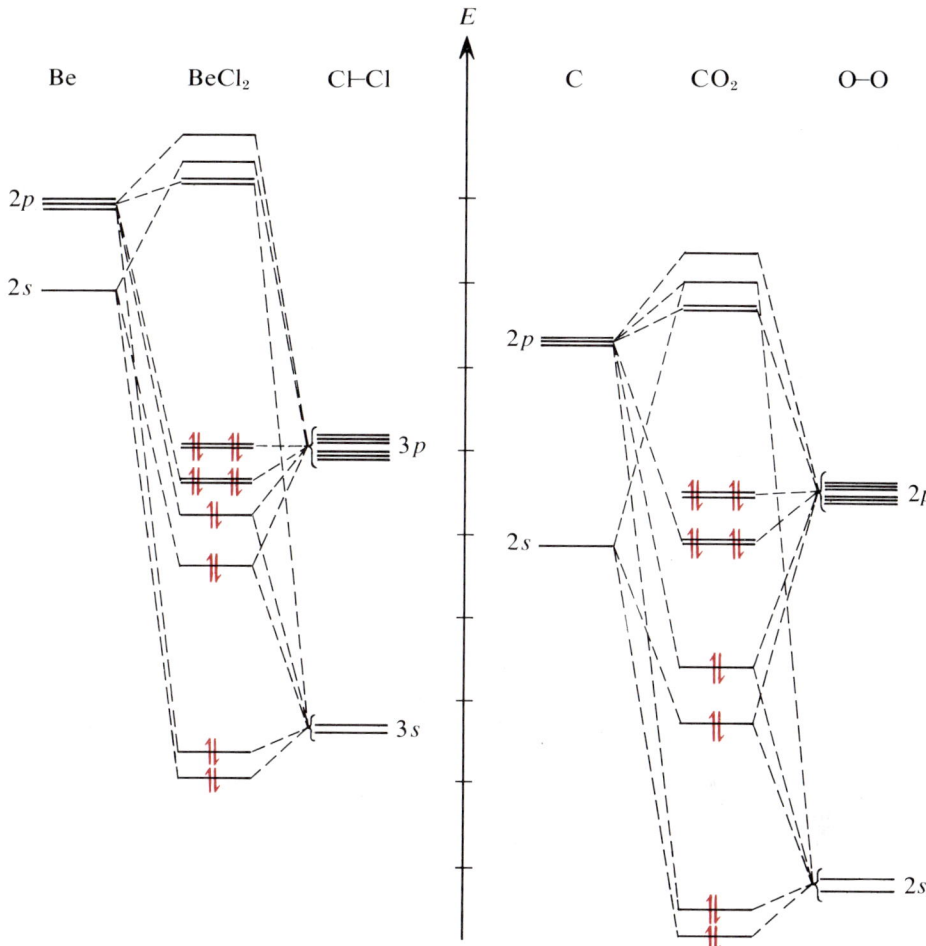

Figure 6-40 Comparison of BeCl$_2$ and CO$_2$ MO energies.

dissociation process with the accompanying **bond-dissociation energy D**, the total bond order is changing from about 4 for CO$_2$ (taking a pair of electrons in a bonding orbital as representing one bond, but discounting σ_{ss}^b and σ_{zs}^b in Fig. 6-40 because the covalent-bonding energy effect is so small) to about 3 for CO (since the σ_s^b and σ_s^* in Fig. 6-28 cancel each other). So approximately one bond is being undone in this process. On the other hand, in the second reaction, which is the **atomization** process with the accompanying **atomization energy E**, all four bonds in CO$_2$ are being broken. It is not surprising, then, that the second process requires much more than twice the energy of the first. Bond-energy terms are described in two ways, corresponding to these two processes. Tables exist in which bond-dissociation energies (D) are given for specific bonds in specific molecules; other tables give **thermochemical bond energies** (E) averaged over the complete atomization of a given molecule. It is apparent that these may be quite different and that it is necessary to distinguish between them.

To return to the general discussion of the MO model for triatomic molecules, there are two reasons why a given triatomic molecule might not be described by the energy levels of Fig. 6-39 or something close to it. Either the atoms involved might not all have s and p valence orbitals or the geometry might not be linear. Let us look at the first possibility.

Linear Triatomic Molecular Orbitals Using Only s Atomic Orbitals on Outer Atoms

Beryllium hydride, BeH_2, would be expected in the gas phase to have a linear geometry (checking this through the VSEPR procedure is a good idea), but the H atoms, unlike the Cl atoms, have only the $1s$ valence orbital. Of course there can be no $1p$ AO. We shall have to look at the overlaps of the various AO's again. Figure 6-41 shows these AO's; because there are now only six AO's involved in bonding, the situation is simpler than for $BeCl_2$. There will be a bonding MO resulting from the combination of the Be $2s$ with the H $1s$'s, and also an antibonding version of the same thing. Similarly, there will be a bonding and an antibonding version of the Be $2p_z$ overlapping the H $1s$'s. The Be $2p_x$ and $2p_y$ have π symmetry with respect to the bond axes; since the H $1s$'s have only σ symmetry, they cannot combine (see Fig. 6-23) and these two Be AO's

Figure 6-41 Overlaps of AO's for BeH_2.

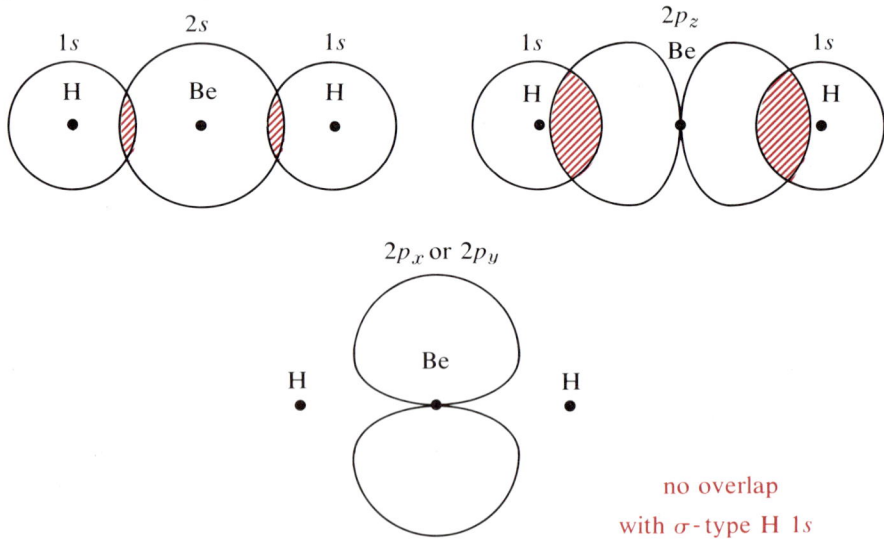

will be nonbonding MO's. The result is the energy levels of Fig. 6-42. There are four valence electrons in BeH$_2$, which just fill the bonding orbitals, as shown.

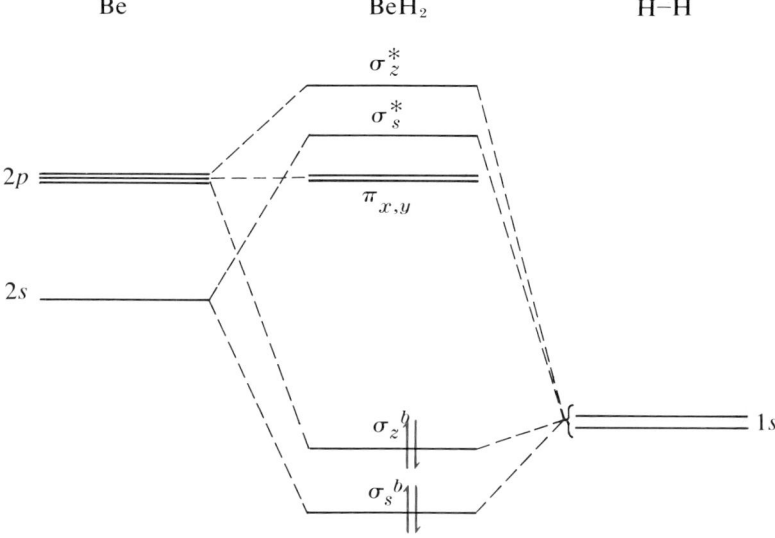

Figure 6-42 Energy-level diagram for BeH$_2$.

Bent Triatomic Molecular Orbitals Using Only *s* Atomic Orbitals on Outer Atoms

The other reason for a significant difference from the MO energy levels of Fig. 6-39 is the possibility of a bent geometry for the triatomic molecule. The simplest example of this is water, because H$_2$O, like BeH$_2$, has a very simple MO description with no π overlap.

Working through the VSEPR approach again, we estimate the geometry of H$_2$O as bent, with a bond angle near the tetrahedral angle of 109°. The most convenient coordinate system is shown in Fig. 6-43. The possibilities for orbit-

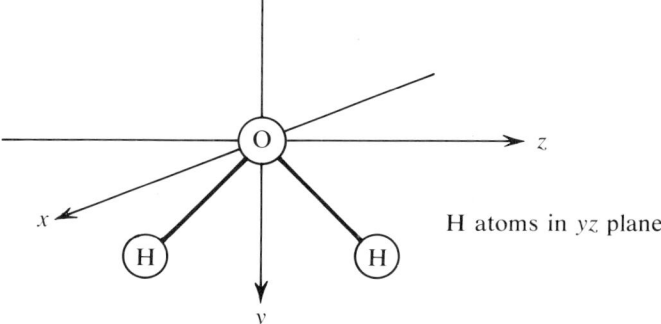

Figure 6-43 Coordinate axes for H$_2$O or other bent triatomic hydride, XH$_2$. No coordinates are needed for the H atoms because the 1*s* orbitals are spherical.

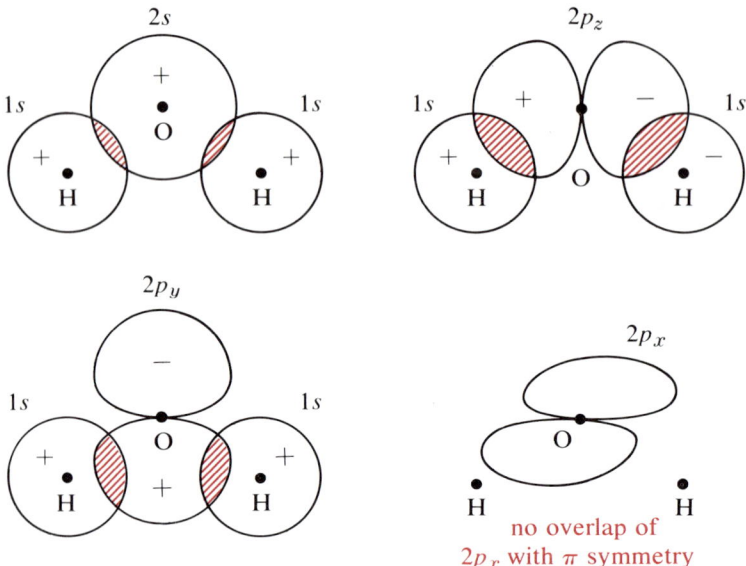

Figure 6-44 AO overlaps in H₂O.

al overlap are shown in Fig. 6-44; notice that the combination of the H $1s$ orbitals with the same sign favorably overlaps not only the O $2s$ but also the O $2p_y$. Since the combination of H $1s$ orbitals with opposite signs overlaps only the O $2p_z$, these five AO's must be combined to produce five MO's. The O $2p_z$ will combine with the H $1s$'s to give a bonding and an antibonding orbital, so three MO's must be formed from the O $2s$, $2p_y$, and the H $1s$'s. As in the BeCl₂ case, there will be a combination with all overlap favorable (bonding) and another with all overlap unfavorable (antibonding). The third MO will have favorable overlap with the O $2p_y$ but unfavorable overlap with the O $2s$, resulting in an intermediate energy between those two O orbitals. The resulting MO energies are shown in Fig. 6-45a; when we take into account the fact that the O $2p_x$ orbital has only π symmetry with respect to the bond axes, which renders it nonbonding, we have the complete energy-level diagram of Fig. 6-45b. Since there are eight valence electrons in a water molecule, all the energy levels are filled up through the nonbonding π MO. This suggests that water should be a poor electron acceptor (vacant MO's are all high-energy antibonding types), but given a sufficiently low-energy acceptor, might donate the nonbonding pair. We shall see that this frequently occurs.

The algebraic representations of these MO's are given below; note that the symmetry of the combinations is, as before, complete in that every combination of orbitals with the same sign is balanced by a combination with opposite sign.

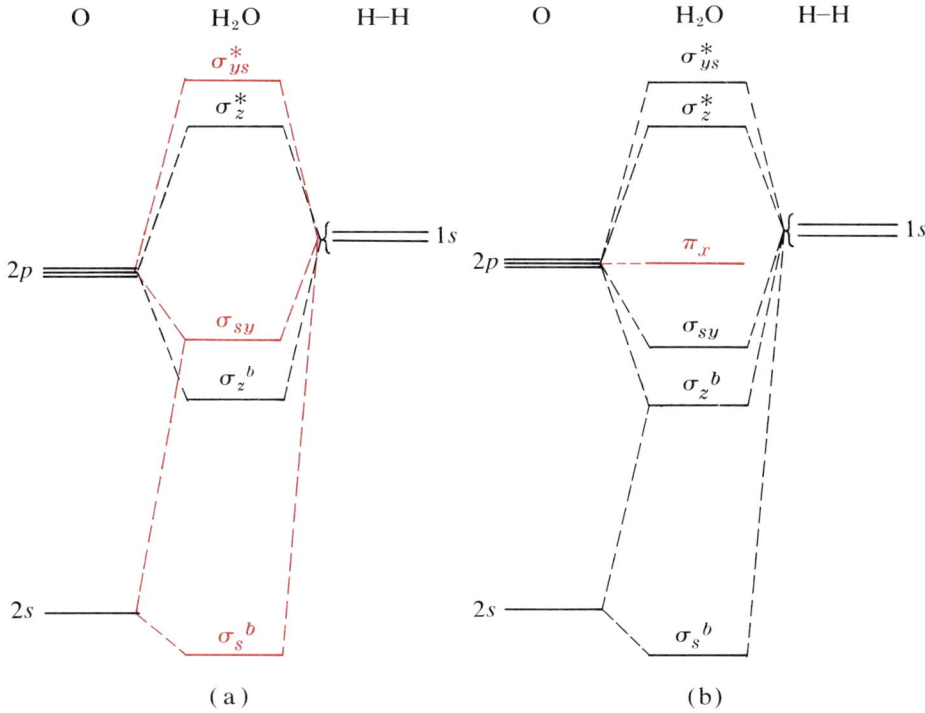

Figure 6-45 Development of the energy-level diagram for H_2O. (a) Bonding and antibonding overlap of both O 2s and $2p_y$ with H 1s's, plus mixed (nonbonding) overlap of same AO's. (b) Nonbonding π symmetry of O $2p_x$.

$$\sigma_s^b = c_1[2s(O)] + c_2[1s(H_1) + 1s(H_2)]$$
$$\sigma_z^b = c_3[2p_z(O)] + c_4[1s(H_1) - 1s(H_2)]$$
$$\sigma_y = c_5[2p_y(O)] - c_6[2s(O)] + c_7[1s(H_1) + 1s(H_2)]$$
$$\pi_x = 2p_x(O)$$
$$\sigma_z^* = c_8[2p_z(O)] - c_9[1s(H_1) - 1s(H_2)]$$
$$\sigma_y^* = c_{10}[2p_y(O)] - c_{11}[1s(H_1) + 1s(H_2)]$$

Again, you should sketch the graphical representations of these to show the bonding, nonbonding, or antibonding nature of the LCAO's.

It is also instructive to compare these MO's with those of BeH_2 to see what kinds of differences result from the change from linear to bent geometry. Figure 6-46 shows these two sets of energy levels plotted so as to correspond as nearly as possible to each other, although of course some differences arise from the fact that the Be AO's are higher energy than the H $1s$, while the O AO's are lower energy. The two most stable bonding orbitals are separated in energy more for H_2O than for BeH_2, but this results primarily from the fact that the

310 | Molecular Structure

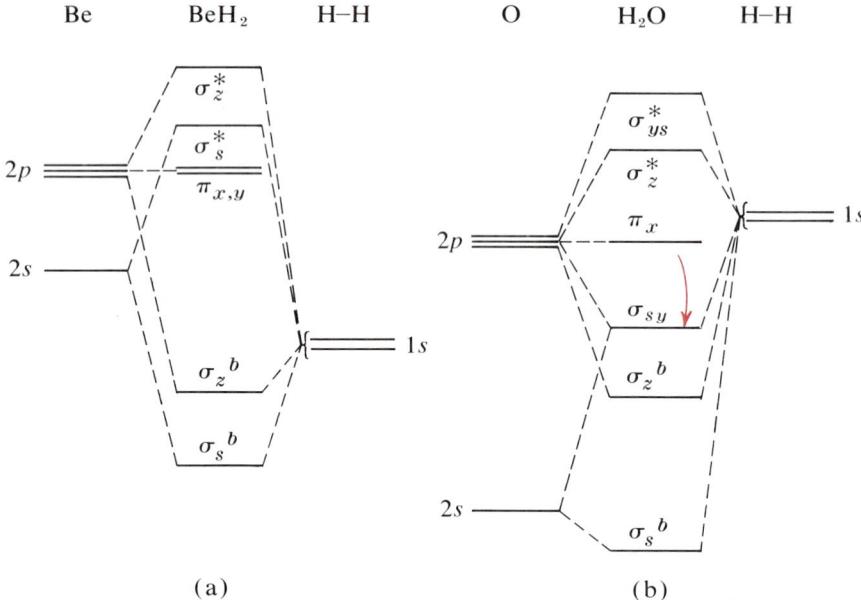

Figure 6-46 Energy differences between (a) linear and (b) bent forms of triatomic molecules.

s–p energy separation is much greater for O than for Be. The only significant difference due to the changed geometry is the change from two π nonbonding orbitals to one π nonbonding orbital and one σ bonding orbital, because bending the molecule keeps the O $2p_y$ nodal plane from falling along the bond axes any more. Otherwise the energy levels are quite similar.

Bent Triatomic Molecular Orbitals from Both s and p Atomic Orbitals

Finally, let us look at the case of a bent triatomic molecule with both σ and π bonds. A good example is oxygen difluoride, OF_2. The VSEPR approach suggests a bond angle of somewhat less than the tetrahedral angle of 109°, and the experimental angle is 103°. Using this geometry we set up the coordinate system of Fig. 6-47a, which gives the AO overlaps shown in Fig. 6-47b. In this figure we have neglected the overlap of the F $2s$ orbitals since they are so far removed in energy from the other AO's as not to produce very effective bonding; obviously there would be two nearly nonbonding MO's slightly lower energy than these two AO's, as shown in Fig. 6-48a. There are thus a total of

Figure 6-47 Coordinate axes and AO overlaps for OF_2 or any bent triatomic molecule with s and p AO's on each atom.

Triatomic Molecules | 311

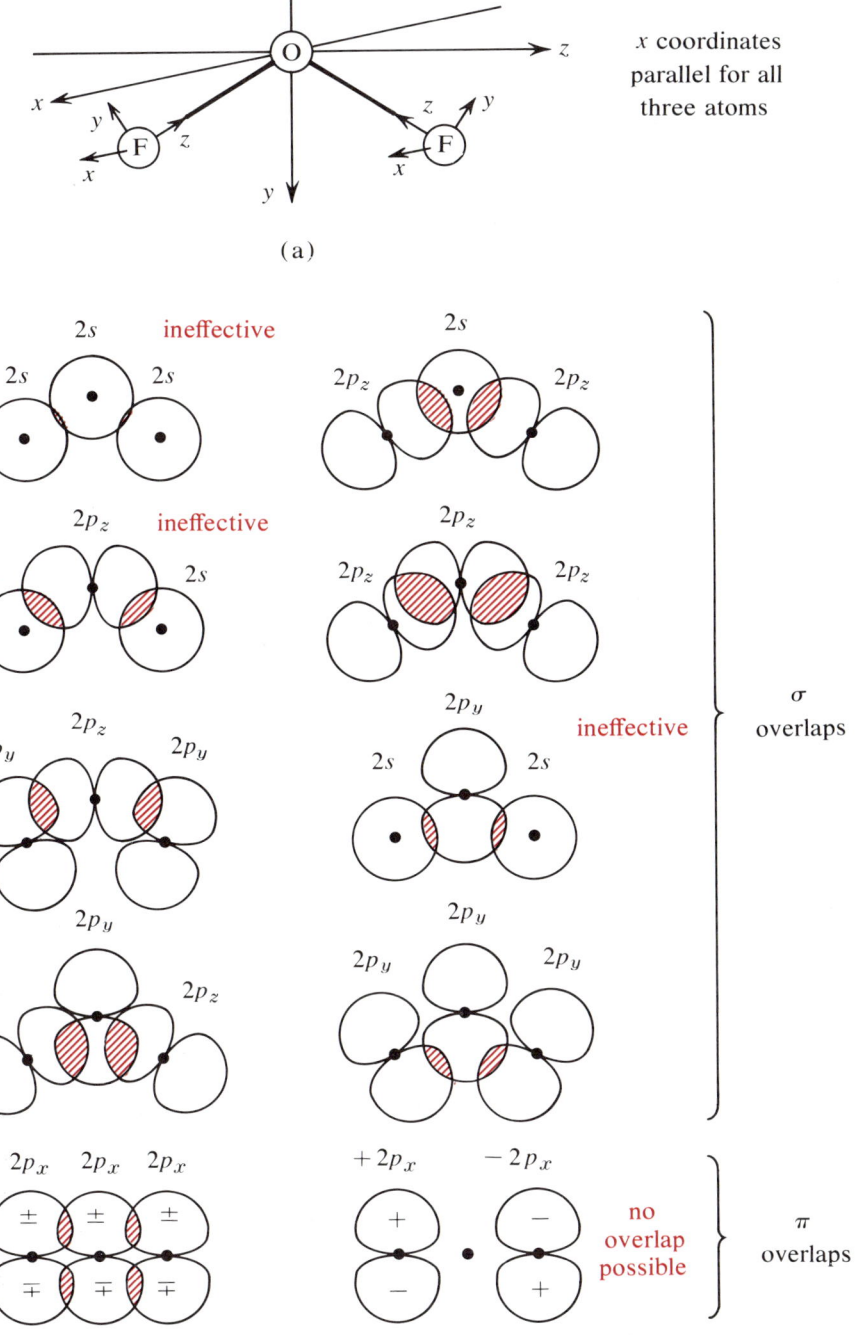

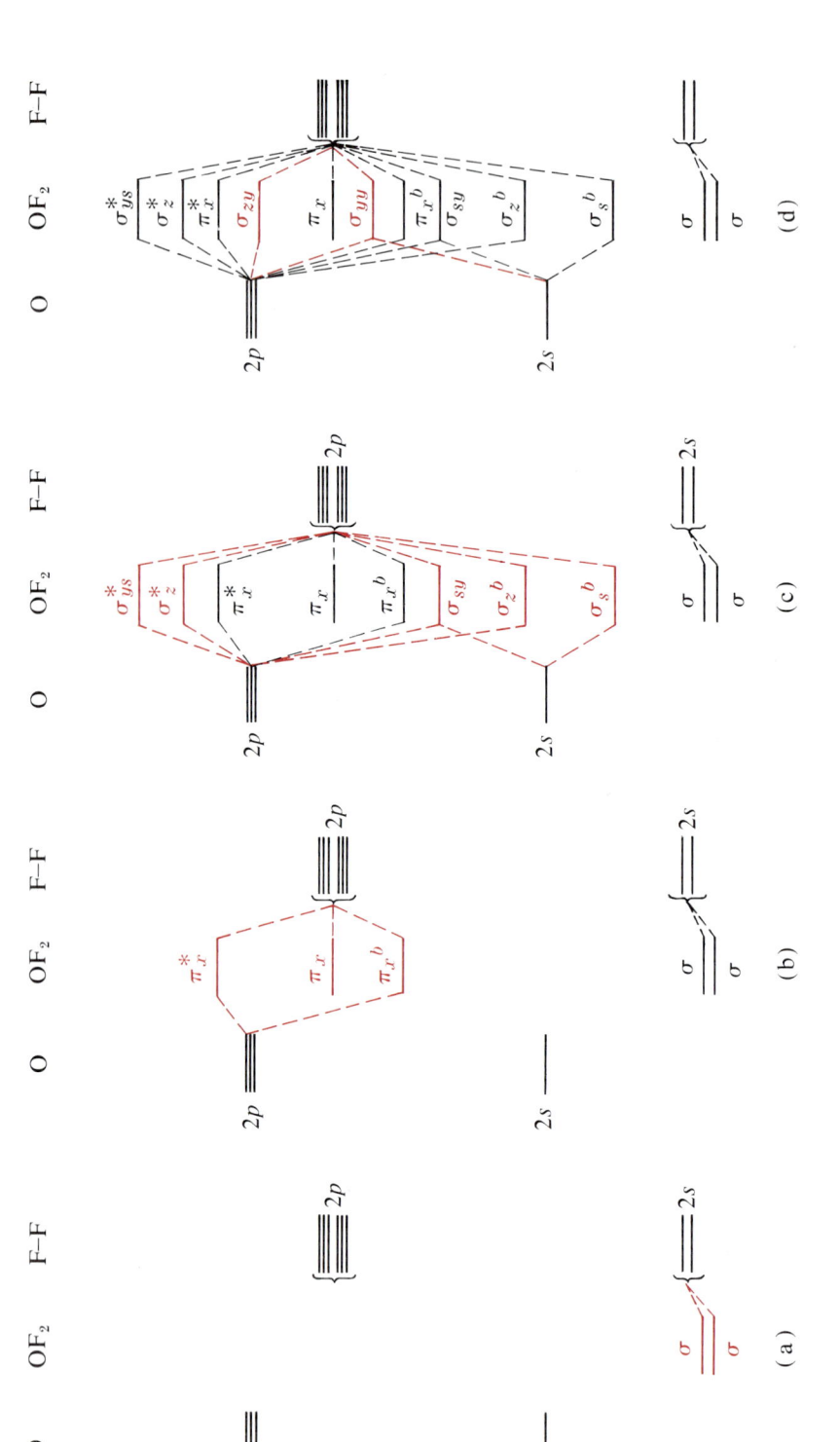

Figure 6-48 Development of the energy-level diagram for a bent triatomic molecule. (a) The nonbonding nature of F $2s$ AO's. (b) Pi bonding, nonbonding, and antibonding overlap of $2p_x$ AO's. (c) Bonding and antibonding overlap of $2p_z$ AO's, and bonding and antibonding overlap of O $2s$ and $2p_y$ with F $2p$'s plus mixed (nonbonding) overlap of the same AO's. (d) Sigma overlap of F $2p_y$'s.

10 AO's that will be considered to overlap, producing a total of 10 MO's. Since all three atoms now have p valence orbitals, π overlap will be possible; the p_x orbitals on the three atoms can combine to produce a π-bonding, a π-antibonding, and a π-nonbonding MO just as they did for $BeCl_2$ (see pp. 300–303). The p_y orbitals do not combine as they did for $BeCl_2$ because of the bent geometry—they now allow some σ overlap, and we shall come back to them. Allowing for the three π MO's as in Fig. 6-48b leaves seven σ MO's to be created. The p_z AO on each F overlaps all of the O AO's in exactly the same fashion that the H $1s$ did, so we can immediately write five of the seven σ MO's simply by substituting the overlap of the F $2p_z$'s for that of the H $1s$'s. These five σ MO's will have the same general sort of energy distribution as they did in the H_2O energy-level diagram, as shown in Fig. 6-48c. Finally, there are two more σ MO's to be created, which correspond to the two new types of σ overlap that can occur involving the F $2p_y$ orbitals, as shown at the bottom of Fig. 6-47b. Because there is a good deal of duplication in the types of overlap that can occur, there is quite a bit of "mixing" of AO's in forming these σ MO's, and other types of AO overlap contribute to these two MO's as well as to the very favorable σ bonding MO's and the very unfavorable σ antibonding MO's. The MO's formed from these two types of overlap can be expected to have roughly intermediate energies near those of the p AO's on the individual atoms; they will be more or less nonbonding, as shown in Fig. 6-48d.

Figure 6-49 shows the complete energy diagram for OF_2, with the 20 valence electrons of the system inserted with due regard for the exclusion principle. Notice that the lowest-energy vacant MO is much higher in energy than any of the individual AO's. This indicates that OF_2 will not be a particularly good electron acceptor, not nearly so good as the valence atomic-orbital ionization energies would suggest. On the other hand, OF_2 would be a very poor electron donor, since even the highest-energy electrons are still more stable than most atoms' and molecules' vacant orbitals. OF_2 should be fairly inert chemically, then, serving as an acceptor only under some provocation. Experimentally, it can be mixed with hydrogen gas or natural gas without any reaction occurring; but if the mixture is ignited by an electric spark a very violent explosion occurs in which HF and H_2O are formed. This sort of behavior is characteristic of OF_2, and we see that it is consistent with the MO model.

Again it is interesting to compare the energy levels of the bent triatomic structure with those of the linear triatomic structure. Figure 6-50 shows the $BeCl_2$ and OF_2 energies arranged to coincide as nearly as possible. We see here as for the simpler case of the hydrides that the principal difference is that three of the π MO's in the linear case (those formed from the p_y AO's) become σ MO's of somewhat different but roughly equivalent energy in the bent case. Sometimes this difference can be important in causing a normally linear molecule to be more reactive when it is bent, but the effect is usually small. The

314 | Molecular Structure

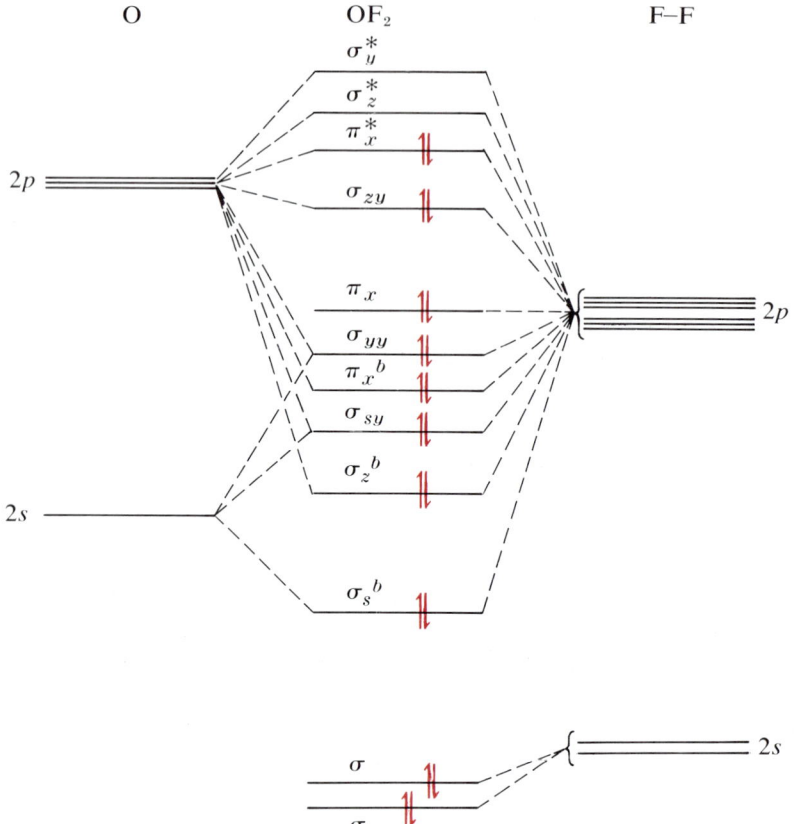

Figure 6-49 Energy-level diagram for a bent triatomic molecule, populated by OF$_2$ valence electrons.

energy difference shows up in the higher-energy MO's and usually tends to make molecules with these MO's occupied (i.e., those with many valence electrons) slightly more stable in a bent configuration; of course, this is just what the VSEPR approach also indicates.

6-10 Tetraatomic Molecules

If we restrict our attention to molecules having the formula AB$_3$ (i.e., without considering A$_4$ or A$_2$B$_2$ molecules) we find that only two geometries occur: trigonal planar and trigonal pyramidal. Examples of the planar geometry are BCl$_3$, SO$_3$, NO$_3^-$ (ion), and BH$_3$; the pyramidal structure is exemplified by NH$_3$, PF$_3$, SO$_3^{2-}$ (ion), and AsCl$_3$. We could approach any of these species in the same way we have constructed MO's for the triatomic systems, but with

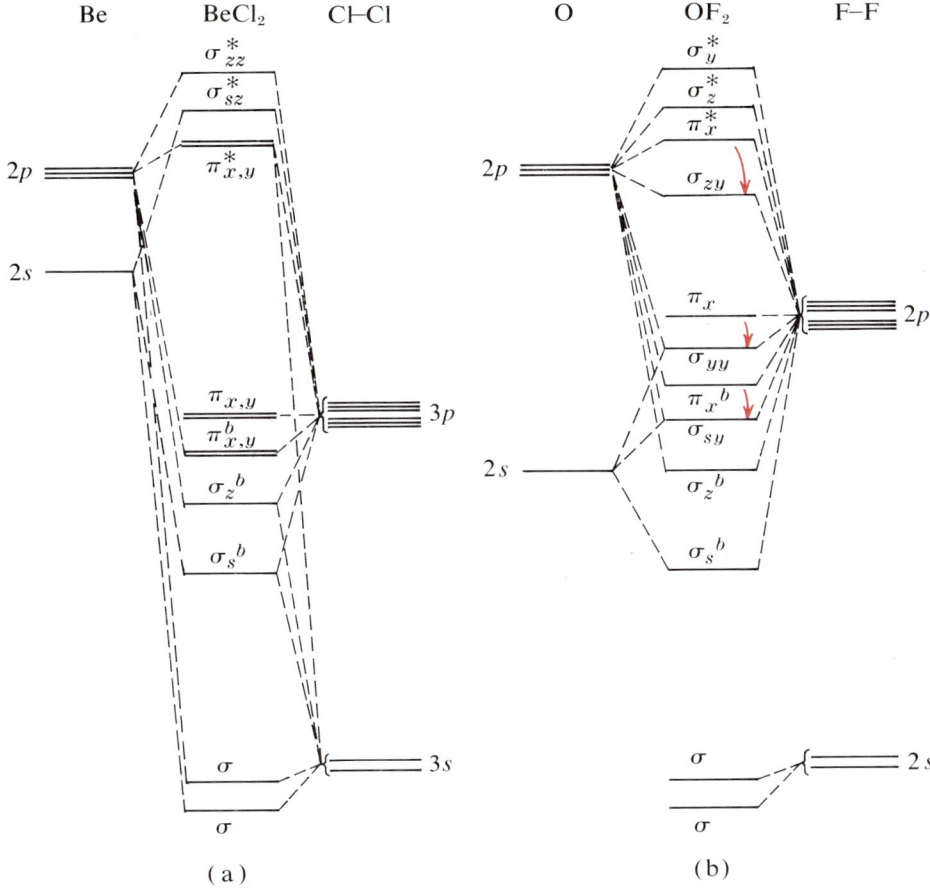

Figure 6-50 Comparison of energy differences between (a) linear and (b) bent forms of triatomic molecules.

the increasing number of atoms the total number of orbitals becomes inconvenient if we try to deal with both s and p orbitals on all of the atoms in the molecule. Accordingly, we shall restrict our examination to the simpler cases of the hydrides.

Planar Systems of Four Atoms

Using the VSEPR technique we can predict that BH_3 ought to be trigonal planar since it will have three bonding electron pairs around the central B atom. Virtually no experimental data are available for this species, however, since it has a very strong tendency to accept electrons from any available source or to **dimerize** to diborane, B_2H_6. The reasons for this are essentially the reasons why $BeCl_2$ is not linear in the solid phase—which we shall discuss in Section 6-11.

316 | Molecular Structure

For the planar BH_3 system we can construct the convenient coordinate system shown in Fig. 6-51a, which indicates overlaps of AO's as shown in Fig. 6-51b. It is clear from a consideration of these overlaps that all but one of the B AO's can effectively overlap the H $1s$ AO's; that one is the B $2p_z$, which has only π symmetry toward the H atoms and so cannot combine with the σ H $1s$ orbitals. The $2p_z$ orbital of the B atom, then, will be a π nonbonding MO. Since we are dealing with a total of seven AO's, which means a total of seven MO's, there are six σ bonding and antibonding orbitals to be constructed from

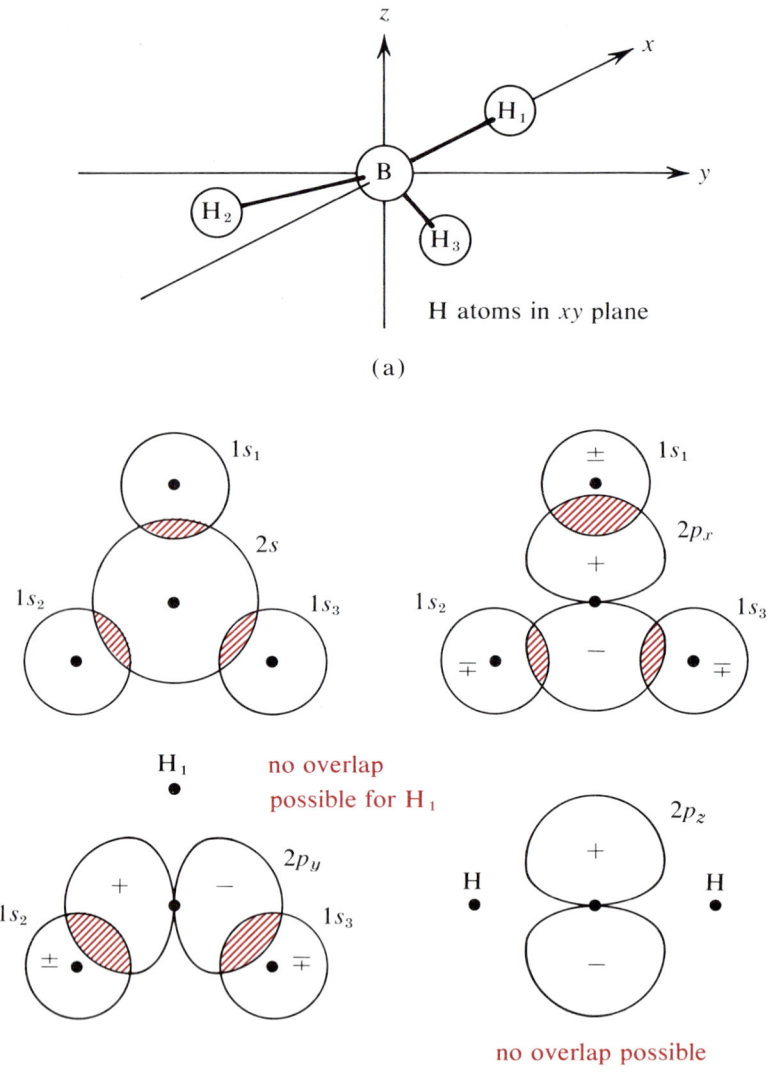

Figure 6-51 Coordinate axes and AO overlaps for BH_3.

the overlaps. Two of these will just correspond to a bonding and an antibonding combination of the B 2s and the H 1s's, all with the same sign, as shown in Fig. 6-52a. The two B orbitals $2p_x$ and $2p_y$ overlap the H 1s orbitals in ways that appear quite different, but that can be shown to be equivalent (for bond angles of 120°, anyway).

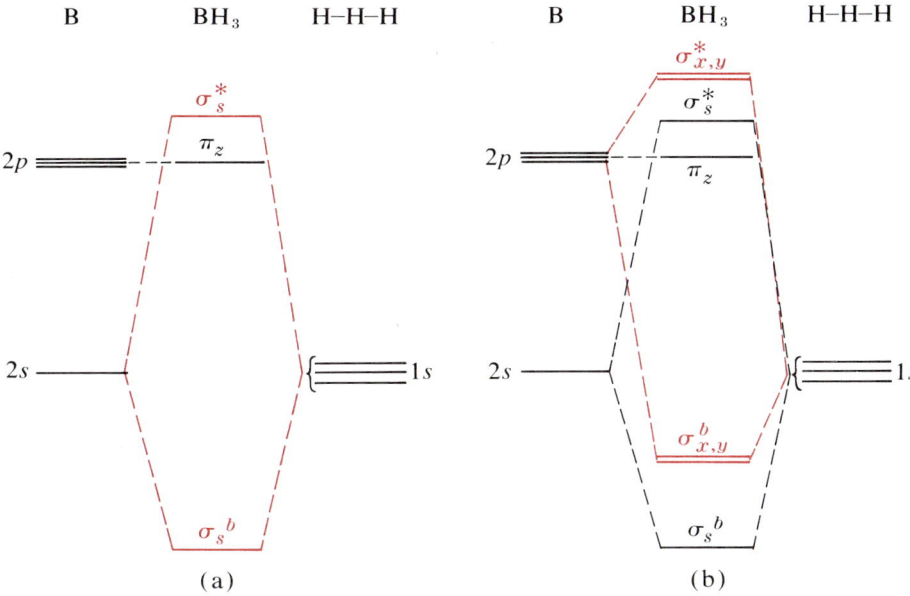

Figure 6-52 Development of the energy-level diagram for BH_3. (a) Bonding and antibonding overlap of B 2s with H 1s AO's. (b) Bonding and antibonding overlap of B $2p_x$ and $2p_y$ with H 1s AO's.

We can establish the equivalence as follows: The three H atoms are all at the same distance from the B nucleus and their 1s orbitals are spherically symmetrical. Any variation in overlap, then, will have to come from the angular dependence of the B 2p orbitals. Looking back at the wave functions tabulated on p. 220, we see that the p orbitals' magnitudes vary as the cosine of the angle from their direction of maximum extension. In Fig. 6-53 we see that H_1 overlaps the B $2p_x$ in exactly this direction, in which the $2p_x$ may be assumed to have an electron population density (proportional to the square of its AO magnitude) of 1 unit. But H_2 and H_3 overlap at 60° from the direction of maximum extension, so the B $2p_x$ magnitude in these directions is proportional (in the same units) to cos 60°, or 0.500, and the $2p_x$ electron population density in these directions is proportional to the square of these magnitudes, or $(0.500)^2 = 0.250$ each. So for the overlap of the B $2p_x$ with the three H orbitals, the total electron population density in the overlap areas is equal to $1 + 0.250 + 0.250 = 1.500$ unit. Now what about the B $2p_y$ overlap? Again referring to Fig. 6-53 we

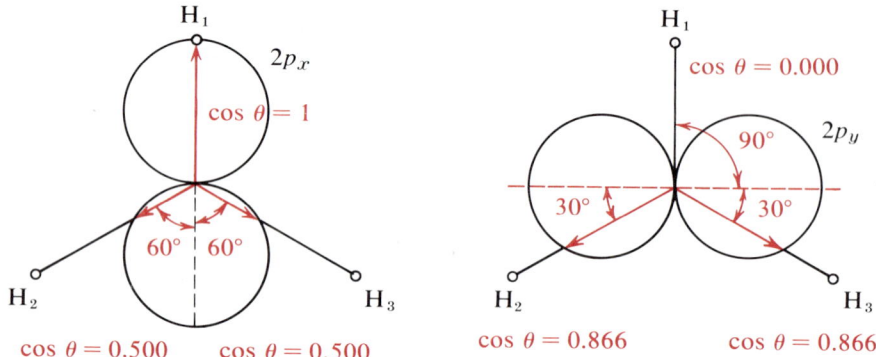

Figure 6-53 Equivalence of $2p_x$ and $2p_y$ overlap for 120°-bond angles.

see that H_1 has no overlap at all, since it lies on the $2p_y$ nodal plane. H_2 and H_3 both lie at 30° to the direction of maximum extension of the B $2p_y$, so the electron population densities in those directions for the $2p_y$ orbital will be proportional to $(\cos 30°)^2 = (0.866)^2 = 0.750$. So for the B $2p_y$ overlap, the total electron population density in the overlap directions equals $0 + 0.750 + 0.750 = 1.500$ unit, which is exactly the same as for the B $2p_x$ overlap.

This equivalence of overlap means that the bonding and antibonding MO's formed from these two B AO's will be degenerate in energy, as shown in Fig. 6-52b. Figure 6-54 shows the final MO energy-level diagram for BH_3, popu-

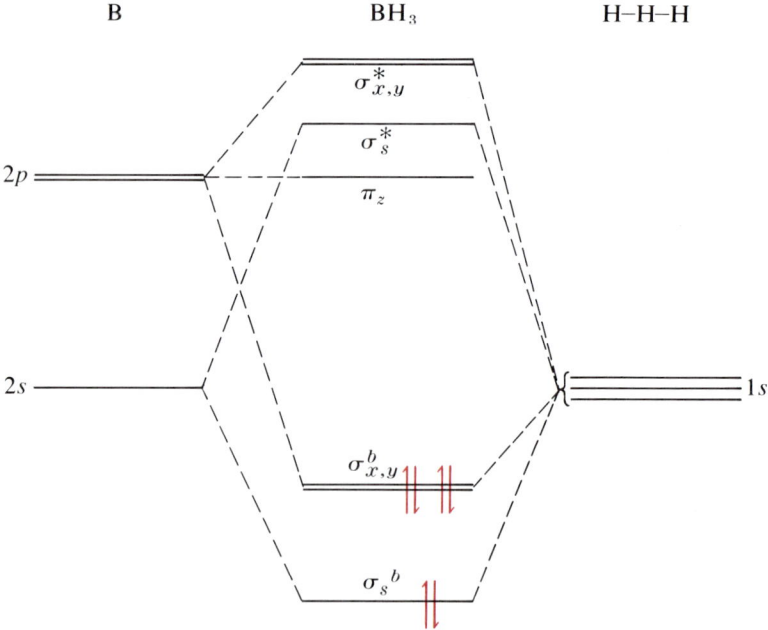

Figure 6-54 Energy-level diagram for BH_3, populated with valence electrons.

lated with the six valence electrons of the system. All the bonding orbitals are filled, but it is not very far up to the empty nonbonding MO. We shall see that this influences the chemistry of BH_3.

Pyramidal Systems of Four Atoms

The other geometry we need to consider is the trigonal pyramidal geometry of NH_3. Again we can start by predicting this experimentally verified geometry through the VSEPR approach; if we then set up the same coordinate system as for the planar case, as in Fig. 6-55a, we can expect the overlaps shown in Fig.

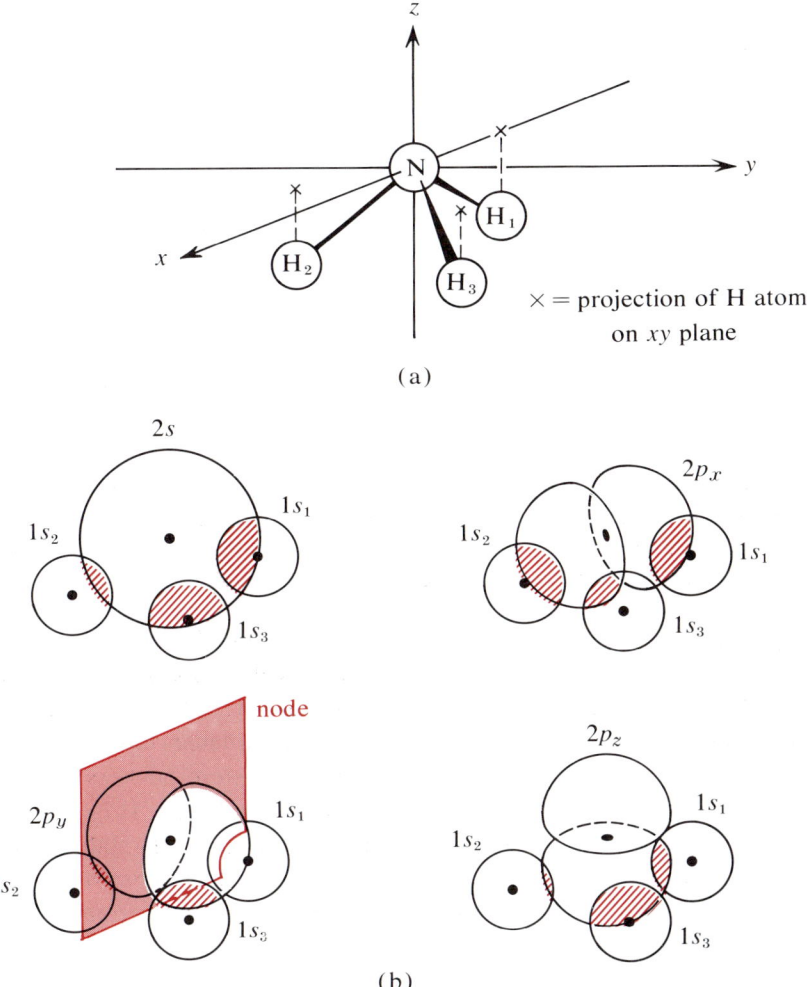

Figure 6-55 Coordinate axes and AO overlaps for NH_3 (trigonal pyramidal).

6-55b. These have exactly the same symmetry as the overlaps in the planar system, except that the overlaps now lie below the N xy plane. There is one difference in that the H orbitals can now overlap the N $2p_z$ orbital's bottom lobe, while they could not overlap the corresponding orbital at all in the planar case. Since this overlap has the same symmetry as that of the N $2s$, it will be mixed with the $2s$ bonding and antibonding MO's and three MO's will result, as shown in Fig. 6-56a. The other N $2p$ orbitals' overlap with the H orbitals will be exactly analogous to those of the planar geometry, so again we get the degenerate

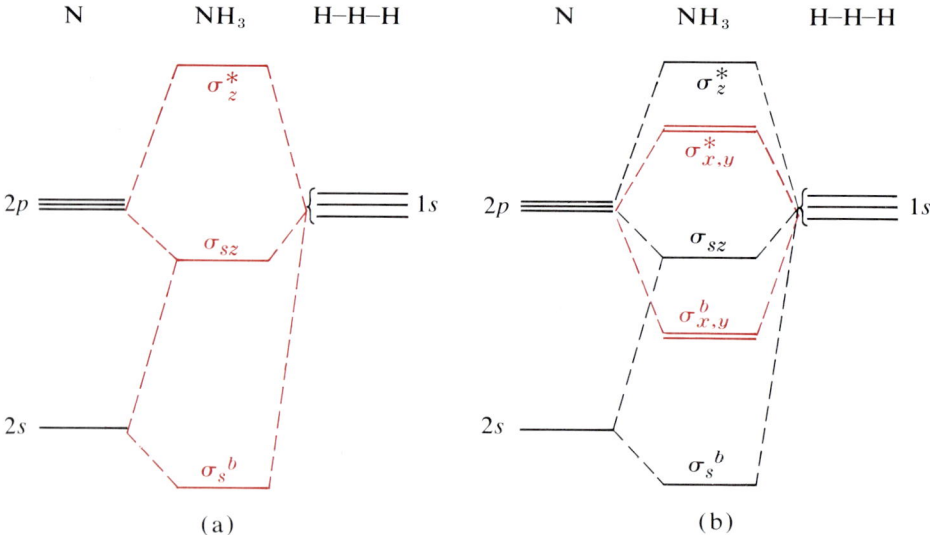

Figure 6-56 Development of the energy-level diagram for NH_3. (a) Bonding and antibonding overlap of both N $2s$ and $2p_z$ with H $1s$ AO's, plus mixed (nonbonding) overlap of the same orbitals. (b) Bonding and antibonding overlap of N $2p_x$ and $2p_y$ with H $1s$ AO's separately but with equivalent energies.

bonding and antibonding MO's shown in Fig. 6-56b. Figure 6-57 shows the final MO energy-level diagram for pyramidal NH_3, populated with its eight valence electrons, and compares it with the planar BH_3 energy levels. The principal difference between the planar and nonplanar energy levels is a stabilization of the π nonbonding level when the system puckers; this is comparable to the effect of bending a linear triatomic molecule. It is still true that the two highest-energy electrons in NH_3 are only moderately stable, and might very well be donated to a good acceptor molecule. The chemistry of NH_3 is largely a reflection of this capability, as we shall see in Section 6-11 and in Chapters 10 and 16.

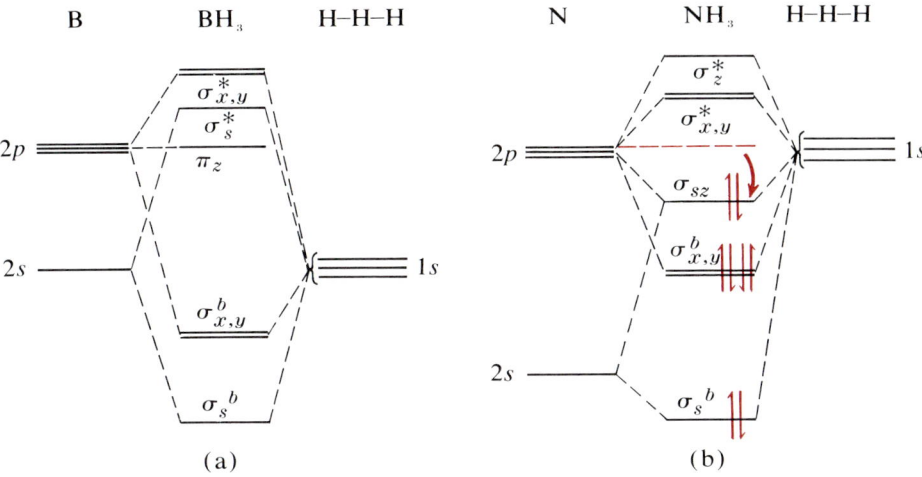

Figure 6-57 Comparison of energies of MO's for (a) planar and (b) pyramidal forms of XH_3.

6-11 Tetrahedral Molecules, Donor–Acceptor Interactions, and Hybridization

The tetrahedral molecule has the most complex geometry (the largest number of ligand atoms) of any central-atom type of molecule that does not have d valence orbitals on the central atom. This geometry is characteristic of molecules with the stoichiometry AB_4, the exclusive exceptions being d-orbital systems. Because the results can be reasonably extended to the enormous number of organic molecules that do not have a central-atom type of geometry, we shall take CH_4 as an example.

Tetrahedral-Molecule Molecular Orbitals

After establishing the tetrahedral geometry of the CH_4 system we can draw the coordinates of Fig. 6-58a, with the overlaps shown in Fig. 6-58b. The cube is indicated in each case because the four apexes of a tetrahedron lie at the alternate corners of a cube, as shown. There are now eight AO's in the system, so we must produce eight MO's. Two of these will obviously be the bonding and antibonding versions of the overlap of the C $2s$ with the H $1s$ orbitals as seen

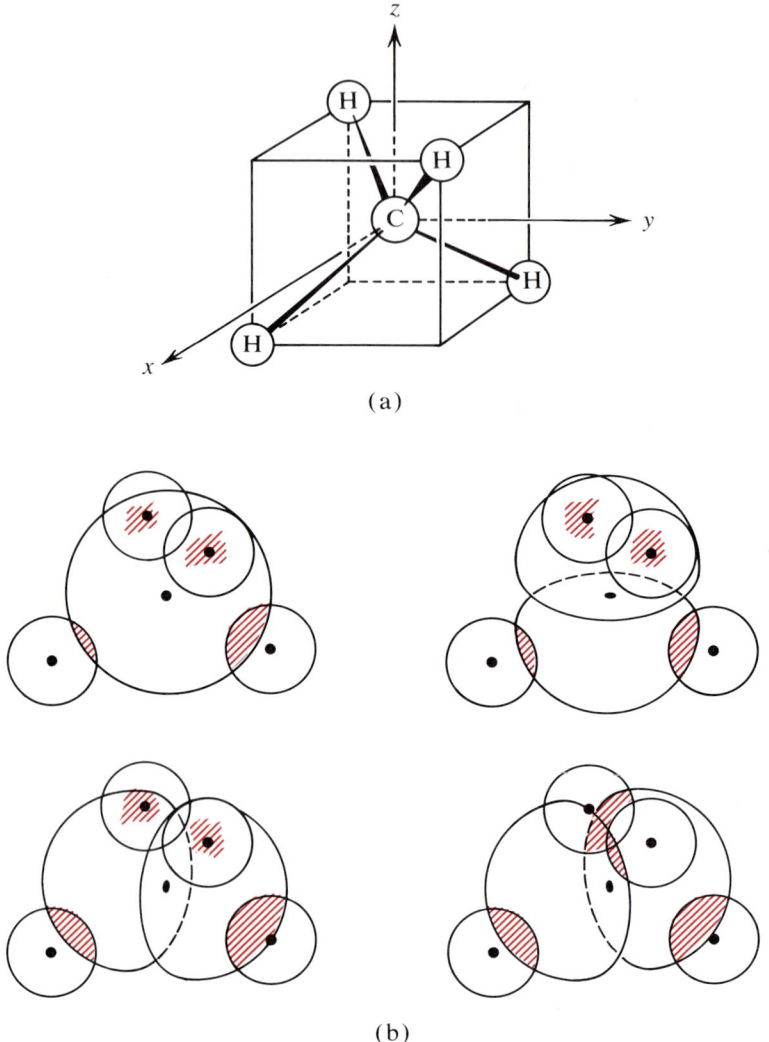

Figure 6-58 Coordinate axes and AO overlaps for tetrahedral CH_4.

in Fig. 6-59a. With respect to the C $2p$ overlaps, all four H $1s$ orbitals overlap a given p orbital in exactly the same way; and further, the overlaps of all three C $2p$ orbitals are exactly the same (since they stick out the centers of the faces of the cube and the H atoms are at the corners of the cube). The bonding and antibonding MO's resulting from these overlaps, then, will be threefold degenerate, as seen in Fig. 6-59b. The resulting energy-level diagram for CH_4 is

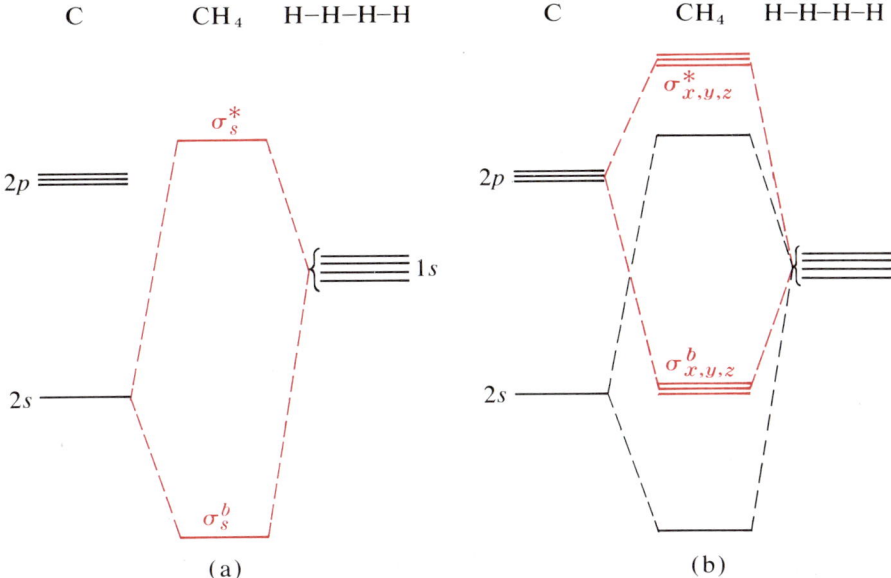

Figure 6-59 Development of the energy-level diagram for CH$_4$. (a) Bonding and antibonding overlap of C 2s with H 1s AO's. (b) Bonding and antibonding overlap of C $2p_x$, $2p_y$, and $2p_z$ with separate combinations of H 1s AO's.

shown in Fig. 6-60, populated with the eight valence electrons of the molecule. There the tetrahedral energy levels are shown in comparison with the corresponding energies of the trigonal pyramidal and trigonal planar systems, also, and it is seen that the two highest-energy electrons of the eight are considerably

Figure 6-60 Comparison of MO energies for trigonal forms of XH$_3$ and the tetrahedral form of CH$_4$. Note the energy lowering on approaching tetrahedral geometry.

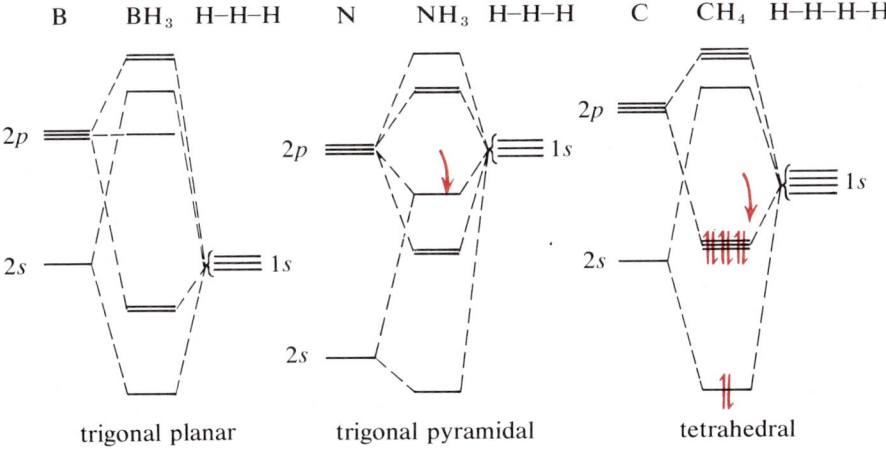

stabilized in the tetrahedral arrangement over their energies in the other two geometries. The drop of the fourth MO from trigonal planar to trigonal pyramidal has already been noted; its further drop, a very substantial one, from trigonal pyramidal to tetrahedral is due, of course, to the addition of another AO with which a bonding combination can be formed. It is true that an antibonding combination must also be formed, but if it need not be populated with electrons, its high energy does not matter.

The Approach of Electron Acceptors to Tetrahedral Geometry

This energy is the key to the very great reactivity of BH_3. Since it has only six valence electrons, it is **electron deficient**. In its planar structure it cannot accommodate any more electrons in low-energy orbitals, but if it puckers into a pyramidal structure it becomes a better acceptor; if another atom with a pair of electrons appears to become a fourth ligand atom, that pair of electrons can be placed in a distribution corresponding to a very low, stable energy in the tetrahedral configuration. Thus BH_3 is very unstable with respect to the formation of, for instance, the tetrahedral ion BH_4^-. This ion is quite stable, and its salts are well known. Another possibility is that one of the hydrogens on another BH_3 can become the fourth ligand atom; this is what happens in the dimerization of BH_3 to diborane, B_2H_6, whose structure is shown in Fig. 6-61. Each B is tetrahedrally surrounded by H atoms, and the central B—H—B form a three-center, two-electron bond that differs from a hydrogen bond in having only two electrons for the three atoms, so the exclusion principle does not prevent extensive overlap of the three AO's.

The relationship between the energies in Fig. 6-60 also accounts for the electron-donor properties of NH_3. Just as for BH_3, the seventh and eighth electrons

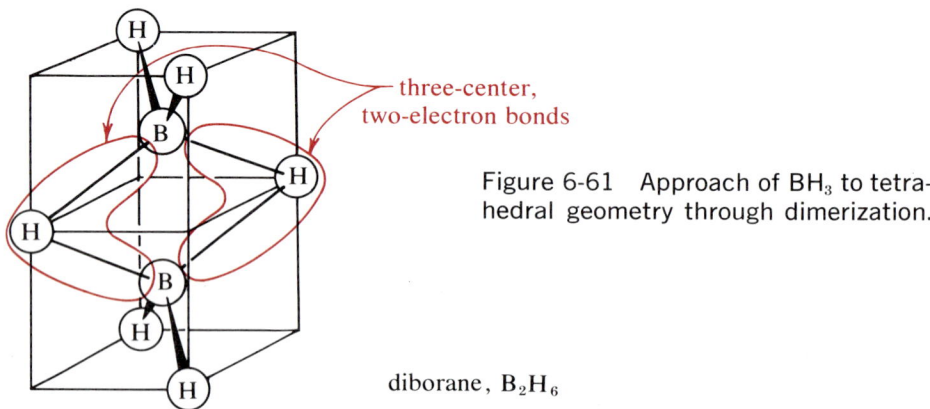

Figure 6-61 Approach of BH_3 to tetrahedral geometry through dimerization.

diborane, B_2H_6

can be greatly stabilized by the adding of another ligand atom (but no more electrons) to form a tetrahedral system. An interesting possibility is that NH_3 ought to be able to donate to BH_3, thereby stabilizing both systems. This does occur, although diborane must be used as the starting material, and the resulting product has the empirical formula $BH_3 \cdot NH_3$ but is actually $[BH_2(NH_3)_2]^+$-BH_4^-. A more familiar result of the donor properties of NH_3 is its formation of ammonium ion, NH_4^+, in acids (with the electron acceptor being an N^+).

Extending these MO results somewhat, we can begin to appreciate why the molecule $BeCl_2$, which is presumably linear in the gas phase, is nonlinear in a crystal. It is also electron deficient, and in forming a crystal with molecules in close contact the more favorable structure turns out to be one in which the Be is surrounded tetrahedrally by Cl atoms; the reasoning from the MO model is analogous to that for BH_3 forming B_2H_6.

The perceptive student may ask why, if forming a fourth ligand-atom bond is so good, molecules do not add a fifth ligand atom, thereby being able to accommodate 10 electrons in bonding orbitals, and so on. The answer, as Study Problem 11 indicates, is that if only four AO's are available on the central atom, then only four bonding MO's can be formed from any molecular geometry, no matter how many ligand atoms it may have. This means that the tetrahedral geometry achieves the maximum number of bonding MO's and the minimum amount of ligand-atom repulsion, which is why it is so commonly found experimentally.

Hybrid Orbitals

There is another approach to the description of orbital energies, which we have already used to some extent without naming it. When two AO's on the same atom both overlap another AO in the same way, we have described them as being "mixed" in the sense that MO's are formed in which the AO's may contribute differently to the overall bonding or antibonding effect. One forthright variation of this is to assume that the two AO's on the same atom are already "mixed" or **hybridized** before any combination into MO's occurs. Thus for an s and a p orbital in a linear triatomic molecule we could form two **sp hybrids** as shown in Fig. 6-62, one of which would strongly overlap the left ligand atom, the other of which would strongly overlap the right ligand atom. From these four AO's we get four MO's, a bonding and an antibonding combination for each hybrid; the only difference between this result and the energy levels of Fig. 6-42 is that the two bonding levels are now degenerate since the hybrids are identical. It is also possible, although not as easy to see graphically, to combine an s and two p AO's to form three **sp²** hybrids that are very directional, like the sp hybrids, but are directed at 120° to each other. This makes them ideal for bonding

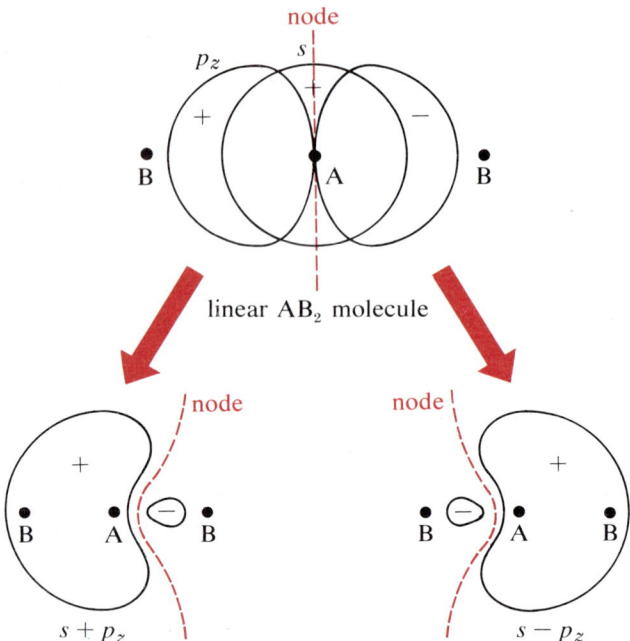

Figure 6-62 Formation of sp hybrid orbitals in a molecule with linear geometry.

in the trigonal planar case, since they give ideal overlap with the ligand orbitals. For the tetrahedral case, the hybridization approach is to form four hybrids from the s and all three p orbitals to give **sp³** hybrids. These turn out to be tetrahedrally arranged and very directional in their extension. Organic chemists usually describe bonding in compounds having more than one carbon atom in terms of these hybrid orbitals; for instance, if methane is considered to have a H atom at the end of each sp^3 hybrid orbital, then ethane, C_2H_6, can be considered as a carbon atom replacing a hydrogen in the tetrahedral ligand-atom arrangement around one carbon, and so on. See Fig. 6-63. Note, however, that the present discussion of hybrid orbitals does not include any possibility of π bonding; we shall return to this subject in Chapter 15.

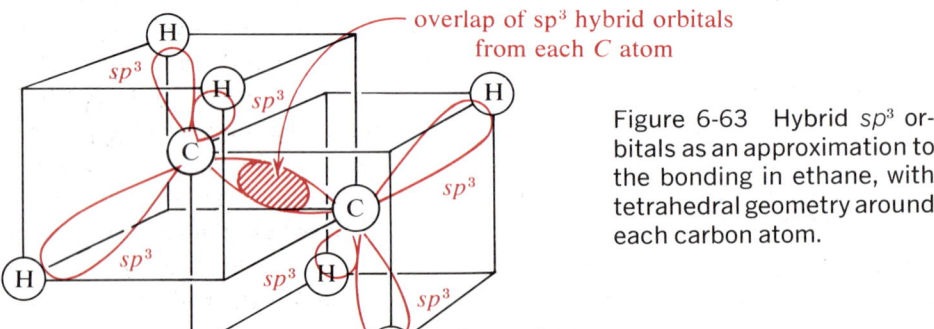

Figure 6-63 Hybrid sp^3 orbitals as an approximation to the bonding in ethane, with tetrahedral geometry around each carbon atom.

6-12 Many-Atom Systems and Metals

So far in this chapter we have examined the process of applying the MO model to small molecules involving only s and p valence orbitals, with the number of ligand atoms and geometry growing progressively more complicated. At the beginning of the chapter we indicated that there were other kinds of polyatomic systems than molecules, and this is an appropriate point at which to expand on that remark. An important class of polyatomic systems that are *not* molecules in the conventional sense is that of the crystals of the elements. Some of these, like iodine (I_2), are composed of discrete molecules held together by van der Waals forces and described using the Lennard-Jones potential. We are not particularly interested in those at this point. There is a much greater number of crystalline elements that consists of single atoms bonded in a more or less close-packed structure. All of the metals (and this includes everything to the left of Group IV in the periodic table plus a few others) fall in this class, plus some other elements such as carbon, silicon, and arsenic. In general, we observe that there must be a significant chemical bonding force operating in these crystals, since otherwise the weak van der Waals forces would allow them to melt and boil at quite low temperatures — while actually some of these elements' melting points are as high as 3000°K.

Following our previous approach, we expect that this bonding force originates in the bonding overlap of the AO's involved. We can check this by calculating the degree of overlap of the Slater valence orbitals for some metal at the internuclear separation of the atoms in that metal (which X-ray data can give us). For aluminum, for instance, the degree of overlap of $3s$ valence orbitals on two Al atoms at the metallic separation from each other is about $S_{Al\text{-}Al} = 0.3$, so we are justified in considering a rather startling possibility: the valence electrons in these elements may be described by MO's spread out over Avogadro's number of atoms, so that a block of metal may be equivalent to one enormous molecule. What can we say about the energy levels of these many-atom MO's?

Molecular Orbitals for Extended Systems as a Function of Atomic Separation

Suppose we first consider two atoms at a very great distance from each other. Their AO's do not overlap, so there is no effective bonding between them. If we now bring these two atoms closer together, the AO's begin to overlap and some bonding effect (and, of course, antibonding effect) begins to exist. If the two atoms both have s and p valence orbitals, they will both interact to give MO's like those in Fig. 6-24. If we bring them closer and closer together, the overlap increases continuously, and the bonding effect also increases (since it is proportional to the degree of overlap). However, at very small separations the inner cores of nonvalence electrons begin to repel each other strongly, as do the nuclei, and this effect eventually swamps the bonding effect for sufficiently small atomic separations. The radial dependence of the MO energies is shown in Fig. 6-64, and these effects can be seen. If the number of atoms in the system increases, the separation between bonding and antibonding levels increases, but not very much. We have already seen that the MO's for more

Figure 6-64 Dependence of the energies of diatomic MO's on interatomic distance.

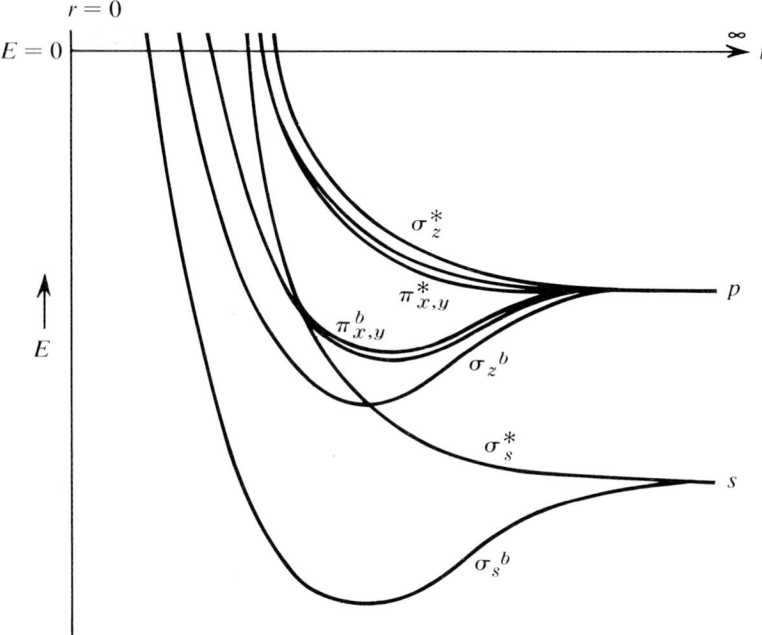

complicated systems tend to be more thickly crowded in the energy-level diagram, and if we go to a very large number of atoms in the bonded system— such as Avogadro's number— the result is Avogadro's number of MO's for each type of interacting AO, but they are very densely packed energetically, and the separation between the lowest and highest levels in such a **band** of energy levels is not much greater than for only two atoms. Figure 6-65 shows the radial dependence of these MO energies; it is very much like Fig. 6-64 except that instead of only two MO's arising from each atom's AO types, there are now so many as to provide almost a continuous band of energy levels.

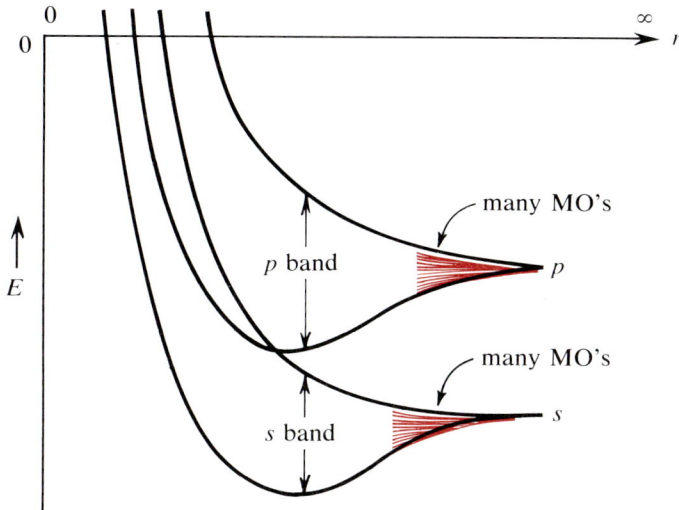

Figure 6-65 Dependence of the energies of many-atom MO's (metals) on interatomic distance.

Electrical Conductivity

This approach, when done with more mathematical elegance, is known as the **band theory of solids**. It permits us to account for many interesting experimental properties of crystals. For instance, suppose we have a crystal of sodium metal, each atom of which has one valence electron. If we consider only the overlap of the 3s valence orbitals, the energy band for the sodium crystal made up of N atoms will contain N energy levels. Since these N orbitals can contain $2N$ electrons but the atoms provide only N electrons, the band is only half full. What does this mean experimentally? It will require only a very tiny amount of energy to put another electron into one side of the sodium crystal, and the MO it enters is delocalized over the whole many-atom system, so it will be very easy to insert a new electron into the crystal and have it pass through the crystal. This is exactly the process of passing an electrical current through the crystal. All of the elements with one s valence electron (Li, Na, K, Rb, Cs, Cu, Ag, and Au) are excellent electrical conductors, which, as we can see, is exactly what the band theory predicts.

What about elements with two s valence electrons? Presumably the s band of energy levels is now full, so there should not be any vacant MO's in which to place a new electron, and these elements (Be, Mg, Zn, etc.) should be electrical insulators. However, if the interatomic distance is small enough in the solid, the s band and the p band (which is empty) overlap; some of the s electrons in effect "spill over" into the p band, and good electrical conduction is still possible since vacant MO's are only a very small amount of energy above the filled ones. However, it is experimentally true that the electrical conductivity of these metals, while good, is distinctly less than corresponding metals with only one valence electron. Figure 6-66 shows the filled and empty MO's for both the sodium and magnesium cases.

Figure 6-66 Filling of energy bands by Na with one valence electron per atom and by Mg with two valence electrons per atom.

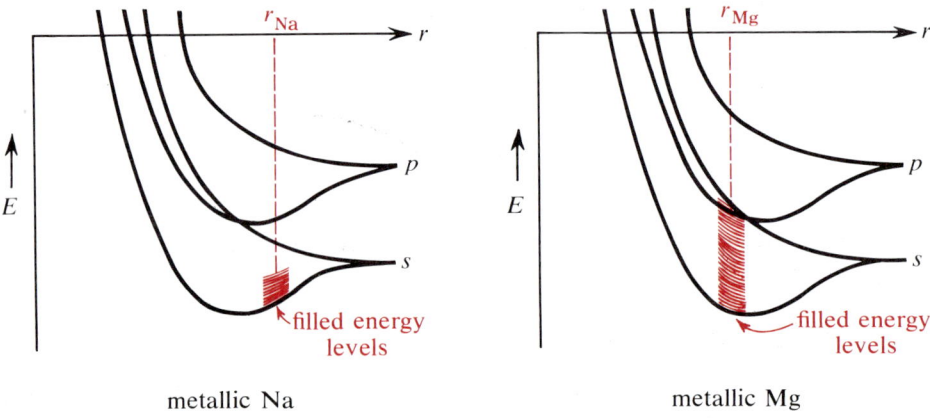

metallic Na   metallic Mg

In general, the three p orbitals on each atom that are degenerate for the free atom will form separate bands in an extended crystal. These bands may or may not overlap in energy, depending on the precise crystal structure for a particular case. In the case of crystalline carbon (diamond) there are four valence electrons and the appropriate energy bands will accommodate exactly four electrons per atom. The next higher band does not overlap at the C—C distance in diamond; in fact it is about 5 eV away, so diamond is an excellent electrical insulator. It is characteristic of electrical insulators (including ionic crystals, which can be treated by the same scheme) that there is a substantial energy gap between the topmost filled energy level and the lowest empty one. This pattern is seen in Fig. 6-67 for a typical insulator.

Also shown in Fig. 6-67 is an interesting intermediate case. We have said that a good electrical conductor has unfilled energy bands or a filled band overlapping an empty one, and that a good insulator has a large separation between filled and empty bands. There is a possibility that a filled band and an empty band might be separated by an energy gap so small that it was about the same size as thermal energy at room temperature (where $kT = 0.025$ eV). In this case there would be a few electrons that had enough thermal energy to be excited up to the empty level at any given time, and at least a small amount of

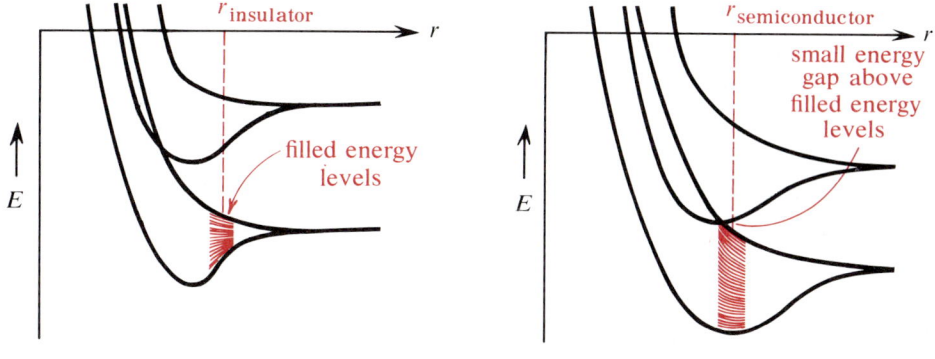

Figure 6-67 Energy gaps affecting conductivity of electrical insulators and semiconductors.

electrical conduction would be possible. Such a system is called a **semiconductor**. Since the application of a small electrical field in such a marginal case can sharply change the electrical conductivity, the electronic effects of semiconductors are most important. Solid-state electronic devices, such as the transistor, germanium diode, and FET (field-effect transistor), are practical applications of the utility of this effect. Since we are seeking a very sharply defined small energy gap, it would be fortuitous to expect many naturally occurring semiconductors; elemental silicon and germanium are the two best-known cases. But it is quite possible, as we have seen all through this chapter, to modify atomic energy levels by forming compounds. What is normally done to produce "tailored" semiconductors is to introduce a small amount of some other element with different electronic properties. which alters the precise positioning of the band edges as desired. This is the reason for the extreme interest in purification of crystals shown by electronic manufacturers; the ultra-pure base material is "doped" with reproducible trace amounts of other elements to reproducibly achieve the desired effect.

6-13 Summary

A sound understanding of chemical structure—crystal structure or molecular structure—is necessary to understand chemical reactivity. In Chapters 3–6 we have established the currently accepted views of chemical structure. Chapters 3 and 4 developed a description of the physical properties of two phases of matter into the atomic/molecular model of matter, and established some of the features that model must have. Chapter 5 developed the modern theory of the atom as a model for these features and others that result from the key experiments of atomic physics and chemical physics. In this chapter we have taken the modern quantum mechanical theory of the atom and developed a further model for molecular structure that is compatible with the one for atomic structure, and that can explain a wide array of chemical facts.

The valence-shell electron-pair repulsion (VSEPR) technique for predicting molecular geometry was shown to be a versatile means of obtaining the information for any molecule that is necessary to establish molecular energy levels. With the molecular geometry known, we developed the LCAO approximation to the distribution of electrons in molecules and showed that it leads to the physically reasonable prediction that bonding electrons are localized between the nuclei they are bonding together. However, we also observed the possibility of forming antibonding orbitals in which the electron is actually prevented from being in the energetically favorable location between the nuclei. We discussed, algebraically, the properties of normalization and orthogonality that proper atomic and molecular orbitals possess and showed that when MO's are *not* orthogonal to each other they can be expected to influence each other's energy.

We then developed MO energy-level diagrams for several of the common molecular geometries shown by systems having only s and p valence AO's, and showed, ultimately, that a molecule's geometry has a distinct effect on its chemical reactivity, since it dictates the MO energies. We saw that electron donor–acceptor properties could be explained from this model. We saw that properly constructed MO's predict that some electron transfer will occur within a heteronuclear molecule, and took an initial look at the resulting ionic bonding. Finally we took a short look at the band theory of solids and its prediction of the electrical conduction properties of solid elements.

The result of all this has been to give us a very versatile and powerful tool for predicting electronic structures in molecules; from this, in succeeding chapters, we shall develop our ideas of chemical reactivity. There will be places at which we shall return for a closer look at some particular aspect of MO electron structures, but all of the later discussion will be built very specifically on the foundation of this chapter.

It is important to realize that this area of chemistry is rapidly growing and changing. It is almost certain that some details of this approach to the description of bonding will be modified or replaced as our understanding of chemical principles grows. For instance, in doing detailed numerical calculations of energies—which we have not attempted—the results, at the present time, are suggestive but numerically imprecise in most cases. Changes that will improve the quantitative accuracy of these calculations are sure to come. However, the qualitative (and in some cases, the quantitative) successes of the LCAO–MO model have been so impressive, and so widely distributed over diverse chemical phenomena, that the general structure of the model is sure to stand as a reliable approximation method, at the very least. Molecular orbitals are here to stay.

We have now finished Part II on chemical structure. We are about to move into Part III on chemical dynamics, in which we shall explore the ways chemical change occurs and the reasons *why* it occurs. Initially we shall go back to a chapter very much like the two at the start of this section, one which describes the structure of liquids. This chapter is placed in Part III because one of the most interesting things about liquids is the nature of their transformations into gases (when they boil) and solids (when they freeze); the principles governing the spontaneous boiling and freezing of liquids provide an excellent simple model for spontaneous chemical change. Following the discussion of liquids, we shall take a hard look at the nature of spontaneity from a purely experimental point of view. In all of this discussion, there will be little mention of MO's or energy levels. But with a sound experimental foundation for the understanding of spontaneous change, we shall return to the molecular model and find it most adaptable to the prediction of spontaneous chemical change—just as we have already predicted.

Study Problems

1. Predict the geometry of the following:
    (a) BeH_4^{2-}
    (b) H_2O
    (c) NO_3^-
    (d) NF_3
    (e) PCl_4^+
    (f) ClO_2F
    (g) XeF_4
    (h) AsH_3
    (i) HCN
    (j) S_3^{2-}

2. Discuss the geometry of the sulfur dichloride (SCl_2) molecule from the viewpoint of both the VSEPR model and the MO model (linear vs. bent). Do the predictions agree with the experimental bond angle of 100.3°?

3. Set up and populate the MO energy-level diagram for HF. Are the electrons polarized more toward H or F?

4. Discuss the partial ionic character that might be present in the B—H, C—H, and N—H bonds in BH_4^-, CH_4, and NH_4^+. Is there a clear trend?

5. Derive the algebraic expression for the normalization constant N^* in the antibonding MO, $\psi(\sigma^*) = N^*(1s_a - 1s_b)$ for the H_2 molecule. Allow for a finite value of the overlap integral S_{ab}.

6. For the two MO's

$$\psi(\pi_y^b) = \frac{1}{\sqrt{2}}(2p_{y_a} + 2p_{y_b})$$

$$\psi(\sigma_s^b) = \frac{1}{\sqrt{2}}(2s_a + 2s_b)$$

show algebraically, by substitution into the orthogonality condition, that they are orthogonal. See Fig. 6-21.

7. For a heteronuclear diatomic molecule AB, $\psi^{MO} = c_1(AO_A) + c_2(AO_B)$, where c_1 may not equal c_2 as it does for a homonuclear A_2 or B_2 molecule. It can be shown that for the lowest energy E the following conditions both apply:

$$(\alpha_A - E) + \frac{c_2}{c_1}(\beta) = 0$$

$$(\beta) + \frac{c_2}{c_1}(\alpha_B - E) = 0$$

Combine these to eliminate the ratio c_2/c_1. Write the resulting equation as a quadratic in E, and solve for E. Simplify this expression as much as possible for the case in which the energies of AO_A and AO_B are quite different, so that $(\alpha_A - \alpha_B) \gg \beta$. (Use the approximation $\sqrt{1+x} \cong 1 + \frac{1}{2}x$ for small x.) Obtain two E expressions. What do these expressions say about the MO energies relative to the AO energies in this case?

8. On graph paper, construct an energy-level diagram for O_2, using the VOIP's from the periodic table inside the front cover and the approximate β expression from p. 274. Assume σ_s overlap of 0.25, σ_z overlap of 0.5, and $\pi_{x,y}$ overlap of 0.2, and ignore σ_s-σ_z interaction. What should the molecular ionization potential be? How does it compare with the experimental value of 281 kcal/mole? What rare gas (He through Rn) does it compare with?

9. If we allow H^+ to react with OH^- to produce water there is considerable evolution of heat and the reaction is very favorable. Draw MO energy-level diagrams to explain the added stability of H_2O over $H^+ + OH^-$. Where does the evolved heat come from?

10. What bond order do you expect for the NO molecule? For the NO_2 molecule? For the NO_2^- ion?

11. Consider a hypothetical molecule AH_5 in which the A atom has only $2s$ and $2p$ valence orbitals. Assuming its geometry to be trigonal bipyramidal and using the BH_3 coordinate system, sketch the overlaps of the five H $1s$ orbitals with the individual A orbitals. Note the sign combinations necessary for favorable overlap (i.e., forming bonding MO's); in all but the σ_s^b case, some H $1s$ AO's are nonbonding so that their signs are undetermined. There will be a σ^b LCAO and a σ^* LCAO for each of these combinations, making a total of eight MO's. The ninth MO (which must be formed since we are combining nine AO's) must correspond to the only other sign combination not yet used. Show that this involves taking the signs of the polar-position H $1s$ AO's opposite to the signs of the equatorial H $1s$ AO's, and that this ligand AO combination can only be nonbonding in overlapping any central atom orbital. There is thus a maximum of four bonding MO's for four central-atom valence orbitals.

Some Further Reading

Gray, H. B., *Electrons and Chemical Bonding*, New York: Benjamin, 1964. Pitched at about the same level as our discussion, with lots of good pictures. Essentially equivalent coverage.

Companion, A. L., *Chemical Bonding*, New York: McGraw-Hill, 1964. This is at a lower level, but quite well written and easy to read. If you had trouble with this chapter, you might fortify yourself with Companion and try again.

Coulson, C. A., *Valence*, Oxford: Clarendon Press, 1961 (2nd ed.). Sooner or later every chemist reads Coulson and wonders how the man says so many hard things so easily. Almost nonmathematical. A classic treatment.

Murrell, J. N., Kettle, S. F. A., and Tedder, J. M., *Valence Theory*, New York: Wiley, 1970 (2nd ed.). Read this after you finish Coulson; it tries to put the numbers to the physical discussion. Chapters 1–5 are review and pretty well adapted to our purposes. Excellent problems with a complete section of hints and complete answers.

III | CHEMICAL DYNAMICS

7

The Behavior of Liquids

What we really want to know as we approach the study of chemistry is how to use it for humane purposes. Inevitably, this requires that we understand how and why chemical reactions occur. It is through chemical reactions that we make the chemical artifacts that are so much a part of our society: fertilizer, steel, plastics, dyestuffs, synthetic blood expanders, color TV phosphors... the list is almost endless. We have not approached the study of chemical reactions—more generally called **chemical dynamics**—until now because all our models for the processes of chemical dynamics involve an appreciation of the features of the atomic/molecular model for the structure of matter. This is, of course, exactly what we said in introducing Part II on chemical structure. In Part II we examined the measurable properties of gases and solids and then constructed a model that was compatible with those properties; with the external features of the atom established, we spent some time discussing the internal structure of atoms and molecules. This background in chemical structure is adequate for the discussion we need to enter now. In the next chapter we shall take a strictly macroscopic look at how we can define the spontaneity of a process in which we are interested. Here, as a sort of bridge between chemical structure and chemical dynamics, we shall investigate the behavior of liquids. One of our purposes, as it was for gases and solids, will be to correlate the properties of liquids with their structure. We shall have another purpose, however, which is to begin looking into the way changes occur spontaneously. Liquids offer a simple model for chemical reactions in their transformations to and from the solid state and the gaseous state. Accordingly, once we have discussed the structure of liquids we shall look at the processes of melting, crystallizing, boiling, and condensing. We shall begin to formulate a model for a process rather than simply for a structure, and this will take us farther toward our goal.

7-1 Experimental Properties of Liquids

We found earlier that we could largely account for the properties of gases by assuming a gas to be a collection of a very large number of molecules moving through space and colliding in a completely random fashion. There was no correlation at all between the positions of the molecules. Conversely, our understanding of crystalline solids rests on a *complete* correlation between the positions of the atoms or molecules in the crystal; regular, completely ordered geometry is the fundamental property of a crystal. Some of the properties of liquids are reminiscent of the properties of gases, others of the properties of solids. For convenience, we shall divide the properties of liquids roughly into these two camps. Let us look first at the solid-like properties of liquids to see not only how they resemble solids but how they differ from solids.

Liquid Density

Our first instinct on seeing a mass of some new form of matter is to pick it up and heft it. The physical property we are testing, of course, is its density. In this respect, liquids are much more like solids than like gases. Solids are quite dense and so are liquids. Table 7-1 shows the relationships between the densities of these three states for several elements or compounds. Both solids and liquids are roughly 1000 times as dense as gases at ordinary pressures.

Table 7-1
Densities of Different States of Matter

Substance	Crystal Density (g/cm³)	Liquid Density (g/cm³)	Vapor Density at 0°C, 1 atm (g/cm³)
mercury	14.193	13.546	0.00895
aluminum	2.699	2.382	0.00120
sodium chloride	2.165	1.550	0.0026
sulfur	2.07	1.803	0.0114
acetone	0.9686	0.792	0.00259
water	0.917	1.000	0.000596
hydrogen	0.0771	0.0710	0.000090

Note, however, that there is a distinct and characteristic difference between the densities of solids and of liquids; in general, liquids are 5–10% less dense than the corresponding solids (but note the exception in the case of water).

Table 7-1 is somewhat misleading because it quotes densities at different temperatures, but the phenomenon is real and says something about the structure of liquids on the molecular scale. To achieve the high densities that are observed, liquids must be composed of nearly close-packed, touching molecules. The modest reduction in density on melting suggests that in a liquid the molecules either must be spaced slightly farther apart or must have some vacancies interspersed in the structure; we shall examine this idea shortly.

Liquid Compressibility

Another property of liquids that resembles the equivalent property of solids is their compressibility. As we noted at the beginning of Chapter 4, gases are about 100,000 times as compressible as solids; liquids are only about 10 times as compressible as solids. Taking the definition of compressibility as $(1/V)(dV/dP)$, the rate of change of volume with increasing pressure, we obtain characteristic values as in Table 7-2. This is sometimes a surprising discovery, because we are so used to observing the fluidity of liquids that we forget that liquid flow means only that the shape of the liquid sample is changing, not its size. There are several familiar examples of the incompressibility of liquids, among them the bursting of an overheated thermometer and the explosion of a plastic bag filled with water and dropped from a window.

Table 7-2
Compressibilities of Liquids and Solids

Substance	Compressibility (atm^{-1})
liquids	
mercury	3.9×10^{-6}
acetone	82×10^{-6}
water	49×10^{-6}
solids	
potassium chloride	5.6×10^{-6}
lead	2.24×10^{-6}
tungsten	0.322×10^{-6}

Again we should note that the difference in compressibility between solids and liquids bears on the physical model we construct; since liquids are slightly more compressible we anticipate that their structure must contain more open space. This is compatible with the reduced density we observe, since it seems to indicate that the atoms are more widely spaced or that there are holes in the array of atoms.

X-Ray Diffraction Patterns of Liquids

A third physical property of liquids resembling that of solids is that they show some order in their diffraction of X rays. Figure 7-1 compares the X-ray powder pattern for solid potassium (a) with the pattern for molten potassium (b) and the absence of any pattern, which is characteristic of gases (c). The sharp lines due to diffraction by the crystalline powder originate in the relatively precise spacing and geometry of atoms in the solid crystal. The interesting feature of

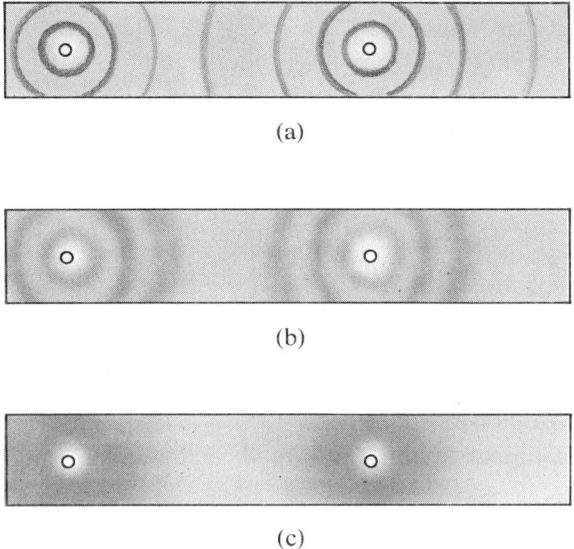

Figure 7-1 X-ray diffraction patterns for potassium. (a) Metal crystals. (b) Liquid metal. (c) Metal vapor.

the liquid diffraction pattern is that it shows the retention of some, but not all, of the order of the crystal. Only an orderly array with regular spacing between atoms can give sharp diffraction lines; since some lines are seen in the liquid pattern some order must be present in the spacing, but since the lines are blurred we assume the order is breaking down. In Fig. 7-2 the liquid pattern is analyzed further. Here the dotted lines correspond to the intensity of X rays or number of atoms that would be characteristic of a completely random distribution; we see that at about 4–4.5 Å from any given atom there are substantially more atoms than a random distribution would suggest, while at about 6 Å there are significantly fewer, and so on. This is entirely reasonable, since the radius of a single potassium atom is 2.23 Å and the center of a nearest-neighbor atom

Experimental Properties of Liquids | 343

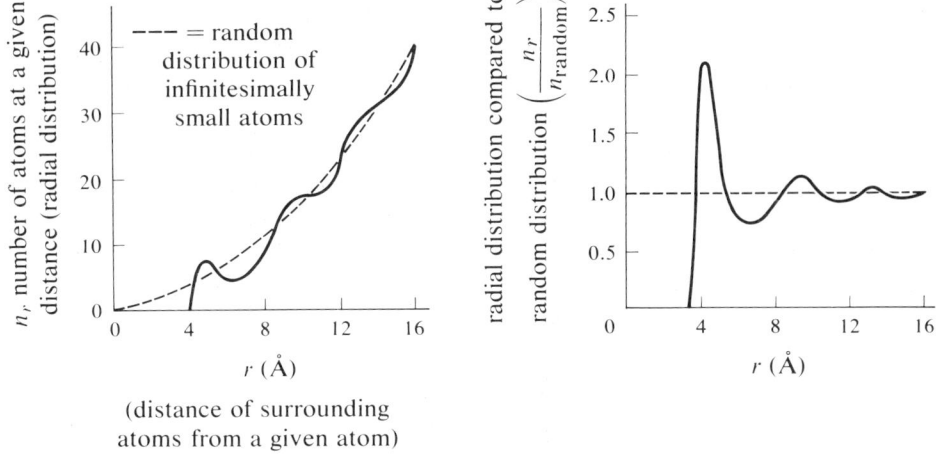

Figure 7-2 Interpretations of the X-ray diffraction pattern of liquid potassium in terms of the average surroundings of a given atom. (Modified from Thomas, C. D. and Gingrich, N. S., *J. Chem. Phys.* 19 (411), 1938.)

would be twice that far away, or 4.46 Å. The fact that there are at least some atoms with spacings intermediate between 4.46 Å and 6.3 Å (the next crystal spacing), however, indicates that some atoms are located in less orderly, more random positions. This is expectable because liquids are isotropic—they do not have the angular dependence of physical properties that we have seen is characteristic of completely ordered systems of atoms.

The partial order that does exist about any given atom in a liquid cannot extend more than about three or four atom diameters in any direction, however, because the atom-distribution curve rapidly approaches the dashed line representing the random distribution. It is also interesting to compare the number of nearest neighbors or coordination number in the solid crystal with that in the liquid. Solid potassium is cubic close packed and has eight nearest neighbors; by comparison, the innermost peak in the radial distribution curve for the liquid indicates slightly less than eight, perhaps an average of 7.5 neighbor atoms. This suggests that a reasonable simple model for liquid structure might be an approximately close-packed array of atoms or molecules with vacancies scattered through it, and indeed this is the model we shall discuss in Section 7-2.

Liquid Viscosity

On the other hand, there are some properties characteristic of liquids that more nearly remind us of gases than solids. The most obvious property of a liquid is that it flows—that is, it has a relatively low viscosity. There are, of course, variations in the viscosity of liquids, with behavior ranging all the way from liquid helium with nearly zero viscosity to more familiar "liquids" such as tar. The measurement of viscosity thus gives us an important physical quantity. In discussing gas viscosity we noted that measurements are usually made by allowing the fluid to flow through a small tube and measuring the rate; Fig. 7-3 shows an **Ostwald viscometer** for liquids. The left arm of the viscometer is filled with liquid to the upper mark, above the bulb. The liquid is allowed to drain under the influence of its own weight through the capillary tube until its surface reaches the lower mark, and the time required for this drainage of a standard volume of liquid is measured. Empirically we find that the time is

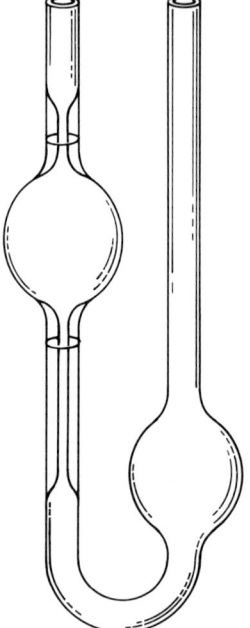

Figure 7-3 Ostwald viscometer.

proportional to the volume of liquid and to the length of the capillary tube, but inversely proportional to the pressure difference (due to gravity) forcing the liquid through the tube and also inversely proportional to the fourth power of the radius of the tube:

$$t \propto \frac{Vl}{Pr^4}$$

Using the coefficient of viscosity η (see p. 119), we can specify the proportionality constant, giving **Poiseuille's equation**:

$$t = \frac{8\eta}{\pi}\left(\frac{Vl}{Pr^4}\right) \quad \text{or} \quad \eta = \frac{\pi r^4 Pt}{8Vl} \tag{7-1}$$

In principle absolute viscosities can be measured this way, but it is a difficult process. A much simpler procedure is to measure the absolute viscosity of a standard liquid such as water, then obtain a relative viscosity for all other liquids. Since the pressure forcing the liquid through the capillary is due to its weight, which is proportional to its density, we can simplify our equation:

$$\frac{\eta_{unknown}}{\eta_{H_2O}} = \frac{(\pi r^4 P_{unkn} t_{unkn})/8Vl}{(\pi r^4 P_{H_2O} t_{H_2O})/8Vl} = \frac{P_{unkn} t_{unkn}}{P_{H_2O} t_{H_2O}} = \frac{\rho_{unkn} t_{unkn}}{\rho_{H_2O} t_{H_2O}}$$

When viscosity is measured by this or similar means, the dimensions of the coefficient of viscosity are seen from Poiseuille's equation to be (distance)^4 × (force/area)(time) divided by (distance)^3(distance), or units of dyne-sec/cm^2. This latter unit is called a **poise** in Poiseuille's honor. The absolute viscosity of water at 20°C is 0.01002 poise (P), or 1.002 centipoise (cP), a convenient unit to use for many liquid viscosities. By way of contrast, the viscosity of ethyl ether is 0.233 cP at the same temperature, and the viscosity of glycerine is 1490 cP or 14.90 P. Gas viscosities are smaller; for instance, the viscosity of dry air at 18°C is 0.01827 cP. Nevertheless, the viscosity of familiar liquids is clearly much closer to that of gases than to any equivalent property of solids; for glacier ice, which "flows" in a fashion, a viscosity of 10^{14} P has been estimated!

Liquid Heat Conductivity

It is appropriate here to look at the liquid constants corresponding to the other transport properties of gases, thermal conductivity and diffusion. In each case we find that the numerical values of the constants are intermediate between the values we might expect for a typical gas and the values for the appropriate type of solid. Thermal conductivity varies among solids just as electrical conductivity does, but if we restrict ourselves to metals we find typical values for solid metals on the order of 0.1 cal/sec conducted across a slab of metal 1 cm thick with a cross-sectional area of 1 cm^2 and a temperature difference of 1° between faces. For a gas with the molecular weight of an average metal, we can calculate (see p. 124) a thermal conductivity of about 10^{-5} cal/deg-sec-cm. Liquid mercury falls between these values, having a thermal conductivity of 0.02 cal/deg-sec-cm; in this case, the liquid value is closer to that for the solid than to that for the gas.

Liquid Diffusion

The other transport property of gases is diffusion. We readily recognize that diffusion is possible in gases and liquids, but it can also be measured between two solids in contact with each other. However, solid–solid diffusion is very slow. The most convenient way to describe the diffusive quality of any substance is to measure its **self-diffusion** coefficient. On p. 123 we wrote an equation governing diffusion:

$$\text{mass flow rate} \left(\frac{dm}{dt}\right) = DA \frac{d(\text{concentration})}{dr}$$

If we are dealing with self-diffusion, the quantity D is a measure of how fast an atom or molecule of a given substance diffuses through a lattice of molecules just like it. Experimentally this is approached by using radioactive atoms or isotopes that are chemically identical to the atoms being investigated; the diffusion rate can be measured by monitoring the spread of radioactivity through the specimen. The self-diffusion coefficient D for gases is characteristically about 0.1 cm²/sec; for liquids it is about 10^{-5} cm²/sec, and for solids it is about 10^{-10}–10^{-15} cm²/sec, all values being taken at room temperature. This gives us an idea of the relative rates involved, but the difference is somewhat exaggerated because the same factors (strong interatomic forces) that make a solid high melting tend to prevent its atoms from diffusing through it; a safer comparison is made between the solid, liquid, and gaseous states of the same substance at the same temperature. Metallic sodium melts at 97°C; at that temperature we can measure the self-diffusion coefficients for the solid and liquid states and extrapolate from measured values for the vapor. The coefficient D for solid sodium is about 1.7×10^{-7} cm²/sec, for liquid sodium D is 4.2×10^{-5} cm²/sec, and for sodium vapor D is about 0.7 cm²/sec. Again we see a sort of averaging of the properties of solids and gases.

This averaging of properties sometimes causes difficulty in deciding when a substance is really a liquid. We shall see later in dealing with liquid–vapor equilibrium that under certain circumstances a liquid cannot be distinguished from a gas, but under more ordinary circumstances we can safely make this distinction—if we can see a surface bounding a specimen, it is a liquid. The properties of this surface thus become important, and we shall examine those properties shortly. However, there is more uncertainty in deciding between the solid and liquid state for certain substances such as tar, hot glass, and soft ice cream. Our intuitive definition is usually based on the viscosity of the substance, that is, its resistance to plastic deformation. If it is very resistant we think of it as a solid; if not, we think of it as a liquid. But there is no sharp dividing line that can be defined unequivocally, and in fact we usually have to content ourselves with a good deal of imprecision in separating noncrystalline or microcrystalline solids from liquids.

Surface Tension

We have now mentioned a number of properties of liquids, some of which are very like those of solids, others of which are very like those of gases. We have not mentioned one of the prominent properties that we can measure for a liquid, but that has no direct counterpart in any measurable property we have previously discussed for gases or solids, **surface tension**. Surface tension as a force is familiar to all of us. It is what makes water bead up on a newly waxed automobile and what makes soap bubbles round. Its magnitude can be measured readily for most liquids by several methods. One of the most widely used is the **du Noüy** ring method, shown in Fig. 7-4. A wire ring suspended from a balance is slowly raised out of the liquid, bringing some liquid with it since the liquid

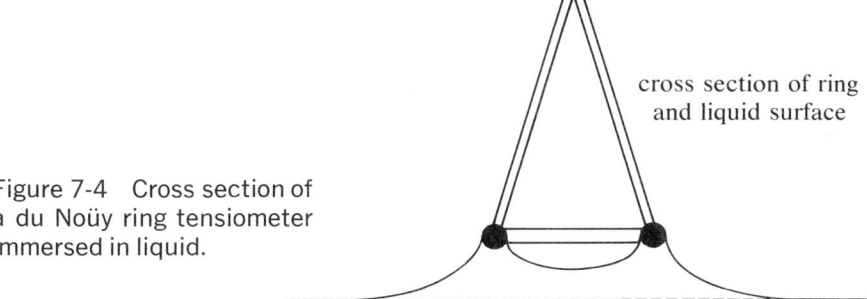

cross section of ring and liquid surface

Figure 7-4 Cross section of a du Noüy ring tensiometer immersed in liquid.

surface hangs from the wire ring. Since surface tension is what holds the liquid to the ring, it is proportional to the maximum weight of liquid the ring can support, wg, where w is the weight of the liquid in gram-mass units and g is the acceleration due to gravity. Surface tension is inversely proportional to the length of liquid surface required to support a given weight of liquid, which is the inner circumference of the ring plus the outer circumference of the ring. Finally, the proportionality constant is related to the volume of liquid contained in the complex shape of the hanging liquid, which can be calculated with some effort. We thus have

$$\gamma = \frac{wg}{2\pi(R_i + R_o)} F \tag{7-2}$$

where R_i and R_o are the inner and outer radii of the wire ring, respectively, F is the proportionality constant, and γ is the coefficient of surface tension or, more often, simply the surface tension of the liquid.

Another technique is the **capillary-rise** method shown in Fig. 7-5. If a liquid wets glass, as water does, there is a considerable force of adhesion holding the liquid to the glass. In a small tube this force attracts the liquid very near the edge of the liquid surface inside the tube, so that the liquid surface adopts a characteristically curved shape; this surface (concave in this case) is called the **meniscus**. In a very small tube the meniscus is spherical and we can again think of the liquid as hanging from a line around the inside of the tube. The surface tension force pulling the liquid up is equal to γ times the inner circum-

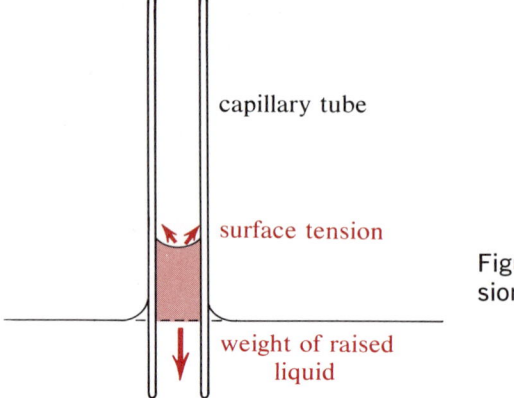

Figure 7-5 Capillary-rise tensiometer.

ference of the tube, $2\pi r$. So the liquid will rise inside the tube until the weight of supported liquid equals the surface-tension force. Since the weight of supported liquid equals the volume of liquid in the capillary, $\pi r^2 h$, times the density ρ of the liquid, times g, we can write

$$2\pi r\gamma = \pi r^2 h\rho g \quad \text{or} \quad \gamma = \tfrac{1}{2}rh\rho g \qquad (7\text{-}3)$$

If the liquid does not wet the tube (as mercury does not wet glass), the surface tension causes the meniscus to be convex instead of concave—the force is trying to minimize the surface area of the liquid. In this case the meniscus will actually be lower than the level outside the capillary, but Eq. (7-3) still holds. Some characteristic surface-tension values for a variety of liquids are given in Table 7-3.

Table 7-3
Surface Tension of Liquids

Liquid	Surface Tension Near 25°C (dynes/cm)
acetone	23.7
benzene	28.8
ethyl alcohol	22.7
glycerin	63.4
mercury	484.0
water	72.0
water with 1% gelatin	8.3

What is surface tension, and why does it occur? Surface tension is a force that tries to minimize the surface area of any body of liquid. All of our other evidence on the structure of liquids suggests that a liquid's molecules are still fairly closely packed, although in a rather irregular fashion; so if a given sample of liquid is to increase its surface area, it must increase the number of surface molecules — they don't just spread out. In the interior of the liquid each molecule experiences the attraction of all its neighbors, which is the reason for the regularity and close-packed nature of most crystal structures. In a close-packed crystal structure there are 12 nearest neighbors, and even if we make allowance for vacancies and irregularities in a liquid structure we can still expect 10 or 11 nearest neighbors (on the average) around a liquid molecule. A molecule on the surface, however, has only about half as many nearest neighbors, because the gas above it is mostly empty space; the surface molecule is thus not stabilized by as many intermolecular attractions as an interior molecule, and if we are to create new surface molecules we must supply energy to the interior molecules that are becoming surface molecules as the surface area expands. This is the origin of the surface tension, which we define as the work required to create a unit area of new liquid surface:

$$\gamma = \frac{\text{energy required to move interior molecules to surface}}{\text{new surface area created}}$$

Expressed this way, in ergs per square centimeter, surface tension has the dimensions of a surface energy. However, inspection of the units shows that ergs per square centimeter is equivalent to dynes per centimeter, which has the dimensions of force per unit length. Figure 7-6 shows the force relationships we have been discussing. Although we have not talked about a surface tension for gases because they expand to fill any container, it should be clear that the surface tension in a liquid is equivalent to the pressure correction term $a(n/V)^2$ in the van der Waals equation of state for gases. Compare Fig. 7-6 with Fig. 3-23.

This rationale also suggests that a force equivalent to surface tension ought to exist at any interface between two phases, not just at the interface between a liquid and a gas. If the forces between molecules are different in the two phases, either because the molecules are more widely separated, as in the case of a

Figure 7-6 Comparison of forces acting on a surface molecule and an interior molecule in a liquid.

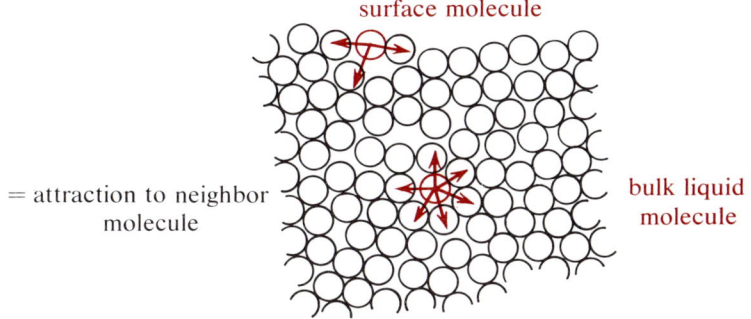

liquid and a gas, or because the molecules are of different chemical natures, as in an oil–water interface, a macroscopic force should result that we can call an **interfacial tension**. This sort of force does exist, as we see when we shake up the oil and vinegar in a salad dressing bottle. The oil forms spherical droplets in the vinegar under the influence of its interfacial tension with the aqueous solution (vinegar).

7-2 A Theory of Liquid Structure

Where does all this experimental information leave us with respect to the formulation of a workable theory of liquid structure? We would like, for instance, to propose an equation of state for liquids, analogous to one of the equations of state we discussed for gases, and we would like to rationalize this equation in terms of the behavior of the molecules in the liquid. Unfortunately, a comprehensive and quantitative theory of liquids does not yet exist. The liquid state presents forbidding theoretical difficulties because of the "halfway" nature of liquids. That is, we can formulate a workable model for gases because we can rely on the complete randomness and lack of correlation of position of their molecules, and correspondingly we can formulate a workable (if more complicated) model for crystalline solids because of their complete order, symmetry, and correlation of position of their molecules. What is difficult is to formulate a mathematical model for the intermediate state, which is disorderly and quite random, yet has considerable correlation of position of its molecules.

The Hole Theory of Liquids

We can say some things about the liquid state, however. In discussing X-ray diffraction by liquids we implied that the distance between nearest neighbors is the same for a liquid as for the solid crystal of the same substance. This is in fact true for most liquids, and we account for the decrease in density by suggesting that to a good approximation the additional "free volume" shows up as vacancies in the more-or-less close-packed liquid lattice. This means that the coordination number of a liquid molecule must decrease as the liquid expands on warming, and this is exactly the inference we draw from the area under the nearest-neighbor hump in the radial distribution graph in Fig. 7-2. Liquid argon, for instance, which is a good model liquid because it has simple spherical monatomic molecules, has a coordination number of 12 in the solid state but 10–11 in the liquid at the melting point. This explains why liquid argon is about

12% less dense than solid argon. As the liquid is warmed its density decreases, and the apparent coordination number from X-ray data decreases to about 4 just below the critical temperature. However, in all this range the nearest-neighbor argon–argon distance is nearly unchanged. This, together with the lack of order that we infer from X-ray data and the isotropic nature of liquids, leads us to a picture of a liquid's structure rather like Fig. 7-7. At relatively low temperatures, any given molecule has its maximum possible number of nearest neighbors (six in this two-dimensional case) since intermolecular attractions cause it to crystallize with the largest possible stabilization due to the nearest-

| crystalline solid | liquid at melting point | liquid near critical temperature |

Figure 7-7 Liquid structure changes as increasing temperature produces more defects in the "close-packed" array of molecules.

neighbor interactions. If we warm such a system, however, the molecules acquire, on the average, an amount kT of thermal kinetic energy (e.g., vibrating about their equilibrium position). When this kinetic energy becomes high enough, some molecules can migrate through the lattice, forcing other molecules aside and leaving behind a vacancy. This is what we implied in the discussion of the high-temperature heat capacity of crystalline silver bromide in Section 4-6, where we saw that the heat capacity increases because the formation of vacancies or defects requires energy. When the concentration of vacancies and interstitial atoms becomes high enough, the orderly structure of the solid can no longer be maintained and the crystal melts. In the liquid, the average coordination number of the molecules is smaller (shown as five in Fig. 7-7), and as the temperature continues to increase and more and more molecules acquire sufficient thermal kinetic energy to migrate through the liquid "lattice," the average coordination number decreases still further. We view a liquid as containing some molecules that can express their kinetic energy by traveling (or **translating**) through the liquid, and others that are confined by their nearest neighbors and have only vibrational (and perhaps rotational) kinetic energy (although the nearest neighbors are no longer in the geometrically orderly arrangement found in the crystal).

This gives us a possible approach to a theory of liquids. In solids, a monatomic molecule can have *only* vibrational kinetic energy; in a gas, the same molecule can have only translational kinetic energy. Suppose we make the approximation that all the "free volume" in a liquid is arranged in molecule-sized vacancies in what is otherwise essentially a close-packed lattice. Then any molecule next to a vacancy can move into it, thereby displaying translational kinetic energy and acting like a gas molecule. On the other hand, any molecule that is completely surrounded (i.e., is *not* next to a vacancy) can only vibrate and is acting like a solid molecule. So the behavior of the liquid as a whole is a composite of the behavior of a fraction of its molecules that are gas-like and the behavior of the remaining molecules that are solid-like. This model of a liquid as a composite of two significant structures is due to Henry Eyring and coworkers and is sometimes called the **Eyring hole theory** of liquids.

How can we calculate the quantitative properties of liquids from this model? We need first an estimate of the number of gas-like molecules in a system. If the molar volume of a liquid is V_l and the molar volume of the corresponding solid is V_s, then the free volume in the liquid is $V_l - V_s$. If all this free volume consists of molecule-sized holes, then the ratio of vacancies to molecules is $(V_l - V_s)/V_s$ at any given temperature. There is some chance that next to a hole will be another hole, but holes have no properties; most of the holes are surrounded by molecules that we define as gas-like because they *are* next to a hole. If holes are distributed randomly through the liquid, then the probability that a given position next to a hole will be filled by a molecule is V_s/V_l. So the fraction of gas-like molecules in the liquid is equal to the ratio of vacancies to molecules, multiplied by the relative probability that a position next to a hole will be occupied by a molecule:

$$\text{fraction of gas-like molecules} = \frac{(V_l - V_s)}{V_s} \frac{V_s}{V_l} = \frac{V_l - V_s}{V_l}$$

The remaining fraction of a mole, V_s/V_l, is taken as the fraction of solid-like molecules. What we propose is to predict any physical property of a liquid to be the sum of a contribution from the solid-like molecules and a contribution from the gas-like molecules. For instance, suppose we wish to explain the temperature dependence of the heat capacity of liquid argon. Since argon is monatomic, its heat capacity as a gas should be $\frac{3}{2}R$ or 2.98 cal/°K (see p. 125). For an idealized solid lattice of argon, the heat capacity would be $3R$, or 5.96 cal/°K (if we ignore the quantum effects shown in Fig. 4-39). Then if we wish to predict the heat capacity of liquid argon at any given temperature, we need only measure the molar volume, V_l, of the liquid at that temperature and add the solid and gas contributions:

$$C_v = 5.96(\text{solid-like fraction}) + 2.98(\text{gas-like fraction})$$
$$= 5.96\left(\frac{V_s}{V_l}\right) + 2.98\left(\frac{V_l - V_s}{V_l}\right)$$

Figure 7-8 shows the results of this approach. The match is obviously quite good. By using more advanced statistical methods than we have discussed,

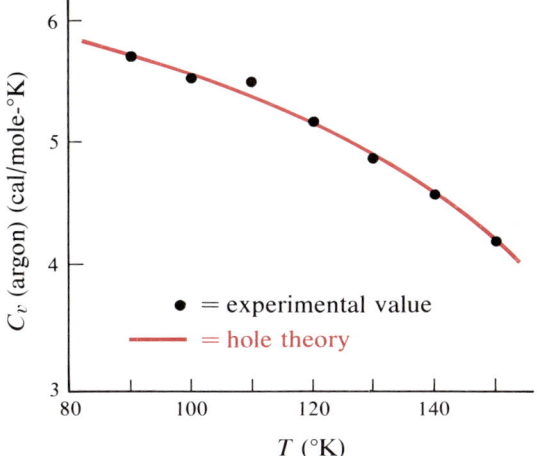

Figure 7-8 Comparison of experimental values of the heat capacity of argon with theoretical values from the "hole theory." (Modified from Eyring, H. and Marchi, R. P., *J. Chem. Ed. 40* (562), 1963.)

Eyring has produced equally good predictions of all the transport properties of liquids in a number of cases as diverse as the rare gases, molten metals, molten salts such as NaCl, and organic liquids such as methane (CH_4). The theory is neither complete nor free of empirical quantities, but it is an interesting and promising example of the power of physical models.

The Tait Equation

Let us apply our theory in an intuitive way to the problem of an equation of state for liquids. For many years an empirically satisfactory equation of state called the **Tait equation** has been known:

$$\frac{dV}{dP} = -\frac{K}{L(T)+P} \qquad (7\text{-}4)$$

where dV/dP is the rate of change of volume of the liquid with increasing pressure, or the compressibility, K is a constant, and $L(T)$ is a function only of the temperature. The Tait equation is an excellent description of the pressure–volume behavior of liquids of several kinds over a wide range of temperatures and pressures, but it has not been derived in the sense that we were able to derive the ideal gas law from the kinetic theory of gases. We can put the Tait equation in a form that is reminiscent of the hole theory, however, by multiplying both sides by the denominator of the right side:

$$\frac{dV}{dP}[L(T)] + \frac{dV}{dP}(P) = -K \qquad (7\text{-}5)$$

To be dimensionally consistent, the constant K must have units of volume. This suggests that we may be able to identify the two terms on the left side as being volumes associated with the gas-like molecules on the one hand and the solid-like molecules on the other. If we assume that the second term represents the volume of the gas-like molecules and that they are ideal gases, we can write:

$$V = \frac{nRT}{P} = nRTP^{-1}$$

$$\frac{dV}{dP} = -1 \cdot nRTP^{-2} = -\frac{nRT}{P^2}$$

$$\frac{dV}{dP}(P) = -\frac{nRT}{P^2}P = -\frac{nRT}{P} = -V$$

So the second term in Eq. 7-5 can be interpreted as the contribution from gas-like molecules.

The first term is a little more difficult. If it is to be interpreted as a volume characteristic of solid-like molecules, we need to show that there is a volume quantity associated with dV/dP and a function of temperature for a crystalline solid. From solid state physics we can obtain a relationship between the coefficient of thermal expansion of a solid and its relative compressibility, $(1/V)(dV/dP)$:

$$\text{thermal expansion} \equiv \alpha = cC_v \frac{1}{V}\frac{dV}{dP}$$

where c is a constant for any given solid and does not depend on temperature. We can solve this expression for V:

$$V = c\frac{C_v}{\alpha}\frac{dV}{dP}$$

If in this expression we identify $c(C_v/\alpha)$ with the function $L(T)$, then the volume of the solid is expressed as a function exactly analogous to the first term in Eq. 7-5. In fact, however, $c(C_v/\alpha)$ does not have the identical dependence on temperature that $L(T)$ specifies, although it is qualitatively correct; this treatment should be regarded only as a qualitative application of the hole theory, not as a derivation of the Tait equation. Nonetheless, it is useful to see that we can write and rationalize an equation of state for liquids that is in some manner analogous to those we developed for gases.

7-3 Liquids in Chemical Mixtures — Solutions

One of the principal chemical features of liquids is their role as solvents. In Chapter 1 we defined solutions as homogeneous mixtures and pointed out that solutions can exist in all three states of matter—solid, liquid, and gaseous. Solid solutions, while useful and interesting, are a topic we shall not discuss except in a very general fashion in Section 7-5 and several places in Part IV. Gaseous solutions are also important (e.g., air), but are not usually thought of as solutions since all gases are miscible with one another. When we think of solutions, we tend instinctively to refer to the liquid state, which is the nature of most familiar solutions.

Solution Composition

The first property of solutions we shall examine is their stoichiometry. Any solution, being a mixture, has at least two components. When solids or gases are dissolved in liquids, the liquid is said to be the **solvent** and the solid or gas the **solute**. When one liquid is dissolved in another, the liquid that is present in the greater amount is usually taken as the solvent, making the other liquid the solute. Since solutions have, by our earlier definition, variable composition, we cannot express their composition as ratios of integers as we did for compounds. A solution is homogeneous; that is, the properties of a solution are the same whether we have a lot or a little, and so we need only the proportions of each component, not the absolute amounts. In other words, we need the concept of **concentration**—not how much A and how much B are present, but how much A *per unit amount of B*. There are several ways to define concentration, but all of them have dimensions that are a ratio of quantities: moles per liter, grams per liter, moles per mole, and so on.

The simplest concentration is percent composition, which usually means **weight percent** or the ratio of grams solute to grams total weight of solution, times 100:

$$\text{weight percent} = \frac{\text{grams solute}}{\text{grams solution}} 100 \tag{7-6}$$

Since mole units are often more convenient than weight units, solution concentrations are sometimes expressed in **mole percent**, with an analogous definition:

$$\text{mole percent} = \frac{\text{moles solute}}{\text{moles solute} + \text{moles solvent}} 100 \tag{7-7}$$

More commonly, however, when proportions in moles of solvent and solute are important, the **mole fraction** is used:

$$\text{mole fraction A} \equiv x_A = \frac{\text{moles A}}{\text{moles A} + \text{moles B}} \tag{7-8}$$

It is sometimes convenient to work with standard weights of solvent even though mole quantities of the solute and solvent are important; in this case we frequently use **molality**, the number of moles of solute per kilogram of solvent (not solution):

$$\text{molality} \equiv m = \frac{\text{moles solute}}{\text{kilograms solvent}} \tag{7-9}$$

A variation of molality used principally with aqueous solutions and extremely convenient in terms of the practical operations involved in the laboratory is **molarity**, the number of moles of solute present per liter of solution:

$$\text{molarity} \equiv M = \frac{\text{moles solute}}{\text{liters solution}} \tag{7-10}$$

One sometimes sees the term **formality** used for solutions of ionic salts that do not really exist as molecules either in the solid or in solution; operationally, formality is identical to molarity:

$$\text{formality} \equiv F = \frac{\text{gram-formula weights solute}}{\text{liters solution}} \tag{7-11}$$

Finally, there is a concentration formalism that is passing out of use but should not be overlooked: **normality**. Normality is the number of moles of chemically reactive units, defined for a particular reaction, present per liter of solution. For instance, if the reaction of interest is the neutralization of barium hydroxide, $Ba(OH)_2$, by acid, the equation is

$$OH^-_{(aq)} + H_3O^+_{(aq)} \rightarrow 2H_2O_{(l)}$$

Suppose we have a solution that is $1\,M$ (one molar) in $Ba(OH)_2$. Since there are two OH^- ions present per formula quantity of $Ba(OH)_2$, the normality of the solution is 2; we write $2N$ (two normal). Normality is molarity defined with respect to a particular reaction, and since the reaction stoichiometry must be known to establish normality, it is usually easier and less confusing to use molarity, taking the stoichiometry from the balanced equation for the reaction.

Levels of Solubility

If, then, we can describe the composition of solutions, we can inquire into the nature of the process of dissolving. If we bring a liquid solvent and some other substance together (whether solid, liquid, or gas) there are three possibilities.

Liquids in Chemical Mixtures — Solutions | 357

The potential solute may dissolve completely; in this case the solute and solvent are said to be completely **miscible**, just as gases are. The solute may dissolve partially, reaching a maximum concentration called the **solubility** of the solute in that solvent. Finally, the potential solute may not dissolve to any measurable extent, in which case it is said to be **insoluble**. For instance, water and acetone are completely miscible; sodium chloride has a solubility of 6.1 moles/l in water; diamonds are insoluble in water. Another complication is that a compound that is quite soluble in one solvent may be nearly or completely insoluble in another. Thus sodium chloride, which is quite soluble in water, is essentially insoluble in gasoline. What relationships between solvent and solute molecules govern solubility?

Favorable Energy Changes on Dissolving

We should first ask why a solid compound dissolves at all. If the compound is a solid at the temperature of the solution, it means that the attractions between atoms or molecules in the solid are strong enough that thermal energy alone cannot destroy the stabilization provided by the geometrically orderly nearest neighbors. But when the compound dissolves the symmetrical, stable crystal lattice is broken up into a disorderly solution just as it would be broken up into a disorderly liquid by melting. Our initial instinct, based on the ideas of the last two chapters, is to say that the low-energy solid is going to an even lower-energy condition by dissolving, in that it has some energy of attractive interaction with the solvent molecules (called the **solvation energy**). In many cases this is true. When iron(III) chloride, $FeCl_3$, dissolves in water a great deal of heat is given off, which must come from somewhere; we say that the iron(III) and the chloride ions are being solvated (in the case of water solvent, **hydrated**), and the stabilization provided by the total **hydration energy** is greater than that provided by the neighbors in the crystal lattice. See Fig. 7-9. The $FeCl_3$ releases

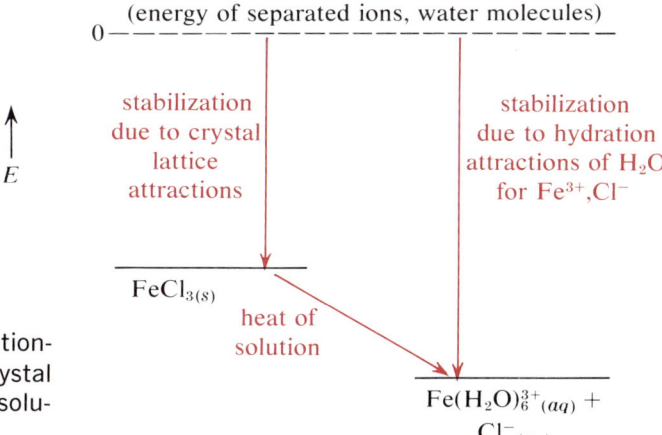

Figure 7-9 Energy relationships for $FeCl_3$ in its crystal lattice and in aqueous solution.

energy as heat in going to the more stable hydrated condition in solution. In Chapter 13 we shall look at the energies involved in the solution process in detail. Here it is only necessary to point out that a crystal that is very strongly bonded together, either by covalent bonds (diamond), metallic bonds (chromium), or ionic bonds (sodium chloride), will dissolve only in a solvent that offers a very high solvation energy, which may actually be the energy of a chemical reaction. This is why sodium chloride, which is ionic, will dissolve in water (which as a very small, very polar molecule has a strong electrostatic interaction with the ions) but not in gasoline (which consists of molecules like octane,

$$\text{H}-\underset{\underset{\text{H}}{|}}{\overset{\overset{\text{H}}{|}}{\text{C}}}-\underset{\underset{\text{H}}{|}}{\overset{\overset{\text{H}}{|}}{\text{C}}}-\underset{\underset{\text{H}}{|}}{\overset{\overset{\text{H}}{|}}{\text{C}}}-\underset{\underset{\text{H}}{|}}{\overset{\overset{\text{H}}{|}}{\text{C}}}-\underset{\underset{\text{H}}{|}}{\overset{\overset{\text{H}}{|}}{\text{C}}}-\underset{\underset{\text{H}}{|}}{\overset{\overset{\text{H}}{|}}{\text{C}}}-\underset{\underset{\text{H}}{|}}{\overset{\overset{\text{H}}{|}}{\text{C}}}-\underset{\underset{\text{H}}{|}}{\overset{\overset{\text{H}}{|}}{\text{C}}}-\text{H}$$

which are neither small nor polar).

Dissolving in Equivalent-Energy Environments

This is not the complete explanation of the solubility of solids in liquids, however. There is an old chemical adage that "like dissolves like," which is simply a summary of our experience: liquid benzene,

dissolves solid naphthalene,

liquid mercury dissolves many other metals, molten salts dissolve other ionic salts, and so on. In all of these cases the forces acting on a particular atom or molecule in the crystal are replaced by essentially identical forces in the liquid. That is, in the benzene/naphthalene case the van der Waals forces between a naphthalene molecule and the benzene molecules surrounding it in solution are essentially identical to those between a naphthalene molecule and other naphthalene molecules in the crystal. Similarly, the metallic bonding forces on a metal atom in a mercury solution (an **amalgam**) are very like those in the metal itself, and the electrostatic attractions on an ion dissolved in a molten ionic salt are quite similar to those acting on the ion in its original crystal. In these cases there can be no strong solvation-energy effect operating, and yet the solids dissolve spontaneously. Why should this be true?

We can reduce the complexity of the discussion by imagining a hypothetical case in which there is no energy difference between the crystal environment and the solution environment for a given molecule; this is very nearly true for naphthalene in benzene. This gives us the molecular situation shown in Fig. 7-10 at the solid surface immersed in liquid solvent, with no energy preference for either phase as far as a molecule at the solid surface is concerned. Consider

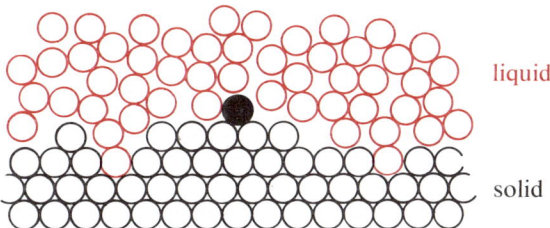

Figure 7-10 Environment of a surface molecule on a crystal immersed in a solvent of similar molecular structure.

the surface molecule shown shaded in the figure. It is only partly surrounded by geometrically regular crystal molecules; part of its environment consists of irregularly arranged liquid molecules, with vacancies being continually created and destroyed as the liquid molecules tumble about. Occasionally a vacancy will be created next to the shaded molecule. When this occurs, the shaded molecule can (in terms of the hole theory) become gas-like by converting some of its vibrational kinetic energy into translational kinetic energy, moving into the hole. A moment later it may move back as other liquid molecules collide with it and change the direction of its momentum. What is more likely, however, is that another liquid molecule will move into the vacancy left at the solid surface, since there are more liquid molecules with a shot at the vacancy; see Fig. 7-11. In this case the former solid molecule finds itself isolated in the solution — it has dissolved.

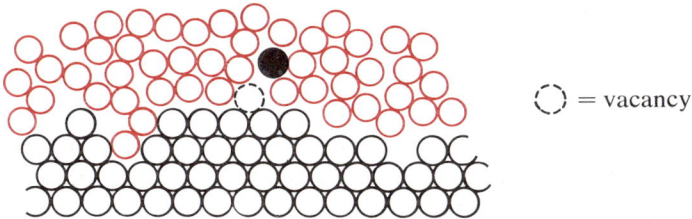

Figure 7-11 Movement of a "solid" molecule into a liquid vacancy.

This whole process rests on the tendency of a large number of random individual events to occur so as to provide the most probable result. Given that liquids have a large number of vacancies, it would be very improbable if none of them were ever to appear next to the solid surface. With vacancies present and no energy difference, it would be very improbable if no solid molecule ever moved into a liquid vacancy. With a solid molecule and several liquid molecules arranged about a vacancy at the solid surface, it would be very improbable for the solid molecule but none of the liquid molecules to move back into it. Of course, all of these improbable events could very well occur in the case of any individual molecule; what becomes staggeringly improbable is that they should *all* occur for *all* of the, say, 10^{15} molecules on the surface of a crystal. Another way of looking at the same idea is to note that there are countless millions of ways to arrange a mole of solid molecules in an irregular solution, all of which are equivalent — but there is only one way to stack molecules to produce a perfect crystal. If all of these arrangements are possible and have equal energy, what is the probability that the ideal crystal will form out of the solution as the moving molecules rearrange themselves? The probability is one in countless millions; in other words, it simply is not going to happen. Of course, all this is true only if there is no energy difference between the crystal environment and the solution environment for a solid molecule. If very strong chemical bonds hold an atom or molecule in its crystal lattice position and no such bonds are possible in the solution environment, then there is no source for the energy required to pull the molecule or atom out of the crystal. In that case even though the liquid vacancies may be present at the solid surface, the solid atoms or molecules will never move into them because we have in effect loaded the odds against such motion; although the solid molecules are there, none has sufficient energy to move away from the crystal into the vacancies.

Thus there are two factors, on a molecular scale, affecting the solubility of a substance (whether solid, liquid, or gas) in a liquid. One is the energy relationship between the molecules in their original environment and the solution environment. The other is the relative probability of the individual events in the process of dissolving a molecule. We shall have more to say about relative probabilities of molecular events later in this chapter and in the next.

The Choice of Solvent Liquids

In Chapter 13 we shall discuss hydration energies in some detail, but that discussion deals with ionic compounds. The question of choosing an effective solvent often arises, particularly in organic chemistry. The relative probability is always favorable for dissolving, but some adjustment of energy relationships usually must be made by choosing an appropriate solvent liquid. Besides the van der Waals forces, which are quite small, two other kinds of energy stabilization are possible for many organic solid molecules. Organic molecules are those that have a basic structure of carbon atoms, but most also have an electronegative atom such as N, O, F, or Cl somewhere in the structure. Since this atom attracts electrons, the molecule will tend to have an **electric dipole moment**; one end of the molecule will be more electrically negative than the other end. If this is true for the solid molecules and also for the liquid molecules of the solvent, then considerable stabilization can be achieved by arranging the positive ends of the solvent molecules near the negative end of the solute molecule, and vice versa, as in Fig. 7-12. The stabilization energy, of course,

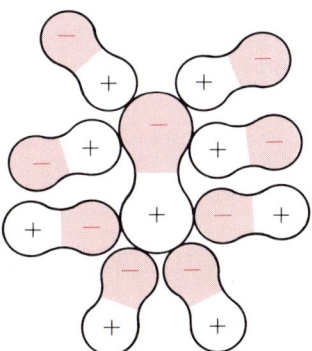

Figure 7-12 Attraction of polar solvent molecules for a polar solute molecule (dipole–dipole interaction).

is simply the Coulomb electrostatic attraction of the opposite charges. As an example of this effect, consider the solid 2,6-dinitroaniline, which has a substantial dipole moment tending to stabilize it in its crystalline environment:

It is insoluble in hydrocarbon solvents such as hexane,

which has no dipole moment, but is reasonably soluble in liquids such as ethyl ether, which has a small dipole moment:

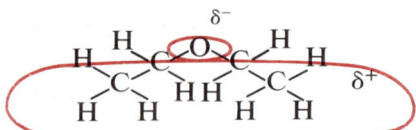

This attraction through the use of a polar solvent for polar compounds is a favorable influence, but only a small one. The other, and more important one, is the possibility of having the solvent hydrogen bond to the solute. As we noted in Chapter 6, hydrogen bonds are formed between hydrogens normally bonded to a very electronegative atom such as O or F (to a small extent N) and another very electronegative atom having a nonbonding pair of electrons. This makes alcohols (which have an —O—H group) good hydrogen-bonding solvents for other molecules having N, O, or F atoms. For instance, 2,6-dinitroaniline is quite soluble in ethyl alcohol, which forms hydrogen bonds to the O atoms on the 2,6-dinitroaniline and also has a small dipole moment. The possibility (or impossibility) of taking advantage of either or both of these factors should be considered whenever a solvent must be chosen for an organic compound.

7-4 Liquids in Physical Mixtures — Liquid–Vapor Equilibrium

Whenever a liquid does not completely fill its container a surface exists which is, to use our former terminology, a liquid–gas interface. Under any practical conditions, the gas phase will contain at least a few molecules of the liquid species. These molecules are said to have passed into the **vapor phase**; since they form a part of the gas mixture above the liquid, they can be said to exert a partial pressure in the gas phase. If the container is closed, this partial pressure will increase spontaneously until some characteristic maximum value is reached, a value that depends on the chemical identity of the liquid and on the temperature. This characteristic maximum value of the partial pressure of the liquid vapor is called the **vapor pressure** of the liquid. Before going any farther into the nature of the liquid–vapor transition, let us comment on how vapor pressure is measured.

Measurement of Vapor Pressure

One fairly obvious method of measuring vapor pressure is shown in Fig. 7-13; the liquid is placed in the bulb, all excess gas is removed by a vacuum pump, and the partial pressure of the vapor (which is now the total pressure) is read on the manometer. This direct procedure is not often used because of the diffi-

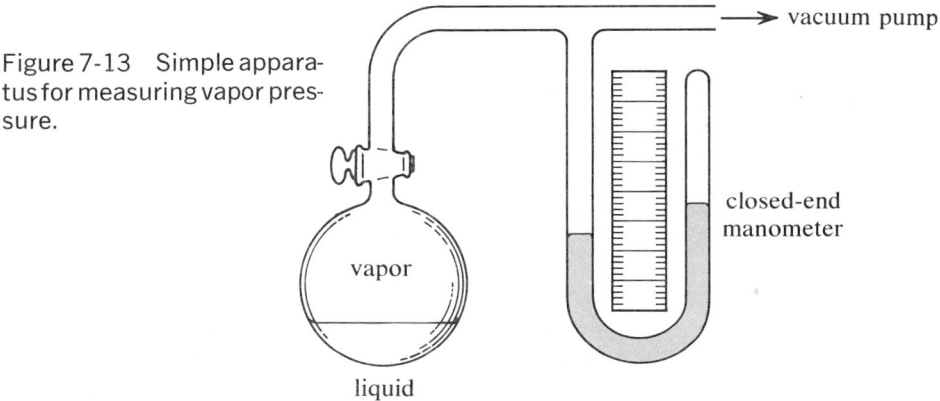

Figure 7-13 Simple apparatus for measuring vapor pressure.

culty of keeping the whole system, including the manometer, at a constant temperature — as we shall soon see, the temperature dependence of vapor pressure is a most interesting property. If we wish to vary the temperature, we usually modify the device simply by placing a trap in the plumbing, as in Fig. 7-14. This device, which is very commonly used, is called an **isoteniscope**.

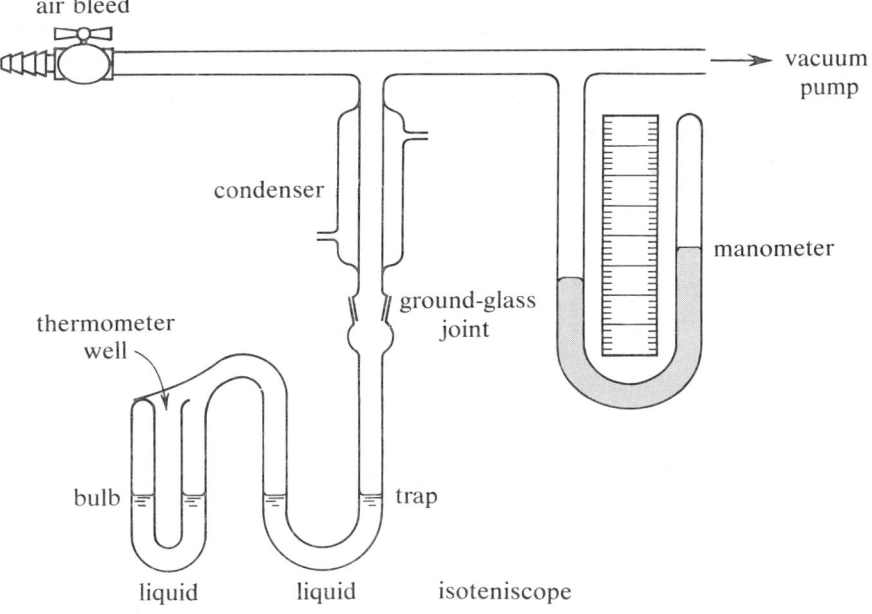

Figure 7-14 Apparatus for measuring the temperature dependence of vapor pressure.

The bulb is partly filled with liquid, attached to the condenser, and only the bulb portion of the apparatus is placed in a constant-temperature bath. When the bulb and liquid are heated, some vapor condenses in the condenser and runs back into the U trap, filling it. This isolates the vapor in the top part of the bulb, so that its pressure can be measured by adjusting the pressure in the rest of the apparatus until the liquid level is at the same height in the two arms of the U trap. When this is done, the bulb (vapor pressure) and the rest of the apparatus are at the same pressure, and the manometer will read the vapor-pressure equivalent.

Molecular Energies and Vaporization

Vapor pressures measured in this way or in other ways vary greatly from liquid to liquid. Hydrogen cyanide, HCN, has a vapor pressure of about 750 torr at room temperature (25 °C); water has a vapor pressure of 24 torr at 24 °C, mercury has a vapor pressure of 1.8×10^{-3} torr, and some silicone oils have vapor pressures as low as 10^{-8} torr. It should be obvious that this list could be extended to include substances whose equilibrium vapor pressures at 25 °C are greater than 760 torr; these substances, however, if kept at a total pressure of 760 torr (in other words, open to the atmosphere) will vaporize spontaneously and completely. This is the process of boiling, and these substances are gases at room temperature and 1 atm pressure. The wide variation in vapor pressures can be understood on a molecular scale by considering Fig. 7-15 and comparing it with Fig. 7-10. Since a gas is mostly empty space, all of the molecules in the surface layer of the liquid are free to move up into the gas phase; the reason all of them do not is that there are intermolecular forces stabilizing them in the

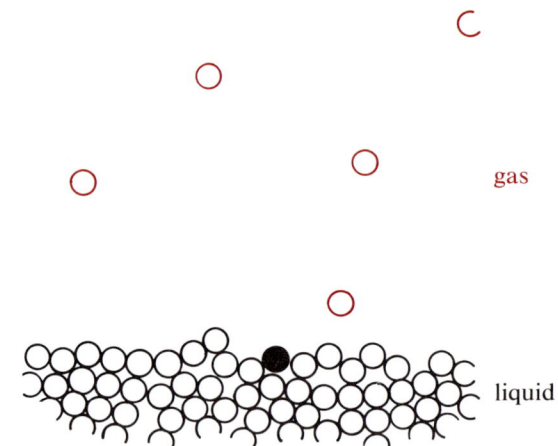

Figure 7-15 Comparison of molecular environments in the liquid and vapor phases.

liquid but no such forces in the gas (or almost none, since at ordinary pressures the molecules are so far apart). Most common liquids have van der Waals and hydrogen bonding as the only intermolecular forces. In Fig. 7-16 is shown the

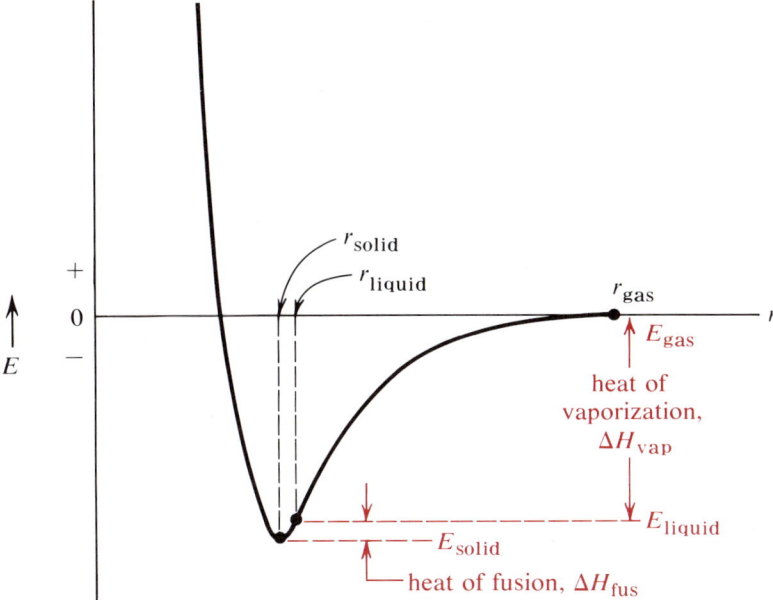

Figure 7-16 Energy relationships between phases as related to intermolecular separation by the Lennard-Jones potential (or another attractive potential). The diagram is for a bimolecular system; to be numerically correct, allowance must be made for the changing number of nearest-neighbor molecules.

Lennard-Jones potential, which we have used to represent the van der Waals forces. Close-packed molecules in the solid are very nearly at the potential minimum; the slight increase in spacing in liquids does not change the potential energy of a pair of molecules much, but spreading the molecules a great distance apart (in vaporizing to a gas) requires a considerable energy input, as shown on the graph. This energy, called the **heat of vaporization**, ΔH_{vap}, is what must be supplied for the molecule to escape the liquid. The only source of energy is the thermal energy, kT, which the molecules of the liquid possess on the average. Of course, some molecules possess more than the average, and the fraction having thermal kinetic energy greater than the potential energy represented by the heat of vaporization is given by the Boltzmann distribution:

$$\frac{\text{number of molecules with KE} > \text{heat of vaporization}}{\text{total number of liquid molecules}} = e^{-(\text{heat of vaporization})/kT}$$

(Review Section 3-2.) Figure 7-17 shows the Boltzmann energy-distribution curve for a given temperature with the energy equivalent of the heat of vaporization indicated. Clearly, if the heat of vaporization is very small due to small van der Waals forces, the fraction of molecules having sufficient energy to escape will be large; while if intermolecular forces are strong, as in a molten

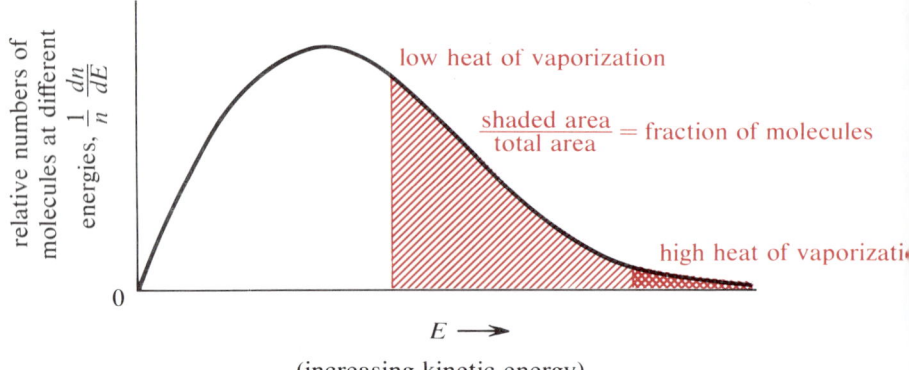

Figure 7-17 Fraction of liquid molecules with sufficient energy to vaporize, for different heats of vaporization.

salt, the fraction will be very small. As in the case of a solid dissolving in a liquid, probability considerations favor vaporizing, but energy considerations limit the number of molecules to which the "equal-energy" probability treatment applies. Thus a small fraction of molecules having kinetic energy equal to the heat of vaporization means a small vapor pressure.

Since the highest-energy, "hottest" molecules are escaping, it is not surprising that a liquid cools as evaporation occurs at its surface. This is a very useful phenomenon; air conditioners and refrigerators operate by blowing air over metal tubing in which a liquid boiling below room temperature is evaporating very rapidly. To keep from running out of liquid, the escaping gas is caught and compressed until it recondenses into a liquid, which is supplied to the evaporation chamber again. The condensation liberates heat, of course (intermolecular distance is moving to the left in Fig. 7-16) but another blower or heat exchanger carries away this heat to what used to be considered a harmless location. Very large heat exchangers, however, such as those associated with power plants, sometimes discharge so much heat into rivers or lakes as to cause what has been called "thermal pollution."

Temperature Dependence of Vapor Pressure of Liquids and Solids

How does the vapor pressure of a given liquid vary with temperature? We can establish an experimental function relating vapor pressure and temperature by using an isoteniscope or other means; in general it will look like Fig. 7-18. This curve rising at an ever-increasing rate looks like an exponential function; if we undo the exponential feature by taking the logarithm of the pressure, we get

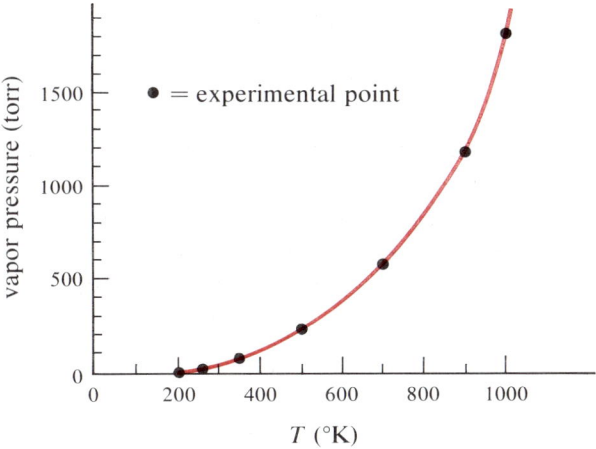

Figure 7-18 Experimental dependence of vapor pressure on temperature.

Fig. 7-19, in which we have sort of overcorrected. Taking a clue from the fact that this new curve looks like a hyperbola with respect to the dotted line, we

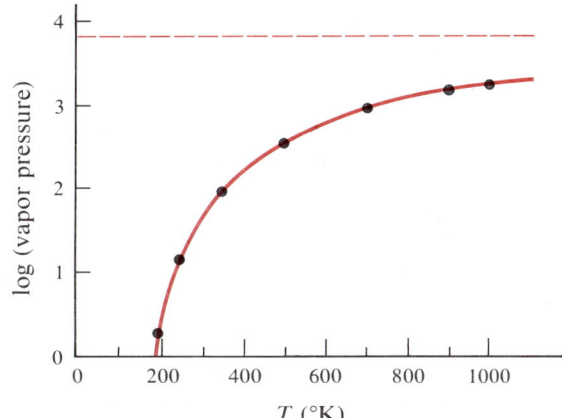

Figure 7-19 Logarithmic plot of vapor pressure against temperature.

can plot log P against the reciprocal of the temperature and get quite a nice straight line, as in Fig. 7-20.

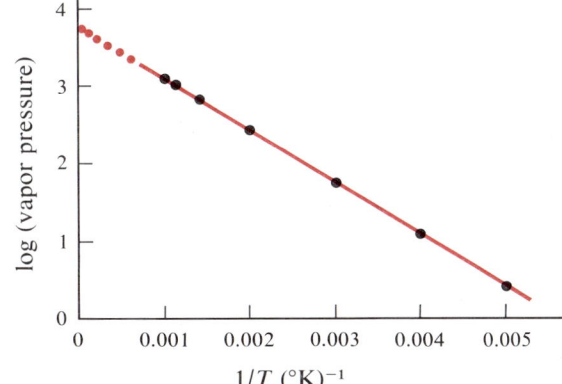

Figure 7-20 Plot of log P vs. $1/T$.

Thus we have a linear equation in log P and $1/T$:

$$\log P = a\left(\frac{1}{T}\right) + b \tag{7-12}$$

In Chapter 9 we shall see that this relationship can be derived theoretically, and in fact gives us a convenient way to measure heats of vaporization. Note that the exponential behavior of the function is entirely consistent with the Boltzmann distribution. If we say that the pressure is linearly proportional to the fraction of molecules having sufficient energy to escape the liquid, then

$$P \propto \frac{n_{\text{escaping}}}{n_{\text{total}}} = e^{-\Delta H_{\text{vap}}/kT}$$

$$\log_e P \propto -\frac{\Delta H_{\text{vap}}}{kT}$$

$$\log_{10} P = K(\Delta H_{\text{vap}})\frac{1}{T} \tag{7-13}$$

Figures 7-15–7-20 and the associated discussion apply equally well to a solid–gas interface. The process of vaporizing molecules from a solid surface is called **sublimation**, and there are some solids, such as iodine, dry ice, and mothballs, that have a considerable vapor pressure. Because a crystalline solid has fewer vacancies than a liquid, all the molecules on the surface of a solid are more tightly bound than those on the surface of a liquid; this follows from the greater average number of neighbors. Accordingly, a solid always has a smaller vapor pressure than the liquid phase of the same substance, except at the melting point where the solid and liquid are in equilibrium with each other and have the same vapor pressure. But just below the melting point the vapor pressure of the solid is increasing with temperature faster than that of the liquid would at the same temperature, because a small increase in temperature causes a marked increase in vacancies in the solid just below the melting point. We can draw a graph of the vapor pressure of a substance over the solid and liquid phases, as in Fig. 7-21. We can look at this graph in two ways: it obviously represents the

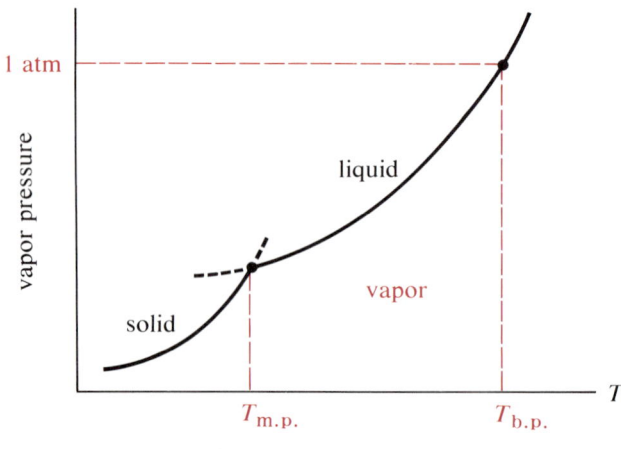

Figure 7-21 Vapor pressure over both solid and liquid as a function of temperature.

vapor pressure over the solid and liquid phases at any given temperature, but it also means that if we apply an external pressure that is *less* than that indicated by the curve, the substance will all vaporize. This is why we have indicated the region below the curve as being the vapor region of the plot.

The Phase Diagram and Critical Phenomena

We can add to this diagram the behavior of the solid–liquid equilibrium as the pressure is increased—that is, the effect of pressure on the melting point. We can introduce here a valuable generalization that applies not only to phase equilibria but to all the equilibria, physical and chemical, we shall encounter. The generalization is called **Le Chatelier's principle**: If a stress of any kind is laid on an equilibrium, the equilibrium conditions will change to minimize the effects of the stress. For those who are electronically minded, this is the equivalent of negative feedback. With respect to solid–liquid equilibrium, Le Chatelier's principle summarizes our experience in that if a liquid is less dense (occupies more volume) than the solid, then increasing the pressure tends to freeze the liquid into the solid, which occupies less volume. Conversely, if the liquid is more dense than the solid, which is not usually the case but is true for the very important case of water, increasing the pressure will tend to melt the solid. So we have the situation shown in Fig. 7-22, where the diagrams are

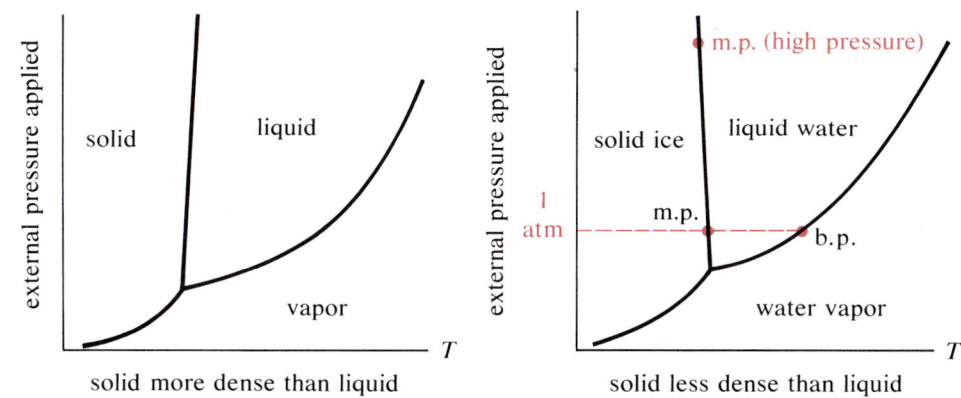

Figure 7-22 Phase diagram for systems in which the solid is more dense than the liquid, compared to that for water in which the solid is less dense than the liquid.

not drawn to scale but are exaggerated to show the essential features. Ice skaters depend on the unusual nature of the water solid–liquid equilibrium, because their skates are able to glide over the ice by the lubrication produced when the skate blade's pressure melts the ice directly under it.

The exponentially increasing function of Fig. 7-18 shows that the liquid has an ever greater tendency to vaporize as its temperature is raised, and to maintain the presence of any liquid we must oppose the vapor pressure by an ever greater external pressure. If the liquid is in an open container, however, the atmospheric pressure of about 760 torr is the maximum available pressure. If the necessary heat of vaporization is supplied to a liquid at the temperature at which its vapor pressure is 760 torr, it will spontaneously form bubbles of pure vapor, since the vapor at 760 torr inside the bubble can support the bubble against the external atmospheric 760 torr. This defines the normal boiling point of the liquid. If we increase the total pressure appropriately, however, as indicated by the curve, we can retain some liquid and keep a well-defined liquid–vapor equilibrium. This is not true indefinitely, however. For any liquid there is a temperature and corresponding pressure above which no liquid–vapor equilibrium can be defined. This is the **critical point**. At temperatures above the critical temperature, no amount of pressure will produce a recognizable liquid–vapor surface or meniscus. In terms of our **phase diagram**, we indicate the critical point by simply letting the liquid–vapor equilibrium line stop, as in Fig. 7-23. This point usually corresponds to a pressure of 30–

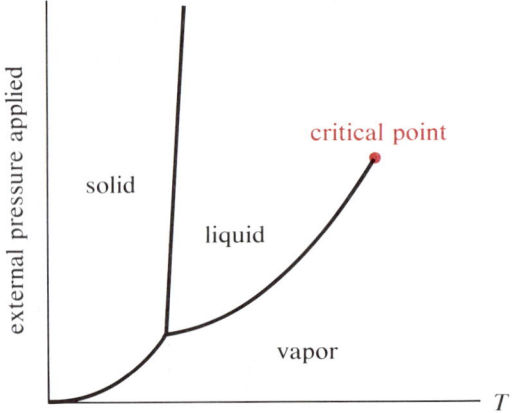

Figure 7-23   Phase diagram including the critical point.

60 atm, although there is considerable variation. Table 7-4 gives some characteristic values of the critical constants.

What do we observe when a substance above its critical temperature cools to the critical point? Well above the critical temperature, the substance is indistinguishable from an ordinary gas except for the greater density due to the very high pressure. As the substance is cooled to near the critical temperature, substantial density fluctuations from point to point occur in the "gas" and it becomes cloudy or opalescent. Just above the critical temperature, gravita-

Table 7-4
Critical Temperatures and Pressures

Substance	T_c (°C)	P_c (atm)
acetone	235	47
air	−141	37
ammonia	132	111
carbon dioxide	31	73
Freon refrigerant (CCl_2F_2)	111	39
helium	−267.9	2.2
n-octane	296	25
propane (LPG)	96	43
water	374	218

tional attraction begins to bring the more dense regions of the gas toward the bottom of the container, as shown in Fig. 7-24. At the critical temperature, the

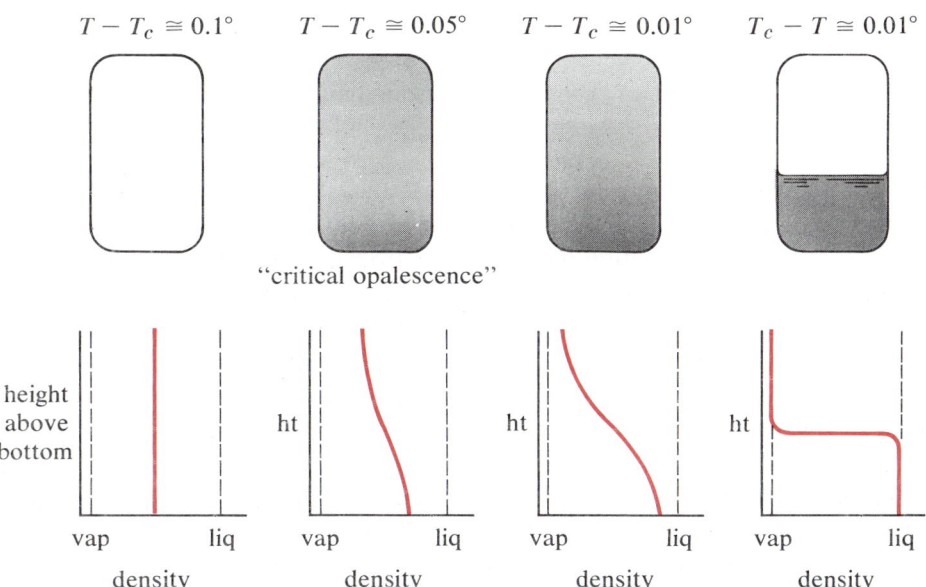

Figure 7-24 Nature of liquid–vapor mixtures near the critical point.

division of densities between the liquid at the bottom of the container and the gas at the top becomes well defined (as seen in Fig. 7-24) and a meniscus appears, allowing us to recognize the liquid phase.

In terms of the hole theory we can rationalize the existence of a critical point by noting that if the number of nearest neighbors around a liquid molecule averages 10–11 at the melting point, this is quite enough to ensure that some of the molecules in the liquid will remain solid-like (held in position by a cage of neighbors so that the molecule can have vibrational, but not translational, kinetic energy). Even as we heat the liquid and the average number of nearest neighbors decreases as more vacancies are formed, the partially solid-like structure persists down to an average number of neighbors equal to four, which is the number indicated by X-ray measurements on liquid argon about 5° below the critical temperature. Figure 7-25 shows that four neighbors arranged more or less tetrahedrally around a molecule are sufficient to prevent its escape; but

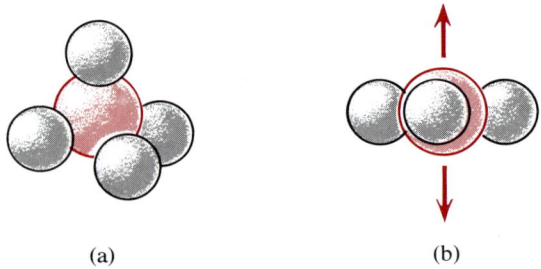

Figure 7-25 Constraint of a molecule by four neighbors (a) but not by three neighbors (b).

there is no geometrical arrangement that will allow three neighbors to constrain another molecule. What this means is that as still more vacancies are created and the average coordination number drops from 4 to 3, none of the molecules are constrained to one position and they all become gaslike (i.e., develop translational kinetic energy). This, of course, is exactly what happens at the critical temperature. The model gives quite a good match for the observed critical volume, since a little geometry on a liquid that was originally cubic close packed suggests the critical volume should be $3.19\sigma^3$, and the observed critical volume for several spherical molecules lies in the range 3.0–$3.3\sigma^3$. Here σ is the Lennard-Jones "diameter" of the molecules—see Fig. 3-22. This result is obtained in Study Problem 3. A good deal of experimental and theoretical work has been done on critical phenomena, and quite sophisticated mathematical models have been constructed. A complete and precise theory of liquids is still in the future, however.

The Vapor Pressure of Solutions — Raoult's Law

We also need to inquire into the effect of a dissolved solute on the vapor pressure of the solvent—that is, how does the vapor pressure of a solution vary with the solute's concentration? Suppose first we consider a solution of a nonvolatile solid, so that the liquid solvent is the only species contributing to the vapor pressure. For at least some solutions of this type, and for all that are sufficiently dilute, we find experimentally that the vapor pressure of the solvent in the solution is equal to that for pure liquid solvent at the same temperature, multiplied by the mole fraction of the solvent:

$$P_{solvent} = P^0_{solvent} x_{solvent}$$

where $P_{solvent}$ refers to the vapor pressure of the solvent in the solution, $P^0_{solvent}$ is the vapor pressure of the pure liquid, and $x_{solvent}$ is its mole fraction. This relationship is called **Raoult's law**, after the French chemist François Marie Raoult, who discovered it. By considering Fig. 7-26 we can see that Raoult's law is consistent with our model; the number of liquid molecules on the surface

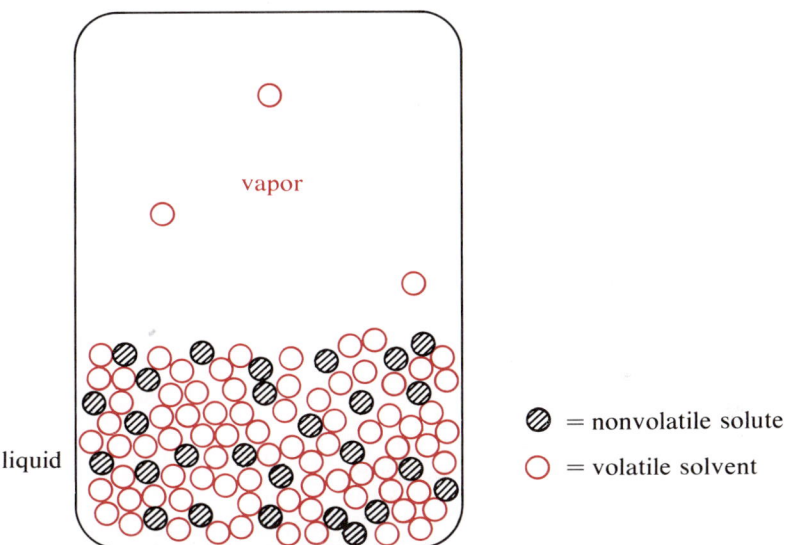

Figure 7-26 Molecular-scale diagram of liquid–vapor equilibrium with dissolved solute in liquid.

that can vaporize is proportional to the fraction of the surface they occupy, and this fraction is the mole fraction. Since mole fractions are involved, it is possible to use Raoult's law to determine molecular weights of solids. For instance, suppose we want the molecular weight of table sugar, sucrose. If we dissolve 10.00 g of sucrose in 50.00 g of H_2O, the vapor pressure of the water will be lowered since its mole fraction is no longer equal to 1. Raoult's law gives

$$P_{H_2O} = P^0_{H_2O} x_{H_2O}$$

but since we are interested in the lowering of the vapor pressure, suppose we define the new quantity ΔP:

$$\Delta P \equiv P^0_{H_2O} - P_{H_2O} = P^0_{H_2O} - (P^0_{H_2O} x_{H_2O}) = P^0_{H_2O}(1 - x_{H_2O})$$

Since the solution contains only water and sucrose, the quantity $1 - x_{H_2O}$ is just $x_{sucrose}$:

$$\Delta P = P^0_{H_2O}(1 - x_{H_2O}) = P^0_{H_2O} x_{sucrose} = P^0_{H_2O} \frac{n_{sucrose}}{n_{sucrose} + n_{H_2O}}$$

where $n_{sucrose}$ and n_{H_2O} are the numbers of moles of each component of the solution. For fairly dilute solutions, including this one, $n_{sucrose} + n_{H_2O}$ is approximately equal to n_{H_2O}, because $n_{sucrose}$ is a very small number relative to n_{H_2O}. This means that to a fairly good approximation we can write

$$\frac{\Delta P}{P^0_{H_2O}} \cong \frac{n_{sucrose}}{n_{H_2O}} = \frac{\text{wt sucrose/MW sucrose}}{\text{wt } H_2O/\text{MW } H_2O}$$

or

$$\text{MW sucrose} = \frac{\text{wt sucrose} \times P^0_{H_2O} \times \text{MW } H_2O}{\text{wt } H_2O \times \Delta P}$$

Writing this in a more general fashion,

$$\text{MW solute} = \frac{\text{wt solute} \times P^0_{solvent}}{\text{wt solvent} \times \Delta P} \times \text{MW solvent}$$

In this particular experiment we have 10.00 g of sucrose and 50.00 g of H_2O. The vapor pressure of pure water at 20°C is 17.535 torr, and we observe a vapor-pressure lowering of 0.185 torr at that temperature. We can get the molecular weight of water from the atomic weights: 18.02 g/mole. So we have

$$\text{MW sucrose} = \frac{10.00 \times 17.535}{50.00 \times 0.185} \times 18.02 = 342 \text{ g/mole}$$

Since the weight of the solvent appears in the above expressions, it should be clear that molality (moles of solute per kilogram of solvent) is an appropriate concentration unit. The vapor-pressure-lowering expression can be rewritten, using molality, as

$$\frac{\Delta P}{P^0_{solvent}} = \frac{n_{solute} \times \text{MW solvent}}{\text{wt solvent}} = \frac{m \times \text{MW solvent}}{1000} \qquad (7\text{-}15)$$

where m is the molality of the solution. So the vapor-pressure lowering is proportional to the molality of the solution. There are several other **colligative properties** that are similarly proportional to molality. A colligative property of a solution is one that depends solely on the number of solute particles (ions or molecules) present in the solution per mole of solvent. All of the colligative properties can be used to establish molecular weights by mathematical procedures analogous to those above.

Boiling-Point Elevation

Consider, for example, what happens to the pressure/temperature plot of Fig. 7-22 when the liquid phase is no longer a pure liquid but a solution. The vapor pressure is lowered at every temperature by a factor given by Raoult's law, resulting in the red line in Fig. 7-27. Since the vapor pressure is lower than it

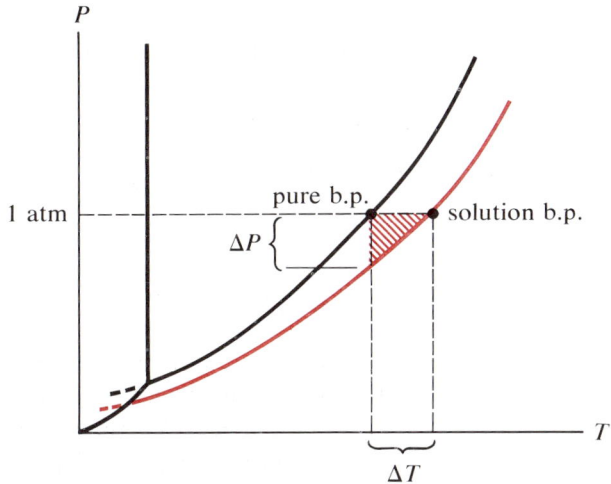

Figure 7-27   The origin of boiling-point elevation as a colligative property.

would have been for the pure liquid, the temperature necessary to give a vapor pressure of 1 atm is higher than for the pure liquid by an amount ΔT, the **boiling-point elevation**. We can show that the boiling-point elevation is proportional to molality by considering the relationship between the sides of the triangle that is shaded red. The amount ΔT times the slope of the red line must be just enough pressure increase to make up for the ΔP pressure decrease:

$$\Delta T \times \text{slope of } P \text{ vs. } T \text{ line} = \Delta P = \frac{m \times \text{MW solvent} \times P^0_{\text{solvent}}}{1000}$$

From Chapter 2 we recall that the slope of the P vs. T line is the derivative of P with respect to T, or dP/dT. To get dP/dT we need a function between vapor pressure and temperature; we use Eq. 7-13:

$$\ln P_{\text{solvent}} = -\frac{\Delta H_{\text{vap}}}{kT} \quad \text{(for 1 molecule)}$$

$$= -\frac{\Delta H_{\text{vap}}}{RT} \quad \text{(for 1 mole)}$$

Differentiate both sides of this expression, remembering that the differential of $\ln P$ is $(1/P) \, dP$, and the differential of $-1/T$ is $(+1/T^2) \, dT$:

$$\frac{d}{dT}(\ln P_{\text{solvent}}) = \frac{d}{dT}\left(-\frac{\Delta H_{\text{vap}}}{RT}\right)$$

$$\frac{1}{P_{\text{solvent}}} \frac{dP}{dT} = \frac{\Delta H_{\text{vap}}}{RT^2}$$

or

$$\frac{dP}{dT} = \frac{P_{\text{solvent}} \Delta H_{\text{vap}}}{RT^2} = \text{slope of } P \text{ vs. } T \text{ line}$$

Substituting this in the first expression we have

$$\Delta T \frac{P_{\text{solvent}} \Delta H_{\text{vap}}}{RT^2} = \frac{(\text{MW solvent}) P_{\text{solvent}}}{1000} m$$

or

$$\Delta T = \frac{(\text{MW solvent}) RT^2}{1000 \, \Delta H_{\text{vap}}} m$$

For dilute solutions, where the red-shaded area in Fig. 7-27 is very small, the red line is nearly straight over the small temperature range ΔT, and the temperature T in this last expression can be taken as T_b, the normal boiling point of the pure liquid. When this is done, all of the terms in the fraction become constants relating only to properties of the solvent, and we can define a boiling-point elevation constant K_b:

$$K_b \equiv \frac{(\text{MW solvent}) RT_b^2}{1000 \, \Delta H_{\text{vap}}}$$

and

$$\Delta T = K_b m \tag{7-16}$$

So the boiling-point elevation is another colligative property, and we can use Eq. 7-16 to determine molecular weights, since values for K_b have been tabulated for many solvents. For instance, if a solution of 3.00 g of an unknown organic solid in 100 g of benzene ($K_b = 2.53$) causes the boiling point of the benzene to increase from 80.10°C to 80.51°C, what is the molecular weight of the unknown solid? Substituting in Eq. 7-16, we have

$$\Delta T = K_b \, m$$

$$80.51 - 80.10 = 0.41 = 2.53 \, m$$

The molality is the number of moles of solute per 1000 g of solvent:

$$m = \frac{(3.00 \text{ g}/\text{MW})(\text{g}/\text{mole})}{100 \text{ g}/1000 \text{ g}}$$

or

$$\text{MW} = \frac{3.00(1000)}{100\ m}$$

But if $0.41 = 2.53\ m$, then

$$m = \frac{0.41}{2.53}$$

$$\text{MW} = \frac{3.00(1000)}{100(0.41/2.53)} = 185\ \text{g/mole}$$

Freezing-Point Depression

Figure 7-28 is the same vapor pressure/temperature plot as Fig. 7-27, but with attention focused on the region of the graph in the vicinity of the freezing point of the liquid. Note that the vapor pressure of the liquid is lowered, but not that

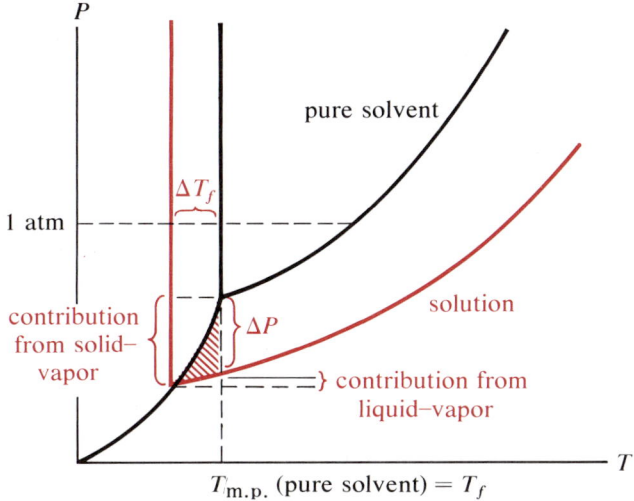

Figure 7-28 The origin of freezing-point depression as a colligative property.

of the solid. This is because when solvent freezes out of solution it is very nearly pure, since the orderly crystal arrangement allows very little size difference between solvent and solute molecules. Thus the freezing solvent normally rejects molecules of the solute in forming its crystal lattice, and the vapor pressure over the solid formed by freezing the solution is no different from that over the solid formed by freezing pure liquid solvent. By an argument analogous to that for the boiling-point elevation, relating the sides of the red-shaded triangle in Fig. 7-28, we can obtain (as in Study Problem 4) an expression for the **freezing-point depression**:

$$\Delta T_f = -\frac{(\text{MW solvent})RT_f^2}{1000 \Delta H_{\text{fus}}} m$$

or

$$\Delta T_f = K_f m \qquad (7\text{-}17)$$

In this expression the **heat of fusion** ΔH_{fus} is the heat that must be put into a solid to melt it (reducing the average coordination number from 12 to 10 or 11). The form of this expression indicates that the freezing-point depression is also a colligative property.

Osmosis

There is one more colligative property, which we cannot treat theoretically at this point but which has considerable significance in polymer chemistry and biological systems. The phenomenon of **osmotic pressure** or **osmosis** occurs whenever a pure liquid solvent and a solution in that solvent are separated by a membrane or barrier that permits solvent molecules but not solute to pass through. In practice, solvent will pass through the **semipermeable membrane** from the pure liquid side to the solution side, diluting the solution. If the experimental arrangement is that of Fig. 7-29, the solution will rise in the capillary tube until the pressure difference due to the height of the liquid column — the hydrostatic pressure — reaches some equilibrium value Π, called the osmotic

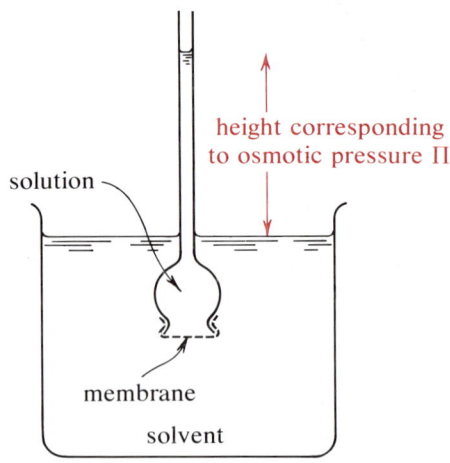

Figure 7-29 Experimental arrangement for observing osmotic pressure.

pressure. It is both experimentally true and theoretically derivable that

$$\Pi V = nRT \tag{7-18}$$

where V is the volume of the liquid solution, n is the number of moles of solute particles present, and RT is the gas constant times the absolute temperature. The analogy to the ideal gas law is apparent and is not accidental. Suppose we had two gas bulbs separated by a stopcock as in Fig. 7-30, but with the stop-

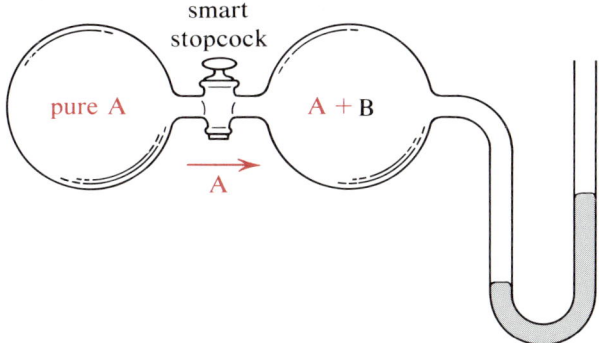

Figure 7-30 Hypothetical apparatus for observing an osmotic effect in gas mixtures.

cock having the unusual property that it permits only molecules of gas A to pass through it. If the two bulbs are initially filled with gas A at equal pressures, no gas will flow through the stopcock. But if half the gas A in the right bulb is removed and replaced by gas B, gas A will flow through the stopcock until the partial pressure of A in the right bulb equals the total pressure of A in the left bulb. The manometer on the right side is sensitive to the pressures of both A and B, so it indicates an increased pressure in the right bulb after A has flowed into it. This is exactly analogous to the solution phenomenon of osmotic pressure; the additional pressure in the right bulb is proportional to the number of moles of B present, and inversely proportional to the volume of the bulb. In solution, the formation of the solution lowers the vapor pressure of the solution. The pure solvent consequently "evaporates" through the membrane and condenses in the solution to increase the rate of escape of its solvent vapor through the membrane (i.e., its vapor pressure) to that of the pure solvent.

In biological systems, cell walls are semipermeable membranes toward aqueous solutions. Perhaps the most familiar example of this is the fact that salt kills grass; the cell fluids in the grass flow out through the cell walls in an attempt to dilute the salt solution outside. There are many more such examples, but perhaps the chief usefulness of osmosis to chemists is, like the other colligative properties, in the establishment of molecular weights. Very high molecular weights can be measured using osmotic pressure, since pressures of very small fractions of an atmosphere can be measured readily. For instance, a sample of synthetic rubber is dissolved in toluene and placed in a cell for osmosis against pure toluene. When equilibrium is attained at 30°C, the solution has risen 11.7 mm in a tube above the height of the pure toluene; the solution has a density of 0.857 g/ml and contains 3.65 g of synthetic rubber per liter of solution. What is the average molecular weight of the rubber? First we need to convert the pressure represented by the solution height to atmospheres. This is most easily done by remembering that 1 atm is equivalent to a height of 760 mm of mercury. Since the density of mercury is 13.54 g/ml, a mercury column of the same weight as the toluene column would be shorter by the ratio of the densities:

$$\text{mercury column height} = \frac{0.857}{13.54} 11.7 = 0.740 \text{ mm}$$

Converting to atmospheres,

$$\Pi = \frac{0.740}{760} = 9.74 \times 10^{-4} \text{ atm}$$

Since the number of moles present equals the weight divided by the molecular weight,

$$\Pi V = nRT = \frac{\text{wt}}{\text{MW}} RT$$

$$9.74 \times 10^{-4} \times 1 = \frac{3.65}{\text{MW}} 0.08205 \, (273 + 30)$$

$$\text{MW} = \frac{3.65 \times 0.08205 \times 303}{1 \times 9.74 \times 10^{-4}}$$

$$= 9.08 \times 10^4 \quad \text{or about } 90{,}000 \text{ g/mole}$$

One last interesting application of the phenomenon of osmosis arises from the fact that if a pressure greater than the osmotic pressure is exerted on the solution side of the membrane, the pure solvent will actually flow back from the solution into the pure liquid side. This phenomenon, called **reverse osmosis**, is being explored as a means of desalinating sea water cheaply, since the only energy requirement is that necessary to pressurize the sea water and circulate it—much less than distillation or freezing. The problems lie in constructing membranes sturdy enough to withstand high pressure and yet porous enough to allow a substantial flow of fresh water.

Henry's Law

Now consider the opposite extreme, a solution of a gas in a liquid (as in a carbonated drink). This situation is governed by **Henry's law**, which was discovered by William Henry 85 years before the formulation of Raoult's law. Henry's law says that the concentration of a dissolved gas is proportional to its partial pressure in the vapor phase, which is obviously similar to Raoult's law:

$$\text{Henry's law:} \quad P_{\text{solute}} = Kx_{\text{solute}} \tag{7-19}$$

$$\text{Raoult's law:} \quad P_{\text{solvent}} = P^0_{\text{solvent}} x_{\text{solvent}}$$

However, the constant in Henry's law is *not* generally equal to P^0_{solute}, the vapor pressure of the pure gas solute. Like Raoult's law, Henry's law is followed by all solutions only at the limit of extreme dilution. From this fact we can see why the constant in Raoult's law is the vapor pressure of the pure solvent but that in Henry's law is not the vapor pressure of the pure solute — in a dilute solution the solvent molecules are in an environment essentially identical to that in the pure liquid, but the solute molecules are in an entirely different environment from the pure solute, being surrounded by solvent molecules instead of other solute molecules.

Extraction and the Distribution Coefficient

Just as in dealing with the property of surface tension or interfacial tension we drew an analogy between a liquid–gas interface and the interface between two liquids, we can apply the analogy to the situation described by Henry's law. From Chapter 3 (see p. 130) we know that the partial pressure of a gas in a gas mixture is proportional to its mole fraction in the mixture, and Henry's law says that the partial pressure is also proportional to its mole fraction in the liquid solution. This means that the ratio of the mole fractions in the gas and liquid phases is constant at constant pressure:

$$P^{\text{gas}}_{\text{solute}} = P^{\text{gas}}_{\text{total}} x^{\text{gas}}_{\text{solute}} \quad \text{and} \quad P^{\text{gas}}_{\text{solute}} = K x^{\text{liquid}}_{\text{solute}}$$

$$\frac{x^{\text{gas}}_{\text{solute}}}{x^{\text{liquid}}_{\text{solute}}} = \frac{K}{P^{\text{gas}}_{\text{total}}}$$

Experimentally we find that we can write an entirely analogous expression for the concentration ratio of a solute — even if it is not a gas — between two immiscible liquids:

$$\frac{x^{\text{solvent A}}_{\text{solute}}}{x^{\text{solvent B}}_{\text{solute}}} = K_d$$

where K_d is the **distribution constant** of the solute between the two solvents. This relationship holds even if the concentrations are expressed in different units, such as molarity or grams per liter, although of course the numerical value of K_d will change from unit to unit. This phenomenon is very useful to the chemist since it enables him to transfer a solute from one solvent to another, which is often desirable in synthetic processes. For instance, suppose K_d has a numerical value of 4 and we have a solution of 10.00 g of solute in 50 ml of solvent B. After shaking the B solution together with 50 ml of solvent A, we have **extracted** 80% of the solute into the A solution:

$$\frac{(x)_{\text{A solution 1}}}{(10.00-x)_{\text{B solution}}} = 4.00$$

$$x = 8.00$$

If we then shake the B solution (now containing only 2.00 g of solute) with a fresh 50-ml portion of solvent A, and combine the two portions of solvent A, we have 96% extraction:

$$\frac{(x)_{\text{A solution 2}}}{(2.00-x)_{\text{B solution}}} = 4.00$$

$$x = 1.60$$

$$1.60 + 8.00 = 9.60 \text{ g out of } 10.00 \text{ g}$$

Repeating the process gives still more complete extraction (see Study Problem 5). Multiple extractors have been devised that give the effect of several hundred extractions without the need of repeated manual solvent handling; these are often used in organic and pharmaceutical preparations.

Liquid–Liquid Solutions and Distillation

We can now move to the intermediate case of solutions of liquids in liquids. From all that we have said previously, we expect these to resemble both solid–liquid and gas–liquid solutions, and indeed this is true. Consider first the condition in which both liquids obey Raoult's law over the entire composition range from pure liquid A to pure liquid B. This situation is represented by the graph in Fig. 7-31, which is valid for a given temperature. The total vapor pressure is the sum of the two individual component vapor pressures, which in turn are dictated by the composition *of the liquid*. The vapor-pressure diagram has two lines, however, representing the fact that the composition of a mixed vapor over a liquid–liquid solution will not be the same as the composition of the solution, because the more volatile component evaporates preferentially from the liquid. Using the diagram, as in Fig. 7-32, we can predict the composition of the vapor over a liquid–liquid solution from the composition of the solution. Drawing a vertical line at the liquid composition, we get the vapor pressure of that mixture; drawing a horizontal line at that vapor pressure, we find the vapor composition that would give that total pressure. Of course, we can do the same thing algebraically, using Raoult's law: consider the two organic liquids *n*-hexane and *n*-octane, which form a nearly ideal solution. Pure

Liquids in Physical Mixtures — Liquid–Vapor Equilibrium | 383

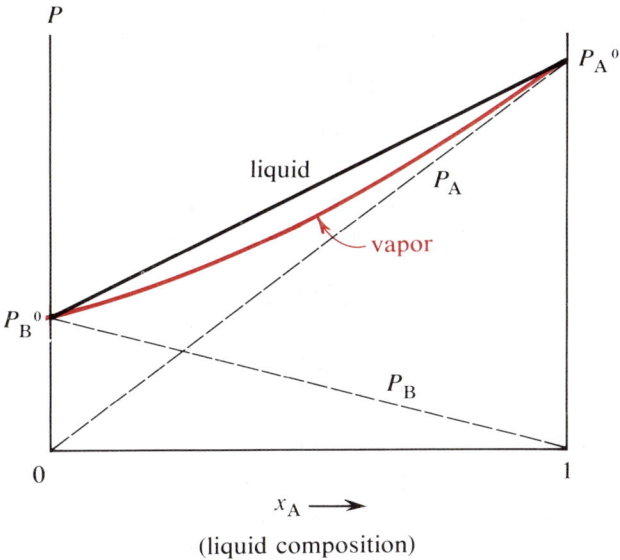

Figure 7-31 Pressure–composition curve for an ideal solution.

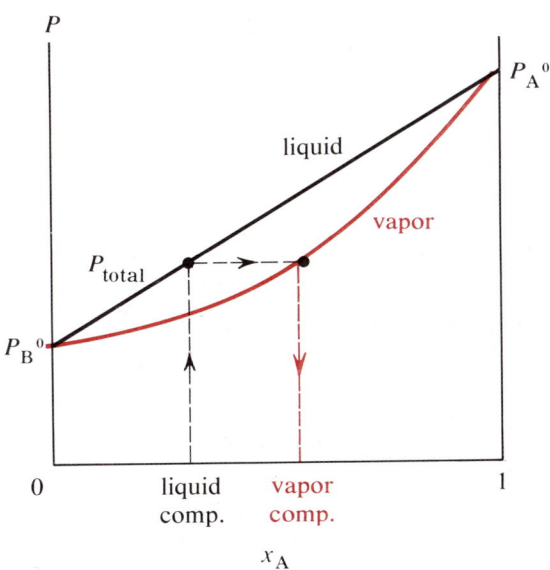

Figure 7-32 Use of pressure–composition curves to establish liquid and vapor compositions.

hexane has a vapor pressure of 188.6 torr at 30°C, and pure octane has a vapor pressure of 18.4 torr at 30°C. What is the composition of the vapor mixture over a 50–50 liquid mixture? Using Raoult's law,

$$30°C \begin{cases} P_{hexane} = 188.6 \, (0.5) = 94.3 \text{ torr} \\ P_{octane} = 18.4 \, (0.5) = 9.2 \text{ torr} \end{cases}$$

The total vapor pressure over the mixture is the sum of these two partial pressures, or 103.5 torr. From Dalton's law of partial pressure, however, we have

$$x_{hexane}^{vapor} = \frac{94.3}{103.5} = 0.911$$

$$x_{octane}^{vapor} = \frac{9.2}{103.5} = 0.089$$

So a mixture that was 50 mole-percent hexane in the liquid phase becomes 91 mole-percent hexane when it vaporizes at 30°C. Since vapor pressures increase with temperature, the relative numbers change if the vaporization occurs at another temperature.

This change of composition on vaporizing is what makes distillation an effective means of purifying many liquids. If the above mixture of hexane and octane is heated, it will begin to boil at about 65 °C, a temperature at which the total vapor pressure of the two components equals 760 torr. At this temperature, the vapor pressure of pure hexane is about 670 torr, and that of octane is about 90 torr. Calculating as before:

$$65°C \begin{cases} P_{hexane} = 670 \, (0.5) = 335 & \frac{335}{380} = 0.88 = x_{hexane}^{vapor} \\ P_{octane} = 90 \, (0.5) = \frac{45}{380} & \frac{45}{380} = 0.12 = x_{octane}^{vapor} \end{cases}$$

So the 0.5-mole-fraction liquid mixture, when boiled, initially distills a vapor that is 0.88-mole-fraction hexane; this obviously represents a substantial change in composition or purity of the hexane. Figure 7-33 shows a simple

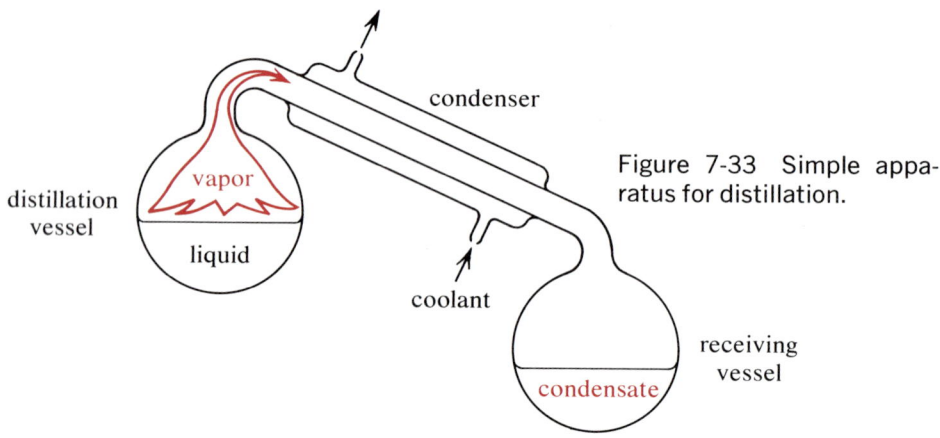

Figure 7-33 Simple apparatus for distillation.

distillation apparatus, in which the vapor distilling over is condensed into a liquid again by the heat exchange or condenser at the right of the diagram. If this condensate were to be redistilled, a further purification would occur, as we can see from Fig. 7-34, which is a pressure–composition diagram for this

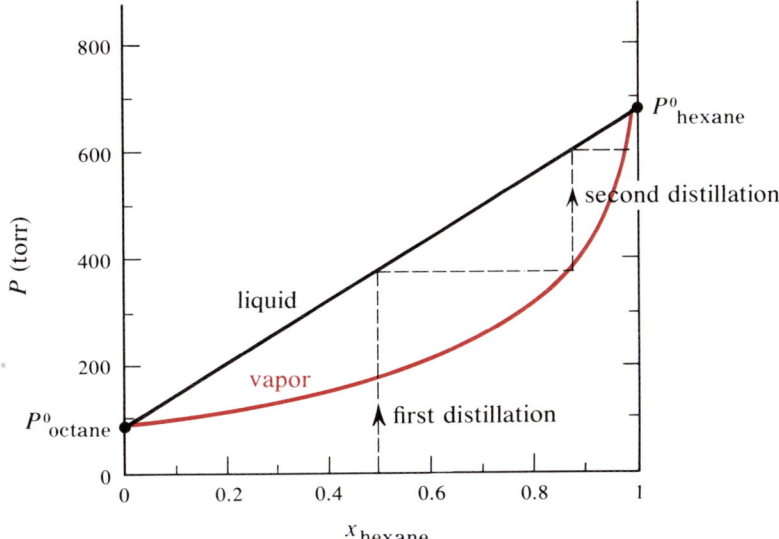

Figure 7-34   Use of the pressure–composition curve to analyze the effect of successive distillations of a 1:1 hexane–octane mixture.

system. Each successive distillation represents another "stair-step" within the lens-shaped liquid–vapor pair of lines. Some effort has gone into designing distillation apparatus that will in effect condense and reevaporate a mixture many times in the process of a single pass through the apparatus; Fig. 7-35 shows an early and reasonably effective version called a **bubble-cap column**, which condenses and reevaporates the distillate at each horizontal plate in the column. A column with 20 of these plates corresponds to an apparatus for 20 successive simple distillations. Other types of distillation columns have been devised that are even more effective, providing separations by distillation equivalent to a bubble-cap column with hundreds of plates; the efficiency of these columns is usually quoted in terms of the number of **theoretical plates** they offer. A very efficient column will offer many theoretical plates but will not occupy much space — if we divide the height of the column by the number of theoretical plates it offers, a number results called the **height equivalent of a theoretical plate (HETP)**, which is small for a good column, large for a poorer

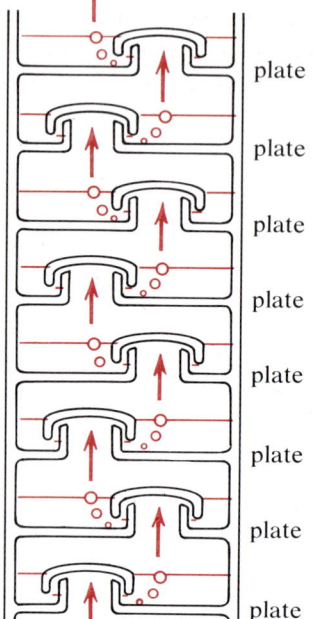

Figure 7-35 Bubble-cap column.

one. For instance, an open column (Fig. 7-36) has a HETP of about 15 cm, a column packed with small glass rings has a HETP of about 4 cm, a popular bubble-cap design has a HETP of about 2 cm, and the best designs have HETP values in the vicinity of 0.5 cm.

Nonideal Solutions

Liquid–liquid solutions in which both components obey Raoult's law at all compositions are called **ideal solutions**. These are relatively rare, however, and we usually have to work with systems for which the pressure–composition diagram shows substantial deviations from the straight lines that represent ideal behavior. The deviations may be either positive or negative, as in Fig. 7-37, but if one component shows a positive deviation then both will. Negative deviations mean that the vapor pressure of each component is lower than would be expected — this usually results when, due to polarity or hydrogen bonding, the molecules of the two components attract each other more strongly than themselves. Positive deviations, for which the vapor pressure of each component is higher than would be expected, result when the intermolecular attractions in each pure liquid are greater than those between the two unlike molecules; for example, the hydrogen bonding in water is very strong, and many solutions of

Liquids in Physical Mixtures — Liquid–Vapor Equilibrium | 387

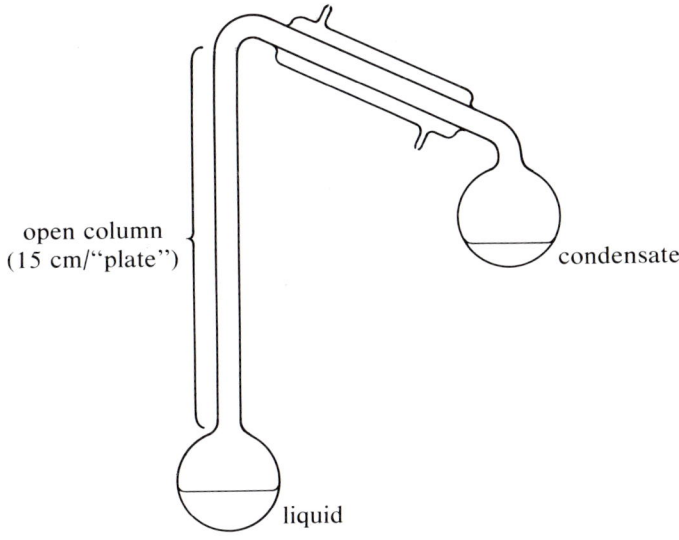

Figure 7-36 Distillation apparatus using a distillation column (often packed with high-surface-area inert solid).

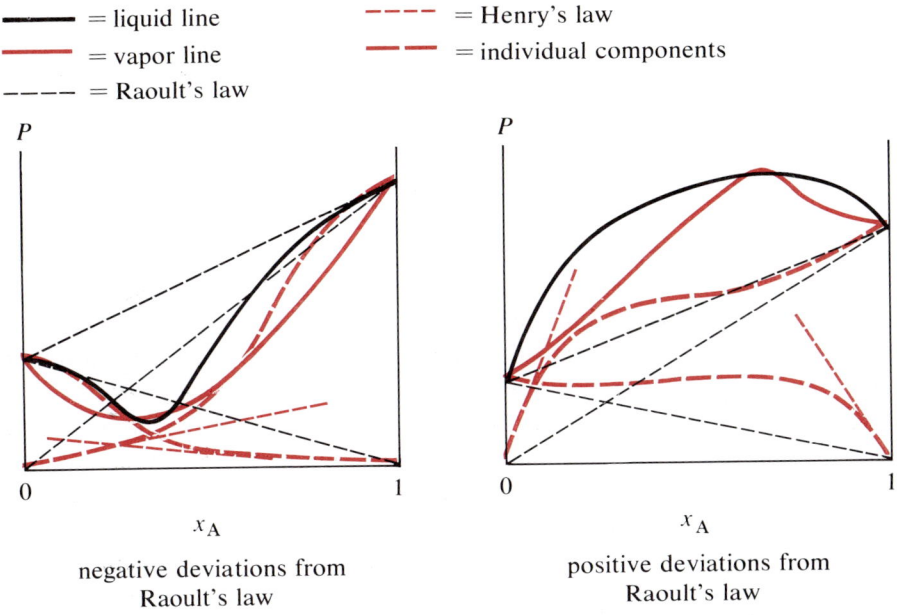

Figure 7-37 Pressure–composition curves for nonideal solutions.

water in other liquids show positive deviations because the hydrogen bonding of the water breaks down, making the vapor pressure of the water higher than Raoult's law suggests. If the negative deviations are great enough, there will actually be a minimum in the pressure–composition curve; if the positive deviations are large enough, there will be a maximum. Positive (or negative) deviations affect both liquid and vapor lines, and at any maximum or minimum in the pressure–composition curve the two lines will touch, as Fig. 7-37 shows. The solution having the composition corresponding to the maximum or minimum is called an **azeotrope** (from the Greek "boiling unchanged"), since the liquid and vapor compositions are the same at that point. Following the "stair-step" treatment of successive distillations, it can be seen that repeated distillations on an azeotropic system always yield the azeotrope, no matter which side of the azeotropic composition is taken initially. Ethyl alcohol, for instance, is usually sold (for chemical, as opposed to social purposes) as 95% alcohol because it is purified by distillation and the alcohol–water azeotrope occurs at that composition.

How can azeotropes be separated? We have two chemical compounds showing a common physical property, volatility. To separate them we must either add another chemical compound or take simultaneous advantage of another physical property, solubility. If we use another chemical compound, it can either react with one component of the azeotrope to remove it, or form a ternary azeotrope with both components and distill away some of the major component but all of the minor component. The ethyl alcohol–water azeotrope, for example, can be taken to 100% ethyl alcohol either by adding CaO, which reacts with water to give solid $Ca(OH)_2$, or by adding benzene, which distills out as the benzene–alcohol–water azeotrope and the benzene–alcohol azeotrope, leaving behind pure alcohol.

Gas Chromatography

We can take simultaneous advantage of the solubility property by using a technique called gas–liquid partition chromatography, or just **gas chromatography**. Suppose we allow a moving stream of an inert gas to pass over the surface of an azeotrope as in Fig. 7-38a. The vapor entering the moving stream of gas (the **moving phase**) will have the same composition as the liquid azeotrope, and no separation has been achieved. But suppose both components of the azeotrope are dissolved in an inert, nonvolatile liquid, as in Fig. 7-38b. If the nonvolatile liquid is properly chosen, the Henry's-law constants for the two azeotrope components dissolved in the nonvolatile liquid (the **stationary phase**) will be different. Then one of the components will preferentially reevaporate into the moving phase (inert-gas stream). If the two vapor components re-

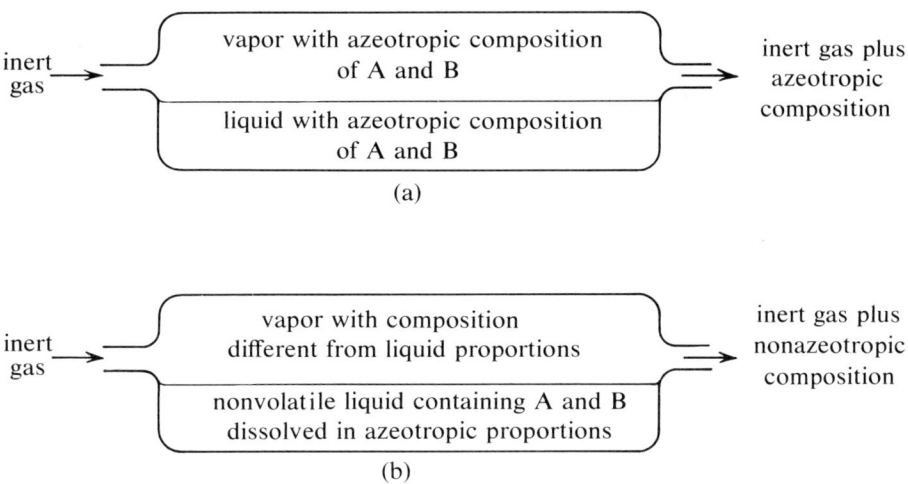

Figure 7-38 Liquid–vapor equilibrium applied to gas chromatography. (a) Binary system; no separation of azeotrope. (b) Ternary system; separation of azeotrope components through different Henry's-law behavior.

dissolve later in the stationary phase and then reevaporate into the moving phase, and if this is repeated many times, a complete separation can eventually be achieved. In practice, gas chromatography is done by packing a small-diameter tube with small granules of solid coated with a very thin film of the nonvolatile liquid, to provide a large surface area. The liquid of interest is injected into the moving inert-gas stream at one end of the tube (or **column**) and passes through the column, dissolving and reevaporating many times. The temperature of the column is controlled in order to keep the volatility of the components at a convenient level; that is, so they come through the column in a reasonable time. In passing through the column they are completely separated, and a detector (often a sensor of the thermal conductivity of the moving stream of gas) indicates the presence of the two separate components, as diagrammed in Fig. 7-39. Gas chromatography is a particularly important tool of the synthetic chemist, both on a very small scale (quantities of a microliter of liquid) for analytical purposes, and on a relatively large scale (injections of 0.5–5 ml) for preparative purposes.

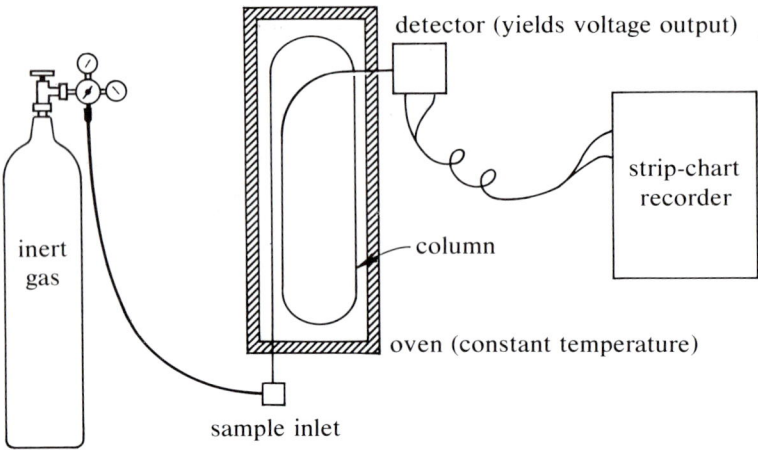

Figure 7-39 Schematic representation of a gas chromatograph.

7-5 Liquids in Physical Mixtures – Liquid–Solid Equilibrium

Besides the phase transition that occurs when a liquid evaporates, there is another phase transition of interest, that between the liquid and solid phases — the phenomena of freezing and melting. These processes are characterized by a temperature called the **melting point**, at which the liquid and solid phases are stable in each other's presence if no heat energy is added or removed. At higher temperatures the solid will melt spontaneously, and at lower temperatures the liquid will freeze spontaneously. Our first interest, then, is the experimental establishment of this characteristic temperature, with the understanding that the terms melting point and freezing point both refer to it.

The earlier discussion has indicated that the melting point is somewhat sensitive to pressure, and consequently it is necessary to specify a pressure at which a compound's melting point is quoted. For laboratory convenience, we normally use the melting point at 1 atm total pressure, although in dealing with the crystallization of rocks in the earth's crust high-pressure freezing points become important. All of the melting/freezing phenomena discussed here, however, will be treated at atmospheric pressure.

Measurement of Melting Points

The melting point is a valuable characterization of solid compounds, particularly organic compounds, and to measure it conveniently an apparatus resembling Fig. 7-40 is often used. A few small crystals of the solid compound are pressed into a very thin-walled glass capillary tube, which is immersed in a heated oil bath near a thermometer. The temperature of the bath is increased slowly, and the melting point is read from the thermometer when the crystals melt.

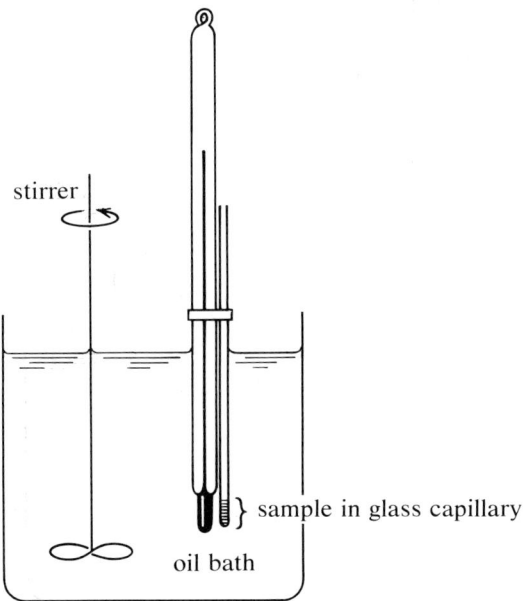

Figure 7-40   Simple apparatus for determining melting points.

For more accurate work, the method of cooling curves is usually employed. Using an apparatus similar to that in Fig. 7-41, a pure liquid compound is allowed to cool slowly, with gentle stirring. If the temperature difference across the insulation barrier is fairly large, the rate of heat transfer through the insulation will be an approximately constant number of calories per minute; we do not need to know how many. The thermometer (or other temperature-sensitive device, such as a thermocouple) is read at regular time intervals—in other words, at regular increments of heat transferred away from the liquid—and the temperatures are plotted as a function of time, as in Fig. 7-42. In the beginning, the temperature goes down fairly sharply as increments of heat are removed, since the liquid is cooling at a rate determined by its heat capacity; this con-

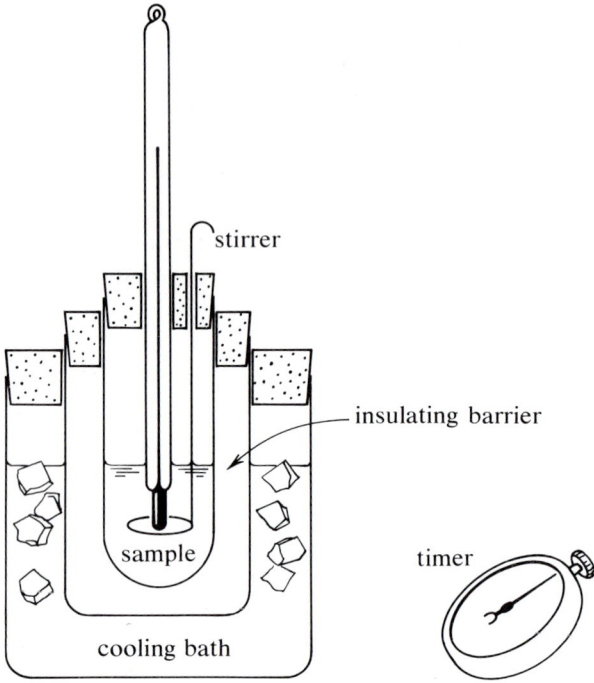

Figure 7-41 Precise freezing-point determination through cooling-curve measurement.

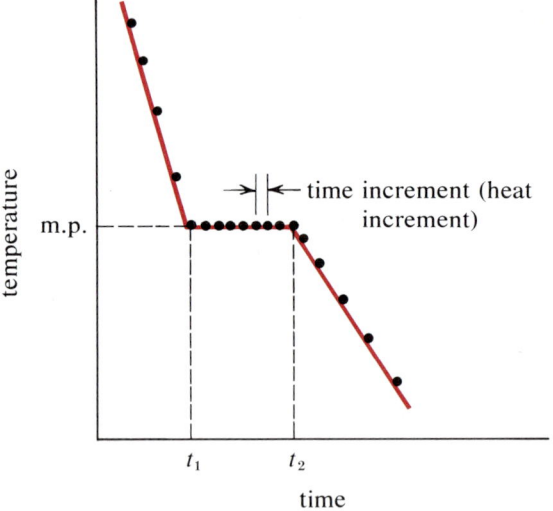

Figure 7-42 Cooling curve for an idealized pure liquid.

tinues until time t_1. At that point the freezing temperature has been reached and some solid begins to form. Heat is released when the disorderly liquid molecules crystallize into the solid lattice, with a higher coordination number and more intermolecular stabilization, so that while the liquid is crystallizing heat is being transferred out even though the temperature does not change. The temperature remains constant at the freezing point while the liquid is crystallizing, from t_1 to t_2. At time t_2 all of the liquid has frozen, and beyond that time the temperature drops at a rate determined by the heat capacity of the solid and the bath temperature. From the horizontal portion of the line a very accurate freezing point can be established.

The Molecular Model for Melting

How can we relate this well-defined freezing point or melting point to the model we have proposed? In Section 7-2 an approach has already been made to the process of melting in which heat energy added to the solid causes increased numbers and densities of crystal defects. At some particularly significant density of defects, the three-dimensional stability of the crystal lattice can no longer be maintained against the increasing vibrational energy of the molecules, and the whole lattice breaks down at once, inside as well as outside. Notice that two effects are operating: first, the increasing vibrational energy makes the constraints on the position of the molecule imposed by the orderly structure less and less effective; and second, the increasing number of vacancies and interstitial atoms reduces the constraints themselves. Melting in a crystal whose temperature is uniform thus cannot occur gradually but must be uniform and sudden, since the statistical behavior of the large number of atoms in a crystal is so reliable as to ensure that disorder will be created all through the crystal at the same rate. Of course, if the temperature varies through a crystal and if only enough heat energy is supplied to provide the heat of fusion for part of the crystal, then only part of the crystal will melt (usually the outside, since it receives the incoming heat first). The rate at which ice melts in ice water is essentially governed by the rate at which heat is supplied to the mixture and conducted through the ice crystals.

The Molecular Model for Freezing

Freezing as a process is a quite different problem. The cooling curve tells us that freezing does *not* occur simultaneously all through a liquid, and since the freezing point and melting point are the same temperature (if measured under equilibrium conditions) we ask why. To produce an answer from our model we must return to probability and statistics. If a sample of material is a large perfect single crystal, there is only one possible geometrical arrangement of the molecules inside it, that dictated by the intermolecular-attraction energy factors. On heating it, disorder is gradually created (in the form of vacancies, etc.), and there are trillions of possible disorderly arrangements, all equally effective in contributing toward the breakdown of the single orderly structure. The chances are thus only one in trillions that the orderly structure would be maintained above the temperature at which the statistical-average energy of the molecules is sufficient to destroy the long-range three-dimensional order of the crystal structure. However, if the sample of material is a pure liquid, there are again trillions of disorderly structures all corresponding equally well to a possible liquid structure—but there is only one possible orderly arrangement corresponding to the crystal structure. The chances are again only one in trillions that crystallization would occur simultaneously throughout the liquid, even though the temperature might be low enough that the energy of the molecules would not be sufficient to destroy the crystal structure if it did exist.

How, then, does freezing occur at all? It appears that we have demonstrated that no liquid can ever freeze. The answer lies in **nucleation**, the process of forming crystal nuclei on which other molecules can land from the liquid phase and stick. Impurities, dust motes, or even the walls of the container usually serve as nuclei by holding surrounding molecules in the proper geometry, as in Fig. 7-43, and freezing begins to occur at or only slightly below the equilibrium

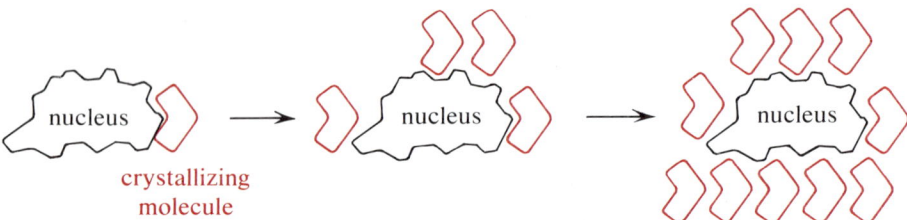

Figure 7-43   The role of dust particles in nucleating crystals.

freezing point. If the liquid is very pure, however, it can usually be **supercooled** by a considerable amount before any crystallization occurs, for the reasons just given. Indeed, some organic compounds are very difficult to crystallize at all, presumably because dust motes are not the right shape to nucleate them. However, when crystallization does begin heat is liberated, as we have noted, and the temperature of the supercooled liquid rises to the equilibrium freezing point, at which temperature crystallization continues until it is complete. Figure 7-44 shows the cooling curve that results when the freezing liquid supercools.

Even a very pure liquid will in many cases form nuclei spontaneously if sufficiently supercooled, however, as for example water, which apparently

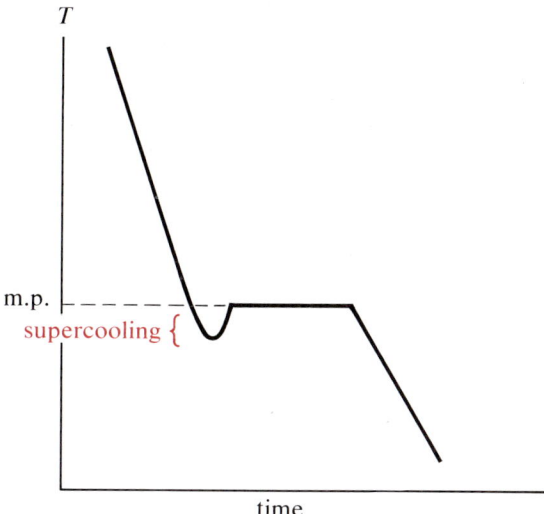

Figure 7-44 Cooling curve showing supercooling due to the lack of appropriate nuclei in the original liquid.

cannot be supercooled below $-40\,°C$. Within a mobile liquid sample that contains a statistically large number of molecules, some very small regions of order containing two, five, or ten molecules are always forming through chance, but at high temperatures the impact of other molecules colliding with them knocks them apart again. The lower the temperature goes below the equilibrium freezing point, however, the less energy the colliding molecules have, and the more likely it becomes that the colliding molecules will fit into the orderly cluster so that it grows instead of being knocked apart. Once it has reached a certain critical size additional impacts cannot knock it apart, and the cluster has become a nucleus for crystallization. For very viscous liquids such as glass or glycerine, the molecules cannot move freely enough to form and enlarge the cluster, and such liquids remain noncrystalline even far below the freezing point for very long times.

Once nuclei exist, molecules striking the nucleus can land in several different types of locations, shown in Fig. 7-45 for a crystal forming from idealized cubic molecules. This picture is really valid only for a crystal forming from vapor, where the crystal surface is mostly surrounded by empty space, but it has some application to growth from liquids. Molecules striking this surface will be most likely to fit into it and be bound into the crystal if they land in a location having a large number of neighbors, such as the hole labeled position 1, in which a molecule has five neighbors. Next most favorable for crystallization is position 2, with four neighbors, and then position 3 with three neighbors. Positions 4 and 5, with two and one neighbors, respectively, are not sufficiently

396 | The Behavior of Liquids

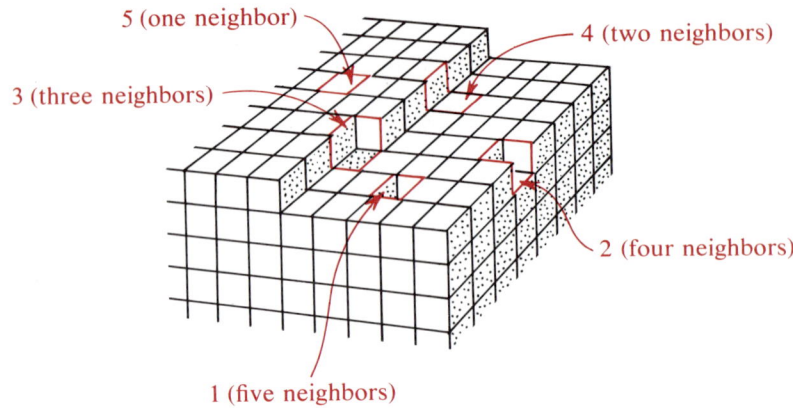

Figure 7-45   Types of idealized sites on a growing crystal.

favorable to lead to growth of the crystal. Since the first three positions all tend to eliminate themselves when one complete plane of molecules has been added to the crystal, this does not seem to be an adequate mechanism for crystal growth. Crystal imperfections play a key role in providing for continued growth, however. The particular type that is known best is called a **screw dislocation**, illustrated in Fig. 7-46. The cubic crystal has been cut through (figuratively) to line AB; then the left side of the split part has been raised relative to the right side by exactly one lattice spacing. The front face of the crystal

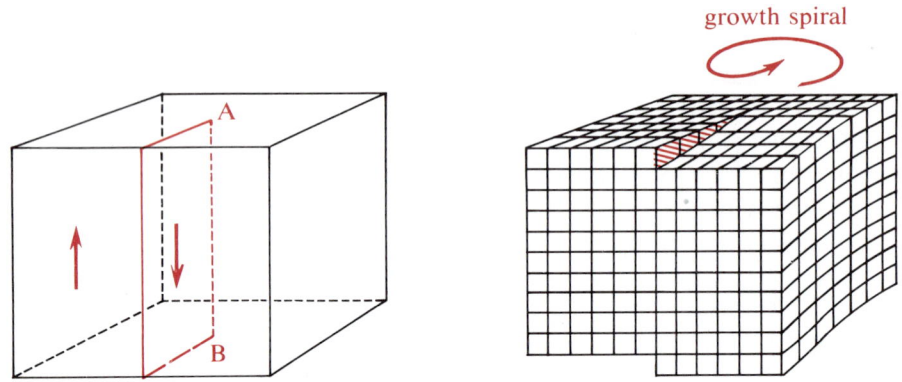

Figure 7-46   Schematic representation of the origin of screw dislocations.

is still perfect, but the top and bottom faces each have a step in them that cannot be removed by crystal growth. Growth on these faces will tend to occur in a spiral fashion, with molecules being deposited at the step. The step will circle around the crystal face but will never reach an edge, since one end of it is anchored at line AB. The resulting growth pattern is shown in Fig. 7-47, both schematically and in a photomicrograph of a real crystal. The net result of this mechanism of growth is to force the crystal to grow one atom at a time at the surface, rather than to have the whole liquid suddenly freeze into a static,

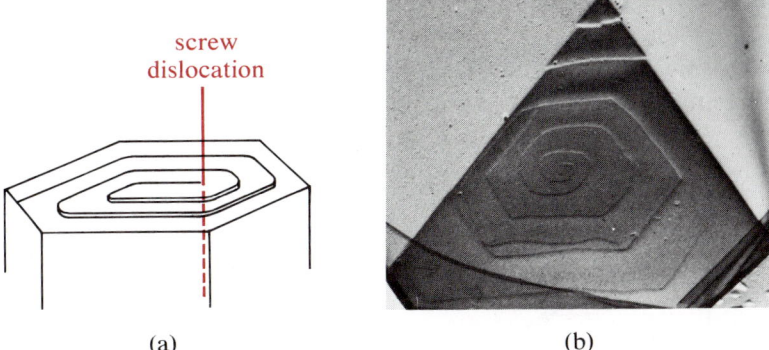

Figure 7-47 Crystal growth around screw dislocations. (Photomicrograph courtesy of Dr. K. R. Lawless, University of Virginia.)

orderly pattern. We can thus rationalize the fact that melting occurs suddenly and a crystal cannot be superheated and remain crystalline, but liquids can be supercooled and crystallize with a finite and relatively slow speed.

Solid–Liquid Phase Diagrams

We can now inquire into the effects of solution formation on melting and freezing. Let us extend our discussion of freezing-point depression by looking at the concentration dependence of the depression for a two-component system. Earlier phase diagrams, such as Fig. 7-28, involved the variables pressure and temperature. Since a third variable — concentration — is being added, a three-dimensional graph is necessary but inconvenient. A great deal can be learned, however, from a two-dimensional plot of temperature vs. composition, taking the pressure as a constant 1 atm. Figure 7-48 shows such a two-component

Figure 7-48 Temperature–composition curve showing freezing-point depression by both components of a mixture.

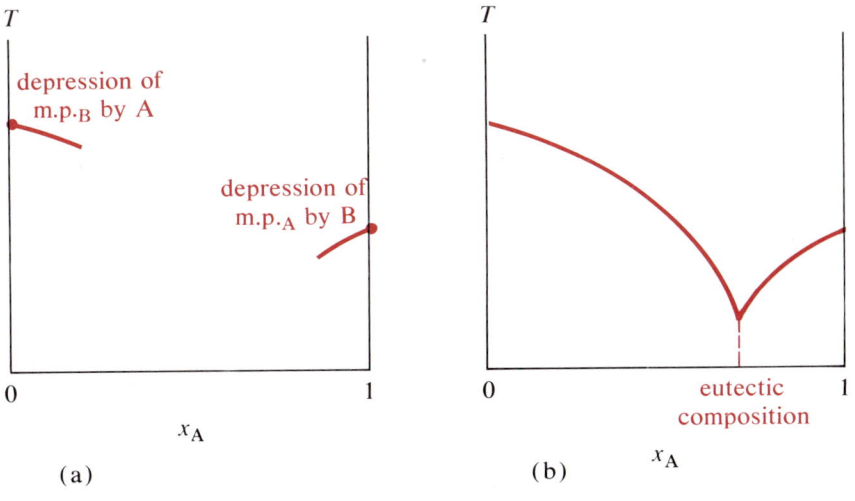

phase diagram for a simple case. In Fig. 7-48a the depression of the freezing point is shown for either pure component when a little of the other is added, and in 7-48b the lines are extended to a composition called the **eutectic**, at which the freezing point calculates the same whichever component is chosen as the solvent. As an example of the use of these diagrams, consider Fig. 7-49 for the system KCl–AgCl. If a mixture with composition (0.3 mole fraction AgCl,

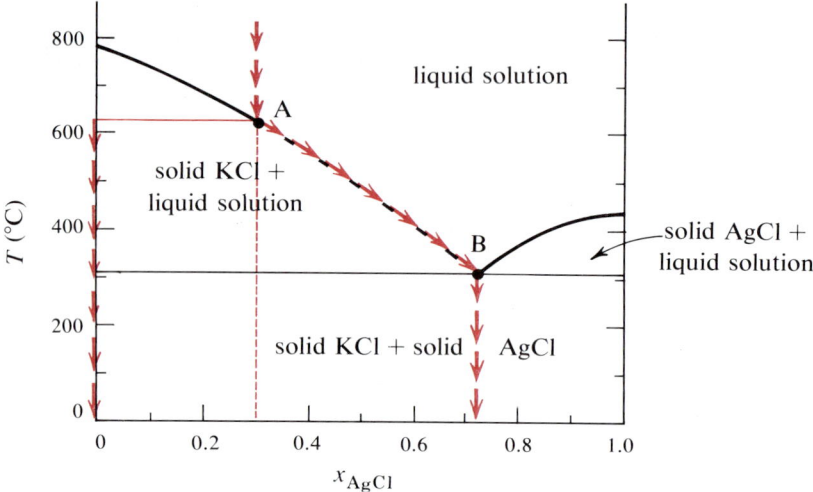

Figure 7-49 Species that freeze out on cooling a liquid solution with noneutectic composition. (Redrawn from Findlay, A., Campbell, A. N., and Smith, N. O., *The Phase Rule*, New York: Dover, 1951.)

0.7 mole fraction KCl) is cooled slowly from an initial temperature of 800°C, as shown by the red line, no change occurs in the liquid until about 650°C (at A), when solid KCl begins to crystallize out of the liquid. With the concentration of KCl thus depleted in the liquid, it moves to a composition with higher mole fraction AgCl, *following the freezing-point-depression line*. As the cooling progresses, more KCl crystallizes out, and the liquid phase represents a smaller and smaller fraction of the total KCl–AgCl system, but is progressively richer and richer in AgCl. Finally, at 306°C, the last liquid (with the eutectic composition B) crystallizes as mixed KCl crystals and AgCl crystals, and below that temperature only the crystals of the two individual compounds exist. Notice that the eutectic composition is somewhat closer to the pure component having the lower melting point, as might be expected from the freezing-point-depression interpretation of the diagram. Another legitimate interpretation of the diagram is as a representation of the solubility of the two individual components in a mixed solvent composed of the two molten salts; thus cooling the liquid reduces the solubility of the individual salts, and when saturation is reached one begins to crystallize out.

Solution Cooling Curves

Having a solution instead of a pure liquid freezing also changes the nature of the cooling curves in a way that is predictable from the phase diagram. Figure 7-50 shows cooling curves for the KCl–AgCl system of Fig. 7-49. Note that instead of a horizontal line indicating crystallization but no temperature change,

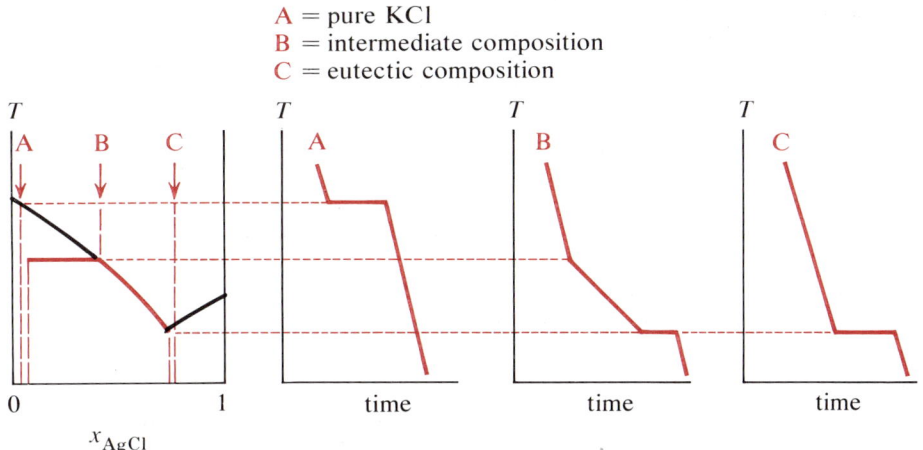

Figure 7-50 Cooling curves for solutions of different compositions.

the solutions give a sloping line over the time span when KCl is crystallizing out and the liquid is diminishing in volume and approaching the eutectic composition; after the eutectic composition is reached by the liquid, however, a horizontal line is observed, indicating that crystallization is occurring with no change in composition. It should be clear from this cooling curve that to define a freezing point for a solution the temperature at which crystals first appear must be used; any other temperature corresponds to a solution of different concentration because one component is freezing out. Only the two pure components and the eutectic mixture show the horizontal portion of the cooling curve exactly like Fig. 7-42.

If the two components of the system form a compound with a definite stoichiometry, the compound itself will be a third component with its own well-defined melting point, and, as in Fig. 7-51, each half of the phase diagram will be a little phase diagram all by itself. The result is that, if two components form a compound, a maximum will appear in the phase diagram of the two components. If more than one compound is formed (at different stoichiometries such as AB, AB_2, A_2B_3, etc.) there will be more than one maximum.

It is also possible to form solid solutions, although these are somewhat rarer than liquid solutions because of the stringent size requirements placed on the

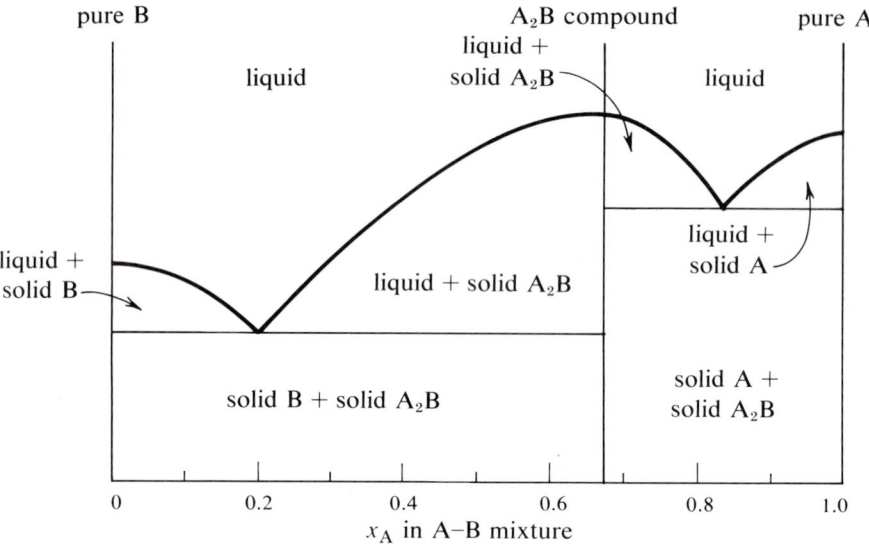

Figure 7-51 Temperature–composition diagram for two components that form a compound.

solute by the crystal structure of the solvent. In favorable cases, however, such as copper and nickel, a continuous series of solid solutions can be formed all the way from one pure component to the other, and the phase diagram looks very much like the pressure–composition diagram for the vapor over a liquid–liquid solution, as shown in Fig. 7-52; maxima and minima can exist in these curves

Figure 7-52 Species present at different stages in cooling a liquid mixture that forms solid solutions.

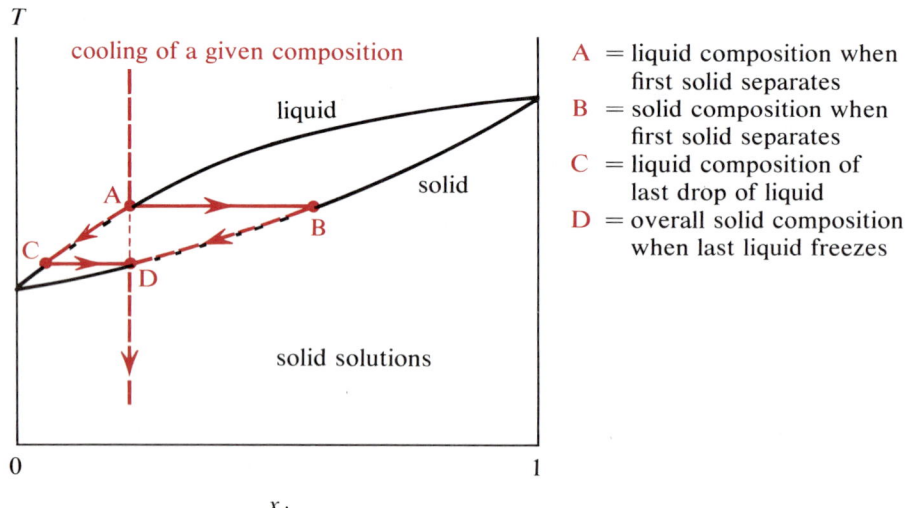

for certain kinds of molecular interactions, just as they do for liquid–liquid azeotropes. More commonly, one component is only slightly soluble in the other (although usually more soluble at higher temperatures) and the solid solutions will occupy only a thin strip on either side of an otherwise normal phase diagram, as in Fig. 7-53. This is true for many metals, and in particular

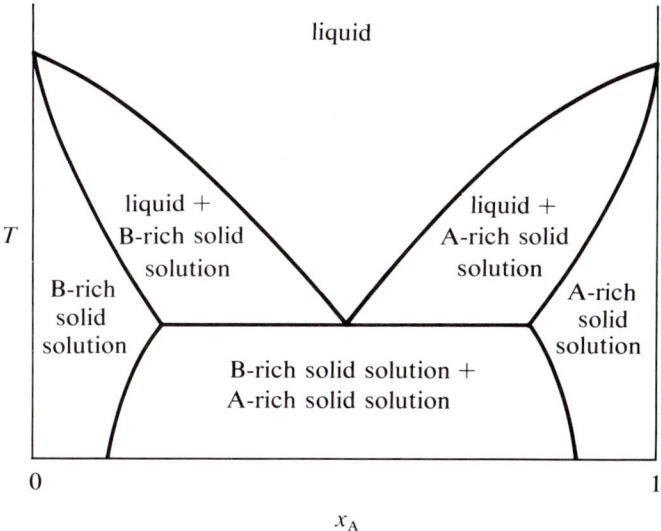

Figure 7-53 Temperature–composition diagram for two components that form limited ranges of solid solutions.

the partial solubility of carbon in iron gives rise to several different phases representing steels with different properties, although the carbon content is usually less than 5%.

7-6 Statistics and Spontaneity – A Summary

In this chapter we have begun laying the groundwork for an understanding of chemical reactivity by examining phase transitions between liquids and their vapors, and between liquids and their crystalline solids. Whereas all our previous discussion has dealt with relatively static structures, this chapter has focused on the ways in which spontaneous change occurs. Spontaneity is important to us, because in chemical reactions as well as phase transitions we cannot pick out molecules with tweezers and alter them. It is always necessary to find a means of causing the change we desire to be spontaneous. The next chapter will survey the spontaneity of macroscopic processes, without any reference (initially, at least) to molecules and their structure. But here we have developed the molecular model to deal specifically with certain kinds of spontaneous processes: melting, freezing, evaporating, and dissolving. What common features have appeared?

One of the ideas we developed for the dissolution process and returned to for the melting/freezing process was the relative probability of different arrangements of the many atoms or molecules in a liquid or solid sample of matter. Since crystalline solids are intrinsically orderly, there is only one possible geometrical arrangement of the molecules that corresponds to the ideal solid phase. On the other hand, since liquids are intrinsically quite disorderly, there are many geometrical arrangements of the molecules in a liquid that are all equivalent and all acceptable liquid structures. In the absence of any energy factors we must assume that all of these are equally probable; this leads us to the conclusion that the spontaneous formation of a solid from a liquid is very improbable, and that all solids should melt or dissolve.

The reason solids are stable under at least some conditions, of course, is that there *are* energy factors that affect the relative probability of the different possible arrangements of molecules. If there is an energy difference ΔE between one arrangement and another, then the relative number of molecules having sufficient thermal kinetic energy to adopt the high-energy arrangement is governed by the Boltzmann distribution $e^{-\Delta E/kT}$. Since not all molecules have sufficient energy to make all the arrangements possible to them, the odds have been loaded in favor of the low-energy arrangements.

In summary, there seem to be two influences at work in determining the macroscopic configuration—the direction of spontaneous transition—of a statistically large number of atoms or molecules. The first is high probability, which we have seen means high disorder or randomness of structure. The second, operating in great degree to cancel the first, is low energy, which is normally achieved in an orderly structure. In surveying macroscopic spontaneity in the next chapter we shall begin in a manner seemingly unrelated to this discussion, but shall arrive eventually at a criterion for spontaneity that combines these two influences.

Study Problems

1. Why does C_v for liquids decrease as the liquid's temperature increases?
2. Calculate the molality, molarity, and mole fraction for
   (a) 10.0 g of NaCl in 100.0 g of H_2O (solution density 1.0707 g/ml).
   (b) 85% H_3PO_4 in water (solution density 1.689 g/ml).
   (c) 35 ml of acetone, $(CH_3)_2CO$, (density 0.792 g/ml) in 65 ml of water; the volume of the solution is 97 ml.

Did solution (c) become warmer or cooler when the two pure liquids were mixed?
3. Consider two ccp cubes of atoms or spherical molecules sharing a face, as shown in Fig. 7-54.

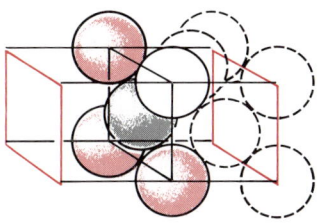

Figure 7-54   Two ccp cubes of atoms sharing a face.

(a) Since each corner molecule is one-eighth inside these cubes, each edge molecule one-fourth inside, and each face molecule one-half inside, show that eight molecules are inside the cubes, although 23 molecules contribute to them.
(b) If the shaded molecule has only three neighbors (shaded red) show that except for the two outer faces shown in red only $1\frac{3}{4}$ molecules are inside the cubes. This means that only four of the original 13 molecules between the two outer faces are still present.
(c) If the same ratio of vacancies applies to the 10 molecules originally in the two outer faces, show that the cubes contain $2\frac{1}{8}$ molecules.
(d) Show that there are three other ways to give the shaded molecule three and only three neighbors, corresponding to $2\frac{3}{8}$, $2\frac{5}{8}$, and $2\frac{7}{8}$ molecules in the cubes.
(e) Taking $2\frac{1}{2}$ molecules as the average occupancy of the cube when the center molecule has only three neighbors, calculate the volume of the cubes in units of d, the molecular diameter, and show that each molecule still present occupies $2.26\,d^3$ volume units.
(f) Assuming that d corresponds to the distance out to the Lennard-Jones potential minimum, $(2)^{1/6}\sigma$, show that the critical volume should be $3.19\sigma^3$.
4. Derive Eq. 7-17 for K_f by considering Fig. 7-28.
(a) Note that the pressure decrease due to the ΔT_f change and the slope of the solid–vapor equilibrium line (sublimation) must equal ΔP due to the Raoult's-law lowering of the vapor pressure plus the additional small pressure decrease due to ΔT_f operating on the liquid–vapor equilibrium line. Write an algebraic statement of this condition, using the symbols $(dP/dT)_{\text{sub}}$ and $(dP/dT)_{\text{vap}}$ for the sublimation and vaporization slopes, respectively.
(b) As in the case of the boiling-point elevation, obtain $(dP/dT)_{\text{sub}}$ by differentiating the expression for the vapor pressure of a solid as a function of temperature, which is

$$\ln P = -\frac{\Delta H_{\text{sub}}}{RT}$$

Note from Fig. 7-16 that the energy difference between a solid and a gaseous molecule is equal to the difference between solid and liquid plus that between liquid and gas:

$$\Delta H_{\text{sub}} = \Delta H_{\text{fus}} + \Delta H_{\text{vap}}$$

Substitute this into the differentiated expression, writing it as the sum of two fractions.
(c) Substitute the expression from (b) and the $(dP/dT)_{vap}$ expression from the earlier derivation of the boiling-point elevation constant into the expression from (a), cancel terms wherever possible, and rearrange the result to the form $\Delta T_f = K_f m$.

5. Show that the fraction of solute remaining in solvent B after n extractions with equal volumes of solvent A is $1/(K_d+1)^n$. How many extractions would be required to get better than 99.9% extracted if $K_d = 4$; if $K_d = 0.1$?

6. Why doesn't mold grow on honey?

7. On p. 345 the viscosities of water, ethyl ether, and glycerine are quoted. Discuss this pattern of numerical viscosity values in terms of the following molecular structures:

water: H—O—H

ethyl ether: CH_3—CH_2—O—CH_2—CH_3

glycerine: HO—CH_2—CH(OH)—CH_2—OH

8. The surface tension of carbon tetrachloride, CCl_4, shows the following temperature dependence (critical temperature is 283 °C):

T(°C)	γ (dynes/cm)
20	26.95
100	17.26
200	6.53

Interpret this trend on a molecular basis.

9. A volatile pure liquid is in a closed container and is in equilibrium with its vapor. If the total number of molecules in the vapor phase is doubled by adding an inert gas that is insoluble in the liquid, what change will occur in the vapor concentration?

10. Using an isoteniscope, a student collected the following vapor-pressure data for benzene:

T(°C)	P(torr)
10	45.4
20	74.7
30	118.0
40	183.1
50	269.2
60	396.6
70	553.0

What value will he calculate for ΔH_{vap} of benzene?

11. Copper(I) chloride, CuCl, forms many coordination compounds in which the copper atom accepts electrons from a donor atom or molecule (called a ligand, L), giving the empirical formula CuClL, with two atoms bonded to the copper (which is said to be two-coordinate). Some ligand molecules form four-coordinate compounds, $CuClL_3$. A ligand in this latter group with a molecular weight $MW_L = 100.0$ g/mole formed a compound that was soluble in chloroform, $CHCl_3$, whose K_b constant is 3.63 °C/molal concentration unit near its normal boiling point of 61 °C. The vapor pressure of pure chloroform is 199.1 torr at 25 °C. When a solution of 0.523 g of $CuClL_3$ in 10.00 g of $CHCl_3$ was heated to boiling, its boiling-point elevation was found to be 1.43 °C. On the other hand, when the change in vapor pressure of a similar solution was measured at 25 °C it proved to be 1.56 torr. What values of the molecular weight for the compound do these two determinations yield? What is happening?
12. If the osmotic-pressure molecular-weight determination on p. 380 were done with the same solution but using the boiling-point elevation technique, what would ΔT_b be? ($K_b = 3.33$ °C/molality unit for toluene.) Comment on the appropriateness of the two techniques for the measurement of high molecular weights.

Some Further Reading

Moore, W. J., *Physical Chemistry*, Englewood Cliffs, N.J.: Prentice-Hall, 1962 (3rd ed.). Compact treatments of liquid structure and phase equilibria are difficult to find, but Chapter 17 deals nicely with liquid structure and Chapter 5 with phase equilibria. Unfortunately, some of Chapter 5 relies on thermodynamics, which we have not yet discussed, but most of it is readable without thermodynamics.

Knight, C. A., *The Freezing of Supercooled Liquids*, New York: Van Nostrand Reinhold, 1967. An interesting and readable little book on one particular aspect of our discussion.

Findlay, A., *The Phase Rule*, New York: Dover, 1951 (9th ed.). This is actually a reprint of a much older book (originally 1903), but phase diagrams have not changed much. There is more here than you need.

8 Spontaneity and Equilibrium

When two chemicals are mixed together a reaction may occur. If chemistry is to be useful we must be able to predict whether or not a reaction will occur, what the reaction will be, and how to guide it in a desired direction. To answer the last implied question, we must have a clear idea of the answers to the first two. By this time the principle should be well established that in approaching such questions as these, the chemist surveys the experimental information available, codifies and arranges it into laws, and formulates a theory or model that is consistent with experimental observations. For spontaneous chemical reactions, the model will need to answer two questions: Will a given chemical reaction occur to any substantial extent? How will it occur? Although obviously linked, these are not the same question and need not be answered together. We shall defer trying to answer the second question until Chapter 11, and limit this discussion to the first.

What does spontaneity mean on a macroscopic, experimental scale? What kinds of processes always occur spontaneously? If such processes can be defined in general terms, the foundation is laid for a law of spontaneity, which must be the first goal. This discussion is the province of chemical **thermodynamics**, which will be the general topic of the next three chapters. Chemical thermodynamics will prove an elegant and powerful tool for predicting the direction of chemical change; it rests, however, on careful definitions of terms, which must be our first concern.

8-1 Thermodynamic Definitions

Early in Chapter 1 we noted that when we wish to describe or deal with a system showing diversity, we subdivide it into smaller systems having only the degree of diversity in which we are interested. Thermodynamics deals quite generally with the whole universe, and consequently it is necessary to define for any particular experiment how much of the universe we are interested in. A **thermodynamic system**, then, is the portion of the universe involved in our experiment, whose properties we are measuring or working with. We choose its boundaries for our own convenience; it may not be surrounded by definite physical boundaries at all, but in some cases only by a theoretical boundary or a dotted line on a diagram. The system as we define it may have certain specified interactions with its **surroundings**, meaning the rest of the universe. For instance, suppose a beaker full of liquid is the system in which we are interested; if it is hot and we place it on a lab bench, heat will flow into the bench as the beaker warms it. This is an example of a thermal interaction of the system with its surroundings.

Possible Spontaneous Interactions

For most chemical systems there are only a few possible interactions. Consider a closed container in which a liquid and its vapor are at equilibrium. Working from the ideas of Chapter 7, how can the equilibrium be altered? The division of matter between the liquid and vapor phases can be changed by changing the temperature; if the container is heated the vapor pressure increases. This, too, is a **thermal interaction**. But we could also enlarge the container, which would allow the vapor to expand and thus do pressure–volume work. As the pressure is decreased, more liquid will vaporize, thereby changing the division of matter between phases again, so **mechanical interaction** is another capability of the system. To be more general we define mechanical interactions to include any interaction involving force, such as electrical interaction in batteries and gravitational interaction in centrifuges. We also saw that the vapor pressure could be lowered by adding another component to the liquid as a solution. This means that a **compositional interaction** is possible for the system. No other interactions are possible, and indeed this is true in general for chemical systems. So a thermodynamic system is a selected object or system of objects having definable thermal, mechanical, and compositional properties. For the purpose of defining these properties we introduce **thermodynamic variables** that describe the contents of a system without reference to anything outside the system. The thermodynamic variables account numerically for these properties—for a gas we have the thermodynamic variables P, V, n, and T, the variables in the ideal gas **equation of state.** If all the thermodynamic variables are specified at certain numerical values for a given system, the **state** of the system is said to be specified. For this reason the thermodynamic variables are sometimes said to be **state variables**, and the relationship between them for any given phase in a system is called the equation of state. An equation of state usually allows some minimal set of state variables to specify the system's state, thereby fixing values for all other state variables. Mechanical variables that are *not* influenced by thermal and compositional variables in an equation of state, such as gravitational potential energy in most cases, are not considered to be state variables. There are other thermodynamic variables, some of which we shall encounter as the discussion continues, but these form an adequate set with which to begin.

Capacity and Intensity Variables

Looking at the state variables for a gas we see that the pressure and volume essentially define the mechanical properties of a gas, the number of moles of each component defines the compositional properties, and the temperature defines the thermal properties of the gas. So, if we know the values of all these variables we ought to know exactly what the gaseous system will do if it is allowed to undergo some specified kind of interaction with its surroundings. That is, we know the initial state of the system, and we should be able to define the state it will reach by the interaction it undergoes. When a system changes from one state to another by such an interaction it is said to have undergone a **thermodynamic process**. Because the equation of state exists for a gas, not all four of the state variables are independent; if the pressure, the number of moles, and the temperature are fixed, the equation of state necessarily requires that the volume take some particular value. Any three of the four variables are equally acceptable, but the three just mentioned are most frequently taken. These four variables contrast with each other in an interesting way that bears on why the volume is usually the variable considered not to be independent. Two of the variables, number of moles and volume, are proportional to the size of the system—a bulky system has more volume than a compact system, and a heavy one has more moles than a light one of the same material. If we brought two identical systems together, the volume and the number of moles of the new system would be doubled. But the other two variables are quite different; pressure and temperature do not depend on the size of the system. Bringing two identical systems together would *not* double the pressure or the temperature. The pressure inside a soap bubble is the same as the pressure inside a blimp. To distinguish between these pairs of variables we say that P and T are **intensity** (or **intensive**) variables, but n and V are **capacity** (or **extensive**) variables. Pressure and temperature are the intensity variables for mechanical interaction and thermal interaction, respectively. Can we construct an intensity variable for compositional interaction? In any given system the "intensity" of a particular kind of matter is represented by its concentration, which does not depend on its size. And concentration can be very well represented by n/V, the number of moles per unit volume. So if the volume is absorbed into the number-of-moles variable we are left with three intensity variables governing the three kinds of interaction usually available to the system. In the next section the particular spontaneous changes that occur reliably for these three interactions will be explored in some detail.

The Nature of Temperature

Let us first, however, comment on the temperature variable. Temperature is particularly difficult to define satisfactorily as a quality, even though it is quite familiar. It is a peculiarly relative and elusive quality:

> I test my bath before I sit,
> And I'm always moved to wonderment
> That what chills the finger not a bit
> Is so frigid upon the fundament.
>
> Ogden Nash[1]

This relative quality means that all we can reliably do is to establish by direct comparison that one body is warmer than another, or that they are at the same temperature. We cannot *prove* that there is a uniform sequence such that if Miami is warmer than Atlanta, and Atlanta is warmer than New York, then Miami is warmer than New York. However, as a result of all our experience we can say — as a law, not a theory — that there *is* a uniform, one-dimensional sequence that describes the hotness or coldness of a sample of matter. We can invent a number to be associated with objects to describe their relative position in this sequence; this number is what we call temperature. The association of temperature with objects gives us a mathematical representation of the sequence $T_{\text{Miami}} > T_{\text{New York}}$, and so on. The concept of temperature alone does not give us a temperature *scale*, however; any sort of function that kept Miami

---

1. From *Verses from 1929 On* by Ogden Nash, copyright 1942 by The Curtis Publishing Company, by permission of Little, Brown and Co. and J. M. Dent & Sons Ltd.

warmer than New York would be satisfactory. Our earlier definition of a temperature scale depended on the properties of a particular substance — water — and as such seems not to be very general. In Section 8-2 we shall derive a temperature scale that does *not* depend on the properties of any particular substance and show that it coincides with the "absolute" temperatures we have been using all along. This is not a trivial accomplishment, since otherwise it would be necessary to choose a reference substance completely arbitrarily. It is a readily verifiable fact that a mercury thermometer and an alcohol thermometer cannot be adjusted so as to read the same temperature at every point along their scale; which is better? Fortunately we need no longer choose.

8-2 The Potential of Spontaneous Transition[2]

When we discussed gases we derived the equation of state and spent some time correlating what we now call thermodynamic variables. But we also spent some time on three other properties, the transfer properties of heat conductivity, viscosity, and diffusion. These properties correspond to mechanisms for moving from one equilibrium state of a gaseous system to another through the thermal, mechanical, and compositional interactions, respectively. They are important to us in considering the possible interactions separately.

The Description of Thermal Interaction

Suppose we have two very simple systems that we shall allow to interact. Each system is a gas, in the states (P_1, V_1, n_1, T_1) and (P_2, V_2, n_2, T_2), respectively. Each system is at equilibrium within itself while isolated, but we can place them together (making one system the surroundings for the other system) in ways that allow them to change their thermodynamic states spontaneously by each of the three possible interactions. In Fig. 8-1 the two systems are connected by a thin, rigid, heat-conducting wall (with no appreciable mass or heat capacity) but are perfectly insulated by mechanically rigid walls from the rest of the universe, thus allowing only thermal interaction between the two systems — the flow of heat. Any heat flow into one system must be equal to the heat flow out of the other. If we define dQ as the heat flow *into* a system, then $dQ_1 = -dQ_2$. What does our experience tell us the spontaneous thermal interaction

---

2. This approach to a function defining a thermodynamic variable measuring spontaneity is adapted from a novel treatment by Peter Fong in *Foundations of Thermodynamics* (New York: Oxford University Press, 1963). The interested student is urged to read this treatment.

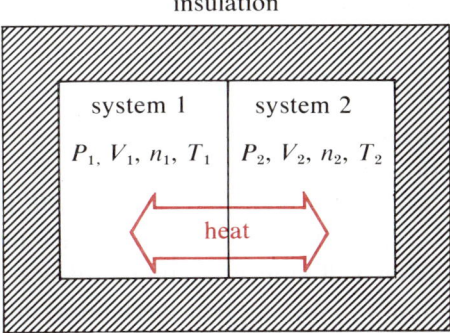

Figure 8-1 Spontaneous thermal interaction between two systems at different temperatures.

will *always* be? We always observe that heat flows from a hot object to a cold object, never the other way. This occurs quite spontaneously and continues until the two temperatures T_1 and T_2 become equal; then no further change occurs. We say that **thermal interaction** leads to **thermal equilibrium**, subject to the **equilibrium condition** that $T_1 = T_2$ (the intensity variables for thermal interaction become equal).

The Description of Mechanical Interaction

The two gases can also have a mechanical interaction, in which they exchange work. Suppose the gases, as in Fig. 8-2, are again insulated from the rest of the universe, but this time instead of being separated by a rigid heat-conducting wall are separated by a perfectly insulated frictionless piston, so that only mechanical interaction can occur. What does our experience tell us the spontaneous mechanical interaction will always be? We always observe that the piston moves away from the high-pressure end and toward the low-pressure

Figure 8-2 Spontaneous mechanical interaction between two systems at different pressures.

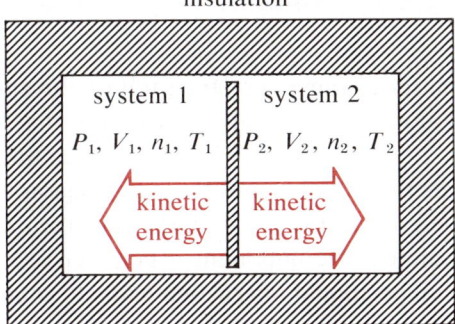

end, never the other way. Then the work done by system 1 in expanding, dW_1, is equal to its pressure times its volume change:

$$dW_1 = P_1\,dV_1 \quad \text{and} \quad dW_2 = P_2\,dV_2$$

By the construction of the piston $dV_1 = -dV_2$, but since in general P_1 is not equal to P_2, dW_1 is not equal to $-dW_2$. Instead, $dW_1 + dW_2$ equal the kinetic energy of the piston, which will bounce back and forth from near one end to near the other. But the viscosity of the gases will use up this kinetic energy eventually, and the piston will stop even though it is frictionless. When the piston stops, P_1 will equal P_2. We say that **mechanical interaction** leads to **mechanical equilibrium** subject to the **equilibrium condition** that $P_1 = P_2$ (the intensity variables for mechanical interaction become equal).

The Description of Compositional Interaction

Gases can also have a compositional interaction, in which they exchange matter. Suppose the systems consist of two different gases, insulated from the rest of the universe as in Fig. 8-3 but separated by a removable panel. If the

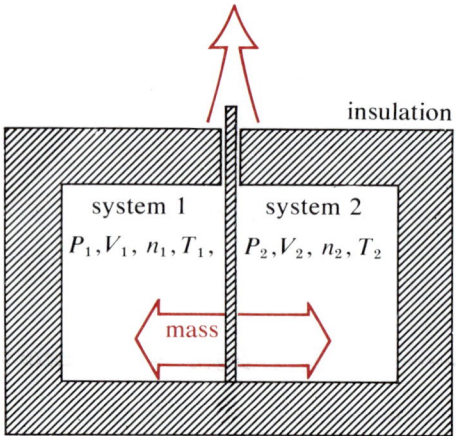

Figure 8-3 Spontaneous compositional interaction between two chemically different systems.

panel is removed, what does our experience tell us the spontaneous compositional interaction will always be? We always observe that the two gases diffuse into each other; they never stay put without mixing. Mass that flows out of one region, dm_1, equals that flowing into the other region, dm_2. The net flow of gas will stop when the concentrations c_1 and c_2 are the same on both sides, c_a being the number of moles of component a per liter of volume in the region it is in. We say that **compositional interaction** leads to **composition equilibrium**, subject to the **equilibrium condition** that $c_1 = c_2$ (the intensity variables for compositional interaction become equal).

In many thermodynamic processes all three types of interactions occur—for instance, many chemical reactions get hot, generate gas, and change composition all at the same time. Change stops—equilibrium is established—only when all three types of equilibria are established. Conversely, we can always analyze general equilibria in terms of these three individual types.

A Potential Function for Spontaneity

All three types of interaction show a very significant common feature. Each change of state proceeds in only one direction. Once a system has passed a particular thermodynamic state it will never reach it again in that process. So in each case we can set up a quantity that *always increases* in a spontaneous interaction. Such a quantity would be a measure of spontaneity, or of the capacity of a system for spontaneous change. If this quantity increased (by calculation) for a certain proposed change, we would know that the change should indeed occur spontaneously. Of course, a decreasing quantity would be just as good—with the rule for spontaneity reversed—but in order to follow convention we shall stick to the formulation of an increasing quantity. For instance, in a thermal interaction the quantity $-(T_{hot} - T_{cold})$ always increases until it reaches zero. We could imagine many—actually an infinite number of—functions of this variable that would always increase. Any of these functions would serve as a guide to what was going to happen next—the next state would have a higher value of the function. We could call the function or variables S_t, the **potential for spontaneous thermal interaction**. In the same way we could construct an infinite number of functions (involving the pressure) that would always increase in any spontaneous mechanical interaction; any of these would be a satisfactory S_m, the **potential for spontaneous mechanical interaction**, but in particular one would have a form identical to the previous S_t: $-(P_{high} - P_{low})$. And using the concentration we could construct an infinite number of functions that would always increase in any spontaneous composition interaction, and one could be $-(c_{concentrated} - c_{dilute})$. This or any of the others would be a satisfactory S_c, the **potential for spontaneous composition interaction**.

These functions or variables S_t, S_m, and S_c resemble each other so much that a unifying physical principle becomes apparent, or at least appealing. There might be a single general variable, S, called the **potential for spontaneous transition**, which takes the form of each of these three variables for that type of interaction. In a general thermodynamic process in which all three types of interaction were occurring, the variable or physical property S would always increase because it is increasing for each of the three types of interaction. Since we are talking here only about physical laws, which are summaries of experimental results, we can justify (or reject) the formulation of the general property S by comparing the results of the deductions from this assumption with genuine experimental results.

The Law of Spontaneity: The Second Law of Thermodynamics

Using the proposed physical property S, we can formulate a basic **law of spontaneous processes**:

> For two interacting thermodynamic systems specified by thermodynamic variables $P_1, V_1, \ldots$, and $P_2, V_2, \ldots$, respectively, there exists a physical property of the composite system represented by a function of the thermodynamic variables $S(P_1, V_1, \ldots, P_2, V_2, \ldots)$ such that a spontaneous process always takes place to change a state of the composite system with a low value of S to a neighboring state with a higher value of S, which is accessible from the first state by the interactions of the process.

There is a fairly obvious corollary to this that defines equilibrium:

> The equilibrium state is that state having the maximum value of S that is consistent with the constraints on the possible variations of dQ, dV, and dm for the composite system.

This follows from the previous law because if S were not a maximum there would be some possibility of spontaneous change remaining; only if it has reached its maximum will the spontaneous change stop.

Reversible and Irreversible Processes

The trouble with real thermodynamic processes is that they do not consist of a sequence of genuine equilibrium thermodynamic states as the basic law of spontaneous processes implies they should. In a thermodynamic state, the state variables are well defined and uniform throughout the system. But when a real gas expands there is always a pressure gradient in front of it, so that the state variable P is not well defined throughout the system. As a matter of fact, this sort of problem arises in all real processes, and in order to counter it we must create a hypothetical idealized spontaneous process for getting from the initial state to the final state, called a **reversible process**. Just as a derivative in calculus is the limit of a sequence of ratios of finite increments of the variables, taken as the increments shrink to zero, a reversible process is the limit of a sequence of real processes involving spontaneous changes, taken as the spontaneous changes become smaller and smaller tending toward zero. For instance, if the spontaneous process were heat transfer between two systems of 0 °C and 100 °C, respectively, an approximation to a reversible path for the process would be heat transfer between the 100 °C system and another system at 90 °C, followed by heat transfer between the second and a third system at 80 °C, and so on to 0 °C. A better approximation would involve 101 systems at 100°, 99°, 98°, ..., 1°, 0°. A still better approximation would involve 1001 systems at 100.0°, 99.9°, 99.8°, ..., 0.1°, 0.0°. The limit of this approximation series as the temperature difference approaches zero is a reversible process. In the idealized case each of the two interacting systems is at the same temperature (the systems are at equilibrium), and so a reversible process consists of an infinite series of infinitesimal steps, each moving from one equilibrium state to another equilibrium state until the whole distance from the initial state of the process to the final state has been traversed. Each of the individual steps could also be called a reversible process, because the least external influence could send it in either direction at any point in the process. If system A is 100° hotter than system B, heat will flow from A to B spontaneously, but the direction of heat transfer could not be reversed (flowing from B to A) by a minute temperature adjustment in either A or B. If the two systems are at the same temperature, however, the slightest temperature increase in A would transfer heat in one direction, but the slightest temperature increase in B would transfer heat in the other direction. Of course, the transfer of heat between two systems takes longer the smaller the temperature difference, and a really reversible process would take an infinitely long time. Real spontaneous processes do not meet this criterion and are thus **irreversible**.

Since the reversible process consists of a linked series of equilibrium states, the property S, the potential for spontaneous transition, must be constant throughout a reversible process—otherwise the process would be a real spontaneous process instead of a reversible one. So in terms of the basic law of spontaneous processes, a reversible process is one in which S is constant for the whole composite system, that is, constant for the system plus its surroundings. This suggests that it would be nice to be able to examine the system and surroundings separately, and so we make the following assumption about the property $S(P_1, V_1, \ldots, P_2, V_2, \ldots)$ of the composite system:

$$S(P_1, V_1, \ldots, P_2, V_2, \ldots) = S(P_1, V_1, \ldots) + S(P_2, V_2, \ldots)$$

In other words, S is an additive property for all the systems in the composite system; it must therefore be a capacity quantity, not an intensity quantity, under this assumption. For a reversible process, then, between two systems A and B that are isolated from the rest of the universe, $dS_{A+B} = dS_A + dS_B = 0$ or $dS_A = -dS_B$, where dS is the change in S through the process. We shall need this property as we establish the detailed form of the S function.

The Change in S During a Reversible Adiabatic Process

To be able to obtain useful predictions from the potential for spontaneous transition, we need to be able to express S in terms of the other, measurable physical properties of the composite system. Toward this end, let us look at a particular type of reversible process called an **adiabatic reversible process**, in which no heat transfer occurs. In Fig. 8-4 two systems can interact with each

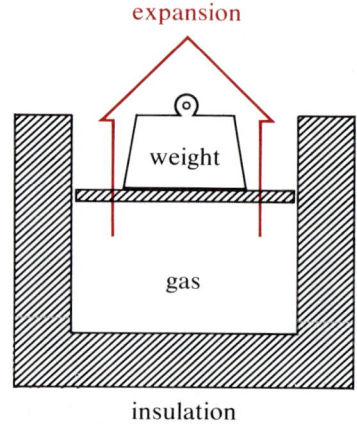

Figure 8-4 A reversible adiabatic process for which $dS_{wt} = -dS_{gas} = 0$.

other and with nothing else: a gas that expands but does not change its composition or transfer heat (since it is insulated), and a weight that is raised by the expansion and weighs just enough to balance the gas pressure and make the process reversible. Since in the expansion nothing changes about the weight except its vertical position, which is not a thermodynamic variable since it is not influenced by the thermal and compositional properties of the weight, all its thermodynamic or state variables remain unchanged, meaning that its potential for spontaneous transition remains unchanged:

$$dS_{wt} = 0 \quad \text{(adiabatic)}$$

And yet, because the adiabatic process is reversible, $dS_{gas} = -dS_{wt}$. So $dS_{gas} = -dS_{wt} = 0$ and we see that in a reversible adiabatic process there is no change of S. All the equilibrium states of a gas that can be connected by a reversible adiabatic process with its surroundings have the same value of S_{gas}. Let us look at what happens in the adiabatic reversible expansion of an ideal gas. Boyle's law tells us that if we plot P vs. V for an ideal gas at constant temperature, we get a hyperbola as in Fig. 8-5 and Eq. 3-1. However, if the gas is expanding it is doing work in lifting the weight, and this energy has to come from somewhere. The only possible source is from the thermal energy of the gas, so the gas cools

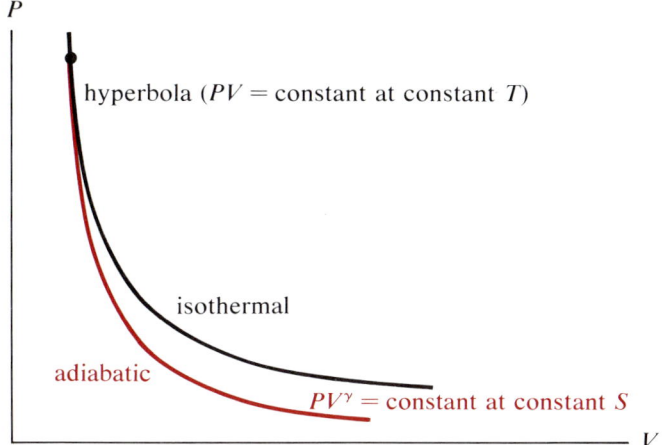

Figure 8-5 Pressure–volume relationships for an isothermal system (Boyle's law) and an adiabatic system.

off in an adiabatic expansion. If it cools off its volume is not as large as it would have been at a constant temperature (where, from Boyle's law, $PV =$ constant), and so the volume is smaller at the end of an adiabatic expansion through a certain pressure change than it would have been at constant temperature. This is illustrated by the red line in Fig. 8-5. Experimentally, instead of PV being constant for a given adiabatic expansion, it is found that $PV^\gamma =$ constant, where γ is a number greater than 1; this also expresses the fact that the volume is smaller than Boyle's law would predict. The adiabatic (red) curve in Fig. 8-5, represented algebraically by $PV^\gamma =$ constant, connects states with equal S.

The Change in S During a Reversible Isothermal Process

Now suppose we look at an **isothermal reversible process**, in which the temperature of the system does not change. This is shown in Fig. 8-6, in which the

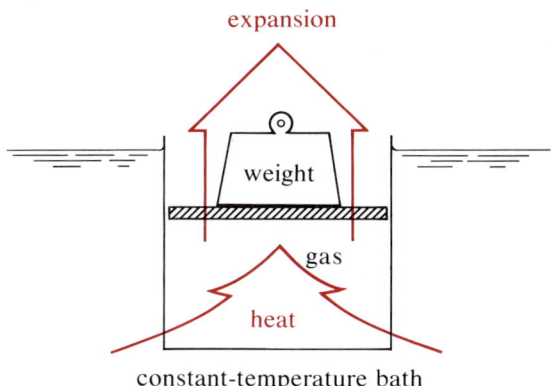

Figure 8-6 A reversible isothermal process, for which $dS_{gas} = -dS_{bath}$.

weight is exactly the same, but the gas is now in a heat-conducting container immersed in an infinitely large constant-temperature bath. Since it is infinitely large it can supply heat without changing its temperature, but it has no other interactions with the gas. It can be seen that what will happen in an isothermal expansion will be that the gas will interact mechanically with the weight and, at the same time, thermally with the bath to absorb the energy (heat) necessary to raise the weight (work). There are thus three systems in the composite system, and

$$S_{\text{gas + bath + wt}} = S_{\text{gas}} + S_{\text{bath}} + S_{\text{wt}}$$

But because the process is reversible, S is constant; and because the weight's thermodynamic variables are not affected by the expansion, S_{wt}, which is a thermodynamic variable, is constant (just as for the adiabatic expansion):

$$\text{constant} = S_{\text{gas}} + S_{\text{bath}} + \text{constant}$$

How does S change in this process? Looking at the above expression we can see that for a reversible isothermal expansion of a gas, $dS_{\text{gas}} = -dS_{\text{bath}}$.

The Relationship Between S and Q

In this reversible isothermal expansion, the gas went from an equilibrium state characterized by $P_1, V_1, T_1, \ldots$ to an equilibrium state characterized by $P_2, V_2, T_1, \ldots$. Suppose we prepare a whole series of gas/weight systems having identical composition, all at P_1, V_1, T_1, and plug one after another into this infinite bath. Each in its turn will expand to P_2, V_2, and each will end up with exactly the same set of thermodynamic variable values. Since the changes in the thermodynamic variables are the same for each gas, and since S is a function of the thermodynamic variables only,

$$dS_{\text{gas 1}} = dS_{\text{gas 2}} = dS_{\text{gas 3}} = \cdots$$

For each of these expansions $dS_{\text{bath}} = -dS_{\text{gas}}$, so for the whole series of expansions

$$dS_{\text{bath}}(\text{total}) = -n\, dS_{\text{gas}}$$

where n is the number of gas systems that have absorbed heat from the bath in expanding. Since heat transfer is the only interaction the gases had with the bath, and since each gas absorbed the same amount of heat from the bath, dS_{bath} for each expansion must relate only to dQ, the amount of heat transferred, and to T, the temperature at which the transfer occurred. Since S is an extensive or capacity quantity it must be proportional to dQ, the capacity quantity for the heat transfer. Then for the whole series of expansions

or
$$dS_{\text{bath}}(\text{total}) \propto dQ$$

$$dS_{\text{bath}}(\text{total}) = K\,dQ$$

The linear relationship is required by the linearity in dS_{gas} from the previous expression. The constant K is a proportionality constant that can involve only the temperature, since in effect the bath has no other physical properties, at least in terms of the interactions we allow it. It wouldn't matter what we built the bath out of; dS_{bath} would be the same for any reversible isothermal expansion as long as its temperature were the same. So K is some function of the temperature, $K(T)$. For the gas itself, in which we are usually more interested than the bath,

$$dS_{\text{gas}} = K(T)\,dQ_{\text{gas}}$$

since $dS_{\text{gas}} = -dS_{\text{bath}}$ and $dQ_{\text{gas}} = -dQ_{\text{bath}}$.

Notice that in this expression we have not specified any particular temperature scale. Any empirical scale would be acceptable, since the only effect of changing the scale would be to change the detailed form of the function $K(T)$. We could even let this relationship define a temperature scale for us; if we set

$$K(\theta) \equiv \frac{1}{\theta}$$

so that

$$dS = \frac{dQ_{\text{rev}}}{\theta}$$

(where rev stands for reversible), the scale of temperature θ is called the **thermodynamic temperature scale** or the **Kelvin scale**. Let us see how it correlates with the ideal gas temperature scale that we have previously called the absolute temperature.

The Thermodynamic Temperature Scale

Consider the graph in Fig. 8-7, in which two isothermal reversible expansions (having constant T) are shown, along with two adiabatic reversible expansions (having constant S). How much work does a gas do in an isothermal reversible expansion? The work done in a differential expansion is $dW = P\,dV$, so for a finite expansion from 1 to 2

$$W = \int_{V_1}^{V_2} dW = \int_{V_1}^{V_2} P\,dV = \int_{V_1}^{V_2} \frac{nRT}{V}\,dV = nRT(\ln V_2 - \ln V_1) = nRT_1 \ln\frac{V_2}{V_1}$$

420 | Spontaneity and Equilibrium

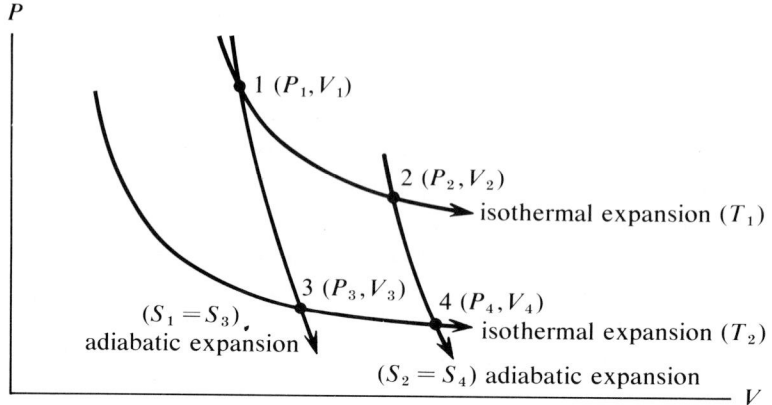

Figure 8-7 Two possible isothermal expansions and two possible adiabatic expansions for the same system.

If the expansion is really isothermal the work done (mechanical energy expended) must equal the heat energy absorbed by the gas, since no heat goes into raising the temperature of the gas:

$$dQ = dW = nRT_1 \ln \frac{V_2}{V_1} \tag{8-1}$$

Then dS, or what we call ΔS for a finite process, for the isothermal reversible expansion of a gas is given by

$$\Delta S_{1 \to 2} = \frac{dQ_{\text{rev}}}{\theta} = \frac{nRT_1}{\theta_1} \ln \frac{V_2}{V_1} \tag{8-2}$$

This applies, of course, to the expansion from point 1 to point 2 on the graph. For the other isothermal reversible expansion, from 3 to 4,

$$\Delta S_{3 \to 4} = \frac{nRT_2}{\theta_2} \ln \frac{V_4}{V_3} \tag{8-3}$$

Now, considering the two adiabatic reversible expansion curves (from 2 to 4 and from 1 to 3), we have $S_1 = S_3$ and $S_2 = S_4$ from the fact that S is constant for an adiabatic expansion. Subtracting these from each other,

$$S_2 - S_1 = S_4 - S_3 \quad \text{or} \quad \Delta S_{1 \to 2} = \Delta S_{3 \to 4}$$

Using the isothermal results, then,

$$\frac{nRT_1}{\theta_1} \ln \frac{V_2}{V_1} = \frac{nRT_2}{\theta_2} \ln \frac{V_4}{V_3}$$

or

$$\frac{T_1}{\theta_1} \ln \frac{V_2}{V_1} = \frac{T_2}{\theta_2} \ln \frac{V_4}{V_3}$$

We can use the ideal gas law and the earlier adiabatic experimental result, PV^γ = constant, to get $TV^{\gamma-1}$ = constant, so for the adiabatic expansion 1 → 3,

$$T_1 V_1^{\gamma-1} = T_2 V_3^{\gamma-1}$$

and for the adiabatic expansion 2 → 4,

$$T_1 V_2^{\gamma-1} = T_2 V_4^{\gamma-1}$$

Dividing the first of these by the second,

$$\frac{T_1 V_1^{\gamma-1}}{T_1 V_2^{\gamma-1}} = \frac{T_2 V_3^{\gamma-1}}{T_2 V_4^{\gamma-1}}$$

so that

$$\left(\frac{V_1}{V_2}\right)^{\gamma-1} = \left(\frac{V_3}{V_4}\right)^{\gamma-1} \quad \text{or} \quad \frac{V_2}{V_1} = \frac{V_4}{V_3}$$

Finally, inserting this result in the earlier expression

$$\frac{T_1}{\theta_1} \ln \frac{V_2}{V_1} = \frac{T_2}{\theta_2} \ln \frac{V_4}{V_3}$$

we get

$$\frac{T_1}{\theta_1} = \frac{T_2}{\theta_2}$$

So the Kelvin or thermodynamic temperature θ is proportional to the ideal gas temperature T, and if we define the degrees to be the same size, the scales are identical. Thus the change in S for an isothermal reversible process is

$$\Delta S_{\text{isothermal}} = \frac{dQ_{\text{rev}}}{T}$$

For any random reversible process, such as the one from M to N in Fig. 8-8, we can calculate ΔS by approximating the curve as a series of small isothermal and adiabatic expansions (the zigzag curve). The total ΔS will be the sum of all the little isothermal ΔS's, and as the zigzags get smaller and smaller (making the approximation better and better), we have

$$\Delta S_{M \to N} = \sum \Delta S_{\text{isothermal}} = \int_M^N dS = \int_M^N \frac{dQ_{\text{rev}}}{T}$$

The Entropy Quantity

Written in this form the quantity S is called the **entropy** of the system to which it refers, and we can use it to measure the spontaneity of a possible physical change or chemical reaction. We must remember, however, that S is a property

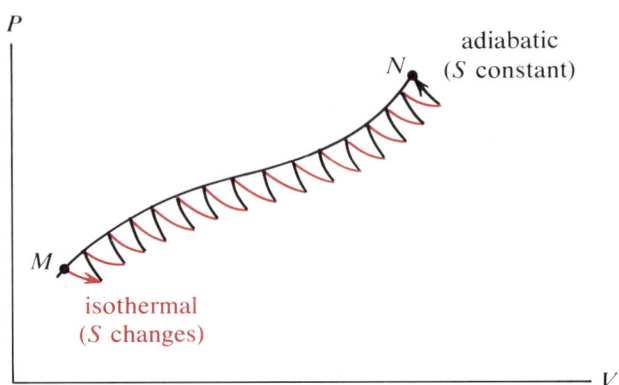

Figure 8-8 A sequence of small adiabatic and isothermal processes approximating a general thermodynamic process.

of the composite system—the system plus its surroundings—in our earlier statement of the law of spontaneous processes. So we cannot simply calculate ΔS for the system alone, because the measure of spontaneity is the total entropy change of the universe! If we can calculate ΔS for the surroundings as well as the system, though, we have a measure of the potential for spontaneous transition, or the spontaneity of the reversible process.

Where does this leave us in dealing with real spontaneous processes? These are never reversible, although they may approximate reversibility. If they are not reversible, then the total entropy will not be constant and our whole derivation will be invalid. Fortunately, the entropy is a function only of the thermodynamic variables of a system. For any given state of a system, specified by definite values of these variables, the entropy of the system must be the same *no matter how the state was reached*. So all we have to do is to devise a path consisting solely of reversible steps that will get us from the initial state of the system to the final state of the system, and the entropy change for these steps (for the system) will be the same as the entropy change of the irreversible process. Notice that we cannot get ΔS directly for the irreversible process by dropping the "reversible" subscript on dQ_{rev} in our differential-entropy expression, because the reversibility of the process is necessary for our derivation. But we need only find a reversible pathway from the initial to the final state in order to calculate the entropy change for an equivalent process.

Let's review what we have so far and work out some examples to see how the spontaneity criterion performs. We set out to develop a function of the state variables that would be a measure of spontaneity, beginning by looking at two simple systems that interacted in each of three possible ways: by thermal, mechanical, and compositional interaction. For each of these three kinds of interaction we established a measure of spontaneity based on the intensity variable of the interaction. Because the three potentials for spontaneous interaction resembled each other so strongly, we proposed that there should be a single potential, S, characteristic of any system plus the surroundings with which it was interacting, which embodied all three and reduced to each under the appropriate circumstances. Working with gaseous systems but thinking in the more general terms of heat and work, it proved to be true that dS for any reversible change was proportional to the heat flow dQ occurring in the change. The proportionality constant could relate only to the temperature at which the heat was transferred, and by considering the properties of expanding ideal gases it proved possible to identify the thermodynamic temperature scale with the ideal gas temperature scale, so that the proportionality constant using the ideal gas temperature scale was $1/T$. We thus obtained, for reversible changes only, an expression governing the change in the measure of spontaneity, S, which we called entropy:

$$dS = \frac{dQ_{\text{rev}}}{T} \tag{8-4}$$

For a finite process instead of a differential one, we need only add up the sum of the infinitesimals in the above relationship:

$$\Delta S_{\text{state 1} \to \text{state 2}} = \int_1^2 dS = \int_1^2 \frac{dQ_{\text{rev}}}{T} \tag{8-5}$$

For a real, irreversible process we can discover whether or not it is spontaneous by calculating ΔS not for the irreversible process but for a reversible pathway from the same initial state to the same final state. Of course, the spontaneity of a process depends not on the entropy change of the system alone but on the entropy change of the system plus that of the surroundings, so there must be two parts to our calculation.

Irreversibility and Spontaneity

Let us look again at the example of an isothermal expansion of a gas. We devised a means of carrying out such an expansion reversibly, but what about an irreversible expansion? In particular, we know that a gas always expands spontaneously into a vacuum. The entropy change for such an expansion, then, should be positive. Consider the system and surroundings shown in Fig. 8-9, in which a gas in the left bulb expands into the right bulb when the stopcock is opened. We can break this down into an imaginary reversible pathway in which the gas expands reversibly and isothermally from V_1 to the new volume,

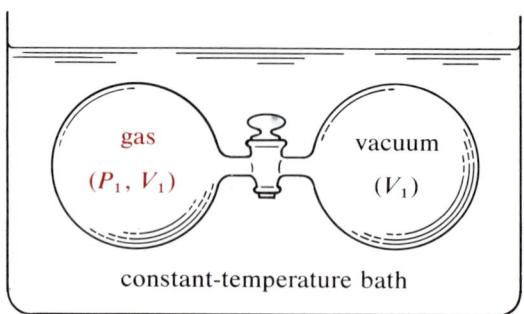

Figure 8-9 A free expansion into a vacuum.

$2V_1$ (which are the initial and final states of the gas in the real expansion), and another reversible pathway in which the bath supplies, reversibly and isothermally, an amount of heat equivalent to the amount of work done in the real expansion. For the first of these (the gas) we return to an expression developed in Eqs. 8-2 and 8-3:

$$\Delta S = \frac{nRT}{\theta} \ln \frac{V_2}{V_1}$$

But we showed subsequently that $\theta = T$, and in this case V_2, the new volume, equals $2V_1$, so we have

$$\Delta S_{\text{gas}} = \frac{nRT}{T} \ln \frac{2V_1}{V_1} = nR \ln 2$$

Since the number of moles must be positive, R is positive, and the logarithm of any number greater than 1 is positive, ΔS_{gas} is positive. For the second part of the calculation (the bath), we need to reversibly supply an amount of heat equivalent to the work done in the real expansion. For any expansion

$$W = \int_{V_1}^{V_2} P\, dV$$

where P is the pressure being pushed back, the force the expansion must overcome or overbalance. In the case of a vacuum, this pressure is zero, so

$$W = \int_{V_1}^{2V_1} 0\, dV = 0$$

In other words, an ideal gas expanding into a vacuum is not doing any work. So it is not necessary for the bath to supply any heat and

$$\Delta S_{\text{bath}} = \frac{dQ_{\text{rev}}}{T} = \frac{0}{T} = 0$$

So for the overall irreversible process,

$$\Delta S_{\text{universe}} = \Delta S_{\text{gas}} + \Delta S_{\text{bath}}$$
$$= nR \ln 2 + 0$$
$$= nR \ln 2$$
$$= \text{positive quantity}$$

This is, of course, exactly what we ought to get, since the gas *is* going to expand into the vacuum.

For another example, consider the system shown in Fig. 8-10, in which thermal interaction is occurring. Two copper bars of equal size and mass are placed in contact with each other, but are insulated from everything else. One is initially at 100 °C, and the other at 0 °C. What will happen spontaneously? Of

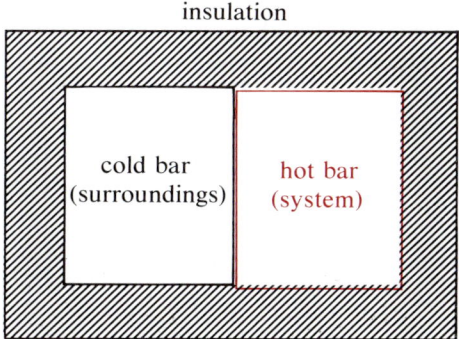

Figure 8-10   Spontaneous heat transfer between two copper bars.

course, we know that they will both wind up at 50°C, but is the entropy change positive for such a process? Let us call the hot bar the system and the cold bar the surroundings. The total heat leaving the hot bar is equal to the total heat entering the cold bar, since there is nowhere else for it to go:

$$-dQ_{\text{hot}} = dQ_{\text{cold}}$$

This is true regardless of whether the heat transfer is reversible or not, so we can perform the process reversibly by imagining a bath that reversibly accepts dQ of heat from the hot bar and reversibly donates it to the cold bar. Since the bath does not appear in the irreversible real process, its entropy change does not appear in the total entropy change, and we need not consider it in our calculation. We can do this because entropy is a state function as applied to each individual bar. In Chapters 3, 4, and 7 we had occasion to examine the process of putting heat into matter to raise its temperature. At constant volume,

$$C_v = \frac{dE}{dT}$$

which defined the heat capacity (see p. 124). The energy change involved is dQ, so we can rearrange the heat-capacity definition to read

$$dQ = C_v \, dT$$

The entropy change for the hot bar (the system), then, is

$$\Delta S_{\text{hot}} = \int_{T_1}^{T_2} \frac{dQ_{\text{rev}}}{T} = \int_{T_1}^{T_2} \frac{C_v \, dT}{T} = C_v(\ln T_2 - \ln T_1)$$

$$= C_v \ln \frac{T_2}{T_1} = C_v \ln \frac{323}{373}$$

Since the temperature ratio is less than 1 its logarithm is negative, which makes the entropy change negative. Taken alone, this result suggests that the hot bar should not cool off, which is certainly not in accord with experiment and should be convincing evidence that the surroundings must be included:

$$\Delta S_{\text{universe}} = \Delta S_{\text{hot}} + \Delta S_{\text{cold}}$$

$$= C_v \ln \frac{323}{373} + C_v \ln \frac{323}{273}$$

$$= C_v \left[\ln \frac{323(323)}{373(273)} \right]$$

$$= C_v \ln (1.025)$$

Since C_v is a positive number and the logarithm of a number greater than 1 is, again, positive, the entropy change for the universe is positive, and the entropy prediction is again in agreement with experiment.

A number of examples of different kinds could be added to the two above, all of which would demonstrate that the entropy function is a reliable measure of spontaneity. In developing the entropy function no mention has been made of the molecular model of matter; entropy thus stands independent of the model we choose to represent the structure of matter. This is at the same time one of the strengths of entropy and thermodynamics and one of the weaknesses. The deficiencies and quantitative inaccuracies of quantum mechanics, for example, do not in any way affect the accuracy of thermodynamic calculations. However, in terms of the structure of matter we have no reference point, no way of uniting molecular properties and molecular motion with the criterion of spontaneity that has proved macroscopically reliable. How, exactly, does entropy relate to the structure of matter? How can we predict the direction of spontaneous change from molecular properties? In Section 8-3 we shall find part of the answer.

8-3 Entropy and Randomness

Part of the answer, indeed, has already appeared in discussion of liquid–gas and liquid–solid phase transitions in Chapter 7. If a system consists of a large number of molecules, the most probable behavior for each molecule becomes, overwhelmingly, the most probable behavior for the system of molecules. A solid melts because its structure becomes more disorderly when heat energy is added to it; the structure becomes more disorderly because the added energy gives the molecules access to many possible geometric distributions, all but one of which are disorderly by comparison with the crystal structure. Presumably the geometric distributions are adopted by chance, and if so the chance that the molecules will select the single orderly structure out of the trillions of possible structures available is vanishingly small. Speaking on the molecular scale, then, there is something that tends to happen spontaneously: when molecular motion can occur, a system of molecules tends to change from a condition that can be represented by fewer equivalent geometrical distributions or other patterns of energy-level occupation, called **configurations**, to a condition that can be represented by more configurations. The more equivalent configurations there are, the more probable that condition becomes. The entropy of a system should increase as the number of equivalent configurations allowed it by its physical condition increases.

The Logarithmic Relation Between Entropy and the System's Configurations

The entropy should be some kind of permanently increasing function of the number of possible configurations: $S = f(W)$, where W is the number of equivalent configurations for some given condition of the system. The entropy of a system is a capacity quantity, not an intensity quantity; so if two identical systems A and B are brought together into a single new system, the new system's entropy must be twice that of A or B. But when the two systems are brought together, the total number of configurations does not add but multiplies — remember that each configuration of A connects equally well to all the configurations of B. If the entropy must add but the numbers of configurations must multiply, we have

$$S_{A+B} = S_A + S_B$$
$$f(W^2) = f(W) + f(W) = 2f(W)$$

What kind of function behaves this way? The logarithmic function, of course:

$$\log W^2 = 2 \log W$$

We propose that the entropy of a system is linearly proportional to the natural logarithm of the number of possible configurations:

$$S = k \ln W + S_0 \tag{8-6}$$

where k is a proportionality constant and S_0 is an additive constant that drops out of any calculation of ΔS by subtraction. Nevertheless, it is convenient (and legitimate) to set it equal to zero:

$$S = k \ln W$$

What is the proportionality constant k? If this formulation of the entropy is to have the success of the macroscopic version, it will have to yield the same prediction for, say, the isothermal expansion of a gas. So we can work the problem out both ways and set k equal to the constants that show up in the macroscopic result. Suppose we have a gas consisting of one molecule. If it "expands" from volume V_1 to V_2, the number of its configurations increases because there are more places it can be in a big volume than in a small volume. That is, the number of configurations is proportional to the volume. We can write

$$\Delta S = S_2 - S_1 = k(\ln W_2 - \ln W_1) = k \ln \frac{W_2}{W_1}$$

and if

$$W = \alpha V$$

$$\Delta S = k \ln \frac{\alpha V_2}{\alpha V_1} = k \ln \frac{V_2}{V_1}$$

The macroscopic result for this case is

$$\Delta S = nR \ln \frac{V_2}{V_1}$$

and n, the number of moles, is in this case $1/N$ — the reciprocal of Avogadro's number — since there is only one molecule present. So

$$n = \frac{1}{N}$$

and

$$k = \frac{1}{N} R = \frac{R}{N} = \text{Boltzmann's constant}$$

Boltzmann's constant is well established numerically, so we have a particularly convenient expression to use for entropy changes:

$$\Delta S = k \ln \frac{W_2}{W_1} \text{ per molecule}$$

or

$$\Delta S = R \ln \frac{W_2}{W_1} \text{ per mole}$$

Detailed calculations using this statistical formulation of the entropy are quite difficult, and are reliable only for ideal gases and perfect crystals. Some interesting approximations have been made for nonideal systems, but in general this is still an area open to development. This contrasts sharply with the macroscopic development, in which numerical agreement of calculations is in general as good as the experimental accuracy of the data involved; the difference is that the macroscopic development is essentially a law, a summary of experimental results, while the molecular approach is a theory involving a model that may or may not be adequate to the task. Nevertheless, the concept of entropy as being (logarithmically) proportional to the number of equivalent configurations of a system — also called **microstates** of the system — is very helpful in developing an intuitive feel for the approximate size of the entropy change associated with a given chemical reaction or physical change.

The Entropy of Various States of Matter

For instance, a crystalline solid will have a relatively small entropy, because even with some crystal defects and vibration of the atoms or molecules about their lattice sites, there will still be only a fairly small number of equivalent configurations the system can adopt. A liquid will have a considerably greater entropy, because the large number of defects and the translational motion of the molecules mean that there are many more possible distributions of the molecules that correspond to acceptable liquid structures, that is, more equivalent configurations or microstates. A gas will have a still higher entropy, because it normally occupies a much larger volume than either a liquid or a solid, giving it many more possible locations in which its molecules can be distributed. In Table 8-1 some absolute entropies, calculated by a method to be described in Section 8-4, are given, and it can be seen that the entropies of solids are very roughly 15 cal/mole-°K, liquids are 30 cal/mole-°K, and gases are 50 cal/mole-°K. Within each individual class the entropies of heavier, more complicated molecules are higher, since the increasing mass gives the molecules a greater variety of states of motion and (for molecules) more bonds to bend and stretch as the molecule flexes; all of these processes lead to increasing disorder and randomness, and in effect a greater number of possible microstates of the system of molecules.

Table 8-1
Absolute Molar Entropies[a] (cal/mole-°K at 298 °K)

Solids

Al	6.77	Cr_2O_3	19.4	$MgCl_2$	21.4		
$AlCl_3$	40	$CoCl_2$	25.4	MgO	6.4		
Al_2O_3	12.19	CoO	10.5	$NiCl_2$	25.6		
NH_4Cl	22.6	Cu	7.96	NiO	9.22		
$BaCl_2$	30	CuCl	20.2	KCl	19.76		
BaO	16.8	Fe	6.49	NaCl	17.30		
BeO	3.37	$FeCl_2$	28.6	$SrCl_2$	28		
$CaCl_2$	27.2	Fe_2O_3	21.5	SrO	13.0		
CaO	9.5	Pb	15.51	S (rhombic)	7.62		
$CaCO_3$	22.2	$PbCl_2$	32.6	Zn	9.95		
C (diamond)	0.58	PbO	16.2	$ZnCl_2$	25.9		
$CrCl_3$	30.0			H_2O (ice, 0 °C)	9.87		

Liquids

$AsCl_3$	55.8	H_2O	16.72	$SiCl_4$	57.2
BCl_3	50.0	Hg	18.5	$SnCl_4$	61.8
Br_2	36.4	HNO_3	37.2	$TiCl_4$	60.4
CH_3OH (methyl alcohol)		30.3		CCl_4 (carbon tetrachloride)	51.2
CH_3CH_2OH (ethyl alcohol)		38.4		C_6H_{12} (cyclohexane)	47.3
CH_2Cl_2 (dichloromethane)		42.7		C_6H_6 (benzene)	37.4
$CHCl_3$ (chloroform)		48.5			

Gases

NH_3	46.01	CO	47.30	F_2	48.6
Ar	36.98	Cl_2	53.29	N_2	45.77
Br_2	58.64	H_2	31.21	O_2	49.00
CO_2	51.06	HCl	44.62	NO_2	57.47
		H_2O	45.11		
CH_4 (methane)		44.50		$CH_2{=}CH_2$ (ethylene)	52.54
CH_3Cl (chloromethane)		55.97		$CH{\equiv}CH$ (acetylene)	48.00
CH_3CH_3 (ethane)		54.85		$CH_3CH_2CH_3$ (propane)	64.51

[a] NBS Circular 500.

Entropy Changes During Chemical Reaction

For chemical reactions the application is fairly obvious. If a chemical reaction generates a liquid (or solution) or a gas from solid reactants, the entropy change *for the system* in the reaction will be positive; likewise if liquid or solution reactants generate a gaseous product. Conversely, if two gases condense to a solid, the entropy change for the system will be quite negative, and only if the entropy change of the surroundings is strongly positive will the reaction occur. For instance, consider the two reactions

$$NH_{3(g)} + HCl_{(g)} \rightarrow NH_4Cl_{(s)} \quad \Delta S^0 = -68.0 \text{ cal/mole reaction-deg}$$

and

$$2H_3O^+_{(aq)} + CaCO_{3(s)} \rightarrow Ca^{2+}_{(aq)} + CO_{2(g)} + 3H_2O_{(l)} \, \Delta S^0 = +20.2 \text{ cal/mole reaction-deg}$$

Both of these are spontaneous at room temperature. The first has a very negative entropy change associated with the system, however, and occurs only because the surroundings undergo a very substantial positive entropy change. The negative value for the system arises, of course, because two disorderly gases with many available microstates are condensing into an orderly crystalline solid with very few microstates. The second reaction involves a positive entropy change for the system, due primarily to the formation of gaseous CO_2.

How can the surroundings increase in entropy? This question becomes important because such an increase is necessary for many reactions, such as the formation of solid NH_4Cl above. If the reaction is carried out in the laboratory open to the atmosphere, the entropy of the atmosphere will increase somewhat as it expands back into the space left by the condensing gas. A more important entropy increase occurs, however, when the surroundings are heated. In the example on p. 425 involving the thermal equilibrium between two copper bars, the expression

$$\Delta S = C_v \ln \frac{T_2}{T_1} \qquad (8\text{-}7)$$

was developed. Since C_v is always a positive number, we see that the entropy change of a substance is always positive when it is heated, that is, when T_2/T_1 is greater than 1. In Chapter 9 we shall use this fact to develop a more convenient criterion for spontaneity, one that involves only the system and not the surroundings. Here we note that a chemical reaction which generates a lot of heat and transfers it to the surroundings may be spontaneous even if ΔS for the system itself is negative, since the surroundings are being strongly heated. In other words, if dQ into the surroundings is positive and T is positive, dS for the surroundings will be positive.

8-4 Entropy and Temperature

The Third Law of Thermodynamics

Section 8-3 has shown that the entropy of a substance is logarithmically proportional to the number of microstates or equivalent configurations available to it. Also, the entropy of a substance increases on warming and decreases on cooling. What will happen to the entropy of a substance as it is cooled to very low temperatures, approaching absolute zero? Even if the substance is a gas, it will eventually liquefy, then solidify as the temperature drops and the thermal energy of the molecules decreases. If it solidifies in a crystalline form, it will still have a significant entropy due to the crystal defects and vibrations previously mentioned. But as the temperature of the crystal is lowered, fewer and fewer molecules will have sufficient thermal energy to translate through the lattice, and the crystal will become more nearly perfect. Eventually, at least in an ideal sense, at the absolute zero of temperature the crystal would be perfect, with no defects whatever. At that point there would be only one possible configuration for the system, and its entropy would be given by

$$S = k \ln W$$
$$= k \ln 1$$
$$= 0$$

In other words, **the entropy of a perfect crystal at the absolute zero of temperature would be zero**. This appears to be a general physical law even though many substances do not form perfect crystals even as the absolute zero of temperature is approached. In every such case, an analysis of the solid's structure reveals some form of disorder that represents a multiplicity of possible configurations and leaves the solid with a residual entropy.

For instance, ice appears experimentally to have a residual entropy of 0.82 cal/mole-°K even as 0°K is approached. Its structure is such that each oxygen in the water molecule is surrounded tetrahedrally by four other water molecules, with hydrogen atoms forming hydrogen bonds between the oxygens, as in Fig. 8-11. If the hydrogens were in the middle of the O—O axis, the crystal would be symmetrical and there would be no residual entropy, so there must be some asymmetry in that the H must be at any given time closer to one O than the other. Presumably in this hydrogen bond the hydrogens can transfer from being near one oxygen to being near the other, but in order to keep two H's on every O, these transfers must be concerted among several molecules. Now, if the $2N$ hydrogens in a mole of ice could each be in either of two positions (near

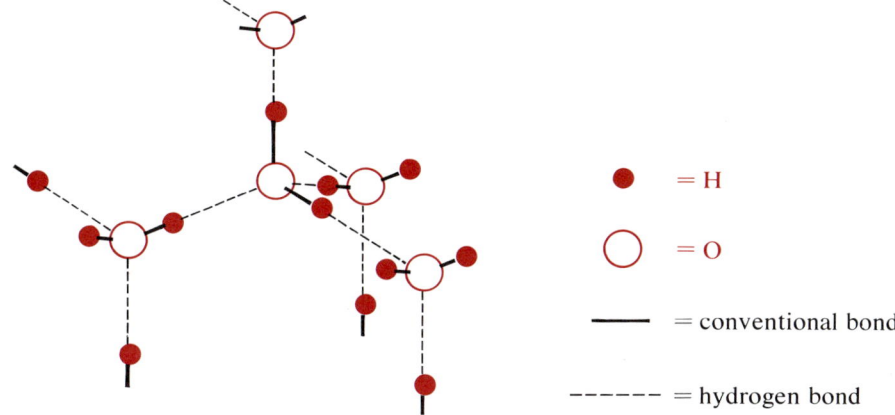

Figure 8-11 Structure of ice.

one oxygen or near the other), there would be 2^{2N} configurations. But not all of these keep the stoichiometry properly at two H close to every O; for a given oxygen with its four neighboring hydrogens, Fig. 8-12 shows that there are

Figure 8-12 Possible arrangements of the four neighbor H atoms around a given O atom in ice.

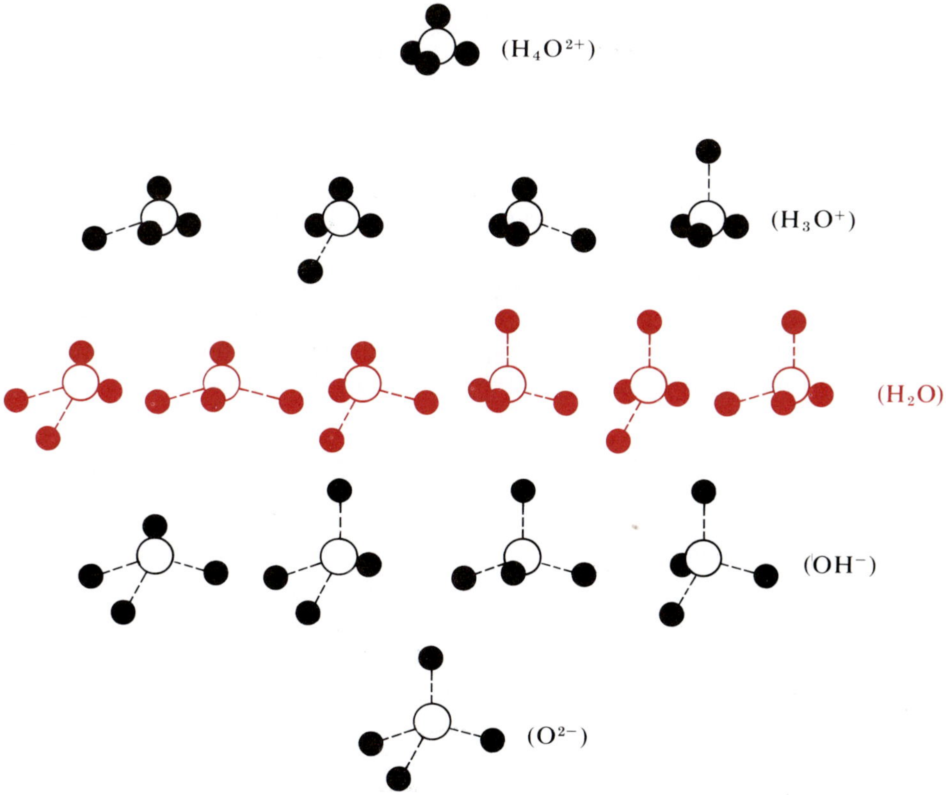

16 possible arrangements of the hydrogens, only six of which correspond to two H's close to the O. So $\frac{6}{16}$ or $\frac{3}{8}$ of the configurations around each oxygen keep the stoichiometry right, and presumably they are the only ones that can occur. Of these configurations, only $\frac{3}{8}$ are right for a neighbor oxygen, and so for N oxygens, having 2^{2N} total configurations of the surrounding hydrogens, only $(\frac{3}{8})^N$ of that number are stoichiometrically correct. The number of available configurations, then, is

$$W = 2^{2N} (\tfrac{3}{8})^N$$
$$= (2^2)^N (\tfrac{3}{8})^N$$
$$= (4 \times \tfrac{3}{8})^N$$
$$= (\tfrac{3}{2})^N$$

We can immediately calculate the entropy of ice from this number as

$$S = k \ln (\tfrac{3}{2})^N$$
$$= Nk \ln \tfrac{3}{2}$$
$$= R \ln \tfrac{3}{2}$$
$$= (1.987 \text{ cal/mole-}°K)(0.405)$$
$$= 0.806 \text{ cal/mole-}°K$$

The agreement is obviously excellent with the observed value of 0.82 cal/mole-°K. This calculation, originally made by Linus Pauling, is a good example of the approach used in performing statistical thermodynamic calculations.

Establishing Absolute Entropies

If we can be certain that the entropy of a perfect crystal at 0°K is zero, we can calculate absolute entropies at any finite temperature. We first calculate the entropy change of heating the solid up to its melting point (or temperature of transition to another crystal structure) by using the relation

$$\Delta S_{0 \to m.p.} = \int_0^{m.p.} dS = \int_0^{m.p.} \frac{C_v(s)}{T} dT$$

Having $T = 0°K$ in the denominator makes it impossible to perform this integration by direct algebraic means, at least for certain algebraic forms of the C_v function, but we can perform the integration graphically by plotting C_v/T as a function of T and remembering that the area under the curve is the integral of the function represented by the curve; see Fig. 8-13. At the melting point, as the cooling curves in Chapter 7 indicated, heat is being put into the solid (the

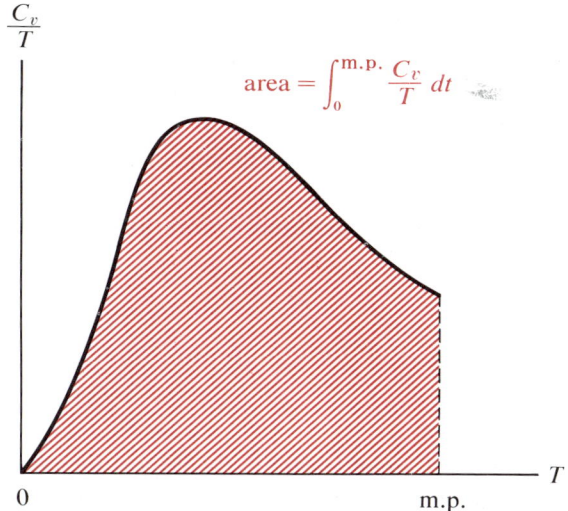

Figure 8-13 Graphical integration to obtain absolute entropies.

heat of fusion), but the temperature is not changing. For such an isothermal change we have

$$\Delta S_{\text{melt}} = \int \frac{dQ_{\text{melting}}}{T_{\text{m.p.}}} = \frac{\text{heat of fusion}}{T_{\text{m.p.}}}$$

which is another contribution to the total entropy of the molten sample. Heating the liquid to the boiling point, we have

$$\Delta S_{\text{m.p.}\to\text{b.p.}} = C_v(\text{l}) \ln \frac{T_{\text{b.p.}}}{T_{\text{m.p.}}}$$

an expression we would have used in heating the solid except for the appearance of the zero temperature in the denominator; here there is no such problem. When the liquid boils, another isothermal process is occurring:

$$\Delta S_{\text{evap}} = \frac{\text{heat of vaporization}}{T_{\text{b.p.}}}$$

Finally, we calculate the entropy change for heating the gas from the boiling point to whatever is the desired temperature:

$$\Delta S_{\text{b.p.}\to T} = C_v(\text{g}) \ln \frac{T}{T_{\text{b.p.}}}$$

The total absolute entropy of the substance at any given temperature is the sum of all these contributions:

$$S = \text{graphical integral of } \frac{C_v(\text{s})}{T} \text{ to m.p.} + \frac{\Delta H_{\text{fus}}}{T_{\text{m.p.}}}$$

$$+ C_v(\text{l}) \ln \frac{T_{\text{b.p.}}}{T_{\text{m.p.}}} + \frac{\Delta H_{\text{vap}}}{T_{\text{b.p.}}} + C_v(\text{g}) \ln \frac{T}{T_{\text{b.p.}}}$$

436 | Spontaneity and Equilibrium

The absolute entropies in Table 8-1 were obtained by this procedure, although they are tabulated at 298 °K and 1 atm, meaning that the pressure has been kept constant throughout the process rather than the volume. The only change in the procedure is that instead of C_v, which refers to the rate of change of heat content at constant volume, we use C_p, a heat capacity that refers to the rate of change of heat content of a system at constant pressure; these values are experimentally established, and we shall have more to say about the concept in the next chapter.

8-5 Summary

This chapter has produced some particularly important concepts relating to chemical reactivity, and we shall need to build on it throughout this part of the book. The initial emphasis was on how spontaneity displayed itself on a macroscopic basis. Spontaneous interactions were broken down into thermal, mechanical, and compositional types, and the patterns by which these changes occur were shown to be very similar. We accordingly formulated a single hypothetical physical property as a measure of spontaneity of all types. This property should increase during any spontaneous change, providing a law for spontaneous changes. For a differential change in this physical property, the entropy, we were able to derive the expression

$$dS = \frac{dQ_{rev}}{T}$$

For different types of interactions, this expression can be integrated to the following expressions for finite entropy changes:

$\Delta S = 0$ (adiabatic process)

$ = nR \ln \frac{V_2}{V_1}$ (isothermal expansion)

$ = C_v \ln \frac{T_2}{T_1}$ (heat transfer at constant volume)

$ = C_p \ln \frac{T_2}{T_1}$ (heat transfer at constant pressure)

$ = \dfrac{\text{heat of phase transition}}{T_{transition}}$ (phase transition)

Unfortunately, each of these expressions refers only to a system or to the surroundings, never to the sum of the two that is the real measure of spontaneity. Accordingly, for a proposed process we must do each calculation twice: once for the system, and again for the surroundings. It is often quite difficult to deal quantitatively with the surroundings (how big is the atmosphere into which a gas is being released?). So ideally we would like to construct a measure for spontaneity that does not depend on the specific nature of the surroundings at all, but only on the nature of the system itself. This is one of the areas we shall examine in Chapter 9.

We also tried to relate the entropy concept to the molecular model of the past few chapters by relating ΔS, the spontaneity measure, to the statistical ideas of melting, dissolving, vaporizing, and so on from Chapter 7. It quickly developed that macroscopic entropy changes and microscopic molecular motion are compatible if

$$S = k \ln W \quad \text{or} \quad \Delta S_{1 \to 2} = k \ln \frac{W_2}{W_1}$$

where S is the absolute entropy of each molecule in the system, k is Boltzmann's constant (the gas constant for a single molecule), and W is the number of possible positions or states of motion available to a single molecule.

From this relationship several important ideas developed. One development was the capacity to predict qualitatively the entropy change for a system undergoing a given process such as a chemical reaction. If a reactant with a relatively small number of equivalent configurations gives rise to a product with a relatively large number, then ΔS will be positive for that system and vice versa. Another development was the general pattern of the temperature dependence of the entropy of a given system, with the theoretical result that a perfect crystal should have zero entropy at the absolute zero of temperature; this is quite generally verified by extrapolations from experiment, and is stated as a law rather than as a theoretical result. Extrapolations are necessary because the absolute zero of temperature cannot be reached. Any spontaneous process taking a substance to absolute zero would lose heat, making dQ negative for the system; if $T = 0°K$ in the denominator of the ΔS expression, this would correspond to an infinite negative entropy change that no positive entropy change in the surroundings could overcome.

There is another development from the $S = k \ln W$ relationship that has not been mentioned explicitly, but that is important to our future discussion. If the entropy change of a system is related to the change in the number of its possible configurations of microstates, it should be clear that it is very specifically related to the increasing disorder that was one of the two competing factors in general processes involving liquids. The other competing factor was the energy of the system of molecules, and this is an area we have carefully avoided in this chapter. How do we deal with the low-energy tendency? How does a tendency toward a low-energy configuration mesh with the general derivation of entropy on a macroscopic scale? The description of the energy relationships in thermodynamic processes will be the initial task of Chapter 9, and from that discussion will arise a more convenient measure of spontaneity, one that combines the two observable tendencies toward high disorder on the one hand and low energy on the other.

Study Problems

1. Under ideal conditions, gaseous CO_2 and H_2O are formed when acetylene burns in oxygen (as in an oxyacetylene torch). Write a balanced equation for this combustion reaction. Remembering that entropy is proportional to the number of moles of each species present, what is the total entropy of the products of the reaction; of the reactants? What is the net change in entropy, ΔS, for the reaction? Why is this reaction spontaneous?
2. Predict the absolute molar entropy of butane, CH_3—CH_2—CH_2—CH_3, by analogy from Table 8-1. The experimental value is 74.10 cal/mole-°K. What does this suggest about the pattern of entropies for chains of —CH_2— groups, as in CH_3—$(CH_2)_n$—CH_3?
3. Compare the numerical entropy changes for the melting and the vaporization of 1 mole of water. Do the relative values seem compatible with our understanding of the structures of solid, liquid, and gaseous water?
4. What factors might be responsible for the fact that the chloride of a given element seems to have a much higher entropy than the corresponding oxide? Why are carbon and hydrogen exceptions?
5. Judging strictly by the discussion of this chapter, which of the following reactions would you expect to go spontaneously to the right?
    (a) $C_8H_{18(g)} + \frac{25}{2} O_{2(g)} \rightarrow 8CO_{2(g)} + 9H_2O_{(g)}$
    (b) $BaO_{(s)} + CO_{2(g)} \rightarrow BaCO_{3(s)}$
    (c) $FeS_{(s)} + 2H_3O^+_{(aq)} + 4H_2O_{(l)} \rightarrow Fe(OH_2)^{2+}_{6(aq)} + H_2S_{(g)}$
    (d) $HCl_{(g)} + H_2O_{(l)} \rightarrow H_3O^+_{(aq)} + Cl^-_{(aq)}$
    All four reactions *do* proceed spontaneously; which ones must evolve heat?
6. Figure 8-7 has some surprising consequences if we consider its four processes as a *cycle*: isothermal expansion from 1 to 2, adiabatic expansion from 2 to 4, isothermal compression from 4 to 3, and adiabatic compression from 3 to 1. This is an approximation to the process involved in a piston engine; how much work can we get from such an engine? If the engine really returns to its original condition at point 1, then its temperature will be the same, no heat will have gone into warming the gas, and the total work done will equal the total heat flow into the system. In the first process of the cycle (from 1 to 2), what is the heat flow Q? What is Q for the second process; the third; the fourth? Use our earlier result for this system, $V_2/V_2 = V_4/V_3$, to combine and simplify these four Q values into an expression for the total heat flow in the cycle.

If the total work done, W_{total}, equals the total heat flow, show that the ratio of the work done to the heat flow Q_1 for the first process only is

$$\frac{W_{total}}{Q_1} = \frac{T_1 - T_2}{T_1}$$

This is the **thermodynamic efficiency** of the engine. Show that a simple steam engine operating at atmospheric pressure and condensing the steam at room temperature (25 °C) is only 20% efficient at converting heat into work. Friction in the engine and other losses would reduce this value still further.

7. The steam turbines in electrical power plants operate under the same limitations as the cycle of Study Problem 6 (a **Carnot cycle**). In a nuclear power plant, steam is generated at about 290 °C, a temperature that is limited by the mechanical properties of reactor components, and is condensed at about 40 °C. What is the plant's thermodynamic efficiency? A coal-fired plant can be designed to operate at higher temperatures, generating steam even above the critical temperature of water, 374 °C. If such a plant operates at 350 °C, what is its thermodynamic efficiency? If both of the plants in this problem were built to produce 1000 megawatts (MW) of electric power (1 W = 1 J/sec), how much heat would each plant return to the surroundings at the point where the steam is condensed? If each plant is cooled by a river that flows at 4000 ft³/sec, how much does the temperature of the river rise at each plant? Think about the environmental consequences of power generation.

8. If the coal-fired plant of Study Problem 7, which produces less thermal pollution than the nuclear plant, burns coal that releases 6.00 kcal/g, how much coal does the plant burn per second? If the coal is 2% sulfur and burns to SO_2, how much SO_2 is released into the atmosphere by the plant in kilograms per second; in tons per hour; in cubic feet per hour at 1 atm and 30 °C?

9. How many ways can three indistinguishable red balls be placed in three boxes, each large enough for one ball? How many ways can two red balls and a green ball be placed in three boxes? How many ways can a red, a green, and a blue ball be placed in three boxes? Briefly discuss the relative entropies of the three molecules shown in Fig. 8-14, which are shown looking down the C—C bond axis, if the substituent atoms on one carbon can rotate about the C—C axis but are always staggered with respect to the atoms on the other carbon.

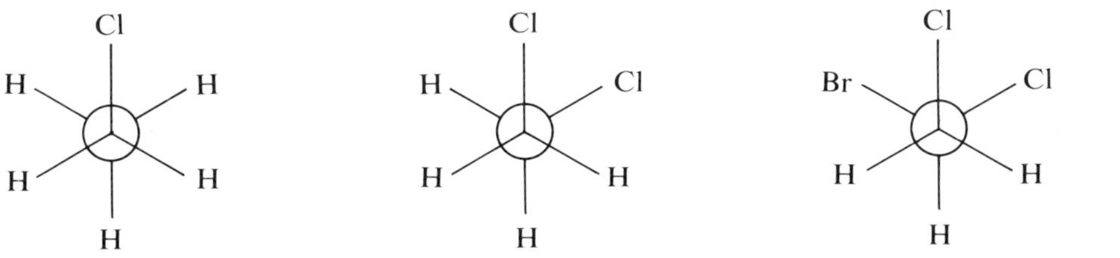

CH_3—CH_2Cl
chloroethane

CH_2Cl—CH_2Cl
1,2-dichloroethane

CH_2Cl—$CHBrCl$
1,2-dichlorobromoethane

Figure 8-14

Some Further Reading

Fong, P., *Foundations of Thermodynamics*, New York: Oxford University Press, 1963. Most of the present chapter is based on Fong's approach, which is unusual but particularly well suited to studies of chemical spontaneity. His discussion is a little more elegant but not much more mathematical.

Bent, H. A., *The Second Law*, New York: Oxford University Press, 1965. About halfway between Fong's approach and the classical one, for which some references are given in the list at the end of Chapter 9. Very readable and even entertaining.

Van Ness, H. C., *Understanding Thermodynamics*, New York: McGraw-Hill, 1969. A very intuitive, nonmathematical approach. Particularly lucid and readable. Some applications to engineering problems.

9 | Energy in Chemical Processes

Our discussion of spontaneity has been somewhat artificial in that it has ignored a very important spontaneous tendency on which we relied in discussing atomic and molecular structure. In Chapters 5 and 6 we postulated the tendency of electrons in atomic systems to distribute themselves so as to have the lowest possible energy. If we had tried at that time to define a criterion for spontaneity, the criterion would have been a tendency to move toward a configuration having lower energy; but in Chapter 8 the criterion proved to be a tendency to move toward a configuration having higher disorder and probability. In this chapter we shall see that these are not incompatible and are, indeed, complementary parts of the general description of spontaneity. To do this it will be necessary to discuss energy as it appears in chemical processes. In what forms does energy appear? How is it transferred between interacting systems? How is the "bookkeeping" of energy transfer done? To answer these questions, we need to return to the energy quantities transferred in the examples from which we derived entropy as a potential for spontaneous interaction, heat and work.

9-1 Heat, Work, and Internal Energy

Heat and work are characteristic of processes, not systems. A system has a specified pressure and volume (and the other thermodynamic variables) but it does not have a specified heat or a specified work. Heat shows up only as heat *transfer* in a process involving the system; work shows up only as mechanical-energy *transfer* in a process involving the system. As a matter of fact, in Chapter 1 work was *defined* as a means of transferring energy from one part of a system to another. Just as different states of a system have different values of the state variables, different processes involving a system and its surroundings have different values of the total amount of heat absorbed and the total amount of work done by the system. If there are two different states of a system, A and B, and several possible processes leading from state A to state B, the total heat flow and the total work performed will be different for different processes. If the heat flow and the work performed are the same for two different processes between the same initial and final states, it is coincidental. In other words, heat and work are not state functions of a system.

Heat Flow and Work Performed for Two Processes Between the Same States

For example, consider the two processes shown in Fig. 9-1. The two states A and B of an ideal gas have the same temperature T (i.e., they lie along the path

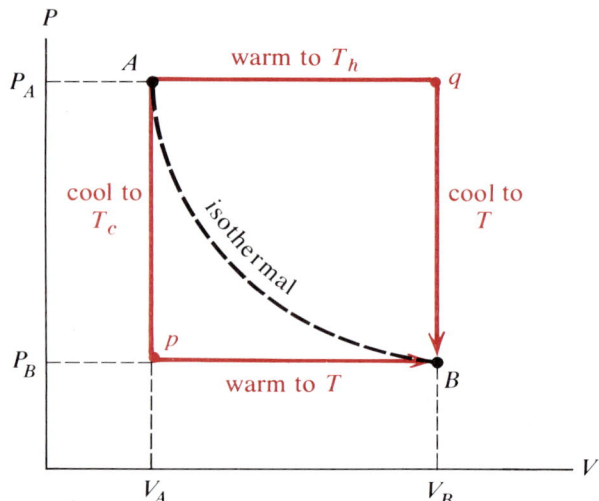

Figure 9-1 Two possible processes between states A and B.

of an isothermal expansion), but are connected by two different reversible processes, 1 and 2. In process 1 the gas is cooled reversibly to a temperature T_c at constant volume until its pressure has fallen from P_A to P_B; then it is heated reversibly at constant pressure (P_B) until its volume has expanded to V_B. In this process the heat flow into the gas, dQ_1, is given for the first part by $C_v(T_c - T)$ and for second part by $C_p(T - T_c)$:

$$dQ_1 = C_v(T_c - T) + C_p(T - T_c) = (C_p - C_v)(T - T_c)$$

The work done by the gas, dW_1, equals $\int P\, dV$; for the first part of the process this is zero since the volume is not changing, and for the second part of the process it is equal to the constant P_B times the volume change:

$$dW_1 = \int_A^B P\, dV = \int_A^P P\, 0 + \int_P^B P_B\, dV = 0 + P_B \int_P^B dV = P_B(V_B - V_A)$$

For process 2, in which the gas is first heated at constant pressure P_A until it has expanded to V_B at temperature T_h, then cooled at constant volume (V_B) until the pressure has dropped to P_B, the heat flow dQ_2 into the gas is given by

$$dQ_2 = C_p(T_h - T) + C_v(T - T_h) = (C_p - C_v)(T_h - T)$$

Similarly, the work done by the gas (by analogy to process 1) is

$$dW_2 = \int_A^B P\, dV = \int_A^q P_A\, dV + \int_q^B P\, 0 = P_A \int_A^q dV + 0 = P_A(V_B - V_A)$$

We can see immediately that dW_1 is smaller than dW_2, since P_B is smaller than P_A. The comparison of heat flows is not as simple, but Study Problem 1 shows that dQ_1 cannot equal dQ_2 either. We shall return to this example later.

If a system and its surroundings are to interact in a spontaneous process between two defined states, the amount of work we can get out of the process depends on the way the process is carried out, not on the two states themselves. How can the maximum work be extracted from the process? There *is* a maximum or limitation of some kind, because otherwise a finite process would produce an infinite amount of work. Let us return to an expanding gas, this time one that is expanding against a piston as in Fig. 9-2. How much work is it

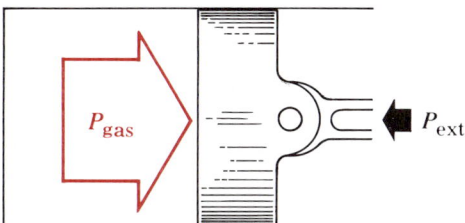

Figure 9-2 A piston-cylinder apparatus by which work can be done on some external object.

doing during the process? The work done by an expanding gas is given by

$$W = \int_{\text{state 1}}^{\text{state 2}} P\, dV$$

but the appropriate pressure is not the pressure of the expanding gas but rather the external pressure it is overcoming; from the last chapter we recall that a gas expanding into a vacuum does no work at all. So for a given length of travel of the piston (which ensures that the initial and final states are always the same if the temperature does not change), the work done by the expanding gas will be larger the larger the external opposing pressure, P_{ext}, is. However, the external pressure cannot increase indefinitely, because as soon as it exceeds P_{gas} the expansion will stop, compression will begin, and the gas will be absorbing work instead of producing it. The limiting maximum work done by the system for a given piston travel will be done when $P_{ext} = P_{gas}$. But this is the equilibrium condition, so the maximum work is obtained from a process that is a succession of linked equilibrium states, or a reversible process. There is, then, some practical justification for an interest in reversible processes, even though we can only approximate them because of the time factor.

The First Law of Thermodynamics

We have still not produced any correlation between dQ and dW for a given process, however. In the discussion of gases in Chapter 3, we introduced the concept of potential energy *to keep the energy bookkeeping simple by making the total energy of an isolated system constant.* (See p. 140.) Here we introduce another concept for exactly the same reason. That is, if two systems interact with each other but are otherwise isolated from the rest of the universe, there should be a state function for each system that correlates the heat and work flows between the systems even though they are not state functions. These two state functions should be energy quantities, and they should add up to a constant. We define a quantity called the **internal energy** of a system, symbolized by E, and add up the initial internal energies of the two systems together with the energy that flows in and out of each one:

$$\text{total energy} = \text{constant} = E_1^i + E_2^i + dQ_1 + dQ_2 - dW_1 - dW_2$$
$$= (E_1^i + dQ_1 - dW_1) + (E_2^i + dQ_2 - dW_2)$$
$$= E_1^f + E_2^f$$

In these expressions E^i refers to the internal energy of the system in its initial state and E^f to the internal energy of the final state. The change in internal energy, ΔE, is given by

$$\Delta E_1 = E_1^f - E_1^i = dQ_1 - dW_1 \quad \text{and} \quad \Delta E_2 = dQ_2 - dW_2$$

More generally, we say that for *any* change between two thermodynamic states of a system,

$$\Delta E = dQ - dW \tag{9-1}$$

This is a general statement of the **first law of thermodynamics**. As a law it is not a theoretical model, but a summary of experimental observation, and its importance lies not in the definition of an internal-energy quantity for bookkeeping purposes but in the fact that ΔE is a state function even though dQ and dW are not. It is quite generally observed that for any process between two thermodynamic states of a system, the heat flow in minus the work flow out is a constant quantity regardless of the process—even though the heat flow and work flow depend quite strongly on the process.

Let us return to the system in Fig. 9-1 for an example. For process 1,

$$\Delta E = dQ_1 - dW_1 = (C_p - C_v)(T - T_c) - P_B(V_B - V_A)$$

To define T_c in terms of states A and B, we need only note that it is the temperature at point p in the diagram, where

$$P_B V_A = RT_c \quad \text{or} \quad T_c = \frac{P_B V_A}{R}$$

Similarly,

$$T = \frac{P_B V_B}{R}$$

Inserting these temperatures in the ΔE expression,

$$\Delta E = \left(\frac{C_p - C_v}{R}\right)(P_B V_B - P_B V_A) - P_B V_B + P_B V_A$$

$$= \left(\frac{C_p - C_v}{R} - 1\right) P_B \Delta V$$

For process 2, we need an expression for T_h. At point q,

$$P_A V_B = RT_h \quad \text{or} \quad T_h = \frac{P_A V_B}{R}$$

and

$$T = \frac{P_A V_A}{R}$$

The internal energy change for process 2 is

$$\Delta E = dQ_2 - dW_2$$
$$= (C_p - C_v)(T_h - T) - P_A(V_B - V_A)$$
$$= \left(\frac{C_p - C_v}{R}\right)(P_A V_B - P_A V_A) - P_A V_B + P_A V_A$$
$$= \left(\frac{C_p - C_v}{R} - 1\right) P_A \Delta V$$

Although this differs from the ΔE expression for the first process in the pressure factor, both ΔE expressions could assume the same value, zero, if $(C_p - C_v)/R = 1$. This would make the first pair of parentheses vanish in each expression. We shall see shortly that for an ideal gas it is indeed true that

446 | Energy in Chemical Processes

$C_p - C_v = R$, so that the internal energy change is the same for both the processes in Fig. 9-1. No such possibility exists for the dQ values or the dW values individually, so for this system at least we can say that the internal energy change is a state function but the heat and work flows are not.

9-2 The Enthalpy Function and Heat Capacities

Suppose we examine the way the first law of thermodynamics applies to processes as they generally occur in the laboratory. The vast majority of chemical processes, both chemical reactions and physical changes (such as distillation), are carried out in vessels that are open to the atmosphere. This means that the total pressure exerted on the components of the systems involved is constant at about 1 atm. If in addition we specify that the chemical system does only pressure–volume work (e.g., evolution of a gas in a reaction), and no electrical or gravitational work, for example, we have introduced some important restrictions by which we can simplify the treatment of the heat flow and work flow. The restrictions of (1) constant pressure and (2) only PV work leave virtually all the ordinary chemical processes unaffected, however.

The Enthalpy Quantity

If only PV work is done by the system, and if it is done at constant pressure, then for a finite change of state dW is very simple:

$$dW = \int_{\text{state 1}}^{\text{state 2}} P\, dV = P_{\text{constant}} \int_{\text{state 1}}^{\text{state 2}} dV = P(V_2 - V_1) = P\Delta V$$

We can insert this in the first law and rearrange it a little:

$$\Delta E = dQ - dW$$
$$= dQ - P(V_2 - V_1)$$
$$E_2 - E_1 = dQ - PV_2 + PV_1$$
$$dQ = (E_2 + PV_2) - (E_1 + PV_1)$$

Since all the variables on the right of this last expression are state variables, it follows that dQ is itself a state variable under these circumstances. The change in this new state variable is represented by the difference between the two quantities in parentheses. The new state variable is called **enthalpy**, symbolized by H:

$$H \equiv E + PV \tag{9-2}$$

The enthalpy change of a process occurring at constant pressure, ΔH, is equal to the heat flow occurring in the process:

$$dQ = \Delta H \quad \text{at constant pressure}$$

Thus the heat input required to evaporate 1 mole of a liquid at a constant pressure is called either the heat of vaporization or the enthalpy of vaporization, with the latter name being less vivid but more precise.

The Relation Between C_v and C_p

We are now in a position to discuss the difference between C_p and C_v, the heat capacities at constant pressure and constant volume, respectively. Experimentally, the heat capacity is evaluated as $dQ/\Delta T$, the measured heat flow divided by the observed temperature change. If only PV work can be done, and if the heating is performed on a system held at constant volume,

$$\Delta E = dQ - \int P \, dV \quad \text{but} \quad dV = 0$$

therefore

$$\Delta E = dQ$$

C_v, then, is in thermodynamic terms $\Delta E/\Delta T$; if a limiting approach is made through smaller and smaller measured heat flows and temperature changes,

$$C_v = \frac{dE}{dT} \quad \text{(constant volume)} \tag{9-3}$$

On the other hand, if the heating is performed on a system held at constant pressure,

$$\Delta E = dQ - P \Delta V$$
$$dQ = \Delta E + P \Delta V$$
$$= \Delta H$$

So the heat capacity at constant pressure, C_p, thermodynamically means $\Delta H/\Delta T$, or in the limit of small heat increments,

$$C_p = \frac{dH}{dT} \quad \text{(constant pressure)} \tag{9-4}$$

How are C_p and C_v related for an ideal gas? We need only write out the C_p relationship in terms of the definition of enthalpy and substitute for PV:

$$C_p = \frac{dH}{dT} = \frac{d}{dT}(E + PV)$$

$$= \frac{d}{dT}(E + RT) \quad \text{(for 1 mole)}$$

$$= \frac{dE}{dT} + \frac{d}{dT}(RT)$$

$$= C_v + R \qquad (9\text{-}5)$$

In other words, $C_p - C_v = R$, which is exactly the relationship we required for ΔE to be a state function in the example of the previous section. The change of the internal energy in that example thus proves to be zero for any process between two states of an ideal gas at the same temperature. With respect to the heat capacities themselves, Chapter 3 (p. 125) has shown that for an ideal monatomic gas,

$$C_v = \tfrac{3}{2}R$$

and for an ideal diatomic gas

$$C_v = \tfrac{5}{2}R$$

if it is considered that the temperature is low enough for kT to be too small to provide the vibrational energy quantum, yet large enough to allow free rotation. This seems to be true for most diatomic gases near room temperature. For heating at constant pressure,

$$C_p = \tfrac{3}{2}R + R = \tfrac{5}{2}R \qquad \text{(monatomic)}$$

$$C_p = \tfrac{5}{2}R + R = \tfrac{7}{2}R \qquad \text{(diatomic)}$$

Adiabatic Expansion of Gases

This is a good place to look at the adiabatic expansion of gases, a process for which, in the last chapter, we took the expansion law to be $PV^\gamma = $ constant. In an adiabatic process, $dQ = 0$, which has the following effect on the first law:

$$dE = dQ - dW = 0 - P\,dV \qquad \text{(for a differential process)}$$

Since $C_v = dE/dT$, we can equally well write $dE = C_v\,dT$:

$$C_v\,dT = -P\,dV$$

Substituting for P from the ideal gas law:

$$C_v\,dT = -RT\frac{dV}{V} \qquad \text{(for 1 mole)}$$

$$C_v\frac{dT}{T} = -R\frac{dV}{V}$$

Integrating,

$$C_v \int \frac{dT}{T} = -R \int \frac{dV}{V}$$

$$C_v \ln T + \text{constant} = -R \ln V + \text{constant}$$

$$\ln T + \frac{R}{C_v} \ln V = \text{constant}$$

$$\ln T + \ln (V^{R/C_v}) = \text{constant}$$

$$\ln (TV^{R/C_v}) = \text{constant}$$

$$TV^{R/C_v} = e^{\text{constant}} = \text{constant}$$

Substituting for T from the ideal gas law:

$$\left(\frac{PV}{R}\right)V^{R/C_v} = \text{constant}$$

$$PV^{(R/C_v+1)} = R(\text{constant}) = \text{constant}$$

$$PV^{[(C_v+R)/C_v]} = \text{constant}$$

$$PV^{C_p/C_v} = \text{constant}$$

The constant γ is thus seen to be the ratio of C_p to C_v; for a monatomic gas this is

$$\gamma = \frac{C_p}{C_v} = \frac{\frac{5}{2}R}{\frac{3}{2}R} = \frac{5}{3} = 1.667$$

For a diatomic gas (under the conditions previously mentioned),

$$\gamma = \frac{C_p}{C_v} = \frac{\frac{7}{2}R}{\frac{5}{2}R} = \frac{7}{5} = 1.400$$

These values are experimentally verified for real gases within about 1% near room temperature.

The Production of Low Temperatures

The relation $TV^{R/C_v} = \text{constant}$ can be transformed into one involving pressure and temperature:

$$TV^{R/C_v} = \text{constant}$$

$$T\left(\frac{RT}{P}\right)^{R/C_v} = \text{constant}$$

Taking the R/C_v root of both sides:

$$\frac{T^{C_v/R} RT}{P} = \text{constant}$$

Dividing by R and combining powers of T,

$$\frac{T^{(C_v+R)/R}}{P} = \text{constant} = \frac{T^{C_p/R}}{P}$$

For a diatomic gas, this means

$$\frac{T_1^{7/2}}{P_1} = \frac{T_2^{7/2}}{P_2}$$

A little thought will show that an adiabatic expansion of an ideal gas provides a good deal of cooling effect. For instance, suppose air at 300°K (27°C) and 10 atm is allowed to expand adiabatically to 1 atm:

$$\frac{300^{7/2}}{10} = \frac{T_2^{7/2}}{1}$$

$$T_2 = \frac{300}{10^{2/7}} = \frac{300}{1.92} = 156°K \quad \text{or} \quad -117°C$$

This substantial degree of cooling means that if a device could be built to precool the gas that is expanding adiabatically, very low temperatures could be achieved. An experiment done under these conditions is known as a **Joule–Thomson expansion**, and liquid nitrogen and liquid oxygen are produced from the atmosphere in exactly this way. Due to the heavy use of liquid oxygen in space-exploration rockets, the liquefaction of gases has become a major industry in the United States; the cost of these once exotic liquids is, for large quantities, lower than that of almost any common solvent.

9-3 Heat Flow in Chemical Reactions

In pursuing the nature of the enthalpy function we have ascertained that, *for a reaction occurring at constant pressure*, the heat flow into the reacting system is equal to the enthalpy change, ΔH. Enthalpy is a state function, meaning that the magnitude of ΔH depends only on the initial and final states of the reaction process and not on the details of the process itself. So for any chemical reaction carried out at constant pressure, it ought to be true that the heat flow can be predicted solely from the composition and other thermodynamic variables of the reactants, and the composition and other thermodynamic variables of the products. This is important because the heat flow (as dQ_{rev}) and the temperature of the reaction determine the entropy change of the system, and this entropy change together with that of the surroundings of the reaction system determine whether or not it will occur spontaneously. By diligent work with a calorimeter, we could presumably measure the heat flow of any existing reaction, although in some cases it would be extremely difficult. This would be a most laborious process, however, and the fact that the enthalpy change is a state function makes it unnecessary.

Enthalpy of Formation

It is only necessary to have a standard enthalpy of some sort for each compound to be able to predict its contribution to the heat flow in a constant-pressure reaction. Since the enthalpy change is a state function, it can be calculated for any reaction by subtracting the standard enthalpy of the reactants (summed up) from the standard enthalpy of the products. Unfortunately, classical thermodynamics is not capable of establishing absolute enthalpies, which means that the obvious, simple procedure,

$$\Delta H = \sum H_{absolute}(\text{products}) - \sum H_{absolute}(\text{reactants})$$

is not available to us. Some means must be found for establishing a standard enthalpy of each compound appearing in a reaction. Since only changes in enthalpy can be measured, we must define a beginning state from which the enthalpy change associated with the formation of any given compound can be measured. This beginning state will itself be assigned a zero enthalpy change of formation; it can be chosen arbitrarily as long as it is consistently adhered to. The beginning state conventionally chosen is the elements in their **standard states** at 298°K. Thus the standard state of sodium is the metallic solid element, the standard state of carbon is graphite, the standard state of bromine is the liquid diatomic molecule Br_2, and the standard state of hydrogen is the gaseous diatomic molecule H_2. From these standard states the enthalpy change of formation, ΔH_f, of any compound (sometimes called the **enthalpy of formation** or the **heat of formation**) can be calculated:

$$Na_{(s)} + \tfrac{1}{2}Cl_{2(g)} \rightarrow NaCl_{(s)} - \Delta H_f$$

The equation is written with $-\Delta H_f$ on the right because this is a convenient mnemonic device for calculations. That is, we can replace each chemical entity in the chemical equation by its enthalpy of formation and get an algebraic equation that correctly evaluates ΔH:

$$Na_{(s)} + \tfrac{1}{2}Cl_{2(g)} \rightarrow NaCl_{(s)} - \Delta H_f$$
$$\Delta H_f(Na) + \tfrac{1}{2}\Delta H_f(Cl_2) = \Delta H_f(NaCl) - \Delta H_f$$
$$0 + \tfrac{1}{2}(0) = \Delta H_f(NaCl) - \Delta H_f(NaCl)$$
$$0 = 0$$

For this case the relationship is trivial, but for general calculations it is very useful:

$$2AB + CD \rightarrow A_2D + CB_2 - \Delta H$$
$$2\Delta H_f(AB) + \Delta H_f(CD) = \Delta H_f(A_2D) + \Delta H_f(CB_2) - \Delta H$$
$$\Delta H = \Delta H_f(A_2D) + \Delta H_f(CB_2) - 2\Delta H_f(AB) - \Delta H_f(CD)$$
$$\Delta H = \sum \Delta H_f(\text{products}) - \sum \Delta H_f(\text{reactants})$$

In this case it would be necessary only to place 1 mole of sodium and $\frac{1}{2}$ mole of chlorine gas together, allow them to react, and measure the heat evolved; since we have defined the standard enthalpies of the reactants to be zero, the molar enthalpy of formation of the product would just be the measured heat flow.

Hess' Law

Not very many compounds can be formed in the laboratory by allowing the elements in their standard states to react, a fact that seems to severely limit obtaining data for enthalpies of formation. From the fact that the enthalpy change is a state function, however, we see that we can obtain enthalpies of formation by algebraically combining reactions and their observed enthalpy changes so as to add up to the hypothetical reaction in which the desired compound is formed from the elements in their standard states. As long as the reactants and products are in their proper thermodynamic condition, the chemical route from one to the other makes no difference. For instance, benzene, C_6H_6, cannot be formed by mixing graphite and hydrogen gas. However, benzene will burn in oxygen, and so will graphite and hydrogen gas:

$$C_6H_{6(l)} + \tfrac{15}{2} O_{2(g)} \rightarrow 6CO_{2(g)} + 3H_2O_{(l)} - \Delta H_1$$

$$C_{\text{graphite}(s)} + O_{2(g)} \rightarrow CO_{2(g)} - \Delta H_2$$

$$H_{2(g)} + \tfrac{1}{2} O_{2(g)} \rightarrow H_2O_{(l)} - \Delta H_3$$

Since all three reactions will proceed spontaneously if ignited, all three enthalpy changes can be measured. Then we need only combine them properly, along with the enthalpy changes, to get the hypothetical reaction of formation of benzene from the elements along with that reaction's enthalpy change:

$$6 \times [C_{\text{graphite}(s)} + O_{2(g)} \rightarrow CO_{2(g)} - \Delta H_2]$$

$$+ 3 \times [H_{2(g)} + \tfrac{1}{2} O_{2(g)} \rightarrow H_2O_{(l)} - \Delta H_3]$$

$$\underline{-1 \times [C_6H_{6(l)} + \tfrac{15}{2} O_{2(g)} \rightarrow 6CO_{2(g)} + 3H_2O_{(l)} - \Delta H_1]}$$

$$6C_{\text{graphite}(s)} + 3H_{2(g)} - C_6H_{6(l)} + \cancel{6O_{2(g)}} + \cancel{\tfrac{3}{2}O_{2(g)}} - \cancel{\tfrac{15}{2}O_{2(g)}} \rightarrow$$

$$\cancel{6CO_{2(g)}} + \cancel{3H_2O_{(l)}} - \cancel{6CO_{2(g)}} - 3H_2O_{(l)} - 6\Delta H_2 - 3\Delta H_3 + \Delta H_1$$

Remembering that we can transpose chemical species across a chemical equation by changing their sign, just as in any algebraic equation, we have

$$6C_{\text{graphite}(s)} + 3H_{2(g)} \rightarrow C_6H_{6(l)} - (6\Delta H_2 + 3\Delta H_3 - \Delta H_1)$$

Since this equation is the same as the hypothetical formation equation, the ΔH quantities within the parentheses must be — because of the state-function nature of the enthalpy change — identical to the enthalpy of formation of benzene:

$$\Delta H_f(C_6H_6) = 6\Delta H_2 + 3\Delta H_3 - \Delta H_1$$

The observation that chemical reactions and their associated heat flows can be combined algebraically to yield the same heat flow as that of the combined reaction was made in the middle of the nineteenth century by G. H. Hess. To restate **Hess' law**:

> The enthalpy change of a chemical reaction or physical process is independent of whether it is carried out in one step or many.

Of course, this is only another way of saying that enthalpy is a state function. Table 9-2 at the end of the chapter gives ΔH_f (and other values we shall discuss later) for a number of common species.

The Calculation of Enthalpy Changes

With enthalpies of formation established, either directly or via Hess's law, we can calculate the enthalpy change for any reaction simply by subtracting the total enthalpy of formation of the reactants from the total enthalpy of formation of the products:

$$\Delta H = \Sigma \Delta H_f(\text{products}) - \Sigma \Delta H_f(\text{reactants})$$

For example,

$$CH_3CN_{(l)} + H_2O_{(l)} \rightarrow CH_3CONH_{2(s)} \quad -\Delta H_{\text{hydrolysis}}$$

$\Delta H_f = 12.7$ kcal/mole $\Delta H_f = -68.3$ kcal/mole $\Delta H_f = -76.6$ kcal/mole

$$\Delta H_{\text{hydrolysis}} = -76.6 - (12.7 - 68.3) = -21.0 \text{ kcal/mole reaction}$$

This reaction, the hydrolysis of acetonitrile to acetamide, is a fairly typical example of ordinary chemical reactions in terms of heat flow. We observe from the fact that ΔH is negative for the reaction that it must be liberating heat rather than absorbing it. A chemical reaction that liberates heat is **exothermic**; most spontaneous reactions are exothermic. Not all are, however:

$$Li_2CO_{3(s)} \rightarrow Li_2O_{(s)} + CO_{2(g)} \quad -\Delta H$$

$\Delta H_f = -290.5$ kcal/mole $\Delta H_f = -142.4$ kcal/mole $\Delta H_f = -94.1$ kcal/mole

$$\Delta H = (-142.4 - 94.1) - (-290.5) = 54.0 \text{ kcal/mole reaction}$$

The positive value of ΔH for the decomposition of lithium carbonate tells us that the reaction absorbs heat—that it is **endothermic**. Endothermic reactions usually have ΔH values smaller than 100 kcal/mole. They are somewhat rarer than exothermic reactions, which can have ΔH values as large as -400 kcal/mole (of product). The reasons for this will become apparent in Section 9-5.

The Measurement of Enthalpy of Combustion

It has been tacitly suggested in the benzene example that most enthalpies of formation are determined by the use of Hess' law in combining the reactions of elements or compounds with oxygen. This is indeed the case, because most elements and compounds will react conveniently with oxygen, either at room temperature or at higher temperatures. The enthalpy changes thus obtained are called **enthalpies of combustion** or **heats of combustion**; they are quite useful both for the establishment of enthalpies of formation and also for the estimation of the energy yield from the burning of fuels. A device called a **bomb calorimeter**, commonly used for heat-of-combustion determinations, is shown in Fig. 9-3. A weighed sample of the substance to be burned is placed in a cup

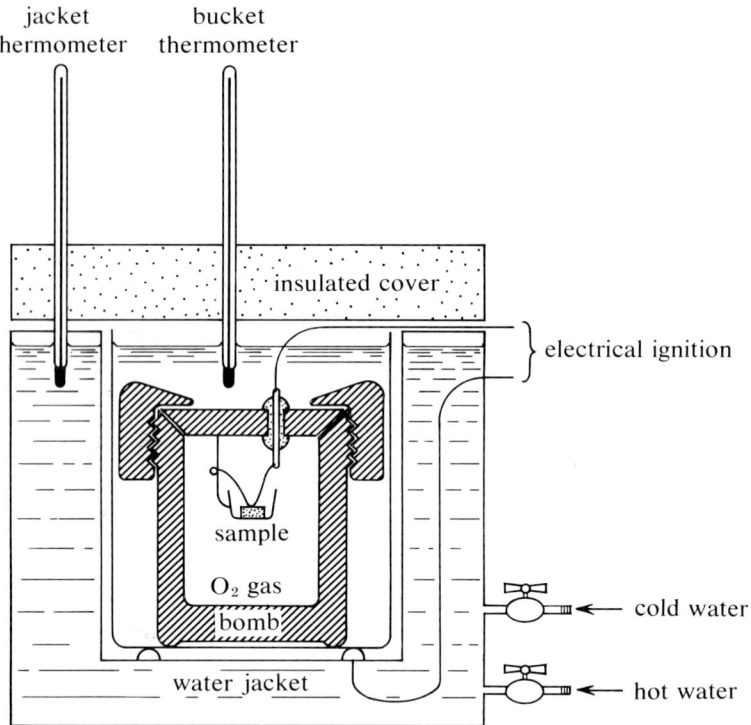

Figure 9-3   Bomb calorimeter for determining heats of combustion.

inside the bomb, with an electrical ignition wire touching it. The bomb is pressurized with pure oxygen gas and placed inside a weighed bucket of water. The bucket containing the bomb is then placed inside a water-jacketed outer container, whose temperature can be adjusted by means of hot- and cold-water taps. The jacket's temperature is adjusted to match that of the bomb and bucket, which is recorded. The sample is ignited, and the resulting reaction releases heat, raising the temperature of the bomb and bucket. Using the hot-water tap on the jacket, the jacket's temperature is raised at the same rate as the bucket's temperature, so no heat flows into or out of the bucket. From the heat capacity of the bomb-and-bucket combination, together with the overall temperature rise, the heat evolved by the chemical reaction can be measured.

This heat flow is not the enthalpy change, however, because it is measured at constant volume (in the bomb), not at constant pressure. At constant volume, the heat flow is equal to ΔE:

$$\Delta E = dQ - dW = dQ - \int P\, dV = dQ - 0 \quad \text{(if } dV = 0\text{)}$$
$$= dQ \quad \text{(constant volume)}$$

Since the enthalpy is defined in terms of the internal energy, however, we need only adjust for the PV change:

$$\Delta H = \Delta E + \Delta(PV)$$

For solids or liquids producing other solids or liquids, the PV change is so small that it may almost always be neglected. For the production or consumption of gases a considerable change may be involved, which is usually approximated by assuming the gases to be perfect and the volume of the solid to be negligible:

$$\Delta(PV) = \Delta(nRT)$$
$$\Delta H = \Delta E + \Delta(n_{gas}RT) = \Delta E + (\Delta n_{gas})RT$$

In other words, the enthalpy change for any reaction is equal to the internal energy change plus RT times the *change* in the number of moles of gas present in the system. Even this change is usually not too large a fraction of the resulting enthalpy change; going back to the burning of benzene to obtain its heat of combustion:

$$C_6H_{6(l)} + \tfrac{15}{2} O_{2(g)} \rightarrow 6CO_{2(g)} + 3H_2O_{(l)}$$

0 moles gas $+ \tfrac{15}{2}$ moles gas $\rightarrow$ 6 moles gas $+ 0$ moles gas

$$\Delta n_{gas} = -\tfrac{3}{2} \text{ moles gas}$$

At 300°K,

$$(\Delta n_{gas})RT = -1.5 \text{ moles} \times 1.987 \text{ cal/mole-°K} \times 300°K$$
$$= -894 \text{ cal} \quad \text{or} \quad -0.894 \text{ kcal}$$

But $\Delta H_{combustion}$ for benzene is -782.3 kcal/mole, which is very large by comparison.

Adiabatic Flame Temperatures

An interesting application of heats of combustion is the calculation of flame temperatures. Suppose we wish to know the temperature of a flame in which butane, C_4H_{10}, is burning in pure oxygen. From the heat of combustion we know the amount of heat being liberated per mole of butane consumed; we assume that the flame is adiabatic, in that it uses all of the released heat in heating the products of the reaction rather than transferring any heat outside the flame. Then for every mole of butane burning, 4 moles of CO_2 and 5 moles of H_2O are produced:

$$C_4H_{10(g)} + \tfrac{13}{2} O_{2(g)} \rightarrow 4CO_{2(g)} + 5H_2O_{(g)} - \Delta H_{combustion}$$

$$\Delta H_{combustion} = -682.5 \text{ kcal/mole reaction}$$

As a crude approximation we may consider the C_p values of CO_2 and water to be constant with temperature, at 10 and 9 cal/deg, respectively. The temperature reached inside the flame will correspond to an increase sufficient to use up all the heat of combustion:

$$C_p \text{(total)} = \frac{\text{heat absorbed}}{\Delta T}$$

$$\Delta T = \frac{\Delta H_{combustion}}{C_p \text{(total)}}$$

$$= \frac{682,500 \text{ cal}}{(4 \times 10) + (5 \times 9) \text{ cal/deg}}$$

$$= 8030 \text{ deg}$$

If the products were initially at room temperature (25 °C), the flame would be expected to achieve a temperature of 8055 °C. However, this is an unrealistic estimate, because at the high temperatures inside flames polyatomic molecules partially atomize (thermal energy is nearly as great as bond energy). As Study Problem 2 shows, if it is assumed that 20% of the CO_2 and H_2O produced atomize to free C, H, and O atoms, the calculated flame temperature is only 1840 °C. And if air is used instead of oxygen to burn the butane, the 4 moles of nitrogen that come along with every mole of oxygen in the atmosphere must also be heated, which increases the total heat capacity of the gases in the flame. Temperatures are thus lower for air flames than for oxygen flames.

Standard States for Solutions: Fugacity and Activity

While we can expect to measure enthalpy changes quite accurately within the framework of thermodynamic theory, there is no way to establish absolute enthalpies. The changes must all be measured from some arbitrarily set zero, which we normally decide on either for conceptual simplicity or for convenience in calculation. The standard used by convention is that already described for enthalpies of formation: the enthalpy of formation of elements in their most stable physical form at 298 °K is defined as zero (solid rhombic sulfur, liquid mercury metal, diatomic oxygen gas at 1 atm, etc.). Since all compounds are formed from elements, enthalpies of formation for all compounds can be established unequivocally with respect to the elements. The standard state for compounds is, again, the most stable physical state at 298 °K, such as the crystalline state for benzoic acid, the liquid state for acetone, and the gaseous state for chlorine trifluoride. Purity is, of course, assumed. A somewhat more difficult situation arises with respect to solutions; since their composition is variable, no one composition can be said to be ideal or "pure." We could elect in the interest of numerical simplicity to take a 1.000-m solution as the standard, and this would be legitimate (since any standard state is chosen arbitrarily). However, as we have seen in Chapter 7, solutions are frequently nonideal and fail to follow Raoult's or Henry's law at any but the most dilute concentrations. If we chose a 1.000-m standard state for a solution, we would frequently find that the properties at other concentrations would bear no reasonable relationship to the properties of the standard state. We need to define a couple of quantities to help us out. In the first place, we have throughout the discussion of thermodynamics been taking gases as being ideal gases, and this is analogous to calling all solutions ideal solutions. At sufficiently low pressures, however, all gases behave as ideal gases (when the molecules are too far apart to have any interaction). So we define a quantity to be used in thermodynamic expressions called the **fugacity**, f, which is an effective pressure for a real gas corresponding to the true pressure for an ideal gas; approaching zero pressure, the fugacity approaches the pressure. The fugacity is that numerical value of the pressure that makes the thermodynamic relationships derived for ideal gases yield values agreeing with experiment. For the solvent in a solution we define another quantity called the **activity**, a, which is the ratio of the fugacity of the solvent in the solution to the fugacity of the pure solvent. Raoult's law says that the vapor pressure (read fugacity) of the solvent in a solution is equal to its mole fraction times the vapor pressure (fugacity) of the pure solvent:

$$a_1 \equiv \frac{f_1(\text{solution})}{f_1^0} \quad \text{and} \quad \left.\begin{array}{l} P_1 = P_1^0 x_1 \\ \text{or} \\ f_1 = f_1^0 x_1 \end{array}\right\} \text{for component 1} \qquad (9\text{-}6)$$

Looking at these two relationships we can see that the activity of the solvent in an ideal solution is just equal to the mole fraction. In a nonideal solution the activity of the solvent is taken as whatever value is necessary to make mole-fraction relationships derived for ideal solutions work for the nonideal solution.

We can express the deviation of the activity from the mole fraction through the **activity coefficient** γ:

$$\gamma_1 \equiv \frac{a_1}{x_1} \qquad \text{or (for molality)} \qquad \gamma_1 \equiv \frac{a_1}{m_1} \qquad (9\text{-}7)$$

The latter is called a practical activity coefficient, because most solution manipulations involve molality rather than mole fraction. The activity coefficient is always 1.000 for an ideal solution, but may be greater or less than 1 for solutions showing positive or negative deviations (respectively) from Raoult's

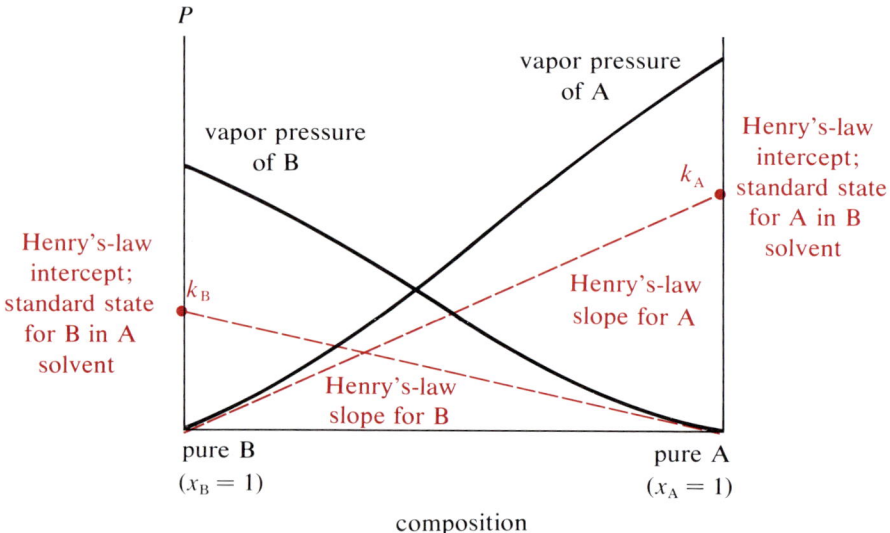

Figure 9-4 Standard states for solutions on a mole-fraction concentration basis.

law. Henry's law is used for the solute just as Raoult's law was for the solvent; the standard state is taken as the intersection of the Henry's-law slope with the concentration axis corresponding to pure solute (see Fig. 9-4 and compare Fig. 7-37). This is not the same as the pressure of the pure solute itself unless the solution is ideal, but it is used because for many solid solutes the solubility limits us to the region of the pressure–composition diagram near where Henry's law applies. The activity of the solute, using this standard state, is the ratio of the vapor pressure of the solute (fugacity) to that of the hypothetical Henry's-law pure solute (represented by the Henry's-law constant k). Moving to practical (i.e., molality-based) activities, the standard state of the solute is entirely analogous: instead of a hypothetical condition corresponding to Henry's-law behavior at $x_2 = 1.000$, the practical standard state is a hypothetical condition, as in Fig. 9-5, corresponding to Henry's-law behavior at $m_2 = 1.000$ — that is, a 1.000-m solution that behaves as if it were infinitely dilute.

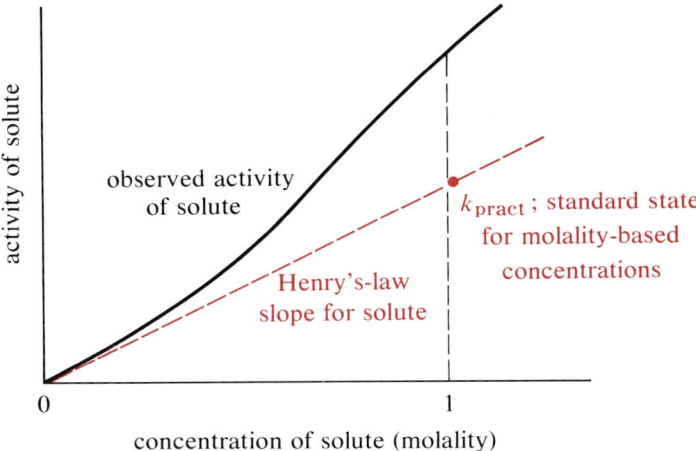

Figure 9-5 Standard state for a solution on a molality basis.

Whatever the student may think of this dodge at this point, it at least has the advantage of allowing us to compare the behavior of any solution with the behavior we might have expected for an ideal solution. We shall need activities shortly, when we consider how concentration changes in a reaction mixture affect the spontaneity of the reaction.

9-4 Enthalpies and Molecular Properties

Although the laws of thermodynamics, by their nature as laws rather than theoretical models, do not require the use of any particular model for the structure of matter, it is worthwhile to try to relate these energy changes to those involved in changing the structures and orbital energies of Chapters 5 and 6. Using either atomic or molecular orbitals for a particular structure, a total electronic energy can be calculated to within the accuracy of the model, which would presumably correspond to an absolute enthalpy for an isolated atom or molecule (at least if considered together with translational, vibrational, and rotational energies for the molecule). But this does not correspond to any quantity in classical thermodynamics, which is unable to establish absolute enthalpies. We can, however, calculate enthalpy changes for chemical reactions by considering the free atoms in their valence electronic state as an arbitrary energy zero, and relating the MO energies of the reactants and products to this zero. Note that this is different from the standard thermodynamic state for elements, which would have been an equally acceptable zero for MO energy comparisons except that the structures of the elements in their standard thermodynamic states are often extremely difficult to get MO energies for. So this is a choice of convenience.

Approximate Enthalpy Changes from Bond Energies

The energies that we want to compare to observed enthalpies, then, are those represented by the resonance integral β between two AO's in a given molecule, since the Coulomb integral α is the energy of the free atom's orbital, which we are taking as a zero point. Refresh your memory from Section 6-4 for definitions of these theoretical concepts. We have seen that a common approximation to the resonance integral between two AO's is to assume that it is proportional to the average energy of the AO's (presumably fixed for any given two orbitals), and also proportional to the overlap integral (presumably fixed *if* the two atoms are at a constant distance from each other and the orbitals are at a constant angle, as for instance pure σ or pure π overlap). So for specified atoms and distances we could hope to define more or less constant bond energies that could be transferred from one molecule to another, a topic that has already come up in Chapter 6 (see pp. 304–306).

As an approximation to enthalpy changes, then, we shall add up the total bond energies of the products and subtract the sum of the bond energies of the reactants:

$$\Delta H = \sum \text{bond } E(\text{products}) - \sum \text{bond } E(\text{reactants})$$

Of course, this is very similar to the Hess'-law addition of enthalpies of formation, using a different zero point for the numbers; the difference is that observed enthalpies of formation appear to be *exactly* additive to within the very small experimental error, while this sum will be good only to within the constancy of the bond energies, which are only moderately reliable. Table 9-1 gives some of the best approximations available for observed bond energies, which are usually inferred from calorimetric data or from vibrational spectroscopy. Using these values, let us look at an example. Suppose we wish the heat of formation of gaseous H_2S:

$$H_{2(g)} + S_{(s, rhombic)} = H_2S_{(g)} - \Delta H_f$$

The bond energies involved in forming the product are $2 \times$ S—H bond energy, or $2(83) = 166$ kcal/mole; this will have a negative sign since it refers to a binding energy relative to the free atoms. The bond energies involved in forming the reactants are, first, the bond energy of diatomic hydrogen gas at 104.2 kcal/mole bond (remember that putting the bond energy in *once* yields *two* atoms), and second, the energy of the bonds, both covalent and van der Waals, holding the S atom in the solid crystal. We need not evaluate the S atom energies

Table 9-1
Bond Energies (kcal/mole)[a]

	H	Li	Be	B	C	N	O	F	Cl	Br	I	S
H	104.2	58	53	70	98.7	93.4	117.5	134	102.2	86.5	70.5	83
Li		25			57		84	137	115	101	81	
Be			17				124	92	109	89	69	
B				69	89	92	128	154	109	90		
C					82.6—	72.8—	85.5—	107	81	65	51	65—
					145.8=	147 =	176 =					128=
					199.6≡	212.6≡						
N						60 —	48 —	65	46	28		
						100 =						
						225 ≡						
O							51 —	45	60	56	56	119=
							118 =					
F								36	60	55	46	68
Cl									57	52	50	61
Br										45	42	52
I											36	
S												54

[a] Principally from Cottrell, T. L., *The Strengths of Chemical Bonds*, London: Butterworths, 1958, 2nd ed.).

specifically, because the experimental heat of sublimation and atomization is available: 53.3 kcal/mole S atoms. So for ΔH_f of gaseous H_2S we have

$$\Delta H_f = \sum \text{bond } E(\text{products}) - \sum \text{bond } E(\text{reactants})$$
$$= 2E_{S-H} - (E_{H-H} + E_S \text{ in crystal})$$
$$= -166 - (-104.2 - 53.3) \text{ kcal/mole}$$
$$= -8.5 \text{ kcal/mole}$$

The experimental value for this enthalpy of formation is -4.8 kcal/mole, so the estimate is not too bad. To take another example with larger numbers, let us calculate the enthalpy of formation of sulfur hexafluoride, SF_6:

$$S_{(s,\,rhombic)} + 3F_{2(g)} \rightarrow SF_{6(g)} - \Delta H_f$$
$$\Delta H_f = \sum \text{bond } E(\text{products}) - \sum \text{bond } E(\text{reactants})$$
$$= 6E_{S-F} - (3E_{F-F} + E_S \text{ in crystal})$$
$$= 6(-68) - 3(-37) - (-53.3)$$
$$= -408 + 111 + 53.3$$
$$= -244 \text{ kcal/mole}$$

The experimental value is -262 kcal/mole, which represents a reasonable agreement although hardly a striking one. Note that all the energies of individual products and reactants are negative, because the standard thermodynamic state of any existing compound is more stable than the free gaseous atoms of its elements. This is not to say that every compound is more stable than other molecules it could form, which is obviously not true; it only says that every compound is more stable than its free gaseous atoms.

The only trouble with this scheme of approximation is that it requires an enormous table of bond-energy values if all the possible compounds are to be dealt with, since presumably all the elements can bond to all the other elements. In Chapter 12 we shall return to this question and see that the scheme can be simplified somewhat further while still yielding useful results.

9-5 Low Energy, High Probability, and Spontaneity

In Chapter 7 we developed the idea that, at least for changes of physical state, the macroscopically observable process that occurs spontaneously is the result of two generally conflicting influences on the system: a high statistical probability of the product state, and a low energy of the product state. In

Chapter 8 we examined the phenomenon of spontaneity and developed the physical property entropy to measure it, where entropy is defined as always increasing in a spontaneous process. On a molecular scale we saw that the entropy of a system can be identified with and even calculated from its level of statistical probability. This left us with two areas of concern. First, can the entropy change of the universe, which is difficult to calculate in most cases, be expressed somehow in terms of properties of the system only in order to make meaningful calculations possible? Second, if the entropy change of a system is equivalent to the change in its degree of probability, what about the low-energy factor? Without it, everything would vaporize spontaneously and no solid or liquid states could even exist. In Sections 9-1 and 9-2 we obtained the enthalpy quantity as a measure of the energy flow in chemical reactions, so it is possible to begin work toward combining the entropy quantity as a measure of high probability, and the enthalpy quantity as a measure of the tendency toward low energy, into a single measure of spontaneity, based on the properties of the system alone.

We can begin by writing the expression that we have said governs the spontaneity of any physical or chemical transformation, that for the entropy change of the universe:

$$\Delta S_{universe} = \Delta S_{system} + \Delta S_{surroundings}$$

In this expression, we can presumably identify ΔS_{system} by reasonable experimental means; it is $\Delta S_{surroundings}$ that is so difficult to deal with. How can we eliminate it, leaving only thermodynamic variables that refer to the system, and at the same time include the tendency of the system toward low energy? By the definition of the entropy change that we have developed,

$$\Delta S_{surroundings} = \int \frac{dQ_{rev}(surroundings)}{T}$$

Now suppose that we can arrange, first, a reversible pathway for the transformation, and second, one that operates at constant pressure. We need only imagine this, not perform it, because since $\Delta S_{surroundings}$ is a state function (of the surroundings, whatever they are), it will depend only on the initial and final states of the surroundings. If we can imagine such a pathway for the transformation, which is easy for most cases, then

$$dQ(surroundings) = -dQ(system)$$

so that

$$\Delta S_{surroundings} = \int \frac{-dQ_{rev}(system)}{T}$$

At constant pressure,

$$dQ = dH$$

or
$$\Delta S_{\text{surroundings}} = \int \frac{-dH_{\text{rev}}(\text{system})}{T}$$

If we further assume an isothermal process,
$$\Delta S_{\text{surroundings}} = -\frac{1}{T} \int dH_{\text{rev}}(\text{system})$$
$$= \frac{-\Delta H_{\text{rev}}(\text{system})}{T}$$

Thus
$$\Delta S_{\text{universe}} = \Delta S_{\text{system}} - \frac{\Delta H_{\text{system}}}{T}$$

or
$$-T \Delta S_{\text{universe}} = \Delta H_{\text{system}} - T \Delta S_{\text{system}}$$

At this point, for special circumstances, we have successfully expressed the entropy change of the universe in terms of the properties of the system alone. Therefore, in future expressions, the subscript "system" will be omitted, and it will be assumed that thermodynamic variables refer exclusively to the system unless specifically indicated otherwise.

The Free-Energy Quantity

In the expression above, the quantity $-T \Delta S_{\text{universe}}$ will always be negative for a spontaneous change, since T (the absolute temperature) is always positive and $\Delta S_{\text{universe}}$ is defined as being positive for a spontaneous change. This expression is so important in its potential use in predicting physical and chemical transformations that we define a new quantity, G,

$$\Delta G \equiv -T \Delta S_{\text{universe}} \tag{9-8}$$

so that
$$\Delta G = \Delta H - T \Delta S \quad \text{or} \quad G = H - TS \tag{9-9}$$

The quantity G has the dimensions of energy and is called the **Gibbs free energy**, or more commonly, just the **free energy**. Since $-TS_{\text{universe}}$ always decreases in a spontaneous change, the free energy always decreases in a spontaneous change (decreasing means becoming more negative). The free-energy change is defined exclusively in terms of thermodynamic state functions of the system, so it is itself a thermodynamic state function of the system alone, but it is also a complete guide to the possibility of spontaneous change by the system. Just as $\Delta S_{\text{universe}}$ is zero for a reversible change, ΔG is zero (since it contains ΔS), and in particular $\Delta G = 0$ at equilibrium. So if we calculate ΔG for some proposed change and it is negative, the change should occur; if it is zero, no change will occur because the system is at equilibrium; if it is positive, the proposed change will not occur, but the reverse of it may (and will if some mechanism for the reverse change exists). In practice, some changes that should occur (ΔG negative) do not at any measurable rate, because no mechanism seems to exist by which the required molecular collisions, and so on, can occur—but no change that should not occur (ΔG positive) has ever been observed. As we have developed it, the free-energy criterion refers exclusively to constant T and P, which seems rather restrictive—but these are precisely the conditions under which most chemical reactions occur, so ΔG will still be quite useful to us.

Notice that the two terms in the expression $\Delta G = \Delta H - T\Delta S$ are, respectively, the low-energy condition (ΔH) and the high-probability condition ($T\Delta S$). The low-energy condition exists because the energy released in moving to a low-energy state of the system is used in heating up the surroundings and increasing their entropy; so even if the entropy of the system should decrease for some transformation, the overall entropy of the universe will increase if the energy yield of the system is converted to a large-enough increase in the surroundings' entropy. Thus, for instance, a liquid freezes when the temperature drops to a point low enough for the entropy increase in the surroundings due to the release of the heat of fusion (ΔH_{fus}) to make up for the unfavorable entropy change in the liquid itself ($T\Delta S_{fus}$). Conversely, a reaction will occur even if it absorbs heat (and lowers the entropy of its surroundings) if its own entropy increases enough. In other words, we have successfully combined the low-energy and high-probability conditions into a single function, while preserving the total entropy of the universe as a measure of spontaneity.

The Free Energy as a Criterion for Spontaneity

Let us see how the free energy of a system changes for a simple example of a physical transformation, more specifically, the one suggested above.

Consider a mole of water that has been supercooled to $-5°C$, or $268°K$. Does the free-energy function predict that it will freeze spontaneously? We have a transformation that we can represent in the form of a chemical equation:

$$-\Delta H_{fus} + H_2O_{(l)} \rightarrow H_2O_{(s)}$$

($-\Delta H_{fus}$ is written on the left because it refers to the transformation going to the left.) We write the expression defining ΔG:

$$\Delta G = \Delta H - T\Delta S$$

For the freezing transformation ΔH is equal to $-\Delta H_{fus}$, if we make the assumption that ΔH does not vary appreciably over a small temperature range. If we make the same assumption about ΔS we can obtain a value for it from the definition of ΔS:

$$\Delta S = \frac{dQ_{rev}}{T} \quad \text{at constant pressure and the temperature at}$$
$$= \frac{-\Delta H_{fus}}{273} \quad \text{which freezing is reversible}$$

Note that we must choose the temperature at which the process is reversible, because the entropy change is defined only for a reversible process. Inserting this in the ΔG expression:

$$\Delta G = \Delta H - T\Delta S$$

$$= -\Delta H_{fus} - 268\left(\frac{-\Delta H_{fus}}{273}\right)$$

$$= -\Delta H_{fus} + \frac{268}{273}(\Delta H_{fus})$$

$$= -\frac{5}{273}\Delta H_{fus}$$

The enthalpy of fusion of any substance is always positive; heat energy must be put in to overcome the interatomic attractions that are maximized in the crystal. So the freezing of water at $-5\,°C$ has associated with it a negative free energy change, indicating that freezing ought to be spontaneous. The same argument would apply for any other liquid supercooled below its normal freezing point. Of course, this corresponds to our experience, provided some means of nucleating the ice crystals exists. The necessity of having a workable mechanism or pathway by which a transformation can occur (nucleation in this case) always limits thermodynamic predictions of spontaneity, since thermodynamics deals only with equilibrium states, not with processes. Again, if ΔG is positive the transformation will *not* occur; but if ΔG is negative the transformation will occur *if* there is a mechanism available. In Chapter 11 we shall look into some of the mechanisms for chemical reactions, developing our ideas of spontaneity still further.

We can calculate the free-energy change for many kinds of physical changes, as Study Problems 3 and 4 suggest. Of more direct interest to us chemically is the calculation of free-energy changes for chemical transformations involving a change of composition. Let us look at a simple example of composition change: the dilution of an ideal solution.

The Dependence of Free Energy on Concentration

Suppose the solution consists of two components, A dissolved in B; the solution is initially $2\,m$ in A and an equal volume of B is added, diluting the solution to $1\,m$:

Low Energy, High Probability, and Spontaneity | 467

$$A_{(2\ m\ in\ B)} + B_{(l)} \rightarrow A_{(1\ m\ in\ B)} - \Delta H_{dil}$$

If the solution is really ideal, ΔH_{dil} will be zero, since the attractions between solvent and solute molecules will be identical to those between solvent and solvent or those between solute and solute. In the expression for ΔG we can thus write

$$\Delta G = \Delta H - T\Delta S$$
$$= 0 - T\Delta S$$

What is ΔS for the dilution? The volume of the solution has doubled, and in terms of the microstates of the system, there are twice as many possible locations for the solute molecules of A. From Chapter 8 we have

$$\Delta S = nR \ln \frac{W_2}{W_1}$$

where W is the number of microstates available to the system. In this case,

$$\Delta S = n_A R \ln \frac{\text{volume}_2}{\text{volume}_1}$$
$$= n_A R \ln 2$$

So the free-energy change for the dilution is given by

$$\Delta G = -T\Delta S$$
$$= -300R \ln 2 \quad \text{for 1 mole}$$
$$= -300 \times 1.987 \times 0.693$$
$$= -413 \text{ cal/mole}$$

This negative value for ΔG tells us that if pure liquid B and an ideal solution of A in B are placed together, mixing and dilution will occur spontaneously. Conversely, it tells us that the reverse reaction — the spontaneous separation of 1 kg of 1-m solution into a 2-m solution and a portion of pure solvent — will not occur, since it would have $\Delta G = +413$ cal/mole. And of course this accords with our experience.

But this expression tells us more than that. Since concentration has dimensions of moles divided by volume, we can write a general expression for the dependence of free energy on concentration (for a dissolved species):

$$\Delta G = -T\Delta S = -TR \ln \frac{\text{volume}_2}{\text{volume}_1}$$
$$= RT \ln \frac{\text{moles A/volume}_2}{\text{moles A/volume}_1}$$
$$= RT \ln \frac{\text{conc}_2}{\text{conc}_1}$$

This applies only to an ideal solution, but for a nonideal solution we can immediately write an analogous expression, since we have already defined solution activities so that the activity of a species in a real solution can be substituted directly for concentration in thermodynamic expressions:

$$\Delta G = RT \ln \left(\frac{a_{\text{final}}}{a_{\text{initial}}} \right)$$

where a_{initial} and a_{final} refer to the activity of the species in which we are interested before and after the concentration change. So the free energy of a substance does depend on its concentration (or activity); we can write

$$G_A = G_A^0 + RT \ln \frac{a_A}{a_A^0}$$

letting G_A^0 be the free-energy content of the standard state of a substance in solution and a_A^0 be its activity in the standard state. But we took the standard state of a solute to be a hypothetical 1.000-m solution following Henry's law, which would make $a_A^0 = 1.000$. So

$$G_A = G_A^0 + RT \ln a_A \qquad (9\text{-}10)$$

for 1 mole of any dissolved species.

Free Energy of Formation and Standard Free-Energy Changes

Equation 9-10 implies that absolute free energies can be established, but this is not true. The free energy is indeterminate because it contains the enthalpy, which is indeterminate. So just as for enthalpies, we arbitrarily establish free energies of formation for compounds in their standard states, relative to the elements in their standard states:

$$\tfrac{1}{2} N_{2(g)} + \tfrac{3}{2} H_{2(g)} \rightarrow NH_{3(g)} - \Delta G_f^0$$

where ΔG_f^0 is the **standard free energy of formation** for ammonia. Since the free energy is a state function, it can be combined in Hess' law just as enthalpies of formation can. So for the purpose of calculation, the free energy of formation can be taken as the absolute free energy to be combined in order to give the free-energy change of a desired reaction:

$$2 CCl_{4(l)} + TiO_{2(s)} \rightarrow TiCl_{4(l)} + 2 COCl_{2(g)} - \Delta G^0$$

ΔG_f^0: $2(-16.4)$ -203 -161.2 $2(-50.3)$

that $\Delta G^0 = \Sigma \Delta G_f^0(\text{products}) - \Sigma \Delta G_f^0(\text{reactants})$

$= -161.2 - 100.6 - (-32.8 - 203.8)$

$= -261.8 + 236.6$

$= -25.2$ kcal/mole reaction

In other words, we can calculate a standard free-energy change (indicated by the superscript zero) for any reaction; in this case ΔG^0 is negative and we expect the reaction to proceed spontaneously, which it does at elevated temperatures. However, this standard free-energy change refers to all substances at unit activity in their standard states, and actually the $TiCl_4$ is being formed as a solution in CCl_4 — in other words, not in its standard state as a pure liquid, and not even as an idealized 1-m solution. For predictive purposes, what we really need is ΔG for the reaction as it actually proceeds, but all we can get from tables such as Table 9-2 (see p. 476) are ΔG^0 values. How can we relate the real ΔG value for a laboratory reaction to the ΔG^0 value we can calculate from tabulated ΔG_f^0 data?

9-6 Free Energy, Equilibrium, and the Equilibrium Constant

Many — perhaps most — of the chemical reactions a chemist runs occur in solution, which makes the relationship of the free-energy change to solution concentrations particularly interesting and useful. In Section 9-5 we have seen that ΔG^0 values for chemical reactions can be calculated, and that the free energy of any individual solution component can be related to its free energy in its standard state. All that is necessary here is to combine these two ideas.

Standard Free-Energy Changes and Real Free-Energy Changes

Suppose we have the following reaction:

$$a\text{A} + b\text{B} \rightarrow c\text{C} + d\text{D} - \Delta G^0$$

in which a moles of A react with b moles of B to give c moles of C product and d moles of D product. From free energies of formation we could, using Hess' law, calculate ΔG^0; but what about ΔG? We can use Eq. 9-10,

$$G_A = G_A^0 + RT \ln a_A$$

$$\Delta G_f(A) = \Delta G_f^0(A) + RT \ln [A]$$

where the free energies of formation are used as the absolute free energies and the square brackets represent activities in solution (concentrations if the solution is ideal). Free energy is an extensive, not intensive, quantity—if we have 2 moles of something we have access to twice as much free energy as if we had only 1 mole. Therefore, for a moles of A, we have

$$a\,\Delta G_f(A) = a\,\Delta G_f^0(A) + aRT\ln[A]$$
$$= a\,\Delta G_f^0(A) + RT\ln[A]^a$$

Of course, analogous expressions can be written for molecules B, C, and D. From Hess' law we have

$$\Delta G^0 = c\,\Delta G_f^0(C) + d\,\Delta G_f^0(D) - a\,\Delta G_f^0(A) - b\,\Delta G_f^0(B)$$

and

$$\Delta G = c\,\Delta G_f(C) + d\,\Delta G_f(D) - a\,\Delta G_f(A) - b\,\Delta G_f(B)$$

Let us substitute in this last equation for $a\,\Delta G_f(A)$ and the other similar quantities:

$$\Delta G = \{c\,\Delta G_f^0(C) + RT\ln[C]^c\} + \{d\,\Delta G_f^0(D) + RT\ln[D]^d\}$$
$$- \{a\,\Delta G_f^0(A) + RT\ln[A]^a\} - \{b\,\Delta G_f^0(B) + RT\ln[B]^b\}$$
$$= \{c\,\Delta G_f^0(C) + d\,\Delta G_f^0(D) - a\,\Delta G_f^0(A) - b\,\Delta G_f^0(B)\}$$
$$+ RT\{\ln[C]^c + \ln[D]^d - \ln[A]^a - \ln[B]^b\}$$

or

$$\Delta G = \Delta G^0 + RT\ln\frac{[C]^c[D]^d}{[A]^a[B]^b} \tag{9-11}$$

It should be clear that with this equation we can calculate ΔG for any existing concentration conditions, since all we need are free energies of formation to get ΔG^0, and the stoichiometry and concentrations. The expression also shows why Le Chatelier's principle works in chemical reactions; if ΔG^0 is slightly positive, so that the reaction should not go, we can make ΔG negative anyway by making the logarithmic term negative. This simply means making the fraction smaller than 1, which means loading the reaction mixture up with reactants A and B while keeping the concentration of products C and D very small.

The Equilibrium Constant

There is another interesting feature of Eq. 9-11. No matter what ΔG^0 is, positive or negative, there is some value of the concentration (activity) fraction that, multiplied by RT, has the same magnitude and the opposite sign. When the

Free Energy, Equilibrium, and the Equilibrium Constant | 471

concentrations meet this condition, we have an interesting relationship:

$$\Delta G = \Delta G^0 + (-\Delta G^0)$$
$$= 0$$

Now $\Delta G = 0$ is the condition for equilibrium, when no further net change will occur spontaneously. So there is a specified condition on the concentration fraction for chemical equilibrium to exist in a solution:

$$RT \ln \frac{[C]^c[D]^d}{[A]^a[B]^b} = -\Delta G^0$$

$$\ln \frac{[C]^c[D]^d}{[A]^a[B]^b} = -\frac{\Delta G^0}{RT}$$

$$\frac{[C]^c[D]^d}{[A]^a[B]^b} = e^{-\Delta G^0/RT}$$

At any given temperature the exponential on the right side of this expression has a fixed, constant value whose numerical magnitude (it has no dimensions) is called the **equilibrium constant**, K_{eq}:

$$K_{eq} = \frac{[C]^c[D]^d}{[A]^a[B]^b} \tag{9-12}$$

This expression is valid *only when all concentrations or activities are at their equilibrium values.* If the concentrations are such as to make the fraction larger than K_{eq}, the reaction will proceed in reverse to reduce [C] and [D] (and increase [A] and [B]) to make the concentration fraction match K_{eq} again. Conversely, if the fraction is smaller, the reaction will proceed forward until the fraction grows to match K_{eq}. We can calculate K_{eq} values very simply by substituting the definition of K_{eq} back into the free energy expression:

$$\Delta G^0 = -RT \ln K_{eq}$$
$$= -2.303 RT \log_{10} K_{eq}$$

If a chemical reaction is written in reverse, the free-energy change associated with it simply has its sign changed, as can be seen from the relationship by which ΔG^0 is calculated:

$$\Delta G^0 = \sum \Delta G_f^0 \text{(products)} - \sum \Delta G_f^0 \text{(reactants)}$$

A little consideration of the derivation of the equilibrium constant will show that the concentration fraction for the reverse reaction is just the inverse of that for the forward reaction:

$$aA + bB \rightleftharpoons cC + dD \qquad cC + dD \rightleftharpoons aA + bB$$

$$K_{\text{forward}} = \frac{[C]^c[D]^d}{[A]^a[B]^b} \quad \text{but} \quad K_{\text{reverse}} = \frac{[A]^a[B]^b}{[C]^c[D]^d}$$

In other words, in an equilibrium-constant expression the concentrations of the products are always in the numerator of the concentration fraction, and those of the reactants in the denominator.

The Relative Importance of Enthalpy and Entropy

In the next chapter we shall take an extended look at equilibrium-constant calculations, and see that many varied calculations are possible. At this point we shall only make some general observations on spontaneity in chemical reactions. Going back to the definition of free energy we see again that it combines the low-energy and high-probability criteria for spontaneity:

$$\Delta G^0 = \Delta H^0 - T \Delta S^0$$

Of these two factors, only the entropy factor is strongly temperature dependent. The enthalpy factor changes with temperature only by the addition of the difference in heat capacities between products and reactants, which is usually not large; Study Problem 5 gives an example of this dependence. The temperature factor in the entropy term, however, means that the spontaneity of reactions will often be quite temperature dependent. A casual examination of the data in Table 9-2 suggests that ΔH^0 values for reactions can be very small but often are in the range -20 to -50 kcal/mole reaction. Similarly, ΔS^0 values can be very small but are often in the range 2–5 cal/deg-mole reaction. Since 1 kcal = 1000 cal, it is clear that the enthalpy term will be the more important at low temperatures, particularly near room temperature (300 °K). This is why most reactions that proceed spontaneously at room temperature are exothermic; a negative ΔH^0 is required to make the free-energy change negative. In a few cases endothermic reactions will actually proceed because of a very large increase in the randomness of the system (ΔS^0). As the temperature is raised, however, the entropy term becomes more and more important, and at very high temperatures endothermic reactions are the rule rather than the exception, because they are very frequently associated with increases in the randomness of the system (see Chapter 8). Of course, this generalization is subject to many exceptions, because everything depends on the relative values of ΔH^0 and ΔS^0 for the particular reaction in which we are interested.

Temperature Dependence of the Equilibrium Constant

We can be a little more specific about the dependence of the equilibrium constant on temperature. We can write

to give
$$-RT \ln K_{eq} = \Delta G^0 = \Delta H^0 - T \Delta S^0$$

$$\ln K_{eq} = \left(\frac{\Delta H^0}{R}\right)\frac{1}{T} + \frac{\Delta S^0}{R}$$

This is a linear equation in $\ln K_{eq}$ and $1/T$, provided that both ΔH^0 and ΔS^0 are independent of temperature. The actual dependence involves only ΔC_p between reactants and products (see Study Problem 5), which is usually small enough to make this quite a good approximation. Let us differentiate this expression to give a relation for the rate of change of K_{eq} with temperature:

$$\frac{d}{dT}(\ln K_{eq}) = \frac{d}{dT}\left(-\frac{\Delta H^0}{R}\frac{1}{T} + \frac{\Delta S^0}{R}\right)$$

$$\frac{d \ln K_{eq}}{dT} = -\frac{\Delta H^0}{R}\left(-\frac{1}{T^2}\right)$$

$$\frac{d \ln K_{eq}}{dT} = \frac{\Delta H^0}{RT^2}$$

A more elegant derivation would show that this is true even if ΔH^0 and ΔS^0 are temperature dependent. From this relationship we can see that K_{eq} will increase with temperature if ΔH^0 is positive (endothermic) and decrease with rising temperature if ΔH^0 is negative (exothermic), which is what Le Chatelier's principle would tell us. Often, however, we do not need the rate of change of the equilibrium constant, but rather the value of the constant itself at a different temperature from the one for which a value is tabulated, normally 298 °K. If we call the tabulated value K_{298} and the other value we need K_T, we can write an expression for the difference between the two $\ln K$ values:

$$\ln K_T - \ln K_{298} = -\frac{\Delta H^0}{RT} + \frac{\Delta S^0}{R} - \left(-\frac{\Delta H^0}{R(298)} + \frac{\Delta S^0}{R}\right)$$

$$\ln \frac{K_T}{K_{298}} = -\frac{\Delta H^0}{R}\left(\frac{1}{T} - \frac{1}{298}\right)$$

For the completely general case involving two temperatures T_1 and T_2, with constants K_1 and K_2, the expression reads

$$\ln \frac{K_2}{K_1} = -\frac{\Delta H^0}{R}\left(\frac{1}{T_2} - \frac{1}{T_1}\right) \tag{9-13}$$

Notice that the linear relation between $\ln K_{eq}$ and $1/T$ often allows us to evaluate ΔH^0 and ΔS^0 for reactions. If we measure concentrations to establish the value of the equilibrium constant at several different temperatures, we can plot the result as in Fig. 9-6. The slope of the straight line is $-\Delta H^0/R$, and its intercept with the $\ln K_{eq}$ axis is $\Delta S^0/R$. The extrapolation may be done graphi-

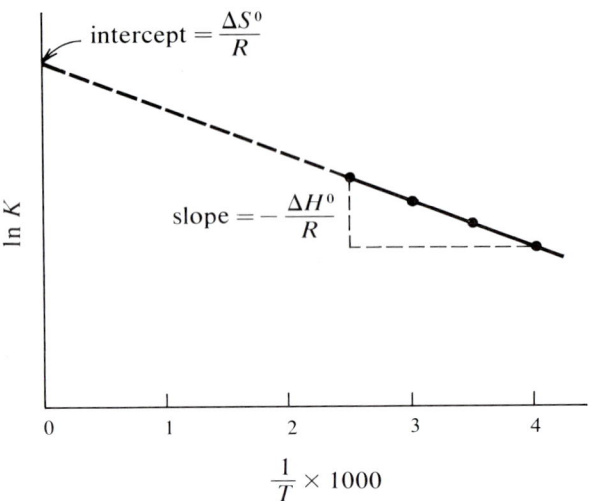

Figure 9-6 Establishing thermodynamic values from equilibrium-constant measurements at different temperatures.

cally to obtain these values, or curve-fitting procedures may be used, such as the **least-squares** procedure, which establishes the values of the slope and intercept of a straight line giving the smallest deviations of the experimental points from a true straight line.

9-7 Summary — The Laws of Thermodynamics

In the past two chapters we have been trying to get at the principles governing spontaneous chemical change. The problem can be resolved into two questions: What can and what cannot be done? How can we do it? Thermodynamics deals only with the first of these questions, but it deals with it fully and accurately; there are no exceptions to the thermodynamic laws we have developed. The second question deals with the detailed mechanisms for physical and chemical change, which we shall discuss in Chapter 11. In Chapter 8 we surveyed spontaneous interaction, then showed that it was entirely compatible with the molecular model for the structure of matter that had already been proposed. It did not appear to apply explicitly to the low-energy criterion that had been applied to electronic transitions in the discussion of molecular structure, however, and in this chapter we have examined the energy relationships that apply to chemical reactions. In proposing the internal-energy state function to eliminate the process dependence of the heat and work quantities, we have in effect stated the *first law of thermodynamics*:

the energy of the universe is constant.

In developing the free-energy function as a single criterion of spontaneity, we combined the energy relation with the probability relation, entropy, from Chapter 8. The definition of entropy in empirical terms allows us to state the *second law of thermodynamics*:

the entropy of the universe continually increases.

From the molecular (statistical) interpretation of entropy we made an interpretation that proved compatible with observed entropies of substances. The interpretation is called the *third law of thermodynamics*:

an ideal crystal would have zero entropy at 0 °K, but 0 °K is experimentally unattainable.

The free-energy function satisfactorily combines the tendency toward states of high probability with the tendency toward states of low energy, and leads further toward the equilibrium constant as a useful shorthand application of the laws of thermodynamics to chemical reaction mixtures. In Chapter 10 we shall take an extended look at the uses of equilibrium constants. The discussion will give us a chance to introduce one of the most fundamental chemical interactions, that between acids and bases. Also in the context of the equilibrium-constant discussion, we shall see that it is possible to measure some free-energy changes directly by electrical means, which will serve as an introduction to electrochemistry. It should be clear that we are at last in a position to discuss with some thoroughness the principles of at least some aspects of chemical dynamics.

Ever since the formulation of the laws of thermodynamics in the nineteenth century their philosophical implications have fascinated chemists and physicists. If entropy is to be correlated with disorder, how is it possible for the fantastically ordered structure of a living cell to arise from the initial chaos? Does the universe face an "entropy death" in which it has uniformly achieved maximum disorder and nothing else can happen? How do error and uncertainty in communicating messages correlate with the physical disorder of entropy? Several good books are available on these and related topics, and the philosophically inclined student may enjoy some or all of them from the bibliography below. As a closing note the author will quote a cynical version of the three laws of thermodynamics (which is not original):

1. You can't win.
2. You can't break even.
3. You can't get out of the game.

Table 9-2
Standard Enthalpies and Free Energies of Formation (298 °K)[a]

Compound or Atom	ΔH_f^0 (kcal/mole)	ΔG_f^0 (kcal/mole)
$AlCl_{3(s)}$	−166.2	−152.2
$Al_2O_{3(s)}$	−399.1	−376.8
$AsCl_{3(l)}$	−80.2	−70.5
$BCl_{3(l)}$	−100.0	−90.6
$B_2O_{3(s)}$	−297.6	−280.4
$BaCl_{2(s)}$	−205.6	−193.8
$BaO_{(s)}$	−133.4	−126.3
$BeO_{(s)}$	−146.0	−139.0
$CaCO_{3(s)}$	−288.5	−269.8
$CaCl_{2(s)}$	−190.0	−179.3
$CaO_{(s)}$	−151.9	−144.4
$CoCl_{2(s)}$	−77.8	−67.5
$CoO_{(s)}$	−57.2	−51.0
$CrCl_{3(s)}$	−134.6	−118.0
$Cr(OH_2)_6Cl_{3(s)}$	−579.0	
$Cr_2O_{3(s)}$	−269.7	−250.2
$CrO_{3(s)}$	−138.4	
$CuCl_{(s)}$	−32.5	−28.2
$CuCl_{2(s)}$	−49.2	
$FeCl_{2(s)}$	−81.5	−72.2
$Fe_2O_{3(s)}$	−196.5	−177.1
$HCl_{(g)}$	−22.06	−22.77
$H_2O_{(g)}$	−57.80	−54.64
$H_2O_{(l)}$	−68.32	−56.69
$HNO_{3(l)}$	−41.4	−19.1
$KCl_{(s)}$	−104.2	−97.6
$MgCl_{2(s)}$	−153.4	−141.6
$MgO_{(s)}$	−143.8	−136.1
$NH_{3(g)}$	−11.04	−3.98
$NH_4Cl_{(s)}$	−75.4	−48.7
$NO_{(g)}$	21.60	20.72
$NO_{2(g)}$	8.09	12.39
$N_2O_{5(s)}$	−10.0	
$NaCl_{(s)}$	−98.2	−91.8
$NiCl_{2(s)}$	−75.5	−65.1
$NiO_{(s)}$	−58.4	−51.7
$PbCl_{2(s)}$	−85.9	−75.0

Table 9-2 (Continued)

Compound or Atom	ΔH_f^0 (kcal/mole)	ΔG_f^0 (kcal/mole)
$PbO_{(s)}$	−52.4	−45.2
$SOCl_{2(l)}$	−49.2	
$SO_{2(g)}$	−71.0	−71.8
$SiCl_{4(l)}$	−153.0	−136.9
$SiO_{2(s)}$	−205.4	−192.4
$SnCl_{4(l)}$	−130.3	−113.3
$SrCl_{2(s)}$	−198.0	−186.7
$SrO_{(s)}$	−141.1	−133.8
$TiCl_{4(l)}$	−179.3	−161.2
$TiO_{2(s)}$	−218.0	−203.8
$ZnCl_{2(s)}$	−99.4	−88.3
$ZnO_{(s)}$	−83.2	−76.1
carbon tetrachloride, $CCl_{4(l)}$	−33.3	−16.4
chloroform, $CHCl_{3(l)}$	−31.5	−17.1
dichloromethane, $CH_2Cl_{2(l)}$	−28	−15.1
methyl alcohol, $CH_3OH_{(l)}$	−57.0	−39.7
ethyl alcohol, $CH_3CH_2OH_{(l)}$	−66.4	−41.8
acetone, $(CH_3)_2C{=}O_{(l)}$	−60.5	−37.2
methane, $CH_{4(g)}$	−17.9	−12.1
ethane, $CH_3CH_{3(g)}$	−20.2	−7.9
ethylene, $CH_2{=}CH_{2(g)}$	12.5	16.3
acetylene, $CH{\equiv}CH_{(g)}$	54.2	50.0
benzene, $C_6H_{6(l)}$	19.8	31.0
carbon monoxide, $CO_{(g)}$	−26.4	−32.8
carbon dioxide, $CO_{2(g)}$	−94.0	−94.3
H	52.1	48.6
Li	37.1	29.2
C	171.7	160.9
N	112.5	108.4
O	59.2	55.0
F	18.3	14.2
Cl	29.0	25.2
Br	26.7	19.7
I	25.5	16.8
S	53.2	43.6

^aFrom NBS Circular 500.

Study Problems

1. Obtain ΔS for process 1 in Fig. 9-1, considering it as a cooling to T_c followed by a warming to T. Obtain ΔS for process 2 in the same general way. What must be the relationship between T, T_c, and T_h if the two ΔS values are the same (as they must be since S is a state variable and the two states are the same)? Can $T_h - T$ equal $T - T_c$?
2. Using the enthalpies of formation in Table 9-2, show that every mole of CO_2 that atomizes uses up 384.0 kcal, and every mole of H_2O that atomizes uses up 221.1 kcal. If 20% of each product from the burning of 1 mole of butane atomizes, how many moles of each product atomize? How much heat is removed from the flame by this atomization, and how much remains from the combustion reaction? To what temperature does this correspond (ignoring the change in C_p due to atomization)?
3. If an ideal gas expands isothermally from V_1 to V_2 with its internal energy remaining constant, show from the definition of ΔG and the entropy relationships of the previous chapter that

$$\Delta G = -nRT \ln \frac{V_2}{V_1}$$

What does this say about the possibility of a spontaneous compression of a gas?
4. What is the free-energy change for the vaporization of 1 mole of liquid at its normal boiling point? Why? Show that vaporization of a liquid at a temperature above its boiling point is always spontaneous.
5. What is ΔH for the heating of 1 mole of butane, C_4H_{10}, from 298 °K to 2100 °K, if its heat capacity is $C_p = 0.47$ cal/g-°K; for heating 1 mole of O_2, with $C_p = 0.22$ cal/g-°K, the same amount; for heating 1 mole of CO_2, with $C_p = 0.25$ cal/g-°K; for heating 1 mole of H_2O vapor, with $C_p = 0.50$ cal/g-°K? For the reaction

$$C_4H_{10} + \tfrac{13}{2} O_2 \rightarrow 4CO_2 + 5H_2O \qquad \Delta H_{298} = -682.5 \text{ kcal}$$

write an expression for ΔH^0_{298} in terms of ΔH_f^0 values at 298 °K. How do the ΔH_f^0 values change for the formation of the species at 2100°K? Show that ΔH^0_{2100} for the reaction is -679.6 kcal. How much effect would this have on the flame temperature calculated on p. 456 and in Study Problem 2?
6. K_w is the equilibrium constant for the reaction

$$H_2O + H_2O \rightleftharpoons H_3O^+ + OH^-$$

At 0°C, $K_w = 1.36 \times 10^{-15}$, while at 100°C, $K_w = 5.50 \times 10^{-13}$. Calculate ΔG^0 and ΔS^0 for the reaction. Is the liquid becoming more or less orderly as the reaction proceeds to the right? What could be causing this?
7. Write a balanced equation for the reaction of ethylene with hydrogen gas to give ethane. What is the equilibrium constant for this reaction at room temperature? What is the equilibrium constant at 1200°C, assuming that the enthalpy and entropy

changes remain the same at the higher temperature? Why does this substantial change in the equilibrium constant occur?

8. The compound 1,2-dichloroethane, $CH_2Cl\text{—}CH_2Cl$, always has the six ligand atoms arranged as shown in Fig. 9-7a. The other possibility, shown in Fig. 9-7b, is not observed because of interatomic repulsion (represented by arrows). Viewed along the C—C axis, the form that does exist can have three specific conformations, as shown in Fig. 9-7c, where "anti" refers to the Cl atoms being opposite each other. We expect the gauche forms to be of somewhat higher energy, since the Cl—Cl repulsion will be more important. We can expect to see an equilibrium established:

$$C_2H_4Cl_{2(anti)} \rightleftharpoons C_2H_4Cl_{2(gauche)}$$

At room temperature, which may be taken as 300°K, experimental measurements show that 80% of the $C_2H_4Cl_2$ is in the anti conformation. Set up the equilibrium-constant expression for the reaction above and calculate the value of the constant and the free-energy change. Taking into account the fact that there are two gauche conformations but only one anti, calculate ΔS^0 for the interconversion. Finally, calculate ΔH^0 for the interconversion. Which of these quantities best corresponds to a measure of the repulsion between gauche Cl's?

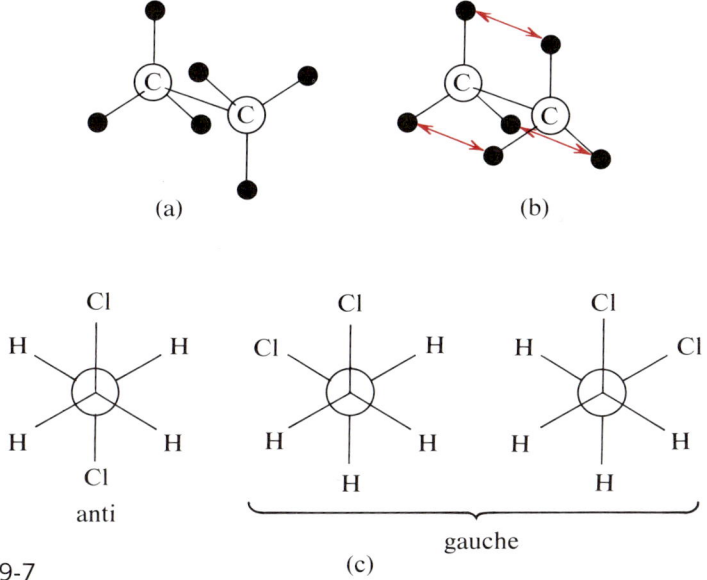

Figure 9-7

9. An interesting proposal for the reduction of air pollution involves having power plants burn ammonia, since the combustion products are N_2 and water vapor. Write a balanced equation for this reaction, and calculate ΔH^0 per gram of ammonia burned. Compare this value with that for the burning of ethane, a component of natural gas.

10. Iron is produced by the reaction of Fe_2O_3 with elemental carbon (coke) at high temperatures to give molten iron and CO_2. Should this reaction be spontaneous at

25 °C? Calculate ΔG^0, ΔH^0, and ΔS^0 at that temperature (but assuming formation of liquid iron). If ΔH^0 and ΔS^0 are taken as constant over the necessary temperature range, at what temperature would ΔG equal zero for the reaction (the break-even point)? ΔH_{fus} for Fe is 3.6 kcal/mole.

11. When anhydrous $CrCl_3$ is added to water it shows no sign of reacting to give the hydrate. Is this because the reaction is thermodynamically forbidden or because no mechanism exists for the reaction?

Some Further Reading

Mahan, B. H., *Elementary Chemical Thermodynamics*, New York: Benjamin, 1964. Covers all the material from this and the last chapter in a very classical and precise fashion, but is easy to read. The approach is quite different, and it is difficult to see what the goal is until you are well along; nevertheless, it is the approach most commonly used.

Klotz, I. M., *Introduction to Chemical Thermodynamics*, New York: Benjamin, 1964. This is somewhat more mathematically rigorous than Mahan, but covers much the same ground. A standard treatment, very solid and lucid but more formal than either Mahan or our discussion.

Blum, H. F., *Time's Arrow and Evolution*, New York: Harper & Row, 1962. The biological implications of the increasing entropy of the universe; the problem of the origin of highly ordered biological molecules and cells. Somewhat outdated by the rapid pace of biochemical research but still very much worthwhile.

Pierce, J. R., *Symbols, Signals, and Noise*, New York: Harper & Row, 1961. A rather informal approach to communication theory and message content. Communication is viewed as the removal of uncertainty by a message; the degree of uncertainty is the "information entropy." The parallels to thermodynamic entropy are sometimes surprising.

10 | Equilibrium Constants and Electrochemistry

The maintenance of solution equilibrium is, in the most literal sense, vital to all of us. Man consists largely of an aqueous solution contained by membranes, and the proper functioning of, for instance, enzymes requires close control of the acidity equilibrium in this solution. For this and many other reasons we shall examine the processes of equilibrium in aqueous solutions. In so doing, however, we must not forget that we have still not attempted to deal with the detailed mechanism of a chemical reaction, nor with the speed with which it reaches its equilibrium; these we defer until Chapter 11. The absence of a satisfactory mechanism can cause a thermodynamically favorable reaction not to occur, or to occur so slowly that it is very difficult to tell when equilibrium has been reached. Fortunately, this difficulty rarely arises for reactions involving ions in aqueous solution. Most ionic reactions, because of the strong Coulomb attraction between opposite charges, occur very rapidly. The collisions between molecules that constitute the mechanism of the reaction are enhanced by this attraction, and the rate of the reaction is accordingly so great that we need not worry about the establishment of equilibrium—it exists as soon as we mix the solution. Thus we need not be limited by our present ignorance of reaction mechanisms. To begin our consideration of chemical-reaction equilibrium, however, let us look at a simple model system involving not solutions, but gases—it is simple because gas mixtures are usually much more nearly ideal gases than aqueous mixtures are ideal solutions. The principles we develop can be transferred readily to ions in solution, where activities may not correspond very well to concentrations.

10-1 Equilibrium Constants for Gaseous Reactions

In the last chapter we developed the concept of the equilibrium constant for a liquid solution; our first concern now will be to show that an analogous expression applies to reactions involving gaseous mixtures. The first step in the derivation for liquids was to obtain an expression for the change in free energy when a solution was diluted; we now ask what the change is when a given quantity of gas expands to a new volume in being diluted by another gas.

Free-Energy Changes on Gas Dilution

For the expansion process we calculate ΔG by starting with its definition:

$$\Delta G = \Delta H - T\Delta S$$

Inserting the definition of enthalpy, we have

$$\Delta G = \Delta E + \Delta(PV) - T\Delta S$$

For ideal gases the internal energy is determined only by the temperature; if the expansion is isothermal $\Delta E = 0$ and $\Delta(PV) = \Delta(nRT) = 0$. For the entropy change on expansion we have from Chapter 8 the expression

$$\Delta S = nR \ln \frac{V_2}{V_1}$$

and inserting all these values in the free-energy expression we get

$$\Delta G = 0 + 0 - TnR \ln \frac{V_2}{V_1}$$
$$= -nRT \ln \frac{V_2}{V_1}$$

This is exactly like the expression derived on p. 467 for a single component of a liquid solution, so we can proceed to write the subsequent expression (see p. 468):

$$G_A(\text{gaseous}) = G_A^0(\text{gaseous}) + RT \ln \frac{P_A}{1 \text{ atm}}$$

or

$$G_A = G_A^0 + RT \ln P_A$$

For a generalized gaseous reaction

$$a\text{A}_{(g)} + b\text{B}_{(g)} \rightleftharpoons c\text{C}_{(g)} + d\text{D}_{(g)}$$

a derivation like that in Chapter 9 can be pursued, eventually giving a dimensionless gas equilibrium constant:

$$K_P = \frac{P_C{}^c P_D{}^d}{P_A{}^a P_B{}^b} \qquad (10\text{-}1)$$

The constant is dimensionless because all the pressures in the fraction represent the dimensionless ratio of the experimental pressure to the standard state of pressure, 1 atm. So in working with gas equilibrium problems we must remember to express the pressures in atmospheres or to write each pressure as a ratio to a standard-state pressure.

Another point to remember is that any equilibrium-constant expression, whether gaseous or solution, is strongly dependent on the way we write the equation for the reaction. In Chapter 1 we noted that an equation can be balanced in more than one way, depending on whether we wish integral numbers of moles to appear in the coefficients or whether we wish the equation to represent the formation of a single mole of product. Take the gaseous reaction

$$N_{2(g)} + 3H_{2(g)} \underset{}{\overset{K_1}{\rightleftharpoons}} 2NH_{3(g)}$$

which can be written equally well as

$$\tfrac{1}{2}N_{2(g)} + \tfrac{3}{2}H_{2(g)} \underset{}{\overset{K_2}{\rightleftharpoons}} NH_{3(g)}$$

All the coefficients in the second equation have been divided by 2; the effect this has on the equilibrium-constant expressions is apparent:

$$K_1 = \frac{(P_{NH_3})^2}{(P_{N_2})(P_{H_2})^3} \qquad K_2 = \frac{(P_{NH_3})}{(P_{N_2})^{1/2}(P_{H_2})^{3/2}} = \left[\frac{(P_{NH_3})^2}{(P_{N_2})(P_{H_2})^3}\right]^{1/2} = K_1^{1/2}$$

In other words, dividing the coefficients by 2 means taking the square root of the equilibrium constant. If the reaction is written in reverse, $K_{\text{reverse}} = 1/K_{\text{forward}}$. So *it is crucial to specify at the very beginning of any equilibrium problem what the balanced chemical equation is for the equilibrium.*

The Determination of K for the H₂/I₂/HI System

Let us start by looking at what has become a classic example of a careful experimental determination of an equilibrium constant: the hydrogen/iodine/hydrogen iodide equilibrium determination by Taylor and Crist [*J. Amer. Chem. Soc.* **63**, 1377 (1941)]. They used the balanced equation

$$2HI_{(g)} \rightleftharpoons H_{2(g)} + I_{2(g)}$$

for which the equilibrium-constant expression is

$$K_p = \frac{P_{H_2} P_{I_2}}{(P_{HI})^2}$$

They approached the equilibrium from both directions; that is, they allowed H_2 and I_2 to combine, forming HI, and they also allowed HI to decompose, forming H_2 and I_2. The equilibrium was established at relatively high temperatures by placing sealed quartz bulbs in a thermostated furnace, each bulb containing the starting materials. The equilibrium was "quenched" by chilling each bulb in ice water; then the bulb was opened under water and the equilibrium I_2 and HI content analyzed by taking advantage of the following quantitative (titration) reactions:

$$I_{2(aq)} + 2S_2O_{3(aq)}^{2-} \rightarrow 2I_{(aq)}^- + S_4O_{6(aq)}^{2-}$$

$$HI_{(aq)} + H_2O_{(l)} \rightarrow H_3O_{(aq)}^+ + I_{(aq)}^- \qquad H_3O_{(aq)}^+ + OH_{(aq)}^- \rightarrow 2H_2O_{(l)}$$

In decomposition reactions the number of moles of iodine formed is the same as the number of moles of hydrogen gas formed, by the stoichiometry of the reaction, so analysis for H_2 was unnecessary. In combination reactions, the H_2 content at equilibrium is determined by accurately measuring the starting H_2 content and subtracting the H content of the HI formed. Since the temperatures were quite high (600–800°K) and the pressures were well below 1 atm, the gases behaved very nearly ideally.

Table 10-1 gives Taylor and Crist's results, expressed not in atmospheres but in moles/l; these could be corrected to atmospheres by using a conversion factor, but since the factor would appear twice in the numerator and twice in the denominator of the equilibrium-constant expression, it would cancel and we need not multiply the values out explicitly. The results are very consistent,

Table 10-1
Hydrogen Iodide/Hydrogen/Iodine Equilibrium[a]

$2HI \rightleftharpoons H_2 + I_2$		(763.8°K)		
		(pressures expressed in moles per liter)		
HI decomposition				
$[I_2]_{equil}$	$[H_2]_{equil}$	$[HI]_{equil}$	$K_{eq} = [H_2][I_2]/[HI]^2$	
0.25792×10^{-2}	0.25792×10^{-2}	1.7632×10^{-2}	2.167×10^{-2}	
0.18961×10^{-2}	0.18961×10^{-2}	1.2835×10^{-2}	2.180×10^{-2}	
0.19806×10^{-2}	0.19806×10^{-2}	1.3417×10^{-2}	2.168×10^{-2}	
0.24237×10^{-2}	0.24237×10^{-2}	1.6406×10^{-2}	2.172×10^{-2}	
$H_2 + I_2$ combination				
$[H_2]_{start}$	$[HI]_{equil}$	$[H_2]_{equil}$	$[I_2]_{equil}$	K_{eq}
1.0851×10^{-2}	1.7151×10^{-2}	0.22754×10^{-2}	0.28396×10^{-2}	2.185×10^{-2}
1.0833×10^{-2}	1.7804×10^{-2}	0.19350×10^{-2}	0.36341×10^{-2}	2.203×10^{-2}
1.0613×10^{-2}	1.7788×10^{-2}	0.17207×10^{-2}	0.40569×10^{-2}	2.193×10^{-2}
1.0645×10^{-2}	1.6900×10^{-2}	0.21930×10^{-2}	0.28702×10^{-2}	2.188×10^{-2}

[a]Taylor and Crist, *J. Amer. Chem. Soc.* **63**, 1377 (1941); see also Bodenstein, *Z. Phys. Chem.* **13**, 56 (1894).

and the direction from which equilibrium is approached does not matter. The calculation of K is made quite simply by multiplying the equilibrium concentration (pressure) of H_2 by that of I_2 and dividing the product by the square of the pressure of HI; for the experiment represented by the top line of Table 10-1 we have

$$K_P = \frac{(0.25972 \times 10^{-2} \text{ mole/l})(0.25972 \times 10^{-2} \text{ mole/l})}{(1.7632 \times 10^{-2} \text{ mole/l})^2}$$
$$= 0.02167$$

Problems in Which One Component Is Present in Large Amounts

In working with any equilibrium-constant problem, the first step is to write the applicable chemical reaction equation and balance it. Next, write the equilibrium-constant expression that is applicable to the reaction equation. Then decide what information—what numbers—are available to plug into the equilibrium-constant expression. Sometimes the object will be the value of the equilibrium constant itself, in which case values must be ascertained for all of the concentrations or pressures in the expression; more often, the value of the constant will be known or can be looked up, and one of the pressures or concentrations will be the desired quantity. Occasionally all of the necessary constants or concentrations will be available directly, but usually they will not and it will be necessary to take advantage of the symmetry in concentrations provided by the stoichiometry of the reaction as it occurs. For example, in the case discussed above, when the reaction consisted of pure HI dissociating it was possible to assume that the pressure of H_2—although unmeasured—was equal to the pressure of I_2, because the only possible source of H_2 gave exactly one molecule of H_2 for every molecule of I_2 it yielded. This sort of substitution is valid only for reactions in which there is only one component present initially, or in which two or more components have been mixed in exactly their stoichiometric mole ratio. Let us look at a few examples of this sort of calculation.

A 2-l bulb is filled at 77°C with 0.100-atm nitrosyl bromide, NOBr, which then dissociates to NO and Br_2. The equilibrium pressure of Br_2 is 0.0342 atm. What is the value of the equilibrium constant at that temperature?

Writing a balanced equation for the dissociation, we have

$$2NOBr_{(g)} \rightleftharpoons 2NO_{(g)} + Br_{2(g)}$$

The equilibrium-constant expression for this reaction equation is

$$K_P = \frac{(P_{NO})^2 (P_{Br_2})}{(P_{NOBr})^2}$$

There does not seem to be enough information available to solve for K_P, because two equilibrium pressures are unspecified; this is where the reaction stoichiometry will help us. Every time the reaction forms a molecule of Br_2, it forms *two* molecules of NO, so the pressure of NO present must always be twice the pressure of Br_2: $P_{NO} = 0.0684$ atm. Notice that this is only true if there is no other source of NO or Br_2 present. To go on, every time a Br_2 molecule forms two NOBr molecules must disappear, so the NOBr pressure *decreases* by twice as much as the Br_2 pressure *increases*: $P_{NOBr} = 0.100 - (2 \times 0.0342) = 0.0316$ atm. So the stoichiometry of the reaction provides the additional information necessary, and we can now substitute directly in the equilibrium-constant expression:

$$K_P = \frac{(0.0684)^2(0.0342)}{(0.0316)^2} = \frac{(4.68 \times 10^{-3})(3.42 \times 10^{-2})}{9.99 \times 10^{-4}}$$

$$= 1.60 \times 10^{-1} \quad \text{or} \quad 0.160$$

To make the necessary substitutions in this equilibrium-constant expression we had to have three numbers, of which only one was given directly. The other two numbers were specified indirectly by two conditions of the equilibrium: the stoichiometry of the reaction and the initial pressure of the reactant. This sort of situation will arise frequently in equilibrium-constant problems, and the use of the reaction stoichiometry in an imaginative fashion is necessary.

Mole Fractions and Fraction Reacted

A constituent of natural gas, ethylene (C_2H_4), can be made to react with water vapor to yield gaseous ethanol, C_2H_5OH. At 400°K the equilibrium constant for the reaction written to produce 1 mole of ethanol is 0.306. What percentage of the ethylene is converted to ethanol if a mixture of ethylene and water having exactly stoichiometric proportions is allowed to reach equilibrium at 10 atm total pressure and 400°K?

We are being asked to calculate a mole quantity, not a pressure quantity. This means that we shall need to work with mole fractions and Dalton's law of partial pressures: $P_a = X_a P_{total}$. Our first step is to write the balanced equation for the reaction:

$$C_2H_{4(g)} + H_2O_{(g)} \rightleftharpoons C_2H_5OH_{(g)}$$

This is the correct equation for the equilibrium constant since it produces 1 mole of ethanol. For this equation the appropriate equilibrium-constant expression is

$$K_P = 0.306 = \frac{P_{C_2H_5OH}}{P_{C_2H_4} P_{H_2O}}$$

What are the equilibrium pressures that must be substituted? For convenience, suppose we start with 1.000 mole of ethylene; since a fractional conversion is sought, this will not affect the answer. If x mole of ethanol is formed at equilibrium, only $(1-x)$ mole of ethylene must remain, since 1 ethylene molecule disappears when 1 ethanol molecule is formed. Similarly, the stoichiometric mixture of ethylene and water must initially have contained 1.000 mole of water, but at equilibrium will contain only $(1-x)$ mole of water. So we have the following situation:

$$C_2H_4 \;+\; H_2O \;\rightleftharpoons\; C_2H_5OH$$

$$1-x \text{ mole} \quad 1-x \text{ mole} \quad\quad x \text{ mole}$$

$$\text{total moles} = (1-x) + (1-x) + x = 2-x$$

mole fraction $C_2H_4 \equiv X_{C_2H_4} = \dfrac{1-x}{2-x} \quad X_{H_2O} = \dfrac{1-x}{2-x} \quad X_{C_2H_5OH} = \dfrac{x}{2-x}$

$P_{C_2H_4} = X_{C_2H_4} P_{\text{total}} \quad\quad P_{H_2O} = X_{H_2O} P_{\text{total}} \quad\quad P_{C_2H_5OH} = X_{C_2H_5OH} P_{\text{total}}$

$= \dfrac{1-x}{2-x} 10 \quad\quad\quad\quad = \dfrac{1-x}{2-x} 10 \quad\quad\quad\quad = \dfrac{x}{2-x} 10$

$$K_P = 0.306 = \dfrac{[x/(2-x)]10}{[(1-x)/(2-x)]10[(1-x)/(2-x)]10} = \dfrac{x(2-x)}{10(1-x)^2}$$

$$= \dfrac{2x - x^2}{10 - 20x + 10x^2}$$

Simplifying to a form suitable for the quadratic formula,

$$4.06x^2 - 8.12x + 3.06 = 0$$

$$x = \dfrac{8.12 \pm \sqrt{(-8.12)^2 - 4(4.06)(3.06)}}{2(4.06)}$$

$$= \dfrac{8.12 \pm 4.04}{8.12}$$

A little common sense is necessary here; x must be less than 1.000, since it represents the number of moles of ethanol formed from 1.000 mole of ethylene. Accordingly, we must choose the minus sign in solving for a physically reasonable answer:

$$x = \dfrac{8.12 - 4.04}{8.12} = \dfrac{4.08}{8.12} = 0.503 \text{ mole}$$

If 1.000 mole of ethylene yields 0.503 mole of ethanol, the conversion has clearly been 50.3%, which is the answer we sought. This problem is typical of many gas-equilibrium problems in specifying a total pressure rather than individual partial pressures and asking for an answer based on moles rather than pressures. In such problems the partial pressures must be obtained by calculating the total number of moles in the reaction mixture, using the stoichiometry of the reaction as we have here, and establishing the mole fraction of each component. Then the law of partial pressures will give the partial pressure from the mole fraction and the total pressure.

The Influence of Relative Quantities of Reactants on the Position of Equilibrium

Suppose we approach the same problem again, with all conditions being the same except that there are now 10 moles of water for every mole of ethylene in the starting mixture. We shall approach the problem the same way, except that the stoichiometry of the reaction no longer makes the pressures of ethylene and water equal, although the stoichiometry is still useful. We first write the balanced equation and the equilibrium-constant expression for it:

$$C_2H_{4(g)} + H_2O_{(g)} \rightleftharpoons C_2H_5OH_{(g)}$$

$$(400°K) \qquad K_P = 0.306 = \frac{P_{C_2H_5OH}}{P_{C_2H_4} P_{H_2O}}$$

We next calculate the total number of moles present in the reaction mixture, taking 1.000 mole of ethylene for convenience again and allowing x mole of ethanol to form.

$$C_2H_4 \quad + \quad H_2O \quad \rightleftharpoons \quad C_2H_5OH$$

$$1-x \text{ moles} \quad 10-x \text{ moles} \quad x \text{ moles}$$

$$\text{total moles} = 11 - x$$

We can now formulate the mole fractions for each component and their partial pressures:

$$X_{C_2H_4} = \frac{1-x}{11-x} \qquad X_{H_2O} = \frac{10-x}{11-x} \qquad X_{C_2H_5OH} = \frac{x}{11-x}$$

$$P_{C_2H_4} = \frac{1-x}{11-x} 10 \text{ atm} \qquad P_{H_2O} = \frac{10-x}{11-x} 10 \text{ atm} \qquad P_{C_2H_5OH} = \frac{x}{11-x} 10 \text{ atm}$$

Inserting these quantities into the equilibrium-constant expression,

$$K_P = 0.306 = \frac{[x/(11-x)]10}{[(1-x)/(11-x)]10[(10-x)/(11-x)]10} = \frac{x(11-x)}{10(1-x)(10-x)}$$

$$= \frac{11-x-x^2}{100-110x+10x^2}$$

$$4.06x^2 - 44.66x + 30.6 = 0$$

$$x = \frac{44.66 \pm \sqrt{(-44.66)^2 + 4(4.06)(30.6)}}{2(4.06)}$$

$$= \frac{44.66 - 38.70}{8.12}$$

$$= \frac{5.96}{8.12}$$

$$= 0.735 \quad \text{or} \quad 73.5\% \text{ conversion}$$

Notice that although the equilibrium-constant expression has not changed, the specific pressures have; we can load up a reaction mixture with a cheap component in order to use up a greater fraction of a more expensive one.

Approximation by Neglecting Small Quantities

Carbon dioxide can be converted to carbon monoxide by the following reaction, for which the equilibrium constant is 1235 at 800°C:

$$CO_{2(g)} + C_{(s)} \rightleftharpoons 2CO_{(g)}$$

Compare the percentage conversion of CO_2 to CO at 1 atm total pressure and 100 atm total pressure.

This problem again involves mole relationships and total pressure rather than partial pressures, so to get the partial pressures we shall again use mole fractions and Dalton's law. With a balanced equation written, as given in the problem, our next move is to write the appropriate equilibrium-constant expression for it. This reaction, however, presents a novel feature in that one of the components that should be included in the pressure fraction is a solid, for which the definition of a pressure would be difficult. What should be written in the equilibrium-constant expression for carbon as a solid? For guidance we return to the derivation of the equilibrium-constant expression from the free energy; each of these pressures (or concentrations) is not really a pressure with pressure dimensions, but rather a ratio of the experimental pressure to the pressure of the 1-atm standard state—in other words, the activity of the gas. By analogy, we use the activity of the solid carbon in the equilibrium-constant expression. The activity is the ratio of the concentration of the species to its concentration in the standard state, which is the pure species at 1 atm pressure. The properties of pure solids and liquids, however, are not affected appreciably by changes in the external pressure, so **the activity of any pure solid or liquid in an equilibrium-constant expression should be taken as 1 exactly**. This is true regardless of whether the equilibrium is a gaseous one or a solution equilibrium. With this principle established, we write the equilibrium-constant expression for this reaction equation as

(800°C) $\quad K_P = 1235 = \dfrac{(P_{CO})^2}{P_{CO_2}(\text{activity of pure solid C})} = \dfrac{(P_{CO})^2}{P_{CO_2}(1)} = \dfrac{(P_{CO})^2}{P_{CO_2}}$

This expression, of course, holds no matter what pressure we choose; let us solve for the 1-atm total pressure first. A little chemical judgment will help simplify the way we set up the mole-fraction expressions. Since the equilibrium constant is quite large, the reaction goes substantially to the right as written; this means that nearly all the CO_2 will be converted to CO. If we start with exactly 1 mole of CO_2, then, very nearly 2 moles of CO will form, leaving only x moles of CO_2 behind. We assign x to small quantities whenever the conditions of the problem make it clear what quantities are small. By the stoichiometry of the reaction, there will be $(2-2x)$ moles of CO present:

$$CO_2 \quad + \quad C \quad \rightleftharpoons \quad 2CO$$

x mole gas $\quad$ 0 mole gas $\quad$ $2-2x$ moles gas

total moles in gas mixture $= 2 - 2x + x = 2 - x$

The resulting gas mole fractions are

$$X_{CO_2} = \dfrac{x}{2-x} \qquad X_{CO} = \dfrac{2-2x}{2-x}$$

and the resulting partial pressures (for 1 atm total pressure) are

$$P_{CO_2} = X_{CO_2} P_{total} = \dfrac{x}{2-x}(1\ \text{atm}) \qquad P_{CO} = \dfrac{2-2x}{2-x}(1\ \text{atm})$$

At 1 atm, then, the equilibrium constant reads

$$K_P = 1235 = \dfrac{\{[(2-2x)/(2-x)]1\}^2}{[x/(2-x)]1} = \dfrac{(2-2x)^2}{x(2-x)}$$

We could multiply this out and solve it just as in the preceding problems. However, it would prove rather unsatisfactory, because the numerator of the quadratic formula would be the difference between two large numbers, one of which would require an electronic calculator to obtain. In this problem, as in many others involving gas and solid equilibria, it is advantageous to simplify the math by making an approximation: when a relatively small number is added to or subtracted from a constant, it may be ignored. Of course, if it is multiplying or dividing the constant it must be kept. How small is small? We cannot know until the problem is solved, but at that point it is imperative that we check the relative size of x and any constants that we have taken as being much larger than x. As a rule of thumb, if x is no more than 5% of the constant we may consider it satisfactorily small; this corresponds to the range of deviation of most gases from ideality at low pressures (as in Fig. 3-17). In other words, this is about the limit of accuracy of the replacement of fugacity by pressure (see

p. 457, which in turn limits the accuracy of the equilibrium-constant value. To return to our problem, we have said that the reaction proceeds substantially to the right as written, leaving only a small number of moles of CO_2. Let us assume that x, the number of moles of CO_2 remaining, is small compared to 2 in the equilibrium-constant expression, so $2-2x \cong 2$. Of course we shall have to check this assumption after we solve for x:

$$K_P = 1235 = \frac{(2-2x)^2}{x(2-x)} \cong \frac{4}{x(2)} \cong \frac{2}{x}$$

$$x \cong \frac{2}{1235} \cong 0.001620 \quad \text{or} \quad 1.620 \times 10^{-3} \text{ mole } CO_2$$

This means that all but 0.001620 mole of the original 1 mole of CO_2 has been converted to CO, corresponding to a percentage conversion of $(1-0.001620)100$, or 99.84% conversion. Checking the assumption we made about the magnitude of x, it is certainly true that $2x$ is less than 5% of 2, so the assumption was valid within the limits of accuracy of the equilibrium constant.

The Influence of Total Pressure on the Position of Equilibrium

Now let us solve the problem for 100 atm total pressure. The mole fraction expressions will be the same, and the only change will be the partial pressures:

$$P_{CO_2} = X_{CO_2} P_{total} = \frac{x}{2-x} 100 \qquad P_{CO} = \frac{2-2x}{2-x} 100$$

Inserting these into the equilibrium-constant expression,

$$1235 = \frac{\{[(2-2x)/(2-x)]100\}^2}{[x/(2-x)]100} = \frac{100(2-2x)^2}{x(2-x)}$$

Making the same approximation as before, that $2x \ll 2$, we write

$$1235 = \frac{100(2-2x)^2}{x(2-x)} \cong \frac{100(2)^2}{x(2)} \cong \frac{400}{2x} \cong \frac{200}{x}$$

$$x \cong \frac{200}{1235} \cong 0.1620 \text{ mole } CO_2$$

This is the amount remaining after conversion to CO, corresponding to 83.8% conversion. Checking the validity of the approximation we made, however, we find that $2x$ at 0.324 is substantially *greater* than 5% of 2 ($2 \times 0.05 = 0.10$). Therefore, the approximation is invalid and the problem must be worked out in full:

$$1235 = \frac{100(2-2x)^2}{x(2-x)} = \frac{400-800x+400x^2}{2x-x^2}$$

$$1635x^2 - 3270x + 400 = 0$$

$$x = \frac{3270 \pm \sqrt{(-3270)^2 - 4(1635)(400)}}{2(1635)}$$

$$= \frac{3270 - 2842}{3270}$$

$$= \frac{428}{3270}$$

$$= 0.1309 \text{ mole } CO_2$$

This corresponds to 86.9% conversion of the CO_2 to CO, which is noticeably different from the approximated value. This problem brings out some interesting points concerning equilibria. First, whenever a pure solid or liquid appears in an equilibrium-constant expression, its activity is taken as 1.000 and it is dropped from the expression, leaving behind only species whose pressure or concentration are subject to variation in the process of the reaction. Second, the approximation that an undetermined but small pressure, concentration, or number of moles may be neglected when added to or subtracted from a constant is often useful. But the approximation may or may not be valid from one equilibrium to another, or even when conditions are changed within a single equilibrium reaction, so the validity of the approximation must be checked whenever it is applied.

As a final note to the discussion of gas equilibria, it is occasionally convenient to express pressures in units other than atmospheres, which is perfectly legitimate as long as each quantity in the equilibrium-constant expression is still given as a ratio of experimental pressure to standard-state pressure (such as 760 torr). Since this standard-state pressure will appear as a factor in any given expression some constant number of times, it is sometimes absorbed into the numerical value of the equilibrium constant to allow pressures in torr (for instance) to be used directly in the equilibrium-constant expression. This can lead to confusion but is acceptable as long as the appropriate units are clearly specified along with the numerical value of the constant. The problems at the end of this chapter will be expressed only in atmospheres—but see Study Problem 1.

10-2 Ionic Equilibrium in Solution—Stability Constants

When we go from gases to solutions, the departures of the solution activities from ideal concentrations become substantial, and the equilibrium-constant expression is numerically reliable only at very low concentrations (remember that very dilute solutions always approach ideality). Of course, this applies only to the use of raw concentrations in the equilibrium-constant expression, since by definition activities would satisfy the equilibrium-constant expression at any concentration. The trouble with the direct use of concentrations in lieu of activities is that the activity depends on, among other things, the strength of the interaction between a dissolved ion and the solvent molecules (normally water, for the purposes of this chapter). The "infinitely dilute" behavior of the ion expresses this well enough, but at any finite concentration the interaction between the ion and the water molecules is modified somewhat by the interaction between the ion and other nearby ions, and by the interactions between the ion's neighbor water molecules and other ions; this causes the true activity, which is the sum of all these effects, to be different from the concentration, which is an extrapolation of the ion's infinitely dilute behavior. Theoretical treatments exist that are moderately successful in predicting the activity coefficient as a function of **ionic strength**, a measure of the total concentration of all ions weighted for the multiply charged ions. However, for our purposes we shall consistently assume that activities can be replaced directly by concentrations and that in water molality = molarity; in practice these are quite good assumptions up to concentrations of $0.01\ M$ and at least suggestive to $0.1\ M$.

One class of ionic equilibria involves the reaction by which a metal ion in solution exchanges its chemically bonded nearest-neighbor atoms or molecules for others in the solution. Molecular systems consisting of a metal ion (positively charged), surrounded by atoms or molecules that are more or less permanently bonded to the metal ion, are called **complexes**. Most metal ions in water solution form **aquo** complexes with H_2O molecules, such as $Mg(H_2O)_6^{2+}$; these are also known as **hydrated ions**. One of these aquo complexes can react with other ions or molecules in solution to give a new complex:

$$Co(H_2O)_{6(aq)}^{2+} + 6NH_{3(aq)} \rightleftharpoons Co(NH_3)_{6(aq)}^{2+} + 6H_2O_{(l)}$$

Since the equilibrium constant for this reaction equation would represent the stability of the new complex in the presence of the complexing water molecules, it is known as a **stability constant** for the complex. Writing the stability-constant expression for the above reaction equation, we have

$$K_{st} = \frac{[Co(NH_3)_6^{2+}][H_2O]^6}{[Co(H_2O)_6^{2+}][NH_3]^6}$$

where the square brackets, as in Chapter 9, indicate the activity or concentration of the species inside the brackets relative to the standard state. The concentration of water molecules in a reasonably dilute aqueous solution is essentially constant; we can to quite a good approximation treat it as a pure liquid, for which we take the activity ratio as 1.000 and omit it from the equilibrium-constant expression. After all, since water's molecular weight is

18 g/mole and there are 1000 g of water in 1 l, pure water is $1000/18 = 55.5\ M$ in water; if the formation of a dilute solution lowers this to $55.4\ M$ the concentration has remained essentially constant. The stability constant for this equilibrium is thus

$$K_{st} = \frac{[Co(NH_3)_6^{2+}]}{[Co(H_2O)_6^{2+}][NH_3]^6}$$

Stepwise and Overall Stability Constants

It is reasonable to expect the reaction to proceed one ligand at a time, so that the overall reaction would consist of a series of reactions of the type

$$Co(H_2O)_6^{2+} + NH_3 \rightleftharpoons Co(H_2O)_5NH_3^{2+} + H_2O$$

The equilibrium constant for the first of these replacements is called K_1, the second K_2, and so on; sometimes the overall constant is called β_i, where i is the total number of ligands on the complex whose stability is being measured. This successive replacement is observed experimentally, and the free-energy changes of the individual steps add up to give the overall free-energy change corresponding to β_i. Since $\Delta G^0 = -nRT \ln K$,

$$\ln \beta_i = \frac{-\Delta G^0}{nRT} = \ln K_1 + \ln K_2 + \cdots + \ln K_i$$

$$\beta_i = K_1 K_2 \cdots K_i$$

The equilibrium constant for a reaction that is a sum of individual equilibria is thus the product of the individual equilibrium constants. Often these overall equilibrium constants are quite large, corresponding to a reaction that is very nearly complete. The following example is typical.

Problems in Which One Component is Present in Large Amounts

Sodium cyanide is an ionic salt that dissociates completely in aqueous solution to give the $Na^+_{(aq)}$ ion and the $CN^-_{(aq)}$ ion. If 1 mole of $Zn(NO_3)_2$ is dissolved in 1 l of such a solution that is initially $1\ M$ in CN^-, what is the equilibrium concentration of CN^-? The overall equilibrium constant β_4 for the cyano complex of zinc ion is 6×10^{16}.

Our first task, as it was for gas equilibria, is to write the balanced equation corresponding to the equilibrium constant:

$$Zn(H_2O)_4^{2+} + 4CN^- \rightleftharpoons Zn(CN)_4^{2-} + 4H_2O$$

The equilibrium-constant expression for β_4 is thus

$$\beta_4 = 6 \times 10^{16} = \frac{[Zn(CN)_4^{2-}]}{[Zn(H_2O)_4^{2+}][CN^-]^4}$$

What numbers can be inserted for the needed concentrations? Since the equilibrium constant is large, the reaction clearly goes far to the right. We assume that nearly all the cyanide ion is used up, leaving only x concentration behind. From the fact that initially only 1 mole of cyanide was present, and from the stoichiometry of the reaction, we can see that nearly 0.25 mole of $Zn(CN)_4^{2-}$ has formed, specifically $0.25-0.25x$ since it takes only 0.25 mole of complex to generate 1 mole of cyanide. This number of moles is also the concentration since the volume is 1 l. The concentration of hydrated zinc ion is also equal to the number of moles of it remaining, $1-[Zn(CN)_4^{2-}]$, or $0.75+0.25x$. We can now substitute these in the equilibrium-constant expression:

$$6 \times 10^{16} = \frac{0.25 - 0.25x}{(0.75 + 0.25x)(x)^4}$$

Since the reaction apparently goes far to the right, x should be small, and we shall assume that $0.25x$ can be neglected compared to 0.25; of course, we shall have to check this after solving for x. Using this approximation,

$$6 \times 10^{16} \cong \frac{0.25}{(0.75)(x)^4} \cong \frac{1}{3x^4}$$

$$3x^4 \cong \frac{1}{6 \times 10^{16}}$$

$$x^4 \cong \frac{1}{18 \times 10^{16}} \cong 0.055 \times 10^{-16} \cong 550 \times 10^{-12}$$

$$x \cong \sqrt[4]{550} \times 10^{-3} = 4.8 \times 10^{-3} \, M$$

This is the remaining concentration of cyanide in the solution; it has obviously been reduced to a low level. Was the approximation we made a valid one? $0.25x$ is 1.2×10^{-3}, which is less than 5% of 0.25 since $0.25 \times 0.05 = 0.0125$, and the approximation was indeed valid.

Problems in Which Two Components Are Present in Large Amounts

We can make a useful distinction between equilibria in which only one component has been added to the solution in macroscopic amounts and those in which two components have been added (as in the previous example). Consider the mercury(II) complex $HgCl_4^{2-}$; if 0.1 mole of it is dissolved in 1 l of water, what will be the concentration of hydrated mercury ions, $Hg(H_2O)_4^{2+}$? What will it be if the complex ion is dissolved in 0.1-M Cl^- instead of in pure water? The overall stability constant for $HgCl_4^{2-}$, β_4, is 1×10^{16}.

The balanced equation corresponding to β_4 is

$$Hg(H_2O)_4^{2+} + 4Cl^- \rightleftharpoons HgCl_4^{2-} + 4H_2O$$

from which we write

$$\beta_4 = 1 \times 10^{16} = \frac{[HgCl_4^{2-}]}{[Hg(H_2O)_4^{2+}][Cl^-]^4}$$

for the equilibrium constant. Taking first the case in which the complex ion is dissolved in pure water, we see from the large value of β_4 that the complex ion will remain largely undissociated, so most of the 0.1 mole will remain at equilibrium as a solution of concentration $0.1 - x$, where x is the concentration of hydrated mercury ions formed. The stoichiometry of the reaction tells us that for each hydrated mercury ion formed four chloride ions will be formed, so the concentration of chloride ion at equilibrium will be $4x$. Substituting these values into the equilibrium-constant expression,

$$1 \times 10^{16} = \frac{0.1 - x}{(x)(4x)^4}$$

Assuming x to be negligible compared to 0.1, we write

$$1 \times 10^{16} \cong \frac{0.1}{256 x^5}$$

$$x^5 \cong \frac{0.1}{256 \times 10^{16}} \cong 3.9 \times 10^{-20}$$

$$x \cong 1.4 \times 10^{-4} \, M = [Hg(H_2O)_4^{2+}]$$

This value of the hydrated mercury ion concentration is indeed small compared to 0.1, so the approximation is valid. This corresponds to a small but easily measurable concentration. When we go to the second part of the problem, however, we find quite a different result. Using the same equilibrium-constant expression, we can still call the hydrated mercury ion concentration x and the $HgCl_4^{2-}$ concentration $0.1 - x$. However, the chloride ion concentration is now primarily governed by the 0.1-M concentration initially present: the total concentration at equilibrium is $0.1 + 4x$. Substituting in the equilibrium-constant expression again and assuming $4x$ to be small compared to 0.1,

$$1 \times 10^{16} = \frac{0.1 - x}{(x)(0.1 \times 4x)^4}$$

$$\cong \frac{0.1}{10^{-4} x}$$

$$x \cong \frac{0.1}{10^{-4} \times 10^{16}} \cong \frac{0.1}{10^{12}} \cong 10^{-13} \, M = [Hg(H_2O)_4^{2+}]$$

This is a very striking difference; the solution in pure water dissociates over a billion times as much as the solution in 0.1-M Cl^-! Checking the validity of our approximation, 10^{-13} is small by comparison with 0.1. This is a good example of the distinction between equilibrium problems having only one component added to the solution and those having two components added. The dramatic change in $[Hg(H_2O)_4^{2+}]$ when Cl^- is added is called the **common-ion effect**.

Complexometric Titrations

A very important practical application of the formation of complexes in aqueous solution lies in the analysis of many metals through the formation of very stable complexes with ligands such as ethylenediamine-N,N,N',N'-tetraacetic acid (EDTA or H_4Y), whose structure is shown in Fig. 10-1a. The molecule con-

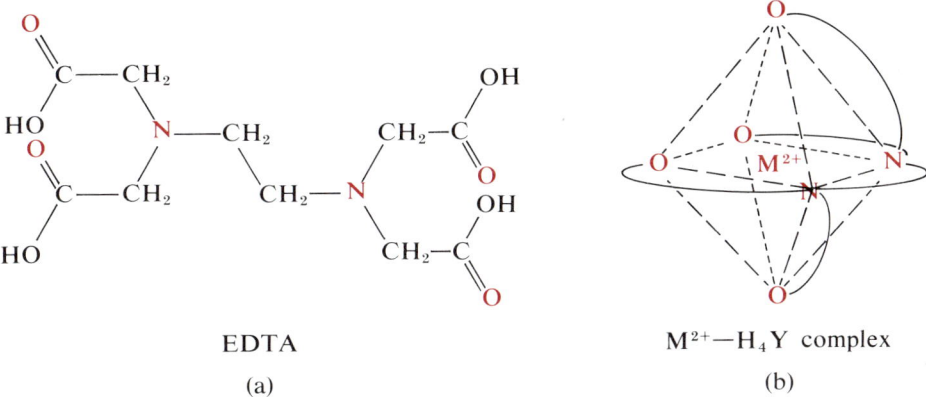

Figure 10-1 Ethylenediaminetetraacetic acid (EDTA). (a) Molecular structure. (b) Formation of octahedral complex with metal ion.

tains six atoms (indicated in red) that by virtue of their atomic electron/orbital ratio have pairs of nonbonding electrons and can serve as electron donors to positively charged metal ions. Furthermore, these donor atoms are separated by chains of atoms of the proper length and geometric configuration to enable all six of them to surround the same metal ion, as shown in a schematic fashion in Fig. 10-1b. EDTA complexes of most metal ions are quite stable, as Table 10-2 indicates. Furthermore, in most instances they are rapidly formed in

Table 10-2
Stability Constants for Metal–EDTA Complexes
(EDTA as Y^{4-} Species)

Metal	K_{st}	Metal	K_{st}
Mg^{II}	5×10^8	Ni^{II}	4×10^{18}
Ca^{II}	3×10^{10}	Cu^{II}	6×10^{18}
Cr^{III}	10^{23}	Zn^{II}	1×10^{16}
Mn^{II}	7×10^{13}	Cd^{II}	3×10^{16}
Fe^{II}	2×10^{14}	Hg^{II}	5×10^{21}
Fe^{III}	1×10^{25}	Pb^{II}	8×10^{17}
Co^{II}	2×10^{16}		

solution. This allows a metal-containing solution to be analyzed by **titration with EDTA** — adding a solution of EDTA of known concentration to another solution containing a given weight of the metal until all the metal has been complexed. From the volume of EDTA solution used and its concentration, along with the stoichiometry of the EDTA—metal ion reaction, the number of moles of metal present can be ascertained:

$$\text{liters EDTA used} \times \frac{\text{moles EDTA}}{\text{liter}} = \text{moles EDTA used}$$

$$M(H_2O)_6^{n+} + H_4Y \rightleftharpoons MY^{(n-4)+} + 4H_3O^+ + 2H_2O$$

$$1 \;:\; 1 = \text{stoichiometric ratio}$$

Therefore,

$$\text{moles EDTA used} = \text{moles metal present}$$

$$\text{liters EDTA} \times \frac{\text{moles EDTA}}{\text{liter}} = \text{moles metal present}$$

If a known weight of metal-containing sample was taken and the atomic weight of the metal is known, the percentage metal present in the sample can be readily calculated:

$$\text{moles metal} \times \frac{\text{grams metal}}{\text{mole}} = \text{grams metal}$$

$$\text{wt \% metal} = \frac{\text{grams metal}}{\text{grams sample}} \times 100 = \frac{\text{liters EDTA} \times M \text{ EDTA} \times \text{AW metal}}{\text{wt sample}}$$

To analyze in this way for metals, we must have some way of knowing when the amount of EDTA that has been added is exactly enough to react with all the metal; the volume for which this is true is called the **equivalence point**. Often an **indicator** is used — another metal-complexing molecule that is one color as a free molecule and another color when complexed to a metal (because of the change in the electronic energy levels when the metal ion enters the molecular system). If a small amount of an indicator compound is added to the initial metal solution it will show the metal-containing color and will continue to do so until the EDTA has removed all of the metal ions except those in the indicator–metal complex. Then if its complex is less stable than the EDTA complex it will lose its metal to the next small amount of EDTA that is added and will change color in the process. The color change, of course, signifies that the equivalence point is very near and is itself known as the **end point** since it is usually taken as the end of the titration.

To show that proper choice of an indicator can indeed give a color change very near the equivalence point, consider the EDTA titration of $Cd(H_2O)_6^{2+}$:

$$Cd(H_2O)_6^{2+} + HY^{3-} \rightleftharpoons CdHY^- + 6H_2O \qquad K_{st} = 1.2 \times 10^9$$

We can use the indicator xylenol orange, for which the effective equilibrium constant is about 10^5:

$$\text{Cd(H}_2\text{O})_6^{2+} + \underset{\text{yellow}}{\text{XO}} \rightleftharpoons \underset{\text{dark red}}{\text{CdXO}} + 6\text{H}_2\text{O} \qquad K_{st} = 10^5$$

We have the equilibrium-constant expressions obeyed throughout the titration:

$$1.2 \times 10^9 = \frac{[\text{CdHY}^-]}{[\text{Cd(H}_2\text{O})_6^{2+}][\text{HY}^{3-}]} \qquad 10^5 = \frac{[\text{CdXO}]}{[\text{Cd(H}_2\text{O})_6^{2+}][\text{XO}]}$$

If we solve the xylenol orange expression for $[\text{Cd(H}_2\text{O})_6^{2+}]$ and insert it into the EDTA expression we have

$$1.2 \times 10^9 = \frac{[\text{CdHY}^-][\text{XO}]}{[\text{HY}^{3-}][\text{CdXO}]} 10^5 \quad \text{or} \quad 1.2 \times 10^4 = \frac{[\text{CdHY}^-][\text{XO}]}{[\text{HY}^{3-}][\text{CdXO}]}$$

Now presumably the color change of the xylenol orange occurs when the concentrations of XO and CdXO are equal:

$$1.2 \times 10^4 = \frac{[\text{CdHY}][\cancel{\text{XO}}]}{[\text{HY}^{3-}][\cancel{\text{CdXO}}]} \quad \text{or} \quad [\text{HY}^{3-}] = \frac{[\text{CdHY}^-]}{12{,}000}$$

so there is no significant excess of HY^{3-} (EDTA) present yet. On the other hand, because of the xylenol orange equilibrium:

$$10^5 = \frac{[\cancel{\text{CdXO}}]}{[\text{Cd(H}_2\text{O})_6^{2+}][\cancel{\text{XO}}]} \quad \text{or} \quad [\text{Cd(H}_2\text{O})_6^{2+}] = 10^{-5}\,M$$

there is not much excess hydrated cadmium ion either (assuming a substantial initial concentration of $\text{Cd(H}_2\text{O})_6^{2+}$; if it were initially $0.01\,M$, only 0.1% of it would remain). And if the total amount of xylenol orange added was small compared to the total amount of cadmium present, there cannot be much CdXO. So the color change occurs just when the cadmium has all been titrated by EDTA, but no excess EDTA exists yet. Using any of a variety of indicators, most metal ions can be titrated by taking advantage of such equilibria as these.

10-3 Proton Transfer in Solution

Another extremely important type of solution is that involving acids. Without yet attempting any definitions, let us consider what is going on in an aqueous solution containing ions. Water is a polar solvent, and the oxygen atom has two nonbonding pairs of electrons. This means that in the presence of a positively charged ion the partially negative, electron-rich end of the water molecule will be attracted to the positive ion monopole. If the charge on the ion is high or its radius is small the overall Coulomb attraction will form a permanent bond between the oxygen of the water molecule and the metal. In this bond the oxygen atom is serving as an electron donor and the metal atom as an acceptor. The resulting species is, of course, the hydrate ion. When this occurs the electrons in the O—H bonds in the hydrated water molecules are attracted back toward the oxygen since that atom, having donated electrons to the metal, is developing a partial positive charge of its own. Consequently the hydrogen atom, in losing its share of the electron that it contributed to the O—H bond, becomes itself partially positively charged. The effect of all this is to spread the positive charge of the metal ion throughout the whole hydrate molecule.

In the liquid state, the hydrate ion is floating about undergoing repeated collisions with water molecules, and an interesting competition arises. If electron withdrawal from the hydrogens on the hydrated water molecule were complete, each would be H^+ (since each contributed one electron to the molecule's electron distribution). But an H^+ ion or **proton** would be very small, since there is no inner core of nonvalence electrons, and it could as readily bury itself in another water molecule's nonbonding electrons as stay on the hydrated water molecule. We thus have the possibility of an unforeseen chemical reaction going on in the water solution, as indicated in Fig. 10-2. Of course, at no time is a *free* proton, H^+ with no electrons, produced in the reaction. To strip the electrons from a proton would require an energy essentially equivalent to the ionization potential of the H atom, or over 300 kcal/mole. The proton transfer occurs by the atom's passing from one electron distribution to another nearby, without ever being entirely free of both. Accordingly, whenever we deal with proton-transfer reactions in water, the species that is sometimes written H^+ will appear as $H_3O^+_{(aq)}$, which is still imprecise because of the possible further hydration of this species (the **hydronium** ion) but at least suggests the true existing species.

Clearly the crux of the existence of this reaction (which is quite an important one as we shall see in Chapter 12) is the capacity of the molecules in solution to serve as electron donors; if no nonbonding pairs were available, the reaction could not occur. We can see that the reaction could be generalized beyond the water solution to essentially any reaction in which an atom is transferred from one atom to another through donor–acceptor interaction. For example, quite a reasonable comparison can be made between all the reactions below and the water reaction:

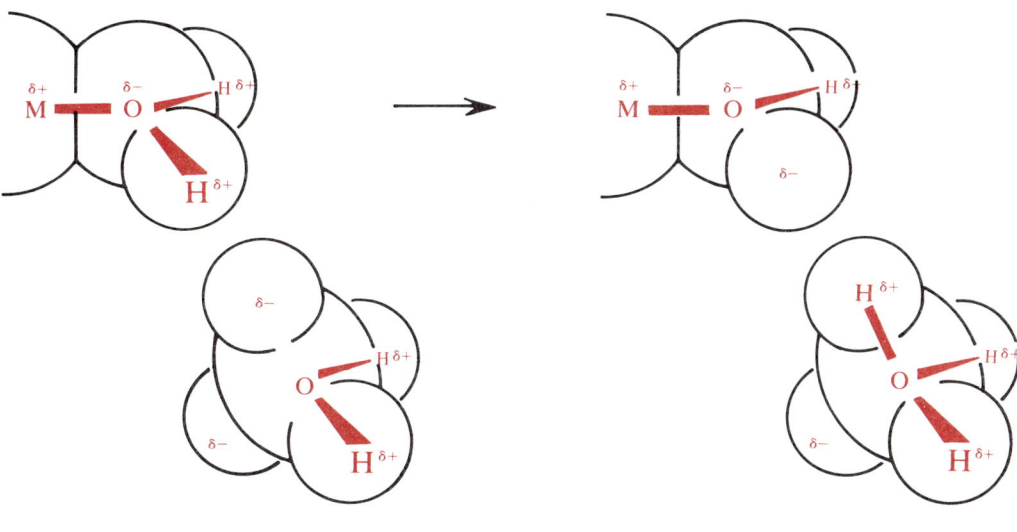

$$M(OH_2)^{n+} + H_2O \rightarrow M(OH)^{(n-1)+} + H_3O^+$$

Figure 10-2 Proton-transfer process between a hydrated water molecule and a free water molecule.

water	$M(OH_2) + H_2O \rightleftharpoons M(OH)^- + H_3O^+$	(H^+ transfer)
compare	$SO_3^{2-} + SOCl_2 \rightleftharpoons 2Cl^- + 2SO_2$	(O^{2-} transfer)
	$HSO_3OH + NO_3^- \rightleftharpoons HSO_4^- + HNO_3$	(H^+ transfer) (10-2)
	$BrF_3 + SbF_5 \rightleftharpoons BrF_2^+ + SbF_6^-$	(F^- transfer)
	$ClO_3OH + NH_3 \rightleftharpoons ClO_4^- + NH_4^+$	(H^+ transfer)

In each of these cases an atom is being transferred from the first species to the second. Looking at the reactions from the electron-donor/acceptor viewpoint, the species H_2O, SO_3^{2-}, NO_3^-, and NH_3 are serving as electron donors and the other species on the left side of the equations as acceptors. All of these reactions occur, although not in water solution for reasons that will become apparent later.

Acid–Base Definitions

The similarity of these reactions and the importance of the donor/acceptor feature led G. N. Lewis to propose that they all be regarded as **acid–base reactions** under the following definitions, which can be applied quite generally:

An acid is an electron acceptor.
A base is an electron donor.

This is the most general theoretical definition of acids and bases, and we shall return to it in describing chemical synthesis.

In proposing it, however, Lewis was appropriating two terms hallowed by antiquity. Acids and bases had always been thought of as peculiarly the property of water solutions. The terms stretch back into the mists of alchemy, but their first systematic definition was due to Svante Arrhenius in the late nineteenth century. He proposed that an acid was a proton source and a base was a hydroxide-ion source. This was satisfactory for simple neutralization reactions:

$$\underbrace{H_3O^+_{(aq)} + Cl^-_{(aq)}}_{\text{aqueous HCl}} + \underbrace{Na^+_{(aq)} + OH^-_{(aq)}}_{\text{aqueous NaOH}} \rightarrow Na^+_{(aq)} + Cl^-_{(aq)} + 2H_2O_{(l)}$$

For other characteristic reactions in water, however, such as the one with which we introduced this section, the definition is inadequate since no hydroxide ion is produced. In 1923, at about the same time that the Lewis definition appeared, two other chemists, Lowry and Brönsted, generalized the Arrhenius definition by proposing the following definitions, which are usually applied to aqueous solutions:

An acid is a proton donor.
A base is a proton acceptor.

These definitions cover the reaction at the beginning of this section, and in addition are completely compatible with, although more narrow than, the Lewis definition, since a proton accepts electrons in the interaction we are describing. We shall use the Lowry-Brönsted definition in discussing acid-base equilibria, but before we do so some more comments on generalized (i.e., Lewis) acids and bases may be desirable.

A Lewis acid or base need not contain protons at all, as is evident from Eq. 10-2. Accordingly, this definition can describe and predict reactions in any of a number of solvents. Just as we write

$$H_3O^+_{(aq)} + OH^-_{(aq)} \rightleftharpoons 2H_2O_{(l)}$$
$$\text{acid} \quad\quad \text{base} \quad\quad \text{solvent}$$

for the neutralization reaction in water, we can in an entirely analogous fashion write

$$SO^{2+}_{(solv)} + SO^{2-}_{3(solv)} \rightleftharpoons 2SO_{2(l)}$$

$$NH^{+}_{4(solv)} + NH^{-}_{2(solv)} \rightleftharpoons 2NH_{3(l)}$$

$$H_3SO^{+}_{4(solv)} + HSO^{-}_{4(solv)} \rightleftharpoons 2H_2SO_{4(l)}$$

acids   bases   solvents

All of these reactions occur and are quite well characterized; in terms of them we can give the operational definition of generalized acids and bases due to Cady:

> An acid is a substance that when dissolved in a solvent gives positive ions (cations) identical to those yielded by the solvent; a base is a substance that yields negative ions (anions) identical to those yielded by the solvent.

Hard and Soft Acids and Bases

Acids differ in strength, as do bases. Toward a given acceptor, various possible electron donors have different capacities for donation; the electrons undergo different degrees of stabilization in being donated or shared. In what classifications can we place acids and bases to account for their patterns of reactivity? In Section 6-8 we briefly discussed the relative patterns of stability between two bonded atoms, concluding that "soft" or readily polarizable atoms tend to bond strongly to each other because the charge separation is small and the covalent energy of bonding is large; "hard" or slightly polarizable atoms tend to bond strongly to each other because the charge separation is great and the ionic attraction energy is large. A contemporary chemist, Ralph Pearson, has proposed that the relative reactivity of acids and bases be considered on this basis:

> Hard acids tend to bond strongly to hard bases.
> Soft acids tend to bond strongly to soft bases.

This pattern is followed with few exceptions if the factors tending to maximize covalency are considered and balanced against those tending to maximize charge separation. Remember that an ion's polarizability is high (it is "soft") if it has a large diameter and a negative or very low positive charge; if it has either a small diameter with any positive charge at all or a high positive charge no matter how large, it is "hard." Besides the polarizability, however, there is the fact that covalent bonding is promoted if the acid or acceptor also has loosely held outer d or f electrons that can form π bonds using vacant orbitals on the base; for this reason cations (acids) on the right of the periodic table having such d electrons tend to be "soft" acids even with a charge of 2+, such as Hg^{2+}. There are other, smaller factors affecting the "hard"/"soft" classification, and the concept is not quantitative, but it is useful in predicting the relative stability of compounds that can reasonably be thought to have formed from a Lewis acid–Lewis base interaction. Table 10-3 indicates the division on empirical grounds, but is consistent with the above discussion.

Table 10-3
Hard and Soft Acids and Bases

Hard Acids (electron acceptors)	Intermediate Acids	Soft Acids
H^+, Li^+, Na^+, K^+	Fe^{2+}, Co^{2+}, Ni^{2+}, Cu^{2+}, Zn^{2+}, Pb^{2+}	Cu^+, Ag^+, Au^+, Tl^+, Cs^+
Be^{2+}, Mg^{2+}, Ca^{2+}, Sr^{2+}, Ba^{2+}	$B(CH_3)_3$, SO_2	Pd^{2+}, Pt^{2+}, Cd^{2+}, Hg_2^{2+}, Hg^{2+}
group IIIa 3+ ions		Tl^{3+}, $Tl(CH_3)_3$, BH_3
any 4+ ion or higher charge		I^+, Br^+
$Be(CH_3)_2$, BF_3, BCl_3		I_2, Br_2, ICN
$Al(CH_3)_3$		M^0 (neutral metal atoms)
R_3C^+ (carbonium ions)		

Hard Bases	Soft Bases
N (electron-donor atom in ligand)	P, As, Sb, Bi (donor atoms in ligand)
O (electron-donor atom in ligand)	S, Se, Te (donor atoms in ligand)
F (electron-donor atom in ligand)	Cl, Br, I (donor atoms in ligand)
	CO, NO, $CH_2{=}CH_2$, CN^-

To give an example of the utility of the concept, consider the variety of ligand atoms necessary to stabilize different degrees of positive charge on cations. For metal compounds to be stable with the metal having no charge or a very low positive charge, the ligands must be relatively large, have a low negative charge, and have a low electronegativity: thus we can make $Mn(CO)_5I$ and $Mn(CN)_6^{5-}$, which contain Mn(I), but not MnF or MnO_3^{5-}, which also would. It can be seen from Table 10-3 that this is a case of soft bases bonding well to soft acids. On the other hand, to stabilize a very high formal positive charge on an atom, ligands such as fluoride and oxide are needed: MnO_4^- and IF_7, both containing atoms with a formal charge of 7+, exist, but not MnS_4^- or $I(CN)_7$. Again, this is simply a case of hard bases bonding well to hard acids, as Table 10-3 indicates.

Relative Degree of Electron Withdrawal from Protons

Another useful classification is an ordering of proton-donor acids according to the strength of their tendency to withdraw electrons from their protons, thereby making them available for transfer to other bases (electron donors). This depends somewhat, but not strongly, on which base is being used; a reasonable average order is given in Table 10-4. In this table $HClO_4$ with-

Table 10-4
Relative Acid Strengths (degree of electron withdrawal from protons on acid molecule)

$$HClO_4 + s \text{ (solvent)} \rightleftharpoons ClO_4^- + Hs^+ \text{ (strongest electron withdrawal)}$$
$$HBr + s \rightleftharpoons Br^- + Hs^+$$
$$H_2SO_4 + s \rightleftharpoons HSO_4^- + Hs^+$$
$$HCl + s \rightleftharpoons Cl^- + Hs^+$$
$$HNO_3 + s \rightleftharpoons NO_3^- + Hs^+$$
$$H_3O^+ + s \rightleftharpoons H_2O + Hs^+$$
$$HSO_4^- + s \rightleftharpoons SO_4^{2-} + Hs^+$$
$$HF + s \rightleftharpoons F^- + Hs^+$$
$$Al(OH_2)_6^{3+} + s \rightleftharpoons Al(OH_2)_5OH^{2+} + Hs^+$$
$$H_2S + s \rightleftharpoons HS^- + Hs^+$$
$$NH_4^+ + s \rightleftharpoons NH_3 + Hs^+$$
$$HS^- + s \rightleftharpoons S^{2-} + Hs^+$$
$$H_2O + s \rightleftharpoons OH^- + Hs^+$$
$$OH^- + s \rightleftharpoons O^{2-} + Hs^+$$
$$NH_3 + s \rightleftharpoons NH_2^- + Hs^+$$
$$CH_4 + s \rightleftharpoons CH_3^- + Hs^+ \text{ (weakest electron withdrawal)}$$

draws electrons most strongly from its proton and thus is the strongest acid; others follow in declining order of acidity. Because of the substantial electron withdrawal by $HClO_4$ it will transfer its proton to water rapidly and completely when it is dissolved in water:

$$HClO_{4(l)} + H_2O_{(l)} \rightarrow ClO_{4(aq)}^- + H_3O_{(aq)}^+$$

In water, this will happen to any acid that is stronger — higher on the list — than H_3O^+. There is thus what is called a **leveling effect** in that no acid stronger than

H_3O^+ can exist in water solution. Conversely, on the other end of the scale, an equivalent reaction will occur with CH_3^-:

$$CH_{3(s)}^- + H_2O_{(l)} \rightarrow CH_{4(g)} + OH_{(aq)}^-$$

where the leveling effect prevents any base stronger than OH^- from existing in water solution. In other words, a sufficiently generous electron donor will strip protons off a neutral water molecule. The acids and bases that can exist in water solution, then, are those lying between H_3O^+ and OH^- on the list. In an entirely equivalent fashion, the acids and bases that can exist in ammonia solution are those that lie between NH_4^+ and NH_2^- on the list, and numerous other such comparisons could be made — some solvents are intrinsically more acidic than others. This is often an important consideration in choosing a solvent for a synthetic reaction, since the properties of the solvent must be matched to those of the desired product.

Restricting ourselves to aqueous solutions, however, the leveling effect gives us a range of acid/base strengths for which the species can exist in water. Within this range, it is useful and important to describe the acid or base strength quantitatively by writing the equilibrium constant for the proton-transfer reaction between water and the acid or base species. This is an area we shall consider carefully.

10-4 Acid–Base Equilibria

For a proton-transfer reaction between species that can exist in aqueous solution (so that a genuine equilibrium can be said to exist), the typical reaction equation is

$$HA + H_2O \rightleftharpoons A^- + H_3O^+ \qquad (10\text{-}3)$$
$$\text{acid} \quad \text{base} \qquad \text{base} \quad \text{acid}$$

This reaction can legitimately be said to run in either direction, since all four species can exist in water solution. Therefore, going back to the Lowry–Brönsted definition of acids and bases, we have indicated the proton donor and proton acceptor for the reaction in both directions. This raises an interesting point. Any species that can exist in water solution in both a protonated form (like HA) and a deprotonated form (like A^-) can serve either as an acid or as a base. The system HA and A^- is called a **conjugate acid–base pair**. If the acid form is a relatively strong acid, the base form will be a weak base, and vice versa. This reciprocal nature means that in tabulating the strengths of acids and bases we need not consider them separately, but can give values for the equilibrium constants (K_a) of the reactions all written in the above form. Table 10-5

Table 10-5
Equilibrium Constants (K_a) for Aqueous Proton-Donor Reactions

Reaction	K_a (25°C)
Cl_3CCOOH(trichloroacetic acid) $+ H_2O \rightleftharpoons Cl_3CCOO^- + H_3O^+$	2.0×10^{-1}
$HOOCCOOH$(oxalic acid) $+ H_2O \rightleftharpoons HOOCCOO^- + H_3O^+$	5.9×10^{-2}
$HSO_4^- + H_2O \rightleftharpoons SO_4^{2-} + H_3O^+$	1.2×10^{-2}
$H_3PO_4 + H_2O \rightleftharpoons H_2PO_4^- + H_3O^+$	7.5×10^{-3}
$Fe(OH_2)_6^{3+} + H_2O \rightleftharpoons Fe(OH_2)_5OH^{2+} + H_3O^+$	3.4×10^{-3}
$HF + H_2O \rightleftharpoons F^- + H_3O^+$	3.5×10^{-4}
$HCOOH$(formic acid) $+ H_2O \rightleftharpoons HCOO^- + H_3O^+$	1.8×10^{-4}
$Cr(OH_2)_6^{3+} + H_2O \rightleftharpoons Cr(OH_2)_5OH^{2+} + H_3O^+$	1.6×10^{-4}
C_6H_5COOH(benzoic acid) $+ H_2O \rightleftharpoons C_6H_5COO^- + H_3O^+$	6.5×10^{-5}
$HOOCCOO^-$(hydrogen oxalate ion) $+ H_2O \rightleftharpoons$ $OOCCOO^{2-} + H_3O^+$	6.1×10^{-5}
$HN_3 + H_2O \rightleftharpoons N_3^- + H_3O^+$	1.9×10^{-5}
CH_3COOH(acetic acid) $+ H_2O \rightleftharpoons CH_3COO^- + H_3O^+$	1.8×10^{-5}
$Al(OH_2)_6^{3+} + H_2O \rightleftharpoons Al(OH_2)_5OH^{2+} + H_3O^+$	1.0×10^{-5}
$C_5H_5NH^+ + H_2O \rightleftharpoons C_5H_5N$(pyridine)$+ H_3O^+$	6.6×10^{-6}
$CO_{2(aq)} + 2H_2O \rightleftharpoons HCO_3^- + H_3O^+$ $(H_2CO_3 + H_2O \rightleftharpoons HCO_3^- + H_3O^+: 2 \times 10^-$,)	4.3×10^{-7}
$H_2S + H_2O \rightleftharpoons HS^- + H_3O^+$	1.0×10^{-7}
$N_2H_5^+ + H_2O \rightarrow N_2H_4$(hydrazine)$+ H_3O^+$	3.3×10^{-9}
$NH_4^+ + H_2O \rightleftharpoons NH_3 + H_3O^+$	5.6×10^{-10}
$HCO_3^- + H_2O \rightleftharpoons CO_3^{2-} + H_3O^+$	5.6×10^{-11}
$(C_2H_5)_2NH_2^+ + H_2O \rightleftharpoons (C_2H_5)_2NH$(diethylamine)$+ H_3O^+$	8.0×10^{-12}
$HS^- + H_2O \rightleftharpoons S^{2-} + H_3O^+$	1.1×10^{-13}
$H_2O + H_2O \rightleftharpoons OH^- + H_3O^+$	1.0×10^{-14}

gives some of these values. The concentration of water is, as usual, omitted from the equilibrium-constant expression.

Some examples of the use of these constants will be helpful; let us begin by considering reactions in which only one component of the system is added to water, then go on to cases in which two components are added.

Problems in Which One Component Is Present in Large Amounts

If 3 g of acetic acid, CH_3COOH, are added to 1.00 l of water, what is the equilibrium concentration of H_3O^+?

Since the reaction equation and the equilibrium-constant expression involve mole quantities, our first step must be to convert the weight of acetic acid to moles:

$$\frac{3 \text{ g acetic acid}}{60 \text{ g acetic acid/mole}} = 0.05 \text{ mole acetic acid}$$

The volume of the solution is 1.00 l, so the molar concentration of acetic acid before proton transfer is 0.05 M. Writing the balanced reaction equation and the appropriate equilibrium-constant expression, we have

$$HAc + H_2O \rightleftharpoons Ac^- + H_3O^+ \qquad K_a = 1.8 \times 10^{-5} = \frac{[Ac^-][H_3O^+]}{[HAc]}$$

where Ac represents the rest of the acetic acid molecule. What numbers can we insert in order to solve for $[H_3O^+]$? Since the constant is quite small, the transfer will occur only to a slight extent; most of the acetic acid molecules will stay acetic acid molecules. Then we can call $[H_3O^+]$ x, expecting it to be fairly small, and the protonated acetic acid concentration, [HAc], will be $0.05 - x$. For $[Ac^-]$ we can rely on the stoichiometry of the reaction: there is no source of acetate ion except the proton-transfer reaction, which forms one Ac^- for every H_3O^+, so $[Ac^-] = [H_3O^+] = x$. We thus have the equilibrium-constant expression

$$K_a = 1.8 \times 10^{-5} = \frac{(x)(x)}{0.05 - x}$$

We have already said that the small value of K_a leads us to believe x will be small; if we assume it to be negligible compared to 0.05 (which we shall have to check at the end), we have

$$1.8 \times 10^{-5} \cong \frac{x^2}{0.05}$$

$$x^2 \cong 9 \times 10^{-7} = 90 \times 10^{-8}$$

$$x \cong 9.5 \times 10^{-4} M = [H_3O^+]$$

This value of x, 0.00095, is indeed smaller than 5% of 0.05 ($0.05 \times 0.05 = 0.0025$), so the approximation was satisfactory. Note that we followed the same procedure as in previous problems: writing a chemical equation and equilibrium-constant expression, judging which quantities are likely to be present in relatively large quantities from the magnitude of the equilibrium constant, using the reaction stoichiometry to establish relationships between concentrations, and simplifying the resulting expression by neglecting small quantities in addition or subtraction.

If 0.100 mole of solid hydrazinium sulfate, $(N_2H_5)_2SO_4$, is dissolved in 500 ml of H_2O, how much free hydrazine, N_2H_4, is present at equilibrium?

The reaction equation and equilibrium-constant expression corresponding to the constant from Table 10-5 are

$$N_2H_{5(aq)}^+ + H_2O_{(l)} \rightleftharpoons N_2H_{4(aq)} + H_3O_{(aq)}^+$$

$$K_a = 3.3 \times 10^{-9} = \frac{[N_2H_4][H_3O^+]}{[N_2H_5^+]}$$

Since the equilibrium constant is a small number, most of the hydrazinium ion originally added to the solution will remain in that form and we can take $[N_2H_4]$ as a small number x. By the stoichiometry of the reaction $[H_3O^+] = [N_2H_4] = x$, and we need only the concentration of $N_2H_5^+$. Now, 0.100 mole of the salt was added to the solution, but each mole of salt contains 2 moles of $N_2H_5^+$ ion, so 0.200 mole of the ion is in solution. Since the volume of the solution is only 500 ml we see that the concentration of $N_2H_5^+$ is initially 0.400 M:

$$\frac{0.200 \text{ mole } N_2H_5^+}{500 \text{ ml}} \; 1000 \frac{\text{ml}}{\text{l}} = 0.400 \text{ mole/l}$$

At equilibrium, the reaction stoichiometry tells us that the concentration will be $0.400 - x$. Inserting these quantities in the equilibrium-constant expression and assuming x to be small compared to 0.400, we have

$$3.3 \times 10^{-9} = \frac{(x)(x)}{0.400 - x} \cong \frac{x^2}{0.400}$$

$$x^2 \cong 1.32 \times 10^{-9}$$

$$x \cong 3.6 \times 10^{-5} \, M = [N_2H_4]$$

To get the desired answer, we need note only that 1 l of solution would contain 3.6×10^{-5} moles of hydrazine, so 500 ml will contain 1.8×10^{-5} moles. Checking the validity of our assumption, 3.6×10^{-5} is certainly less than 5% of 0.400, so the approximation is adequate.

The Water Equilibrium

Looking at the hydrazine system in another way, another feature of aqueous acid–base equilibrium becomes apparent. Suppose that instead of adding $N_2H_5^+$ ion to the solution, we had added N_2H_4 to pure water. What reaction would occur and what kind of equilibrium would it reach? The reaction indicated in Table 10-5 suggests that N_2H_4 would react with H_3O^+ if any were present; is there any? There is, because water itself can serve as a proton donor to another water molecule:

$$H_2O_{(l)} + H_2O_{(l)} \rightleftharpoons H_3O^+_{(aq)} + OH^-_{(aq)} \tag{10-4}$$
$$\text{base} \quad \text{acid} \quad\quad \text{acid} \quad\; \text{base}$$

This reaction goes on in any water solution, and the equilibrium for it is always satisfied in any aqueous equilibrium. For the equilibrium-constant expression we write

$$K_{eq} \equiv K_w = \frac{[H_3O^+][OH^-]}{[H_2O][H_2O]} = [H_3O^+][OH^-] \tag{10-5}$$

The equilibrium constant for this expression at 25 °C is called K_w and is equal to 1.0×10^{-14}. So in any aqueous solution at equilibrium it is always true that

$$[H_3O^+] = \frac{10^{-14}}{[OH^-]} \quad \text{or} \quad [OH^-] = \frac{10^{-14}}{[H_3O^+]} \tag{10-6}$$

This means that there is some H_3O^+ present for the N_2H_4 to react with; what happens to the water equilibrium when some of this H_3O^+ disappears? Since the product of $[H_3O^+]$ and $[OH^-]$ is then less than 10^{-14}, more water reacts to produce H_3O^+ and OH^- and raise the product of their concentrations to 10^{-14} again. But this provides more H_3O^+ for the N_2H_4 to react with, so the reaction goes on until *both* equilibria are satisfied:

$$3.3 \times 10^{-9} = \frac{[N_2H_4][H_3O^+]}{[N_2H_5^+]} \quad \text{and} \quad 1.0 \times 10^{-14} = [H_3O^+][OH^-]$$

Combining these into a single expression we have

$$3.3 \times 10^{-9} = \frac{[N_2H_4]}{[N_2H_5^+][OH^-]} \times 10^{-14} \quad \text{or} \quad \frac{[N_2H_5^+][OH^-]}{[N_2H_4]} = 3 \times 10^{-6}$$

The Base Constant K_b

This last expression is just the equilibrium constant for the reaction viewed a different way: hydrazine reacting as a base with water to produce hydroxide ion. That is,

$$N_2H_{4(aq)} + H_2O_{(l)} \rightleftharpoons N_2H_{5(aq)}^+ + OH_{(aq)}^-$$

which expresses the same result as the previous pair of reactions: hydrazine and water are used up and $N_2H_5^+$ ion and OH^- ion are produced. So the reaction of hydrazine with water can be written so as either to rely on the K_a expression for $N_2H_5^+$, or to express directly the base property of N_2H_4. An equilibrium constant for a reaction in which a base reacts with water is called K_b, and from the last equilibrium-constant expression above we can see that K_b and K_a for a given conjugate acid/base pair are related by

$$K_a = \frac{1}{K_b} K_w \quad \text{or} \quad K_a K_b = K_w \tag{10-7}$$

In general, it is important to analyze an equilibrium problem to see which form of the equilibrium expression is more convenient for the purposes of the problem. Usually the more convenient form will describe the reaction of the principal solute with water, as the following example will show.

Is it possible to reduce the H_3O^+ concentration of water to 10^{-9} M by adding pyridine?

In this problem the reaction is clearly one between pyridine and water to use up H_3O^+, since in pure water $[H_3O^+]$ is given by

$$H_2O_{(l)} + H_2O_{(l)} \rightleftharpoons H_3O^+_{(aq)} + OH^-_{(aq)}$$

$$10^{-14} = [H_3O^+][OH^-] = (x)(x) = x^2$$

$$x = [H_3O^+] = 10^{-7} M$$

Pyridine is thus acting as a base (proton acceptor), and the appropriate reaction is

$$py_{(aq)} + H_2O_{(l)} \rightleftharpoons pyH^+_{(aq)} + OH^-_{(aq)}$$

where py stands for the pyridine molecule. The equilibrium-constant expression for this reaction equation is

$$K_b = \frac{K_w}{K_a} = \frac{10^{-14}}{6.6 \times 10^{-6}} = 1.5 \times 10^{-9}$$

$$= 1.5 \times 10^{-9} = \frac{[pyH^+][OH^-]}{[py]}$$

If we know the concentration of H_3O^+ in the solution at equilibrium, we know the concentration of OH^- from the water equilibrium and K_w:

$$[OH^-] = \frac{K_w}{[H_3O^+]} = \frac{10^{-14}}{10^{-9}} = 10^{-5} M$$

This concentration is large compared to the amount that could come from proton exchange in water, which is indicated by $[H_3O^+]$: 10^{-9} M. Accordingly we assume that all the OH^- comes from the pyridine reaction, which means that the concentration of pyH^+ (looking back at the reaction stoichiometry) equals the OH^- concentration, at 10^{-5} M. We thus have the K_b expression

$$K_b = 1.5 \times 10^{-9} = \frac{[pyH^+][OH^-]}{[py]} = \frac{(10^{-5})(10^{-5})}{[py]}$$

$$[py] = \frac{10^{-10}}{1.5 \times 10^{-9}}$$

$$= 6.7 \times 10^{-2} \quad \text{or} \quad 0.067 M$$

From this result we can see that the answer is "yes," since $[H_3O^+] = 10^{-9}$ can be attained using pyridine at a reasonable concentration.

pH and pK

In the previous problem we encountered the idea that very low concentrations of H_3O^+ (or OH^-) may be of interest to us. The power-of-ten notation for these concentrations is compact and reasonably convenient, but a further abbreviation and convenience can be achieved by introducing the concept of **pH**. The pH of an aqueous solution is defined as the negative of the logarithm of the H_3O^+ activity (or in an approximate sense, concentration). It has only an operational definition, based on several standard solutions that have defined pH values to 0.001 pH unit. Using the pH notation, we say that the solution in the previous example has a pH of 9.00:

$$[H_3O^+] = 1 \times 10^{-9}$$

$$\log [H_3O^+] = -9.00$$

$$-\log [H_3O^+] \equiv pH = 9.00$$

The pH notation is also convenient in other respects. Hydroxide ion concentration is sometimes quoted as pOH, which has the obvious definition $-\log[OH^-]$. Equilibrium-constant values that differ substantially from 1, particularly if they are very small, are often quoted as pK values, which is $-\log K$. The pH and pK notation is quite commonly used, the pOH notation somewhat less so because of the ease of conversion from pH:

$$K_w = 10^{-14} = [H_3O^+][OH^-]$$

$$\log K_w = -14 = \log [H_3O^+] + \log [OH^-]$$

$$-\log K_w \equiv pK_w = 14 = -\log [H_3O^+] - \log [OH^-]$$

$$14 = pH + pOH$$

Hydrolysis Problems

Another topic that is usually considered separately from the previous types of problems is **hydrolysis**, the reaction of an ion in a salt with water. A salt that consists of a cation whose hydroxide is a strong base, and an anion whose protonated form is a strong acid, will show no hydrolysis by either ion and will not change the pH of pure water from 7.00. For instance, NaCl is such a salt, since sodium hydroxide is a very strong base and hydrogen chloride in water is a very strong acid. On the other hand, ammonium salts (no matter what the anion) always show some hydrolysis due to the reaction of the ammonium ion with water, and so do acetate salts (no matter what the cation):

$$NH_{4(aq)}^+ + H_2O_{(l)} \rightleftharpoons NH_{3(aq)} + H_3O_{(aq)}^+$$

$$Ac_{(aq)}^- + H_2O_{(l)} \rightleftharpoons HAc_{(aq)} + OH_{(aq)}^-$$

The first of these is the K_a reaction for the ammonium ion, but the second differs from the K_a reaction for acetic acid in the same way that the reaction for N_2H_4 on p. 510 differed from the K_a reaction for $N_2H_5^+$. In other words, the second reaction above is the K_b reaction for the acetate ion, and all the relationships that apply to K_b expressions apply to it. The constant is sometimes called K_h to indicate the hydrolysis process, but it is really only the acetate ion serving as a base in a familiar K_b reaction. To be more explicit, when an anion acts as a base in hydrolyzing, we have

$$K_h = \frac{K_w}{K_a} = \frac{[HA][OH^-]}{[A^-]}$$

As an example of hydrolysis calculations, what is the pH of a solution of 0.15 mole of sodium azide, NaN_3, in 750 ml of H_2O?

The reaction that occurs in establishing this equilibrium is

$$N_3^- + H_2O \rightleftharpoons HN_3 + OH^-$$

$$K_h = \frac{K_w}{K_a} = \frac{10^{-14}}{1.9 \times 10^{-5}} = \frac{[HN_3][OH^-]}{[N_3^-]}$$

The calculation proceeds exactly like the previous ones. The initial concentration of azide ion is 0.15 mole/0.750 l = 0.200 M; if x concentration of OH^- forms, then the stoichiometry of the reaction tells us that the equilibrium concentration of azide ion will be $0.200 - x$. From the stoichiometry we also know that the concentration of hydrazoic acid, HN_3, will be x. If we insert all these values into the equilibrium-constant expression and apply our usual approximation,

$$\frac{10^{-14}}{1.9 \times 10^{-5}} = 5.3 \times 10^{-10} = \frac{(x)(x)}{0.200 - x}$$

$$5.3 \times 10^{-10} \cong \frac{x^2}{0.200}$$

$$x^2 \cong 1.06 \times 10^{-10}$$

$$x \cong 1.03 \times 10^{-5} = [OH^-]$$

The approximation was valid: 1×10^{-5} is less than 5% of 0.200. So we can go on to the calculation of the pH; this is one of the places where pOH is convenient:

$$pH = 14 - pOH = 14 - [-\log(1 \times 10^{-5})]$$
$$= 14 - [-(-5)]$$
$$= 14 - 5 = 9.00$$

Any hydrolysis of a single ion can be calculated in just this fashion.

The Exact Treatment for Two Hydrolyzing Ions

A salt may contain more than one hydrolyzing ion—for instance, ammonium acetate. Because the number of unknown concentrations is now larger (HAc, NH_3, H_3O^+, and OH^-), we must use one of a hitherto-neglected pair of relationships to establish all the unknown concentrations. There is a **mass-balance** condition: the total concentration of species arising from the positive ion is equal to the total concentration of species arising from the negative ion. In this case,

$$[NH_4^+] + [NH_3] = [Ac^-] + [HAc]$$

There is also a **charge-balance** condition: the total positive charge in the solution must equal the total negative charge in the solution. Here we write

$$[NH_4^+] + [H_3O^+] = [Ac^-] + [OH^-]$$

This makes a total of five relationships that we can apply in solving for the four unknown concentrations in the ammonium acetate example; in a generalized form, they are

K_a (cation)
K_a (anion)
K_w
mass balance
charge balance

We generally use only as many as necessary for a given problem. By taking full advantage of them, the more involved problem of a salt of two hydrolyzing ions can be solved, but this is an area we shall not discuss. The interested student may want to consider Study Problem 3, however.

Problems in Which Two Components Are Present in Large Amounts

Having fortified ourselves with some additional concepts and notation that are important to aqueous solution equilibrium, let us go on to problems in

Acid–Base Equilibria | 515

which two components of the equilibrium are present in substantial quantities. In the typical acid-equilibrium expression, three concentrations appear:

$$HA_{(aq)} + H_2O_{(l)} \rightleftharpoons A^-_{(aq)} + H_3O^+_{(aq)}$$

$$K_a = \frac{[A^-][H_3O^+]}{[HA]}$$

If none of these three are specified, the stoichiometry of the reaction cannot help us; all three remain undetermined and only their ratio is known. If one of the three is specified, we can usually get a relationship between the other two by applying the concentration relationships arising from the stoichiometry of the reaction equation, so that in solving for one we are also getting the other. If two of the three are specified, the stoichiometry of the reaction does not matter; the third concentration is up against the wall, so to speak, and must adjust itself through the chemical reaction to conform to the equilibrium constant. A couple of examples will be helpful.

What is the pH of 1 l of water to which has been added 0.05 mole of formic acid, HCOOH, and 0.20 mole of sodium formate, $NaCHO_2$?

The balanced reaction equation and K_a expression are

$$HFo_{(aq)} + H_2O_{(l)} \rightleftharpoons Fo^-_{(aq)} + H_3O^+_{(aq)}$$

$$K_a = 1.8 \times 10^{-4} = \frac{[Fo^-][H_3O^+]}{[HFo]}$$

where Fo is the rest of the formic acid molecule. We can call the concentration of H_3O^+ x; it will be small since K_a is small. Since we are dealing with 1 l of solution, the concentration of formic acid will be 0.05 M initially, and by the stoichiometry of the reaction will be $0.05 - x$ at equilibrium (if we assume that the H_3O^+ contribution from water is negligible). Similarly, the concentration of formate ion, CHO_2^- or Fo^-, will be $0.20 + x$ at equilibrium. We thus have the equilibrium-constant expression

$$K_a = 1.8 \times 10^{-4} = \frac{(0.20 + x)(x)}{0.05 - x}$$

Approximating that x is negligible compared to 0.05, we write

$$1.8 \times 10^{-4} \cong \frac{0.20x}{0.05} = 4x$$

$$x \cong 0.45 \times 10^{-4} = 4.5 \times 10^{-5} \, M \cong [H_3O^+]$$

We can see that the approximation is satisfactory since 4.5×10^{-5} is indeed small compared to 0.05. To get the pH of the solution, we need only take the logarithm:

$$pH \equiv -\log[H_3O^+] = -\log(4.5 \times 10^{-5})$$
$$= -\log(4.5) - \log(10^{-5})$$
$$= -(0.65) - (-5)$$
$$= 5 - 0.65$$
$$= 4.35$$

This problem was solved by essentially the same procedure as earlier ones, but the only contribution from the reaction stoichiometry was the additive values of x that we disregarded. Of course, this will not always be permissible.

If equal volumes of two solutions that are, respectively, 0.02 M in trichloroacetate ion, Cl_3CCOO^-, and 0.04 M in trichloroacetic acid, Cl_3CCOOH, are mixed, what is the pH of the resulting solution at equilibrium?

Using the abbreviation HTC for the acid, we have the reaction equation and equilibrium-constant expression as follows:

$$HTC_{(aq)} + H_2O_{(l)} \rightleftharpoons TC^-_{(aq)} + H_3O^+_{(aq)}$$

$$K_a = 2 \times 10^{-1} = \frac{[TC^-][H_3O^+]}{[HTC]}$$

When equal volumes of the solutions are mixed, their initial concentration is halved in the resulting solution (at least before any reaction occurs). In the mixed solution, then, if we call $[H_3O^+]$ x, the concentration of trichloroacetic acid, HTC, will be not $0.04 - x$ but $0.02 - x$ (since x moles of it disappear when x moles of H_3O^+ are formed). Also, the concentration of trichloroacetate ion, TC^-, will be $0.01 + x$ by the same reasoning. We thus have

$$K_a = 2 \times 10^{-1} = 0.2 = \frac{(0.01 + x)(x)}{0.02 - x}$$

an expression exactly analogous to those in previous problems. If we make the standard approximation that x is negligible compared to 0.01, we write

$$0.2 \cong \frac{0.01x}{0.02} = \frac{x}{2} \quad \text{or} \quad x \cong 0.4$$

Checking the validity of the approximation, however, we find that 0.4 *is not* small compared to 0.01. We shall have to reject this answer and work out the quadratic version of the equilibrium-constant expression:

$$0.2 = \frac{0.01x + x^2}{0.02 - x} \quad \text{or} \quad 0.004 - 0.2x = 0.01x + x^2$$

$$x^2 + 0.21x - 0.004 = 0$$

$$x = \frac{-0.21 \pm \sqrt{(0.21)^2 - 4(1)(-0.004)}}{2(1)}$$

$$= \frac{-0.21 \pm \sqrt{0.0602}}{2}$$

Now x, a concentration, must be a positive number, so only the plus sign in the numerator will be physically acceptable:

$$x = \frac{-0.21 + 0.245}{2} = \frac{0.035}{2}$$

$$= 0.018\,M = [H_3O^+]$$

This is quite a different result from the approximate answer, and shows the necessity of checking the validity of any approximation that is used. The usual approximation breaks down only when K_a itself is larger than about 0.01 of the concentrations involved; any problem involving normal concentrations (0.01 M up) and K_a no larger than 10^{-4} will usually allow the approximation. But it is necessary to check it in every case, anyway. Using the exact result in this case, the pH of the solution is given by

$$\text{pH} = -\log(0.018) = -\log 1.8 - \log 10^{-2}$$
$$= -(0.25) - (-2)$$
$$= 1.75$$

Reactions involving a base, in which both components of the conjugate acid–base pair are present in macroscopic amounts, can be solved as simply (or perhaps even more simply) from the acid equilibrium reaction of Table 10-5 as they can by converting to the base reaction. There is no need to convert, since all species necessary for the reaction in either direction are present.

What is the pH of 300 ml of water in which are dissolved 10 g of diethylamine, $(CH_3CH_2)_2NH$ or DEA, and 10 g of diethylammonium chloride, $(CH_3CH_2)_2NH_2{}^+Cl^-$ or $DEAH^+Cl^-$?

The reaction equation and equilibrium-constant expression, from Table 10-5 for the acid reaction are

$$DEAH^+_{(aq)} + H_2O_{(l)} \rightleftharpoons DEA_{(aq)} + H_3O^+_{(aq)}$$

$$K_a = 8.0 \times 10^{-12} = \frac{[DEA][H_3O^+]}{[DEAH^+]}$$

To substitute in the equilibrium-constant expression we need to establish the initial concentrations from the data given in the problem:

DEA: MW = $4(12.01) + 11(1.008) + 14.01 = 73.1$ g/mole

$$\frac{10 \text{ g DEA}}{73.1 \text{ g/mole}} \frac{1}{300 \text{ ml}} 1000 \frac{\text{ml}}{1} = 0.45 \frac{\text{mole}}{1} \quad \text{or} \quad 0.45 \text{ M}$$

DEAH$^+$Cl$^-$: MW $= 73.1 + 1.008 + 35.46 = 109.6$ g/mole

$$\frac{10 \text{ g DEAH}^+\text{Cl}^-}{109.6 \text{ g/mole}} \frac{1}{300 \text{ ml}} 1000 \frac{\text{ml}}{1} = 0.30 \text{ M}$$

If we set $[H_3O^+] = x$, the DEA concentration at equilibrium will be $0.45 - x$; the minus sign arises because K_a is so small that we expect DEA to be a fairly good base, actually removing H_3O^+ from the solution. Pursuing the same reasoning, the DEAH$^+$ concentration at equilibrium will be $0.30 + x$. Inserting these values and making the approximation that $x \ll 0.30$, we have

$$K_a = 8.0 \times 10^{-12} = \frac{(0.45 - x)(x)}{0.30 + x} \cong \frac{0.45}{0.30} x$$

$$x \cong 5.3 \times 10^{-12} \cong [H_3O^+]$$

This is such a small number that our approximation was certainly justified, and we can proceed to calculate the pH:

$$pH = -\log 5.3 \times 10^{-12} = -\log 5.3 - \log 10^{-12}$$
$$= -(0.73) - (-12)$$
$$= 11.27$$

This solution is quite basic; any solution having more H_3O^+ than pure water is acidic, of course, corresponding to a pH less than 7.00, while any solution with a pH greater than 7.00 is basic. Our suspicion that the principal reaction would be that of a base is thereby verified. If it had not been, we would have had to rework the problem using concentrations $0.45 + x$ and $0.30 - x$ to get a different result. Of course, if x is negligible it will not matter. Generally, when both components of the acid–base conjugate pair are present we expect a basic solution if K_a is smaller than 10^{-7}, and an acidic solution if it is larger.

Polyprotic Acids

A few of the acids for which the equilibrium reactions are given in Table 10-5 can transfer more than one proton to a surrounding water molecule; they are **polyprotic**. Carbonic acid (H_2CO_3) and oxalic acid ($H_2C_2O_4$) are good examples of this behavior. Taking carbonic acid as an example, there are two possible proton-transfer reactions:

$$H_2CO_{3(aq)} + H_2O_{(l)} \rightleftharpoons HCO_{3(aq)}^- + H_3O_{(aq)}^+$$

$$HCO_{3(aq)}^- + H_2O_{(l)} \rightleftharpoons CO_{3(aq)}^{2-} + H_3O_{(aq)}^+$$

K_a for the transfer of the first proton is called K_1; K_a for the second proton is called K_2. The value of K_1 is always larger than K_2, because the strain on the electron distribution in the original molecule that makes the proton transfer possible is somewhat relieved by the departure of the first proton, so that the second is harder to remove. It should be intuitively clear that an equilibrium in which both the K_1 and K_2 reactions are significant sources of H_3O^+ will be difficult to treat for the same sort of reasons that the hydrolysis of ammonium acetate (see p. 514) was; there are too many variables and not enough interrelationships. The exact treatment follows the same lines as that for ammonium acetate hydrolysis, and we shall not go into it. What we can often do is to make the simplifying approximation that the only significant source of H_3O^+ is the K_1 reaction—of course, in any such problem, this approximation must be checked at the end. An argument can be presented, however, to guarantee the validity of the approximation for certain limiting values of K_1 and K_2. Consider a solution of the acid H_2A, in which the two reactions

$$H_2A + H_2O \rightleftharpoons HA^- + H_3O^+ \quad K_1 = \frac{[HA^-][H_3O^+]}{[H_2A]}$$
$$HA^- + H_2O \rightleftharpoons A^{2-} + H_3O^+ \quad K_2 = \frac{[A^{2-}][H_3O^+]}{[HA^-]}$$
(10-8)

are occurring. If K_2 is small enough that reaction's contribution to $[H_3O^+]$ will be negligible; how small does it have to be? Using our 5% criterion, the largest allowable contribution corresponds to $[A^{2-}] = 0.05\,[HA^-]$, since the proton transfer from HA^- must be no more than 0.05 of that which generates HA^-. So we have the condition

$$K_2 = \frac{0.05\,[HA^-][H_3O^+]}{[HA^-]} \quad \text{or} \quad \frac{K_2}{[H_3O^+]} = 0.05$$

That is, K_2 must not be larger than 5% of the actual hydronium ion concentration in the solution. If K_1 is relatively large, so that the hydronium ion concentration is fairly substantial, then K_2 can also be relatively large without destroying the usefulness of the approximation. If only the pure acid is added to the solution, $[H_3O^+]$ will always be greater than 10^{-7}, however, and so—for a solution of the pure acid only—K_2 can always be neglected if

$$\frac{K_2}{(>10^{-7})} = 0.05 \quad \text{or} \quad K_2 < 5 \times 10^{-9}$$

Carbonic acid meets this criterion, but oxalic acid does not. Suppose 0.1 mole of oxalic acid is dissolved in 1 l of water. Is the K_2 proton transfer important in this equilibrium?

Assuming that it is not, we solve the K_1 equilibrium-constant expression by familiar methods, although the exact quadratic solution is necessary:

$$K_1 = 6.5 \times 10^{-2} = \frac{[HOx^-][H_3O^+]}{[H_2Ox]} = \frac{(x)(x)}{0.1-x}$$

$$x^2 + 0.065x - 0.0065 = 0$$

$$x = \frac{-0.065 + \sqrt{(0.065)^2 - 4(1)(-0.0065)}}{2}$$

$$= \frac{-0.065 + \sqrt{0.03023}}{2} = \frac{-0.065 + 0.1738}{2}$$

$$= 0.0544 \, M = [H_3O^+]$$

If we now insert this value into the K_2 expression (since the same hydronium ion concentration governs both steps) we have

$$K_2 = 6.1 \times 10^{-5} = \frac{[H_3O^+][Ox^{2-}]}{[HOx^-]}$$

$$= 0.054 \frac{[Ox^{2-}]}{[HOx^-]}$$

$$\frac{[Ox^{2-}]}{[HOx^-]} = 1.1 \times 10^{-3}$$

In other words, much less than 5% of the HOx^- ion has transferred another proton — only about 1 in 1000 — and we are still justified in ignoring the K_2 reaction. If we needed the concentration of oxalate ion, Ox^{2-}, we could solve the last expression above for it. Alternatively, we could add the K_1 and K_2 reactions together, producing a new reaction whose overall equilibrium constant, like β_i for complexes, is the product of the individual steps:

$$H_2A + H_2O \rightleftharpoons HA^- + H_3O^+ \quad K_1$$
$$HA^- + H_2O \rightleftharpoons A^{2-} + H_3O^+ \quad K_2$$

$$H_2A + 2H_2O \rightleftharpoons A^{2-} + 2H_3O^+ \quad K_a = K_1 K_2$$

It is not legitimate to solve this reaction equation's equilibrium-constant expression directly for both $[H_3O^+]$ and $[A^{2-}]$ because the equilibrium solution does not really have A^{2-} as the principal product, but if we know $[H_3O^+]$ we can use this expression to solve for $[A^{2-}]$.

Under certain circumstances the limiting relationship for this approximation,

$$\frac{K_2}{[H_3O^+]} \leq 0.05$$

will not be true, either because of the relative size of K_1 and K_2 or because the pH of the solution is controlled at a value corresponding to too low a concentration of hydronium ion. When this happens the calculation of the pH becomes very difficult, but the pH range over which the approximation breaks down is not very wide. We have seen that if $K_2 = 5 \times 10^{-9}$ the approximation is good all the way to $[H_3O^+] = 10^{-7}$, at which point the second proton transfer is occurring 5% (or $\frac{1}{20}$) as much as the first. If we reduce the H_3O^+ concentration still further, the second proton transfer becomes still more important and there is less of the original H_2A species left to contribute protons. When the concentration ratio of 1:20 is reversed to 20:1 in favor of A^{2-}, we can to the same 5% degree of accuracy ignore any contribution from the K_1 reaction. This occurs when $[A^{2-}]/[HA^-] = 20$, or for our model system at $[H_3O^+] = 2.5 \times 10^{-10}$:

$$K_2 = 5 \times 10^{-9} = \frac{[A^{2-}]}{[HA^-]}[H_3O^+] = 20[H_3O^+]$$

$$[H_3O^+] = \frac{5 \times 10^{-9}}{20} = 2.5 \times 10^{-10}$$

In other words, from pH 0 to 7 we can assume that only the K_1 proton transfer is important, and from pH 9.6 (10 minus log 2.5) up we can assume that only the K_2 proton transfer is important. Only in the relatively short range from pH 7 to 9.6 is neither approximation valid. If a problem is phrased so that we cannot be sure what the resulting pH will be, we can make one assumption or the other and check the resulting pH to see whether the assumption was justified.

Concentration Control Through pH Adjustment

The argument of the last paragraph raises the possibility of controlling the concentration of certain ions by controlling the pH of the solution in which they are found. Suppose we go back to the EDTA determination of metals, in which EDTA or H_4Y is an acid capable of transferring four protons; it has $K_1 = 1.02 \times 10^{-2}$, $K_2 = 2.14 \times 10^{-3}$, $K_3 = 6.91 \times 10^{-7}$, and $K_4 = 5.50 \times 10^{-11}$. The five species H_4Y, H_3Y^-, H_2Y^{2-}, HY^{3-}, and Y^{4-} have different complexing abilities, since their complexing depends on their ability to donate electrons, which is greatest for the most negatively charged species. This means that the species H_2Y^{2-} may form a complex with one metal but not another (depending on the different stability constants), while HY^{3-} may complex both. If we wish to analyze for both metals in a mixture, we need only adjust the pH to make sure that the EDTA is in the H_2Y^{2-} form and titrate the one metal that it will complex, then readjust the pH to put the EDTA in the HY^{3-} form and titrate the other metal.

What pH would we need to maintain to keep EDTA in the HY^{3-} form to the maximum possible extent?

There are two equilibria that concern us: one creates HY^{3-} from H_2Y^{2-}, and one turns it into Y^{4-}. For these two we write

$$H_2Y^{2-} + H_2O \rightleftharpoons HY^{3-} + H_3O^+ \qquad K_3 = 6.91 \times 10^{-7} = \frac{[HY^{3-}][H_3O^+]}{[H_2Y^{2-}]}$$

$$HY^{3-} + H_2O \rightleftharpoons Y^{4-} + H_3O^+ \qquad K_4 = 5.50 \times 10^{-11} = \frac{[Y^{4-}][H_3O^+]}{[HY^{3-}]}$$

From these equilibrium constants we can obtain the two relationships

$$\frac{[HY^{3-}]}{[H_2Y]} = \frac{6.91 \times 10^{-7}}{[H_3O^+]} \quad \text{and} \quad \frac{[HY^{3-}]}{[Y^{4-}]} = \frac{[H_3O^+]}{5.50 \times 10^{-11}}$$

both of which must be maximized to get the maximum possible amount of HY^{3-}. In the first $[H_3O^+]$ must be as small as possible; in the second it must be as large as possible. Since it must take an intermediate value, we set the ratios equal to each other to solve for the most mutually favorable $[H_3O^+]$:

$$\frac{6.91 \times 10^{-7}}{[H_3O^+]} = \frac{[H_3O^+]}{5.50 \times 10^{-11}}$$

$$[H_3O^+]^2 = (6.91 \times 10^{-7})(5.50 \times 10^{-11})$$

$$[H_3O^+] = 6.18 \times 10^{-9} \, M$$

This means that less than 1% of the EDTA is in either of the other forms:

$$\frac{6.91 \times 10^{-7}}{6.18 \times 10^{-9}} = \frac{6.18 \times 10^{-9}}{5.50 \times 10^{-11}} = 112 \qquad \text{(concentration ratio)}$$

The student may verify that this corresponds to a pH of 8.21. It can be seen from Table 10-5 that there are many ions whose concentration can be controlled by controlling the pH; in the next section we shall need to use this technique.

pH Control Through Concentration Adjustment: Buffer Solutions

The other side of the coin is that by controlling the concentration of certain species in solution, the pH of the solution can be controlled. For this purpose we return to the idea that if both components of an acid–base conjugate pair are present in substantial quantities, the pH is essentially fixed. We realize that this is true for a given solution, because two of the three variables in the equilibrium-constant expression are given; what is not immediately clear is that if properly chosen, these concentrations will allow only a small pH change for a substantial addition of strong acid or strong base. Such a conjugate-pair solution is called a **buffer solution**. The diethylamine solution on p. 517 was an example of a buffered solution.

If 0.01 mole of HCl, a strong acid, is added to 1 l of water, how much does the pH change? If it is added to 1 l of solution that is 1 M in both acetic acid and sodium acetate, how much does the pH change?

Pure water has a pH of 7.00; if 0.01 mole of a monoprotic strong acid is added to 1 l of it, $[H_3O^+] = 0.01$, corresponding to a pH of 2.00. For the pure water, then, the pH change is 5.00 units.

Using our standard procedure, the pH of the acetic acid–acetate buffer solution is seen initially to be 4.76:

$$HAc + H_2O \rightleftharpoons Ac^- + H_3O^+ \qquad K_a = 1.75 \times 10^{-5} = \frac{[Ac^-][H_3O^+]}{[HAc]}$$

$$1.75 \times 10^{-5} = \frac{(1+x)(x)}{1-x} \cong \frac{1x}{1}$$

$$x = [H_3O^+] \cong 1.75 \times 10^{-5} \, M \qquad (1.75 \times 10^{-5} \ll 1)$$

$$pH = -\log(1.75 \times 10^{-5}) = 5 - 0.24 = 4.76$$

If we add 0.01 mole of H_3O^+ to this, the concentration fraction will exceed the equilibrium constant and the reaction will run in reverse until enough H_3O^+ and Ac^- have been used up to reduce the concentration fraction to the equilibrium value. Since there is a lot more acetate present than H_3O^+, even after the addition, the H_3O^+ will be almost completely used up, shrinking from 0.01 M to some new concentration y, but the acetate ion will still be present in the reaction, and its concentration will become $1 + 0.01 - y$. We thus have the equilibrium-constant expression

$$1.75 \times 10^{-5} = \frac{(1 - 0.01 + y)(y)}{1 + 0.01 - y} \cong \frac{0.99}{1.01} y$$

$$y = [H_3O^+] \cong 1.78 \times 10^{-5} \, M \qquad (1.78 \times 10^{-5} \ll 0.99)$$

$$pH = -\log(1.78 \times 10^{-5}) = 5 - 0.25 = 4.75$$

The pH change in the buffer solution is only 0.01 unit, compared to a change of 5.00 units in pure water. Even if the pH of the pure water had been adjusted with strong acid to 4.76 before adding the 0.01 mole of strong acid, it would still have changed 2.75 units rather than 0.01 unit. So buffer solutions have a considerable resistance to changing their pH. It can be seen from the form of the equilibrium-constant expression that the effectiveness of a buffer depends on having large concentrations of *both* acid and base present; if either falls below roughly 0.01 M the buffer loses its pH stability. A corollary is that a conjugate acid–base pair can serve as a buffer only over a short pH range on either side of its pK_a value, perhaps one pH unit either way. Within this range, of course, the pH of the buffer can be adjusted by adjusting the relative amounts of each component.

pH Titrations

As a last example of the use of acid–base equilibria and the pH concept, let us look at acid–base or pH titrations. For many years, acid–base titrations have been done using an unknown acid and a standard solution of hydroxide ion, or an unknown base and a standard solution of hydronium ion. These are quite analogous to EDTA titrations in that a "metal ion" (H_3O^+) is being "complexed" by an electron donor, OH^-. Just as in the EDTA case, we can use an indicator, which for acid–base titrations is a donor molecule that is one color in its protonated form and another color in its deprotonated form; the same considerations apply to the "stability constant" or K_a for the indicator molecule as to the EDTA indicator. However, in recent years, an instrument called a pH meter has become widely available, which by electrical means reads the pH of a solution directly on an electrical meter or digital readout; we shall delay a discussion of its operating principles until Section 10-7. In monitoring an acid–base titration with a pH meter, we observe for all but the very weakest acids a sharp jump in the pH of the solution when the addition of a strong base solution passes the equivalence point. The exact shape of the titration curve (in a graphical representation) depends on the K_a of the acid, as in Fig. 10-3.

We can best appreciate the reasons for the break in the pH titration curve by calculating the pH for several points during a sample titration. In so doing, we

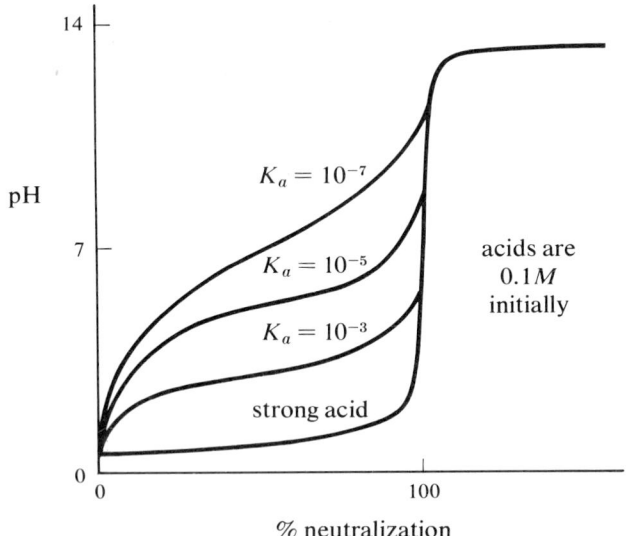

Figure 10-3 Titration curves for acids of decreasing strength. Reducing acid concentration has the same effect.

are again controlling pH by controlling concentrations of the other species present. Consider a solution that is $0.1000\,M$ in acetic acid, and is being titrated with sodium hydroxide with no dilution effect. The pH of the solution is initially 2.88:

$$1.75 \times 10^{-5} = \frac{[H_3O^+][Ac^-]}{[HAc]} = \frac{(x)(x)}{0.1000-x}$$

$$x^2 \cong 1.75 \times 10^{-6}$$

$$x \cong 1.32 \times 10^{-3} \cong [H_3O^+] \qquad (0.00132 \ll 0.1000)$$

$$pH = -\log(1.32 \times 10^{-3}) = 3 - 0.12 = 2.88$$

When 10% of the acetic acid has been neutralized, the acetate concentration has climbed to $0.0100+x$, and the acetic acid concentration has shrunk to $0.0900-x$:

$$1.75 \times 10^{-5} = \frac{(x)(0.0100+x)}{0.0900-x} \cong \frac{0.01}{0.09}x$$

$$x \cong 9 \times 1.75 \times 10^{-5} = 1.57 \times 10^{-4} \cong [H_3O^+] \qquad (1.57 \times 10^{-4} \ll 0.01)$$

$$pH = -\log(1.57 \times 10^{-4}) = 4 - 0.20 = 3.80.$$

When 50% of the acetic acid has been neutralized, the acetate concentration is $0.0500+x$, and the acetic acid concentration is $0.0500-x$:

$$1.75 \times 10^{-5} = \frac{(x)(0.0500+x)}{0.0500-x} \cong \frac{0.0500}{0.0500}x$$

$$x \cong 1.75 \times 10^{-5} \cong [H_3O^+] \qquad (1.75 \times 10^{-5} \ll 0.05)$$

$$pH = -\log(1.75 \times 10^{-5}) = 5 - 0.24 = 4.76$$

When 90% of the acetic acid has been neutralized, the acetate concentration has risen to $0.0900+x$ and the acetic acid concentration has diminished to $0.0100-x$:

$$1.75 \times 10^{-5} = \frac{(x)(0.0900+x)}{0.0100-x} \cong \frac{0.09}{0.01}x$$

$$x \cong \frac{1.75}{9} \times 10^{-5} = 1.95 \times 10^{-6} \cong [H_3O^+] \qquad (1.95 \times 10^{-6} \ll 0.01)$$

$$pH = -\log(1.95 \times 10^{-6}) = 6 - 0.29 = 5.71$$

Close to the equivalence point, when 99% of the acetic acid has been titrated, the acetate concentration is $0.0990+x$ and the acetic acid concentration is $0.0010-x$:

$$1.75 \times 10^{-5} = \frac{(x)(0.0990 + x)}{0.0010 - x} \cong \frac{0.099}{0.001} x$$

$$x \cong \frac{1.75}{99} \times 10^{-5} = 1.77 \times 10^{-7} \cong [H_3O^+] \quad (1.77 \times 10^{-7} \ll 0.001)$$

$$pH = -\log(1.77 \times 10^{-7}) = 7 - 0.25 = 6.75$$

When exactly enough sodium hydroxide has been added to reach the equivalence point, there is no acetic acid left—except for what is formed through the hydrolysis of the acetate ion (whose concentration is now $0.1000 - y$):

$$K_h = \frac{K_w}{K_a} = \frac{10^{-14}}{1.75 \times 10^{-5}} = 5.71 \times 10^{-10} = \frac{[HAc][OH^-]}{[Ac^-]} = \frac{(y)(y)}{0.1000 - y}$$

$$5.71 \times 10^{-10} \cong \frac{y^2}{0.1000}$$

$$y^2 \cong 57.1 \times 10^{-12}$$

$$y \cong 7.56 \times 10^{-6} \cong [OH^-] \quad (7.56 \times 10^{-6} \ll 0.1000)$$

$$pH = 14 - pOH = 14 + \log(7.56 \times 10^{-6}) = 14 - 6 + 0.88 = 8.88$$

At 1% past the equivalence point, the excess OH^- concentration has risen to $0.0010\ M$ (1% of the original acetic acid concentration), plus whatever has formed through hydrolysis of the acetate ion:

$$5.71 \times 10^{-10} = \frac{[HAc][OH^-]}{[Ac^-]} = \frac{(y)(0.0010 + y)}{0.1000 - y}$$

It can be seen that y must be a very small number, so $[OH^-] = 0.0010\ M$. Then

$$pH = 14 - pOH = 14 + \log(0.0010) = 14 - 3 = 11.00$$

At 10% past the equivalence point, a similar argument shows that the pH is 12.00, at which point we may consider the titration to be over. Figure 10-4 shows these points plotted to emphasize the break in the curve at the equivalence point. The low slope in the vicinity of 50% titration is simply a reflection of the buffer capacity of the acetate–acetic acid conjugate pair, both of which are present in substantial quantity in that range of the titration. Such titrations as this are extremely useful in analysis, although very weak acids give a poorly defined break at the equivalence point; this problem can be overcome by going to a more basic solvent than water, such as liquid ammonia (see p. 505 and the discussion of the leveling effect). Furthermore, with a little electronics and a motor-driven buret, an automatic titrator can be constructed that cuts itself off when the pH reaches the value characteristic of the equivalence point.

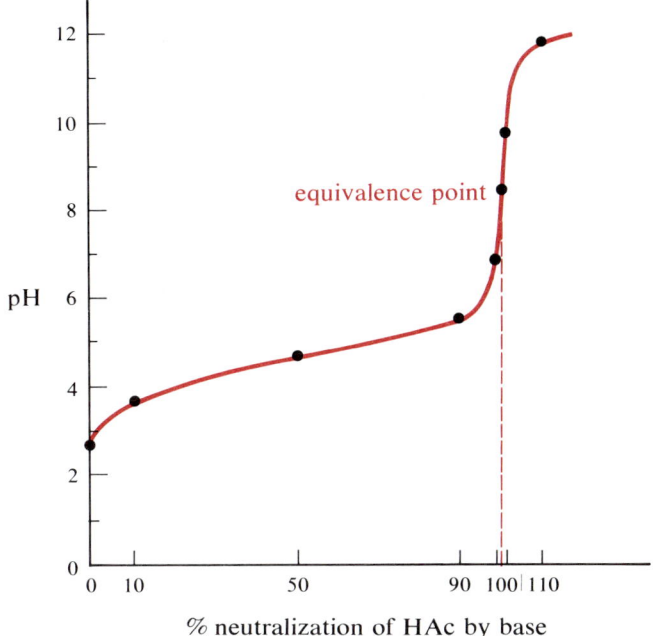

Figure 10-4 The pH titration curve for acetic acid.

10-5 Solubility Equilibria in Aqueous Solution

Another type of reaction in aqueous solution is that in which a solid compound, in which some degree of charge separation into ions exists, dissolves to form aqueous ions. The positive ions are usually hydrated in a stoichiometric fashion, with four or six water molecules firmly bonded to the positive ion; but all of the ions at least exert an orienting effect on the polar water molecules that surround them. The dissolving of ionic compounds thus represents a form of chemical reaction. We can describe the solubility of the ionic or partially ionic compound in terms of the equilibrium constant for this "dissolving reaction." We write the reaction

$$A_m B_{n(s)} \rightleftharpoons m A^{n+}_{(aq)} + n B^{m-}_{(aq)}$$

and note that the pure solid will have an activity ratio of 1, so that it need not appear in the equilibrium-constant expression:

$$K_{\text{solubility}} = [A^{n+}]^m [B^{m-}]^n$$

Because the solid's activity ratio disappears, the equilibrium-constant expression has the appearance of a product rather than a fraction; it is often called the **solubility product**, with a **solubility-product constant**, K_{sp}. In principle such a solubility product exists for all such compounds, since they all have a limited (or equilibrium) solubility. In practice, however, the expression is applied only to sparingly soluble compounds, since when the concentrations at equilibrium are high the departure of the activity from the measured concentrations is great and the expressions are quite difficult to apply. For compounds that we normally think of as quite soluble, such as sodium chloride, the solubility is usually limited by the use of all the water molecules in forming hydrated ions. Since water is 55 M in water and most hydrated ions involve roughly four to six water molecules per ion, concentrations exceeding 10 M are most unusual. Most such solubility limits are in the range 1–5 M.

Problems in Which One Component Is Present in Large Amounts

For sparingly soluble compounds, however, the solubility-product expression is very useful and accurate, since for low concentrations the activity is very near the concentration. As for other types of equilibria, we can subdivide solubility-product problems into those involving only one component and those involving two. Taking first problems involving only one component present in large quantity, consider the following example.

Fifteen milligrams of calcium fluoride, CaF_2, dissolve in 1 l of water at equilibrium. What is the value of the solubility-product constant?

CaF_2 has a molecular weight (or formula weight) of 78.1. In dissolving, it follows the "reaction" equation

$$CaF_{2(s)} \rightleftharpoons Ca^{2+}_{(aq)} + 2F^-_{(aq)}$$

For this reaction we write the solubility-product expression

$$K_{sp} = [Ca^{2+}][F^-]^2$$

Again, as in all the other equilibrium problems for which only one component is specified, we must rely on the stoichiometry of the reaction. From the weight of solid calcium fluoride dissolving we can get the number of moles dissolving per liter, from which the stoichiometry gives us the ion concentrations necessary. To get the moles of solid calcium fluoride dissolving, we use the molecular weight:

$$\text{moles dissolving} = \frac{0.015 \text{ g}}{78.1 \text{ g/mole}} = 1.9 \times 10^{-4} \text{ mole (per liter)}$$

Every time 1 mole of solid salt dissolves, it forms 1 mole of Ca^{2+} in solution, so the concentration of Ca^{2+} is also 1.9×10^{-4} M. But for each mole of solid salt dissolving we get 2 moles of fluoride ion (see the reaction equation), so the fluoride-ion concentration will be *twice* the calcium ion concentration — $3.8 \times 10^{-4}\ M$. This will be true only if solid CaF_2 is the only source of both ions, or if we make up the solution in this special way. With these concentrations, we can quickly get the value of the solubility-product constant:

$$K_{sp} = [Ca^{2+}][F^-]^2$$
$$= (1.9 \times 10^{-4})(3.8 \times 10^{-4})^2$$
$$= (1.9 \times 10^{-4})(14.8 \times 10^{-8})$$
$$= 2.8 \times 10^{-11}$$

Some K_{sp} values obtained in this and other ways are given in Table 10-6.

Table 10-6
Solubility-Product Constants for the Aqueous Reaction $M_mX_{n(s)} \rightleftharpoons mM^{n+}_{(aq)} + nX^{m-}_{(aq)}$ at 25°C

Solid Species	K_{sp}	Solid Species	K_{sp}
$Mg(OH)_2$	1.1×10^{-11}	AgBr	5.0×10^{-13}
$Ca(OH)_2$	9×10^{-6}	AgI	8.3×10^{-17}
$Al(OH)_3$	1×10^{-32}	MnS	2×10^{-13}
$Cr(OH)_3$	1×10^{-30}	FeS	5×10^{-18}
$Fe(OH)_3$	2×10^{-39}	CoS	4×10^{-21}
$Co(OH)_2$	2×10^{-15}	NiS	3×10^{-19}
$Ni(OH)_2$	1×10^{-15}	CuS	6×10^{-36}
$Cu(OH)_2$	1.6×10^{-19}	ZnS	2×10^{-24}
$Zn(OH)_2$	3×10^{-17}	$CaSO_4$	1×10^{-5}
CaF_2	2.8×10^{-11}	$BaSO_4$	1.0×10^{-10}
PbF_2	3×10^{-8}	$PbSO_4$	1.7×10^{-8}
AgCl	1.8×10^{-10}	$BaCrO_4$	1×10^{-10}
$PbCl_2$	1.7×10^{-5}	$PbCrO_4$	3×10^{-13}
Hg_2Cl_2 (to Hg^{2+} ion)	1.6×10^{-18}	Ag_2CrO_4	2.5×10^{-12}

Since these compounds are only slightly soluble, the equilibrium constants are all smaller than 1.0, and some of them are very small indeed. In general, the smaller an equilibrium constant is, the less reliable the value is, even sometimes by several orders of magnitude (powers of 10). But even at their worst, the values are useful approximations from which we can calculate solubilities when needed.

How much lead(II) chloride will dissolve in 100 ml of water? We assume that only one component, the solid lead(II) chloride, is present in any substantial quantity. Writing the dissolution equation and solubility product, we have

$$PbCl_{2(s)} \rightleftharpoons Pb^{2+}_{(aq)} + 2Cl^-_{(aq)} \qquad K_{sp} = [Pb^{2+}][Cl^-]^2$$

From Table 10-6 we have $K_{sp} = 1.7 \times 10^{-5}$. If we allow the concentration of Pb^{2+} to be x, then the concentration of Cl^- can only be $2x$, since the dissolution is the only source of both ions. This immediately gives us the molarity of a a saturated solution:

$$K_{sp} = 1.7 \times 10^{-5} = [Pb^{2+}][Cl^-]^2$$
$$= (x)(2x)^2$$
$$= 4x^3$$
$$x^3 = 4.3 \times 10^{-6}$$
$$x = 1.6 \times 10^{-2} \, M = [Pb^{2+}] \quad \text{or} \quad 0.016 \, M$$

The molecular weight of $PbCl_2$ is 278.1, which gives us the weight solubility:

$$0.016 \frac{\text{moles}}{1} \times 278.1 \frac{\text{g}}{\text{mole}} \times 0.1 \frac{1}{100 \text{ ml}} = 0.45 \text{ g}/100 \text{ ml}$$

Note that these problems are considerably simplified by the use of dimensional analysis.

Problems in Which Two Components Are Present in Large Amounts

For cases in which two components of the equilibrium are present in substantial quantities—usually the solid and one of the ions—we find a considerable analytical application for the calculations. For instance, compounds containing Cl, Br, or I are usually analyzed by precipitating these elements using silver ion, since AgCl, AgBr, and AgI are nearly insoluble (see Table 10-6). Suppose we have a solution whose volume is 125 ml in which a chloride-containing unknown compound has been dissolved (and we assume that the separation into aqueous ions is complete). If silver nitrate is added, AgCl precipitates. Suppose enough silver nitrate is added to make the solution $0.05 \, M$ in Ag^+; how much chloride remains unprecipitated?

The precipitation reaction is just the reverse of the dissolution reaction, for which we write

$$AgCl_{(s)} \rightleftharpoons Ag^+_{(aq)} + Cl^-_{(aq)} \qquad K_{sp} = 1.8 \times 10^{-10} = [Ag^+][Cl^-]$$

In this case we know the concentration of the silver ion, and we need only substitute to get the concentration of the remaining chloride:

$$1.8 \times 10^{-10} = [\text{Ag}^+][\text{Cl}^-]$$
$$= (0.05)(x)$$
$$x = 3.6 \times 10^{-9} \, M$$

In a volume of 125 ml, this corresponds to a very small concentration of chloride that has escaped our clutches:

$$3.6 \times 10^{-9} \frac{\text{mole}}{\text{l}} \times 0.125 \, \text{l} = 4.5 \times 10^{-10} \, \text{mole Cl}^- \text{ remaining}$$

$$4.5 \times 10^{-10} \, \text{mole} \times 35.46 \, \text{g Cl/mole} = 1.53 \times 10^{-8} \, \text{g Cl}^- \text{ remaining}$$

This would only amount to a 1% error if we had weighed out a sample containing just one and a half micrograms of chloride. To prevent overconfidence, however, it should be pointed out that analysis of foodstuffs for residues of chlorinated pesticides is presently being carried out at the nanogram level— 10^{-9} g—by gas chromatography and other more sensitive techniques. This amount, of course, would not be precipitated at all.

Solubility Control Through Solution Equilibria

As we have seen, it is possible to control ion concentrations through solution equilibria involving the ions. This technique is quite important in analytical separations involving precipitation. For example, consider the previous solution, which before precipitation was 0.01 M in the chloride. However, suppose the solution is made 1 M in NH_3, which complexes silver ion to $Ag(NH_3)_2^+$ with a stability constant of 1.6×10^7. This has the effect of controlling the concentration of free (i.e., hydrated) silver ion as it is added. If 0.00125 mole of silver ion is added to the 125 ml of solution (the same number of moles as the chloride), will the chloride ion precipitate?

We need to find out what concentration of hydrated silver ion is allowed by the ammonia. If this is greater than the minimum needed for precipitation of chloride ion, precipitation of solid silver chloride will occur; otherwise, it will not. The reaction equation and the equilibrium-constant expression for the formation of the complex are as follows:

$$\text{Ag}^+_{(aq)} + 2\text{NH}_{3(aq)} \rightleftharpoons \text{Ag(NH}_3)^+_{2(aq)}$$

$$K_{st} = 1.6 \times 10^7 = \frac{[\text{Ag(NH}_3)_2^+]}{[\text{Ag}^+][\text{NH}_3]^2}$$

Calculating as in Section 10-2, we take [Ag$^+$] as x and from the large value of the stability constant assume it to be small compared to [Ag(NH$_3$)$_2^+$]:

$$1.6 \times 10^7 = \frac{[\text{Ag(NH}_3)_2^+]}{[\text{Ag}^+][\text{NH}_3]^2} = \frac{0.01 - x}{(x)(1 - 2 \times 0.01 + 2x)}$$

$$1.6 \times 10^7 \cong \frac{0.01}{(x)(0.98)}$$

$$x \cong \frac{0.01}{1.6 \times 10^7 \times 0.98} = 6.3 \times 10^{-10}\, M \cong [Ag^+] \qquad (6.3 \times 10^{-10} \ll 0.01)$$

Now we can ask whether this concentration of hydrated silver ion is sufficient to precipitate the chloride present. The silver concentration just sufficient to start precipitation is that given by the substitution of the actual chloride-ion concentration into the solubility product:

$$K_{sp} = 1.8 \times 10^{-10} = [Ag^+][Cl^-]$$
$$= [Ag^+](0.01)$$

$$[Ag^+] = \frac{1.8 \times 10^{-10}}{0.01} = 1.8 \times 10^{-8}\, M$$

In other words, no precipitation of AgCl is going to occur until the concentration of silver ion builds up to $1.8 \times 10^{-8}\, M$. But we previously calculated that the presence of 1-M ammonia limited the presence of hydrated silver ion to $6.3 \times 10^{-10}\, M$, a much smaller concentration. Consequently, no solid AgCl will be formed.

This result suggests that silver ion will not precipitate halides from strongly ammoniacal solutions—but suppose the same experiment were done on a solution identical in every respect except that the halide ion was I^- instead of Cl^-. The resulting concentration of hydrated silver ion would be the same: $6.3 \times 10^{-10}\, M$. In this case, however, the concentration necessary to begin precipitation is much smaller:

$$K_{sp} = 8.3 \times 10^{-17} = [Ag^+][I^-]$$
$$= [Ag^+](0.01)$$

$$[Ag^+] = \frac{8.3 \times 10^{-17}}{0.01} = 8.3 \times 10^{-15}\, M$$

Since this threshold for precipitation is smaller than the existing concentration of silver ion, a precipitate will form in this case. The smallness of the numbers may suggest that only a tiny amount of precipitate would form, but we can see that this is not the case by calculating the equilibrium concentration of iodide in the presence of this controlled concentration of silver ion:

$$K_{sp} = 8.3 \times 10^{-17} = [Ag^+][I^-]$$
$$= (6.3 \times 10^{-10})[I^-]$$

$$[I^-] = \frac{8.3 \times 10^{-17}}{6.3 \times 10^{-10}} = 1.3 \times 10^{-7}\, M$$

So of the original 0.01 M solution of iodide, only a very small amount remains—all the rest has precipitated out. This can happen because, as the hydrated silver ion is used up in precipitation, the silver ammine complex maintains its equilibrium by releasing some silver into the hydrated form:

$$Ag(NH_3)_{2(aq)}^+ \rightleftharpoons Ag_{(aq)}^+ + 2NH_{3(aq)}$$

Although not very much of the silver in the reaction mixture exists as the hydrated ion at any given time, nearly all of it passes through that form on its way to precipitate the iodide. When the relative concentrations of the two halide ions are roughly of the same order of magnitude, this difference in behavior toward the silver ammine complex ion offers a way to selectively precipitate the iodide while leaving the chloride in solution. Such selective precipitation is important to many analytical methods, and we shall look at other means of achieving it.

Solubility Control Through pH Adjustment

One of the most useful approaches to selective precipitation involves controlling the concentration of a negative ion through protonation. The anions of many weak acids form quite insoluble salts with many metals, for reasons we shall go into in Chapter 13. One of the most widely used anions is the sulfide ion, S^{2-}, which can be regarded as the base conjugate with the weak acid HS^- or with the diprotic weak acid H_2S. For any pH lower than about 7, it can be assumed that the largest portion of the sulfide is present as H_2S, and the combined equilibrium constant can be used for calculating the S^{2-} concentration:

$$H_2S_{(aq)} + 2H_2O_{(l)} \rightleftharpoons S_{(aq)}^{2-} + 2H_3O_{(aq)}^+$$

$$K_a = K_1 K_2 = 1.1 \times 10^{-20} = \frac{[S^{2-}][H_3O^+]^2}{[H_2S]}$$

The sulfide is commonly added as gaseous H_2S, by bubbling it into the solution until it is saturated at about 0.10 M. It should be apparent that under these circumstances the concentration of S^{2-}, and hence the precipitation of metal sulfides, can be controlled by controlling the acidity of the solution.

A solution is 0.100 M in both $Mn(H_2O)_6^{2+}$ and $Ni(H_2O)_6^{2+}$. What pH must be maintained in the solution if only one of the metals is to be precipitated (as completely as possible) when the solution is saturated with gaseous H_2S?

The solubility reactions for these two sulfides, and the values of the solubility-product constants, are found in Table 10-6:

$$MnS_{(s)} \rightleftharpoons Mn(H_2O)_{6(aq)}^{2+} + S_{(aq)}^{2-} \qquad K_{sp} = 2 \times 10^{-13}$$

$$NiS_{(s)} \rightleftharpoons Ni(H_2O)_{6(aq)}^{2+} + S_{(aq)}^{2-} \qquad K_{sp} = 3 \times 10^{-19}$$

If only one of the two metals is to precipitate it will have to be nickel, since nickel sulfide has the smaller solubility, as represented by the smaller number given as the solubility-product constant. To precipitate nickel as completely as possible we must have the sulfide ion concentration as high as possible, but we are limited by the fact that if it gets very concentrated the manganese will begin to precipitate. First we must see what S^{2-} concentration just suffices to start precipitation of the manganese, and then use a little less than that in precipitating the nickel:

$$K_{sp} = 2 \times 10^{-13} = [Mn(H_2O)_6^{2+}][S^{2-}]$$
$$= (0.100)[S^{2-}]$$

$[S^{2-}] = 2 \times 10^{-12}\ M$ for incipient precipitation of MnS

To stay slightly under this value, we might select $[S^{2-}] = 1 \times 10^{-12}\ M$ for the optimum precipitation of NiS. Assuming that H_2S will be the predominant solution species for sulfide, we can calculate the desired pH from the combined acid-equilibrium constant for H_2S:

$$1.1 \times 10^{-20} = \frac{[S^{2-}][H_3O^+]^2}{[H_2S]}$$
$$= \frac{(1 \times 10^{-12})[H_3O^+]^2}{0.1}$$

$$[H_3O^+]^2 = \frac{(1.1 \times 10^{-20})(0.1)}{1 \times 10^{-12}} = 1.1 \times 10^{-9}$$

$$[H_3O^+] = 3.3 \times 10^{-5} \qquad pH = 5 - 0.52 = 4.48$$

This pH is low enough to ensure that the H_2S has not transferred protons substantially, as we could show by using K_1 for H_2S. What concentration of $Ni(H_2O)_6^{2+}$ is left in solution when equilibrium is reached at this pH; that is, how complete is the precipitation? We need only substitute the existing $[S^{2-}]$ into the NiS solubility product to calculate it:

$$3 \times 10^{-19} = [Ni(H_2O)_6^{2+}][S^{2-}]$$
$$= [Ni(H_2O)_6^{2+}](1 \times 10^{-12})$$

$$[Ni(H_2O)_6^{2+}] = \frac{3 \times 10^{-19}}{1 \times 10^{-12}} = 3 \times 10^{-7}\ M$$

This very low value assures us that the separation of nickel from manganese in solution can be made quantitative by proper selection of the solution conditions.

Hydroxide Precipitation by Hydrated Ions

Acid-base equilibria and solubility equilibria interact in other ways that have important effects on solution chemistry. What limits exist on the pH of a solution of iron(III)?

In analyzing the problem, we see from Table 10-6 that the difficulty is that $Fe(OH)_3$—a gloppy, gelatinous material—is quite insoluble. As a measure of its insolubility we can write the solubility reaction and the solubility product:

$$Fe(OH)_{3(s)} \rightleftharpoons Fe(H_2O)_{6(aq)}^{3+} + 3OH^-_{(aq)}$$

$$K_{sp} = 2 \times 10^{-39} = [Fe(H_2O)_6^{3+}][OH^-]^3$$

From this very low value it is clear that if the concentration of OH^- is anything other than very small, the solution will be unstable because $Fe(OH)_3$ will precipitate. This situation is described equally well by considering the proton transfer away from $Fe(H_2O)_6^{3+}$ to water molecules or OH^- ions in solution; there is a stepwise formation of $Fe(H_2O)_5(OH)^{2+}$, then $Fe(H_2O)_4(OH)_2^+$, then $Fe(H_2O)_3(OH)_3^0$, which in a less stoichiometric form is the **hydrous oxide** precipitate that appears. This proton transfer is influenced by the concentration of OH^-, a good base or proton acceptor, in solution, and we could approach the problem in this way. But since in our problem we only wish to avoid the formation of a precipitate, we use the solubility product:

$$K_{sp} = 2 \times 10^{-39} = [Fe(H_2O)_6^{3+}][OH^-]^3$$

$$= (x)(3x)^3 \quad \text{(since only one component)}$$

$$x^4 = \frac{2 \times 10^{-39}}{27} = 7.4 \times 10^{-41} = 7400 \times 10^{-44}$$

$$x = 9.3 \times 10^{-11} \, M = [Fe(H_2O)_6^{3+}]$$

$$[OH^-] = 3x = 2.8 \times 10^{-10} \, M$$

$$pH = 14 - pOH = 14 + \log 2.8 + \log 10^{-10} = 14 + 0.45 - 10 = 4.45$$

This pH of 4.45, which is moderately acidic, is that which is developed by $Fe(OH)_3$ dissolving. It is thus approximately the most basic solution of Fe^{III} that can exist. If we insist on having the iron present in substantial concentration, the answer is quite different. Suppose the solution is 0.1 M in iron:

$$2 \times 10^{-39} = [Fe(H_2O)_6^{3+}][OH^-]^3$$

$$= (0.1)[OH^-]^3$$

$$[OH^-]^3 = \frac{2 \times 10^{-39}}{0.1} = 20 \times 10^{-39}$$

$$[OH^-] = 2.7 \times 10^{-13}$$
$$pH = 14 - pOH = 14 - 13 + 0.43 = 1.43$$

In other words, for a 0.1-M solution of iron(III) to be stable the pH must be *less* than 1.43, since any greater concentration of OH^- will begin to reduce the iron concentration through precipitation of $Fe(OH)_3$ [or $Fe(OH)_3 \cdot nH_2O$].

Of course we can also calculate the conditions under which a precipitated solid will dissolve. An interesting example is $Cu(OH)_2$, which can be dissolved in two ways: either by the addition of acid, which through the water equilibrium takes OH^- ions out of solution; or by the addition of ammonia, which makes the solution even more basic but also takes $Cu(H_2O)_4^{2+}$ ions out of solution through the formation of the complex $Cu(NH_3)_4^{2+}$. What pH would be necessary to dissolve $Cu(OH)_2$ up to a $Cu(H_2O)_4^{2+}$ concentration of 0.1 M? Using ammonia, what solution conditions (including pH) would be necessary to dissolve $Cu(OH)_2$ up to a $Cu(NH_3)_4^{2+}$ concentration of 0.1 M?

The first part of the problem is quite straightforward; we need only write the reaction equation and solubility-product expression, and solve the latter for the OH^- concentration, from which we can get the pH:

$$Cu(OH)_{2(s)} \rightleftharpoons Cu(H_2O)_{4(aq)}^{2+} + 2OH^-_{(aq)}$$

$$K_{sp} = 1.6 \times 10^{-19} = [Cu(H_2O)_4^{2+}][OH^-]^2$$

$$1.6 \times 10^{-19} = [Cu(H_2O)_4^{2+}][OH^-]^2$$

$$= (0.1)[OH^-]^2$$

$$[OH^-]^2 = 1.6 \times 10^{-18}$$

$$[OH^-] = 1.25 \times 10^{-9}\ M$$

$$pH = 14 - pOH = 14 + 0.10 - 9 = 5.10$$

Such a solution is only slightly acidic, so $Cu(OH)_2$ is readily dissolved in acid. The second part of the problem is somewhat more complicated. In dissolving, the tetramminecopper(II) complex is being formed, but it is not being formed from the aqueous ion, so the ordinary stability-constant reaction equation does not apply. Consideration of the reacting species gives us the balanced equation

$$Cu(OH)_{2(s)} + 4NH_{3(aq)} \rightleftharpoons Cu(NH_3)_{4(aq)}^{2+} + 2OH^-_{(aq)}$$

The equilibrium-constant expression for this reaction equation would be

$$K_{eq} = \frac{[Cu(NH_3)_4^{2+}][OH^-]^2}{[NH_3]^4}$$

for which we have no value of the constant. However, we can generate an equivalent expression using the overall stability constant β_4 and the solubility-product constant:

$$K_{eq} = \frac{[Cu(NH_3)_4^{2+}][OH^-]^2}{[NH_3]^4} = \frac{[Cu(NH_3)_4^{2+}]}{[Cu(H_2O)_4^{2+}][NH_3]^4}[Cu(H_2O)_4^{2+}][OH^-]^2$$

$$= \beta_4 K_{sp}$$

The stability constant β_4 for the tetramminecopper(II) complex is available; it is 1×10^{13}. The equilibrium constant for the dissolving reaction is thus

$$K_{eq} = \beta_4 K_{sp} = (1 \times 10^{13})(1.6 \times 10^{-19}) = 1.6 \times 10^{-6}$$

When the reaction occurs it is because ammonia (*not* ammonium ion) is being added to the suspension of solid $Cu(OH)_2$ in water. For every mole of $Cu(OH)_2$ that dissolves, 1 mole of $Cu(NH_3)_4^{2+}$ is formed in solution, and 2 moles of OH^-. This OH^- is not consumed by NH_3 because the corresponding base (to NH_3 as an acid) is the amide ion, NH_2^-, which is a stronger base than OH^- and thus cannot exist in water solution. So if the solution is ultimately to be 0.1 M in $Cu(NH_3)_4^{2+}$, it must also be 0.2 M in OH^-—a surprising result for an ammonia solution since ammonia is a weak base. Of course, a similar OH^- concentration could be obtained by adding NaOH to an ammonia solution, which as a strong base would raise the OH^- concentration at the expense of the NH_4^+ concentration. By forming a complex with the copper, we are forcing the $Cu(OH)_2$ to act as a strong base, yielding a pH of $14 + \log(0.2) = 13.20$. We can solve the equilibrium-constant expression for the concentration of free NH_3 that must be present:

$$1.6 \times 10^{-6} = \frac{[Cu(NH_3)_4^{2+}][OH^-]^2}{[NH_3]^4}$$

$$= \frac{(0.1)(0.2)^2}{[NH_3]^4}$$

$$[NH_3]^4 = \frac{(0.1)(0.04)}{1.6 \times 10^{-6}} = 2.5 \times 10^3 = 2500$$

$$[NH_3] = 7.1\ M$$

So in order to get 0.1-M copper complex the solution must be made 7 M in ammonia, thereby driving the dissolving reaction to the right. We have tacitly assumed that there was no change in either $[OH^-]$ or $[NH_3]$ due to formation of NH_4^+ from the existing NH_3, and we can check on this by calculating the equilibrium NH_4^+ concentration, since the NH_3 consumed and the OH^- formed in the hydrolysis of NH_3 are both equivalent to the NH_4^+ formed:

$$NH_{3(aq)} + H_2O_{(l)} \rightleftharpoons NH_{4(aq)}^+ + OH_{(aq)}^-$$

$$K_h = \frac{K_w}{K_a} = \frac{10^{-14}}{5.8 \times 10^{-10}} = \frac{[NH_4^+][OH^-]}{[NH_3]}$$

$$\frac{10^{-14}}{5.8 \times 10^{-10}} = 1.7 \times 10^{-5} = \frac{[NH_4^+][OH^-]}{[NH_3]}$$

$$= \frac{[NH_4^+](0.2)}{7.1}$$

$$[NH_4^+] = \frac{7.1 \times 1.7 \times 10^{-5}}{0.2} = 6.0 \times 10^{-4} \, M$$

This is indeed negligibly small compared to the NH_3 and OH^- concentrations present at equilibrium, and we need not change any of the results.

This problem illustrates an important principle of equilibrium calculations, which is the necessity of writing an equation for the reaction as it actually occurs, even though it may be different from the standard forms in our tables. When this is done, a value for the appropriate equilibrium constant can be produced by combining standard equilibrium-constant expressions and their constants. The resulting equilibrium-constant expression will accurately describe the concentration relationships in the solution equilibrium of interest, whereas a standard form that neglects one or more of the several equilibria that must be satisfied in some solutions will be entirely misleading.

10-6 Electron-Transfer Reactions in Solution

Throughout this chapter we have been using the thermodynamic concept of the equilibrium constant, derived from free energies, to quantitatively measure the extent to which a reversible reaction occurs. There is another class of reaction that lends itself particularly well to the experimental measurement of the extent of reaction, or, more accurately, to the direct measurement of the free energy of reaction. This class comprises the electron-transfer reactions, in which one species donates electrons to another so completely that its overall net electrical charge changes. A model equation for this type of reaction is

$$A^0_{(aq)} + B^0_{(aq)} \rightleftharpoons A^+_{(aq)} + B^-_{(aq)}$$

In this reaction an electron has been transferred completely from A to B. What governs the extent to which this reaction occurs? Of course, the free energy does, but what specific influences affect the free-energy change? Let us look at what happens to the electron-donor atom (or molecule) only. It is releasing an electron:

$$A^0_{(aq)} \rightleftharpoons A^+_{(aq)} + e^-$$

This process absorbs a good deal of energy, because it strongly resembles the ionization process (Chapter 5):

$$A^0_{(g)} \xrightarrow{IP_1} A^+_{(g)} + e^-_{(g)}$$

For any stable, storable system, whether atom or molecule, this process requires an energy input called the ionization energy. We also used the name ionization potential, however, since we could measure the energy of the process by using an electrical potential or voltage to strip away the electron; the ionization energy is the charge on the electron times the voltage difference necessary to remove it, which is what we call the ionization potential. (Refresh your memory from pp. 246–249.) The comparison is not exact, though, because the system we are now discussing is not gaseous. Since the species are now in aqueous solution, there are hydration energies as well as entropy effects arising from the different degree of ordering of water molecules around the positive ion and the neutral atom. We indicate the relationship by the following diagram:

$$\begin{array}{ccc} A^0_{(g)} & \xrightarrow{\text{first ionization potential}} & A^+_{(g)} + e^- \\ \uparrow {\scriptstyle -A^0 \text{ hydration energy}} & & \downarrow {\scriptstyle A^+ \text{ hydration energy}} \\ A^0_{(aq)} & \xrightarrow{\text{electron-transfer energy}} & A^+_{(aq)} + e^- \end{array}$$

Since the free-energy quantity includes both energy and entropy effects and is a thermodynamic state variable, the free energy of the bottom reaction in the diagram is equal to the sum of the free energies of the three top reactions. What is common to this electron-transfer reaction in solution and the ionization reaction in the gas phase is that both liberate exactly one electron from its bound condition. They differ in that the solution reaction is more complicated due to the ion–dipole attractions, and so on, appearing in the hydration processes. Nonetheless, since we can measure the ionization energy by measuring the electrical potential necessary to strip the electron from the gaseous species, we ought to be able to measure the solution electron-transfer energy—a free-energy quantity—by measuring the electrical potential necessary to strip the electron from the solution species. In either case, exactly one electron is being moved through space from one bound system to another location; an electrical voltage applied over that space of movement should be sufficient to cause that movement. Since the free energy of the ionization process is equal to the number of electrons being moved times the voltage difference through which they move, the free energy of the solution electron-transfer process should also be equal to the number of electrons being moved times the voltage difference through which they move. In other words, for the slightly more general reaction

$$A^0_{(aq)} \rightarrow A^{n+}_{(aq)} + ne^- \quad -\Delta G^0$$

we ought to be able to write the general relationship

$$\Delta G^0 = Cn\mathscr{E}^0$$

where $\mathscr{E}^0$ is the standard electrical potential or voltage difference associated with the reaction and C is a conversion factor from electrically based energy units to mechanical or heat-based energy units; it would be 1.00 if ΔG^0 were taken for a single molecule and measured in electron volts. It is customary, however, to give ΔG^0 for 1 mole of A, and to express it in electrical energy units of joules. Looking back at p. 12, we see that 1 eV/atom = 96,487 J/mole; since a coulomb-volt is a joule, 1 electron charge/atom = 96,487 C/mole, a quantity called the **faraday** after Michael Faraday, who did the pioneering experiments on the chemical effects of electricity (see Chapter 5). Using the symbol $\mathscr{F}$ for the faraday, we have

$$\Delta G^0 = -n\mathscr{F}\mathscr{E}^0 \tag{10-9}$$

where $\mathscr{E}^0$ is, again, a voltage associated with the electron-transfer reaction, which we could presumably measure, and the minus sign arises from the standard convention about measuring electrical potential. That is, a positive voltage corresponds to a spontaneous electrical change, and spontaneity corresponds to a negative ΔG^0.

Electric Potential Differences and Their Measurement

To be able to measure in the laboratory the free-energy change for an electron-transfer reaction involving only a single atom or molecule, we need to be able to measure a voltage *difference*, and because the electron is not really free but has gone to some other bound system, it is not at all clear what difference is to be chosen. The answer is that we cannot measure a characteristic potential $\mathscr{E}^0$ associated with the undefined reaction

$$A^0_{(aq)} \rightleftharpoons A^{n+}_{(aq)} + ne^-(?)$$

To make the measurement in the laboratory, we must consider the whole electron-transfer reaction

$$A^0_{(aq)} + B^0_{(aq)} \rightleftharpoons A^{n+}_{(aq)} + B^{n-}_{(aq)}$$

and measure the voltage difference between the **couple** A^0/A^{n+} and the other couple B^0/B^{n-}. (An electron-transfer couple is analogous to a conjugate acid–base pair.) Figure 10-5 shows one way this can be done experimentally. The electron-transfer reaction under consideration is

$$Cu^0_{(s)} + 2Ag^+_{(aq)} + 4H_2O_{(l)} \rightleftharpoons Cu(H_2O)^{2+}_{4(aq)} + 2Ag^0_{(s)}$$

Metallic copper is placed in the left beaker, in contact with a solution containing $Cu(H_2O)^{2+}_4$ ions, and metallic silver is placed in the right beaker, in

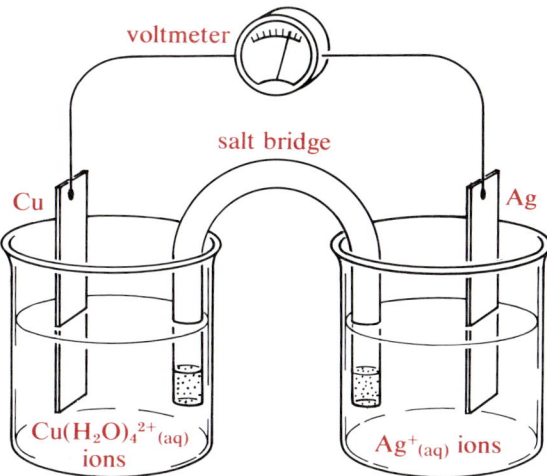

Figure 10-5 Experimental arrangement for observing differences in half-reaction potential. The salt-bridge tube contains a salt solution (such as K^+ or NO_3^-) and is given a porous plug at each end to allow diffusion of ions but not bulk flow of the solution.

contact with a solution containing $Ag^+_{(aq)}$ ions. The electron transfer from copper to silver (going to the right in the reaction equation above) occurs by having the electrons flow through a wire connecting the metallic copper and metallic silver, and overall electrical neutrality in each beaker is maintained by allowing ions to flow through the middle tube called a **salt bridge**. We can measure the free-energy change of the reaction by measuring the voltage difference between the two metal **electrodes**, as shown. In fact, a voltage difference does exist, and the two beakers plus salt bridge are said to form an electrical **cell**.

Free-energy changes can be very conveniently measured for electron-transfer reactions in solution by measuring the potential difference between the donor couple and acceptor couple, after arranging the geometry of the reaction mixture so that the electrons being transferred must flow through an electrical circuit external to the cell. But there is no way to establish an absolute potential for any single electrical couple. However, by making a series of comparative measurements with different couples it is possible to place the various couples in an order of relative ease of donation or vigor of acceptance of electrons. That is, if A always donates electrons to B and B always donates electrons to C, then A also always donates electrons to C, and we can establish a little table:

$$C^+ + e^- \rightleftharpoons C^0$$

$$B^+ + e^- \rightleftharpoons B^0$$

$$A^+ + e^- \rightleftharpoons A^0$$

542 | Equilibrium Constants and Electrochemistry

in which the higher a couple is in the table, the more strongly it tends to go to the right when in a cell with another couple.

The Standard Electron-Transfer Couple and Relative Reduction Potentials

There is a sense, then, in which the individual couples or **half-cell reactions** can be considered individually—in a table such as the above. No absolute potential can be assigned to a given couple, but since all we need is internal consistency in the table, we can arbitrarily choose a single one of the couples to have a defined potential of 0.0000 V and relate all the others to that one. By agreement, the couple

$$2H_3O^+_{(aq)} + 2e^- \rightleftharpoons H_{2(g)} + 2H_2O_{(l)} \tag{10-10}$$

has been defined to have an individual half-cell potential of exactly 0 V with all of the components in their thermodynamic standard states: 1 atm pressure, 298 °K temperature, 1 m (hypothetical Henry's law) concentration H_3O^+. By careful measurement, a substantial series of other relative (not absolute) half-cell potentials have been established under this arbitrary convention; some of these are given in Table 10-7. All of the half-reactions in this table are written so that they are accepting electrons, which is again a convention. When an atom or molecule accepts electrons it is being **reduced**, and these half-cell potentials are **reduction potentials**. It would be equally possible to write the half-reactions so that they all represented donation of electrons, and this was done for many years with attendant confusion. When an atom or molecule donates electrons it is being **oxidized**, and such a table of half-cell potentials would be **oxidation potentials**. The student may encounter such a table, and it is quite workable as long as the two types of potentials are not mixed; but reduction potentials are more consistent with other electrical sign conventions.

From Table 10-7 we can immediately predict the voltage that would be developed by any given cell—it is simply the difference between the two individual reduction potentials. For instance, the potential developed by the copper/silver cell in the earlier example would, from Table 10-7, be $0.800 - 0.346 = 0.454$ V. We can tell the direction in which the reaction would go because, as in the little table on the previous page, the half-reaction that is higher in the table—has a more positive potential—will go in the direction it is written, and the other half-reaction will go in the reverse direction. One of the two electrodes in any cell has a surplus of negatively charged electrons and consequently is said to be the **negative electrode**. The other has a deficiency of electrons and is consequently positively charged and is called the **positive electrode**. The electrode from which electrons leave the cell to go into the wire is known as the **anode**; the other, where electrons enter the cell from the wire,

Table 10-7
Reduction Half-Cell Potentials[a]

Half-Reaction	$\mathscr{E}^0$ (V at 25°C)
$H_4XeO_6 + 2H_3O^+ + 2e^- \rightarrow XeO_3 + 5H_2O$	3.0
$F_2 + 2e^- \rightarrow 2F^-$	2.87
$O_3 + 2H_3O^+ + 2e^- \rightarrow O_2 + 3H_2O$	2.07
$PbO_2 + SO_4^{2-} + 4H_3O^+ + 2e^- \rightarrow 6H_2O + PbSO_4$	1.686
$MnO_4^- + 8H_3O^+ + 5e^- \rightarrow Mn(OH_2)_6^{2+} + 6H_2O$	1.491
$ClO_4^- + 8H_3O^+ + 8e^- \rightarrow Cl^- + 12H_2O$	1.36
$Cl_2 + 2e^- \rightarrow 2Cl^-$	1.358
$Cr_2O_7^{2-} + 14H_3O^+ + 6e^- \rightarrow 2Cr(OH_2)_6^{3+} + 9H_2O$	1.33
$O_2 + 4H_3O^+ + 4e^- \rightarrow 6H_2O$	1.229
$Br_2 + 2e^- \rightarrow 2Br^-$	1.066
$2Hg^{2+} + 2e^- \rightarrow Hg_2^{2+}$	0.910
$ClO^- + H_2O + 2e^- \rightarrow Cl^- + 2OH^-$	0.90
$Hg^{2+} + 2e^- \rightarrow Hg^0$	0.854
$Ag^+ + e^- \rightarrow Ag^0$	0.800
$Hg_2^{2+} + 2e^- \rightarrow 2Hg^0$	0.798
$Fe(OH_2)_6^{3+} + e^- \rightarrow Fe(OH_2)_6^{2+}$	0.770
$I_2 + 2e^- \rightarrow 2I^-$	0.535
$Cu(OH_2)_4^{2+} + 2e^- \rightarrow Cu^0 + 4H_2O$	0.346
$Hg_2Cl_2 + 2e^- \rightarrow 2Hg^0 + 2Cl^-$ (in saturated KCl solution; saturated calomel electrode)	0.242
$AgCl + e^- \rightarrow Ag^0 + Cl^-$	0.222
$Cu(OH_2)_4^{2+} + e^- \rightarrow Cu(OH_2)_4^+$	0.153
$2H_3O^+ + 2e^- \rightarrow H_2 + 2H_2O$	0 (defined)
$Pb(OH_2)_4^{2+} + 2e^- \rightarrow Pb^0 + 4H_2O$	−0.125
$Ni(OH_2)_6^{2+} + 2e^- \rightarrow Ni^0 + 6H_2O$	−0.250
$Co(OH_2)_6^{2+} + 2e^- \rightarrow Co^0 + 6H_2O$	−0.277
$PbSO_4 + 2e^- \rightarrow Pb^0 + SO_4^{2-}$	−0.356
$Cd(OH_2)_4^{2+} + 2e^- \rightarrow Cd^0 + 4H_2O$	−0.402
$Cr(OH_2)_6^{3+} + e^- \rightarrow Cr(OH_2)_6^{2+}$	−0.41
$Fe(OH_2)_6^{2+} + 2e^- \rightarrow Fe^0 + 6H_2O$	−0.440
$Cr(OH_2)_6^{3+} + 3e^- \rightarrow Cr^0 + 6H_2O$	−0.74
$Zn(OH_2)_4^{2+} + 2e^- \rightarrow Zn^0 + 4H_2O$	−0.763
$Cr(OH_2)_6^{2+} + 2e^- \rightarrow Cr^0 + 6H_2O$	−0.91
$V(OH_2)_6^{2+} + 2e^- \rightarrow V^0 + 6H_2O$	−1.18
$Mn(OH_2)_6^{2+} + 2e^- \rightarrow Mn^0 + 6H_2O$	−1.182
$Al(OH_2)_6^{3+} + 3e^- \rightarrow Al^0 + 6H_2O$	−1.66
$Al(OH_2)_2(OH)_4^- + 3e^- \rightarrow Al^0 + 4OH^- + 2H_2O$	−2.36
$Mg(OH_2)_6^{2+} + 2e^- \rightarrow Mg^0 + 6H_2O$	−2.37
$Na(OH_2)_n^+ + e^- \rightarrow Na^0 + nH_2O$	−2.71
$Ca(OH_2)_6^{2+} + 2e^- \rightarrow Ca^0 + 6H_2O$	−2.87
$K(OH_2)_n^+ + e^- \rightarrow K^0 + nH_2O$	−2.92
$Li(OH_2)_4^+ + e^- \rightarrow Li^0 + 4H_2O$	−3.05

[a] In acid solution unless OH^- appears in the half-reaction

544 | Equilibrium Constants and Electrochemistry

is the **cathode**. The half-reaction that goes on at the anode is always the giving up of electrons by the cell, or an oxidation; at the cathode the half-reaction is always accepting electrons in a reduction. In the copper/silver example, the silver couple has the more positive reduction potential and thus will proceed as a reduction when the electrical circuit is completed, and the copper couple will proceed in reverse, as an oxidation. Since reduction occurs at the silver electrode it is the cathode and the copper electrode is the anode. Since at the silver electrode silver ions are accepting electrons and draining them out of the metal, that electrode is the positive electrode, which agrees with the sign of the reduction potential relative to the copper electrode; this is one of the virtues of reduction potentials — the electrode with the more positive reduction potential is also the positive electrode in the cell.

Balancing Redox Reaction Equations

We shall have a good deal more to say about electrical cells and electrochemistry but let us take time out here to discuss the stoichiometry of electron-transfer reactions (also called **oxidation–reduction reactions** or **redox reactions**). Redox reactions occur in any electrical cell, but they may also occur in a conventional reaction mixture; the stoichiometry is the same. We have already written one balanced redox equation:

$$Cu^0_{(s)} + 2\,Ag^+_{(aq)} + 4H_2O_{(l)} \rightleftharpoons Cu(H_2O)^{2+}_{4(aq)} + 2\,Ag^0_{(s)}$$

Why does the coefficient 2 appear? The number of atoms would be equally well balanced without it. The answer is that the total electrical charge on each side of the equation must also be the same. These are the two basic rules for the balancing of redox equations: balance the number of atoms (or moles) of each element, and balance the charge. The reason for the first rule is simply the conservation of matter. To understand the reason for the second rule, we need to think about the nature of the electron-transfer reaction and look at it in terms of its individual half-reactions. The half-reactions, as they appear in Table 10-7, are

$$Ag^+_{(aq)} + e^- \rightleftharpoons Ag^0_{(s)}$$

$$Cu(OH_2)^{2+}_{4(aq)} + 2e^- \rightleftharpoons Cu^0_{(s)} + 4H_2O_{(l)}$$

The silver half-reaction goes as it is written, accepting electrons, and the copper half-reaction goes in reverse (to the left), donating electrons. Now there are no free electrons in a reaction mixture, any more than there are free protons in an acid solution. The electrons are transferred from a bound condition on one atom or molecule to a bound condition on another. In terms of the stoichiometry of the reaction equation this means that the total number of electrons donated must equal the total number accepted. Since the silver half-reaction as written accepts only one electron, we must take two of them for every copper half-reaction that donates two electrons. We can develop this idea into a systematic procedure for balancing redox equations, as follows.

When we are faced with establishing the stoichiometry of a redox reaction in aqueous solution, we need to remember that in many cases the water enters the reaction either as H_2O, H_3O^+, or OH^-, and we frequently need to use these species to account for all the atoms that may be present in the two existing half-reactions. We shall come to the use of these in a moment. Perhaps the most convenient approach to the balancing of these equations is a stepwise list of operations:

1. Pick out the two half-reactions and write each separately, using only the chemical species in that half-reaction without worrying about H_2O species or electrons. For instance, suppose we have the skeleton reaction

$$Cr_2O_{7(aq)}^{2-} + H_2CO_{(aq)} \rightleftharpoons Cr(H_2O)_{6(aq)}^{3+} + HCOOH_{(aq)}$$

Identifying each half-reaction as a couple involving like atoms or molecules, we write:

$$Cr_2O_7^{2-} \rightleftharpoons Cr(H_2O)_6^{3+}$$

$$H_2CO \rightleftharpoons HCOOH$$

as the skeleton half-reaction equations.

2. Take one of these skeleton half-reactions and insert coefficients to balance the number of atoms of each element other than H and O, which may need to be balanced using water species. In our example:

$$Cr_2O_7^{2-} \rightleftharpoons 2Cr(H_2O)_6^{3+}$$

balances the number of Cr atoms on each side.

3. If all the atoms are balanced, go to step 4; if O or H is still unbalanced, add enough H_2O molecules to the side that is short in O to make the O atoms balance, then add enough H^+ (we'll hydrate it in a minute) to the side that is short in H to balance the H atoms including those on the H_2O just added:

$$5H_2O + Cr_2O_7^{2-} \rightleftharpoons 2Cr(H_2O)_6^{3+}$$

[10H, 12O]   [24H, 12O]

$$14H^+ + 5H_2O + Cr_2O_7^{2-} \rightleftharpoons 2Cr(H_2O)_6^{3+}$$

[24H, 12O]   [24H, 12O]

4. With all the atoms balanced, balance the total electrical charge on each side by adding electrons to the side that is too positive (or not negative enough):

$$6e^- + 14H^+ + 5H_2O + Cr_2O_7^{2-} \rightleftharpoons 2Cr(H_2O)_6^{3+}$$

[total charge 6+]   [total charge 6+]

This completes the balancing process for one half-reaction.

5. Repeat steps 2, 3, and 4 for the other half-reaction. Sometimes the reaction will be a disproportionation, in which one element serves both as a donor and an acceptor ($A^{2+} \rightleftharpoons A^0 + A^{4+}$), or the reverse reaction in which both half-reactions have a common product. Either way, the two half-reactions that are written will have a common species either as a reactant or as a product, but they should be balanced individually in exactly the same way. In our example, the other half-reaction is balanced as follows:

(2) $\qquad\qquad H_2CO \rightleftharpoons HCOOH \qquad$ C already balanced

(3) $\qquad\qquad H_2O + H_2CO \rightleftharpoons HCOOH$

$\qquad\qquad\qquad$ [4H, 2O] $\qquad$ [2H, 2O]

$\qquad\qquad H_2O + H_2CO \rightleftharpoons HCOOH + 2H^+$

$\qquad\qquad\qquad$ [4H, 2O] $\qquad$ [4H, 2O]

(4) $\qquad\qquad H_2O + H_2CO \rightleftharpoons HCOOH + 2H^+ + 2e^-$

$\qquad\qquad\qquad$ [0 charge] $\qquad$ [0 charge]

6. Now that both half-reactions are balanced, check to see that one is producing and the other consuming electrons. Multiply each half-reaction through by a factor such that the number of electrons produced equals the number consumed; this number should be the least common multiple of the two factors. Here we have

$$6e^- + 14H^+ + 5H_2O + Cr_2O_7^{2-} \rightleftharpoons 2Cr(H_2O)_6^{3+}$$

and

$$3(H_2O + H_2CO \rightleftharpoons HCOOH + 2H^+ + 2e^-)$$

$$3H_2O + 3H_2CO \rightleftharpoons 3HCOOH + 6H^+ + 6e^-$$

7. Add these two half-reactions together to get the full reaction, canceling the equal numbers of electrons on each side of the sum and also canceling and combining H_2O or H^+ wherever possible. Free electrons MUST NOT APPEAR in the balanced equation.

$$\cancel{6e^-} + \overset{8}{\cancel{14}}H^+ + \overbrace{5H_2O + Cr_2O_7^{2-} + 3H_2O}^{8} + 3H_2CO \rightleftharpoons 2Cr(H_2O)_6^{3+}$$

$$+ 3HCOOH + \cancel{6}H^+ + \cancel{6e^-}$$

$$8H^+ + 8H_2O + Cr_2O_7^{2-} + 3H_2CO \rightleftharpoons 2Cr(H_2O)_6^{3+} + 3HCOOH$$

8. Finally, convert any H^+ in the resulting equation to the proper species as follows: (a) If the reaction mixture is acid or neutral (in a pH sense) add H_2O to

both sides to make all H^+ H_3O^+, and cancel with previous H_2O wherever possible; (b) if the reaction mixture is basic or alkaline, add OH^- to both sides to make all H^+ H_2O, and cancel with previous H_2O wherever possible. In our example, the solution is acidic (we could assume that if not otherwise specified), and we have

$$8H_3O^+ + 8H_2O + Cr_2O_7^{2-} + 3H_2CO \rightleftharpoons 2Cr(H_2O)_6^{3+} + 3HCOOH + 8H_2O$$

or

$$8H_3O^+ + Cr_2O_7^{2-} + 3H_2CO \rightleftharpoons 2Cr(H_2O)_6^{3+} + 3HCOOH$$

as our final balanced equation.

Since the stoichiometry of any reaction is the most basic element in understanding it, practice applying this method (or an equivalent one) to some of the skeleton equations in the study problems until you are certain of your ability to balance redox equations; the test of balance is, very simply, atom balance *and* charge balance in the final equation.

10-7 Electrochemistry and Thermodynamics

With an understanding of the stoichiometry of redox reactions, let us go on to a further examination of the ways they occur and a little more of the theoretical background. Any redox reaction that can run as an electrical cell can also occur in the simpler manner when the reactants are mixed, but the reverse is not true. Some redox reactions or half-reactions that are conceptually straightforward cannot be made to occur in an aqueous solution half-cell for various reasons: sometimes because one of the species in the couple reacts vigorously and irreversibly with the water molecules, and sometimes because there is no available mechanism for the electrode reaction. For those systems that can be made to function in a cell, however, there are a number of different kinds of electrodes that have been developed.

Types of Cell Electrodes

The most obvious type is that used in the cell of Fig. 10-5. In that cell each electrode is a metal plate that actually takes part in the cell reaction—copper dissolves away from the copper plate and silver plates out onto the silver plate. Such electrodes are called **active** or **metal-solution** electrodes, the names being fairly obvious. A variation of this electrode that is sometimes useful with very active metals or to eliminate surface effects is the **amalgam electrode** shown in Fig. 10-6. An amalgam is a solution of a metal in liquid mercury; as an elec-

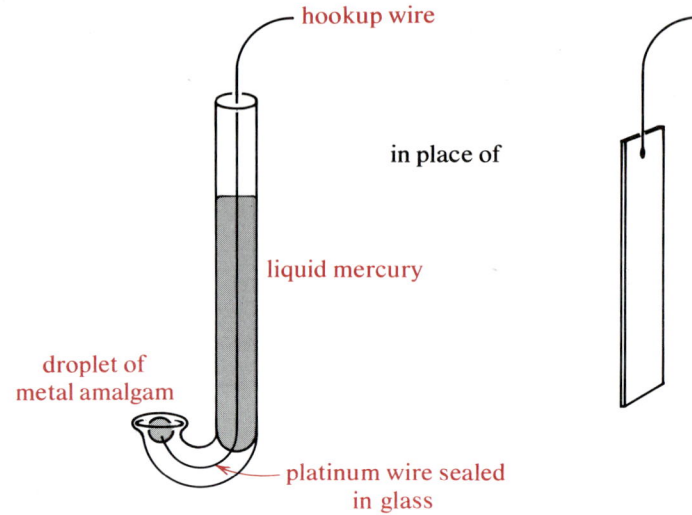

Figure 10-6 Metal amalgam (J-tube) electrode.

trode it is desirable because, as a liquid, it has a very reproducible surface exposed to the aqueous solution. If it is dilute (which is usually the case) it also prevents the reaction of its solute metal from being too vigorous in case the metal is a very good electron donor.

A case in which the electrode serves a different purpose is shown in Fig. 10-7. Here the redox couple is a pair of ions in solution, as for example iron(II) and iron(III). The two ions can transfer electrons between themselves just as

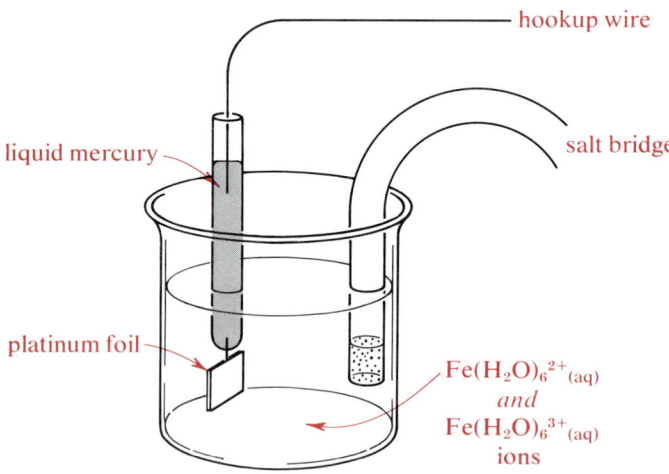

Figure 10-7 Inert electrode in a half-cell having both redox-active species in solution.

the copper plate and the copper(II) ion did, but the metal of the electrode—usually platinum because of its chemically inert nature—does not take part in the electrode reaction. Instead, it serves only as a carrier for the electrons flowing into or out of the redox couple, and in effect adjusts its potential to match that of the couple. Such an electrode is variously called an **inert**, **sensing**, or **redox electrode**. A variation of this is the **gas electrode** shown in Fig. 10-8. Here

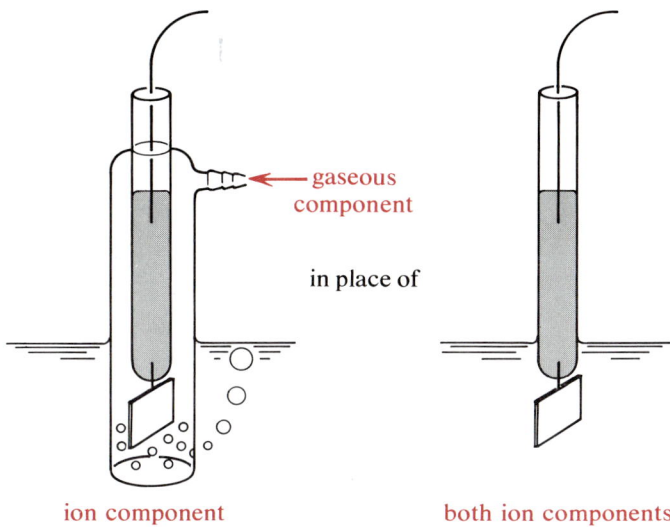

Figure 10-8   Gas electrode compared to inert electrode.

the platinum serves exactly the same purpose, but instead of having both components of the redox couple as ions in solution, one is a gas that is bubbled over the platinum strip while it is wet by the solution containing the other component. The most important gas electrode is the hydrogen electrode, which serves as the reference or zero point for the arbitrary assignment of individual half-reaction potentials.

A recent development of potentially great usefulness in analytical chemistry is the **ion-selective** or **membrane** electrode. All of the electrodes mentioned thus far are sensitive to changes in the concentration of the dissolved ion or ions that are represented in the redox couple, for reasons we shall discuss shortly. By choosing the couple properly it is possible to produce an electrode whose potential is not concentration sensitive, because the concentration is controlled by a solubility equilibrium; for instance, silver metal and silver ion form a concentration-sensitive electrode, but not if the silver ion comes from insoluble AgCl in the presence of a fairly concentrated Cl⁻ solution. Such electrodes as Ag/AgCl are called **reference electrodes** because they provide a constant reference potential against which the potential of some other concentration-sensitive electrode may be checked. However, if a silver/silver chloride electrode is surrounded by a membrane that can absorb and exchange ions of a particular type from solution (by virtue of its molecular or crystal structure providing "sites" for the solution ions to fit into), and if this membrane can conduct at least small currents of electricity by diffusion of ions through it, the potential of the internal silver/silver chloride electrode will depend on the concentration (or activity) of the ion that is being absorbed from the solution. Depending on the structure of the membrane, the ion-selective electrode can be tailored to be selectively sensitive to any of a wide variety of ions, both cations and anions. The oldest and most familiar of the ion-selective electrodes is the **glass electrode**, which is sensitive to the solution's H_2O^+ concentration because its thin glass membrane exchanges protons with the solution's H_3O^+ ions. Fig. 10-9 illustrates this and some newer ion-selective electrodes for other ions.

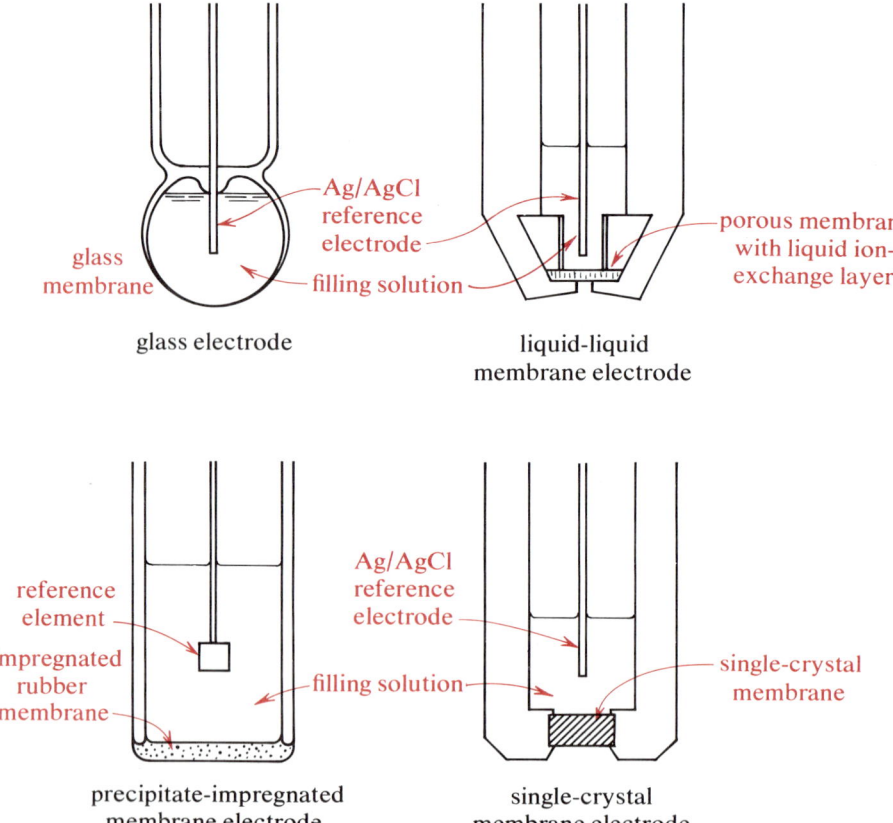

Figure 10-9 Ion-sensitive electrodes.

Cells for Power Generation

For analytical purposes, where we want to know what the concentration of some ion is, a concentration dependence of the electrode's potential is desirable. On the other hand, for the generation of electrical potential by a cell for external use, we would like the potential to be constant from the time the cell is fresh until it is completely discharged. This ideal cannot be achieved exactly, but several cells approximate it reasonably well. Most practical cells develop a rather low voltage, and indeed we can see from Table 10-7 that no conceivable electrochemical cell could develop more than about 6 V as its $\mathscr{E}^0$. Within reasonable limits, however, higher voltages can be achieved by connecting cells together in series, that is, with the positive electrode of one cell connected to the negative electrode of the next as in Fig. 10-10, to form a **battery**. An auto-

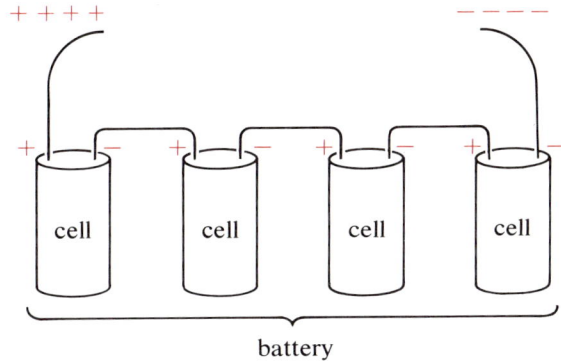

Figure 10-10 Series connection of cells to form a battery.

mobile battery develops about 12 V by having six cells, each of which develops approximately 2 V between the two electrodes:

$$Pb^0_{(s)} + SO^{2-}_{4(aq)} \rightleftharpoons PbSO_{4(s)} + 2e^-$$

and

$$PbO_{2(s)} + 4H_3O^+_{(aq)} + SO^{2-}_{4(aq)} + 2e^- \rightleftharpoons PbSO_{4(s)} + 6H_2O_{(l)}$$

Notice that these half-reactions are only moderately concentration sensitive since most of the components are solids. The other familiar use of electrochemical cells as a power source is the flashlight cell, which is a cell rather than a battery, since it needs to yield only about 1.5 V. The flashlight cell or dry cell does not use liquid solutions for ion transport within the cell, but rather a heavy paste of solid MnO_2 held together by a small amount of ammonium chloride-zinc chloride solution. An electrically conducting carbon rod is used in the same sense as a platinum inert electrode, and the whole thing is encased in a zinc can that serves as the other electrode:

$$Zn_{(s)} \rightleftharpoons Zn^{2+}_{(paste\ matrix)} + 2e^-$$

$$2MnO_{2(s)} + Zn^{2+}_{(paste)} + 2e^- \rightleftharpoons ZnMn_2O_{4(s)}$$

Again the preponderance of solids in the redox couples serves to help keep the voltage of the cell constant as it is used.

An active area of engineering research is the development of **fuel cells**, a few of which have reached considerable refinement. The goal is to achieve greater efficiency in the use of conventional fuels such as gasoline by allowing the reaction between gasoline and air, which involves electron transfer, to occur in an electrochemical cell rather than in a flame. The burning of gasoline in any type of internal combustion engine is relatively inefficient, besides leading to severe air pollution in urban areas. If the gasoline burned by an ordinary automobile could be allowed to react with air in an electrochemical cell instead, a considerable electrical current could be drawn to run the car through electrical motors with greatly improved gasoline mileage—and the chemical reaction to CO_2 and H_2O would be so complete as to yield very little smog. Unfortunately the technical problems in providing suitable electrode surfaces and getting rid of the reaction products are severe, and the most satisfactory fuel cells built so far involve chemically simpler systems such as hydrogen gas–oxygen gas cell reactions.

Electroplating

One of the most interesting features of reversible cell reactions is that their voltage is completely characteristic of the driving force of the reaction; if the two electrodes are connected to a larger potential in the opposite sense, the cell reaction is driven in reverse. Thus in the Cu/Ag case we examined initially, the spontaneous cell reaction caused copper to dissolve and silver to plate out. The characteristic cell potential was 0.454 V. If we connected a cell developing 0.5 V to the Cu/Ag cell with the positive terminal of the new cell at the negative terminal of the Cu/Ag, the reaction that would proceed in the Cu/Ag cell would dissolve silver and plate out copper. In practice, electroplating is carried out a little more simply, by constructing a cell having the plating metal as one electrode as in Fig. 10-11. An external potential is imposed such that the article to be electroplated is the cathode of the cell, and reduction of the metal ions—silver, for example—to the pure metal occurs on the surface of the article. It should be clear from our previous discussion that a characteristic voltage is necessary to plate out any given metal, so that in a mixture of metal ions only the most easily reduced should be plated out. However, by controlling the concentrations of the different ions in the cell solution, it is possible to plate two

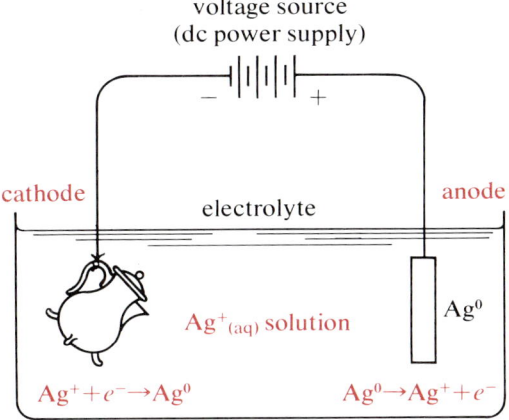

Figure 10-11   Electroplating cell with external voltage imposed.

metals simultaneously. Thus, although brass is an alloy of two metals, copper and zinc, by controlling the concentration of the two metals through the formation of complexes (with different stability constants for the two metals) it is possible to plate brass, and even to choose with some precision the relative composition of the plated metal.

The quantity of the metal plated is very simply related to our initial thermodynamic expression $\Delta G^0 = -n\mathscr{F}\mathscr{E}^0$. If the necessary $\mathscr{E}^0$ is provided, one faraday of electricity will plate out exactly $1/n$ mole of metal. Since a faraday is 96,487 coulombs and a coulomb is an ampere-second, it is easy to calculate the quantity of metal that would be deposited by a given current flowing for a given length of time. For instance, suppose a current of 1.30 amp passes through a copper plating solution for 10 min; what weight of copper metal is plated out? From the half-reaction for copper we can identify n as 2:

$$Cu(H_2O)_{4(aq)}^{2+} + 2e^- \rightleftharpoons Cu_{(s)}^0 + 4H_2O_{(l)}$$

The number of coulombs of charge that passes through the solution (the **electrolytic cell**) is equal to the current in amperes times the time of passage in seconds:

$$\text{Coulombs} = 1.30 \text{ amp} \times 10 \text{ min} \times 60 \frac{\text{sec}}{\text{min}}$$

$$= 780 \text{ amp-sec}$$

$$\text{faradays} = \frac{780}{96487}$$

$$= 8.10 \times 10^{-3}$$

Since $n = 2$, each faraday quantity of electrical charge will plate out 0.5 mole of copper metal, so the amount of metal plated is given by

$$\text{moles Cu} = \frac{1}{2}\frac{\text{mole}}{\text{faraday}} \times 8.10 \times 10^{-3} \text{ faraday}$$

$$= 4.05 \times 10^{-3} \text{ mole}$$

$$\text{wt Cu} = 4.05 \times 10^{-3} \text{ mole} \times 63.54 \text{ g/mole}$$

$$= 0.258 \text{ g}$$

The electrolytic deposition of metals is important to the chemist for another reason. The more electropositive metals such as sodium, magnesium, and aluminum are very difficult or impossible to obtain by chemical means from their compounds. But only a modest voltage is required to reduce the positive ions of these metals to the metallic state, so that electrolytic cells are very powerful tools for the preparation of very strong reducing (or oxidizing) agents. In aqueous solution the electrolysis of these ions produces only decomposition of water at the electrode surface. However, molten ionic salts can conduct electricity, and all of the above-mentioned metals are produced commercially by massive electrolysis of their fused salts. Such commercial electrolytic cells operate at currents on the order of 10^5 amp, which ensures rapid production of the metals but also explains why aluminum plants are always located near large power dams.

Cell Potentials and Equilibrium Constants

With this brief excursion into the experimental and practical aspects of cells, we can go on to examine more closely the thermodynamics of the operation of a cell and the nature of electron-transfer equilibrium. From the fundamental relationship

$$\Delta G^0 = -n\mathscr{F}\mathscr{E}^0 \quad \text{(for 1 mole)}$$

we can immediately move to an expression for the equilibrium constant for the cell reaction:

$$\Delta G^0 = -n\mathscr{F}\mathscr{E}^0 \quad \text{and} \quad \Delta G^0 = -RT \ln K_{eq}$$

$$-n\mathscr{F}\mathscr{E}^0 = -RT \ln K_{eq}$$

$$\ln K_{eq} = \frac{n\mathscr{F}\mathscr{E}^0}{RT}$$

or

$$\log_{10} K_{eq} = \frac{n\mathscr{F}}{2.303RT}\mathscr{E}^0 \quad (10\text{-}11)$$

Thus the cell reaction between copper and silver of our initial example, with a voltage ($\mathscr{E}^0$) of 0.454 V, would have an equilibrium constant of

$$\log K = \frac{n\mathscr{F}}{2.303RT}\mathscr{E}^0$$

$$= \frac{(2 \text{ moles electrons/mole reaction})(96487 \text{ C/mole electrons})}{2.303(8.314 \text{ J/mole-}°\text{K})298°\text{K}} 0.454 \text{ V}\left(\frac{J}{C}\right)$$

$$= 15.34$$

$$K = 2.2 \times 10^{15}$$

From this example we can see that it is possible to express all of the electron-transfer equilibria derivable from Table 10-7 in terms of equilibrium-constant expressions if we choose; however, the table of individual reduction potentials is a much more compact and convenient way of expressing the same data. An interesting and useful sidelight of this equivalence involves the obtaining of more conventional equilibrium constants through the use of cell potentials. The process consists of using Hess' law on the free energies of cell half-reactions to produce a free-energy change for an overall reaction that does *not* involve electron transfer; the electron transfer is avoided by choosing systems involving the same ions and subtracting the electrons. For instance, if we wish to establish the solubility-product constant for AgCl, we need only determine $\mathscr{E}^0$ for the following half-reactions and combine them:

$$\text{AgCl}_{(s)} + e^- \rightleftharpoons \text{Ag}^0 + \text{Cl}^-_{(aq)} \qquad \text{(Ag/AgCl electrode)} \qquad \mathscr{E}^0 = 0.2221 \text{ V}$$

$$\frac{(\text{Ag}^+_{(aq)} + e^- \rightleftharpoons \text{Ag}^0)}{\text{AgCl}_{(s)} \rightleftharpoons \text{Ag}^+_{(aq)} + \text{Cl}^-_{(aq)}} \qquad \text{(Ag active electrode)} \qquad \frac{-\mathscr{E}^0 = -0.7996 \text{ V}}{-0.5775 \text{ V}}$$

$$\log K_{sp} = \frac{n\mathscr{F}(-0.5775)}{2.303RT}$$

$$= \frac{1(96487)(-0.5775)}{2.303(8.314)(298)}$$

$$= -9.76 = \overline{10}.24$$

$$K_{sp} = 1.7 \times 10^{-10}$$

Many of the tabulated solubility-product constants and metal-complex stability constants have been determined in this way.

Combining Half-Cell Potentials

The application of Hess' law to electrode potentials is also convenient in predicting other electrode potentials; the process is similar to the previous example except that the total electron change does not cancel out. This causes an important difference. Hess' law applies to thermodynamic energies, not to potentials, and the free-energy change for a half-reaction is proportional to its potential *times the number of electrons being transferred*. For cases in which the electrons are to cancel out, n is the same for each half-reaction and we have

$$A^{n+} + ne^- \rightleftharpoons A^0 \qquad \Delta G^0 = n\mathscr{F}\mathscr{E}_A^0$$

$$\underline{-(B^{n+} + ne^- \rightleftharpoons B^0) \qquad \Delta G^0 = n\mathscr{F}\mathscr{E}_B^0}$$

$$A^{n+} + B^0 \rightleftharpoons B^{n+} + A^0 \qquad \Delta G^0 = n\mathscr{F}\mathscr{E}_{A,B}^0$$

or

$$\not{n}\not{\mathscr{F}}\mathscr{E}_A^0 - \not{n}\not{\mathscr{F}}\mathscr{E}_B^0 = \not{n}\not{\mathscr{F}}\mathscr{E}_{A,B}^0$$

$$\mathscr{E}_{A,B}^0 = \mathscr{E}_A^0 - \mathscr{E}_B^0$$

That is, the potentials can be combined directly since n cancels out of the Hess'-law expressions. But for cases in which the electrons do not cancel out, the number of electrons must be kept:

$$A^{n+} + ne^- \rightleftharpoons A^0 \qquad \Delta G^0 = n\mathscr{F}\mathscr{E}_A^0$$

$$\underline{-(B^{m+} + me^- \rightleftharpoons B^0) \qquad \Delta G^0 = m\mathscr{F}\mathscr{E}_B^0}$$

$$A^{n+} + B^0 + (n-m)e^- \rightleftharpoons A^0 + B^{m+} \qquad \Delta G^0 = (n-m)\mathscr{F}\mathscr{E}_{A,B}^0$$

so that

$$n\mathscr{F}\mathscr{E}_A^0 - m\mathscr{F}\mathscr{E}_B^0 = (n-m)\mathscr{F}\mathscr{E}_{A,B}^0$$

To give a concrete example, consider the prediction of $\mathscr{E}^0$ for the electrode half-reaction:

$$ClO_{4(aq)}^- + 2H_3O_{(aq)}^+ + 2e^- \rightleftharpoons ClO_{3(aq)}^- + 3H_2O_{(l)} \qquad \mathscr{E}^0 = ?$$

We can combine the two half-reactions

$$ClO_{4(aq)}^- + 8H_3O_{(aq)}^+ + 8e^- \rightleftharpoons Cl_{(aq)}^- + 12H_2O_{(l)} \qquad \mathscr{E}^0 = 1.36 \text{ V}$$

$$\underline{-(ClO_{3(aq)}^- + 6H_3O_{(aq)}^+ + 6e^- \rightleftharpoons Cl_{(aq)}^- + 9H_2O_{(l)}) \qquad \mathscr{E}^0 = 1.48 \text{ V}}$$

$$ClO_{4(aq)}^- + 2H_3O_{(aq)}^+ + 2e^- \rightleftharpoons ClO_{3(aq)}^- + 3H_2O_{(l)}$$

Now, simply subtracting the $\mathscr{E}^0$ values would indicate that the standard reduction potential for the desired half-reaction would be -0.12 V. But subtracting free energies instead—which is the only legitimate procedure—gives

$$\Delta G^0_{ClO_4^-, Cl^-} = 8\mathscr{F}(1.36)$$

$$\underline{-[\Delta G^0_{ClO_3^-, Cl^-} = 6\mathscr{F}(1.48)]}$$

$$\Delta G^0_{ClO_4^-, ClO_3^-} = 2\mathscr{F}\mathscr{E}^0_{ClO_4^-, ClO_3^-}$$

$$2\mathscr{F}\mathscr{E}^0_{ClO_4^-, ClO_3^-} = (8\mathscr{F}\,1.36) - (6\mathscr{F}\,1.48)$$

$$\mathscr{E}^0_{ClO_4^-, ClO_3^-} = \frac{10.88 - 8.88}{2} = 1.00 \text{ V}$$

This remarkably different result should dramatize the dangers of combining $\mathscr{E}^0$ values indiscriminately.

Thermodynamic Quantities from Electrical Measurements

Since electrical potential can be measured easily and very accurately, it is often desirable to use potential measurements to establish other thermodynamic values besides ΔG^0 and K_{eq}. We can readily show that entropy and enthalpy changes are accessible from $\mathscr{E}^0$ values. Writing the definition of ΔG^0,

$$\Delta G^0 = \Delta H^0 - T \Delta S^0$$

Differentiating this expression with respect to the temperature (but considering ΔH^0 and ΔS^0 to be constant over a small temperature range), we have

$$\frac{d(\Delta G^0)}{dT} = 0 - 1 \Delta S^0 = -\Delta S^0$$

Substituting $-n\mathscr{F}\mathscr{E}^0$ for ΔG^0, we have

$$\frac{d(-n\mathscr{F}\mathscr{E}^0)}{dT} = -\Delta S^0$$

$$n\mathscr{F}\frac{d\mathscr{E}^0}{dT} = \Delta S^0$$

The temperature dependence of the standard potential from a cell thus gives us a value for the standard entropy change for the cell reaction. Going back to the ΔG^0 definition, we can rearrange it to yield ΔH^0:

$$\Delta H^0 = \Delta G^0 + T \Delta S^0$$

$$= -n\mathscr{F}\mathscr{E}^0 + n\mathscr{F}T\frac{d\mathscr{E}^0}{dT}$$

$$= n\mathscr{F}\left(T\frac{d\mathscr{E}^0}{dT} - \mathscr{E}^0\right)$$

Study Problem 4 gives a typical example of the use of these relationships.

The Nernst Equation

We have observed that most electrodes are concentration dependent, without indicating the nature of the dependence. The quantity $\mathscr{E}^0$ refers to the voltage developed by a cell when all the components of both redox couples are in their standard states, meaning unit activity and in particular, for dissolved species, a "Henry's-law" 1-m activity or concentration. If the concentrations differ from this condition, we still measure a voltage, but it is not $\mathscr{E}^0$. Rather, it is $\mathscr{E}$, the voltage developed by a cell in which not every component is in its standard state. This corresponds to ΔG, the free-energy change for a reaction in which not all reactants and products are in their standard states. From p. 470 we have, for the reaction

$$aA + bB \rightleftharpoons cC + dD$$

the expression

$$\Delta G = \Delta G^0 + RT \ln \frac{[C]^c[D]^d}{[A]^a[B]^b}$$

If the reaction is a cell reaction, we can substitute for both ΔG and ΔG^0:

$$-n\mathscr{F}\mathscr{E} = -n\mathscr{F}\mathscr{E}^0 + RT \ln \frac{[C]^c[D]^d}{[A]^a[B]^b}$$

$$\mathscr{E} = \mathscr{E}^0 - \frac{RT}{n\mathscr{F}} \ln \frac{[C]^c[D]^d}{[A]^a[B]^b}$$

or

$$\mathscr{E} = \mathscr{E}^0 - \frac{2.303 RT}{n\mathscr{F}} \log_{10} \frac{[C]^c[D]^d}{[A]^a[B]^b} \qquad (10\text{-}12)$$

This expression is known as the **Nernst equation**, after its formulator, Walther Herman Nernst, a brilliant electrochemist and early Nobel laureate. It gives us a tool for predicting the voltage developed by a cell under any concentration conditions, subject only to the validity of replacing activity ratios by concentration ratios. The equation can be applied equally well to a cell voltage or to the arbitrary but consistent half-cell potential. For convenience, note that the factor $2.303RT/\mathscr{F}$, at 25°C, is equal to 0.05916, so that the Nernst equation at that temperature reads

$$\mathscr{E} = \mathscr{E}^0 - \frac{0.05916}{n} \log_{10} \frac{[C]^c[D]^d}{[A]^a[B]^b} \qquad (10\text{-}13)$$

As an example of the use of the Nernst equation, suppose the Cu/Ag cell of Fig. 10-4 is initially set up in its standard state; that is, with $Cu(H_2O)_4^{2+}$ and Ag^+ both present at 1-m concentration (activity) and equal volumes. When the cell is connected to some electrical device and allowed to run down completely, what are the concentrations of the two ions?

The cell reaction follows the equation

$$Cu^0_{(s)} + 4H_2O_{(l)} + 2Ag^+_{(aq)} \rightleftharpoons Cu(H_2O)_4^{2+}{}_{(aq)} + 2Ag^0_{(s)}$$

$$n = 2 \qquad \mathscr{E}^0 = 0.454 \text{ V}$$

When a cell has completely run down it no longer develops any voltage, so in the Nernst equation we write

$$\mathscr{E} = \mathscr{E}^0 - \frac{0.05916}{n} \log \frac{[\text{Cu}(\text{H}_2\text{O})_4^{2+}]}{[\text{Ag}^+]^2}$$

$$0 = 0.454 - \frac{0.05916}{2} \log \frac{[\text{Cu}(\text{H}_2\text{O})_4^{2+}]}{[\text{Ag}^+]^2}$$

$$\log \frac{[\text{Cu}(\text{H}_2\text{O})_4^{2+}]}{[\text{Ag}^+]^2} = \frac{0.454 \times 2}{0.05916} = 15.35$$

$$\frac{[\text{Cu}(\text{H}_2\text{O})_4^{2+}]}{[\text{Ag}^+]^2} = 2.2 \times 10^{15}$$

As the reaction proceeds silver ions are consumed and copper ions are formed, but since the silver solution is initially only 1 m and the reaction of 1 mole of silver ion produces only 0.5 mole of copper ion, the final copper ion concentration is very near 1.50 m. We thus have

$$\frac{1.50}{[\text{Ag}^+]^2} \cong 2.2 \times 10^{-15}$$

$$[\text{Ag}^+]^2 \cong \frac{1.50}{2.2 \times 10^{15}} = 6.8 \times 10^{-16}$$

$$[\text{Ag}^+] \cong 2.6 \times 10^{-8}\ M\ (\text{or}\ m)$$

For any redox couple having a reduction potential greater than about 1 V, the oxidized form of the couple is usually considered a strong oxidizing agent; inspection of Table 10-7 will show that the perchlorate ion is such an oxidizing agent. However, $\mathscr{E}^0 = 1.36$ V for $\text{ClO}_4^-/\text{Cl}^-$ refers to a system in which all ions including H_3O^+ are at 1-m concentration. What is $\mathscr{E}$ for the $\text{ClO}_4^-/\text{Cl}^-$ couple if ClO_4^- and Cl^- are still 1 m, but the solution is neutral (i.e., pH 7)?

For the half-reaction

$$\text{ClO}_{4(aq)}^- + 8\text{H}_3\text{O}_{(aq)}^+ + 8e^- \rightleftharpoons \text{Cl}_{(aq)}^- + 12\text{H}_2\text{O}_{(l)}$$

we write the Nernst equation

$$\mathscr{E} = \mathscr{E}^0 - \frac{0.05916}{8} \log \frac{[\text{Cl}^-]}{[\text{ClO}_4^-][\text{H}_3\text{O}^+]^8}$$

ignoring the concentration of electrons. Since $[\text{H}_3\text{O}^+]$ now equals 10^{-7}, we have

$$\mathscr{E} = 1.36 - \frac{0.05916}{8} (56)$$

$$= 1.36 - 0.414$$

$$= 0.95\ \text{V}$$

This striking change illustrates a common feature of oxidizing ions such as perchlorate, chlorate, and nitrate; they are much stronger oxidizing agents in strongly acidic solutions than in plain water. Perchloric acid is dangerously explosive in the presence of any organic material, but the perchlorate ion in neutral aqueous solution is almost completely chemically unreactive.

pH Meters and Potentiometric Titrations

A rather direct application of the Nernst equation is the pH meter. The glass electrode (see Fig. 10-9) develops a potential, against any reference electrode, which is sensitive to the concentration of H_3O^+ in the solution; more specifically, its response to $[H_3O^+]$ is Nernstian:

$$\mathscr{E}_{\text{glass electrode}} = \mathscr{E}^0_{\text{glass electrode}} - \frac{0.05916}{1} \log [H_3O^+]$$

$$[H_3O^+] = m \quad [H_3O^+] = 1$$

$$= \mathscr{E}^0 + 0.05916 \text{ pH}$$

A pH meter is thus just a voltmeter with a scale calibrated so that 59 mV equals one pH unit. The difficulty in measurement lies in the very high electrical resistance of the glass electrode; the meter must not draw more than about 10^{-12} amp of current while it is measuring the voltage. This has traditionally made pH meters expensive, but the advent of solid state electronics has revolutionized this field along with many others; satisfactory electrometer operational amplifiers are now available for about $12.

Using a pH meter, or even an ordinary vacuum-tube voltmeter and any concentration-sensitive electrode, whether a glass electrode for H_3O^+, another ion-selective electrode, or any other kind, a **potentiometric** endpoint can be obtained for many titrations involving the quantitative consumption of one ion in a redox couple. The reaction need not be a redox reaction; for instance, both of the following titration reactions can be followed potentiometrically:

$$Fe(H_2O)_6^{2+} + Ce(H_2O)_6^{4+} \rightleftharpoons Fe(H_2O)_6^{3+} + Ce(H_2O)_6^{3+} \quad \text{(redox)}$$

$$H_3O^+ + OH^- \rightleftharpoons 2H_2O \quad \text{(acid-base)}$$

To demonstrate the nature of a potentiometric endpoint, let us assume that a 0.01-M solution of $Cd(H_2O)_6^{2+}$ is being titrated with EDTA (1.00 M, to make dilution negligible) as on p. 498. A cadmium electrode is being used:

$$Cd(H_2O)_{6(aq)}^{2+} + 2e^- \rightleftharpoons Cd^0_{(s)} + 6H_2O_{(l)} \quad \mathscr{E}^0 = -0.4021 \text{ V}$$

$$\mathscr{E} = -0.4021 - \frac{0.05916}{2} \log \frac{1}{[Cd(H_2O)_6^{2+}]}$$

$$= -0.4021 - \frac{0.05916}{2} (2)$$

$$= -0.4613 \text{ V initially}$$

Here we have used the Nernst equation to calculate the initial potential of the cadmium electrode in the 0.01-M solution. We can similarly calculate the potential at other points in the progress of the titration. Halfway to the equivalence point, the hydrated cadmium ion concentration is still 0.005 M:

$$\mathscr{E} = -0.4021 - \frac{0.05916}{2} \log \frac{1}{0.005}$$

$$= -0.4021 - \frac{0.05916}{2} (2.30)$$

$$= -0.4701 \text{ V}$$

Ninety percent of the way to the endpoint, the hydrated cadmium ion concentration, which is the only species to which the electrode is sensitive, is still further reduced to 0.001 M:

$$\mathscr{E} = -0.4021 - \frac{0.05916}{2} \log \frac{1}{0.001}$$

$$= -0.4021 - \frac{0.05916}{2} (3)$$

$$= -0.4907 \text{ V}$$

Ninety-nine percent of the way to the endpoint, the hydrated cadmium ion concentration is only 0.0001 M:

$$\mathscr{E} = -0.4021 - \frac{0.05916}{2} \log \frac{1}{0.0001}$$

$$= -0.4021 - \frac{0.05916}{2} (4)$$

$$= -0.5203 \text{ V}$$

At the equivalence point, the only hydrated cadmium ion comes from hydrolysis of the CdHY complex, which we can calculate from the stability constant:

$$Cd(H_2O)_{6(aq)}^{2+} + HY_{(aq)}^{3-} \rightleftharpoons CdHY_{(aq)}^{-} + 6H_2O_{(l)}$$

$$K_{st} = 1.2 \times 10^9 = \frac{[CdHY^-]}{[Cd(H_2O)_6^{2+}][HY^{3-}]}$$

$$= \frac{0.01 - x}{(x)(x)}$$

$$x^2 \cong \frac{0.01}{1.2 \times 10^9} = 8.3 \times 10^{-12}$$

$$x \cong 2.9 \times 10^{-6}\ M \cong [\text{Cd}(\text{H}_2\text{O})_6^{2+}] \quad (2.9 \times 10^{-6} \ll 0.01)$$

$$\mathscr{E} = -0.4021 - \frac{0.05916}{2} \log \frac{1}{2.9 \times 10^{-6}}$$

$$= -0.4021 - \frac{0.05916}{2}(5.54)$$

$$= -0.5661\ \text{V}$$

We could continue calculating cadmium electrode potentials past the equivalence point in the same way that the last calculation was done; the student may profit by verifying that at 1% excess EDTA $\mathscr{E} = -0.6121$ V, at 10% excess $\mathscr{E} = -0.6413$ V, and at 50% excess $\mathscr{E} = -0.6621$ V. Figure 10-12 shows these voltages, plotted as a function of volume of EDTA delivered. Clearly the equivalence point is marked by the most rapid change of potential, and such a graph can be used in the absence of an indicator to establish the equivalence point for analytical purposes. When a glass electrode is used for acid–base titrations, the procedure is usually called a pH titration, since the vertical scale can be constructed in pH units. Other applications include genuine redox titrations such as the $\text{Fe}^{\text{II}}/\text{Ce}^{\text{IV}}$ mentioned earlier, complexation titrations such as this one, precipitation titrations such as Ag^+/Cl^-, and others using all sorts of concentration-sensitive or **indicating electrodes**. All that is necessary, again, is that the titration use up one component of a redox couple for which an electrode exists.

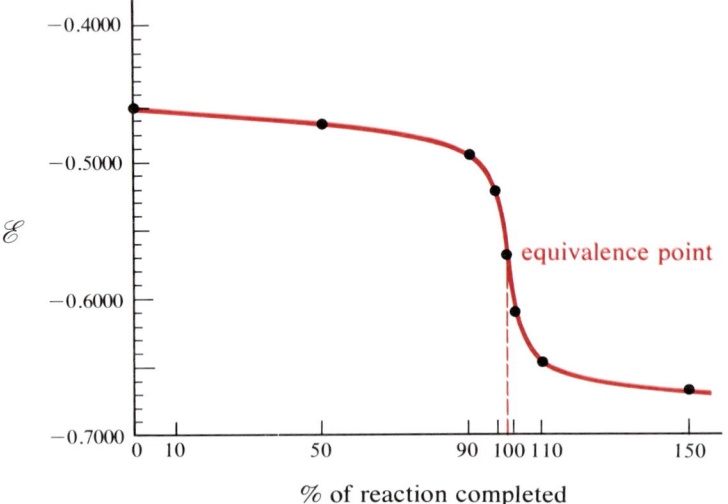

Figure 10-12 Titration curve for potentiometric titration of Cd^{2+} by EDTA.

10-8 Summary

The length and varied nature of this chapter make it difficult to summarize. Its principal purpose has been to introduce the student to some of the quantitative applications of the free-energy function, principally through the equilibrium-constant expression. The discussion has, of course, been abbreviated and incomplete; in particular, no discussion has been given to the use of the charge-balance and material-balance expressions in the exact solution of multiple equilibria, although Study Problem 3 tentatively explores this area. What *has* been done is to explore the simple applications of several areas to which equilibrium constants are commonly applied: gas equilibria, metal complexes, acid–base equilibria, solubility equilibria, and mixtures of these. Electron-transfer or redox equilibria have also been considered; they could be described by equilibrium constants, but because of the ease of measurement of electrical potential are usually characterized by the reduction potentials of the redox couples in the reaction. All of these applications rest on the free-energy function as a measure of spontaneity, and the derivation of the equilibrium-constant expression should be thoughtfully considered until the nature of the expression is understood.

What we have *not* done in this or either of the preceding two chapters is to look at the details of *how* a reaction proceeds when its thermodynamics make the reaction spontaneous. Several times we have mentioned the role of reaction mechanisms in governing the rate at which a reaction occurs, but have always deferred their consideration until the measure of spontaneity, the go/no-go aspect of chemical dynamics, had been put in its proper perspective. With the feeling that for some classes of reactions, at least, we can confidently predict the extent of reaction, it is now time to change from chemical thermodynamics to chemical kinetics, the study of reaction mechanisms and reaction rates.

Study Problems

1. Consider the gaseous reactions

$$N_2 + 3H_2 \rightleftharpoons 2NH_3 \quad \text{and} \quad N_2 + O_2 \rightleftharpoons 2NO$$

Write the equilibrium-constant expression for each reaction. Rewrite these expressions giving each pressure as an activity ratio in torr. Which numerical K_p value changes when the 1/760 factors are absorbed into it? What is the condition for a reaction's K_P to be unchanged on changing pressure units?

2. Which of the following compounds should be unstable, viewing each as a combination of a positively charged acid and a negatively charged base?

AuF	$Fe_3(CO)_9$	CdO_4^{6-}	BF_3Br^-
Au_2S	FeO_4^{2-}	Cr_2O	BF_4^-
SF_6	$FeSe_4^{2-}$	$MoCl_6$	SI_6

3. To solve for the pH of an ammonium acetate solution, as p. 514 has pointed out, an additional condition on the solution equilibria is necessary. If we use the mass-balance condition, we can eliminate one unknown concentration from our calculations since $[OH^-]$ does not appear. Solve the $K_a(NH_4^+)$ expression for $[NH_3]$ and the $K_a(HAc)$ expression for $[HAc]$, and substitute these into the mass-balance expression. Rearrange the result to a quadratic in $[H_3O^+]$ and show that $[H_3O^+] = \sqrt{K_a(NH_4^+)K_a(HAc)}$ if enough NH_4Ac salt is dissolved to make $[NH_4^+] \gg [NH_3]$ and $[Ac^-] \gg [HAc]$, so that $[NH_4^+]/[Ac^-] \cong 1$.

4. At 0°C each cell of a lead storage battery develops 2.0374 V; at 25°C each develops 2.0420 V, and at 50°C each develops 2.0476 V. Calculate ΔG^0, ΔH^0, and ΔS^0 for the cell reaction at 25°C, using graphical methods to get $d\mathscr{E}^0/dT$ at 25°C.

5. Balance the following skeleton equations for aqueous redox reactions:
   (a) $Cu + NO_3^- \rightarrow Cu(OH_2)_4^{2+} + NO$ (dilute)
   (b) $Cu + NO_3^- \rightarrow Cu(OH_2)_4^{2+} + NO_2$ (concentrated)
   (c) $ClO_4^- + SnS \rightarrow Cl^- + Sn(OH_2)_2(OH)_2^{2+} + SO_4^{2-}$
   (d) $Cr_2O_7^{2-} + (CH_3)_2CHCOOH \rightarrow Cr(OH_2)_6^{3+} + CO_2$
   (e) $S_2O_8^{2-} + NH_3 \rightarrow N_2 + SO_4^{2-}$ (basic solution)
   (f) $S_2O_8^{2-} + NH_4^+ \rightarrow NO_3^- + SO_4^{2-}$
   (g) $Se_2Cl_2 \rightarrow H_2SeO_3 + Se + Cl^-$
   (h) $P_4 \rightarrow PH_3 + H_2PO_2^-$ (basic solution)
   (i) $(CH_3)_2As(O)OH + H_3PO_2 \rightarrow (CH_3)_2AsCl + H_2PO_4^-$ (aqueous HCl)
   (j) $MoCl_4 \rightarrow Mo(OH_2)_6^{3+} + MoOCl_3 + Cl^-$
   (k) $Cl_2 + Br^- \rightarrow BrO_3^- + Cl^-$ (basic solution)
   (l) $HN_3 + Zn \rightarrow N_2H_5^+ + Zn^{2+}$

6. Calculate the pH of the solution that results when 50.0 ml of 0.100-M hydrazine are neutralized by 25.0 ml of 0.200-M HCl.

7. If the solubility-product constant for CaC_2O_4 is 2×10^{-9}, what $C_2O_4^{2-}$ concentration is necessary to start precipitation from a saturated solution of $CaSO_4$? What is the lowest pH at which such precipitation can occur from a 0.0100 M solution of $Na_2C_2O_4$?

8. At 2000°C the equilibrium constant for the dissociation of gaseous Cl_2 into atoms is 0.56. If the total pressure is 0.10 atm, what fraction of the Cl_2 is dissociated?

9. Water vapor reacts with liquid sulfur at its boiling point, 718°K, to give H_2S and SO_2:

$$2H_2O_{(g)} + 3S_{(l)} \rightarrow 2H_2S_{(g)} + SO_{2(g)}$$

Since the sulfur is at its normal boiling point there is a constant 1-atm pressure of S vapor, and the reaction's progress can be followed by adding a known pressure P^0 of H_2O and measuring the pressure change ΔP at constant volume. Will the pressure increase or decrease as the reaction proceeds? Derive an expression for K involving only P^0 and ΔP.

10. What weight of Pb^{2+} could remain dissolved in 1 l of 0.500-M HF?

11. Ocean water has an average pH of about 8.1. A simple model for its pH control proposes that this is due to dissolved atmospheric CO_2 reacting with carbonate from rock and the sea bed:

$$CO_{2(aq)} + CO_{3(aq)}^{2-} + H_2O_{(l)} \rightleftharpoons 2HCO_{3(aq)}^-$$

Show that the equilibrium constant for this reaction equals K_1/K_2, the ratio of the first and second proton-transfer constants for $CO_{2(aq)}$. Since $K_1/K_2 = 7700$, this reaction goes well to the right. If we pursue this and assume that the ocean is a pure HCO_3^- solution, the reverse reaction's stoichiometry makes $[CO_2] \cong [CO_3^{2-}]$. Substitute for $[CO_2]$ from the K_1 expression and for $[CO_3^{2-}]$ from the K_2 expression in this equality and show that $[H_3O^+] \cong \sqrt{K_1K_2}$. How does this predicted pH check with the experimental value?

12. The reduction potential for the couple Hg^{2+}/Hg is measured as 0.654 V in a solution that is 0.100 M in HgI_4^{2-} ion but has no other source of Hg or I. Assuming that this $\mathscr{E}$ value represents a concentration dependence of the tabulated $\mathscr{E}^0$ value through the Nernst equation, calculate β_4 for HgI_4^{2-}.

13. Can lead chromate be precipitated from a 0.100-M solution of lead-EDTA complex by adding 0.100-M CrO_4^{2-}?

14. The pH quantity is defined only to ± 0.001 unit; a good pH meter is sensitive to a change in pH of 0.002 unit. If a meter indicates a change from pH = 12.002 to pH = 12.000, how many hydronium ions have been added to each milliliter of solution?

15. K_a for iodic acid, HIO_3, is 1.9×10^{-1}, and K_{sp} for copper(II) iodate is 1.4×10^{-7}. Will 1 l of a solution that is 1 M in HCl and 1 M in HIO_3 precipitate any copper(II) iodate if 1 mg of Cu^{2+} is added to it?

16. In physiological chemistry it is convenient to have a buffer at the pH of blood, 7.38. If the dihydrogen phosphate–monohydrogen phosphate system is used to form the buffer,

$$H_2PO_4^- + H_2O \rightleftharpoons HPO_4^{2-} + H_3O^+ \qquad K_a = 6.3 \times 10^{-8}$$

and 0.1000 mole of $Na_2HPO_4 \cdot 2H_2O$ (17.80 g) is used for 1 l of buffer solution, what weight of $NaH_2PO_4 \cdot H_2O$ should be used?

17. Calculate $\mathscr{E}^0$ for $Fe(OH_2)_6^{3+} + 3e^- \rightleftharpoons Fe + 6H_2O$, using Table 10-6.

18. A saturated solution of $CuSO_4$ is about 2.0 M at room temperature. If a 0.01-M solution of Cu^{2+} is allowed to stand without stirring in a beaker over solid $CuSO_4$ the bottom layer of the solution will dissolve more Cu^{2+} until it is saturated. Suppose a copper-foil electrode is then inserted into the top part of the solution and another is inserted into the bottom layer. What voltage would be measured between the two electrodes? Which electrode is supplying electrons to the solution? Does the electrochemical reaction that is going on work against or assist the normal diffusion process?

19. Calculate the equilibrium constant for the reaction

$$4Cr(OH_2)_6^{2+} + O_2 + 4H_3O^+ \rightleftharpoons 4Cr(OH_2)_6^{3+} + 6H_2O$$

20. Given the two half-reactions

$$Fe(OH_2)_6^{3+} + e^- \rightleftharpoons Fe(OH_2)_6^{2+} \quad \mathscr{E}^0 = 0.770 \text{ V}$$

$$PtCl_6^{2-} + 2e^- \rightleftharpoons PtCl_4^{2-} + 2Cl^- \quad \mathscr{E}^0 = 0.758 \text{ V}$$

if a solution of both the iron species is mixed with a solution of Cl^- and both the platinum species to form a solution initially $1\,M$ in all these species, will the iron be oxidized or reduced? Should the Cl^- concentration be increased or decreased if we wish to reverse the direction of the iron reaction? What value of the Cl^- concentration would cause no net reaction on mixing?

21. How long would it take to silverplate a coffee pot with a layer of plate 0.0100 in. thick if the surface area of the pot is 250 in.² and a current of 20.0 amp is used? The density of silver is 10.5 g/cm³.

22. Calculate the titration curve for 0.1000-M HCl with 0.1000 M NaOH and plot it on graph paper. Repeat for 10^{-3}-M and 10^{-5}-M solutions of HCl. What happens to the sharpness of the break in the curve at the equivalence point?

Some Further Reading

Blackburn, T. R., *Equilibrium—A Chemistry of Solutions*, New York: Holt, Rinehart & Winston, 1969. Equilibrium calculations tend to be messy and hard to describe, but this book is thorough, clear, helpful, and sometimes funny.

Butler, J. N., *Ionic Equilibrium: A Mathematical Approach*, Reading, Massachusetts: Addison-Wesley, 1964. Beautifully crisp mathematical discussion of this same area. Somewhat more sophisticated, but still very clear.

Barrow, G. M., *Physical Chemistry*, New York: McGraw-Hill, 1966 (2nd ed.). This is another widely used text; Chapter 23 is a good discussion of the electrochemistry in the current chapter.

Lyons, E. H., *Introduction to Electrochemistry*, Boston: Heath, 1967. Covers pretty much the same ground as Barrow; a little bit more mathematical than our discussion here, but along the same general lines.

11

The Mechanism of Chemical Change

When we run a chemical reaction, it is not enough to know that the equilibrium composition of the reaction mixture ought to be some calculable set of concentrations. These will not do us much good if the reaction requires a thousand years to reach equilibrium. As we have pointed out before, this is where thermodynamics fails as a complete guide to chemical dynamics. Nowhere does time enter into its considerations—yet a reaction mixture that is essentially unchanged after a month because the attainment of equilibrium would take a thousand years is just as unproductive as one that is unchanged after a month because its associated free-energy change is large and positive. We must therefore inquire into the time dependence of chemical reactions: what are the rates of chemical reactions, and why are some faster than others? Furthermore, we do not understand the fundamental nature of a chemical reaction unless we can relate it to our atomic/molecular model; what is happening to the molecule while it is in the process of being transformed? These questions, and the partial answers that exist, are the province of **chemical kinetics**. Any complete discussion of a chemical reaction process must include both the thermodynamics of the reaction and its kinetics. It may already be intuitively obvious that the questions of rates and molecular processes are intimately related. This is indeed the case, and we shall devote a good portion of this chapter to the relationship. Our first task, however, is to see what is meant, experimentally, by the rate of a reaction.

11-1 Rates and Rate Constants

Not all reactions that occur spontaneously are part of any meaningful equilibrium, as for instance explosions. Many are, however, and consistency requires us to define the rate of a reaction in a manner compatible with the way we define or describe equilibrium.

The Representation of Rates

Equilibrium is determined solely by concentrations (as an approximation to activities), so we can define the rate of a reaction as the rate of change of the concentrations of the components of the reaction mixture. If we have the reaction

$$A \rightleftharpoons B + C$$

we can define the rate of the reaction as the negative derivative of the concentration of A with respect to time,

$$\text{reaction rate} \equiv -\frac{d[A]}{dt}$$

since derivatives are the conventional representation of rates of change. However, we could equally well represent the reaction rate in terms of B or C:

$$\text{reaction rate} \equiv -\frac{d[A]}{dt} = +\frac{d[B]}{dt} = +\frac{d[C]}{dt}$$

Notice the signs; that for A, the reactant, is negative because it is disappearing as the reaction progresses, while those for B and C are positive because as products their concentrations are increasing as the reaction progresses. The stoichiometry of the reaction guides us here, in telling us that B and C must appear at equal rates. If the reaction produced six C for every B, we would obviously have

$$A \rightleftharpoons B + 6C$$

$$\frac{d[B]}{dt} = \frac{1}{6}\frac{d[C]}{dt}$$

Guldberg and Waage, in an early investigation of reaction rates (1867) studied the reaction between ethyl alcohol and acetic acid to give ethyl acetate and water:

$$CH_3CH_2OH_{(l)} + CH_3COOH_{(l)} \rightleftharpoons CH_3COOCH_2CH_{3(l)} + H_2O_{(l)}$$

This reaction is a true equilibrium and can be run in either direction. For the forward reaction they found that the rate was proportional to the concentration of *each* reactant:

$$-\frac{d[CH_3COOH]}{dt} = k_f[CH_3COOH][CH_3CH_2OH] \qquad \text{forward reaction}$$

Similarly, for the reverse reaction they found that the rate was proportional to the concentration of each reactant:

$$\frac{d[CH_3COOH]}{dt} = k_r[CH_3COOCH_2CH_3][H_2O] \qquad \text{reverse reaction}$$

In these two expressions the two proportionality constants k_f and k_r are not numerically the same; the reactions are, after all, different. Each number is a valuable characterization of the rate of the corresponding reaction, and is called a **rate constant**. The expression in which it appears is called a **rate law**. The form of the two rate laws obtained by Guldberg and Waage suggests that for any reaction the rate law should involve the product of the concentrations of the reactants, in much the same fashion as did the equilibrium constant. This is only rarely true. Often some form of product does appear, but it is never raised to a very high power, and there are sometimes additive terms, concentration ratios, and other such complications, which we shall return to later. The rate laws, in other words, do not very often express the overall stoichiometry of the reaction.

Measuring Reaction Rates

Before we try to account for the lack of correspondence between the reaction stoichiometry and its rate law, let us be certain we appreciate how rates are measured in the laboratory. To some degree it is possible to categorize the methods of measurement, but they are many and varied and the categories are somewhat arbitrary. The most obvious means is by direct chemical analysis of the reaction mixture. In Guldberg and Waage's reactions, the acetic acid concentration at any time could be measured by removing a portion and titrating with a strong base. This sort of thing can be done in many cases, but it must be true that the reaction is slow under the conditions of the analysis; otherwise, in the relatively slow process of titration or precipitation, the reaction would continue to some extent and confuse the result. Sometimes direct analysis can be applied to high-temperature reactions that are **quenched** (i.e., stop reacting) when they are cooled to room temperature, even though a sequence of analyses would take too long at the higher temperature. One plots the concentrations (as obtained at several times during the reaction) against time as in Fig. 11-1.

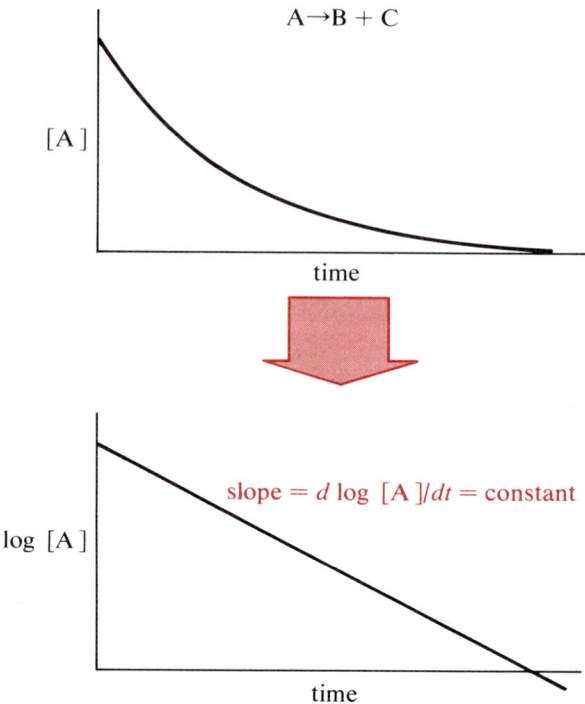

Figure 11-1 Experimental concentration-vs.-time curve and its conversion to a straight-line plot.

The result is a smooth curve, which can usually be transformed into a straight line by taking some more complicated function of the concentration as the vertical axis. When a straight line can be obtained its slope (or rate of change) is the rate constant in a form of the rate law given by the concentration function on the vertical axis. Thus in Fig. 11-1, for the straight-line plot $d(\log[A])/dt = k$, or, using the chain rule,

$$-\frac{d}{dt}(\log[A]) = \frac{-d}{d[A]}(\log[A])\frac{d[A]}{dt} = \frac{-1}{A}\frac{d[A]}{dt} = k$$

$$-\frac{d[A]}{dt} = k[A] \qquad (11\text{-}1)$$

There are also many indirect methods for monitoring the progress of a reaction. Any physical property that can be related to concentration can be the basis of a kinetic determination. Perhaps the most fundamental are the thermodynamic variables pressure and volume; temperature is not used because the temperature dependence of reaction rates is itself of great significance, as we shall see. For gaseous systems in which the total number of moles of gas present is changing, one can very readily measure the total pressure at constant volume, or in some cases the total volume at constant pressure. Similarly, for liquid systems the total volume at constant pressure can be readily measured by using a **dilatometer**, as in Fig. 11-2—if the volume expands as the reaction progresses, the advance of the meniscus up the capillary gives the time dependence of the reaction's progress. Of course, with either of these methods or with any indirect method, a preliminary calibration curve of the physical property's dependence on concentration must be obtained.

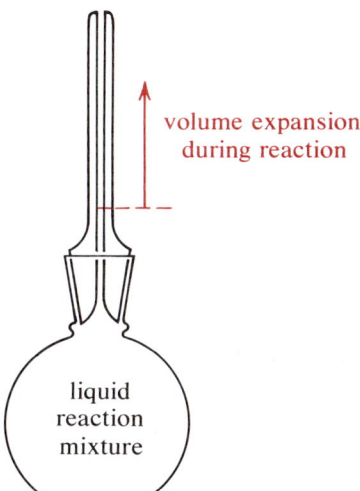

Figure 11-2 Dilatometer.

Electrical properties can also be used in indirect methods. The potential of a concentration-sensitive electrode, the electrical conductivity of the reaction mixture, the dielectric constant of the mixture, its degree of paramagnetism, and other related properties can all be used in exactly the same way as pressure or volume; some of these methods are even applicable to extremely fast reactions in which equilibrium is reached within milliseconds.

Optical properties are widely used to monitor kinetics, usually by observing the changes in the visible, ultraviolet, or infrared spectrum of the reaction mixture as a function of time. Other optical properties that are used include the refractive index of the reaction mixture, the degree to which it may rotate a polarized beam of light, and light scattering by a turbid reaction mixture. Again, all of these methods require interpretation through a calibration curve.

Forms of Rate Laws

Using one or more of these techniques we can usually produce a rate law for any reaction in which we are interested, unless it is either too fast or too slow for any of the applicable properties. While there are many obscure forms that rate laws can take, some are sufficiently common to deserve individual examination. The simplest type is represented by Eq. 11-1:

$$-\frac{d[A]}{dt} = k[A]$$

This might occur for the simple reaction

$$A \rightarrow B + C$$

or for others whose stoichiometry is somewhat more complicated:

$$A + B \rightarrow C + D \quad \text{or} \quad 2A \rightarrow B + C$$

This rate law, since it contains the concentration of A to the first power, is said to represent a **first-order reaction**. Higher orders are possible:

$$-\frac{d[A]}{dt} = k[A]^2$$

$$-\frac{d[A]}{dt} = k[A][B]$$

Both of these rate laws refer to **second-order** reactions; the first is second order in A and also second order overall, while the second is first order *in component A* but second order overall. In general, for rate laws involving only a product of concentrations,

$$-\frac{d[A]}{dt} = k[A]^m[B]^n \tag{11-2}$$

the reaction is mth order in A, nth order in B, and $(m+n)$th order overall. As a matter of experimental fact, neither m, n, nor $m+n$ is found to exceed 3, even though the reaction's stoichiometry may involve quite large integers. Most common solution reactions follow second-order kinetics, first order in each of two reactants, as in Guldberg and Waage's reaction. The fact that the stoichiometry does not seem to govern the kinetics of the reaction arrests our attention, but it is unquestionably true. There is no way we can predict the nature of the rate law for a reaction from the balanced stoichiometric equation alone. To understand why this is true, let us look at what is happening to the molecules themselves.

11-2 Molecular Mechanisms and Macroscopic Rates

The chemical transformation of a molecule is paradoxical, in a sense. The electrons that do the bonding within the molecule are normally arranged so as to provide the ground state of the molecule, which has a negative total energy relative to the separated atoms. Even when some of the electrons are antibonding, there must be a net positive bond order or the molecule would not exist in the first place. Then where does the energy come from to break the bonds? It is not enough to say that the electrons would have a lower energy in the new configuration, because they don't know that.

Molecular Interaction via Collision

The condition our model requires to account for this is that molecules must collide with something in order to react:

$$A + B \rightarrow C + D$$

or

$$A + M \rightarrow A^* \rightarrow B + C$$

The first of these is a straightforward collision between two molecules that react in the collision to give two new molecules; the energy necessary to break the bonds in A and B (to form new bonds in C and D) comes from the trans-

Figure 11-3 Energy transformation in a molecular collision

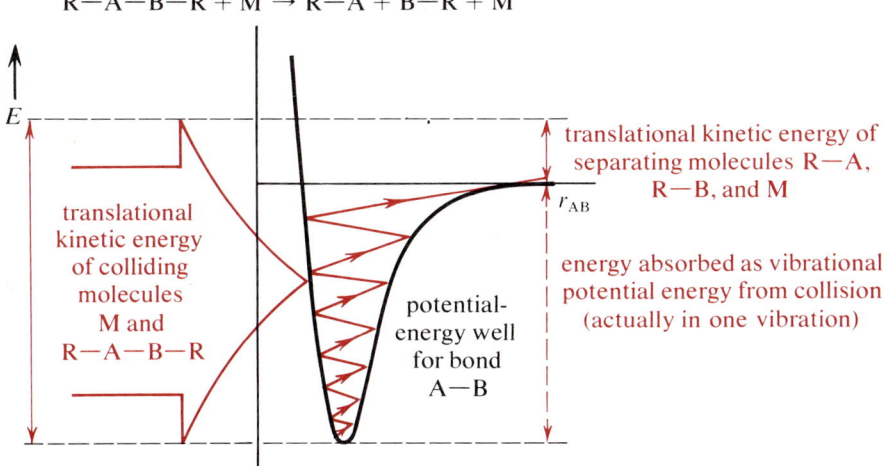

lational energy A and B had before they collided. In effect, as Fig. 11-3 indicates, the two molecules were moving rapidly toward each other, collided, transferred their translational energy to bond vibrational energy, which was then great enough to vibrate one or more of the atoms all the way out of the potential-energy well representing the bond energy, and moved apart slowly with the atoms in a new arrangement. The second collision reaction above indicates how the energy to break the bonds in A can come from a collision even if A is the only reactant. M could be another, unreactive molecule, or it could be the wall of the beaker; the only important feature is that A* is a high-energy **activated** molecule after it has collided with M and transformed some of its translational energy into vibrational energy. It is this A* that reacts to give B and C, not the ground-state A molecule. There is a class of reactions, known collectively as **photochemistry**, in which the activation occurs by having a photon with energy $h\nu$ hit the ground-state molecule, causing electronic excitation, but these are similar in principle.

Suppose the reaction in which we are interested occurs in the gas phase. We could calculate the collision frequency using the kinetic theory of Chapter 3, and we could construct a rudimentary theory of reaction rates out of the collision frequency. We will do just this in the next section. At this point, however, we need only observe that there are some statistical limitations on the kinds of collisions that can occur. There will be many **two-body** collisions; these are the ones we normally think of. It is possible to have a **three-body** collision, but it will occur only very rarely, since it would be quite unusual for a third molecule to strike two others while they were in the act of colliding. Nevertheless, since the molecules are not billiard balls but are somewhat soft and sticky, as witness the Lennard-Jones potential of p. 141, such three-body collisions will sometimes occur. Collisions involving four or more molecules have, by an extension of the same reasoning, a finite probability of occurrence, but it is so small that any chemical reaction that relied on them would take a fantastically long time to occur. The upshot of this is that any chemical reaction, even if its stoichiometry involves many molecules, must occur as one or a series of collisions involving usually two but at most three bodies. The detailed pathway of molecular motion and interaction by which a reaction occurs — the reaction **mechanism** — must consist of a series of steps, each of which represents a single collision or decomposition of an activated molecule. Each of the steps in the series is called an **elementary process**; because of the limits on the number of bodies that can collide, the stoichiometry of an elementary process is always very simple.

The number of molecules that react in an elementary process is called the **molecularity** of the process. If a single molecule is decomposing, the elementary process is **unimolecular**; if two molecules collide, the process is **bimolecular**, and if three molecules collide, the process is **termolecular**. The molecularity of an elementary process is the same as the order of that process, but it is not necessarily the same as the order of the overall reaction. The molecularity and order are the same for an elementary process because the rate of any reaction is governed by the rate at which collisions of the right sort can occur, and the collision rate is proportional to the concentration of each of the two (or three) colliding molecules. Thus if A and B must collide in an elementary process,

$$A + B \rightarrow C$$

the frequency of A–B collisions is proportional to the concentration of A, and also to the concentration of B:

$$\text{collision frequency} = K[A][B]$$

Since we think the collision frequency limits the reaction rate, this is equivalent to the rate-law expression as it would apply to the elementary process. But this elementary process is part of our molecular model; the observed macroscopic reaction rate and order is independent of the model we propose, and the term "molecularity" may not even apply to it. Even if we do apply this model, the molecularity of the elementary processes may not correspond to the order of the reaction.

Mechanisms as Sequences of Simple Collision Steps

As a generalized example, suppose we have a chemical reaction whose overall stoichiometry is given by the equation

$$A + 3B \rightarrow C + 2D$$

A mechanism for this reaction might involve the following elementary processes:

$$\left.\begin{array}{ll} A + B \rightarrow E & \text{bimolecular elementary process} \\ B + E \rightarrow F & \text{bimolecular elementary process} \\ B + F \rightarrow C + G & \text{bimolecular elementary process} \\ G \rightarrow 2D & \text{unimolecular elementary process} \end{array}\right\} \text{mechanism}$$

Supposing this to be the true mechanism, what is the rate of the reaction? If all the elementary processes proceeded at very nearly the same rate, the overall reaction rate would be a complicated combination of the rate expressions for each elementary process. But this is not usually true, and we should not expect it to be, since the four elementary processes are quite different reactions. One ordinarily expects the various elementary processes in a mechanism to show quite different rates, some being fast and others slow. If one is distinctly slower than the others, it will in effect control the overall reaction rate; the products from all the steps before it will pile up waiting to react, and its products, when they are finally produced, will quickly finish off the overall reaction. Such an elementary process is said to be the **rate-determining step** in the proposed mechanism. Whenever such a rate-determining step exists, the overall reaction rate law will be reasonably simple; a complicated rate law guarantees that several elementary processes share in the controlling of the rate. Suppose the first step in the above mechanism is rate determining; then its rate law is in effect that for the whole reaction—second order overall, first order in both A and B:

$$\text{rate} = k[A][B]$$

We can see that the rate law for a reaction is governed essentially by the kinds of collisions that can occur, not by its overall stoichiometry. Of course, if a reaction involves only one step, as many do, its overall stoichiometry will be the same as that of its only elementary process, whose rate law does correspond to its molecularity or stoichiometry. If the rate law does not correspond to the overall stoichiometry of the reaction, it shows that more than one elementary process is involved in the mechanism.

The order of a reaction may not correspond to that of its rate-determining step, even when the latter exists. Consider the following reaction and mechanism:

$$A + B + C \rightarrow D$$

$$\text{mechanism} \begin{cases} A + B \rightleftharpoons E & \left(\text{rapid equilibrium: } K_{eq} = \frac{[E]}{[A][B]}\right) \\ C + E \rightarrow D & \text{slow (rate determining)} \end{cases}$$

The law for the rate-determining step is

$$\text{rate} = k[C][E]$$

But because of the preceding rapid equilibrium, we can substitute for [E] its equivalent from the equilibrium-constant expression:

$$[E] = K_{eq}[A][B]$$

$$\text{rate} = k[C][E] = kK[A][B][C]$$

This rate law is third order, while the rate-determining step is bimolecular. It is impossible to distinguish this mechanism from a one-step mechanism involving a termolecular elementary process unless we have some evidence of the existence of the **reaction intermediate E**. Frequently several possible mechanisms account equally well for the observed rate law; to distinguish between them one must discover at least transient evidence of an intermediate that one mechanism has and the others do not.

Thus we attempt to elucidate reaction mechanisms by first establishing a rate law for the overall reaction. Then, using analogy, chemical intuition, or just gall, we propose a mechanism that will account satisfactorily for the observed rate law. If more than one can be proposed, we attempt to find experimental evidence for intermediates that will distinguish between the possible mechanisms. Finally, we test the most satisfactory mechanism by investigating the kinetics of analogous reactions to see if the changes are those we might expect for the given changes in structure. This last step requires that we apply our ideas of chemical structure to species present in elementary processes, which is the subject of Section 11-4. In the meantime, however, let us ask the other question from the beginning of the chapter: why are some reactions faster than others?

11-3   Gaseous Reaction Rates — Collision Theory

Having gone this far through the book, the student will probably suspect — correctly — that we shall attempt to construct a mathematical model that will predict reaction-rate constants to some degree of accuracy. This is more difficult than might be supposed, however, and no complete, quantitative theory of reaction dynamics exists. The present state of the theory is such that we shall have to be content with order-of-magnitude agreement (within one power of 10) when we can get it, which will not be very often. However, we can make a start in what seems an obvious direction, and try to refine our ideas from there.

The **collision theory** is a fairly obvious name for a mathematical model that we can build from our intuitive ideas about reaction from molecular collision. We shall restrict our attention to gaseous reactions, since it is only for gases that we have a satisfactory kinetic molecular theory to serve as a tool in discussing molecular collisions. We hope that the results will in some sense transfer to solution reactions, but the differences will be important; we shall try to make the transfer in Section 11-5.

From Chapter 3 we have an expression derived under fairly reasonable approximations for the collision frequency or **collision number**, Z_c, of a gaseous molecule,

$$Z_c = \frac{4\pi D^2 v_{\text{rms}}}{3} \frac{N}{V}$$

where v_{rms} is the root-mean-square average velocity of the molecule, D^2 (which we shall call σ^2 from now on to agree with conventional notation) is the square of the distance between the centers of two molecules when the molecules are touching, and N/V is a concentration factor (number of molecules per unit volume). If we insert an expression for v_{rms} from Chapter 3 (p. 111) in this relation, we have

$$v_{\text{rms}} = \sqrt{\frac{3RT}{M}}$$

$$Z_c = \frac{4}{3}\pi\sigma^2 \left(\frac{3RT}{M}\right)^{1/2} \frac{N}{V}$$

This collision rate can presumably be turned into a reaction rate for comparison with experiment, but first we need to define a reacting system. We can take the reaction

$$HI_{(g)} + HI_{(g)} \rightleftharpoons H_{2(g)} + I_{2(g)}$$

which has a complex mechanism but to a good approximation can be regarded as being a one-step bimolecular elementary process:

$$\frac{d[H_2]}{dt} = k[HI][HI]$$

(Check that the units of k must be liters per mole-second to keep this rate law dimensionally consistent with concentrations in moles per liter.) This reaction has been studied extensively, and at 400°C (673 °K), for instance, $k = 8.0 \times 10^{-4}$ l/mole-sec. How close can we come to predicting this number?

Simple Collision Number — Reaction at Every Collision

If we take the collision-number expression above and eliminate N/V, we shall have an expression for the collision frequency per concentration unit, which as a first approximation we can say ought to equal the rate constant (which does not contain concentration): k = reaction frequency at unit concentration = collision frequency at unit concentration = $\frac{4}{3}\pi(3RT/M)^{1/2}\sigma^2$. However, if we count all the HI collisions using this expression we shall count each one twice, since HI is colliding with HI. Since a reaction involves *two* HI molecules, we need

$$k = \frac{1}{2}\left(\frac{4}{3}\right)\pi\left(\frac{3RT}{M}\right)^{1/2}\sigma^2$$

$$= 2\pi\left(\frac{RT}{3M}\right)^{1/2}\sigma^2$$

For HI the molecular weight, M, is 127.9 g/mole, and σ^2 (established by viscosity measurements as in Chapter 3) is $(3.5 \text{ Å})^2$, or $(3.5 \times 10^{-8})^2$ cm²/molecule. Substituting these numbers into the above expression we have

$$k = 2 \times 3.142 \left(\frac{8.314 \times 10^7 \text{ erg/mole-deg} \times 673 \text{ deg}}{3 \times 127.9 \text{ g/mole}}\right)^{1/2}$$

$$\times ((3.5 \times 10^{-8})^2 \text{ cm}^2/\text{molecule})$$

$$= (6.284)(1.203 \times 10^4 \text{ cm/sec})(1.225 \times 10^{-15} \text{ cm}^2/\text{molecule})$$

$$= 9.23 \times 10^{-11} \text{ cm}^3/\text{molecule-sec}$$

Converting the concentration units,

$$k = 9.23 \times 10^{-11} \text{ cm}^3/\text{molecule-sec} \times \frac{1}{1000} \text{ l/cm}^3 \times 6.023 \times 10^{23} \text{ molecules/mole}$$

$$= 5.57 \times 10^{10} \text{ l/mole-sec}$$

This is a rather disastrous answer when compared with experiment; specifically, it is about 10^{14} times too large! Where have we gone astray?

The Activation Energy and the Boltzmann Limit on Reactive Collisions

We have assumed that every HI–HI collision would result in one molecular reaction. This is clearly not true; let us see why. The collision must transfer enough translational energy into H—I bond vibrational energy to allow the atoms to separate, and this energy may be quite large. If it is large, only a very small fraction of all the collisions that occur will be able to provide it — namely those between molecules far out at the high-energy end of the Maxwell–Boltzmann velocity-distribution function. If the necessary energy, which we call the **activation energy**, E_a, could be specified, the Boltzmann distribution would give us the fraction of collisions having at least this energy:

$$\text{collisions with energy} > E_a = \text{total collisions} \times e^{-E_a/RT} \qquad (11\text{-}3)$$

Unfortunately, there is no reliable way to calculate a theoretical value of E_a, so we cannot carry our purely theoretical treatment any further. But if this activation energy is a genuine physical quantity, we should be able to obtain it by looking at the temperature dependence of the reaction rate, since increasing the temperature will increase the fraction of collisions having the necessary energy and thus should increase the rate. What is the experimental temperature dependence of the rate?

In 1889 Svante Arrhenius, a Swedish chemist (also the author of the early acid–base definition), published a paper dealing with the general temperature dependence of reaction rates. For many reactions, of diverse types, he noted that the logarithm of the rate constant was proportional to the reciprocal of the absolute temperature:

$$\log k = Q \left(\frac{1}{T} \right) + B$$

See Fig. 11-4. If we differentiate Arrhenius' expression with respect to T, we have

$$\frac{d \log k}{dT} = \frac{-Q}{T^2}$$

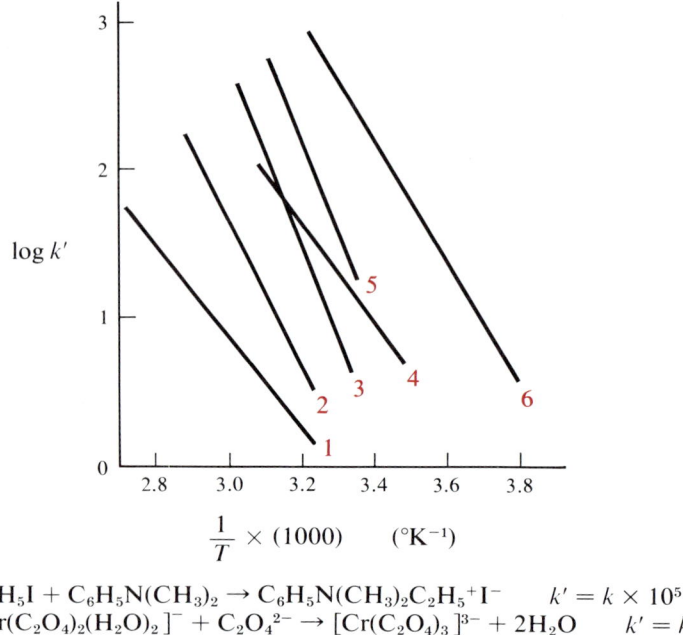

Figure 11-4 Temperature dependence of the rate constant for several reactions. (Redrawn from Harris, G. M., *Chemical Kinetics*, Lexington, Mass.: D.C. Heath, 1966.)

1 $C_2H_5I + C_6H_5N(CH_3)_2 \rightarrow C_6H_5N(CH_3)_2C_2H_5{}^+I^-$ $k' = k \times 10^5$
2 $[Cr(C_2O_4)_2(H_2O)_2]^- + C_2O_4{}^{2-} \rightarrow [Cr(C_2O_4)_3]^{3-} + 2H_2O$ $k' = k \times 10^2$
3 $C_6H_5N_2Cl \rightarrow C_6H_5Cl + N_2$ $k' = k \times 10^3$
4 $OCl^- \rightarrow Cl^- + \frac{1}{2}O_2$ $k' = k \times 10^4$
5 $N_2O_5 \rightarrow 2NO_2 + \frac{1}{2}O_2$ $k' = k \times 10^5$
6 $\alpha - C_6H_{12}O_6 \rightarrow \beta - C_6H_{12}O_6$ $k' = k \times 10^4$

Arrhenius was familiar with the relationship for the temperature dependence of an equilibrium constant that we derived on p. 473:

$$\frac{d \ln K_{eq}}{dT} = \frac{\Delta H^0}{RT^2}$$

The resemblance between these expressions and the idea that equilibrium can be regarded as equality of the forward and reverse reaction rates led him to propose the existence of an activation energy E_a defined by the expression

$$\frac{d \ln k_{rate}}{dT} = \frac{E_a}{RT^2}$$

Integrating this expression,

$$\ln k = -\frac{E_a}{RT} + \text{constant} \quad (\text{constant} \equiv \ln A)$$

$$k = A e^{-E_a/RT} \tag{11-4}$$

This **Arrhenius equation** was in fact the first definition of an activated state, preceding the collision theory. Experimental results are commonly reported in terms of an **Arrhenius activation energy**, which for most common reactions is roughly 20–30 kcal/mole; Study Problem 2 shows that the rate is approximately doubled for a 10° temperature rise at room temperature. We have symbolized this activation energy in the same way as the collision activation energy even though we do not know they are equivalent. In fact, they must differ somewhat, since there is an extra factor of $1/T^{1/2}$ in the collision-theory expression. Making allowance for this, we can plot the results of temperature studies of the rate of the $HI + HI \rightarrow H_2 + I_2$ reaction in the Arrhenius fashion to obtain an empirical value of E_a from the slope of the line: $E_a = 42.5$ kcal/mole. With this empirical value, suppose we return to the collision-theory expression:

$$\text{rate constant} = \text{collision frequency with energy} > E_a$$

$$= \text{total collision frequency} \times e^{-E_a/RT}$$

$$= 2\pi \left(\frac{RT}{3M}\right)^{1/2} \sigma e^{-E_a/RT}$$

$$= (5.57 \times 10^{10} \text{ l/mole-sec}) \, e^{-(42500/1.987(673))}$$

$$= (5.57 \times 10^{10}) e^{-31.70}$$

$$= (5.57 \times 10^{10}) \, 10^{-(31.70/2.303)}$$

$$= (5.57 \times 10^{10})(1.62 \times 10^{-14})$$

$$\text{rate constant} = 9.05 \times 10^{-4} \text{ l/mole-sec} \tag{11-5}$$

The Steric Factor p

Equation 11-5 is a very good match for the observed rate constant of 8.0×10^{-4} l/mole-sec. In fact, it is so good it is suspicious, because when other gaseous reactions are treated in the same way, the rate constants predicted are usually too large still. A few rate constants can be fit in this way, but many are 10^1–10^5 times too large. In other words, not even all the collisions with enough energy are effective in producing reaction. It has been proposed that the necessary factor of 10^{-1}–10^{-5} necessary to convert the theoretical rate constants to the observed values is a **steric factor** p:

$$k = p \times \text{collision frequency} \times e^{-E_a/RT} \tag{11-6}$$

That is, not all the collisions are effective because some of them bring the molecules together facing in the wrong direction, and the steric factor expresses the acceptable angles of approach. Thus for the gaseous reaction between potassium and methyl iodide,

$$K_{(g)} + CH_3-I_{(g)} \rightarrow KI_{(g)} + CH_{3(g)}$$

Fig. 11-5 shows a satisfactory angle of approach and an unsatisfactory one.

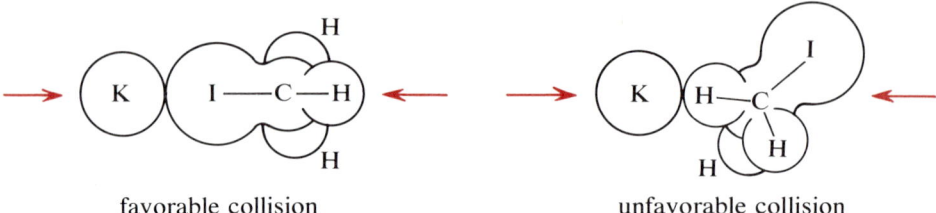

favorable collision unfavorable collision

Figure 11-5 Geometric origin of the steric factor for molecular collisions.

Unfortunately, our concepts of the dependence of molecular energies on molecular geometry do not allow us to predict the magnitude of the steric factor accurately. We can only say that it should be very near 1 for any reaction between single atoms, and quite small for a complicated organic molecule with only a single reactive atom deep inside the molecule. To make any further progress in understanding reaction rates and mechanisms, we must examine what happens when two molecules come together, collide, and form new bonding arrangements. The nature of the "collided" molecule pair or trio is missing from the collision theory, just as the internal nature of molecules is missing from the kinetic theory of gases.

11-4 Gaseous Reaction Rates — Transition-State Theory

Let us begin our consideration of the internal structure of collided pairs of molecules in somewhat the same way we began the study of bonding, by considering a chemical reaction in which two atoms combine to a diatomic molecule:

$$A + B \rightarrow AB$$

As atoms A and B approach each other from a great distance they begin to experience an attraction due to the van der Waals forces of polarization, which increases until overcome by the repulsive force at very close range; this is the behavior described by the Lennard-Jones potential for the approach of non-bonded atoms, as in Fig. 11-6. However, if the Lennard-Jones potential were the only attraction the bonded AB molecule would not exist. There is a sub-

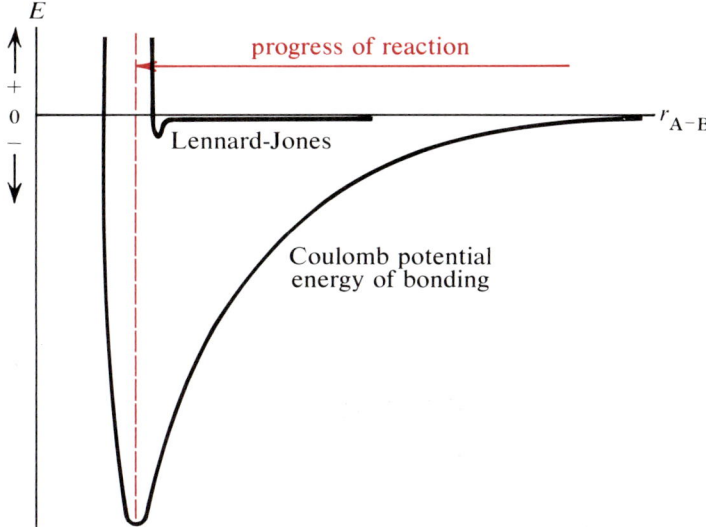

Figure 11-6 Potential-energy curves for nonbonding (Lennard-Jones) and bonding approach of atoms or molecules.

stantially larger attraction due to the Coulomb forces on the bonding electrons, and the figure also shows this bonding potential. The progress of the reaction consists of coming in from the right side of the figure until the internuclear distance equals the bond distance, for which the potential energy is all downhill, so to speak. There should thus be no activation energy for the reaction between two atoms, and this accords with the experimental results for reactions such as

$$H + H + M \rightarrow H_2 + M$$

and

$$H + Br + M \rightarrow HBr + M$$

The rate constant for all such reactions is very high, corresponding essentially to the termolecular collision frequency.

The Calculation of Relative Energies for Colliding Molecules

Now suppose we move on to the next stage of complexity, in which an atom reacts with a diatomic molecule:

$$A + BC \rightarrow AB + C$$

If we suppose that the A atom comes at the BC molecule end-on, there is at the time of collision a linear activated complex A—B—C, in which whatever bonding exists could presumably be described as we did previous molecular orbital structures, in terms of Coulomb integrals, α, and resonance integrals, β. Using the valence bond theory, Fritz London produced an approximate treatment of this system in which the total energy can be described in terms of the energies applying to the possible diatomic molecules AB, AC, and BC. Translating into MO energies, in which the electronic energy of molecule AB is $\alpha_A + \alpha_B + 2\beta_{AB}$, the total electronic energy of the activated-complex system, according to London's treatment, is

$$E_{A-B-C \text{ system}} = 2\alpha_A + 2\alpha_B + 2\alpha_C - \sqrt{2[(\beta_{AB}-\beta_{BC})^2 + (\beta_{BC}-\beta_{AC})^2 + (\beta_{AC}-\beta_{AB})^2]} \quad (11\text{-}7)$$

Potential Surfaces and the Reaction Coordinate

Each of these terms in Eq. 11-7 is strongly influenced by internuclear distance, so if we wished to make a large number of calculations of the appropriate integrals, we could not only calculate E_{A-B-C} at any given distances but we could also construct a **potential surface** like that shown in Fig. 11-7, in which the energy is shown as a topographic map in the third dimension. This figure is worthy of close study, inasmuch as it contains much information about the

Figure 11-7 Potential-energy surface for A + BC → AB + C, with cross sections for B—C interaction (1–2) and A—B interaction (1–4).

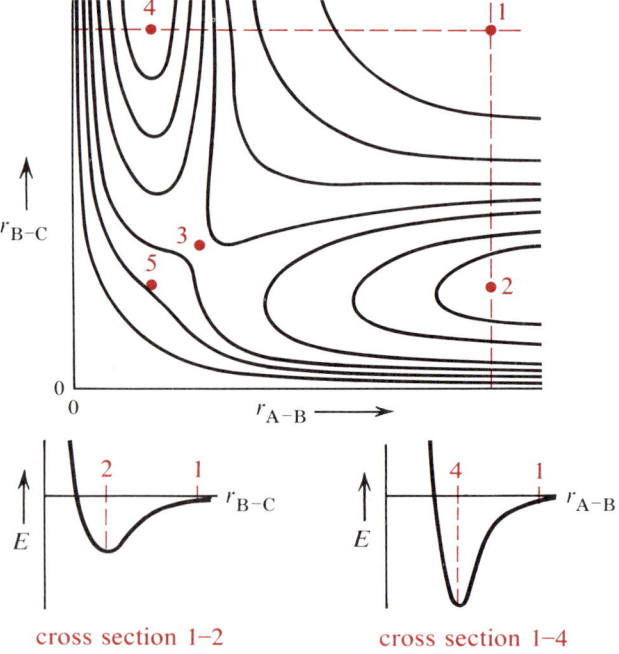

reacting system. First, we need to orient ourselves: at point 1 atoms A, B, and C are all far apart; at point 2 atom A is far from atom B but atoms B and C are close together, as in the reactants of our sample reaction; at point 3 atoms A, B, and C are all close together, as in the collided pair of molecules or activated complex; and at point 4 atoms A and B are close together but far from atom C, as in the products of the reaction. Some feeling for the shape of the surface may be gained by considering the cross sections 1–2 and 1–4, represented by the dashed lines in the figure. The former is simply the potential-energy diagram for the diatomic molecule BC, since A is at all times far away; similarly, the latter is the potential-energy diagram for AB. These cross-section diagrams are shown in the figure. At point 3, called the **transition state**, the energy of the system is significantly higher than at either point 2 or 4, essentially because the B atom has begun to withdraw from the B—C bond but has not yet fully entered the A—B bond. Now suppose we consider the reaction pathway along the **reaction coordinate**, which is the horizontal dimension of the cross section shown in Fig. 11-8. Beginning at point 2, we have atom A and molecule BC;

Figure 11-8 Relationship of the reaction coordinate to the potential-energy surface (as a cross section), showing the activation energy.

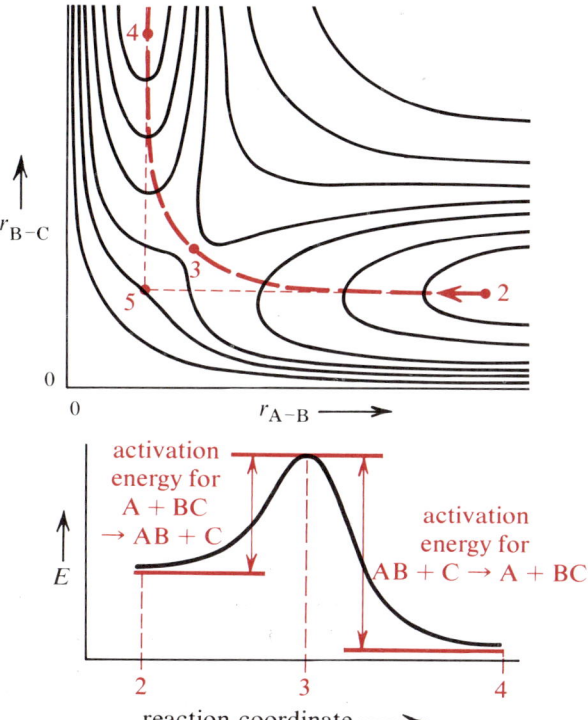

the reaction occurs by having A close on B, moving to the left on the potential-energy surface. Some stretching of the B—C bond has to occur to get to point 3, which accounts for the activation energy; however, the overall energy is still lower than, say, at point 5, because point 3 is a **saddle point**, whose name describes its shape on the potential-energy surface. Finally, after going over the activation-energy hump, the new molecule AB and the atom C separate, moving up on the surface to point 4. Looking at the cross section—the reaction-coordinate diagram—we see clearly the nature of the activation energy, and also the fact that the activation energy for the reverse reaction is in general different from that for the forward reaction.

Figure 11-9 represents a more realistic reaction pathway, in which the zigzags represent simply the vibration of the atoms in the bond before and after the reaction. Notice that the vibration is damped out as the transition state is approached. This may seem to conflict with our earlier idea that translational energy is converted into vibrational energy on collisions, but it does not. The

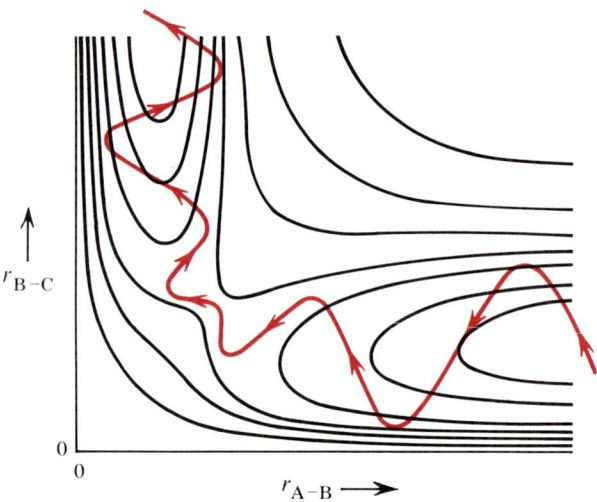

Figure 11-9 A more realistic molecular path across the potential-energy surface.

vibrational energy acquired is potential, not kinetic, energy, and all or most of the existing vibrational energy is absorbed into potential energy to get the molecule over the activation-energy potential barrier. Thus the activated complex is a rather rigid, nonvibrating molecule with stretched bonds. When it decomposes, however, the molecule that is formed moves out into a potential-energy well and converts some of its vibrational potential energy into kinetic energy.

In somewhat the same sense, a collision that is not sufficiently energetic to cause reaction can still convert some translational energy into vibrational energy, as in Fig. 11-10. Study of these figures until a "feel" is acquired will prove extremely helpful in understanding the concepts of reaction dynamics.

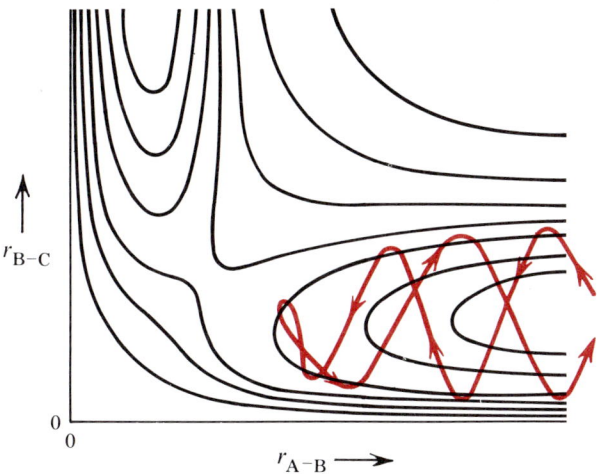

Figure 11-10 Unreactive collision of A and BC.

Two comments should be made about the technique for calculating the potential surface and the surface itself. With respect to the latter, we have only three dimensions in which to draw graphs, so we are limited to systems no more complicated than $A + BC = AB + C$, and even in this case we have limited the geometry to a linear activated complex. If the angle A—B—C changed, the energies would change; at 180° we would have a completely different surface for $A + CB = AC + B$. Although the technique can be applied to other systems, no graphical representation of the energy relationships can be drawn for the results. The other cautionary remark is that since the activation energy appears in the exponent of the Arrhenius expression for the rate constant, a small error in calculating it leads to a very large error in the calculated rate constant. The present level of quantum mechanical calculations is such that the approximations involved in evaluating the necessary integrals for molecules of any complexity lead to errors in E_a of at least one order of magnitude, perhaps two. We are in the frustrating position of knowing what calculations have to be made but not knowing how to do them. However, this situation is not only frustrating but stimulating, and the theory of chemical kinetics is one of the most exciting areas of current research.

Nucleophiles and Electrophiles

By considering these potential-energy surfaces we can introduce another concept which will be useful in considering solution reactions later on. Suppose atom A has a negative charge and there are two kinds of BC molecules available in the reaction mixture: one has both B and C electrically neutral and the other has a partial positive charge on the B atom. For the reaction of A with the neutral molecule we could presumably establish an activation energy and a rate constant, and similarly for the reaction with the molecule having a charged B atom. For the latter, however, there will be an electrostatic attraction between A^- and $B^{\delta+}$ that will deepen the potential-energy well for AB since it is an additional binding influence. Then, as Fig. 11-11 shows, not only will the

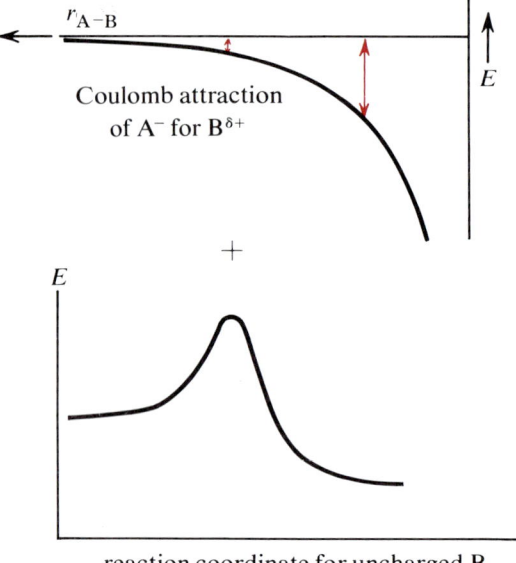

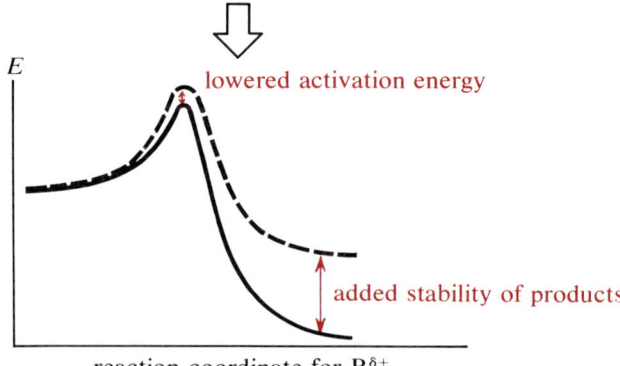

Figure 11-11 Effect of electrostatic attraction in lowering activation energy for reacting atoms or molecules.

product be more stable, but the activation energy will be lower since the Coulomb attraction contributes to the total energy of the system in the activated complex as well as in the product molecule. Since the activation energy is lower than it would have been for a similar reaction with an uncharged B atom (in a BC molecule), the resulting rate is greater, and the result of the overall reaction is that the A^- reacts preferentially with the positively charged B as against the neutral B. We say that A^-, or any reactive atom in a molecule that reacts preferentially with positively charged sites on another molecule, is a **nucleophile**; the name arises from the fact that the nucleus is the source of the positive charge. If the charges on the reactants had been reversed, A^+ would react preferentially with negatively charged sites and would be an **electrophile** (since the electrons provide the negative charge that it seeks).

In Fig. 11-11 both the activation energy and the stability of the products have been changed, but it is possible to imagine cases in which the nucleophile has a choice of a charged or an uncharged atom on the same molecule, giving products with presumably the same stability:

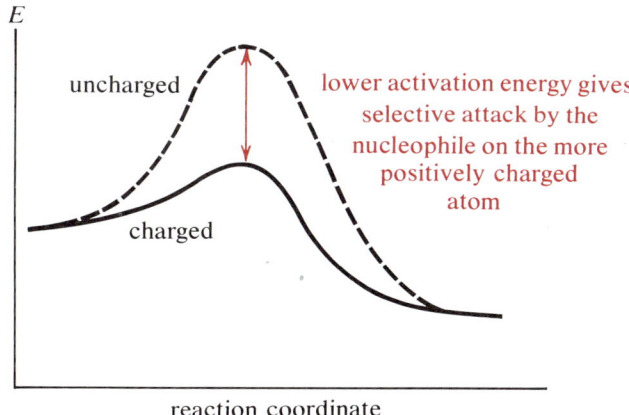

In such a case the reaction coordinate will look like that of Fig. 11-12, from

Figure 11-12 Kinetic control of reaction products by different activation energies.

which it can be seen that the product whose formation requires the lower activation energy will be formed preferentially even though there may be no thermodynamic difference between the stabilities of the reactants and the two possible products. This preference is called **kinetic control** of the reaction products; it is often the determining factor in establishing the principal product of an organic reaction for which several products are possible.

Molecular-Beam Studies of Molecular Collisions

In discussing collisions of molecules we noted that it is somewhat difficult to define collision because molecules are somewhat soft and sticky, as the Lennard-Jones potential implies, and to different degrees. It would be very useful, then, to get some experimental data on how an individual collision occurs. With this sort of data we could make considerable progress toward constructing theories that would yield accurate activation energies and reaction cross sections (the quantity σ^2 that appeared in our collision-theory expressions). Experimentally, however, the problem is difficult. To study the collision between A and BC on an individual-molecule basis requires, first, that no other collisions occur. This means that the whole apparatus must be held to a vacuum such that the mean free path of the A and BC species is much longer than the dimensions of the apparatus; in practice this means pressures on the order of 10^{-7} torr, which is a good vacuum. This already means a very small amount of reaction occurring, on a molar basis, and when we add the restrictions imposed in selecting only those A and BC molecules having the desired velocity (out of all possible velocities), along with other selections of properties we might wish (such as allowing only certain molecular quantum mechanical states), the problem of quantitatively detecting the microscopic amount of reaction becomes very severe. Remarkable progress has been and is being made, using apparatus similar to that in Fig. 11-13. In this figure, the small box labeled A contains gas A at about 0.01 torr, with a hole opening into the high-vacuum chamber of such a size that the A atoms effuse through it rather than streaming through. Inside the chamber the effused molecules are collimated into a reasonably narrow **molecular beam** by passing through slits that block the atoms headed in the wrong direction. The beam has, thanks to the Maxwell–Boltzmann distribution, a wide range of velocities; we would like only a single velocity or at worst a small range of velocities, since defining the velocity gives us a value of the translational kinetic energy that we can turn into information about the collisional activation energy. So we select velocities by aiming the beam at the end of a spiral groove on a rapidly rotating drum. The effect is that the very fast molecules run into the leading edge of the groove, since they move down the groove faster than the drum can turn; the slow molecules do not get all the way down the groove before the trailing edge of the

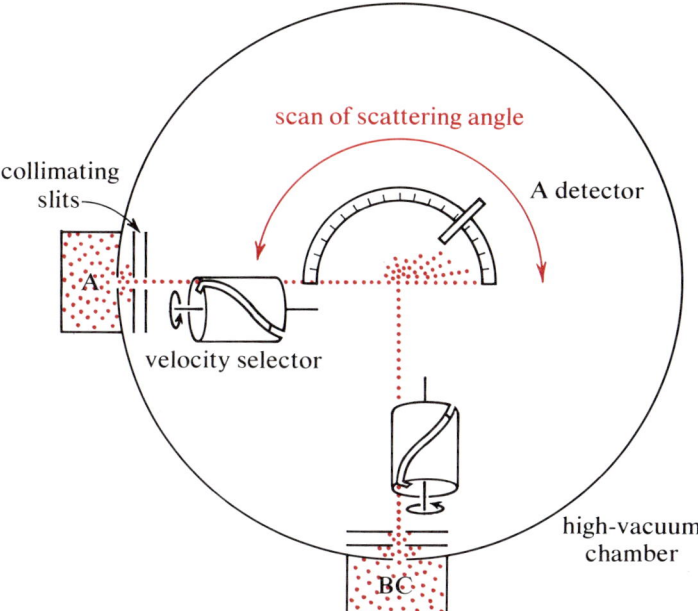

Figure 11-13 Crossed-molecular-beam apparatus.

groove catches them; and only a narrow range of velocities "threads" its way down the groove without hitting either side, giving us a molecular beam with reasonably well-defined velocity. Exactly the same apparatus is set up for the effusion of BC molecules to produce another molecular beam at an angle of 90° to the first, and the two are aligned to cross, so that some collisions occur. Some scattering occurs, both because of the interaction of A with BC's Lennard-Jones potential and because of chemical reaction—which, of course, is what we are interested in. A detector for atom A is mounted on a movable track so as to record the intensity of scattered particles as a function of angle.

Beam Scattering Patterns

Now let us consider what we expect the scattering to be like for various values of the impact parameter b, defined in Fig. 11-14 as the perpendicular distance between the BC molecule's center and the straight-line trajectory of A. As the figure shows, there are two distances at which the A atoms will pass straight through: any large b, meaning substantially greater than the radius of the bottom of the Lennard-Jones potential well, shown in the figures as a red circle, and also b equal to that distance, since it would be attracted while approaching, repelled when it came inside that radius, and attracted again on leaving. All the atoms having b greater than the characteristic Lennard-Jones radius are attract-

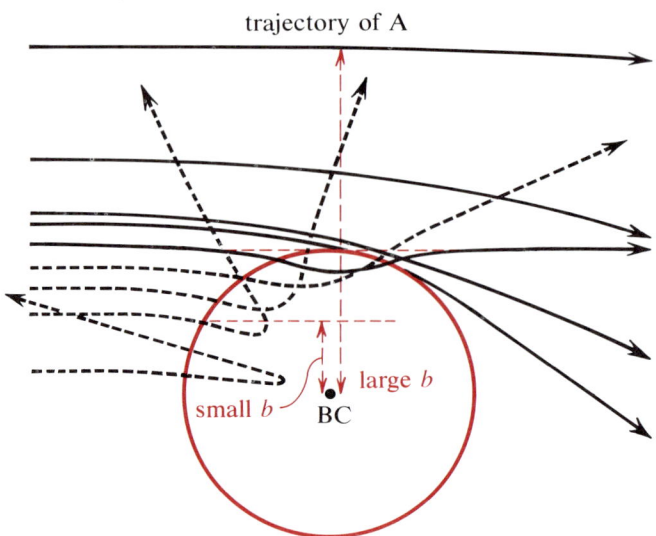

Figure 11-14 Trajectories of A passing BC with different values of the impact parameter b.

ed but never penetrate that radius, and so are scattered toward the bottom of the figure at various angles depending on b. There is a maximum angle of deflection, though, corresponding to the value of b that brings the A atom's actual curved path just to the Lennard-Jones radius but not inside it, and for b smaller than this value but larger than the Lennard-Jones radius the angle of deflection is actually smaller than for larger b, since the atom is now penetrating to the Lennard-Jones repulsion region. On the other hand, for b smaller than the Lennard-Jones radius, the incoming A atom is repelled and deflected up in the figure, as shown by the dotted lines; for very small b the angle of deflection is very large, approaching 180°. In Fig. 11-15 this pattern is extended to the other side of the BC molecule, and it can be seen that between the attracted trajectories (solid lines) from one side of the BC molecule and the repelled trajectories (dotted lines) from the other side, there is a preponderance of scattering more or less parallel to the red lines superimposed on the figure. These red lines correspond to the maximum attracting deflection, and their angle is called the **rainbow angle** since this situation is exactly analogous to the scattering of light that causes rainbows in the atmosphere. From this rainbow angle, then, we can get a good deal of information about the parameters in the Lennard-Jones potential function for the BC molecule. For a system in which no chemical reaction occurs, only elastic scattering, the intensity of scattered A atoms as a function of angle (the experimental result, in other words) is shown

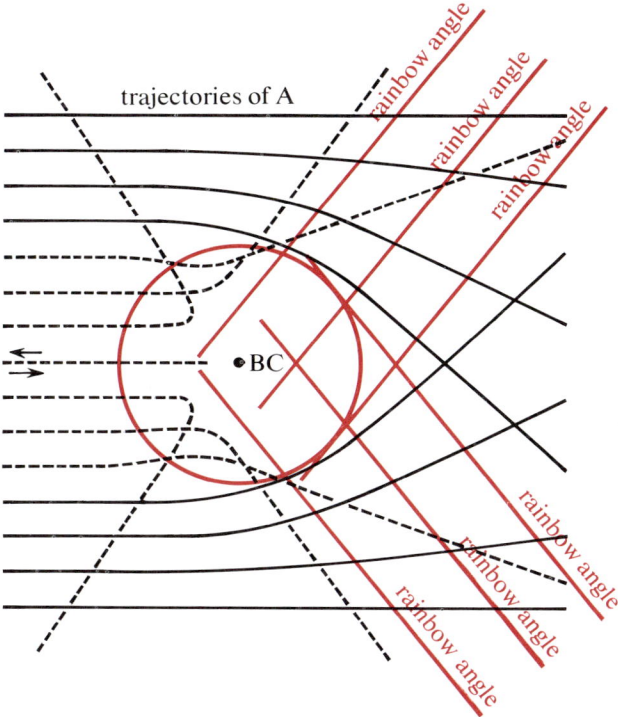

Figure 11-15 The full set of molecular trajectories and the rainbow angle.

in Fig. 11-16a. There are fewer atoms scattered through very large angles because the target is smaller. The rainbow angle is obvious, and from it and the beam velocity we can calculate the Lennard-Jones parameters.

Figure 11-16 Experimental scattering results showing reaction threshold. (a) No reaction, elastic scattering only; (b) chemical reaction occurring.

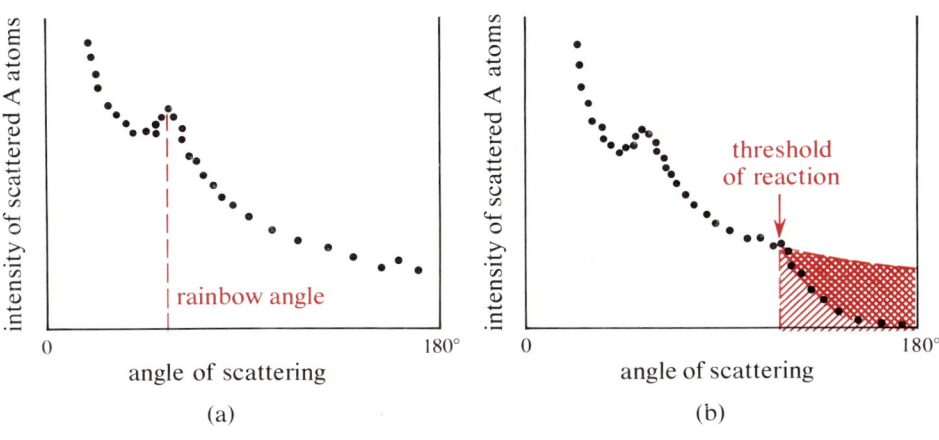

Measurement of Activation Energies

We can get even more interesting information from systems in which some reaction occurs. Figure 11-16b shows the experimental scattering for a system in which the following reaction is occurring:

$$K + CH_3I \rightarrow KI + CH_3$$

As the dotted line suggests, at large scattering angles many fewer K atoms are scattered than if no reaction occurred. The relative extent of reaction is given by the ratio of the area that is "missing" from the curve to the total area that "ought" to be there. From the velocity and the scattering angle at which reaction begins to subtract K atoms one can calculate what b, the impact parameter, must have been, thereby obtaining a value for σ^2, the reaction cross section that appears in the collision-theory equations. This is particularly useful since it differs from the viscosity value in many cases; viscosity, after all, depends on elastic collisions with no possibility of chemical bonding, and the distance of closest approach may well be different for cases in which a bond is being formed. We can go still further: if we measure σ^2 for a range of molecular velocities and plot the resulting function, we get Fig. 11-17. There is a relative kinetic energy (from the two beam's velocities) below which the cross section is zero and no reaction occurs. What does this mean? We have said that a collision has to involve a certain minimum amount of kinetic-energy transfer in order to supply the activation energy. In this case the energy threshold at which the reaction cross section becomes zero is the activation energy, and we thus have a direct means of measuring the activation energy for simple systems. All these results are exciting in their implications, but careful studies are only beginning in this area, and the data so far are only suggestive of what is really happening in reactive molecular collisions.

Figure 11-17 Direct measurement of activation energy.

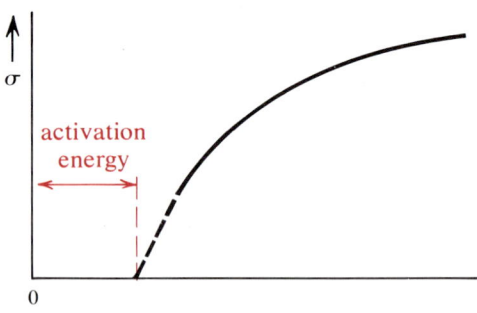

relative kinetic energy from impinging molecular beams

11-5 Reaction Rates in Solution

The preceding sections on the rates of gas reactions have been helpful in understanding the actual molecular processes of chemical reactions as they occur in the gas phase. Most chemical reactions are not run in the gas phase, however, but in solution in a liquid solvent. Even if we considered the same reaction in the gas phase and in a solution, we would expect to see substantial differences in the rate due to the solvent-molecule environment of the dissolved reactants and the dissolved activated complex. More important, very few reactions can be run both in the gas phase and in solvents; in the overwhelming majority of cases, a reaction that proceeds smoothly in solution cannot be observed to occur at all in the gas phase, or else a gas-phase reaction will be replaced in solution by another reaction giving different products. When we stop to think about the profound differences of molecular environment between the gas and the solution this is perhaps not surprising, but what can we say specifically about the comparison? What are the similarities and differences that we must take into account in translating rates and mechanisms for gaseous reactions into those for solution reactions? Most important of all, how can we classify and bring order to the wide array of real reactions in solution, with their diverse rates?

Diffusion, the Solvent Cage, and Encounters

The most important constant feature is that for two molecules to react they still have to collide. In light of the crossed-molecular-beam results, collisions are difficult to define exactly, but the molecules at least have to get close enough together that their centers are separated by not more than, say, 1.5 times the sum of the molecules' radii. In gases under ordinary conditions the mean free path is perhaps 1000 times the radius of a molecule, so that the colliding molecules are essentially completely free of all other influences (meaning attractions, etc., from neighboring molecules) through this distance. In a solution with the same concentration of reactive molecules per liter, however, starting with two molecules that need to collide from 1000 radii apart, millions of collisions with solvent molecules must occur before the molecules can diffuse together. The efficiency of the diffusion process is described by the diffusion constant D for the diffusion of the molecule through the solvent (see p. 123), so we expect the collision frequency in solutions, and hence the rate constant, to be proportional to the diffusion constant. If the theory is worked out in detail and typical values of radii and the diffusion constant are used, the collision frequency between reactive molecules (ignoring the many solvent-molecule collisions) is about 10^9 l/mole-sec, which can be compared to about 10^{11} l/mole-sec for gases (see p. 579). In liquid solutions of moderate concentration, then, collisions between reactive molecules occur only about 1/100 as often as in gases under ordinary conditions. This seems to require solution reactions to be slower than gas reactions, but there is another effect to be considered. If we start with a pair of reactive molecules that have collided and are inside a **solvent cage** of molecules, as in Fig. 11-18, we can calculate through the diffusion constant the time required, on the average, for one of the molecules to diffuse one whole molecular diameter away from the other (i.e., out of the sol-

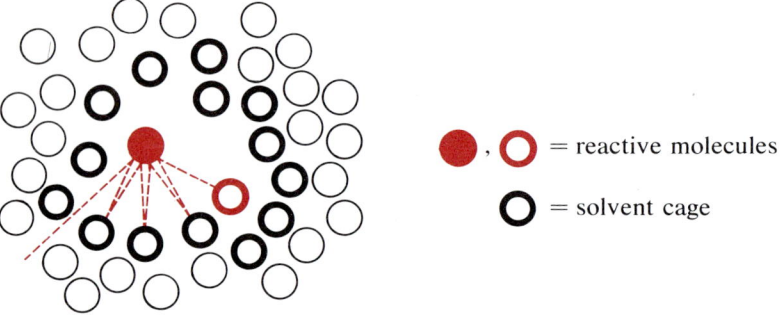

Figure 11-18 Molecular collisions in a liquid environment.

vent cage). If we compare this with the time required for a trip across the solvent cage at the average velocity of the molecule, we find that the average pair of reactive molecules undergoes about 100 collisions with each other while they are both inside the solvent cage (unless their reaction has zero activation energy, which would allow reaction at the first collision); this period is called an **encounter**. So we have the approximate theoretical result that in liquid solutions reactive molecules take 100 times as long to get to each other, but then hit each other 100 times while they are near. In other words, if there are no other forces operating, such as electrical charges, dipole moments, and so on, the rate constant for a reaction in solution ought to be nearly the same as for the same reaction in the gas phase. A nice nonpolar example is the cyclopentadiene hydrocarbon dimerization, one of the few reactions that can be studied under both conditions:

$$C_5H_6 + C_5H_6 \longrightarrow C_{10}H_{12}$$

The rate constants k for this reaction run in various media are shown below; the rate is indeed nearly independent of the medium:

gas reaction	$k = 6 \times 10^{-6}$ l/mole-sec
liquid cyclopentadiene	$k = 6 \times 10^{-6}$ l/mole-sec
benzene solvent	$k = 1 \times 10^{-5}$ l/mole-sec
nitrobenzene solvent	$k = 4 \times 10^{-5}$ l/mole-sec
ethyl alcohol solvent	$k = 4 \times 10^{-5}$ l/mole-sec
acetic acid solvent	$k = 1 \times 10^{-5}$ l/mole-sec

The Thermodynamics of the Transition State

The argument changes dramatically, however, if there are genuine forces acting between the reactive molecules or ions and the surrounding solvent medium. Let us look again at the formation of the transition state. An assumption of the transition-state theory that we have only implied before is that since the transition state can dissociate back to reactants as well as toward products, and since reactants are present and have the necessary activation energy, the transition state (indicated by the superscript‡) is in equilibrium with the reactants:

$$A + B \underset{k_2}{\overset{k_1}{\rightleftarrows}} (A\text{—}B)^{\ddagger} \overset{k_3}{\rightarrow} C + D$$

If the equilibrium is rapid compared with the dissociation into reaction products C and D, the rate-determining step will be that dissociation, and the reaction rate (say $d[C]/dt$) will be given by

$$\frac{d[C]}{dt} = k_3[(A\text{—}B)^{\ddagger}]$$

But, as on p. 576, we can replace $[(A\text{—}B)^{\ddagger}]$ by the equilibrium-constant expression:

$$K_{eq}^{\ddagger} = \frac{[(A\text{—}B)^{\ddagger}]}{[A][B]} \quad \text{or} \quad [(A\text{—}B)^{\ddagger}] = K_{eq}^{\ddagger}[A][B]$$

So

$$\frac{d[C]}{dt} = k_3 K_{eq}^{\ddagger}[A][B]$$

If the observed rate is found to be first order in both A and B, then we have

$$k_{obs} = k_3 K_{eq}^{\ddagger}$$

or, comparing the gas reaction with the solution reaction,

$$\frac{k_{obs}^{gas}}{k_{obs}^{sol}} = \frac{k_3^{gas} K_{eq}^{\ddagger \, gas}}{k_3^{sol} K_{eq}^{\ddagger \, sol}}$$

Now k_3, the dissociation rate constant, is the rate constant for the separation of the activated complex, with all the necessary energy and all the atoms in the proper positions. This will take only as long as one vibration of the bond to be broken, and the time for that vibration will be determined mostly by the structure of the activated complex, not by the surroundings. Therefore k_3 will be nearly the same number whether the reaction is occurring in solution or in the gas phase; to a good approximation we can cancel it and introduce the thermodynamic definition of the equilibrium constant:

$$\frac{k_{obs}^{gas}}{k_{obs}^{sol}} \cong \frac{K_{eq}^{\ddagger gas}}{K_{eq}^{\ddagger sol}} \cong \frac{\exp\left[-\Delta G^{0\ddagger}(gas)/RT\right]}{\exp\left[-\Delta G^{0\ddagger}(sol)/RT\right]} \cong \exp\frac{-[\Delta G^{0\ddagger}(gas) - \Delta G^{0\ddagger}(sol)]}{RT}$$

[Remember that exp( ) is another way of saying $e^{(\ )}$.] In this expression $\Delta G^{0\ddagger}$ is the free-energy change for the production of the activated complex, which is just a careful definition of the activation energy. We see that the ratio of observed reaction rates between the gas phase and a liquid solution depends exponentially on the difference in activation energies between the two phases.

Solvent Effects on $\Delta G^{0\ddagger}$ and Rates

Suppose a reaction occurred between two electrically neutral molecules, but the transition state involved a considerable charge separation so that the activated complex was quite polar. In the gas phase the activation energy would be determined essentially by structural factors within the molecules alone, but if we ran the reaction in a strongly polar solvent the Coulomb forces between the polar solvent molecules and the polar activated complex would stabilize the activated complex, lower the activation energy, and speed up the reaction. The same argument could be applied to a polar solvent and a nonpolar solvent, which, like the gas, would not stabilize the polar activated complex. In practice, this argument is used in two ways: first, the dependence of the rate constant on the polarity of the solvent is taken as evidence for or against a strongly polar or charged transition state; second, if a mechanism seems fairly likely for a proposed new reaction, this argument is used to select a solvent that will favor the desired reaction.

We have seen that the nonpolar reaction in which cyclopentadiene dimerizes is virtually unaffected by choice of solvents; now consider the hydrolysis of t-butyl chloride:

$$(CH_3)_3C{-}Cl + 2H_2O \rightarrow (CH_3)_3COH + H_3O^+ + Cl^-$$

The reaction is allowed to proceed in a mixed solvent of ethanol and water; the more water present, the more polar the solvent mixture is. The rate-controlling step in the mechanism evidently involves only the t-butyl chloride molecule, since, experimentally,

$$\text{rate} = \frac{d[(CH_3)_3COH]}{dt} = k[(CH_3)_3C-Cl]$$

This does not mean, however, that the water concentration did not affect the rate:

90% ethanol–10% water	$k = 0.00616 \text{ hr}^{-1}$
80–20	0.0329
70–30	0.145
60–40	0.453
50–50	1.32
40–60	4.66

Quite obviously the rate increases dramatically as the solvent becomes more polar, in contrast to the rate of the cyclopentadiene dimerization, which was nearly unchanged through a wide range of solvent polarities. This suggests that the mechanism must involve the ionization or at least partial ionization of the t-butyl chloride:

$$(CH_3)_3C-Cl \rightarrow (CH_3)_3C^+ \cdots Cl^- \quad \text{(rate determining)}$$

$$(CH_3)_3C^+ \cdots Cl^- + 2H_2O \rightarrow (CH_3)_3COH + Cl^- + H_3O^+ \quad \text{(rapid)}$$

Why does the water concentration not appear in the rate law, since it strongly affects the rate and is a reactant in the reaction? Its effect cannot be reproduced by any power of $[H_2O]$, so its effect on the rate must not involve participation in the activated complex in any stoichiometric fashion. In addition, the reaction does not occur to any measurable extent in the gas phase even at elevated temperatures, indicating a very strong solvent effect rather than water simply participating in the rate-determining step of the reaction.

Free-Radical and Ionic Mechanisms

This sharp distinction between reactions indifferent to the solvent in which they occur, and other reactions whose rate constant changes by many orders of magnitude for relatively modest changes in the nature of the solvent, suggests that we should distinguish in our mechanism models between reactions having a nearly covalent transition state and those having a strongly polar or perhaps ionic transition state. For bimolecular elementary processes the distinction is fairly simple:

$A^+ + B^- \rightarrow AB$	polar or ionic
$A + B \rightarrow A^+ + B^-$	polar or ionic
$A + B \rightarrow AB$	covalent

For unimolecular elementary processes we can make a distinction in terms of the kinds of bond-breaking processes that occur (see Fig. 6-31):

$$A - B \rightarrow A^0 + B^0 \quad \text{covalent}$$

$$A - B \rightarrow A^+ + B^- \quad \text{ionic}$$

If the bond being broken is a pair of electrons, and if two electrically neutral molecules or fragments are to result, each will have to have one of the two formerly bonding electrons, becoming a **free radical** (commonly written A· or B·) with an odd number of electrons (except in the case of diradicals). Free radicals are usually quite reactive since little activation energy is needed for their recombination:

$$A \cdot + B \cdot \longrightarrow A—B$$

They also readily undergo other reactions such as **abstraction**:

$$A \cdot + R—CH_3 \rightarrow A—H + R—CH_2 \cdot$$

We shall discuss free-radical reactions more extensively in Chapter 15, because they are very important to the production of many polymers and plastics.

The reactions in which a bond breaks with both electrons going with one of the formerly bonded atoms, giving an ionic or at least a polar transition state, are more common than free-radical reactions, because most bonds between unlike elements are already polar due to the different valence-orbital ionization potentials. Many organic and inorganic substitution reactions fall into this category:

$$(CH_3)_3CBr + OH^- \rightarrow (CH_3)_3COH + Br^-$$

$$Co(NH_3)_5Cl^{2+} + OH^- \rightarrow Co(NH_3)_5OH^{2+} + Cl^-$$

The first, of course, is the example in which the rate-determining step is the unimolecular production of the $(CH_3)_3C^+$ ion. The second (to which we shall return in Chapter 16) similarly has as a rate-determining step the production of the $Co(NH_3)_4NH_2^{2+}$ ion; both rate-determining steps involve the departure of the halide ion. The ensuing rapid step in each case is the attack of water or OH^- ion, which donates a pair of electrons to the positively charged species. In this subsequent reaction the water or OH^- is serving both as an electron donor and as a seeker of positive charge. It is thus both a Lewis base and a nucleophile; the terms are in fact synonymous. An abbreviation has been proposed that both organic and inorganic chemists use frequently for such a mechanism: S_N1, meaning a *S*ubstitution that involves the attack of a *N*ucleophile and is kinetically *1*st order. The rate-determining step in these reactions is always the unimolecular separation of an ion or molecule into a more positively charged ion and a negative ion. It is always, for this reason, favored by polar solvents and may not occur at all in nonpolar solvents. It is also favored by an electron distribution in the reactant molecule that places a partial positive charge on the atom from which the negative ion must leave. Some examples of this relative behavior will appear in both Chapters 15 and 16.

Returning to reactions in which the rate-determining step is bimolecular, it is possible to have substitution reactions that involve attack by a nucleophile but (because of the bimolecular step) are second order; these are called S_N2 reactions. Again they apply equally well to both organic and inorganic reactions:

$$\underset{\text{benzene}}{C_6H_6} + NO_2^+ + HSO_4^- \rightarrow \underset{\text{nitrobenzene}}{C_6H_5NO_2} + H_2SO_4$$

$$\begin{cases} C_6H_6 + NO_2^+ \rightarrow C_6H_6NO_2^+ & \text{slow (rate determining)} \\ C_6H_6NO_2^+ + HSO_4^- \rightarrow C_6H_5NO_2 + H_2SO_4 & \text{fast} \end{cases}$$

$$\text{rate} = k[C_6H_6][NO_2^+]$$

$$\underset{\text{pyridine}}{Pt(NH_3)Cl_3^- + C_5H_5N} \rightarrow Pt(NH_3)(C_5H_5N)Cl_2 + Cl^-$$

$$\text{rate} = k[Pt(NH_3)Cl_3^-][C_5H_5N]$$

For a substitution reaction

$$Y^- + R-X \rightarrow R-Y + X^-$$
nucleophile or
Lewis base

there are thus two general categories of mechanism (although gradations in between are possible and often seen):

$$S_N1 \quad R-X \longrightarrow R^+ + X^-$$
$$R^+ + Y^- \longrightarrow R-Y$$
$$S_N2 \quad Y^- + R-X \longrightarrow Y^{\delta-}\cdots R \cdots X^{\delta-} \rightarrow Y-R + X^-$$

A more polar solvent will speed up the S_N1 mechanism, since it stabilizes the ionic R^+ intermediate whose formation is the bottleneck in the whole reaction. On the other hand, a more polar solvent will actually slow down the S_N2 mechanism, since it stabilizes the compact charged reactant Y^- more than it does the transition state $Y^{\delta-}\cdots R \cdots X^{\delta-}$, in which the charge is spread out over a larger molecule; this makes the activation energy greater than it was in the less polar solvent.

Acid–Base Catalysis

By this time it should be clear that the stoichiometry of a chemical reaction does not dictate the concentration dependence of its rate law. We have illustrated this for cases in which not all of the reactants appear in the rate law, or in which they appear raised to smaller powers than their stoichiometric coefficients suggest. Having prepared ourselves in this fashion, we are not surprised to discover that some reactions having polar or ionic transition states in protic solvents (ones that are capable of transferring protons from one molecule to another, like water) can actually contain in their rate laws a stoichiometric dependence on proton-donor or -acceptor concentration, *even when the reactions as written do not generate or consume protons*. That is, the rate of the reaction may be governed by the concentration of an ion that does not appear in the overall stoichiometry of the reaction. This remarkable kind of rate dependence is called **catalysis**; when proton donors or acceptors (Brönsted acids or bases) are the foreign species, the reaction is subject to **acid–base catalysis**. The effect is not mystical, of course, even though like other kinetic phenomena it does not have as yet a completely quantitative interpretation. The reason a foreign molecule appears in the rate law is that it actually takes part in the rate-determining step or steps. It does not appear in the overall reaction stoichiometry because it is regenerated in a step subsequent to the one in which it is used. Let us look at an example. When acetone reacts in aqueous solution with elemental bromine, the overall stoichiometric equation is

$$CH_3\overset{O}{\underset{\|}{C}}CH_3 + Br_2 + H_2O \rightarrow CH_3\overset{O}{\underset{\|}{C}}CH_2Br + Br^- + H_3O^+$$

Whatever the rate law is, our previous discussion suggests that it ought to contain only acetone and/or bromine. Instead, we find experimentally that

$$\text{rate} = \frac{d[Br^-]}{dt} = k\left[CH_3\overset{O}{\underset{\|}{C}}CH_3\right][H_3O^+]$$

To explain this result we propose the following mechanism, in which the slow or rate-determining step does indeed require a prior stoichiometric reaction with H_3O^+:

$$CH_3\overset{O}{\underset{\|}{C}}CH_3 + H_3O^+ \underset{}{\overset{K_{eq}}{\rightleftharpoons}} \left[CH_3\overset{OH}{\underset{|}{C}}CH_3\right]^+ + H_2O \qquad \text{rapid equilibrium}$$

$$\left[CH_3\overset{OH}{\underset{|}{C}}CH_3\right]^+ + H_2O \overset{k_1}{\longrightarrow} CH_3\overset{OH}{\underset{|}{C}}=CH_2 + H_3O^+ \qquad \text{slow (rate determining)}$$

$$CH_3\overset{OH}{\underset{|}{C}}=CH_2 + Br_2 + H_2O \overset{k_2}{\longrightarrow} CH_3\overset{O}{\underset{\|}{C}}CH_2Br + Br^- + H_3O^+ \qquad \text{fast}$$

If the second step is indeed rate determining, we have

$$\text{rate} = k_1\left[\left[CH_3\overset{OH}{\underset{|}{C}}CH_3\right]^+\right][H_2O]$$

But, from the prior equilibrium-constant expression we have

$$K_{eq} = \frac{\left[\left[\begin{array}{c} OH \\ | \\ CH_3CCH_3 \end{array}\right]^+\right][H_2O]}{\left[\begin{array}{c} O \\ \| \\ CH_3CCH_3 \end{array}\right][H_3O^+]}$$

or

$$\left[\left[\begin{array}{c} OH \\ | \\ CH_3CCH_3 \end{array}\right]^+\right] = \frac{K_{eq}\left[\begin{array}{c} O \\ \| \\ CH_3CCH_3 \end{array}\right][H_3O^+]}{[H_2O]}$$

So

$$\text{rate} = k_1 \left\{ \frac{K_{eq}\left[\begin{array}{c} O \\ \| \\ CH_3CCH_3 \end{array}\right][H_3O^+]}{[H_2O]} \right\}[H_2O]$$

$$= k_1 K_{eq}\left[\begin{array}{c} O \\ \| \\ CH_3CCH_3 \end{array}\right][H_3O^+]$$

This agrees with the experimental rate law. Note that although H_3O^+ is a product in the overall reaction, it is a reactant in one step. However, it is produced in the other two steps, which gives a net production of one H_3O^+ by the overall reaction. The combination of a rate-determining step preceded by a rapid equilibrium is frequently found in multistep mechanisms.

The idea that proton transfer plays an important role in some reaction mechanisms not dealing specifically with acid–base reactions suggests that it might be possible to use any proton donor, not just H_3O^+, or conversely any proton acceptor, not just OH^-, for catalytic purposes. This is indeed true, and as a matter of fact if more than one proton-transfer species are in a reaction mixture that is susceptible to acid–base catalysis, the rate law is a sum of individual contributions from each species as catalyst:

$$\text{overall rate constant } k = k_0 + \sum_{i=1}^{n} k_i[\text{acid}_i] + \sum_{j=1}^{m} k_j[\text{base}_j]$$

if there are n acids and m bases present. Which catalyst species is most efficient? Since we are talking about proton donation by the acid, the most obvious correlation is on the basis of relative acid constants in aqueous solution,

K_a. Consider the possibility that the free energy of activation for the rate-determining step might be some fraction α of the free-energy change for the proton-transfer reaction in water (the K_a reaction):

$$\Delta G^{0\ddagger} = \alpha \Delta G^0_{HA}$$

where $\Delta G^{0\ddagger}$ is the free energy of activation and α in effect allows for the possibility that the proton transfer in catalysis may not need to be complete since it is only temporary. We can relate this to the rate constant itself through a modification of the Arrhenius equation:

$$k = A e^{-\Delta G^{0\ddagger}/RT}$$

or

$$\ln k = \ln A - \frac{\Delta G^{0\ddagger}}{RT}$$

Substituting in our proposed relationship,

$$\ln k = \ln A - \frac{\alpha}{RT}(\Delta G^0_{HA})$$

But since ΔG^0_{HA} refers to the K_a reaction in water, we can equally well substitute in the thermodynamic definition of K_a:

$$\ln k = \ln A - \frac{\alpha}{RT}(-RT \ln K_a)$$

$$= \ln A + \alpha \ln K_a$$

$$k = A(K_a)^\alpha \tag{11-8}$$

Within the limits of our approximations we can thus relate the rate constant directly to the acid constant for the particular catalyzing species; this expression is extremely successful in linking kinetic phenomena to thermodynamic quantities. Equation 11-8 is called the **Brönsted equation**; it is one of a series of similar relationships known as **linear free-energy relationships**. There are two parameters involved: A is the collision frequency combined with the steric factor, and is constant only for acids of similar structure; α is presumably characteristic of a particular reaction and solvent. When A and α are properly chosen, the proportionality implied in the Brönsted equation can in favorable cases be shown to hold true over K_a ranges of 10^{15} with quite modest deviations, which is a ray of quantitative hope in the gloom of order-of-magnitude speculations.

The phenomenon of catalysis is of course not limited only to the sort of acid–base interaction we have discussed. In redox reactions, for instance, a foreign ion that can accept both electrons from a two-electron donor and release them one at a time to a one-electron acceptor serves as a catalyst, since its two-body collisions occur much more frequently than the three-body collisions that would otherwise be necessary. Many other kinds of reactions are subject to catalysis of different sorts; all that is necessary is that some species in the reaction mixture be used to provide a mechanism with lowered activation energies and be regenerated after the rate-determining step is over. Indeed, it need not even lose its own chemical identity, as the next section will show.

11-6 Heterogeneous Reactions and Catalysis

We have progressed from gaseous reactions through solution reactions, both of which are homogeneous in that all reaction components, even catalysts, are present in the same phase. Some reactions, however, occur at phase boundaries, either because they involve the reaction of, say, a solid with a liquid, or because they are homogeneous in the gas or liquid phase and are catalyzed by a solid surface. These are heterogeneous reactions, or at least heterogeneous catalysis. We shall consider the two topics together.

The Influence of Catalysts on Reactions with Kinetic Control of Reaction Productivity

Heterogeneous catalysis is one of the most intriguing aspects of chemistry since it has the appearance of providing something for nothing. After all, one simply mixes two chemicals that would not react, throws in a few pellets of something, and shortly draws off a useful product with the pellets left over. What is not clear on such a superficial examination is that for heterogeneous catalysis to succeed it must be true that the failure of the reaction otherwise is due to kinetic factors, not thermodynamic ones. If the catalyst is really unchanged in the reaction (which is sometimes true), its description using thermodynamic variables cancels out of the difference between the initial state and the final state of the reaction system; ΔG^0 is absolutely unchanged by the presence of the catalyst since the catalyst's own thermodynamic properties are unchanged, and the degree of spontaneity of the reaction is exactly the same as it would have been without the catalyst. For any reaction that can be thought of as an equilibrium, the constancy of ΔG^0 means that the equilibrium constant and thus the equilibrium composition are unchanged by the presence of a catalyst. The only thing a catalyst can change is the *rate* at which equilibrium is established; the misleading quality of the use of catalysts is due to the fact that it is very difficult to establish whether lack of reactivity is due to thermodynamic or kinetic factors. Figure 11-19 shows the difference in terms of the reaction coordinate for two reactions: one is unreactive because ΔG is positive; the other is unreactive because $\Delta G^{0\ddagger}$ is not only positive but very large compared to the energy available from molecular collisions, and may also involve very severe steric or entropy requirements on the reacting system. The

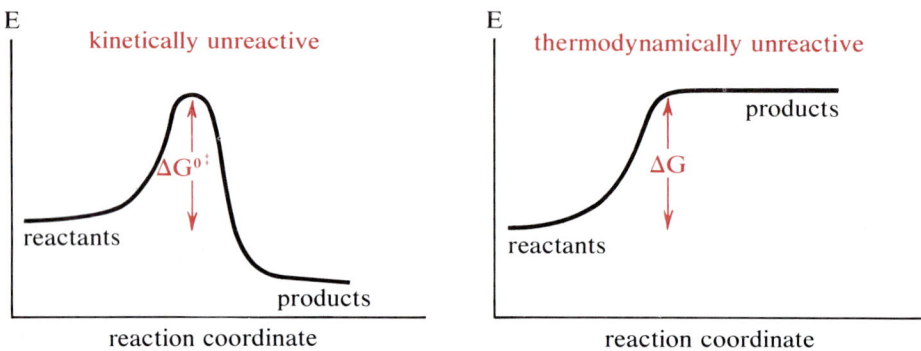

Figure 11-19 Comparison of two systems that are unreactive for different reasons.

only thing a catalyst can do is to reduce the free energy of activation, either through providing a low-energy mechanism or by assisting in modifying the geometry of a reactant to reduce steric hindrance. A catalyst cannot change the position of equilibrium or the overall extent of reaction.

The Langmuir Isotherm

A solid catalyst for a liquid or gaseous reaction will work in much the same way that a homogeneous catalyst does, by entering into a chemical reaction to make the rate-determining elementary process easier. In order for this to happen, it is necessary for the molecule to hit a site on the surface of the solid at which such chemical bonding can occur. Obviously this places a premium on having the maximum possible surface area on a given amount of solid catalyst; most commercial catalysts are very fine powders, having around a million square centimeters of surface area per gram of material. When the molecule hits the catalyst surface it must stick, and there are two rather loosely defined, overlapping kinds of **sorption** or sticking process. The first is **physical adsorption**, in which the surface is gradually coated with a layer of molecules held loosely in place by van der Waals forces. In general this sort of adsorption cannot have a sufficiently large energy effect on the adsorbed molecules to affect their chemical properties significantly, because the van der Waals forces are so much smaller than the chemical binding forces. The other type of sorption is called **chemisorption**. It is similar to physical adsorption except that the forces are much stronger and are apparently equivalent to conventional chemical bonds. Chemisorption occurs in most cases of surface reaction or catalysis, although it is not necessary for all reaction components to be chemisorbed. If all reactants are chemisorbed, the reaction must occur by surface migration until the reactants meet, or perhaps by evaporation from one site and recondensation on another, either of which is relatively slow. For this reason the sorption process is fairly often the rate-controlling factor in the overall reaction.

We can represent the chemisorption as a chemical reaction between the free reactant molecule A and an active site S on the catalyst surface to give the bonded system A·S:

$$A_{(g)} + S_{(s)} \rightleftharpoons A \cdot S_{(surface)}$$

Written in this way the equilibrium can be represented by an equilibrium constant K_s for the sorption process, or K_d for the desorption process (the reverse reaction):

$$K_s = \frac{[A \cdot S]}{[A][S]} \qquad K_d = \frac{[A][S]}{[A \cdot S]} = \frac{1}{K_s}$$

Suppose we designate the initial total number of sites as S^0. Then

$$S^0 = [A \cdot S] + [S]$$

since all the sites are either covered or uncovered. The fraction covered, called θ, is

$$\theta \equiv \frac{[A \cdot S]}{S^0} = \frac{[A \cdot S]}{[A \cdot S] + [S]} = \frac{K_s[A][S]}{K_s[A][S] + [S]} = \frac{K_s[A]}{K_s[A] + 1}$$

$$= \frac{[A]}{[A] + 1/K_s} = \frac{[A]}{[A] + K_d} \qquad (11\text{-}9)$$

This expression is called the **Langmuir isotherm** after its discoverer, Irving Langmuir, the first American scientist to do important industrially sponsored research. Since [A] is proportional to the pressure of A if A is a gas, the Langmuir isotherm predicts, with moderate accuracy, that at low pressures the fraction of sites covered will be proportional to P_A (if $P_A \ll K_d$) and at high pressures the fraction covered will approach 1 (when $P_A \gg K_d$), as in Fig. 11-20.

Let us apply this relationship to an actual case. At elevated temperatures (around 300°C) many alcohols will dehydrate to double-bonded hydrocarbons if catalyzed by the presence of solid metal oxides. The reaction is

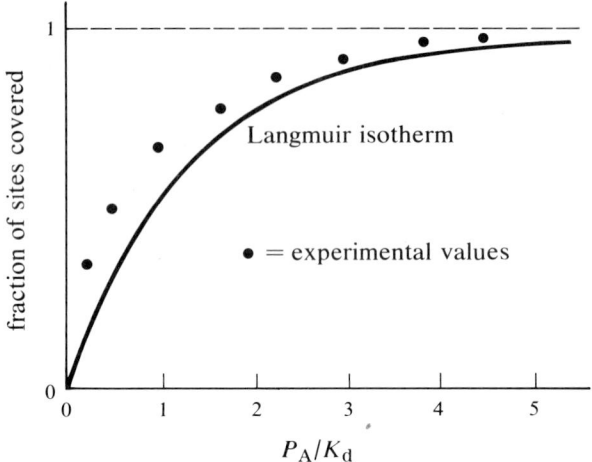

Figure 11-20 Experimental study of catalyst surface coverage compared to Langmuir isotherm prediction. (Redrawn from S. W. Benson, *The Foundations of Chemical Kinetics*, New York: McGraw-Hill, 1960.)

$$RCH_2-CH_2OH \rightarrow RCH=CH_2 + H_2O$$

The experimental rate law is

$$\frac{d[RCH=CH_2]}{dt} = kP_{RCH_2CH_2OH}$$

and it is also found that this is slower than the rate of sorption, so that it must represent a step taking place on the surface. The proposed mechanism is as follows:

(1) $RCH_2-CH_2OH_{(surface)} \xrightarrow{slow} RCH_2-CH_{2(surface)}^+ + OH_{(surface)}^-$

(2) $RCH_2-CH_{2(surface)}^+ \xrightarrow{fast} RCH_2-CH_{2(surface)}^+ + OH_{(surface)}^-$

(3) $RCH=CH_{2(surface)} \xrightarrow{fast} RCH=CH_{2(g)}$ (desorption)

(4) $H_{(surface)}^+ + OH_{(surface)}^- \xrightarrow{fast} H_2O_{(g)}$

The rate-determining elementary process is thus unimolecular:

$$\frac{d[RCH_2-CH_2^+]}{dt} = k_1[RCH_2-CH_2OH_{(surface)}]$$

At low pressures, the Langmuir isotherm says that

$$[RCH_2-CH_2OH_{(surface)}] = CP_{RCH_2-CH_2OH}$$

so

$$\text{rate} = k_1 CP_{RCH_2-CH_2OH}$$

which agrees with the observed rate law. The pressure dependence of the rate would give us an interesting check on this application of the Langmuir isotherm, but side reactions and the difficulty of measuring adsorption make it impossible to explore a very wide pressure range. This is frustrating, because among other things it would be interesting to know whether the metal oxide actually has two kinds of sites, acid and basic, to accommodate the OH^- and H^+ respectively, or whether it represents simply a very polar medium in which the polar transition state for the first elementary process is more stable. Most such studies encounter this or similar difficulties, with the result that kineticists feel reasonably confident only about the broad features of heterogeneous reactions.

Catalyst Poisons

One aspect of heterogeneous catalysis that does relate nicely to the above model is the existence of "catalyst poisons"—substances that in very small quantities prevent the solid surface from acting as a catalyst. In general, the number of sites on a catalyst surface is much smaller than the total number of molecular reactions that the surface catalyzes. Therefore a given site must serve many reacting molecules in the course of the reaction. If it chemisorbs an atom or molecule foreign to the reaction that bonds very strongly to the site so that desorption is impossible under the conditions of the reaction, then that site is lost to further catalysis. "Poisons," then, chemisorb on the particular catalyst so strongly that they use up all the active sites and are not removed. The concept of hard vs. soft acids and bases can be applied with some utility here; for instance, when metal surfaces serve as catalysts, their low charge (zero) makes them very soft bases. Accordingly, they bond strongly to soft acids, such as Cs^+, Ag^+, Hg^{2+}—but not strongly to hard acids such as Na^+, Mg^{2+}, or Cr^{3+}. The former set of ions are poisons for the catalyst surface but the latter are not, since they do not bond strongly enough to prevent desorption.

Enzyme Catalysis of Biochemical Reactions

Many cases of heterogeneous catalysis are extremely important to our society: the synthesis of ammonia from N_2 and H_2, the contact process in which SO_3 for the manufacture of sulfuric acid is made from SO_2 and O_2, and the catalytic cracking of hydrocarbons to maximize the yield of gasoline from crude oil. Perhaps the most interesting example is the functioning of enzymes. Most biochemical processes in living systems are catalyzed by enzymes, which are very large molecules having one or a few sites that catalyze a very specific reaction; these sites often involve a complexed metal atom. For instance, urease catalyzes only the hydrolysis of urea to ammonia:

$$\mathrm{NH_2\overset{O}{\overset{\|}{C}}NH_2 + H_2O \xrightarrow{\text{urease}} 2NH_3 + CO_2}$$

A model exists for the kinetic function of a general enzyme, which is very similar to the Langmuir isotherm. Suppose we have an enzyme E and a molecule whose reaction it specifically catalyzes (the **substrate**, S). We write the same sort of equation for the formation of the enzyme-substrate complex E·S:

$$\mathrm{E + S \underset{2}{\overset{1}{\rightleftharpoons}} E\cdot S}$$

This is an equilibrium, with forward rate constant k_1 and reverse rate constant k_2. The next elementary process in this simplified model is the reaction itself, with rate constant k_3:

$$\mathrm{E\cdot S \xrightarrow{3} E + products}$$

Now suppose that in the process of reaction the intermediate complex E·S reaches a maximum and constant concentration; that is, it is formed by reaction 1 as fast as it is lost by reactions 2 and 3. This assumption is called the **steady-state approximation** and is applied in many kinetic problems. Since the rate of formation must equal the rate of disappearance,

$$\underset{\text{formation}}{k_1[\mathrm{E}][\mathrm{S}]} = \underset{\text{disappearance}}{k_2[\mathrm{E\cdot S}]} + \underset{\text{disappearance}}{k_3[\mathrm{E\cdot S}]}$$

or

$$[\mathrm{E\cdot S}] = \frac{k_1[\mathrm{E}][\mathrm{S}]}{k_2 + k_3}$$

For the total enzyme concentration $[\mathrm{E^0}]$ we have, as we did for the active sites, $[\mathrm{E^0}] = [\mathrm{E}] + [\mathrm{E\cdot S}]$. To get this in a form resembling the above, we multiply by $k_1[\mathrm{S}]$:

$$k_1[\mathrm{S}][\mathrm{E^0}] = k_1[\mathrm{S}][\mathrm{E}] + k_1[\mathrm{S}][\mathrm{E\cdot S}]$$

For the first term on the right we can substitute from the steady-state [E·S] expression:

$$k_1[\mathrm{S}][\mathrm{E^0}] = (k_2 + k_3)[\mathrm{E\cdot S}] + k_1[\mathrm{S}][\mathrm{E\cdot S}]$$

or

$$[\mathrm{E\cdot S}] = \frac{k_1[\mathrm{S}][\mathrm{E^0}]}{k_1[\mathrm{S}] + k_2 + k_3}$$

For the rate of reaction we then have

$$\text{rate} = \frac{d[\text{products}]}{dt} = k_3[\mathrm{E\cdot S}]$$

$$= \frac{k_3 k_1 [S][E^0]}{k_1 [S] + k_2 + k_3}$$

$$= \frac{k_3 [S][E^0]}{[S] + (k_2 + k_3)/k_1}$$

Let $(k_2 + k_3)/k_1 \equiv K_M$:

$$\text{rate} = \frac{k_3 [S][E^0]}{[S] + K_M}$$

$$\frac{1}{\text{rate}} = \frac{1}{k_3 [E^0]} + \frac{K_M}{k_3 [E^0]}\left(\frac{1}{[S]}\right)$$

This is a linear function between 1/rate and 1/[S]. So by taking the observed rate and plotting its reciprocal against 1/[S], we should get a straight line; from its slope and intercept, if we know $[E^0]$, we can get the constants K_M and k_3. Finally, if we measure the temperature dependence of the reaction, we can get the activation energy for the reaction step represented by k_3. In the few cases for which a careful treatment has been made, the activation energy for the actual reaction (step 3) proves to be exceedingly low compared to ordinary solution reactions: perhaps 5–15 kcal/mole reaction. This is, of course, exactly what a catalyst ought to do, namely, reduce the activation energy for the desired reaction. This procedure and the resulting expressions are known as the **Michaelis–Menten** equations, and are the basis for most treatments of enzyme kinetics. Enzymes show many of the properties of other heterogeneous catalysts, but the subject is too vast for us to consider in detail here. In Chapter 15 we shall encounter enzymes again, and consider ways in which we might be able to explain their catalytic properties.

11-7 Summary

After an extended discussion of the principles of spontaneous change we have finally approached the more detailed and more difficult problem of how chemical transformations take place. We have established the distinction between thermodynamic control of a reaction and kinetic control, which means that the free energy of activation for the reaction is so large that molecular collisions, the only source of energy available to the reactants, cannot provide it. We thus have an explanation of the fact that some reactions that have fairly large negative free-energy changes associated with them do not occur; reactants can be prevented from combining by a large potential-energy barrier, whether that barrier represents a free-energy change for the overall reaction or just for the activation step.

Rates and mechanisms are thus intimately linked, and we have seen that the overall rate law for a reaction is the most fundamental piece of information about its mechanism. Since in many cases the mechanism consists of not one but a series of collisions, the rate law may have a concentration dependence quite different from that suggested by the overall stoichiometry, particularly if one step is much slower than the others, so that it is rate-determining. For any individual step or elementary process, however, the rate law corresponds exactly to the stoichiometry of that process. The process of establishing a mechanism for a reaction usually involves establishing its rate law, then proposing on the basis of reasonable chemical structures a series of elementary processes that yield an identical rate law. Since several mechanisms can often be proposed, additional information must be obtained, such as evidence of an intermediate molecule, solvent dependence of the rate, or other physical properties of the reacting system.

We can calculate very roughly what reaction rates ought to be in gases by considering the frequency with which the reactive atoms collide. Since observed rates are much slower than the collision frequency, we must add two moderating factors: first, the concept of activation energy and the Boltzmann distribution of energies to account for the fact that for most molecules only a small fraction of collisions seem to yield reaction; second, a steric factor that, because of restrictions imposed by the geometry of the reacting molecules, prevents reaction even in high-energy collisions when the specific reactive atoms in the molecules do not meet. We can improve on this collision theory at least conceptually if not quantitatively by considering the energy (and entropy) relationships in a collided pair of molecules that are about to react. This is the transition state, and within severe limitations imposed by calculation approximations we can calculate potential-energy wells and surfaces for very simple elementary processes. At present these calculations are still very crude, but they will undoubtedly improve as quantitative procedures in quantum mechanics do—and as more detailed data from molecular-scale experiments, such as those involving molecular beams, become available.

In solution the situation is more complicated but not qualitatively different. The overall rate of individual collisions between A and B is about the same as in the gas phase if there are no interactions between reactive molecules and solvent due to polarity, hydrogen bonding, and so on. The slowing of molecular travel to diffusion rates is just about made up for by the phenomenon of encounters—multiple collisions of reactive molecules while they are held in a solvent cage. However, if interactions of either the reactants or the transition state with the solvent become important, the rate can be altered very dramatically since those interactions directly modify the activation energy. Reactions between charged systems can involve attack by nucleophiles, which seek a positive charge, or electrophiles, which seek a negative charge. Charged or polar transition states are much more common than nonpolar ones (perhaps free radicals), since most chemical bonds are themselves polar to some extent and a polar solvent can usually be found to stabilize a polar transition state.

Catalysis, the speeding up of reactions by species not appearing in the reaction stoichiometry, was discussed for both the homogeneous and heterogeneous cases. Homogeneous catalysis proved no different from ordinary mechanisms and elementary processes, except that a reactant in the rate-determining step was generated by a subsequent step so that it canceled out of the overall reaction. Heterogeneous catalysis by solid surfaces is analogous in that the surface bonds to a reactant so as to reduce the activation energy for the rate-determining step, but is a little more dramatic because the rigid solid structure enables the catalyst to retain its structural identity throughout the reaction. The important case of enzyme catalysis was introduced and a mechanism for general enzyme kinetics developed.

Even though this has been a sketchy survey, and even though chemical kinetics is still far from being fully developed, we have achieved at least a partial feel for what is going on among molecules when they react. This is a desirable end in its own right, but perhaps even more important is that it allows us to make an intelligent choice of conditions that will favor a chemical reaction we wish to carry out. With this background, then, we need to go on to a survey of the ways in which our ideas of chemical structure and chemical dynamics can guide us in understanding the bewildering diversity of chemical compounds that exist, and also in understanding how to change them for our own purposes. This final portion of the book is at last the heart of the matter.

Study Problems

1. A frequently used measure of a reaction's rate is its *half-life*, $t_{1/2}$, the time necessary for half of a reactant to be transformed. The half-life is particularly convenient for first-order reactions, in which the concentration dependence is that shown in Fig. 11-1. Integrate the algebraic expression of the first-order rate law between limits C_A and $C_A/2$ for the concentration of A, and $t=0$ and $t=t_{1/2}$ for time. Show that for this type of rate law $t_{1/2} = \ln 2/k = 0.693/k$.
2. If the Arrhenius activation energy of a chemical reaction is 20 kcal/mole, what will be the ratio of the reaction's rate constants at 300°K and 310°K?
3. In seeking a function of concentration to plot against time that will yield a straight line, we are in effect finding the form of the integrated rate law, as consideration of Study Problem 1 will show. Rearrange the second-order differential rate law, integrate it, and show that the proper function of concentration to plot is $1/C_A$.
4. Copper(II) is reduced by tin(II) in the presence of chloride ion:

$$2Cu^{2+} + Sn^{2+} \rightarrow 2Cu^{+} + Sn^{4+} \quad \text{(all species are complexes)}$$

When solutions of Cu^{2+} and excess Sn^{2+} are mixed and the concentration of Cu^{2+} monitored spectroscopically, the following data result:

t (sec)	Cu^{2+} (mM)	t (sec)	Cu^{2+} (mM)
0	35	50	2.86
10	14.3	60	2.38
20	7.16	90	1.61
30	4.78	120	1.20
40	3.58	180	0.82

Do these data indicate first- or second-order dependence on $[Cu^{2+}]$? What is the indicated rate constant? Further studies indicate that the reaction is also first order in $[Sn^{2+}]$, which is consistent with the reaction's stoichiometry. However, the rate also has an inverse dependence on $[Cu^+]$ but no dependence whatever on $[Sn^{4+}]$. Propose a mechanism that is consistent with these findings. [Nunes, *Inorg. Chem.* **9**, 1325 (1970).]

5. The gas arsine, AsH_3, decomposes when heated:

$$2AsH_{3(g)} \rightarrow 2As_{(s)} + 3H_{2(g)}$$

The progress of the reaction can be followed by monitoring the pressure, since the number of moles of gas is changing. At 350°C the following data were obtained:

t (hr)	P (torr)
0	392.0
4.33	403.0
16.00	436.5
25.50	453.5
37.66	480.5
44.75	488.5

If the initial pressure is P^0, write an algebraic expression for the pressure of arsine, P_{AsH_3}, when a fraction f has reacted, and a corresponding expression for P_{H_2}. What will be the total pressure? For each time datum, calculate the arsine pressure using these expressions. Is the reaction first or second order in arsine? What is the rate constant? [Tamaru, *J. Phys. Chem.* **59**, 777 (1955).]

6. The Arrhenius activation energy can be obtained by plotting $\log k$ vs. $1/T$, since it determines the slope in the linear equation

$$\ln k = \text{constant} - \frac{E_a}{R}\left(\frac{1}{T}\right) \quad \text{or} \quad \log k = \text{constant} - \frac{E_a}{2.303R}\left(\frac{1}{T}\right)$$

For the gaseous reaction

$$NO_2Cl + NO \rightarrow NO_2 + NOCl$$

the following rate data have been observed:

T (°C)	k (cm³/mole-sec)
27	0.79×10^7
38	1.25×10^7
50	1.64×10^7
61	2.56×10^7
71	3.4×10^7

Convert these values to $\log k$ and reciprocal absolute temperature values and plot them on graph paper. Draw the best straight line you can through the points. Get the slope [as $\Delta(\log k)/\Delta(1/T)$] and solve for E_a. Using the expression

$$k = p(\text{collision frequency})e^{-E_a/RT}$$

take the rate constant at 27 °C and solve for p, the steric factor. Follow the procedure on p. 578 for the collision frequency, but note that this k is in cubic centimeters rather than liters. Take the molecular weight as being that of the lightest reactant, and use 3.5 Å for σ. [Freiling et al., J. Chem. Phys. **20**, 327 (1952).]

7. The fact that the half-life of a first-order reaction is independent of concentration (see Study Problem 1) permits radiological dating, since the radioactive decay of nuclei is a first-order process. A particularly useful technique is radiocarbon dating. A small fraction of the atmospheric CO_2 is radioactive, since ^{14}C is continually being formed in the upper atmosphere; the ^{14}C decays with a half-life of 5720 years. Living plants take up this $^{14}CO_2$ at what we assume to be a constant ratio to $^{12}CO_2$, but after a plant dies its ^{14}C content diminishes due to the nuclear reaction. A living plant shows 15.3 nuclear disintegrations per minute per gram of carbon, which may be taken as $[A]^0$ in the first-order integrated rate equation:

$$\ln \frac{[A]^0}{[A]} = kt \quad \text{or} \quad \log \frac{[A]^0}{[A]} = \frac{k}{2.303}t$$

A tree trunk found embedded in volcanic debris on the Greek island of Thera was carbon dated in 1967. Its date is interesting because the enormous volcanic explosion that buried it is thought to have ended the nearby Cretan civilization and given rise to the Atlantis legends. The tree showed 10.0 disintegrations/g C-min. When did Atlantis disappear?

8. The reaction between nitrite ion and oxygen gas,

$$2NO_2^- + O_2 \rightarrow 2NO_3^-$$

proceeds at a rate that is first-order in $[NO_2^-]$ and also in $[O_2]$, to a good approximation. A mechanism has been proposed:

$$NO_2^- + O_2 \xrightarrow{k_1} NO_3^- + O$$
$$NO_2^- + O \xrightarrow{k_2} NO_3^-$$

Assume that the O atom is very reactive so that its concentration quickly reaches a steady state: $d[O]/dt = 0$ (see p. 610). Use this approximation to obtain an expression for [O] and from this show that the mechanism is consistent with experiment and that the rate depends only on k_1. [Anderson and Freeman, *J. Phys. Chem.* **65**, 1648 (1961).]

9. We have pointed out that a mechanism can sometimes be inferred from the rate effect when the polarity of the solvent changes. Consider the reaction

$$OH^- + (CH_3)_3S^+ \rightarrow CH_3OH + (CH_3)_2S$$

run in a mixed water–ethanol solvent. If the mechanism were S_N1 [with $(CH_3)_3S^+$ dissociating to CH_3^+ and $(CH_3)_2S$] what should be the effect on the rate if more water is added to the solvent, making it more polar? What if the mechanism were S_N2? Experimentally it is found that the rate decreases when solvent polarity increases; is this consistent with either mechanism?

10. Another way of looking at S_N1 and S_N2 mechanisms is to regard them as *dissociative* and *associative*, respectively, referring to the principal process going on in the rate-determining step. When the reaction involves a metal complex ion:

$$CoL_4Cl_2^+ + H_2O \rightarrow CoL_4Cl(OH_2)^{2+} + Cl^-$$

an important factor in any mechanism is the degree of crowding around the central atom; if crowding (**steric hindrance**) is substantial a dissociative mechanism will be enhanced, for example. If the four ligand atoms L are provided by the two N atoms on each of two diamine molecules as in the table below, what do the observed rate constants for the analogous reactions indicate about the mechanism, assuming it is the same for all diamines? [Pearson *et al.*, *J. Amer. Chem. Soc.* **75**, 3089 (1953).]

Co(diamine)$_2$Cl$_2^+$	Diamine	k
	$H_2N-CH_2-CH_2-NH_2$	1.9×10^{-3} min^{-1}

	$H_2N-CH_2-CH-NH_2$
	       $\quad\quad\quad\quad\quad\,\,\,|$
	       $\quad\quad\quad\quad\,\,CH_3$
		3.7×10^{-3}

	$H_2N-CH-CH-NH_2$
	       $\quad\quad\,\,\,|\quad\,\,\,|$
	       $\quad\,\,CH_3\,\,CH_3$
		250×10^{-3}

	$\quad\quad\,\,CH_3\,\,CH_3$
	       $\quad\quad\,\,\,|\quad\,\,\,|$
	       $H_2N-C\!-\!-\!-\!C-NH_2$
	       $\quad\quad\,\,\,|\quad\,\,\,|$
	       $\quad\quad\,\,CH_3\,\,CH_3$
		instantaneous

11. The removal of tetraethyllead from gasoline is intended partly to lower the lead content of the atmosphere, but also to allow the use of catalytic afterburners in the automobile enhaust line to ensure complete combustion of hydrocarbons to CO_2 and H_2O. Why does lead-containing gasoline sharply reduce these catalysts' effectiveness?

Some Further Reading

Dence, J. B., Gray, H. B., and Hammond, G. S., *Chemical Dynamics*, New York: Benjamin, 1968. A rather freewheeling survey of much of the material of this chapter plus some uses of mechanisms in descriptive chemistry. Not exactly equivalent to the current chapter but good reading.

King, E. L., *How Chemical Reactions Occur*, New York: Benjamin, 1963. An introductory treatment of this material, very clearly written and widely used.

Harris, G. M., *Chemical Kinetics*, Boston: Heath, 1966. Somewhat more mathematical than this chapter, this is still a very good short account of modern kinetics. The reading may be slow but the emphasis is right.

Trotman-Dickenson, A. F., *Chemical Kinetics*, Oxford: Pergamon Press, 1966. Another very short but authoritative account, comparable to Harris.

Most physical chemistry textbooks have good sections on kinetics, although the emphasis is usually on rate laws and the theoretical calculation of rate constants rather than on mechanisms or synthetic uses of the material.

IV | CHEMICAL SYNTHESIS

12 Periodicity and Electronegativity

We have tried to provide an outline of the basic concepts of chemistry, both with respect to the basic structure of chemical compounds and with respect to the tendency of chemical compounds to change their form or composition. We are now in a position to try to produce a logical structure, a pattern, for the immense array of information known as descriptive chemistry. A consistent application of the basic concepts from the preceding chapters should allow us to predict accurately the vast majority of the chemical properties of the various elements and their compounds. Throughout our treatment of descriptive chemistry—the principles of chemical synthesis—we shall be attempting to unite the material into a logical structure: to replace memorization by logical reasoning. Because chemistry is an unfinished science, this will not always be possible; perhaps the places in which our present understanding is inadequate will stimulate you to creative thought of your own.

12-1 The Periodic Table

The principal instrument chemists use to systematize the chemistry of the elements is the periodic table. In the past hundred years, many forms of the periodic table have been devised, each with some features that made certain relationships between elements apparent. Inside the front and back covers of this book are two versions of a particular form, the "long form" of the periodic table. You have already had occasion, in Chapters 5 and 6, to use some of the information from the table at the front of the book, which deals with the atomic properties of the elements: their atomic numbers and atomic weights, valence electron structures, valence-orbital ionization potentials, electronegativities, and radii. These are largely derived quantities. In the table at the back of the book will be found data on the bulk properties of the elements: their melting and boiling points, densities, enthalpies of vaporization, and electrical conductivities. These are the properties that may readily be established by direct measurement. The original arrangement of the elements into a table emphasizing their periodic properties (principally due to Mendeleev) was based largely on such bulk properties, aided by the atomic weight, which was almost the only atomic property known at that time. In this chapter we shall look at these bulk properties and their periodic character and relate them to the basic electronic structure of the elements.

In Chapter 5 we first encountered the periodic table by deriving its form with the help of the exclusion principle. At that time the criterion for placing sodium in the same column as lithium, for instance, was the similarity of electron distribution in the outer or valence AO's. This is the principle we shall follow here, and it should become apparent that a great many of the periodic properties, and the trends in those properties, follow directly from a consideration of these electron distributions.

12-2 Ionization Potentials

As we have seen in developing the MO description of chemical bonding, the valence-orbital ionization potential is one of the most fundamental properties of an atom in determining the electron distribution in the atom's compounds and the resulting properties of those compounds. Let us look at the factors that affect the valence-orbital ionization potential (VOIP).

Hydrogen

Consider first the hydrogen atom, with an atomic number of 1 and only one electron. The valence orbital in this case is the lowest-energy AO the lone electron can fill, which is the $1s$ orbital. The VOIP is 13.60 electron volts (eV), which represents simply the energy necessary to remove the electron from its average position near a nucleus with a charge of +1. Since the charge on the electron is opposite to that on the hydrogen nucleus, it experiences a strong attraction to the nucleus, which we can express quantitatively by the energy of 13.6 eV necessary to completely remove the electron to an infinite distance from the nucleus, the arbitrarily chosen condition of zero energy.

Helium

When we go to helium, with an atomic number of 2 and two electrons, we might initially expect the nuclear charge, which is now +2, to attract the electron twice as strongly, giving a VOIP of $2 \times 13.6 = 27.2$ eV. This is not far from the true value, but there are now two more effects we must consider. One is that the average distance of the electron from the helium nucleus, while distributed according to a $1s$ orbital pattern, is different from the average distance of a $1s$ electron from a hydrogen nucleus. Specifically, it is smaller because of the greater attractive energy, and since the average radius appears in the denominator of the Coulomb-energy expression this means the attractive energy should be greater yet. On the other hand, we have neglected the electron-repulsion energy; since there are now two electrons in the atom, the energy of either electron we choose to remove will be affected not only by the attraction of a nucleus with a charge of +2, but also by the repulsion of another electron with a charge of -1. This should reduce the net attractive energy of the electron for the system consisting of the nucleus and the other electron, and indeed when these effects are all considered, the VOIP is 24.6 eV – somewhat lower than twice the $1s$ VOIP for hydrogen.

Lithium

Lithium, with an atomic number of 3, markedly displays the effects of changed electron radius and repulsion energy. Again our first guess might be that the electron would be held three times as strongly by the Li+3 nucleus as by the H+1 nucleus, but this would be dramatically wrong. Remembering the sequence in which the electron distribution is built up, we expect the easiest electron to remove in the Li atom to be the third electron put in, which because of the exclusion principle must be distributed according to the $2s$ orbital pattern. On the average, a $2s$ electron in any atom is substantially farther from the nucleus than a $1s$ electron in the same atom. For lithium, the radius at which the $1s$ charge density is greatest is 0.20 Å, while the radius at which the $2s$ charge density is greatest is 1.50 Å. Since the corresponding distance for the hydrogen $1s$ electron is 0.53 Å, the greater distance in the Li atom should lead to substantially less attraction. (It may be helpful here to consider Figs. 5-9 and 5-10 again.) There is also the influence of the repulsion of the $2s$ electron by the other two electrons; we have already discussed the approach chemists usually take to this, which is to regard the repulsion of the inner electrons for an outer valence electron as a simple shielding of the full nuclear charge. Using this approach, the effective nuclear charge for lithium's $2s$ electron is $+1.30$, and it is obvious that the repulsion energy is a very important effect. For Li, both the larger average electron radius and the substantial repulsion due to the other electrons tend to reduce sharply the net attractive energy holding the $2s$ electron in the atom. We expect the VOIP for the Li $2s$ orbital to be approximately that for the H $1s$ orbital, scaled up to account for the somewhat larger effective nuclear charge, but scaled down substantially because of the greater electron radius:

$$\text{VOIP}_{\text{Li}} = 13.6 \left(\frac{1.30}{1.00}\right)\left(\frac{0.53}{1.50}\right) = 6.2 \text{ eV}$$

This compares reasonably well with the spectroscopically observed value of 5.39 eV. Notice that the overall effect is to make the Li $2s$ valence electron very easy to remove, compared to the hydrogen electron, while the helium VOIP was so high as to make He electrons very hard to remove.

Beryllium

Going on across the first row of the periodic table, beryllium, Be, has the electron configuration $1s^2 2s^2$, so the VOIP in which we are interested is that of the $2s$ AO. Again we shall use Slater's screening rules to try to represent the mutual repulsion energy of the electrons by an effective nuclear charge. Following these rules (see Section 5-8), the effective nuclear charge for a Be $2s$ electron is $+1.95$, again a substantial reduction from the full charge of $+4$ that would exist on the bare nucleus. Looking at the radius effect, we see that (because of the increased nuclear charge and attraction) the radius at which the electronic charge density is greatest is 1.19 Å, substantially less than for Li. If we attempt to scale the H $1s$ VOIP again to account for these factors, we have

$$\text{VOIP}_{\text{Be}} = 13.6 \left(\frac{1.95}{1.00}\right)\left(\frac{0.53}{1.19}\right) = 11.8 \text{ eV}$$

The spectroscopically observed value in this case is 9.3 eV, and while the quantitative agreement is not quite as good as before, we would predict correctly that the VOIP for an electron distributed according to the Be $2s$ atomic orbital would be substantially greater than that for Li but smaller than that for a H $1s$ electron. The trouble is that it is a rather imperfect approximation to represent all of the electron–electron repulsions by a simple screening effect. Nonetheless, it is worth noticing that we can predict trends accurately; the Be $2s$ electron will be much harder to remove than the Li $2s$ electron. Incidentally, if you are inclined to worry about whether the $1s$ electrons should be considered as valence electrons, look at what we would predict as their VOIP in the Be atom. The screening rules predict an effective nuclear charge of +3.70, and this large attractive force causes the radius of maximum electronic charge density to be only 0.14 Å. Then our scaling calculation on the H 1s VOIP gives

$$\text{VOIP}_{\text{Be}(1s)} = 13.6 \left(\frac{3.70}{1.00}\right)\left(\frac{0.53}{0.14}\right) = 190 \text{ eV}$$

This is probably not a very reliable number, but we can see that the 1s electrons on Be are certainly tremendously more stable than the 2s electrons.

Boron

The element boron, B, has the electronic configuration $1s^2 2s^2 2p$. Continuing, for the moment, to consider the 2s electrons, the screening rules give an effective nuclear charge of +2.60, with the accompanying decrease in radius of maximum electronic charge density to 0.88 Å we might expect from the larger nuclear charge. The increasing number of electrons may cause some difficulty in representing the repulsion energy by a screening constant, however. If we calculate the VOIP for the B 2s orbital in the same way as before,

$$\text{VOIP}_{\text{B}(2s)} = 13.6 \left(\frac{2.60}{1.00}\right)\left(\frac{0.53}{0.88}\right) = 21.3 \text{ eV}$$

it does not compare favorably with the experimental value of 14.0 eV. Moreover, we must now consider the VOIP for the 2p orbital; since its radius of maximum electronic charge density is slightly smaller than that for the 2s orbital (0.85 Å), and the screening of the nuclear charge is predicted to be the same, we would expect it to be slightly more tightly held than the 2s. Actually it is substantially *less* tightly held (experimental VOIP = 8.3 eV), and the reason for this difference is worth examining. If we look at Fig. 12-1, which

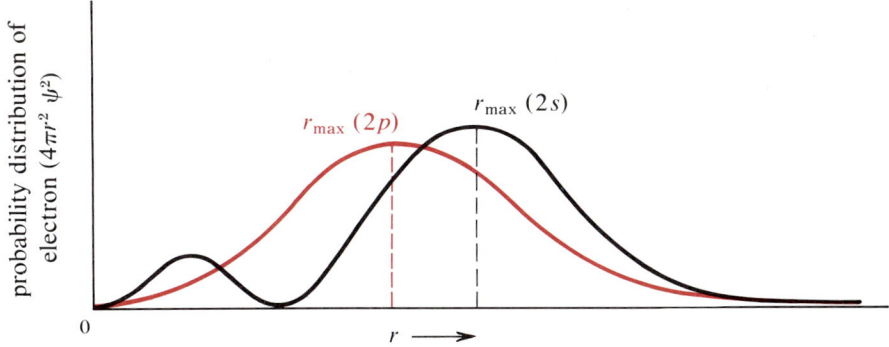

Figure 12-1 Comparison of radial distribution of 2s and 2p electrons on the same atom.

compares the radial distributions of electronic charge for the B $2s$ and $2p$ orbitals, we see that although the *maximum* in the curve is farther out for the $2s$ orbital, a $2s$ electron also has a substantial chance of being found quite near the nucleus, experiencing a very strong attraction by penetrating the shielding electrons, so to speak. Since the $2p$ orbital distribution does not allow the electron this possibility of penetration, a $2p$ electron will be *less* tightly bound to the nucleus, on the whole, than a $2s$ electron. It is necessary to consider the whole pattern of electron distribution — we cannot characterize it reliably by a single radius. This is exactly why it is not reasonable to think of the electron as being in an orbit at some fixed distance from the nucleus as the Bohr model of the atom did.

The Periodic Behavior of the VOIP

The overall effect of this is that the $2p$ VOIP is much smaller than the $2s$ VOIP for the same atom, and that as the atomic number and nuclear charge increase the VOIP for a given orbital increases more or less uniformly. We can see this effect in Fig. 12-2, in which VOIP values are plotted as a function of atomic number for the elements up through neon ($Z = 10$). Notice that the separation between $2s$ and $2p$ orbitals also becomes greater as the atomic number increases. The reason the graph is plotted upside down, so to speak, with the VOIP's increasing going down the scale instead of up, is that plotted in this way they bear the relation to each other that they would in constructing

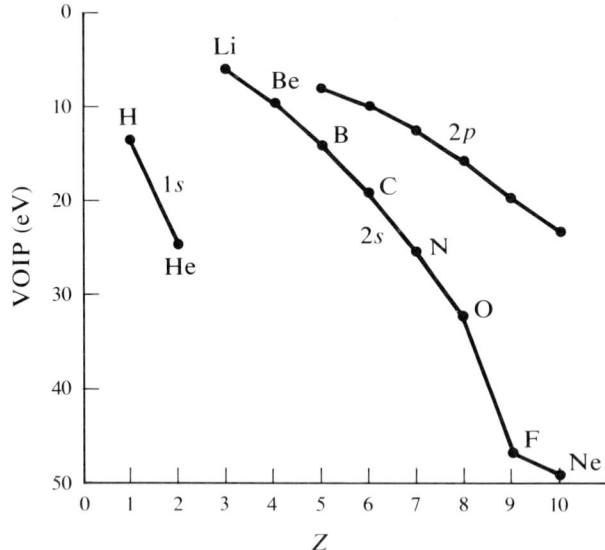

Figure 12-2 Periodic dependence of VOIP values for the first 10 elements.

a MO energy-level diagram; that is, the C $2s$ AO is slightly lower in energy than the H $1s$, while the C $2p$ is slightly higher, and so on. We can use these VOIP's or orbital energies very conveniently and directly to account for the physical and chemical properties of the elements and their compounds, and later in this chapter we shall do this at some length. An example of how this trend in VOIP's affects a particular property may be helpful here, however. In considering the MO description of bonding we saw that if the AO's that are combined to yield the MO's have nearly the same energy the electrons in the MO will be distributed uniformly over the atoms and in general no significant charge separation will occur. That is, MO's formed from AO's with similar VOIP's will correspond to almost completely covalent bonding. On the other hand, if the AO energies differ substantially the electrons in the MO formed will be concentrated around the atom or atoms whose orbital energies are closest to the MO energy, so that a good deal of electron transfer occurs and we expect the bonding to be distinctly ionic. In considering the properties of crystals we saw that ionic crystals are quite high melting because of the very strong Coulomb forces holding the crystal lattice together, while covalent-molecule crystals are very low melting since little thermal energy is needed to break the weak van der Waals forces holding the molecules together. So to a first approximation the melting point of a compound may be taken as an index of its degree of ionicity, and we ought to be able to tie this together with the MO bonding description by comparing VOIP's for the AO's involved. Suppose we consider the fluorides of the elements in the first row of the periodic table, since fluorine, according to Fig. 12-2, will have the lowest-energy valence orbitals. The melting points of the fluorides are summarized in Table 12-1.

Table 12-1

Melting Points of First-Row Fluorides (°K)

LiF	1143
BeF_2	1073
BF_3	146
CF_4	89
NF_3	56
OF_2	49
FF (fluorine)	53

Clearly LiF, which the VOIP relationship suggests should be most ionic, is also the highest melting. Figure 12-3 shows the same melting-point data plotted on the same graph as the difference in VOIP's between the fluorine $2p$ orbitals and the other atom's averaged VOIP's. The correspondence is not perfect—

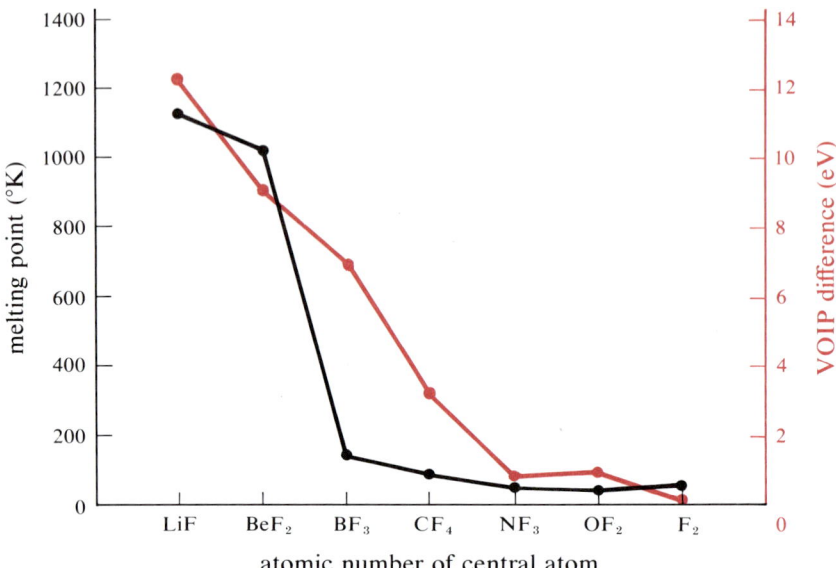

Figure 12-3 Comparison of melting-point trend and VOIP-difference trend for first-row fluorides.

some other factors are also at work — but clearly the VOIP trend is strongly influencing the melting-point trend.

In going from H through He to Li we saw that when n, the principal quantum number, increases, a dramatic change in VOIP occurs due to the much greater radius (on the average) of an electron distributed according to an orbital with higher n. This greater distance, besides decreasing the Coulomb attraction of the nucleus for the electron, also improves the accuracy of the approximation in which all electron–electron repulsion is treated as a screening of the nuclear charge, since most of the time the earlier electrons are all between the last electron and the nucleus. Consequently, when we go beyond neon to sodium, Na, we expect to see the valence electron, which is distributed according to a $3s$ orbital, very loosely held for the same reasons that the Li valence electron was very loosely held. Thus for sodium the radius of maximum electronic charge density is 1.55 Å for the $3s$ electron but only 0.32 Å for the $2s$ electrons. In fact, because the radius is slightly greater for Na, the electron is even less tightly held than on the Li atom, with a $3s$ VOIP for Na of 5.1 eV, compared to the Li $2s$ VOIP of 5.4 eV. On going across the second row of the periodic table from Na to argon, Ar, the same sort of trends in VOIP's ought to be evident as for the first row. Figure 12-4 shows this, with the graph plotted in the same way as before. The trends are less pronounced because the screening of the nuclear charge is somewhat better for the electrons with $n = 3$ than for the first row, in

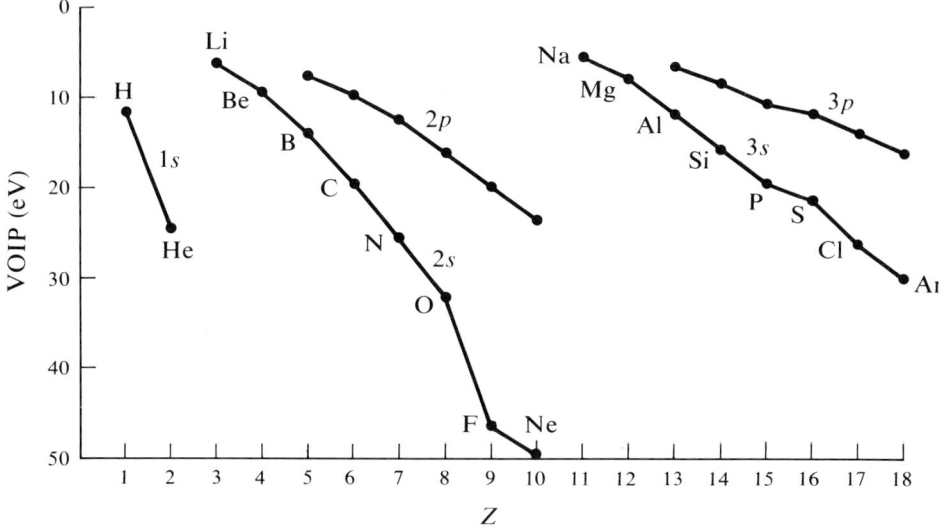

Figure 12-4 Periodic trends in VOIP values for the first two rows of the periodic table.

which $n = 2$. Finally, in Fig. 12-5, the same sort of graph is shown for all the elements up to atomic number 36, krypton (Kr). The relatively flat portions of the $4s$ and $4p$ curves between scandium, Sc, and copper, Cu, are due to the filling of the $3d$ AO's as the atomic number increases through that range. The $4s$ and $4p$ electron energies do not change much with increasing Z because the $3d$ electrons that are also added screen the increased nuclear charge fairly

Figure 12-5 Periodic trends in VOIP values through the first-row transition metals.

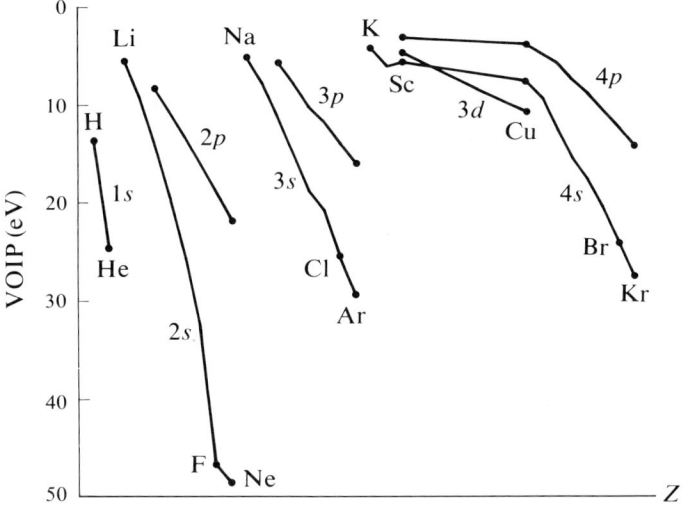

effectively. We shall consider this penetration effect again in Chapter 16, which deals with transition-metal chemistry.

There is one more trend in the VOIP's we ought to note. Figure 12-6 shows a few of the VOIP's plotted exactly as in Fig. 12-5, but with lines added to emphasize the relationships between different elements with analogous electronic configurations, that is, the elements in several columns of the periodic table. The relationships are mostly ones of great similarity, which means simply that elements in the same column of the periodic table will have, in general,

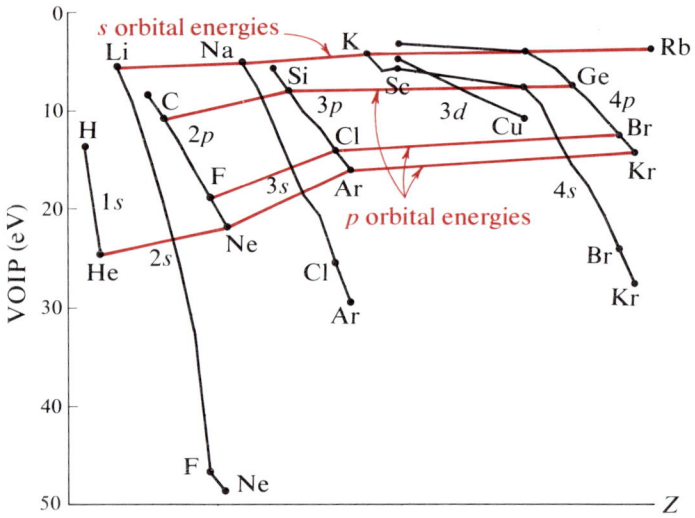

Figure 12-6 Similarities in VOIP values for different elements in the same column of the periodic table.

quite similar tendencies to donate or accept electrons in forming chemical bonds. Just as it is easy to remove an electron from Li, it should be easy to remove an electron from Na, K, and Rb. There is a slight upward slope to the right in all of the curves, however, representing the fact that VOIP's become somewhat smaller—the electron in question becomes easier to remove—as the average radius of the electron in the particular valence orbital distribution increases. The greater radius, of course, means a decreased Coulomb attraction for the nucleus. Later we shall see the chemical results of this trend; for now we note only that Li reacts relatively slowly with water at room temperature, but Na reacts quite rapidly, K almost invariably catches fire, and Rb and Cs explode.

12-3 Electron Affinities

There is another property of atoms that is closely related to the ionization energy: the electron affinity. Just as the ionization energy is the amount of energy that must be put into an atom to *remove* an electron, the electron affinity is the amount of energy that is released when a neutral atom *adds* an electron. Looked at another way, the electron affinity is equal to the ionization energy of an atom with a single negative charge: $Z^- + IP(-) = Z^0 + e^-$ is just the same as $Z^0 + e^- = Z^- + EA$. Since the ionization energy represents energy going into the system while electron affinity represents energy coming out, they will have different algebraic signs when we do any sort of Hess'-law calculation with them. Only a few electronic affinities are known experimentally, since the measurement is extraordinarily difficult to make for most atoms. A representative sampling of values, those for the first row of the periodic table, is shown in Fig. 12-7. They tend to be much smaller than ionization energies for the same atoms, because an electron near a neutral atom (the situation electron affinity describes) does not feel the great attractive force experienced by an electron

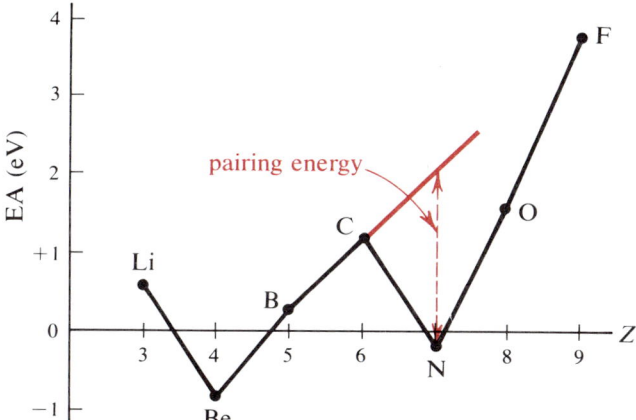

Figure 12-7 Electron-affinity trends and the electron-pairing energy.

near an atom with a net positive charge (the situation ionization energy describes). Furthermore, electron affinities can actually be negative, in the sense that energy may be required to put another electron onto a free atom. In the case of Be, and some other atoms with filled subshells, this seems to be because the new electron, in a much higher-energy AO, experiences greater repulsion from the other electrons than attraction from the nucleus. As atomic number increases through B and C, the effect of penetration of the new electron near the nucleus becomes more important because of the higher charge on the nucleus and the electron–nucleus attraction becomes larger than the net electron–electron repulsion.

Pairing Energy

When we get to nitrogen, however, we find that to put another electron into the atom we must pair it with an existing electron, which means that they must occupy the same AO and be, on the average, quite near each other. This causes a substantial increase in the net electron–electron repulsion, and again gives a net negative electron affinity. As the nuclear charge continues to increase in going from N to O to F, its attraction for the incoming electron is able to overcome this **pairing energy** to give a positive electron affinity again.

This pairing energy will also affect the ionization energies of real, polyelectronic atoms. We have previously considered only the VOIP values, without considering the effect of a possible pairing of electrons in the valence orbitals. Figure 12-8 shows the ionization energies of the atoms through Kr, superimposed on a graph that is identical to the VOIP plot of Fig. 12-5. These

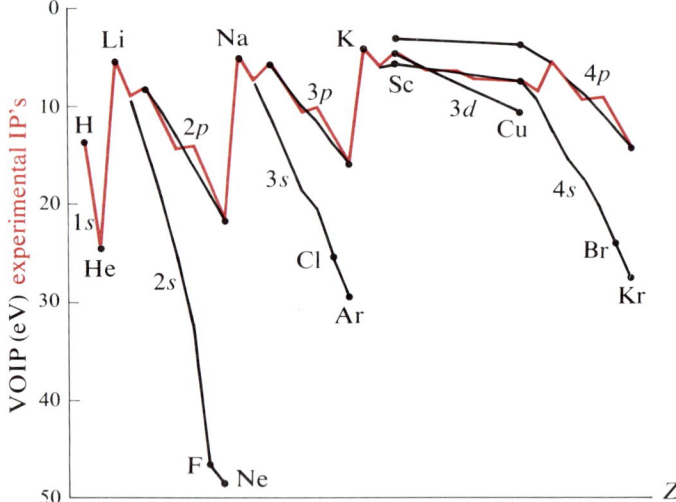

Figure 12-8 Comparison of VOIP trends and observed ionization potentials.

experimental ionization energies follow the VOIP's for the highest-energy electron in each atom fairly closely, except that at O and S, for instance, the value of the experimental IP is higher than the VOIP by an amount roughly equal to the pairing energy indicated in Fig. 12-7. The pairing, of course, is adding repulsive energy that is not considered in arriving at the VOIP, thereby making the last (paired) electron in O, for instance, less stable — higher energy — than the VOIP predicts.

12-4 Differential Ionization Energies

In dealing with ionization energy and electron affinity we are at the very heart of the problem of dealing with chemical reactivity. All of our ideas about predicting the chemical nature of the compounds of an element are based on the electronic structure of the compounds. As we have seen in Chapter 6, we can rationalize the electronic structure of compounds in some detail by using the MO model, which apportions the bonding electrons according to the relative VOIP's of the atoms involved in the compound. The electron affinity is also involved, because if one atom is to lose a relatively high-energy electron, another must accept it in a relatively low-energy orbital. Since any compound will adopt the electronic structure that gives it the lowest energy, whether this means forming ionic or covalent bonds, a reasonable criterion of the general chemical reactivity of an element might be the rate at which its energy changes as it gains or loses electrons. To express this more precisely, we can propose as a criterion of reactivity the rate of change of an atom's energy with respect to its charge: dE/dq, where E is the atom's energy and q is its net charge. This quantity, which is due to a contemporary chemist, Klixbull Jorgensen, is called the **differential ionization energy** (DIE), since we can express the nth ionization energy of an atom, IP_n, as

$$IP_n = \int_{n-1}^{n} \frac{dE}{dq} dq$$

The derivative or differential feature is simply the standard way of expressing the rate of change to which we referred, and if we integrate (over charge) the derivative of energy with respect to charge we shall get energy; since the limits of integration are the two charges on the atom of $+(n-1)$ and $+n$, the definite integral is the energy difference between these two conditions, which is just what we expect the ionization energy to be.

Table 12-2 gives the successive ionization energies for the elements, with values that represent removal of nonvalence electrons in parentheses. Leaving out the ionization of nonvalence electrons, which would not be chemically significant, we can use these values together with such experimental electron affinities as are available to produce a series of functions, one for each atom,

Table 12-2
Ionization Potentials of the Elements[a]

Element		Ionization Potential (eV)							Average to
No.	Symbol	I	II	III	IV	V	VI	VII	VIII	Usual Oxidation State
1	H	13.595								13.595
1	D	13.598								
2	He	(24.580)	(54.40)							
3	Li	5.390	(75.619)	(122.42)						5.39
4	Be	9.320	18.206	(153.85)	(217.657)					13.76
5	B	8.296	25.149	37.920	(259.30)	(340.127)				23.79 (3+)
6	C	11.264	24.376	47.864	64.476	(391.99)	(489.84)			37.00 (4+)
7	N	14.54	29.605	47.426	77.450	97.863	(551.93)	(666.83)		30.52 (3+)
8	O	13.614	35.146	54.934	77.394	113.873	138.080	(739.11)	(871.12)	24.38 (2+)
9	F	17.42	(34.98)	(62.646)	(87.23)	(114.214)	(157.117)	(185.139)	(953.6)	
10	Ne	(21.559)	41.07	64	97.16	126.4	(157.9)	(208.444)	(264.155)	
11	Na	5.138	(47.29)	(71.65)	(98.88)	(138.60)	(172.36)	(225.31)	(265.957)	5.14
12	Mg	7.644	15.03	(80.12)	(109.29)	(141.23)	(186.86)	(241.93)	(285.13)	11.34
13	Al	5.984	18.823	28.44	(119.96)	(153.77)	(190.42)	(246.41)	(303.87)	17.75 (3+)
14	Si	8.149	16.34	33.46	45.13	(166.73)	(205.11)	(263.31)	(309.26)	25.77 (4+)
15	P	11.0	19.65	30.156	51.35	65.007	(220.414)	(280.99)	(328.80)	35.43 (5+)
16	S	10.357	23.4	35.0	47.29	72.5	88.029	114.27	(348.3)	16.88 (2+)
17	Cl	13.01	23.80	39.90	53.5	67.80	96.7	(124.0)	(143.46)	25.57 (3+)
18	Ar	(15.755)	(27.62)	(40.90)	(59.79)	(75.0)	(91.3)	(118)	(155)	
19	K	4.339	(31.81)	(46)	(60.90)		(99.7)	(128)	(147)	4.34
20	Ca	6.111	11.87	(51.21)	(67)	(84.39)			(159)	8.99
21	Sc	6.56	12.89	24.75	(73.9)	(92)	(111.1)	(140.8)		14.73 (3+)
22	Ti	6.83	13.63	28.14	43.24	(99.8)	(120)	(151)	(173.7)	22.96 (4+)/16.20 (3+)
23	V	6.74	14.2	29.7	48	65.2	(128.9)	(161.1)	(185)	16.88 (3+)
24	Cr	6.763	16.49	30.95	49.6	73	90.6	119.24	(196)	18.07 (3+)
25	Mn	7.432	15.64	33.69		76			(151)	18.92 (3+)
26	Fe	7.90	16.18	30.64						18.24 (3+)
27	Co	7.86	17.05	33.49						19.47 (3+)
28	Ni	7.633	18.15	36.16						20.65 (3+)
29	Cu	7.724	20.29	36.83						14.01 (2+)
30	Zn	9.391	17.96	39.70						13.68 (2+)
31	Ga	6.00	20.51	30.70	(64.2)					19.07 (3+)
32	Ge	7.88	15.93	34.21	45.7	(93.4)	(127.5)			25.93 (4+)
33	As	9.81	20.2	28.3	50.1	62.6	81.7	(155)		19.44 (3+)
34	Se	9.75	21.5	32.0	42.9	73.1			(193)	26.54 (4+)
35	Br	11.84	21.6	35.9						
36	Kr	(13.996)	(24.56)	(36.9)						4.18
37	Rb	4.176	(27.5)	(40)						8.36

Z	El	I1	I2	I3	I4	I5	I6	Other
39	Y	6.5	12.4	20.5				
40	Zr	6.95	14.03	24.8	33.97	(77)		13.13 (3+)
41	Nb	6.77	14	28.1	38.3	50		19.94 (4+)
42	Mo	7.383	16.1	27.1	46.4			27.43 (5+)
43	Tc	7.45	15.3					16.86 (3+)
44	Ru	7.7	16.8	28.4				
45	Rh	7.7	18.1	31.1				
46	Pd	8.334	19.9	32.9				
47	Ag	7.574	(21.960)	(36.10)				7.57
48	Cd	8.991	16.904	(38.217)				12.95
49	In	5.785	18.867	28.030	(58.037)	(81.13)		17.56 (3+)
50	Sn	7.332	14.629	30.654	40.740	55.69		23.34 (4+)/10.98 (2+)
51	Sb	8.64	18.6	24.825	44.147	60.27		30.38 (5+)
52	Te	9.007	21.543	30.611	37.817			24.74 (4+)
53	I	10.44	19.010					
54	Xe	12.127	21.204	32.115				
55	Cs	3.893	(32.458)	(35)	46	76	(99)	3.89
56	Ba	5.2097	10.001		(51)	(58)	(103)	7.61
57	La	5.614	11.43	19.17	36.715			12.07 (3+)
58	Ce	6.57		19.70				
62	Sm	5.6	11.4					
63	Eu	5.67	11.24					
64	Gd	6.16						
70	Yb	6.25	12.11					
72	Hf	7.3	14.9					
73	Ta	7.6	16.2					
74	W	7.98	17.7					
75	Re	7.85	16.6					
76	Os	8.7	16.9					
77	Ir	9.2						
78	Pt	9.0	19.3					
79	Au	9.223	20.1					
80	Hg	10.434	18.752	(34.5)	(72)	(82)		14.59 (2+)
81	Tl	6.106	20.423	29.8	(50.8)			18.78 (3+)
82	Pb	7.415	15.04	32.1	38.97	(69.7)	(125)	23.38 (4+)/11.23 (2+)
83	Bi	7.29	16.7	25.56	45.3	56.0		
84	Po	8.3						
85	At	9.4						
86	Rn	10.746						
87	Fr	4.0						
88	Ra	5.278	10.145					
89	Ac	5.5						
90	Th	5.7						
91	Pa	5.7						
92	U	5.7						

[a] Values principally from NBS Circular 467.

representing that atom's differential ionization energy, $DIE(q)$. To a reasonable approximation we can represent this function relating energy and charge as a linear function of the charge: $DIE(q) = a_0 + a_1 q$. Table 12-3 gives the appro-

Table 12-3

Differential Ionization-Energy Constants for Selected Elements[a]

At. No.	Element	Valence Orbital	a_0	a_1	At. No.	Element	Valence Orbital	a_0	a_1
3	Li	s	2.0	6.1	28	Ni	d	−3.5	12.9
4	Be	s	5.5	7.0	29	Cu	s	4.3	6.3
5	B	p	3.4	8.7	30	Zn	s	5.6	7.1
6	C	p	5.0	10.3	31	Ga	p	2.6	6.2
7	N	p	6.8	11.9	32	Ge	p	3.7	6.9
8	O	p	8.6	13.7	33	As	p	5.0	7.7
9	F	p	10.3	15.8	34	Se	p	6.5	8.2
11	Na	s	2.4	5.1	35	Br	p	7.5	9.1
12	Mg	s	4.5	5.8	37	Rb	s	2.4	3.1
13	Al	p	2.6	6.2	38	Sr	s	3.5	3.9
14	Si	p	4.0	7.1	47	Ag	s	4.3	5.7
15	P	p	5.6	8.1	48	Cd	s	5.5	6.5
16	S	p	6.8	9.2	49	In	p	3.2	5.7
17	Cl	p	8.3	10.2	50	Sn	p	4.3	6.1
19	K	s	2.4	3.5	51	Sb	p	5.6	6.3
20	Ca	s	3.7	4.2	52	Te	p	6.5	6.8
23	V	d	−9.2	12.0	53	I	p	7.3	7.6
24	Cr	d	−7.2	12.3	55	Cs	s	2.4	2.5
25	Mn	d	−6.6	12.6	56	Ba	s	3.4	3.2
26	Fe	d	−5.5	12.7	79	Au	s	6.1	6.1
27	Co	d	−4.5	12.8	80	Hg	s	6.7	6.8

[a]From Jorgensen, C. K., *Orbitals in Atoms and Molecules*, New York: Academic, 1962.

priate values of the constants a_0 and a_1 for a number of the elements. Figure 12-9 shows a graph of the energy of several of the first-row elements as a function of the charge; remember that the *slope* of this curve is the differential ionization energy. Notice that the slopes increase more or less uniformly as the atomic number increases; the slope of this curve at zero charge, which is just the quantity $DIE(q=0)$, is the constant a_0. If we now plot a graph of the differential ionization energy, $DIE(q)$, for each of these elements as a function of charge we get Fig. 12-10.

Charge Flow in Compounds

What do these graphs mean physically? If we form a compound from two elements that have quite different slopes of the energy-vs.-charge curve (i.e., quite different DIE's), such as LiF, the element that has a large DIE or slope will tend to attract electrons strongly so as to become negatively charged and move toward lower energy, that is, toward the left from zero charge in Fig.

Differential Ionization Energies | 637

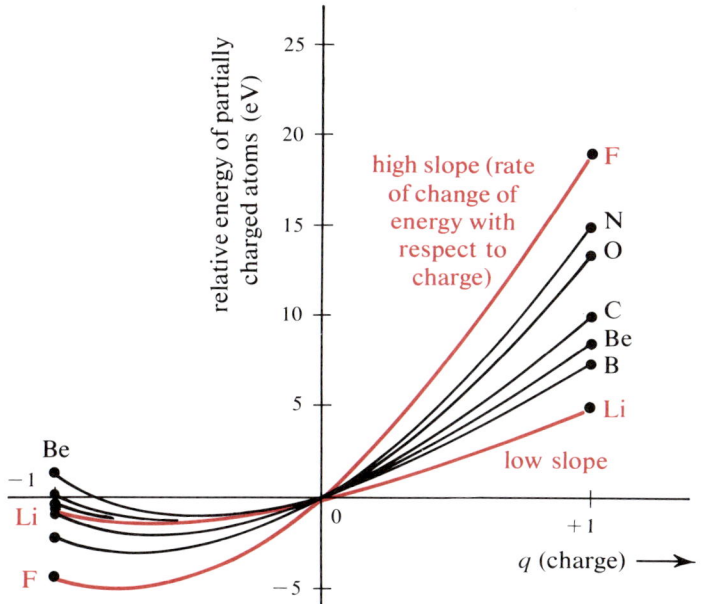

Figure 12-9 The energy of atoms as a function of their electric charge, for first-row elements.

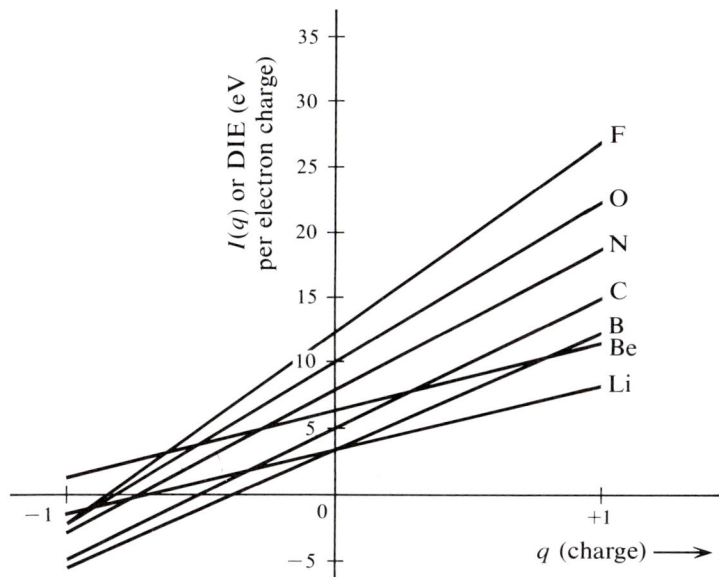

Figure 12-10 The differential ionization-energy curves (nearly linear) for first-row elements; the derivative of Fig. 12-9.

12-9. Since it can do this only by taking electrons away from the atom with the small DIE, that atom will have to become positively charged in the process. It is quite practical to do this since the small DIE means the atom with the small DIE does not go to much higher energy as it becomes positively charged.

Where will the process stop? What will be the final charges on the positive atom and negative atom? The process of transferring electrons will stop lowering the energy when the charge is reached at which F, with a charge of $-q$, and Li, with a charge of q, have equal DIE's. That is, as F becomes more negative its DIE decreases and a further increment of negative charge does less good, and as Li becomes more positive its DIE increases and a further increment of positive charge does more harm. The break-even point will come when the two DIE's are equal; Fig. 12-11 shows the situation. We can then read the charge on the gaseous molecule LiF's atoms as being about 0.3 on the Li, about -0.3 on the F.

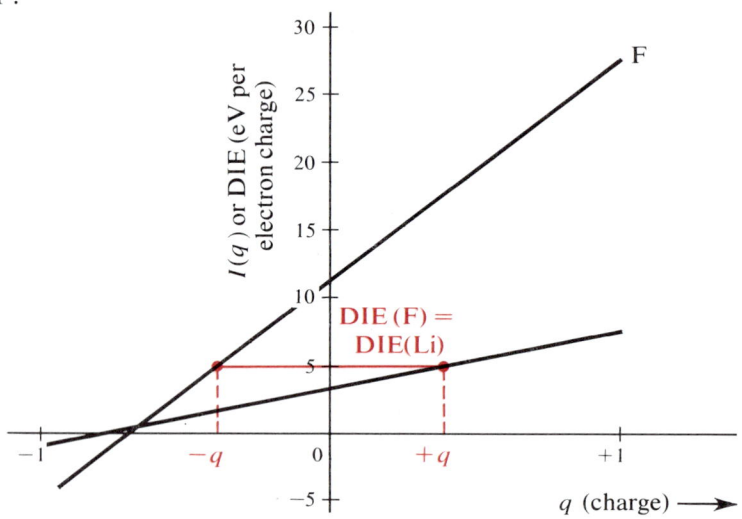

Figure 12-11 Condition for balance in electron flow from one atom to another through a bond; large DIE difference.

On the other hand, if we form a compound between two atoms having nearly the same DIE — nearly the same rate at which the atoms' energy changes with charge — we expect to see the electron-transfer process stop when only a very small amount of charge separation has occurred. Figure 12-12 shows the situation for the compound CO. Since the two curves are so close together no very great charge separation can develop; the charges appear to be about 0.15 on the C and -0.15 on the 0. We shall have an essentially covalent compound, with very little charge separation, if the two DIE's are nearly the same no matter whether both atoms have a high DIE or a low DIE. The substantially ionic case will occur only when the two DIE's are quite different.

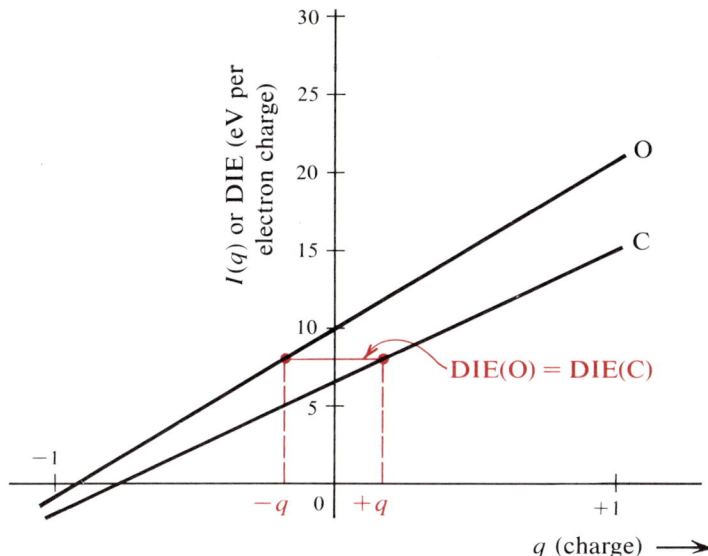

Figure 12-12 Net charges for small DIE difference.

12-5 Electronegativity

Can we come up with a single number that will be a reasonable characterization of the chemical nature of an atom of a particular element? Obviously both the slope and the intercept of the DIE curves influence the charge separation, but if we select the intercept more or less arbitrarily we might be able to characterize an atom by the appropriate quantity a_0. (Remember the definition of intercept for a linear function from Chapter 2.) What is the intercept a_0, physically? It is simply $DIE(q = 0)$, which we can interpret in terms of the ionization energies since $IP_n = \int_{n-1}^{n} DIE(q)\, dq$. Plugging into this expression,

$$IP_n = \int_{n-1}^{n} (a_0 + a_1 q)\, dq$$

$$= \left(a_0 q + \frac{a_1}{2} q^2\right)\Big|_{n-1}^{n} = a_0 n + \frac{a_1}{2} n^2 - a_0(n-1) - \frac{a_1}{2}(n-1)^2$$

$$= a_0 n + \frac{a_1}{2} n^2 - a_0 n + a_0 - \frac{a_1}{2} n^2 + \frac{a_1}{2} 2n - \frac{a_1}{2}$$

$$= a_0 + a_1 n - \frac{a_1}{2} = a_0 + a_1\left(n - \frac{1}{2}\right)$$

If we interpret a_0 in terms of ionization energies, then $a_1(n-\frac{1}{2})$ must equal zero for the particular value of n we are considering, and n must be $\frac{1}{2}$. But what is the "one-halfth" ionization energy? We cannot take half an electron away from an atom; the only interpretation of this quantity that makes physical sense is to say that $IP_{1/2} = (IP_0 + IP_1)/2$. But IP_0 is just the energy necessary to go from an atom with a charge of -1 to the same neutral atom and a free electron, and this is the electron affinity. Thus $a_0 = (IP_1 + EA)/2$. This quantity, the average of the electron affinity and the first ionization energy, is called the **Mulliken electronegativity**. The electronegativity of an atom is the best single guide we have to its reactivity and the nature of the compounds it will form; let us stop here and consider the concept of electronegativity carefully.

12-6 Electronegativity Scales

One of the most curious features of electronegativity is that essentially the same set of numbers can be obtained for the elements in the periodic table in a number of different ways, none of which, on their face, seem to have anything to do with each other. The concept was originated by Pauling, who noticed that bond energies between unlike atoms were always greater than the average of the bond energies of either atom to itself. He proposed that the "extra" bond energy be proportional to the square of the electronegativity difference between the atoms. Using electron volts as energy units, his electronegativity differences range between 0 and 3.3 electronegativity units. Since his definition of electronegativity, like the definition of standard reduction potentials, depended on a difference measurement, he arbitrarily defined the electronegativity of F as 4.0 units and scaled all other atomic electronegativities from that value using measured bond energies. His table of values gained wide acceptance, and it has become conventional to adjust other electronegativity scales to a range of 0–4 units.

To give another example of an electronegativity scale, Sanderson has proposed a scale based on the electron density in the space about the atom. He assumes that the rare gases have ideal electron densities and are unreactive for that reason; then he takes the ratio of the radius of the real atom to the radius a rare gas with that atomic number would have as his "stability ratio" or electronegativity.

The values given in the periodic table inside the front cover represent yet another scale, that of Allred and Rochow, who define the electronegativity as the Coulomb force of attraction on an electron at the distance of the covalent radius, that is, as Z_{eff}/r^2. The significant thing about all of these scales, which seem to differ widely from each other in physical basis, is that when they are placed in the range 0–4 units, they are all very nearly identical, Mulliken as well as Pauling, Sanderson, and Allred and Rochow. This seems to suggest that the property, although perhaps difficult to define unequivocally, is still a very basic one and is a valuable (if not flawless) guide to the basic chemical nature of an atom. For instance, suppose we wanted to compare the aqueous reduction potentials of elemental F, Cl, Br, and I by the half-reaction $X_2 + 2e^- = 2X^-$. The reduction potential is the tendency of an atom to capture and hold an electron in solution. This will be closely related to the electron affinity, obviously, and it might be proportional to the Mulliken electronegativity (or anybody else's). Figure 12-13 is a double graph of both the electronegativity

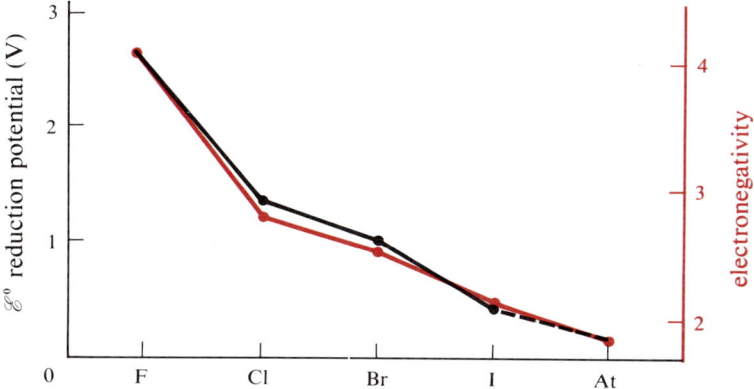

Figure 12-13 Comparison of reduction potential and electronegativity trends for the halogens.

and the experimental reduction potential of the elements F, Cl, Br, and I, and the agreement is quite satisfactory. The fifth halogen, astatine (At), is extremely radioactive and difficult to handle. Using this sort of chart, however, we can predict with some confidence that its reduction potential in aqueous solution would be about +0.3 V, so that it should be able to reduce Cu^{2+} to copper metal, for instance, although probably not quantitatively ($E^0_{Cu(II)=Cu^0} = +0.34$ V).

Even when we are dealing with aqueous reduction potentials, however, which seems to be a particularly favorable property to correlate with electronegativity, we cannot rely absolutely on these values. We expect the alkali metal ions Li^+, Na^+, and K^+ to have quite negative reduction potentials since they are formed so readily from the neutral atoms, and indeed they do. But while the reduction potentials of these three are in the order Na–K–Li, the electronegativities are in the order Na–Li–K. The trouble is simply that chemistry is too complex to be able to represent all its diversity by a single parameter.

12-7 Electronegativity and Bond Energies

Electronegativity is remarkably versatile. Using a table of single-bond energies for an atom with itself (for instance, the bond energy of the Li_2 molecule in the gas phase), it is possible to use Pauling's definition of electronegativity to predict bond energies between unlike atoms. Table 12-4 gives a number of

Table 12-4
Homonuclear Bond Energies Between Like Atoms[a]

Atoms in Bond	Bond Energy (kcal/mole)	Atoms in Bond	Bond Energy (kcal/mole)
H—H	104.2	As—As	35
Li—Li	25	Se—Se	50
B—B	69	Br—Br	46.1
C—C	82.6	Rb—Rb	11
N—N	38	Ag—Ag	39
O—O	35	Cd—Cd	2
F—F	37	Sn—Sn	39
Na—Na	17	Sb—Sb	29
Si—Si	51	Te—Te	53
P—P	48	I—I	36.1
S—S	60	Cs—Cs	10
Cl—Cl	57.9	Au—Au	52
K—K	12	Hg—Hg	3
Cu—Cu	47	Pb—Pb	23
Zn—Zn	6	Bi—Bi	39
Ge—Ge	34		

[a] From Cottrell, T. L., *The Strengths of Chemical Bonds*, London: Butterworths, 1958, 2nd ed.

homonuclear bond energies; let us try to predict a heteronuclear bond energy from these and then come up with an enthalpy of formation. Suppose we are interested in the compounds nitrogen trichloride, NCl_3, and nitrogen trifluoride, NF_3. Considering NCl_3 first, we can estimate the energy of a N—Cl bond (according to Pauling) as $E_{N-Cl} = \sqrt{E_{N-N} \times E_{Cl-Cl}} + \Delta$, where Δ is the additional bond energy above the geometric mean that is due to ionic attraction. If E_{N-N} and E_{Cl-Cl} are expressed in energy units of electron volts, Δ turns out to be just the electronegativity difference between N and Cl. Then, substituting values from Table 12-4,

$$E_{N-Cl} = \sqrt{\left(38\frac{\text{kcal}}{\text{g-bond}}\right)\left(58\frac{\text{kcal}}{\text{g-bond}}\right)\left(\frac{1}{23}\frac{\text{eV/bond}}{\text{kcal/g-bond}}\right)} + (3.07 - 2.83)$$

$$= (47)\left(\frac{1}{23}\right) + 0.24$$

$$= 2.04 + 0.24$$

$$= 2.28 \text{ eV/bond, or } 2.28(23) = 52.5 \text{ kcal/g-bond}$$

Since NCl$_3$ has three N—Cl bonds, its total bond energy (the amount by which the compound is more stable than the free atoms) ought to be just 3(52.5) or 157 kcal/mole. Since this is a "stability energy," the bond energy is taken as a *negative* potential energy; that is, −157 kcal/mole. We can get an enthalpy of formation from this value by applying Hess' law to the reaction that represents the formation of NCl$_3$ from the elements.

$$\tfrac{1}{2}N_2 + \tfrac{3}{2}Cl_2 \rightarrow NCl_3 - \Delta H_f(NCl_3),$$

$$\tfrac{1}{2}E_{N_2} + \tfrac{3}{2}E_{Cl_2} \rightarrow E_{NCl_3} - \Delta H_f(NCl_3),$$

$$\Delta H_f(NCl_3) = E_{NCl_3} - \tfrac{1}{2}E_{N_2} - \tfrac{3}{2}E_{Cl_2}$$

The total bond energy of N$_2$ is just the N≡N triple bond energy, or −225 kcal/g-bond, and the total bond energy of Cl$_2$ is the Cl—Cl single-bond energy of −58 kcal/g-bond. So

$$\Delta H_f(NCl_3) = (-157) - \tfrac{1}{2}(-225) - \tfrac{3}{2}(-58)$$

$$= -157 + 112 + 87$$

$$= +42 \text{ kcal/mole}$$

Obviously NCl$_3$ is a very endothermic compound. What about NF$_3$? We can go through exactly the same sort of calculation:

$$E_{N-F} = \sqrt{E_{N-N}E_{F-F}} + \Delta$$

$$= \sqrt{38(37)}\,(\tfrac{1}{23}) + (4.10 - 3.07)$$

$$= 1.63 + 1.03 = 2.66 \text{ eV/bond}$$

$$= 2.66(23) = 61 \text{ kcal/g-bond}$$

Then $E_{NF_3} = 3(-61) = -183$ kcal/mole, and for

$$\tfrac{1}{2}N_2 + \tfrac{3}{2}F_2 \rightarrow NF_3 - \Delta H_f(NF_3)$$

$$\tfrac{1}{2}E_{N_2} + \tfrac{3}{2}E_{F_2} \rightarrow E_{NF_3} - \Delta H_f(NF_3)$$

$$\Delta H_f(\mathrm{NF_3}) = E_{\mathrm{NF_3}} - \tfrac{1}{2}E_{\mathrm{N_2}} - \tfrac{3}{2}E_{\mathrm{F_2}}$$
$$= (-183) - \tfrac{1}{2}(-225) - \tfrac{3}{2}(-37)$$
$$= -183 + 112 + 56 = -15 \text{ kcal/mole}$$

So we have the curious prediction that $\mathrm{NCl_3}$ ought to be quite endothermic, but $\mathrm{NF_3}$ exothermic. Since the enthalpy change is usually the larger factor in the free-energy change (at room temperature, anyway), we would expect the free energies of formation to follow this same pattern; then $\mathrm{NCl_3}$ should be quite unstable with respect to decomposition into the elements, but $\mathrm{NF_3}$ should be stable. What are the experimental facts? $\mathrm{NF_3}$ is a very stable gas at room temperature, which is chemically reactive only at relatively high temperatures, while $\mathrm{NCl_3}$ is an oily liquid at room temperature and is one of the most treacherous and violent explosives known, decomposing into $\mathrm{N_2}$ and $\mathrm{Cl_2}$. The experimental enthalpies of formation of the two compounds are -26 kcal/mole for $\mathrm{NF_3}$ (compare our value of -15 kcal/mole) and 55 kcal/mole for $\mathrm{NCl_3}$ (compare our value of 42 kcal/mole). We should not expect numerical agreement much better than this, because as we have seen in setting up MO energy-level diagrams, bonding electrons do not all have the same energy. The approximation that bond energies are constant, then, is bound to lead to trouble, even if we try to approximate bonding overlap by adding an electronegativity term. Furthermore, we have tacitly assumed in Pauling's treatment that the ionic contribution, Δ, to the bond energy is independent of the structure of the rest of the molecule. Looking at Fig. 12-11 we can see that its prediction of charge is valid only for the case in which *one* atom is becoming positive and *one* atom negative. That treatment (using differential ionization energies) will apply to polyatomic molecules, say MX_n, but now $q_\mathrm{M} = -nq_\mathrm{X}$. Thus the charge — and the ionic contribution to the bond energy — will depend on n, the number of other atoms, and also on their electronegativity. So we have oversimplified somewhat. Nevertheless, we see that it is possible to produce substantially correct predictions of the chemical nature of compounds by this very simple procedure.

12-8 Electronegativity and Bond Lengths

Another very useful correlation of electronegativity with molecular properties lies in the prediction of bond lengths. We have seen that the average radius of an electron in a given orbital depends on the nuclear charge; the larger this charge is, the more the orbital shrinks toward the nucleus. This accounts for the fact that the covalent radii tabulated in the periodic table in the front of the book get smaller going to the right in any given row. The electrons are being added in orbital distributions having the same principal quantum number n, but the nuclear charge is increasing as the atomic number increases. We should also remember, however, that for any given atom the average radii of its orbitals will be affected by its net electric charge. If the atom has a net positive charge, the electrons will be held closer to the nucleus, the source of the positive charge. If the atom has a net negative charge, the added electron density will force all the electrons to expand outward slightly to reduce the net electron–electron repulsion. This is reflected in the expression on p. 239 for the algebraic form of Slater orbitals, where each wavefunction contains a term in $e^{-Z_{eff}r}$. The larger this Z_{eff} is, the more rapidly the negative exponential drops off, and the smaller the orbital is.

The Origin of Covalent Radii

This effect is going to show up in the lengths of chemical bonds. At first glance we might think that an atom in its customary oxidation state or coordination number would be a standard size, so that we could characterize it by a single radius. Then the length of a chemical bond would be the sum of the two radii of the two atoms in the bond. This approximation is sometimes made; for most atoms and compounds, the radii involved are called **covalent radii** and are derived wherever possible from the length of a single bond between two of the same kind of atom. Thus the Cl covalent radius would be half of the distance between nuclei in the Cl_2 molecule. Not all of these molecules exist, however, and with no further adjustment even the values that have been measured are sometimes misleading when we try to apply them to compounds of the atom with other atoms than itself. Most tables of covalent radii have been adjusted to give the best possible approximation for a variety of compounds, but even so the bond lengths arrived at by simply adding covalent radii are often sadly mistaken. For example, if we wanted to predict the length of the C—N bond in, say, methylamine (CH_3NH_2), we could simply add the covalent radius of C, which is 0.77 Å, to that of N, which is 0.75 Å, and get a predicted C—N bond length of 1.52 Å. This is not a bad approximation, since the actual bond length has been measured as 1.47 Å. If we proceed to SiF_4, though, and try to predict the Si—F bond length, we have 1.17 Å (Si) + 0.72 Å (F) = 1.89 Å. This is most unfortunate – the experimental bond length is only 1.56 Å. Covalent radii simply are not reliably additive.

Bond Contraction Due to Electrostatic Attraction

The trouble, of course, is that not all bonds are genuinely covalent. If some charge separation exists – if the bonding is partly ionic – the atoms will be pulled together by the attraction of their opposite charges, and the bond length will be shorter than the "covalent" bond length. Several attempts have been made to devise relationships that would allow for this effect. One of the best known is that of Schomaker and Stevenson, who in effect shorten the "covalent" bond by an amount proportional to the electronegativity difference between the two atoms. This seems reasonable, since the greater the electronegativity difference the more ionic the compound and the more the bonds should be shortened by the Coulomb attraction of the opposite charges. The Schomaker–Stevenson relationship is:

$$r_{XY} = r_X^{cov} + r_Y^{cov} - 0.09(\chi_X - \chi_Y)$$

where χ represents the electronegativity. A better match of predicted and experimental values is obtained, however, if we use the square of the electronegativity difference:

$$r_{XY} = r_X^{cov} + r_Y^{cov} - 0.07(\chi_X - \chi_Y)^2$$

Using the covalent radii and electronegativities from the periodic table inside the front cover, let us calculate one or two bond lengths using this expression.

The C—N bond length (to try one we have already done), according to this expression, should be $r_{C-N} = r_C + r_N - 0.07(\chi_C - \chi_N)^2$, or

$$r_{C-N} = 0.77 + 0.75 - 0.07(2.50 - 3.07)^2$$
$$= 1.52 - 0.07(0.32)$$
$$= 1.50 \text{ Å} \quad \text{(experimental} = 1.47 \text{ Å)}$$

As a matter of fact, this is slightly better than the value we had predicted using covalent radii alone, even though the C—N bond is very nearly covalent. What about the stickier case of the Si—F bond length?

$$r_{Si-F} = r_{Si} + r_F - 0.07(\chi_{Si} - \chi_F)^2$$
$$= 1.17 + 0.72 - 0.07(1.74 - 4.10)^2$$
$$= 1.89 - 0.39$$
$$= 1.50 \text{ Å} \quad \text{(experimental} = 1.56 \text{ Å)}$$

This is somewhat too small, of course, but it is much better than the sum of the covalent radii alone. Table 12-5 gives a number of bond lengths predicted by this method, with the experimental value given in each case for comparison. The method is seen to be reasonably accurate in almost every case. It should be pointed out that even for a given pair of atoms in a bond, the charges can vary substantially depending on what other atoms are in the molecule or ion, and the bond lengths vary accordingly. For instance, the Ge—Cl distance is about 2.07 Å in the molecule GeF_3Cl, but about 2.15 Å in GeH_3Cl. Usually the effect is smaller than this, perhaps 0.01–0.03 Å, but it is worth noticing that the uncertainty is about as big as the error in the predicted values.

Table 12-5
Comparison of Predicted and Experimental Bond Lengths

Bond	Predicted Length (Å)	Experimental Length (Å)[a]
B—H	1.16	1.10–1.19
C—H	1.13	1.09
N—H	1.05	1.01
O—H	0.97	0.96
Li—Br(gas)	2.26	2.17
Na–Cl(gas)	2.30	2.36
K–I(gas)	3.17	3.05
Rb—Cl(gas)	2.84	2.79
Cs—F(gas)	2.24	2.35
B—O	1.38	1.36
B—Br	1.89	1.87
B—F	1.21	1.29
C—O	1.44	1.43
C—Cl	1.75	1.76
C—As	1.96	1.95
Al—C	1.93	2.00
Al—Cl	2.09	2.06
Al—I	2.52	2.53
Si—O	1.69	1.64
Si—F	1.50	1.56
N—Cl	1.74	1.74
P—Cl	2.05	2.02
As—O	1.81	1.80
I—O	1.94	1.93
Sb—Cl	2.33	2.37
Ge—C	1.94	1.95
Ge—Cl	2.14	2.08–2.15

[a]From Wells, A. F., *Structural Inorganic Chemistry*, London: Oxford, 1962, 3rd ed.

Multiple Bonds

There is another effect on bond length that has nothing to do with electronegativity. All of the values in Table 12-5 are for single bonds—in the VSEPR approach, a single pair of electrons between bonded atoms. As we saw in our discussion of the VSEPR treatment, it is quite possible to have double or triple bonds. Since this means there are more electrons *between* the two atoms involved, their mutual Coulomb attraction for the two nuclei draws the nuclei closer together. Thus double bonds are shorter than single bonds, and triple bonds are shorter yet. Fortunately, this shortening occurs in nearly a constant proportion, and we can get multiple-bond lengths directly from single-bond lengths simply by multiplying the single-bond length by a constant factor. For double bonds, this factor is 0.86 if the elements are in the first row of the periodic table and 0.91 for all elements past fluorine. For triple bonds, the factor is 0.78 for the first row and 0.85 for other elements. In this way we can calculate the length of a C=N double bond by multiplying the single-bond length, which we have calculated as 1.50 Å, by 0.86 (since both C and N are in the first row): $1.50 \times 0.86 = 1.29$ Å; the experimental value is approximately 1.28 Å. For the C≡N triple bond, we calculate $1.50 \times 0.78 = 1.17$ Å, and the experimental value for HCN is 1.16 Å. This is quite straightforward, and we need only watch out for the fact that in many cases it is very difficult to tell what the bond order is. For instance, for the molecule NO the total bond order is $2\frac{1}{2}$, since five more electrons are in bonding MO's than are in antibonding MO's. Sometimes it is not even this easy to tell, particularly in molecules in which there is a substantial opportunity for delocalization of π bonding electrons over a number of atoms. In the simpler cases, however, the procedure outlined here is entirely adequate.

12-9 Periodicity of Density and Atomic Volume

Density Trends

The experimental fact that bond lengths — or, we might say, effective atomic radii in molecules — depend on electronegativity suggests another property that might be influenced by electronegativity. The density of a solid obviously depends on the volume each atom occupies, which in turn is related to the radius cubed: $V(\text{sphere}) = 4/3\pi r^3$. For most types of solids the density will be affected by the shape of the molecules and by the nature of the bonding within the molecule, so that a comparison would have to include many factors. However, there is one kind of crystal that consists of a lattice of single, more or less spherically symmetrical atoms that have a well-defined electronegativity — the metallic crystal. In Chapter 4 we saw that metallic crystals are, in effect, a lattice of positively charged ions held together by a delocalized "sea" of valence electrons. The more electronegative an atom is, the more tightly it holds its electrons, as may be seen from the trend in radii going across any given row of the periodic table. We expect metal atoms to shrink, so that the density goes up, as electronegativity increases within a series of metals. Let us assume as an approximation that the radius is inversely proportional to the electronegativity. Then the volume of the atom will be inversely proportional to the cube of the electronegativity and the density, which is mass/volume, will be proportional to χ^3. Choosing the proportionality constant to give a reasonable fit, we would suggest $\rho = 1.8\chi^3$, where ρ is density. Table 12-6 shows the results of such an approximation for the metals along the first long row of the periodic table. Obviously the trend of metal densities is being reproduced, but there are some rather substantial errors. One of the reasons for this is that we have neglected the fact that the mass of the atom is increasing as we go across the row. We could take care of this by considering a quantity known as **atomic volume**, which is the volume of 1 gram-atomic weight of an element. A little thought will show that atomic volume is equal to the atomic weight of the element divided by the density.

Table 12-6
Density of Metals (g/cm³)

Metal	Predicted Density	Experimental Density
K	1.35	0.86
Ca	2.0	1.5
Sc	3.1	3.0
Ti	4.1	4.5
V	5.5	6.1
Cr	6.8	7.2
Mn	7.4	7.4
Fe	7.9	7.9
Co	8.8	8.9
Ni	9.6	8.9
Cu	9.6	9.0
Zn	8.2	7.1

Atomic Volume Trends

Historically, atomic volume has been one of the most significant properties in establishing the periodicity of the properties of the elements. In 1870 Lothar Meyer pointed out the striking periodicity that this quantity displays, as seen in Fig. 12-14. Since atomic volume is proportional to atomic weight, it places all the elements on an equal weight basis. If we again assume that density is proportional to the cube of the electronegativity, and also proportional to atomic weight, we can establish the following empirical relationship: atomic volume $= 32/\chi^3$. Table 12-7 compares the values predicted by this relationship with the actual values; the agreement is somewhat better than for density alone. The major difficulty lies in the alkali metals Na, K, Rb, and Cs, and at this point it might be well to point out that we might get quite different values for only a slight change in the electronegativity, since we are cubing it. For instance, if we assumed a value of 0.77 for the electronegativity of Cs, its

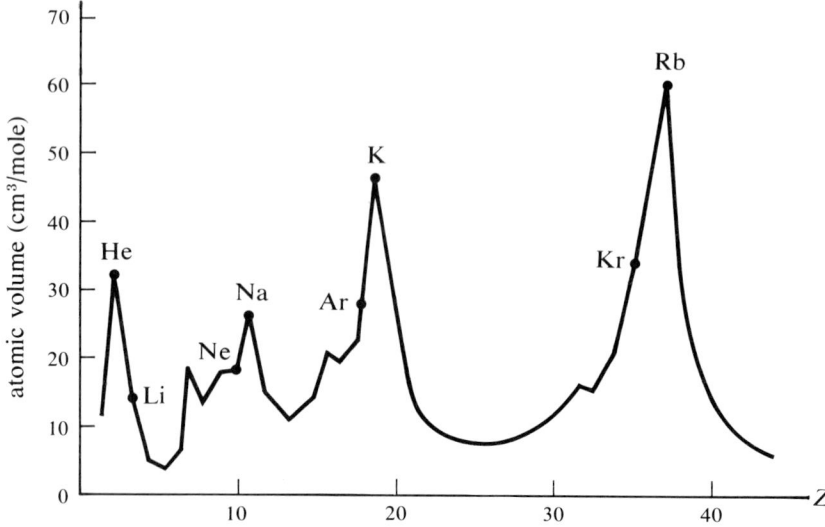

Figure 12-14 Periodic trends in atomic volume.

Table 12-7
Atomic Volumes of Metals (cm³/mole)

Metal	Predicted Atomic Volume	Experimental Atomic Volume
Na	31	23.7
K	42	45.3
Rb	45	55.9
Cs	50	70
Mg	17	14.0
Sr	33	33.7
Ba	35	39
Al	10	10.0
Sc	18	15.0
Y	23	19.8
La	23	22.5
Ti	14	10.6
Zr	17	14.1
Hf	17	13.6
Mn	7.8	7.4
Os	9.1	8.4
Cu	6.0	7.1
Zn	7.0	9.2
Hg	11	14.8

atomic volume would match exactly. There are other factors involved, however, such as the differences in crystal structures between metals, so that we should not expect an exact match from electronegativity alone.

12-10 Periodicity of Other Physical Properties

There are many other properties of the elements and their compounds that show distinct periodic relationships, although most of these relationships are difficult to predict in detail because they are influenced by a number of factors such as electronegativity, polarizability, crystal structure, the total number of valence orbitals, and so on. One of these properties, shown in Fig. 12-15, is the melting point of the elements. In general, melting points tend to be quite high

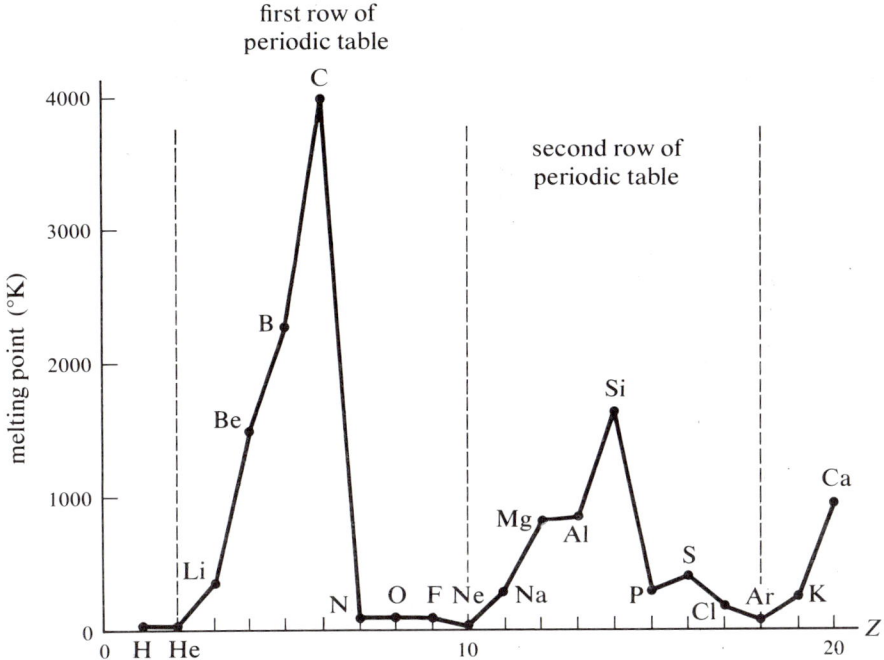

Figure 12-15   Periodicity of elemental melting points.

near the middle of the periodic table because there the number of valence electrons is nearly equal to the number of valence orbitals, so each atom can form a maximum number of bonds to its neighbor atoms. Since the energies of these bonds are much higher than the relatively weak van der Waals forces holding separate molecules together in molecular crystals such as I_2, much more thermal energy must be supplied to disrupt the crystal. However, the substantial irregularities that are evident prevent us from offering a simple correlation.

We have already seen, in Fig. 12-8, the periodic properties of the first ionization potential. Figure 12-16, however, shows a wider range of elements and perhaps a more convincing demonstration of the periodic nature of this property. The cross lines show the group or column relationships, and they evidently have a strong tendency to parallel each other, as we have already noted.

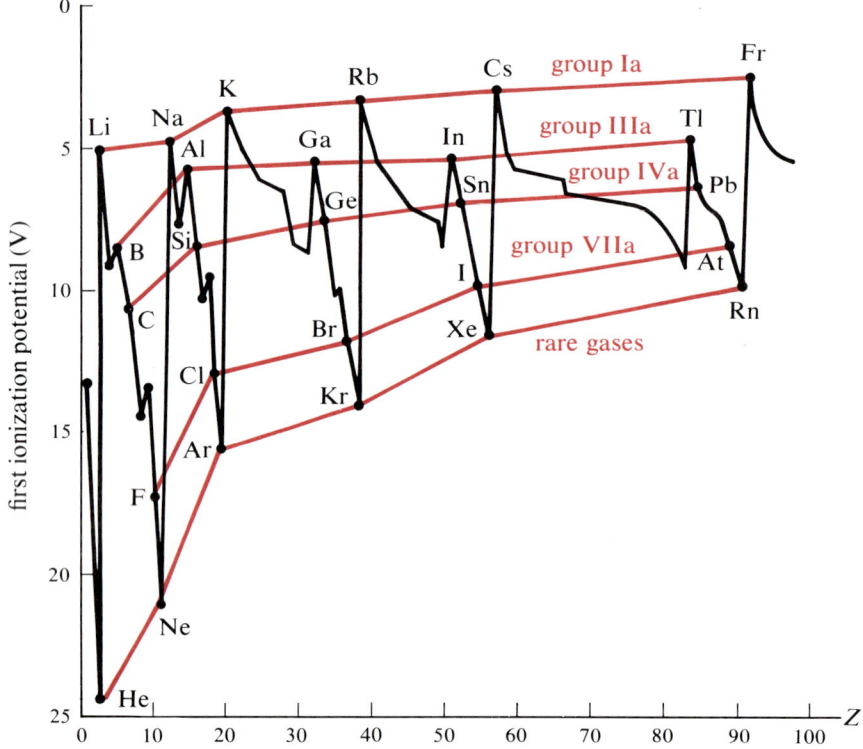

Figure 12-16 Periodicity of first ionization potential, with group relationships indicated.

Figures 12-17, 12-18, and 12-19 show some additional periodic trends in physical properties: the trends in melting points of alkali halides, the trends in boiling points of several groups of halides, and the trends in boiling points of several groups of hydrides. Many graphs of this sort can be constructed for comparable compounds of groups of elements, using all sorts of properties such as the enthalpy of formation or free energy of formation, the dipole moment, the standard reduction potential, and so on. Usually, as in Fig. 12-17,

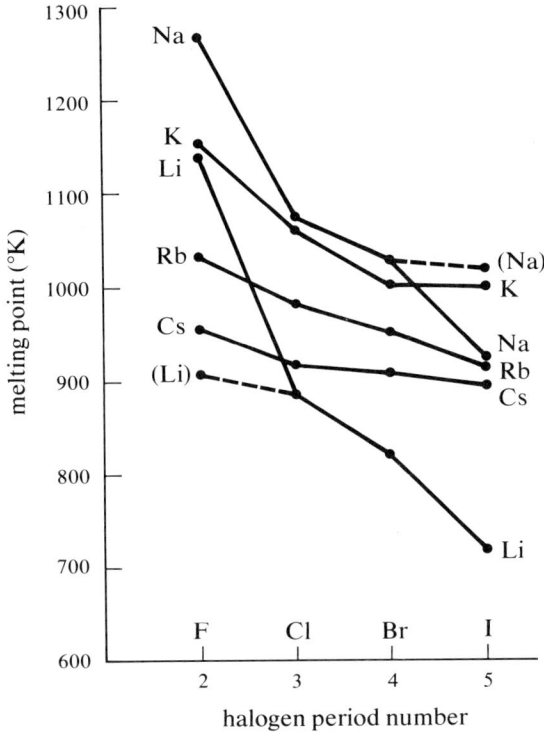

Figure 12-17 Periodicity of alkali halide melting points. (From Tyree, S. Y., and Knox, K., *Textbook of Inorganic Chemistry*, New York: Macmillan, 1961.)

the physical property is plotted as a function of the period number of the atom of interest, but it is interesting to note the correlation of the boiling-point data with the boiling points of the corresponding rare gases, as in Figs. 12-18 and 12-19. This has a limited application since the rare gases lack many of the interesting properties such as meaningful aqueous reduction potentials and free energies of formation. The remarkable deviations of the boiling points of H_2O, HF, and NH_3 from those of their families are worthy of mention here, although we shall return for a fuller discussion of this later. The deviations are such that the molecules are much more firmly held in the liquid phase than might be expected on the basis of their low molecular weight—this is why the boiling points are remarkably high, since more thermal energy is required to remove the molecules from the attraction of their neighbors. This is one of the most striking demonstrations of the effects of hydrogen bonding in these liquids. You may remember the earlier discussion of this phenomenon in connection with polarization of bonds, and also for proton transfer in aqueous acid–base systems. We see from this graph that this phenomenon is apparently possible only for

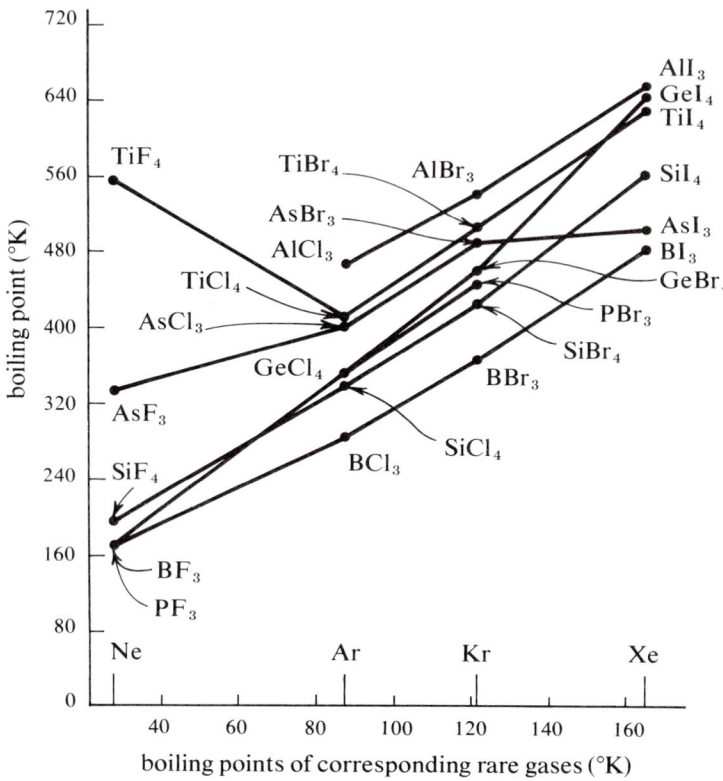

Figure 12-18 Periodicity of boiling points of covalent halides. (From Tyree, S. Y., and Knox, K., *Textbook of Inorganic Chemistry*, New York: Macmillan, 1961.)

hydrides of the small first-row elements that have unshared pairs of electrons, which fits into the previous discussions. Notice that H_2S, with larger, higher-energy, valence orbitals, shows only a very small effect, and CH_4, with no unshared electron pairs, shows no effect at all. But imagine the effect on the world we know if hydrogen bonding did not exist and the boiling point of water fell in a straight line with those of H_2Se, H_2Te, and H_2Po, at about 140°K or −130°C!

12-11 Periodicity of Acid–Base Properties

Finally, we need to examine some of the trends in chemical properties that are seen in the different groups of elements of the periodic table. In general these can be related quite nicely to the numbers of valence electrons, their orbital

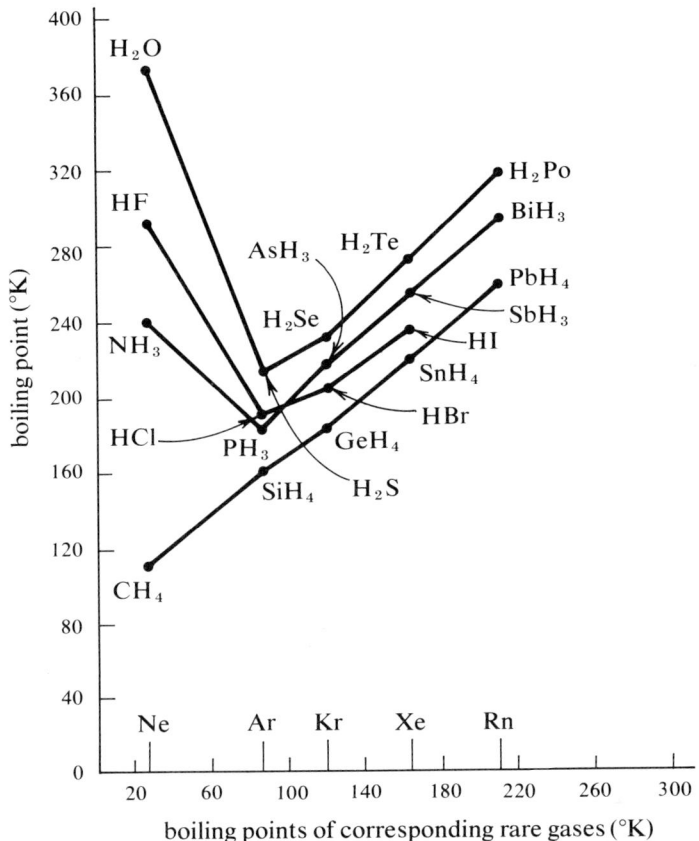

Figure 12-19 Periodicity of boiling points of covalent hydrides. (From Tyree, S. Y., and Knox, K., *Textbook of Inorganic Chemistry*, New York: Macmillan, 1961.)

energies, and the relative charges and radii. Let us begin by trying to account for the acidity or basicity of ions in aqueous solution.

Polarization in Hydrated Ions

To begin, consider the elements in the second row of the periodic table—Na through Cl—and suppose that an atom of each has had all its valence electrons removed by successive ionizations. That is, the ions are Na^+, Mg^{2+}, ..., Cl^{7+}. If we place these bare ions in water, what will happen? The Na^+ ion, with a relatively large radius and low charge, will attract the unshared electron pairs on the nearby water molecules, but will be unable to form permanent bonds with any of the oxygen atoms in the water molecules. The species in solution will be $Na(OH_2)_n^+$, where the water formula is written backward to emphasize that the oxygen electrons are the ones principally attracted to the positive ion.

Since no permanent bond is formed between Na and O, the number n is not constant but varies roughly in the range 6–8, with water molecules drifting in and drifting away again. The Mg^{2+} ion has a greater positive charge, and thus a smaller radius. (Each of these ions has the same number of electrons—this is called an **isoelectronic** series—but the number of protons in the nucleus is increasing, which shrinks the AO's). The effect of the Coulomb attraction of the Mg^{2+} for the O unshared pairs, then, is much greater than for Na^+, and the result is that the solution species is $Mg(OH_2)_6^{2+}$, where the six water molecules are firmly bound to the magnesium ion, and permanent Mg—O bonds can be said to exist. This is an example of increasingly strong covalent bonding due to increasingly strong polarizing power of an electron acceptor toward a donor; see Section 6-8.

Going on to the Al^{3+} ion, the effects of increasing charge and decreasing radius are so great that the oxygen unshared pairs are substantially transferred to the Al^{3+}, which in turn causes the electrons in the water O—H bonds to be attracted toward the oxygen. All this has the result that it is not difficult to transfer a H^+ from one of the water molecules on the aluminum to another water molecule that is free in the solution; after all, this happens anyway in the self-ionization of water. Thus we have

$$Al(OH_2)_6^{3+} + H_2O = Al(OH_2)_5OH^{2+} + H_3O^+$$

A glance at the table of relative acid strengths (Table 10-5) will show that this reaction occurs to some extent, but the predominant solution species is still the hexaaquoaluminum(III) ion. For the Si^{4+} ion, all the same factors are at work, with the charge now still higher and the radius still smaller. The polarization of the oxygen electrons, leaving substantial positive charge on the hydrogen atoms, is now so extensive that the predominant solution species is one that has had several H^+ ions transferred to free water molecules: $Si(OH_2)_2(OH)_4$, an electrically neutral species that is not very soluble in water and that may be considered as the raw material for silica gel.

The very high positive charge on the P^{5+} ion, with its accompanying very small radius, accentuates the process of polarizing the oxygen electrons to such an extent that some of the water molecules (now only four because of the small size of P^{5+}) lose not only one, but two H^+, leaving the solution species $P(OH)_2O_2^-$, which is more recognizable written as $H_2PO_4^-$ ion, left behind when phosphoric acid ionizes in water. By this time it should not be difficult to see that the S^{6+} ion will lead to either $S(OH)O_3^-$—that is, HSO_4^-—or SO_4^{2-}, and the Cl^{7+} ion will give nothing but ClO_4^-. Indeed, the species $Cl(OH)O_3$, anhydrous perchloric acid or hydrogen perchlorate, is difficult to make and reacts very violently with water.

Acidic and Basic Oxides

How does this affect the acidity or basicity of ions in aqueous solution? For any case in which we are interested we can write an equation in which the species in which we are interested reacts with water to produce the stable aqueous ion, balance it, and see whether it produces excess H_3O^+ or OH^-. For instance, suppose we have the two oxides Na_2O and SO_3, and that we dissolve each in water. Will the solutions be acidic or basic? For the Na_2O case we can write

$$Na_2O + 15H_2O = 2Na(OH_2)_7^+ + 2OH^-$$

and for the SO_3 case,

$$SO_3 + 3H_2O = SO_4^{2-} + 2H_3O^+$$

So we can say that Na_2O is a **basic oxide**, while SO_3 is an **acidic oxide**. In general any oxide having a central atom in a high formal oxidation state will tend to transfer several protons from its water molecules of hydration because of the transfer of electrons to reduce the charge; in other words, a high formal oxidation state gives an acidic solution. This is a fairly smooth trend going across the periodic table, so that Na_2O is very basic, SiO_2 is nearly neutral but slightly acidic (why does glass resist attack by acids better than bases?), and Cl_2O_7 is very acidic. Since the net formal charge on the ion is the basic factor influencing the proton transfer from the hydrated water molecules, we can infer correctly that the acidity or basicity of any species will be dictated by the formal oxidation state of its central atom. Thus SO_2, in which the sulfur has a formal oxidation of 4+, has acid–base properties roughly equivalent to those of SiO_2; it is a very weakly acidic oxide.

Acid–Base Properties of Other Compounds in Water Solution

The properties are related to the fact that, in addition to the central ion, each oxide contains, at least formally, the O^{2-} ion, which is a very strong base (note its position in Table 10-4). If a compound in which we are interested does not contain oxygen, however, the same sort of analysis applies. Suppose we are considering the compound titanium tetrachloride, $TiCl_4$. In this compound Ti has a formal oxidation state of 4+, and we would expect it to form the same sort of hydrated ion that Si^{4+} does. Then we would write the equation

$$TiCl_4 + 10H_2O = Ti(OH_2)_2(OH)_4 + 4Cl^- + 4H_3O^+$$

This reaction is quite vigorous, and the electrically neutral Ti species is an insoluble solid, although $TiCl_4$ itself is a liquid at room temperature. Under these circumstances it is perhaps not surprising that TiCl$_4$ is used to generate smoke screens in military and naval maneuvers, simply by spraying it and water into the air together; the dense white smoke is a suspension of very fine particles of the more or less hydrated TiO$_2$ solid. We also see that dissolving TiCl$_4$ in water produces a very acidic solution, which is characteristic of chlorides in which the central atom has a formal oxidation state greater than about 2+. Most soluble compounds also follow the pattern of the chlorides.

12-12 Electronegativity and Periodicity of Oxidation States

Using the periodic table we can also predict the relative stabilities of the various formal oxidation states of an element. Our argument will depend on the valence electrons' being the outermost electrons in the atom, which effectively prevents us from dealing with the transition metals, with d valence electrons. For the time being, then, we shall restrict ourselves to the **characteristic elements**—the two eight-member rows of the periodic table and their congeners.

Considering only the characteristic elements, then, suppose we have an electrically neutral atom of an element, and we want to know what formal oxidation states it can assume in compounds with other elements, and the relative stability of these oxidation states. Is the atom more likely to gain or lose electrons? This tendency is a relative one—it depends on what other element the atom is reacting with. If we look back at Fig. 12-9, we see that, starting from zero charge, all atoms gain stability by gaining electrons. But the electrons must come from somewhere, specifically from another atom, and the question becomes whether the two atoms *together* would have lower energy if one becomes negative or the other does. Electronegativity is a fairly reliable guide to this choice. In any particular compound, the element with the lowest electronegativity will have a positive formal oxidation state, the others, generally, a negative one. In general, it will not be possible for any atom to have a highly positive formal oxidation state unless its electronegativity is very much lower than that of the atoms surrounding it in the molecule or ion. Thus the highest fluorides that can be formed by the other halogens are ClF$_3$, BrF$_5$, and IF$_7$, in which the formal oxidation states of the central atoms fall in the same order as their decreasing (from F) electronegativities. The most prominent exception to this generalization is the perchlorate ion, ClO$_4^-$, for which the electronegativity difference is only about 0.7 unit; this is due to the great capacity of water to stabilize high formal oxidation states (as a ligand) by relieving the strained electron distribution through proton transfer. This is precisely the process involved in the discussion of the previous section.

Positive Oxidation States

What will the oxidation state be? If the electronegativity difference is substantial, as it is for nearly all the compounds of the metals in groups I, II, and III, the greatest stability will be achieved only when the electropositive atom (the metal) has lost all of its valence electrons. This reduces the electron configuration around the atom to that of the previous rare gas. No more electrons are lost because the additional energy of forming more bonds cannot compensate for the tremendous ionization energy of an electron from an inner shell. The result is that we expect the oxidation state of the metals in these groups of the periodic table to be equal to the number of valence electrons, which is the same as the group number.

Going farther to the right in the table, the increasing electronegativity makes it less and less energetically favorable to remove all valence electrons. The smaller electronegativity differences indicate that the bonding electrons will be predominantly shared, and the ionic nature of the bonding decreases. If the electronegativity is high enough — if the electrons are sufficiently tightly bound — there will be reasonably stable oxidation states that are less positive than the maximum. One might suspect that these would tend to be one unit less than the maximum, but if this were the case the atom would have an odd number of electrons. But the existence of an odd number of electrons causes substantial chemical reactivity, since such a molecule or fragment can gain stability by forming another covalent bond in dimerizing. So the stable oxidation states will be those that have an even number of electrons. Thus we have not only SO_4^{2-}, with S formally +6, but SO_3^{2-}, with S formally +4; similarly we have PO_4^{3-} and PO_3^{3-}, and not only ClO_4^- but ClO_3^-, ClO_2^-, and ClO^-. This suggests a valid rule: the higher the electronegativity of the formally positive atom, the more stable the lower oxidation states will be. Phosphite ion, in which P is formally 3+, is a good reducing agent, but chlorite ion, in which Cl is formally 3+, is a good oxidizing agent.

Negative Oxidation States

Of course, there is the possibility of an atom adopting a negative formal oxidation state. This will occur only when the atom is more electronegative than its surrounding atoms, and the compounds tend to be stoichiometrically well defined only if the negative atom's electronegativity is greater than about 2.0; below that value the atom's attraction for electrons is simply not sufficiently great to make such a compound stable. Single-atom negative ions have always acquired enough electrons to fill their valence orbitals to a rare-gas configuration. Examples of this, in addition to the familiar halides—F^-, Cl^-, Br^-, and I^-—are O^{2-}, S^{2-}, Se^{2-}, N^{3-}, and C^{4-}. It must be pointed out, however, that these are formal oxidation states only, and the genuine electron configuration around one of these atoms will rarely have more than one extra electron, the rest being polarized to the extent of being essentially covalently bound.

Redox Properties of Intermediate Oxidation States

Another effect can be noticed on the right side of the table, where multiple oxidation states are possible. These atoms have both s and p valence electrons, and there is an energy difference between these distributions. In each case, then, there tends to be an oxidation state two units less positive than the maximum that corresponds to the retention of the s electrons. The heavier members of these groups have the more stable lower oxidation states: there are no stable halides of C and Si other than the tetrahalides, but $GeCl_2$ exists although it is rapidly oxidized to GeO_2 and $GeCl_4$ by air at room temperature; $SnCl_2$ is only slowly oxidized, and $PbCl_2$ is not oxidized at all. In fact, $PbCl_4$ is a very powerful oxidizing agent, being itself reduced to $PbCl_2$. This does not seem to be the result of a higher ionization energy for the s electrons; indeed, the s electrons are more readily removed (third and fourth ionization potentials) for Pb than for C or Si. Rather, the difference seems to be that the relatively large $5s$ or $6s$ orbital on the heavier elements is not capable of good overlap with ligand orbitals as the $2s$ and $3s$ are (see Fig. 12-20). Thus the formation of a bond involving the $5s$ or $6s$ AO's is not as energetically favorable as for the smaller s orbitals, and the maximum-oxidation-state compounds are not as thermodynamically stable.

Figure 12-20 Overlap comparison for large and small orbitals.

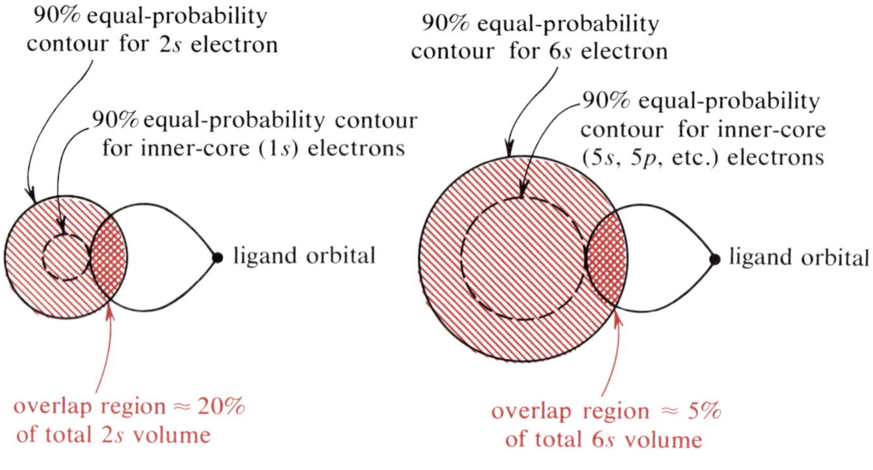

90% equal-probability contour for $2s$ electron

90% equal-probability contour for $6s$ electron

90% equal-probability contour for inner-core ($1s$) electrons

90% equal-probability contour for inner-core ($5s$, $5p$, etc.) electrons

ligand orbital

ligand orbital

overlap region ≈ 20% of total $2s$ volume

overlap region ≈ 5% of total $6s$ volume

12-13 Summary

In the periodic table we have a remarkably flexible and useful guide to the properties of atoms and molecules. We have seen that a consideration of the radii and charges involved allows us to predict the trends in the valence-orbital ionization potentials that we have used to construct MO energy-level diagrams, and from these the trends in atoms' ionization energies. Using the ionization energies and the related electron affinities, we have developed the concepts of differential ionization energy and electronegativity, which allow us to predict bond energies and enthalpies of formation fairly accurately, and bond lengths quite accurately, for example. Other physical properties show very strong correlations with the atom's position in the periodic table, and in general these can be rationalized by considering the numbers of valence electrons and orbitals, and the radii and charges of the atoms. Similarly, we can predict a number of chemical properties, such as acidity and redox behavior, by these considerations, and we see that this characterization of descriptive chemistry by electronegativity differences is a very general one. In the next chapters we shall pursue this pattern by considering first the descriptive chemistry of compounds that have a very great electronegativity difference and are thus predominantly ionic in their bonding, then the chemistry of compounds that are only partially ionic, having only a moderate electronegativity difference, and finally the chemistry of compounds that have a very small electronegativity difference (or none) and are thus almost completely covalent. This pattern of discussion may be unfamiliar, but it allows us to introduce a much greater order to the great mass of descriptive chemistry.

Study Problems

1. If the equilibrium charge separation for a gaseous LiF molecule is appropriately described by the condition $DIE(q)(Li) = DIE(-q)(F)$, as indicated in Fig. 12-11, what is the algebraic expression that is equivalent to Fig. 12-11? Solve this expression for the charge q. (Be careful to keep the signs straight for the $+$ and $-$ charges.)
2. Following the treatment of Study Problem 1, consider a single C—F bond in the CF_4 molecule, remembering that now the positive charge on the C atom will be four times as great as the negative charge on the F atom. What is the appropriate algebraic expression governing the charge separation in this case? Solve for the charges on C

and F. Repeat for a single C—Cl bond in CCl_4. Do the relative charges on the atoms in these two cases agree with our ideas of the relative electronegativities of F and Cl?

3. In the molecule CCl_3F, the charge on the C atom must be $3q_{Cl} + q_F$, since we expect both the Cl and F to have a negative charge. Write an expression governing the charge separation in a single C—Cl bond in CCl_3F, and another for the C—F bond in CCl_3F. Solve these two simultaneously for the charges on C, Cl, and F. How does the C charge compare with its charge in CF_4 and CCl_4? Is this reasonable? Why is Cl less negative than in CCl_4? Why is F more negative than in CF_4?

4. Using Table 12-4, calculate the Zn—I, Zn—C, and C—I bond energies, identifying the covalent and ionic contributions to each. If the three compounds ZnI_2, $Zn(CH_3)_2$, and CH_3I are each placed in water, and if water attacks each of the above bonds by essentially an ionic mechanism, which should hydrolyze most rapidly, judging from the ionic contributions to the bond energies?

5. The molar volume is a quantity analogous to the atomic volume, which can be established from the molecular weight and the compound's density. It is proportional to the molecular volume. Why should the molar volume of AlI_3 be smaller than that of BI_3, when the Al atom has about three times the volume of the B atom?

6. Plot a graph of the enthalpy of vaporization of the elements from cesium ($Z = 55$) to radon ($Z = 86$), leaving out the lanthanide rare earths ($Z = 58$–71), against atomic number. Why should the curve have the double-humped shape it shows?

7. How does the fact that we dealt with the acidity of SO_3 in terms of the hypothetical ion S^{6+} fit in with the earlier demonstration that even a compound with as great an electronegativity difference as LiF develops a charge of only about $0.3/-0.3$ within the gaseous molecule?

8. Lithium nitride, Li_3N, may be considered to contain the N^{3-} ion. Water molecules hydrate negative ions through the more positive hydrogen atoms. Show through appropriate chemical equations that this leads to the prediction that Li_3N solutions in water should be very basic.

9. MgS and NaCl have the same crystal structure, and the ions Na^+/Mg^{2+} and S^{2-}/Cl^- are isoelectronic. Explain the observation that MgS is much more dense than NaCl (2.84 g/cm³ vs. 2.17 g/cm³).

Some Further Reading

Sisler, H. H., *Electronic Structure, Properties, and the Periodic Law*, New York: Van Nostrand Reinhold, 1963. A very good discussion of most of the material of this chapter, except for that on differential ionization energies and the origin of electronegativity. The best single place to look.

Rich, R., *Periodic Correlations*, New York: Benjamin, 1965. A little offbeat and somewhat more sophisticated than Sisler, but some very interesting material is presented.

Jorgensen, C. K., *Orbitals in Atoms and Molecules*, New York: Academic, 1962. Chapter 7 of this book is just about the only place to find out more about differential ionization energies, and it is hard to read. But he adds some interesting material on combining lattice or Madelung energies with the charge prediction; more on this in the next chapter of this book.

Day, M. C. and Selbin, J., *Theoretical Inorganic Chemistry*, New York: Van Nostrand Reinhold, 1969 (2nd ed.). Most inorganic chemistry books have a section on periodicity, and usually a pretty good one. This one is a little more thorough than some others.

13 Ionic Compounds

One of the aims of Chapter 12 was to establish the parameter called electronegativity for each atom from its differential ionization-energy constants, and ultimately from its valence-orbital ionization energies. Because the basic forces governing chemical behavior are electrostatic (Coulomb) forces, the distribution of electrical charge in molecules or other polyatomic systems is of central importance in predicting and rationalizing the chemistry of these systems. The electronegativity of the atoms involved is a very useful guide to this charge distribution. For this reason, it is convenient and helpful to arrange descriptive chemistry in groups according to electronegativity difference between the atoms in the compounds of a particular group; the similar charge distributions produce a considerable chemical similarity. In this discussion we shall restrict ourselves initially to compounds formed from atoms having only s and p valence orbitals, and shall return later, in Chapter 16, to d-electron compounds, which have many similarities but some important differences.

13-1 Charge Separation in Molecular Orbitals and Molecular Energies

We pointed out in Section 6-7 that if the valence-orbital ionization energies of two atoms in a molecule are different, some electron transfer will occur. The coefficients c_1 and c_2 in the LCAO MO will be different:

$$\psi^{MO} = c_1\psi_1^{AO} + c_2\psi_2^{AO}$$

In terms of the molecule's energy-level diagram, as shown in Fig. 13-1, we say that if the bonding MO is much closer in energy to one AO than the other,

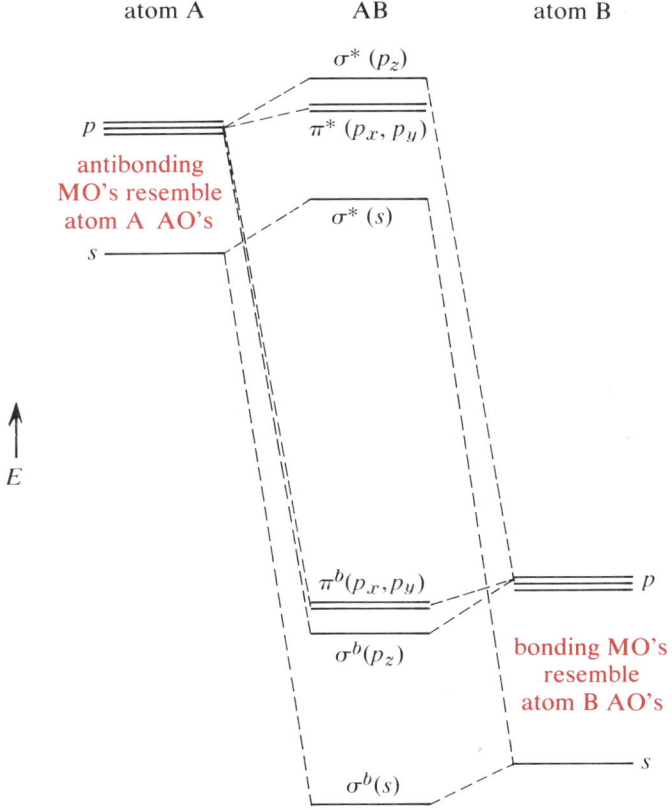

Figure 13-1 Energy-level diagram for a diatomic molecule in which the atoms have quite different VOIP's.

then the bonding electrons will be distributed in the molecule mostly according to the pattern described by the AO that is closer in energy. If one of the two bonding electrons were originally owned by each atom, then there will be some net electrical charge developing on each atom since both electrons are now mostly distributed around one atom—a positive charge on the atom with the higher-energy AO's and a negative charge on the other. Since these opposite charges attract each other according to Coulomb's law, they help hold the molecule together, and we say that there is an ionic contribution to the bonding. Being more precise about the spontaneous change of an electron from a higher energy to a lower one, we say that the free-energy change of this electron transfer is negative. Since the free-energy change is composed of a very large enthalpy change (if the AO's are really quite different in energy) and usually a small entropy change, we normally neglect the entropy change for this type of transition. It becomes important only when the electronic energies are fairly close in energy.

In Chapter 6 we also noted that when this electron-transfer process is accentuated by a large difference in valence-orbital ionization energies, the total covalent bonding energy (represented by the resonance integral β) may be less important than the electron-transfer energy arising from the transfer of an electron from the **electropositive** element with low electronegativity to the electronegative element with high electronegativity. We considered LiF as an example of this sort of significantly ionic bonding. In Chapter 12 we looked at the LiF molecule (still a single molecule in the gas phase) again by using differential ionization energies to predict the degree of charge separation, which we saw to be about 0.3 electron transferred. Of course, this fraction represents a statistical average of the electron's position over a long period of time, not an actual fraction of an electron.

Energy Quantities Involved in Bonding of Gaseous LiF

Let us look into the relative sizes of the energy quantities involved in the formation of a LiF gaseous molecule. If we start with a neutral Li atom and a neutral F atom, so that the chemical reaction reads

$$Li^0_{(g)} + F^0_{(g)} = LiF_{(g)} \tag{13-1}$$

we can break the energies down as follows (see Fig. 6-30). Suppose, for the sake of convenience, that the σ_s^b orbitals have the same energy as the F AO's from which they are formed, and that they are filled by electrons originating on the F atom. Then only the σ_z^b MO will show a bonding energy effect, and it will contain one electron from the Li and one from the F. What energy changes are involved in reaching this situation, starting from the AO's? We need an expression that will allow us to estimate the position of the σ_z^b MO; such an expression resulted from Study Problem 7 in Chapter 6. The expression developed there for the energy of the bonding MO in a significantly ionic molecule was

$$E^b_{MO} = I_F + \frac{\beta^2}{I_F - I_{Li}}$$

where I_F represents the valence-orbital ionization energy of the F orbital involved in the bonding (2p in our approximation), I_{Li} the same quantity for the Li 2p orbital, and β the resonance integral. We have already developed (also in Chapter 6) an approximate expression for β:

$$\beta = \frac{S(I_{Li} + I_F)}{2}$$

where S is the overlap of the two AO's, which for a σ bond is probably near

0.3. So for the energy of the bonding MO in LiF we can write

$$E_{MO}^b \cong I_F + \frac{[0.3(I_{Li}+I_F)/2]^2}{I_F - I_{Li}} \qquad (13\text{-}2)$$

and substituting the energies from inside the front cover into this expression we have

$$E_{MO}^b \cong -18.7 + \frac{[0.3(-18.7-4.0)/2]^2}{-18.7+4.0} = -18.7 - 0.79 = -19.49 \text{ eV}$$

With respect to the energies in Eq. 13-1, we can first note the electron-transfer energy, which can be expressed simply as the difference between the Li$2p$ energy and the F$2p$ energy: $18.7 - 4.0 = 14.7$ eV. With two electrons at the F$2p$ energy, then, there is the bonding-energy effect represented by the energy difference between the F$2p$ energy and the bonding MO energy. This is simply the second term in Eq. 13-2, or 0.79 eV; but two electrons are experiencing this stabilization, so the covalent bonding energy is $2 \times 0.79 = 1.58$ eV. The striking demonstration here is that the covalent bonding energy is only about one-tenth as big as the electron-transfer energy. This verifies numerically a qualitative observation we had made before.

But there is another energy effect we have overlooked. The two ions, which we may take as having 0.3 of the charge on an electron each, are attracting each other according to the Coulomb law. We can write the energy of attraction as

$$\text{ionic attractive energy} = -\frac{q_1 q_2}{r_{12}} = -\frac{(0.3e)(-0.3e)}{r_{LiF}}$$

Unfortunately, no experimental determination of internuclear distance exists for gaseous LiF; but using the method developed in Chapter 12 for predicting radii using electronegativity corrections, we can expect the radius to be near 1.36 Å, or 1.36×10^{-8} cm. The charge on the electron, e, in the proper units is 4.80×10^{-10} esu. Substituting these into the ionic attractive-energy expression,

$$\text{ionic attractive energy} = \frac{(0.3 \times 4.80 \times 10^{-10})^2}{1.36 \times 10^{-8}} = 1.53 \times 10^{-12} \text{ erg/molecule}$$

$$= 0.95 \text{ eV/molecule}$$

This is still smaller than the electron-transfer energy, but it is nearly equal to the covalent bonding energy. So we can divide the energy of Eq. 13-1 into three parts. The first, the electron-transfer energy, is responsible for most of the energy release of Eq. 13-1, but it is not a contribution to the total *bonding* energy. The second, the covalent bonding energy, is significant even for this molecule with atoms of very different electronegativity, but it is only about equal to the third, the ionic attractive energy. The ionic energy is as important as the covalent energy in the total bonding of the gaseous molecule.

13-2 Systems of Charged Particles and Madelung Energies

In the previous calculation we dealt with a molecule having considerable covalent character; we were careful to restrict ourselves to a single gaseous molecule of LiF. But at ordinary temperatures LiF is not a gas. It is a solid, crystalline material, having all the characteristics of the ionic crystals described in Chapter 4. And the ionic crystal differs in several important respects from the gaseous molecule. In the first place, the Li and F atoms are not the same distance apart, by a wide margin; we have estimated $r_{Li-F} = 1.36$ Å for the gaseous molecule, but in the ionic crystal the experimental distance (by X-ray diffraction) is 2.01 Å, which represents an increase of almost 50%. This sort of increase is characteristic of all the compounds of group Ia with group VIIa elements, the **alkali halides**. For instance, using experimental data for CsF, in the gas phase $r_{Cs-F} = 2.35$ Å, while in the ionic crystal $r_{Cs-F} = 3.01$ Å. This is a very large change in the internuclear distance, much larger than the differences of about 0.1 Å or less that we noted in Chapter 12 for different compounds involving the same two elements; such a substantial change must signify a major change in the nature of the bonding in the molecule when it condenses into an ionic solid. There is also the effect of crystallization on the total bonding energy. As we have just seen, the bonding energy is about equally divided for the gaseous molecule between the covalent energy and the ionic energy. As we shall see presently, in the ionic crystal the experimental (thermodynamic) bonding energy is accounted for almost completely by the ionic attraction, and there is hardly any covalent contribution at all.

Promotion of Charge Transfer in Crystal Lattices

What has happened to the bonding when LiF crystallizes? Apparently the process of crystallization promotes the transfer of electrons from the Li to the F, so that the charges approach those corresponding to complete transfer of valence electrons — 1.00+ and 1.00−. This would, of course, increase the importance of the ionic attraction, but it would also decrease the covalent bonding, since if the Li atom no longer has any share in the valence electrons it cannot contribute to a bonding MO. Suppose, then, that we have a crystal lattice composed of ions of alternating opposite charges. What can we say about the total bonding energy of such a system, and how can we predict the change from partial to complete charge separation when a gaseous molecule crystallizes?

A One-Dimensional Crystal

LiF crystallizes in the same lattice geometry as NaCl, whose structure we discussed in Chapter 4. Figure 13-2 shows this lattice; it is composed of alternating Li$^+$ and F$^-$ ions in a three-dimensional arrangement. Let us consider the

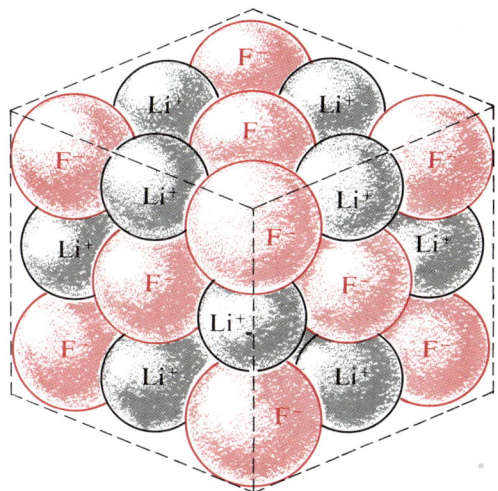

Figure 13-2 Lithium fluoride crystal structure (NaCl lattice).

energy of a single Li$^+$ ion in such a crystal. We can begin by considering a simpler model of an ionic crystal that is one-dimensional, as shown in Fig. 13-3; we shall expand it later to the real three-dimensional structure. If we consider the bonding energy of the lithium ion in the middle of the row of ions, we can see immediately that the two F$^-$ ions next to it, at a distance R away, will exert an attractive Coulomb force on the Li$^+$. The magnitude of this force will be e^2/R^2 for each F$^-$, and the bonding energy will be $-e^2/R$. The total bonding

Figure 13-3 One-dimensional approximation to the LiF structure.

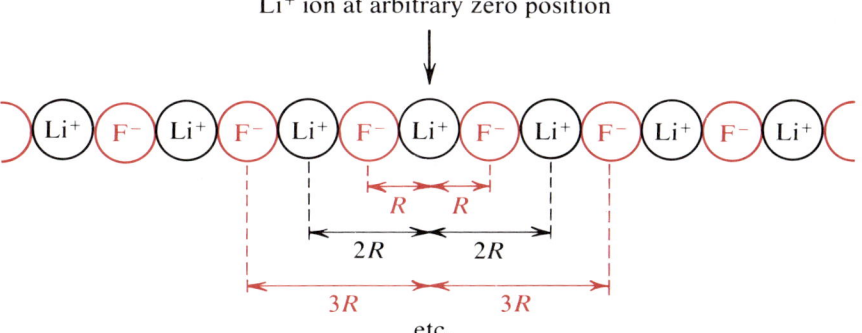

energy due to these two ions will be $-2e^2/R$. Farther away, at a distance $2R$, are two more Li⁺ ions. Since these have the same charge as the Li⁺ we are considering, they will repel it and *reduce* the bonding energy of the central ion. The bonding energy including these will be

$$U = -\frac{2e^2}{R} + \frac{2e^2}{2R}$$

where we have symbolized the bonding energy by U. As we consider the effects of ions farther and farther away from the central ion, the terms in the expression for U will have alternating signs (since the ions in the row do) and the denominator will increase by R in each succeeding term (since the total distance in the row of ions does):

$$U = -\frac{2e^2}{R} + \frac{2e^2}{2R} - \frac{2e^2}{3R} + \frac{2e^2}{4R} - \frac{2e^2}{5R} + \frac{2e^2}{6R} - \cdots$$

We can factor this expression as follows:

$$U = -\frac{2e^2}{R}\left(1 - \frac{1}{2} + \frac{1}{3} - \frac{1}{4} + \frac{1}{5} - \frac{1}{6} + \cdots\right)$$

What does the series of fractions in the parentheses add up to? If we include 10 ions on either side of the central ion, the parenthetical fractions add up to 0.74564. If we include 20 ions on either side, the series sums to 0.71877. If the row of ions were as big as an actual crystal, so that perhaps 10 million ions were on each side of the central ion, we would find that the series summed to 0.69315. The **infinite series** is said to **converge** to 0.69315; we can see the convergence developing below:

> 10 ions on each side = 0.74564
> 20 ions on each side = 0.71877
> 30 ions on each side = 0.71009
> 40 ions on each side = 0.70580
> 50 ions on each side = 0.70325
> ∞ ions on each side = 0.69315

So we can write the bonding energy of a single ion in an essentially infinite one-dimensional crystal as

$$U = -(0.69315)\frac{2e^2}{R}$$

or more generally as

$$U = -\frac{Ae^2}{R} \tag{13-3}$$

where A is the **Madelung constant** for the crystal under discussion; it is characteristic only of the crystal's symmetry, since the interatomic distance is not included in it. In other words, the Madelung constant for a one-dimensional crystal with alternating ions is 0.69315×2, or 1.38630.

A Three-Dimensional Crystal

What about the real three-dimensional LiF crystal structure? We can still adopt the approach of adding up all the contributions to bonding energy from all the ions at an equal distance from the central ion, then going to the next-farther set of ions, and so on. But we shall have to watch out for one thing that is quite different in the three-dimensional case. Since the packing in the crystal is cubic, we shall take successively larger cubes around the central ion. If the series we generate in this way is to converge satisfactorily, however, the cubes must each contain a net charge of zero; otherwise, if the cube is positively charged, say, the central positive ion will have a much smaller bonding energy than in an electrically neutral real crystal. Neutral cubes can be generated rather neatly, as shown in Fig. 13-4, by including (for bookkeeping purposes) only that portion of the spherical ion that is inside the cube. Thus each F^- in the face of the cube is half inside, and half its charge is counted; each Li^+ on the edge of the cube is one-fourth inside, and one-fourth of its charge is counted; and each F^- on the corner of the cube is one-eighth inside the cube, and one-eighth of its charge is counted.

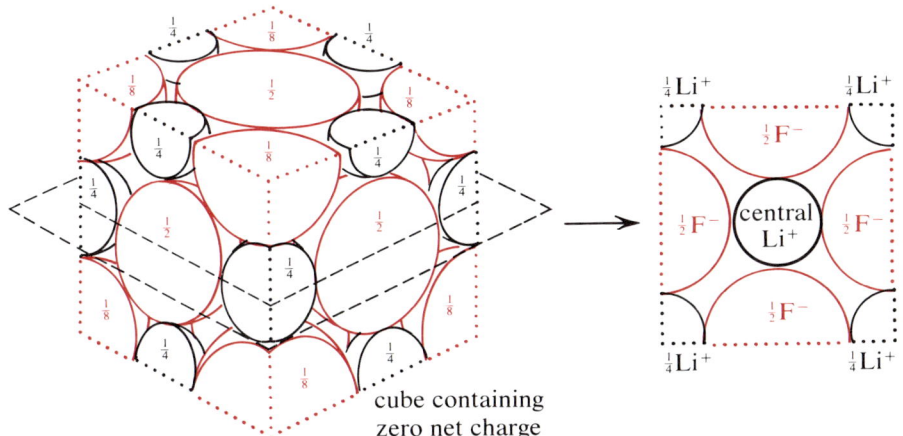

Figure 13-4 Evaluation of the Madelung constant using successively larger cubes containing zero net charge.

Now there are six F^- ions in the faces of the cube a distance R away, and we are counting half of that charge, so the first term in the series is

$$U = -\frac{e^2}{R}\left(\frac{6}{2} + \cdots\right)$$

There are 12 Li⁺ ions on the edges of the cube, which are $R\sqrt{2}$ away, and we shall count one-fourth of their charge:

$$U = -\frac{e^2}{R}\left(\frac{6}{2} - \frac{12}{4(1.414)} + \cdots\right)$$

The last ions in this particular cube are eight F⁻ ions, which are one-eighth inside the cube (at the corners) and which are $R\sqrt{3}$ away:

$$U = -\frac{e^2}{R}\left(\frac{6}{2} - \frac{12}{4(4.414)} + \frac{8}{8(1.732)}\right) = -\frac{e^2}{R}(3.000 - 2.120 + 0.577)$$

So our first approximation to the Madelung constant for LiF (the NaCl crystal structure) is $A = 1.457$. If we go to the next larger cube, all the ions that were fractionally included in this cube will be completely inside the larger one and there will be another set fractionally included. Adding these up we get $A = 1.750$. The third larger cube, treated in the same way, yields $A = 1.714$, and so on. The series rapidly converges to $A = 1.748$ for a crystal of normal dimensions. Of course, if the crystal geometry were different we would get a different Madelung constant; some other constants for other structures are given in Table 13-1.

Table 13-1
Madelung Constants for Different Crystal Lattices

NaCl lattice	(6:6 coordination)	$A = 1.748$
CsCl lattice	(8:8 coordination)	$A = 1.763$
wurtzite lattice	(4:4 coordination)	$A = 1.641$
zincblende lattice	(4:4 coordination)	$A = 1.638$
CaF₂ lattice	(6:3 coordination)	$A = 2.519$ (note 2+ charge)
TiO₂ (rutile) lattice	(6:3 coordination)	$A = 2.408$ (note 4+, 2− charges)

$$U = A\frac{(\text{charge on positive ion})(\text{charge on negative ion})}{\text{positive ion radius} + \text{negative ion radius}}$$

Ionic-Attraction Lattice Energy

If we calculate the energy of crystalline LiF,

$$U = -\frac{Ae^2}{R} = -\frac{1.748(4.803 \times 10^{-10} \text{ esu})^2}{2.009 \times 10^{-8} \text{ cm}} = -2.005 \times 10^{-11} \text{ erg/Li}^+ \text{ ion}$$

$$= -288.2 \text{ kcal/mole}$$

The energy U calculated here is called the **lattice energy**. The lattice energy is the energy absorbed when a mole of positive ions and an equivalent number of negative ions evaporate from a crystalline solid to free gaseous ions:

$$\text{LiF}_{(s)} + U \rightarrow \text{Li}^+_{(g)} + \text{F}^-_{(g)}$$

Born–Haber Cycles

We can apply Hess' law to this hypothetical reaction to get an experimental value for the lattice energy so that we can check our calculated value. We need the enthalpies of formation of each of the three species above. We can get that of $\text{LiF}_{(c)}$ directly. The other two require a little further attention. Since enthalpies of formation are referred to the elements in their standard states at room temperature, we need to break ΔH_f for $\text{Li}^+_{(g)}$ down into two steps:

$$\text{Li}_{(s)} \xrightarrow[\substack{\Delta H_{\text{sub}} \\ (S)}]{\text{absorbs } E} \text{Li}^0_{(g)} \xrightarrow[\substack{\text{ionization energy} \\ (\text{IE})}]{\text{absorbs } E} \text{Li}^+_{(g)} \qquad \Delta H_f = S + \text{IE}$$

Similarly, we need to break ΔH_f for $\text{F}^-_{(g)}$ down into two steps:

$$\tfrac{1}{2}\text{F}_{2(g)} \xrightarrow[\substack{\frac{1}{2} \text{ bond energy} \\ (\frac{1}{2}D)}]{\text{absorbs } E} \text{F}^0_{(g)} \xrightarrow[\substack{\text{electron affinity} \\ (\text{EA})}]{\text{releases } E} \text{F}^-_{(g)} \qquad \Delta H_f = \tfrac{1}{2}D - \text{EA}$$

All of these last four quantities are measurable, and so we can write, following Hess' law:

$$\text{Li}^+_{(g)} + \text{F}^-_{(g)} \rightarrow \text{LiF}_{(c)} - U \text{ (lattice energy)}$$

$$U = \Delta H_f(\text{LiF}) - \Delta H_f[\text{Li}^+_{(g)}] - \Delta H_f[\text{F}^-_{(g)}]$$
$$= \Delta H_f(\text{LiF}) - S - \text{IE} - \tfrac{1}{2}D + \text{EA}$$
$$= -146.3 - 37.1 - 124.3 - 18.3 + 97.8 \text{ kcal/mole}$$
$$= -229.7 \text{ kcal/mole}$$

This particular application of Hess' law is referred to as a **Born–Haber cycle**. Perhaps the "cyclic" nature of the calculation will be more obvious in graphical form:

$$\begin{array}{ccc}
\text{Li}_{(g)} + \text{F}_{(g)} & \xrightarrow{\text{IP} + \text{EA}} & \text{Li}^+_{(g)} + \text{F}^-_{(g)} \\
\uparrow S \quad \uparrow D & & \downarrow U \\
\text{Li}_{(s)} + \text{F}_{2(g)} & \xrightarrow{\Delta H_f} & \text{LiF}_{(c)}
\end{array}$$

There are several other interesting and important applications of Born–Haber cycle calculations; we shall come back to them from time to time.

Addition of Repulsion Energy to the Madelung Expression

There is a significant discrepancy between the "experimental" (Born–Haber) value of the LiF lattice energy and the theoretical value we produced using the

Madelung treatment. Specifically, the lattice energy is not nearly as large as the Madelung ionic attraction energy — this is quite generally true for ionic crystals, although most values match better than this one, with errors being on the order of 10%. What have we overlooked? If the ionic attraction were the only energy quantity we needed to consider, the interionic distance would shrink down to nothing, since the attraction gets larger as the distance gets smaller. But in fact the interionic distance is *not* zero, because the ions have a finite size; that is, the inner-core electrons repel each other very strongly at small distances. So we need to add to our Madelung lattice-energy expression a repulsion term, and we shall use a term with the same form as the repulsion term in the Lennard–Jones potential:

$$U = -\frac{Ae^2}{r} + \frac{B}{r^n} \tag{13-4}$$

Here B is the proportionality constant for the repulsion energy and n is an exponent whose value increases as the number of inner-core electrons in the ions increases. If both ions have the electronic structure of helium, as in LiH, n is usually taken as 4; the neon structure has $n = 7$, argon 9, krypton 10, and xenon 12. If the ions have different electronic structures, the average is used.

Curiously enough, we do not need to know what B is. For any given crystal, we need only know what its equilibrium interionic distance is. At that distance the attraction and repulsion have reached their best compromise, and the energy, as a function of distance, is at a minimum. At a minimum a function has a slope of zero, or a derivative of zero. So the derivative of U with respect to r will be zero at R, the equilibrium distance:

$$\frac{dU(r)}{dr} = 0 = \frac{Ae^2}{R^2} - \frac{nB}{R^{n+1}}$$

We can solve this expression for the value of B:

$$B = \left(\frac{Ae^2}{R^2}\right)\left(\frac{R^{n+1}}{n}\right) = \frac{Ae^2 R^{n-1}}{n}$$

Substituting this in the Madelung-energy expression, we have

$$U = -\frac{Ae^2}{R} + \frac{Ae^2 R^{n-1}}{nR^n} = -\frac{Ae^2}{R} + \frac{Ae^2}{nR} = -\frac{Ae^2}{R}\left(1 - \frac{1}{n}\right)$$

For LiF we take n as the average of the Li$^+$ value of 4 and the F$^-$ value of 7, or 5.5. This gives us 4.5/5.5 of our earlier value for the lattice energy, or

$$U = -288.2\left(\frac{4.5}{5.5}\right) = -236 \text{ kcal/mole}$$

We can see that this is quite a good match for the experimental value of 229.7 kcal/mole.

Partial Charge Transfer

Another approach to the problem of the exaggerated attraction energy is to assume that the charge separation into ions is not really complete. We can try to get the charge separation in the crystal by using differential ionization energies as in Study Problems 2 and 3 of Chapter 12. The condition for charge equilibrium in a bond is that the DIE's of the two atoms be equal. If we take the lithium atom as the central atom and consider the six fluorines nearest it as bonded to it, we can suppose that only the charge on each F due to its interaction with this Li need be considered. Then the condition for charge equilibrium is that $DIE_{Li}(6q) = DIE_F(-q)$. Setting up this expression numerically from Table 12-3, we have

$$1.74 + 39.4q = 10.3 - 15.7q$$

$$55.1q = 8.56$$

$$q = 0.155$$

$$\text{charge on Li} = 6 \times 0.155 = 0.934$$

Because the stoichiometry of the compound LiF is 1:1 the charges must be equal, and we can write an ionic attractive energy using the Madelung expression:

$$U = -\frac{A(\text{fractional charge})^2}{R}$$

$$= -1.748 \frac{(0.934)^2 (4.803 \times 10^{-10} \text{ esu})^2}{2.009 \times 10^{-8} \text{ cm}}$$

$$= -1.745 \times 10^{-11} \text{ erg/Li atom} = -248 \text{ kcal/mole}$$

This method does not give as good a match for the lattice energy as the previous consideration of the ion–ion repulsion, particularly when we realize that if charge separation is not complete the shared electron fraction must also contribute some covalent bonding energy to the lattice energy. For instance, if we consider LiI, which we have previously observed to be much more covalent than LiF, the same sort of calculation gives an ionic charge of 0.67 and an ionic attraction (Madelung) energy of 86 kcal/mole; but the observed lattice energy is 177 kcal/mole. This suggests that the lattice energy in this case is about half due to covalent bonding. On the other hand, all cesium halides have complete charge separation (to 1.00 and −1.00), so that some treatment such as the consideration of repulsions is necessary to reproduce the observed lattice energies. It is interesting to observe for the alkali halides, however, that the ionic charges calculated in this way usually are in the range 0.7–1.0, so the degree of charge separation is in any case much greater than for the isolated diatomic molecule.

Lattice Energy vs. Gaseous-Molecule Bond Energy

The lattice energy is much more negative, at -229.7 kcal/mole, than the binding energy of an individual gaseous molecule of LiF, which we have already calculated as $-1.58 - 0.95 = -2.53$ eV/molecule or -58.3 kcal/mole. It is obvious that the ionic lattice will be the normal condition of compounds of atoms whose electronegativities are so different as to indicate substantial charge separation to be possible. Considering this energy difference, which would have to be made up by thermal energy (kT), we can also see why the boiling points of ionic crystals are so high.

13-3 Born–Haber Cycle Calculations

As we have just seen, Born–Haber cycle calculations offer us a convenient way to account for the stability of ionic lattices. They are much more versatile than that, however; basically a Born–Haber cycle is a form of Hess' law involving atomic or molecular energies rather than strictly descriptive thermodynamic energies. These cycles offer us, then, a way to reconcile our molecular model with the great quantity of existing thermodynamic data for all kinds of physical and chemical changes. Let us look at some of these.

Lattice Energies for Multiply Charged Ions

Let us restrict ourselves initially to an examination of some further properties related to the stabilities of ionic lattices. For instance, we indicated in Chapter 12 that we expect the formal oxidation states of atoms at the left of the periodic table to be equal to the number of valence electrons, or the group number. We have seen that for group Ia metals—the alkali metals—this is true and that the actual electrostatic charge on the atoms approaches that indicated by the formal oxidation state. What can we say about group IIa metals—the **alkaline earth metals**?

Suppose we examine calcium oxide, CaO. Calcium oxide has the NaCl crystal structure, so we can conveniently calculate the lattice energy using the NaCl Madelung constant of 1.748 and the interionic distance of 2.39Å. There is some reason for concern about the stability of CaO, because if we adopt the ionic model it will be necessary to provide, from somewhere, enough energy to ionize the calcium atom up to Ca^{2+}. The Born–Haber cycle for the formation of CaO is shown below.

$$\text{Ca}_{(g)} + \text{O}_{(g)} \xrightarrow[+\text{EA}_1+\text{EA}_2]{\text{IE}_1+\text{IE}_2} \text{Ca}^{2+}_{(g)} + \text{O}^{2-}_{(g)}$$

$$S \uparrow \quad \tfrac{1}{2}D \uparrow \qquad\qquad\qquad\qquad \downarrow U$$

$$\text{Ca}_{(s)} + \tfrac{1}{2}\text{O}_{2(g)} \xrightarrow{\Delta H_f} \text{CaO}_{(c)}$$

We can identify these quantities experimentally:

Ca	O
$S = +46.0$ kcal/mole	$\tfrac{1}{2}D = \tfrac{1}{2}(118.3)$
	$= +59.3$ kcal/mole
$\text{IE}_1 = +6.11$ eV/atom $= +142.4$ kcal/mole	$\text{EA}_1 = -33$ kcal/mole
$\text{IE}_2 = +11.87$ eV/atom $= +275.2$ kcal/mole	$\text{EA}_2 = +189$ kcal/mole

Notice that although it is energetically favorable to add a free electron to an oxygen atom to make O^- it is very unfavorable (positive enthalpy change) to add another free electron to make O^{2-}. Using these quantities, we can express the enthalpy of formation of CaO as follows:

$$\Delta H_f = S + \tfrac{1}{2}D + (\text{IE}_1 + \text{IE}_2) + (\text{EA}_1 + \text{EA}_2) + U$$

$$= 46.0 + 59.2 + 142.4 + 275.2 - 33 + 189 + U$$

$$= 678.8 + U \text{ kcal/mole}$$

There is obviously a heavy responsibility on the lattice energy U. If the solid compound CaO is to be shown stable (as we know experimentally it is), ΔH_f will have to be substantially negative, because the entropy change involved in making an orderly crystal is distinctly unfavorable. But to make ΔH_f negative, U will have to be at least -700 kcal/mole, which is a very large energy by chemical standards. Let us calculate it.

$$U = -\frac{A(2e)^2}{R}\left(1 - \frac{1}{n}\right)$$

$$= -\frac{(1.748)(4)(4.803 \times 10^{-10})^2}{2.39 \times 10^{-8}}\left(\frac{7}{8}\right)\left(1.440 \times 10^{13} \frac{\text{kcal/mole}}{\text{erg/molecule}}\right)$$

$$= -850 \text{ kcal/mole} \quad \text{(averaging } n \text{ values for Ca}^{2+} \text{ and O}^{2-}\text{)}$$

Thus $\Delta H_f = 678.8 - 850 = -171$ kcal/mole. This compares fairly favorably with an experimental value of -152 kcal/mole for ΔH_f. The reason it is somewhat too large is that polarization of the O^{2-} ion is reducing the effective charges somewhat, so that the ionic attractive energy is not as large as we have proposed. We see that the lattice energy is an enormous factor in the stability of ionic solid compounds.

Stability of Oxidation States in Ionic Lattices

How far can we go with this model? If the lattice energy is so large, why doesn't it make a solid ionic compound like LiO stable? The calculation is exactly the same as for CaO, but we must change the numbers to those for Li: $S = 37.1$ and $(IE_1 + IE_2) = 124.3 + 1744$ kcal/mole. Using these,

$$\Delta H_f(\text{LiO}) = S + \tfrac{1}{2}D + (IE_1 + IE_2) + (EA_1 + EA_2) + U$$
$$= 37.1 + 59.2 + (124.3 + 1744) + (-33 + 189) + U$$
$$= 2120.6 + U$$

U, the lattice energy, will be large, because Li^{2+} would be very small. Since its one electron feels three times the nuclear charge of the electron in the hydrogen atom, we can assume its radius is one-third that of the hydrogen atom or 0.176 Å. If we add this to the ionic radius of O^{2-} (from inside the front cover) we have $R = 1.58$ Å, and

$$U = -\frac{A(2e)^2}{R}\left(1 - \frac{1}{n}\right)$$
$$= -\frac{(1.748)(4)(4.803 \times 10^{-10})^2}{1.58 \times 10^{-8}}\left(\frac{4.5}{5.5}\right)\left(1.440 \times 10^{13} \frac{\text{kcal/mole}}{\text{erg/molecule}}\right)$$
$$= -1204 \text{ kcal/mole}$$

So even allowing for rather substantial errors in our calculated value of the lattice energy of LiO, it is not anywhere near large enough to take care of the enormous ionization energy of $Li^+ \rightarrow Li^{2+}$. And, in fact, there is no evidence for the formation of LiO.

We see that lattice-energy calculations have a considerable bearing on the stability of oxidation states. Consider another case. If we can show that group Ia metals cannot be oxidized to a stable 2+ oxidation state, what about the opposite problem: can there be a stable 1+ oxidation state of a group IIa metal? Take the case of calcium fluoride. If we could make CaF, calcium monofluoride, would its lattice energy make it stable against disproportionation?

$$2\text{CaF}_{(c)} = \text{Ca}_{(s)} + \text{CaF}_{2(c)}$$

To decide whether this should be spontaneous we need a value for the free-energy change of this reaction, but if the enthalpy change is large it will dominate the free-energy change. So we can calculate ΔH for the reaction by using ΔH_f for each of the species, and we can get these values from Born–Haber cycle calculations:

$$Ca_{(g)} + F_{(g)} \xrightarrow{IE+EA} Ca^+_{(g)} + F^-_{(g)} \qquad\qquad Ca_{(g)} + 2F_{(g)} \xrightarrow[+2(EA)]{IE_1+IE_2} Ca^{2+}_{(g)} + 2F^-_{(g)}$$

$$S\uparrow \quad \tfrac{1}{2}D\uparrow \qquad\qquad \downarrow U \qquad\qquad\qquad S\uparrow \quad D\uparrow \qquad\qquad \downarrow U$$

$$Ca_{(s)} + \tfrac{1}{2}F_{2(g)} \xrightarrow{\Delta H_f} CaF_{(c)} \qquad\qquad\text{and}\qquad\qquad Ca_{(s)} + F_{2(g)} \xrightarrow{\Delta H_f} CaF_{2(c)}$$

ΔH_f for $Ca_{(s)}$, of course, is zero since it is an element in its standard state. For CaF we have

$$\Delta H_f = S + \tfrac{1}{2}D + IE + EA + U$$
$$= 46.0 + 18.3 + 142.4 - 97.8 + U$$
$$= 108.9 + U$$

Assuming that Ca^+ would have the same radius as K^+, we can estimate U:

$$U = -\frac{Ae^2}{R}\left(1 - \frac{1}{n}\right)$$

$$= -\frac{(1.748)(4.803 \times 10^{-10})^2}{2.70 \times 10^{-8}}\left(\frac{7}{8}\right)(1.440 \times 10^{13})\frac{\text{kcal/mole}}{\text{erg/molecule}}$$

$$= -188 \text{ kcal/mole, and } \Delta H_f = 109 - 188 = -79 \text{ kcal/mole}$$

So CaF would be stable against decomposition into the elements; but what about disproportionation into CaF_2? For CaF_2,

$$\Delta H_f = S + D + (IE_1 + IE_2) + 2(EA) + U$$
$$= 46.0 + 36.6 + (142.4 + 275.2) - 195.6 + U$$
$$= 304.6 + U$$

The Madelung constant for CaF_2 is 5.039 (see Table 13-1), after taking the 2+ charge on the calcium ion into account. Using it:

$$U = -\frac{Ae^2}{R}\left(1 - \frac{1}{n}\right) = -\frac{5.039(4.803 \times 10^{-10})^2}{2.35 \times 10^{-8}}\left(\frac{7}{8}\right)(1.440 \times 10^{13})\frac{\text{kcal/mole}}{\text{erg/molecule}}$$

$$= -622 \text{ kcal/mole}$$

$$\Delta H_f = 304.6 - 622 = -317 \text{ kcal/mole}$$

So for the disproportionation reaction,

$$2CaF \rightarrow Ca^0 + CaF_2 - \Delta H_d$$
$$\Delta H_d = \Delta H_f(CaF_2) + 0 - 2\Delta H_f(CaF)$$
$$= -317 - 2(-79)$$
$$= -159 \text{ kcal} \qquad \text{(favorable for disproportionation)}$$

This result is a typical one. It is not possible to make stable compounds of group IIa metals in an oxidation state lower than 2+ because the lattice energy of the divalent condition is so favorable as to make the monovalent compound disproportionate.

Thresholds of Energy for the Formation of Ionic Compounds

We have seen that when the total ionization energy of the positively charged atom becomes very large, the lattice energy cannot supply it. This does not always mean that the compound will not form, because the bonding energy has not only an ionic contribution but a covalent contribution (which is somewhat more difficult to evaluate numerically from our MO model). It does mean that if the compound forms (with a calculated lattice energy insufficient to provide the total ionization energy) it will not be an ionic crystal but will have strongly covalent bonding. For instance, looking back at the ionization energies in Table 12-2, there is a column representing the average ionization energy per electron of the various atoms up to an actual charge equal to their formal charge or formal oxidation state. We can make a rough generalization; if the total ionization energy of an atom is less than about 400 kcal/mole (or about 17 eV) *per fluoride*, its fluoride will be an ionic solid, while if it is greater than about 450 kcal/mole (or 20 eV) per fluoride, its fluoride will be strongly covalent. In the middle, compounds can be either ionic, partly ionic, or covalent depending on their exact structure, electronegativity, and so on. The corresponding numbers for the chlorides would be about 300 and 350 kcal/mole (or 13 and 15 eV), and for bromides, very roughly, 275 and 325 kcal/mole (or 12 and 14 eV). This decrease reflects the greater radius of the chloride and bromide ions and their smaller electron affinity. The greater the interionic distance, the smaller the ionic lattice energy will be.

Following this generalization, for instance, we can expect BeF_2 to be ionic; the first two ionization energies of Be add up to 27.5 eV/molecule, which is only 13.8 eV per fluoride in the compound. But $BeCl_2$ ought to be a borderline case and so should the bromide. There is a distinct difference, as exemplified by the melting points of the three compounds: 800°, 440°, and 490°, respectively. This generalization should not be pressed too far, since it assumes that all positive ions are the same size and all lattices have the same geometry; it does offer at least a rough guide to the size of the lattice-energy effect, however.

Acid Strengths

Let us digress here to look at a couple of other applications of the Born–Haber cycle, although they are not completely related to the chemistry of ionic compounds. One interesting usage is the prediction of relative acidities (proton-transfer capabilities) of hydrogen halides in aqueous solution. These acidities are normally described by equilibrium constants K_a; but $\Delta G^0 = -RT \ln K_a$, so we can describe acidities in terms of the free energy of the proton-transfer reaction:

$$HX_{(aq)} + H_2O_{(aq)} \rightarrow H_3O^+_{(aq)} + X^-_{(aq)} - \Delta G^0$$

We can reduce this to individual steps involving atoms or molecules in the following Born–Haber cycle:

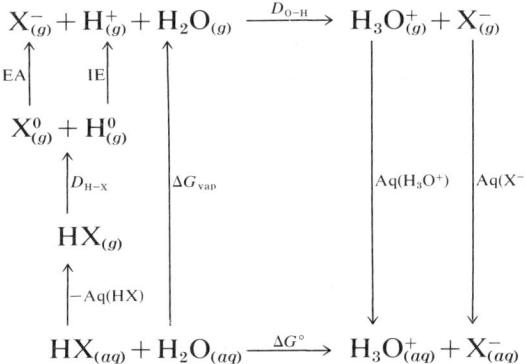

Obviously this is a fairly complicated calculation involving many individual physical quantities. We can write ΔG^0 as the following expression:

$$\Delta G^0 = -Aq(HX) + \Delta G_{vap}(H_2O) + D(H-X) + IE(H) + EA(X) + D(O-H) + Aq(H_3O^+) + Aq(X^-)$$

Some of these quantities are measurable directly, while others are measured indirectly or must be estimated; Table 13-2 gives values of each of these

Table 13-2
Free Energies Related to Hydrogen Halide Proton Transfer[a]

Process	ΔG(HF) (kcal/mole)	ΔG(HCl) (kcal/mole)	ΔG(HBr) (kcal/mole)	ΔG(HI) (kcal/mole)
$-Aq(HX)[HX_{(aq)} \rightarrow HX_{(g)}]$	5.7	-1	-1	-1
$D(HX)[HX_{(g)} \rightarrow H_{(g)} + X_{(g)}]$	127.8	96.5	81.0	65.0
$EA(X)[X_{(g)} + e^- \rightarrow X^-_{(g)}]$	-83.0	-87.6	-82.5	-75.3
$IE(H)[H_{(g)} \rightarrow H^+_{(g)} + e^-]$	315.3	315.3	315.3	315.3
$\Delta G_{vap}(H_2O)[H_2O_{(l)} \rightarrow H_2O_{(g)}]$	2.1	2.1	2.1	2.1
$-D(H^+-OH_2)[H^+_{(g)} + H_2O_{(g)} \rightarrow H_3O^+_{(g)}]$	-206.6	-206.6	-206.6	-206.6
$Aq(H_3O^+)[H_3O^+_{(g)} \rightarrow H_3O^+_{(aq)}]$	-55.7	-55.7	-55.7	-55.7
$Aq(X^-)[X^-_{(g)} \rightarrow X^-_{(aq)}]$	-101.3	-72.6	-65.5	-57.5
ΔG^0 (overall proton transfer)	4.3	-10	-13	-14
$K_a = $ antilog $\dfrac{-\Delta G^0}{2.303RT}$	6.3×10^{-4}	10^8	10^9	10^{10}

[a]Adapted from Cotton, F. A., and Wilkinson, G., *Advanced Inorganic Chemistry*, New York: Wiley-Interscience, 1966, (2nd ed.).

quantities for the acids HF, HCl, HBr, and HI, and the resulting calculated acidities. By comparison, the experimental K_a value for HF is tabulated as 6.7×10^{-4}, which is quite a good match. Too good, in fact—the numbers in the table are simply not that reliable. It is apparent that we are taking the difference of two large numbers to get ΔG^0, and this always has the effect of magnifying relative error. It should be clear that the overall trend of acidities of related compounds such as these is not determined by any one factor and that there is no simple relationship that can establish the trend. To the limited extent this is possible, however, we expect the major factors causing HF to be a weak acid to be the high H—F bond energy $[D(H—X)]$, which reflects the good overlap of the H $1s$ and F $2p$ orbitals, and the relatively large free energy of solution of molecular HF $[-Aq(HX)]$, which reflects the hydrogen-bonding capability of the F atom.

Reduction Potentials

One last application, from many that could be given, of the Born–Haber cycle is the explanation of the order of the reduction potentials of the alkali metals. We have already noted in Chapter 12 that the single parameter of electronegativity is not adequate to predict the order of these potentials. We can combine the various factors that affect the potential in the following cycle, however:

$$\begin{array}{ccc} e^- + M^+_{(g)} & \xrightarrow{-IE} & M^0_{(g)} \\ \uparrow {\scriptstyle -\Delta H_{\text{hydr}}} & & \downarrow {\scriptstyle -S} \\ e^- + M^+_{(aq)} & \xrightarrow{\Delta H^\circ} & M^0_{(s)} \end{array}$$

In the interests of simplicity, we shall deal only with the enthalpy changes for the reduction half-reactions, on the premise that large enthalpy changes will determine the free-energy changes and thus the reduction potentials. Since an isolated half-reaction is experimentally impossible, the numbers will not be meaningful, but will simply reflect a trend in the potential values. Table 13-3 gives the values of the sublimation energies, ionization energies, and hydration

Table 13-3
Energies Related to Aqueous Reduction Potentials of Alkali Metals

Process	ΔH^0(Li) (kcal/mole)	ΔH^0(Na) (kcal/mole)	ΔH^0(K) (kcal/mole)	ΔH^0(Rb) (kcal/mole)	ΔH^0(Cs) (kcal/mole)
$-S[M^0_{(g)} \to M^0_{(s)}]$	-37.1	-26.0	-21.5	-20.5	-18.8
$-IE[M^+_{(g)} \to M^0_{(g)}]$	-125.8	-120.0	-101.5	-97.8	-91.3
$-\Delta H^0_{\text{hydr}}[M^+_{(aq)} \to M^+_{(g)}]$	130.3	104.1	84.0	78.0	70.1
ΔH^0 (overall electron transfer)	-32.6	-41.9	-39.0	-40.3	-40.0
$\mathscr{E}^0$ (experimental, in volts)	-3.045	-2.713	-2.925	-2.924	-2.923

energies that affect the reduction potential, together with the net enthalpy change. We see that the resulting ΔH^0 values lie in the same order as the experimentally established reduction potentials. In this case, which is somewhat simpler than the previous one, the major factor in the effectiveness of lithium as a reducing agent is its high enthalpy of hydration. Usually no single factor can be seen to determine a trend in some physical or chemical property so readily, however, and the great usefulness of Born–Haber cycle calculations is that they allow the encompassing of many different influences and a rather more sophisticated treatment of the origin of chemical differences.

13-4 Solvation Energies and Solubility

The last two examples in Section 13-3 involved hydration energies in a rather crucial way. Since much of common chemistry takes place in water solution, it is very frequently true that hydration energies have a strong influence on experimental chemical behavior. What is the origin of these hydration energies, and what influences do they have?

Ion–Dipole Attraction

Hydration energies, like all the other energies we have discussed, are electrostatic in origin. In treating lattice energies, we added up the total interaction of a central ion with other ions arranged around it. When an ion is placed in a solvent, this specific sort of interaction is not possible, because there is no charge on the solvent molecules. However, if the solvent is polar — having an electric dipole in the sense that one end of the solvent molecule is more electrically positive than the other — it is possible to have an electrostatic interaction between the ion (monopole) and the solvent molecules (dipole). Consider a positive ion with a charge $z+$, which is a distance R from the center of a solvent molecule that is an electric dipole, as in Fig. 13-5. In the solvent

Figure 13-5 Definition of quantities involved in ion–dipole attraction.

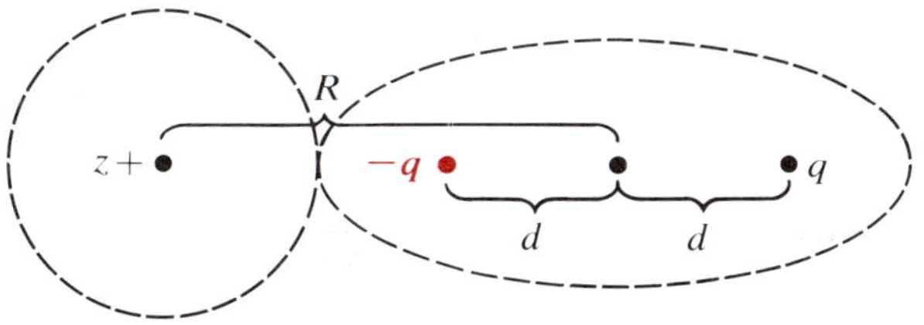

$$U = -\frac{(z+)(-q)}{R-d} - \frac{(z+)(q)}{R+d}$$

molecule, electrons with a total charge of $-q$ have been shifted from one end of the molecule to the end nearest the positive ion — either permanently or as a result of the attraction of the positive ion. The distance between the center of negative charge and the center of positive charge is $2d$. The Coulomb energy of interaction of the ion with these two charges is the sum of the energies of its attraction to each separately:

$$U_1 = -\frac{qz}{R+d} + \frac{qz}{R-d}$$

Taking a common denominator,

$$U_1 = -\frac{qz(R-d)}{R^2-d^2} + \frac{qz(R+d)}{R^2-d^2} = \frac{qz}{R^2-d^2}(-R+d+R+d)$$

$$= \frac{2dqz}{R^2-d^2}$$

The Dipole Moment

But $2d$ is the distance between the centers of charge in the solvent dipole, which is presumably more or less constant; we can lump the dipole properties together into a single quantity called the **dipole moment** μ:

$$\mu = 2dq \quad \text{so} \quad U_1 = \frac{\mu z}{R^2-d^2}$$

So we see that there will be a net attractive energy due to an ion–dipole interaction in a solvent that has a nonzero dipole moment. The bigger the dipole moment, the bigger this interaction will be, and the more the ion will be stabilized. Notice that while the attraction between two ions is just the Coulomb-law energy with r in the denominator, this energy expression has r^2 in the denominator, which means that the attractive force will drop off more quickly for an ion–dipole interaction than for an ion–ion interaction.

Table 13-4 gives some dipole moments for common molecules; since the electron distribution is unequal we expect any heteronuclear diatomic molecule to have a permanent dipole moment — and, indeed, any heteronuclear molecule that did not have ligand atoms arranged around the central atom with complete symmetry. Thus water, which is a bent molecule, has a dipole moment of 1.85×10^{-18} esu-cm (a unit sometimes called a **debye**), but $HgCl_2$, which is linear, has $\mu = 0.0$ esu-cm. Of course, this does not preclude a dipole being associated with a specific atom pair (Hg–Cl, for instance), but means simply that in a symmetrical molecule such as $HgCl_2$ these **bond dipoles** cancel each other out, which precludes their measurement.

Returning to the expression for the energy of attraction between an ion and a single molecule of a polar solvent, we can see that it allows us to calculate an

Table 13-4
Dipole Moments of Common Molecules (in debyes; 1 debye = 10^{-18} esu-cm)

Substance	μ	Substance	μ
HI	0.42	SF_6	0
HBr	0.80	CO	0.10
HCl	1.05	CO_2	0
HF	1.91	$CHCl_3$ (chloroform)	1.02
H_2O	1.85	CCl_4 (carbon tetrachloride)	0
KCl vapor	8.0	CH_3OH (methyl alcohol)	1.70
NO	0.07	CH_3COOH (acetic acid)	1.74
NO_2	0.39	CH_3CH_2OH (ethyl alcohol)	1.70
NH_3	1.47	$CH_3C(O)CH_3$ (acetone)	2.89
PCl_3	0.78	C_6H_6 (benzene)	0

energy that will be related to the hydration energy in which we are interested. For example, take a magnesium ion in water solution. The ion Mg^{2+} forms a 6-hydrate; that is, it is surrounded by six polar water molecules. The ion–dipole attraction in this case, then, is

$$U_1 = -\frac{6\mu_{H_2O}(2e)}{R^2 - d^2}$$

R and d are defined as shown in Fig. 13-6. It will simplify the calculation if we

Figure 13-6 Radii definitions involved in ion hydration. The "effective spherical volume" has a radius equal to half the distance between the oxygen nuclei in ice—it includes the O—H bond length and the hydrogen-bonding distance.

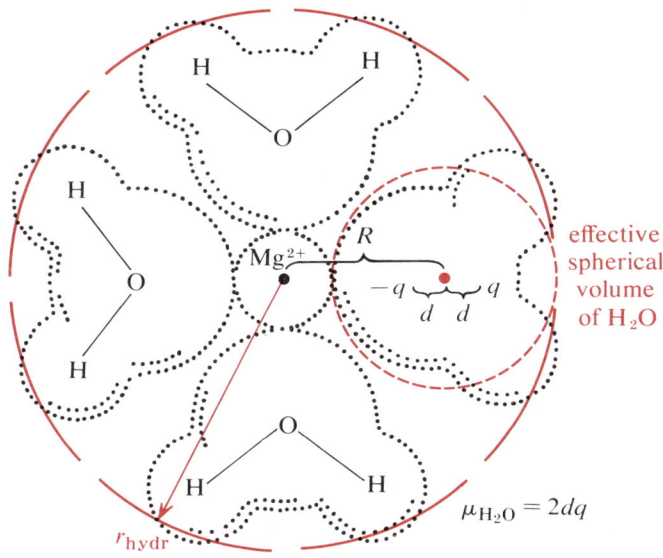

neglect d^2, which is difficult to define numerically anyway:

$$U_1 = -\frac{6\mu_{H_2O}(2e)}{R^2}$$

In a careful treatment we would have to take note of the fact that μ_{H_2O} has two parts, the permanent dipole moment and the induced dipole moment (because the molecule is near a positive charge). However, there are also dipole–dipole repulsion terms because the six waters are near each other, and other repulsions due to the inner electrons on each atom like the ones we included in the Madelung energy expression; all of these repulsions usually add up to a total repulsion that very nearly offsets the attraction of the ion for the *induced* dipole moment. We can make a rough calculation, then, simply by using the permanent dipole moment for water. For R we can use the sum of the crystal radius of the ion (from the inside front cover) and the radius of the water molecule (1.38 Å from the X-ray diffraction pattern of ice). This gives us

$$U_1 = -\frac{6(1.85 \times 10^{-18})(9.60 \times 10^{-10})}{(2.03 \times 10^{-8})^2}(1.440 \times 10^{13})\frac{\text{kcal/mole}}{\text{erg/molecule}}$$

$$= -373 \text{ kcal/mole}$$

Solvation of the Hydrated Ion

This is not all the hydration energy, however, because this charged hydrate ion will still interact with other water molecules in solution. It may look as if this could go on forever, but fortunately for an ion as large as the hydrate ion we can simply average the effects of all the other dipoles in the solution as a bulk **dielectric constant**, ϵ, for water. Using this approximation, the energy of dissolving the hydrate complex can be shown to be

$$U_2 = -\frac{(Ze)^2}{2r_{\text{hydr}}}\left(1 - \frac{1}{\epsilon}\right) = -162\frac{Z^2}{r_{\text{hydr}}}\text{kcal/mole} \qquad (r \text{ in Å})$$

Here r_{hydr} is the quantity shown in Fig. 13-6: the ion's radius plus the *diameter* of the water molecules, 2.76 Å; Z is the number of electron charges on the ion. For Mg^{2+},

$$U_2 = -162\left(\frac{4}{3.41}\right) = -190 \text{ kcal/mole}$$

Origin of the Hydrated Water Molecules

There is one last effect we have forgotten. Forming the initial hydrate effectively involved using free water molecules, so we must add on the energy necessary to remove 6 moles of water from solution — 6 times the enthalpy of vaporization, or 63 kcal/mole:

$$U_3 = (\text{No. } H_2O \text{ in hydrate})[\Delta H_{vap}(H_2O)] = 6 \times 10.5 = 63 \text{ kcal/mole}$$

All this is summarized in the Born–Haber cycle shown below:

$$\begin{array}{ccc}
Mg^{2+}_{(g)} + 6H_2O_{(g)} & \xrightarrow{U_1} & Mg(OH_2)^{2+}_{6(g)} \\
\uparrow \quad \quad \uparrow U_3 & & \downarrow U_2 \\
Mg^{2+}_{(g)} + 6H_2O_{(aq)} & \xrightarrow[\text{hydration energy}]{\text{single-ion}} & Mg(OH_2)^{2+}_{6(aq)}
\end{array}$$

There is only one thing wrong with this cycle. It is not possible to work with isolated ions. We need to replace it with a cycle in which the bottom line is simply the dissolution reaction in water, with a measurable enthalpy of solution:

$$\begin{array}{ccc}
Mg^{2+}_{(g)} + 2Cl^-_{(g)} + 6H_2O_{(g)} & \xrightarrow{U_1} & Mg(OH_2)^{2+}_{6(g)} + 2Cl^-_{(g)} \\
\uparrow \text{lattice energy} & \uparrow 6\Delta H_{vap} & \downarrow U_2 \quad \downarrow 2\Delta H_{hydr}(Cl^-) \\
MgCl_{2(g)} & +6H_2O_{(aq)} \xrightarrow{\Delta H_{sol'n}} & Mg(OH_2)^{2+}_{6(aq)} + 2Cl^-_{(aq)}
\end{array}$$

Overall Hydration Energies

From an examination of the cycle above we can see that it is easy to get accurate experimental values for the *sum* of the single-ion hydration energies but quite difficult to divide this sum up appropriately. Table 13-5 gives some hydration energies for ions, listed in two ways: first with the hydration energy of H^+ chosen arbitrarily as zero, and second with that energy estimated theoretically and other values scaled accordingly.

From Table 13-5 we can see that the estimated "experimental" hydration energy of Mg^{2+} is -452 kcal/mole. The quantities U_1, U_2, and U_3 that we produced add up to $-373 - 190 + 63 = -500$ kcal/mole, which is not too far from the tabulated value. Other ions are roughly comparable: Al^{3+} calculates -1031 kcal/mole vs. tabulated -1102; Cu^{2+} calculates -481 vs. -495; and K^+ calculates -81 vs. -73. The agreement is fairly good — about 10% error or less. We can see that these are very large energies, of the same order of magnitude as the lattice energies, so that an ion in water solution is stabilized to about the same extent as in a crystal.

Table 13-5

Hydration Energies of Ions

Ion	ΔH_{hydr} [if $\Delta H_{hydr}(H^+) = 0$]	ΔH_{hydr} [if $\Delta H_{hydr}(H^+) = -260$ kcal/mole]
H^+	0	−260.0
Li^+	137.7	−122.3
Na^+	163.8	−96.2
K^+	184.0	−76.0
Be^{2+}	−73	−593
Mg^{2+}	62.0	−458.0
Ca^{2+}	140.8	−379.2
Ba^{2+}	209.8	−310.2
Cr^{2+}	79.3	−440.7
Mn^{2+}	80.4	−439.6
Fe^{2+}	62.5	−457.5
Co^{2+}	30.4	−489.6
Ni^{2+}	18.1	−501.9
Cu^{2+}	19.5	−500.6
Zn^{2+}	32.8	−487.2
Pb^{2+}	167.7	−352.3
Al^{3+}	−331.6	−1111.6
Fe^{3+}	−264	−1044
F^-	−366.3	−106.3
Cl^-	−348.8	−88.8
Br^-	−340.7	−80.9
I^-	−330.3	−70.3

Enthalpy Effects on Solubility

This stabilization in water solution has an important influence on solubility. Spontaneous dissolving of a salt in water means that the free-energy change of solution is negative. The entropy change of solution is always positive, which is favorable, since an ionic crystal is very orderly while a solution offers many more microscopic states for the ions. However, if the enthalpy change of solution is quite positive (endothermic) it can overcome the entropy effect and make dissolution unfavorable. Because lattice energies are so large, the hydration energies of the ions must be very substantial in order to prevent a strongly positive enthalpy of solution. We can represent this by a simplified version of the last Born–Haber cycle:

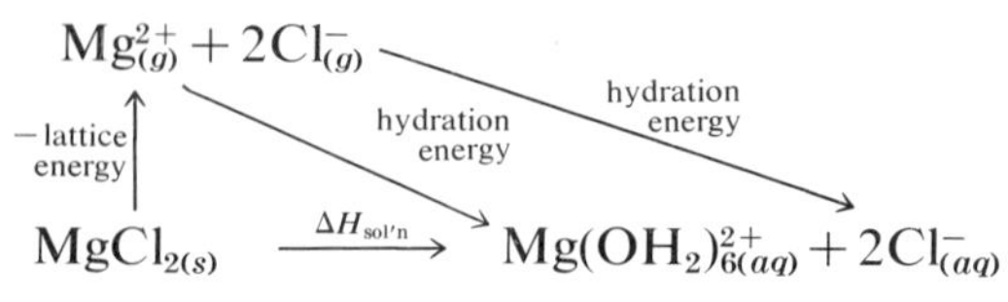

$$Mg^{2+}_{(g)} + 2Cl^-_{(g)}$$

$$\nearrow^{-\text{lattice energy}} \quad \searrow^{\text{hydration energy}}$$

$$MgCl_{2(s)} \xrightarrow{\Delta H_{sol'n}} Mg(OH_2)^{2+}_{6(aq)} + 2Cl^-_{(aq)}$$

A few examples of the use of this procedure may be helpful. Sodium chloride has a lattice energy (which we could calculate) of −183 kcal/mole, Na^+ has a hydration energy of −96 kcal/mole, and Cl^- has a hydration energy of −89 kcal/mole, so

$$\Delta H_{sol'n} = -96 - 89 - (-183) = -185 + 183 = -2 \text{ kcal/mole}$$

This indicates that there ought to be virtually no heat effect when NaCl dissolves, since the lattice energy and hydration energies come so near cancelling. We expect NaCl to be soluble, however, since the entropy of solution will be quite favorable.

Calcium fluoride, CaF_2, has a lattice energy (see p. 679) of −622 kcal/mole, Ca^{2+} has a hydration energy of −379 kcal/mole, and two F^- have a total hydration energy of −212 kcal/mole, so

$$\Delta H_{sol'n} = -379 - 212 - (-622) = -591 + 622 = +31 \text{ kcal/mole}$$

This is a rather substantial positive enthalpy change, and we expect that it will probably outweigh the favorable entropy change, so CaF_2 ought to be insoluble in water. This is exactly the case: K_{sp} for CaF_2 is 2×10^{-10}.

Does the high lattice energy of a divalent ion prevent its ionic crystals from dissolving? Consider ZnI_2. We can calculate its lattice energy as about −519 kcal/mole (although it is probably not really ionic) and compare that with the hydration energies:

$$\Delta H_{sol'n} = -487 - 141 - (-519) = -628 + 519 = -109 \text{ kcal/mole}$$

This is very favorable for dissolution, and it will be aided by the entropy effect. And in fact, we can dissolve over 400 g of ZnI_2 in only 100 ml of water; it is one of the most soluble salts known.

There is an important point to note with respect to lattice energies and solubility, however. The lattice energy depends on the *product* of the charges on the ions, while the hydration energies depend only on the *individual* charges. This means that it will usually be difficult to dissolve a salt of a 2+ ion with a 2− ion, such as CoS or $BaSO_4$, and it will be extremely difficult to dissolve a salt of a 3+ ion with a 3− ion, such as $CrPO_4$. Of course, in the latter case polarization of the ions means that there will be considerable covalent nature to the bonding of the crystal, but the generalization is still valid. Since most metals form ions with charges of 2+ or higher, this means that most carbonates, sulfates, and chromates, for instance, are insoluble; but the alkali metal ions with an oxidation state of 1+ and halides, nitrates, and acetates with a charge of 1− are usually soluble.

All of this discussion has centered on the use of water as a solvent, and we have used hydration energies. The discussion can be extended to other solvents, however, and the theory is the same for **solvation energies**. Ionic compounds are usually not as soluble in nonaqueous solvents as they are in water; the reasons are readily seen. The solvation energy must be large if it is to overcome the lattice energy, and yet the solvation energy is directly proportional to the dipole moment of the solvent. Looking at Table 13-4, we see that water has a much higher dipole moment than other common solvents such as ethanol, benzene, carbon tetrachloride, and ethyl ether. There are some solvents with an even higher dipole moment than water, but they are much larger molecules. For effective solvation the solvent molecule should be small, since the radius of that molecule occurs in the denominator of the energy expression. Water is unique as a solvent for ionic solids because of its combination of high dipole moment and small molecular size. This is quite different from the earlier discussion of the solubility of molecular solids or liquids (Chapter 7), where we saw that the entropy effect leads to solubility of nonpolar compounds in nonpolar solvents if no hydrogen bonding occurs, and that hydrogen bonding leads to solubility of oxygen-substituted compounds in protic solvents but internal hydrogen bonding prevents solubility in completely nonpolar solvents.

All in all, we can see that the solubility of compounds is governed by a rather intricate set of properties. It is extremely important, though, because so many chemical reactions — including, for instance, all of the life processes — take place in solution. There is an alternative approach to the origin of hydration energies, through MO's, which we shall examine in Chapter 16. This leads to essentially the same results by considering water as an electron donor. As we have indicated before, the MO and ion-polarization approaches are compatible, and we use whichever is more convenient in a particular discussion.

13-5 Chemistry of Ionic Compounds — Electronegativity Differences Greater than 2.0 Units

Descriptive chemistry falls into broad categories that are most easily approached by recognizing the similarities in chemical behavior that exist for compounds of approximately equal electronegativity differences. In this chapter we have developed some principles relating to the bonding in compounds in which electron transfer is complete or nearly so — *ionic* compounds. These are usually the simplest to describe; let us begin with them in attempting to unify the diverse patterns of chemistry.

Chemistry of Ionic Compounds | 691

The Range of Ionic-Compound Formation

The first task is to define the range of compounds that we can agree to call ionic. As this section heading indicates, it will be convenient to make the arbitrary division between ionic and partly ionic compounds by saying that compounds are ionic when they possess a bond between two atoms whose electronegativities differ by more than 2.0 units. This choice *is* arbitrary; an electronegativity difference of 1.9 units gives a compound properties that are nearly indistinguishable from those of a compound having an electronegativity difference of 2.1 units. But the generalization is a useful one nonetheless. Figure 13-7 shows the range of elements whose compounds can be described as ionic using this criterion. Note that this does not mean that *all* of an element's compounds will be ionic—only that its electronegativity is low enough or high enough that some of them can be ionic. The figure 2.0 has not been adhered to with complete fidelity in Fig. 13-7 to avoid too great a difference between fluorine and the other halogens; a study of the electronegativities inside the front cover will show that over 90 elements have electronegativities 2.0 units (or more) less than fluorine, while none have electronegativities 2.0 units less than chlorine. Fluorine chemistry is quite different from that of the other halogens, but not that different.

Figure 13-7 Ionic-compound formation in the periodic table.

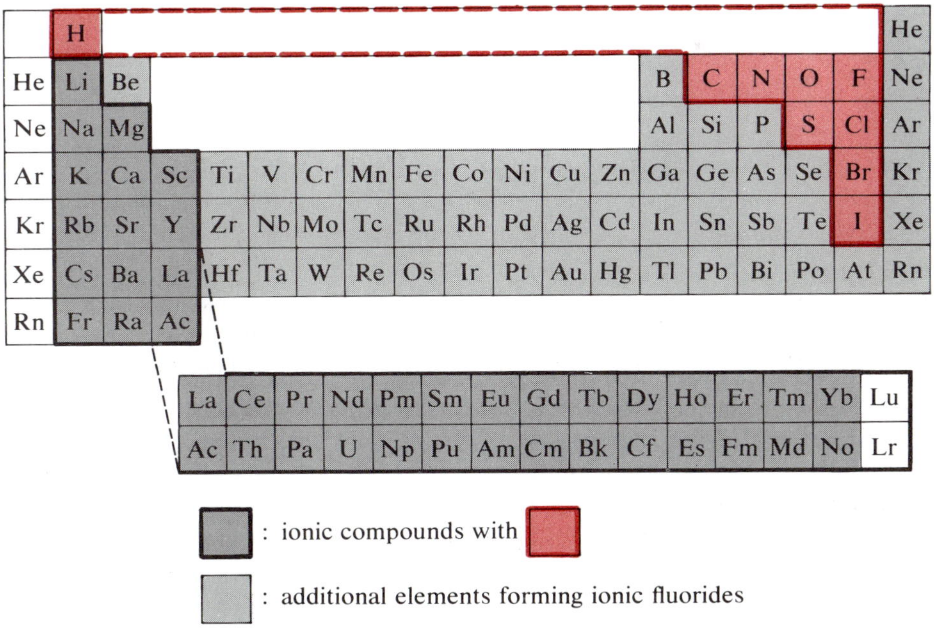

Ionic compounds in general are crystalline solids at room temperature. They melt (or sublime) and boil at relatively high temperatures and have no appreciable vapor pressure at room temperature. This is a result of their high lattice energies. The relative magnitudes of the lattice and hydration energies dictate that most ionic compounds are water soluble if they contain a 1+ or 1− ion, but not otherwise. They form spontaneously (substantial negative free-energy change) when covalent or partially covalent compounds of the potential positive ion and the potential negative ion are mixed, because of the favorable enthalpy change on formation of the ionic lattice; this is a particularly important synthetic technique.

The characteristic reactions of ionic compounds are those in which the positive ion serves as an electron acceptor, the negative ion serves as an electron donor, or a more stable lattice is formed. For instance, calcium chloride is used as a drying agent because the positive calcium ion accepts electrons from water molecules, forming a hydrate; KCl reacts with $TiCl_4$ vapor to form K_2TiCl_6 (which is really $2K^+[TiCl_6]^-$) because the negative chloride ion serves as an electron donor to the titanium; and when calcium carbonate (limestone, which is primarily ionic) is heated, it forms calcium oxide (quicklime), releasing CO_2 because of the greater stability of the lattice involving the smaller oxide ion. These general principles will be useful in rationalizing the chemistry of the various ionic compounds; let us look at these compounds, grouping them by their negative ions and starting with the least electronegative.

Ionic Hydrides

Hydrogen, with an electronegativity of 2.20, might be considered too electropositive to form a negative ion in an ionic lattice, since no metals have electronegativities as low as 0.20; but the experimental fact is that all of the most electropositive metals form ionic-lattice hydrides: Li, Na, K, Rb, Cs, Mg, Ca, Sr, and Ba. X-ray studies of hydride crystals show that the hydride ion has an effective ionic radius about the same as the fluoride ion (1.53 Å vs. 1.36 Å), so the calculated lattice energies will be no larger than those of fluorides. Considering this in a Born–Haber cycle calculation for the formation of these hydrides shows that only the most electropositive metals have ionization energies low enough for the ionic hydride formation to be favorable; thus the group IIIa metals do not form ionic hydrides.

The ionic hydrides are prepared by direct reaction of the molten metal with hydrogen gas at high temperatures:

$$2Na_{(l)} + H_{2(g)} \xrightarrow{375°} 2NaH_{(s)}$$

They are extremely reactive because of the pronounced tendency of the hydride ion to serve as an electron donor. Since hydrogen is not very electronegative, the two electrons on the hydride ion are greatly stabilized by forming a bond with an acceptor atom, and any acceptor atom that has lower-energy orbitals than the group Ia and IIa metals will do. This donor tendency takes two forms: first, the "acid" of this anion, H—H, is an exceedingly weak acid, so that if an ionic hydride is placed in a solution of any protic solvent, complete solvolysis occurs very rapidly, and sometimes violently:

$$LiH + 5HOH \rightarrow Li(OH_2)_4^+ + OH^- + H_2 \uparrow$$

The second aspect of electron donation by H^- is that it is a very powerful reducing agent. It is frequently used by preparative organic chemists for this purpose, either as LiH, CaH_2, $LiAlH_4$, or $NaBH_4$. The last two of these have the hydride ions bound to a central atom that is much less electropositive than the group Ia and IIa metals, so the bonding is only somewhat ionic; but the AlH_4^- and BH_4^- are ions that show the same reducing tendencies as H^-, although somewhat milder. A typical reducing reaction of the hydride ion is

$$CH_3-\overset{\overset{O}{\|}}{C}-CH_3 + LiAlH_4 \rightarrow Li^+[Al(O-CH(CH_3)_2)_4]^-$$

$$\xrightarrow{H_2O} CH_3-\overset{\overset{OH}{|}}{C}H-CH_3 + LiOH + Al(OH)_3$$

In humid air, exothermic hydrolysis can generate enough heat to initiate the reduction of atmospheric oxygen, and several of the ionic hydrides ignite spontaneously if left open to the atmosphere.

Ionic Carbides

Carbon, with an electronegativity of 2.50, appears to be a somewhat better candidate than hydrogen for the preparation of ionic compounds with the most electropositive metals. We expect the physical and chemical properties of ionic carbides to be very similar to those of the ionic hydrides, and this is in fact true. There is an interesting difference, however, in the nature of the carbide ion. Our initial expectation might be that the carbide ion should be C^{4-}, so that (for instance) hydrolysis of the carbide ion should give the very weak acid H_4C, which, of course, is methane. This does occur for two carbides, Be_2C and Al_4C_3, but these have an electronegativity difference so small that the bonding cannot be considered primarily ionic. What we have neglected to consider is the great difficulty of placing four electrons on a neutral atom. Carbon has an electron affinity of 1.25 eV/atom, which is favorable, but this is only for the first electron. No experimental values exist for the energy effect of adding the second, third, and fourth electrons, but from the example of oxygen ($EA_1 = 1.43$ eV, $EA_2 = -8.2$ eV) we can be certain they will be added endothermically, and the lattice energy of the hypothetical C^{4-} crystal would have to overcome not only the ionization energy of the positive ion, but an unfavorable electron affinity as large or larger. It is not surprising, then, that no clearly ionic C^{4-} carbides exist. Instead, the charge is reduced sharply by forming the covalent diatomic C_2^{2-} ion, which is variously called the carbide or **acetylide** ion. The energy levels of Fig. 6-26 for C_2 presumably apply here but now there are two additional electrons, which would be placed in the σ_z^b MO, giving the ion a total of three bonds. This is the basic structure of acetylene, C_2H_2 or H—C≡C—H, without two protons, so that the acetylide ion can be considered the conjugate base of the weak acid acetylene.

All of the positive ions shaded in Fig. 13-7 form ionic acetylides, and many form the intermediate compounds **hydroacetylides**, MHC_2 or $M^+[H-C\equiv C]^-$. Most of the acetylides are prepared by passing acetylene gas over the hot metal, so that the negative free-energy change for the spontaneous reaction results from the formation of the ionic lattice:

$$Mg_{(s)} + C_2H_{2(g)} \xrightarrow{500°} MgC_{2(s)} + H_{2(g)}$$

CaC_2, however, is prepared by heating CaO with carbon (as charcoal or coke):

$$CaO_{(s)} + 3C_{(s)} \xrightarrow{2500°} CaC_{2(s)} + CO_{(g)}$$

Like hydrides, ionic carbides are salts of a very weak acid (methane or acetylene) so that they react vigorously with protic solvents. This is a result of the electron-donor capacity of the ion, which has electrons in relatively high-energy orbitals since C is not very electronegative. Thus acetylides hydrolyze,

$$CaC_{2(s)} + 2HOH_{(l)} \rightarrow Ca(OH)_{2(s)} + C_2H_{2(g)}$$

and react rapidly with ethanol,

$$CaC_{2(s)} + 2CH_3-CH_2-OH_{(l)} \rightarrow$$
$$Ca(OH)_{2(s)} + CH_3-CH_2-C\equiv C-CH_2-CH_{3(l)}$$

The reducing power of the acetylide ion, which we expect to see because of its electron-donor ability, is not as pronounced as for the hydride ion. This is in keeping with carbon's greater electronegativity. The mixed salt of magnesium with acetylide and iodide ions will reduce elemental iodine, however:

$$2IMgC\equiv C-H + I_2 \rightarrow 2MgI_2 + H-C\equiv C-C\equiv C-H$$

Ionic Nitrides

The higher electronegativity of nitrogen (3.07) makes it more likely that its negative ion will be stable. However, the remarks that applied to the hypothetical C^{4-} ion also apply to the N^{3-} ion. Because of the substantial energy necessary to put three electrons on the nitrogen atom, only the most electropositive metals will form an ionic nitride. In addition, the high negative charge on the ion makes it very polarizable, so there is a substantial tendency for its compounds to be partly covalent. The physical and chemical properties of several nitrides resemble those of the hydrides and carbides sufficiently that we may regard them as ionic, however. For instance, in analogy with the ionic hydrides, magnesium nitride, Mg_3N_2, is made by heating magnesium in nitrogen:

$$3 Mg_{(s)} + N_{2(g)} \xrightarrow{800°C} Mg_3N_{2(s)}$$

Lithium even reacts with nitrogen at room temperature:

$$6 Li_{(s)} + N_{2(g)} \xrightarrow{25°C} 2 Li_3N_{(s)}$$

These ionic nitrides are formed by lithium and the group IIa metals; sodium and potassium also form nitrides with the corresponding formula, but they cannot be prepared from molecular nitrogen and are less stable than one might expect. One possible reason for this, which also applies to the C^{4-} ion, is that the lattice energy of a system that must contain many large positive ions for each monatomic negative ion will not be very favorable because the positive ions will repel each other too strongly. Thus the small Li^+ will form a more stable lattice than the larger Na^+ and K^+.

The reactions of nitrides are analogous to those of hydrides and carbides; the electron–donor properties of N^{3-} are sufficiently pronounced that the ion's conjugate "acid," NH_3, is extremely weak. Thus ionic nitrides react vigorously with protic solvents,

$$Mg_3N_{2(s)} + 6HOH_{(l)} \rightarrow 3Mg(OH)_{2(s)} + 2NH_{3(aq)}$$

and have significant reducing properties,

$$Mg_3N_{2(s)} + 3CO_{(g)} \rightarrow 3MgO_{(s)} + N_{2(g)} + 3C_{(s)}$$

Triatomic Ions Just as ionic carbides could be produced more readily by combining two carbon atoms into the acetylide ion, ionic "nitrides" involving the N_3^- ion are more common than those involving the monatomic ion. These are the **azides**, and they are known for all of the group Ia and IIa metals. The N_3^- ion is linear and can presumably be described by MO's resembling those of

CO_2, with which it is isoelectronic (see Fig. 6-40). The least stable electrons in the ion are those distributed according to the $\pi_{x,y}$ MO's that have the same energy as the N $2p$ AO's; they are also not destabilized by a high negative charge on the ion. This means that the ion is not as good an electron donor as the N^{3-} ion, and correspondingly its acid, HN_3, hydrazoic acid, really is an acid (although a weak one; $K_a = 3 \times 10^{-5}$) and its reducing properties are quite modest:

$$2N_3^- + I_2 \rightarrow 2I^- + 3N_2$$

Nearly all the reactions of the azide ion involve the generation of N_2 gas. For instance, heating sodium or barium azide smoothly generates very pure N_2; the reaction is sometimes used for that purpose:

$$Ba(N_3)_{2(s)} \rightarrow Ba_{(s)} + 3N_{2(g)}$$

The driving force for this reaction is the very high covalent bond energy of N_2; looking back at the MO's for this system in Fig. 6-26 we see that it has a bond order of 3, and the experimental bond energy for N≡N is $D = 225.0$ kcal/mole. Obviously the forming of N_2 will be a very exothermic process, and when covalent azides decompose, this tremendous energy release nearly always occurs as a violent explosion. Lead azide, which is much less ionic than sodium or barium azide, is used commercially as a detonator, and organic azides and polyazides are more sensitive yet. The difference between the explosive nature of covalent azides and the smooth, controllable nitrogen release by ionic azides is just the lattice energy of the ionic azides, which uses up the large bond energy release of the N_2 that is formed.

Heteronuclear Ions Another way in which nitrogen-containing negative ions can be formed is to allow the nitrogen to combine with some more electropositive element such as carbon. Thus we find stable lattices containing **cyanide** ions, CN^-, and **cyanamide** ions, NCN^{2-}, which are analogous to (and isoelectronic with) the acetylide and azide ions, respectively. This resemblance is more than coincidence, because the carbon–nitrogen ions are produced commercially by starting with calcium carbide:

$$CaC_{2(s)} + N_{2(g)} \xrightarrow{1100°} CaNCN_{(s)} + C_{(s)}$$

$$CaNCN_{(s)} + C_{(s)} + Na_2CO_{3(s)} \rightarrow CaCO_{3(s)} + 2NaCN_{(s)}$$

These ions have the properties we might expect from comparison with C_2^{2-} and N_3^- when we allow for the change in atom electronegativities and charges; thus CN^- is a good electron donor and its conjugate acid HCN a weak acid, but the lower charge and the more electronegative N atom make it a poorer donor and HCN a stronger acid than C_2^{2-} and C_2H_2. The donor properties of the cyanide ion mean that it is quite frequently found in transition-metal coordination compounds in which a metal atom or ion is stabilized by electron-donor ligands (see Chapter 16). Making the other comparison, the cyanamide ion has a higher negative charge and a less electronegative atom than the azide ion, so we expect it to be a better donor; it does hydrolyze in water to the compound cyanamide, H_2NCN, which is much less acidic than HN_3.

Protonated Ions Still another way in which nitrogen-containing negative ions can be formed is to take up some of the negative charge on the N^{3-} ion with protons: we expect the formation of NH^{2-}, the **imide** ion, and NH_2^-, the **amide** ion. Imides are not well known, although lithium imide, Li_2NH, has been characterized. Amides are much better known and have been shown to have the properties expected from a sort of averaging of the properties of the nitride and hydride ions. They are formed by heating the metal in a stream of ammonia gas,

$$2Na_{(s)} + 2NH_{3(g)} \rightarrow 2NaNH_{2(s)} + H_{2(g)}$$

or by heating the hydride in a stream of ammonia,

$$NaH_{(s)} + NH_{3(g)} \rightarrow NaNH_{2(s)} + H_{2(g)}$$

They react vigorously with protic solvents, reflecting their strong electron-donor properties,

$$NH_2^- + HOH \rightarrow NH_3 + OH^-$$

and are convertible into the other nitrogen negative ions,

$$NaNH_{2(s)} + C_{(s)} \xrightarrow{600°} NaCN_{(s)} + H_{2(g)}$$

$$2NaNH_{2(s)} + CO_{2(g)} \longrightarrow H_2NCN_{(s)} + 2NaOH_{(s)}$$

$$NaNH_{2(s)} + NNO_{(g)} \longrightarrow NaN_{3(s)} + H_2O_{(g)}$$
$$\text{(nitrous oxide)}$$

In these last reactions, notice that the total number of formula weights of ionic-lattice solids does not change, so that the negative free-energy change is a combination of small lattice-energy changes plus entropy terms, which are significant at the high temperatures involved. The amide ion is not a good reducing agent even though it is a good electron donor; this paradox is probably due to the absence of a mechanism for forming a convenient oxidation product such as **hydrazine**, H_2N—NH_2. When an electrical current is passed through molten amides (showing that they are indeed ionic) the metal ion is reduced at the cathode to the element, but only nitrogen gas and hydrogen gas appear at the anode from the oxidation of NH_2^-. It is often true that characteristic patterns of reaction are dictated as much by available mechanisms as by overall thermodynamic stability. Thus it is very easy for covalent azides to explode into N_2 because the two N atoms are already next to each other, but covalent nitrides are not usually explosive because no mechanism exists for getting the two atoms in N_2 together.

Ionic Oxides

Oxygen is quite electronegative and does not have an extremely unfavorable electron affinity for forming the ion O^{2-}, so we expect to find numerous examples of stable ionic systems containing the monatomic ion. These are known for all of the elements in the shaded portion of Fig. 13-7. Even though oxygen is very electronegative, the 2− charge destabilizes the ion's electrons enough to make the ion's conjugate acid, H_2O, very weak; when ionic oxides are placed in protic solvents they react vigorously:

$$BaO + HOH \rightarrow Ba(OH)_2$$

The oxide ion has virtually no reducing properties, which again may represent the absence of a mechanism for generating O_2. Indeed, several unusually high formal oxidation states of metals are stable as the solid oxide without being substantially reduced, such as AgO, MnO_2, and PbO_2. This is not true for the ionic oxides, however, since in all of them the metal is already in its highest chemically possible oxidation state. We have already discussed, in Chapter 12, the designation of oxides as acidic or basic, which is decided by the other atom in the compound, since the oxide ion is always a strong base. All ionic oxides are basic, since these positive ions have too low a charge to strongly polarize the O—H bonds in the water molecules surrounding them in solution.

Polyatomic Ions As in the previous cases, it is possible to make an oxygen-containing ion with lower charge by allowing the oxygen atoms to combine. Just as we saw the formation of C_2^{2-} and N_3^-, we can observe the formation of O_2^{2-}, the **peroxide** ion; O_2^-, the **superoxide** ion; and O_3^-, the **ozonide** ion. All the alkali metals (group Ia) and also Ca, Sr, and Ba form ionic peroxides, but only Na, K, Rb, and Cs form superoxides, and ozonides are known only for K, Rb, and Cs. Most of the salts of these ions can be formed by heating the metal's oxide in either oxygen or **ozone**, O_3:

$$2BaO + O_2 \xrightarrow{500°} 2BaO_2$$

$$4KOH + 3O_2 \xrightarrow[\text{pressure}]{375°} 4KO_2 + 2H_2O$$

$$6KOH + 4O_3 \xrightarrow{-15°} 4KO_3 + 2KOH \cdot H_2O + O_2$$

All of these ions show oxygen in a formal oxidation state between 0 and 2−, so in principle they could serve either as reducing or as oxidizing agents; but in fact the great electronegativity of O and the favorable lattice energies of systems involving the O^{2-} ion cause all of the three polyatomic ions to serve as strong oxidizing agents:

$$3Na_2O_2 + Fe \rightarrow Na_2FeO_4 + 2Na_2O$$

$$2KO_2 + H_2 \rightarrow 2KOH + O_2$$

Only the peroxide ion ever serves as a reducing agent, and then only with the most powerful oxidizing agents:

$$16H_3O^+ + 2MnO_4^- + 5O_2^{2-} \rightarrow 2[Mn(OH_2)_6]^{2+} + 5O_2 + 12H_2O$$

This behavior is understandable if we consider that O^{2-} has acquired two extra electrons per atom, O_2^{2-} one extra electron per atom, O_2^- one-half, and O_3^- only one-third. Since electronegativity is an electron-acquiring tendency, the donor ability (reducing tendency) will be smaller if fewer electrons have been acquired. Accordingly, O_2^{2-} is a poorer donor than O^{2-} and its conjugate acid, H_2O_2, is a stronger acid than H_2O ($K_a = 2 \times 10^{-12}$). The other two ions oxidize water:

$$O_2^{2-} + 2H_2O \rightarrow H_2O_2 + 2OH^-$$

but

$$2O_2^- + 2H_2O \rightarrow H_2O_2 + 2OH^- + O_2$$

and

$$2O_3^- + 2H_2O \rightarrow H_2O + 2OH^- + \tfrac{3}{2}O_2$$

Heteronuclear Ions There is a large number of oxygen-containing ions in which the O atom or atoms are on the outside of the ion and some more electropositive atom on its inside; these are the most familiar polyatomic ions:

BO_3^{3-}	borate (and polyborates)
CO_3^{2-}	carbonate
$CH_3CO_2^-$	acetate
NO_3^-	nitrate
NO_2^-	nitrite
SiO_3^{2-}	silicate (and polysilicates)
PO_4^{3-}, PO_3^{3-}	phosphate, phosphite (and polyphosphates)
SO_4^{2-}, SO_3^{2-}	sulfate, sulfite (and polysulfates)
ClO_4^-, ClO_3^-, ClO_2^-, ClO^-	perchlorate, chlorate, chlorite, hypochlorite
AsO_4^{3-}, AsO_3^{3-}	arsenate, arsenite
SeO_4^{2-}, SeO_3^{2-}	selenate, selenite
BrO_4^-, BrO_3^-	perbromate, bromate
IO_4^-, IO_3^-	periodate, iodate

Even some metals form oxyanions, although not the most electropositive ones:

VO_4^{3-} vanadate (and polyvanadates)
CrO_4^{2-} chromate (and dichromate)
MnO_4^-, MnO_4^{2-} permanganate, manganate
FeO_4^{2-} ferrate

And finally, the electronegativity of O is sufficiently great that it can form XeO_6^{4-}, the perxenate ion.

The preparation of these ions involves a number of techniques, depending on the properties of the ion; perhaps the most common is the reaction of the other element's oxide with the oxide ion or with water. Of course, a number of these ions occur naturally in the earth's crust and require no preparation.

The properties of these oxyanions are determined largely by the overall charge on the ion and by the central atom's electronegativity. The larger the charge and the more electropositive the central atom, the better electron donor the ion will be. The better electron donor the ion is, the weaker its conjugate acid will be; compare this generalization with the K_a values in Table 13-6. In a number of cases, such as nitrate and nitrite, the charge and central atom electronegativity are the same. In this case we need to look at the formal oxidation state of the central atom; although this will not be a realistic estimate of the charge on that atom, it will give a correct idea of the availability of electrons within the ion. The higher this formal (positive) oxidation state, the poorer electron donor the ion will be, and the stronger its conjugate acid will be. This statement should also be compared with Table 13-6.

Table 13-6
Relative Strengths of Oxyanion Acids

Ion	Charge	Electronegativity of Central Atom	Oxidation State of Central Atom	K_a of Conjugate Acid
NO_3^-	-1	3.07	5+	2×10^1
NO_2^-	-1	3.07	3+	5×10^{-4}
ClO_4^-	-1	2.83	7+	large (10^7?)
ClO_3^-	-1	2.83	5+	large (10^3?)
ClO_2^-	-1	2.83	3+	1×10^{-2}
ClO^-	-1	2.83	1+	3×10^{-8}
CO_3^{2-}	-2	2.50	4+	6×10^{-11}
SO_4^{2-}	-2	2.44	6+	1×10^{-2}
SO_3^{2-}	-2	2.44	4+	6×10^{-8}
PO_4^{3-}	-3	2.06	5+	5×10^{-13}
AsO_4^{3-}	-3	2.20	5+	3×10^{-12}

The redox behavior of these ions proceeds from much the same sort of reasoning as before. If the central atom is very electronegative or has a very high positive oxidation state, the ion will be a strong oxidizing agent for the same reasons that the polyatomic oxide ions were. If the central atom is not too electronegative *and* is not in its highest formal oxidation state, the electron-donor properties of the ion will be enhanced and it will be able to serve as a moderate reducing agent, although the electronegativity of the oxygens keeps these ions from ever being extremely strong reducing agents. Table 13-7 gives some reduction potentials for comparison purposes, but remember that all or nearly all of these will be dependent on the acidity of the solution being used, through the Nernst equation. Thus, as we saw in Chapter 10, perchlorate is a powerful oxidizing agent in very acid solution but has virtually no oxidizing properties in neutral solution; the same reasoning applies to many of the others.

Table 13-7
Relative Oxidizing Strengths of Oxyanions

Ion	Electronegativity of Central Atom	Oxidation State of Central Atom	Reduction Potential (V)
NO_3^-	3.07	5+	0.96 (to NO)
NO_2^-	3.07	3+	1.00 (to NO)
ClO_4^-	2.83	7+	1.37 (to Cl^-)
ClO_3^-	2.83	5+	1.45 (to Cl^-)
ClO_2^-	2.83	3+	1.57 (to Cl^-)
ClO^-	2.83	1+	1.49 (to Cl^-)
CrO_4^{2-}	1.56	6+	1.20 (to Cr^{3+})
CO_3^{2-}	2.50	4+	-0.20 (to HCOOH)
SO_4^{2-}	2.44	6+	0.36 (to S)
SO_3^{2-}	2.44	4+	0.45 (to S)
PO_4^{3-}	2.06	5+	-0.28 (to H_3PO_3)
AsO_4^{3-}	2.20	5+	0.56 (to $HAsO_2$)
XeO_6^{4-}	(≈ 3?)	8+	3.0 (to XeO_4^{2-})

Reduction potentials and other thermodynamic data allow us to calculate equilibrium constants for possible redox or acid–base reactions involving these ions. As a very rough approximation, however, nitrate, all of the halogen oxyanions, chromate, the manganates, ferrate, and perxenate are oxidizing agents, with the strongest being perxenate, perbromate, permanganate, and perchlorate; nitrite, phosphite, arsenite, and sulfite are all weak reducing agents.

When one of these oxyanions serves as an oxidizing agent it is reduced, of course. The reduction product may vary under different reaction conditions, in a predictable way. Thus if concentrated nitric acid dissolves (and oxidizes)

Cu, the formal oxidation state of N decreases only from 5 to 4:

$$Cu^0 + 2NO_3^- + 4H_3O^+ \rightarrow Cu(OH_2)_4^{2+} + 2NO_2 + 2H_2O$$

On the other hand, if dilute nitric acid dissolves Cu there are relatively fewer NO_3^- ions present to accept electrons, and each accepts more, reducing its formal oxidation state further:

$$3Cu^0 + 2NO_3^- + 8H_3O^+ \rightarrow 3Cu(OH_2)_4^{2+} + 2NO$$

Pyrolysis Another property shared by many of the oxyanions is that of **pyrolysis** or thermal decomposition to give the ionic oxide lattice and the volatile oxide of the other element. The pyrolysis of oxyanions is still not fully understood, but at least the following ions pyrolyze to an ionic lattice as above: carbonate, acetate, nitrate, nitrite, and sulfate. The best known of these reactions is that of the carbonate, which for the group IIa cations shows an interesting periodic trend. The reaction is

$$MCO_{3(s)} \xrightarrow[\text{heat}]{\Delta\Delta\Delta} MO_{(s)} + CO_{2(g)}$$

When M = Be, the temperature at which the carbonate gives off CO_2 at 1 atm pressure is approximately room temperature (the basic carbonate or oxycarbonate forms) or 25 °C. When M = Mg, the corresponding temperature is 540 °C; for Ca, 900 °C; for Sr, 1290 °C; for Ba, 1360 °C. The steady increase in stability must be exclusively an enthalpy effect, since the entropy change must be comparable for all the reactions (1 mole of gas being generated). Like most of the other properties of ionic solids, it is related to the lattice energy. The carbonate ion is relatively large, the oxide ion small; so on going from the large 2− ion to the small 2− ion, the lattice energy increases (r in denominator). But when the cation is also large, the relative effect of shrinking the anion is much smaller, and the lattice energy changes less. Therefore more thermal energy must be put into the barium carbonate lattice to decompose it than into the beryllium or magnesium carbonate lattices. That is, ΔG^0 is positive at room temperature (since the reaction does not occur at 25 °C) and it can be made negative only by going to a temperature high enough to make $T\Delta S^0$ large; when the lattice-energy effect does not help much, the temperature must be made higher.

Solubility As we commented earlier, the solubility trends in these ionic salts of oxyanions can also be predicted approximately. Nearly all the oxyanions with 1− charge will form soluble salts, since the hydration energy of the ions will usually exceed the product-of-the-charges' effect on the lattice energy. Salts of oxyanions with 2− charge will usually be water soluble if the positive ion has a 1+ charge, but not if it is higher, since the lattice energy increases too much. For 3− oxyanions, even the 1+ cations' salts usually are not too soluble; thus LiCl dissolves up to approximately $1\,M$, but Li_3PO_4 is saturated at $0.003\,M$. For a given oxyanion and cation charge, the lighter cations will usually form the more soluble salts, since the smaller cations have higher hydration energies but the interionic distance in the lattice-energy expression is largely fixed by the large oxyanion at a more or less constant value.

Protonated Ions Finally, there is the possibility, as for nitrides, of forming an oxide–hydride negative ion — OH^-, the hydroxide ion. Although the negative charge is not high, the presence of the relatively electropositive H makes the hydroxide ion a good electron donor, and thus its conjugate acid, HOH, a weak acid. The reducing power of OH^- is slight:

$$4OH^- \rightarrow O_2 + 2H_2O + 4e^- \qquad \mathscr{E}^0_{ox} = -0.40 \text{ V in } 1\text{-}M \text{ } OH^-$$

but it will react with strong oxidizing agents such as MnO_4^-:

$$4OH^- + 4MnO_4^- \rightarrow O_2 + 4MnO_4^{2-} + 2H_2O$$

We have mentioned earlier, in connection with the "leveling effect," that the OH^- ion is the strongest base that can exist in aqueous solution. It follows that the oxygen electrons in OH^- form very stable hydrogen bonds with surrounding water molecules; solid ionic hydroxides such as KOH are fairly good drying agents for this reason, and it is difficult to form crystals of ionic hydroxides without some rather nonstoichiometric amount of hydrated water. This hydrogen-bonding tendency also leads to a capacity for coordinated (i.e., partially covalent) hydroxides to polymerize; we shall look at this in the next chapter. As a strong base (or nucleophile) the OH^- ion often serves as a catalyst in reactions of organic or other covalent compounds with water:

$$\left[\text{R}-\text{C}\begin{array}{c}\nearrow \text{O} \\ \searrow \text{OCH}_3\end{array} \longleftrightarrow \text{R}-\overset{+}{\text{C}}\begin{array}{c}\nearrow \text{O}^- \\ \searrow \text{OCH}_3\end{array} \right]$$

$$+ \text{ OH}^- \longrightarrow \left[\text{R}-\underset{\text{OCH}_3}{\overset{\text{O}^-}{\underset{|}{\overset{|}{\text{C}}}}}-\text{OH} \right] \longrightarrow \text{R}-\underset{\text{OH}}{\overset{\text{O}}{\underset{|}{\overset{\|}{\text{C}}}}} \text{ } + \text{ OCH}_3^- \xrightarrow{H_2O} CH_3OH + OH^-$$

Here its role is to stabilize a positive carbon ion as it forms.

Ionic Sulfides

In contrast to the trend of electronegativities we have been following, sulfur is less electronegative than oxygen. The sulfide ion, S^{2-}, is also larger and correspondingly more polarizable than O^{2-}. Both of these characteristics tend to make the sulfide ion a better electron donor than the oxide ion. Accordingly, ionic sulfides are less common than ionic oxides and occur only for the most electropositive elements, groups Ia and IIa; the sulfide ion resembles the hydride ion in this regard. Because of its donor properties, the sulfide ion hydrolyzes extensively in water:

$$S^{2-} + H_2O \rightleftharpoons HS^- + OH^- \xrightarrow{H_2O} H_2S + OH^-$$

The solution species, H_2S, is a much better reducing agent than the corresponding oxide solution species, H_2O:

$$2Fe(OH_2)_{6(aq)}^{3+} + H_2S_{(g)} \xrightarrow{H_2O} 2Fe(OH_2)_{6(aq)}^{2+} + 2H_3O_{(aq)}^+ + S_{(s)}^0$$

Polyatomic Ions Like the other ions we have discussed (except hydrogen), sulfur forms polyatomic negative ions. These are prepared by dissolving sulfur in solutions of the sulfide ion that are basic enough to prevent complete hydrolysis:

$$S^{2-} + (n-1)S^0 \rightleftharpoons S_n^{2-}$$

As we shall see in Chapter 15, sulfur **catenates**—it forms extended chains of atoms bonded to other atoms of the same element—better than any other element except carbon; thus compounds have been characterized from the above reaction with $n = 1, 2, 3, 4,$ and 5. The **disulfide** ion, S_2^{2-}, which is analogous to the peroxide ion, is probably the best known, although S_4^{2-} is probably more stable. Again by analogy with the peroxide ion (as well as the other polyatomic negative ions), the polysulfides are poorer electron donors and H_2S_n is a progressively stronger acid than H_2S.

Heteronuclear Ions Several ions can be prepared in which one or more of the O atoms on an oxyanion have been replaced by S atoms. These are named by placing *thio-* before the ion's name. Examples are **thiosulfate**, SSO_3^{2-}; **thiocyanate**, SCN^-, which is analogous to **cyanate**, OCN^-; and **trithiocarbonate**, CS_3^{2-}. These have, in general, the properties of the oxyanions but are better reducing agents (electron donors) due to the presence of the less electronegative S atom or atoms. None are strong oxidizing agents. The stability of covalent bonding between O and S and the ease of replacement of O atoms by S as above lead to the formation of a remarkable number of S oxyanions, some of which are shown in Table 13-8. The peroxo ions are strong oxidizing agents, sulfate and its polymers are not very reactive toward either oxidants or reductants, and most of the rest of the ions are either weak or strong reducing agents.

Table 13-8
Sulfur Oxyanions

SO_3^{2-} sulfite	
SO_4^{2-} sulfate	
$S_2O_3^{2-}$ thiosulfate	
$S_2O_4^{2-}$ dithionite	
$S_2O_5^{2-}$ pyrosulfite	
$S_2O_6^{2-}$ dithionate	
$S_2O_7^{2-}$ pyrosulfate	
SO_5^{2-} peroxymonosulfate	
$S_2O_8^{2-}$ peroxydisulfate	

Ionic Halides

We can consider the halogen anions together, although there is a substantial electronegativity difference between fluorine and iodine. Chlorine, bromine, and iodine have electronegativities sufficiently close together that they can be considered as a class, and we shall look at the fluorides along with these and also separately. The monatomic halide ions F^-, Cl^-, Br^-, and I^- are all reasonably electronegative (F^-, of course, extremely so) and they do not have a high negative charge, so they are relatively poor electron donors. Their conjugate acids are all fairly strong or very strong, although the precise nature of these acids is a balance of complex factors, as we have seen. They show little or no reducing power, although iodide and bromide react with moderate oxidizing agents:

$$2Cu(OH_2)_6^{2+} + 4I^- \rightarrow 2CuI + I_2 + 12H_2O$$

$$Cl_2 + 2Br^- \rightarrow Br_2 + 2Cl^-$$

and Cl^- with strong oxidizing agents:

$$4H_3O^+ + 2Cl^- + MnO_2 \rightarrow Cl_2 + Mn(OH_2)_6^{2+}$$

The elements in the shaded portion of Fig. 13-7 form ionic halides, although as might be expected from the variations in cation size and anion electronegativity there are differences. Table 13-9 indicates the trends in melting point of the alkali halides. To the extent that the charges on the ions are reduced by polarization (and resulting covalency), we expect the high melting point that is characteristic of ionic salts with their high lattice energies to be lowered. Notice that the smallest cation, Li^+, polarizes the "softest" anion, I^-, enough to lower its melting point more than 400° below that of the ionic LiF. On the other hand, the large Cs^+ ion has a very small effect on any of the halide ions, and CsI melts nearly as high as CsF.

Table 13-9
Melting Points of Alkali Halides (°C)

	Li	Na	K	Rb	Cs
F	870	997	880	760	684
Cl	613	801	776	715	646
Br	547	755	730	682	636
I	446	651	723	642	621

Polyatomic Ions The halogens also form polyatomic negative ions, but they differ substantially in ability to form these. Only Br and I form homonuclear polyanions: Br_3^- and I_3^-, I_5^-, I_7^-, and I_9^-. Cl_3^- may exist but is very unstable, and there is no evidence for F_3^-. The greater the electronegativity of an atom, the less stable its polyanions will be, since it will seek to add electrons to form the monatomic ion:

$$X_3^- + 2e^- \rightarrow 3X^- \begin{cases} \mathscr{E}^0_{I_3^-} = 0.536 \text{ V} \\ \mathscr{E}^0_{Br_3^-} = 1.06 \text{ V} \end{cases}$$

Heteronuclear Ions There is a greater number of negative ions that involve mixed halogens: ClF_2^-, ClF_4^-, $ClBr_2^-$, BrF_4^-, BrF_6^-, $BrCl_2^-$, IF_4^-, IF_6^-, ICl_2^-, ICl_4^-, IBr_2^-, $IFCl_3^-$, $IFBr^-$, and $IClBr^-$. All of these polyanions are made by adding excess halogen to a crystallizing solution of the simple alkali salt. They decompose when heated, giving off the gaseous halogen or interhalogen compound and leaving the simple halide behind:

$$CsI_3 \rightarrow CsI + I_2 \quad 250°$$
$$CsIBr_2 \rightarrow CsBr + IBr \quad 242°$$
$$CsICl_2 \rightarrow CsCl + ICl \quad 209°$$

The smaller monatomic halide ion provides a more favorable lattice energy for the product, and the greater the favorable change in lattice energy, the lower the temperature required to achieve the change; thus the chloride is formed at a lower temperature than the bromide, which is formed at a lower temperature than the iodide.

A number of other ions form that are analogous to the oxyanions in that they represent the halogen on the outside of the system and a more electropositive atom in the center: BeF_4^{2-}, BF_4^-, SiF_6^{2-}, PF_6^-, PCl_6^-, and HgI_4^{2-}, for instance. Most are prepared by adding halide ion to a covalent halide, analogous to the polyhalide ions. Many other examples exist in which halides are coordinated to transition metals, but we shall defer discussion of these until Chapter 16.

With respect to their acid–base behavior, ionic salts of these haloanions show one of three responses toward hydrolysis: (1) they are insoluble, which is usually a reflection of the ion charge relationships mentioned earlier; (2) they dissolve without hydrolysis to give the anion of a strong acid in solution; or (3) they hydrolyze the halides off, producing an oxyanion or more rarely a hydrated cation. We shall look at the fluoroanions separately; most of the others hydrolyze the halides off if the central atom is sufficiently electronegative or in a sufficiently high formal oxidation state to make an oxyanion stable. Since the bonding within these ions is predominantly covalent, their stability is strongly influenced by the orbital overlap; as we showed in Chapter 12 (Fig. 12-20), large atoms with $n = 5$ or 6 for the valence electrons overlap other AO's poorly, and as Fig. 13-8 shows, they overlap other AO's with large n even more poorly than those with small n. Going from a metal–iodine bond to a metal–oxygen bond, then, improves the overlap and the bonding effect. This change is much

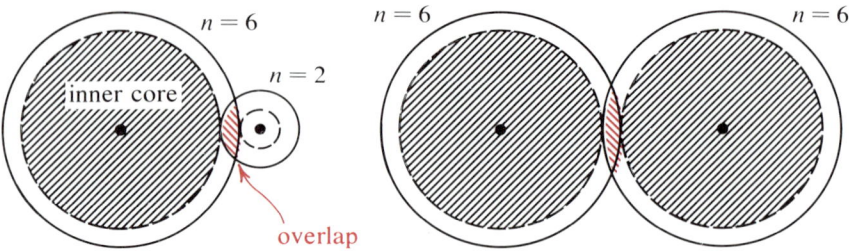

Figure 13-8 Poor overlap by s orbitals with large n and large inner core. Overlap is the fraction of total area covered by the red shading in this schematic representation.

less pronounced for chloride anions and nonexistent for fluoroanions, so we can understand the trend of stability for halostannates:

$$SnF_6^{2-} + H_2O \rightarrow \text{N.R.} \quad \text{(no reaction)}$$

$$SnCl_6^{2-} + H_2O \rightarrow \text{N.R.}$$

$$SnBr_6^{2-} + \begin{cases} H_2O & \rightarrow Sn(OH)_6^{2-} \\ CH_3OH & \rightarrow \text{N.R.} \end{cases}$$

$$SnI_6^{2-} + \begin{cases} H_2O & \rightarrow Sn(OH)_6^{2-} \\ CH_3OH & \rightarrow \text{alcoholate} \end{cases}$$

stability ↑

Mercury, on the other hand, because of its low electronegativity and low charge, will not form a covalent bond with light electronegative elements at all unless they are quite polarizable. Thus it will not form a fluoroanion at all, and the stability of HgX_4^{2-} *increases* with the atomic weight of the halogen; HgI_4^{2-} is most stable because iodide is the most polarizable of the halide ions (and has the lowest hydration energy).

Ionic Fluorides

Fluorine's very great electronegativity places it in a class by itself. As Fig. 13-7 shows, many metals will form ionic fluorides when all their other halides are partially or substantially covalent. Thus $AlBr_3$ melts at 97° and has a completely molecular crystal structure (as Al_2Br_6); $AlCl_3$ melts at 190° and has a layered crystal structure with little attraction between layers; but AlF_3 melts at 1040° and has the infinite-octahedral-network crystal structure typified by NaCl (although different because of the different stoichiometry). Comparable, if less pronounced, differences exist for most of the transition metals in low oxidation states. On the other hand, any element in a very high formal oxidation state has such a high polarizing power that even F will form covalent bonds; thus UF_6 is a gas at 60 °C. At this point we might mention again the generalization that a fluoride will be ionic if the other element's total ionization energy up to its formal oxidation state is less than about 17 eV per F atom in the formula.

Ionic fluorides have no reducing properties at all, since no other element attracts electrons strongly enough to take them away from F^-; they will serve as partly covalent electron donors to other elements that are fairly electronegative or have a high positive charge, as witness the ions BF_4^- and TaF_7^{2-}. When fluorides are water-soluble, F^- hydrolyzes substantially to HF, since as we have seen earlier the energy factors affecting the proton-transfer reaction from HF make it a moderately weak acid. However, since F^- is a small ion its lattice energy with small cations is very high, and with the lighter elements the fluoride tends to be the least soluble of the halides. In several of these cases the fluoride does not dissolve enough to hydrolyze significantly. This is particularly true for cations with a charge of 2+ or greater. For heavier cations with a large radius and small charge (so that the lattice energy is not too large) the fluoride is usually the *most* soluble of the halides, due to its large hydration energy. Thus LiF and CaF_2 are nearly insoluble—much less soluble than LiCl and $CaCl_2$—but TlF is more than a hundred times as soluble as TlCl on a molar basis.

No polyfluorides exist because of the great electronegativity of F. Among halogen fluoride anions the highest chlorine-centered ion is ClF_4^-, but bromine and iodine make BrF_6^- and IF_6^-; to stabilize the molecule-ion a fairly low electronegativity is apparently necessary on the central atom to provide some partially ionic bonding. If the central atom is as electronegative as Cl, it cannot donate electrons to a large number of F's. In keeping with this idea, N and O do not form any fluoro anions, and the SF_5^- ion, like the ClF_4^- ion, is rather unstable and a strong fluorinating agent. Other fluoro anions, such as BF_4^-, SiF_6^{2-}, and PF_6^-, are much weaker fluorinating agents and are presumably stabilized by having less electronegative central atoms that donate to the F's better. When these fluoro anions are exposed to water, they normally dissolve without any significant hydrolysis of the F's, which is quite different from most of the other halo anions; since the F orbitals and O orbitals are essentially the same size, the bonding within the ion is not noticeably improved by replacing F by O in hydrolysis. Usually they can be hydrolyzed by treatment with base, since the hydrolysis reaction generates H_3O^+. The reaction is not very vigorous, however, and partial hydrolysis to such ions as $[SnF_5OH]^{2-}$ is possible.

Polyatomic Positive Ions

The positive ions in all of our discussion so far have been the ions of groups Ia, IIa, and sometimes IIIa, all of which are monatomic and are fairly simply characterized. But just as we have formed polyatomic anions, we can form polyatomic cations. In order for such a cation to be stable, it must not contain any extremely electronegative atoms, since they will not withstand the approach of a negative ion without accepting electrons from it in a partially covalent bond. On the other hand, it cannot contain any extremely electropositive atoms either, because they would achieve greater stability by forming monatomic ions in a lattice. And to be stable, it must be composed of relatively light elements to get favorable overlap of AO's. The most familiar ion meeting these requirements is NH_4^+, the **ammonium** ion. There are others that are less stable or whose compounds are less ionic in nature, however: PH_4^+, PCl_4^+, NO_2^+, VO^{3+}, I_3^+, and others. The H_3O^+ ion can even be found in an apparently ionic lattice in the monohydrate of perchloric acid, $HClO_4 \cdot H_2O$ or $[H_3O]^+[ClO_4]^-$. By far the largest number of these cations arise from the substitution of a carbon-based organic molecular fragment for one or more of the hydrogens on NH_4^+, however, and these are named as substituted ammonium salts: $CH_3NH_3^+$ is **methylammonium**, and so on.

The ammonium ion is related to ammonia and the amide ion in the same sense that the hydronium ion is related to water and hydroxide ion; thus titrations can be performed in liquid ammonia using an ammonium salt as the acid and an amide as the base. Pursuing this analogy, the amide ion reacts vigorously with water, ammonia reacts slightly (to produce a little NH_4^+ and OH^-), but ammonium ion does not react in the same sense at all, and has very little tendency to hydrate. The ammonium ion is quite small, having an ionic radius (1.43 Å) near that of rubidium, and consequently the lattice energies of ammonium salts resemble those of rubidium (or potassium) salts, as do the solubilities of ammonium salts. The ammonium ion in solution shows no strong redox properties, but in an ionic salt with a strongly oxidizing anion it can be oxidized to the very stable N_2 molecule, with considerable energy release. The result is that while ammonium nitrate is a valuable fertilizer, it is also widely used as a mining explosive. For the same reasons, ammonium perchlorate is commonly used in solid rocket propellants, although $NO_2^+ClO_4^-$, **nitronium perchlorate**, is even more energetic.

When ammonium salts are heated, they do not vaporize either molecular species having the formula of the salt or gaseous ions, but rather undergo a chemical reaction to give gaseous ammonia and the hydride of the negative ion—usually an acid. Thus:

$$NH_4Cl_{(s)} \xrightarrow{\Delta\Delta} NH_{3(g)} + HCl_{(g)}$$

$$HN_4NO_{3(s)} \xrightarrow{\Delta\Delta} NH_{3(g)} + HNO_{3(g)}$$

A bottle of ammonium carbonate always smells of ammonia when opened:

$$(NH_4)_2CO_{3(s)} \xrightarrow{\text{room temp.}} 2NH_{3(g)} + CO_{2(g)} + H_2O_{(g)}$$

With care, ammonium salts with oxidizing anions can be caused to oxidize partially on heating:

$$NH_4NO_{3(l)} \xrightarrow{\Delta\Delta} N_2O + 2H_2O$$

This is the usual means of preparing nitrous oxide, N_2O.

Substituted ammonium salts are useful to the chemist in several applications. They offer a convenient way to crystallize organic amines, R—NH_2 compounds, since the substituted ammonium salt R—$NH_3^+Cl^-$ forms when HCl is added to the amine. The very large ammonium ions such as tetrabutylammonium, $(CH_3CH_2CH_2CH_2-)_4N^+$, have very low lattice and hydration energies because of their large radii and are useful ways to get desired anions such as OH^- into relatively nonpolar, nonprotic solvents. Thus, using a solution of tetrabutylammonium hydroxide in benzene, acids can be titrated in solvents such as methyl isobutyl ketone that show no levelling effect (see Chapter 10). And conversely, when it is desired to crystallize some very large anion such as tetraiodocobaltate(II), $[CoI_4]^{2-}$, the bulky tetrabutylammonium ion allows the formation of a stable crystal without undue anion–anion contact, which would occur with smaller cations; it often forms crystalline products with large anions when smaller cations lead only to decomposition of the desired anion.

13-6 Summary

In this chapter we have discussed the models chemists use to describe solids in which complete or nearly complete charge separation occurs. We have developed a simple model for the ionic crystal that allows us to explain and predict the large lattice energies these systems display, and we have also seen that the close-packed feature of ionic crystals leads to the prediction of increased charge separation, using differential ionization energies. We can thus account for the striking differences between crystalline and gaseous ionic compounds such as the alkali halides. We have also used a molecular version of Hess' law, the Born–Haber cycle, to analyze the various energy factors surrounding the reactions of some ionic systems, and have extended this to describe hydration and solvation energies. We have developed a simple technique for predicting approximate values of single-ion hydration energies. It has become apparent that the balance between lattice energies and hydration energies governs the solubility of ionic compounds, and indeed that lattice-energy relationships have a profound effect on nearly all of the chemical and physical properties of ionic compounds.

712 | Ionic Compounds

We have described a little of the descriptive chemistry of ionic compounds, attempting to group the compounds by electronegativities, and have seen substantial regularities develop—patterns of behavior that are consistent from hydrides to carbides to nitrides and so on; and we have seen that the differences between these groups are also patterned and predictable from our earlier background.

We need now to go on to compounds that show only partial charge separation. The fact that covalent bonding is important in these compounds will give rise to geometric differences, and will also lead to some striking differences in chemical properties.

Study Problems

1. Note in Table 13-1 that the Madelung constants for different crystal lattices are approximately constant *per ion* in the compound's formula. Obtain an average value for the Madelung constant per atom from the six values in the table and derive an approximate expression for U in terms of a single numerical constant, the number of ions in the formula, the number of electron charges on each ion (e.g., 2+, 1−), and the ionic radii. Include in the constant an average value for the Born repulsion exponent, and adjust the constant to kilocalories per mole. How well does your approximation match the Born–Haber cycle value of $U = -229.7$ kcal/mole for LiF?
2. Show by using a Born–Haber cycle that ScH_3, as an ionic compound, would have a positive enthalpy of formation. The electron affinity of H is 0.72 eV.
3. Should BH_4^- or BeH_4^{2-} be the better reducing agent?
4. What molecular-structure influences relate to the fact that carbon and oxygen predominantly form diatomic anions, while nitrogen and iodine form triatomic anions, as C_2^{2-} and O_2^{2-}, but N_3^- and I_3^-?
5. The K_{sp} values for the group IIa hydroxides are approximately as follows:

$$Be(OH)_2 \quad K_{sp} = 10^{-26}$$
$$Mg(OH)_2 \quad K_{sp} = 10^{-11}$$
$$Ca(OH)_2 \quad K_{sp} = 10^{-5}$$
$$Sr(OH)_2 \quad K_{sp} = 10^{-3}$$
$$Ba(OH)_2 \quad K_{sp} = 10^{-2}$$

Comment on the reasons for this trend. Is it consistent with the trend in hydration energies?

6. Consider the reaction

$$CH_3-Na + CH_3-I \rightleftharpoons Na-I + CH_3-CH_3$$

Suppose all four species are isolated molecules, so that the difference in bond energies is essentially the only component of ΔH^0. Using Pauling's electronegativity definition, calculate ΔH^0; assuming $\Delta H^0 \cong \Delta G^0$, calculate K_{eq} for the reaction at room temperature. Now calculate U for NaI as an ionic crystal and recalculate ΔH^0 and K_{eq} including this effect. Is the difference important?

7. How should liquid HF compare with water as a solvent for ionic crystalline solids? What about liquid SF_6?

8. Lithium reacts with O_2 to give Li_2O; sodium gives Na_2O_2, and potassium, rubidium, and cesium give KO_2, RbO_2, and CsO_2. How does ionic radius affect this trend in properties?

9. Discuss the following series of reduction potentials in terms of the bonding and electron distribution in the ions involved:

$$S_2O_8^{2-} + 2e^- \rightarrow 2SO_4^{2-} \qquad \mathscr{E}^0 = 2.01 \text{ V}$$

$$SO_4^{2-} + H_2O + 2e^- \rightarrow SO_3^{2-} + 2OH^- \qquad \mathscr{E}^0 = -0.93 \text{ V}$$

$$2SO_3^{2-} + 2H_2O + 2e^- \rightarrow S_2O_4^{2-} + 4OH^- \qquad \mathscr{E}^0 = -1.12 \text{ V}$$

$$2SO_3^{2-} + 3H_2O + 4e^- \rightarrow S_2O_3^{2-} + 6OH^- \qquad \mathscr{E}^0 = -0.58 \text{ V}$$

10. Is the pattern of melting points below consistent with our ideas of the nature of bonding in halides?

KF	CaF_2	GaF_3	GeF_4	AsF_5	SeF_6
846°C	1360°C	800°C (sub)	−37°C (sub)	−80°C	−39°C

KI	CaI_2	GaI_3	GeI_4	AsI_5	
686°C	740°C	212°C	144°C	76°C	

11. HF, in very concentrated solutions on the order of 10 M, is a much stronger acid than its K_a value in dilute solution indicates. What is influencing the proton-transfer equilibrium?

12. In Lavoisier's *Elements of Chemistry* (1789) we learn that "Although the carbonic acid has less affinity with potash than any other acid, yet it is difficult to separate the last portions from it. The most usual method of accomplishing this is to dissolve the potash in water; to this solution add two or three times its weight of quicklime, then filtrate the liquor and evaporate it in close vessels; the saline substance left by the evaporation is potash almost entirely deprived of carbonic acid." Given that potash deprived of carbonic acid is KOH and that quicklime is CaO, what's going on here?

Some Further Reading

Gould, E. S., *Inorganic Reactions and Structure*, New York: Holt, Rinehart, and Winston, 1962 (rev. ed.). Chapter 12 of this book is a very readable introduction to lattice energies; Chapters 4 and 6 have much of the descriptive material on the chemistry of these systems. Particularly good problems.

Cotton, F. A., and Wilkinson, G., *Advanced Inorganic Chemistry*, New York: Wiley-Interscience, 1966 (2nd ed.). Most chemists regard this as somewhere between a text and a reference work. Chapter 2 has a good discussion of ionic systems, in a somewhat terse style, and the other chapters, arranged by elements, have an enormous amount of information about synthetic inorganic chemistry. Cotton and Wilkinson is useful in all the remaining chapters of this book.

Phillips, C. S. G., and Williams, R. J. P., *Inorganic Chemistry*, New York: Oxford, 1965. Chapter 5 has a good treatment of the energies of ionic systems, including hydration energies. In general, a very good book, although sometimes topics are hard to find.

Gurney, R. W., *Ions in Solution*, New York: Dover, 1962. The title describes the limited area of this book. Very readable and helpful in achieving a physical feel for the process of dissolving ionic substances.

Jolly, W. L., *The Chemistry of the Non-Metals*, Englewood Cliffs, N.J.: Prentice-Hall, 1966. A low-level discussion of the descriptive chemistry mentioned here, but not arranged in the same way. Also carries over to the next two chapters.

Bloom, H., *The Chemistry of Molten Salts*, New York: Benjamin, 1967. This is strictly a special topic, which we have scarcely mentioned. Some students may be interested in its expansion on some of the ideas of this chapter, however.

14 Semicovalent Compounds

We began our study of descriptive chemistry by looking at the chemistry of those compounds formed from elements with electronegativities so different that in a solid lattice they spontaneously transfer electrons to give essentially completely ionized, electrostatically bound systems. These are ionic compounds. We observed that the structures tend to be close packed, or as close to it as stoichiometry and ionic radii will allow, and that the chemistry of separate ions is relatively uncomplicated. Ions with the same charge rarely react with each other, since they are held apart by electrostatic repulsions. Ions with opposite charges tend to form lattices unless solvation energies keep them in solution. Ions with either a positive or a negative charge react with neutral molecules by inducing dipoles in them or by aligning permanent dipoles they may possess; this is the origin of solvation energies. In effect, a positive ion serves as an electron acceptor and a negative ion as an electron donor in this type of interaction; depending on the strength of the donor-acceptor interaction the only effect may be the aligning of molecules, or complete disruption of bonds may occur and a new, more-or-less covalent molecule may form.

All of these statements were made about compounds in which there was an electronegativity difference of (roughly) 2.0 units or more. Only if the electronegativity difference is of this order of magnitude will charge separation in a fairly close-packed lattice be essentially complete. We have seen that electronegativity is related to the differential ionization energies that we used to predict the approximate degree of charge separation. Suppose we go now to systems of atoms in which the electronegativity differences are less than 2.0 units. What sort of charge separations do we predict, and how important will the ionic bonding be? What differences in chemical and physical properties can we expect, based on the change in the chemical bonding?

14-1 Charge Separation in Partly Covalent Systems

We can get an idea of the degree of charge separation in these systems by picking out a few and calculating the charge on each positive ion in the same way we did in Study Problems 2 and 3 of Chapter 12 for CCl_4 and CF_4, and in Section 13-2 for LiF. We must remember that for extended crystals this is only a rough approximation, since we are ignoring all the atoms or ions except the nearest neighbors. Suppose we try aluminum nitride, AlN, which has an electronegativity difference of 1.60 units, nearly enough to qualify for our previous definition of an ionic compound. Aluminum nitride has the wurtzite structure shown in Fig. 4-36, with each Al surrounded by four N atoms and vice versa. Using our earlier differential ionization-energy expression, we can say as an approximation that the Al, as the central atom, is influenced by only one-fourth of the charge on each N. Then the condition for charge equilibrium in each Al—N bond is

$$DIE_{Al}(4q) = DIE_N(-q)$$
$$2.6 + 6.2(4q) = 6.8 + 11.9(-q)$$
$$36.7q = 4.2$$
$$q = 0.115$$
$$\text{charge on Al} = 4(0.115) = 0.46$$

This is a noticeably smaller charge than the 0.67 we calculated for LiI, even though the electronegativity difference is actually somewhat larger for AlN. This suggests again that electronegativity is a valuable guide but sometimes an oversimplification. We can go through the same sort of calculation for a number of other compounds with varying electronegativity differences, as shown in Table 14-1. In this table only compounds with 1:1 stoichiometry are considered, to keep electronegativity and stoichiometry effects from being mixed up. Although there is no sharp division it seems to be true that below an electronegativity difference of 1.0 unit the charges are very small, whereas above that value they are fairly substantial. This has an even greater effect on the bond or lattice energy relationships of the system, because the ionic energy requires

Table 14-1
Electronegativity Differences and Calculated Charge Separations

Compound	Electronegativity Difference	Charge (from DIE calc.)	Structure[a]
$NaCl_{(s)}$	1.82	0.80	N
$AlN_{(s)}$	1.60	0.46	W
$AgBr_{(s)}$	1.32	0.44	N
$LiI_{(s)}$	1.24	0.67	N
$CO_{(g)}$	1.00	0.15	G
$HgS_{(s)}$	1.00	0.01	Z
$BeS_{(s)}$	0.97	0.16	Z
$NO_{(g)}$	0.43	0.07	G
$C_{(s)}$	0	0	Z
$O_{2(g)}$	0	0	G

[a] N = NaCl structure (six neighbors), W = wurtzite structure (four neighbors), Z = zincblende structure (four neighbors), and G = gaseous diatomic molecule.

multiplying the two charges together — in effect squaring these fractional charges. Thus a fractional charge of $0.80e$ on each atom leads to a lattice energy 0.8×0.8 or 0.64 times as great as for complete charge separation; but a fractional charge of 0.10 on each atom leads to a lattice energy 0.1×0.1 or only 1% as great as for complete charge separation. Clearly for systems having small charges we can ignore the ionic attraction energies. The presence of a small charge is still quite important, since it often dictates a mechanism for reactions of a compound. However, for our purposes we can create another arbitrary division between partly ionic compounds and covalent compounds by taking the former as those compounds having an electronegativity difference of 1.0–2.0 units between adjacent atoms. These are the compounds we shall discuss in this chapter. It is immediately obvious that there is to be a third category, covalent compounds, having no bonds between atoms with electronegativity differences greater than 1.0 units. The three divisions we have set up, then, are ionic compounds, in which the bond energy is due almost solely to electrostatic attraction of positive ions for negative ions (electronegativity difference greater than 2.0 units); partly ionic compounds, in which this electrostatic attraction is about as important as covalent bonding (electronegativity difference 1.0–2.0 units); and covalent compounds, in which the effect of ionic attraction on the total energy of the molecule is negligibly small (electronegativity difference less than 1.0 unit).

Figure 14-1 shows the range of elements for which this partly ionic bonding can be expected. This is only a rough guide, and ought not to be considered as

Semicovalent Compounds

	H																He
He	Li	Be											B	C	N	O	F	Ne
Ne	Na	Mg											Al	Si	P	S	Cl	Ar
Ar	K	Ca	Sc	Ti	V	Cr	Mn	Fe	Co	Ni	Cu	Zn	Ga	Ge	As	Se	Br	Kr
Kr	Rb	Sr	Y	Zr	Nb	Mo	Tc	Ru	Rh	Pd	Ag	Cd	In	Sn	Sb	Te	I	Xe
Xe	Cs	Ba	La	Hf	Ta	W	Re	Os	Ir	Pt	Au	Hg	Tl	Pb	Bi	Po	At	Rn
Rn	Fr	Ra	Ac															

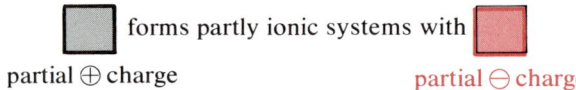 forms partly ionic systems with

partial ⊕ charge   partial ⊖ charge

Figure 14-1 Partly ionic compound formation in the periodic table. Properties vary depending on electronegativity differences.

having been handed down on stone tablets. For one thing, neighboring elements such as Se and Br will have electronegativities much closer together than 1.0 unit, and their compounds will be predominantly covalent. For another thing, it is even possible to have some partly ionic compounds of the alkali metals and alkaline earth metals, as we shall see. But this figure does provide a reasonable guide to the majority of the chemistry of these elements, and the exceptions are the ones we would expect to see. The partly negative elements are the same ones we observed as negative ions in Chapter 13, with the exception of fluorine. On the basis of electronegativities alone, there seems to be no basis for excluding F, but its great electronegativity gives it some unique properties in nonionic compounds that we shall return to in Chapter 15; very briefly, its nature makes its compounds either very ionic or—in effect—very covalent. The partly positive elements are those that were *not* shown as being positive or negative ions in the similar division of the periodic table for ionic compounds, Fig. 13-7. We have already seen that most of these form ionic fluorides, however.

If covalent bonding is to be significant in partly ionic compounds, what difference will this make in the chemistry of these compounds? Chemical reactivity is influenced by structural factors; what structural differences exist between an ionic solid and a partly ionic solid? And what kind of bonding is responsible for the very great thermal stability we observe for many of these compounds, such as SiO_2 (quartz), which melts at 1400–1700°C, depending on its exact crystal structure? These are areas we shall explore.

14-2 Lattice Geometry, Directional Overlap, and Lattice Energies

Sodium chloride is usually taken as the typical ionic crystal. In it, each atom has six nearest neighbors. This is somewhat unusual, because if an atom has only s and p valence orbitals it can form only four bonding MO's. If we insisted on using a MO model for solid NaCl we could perhaps speculate that since both Na and Cl have $3s$ and $3p$ valence orbitals, the $3d$ orbitals on each might be involved in the bonding so that more bonding MO's could be constructed. But LiF also has the NaCl structure, with 6:6 coordination (six neighbors around each kind of atom), and Li and F have only $2s$ and $2p$ valence orbitals; there is no $2d$ orbital, of course, and the $3d$ is too far away in energy to be significantly involved.

Covalence and Coordination Number

How do we account for the 6:6 coordination? The answer is that MO's describe covalent or at least significantly covalent bonding, but not ionic bonding. If charge transfer is complete, the electrons are described by only one atom's AO's and the resonance integral—the attraction of the electron mutually for two nuclei—becomes zero, so that there is no covalent bonding effect. Instead (and this is just the result of our earlier lattice-energy discussion) the atoms are held together in the crystal by their mutual electrostatic forces. Electrostatic attraction is **centrosymmetric**; that is, the Coulomb attraction $-q_1q_2/r$ does not depend on the direction of one charge from the other, but only on the distance separating them. This means that when the bonding is ionic, the number of nearest neighbors will be limited only by the repulsion between neighbors or ligand ions; the best compromise for NaCl and LiF between the favorable energy of a large number of ligand ions and the unfavorable energy of many (ligand ion)–(ligand ion) repulsions is 6:6 coordination. But this is entirely independent of any angular factors. There are no overlap problems to worry about.

In the partly ionic compounds we are now looking at, we cannot afford to take this simplified view; the charge separation is not complete. In Table 14-1 the calculated charge separations range from a high of 0.67 to a low of 0.01 electron. Covalent bond energy is important to the overall lattice energy, since electrons are being shared rather than completely transferred. This means that the structure the system adopts will not represent the same sort of compromise as that described for NaCl. The same factors will still be important, but in addition the lattice geometry will have to be such as to make orbital overlap favorable, since that dictates the size of the covalent bonding energy. If the two bonding atoms have no d valence orbitals, 6:6 coordination will not be possible unless the bonding is essentially completely ionic. Thus boron nitride, BN, which has the same stoichiometry as LiF, has two crystal structures; in one of these the B and N atoms are in a layer with each atom having three neighbors, and in the other each atom is tetrahedrally surrounded by four of the other kind of atom. This is the zincblende structure shown in Fig. 4-35, with 4:4 coordination. This structure, or others having four or fewer neighbors, are very frequently found in partly ionic compounds. Since transition metals have d valence electrons, they can form more than four bonding MO's, and usually have more than four (usually six) ligand atoms; but for elements in groups IIb–Vb, in which the d orbitals are completely filled and strongly bound, coordination numbers above four are much less common.

One of the consequences of the lower coordination number, when it occurs, is that the crystals with 4-coordination are much less closely packed than the 6-coordinate NaCl structure. The tetrahedral geometry is quite open, and the resulting crystals are much less dense than comparable crystals with 6-coordination. Thus ZnO, which is tetrahedrally coordinated, has a density of 5.5 g/ml; but NiO (with nearly the same formula weight and similar metal–oxygen distance) has the NaCl structure and a density of 7.5 g/ml. Zinc sulfide (4-coordinate) has a density of 4.1 g/ml, and NiS (6-coordinate) a density of 5.5 g/ml. This is the same reasoning we gave earlier for the fact that ice is less dense than water, although in that case we were talking about much weaker bonds—hydrogen bonds—between the centers of tetrahedral geometry.

In Chapter 13 we saw that the lattice energy was a very important factor in determining the stability and reactivity of solid ionic compounds. We expect that this will be true again in the case of partly ionic compounds. How is the lattice energy affected by the relatively open structure and by the reduced charges on the atoms? We can answer the first question relatively simply by looking back at Table 13-1, which gives the Madelung constants for various crystal structures. If we restrict ourselves again to compounds with 1:1 stoichiometry to avoid confusion due to the effects of stoichiometry, we can pick out four pertinent examples: the CsCl structure, the NaCl structure, the zincblende structure, and the wurtzite structure. CsCl has coordination of 8:8, as shown in Fig. 4-33; the Madelung constant for this geometry is 1.763. NaCl has 6:6 coordination and a Madelung constant of 1.748, only about 1% lower than the CsCl value even though the number of nearest neighbors has dropped by 25%. Zincblende and wurtzite, pictured in Figs. 4-35 and 4-36, both have 4:4 coordination although the geometries are different. Their Madelung constants are, respectively, 1.638 and 1.641; these values are only about 7% lower than CsCl even though the number of nearest neighbors has dropped by 50%. This is not really as surprising as it might seem at first glance, because the Madelung constant sums up the effects of *all* the ions in an infinite crystal, and although there are fewer nearest neighbors in the 4-coordinate structures there are more second-nearest neighbors, and so on; the net result for the whole crystal is remarkably insensitive to the changing coordination number.

Lattice Energies and Born–Haber Cycles for Species with Partial Charge

The reduced charges on the atoms have a more substantial effect at which we ought to look. Suppose we take zinc sulfide, with the zincblende structure, as our example. If we follow the treatment of Chapter 13 and assume that both the Zn and the S have a charge of 2 (2+ and 2−, respectively), we can use the zincblende Madelung constant and the observed Zn—S distance of 2.58 Å to calculate the ionic-attraction lattice energy:

$$U_{ionic} = -1.638 \frac{(2)^2 (4.803 \times 10^{-10})^2}{2.58 \times 10^{-8}} \left(1 - \frac{1}{9}\right) \left(1.440 \times 10^{13} \frac{\text{kcal/mole}}{\text{erg/molecule}}\right)$$

$$= -748 \text{ kcal/mole}$$

If we compare this with the value obtained from a Born–Haber cycle assuming the formation of Zn^{2+} and S^{2-}, we get a somewhat different number:

$$Zn_{(g)} + S_{(g)} \xrightarrow{(IP_1+IP_2)_{Zn}}_{(EA_1+EA_2)_S} Zn^{2+}_{(g)} + S^{2-}_{(g)}$$

$$\Delta H_{sub} \uparrow \quad \uparrow \Delta H_{sub} \qquad \qquad \downarrow U$$

$$Zn_{(s)} + S_{(s)} \xrightarrow{\Delta H_f} ZnS_{(s)}$$

$$U = \Delta H_f - \Delta H_{sub} - IP - EA = -835 \text{ kcal/mole}$$

The ionic attraction does not account for all of the lattice energy. But there cannot be any covalent contribution to the lattice energy if the valence electrons have been transferred completely as we assumed here, because there are no longer any electrons in what might be called the "overlap population" of the two atoms' orbitals. For a covalent contribution it is necessary that the charges be less than the maximum, so that there are still some electrons described by the valence orbitals of *both* atoms. We can use differential ionization energies to produce a rough estimate of the charges on the atoms in solid ZnS as we have already done for AlN:

$$DIE_{Zn}(4q) = DIE_S(-q)$$

$$5.6 + 28.4q = 6.8 - 9.2q$$

$$37.6q = 1.2$$

$$q = 0.0318$$

$$\text{charge on Zn} = 4q = 0.13$$

Again we must remember that this is only a fairly crude approximation to the charge, since we are considering only the nearest neighbors to the Zn atom in the crystal. To the extent that it is even qualitatively correct in predicting a very low charge in the ZnS crystal, however, we see that our ideas of lattice energy must be changed. We can still calculate an ionic attraction energy using the Madelung expression, but now it will be very small:

$$U_{\text{ionic}} = -1.638 \frac{(0.13)^2 (4.803 \times 10^{-10})^2}{2.58 \times 10^{-8}} \left(1 - \frac{1}{9}\right)\left(1.440 \times 10^{13} \frac{\text{kcal/mole}}{\text{erg/molecule}}\right)$$

$$= -3.0 \text{ kcal/mole}$$

To compare this with an experimental thermodynamic lattice energy, we can set up a Born–Haber cycle again. But now we cannot use the experimentally determined ionization energies and electron affinities, because they refer to the removal or addition of integral numbers of electrons. Here we are using as a model a system in which *partial* electron transfer has occurred. The relevant cycle, then, is

$$\begin{array}{ccc}
Zn_{(g)} + S_{(g)} & \xrightarrow{\text{(partial IP)}_{Zn} + \text{(partial EA)}_S} & Zn^{+0.13}_{(g)} + S^{-0.13}_{(g)} \\
\uparrow \Delta H_{\text{sub}} & \uparrow \Delta H_{\text{sub}} & \downarrow U_{\text{ionic}} + U_{\text{covalent}} \\
Zn_{(g)} + S_{(s)} & \xrightarrow{\Delta H_f} & ZnS_{(s)}
\end{array}$$

We could establish theoretical values for the partial ionization energy of Zn and the partial electron affinity of S by remembering from Chapter 12 that any ionization energy is defined by the integral of the differential ionization energy — in this case we could integrate between the limits of 0 and 0.13. However, in the simple approximation we are using the partial ionization energy of the positive atom (evaluated in this way) and the partial electron affinity of the negative atom are exactly equal and opposite; thus:

$$U_{\text{ionic}} + U_{\text{covalent}} = \Delta H_f(\text{ZnS}) - \Delta H_{\text{sub}}(\text{Zn}) - \Delta H_{\text{sub}}(\text{S}) - \text{partial IP}_{\text{Zn}} - \text{partial EA}_{\text{S}}$$

$$= \Delta H_f(\text{ZnS}) - \Delta H_{\text{sub}}(\text{Zn}) - \Delta H_{\text{sub}}(\text{S}) - 0$$

Inserting experimental numbers for the sublimation energies of Zn and S, and for the enthalpy of formation of ZnS,

$$U_{\text{ionic}} + U_{\text{covalent}} = (-45.3) - (31.2) - (53.2) - 0$$

$$= -129.7 \text{ kcal/mole}$$

$$U_{\text{covalent}} = -129.7 - U_{\text{ionic}} = -129.7 - (-3.0) = -126.7 \text{ kcal/mole}$$

The covalent contribution to the lattice energy, U_{covalent}, is 127 kcal/mole, which compares fairly well with the experimental bond-dissociation energy $D_{\text{ZnS}} = 98$ kcal/mole. The difference is presumably due to our crude estimate of the charges on the atoms.

Another example may be worthwhile. Consider beryllium oxide, BeO, which has the wurtzite structure and a Be—O distance of approximately 1.71 Å. Since the wurtzite structure makes each Be 4-coordinate, we can set up the differential ionization-energy expression for the charge separation as follows:

$$DIE_{\text{Be}}(4q) = DIE_{\text{O}}(-q)$$

$$5.5 + 28.0q = 8.6 - 13.7q$$

$$41.7q = 3.1$$

$$q = 0.0744$$

$$\text{charge on Be} = 4q = 0.30$$

This charge is substantially larger than the charge on the atoms in ZnS; to what ionic lattice energy does it lead? Using our Madelung expression,

$$U_{\text{ionic}} = -\frac{1.641(0.30)^2(4.803 \times 10^{-10})^2}{1.71 \times 10^{-8}}\left(1 - \frac{1}{5.5}\right)\left(1.440 \times 10^{13}\frac{\text{kcal/mole}}{\text{erg/molecule}}\right)$$

$$= -23.5 \text{ kcal/mole}$$

The same sort of Born–Haber cycle applies as for ZnS:

$$Be_{(g)} + O_{(g)} \xrightarrow[\text{(partial EA)}_{\text{O}}]{\text{(partial IP)}_{\text{Be}}^+} Be_{(g)}^{+0.30} + O_{(g)}^{-0.30}$$

$$\Delta H_{\text{sub}} \uparrow \quad \uparrow \tfrac{1}{2}D \quad \quad \downarrow U_{\text{ionic}} + U_{\text{covalent}}$$

$$Be_{(s)} + \tfrac{1}{2}O_{2(g)} \xrightarrow{\Delta H_f} BeO_{(s)}$$

$$U_{\text{ionic}} + U_{\text{covalent}} = \Delta H_f(\text{BeO}) - \Delta H_{\text{sub}}(\text{Be}) - \tfrac{1}{2}D(O_2) - \underbrace{\text{partial IP}_{\text{Be}} - \text{partial EA}_{\text{O}}}_{0}$$

$$= -146.0 - (76.6) - (59.2)$$

$$= -281.8 \text{ kcal/mole}$$

$$U_{\text{covalent}} = -281.8 - U_{\text{ionic}} = -281.8 - (-23.5) = -258.3 \text{ kcal/mole}$$

We see that both the ionic and covalent contributions to the lattice energy are larger than in the ZnS case. The total lattice energy is over twice as large as that of ZnS. What experimental evidence do we have that this is reasonable?

Boiling Points and Lattice Energies

For compounds like this in which the charge separation is relatively low, the boiling point is a fairly good guide to the relative lattice energy, because the energy necessary to break up the lattice is approximately equivalent to the thermal energy needed to take molecules from a condensed phase into the vapor. Zinc sulfide sublimes at about 1185°C or 1460°K, which is in keeping with the approximate lattice energy we have calculated of −130 kcal/mole; by comparison, NaCl, which is presumably almost completely ionic, has a lattice energy of −183 kcal/mole and a boiling point of 1690°K. On the other hand, the lattice energy of BeO is about −282 kcal/mole, so it should boil at a much higher temperature; the boiling point of BeO is experimentally estimated as 4200°K, a temperature so high that measurements are very difficult. We can see that the sublimation or boiling temperatures are roughly proportional to the lattice energies, as shown in Fig. 14-2. This lends support to our statement that boiling points are related to lattice energies, and even suggests that ionic compounds fall in the same approximate order.

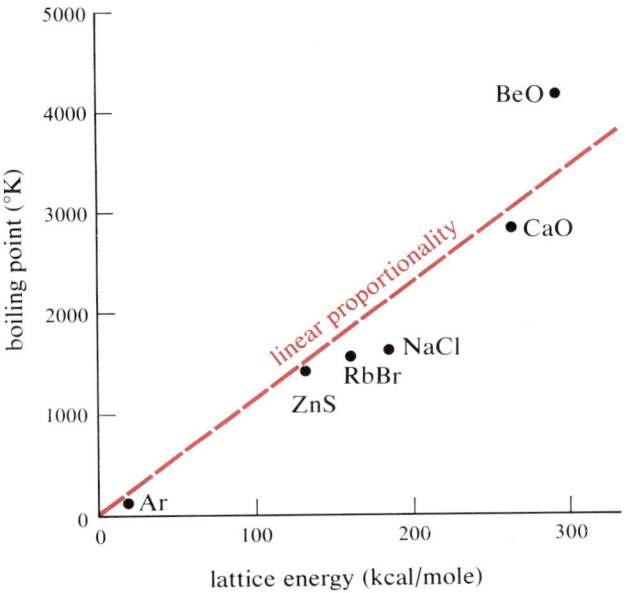

Figure 14-2 Proportionality of boiling points to lattice energies.

Melting Points and Covalent Lattice Energies

High lattice energies are generally characteristic of partly ionic oxides in which the charge separation is fairly significant (so that the O has a charge of approximately 0.2–0.5). These are, generally speaking, the oxides of the elements to the left of the shaded area in Fig. 14-1, because at the far right of that area the elements' electronegativities are too high to give substantial charge separation. The element is also usually in a fairly low oxidation state, because a high formal charge has the same effect as a high electronegativity. Thus this description includes such oxides as Al_2O_3, MgO, La_2O_3, Cr_2O_3, TiO_2, ThO_2, and ZrO_2. These oxides are known as **refractory oxides**, which refers to their particular property of having a very high melting point. They are used in a number of applications such as firebrick, in which a solid material must withstand very high temperatures. This property illustrates one of the major distinctions between ionic and partly ionic compounds. Ionic compounds have high lattice energies but melt far below these refractory oxides; most of the alkali halides melt around 700°C, while these oxides melt around 1900–2400°C. The difference is simply that covalent bonding is directional, depending on the overlap of certain orbital lobes, while ionic bonding is completely nondirectional and depends only on the distance between ions. The process of melting to a liquid changes the "bond angles" between adjacent atoms substantially, but only slightly expands their average internuclear distance, as we saw in Chapter 7. For this reason, compounds in which covalent bonding is important require a great deal of thermal energy (in other words, a high temperature) to melt, while purely ionic compounds, which are rather insensitive to changes in "bond angles," require much less. But if the charge is very small—in a nearly covalent molecule—there is a tendency to form individual molecules such as CO_2, which are bound together only by the very weak van der Waals forces; these compounds melt and boil at very low temperatures. Compounds with very high melting points invariably have covalent bonds extending throughout the crystal. These can even be elements, such as carbon (diamond), with no ionic attraction, but this depends on the special circumstance of having approximately one valence electron per valence orbital, so that the donor-acceptor capacities require the formation of an extended lattice rather than a simple molecule such as Cl_2 or P_4. A wide range of elements, however, form refractory oxides, nitrides, carbides, silicides, borides, and so on. These are partly covalent because the electronegativity difference is too small to give complete charge separation, but the partial charge that does exist is sufficient to ensure bonding throughout the crystal. We shall look at some of these in Section 14-4.

14-3 Polymeric Structures in Solids and Solutions

A polymer is a chemical structure containing repeated units of a single structure (a **monomer**) bonded to one another. We commonly think of polymers in terms of nylon and trademarked commercial products — Teflon, Dacron, Herculon, Kodel, and many others; we shall look at a few of these in Chapter 15. What sort of tendency to polymerize do ionic and partly ionic compounds display? In one sense the ionic lattice structure could be considered a polymer, with the monomer being the unit cell. Completely ionic lattices contain no directional chemical bonds, however, and we cannot really say that this "monomer" is bound into a "polymer" without stretching the definition rather badly.

The refractory oxides we have just been looking at, with their partly covalent character, clearly qualify as polymers under this definition, however. In a very real sense, a crystal of one of these oxides is a giant molecule consisting of units equivalent to the crystal unit cell repeated throughout the crystal. To what degree is this polymeric structure characteristic of partly ionic compounds, and what sorts of structures result?

The answer to the first part of this question is that polymeric structures are very general. The fact that significant charge separation or partial ionization still exists in these compounds means that there are electrostatic attractions or repulsions between neighbor atoms throughout the solid crystal. By the very definition of these compounds as those in which both ionic and covalent bonding are important, we expect ionic bonding to produce this attraction between extended sets of atoms, and covalent bonding to provide the directional-bond framework that we expect in a polymer.

Three-Dimensional Polymeric Structures

The answer to the second part of the question includes some very interesting structures. If we characterize purely ionic compounds as having "bonding" or ionic attractions in all possible directions in a three-dimensional lattice, then it might be reasonable to describe covalent solids built up of isolated individual molecules held together only by van der Waals forces as having zero-dimensional bonding, since there is no bonding between the individual molecules. The partly ionic compounds fall in between. They can have three-dimensional bonding, as in the case of the refractory oxide TiO_2, whose **rutile** structure is shown in Fig. 14-3. Each Ti is surrounded octahedrally by six O

Figure 14-3 Rutile structure of TiO_2. Note the octahedral coordination of the central Ti atom.

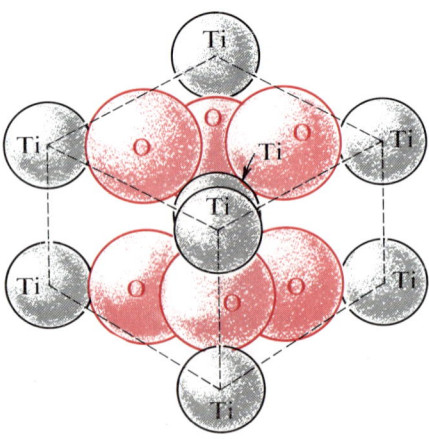

atoms, each O by three Ti atoms, and the structure extends indefinitely in all three directions. Because the large number of nearest neighbors to the Ti tends to promote charge separation, this structure is usually found in compounds having fairly large electronegativity differences and considerable ionic character: the electronegativity difference for TiO_2 is 2.18 units, which ought to make it ionic by our earlier generalization. If we calculate its lattice energy from its Madelung constant assuming it to be completely ionic, however, the energy falls substantially short of the Born–Haber cycle value, showing that there must be some significant covalent bonding. Many other three-dimensional structures exist: a few more are shown in Figs. 4-34, 4-35, and 4-36.

Two-Dimensional Polymeric Structures

These systems can also form two-dimensional polymeric structures or layer structures. Figure 14-4 shows several of these, and makes another interesting point. The CdI_2 and MoS_2 structures are essentially identical, except that in the CdI_2 case each Cd is octahedrally surrounded by I atoms, while in the MoS_2 case each Mo is surrounded by the same number of atoms — six — but the S atoms are at the corners of a trigonal prism. This structure has been found for other systems in which six sulfur atoms surround a metal atom, but for very few nonsulfur systems. It has been suggested that partial bonding between the two superimposed sulfur atoms is responsible for keeping them together, since we would normally expect them to adopt the less hindered octahedral structure.

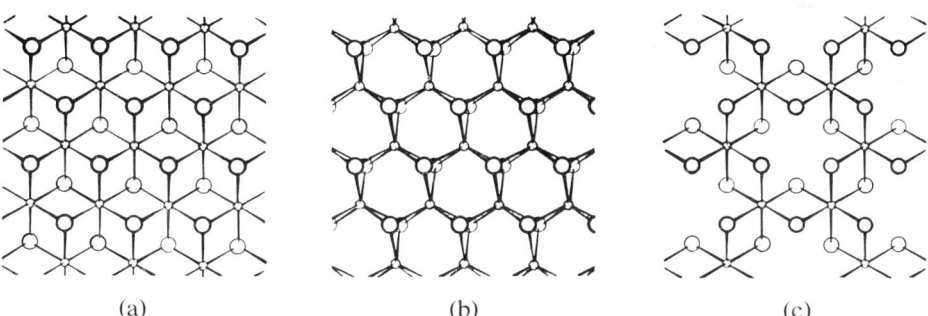

(a)   (b)   (c)

Figure 14-4 Layer structure of several compounds. (a) CdI_2. (b) MoS_2. (c) $CrCl_3$. Small circles represent metal atoms; large circles, nonmetal atoms. (Redrawn from Wells, A. F., *Structural Inorganic Chemistry*, Oxford: The Clarendon Press, 1962.)

The $CdCl_2$ and $CrCl_3$ structures are also interesting. They have the Cl atoms in the same positions and differ only in the locations of metal atoms, which is simply a result of the differing stoichiometry in the two compounds. This suggests that it may be a reasonable approximation to think of these solid compounds as being essentially a close-packed lattice of the negative ion (or partially negative atom), with the metal atoms inserted in the holes between the other close-packed atoms. We remember from Chapter 4 that close-packed lattices have both tetrahedral and octahedral holes—see Figs. 4-29 and 4-30. In these cases the Cd atoms are occupying all the octahedral holes in the double layer of Cl atoms, and the Cr atoms are occupying two-thirds of the octahedral holes. In the crystal these "sandwiches" of metal atoms between electronegative atoms are stacked into a three-dimensional close-packed lattice of electronegative atoms. There are no metal atoms between "sandwiches," so these double layers are held together in most cases only by van der Waals forces. Consequently the layers can slide over each other relatively freely, and some of these compounds are used as lubricants as a result; MoS_2 is a notable example of a layered lubricant (although it is not close-packed).

We also remember from Chapter 4 that there are two ways to pack spherical atoms or layers of atoms into a close-packed solid. If the third layer is superimposed over the first layer *ababababa*, we have a hexagonal close-packed structure—see Fig. 4-26. If the third layer is in the third kind of position *abcabcabc*, the three-dimensional array is cubic close packed. This is the only difference between the $CdCl_2$ and CdI_2 structures; the two layers of halide ions are *ab* in each case, so the two structures differ only in how the double layers pack together. $CdCl_2$ has cubic close-packed Cl's, while CdI_2 has hexagonal close-packed I's. Within each double layer, however, the structures are identical.

One-Dimensional Polymeric Structures

These polymeric, partly ionic structures can also be one dimensional, as seen in Fig. 14-5. These structures are somewhat more rare, since it is energetically favorable for most chemical systems to rearrange into a geometry that provides each atom with more nearest-neighbor bonds. Most of these **chain** structures involve metal atoms whose coordination number is limited, for one reason or another, to four, although there are MX_4 systems in which a chain is formed by octahedral MX_6 units sharing an edge, and MX_5 systems in which the chain consists of MX_6 units sharing a corner. In addition, chains of square 4-coordinate units are sometimes held together by weaker, longer bonds that in effect surround the metal atom by a distorted octahedron of electronegative atoms; Fig. 14-6 shows this behavior as found in $CuCl_2$. Sometimes both chains and layers can be held together by hydrogen bonds; a number of hydroxides form polymers that are stabilized by this kind of interaction.

Polymeric Structures in Solids and Solutions | 729

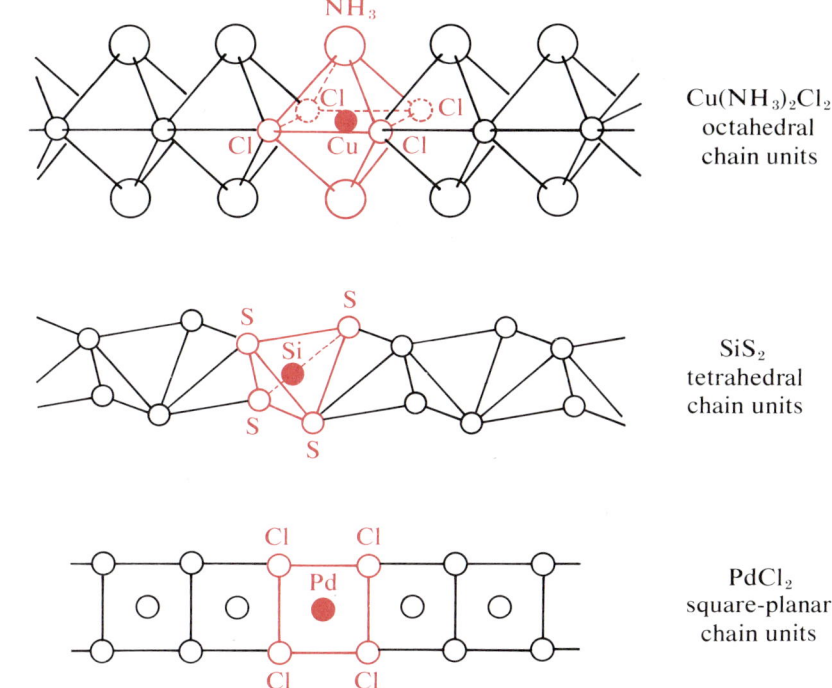

Cu(NH$_3$)$_2$Cl$_2$
octahedral
chain units

SiS$_2$
tetrahedral
chain units

PdCl$_2$
square-planar
chain units

Figure 14-5 One-dimensional or chain structures in partly ionic compounds.

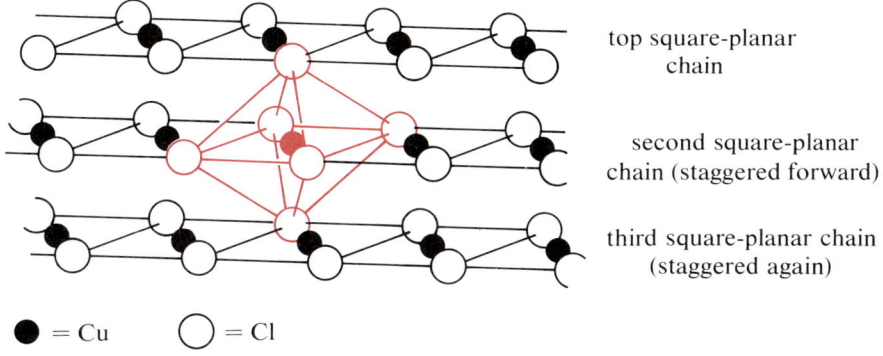

top square-planar
chain

second square-planar
chain (staggered forward)

third square-planar chain
(staggered again)

● = Cu   ○ = Cl

Figure 14-6 Chain stacking in CuCl$_2$ to provide octahedral coordination.

Isopoly and Heteropoly Anions

When the electronegative atoms are oxygens, the formation of these polymers in aqueous solution is facilitated by the hydrogen bonding that occurs between the monomer unit and surrounding water molecules or hydronium ions. For instance, the chromate ion, CrO_4^{2-}, is a monomeric unit that is stable in neutral solution. If the solution is acidified, however, a **condensation** reaction occurs, in which two chromates react with surrounding hydronium ions to dimerize by eliminating a molecule of water:

$$2\,CrO_4^{2-} + 2\,H_3O^+ \rightarrow Cr_2O_7^{2-} + 3\,H_2O$$

The geometry of the resulting dichromate ion is that of two tetrahedral CrO_4 groups linked at a corner. This reaction may be taken as a model for the formation of oxypolymers in general; if the condensation goes on indefinitely, the resulting species will be a solid (either a crystal or a glass), but if it terminates before the molecule has become too large, the polymer will stay in solution as a **polyanion**. If the monomer is the phosphate ion, PO_4^{3-}, condensation gives the **pyrophosphate** ion, $P_2O_7^{4-}$, again with the geometry of two tetrahedra sharing a corner. Further condensation gives the **triphosphate** ion, $P_3O_{10}^{5-}$, which has a central tetrahedron sharing two of its corner O atoms with other tetrahedra; see Fig. 14-7. This ion can condense still further by forming a six-membered ring containing three P and three O atoms, $P_3O_9^{3-}$, the **trimetaphosphate** ion, also shown in Fig. 14-7. Note that three molecules of water have been eliminated from the trimer at this stage, so the empirical formula for the sodium salt, $NaPO_3$ (corresponding to the true structure $Na_3P_3O_9$) is what would be

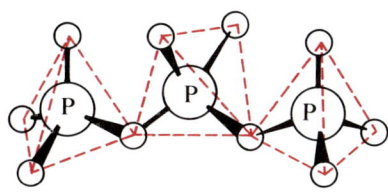
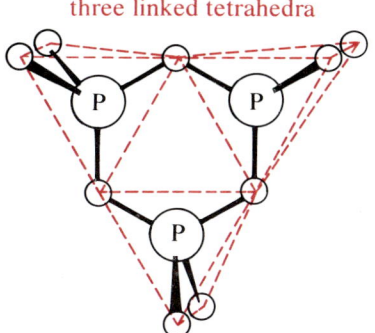

triphosphate ion, $P_3O_{10}^{5-}$ trimetaphosphate ion, $P_3O_9^{3-}$

Figure 14-7 Polyphosphate ions with linked-tetrahedron structures.

expected if one water molecule were lost for each phosphate ion. Up to this point all the polymers are soluble, but long-chain polymers, again with the empirical formula $NaPO_3$, exist that are so large they form a separate solid phase and are insoluble.

Other systems sometimes give even more dramatic polymerization products in solution. For instance, if a solution of the molybdate ion, MoO_4^{2-}, is acidified, proton transfer and further hydration of the Mo occurs, together with very extensive condensation; the main product is the **paramolybdate** ion, $Mo_7O_{24}^{6-}$. In more acid solutions the **octamolybdate** ion is formed ($Mo_8O_{26}^{4-}$). Figure 14-8 shows these ions' structure as determined by X-ray diffraction of crystals of their salts; each octahedron represents a MoO_6 unit, and it can be seen that these units share corners (one O atom) or edges (two O atoms) to form a very compact, symmetrical structure. In a similar way, the tungstate ion, WO_4^{2-}, rapidly polymerizes to form first $HW_6O_{21}^{5-}$, then $W_{12}O_{41}^{10-}$. All of these ions are hydrated to an extent that is difficult to determine, so that some O atoms in these formulas may actually be OH groups, and so on.

Figure 14-8 Isopolymolybdate ions, shown as linked octahedra.

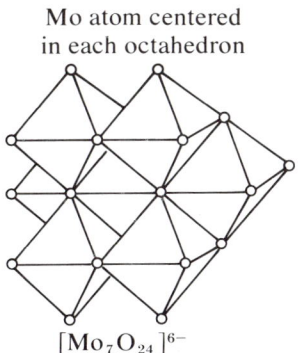
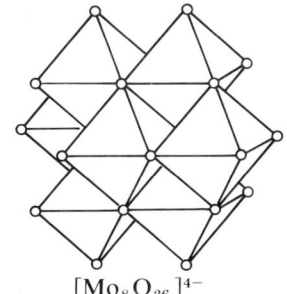

Mo atom centered in each octahedron

symmetrical on other side—one octahedron hidden in rear

$[Mo_7O_{24}]^{6-}$ $[Mo_8O_{26}]^{4-}$

As a natural extension of the formation of these **isopoly** ions, we might expect — correctly — that condensations could occur between oxyanions with different central atoms to give **heteropoly** ions. A great number of these are known, with many different heteroatom oxyanions combining with either MoO_4^{2-} or WO_4^{2-}. Such ions as $[PMo_{12}O_{40}]^{3-}$, **12-molybdophosphate**, and $[Co_2W_{12}O_{42}]^{8-}$, **dimeric 6-tungstocobaltate**, are formed; most of the later transition metals and members of groups IVa and Va will form these species (although not carbon and nitrogen, because they cannot form bonds to an octahedron of ligand atoms).

14-4 Chemistry of Partly Ionic Compounds — Electronegativity Differences Between 1.0 and 2.0 Units

We can organize the discussion in this chapter in the same manner as in the last, by considering partly ionic compounds as being grouped according to the partly negative atom and taking these groups in the approximate order of increasing electronegativity. In all of these groups we shall see distinct changes in the individual compounds' chemical and physical properties as the partly positive atom is changed. In part these changes are due to the changing electronegativity difference between the partly positive and the partly negative atom, but also in part they are due to changing size relationships. We shall explore both of these effects in treating the compounds, but the size effect deserves some immediate attention.

In Chapter 13, in connection with the structures and stabilities of ionic nitrides, we mentioned that a stable solid ionic lattice cannot contain many large positive ions for each negative ion. Size relationships are obviously the key to the stability of these compounds. For partly ionic compounds the stable solid structure is often a close-packed lattice of one kind of atom, with the other atom inserted in the interstices of the close-packed lattice. Clearly this sort of structure will not be stable if one atom is too big to fit into the holes in the other atom's lattice. How big can an atom be, relative to another, and still fit into the other's lattice? In Chapter 4 we saw that in a close-packed host lattice, other atoms must be no larger than about 0.41 of the host atom's radius to fit into octahedral holes. This figure will govern the formation of some of these compounds, but it must be remembered that the atoms are not hard little spheres but have electron distributions trailing off exponentially from the nucleus; this means that there is some flexibility in the radius relationships. In particular, polarization may change some of the observed relationships substantially.

Chemistry of Partly Ionic Compounds | 733

Still, the radius ratio is a useful guide. We also saw in Chapter 4 that for an atom to fit into the tetrahedral holes in another's lattice it must not be larger than about 0.22 of the other atom's size, which is very small indeed. This has some bearing on the fact that many systems are found in which octahedral holes are occupied, but relatively few in which tetrahedral holes are occupied. There simply are not very many atoms small enough to fit in.

Partly Ionic Hydrides

Figure 14-9 shows the range of elements that form this type of hydride. Notice that because hydrogen has a rather low electronegativity the elements to the right of the shaded (positive) portion of Fig. 14-1 have very small electronegativity differences and form covalent hydrides; we shall return to these elements in Chapter 15. Even within the range of partly ionic hydrides indicated in Fig. 14-9, however, there are significant differences in properties.

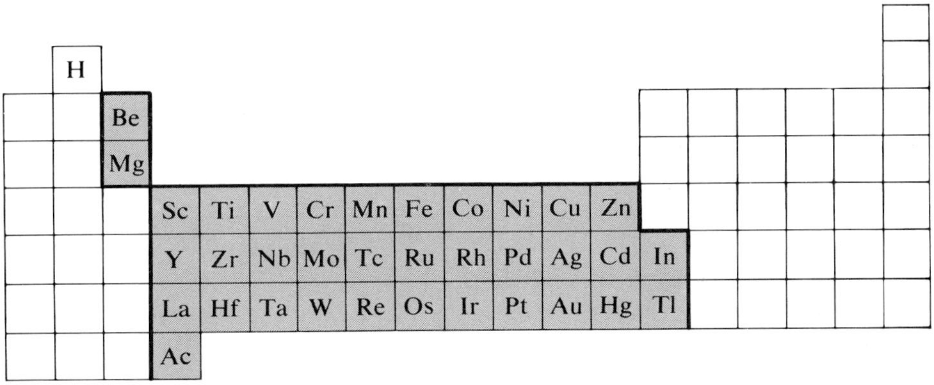

	H																
		Be															
		Mg															
			Sc	Ti	V	Cr	Mn	Fe	Co	Ni	Cu	Zn					
			Y	Zr	Nb	Mo	Tc	Ru	Rh	Pd	Ag	Cd	In				
			La	Hf	Ta	W	Re	Os	Ir	Pt	Au	Hg	Tl				
			Ac														

Figure 14-9 Range of partly ionic hydrides in the periodic table. The shaded elements form partly ionic hydrides.

Beryllium and magnesium hydrides, BeH_2 and MgH_2, are similar to the ionic hydrides in their preparation and properties, and are distinguished from them primarily by their crystal structures.

Polymeric Structures Instead of the characteristic ionic structures with all neighbors to an atom equivalent, these hydrides are polymeric; in particular, BeH_2 has a chain structure in which tetrahedral BeH_4 units share an edge (two H atoms).

The other metals in this group form hydrides by direct combination of the heated metal with the gas. We cannot write a stoichiometric equation for this reaction because the compounds themselves are quite nonstoichiometric.

Interstitial Structures In effect, the compound consists of close-packed metal atoms with hydrogen atoms in the interstices — for this reason, the compounds are sometimes known as **interstitial** hydrides. Sometimes the metal structure remains unchanged except for a slight expansion as hydrogen is added to it, while in other cases it changes to a different structure (e.g., from hexagonal to cubic close packed) but the hydrogens continue to occupy holes in the metal lattice. With this mode of formation, it is not surprising that the exact ratio of H to metal should be variable depending on the past history of the compound — the pressure of hydrogen, and so on. There is also no particular reason for these hydrides to assume a H:metal ratio such that the metal has one of its characteristic formal oxidation states; as a matter of fact, we saw in Chapter 4 that there are only two tetrahedral sites and one octahedral site for each metal atom in the close-packed lattice, which means that H:M cannot exceed 3 no matter what the formal oxidation state of the metal usually is. Usually, since H is a very small atom, it occupies tetrahedral sites, and the limit of stoichiometry for most of these hydrides is MH_2 (although a wide range of values less than this limit is often possible).

There is also another class of metal hydrides that represents an interesting application of the synthetic technique of making the free energy of a reaction negative by having it form an ionic lattice from covalent reactants. A partly ionic or covalent halide is dissolved in ether, and an ionic hydride is added:

$$3\,LiH + AlCl_3 \xrightarrow{\text{ether}} AlH_3(\text{polymer})\cdot n\ \text{ether} + 3\,LiCl$$

$$LiAlH_4 + 4\,CuI \longrightarrow 4\,CuH + LiI + AlI_3$$

These hydrides have conventional stoichiometries but always contain some solvated ether. They show reactions like those of the hydride ion and have the appearance of ordinary salts, in contrast to the interstitial hydrides, which are relatively stable and look almost exactly like the metals from which they were formed.

Partly Ionic Borides, Carbides, Nitrides, Silicides, and Phosphides

We can appropriately consider these compounds together because of a very great resemblance in their properties. It should be pointed out immediately that B, Si, and P fall outside the rather arbitrary range of partly negative atoms indicated in Fig. 14-1; their electronegativities (2.01, 1.74, and 2.06, respectively) indicate that the more electropositive elements should be able to form partly ionic compounds with them, however.

All of these compounds can be prepared by direct combination of the elements, although in most cases very high temperatures, from 1200 to 2200 °C, are required. This is simply a reflection of the very high activation energy needed to break the covalent bonds holding the elements together. The reactions themselves are exothermic:

$$\text{Ti}_{(s)} + \text{C}_{(s)} \xrightarrow{2200\,°C} \text{TiC}_{(s)} \qquad \Delta H_f^0 = -54 \text{ kcal/mole}$$

$$6\text{Ca}_{(s)} + \text{P}_{4(s)} \longrightarrow 2\text{Ca}_3\text{P}_{2(s)} \qquad \Delta H_f^0 = -120 \text{ kcal/mole}$$

Table 14-2 gives the formulas of a number of these compounds. The most common formula is MX, with one electronegative atom per metal atom.

Table 14-2
Partly Ionic Borides, Carbides, Nitrides, Silicides, and Phosphides

K	Ca	Sc	Ti	V	Cr	Mn	Fe	Co	Ni	Cu
	CaB_6	ScB_6	TiB	VB	CrB	MnB	FeB	CoB	NiB	Cu_3B_2
KC_8	CaC_2	Sc_2C_6	TiC	VC	Cr_3C_2	Mn_3C	Fe_3C	Co_3C	Ni_3C	Cu_2C_2
	Ca_3N_2	ScN	TiN	VN	CrN	Mn_4N	Fe_2N	Co_2N	Ni_3N	Cu_3N
KSi_6	$CaSi_2$		$TiSi_2$	VSi_2	CrSi	MnSi	FeSi	CoSi	NiSi	Cu_5Si
	Ca_3P_2			VP	CrP	MnP	FeP	CoP	Ni_3P	Cu_3P

Interstitial Structures This formula results from the interstitial nature of most of these compounds; these electronegative atoms, unlike the hydrogen atom, are too large to fit in tetrahedral holes in the close-packed metal lattice. Each fills an octahedral hole, although there is some distortion of the metal lattice in the process (as for hydrides). Since there is only one octahedral hole per atom in a close-packed lattice, this limits the stoichiometry for many of the systems. As for the hydrides, the compounds are not necessarily stoichiometric, since the filling of holes is a gradual process depending on the migration of electronegative atoms through the metal lattice.

These interstitial compounds tend to be very hard and high melting, like diamond, which melts about 4000 °C (under high pressure) and for many years defined the maximum hardness of 10 on the Mohs scale of hardness (talc = 1, ordinary glass = 6). In these compounds, as in diamond, the hardness and very great thermal stability are the result of an extended network of covalent bonds throughout the crystal. For example, HfC melts at about 4100 °C, and is sometimes used as a liner for the throat of rocket exhaust nozzles. Most of the interstitial borides and carbides melt above 3000 °C, the nitrides slightly lower. Nearly all of these compounds have Mohs hardnesses greater than 8, and a few will scratch (and be scratched by) diamond. Carbide-tipped cutting tools and drill bits are widely used in industry.

Not all of these compounds can be described as interstitial, however. Interatomic distances in the metal lattices range from 3.60 Å for Th to 2.48 Å for Ni, following trends we have discussed. Since an octahedral interstitial atom can be only 0.414 as big as the metal radius—half the above distances—it would presumably be impossible for atoms with a covalent radius bigger than about 0.65 Å to fit into these metal lattices if there were no expansion. All five of the electronegative elements we are discussing are larger than this, and presumably only lattice expansion permits the formation of interstitial compounds. But as the lattice expands it becomes more and more possible for the electronegative atoms to bond to each other.

Polymeric Structures Consequently, for the smaller metal atoms the structures of these binary compounds tend to involve extended polymers—either chains or layers—of the electronegative atom, particularly if there are more X atoms than M atoms.

For instance, isolated boron atoms are found only in borides with M:B at least 2:1. The MB borides have chains of boron atoms, the M_3B_4 borides usually have double chains, the MB_2 borides have layers of boron atoms, and higher B:M ratios (up to 12 B per M) show three-dimensional networks of B atoms. Some of these structures are shown in Fig. 14-10. The layer structure of

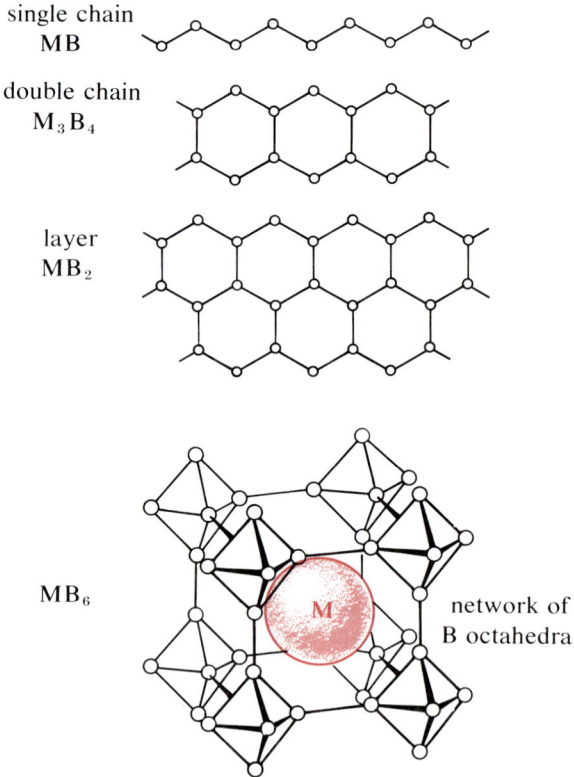

Figure 14-10   Various linkages of B atoms in partly ionic borides.

MB$_2$ borides shows up in modified form in other compounds of this series, such as the **lamellar** (layered) compounds of graphite with both electropositive atoms such as K, Rb, and Cs, and electronegative atoms such as O and F. Only the O and F compounds have even remotely regular stoichiometry, approaching the values $(C_2O)_n$ and $(CF)_n$. Graphite also spontaneously absorbs many molecules, including Cl_2, Br_2, $FeCl_3$, MoO_3, and others, with the layers of C atoms separating to form a host lattice for the incoming atom or molecule, but the nature of these compounds is not well understood.

The layer structure is also present in some silicides. For example, CaSi$_2$ has a structure quite different from the ionic CaC$_2$, with the Si atoms arranged in puckered layers as shown in Fig. 14-11. The differing structure has a profound effect on the compound's reactions; for instance, when CaC$_2$ hydrolyzes, the

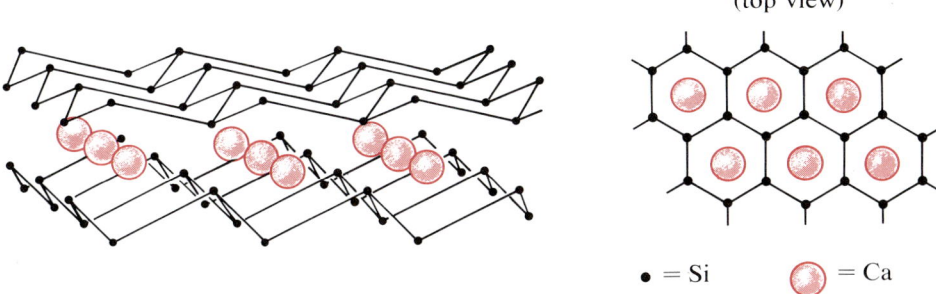

Figure 14-11 Layer structure of CaSi$_2$.

product is acetylene, C_2H_2, but when CaSi$_2$ hydrolyzes, the product maintains the polymeric nature of the silicon layers: **siloxene**, $(Si_6O_3H_6)_n$, has the structure shown in Fig. 14-12, with a H on each Si.

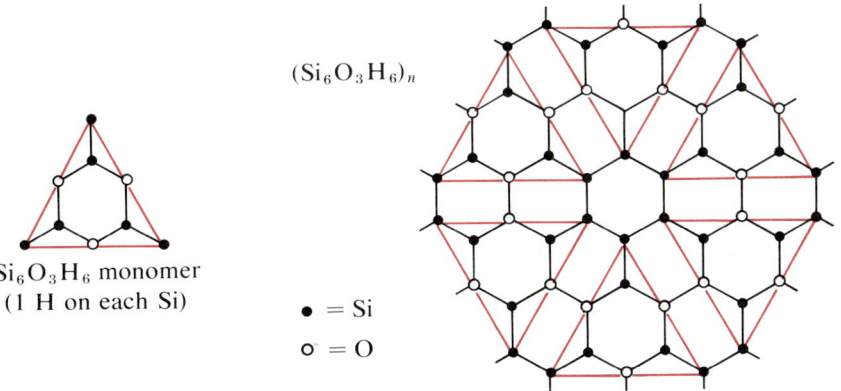

Figure 14-12 Layer structure of siloxene for comparison with CaSi$_2$.

All of these interstitial compounds (in the sense that the nonmetal atoms are isolated from each other in a close-packed metal–atom lattice) are very inert chemically. Thermodynamically, they should hydrolyze to the metal oxide — with a high lattice energy — and the nonmetal hydride:

$$2VN + 5H_2O \rightarrow V_2O_5 + 2NH_3 + 2H_2 \quad \Delta H^0 = -65 \text{ kcal/mole}$$

$$WC + 3H_2O \rightarrow WO_3 + CH_4 + H_2 \quad \Delta H^0 = -37 \text{ kcal/mole}$$

They should also oxidize in air:

$$TiC + 2O_2 \rightarrow TiO_2 + CO_2 \quad \Delta H^0 = -258 \text{ kcal/mole}$$

None of these reactions can be induced to go as written, however, except at very high temperatures. The strong covalent bonding and great delocalization of the bonding electrons in the lattice means that any reaction involving these systems has an enormous activation energy to overcome, since the atoms must be broken loose to react. When sizes (and stoichiometries) prevent the nonmetal atoms in the compound from being interstitial, the compounds become somewhat more reactive. Thus Cr_3C_2 and Fe_3C hydrolyze fairly readily to a rather puzzling mixture of hydrocarbons and even free carbon.

An aspect of these compounds' behavior that we have not discussed is the fact that many of them have a high electrical conductivity — in some cases even higher than that of the bulk metal. This is taken as evidence that the electron structure in the compound is still related to that of the metal. Where delocalization of the bonding electrons in extended π orbitals is possible, as in graphitic compounds, the formation of the compound often contributes electrons to the charge-carrying delocalized system. In other cases, however, the theoretical treatment involves application of the band theory of solids (see Section 6-12) in an advanced form.

Partly Ionic Oxides

Like the hydrides, carbides, and so on, that we have just discussed, partly ionic oxides tend to have a close-packed structure with heteroatoms in the interstices. The greater electronegativity of the oxygen atom causes substantially greater charge separation and ionic character in its compounds, however, and this has an effect on the geometry. As electrons are added to the O atom it becomes larger, because the effective nuclear charge is decreasing due to the additional shielding. For relatively large metal atoms such as Ti it is possible to regard the metal oxide lattice of TiO — which has the NaCl structure — as being approximated by either close-packed metal atoms or close-packed O atoms, since the NaCl structure is essentially two interwoven close-packed lattices. But O has a relatively low formal oxidation state (−2), and many oxides require several O's per metal. Thus the more common oxide of titanium is TiO$_2$, which has the rutile lattice shown in Fig. 14-3 and in Fig. 4-37. Those with good spatial perception may be able to see that in this structure the O's are more or less close-packed, and the Ti atoms are in half the octahedral holes in the close-packed O lattice. The Ti atoms, however, are not in any sense close-packed, and this is customary for partly ionic oxides. The structures are usually those corresponding to close-packed O atoms with interstitial metal atoms, rather than the other way around.

Partly ionic oxides can be nonstoichiometric in the same sense as the hydrides, carbides, and so on. A good example of this is the Ti–O system. Oxygen can be absorbed into the hexagonal close-packed metal lattice of Ti up to a concentration corresponding to an overall composition of about Ti$_2$O, or TiO$_{0.5}$. Further addition of O produces a change of structure to the NaCl structure, but with some O atoms missing—the lower limit of composition is about TiO$_{0.8}$. Further addition of oxygen carries the composition more or less continuously up to about TiO$_{1.2}$, which corresponds to the idealized TiO lattice with some Ti's missing. So we see that this is quite comparable to the previously discussed compounds. However, oxides tend to show more exact adherence to formal oxidation states than the other systems because of their greater ionic character. When considerable electron transfer is occurring, the combination of elements tends to go on until all the electrons in the system have reached their lowest-energy state; if one atom is very electronegative, such as O, there is a considerable tendency to transfer electrons more or less completely, so that the charges in the system are fairly close to the formal oxidation states that are expected. Figure 14-13 shows the range of elements that form

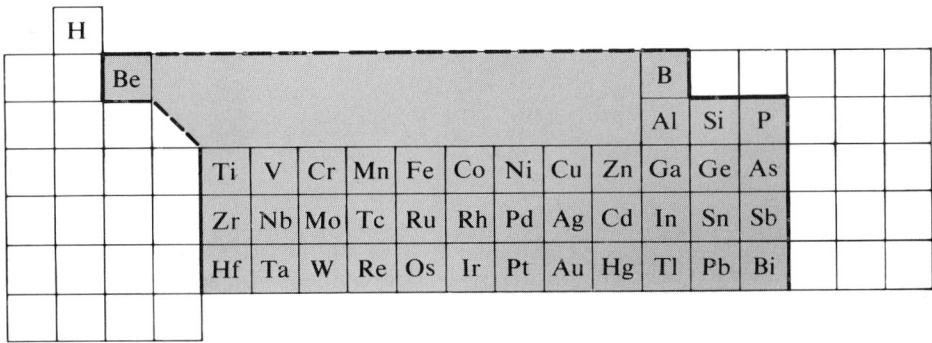

Figure 14-13   Range of partly ionic oxides in the periodic table. The shaded elements form partly ionic oxides.

partly ionic oxides within our definition, and Table 14-3 gives the formulas of these oxides; the formulas correspond much better to conventional oxidation states than those of any of the previously discussed compounds. It is also useful to point out the difference between the partly ionic oxides in Fig. 14-13 and the partly ionic hydrides in Fig. 14-9. The much greater electronegativity of O has the effect of shifting the range of partly ionic compounds to the right—this is why the general division of the periodic table in Fig. 14-1 is only a rough indication.

Table 14-3

Partly Ionic Oxides

IIa	IIIa	IVa	Va	VIa	VIIa	VIII			Ib
BeO	B_2O_3								
	Al_2O_3								
		SiO_2							
		TiO_2	V_2O_5	CrO_3		Fe_2O_3	Co_2O_3	$Ni_2O_3 \cdot H_2O$	CuO
		Ti_2O_3	VO_2	CrO_2	MnO_2	Fe_3O_4	Co_3O_4	NiO	Cu_2O
			V_2O_3	Cr_2O_3	Mn_2O_3	FeO	CoO		
			VO	CrO	MnO				
		ZrO_2	Nb_2O_5	MoO_3					
			NbO_2	MoO_2					
								PdO_2	AgO
						RuO_4	Rh_2O_3	PdO	Ag_2O
		HfO_2	Ta_2O_5	WO_3	Re_2O_7	OsO_4	IrO_3	$PtO_2 \cdot H_2O$	Au_2O_3
			TaO_2	WO_2	ReO_3	OsO_2	IrO_2	PtO	Au_2O
					ReO_2		Ir_2O_3		
					Re_2O_3				

IIb	IIIb	IVb	Vb	
			P_4O_{10}	⎫ covalent
ZnO	Ga_2O_3	GeO_2	$(PO_2)_n$	⎬ molecules
	Ga_2O	GeO	P_4O_6	⎭ in some forms
			As_4O_{10}	
			As_2O_4	
			As_4O_6	
CdO	In_2O_3	SnO_2	Sb_4O_{10}	
	In_2O	SnO	Sb_2O_4	
			Sb_2O_3	
HgO	Tl_2O_3	PbO_2	Bi_2O_3	
	Tl_2O	PbO		

Since oxygen is the most abundant element (by weight) in the earth's crust (almost 50%) it is not surprising that many of the elements that form partly ionic oxides occur naturally primarily as the oxides. Silicon, the next most abundant element, is found exclusively as SiO_2 (quartz and other forms) and silicates, which are essentially mixed oxides of Si and other electropositive atoms. We shall come back to these shortly. Iron is mined primarily as **magnetite**, Fe_3O_4, and **hematite**, Fe_2O_3; phosphorus is normally mined as the phosphate ion, PO_4^{3-}; chromium is mined as **chromite**, $FeCr_2O_4$, and so on. When it is desired to prepare an oxide, however, numerous methods are available. The elements will combine directly, although sometimes quite high temperatures are necessary to overcome the large activation energy (as for carbides, etc.):

$$2Al + 3O_2 \rightarrow Al_2O_3 \quad \Delta H^0 = -399 \text{ kcal/mole}$$

$$2Pb + O_2 \rightarrow 2PbO \quad \Delta H^0 = -104 \text{ kcal/mole}$$

Halides can often be hydrolyzed to the hydroxide or oxide; heating removes the H from the system as water vapor:

$$NiCl_2 + 2OH^- \rightarrow 2Cl^- + Ni(OH)_{2(s)}$$

$$Ni(OH)_2 \xrightarrow{600°} NiO + H_2O_{(g)}$$

This technique is applicable to any element with a high enough charge or electronegativity to have a neutral or acidic oxide, since those are the ones that will hydrolyze extensively in solution. Another technique, as we mentioned in Chapter 13, is simply to heat a carbonate, nitrate, or other oxysalt:

$$ZnCO_3 \rightarrow ZnO + CO_2 \quad \Delta H^0 = +17 \text{ kcal/mole}$$

$$2Bi(NO_3)_3 \cdot 5H_2O \xrightarrow{700°} Bi_2O_3 + 10H_2O + 6NO_2 + \tfrac{3}{2}O_2$$

Acid–Base Properties Partly ionic oxides can be slightly basic, for the more electropositive atoms (such as Ti or Fe), slightly acidic, for the more electronegative atoms (such as B or Si), or neutral or amphoteric. These properties show up, however, as reactivity toward acids or bases, since partly ionic oxides are insoluble in water. The insolubility is due to the extended covalent bond structure combined with the fact that ionic lattice energies are usually larger than the hydration energies released on dissolving. Furthermore, when high acid or base concentrations make it thermodynamically favorable for the oxide to dissolve, the process is usually very slow; thus, for instance, strongly basic solutions can be handled in glass beakers even though the glass (SiO_2 and silicates) ought thermodynamically to be soluble in the solution. The slowness is, of course, due to the very great activation energy needed to disrupt the extended lattice of the oxide. As we might expect from our kinetics discussion of Chapter 11, the rate increases on heating, and hot concentrated OH^- solutions will etch glass vessels fairly quickly. Also, the dissolution process is aided kinetically by grinding the oxide to a very fine powder, which has the effect of partly overcoming the lattice binding and providing more surface atoms not held by the full lattice energy. This is synthetically useful for the formation of hydrated salts of strong acids:

$$NiO_{(s)} + 2H_3O^+_{(aq)} + 3H_2O_{(l)} + 2ClO^-_{4(aq)} \rightarrow Ni(OH_2)_6(ClO_4)_{2(s)} \text{ on evaporation}$$

When aqueous solutions of the positive ions (present in these oxides in partly ionic form) are brought toward pH neutrality, perhaps going from pH 0 or 1 to pH 3 or 4, precipitation occurs but the oxide does not form. The oxide ion is not present in the solution because it is too strong a base, reacting with water to give OH$^-$ because the oxide ion is a better electron donor than the rest of the water molecule on which the proton was initially bound. Instead, the hydrated positive ion is more and more able to transfer its hydrate protons to surrounding water molecules, and eventually all the positive charge is lost. The resulting electrically neutral species collects or **agglomerates** into a rather gloppy semi-solid called a **hydrous oxide**. Figure 14-14 suggests the extent to which the structure of a hydrous oxide is strongly, but nonstoichiometrically, hydrated. As might be inferred from the figure, it is often difficult to know where the hydrous oxide leaves off and the solution begins.

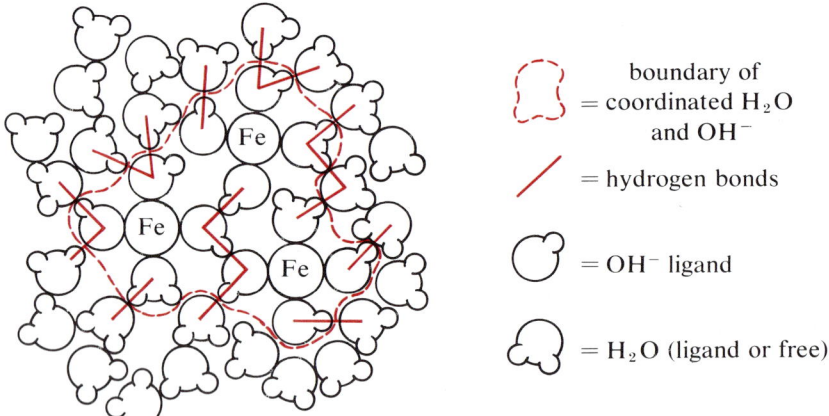

Figure 14-14 Schematic representation of the aggregation of Fe(OH)$_3$ in water, with an indefinite amount of hydrogen-bonded solvent water attached to the structure. Iron is represented here as two-dimensional Fe(OH)$_3$(OH$_2$); these units are held together by hydrogen bonds, which also bind bulk water molecules to the coordinated H$_2$O and OH$^-$.

Redox Properties Partly ionic oxides can usually be reduced to the element by either hydrogen or carbon, but this will not work for the most electropositive elements:

$$Sb_2O_3 + 3C \rightarrow 2Sb^0 + 3CO_{(g)} \qquad \Delta H^0 = +89 \text{ kcal/mole}$$

(gas generation makes entropy term favorable at high temperatures)

$$NiO + C \rightarrow Ni^0 + CO \qquad \Delta H^0 = +32 \text{ kcal/mole}$$

$$Cr_2O_3 + 3H_2 \xrightarrow{1500^\circ C} 2Cr^0 + 3H_2O \qquad \Delta H^0 = +96 \text{ kcal/mole}$$

$$Al_2O_3 + 3H_2 \longrightarrow \text{N.R.} \qquad \Delta H^0 = +225 \text{ kcal/mole}$$

$$BeO + C \longrightarrow \text{N.R.} \qquad \Delta H^0 = +120 \text{ kcal/mole}$$

These last oxides can be reduced by the electrolysis of molten salts such as AlF_6^{3-}, or by the more electropositive metals:

$$6Na + Al_2O_3 \rightarrow 2Al + 3Na_2O$$

$$2Al + Cr_2O_3 \rightarrow 2Cr + Al_2O_3$$

Reduction of oxides with powdered aluminum is known as the **Goldschmidt** or **thermite** reaction. Because of the great thermodynamic stability of Al_2O_3, this is a very versatile reaction for the preparation of elements from partly ionic oxides; Table 14-4 compares the enthalpy of formation of a number of these oxides *per oxygen*, and it can be seen that Al_2O_3 is most exothermic in its formation.

Table 14-4
Enthalpy of Formation of Oxides Per Oxygen Atom (kcal/mole O)

BeO	B_2O_3							
−146	−101							
	Al_2O_3	SiO_2						
	−133	−103						
	Ga_2O_3	TiO_2	Cr_2O_3	Fe_2O_3	NiO	Cu_2O	GeO_2	As_4O_{10}
	−86	−109	−90	−66	−29	−40	−64	−45
			WO_3	OsO_4			PbO_2	
			−64	−23			−33	

A reaction related to the carbon reduction of oxides is the chlorination by C and Cl:

$$TiO_2 + C + 2Cl_2 \xrightarrow{800°} TiCl_4 + CO_2$$

A somewhat simpler reaction, although it must be carried out under pressure at temperatures above the normal boiling point of CCl_4, is

$$2Fe_2O_3 + 3CCl_4 \xrightarrow{200°} 4FeCl_3 + 3CO_2$$

Higher-boiling chlorocarbons will react with partly ionic oxides at reflux:

$$Nb_2O_5 + Cl_2C{=}CCl{-}CCl_3 \xrightarrow{200°} NbCl_5 + Cl_2C{=}CCl{-}COCl$$

The higher electronegativity of oxygen (relative to the hydride, carbide, and other negatively charged atoms) makes higher formal oxidation states possible on the positively charged atom. This is true partly because of the greater electron-withdrawing ability of the oxygen, and partly because the added electron density expands the oxygen so that it becomes the dominant atom in the lattice and need no longer fit in a limited number of holes in the positive-atom lattice. Thus phosphorus, besides forming a covalent gaseous hydride **phosphine** (molecular PH$_3$), also forms a nonstoichiometric solid formulated more or less as P$_2$H. This almost surely represents hydrogen atoms occupying sites in a phosphorus-atom lattice, and shows phosphorus with an average oxidation state of only $\frac{1}{2}+$, whatever that means. The oxide with a similarly extended polymeric structure is P$_2$O$_5$, which is a stoichiometric layer structure having approximately close-packed oxygens (rather than phosphorus atoms) and a formal oxidation state of 5+.

Polymeric Structures The fact that many partly ionic oxides have a basic structure of close-packed O atoms makes it possible for some substitution of positive atoms to occur, either on a stoichiometric or nonstoichiometric basis. We can imagine in the simplest case that a positive atom is replaced by another with the same formal oxidation state, so that electrical neutrality is preserved and no other change is necessary. For instance, if aluminum oxide, Al$_2$O$_3$, is doped in this way with a little Cr^{III} the result is a synthetic (but perfectly authentic) **ruby**. A somewhat more complicated substitution involves a positive atom with a lower formal oxidation state than the one being replaced and another one with a higher oxidation state to make up the electrical charge difference. If a few of the Al atoms in Al$_2$O$_3$ are replaced by Fe^{III}, others by Fe^{II}, and an equal number by Ti^{IV}, the result is synthetic **sapphire**. Ruby and sapphire actually have only trace amounts of foreign atoms, but it is equally possible to form mixed oxides with the positive atoms in a stoichiometric ratio. In this event there is no point in trying to maintain the "host-compound" stoichiometry; if, for example, we take the form of Al$_2$O$_3$ that has approximately cubic close-packed O atoms and insert Mg atoms partly as replacements for Al (giving a deficiency of positive charge) and partly in holes (bringing the positive charge back up to electrical neutrality) we get the compound MgAl$_2$O$_4$, a mineral called **spinel**. Many oxides with the formula AB$_2$O$_4$ have the spinel structure, with half of the octahedral holes occupied and one-eighth of the tetrahedral holes occupied. Any combination of formal oxidation states on A and B adding up to eight is possible: 2+ and two 3+, 4+ and two 2+, 6+ and two 1+. Thus besides MgAl$_2$O$_4$ we find FeCr$_2$O$_4$ and many others in the first category, Zn$_2$TiO$_4$ and SnCo$_2$O$_4$ in the second, and Na$_2$MoO$_4$ in the third. If the spinel structure consists of Fe^{II} and Fe^{III}, the resulting mineral is FeFe$_2$O$_4$, more commonly written Fe$_3$O$_4$, **magnetite**, which as the name implies is the magnetic oxide of iron. In antiquity natural magnetite was the lodestone used as the precursor of the compass.

An interesting variation of close-packed O structures is seen in the **perovskite** structure. Perovskite is the mineral CaTiO$_3$; in it the O's *and* the large positively charged atom Ca form a cubic close-packed lattice, with the Ti in octahedral sites that are surrounded only by O atoms. Many mixed oxides with this and the spinel structure have unusual and valuable magnetic and electrical properties. A striking example of this is the perovskite barium titanate $BaTiO_3$, which is **piezoelectric**, meaning that it develops an electrical voltage between its crystal faces if it is squeezed. If a phonograph needle is attached to a $BaTiO_3$ crystal, the wiggles in the phonograph record drive the needle, which results in a varying electrical voltage, which shortly turns into 400 watts of Missa Solemnis or the Electric Indian, as the case may be. This and other unusual electrical properties result from the asymmetric distribution of the charged atoms in the perovskite structure, but the full theory of the behavior is mathematically quite sophisticated.

By far the most abundant and important class of mixed oxides is the **silicates**, which consist of foreign positively charged atoms introduced into a host lattice of silicon atoms in a more or less close-packed array of O atoms. It is sometimes difficult for the introductory student to appreciate how much of the world he sees around him is composed of silicates. Glass, beach sand, bricks, water demineralizers, granite, talcum powder, concrete, and plain dirt: all of these are silicates, and their widely diverse properties reflect not intrinsic chemical difference as much as structural difference. Perhaps the simplest place to begin is with SiO_2 (silica) itself. The most common form of silica is quartz, which consists of Si atoms surrounded tetrahedrally by O atoms to form SiO_4 groups. For this to add up to an overall composition of SiO_2, all the O atoms must be shared by two tetrahedra, and the structure of this and all other silicates consists of linked SiO_4 tetrahedra, with or without substituted atoms. There is a considerable amount of open space in such a lattice, because there are two tetrahedral holes for each atom in a close-packed lattice, but there is only half a Si atom for each O atom in a system having the composition SiO_2. The open space shows up because O atoms are missing from the hypothetical close-packed lattice to maintain proper stoichiometry; a SiO_4 tetrahedron and the **cristobalite** variation of quartz are shown in Fig. 14-15. Although the three-dimensional structure is still rather difficult to visualize from this figure, it

Figure 14-15   Silica as linked SiO_4 tetrahedra, with O sharing.

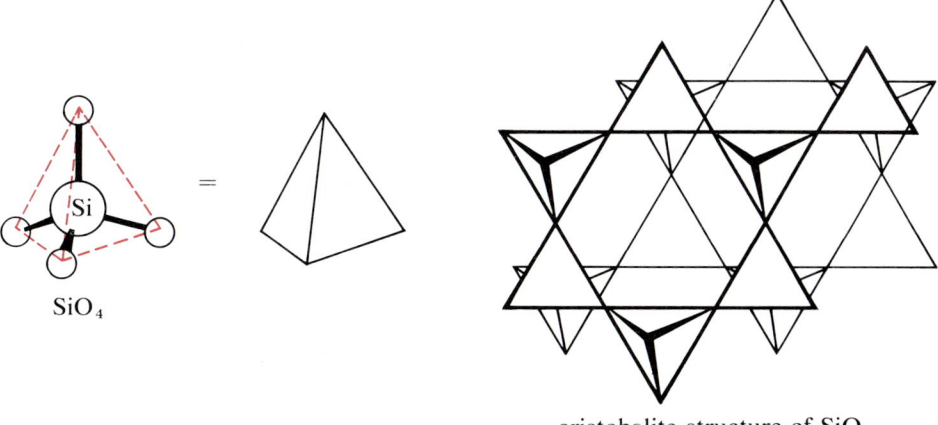

cristobalite structure of SiO_2

should be clear that there is a wide range of possibilities for the combining of these SiO_4 tetrahedra in different geometric symmetries. Indeed, if we melt quartz the result of introducing the disorder characteristic of a liquid structure is usually that the tetrahedra link in a random fashion rather than with a precise geometric symmetry. The result is a glass—a substance that on cooling yields an isotropic solid without the long-range order characteristic of crystals.

Zero-Dimensional SiO_4 Polymers Two kinds of chemical modification of linked-tetrahedra structures are possible: in one, other atoms are placed in the open spaces in the structure, and in the other, a positively charged atom is substituted for Si in some of the SiO_4 tetrahedra. These frequently occur together. When a surplus of positively charged atoms is present, the charge balance is maintained by unlinking some of the corners of the tetrahedra, so that there are more than two O per Si. On this basis we can distinguish several kinds of silicate structures; the simplest are those involving positive ions (or at least substantially charged atoms) and isolated SiO_4^{4-} tetrahedra. The semiprecious stone **zircon**, $ZrSiO_4$, is an example of this kind of structure, as is **Portland cement**, which is basically Ca_2SiO_4. There are four observed structures for Ca_2SiO_4, of which one, the γ structure, occupies a relatively large volume and is completely impervious to water. On the other hand, the β form is denser and is converted to the γ form fairly quickly by the presence of water, so if it is finely ground and mixed with water it swells and changes to a single mass of the waterproof β form. It should be apparent why setting concrete must be kept wet and why it cannot be allowed to freeze—only if the $\beta \rightarrow \gamma$ transformation proceeds to completion is the full strength of the concrete realized.

One-Dimensional SiO_4 Polymers The first step in linking SiO_4 tetrahedra involves only two of them, yielding the $Si_2O_7^{6-}$ ion if the other positively charged atoms in the lattice structure are sufficiently electropositive, but these are relatively rare, an example being $Zn_4(OH)_2Si_2O_7 \cdot H_2O$. The next step in linking the tetrahedra is the formation of either chains or rings, as suggested in Fig. 14-16. The six-tetrahedron ring Si_6O_{18} occurs in **emerald**, which is chemically $Be_3Al_2Si_6O_{18}$, and chain structures are quite common, both as the **pyroxene** single chain and the **amphibole** double chain. A fundamental feature of silicate structures is that the physical properties of the mineral often reproduce the basic symmetry of the elementary linkages of tetrahedra, and in this connection it is interesting that the term **asbestos** refers to a mixture of minerals that have the amphibole double-chain structure (along with others that are "one dimensional" because they are formed from rolled sheets of tetrahedra). The fibrous nature of asbestos, which allows it to be made into cloth, paper, and string, is of course entirely compatible with this structure.

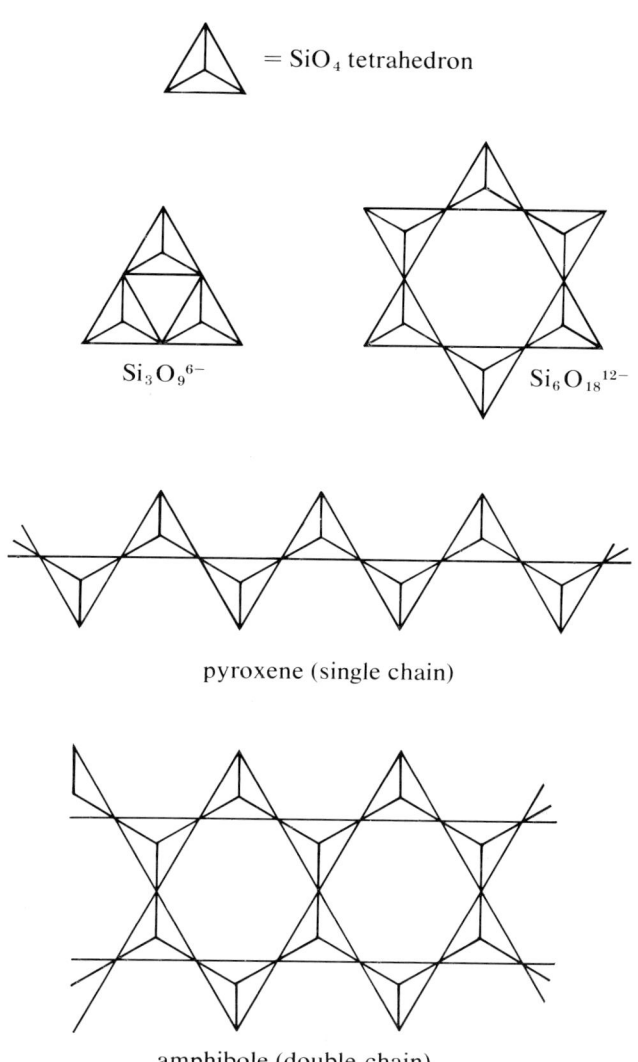

Figure 14-16 Ring and chain networks of SiO_4 tetrahedra.

Two-Dimensional SiO_4 Polymers By this time the student may correctly anticipate that the next conceptual step in developing silicate structures and properties is to form two-dimensional sheets of SiO_4 tetrahedra. Figure 14-17a shows two ways this linkage can occur, but only the one involving six-tetrahedron rings is common. These sheets can have all the tetrahedra pointing either in the same direction or in alternating directions, as in Fig. 14-17b. If we take two layers, each having all tetrahedra pointing in the same direction, and make a sandwich with the tetrahedra pointing toward the inside of the sandwich,

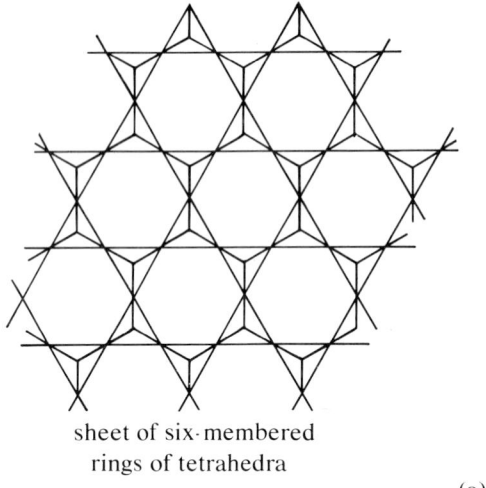
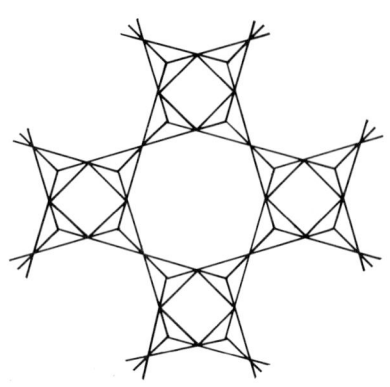

sheet of six-membered rings of tetrahedra

sheet of alternating four- and eight-membered rings of tetrahedra

(a)

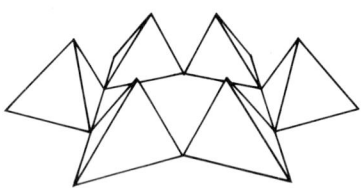
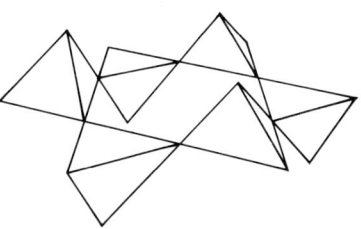

single ring from six-membered sheet in (a), with all tetrahedra pointed the same way

same ring, but with tetrahedra alternating in direction as in cristobalite structure of Fig. 14-15

(b)

Figure 14-17 Layer linkages of SiO$_4$ tetrahedra. (a) Infinite sheets. (b) Linkages of tetrahedra within a single ring.

the result is a negatively charged system, $(Si_4O_{10}^{4-})_n$. The double layer can be rendered electrically neutral by placing Al^{3+} ions in the center of the sandwich, normally accompanied by OH$^-$ ions to fill out the octahedral coordination around the Al. The resulting structure, Al$_2$(OH)$_2$Si$_4$O$_{10}$, and an analogous one, Mg$_3$(OH)$_2$Si$_4$O$_{10}$, **talc**, are both important minerals. As might be expected, the "sandwich" is very strongly bound by typical partly-ionic oxide bonds — but the sandwich itself is bound to the next sandwich only by very weak van der Waals forces, since there is no net charge on the sandwich and the outer layers of atoms which face each other are all the same kind of atom (O). This means that very little force should be necessary to deform the structure, and indeed talc is one of the softest minerals.

Suppose in addition to placing Al^{3+} or Mg^{2+} in the center of the sandwich we replace one of every four Si atoms in the layers of tetrahedra by Al. Since Si^{4+} is being replaced by Al^{3+} (as far as the electron bookkeeping is concerned), the sandwich will now have a net negative charge: $[Al_2(OH)_2Si_3AlO_{10}]^-$. Such sandwiches can be held together in an electrically neutral structure by positive ions such as K^+, as in Fig. 14-18, and the result is **mica**. Mica is harder than

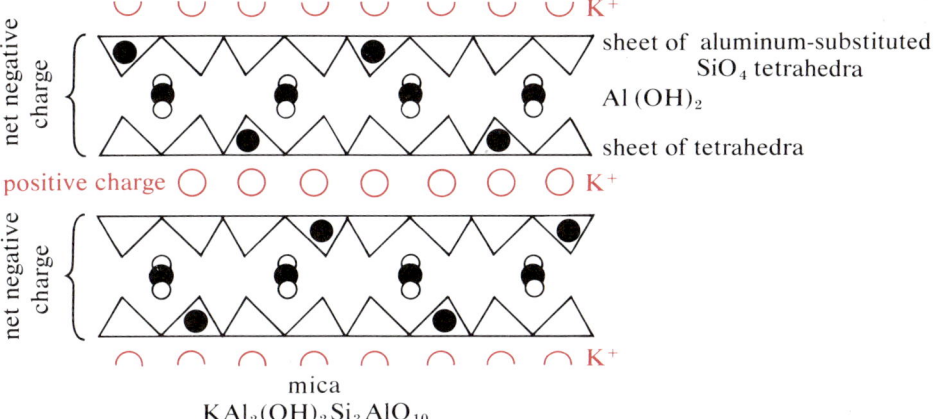

Figure 14-18 Double-layer structure of mica.

talc because the sandwiches are now being held together by ionic or electrostatic attraction, but the net charge on the sandwiches is rather diffuse and the layers can be split apart very easily, which is, of course, quite characteristic of mica. If instead of a simple ion like K^+ we use a hydrated ion such as $Mg(OH_2)_6^{2+}$ to bind the layers together, the result is **vermiculite**. Natural vermiculite is very much like mica, but if it is heated strongly the waters on the hydrated ions are driven off as vapor, and the pressure of the vapor expands the crystal by forcing the layers apart, rather like the operational principle of popcorn. The result, called expanded mica, is very light and compressible and finds considerable use as a packing material for fragile objects and as a soil conditioner. Most of the common clay minerals are layer structures more or less analogous to talc, in which the principal constituents are silicon, magnesium, and aluminum. White clays such as **kaolin** are used in pottery and in glazing paper; the more familiar orange or red clays usually owe their color to small amounts of Fe, whose color-producing properties will be discussed in Chapter 16.

Three-Dimensional SiO_4 Polymers The only remaining general possibility for linking SiO_4 tetrahedra is to connect them in a three-dimensional polymeric structure. If each O on a SiO_4 tetrahedron is shared by two Si's, the result has the formula SiO_2. This is analogous to the quartz structure mentioned earlier. More generally, the tetrahedra have some substitution of Al for Si, which requires that some positive ions be inserted somewhere in the structure. There are both natural and synthetic examples of three-dimensional networks of these tetrahedra, having a wide variety of geometries. Some are quite compact and dense, as in granite, which is a mixture of quartz, mica, and a silicate structure known as a **feldspar**, with a three-dimensional network of alternating four- and eight-membered rings of tetrahedra, as in Fig. 14-17a. Others have large cavities within their structures and the resulting low densities, for example **zeolites**. These have two interesting properties: they take up water and other small neutral molecules quite readily into the cavities, and they readily exchange whatever positive ions are present for others in any water solution that may pass over them. The first property is used to good advantage in artificial silicates called **molecular sieves**, which can be tailored to accept small molecules such as water or small organic molecules but reject larger, more bulky molecules even if there is no appreciable chemical difference between the molecules being separated. The second property is called **ion exchange**. Water softeners and demineralizers rely on ion exchange, although usually with organic polymers rather than silicates; home water softeners have artificial zeolites with Na^+ ions that exchange for Ca^{2+} ions in the incoming hard water, while demineralizers have H_3O^+ ions that exchange for any positive ion in the water being treated. There are also a number of applications in analytical chemistry for ion-exchange methods, since the technique allows all the ions in a very dilute solution to be collected and regenerated on demand.

The number of naturally occurring mixed oxides having partly ionic bonding is enormous, as witness the very large number of minerals, most of which fall into this class. The preponderant composition of the earth's crust is that of an aluminum- and magnesium-substituted silicate, which primarily reflects the relative abundance of these elements and their readiness to form mixed-oxide structures. It is interesting to note that the mineralogy of the moon appears to be noticeably different from that of the earth, because of the differences in elemental abundances.

Biochemical Polymeric Phosphates One last topic under the general heading of partly ionic oxides involves the condensation reaction (eliminating water) discussed on p. 730, where the formation of the triphosphate ion is mentioned. This type of reaction can proceed in either direction, depending on solution conditions, and it occurs with equal readiness even if the phosphate ion is partially substituted. In particular, the reaction shown in Fig. 14-19 is important because it represents the principal source of energy in biological systems. The muscular energy of any animal results from the controlled hydrolysis of

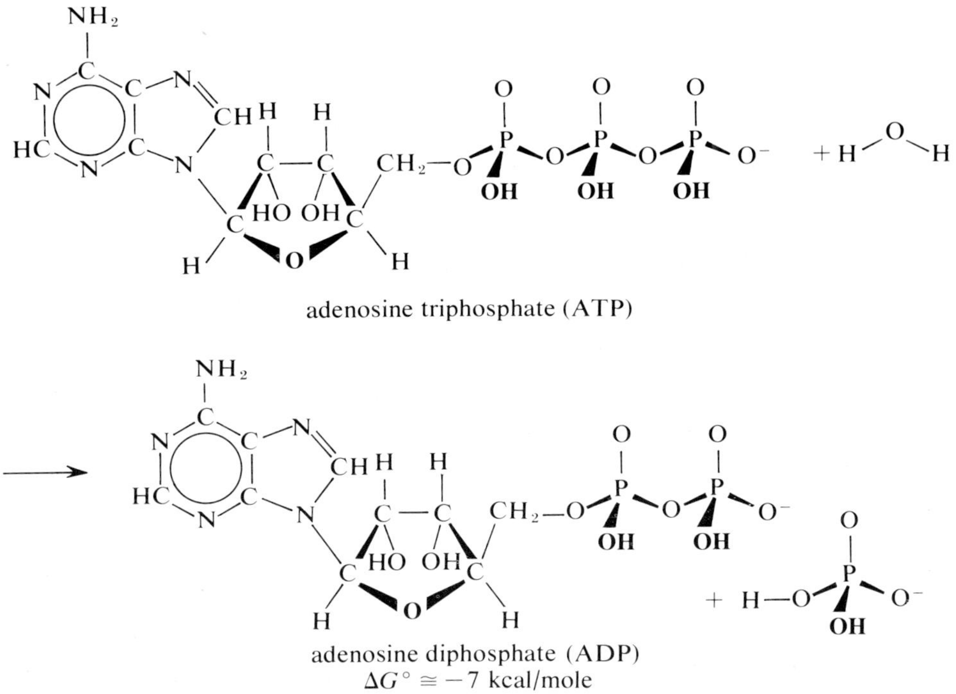

Chemistry of Partly Ionic Compounds | 751

adenosine triphosphate (ATP)

adenosine diphosphate (ADP)
$\Delta G° \cong -7$ kcal/mole

Figure 14-19 Linked-tetrahedra structures of biologically important partly ionic oxides (polyphosphates).

ATP, as shown, and even the energy of the firefly's glow results from this reaction. The molecules ADP and ATP are not exactly oxides, but it is nonetheless true that the energy release in the hydrolysis reaction arises from the partly ionic character of the P—O—P bonds; note that the electronegativity difference between P and O is $3.50 - 2.06 = 1.44$.

If the ATP hydrolysis reaction really represents some kind of equilibrium, its free energy must be influenced by the concentrations of the species in the reaction, and in particular by the pH of the cell solution. At the pH of normal human cell solutions (7.38) the hydrolysis reaction has a free-energy change of about -7 kcal/mole. It is clear that this is a source of energy; basically it is due to the lowering of electronic energy that occurs when nonbonding oxygen electrons in water are shared with a phosphorus atom in forming the free phosphate monomer. The phosphorus is a good acceptor because its electrons have to a substantial extent been withdrawn by the chain of electronegative atoms in the triphosphate part of the ATP molecule. However, the question naturally arises of how the ATP is formed in the first place, since the reverse of the hydrolysis reaction would have a positive ΔG associated with it. There are two ways the free-energy change can be modified for a given reaction: by statistically affecting the odds for a particular molecular transformation through a concentration change (as in pH control), or by changing the energy relationships within the reacting molecule itself through control of its geometry. Both of these are probably involved in the formation of ATP, but the second seems to be a major factor and to be responsible for the very complex molecular organization that is being discovered within the cell—not only with respect to the formation of ATP but in many other cellular processes, such as genetic control through nucleic acid replication. In Chapter 6 we noted that NH_3 was much more stable in a pyramidal conformation than in a planar one, because a pair of electrons that would be nonbonding in the planar geometry become bonding in the pyramidal shape. This is the sort of effect thought to be at work here—if a molecule's environment can force it to depart from its normal geometry to some extent, its electronic energy may change enough to make reactions possible that ordinarily would not occur. Still, you can't get something for nothing! The energy that ATP releases on hydrolysis must be built into it in its formation from some energy source, which in all biochemical systems is the oxidation of food. Such oxidations, using atmospheric oxygen, release substantial amounts of energy, which the cell uses very efficiently (in a thermodynamic sense) by coupling a series of redox reactions having reduction potentials (i.e., free-energy changes) intermediate between the food oxidation and the ATP formation. The many steps in the mechanism provide an approximation to a thermodynamically reversible reaction, which we have seen provides the maximum amount of possible work output for two given initial and final states. A good deal is known about the energy-transport process in cells, but a complete explanation on a molecular basis is still beyond our reach.

Partly Ionic Sulfides

Sulfur is much less electronegative than oxygen (2.44 vs. 3.50) and if we were strictly to continue the practice of defining a partly ionic compound as one having bonded atoms with an electronegativity difference of between 1.0 and 2.0 units hardly any sulfides would qualify, as a survey of the periodic table inside the front cover will indicate. However, the physical properties of many sulfides of elements in the shaded area of Fig. 14-1 are akin to those of the corresponding oxides, in spite of this, and studies of the structures suggest the same sort of extended networks, although with less charge separation and ionic character. Even with less negative charge accumulating on the sulfur atoms in the lattice, they are sufficiently large to form more or less close-packed structures in most cases, with the positively charged atoms in the octahedral sites of the close-packed S lattice. A common example is the structure adopted by many transition-metal sulfides: the nickel arsenide, NiAs, structure of Fig. 14-20, in which the S's are nearly hexagonally close packed and the metal

Chemistry of Partly Ionic Compounds | 753

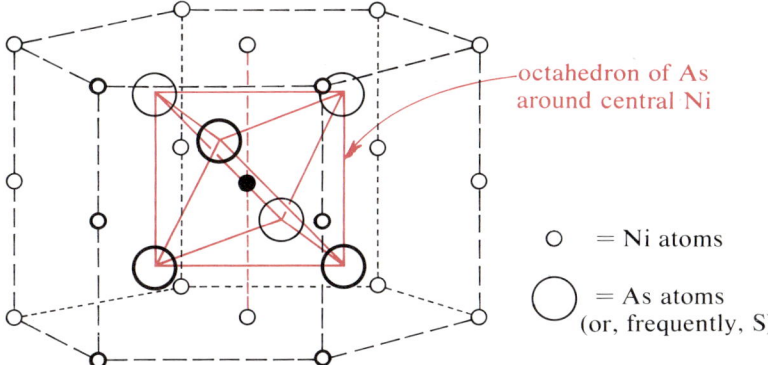

Figure 14-20 Nickel arsenide structure.

atoms are in the octahedral holes. Note that the arrangement is such that the metal atoms have two other metal atom neighbors at about the same distance as the S's. This suggests the possibility of metal–metal bonding as in the metal itself, and indeed many of the properties of these sulfides, such as a substantial electrical conductivity and metallic luster, are reminiscent of metals. In fact, the brilliant yellow metallic luster of a related structure, **pyrites** (FeS_2), has countless times been responsible for the rueful application of its common name, fools' gold.

Like the other partly ionic compounds discussed in this chapter, partly ionic sulfides are often nonstoichiometric. For example, a compound we may approximately call iron(II) sulfide is found in Canada as a mineral called **pyrrhotite** (which is mined because it contains moderate quantities of Ni). The lattice is deficient in Fe atoms, and instead of the simple stoichiometry FeS some rather ludicrous formulas such as Fe_6S_7 and even $Fe_{16}S_{17}$ have been proposed. There is no reason to think that integers govern the relative number of missing Fe atoms, however, and a more reliable guide is simply the statement that it is iron-deficient FeS. Apparently, however, the particular formulation 7Fe:8S can be prepared so as to have the Fe vacancies distributed symmetrically in alternate layers of Fe atoms, with one-eighth missing overall, and the resulting structure (like some of the mixed oxide structures) has unusual magnetic properties. Quite generally, transition-metal sulfides can be prepared over a surprisingly wide variety of nonstoichiometric proportions, for reasons we shall explore further in Chapter 16. The other metals and metalloids having only s and p valence electrons (such as Hg, Ga, and As) show more conventional and more reliable stoichiometry.

Partly ionic sulfides are easily prepared, but even the modest effort can often be avoided, because many of them occur naturally. At least a dozen elements are mined as their sulfide or as a mixed sulfide (analogous to the mixed oxides), including most of the metals and metalloids having p valence electrons, such as lead. This has been true in at least a few cases since before the dawn of history, and the last mists of alchemy hover, perhaps, about the names of these minerals; zinc occurs as ZnS, **zincblende**, mercury as HgS, **cinnabar**, lead as PbS, **galena**, and arsenic as As_4S_4 and As_4S_6, **realgar and orpiment**.

The obvious way to prepare partly ionic sulfides is almost universally effective; one warms elemental S with the other element, either directly or in some inert solvent such as CS_2. Only iridium, platinum, and gold from the shaded area of Fig. 14-1 fail to react directly with sulfur under any conditions thus far explored. The favorable free-energy change for the reaction results from the substantial lattice energy of the resulting sulfide; even so, most of the ΔG^0 values for the reactions are relatively small, ranging from about -10 to -40 kcal/mole reaction:

$$2Sb_{(s)} + 3S_{(s)} \longrightarrow Sb_2S_{3(s)} \qquad \Delta G^0 = -43.5 \text{ kcal}$$

$$Pb_{(s)} + S_{(g)} \xrightarrow{500°} PbS_{(s)} \qquad \Delta G^0 = -22.2 \text{ kcal}$$

$$Cu_{(s)} + S_{(solution\ in\ CS_2)} \xrightarrow{100°} CuS_{(s)} \qquad \Delta G^0 = -11.7 \text{ kcal}$$

Another method commonly used, particularly for the more metallic elements, is the precipitation of the sulfide from an aqueous solution of the positive ion (hydrated). Since the precipitation is an equilibrium, and since H_2S is a weak acid also governed by a proton-transfer equilibrium, sulfides can often be precipitated selectively by controlling the pH of the initial solution; this is the basis of several schemes for identification of metal ions, which have the additional advantage that most of the sulfides are brightly colored, facilitating identification. The formation of the sulfides from ions should not lead the student to think the sulfides themselves are ionic, however. Perhaps the best evidence that they are not comes from their formulas; although they are formed from the hydrated metal ion, the sulfides are never hydrated, which says that the sulfide ion is such a good electron donor as to be able to displace the water donor molecules in every case. A very good electron donor will, of course, leave only a very small positive charge on the acceptor ion, which is the condition described in this chapter generally.

One of the most characteristic reactions of the partly ionic sulfides is that with atmospheric O_2 at elevated temperatures, called "roasting" in commercial refining processes:

$$2CuS_{(s)} + 3O_{2(g)} \rightarrow 2CuO_{(s)} + 2SO_{2(g)} \qquad \Delta G^0 = -181 \text{ kcal}$$

$$Sb_2S_{3(s)} + 5O_{2(g)} \rightarrow Sb_2O_{4(s)} + 3SO_{2(g)} \qquad \Delta G^0 = -316 \text{ kcal}$$

The rather large free-energy changes are due to the formation of the oxide lattice with greater charge separation and the smaller oxide ion both contributing to a greater lattice energy. The oxides that result from the roasting process are subsequently reduced to the elements by heating with coke:

$$2CuO_{(s)} + C_{(s)} \rightarrow 2Cu^0_{(l)} + CO_{2(g)}$$

A more direct method involves simply strongly heating the sulfide with scrap iron:

$$PbS_{(s)} + Fe_{(s)} \rightarrow Pb_{(l)} + FeS_{(s)}$$

$$Sb_2S_{3(s)} + 3Fe_{(s)} \rightarrow 2Sb_{(l)} + 3FeS_{(s)}$$

Iron is, in general, less electronegative than the elements whose sulfides can be reduced in this way. This means that the FeS lattice is more ionic than the other sulfide lattice and generally has a higher lattice energy, which makes the free-energy change favorable for the reaction as written. These reactions are of great importance because of their mining and smelting applications; in particular, all of the myriad electrical applications of copper rely on obtaining the metal by this process. However, the enormous quantities of SO_2 that are produced take their toll in severe pollution of the plant's environment.

Partly Ionic Halides

Essentially all the metals in the shaded area of Fig. 14-1 form partly ionic halides with properties analogous to those of the other partly ionic compounds. Deferring (again) a discussion of the transition-metal compounds until Chapter 16, we can note that the halides generally show somewhat better stoichiometry than the other partly ionic compounds. As with the oxides, this is due to the greater electronegativity of the halogens, which tends to promote charge separation and produce larger negatively charged species that will not fit into sites in a lattice of the positively charged atom. Rather than fitting into an established lattice, then, the halides form a new lattice, and several of the most characteristic partly ionic lattices are formed predominantly by halides (see Section 14-3). The resulting compounds are usually nicely crystalline, with the stoichiometry nearly exact. With the less electronegative halogens (Br and I) and metals having a reasonably stable lower oxidation state, departures from stoichiometry can occur due to the reaction

$$MI_n \rightarrow MI_{n-2} + I_2$$

or its equivalent. If this occurs to a slight extent, the stoichiometry of the compound will show a slight deficiency of Br or I; if it occurs to a substantial extent it may prove impossible to prepare the higher-formal-oxidation-state compound. For instance, Tl has two commonly observed formal oxidation states, (I) and (III); both TlCl and $TlCl_3$ can be prepared, and both have properties corresponding to partly ionic compounds, but $TlCl_3$ decomposes to $TlCl$ and Cl_2 only slightly above room temperature. On the other hand, $TlBr_3$ cannot be prepared or stored at room temperature—and although TlI_3 exists, its properties make it clear that it is really thallium(I) tri-iodide, $Tl^+I_3^-$. This pattern is reflected in other metals' halides, and even in covalent halides, as Chapter 15 will show.

Preparation Several methods are available for the preparation of partly ionic halides. The most obvious is the direct reaction of the elements, which is effective for nearly all metals:

$$Be_{(s)} + Cl_{2(g)} \xrightarrow{heat} BeCl_{2(s)}$$

$$2Al_{(s)} + 3I_{2(g)} \xrightarrow{heat} 2AlI_{3(s)}$$

$$2Bi_{(s)} + 3Br_{2(l)} \xrightarrow{heat} 2BiBr_{3(s)}$$

The lattice energy of the product halide serves to provide the favorable free-energy change for the reaction. An alternative preparation relies on influencing the equilibrium for a gaseous reaction by sweeping away a gaseous product by a gaseous reactant:

$$CoO_{(s)} + HBr_{(g)} \xrightarrow{heat} CoBr_{2(s)} + H_2O_{(g)}$$

$$Al_2O_{3(s)} + 3C_{(s)} + 3Cl_{2(g)} \xrightarrow{1000\,°C} Al_2Cl_{6(g)} + 3CO_{(g)}$$

$$ZnS_{(s)} + Cl_{2(g)} \xrightarrow{700\,°C} ZnCl_{2(g)} + S_{(g)}$$

For the first of these reactions the equilibrium constant involving the pressures of the gaseous reaction components is

$$K_P = \frac{P_{H_2O}}{P_{HBr}}$$

If the reaction is carried out in a heated tube with a fresh current of HBr continually sweeping away any H_2O as it forms, the pressure of H_2O can never get high enough in the reaction area—the oxide surface—to reach the equilibrium value, with the result that the reaction proceeds completely to the right. The other reactions are analogous. In the second, the purpose of the C is to render the overall free-energy change more favorable by forming the strongly bonded molecule CO, but as the third reaction suggests and as other reactions with less stable oxides demonstrate, the C is not always necessary. A related series of reactions has already been described under the reactions of oxides:

$$2BeO_{(s)} + CCl_{4(g)} \xrightarrow{800°} 2BeCl_{2(g)} + CO_{2(g)}$$

$$V_2O_{5(s)} + C_3Cl_{6(l)} \xrightarrow{heat} VCl_{3(s)} + C_3Cl_4O_{(l)}$$

Chemistry of Partly Ionic Compounds | 757

In these reactions a favorable free-energy change is due to the formation of the very strong C—O bond in the product, overcoming the large lattice energy of the metal oxide.

Preparation in the Presence of Water: Hard and Soft Acids (Cations) Many of these halides can be formed in aqueous solution, a process that is particularly favorable with metals that form stronger bonds to halogen atom donors than to water (oxygen donor atom). In Sections 6-8 and 10-3 we introduced the concept of hard and soft acids and bases; since, (from Table 10-2) the halide ions are soft bases, we expect them to displace hydrated water molecules on positive ions if the positive ion is a soft acid. In practice, this means that most of the metal atoms with p valence electrons, a filled inner d shell, and a relatively low formal charge, such as Pb^{2+}, will form anhydrous halides out of water solution:

$$Pb(OH_2)_{6(aq)}^{2+} + 2Cl^-_{(aq)} \rightarrow PbCl_{2(s)} + 6H_2O_{(l)}$$

$$Hg(OH_2)_{2(aq)}^{2+} + 2Cl^-_{(aq)} \rightarrow HgCl_{2(s)} + 2H_2O_{(l)}$$

Not all the metal ions in this group are so obliging; when $ZnCl_2$ is formed by dissolving ZnO_2 in HCl and evaporating away the water and excess HCl, the product contains at least one water of hydration per Zn atom. If the evaporation is carried out under a stream of HCl gas, however, anhydrous $ZnCl_2$ results. In a number of cases the anhydrous halide can be obtained from an aqueous solution or a hydrated salt by heating, but only if the metal ion is reasonably "soft" — has a relatively low charge and, usually, a substantial inner core of d electrons. If the metal ion is "hard" it will bond more strongly to the O in the water than to the halide, and the result of heating will be to drive off HCl, HBr, or HI vapor, not water, leaving behind the metal oxide or oxyhalide:

$$Co(OH_2)_6Cl_{2(s)} \xrightarrow{heat} CoCl_{2(s)} + 6H_2O_{(g)}$$

$$BiCl_3 \cdot 2H_2O_{(s)} \xrightarrow{distill} BiCl_{3(s)} + 2H_2O_{(l)}$$

but

$$SnCl_4 \cdot 5H_2O_{(s)} \xrightarrow{heat} SnO_{2(s)} + 3H_2O_{(g)} + 4HCl_{(g)}$$

$$2Fe(OH_2)_6Cl_{3(s)} \xrightarrow{heat} Fe_2O_{3(s)} + 9H_2O_{(g)} + 6HCl_{(g)}$$

However, in many cases the hydrated halide can be converted into the anhydrous compound even for "hard" positive ions, by allowing the hydrate to react with a compound that preferentially removes water:

$$Cr(OH_2)_6Cl_{3(s)} + 6SOCl_{2(l)} \xrightarrow{heat} CrCl_{3(s)} + 6SO_{2(g)} + 12HCl_{(g)}$$
thionyl
chloride

$$\text{Ni(OH}_2)_6\text{Cl}_{2(s)} + 6\,\text{CH}_3\underset{\underset{\text{OCH}_3}{|}}{\overset{\overset{\text{OCH}_3}{|}}{\text{C}}}\text{CH}_{3(l)} \xrightarrow{\text{heat}} \text{NiCl}_{2(s)} + 6\,\text{CH}_3\overset{\overset{\text{O}}{\|}}{\text{C}}\text{CH}_{3(l)} + 12\,\text{CH}_3\text{OH}_{(l)}$$

2,2-dimethoxypropane
(DMP)   acetone   methanol

In effect we are allowing the hydrated ions to react with compounds that are even "harder" acids than the ions; the S in $SOCl_2$ and the central C in DMP serve this purpose, having a high formal oxidation state and no inner d electrons. Note that protic acids cannot be used, because the result would be the loss of HCl gas — the compound must be a Lewis acid, preferably one that gives gaseous products or neutral solvent molecules. Naturally, if we need to get an anhydrous partly ionic halide from water solution, we use only the necessary degree of force, working up to these latter methods for the tough ones.

Donor–Acceptor Reactions Partly ionic halides have one reaction that is of primary interest, that with electron donors. Remember that, using our rough criterion of electronegativity difference, a partly ionic halide necessarily has a more electronegative metal atom than an ionic halide. This means that it attracts electrons more strongly and is a better electron acceptor, which in turn means generally stronger bonds formed with electron donors and favorable enthalpy changes for reactions with good donors. Hydrolysis is a good example of a donor–acceptor reaction of this type, the donor being, of course, water. For ionic compounds, the reaction of the positive ion with water corresponds only to an electrostatic ion–dipole attraction, which is broken when the ionic lattice is reformed. For partly ionic compounds, the reaction is usually considerably more vigorous, with the O electrons (from H_2O) being drawn into bonds to the positively charged atom to such an extent as frequently to transfer protons to other water molecules, which then acquire their own hydration energies — see Section 12-11.

The concept of hard and soft acids and bases can again be useful in rationalizing the differences within the general grouping of partly ionic halides. The "harder" the acid — the electron-acceptor positively charged atom — is, the more vigorously it reacts with water, which is a "hard" base. Thus $BeCl_2$ hisses when it hits water, the heat of hydration being so great as to boil some of the surrounding water. On the other hand, larger atoms or those with a less positive formal oxidation state are relatively "soft" and their compounds with the relatively soft halogens release much less heat on reacting with water, as when $SnCl_2$ dissolves in water with modest evolution of heat. And the very softest atoms essentially show no reaction with water at all, since the bonding to the soft halogen provides a superior stability; thus the soft halides CuCl, AgCl,

Hg_2Cl_2, and $PbCl_2$ are only slightly soluble in water. Even with the positive formal oxidation state increased in $HgCl_2$, which increases the attraction for the water electrons and the solubility, the dissolved species appears to be essentially molecular $HgCl_2$, with its Hg—Cl bonds intact, since the solution conducts electricity very poorly.

Other donors have analogous reactions with the positively charged atom, and indeed some soft bases such as halide ions will form donor–acceptor compounds or complexes with soft acid atoms even in water solution where statistically the odds favor water by 10 or even 100 to 1. For instance, aqueous solutions of $CuCl_2$ are blue, because of the characteristic light absorption of the $Cu(OH_2)_4^{2+}$ ion, but making the solution very strongly acidic with HCl (thereby raising the concentration of the Cl^- ion) turns it brown, because of the formation of the $CuCl_3^-$ or $CuCl_4^{2-}$ ion. Nearly all metals can form halide complexes such as MCl_4^{n-} or MCl_6^{n-}, but of course only the "softest" metal ions can form such complexes or remain stable in water solution. Many electron-donor atoms and molecules exist, and a very wide variety of donor–acceptor compounds has been described; those that are stable or nearly so in aqueous solution have been discussed in Section 10-2. Of course, providing an environment other than that of a water solution can change the relative stability of a compound completely, and nonaqueous methods are widely used in preparative chemistry.

14-5 Summary

In this chapter we have pointed out the changes in chemical properties that result when some directional covalent bonding is superimposed on the nondirectional electrostatic attraction that binds ionic lattices together. Using a crude approximation to predict charge on atoms bound in a lattice, we have demonstrated that the covalent contribution to the lattice energy is substantial even in compounds with a rather substantial difference in electronegativities. A rather arbitrary but helpful division may be made by taking these partly ionic compounds as those in which the electronegativity difference between adjacent atoms is between 1.0 and 2.0 units. We saw that for these compounds a polymeric structure was characteristic, either in one, two, or three dimensions. Frequently this structure consists of a more or less close-packed lattice of one kind of atom, with the other kind of atom present in the interstices of the original lattice. For small electronegative atoms, this usually means the formation of compounds with metals in which the lattice of the metallic element is changed very little, and indeed the interstitial hydrides and carbides are sometimes called **metallic** hydrides or carbides. However, for larger atoms or for stoichiometries in which the electronegative atom must be present in large excess, the reverse often happens. Oxides, sulfides, and such compounds as CaB_6 often have structures that are nearly close packed in the electronegative atom, with the metal atoms in appropriate interstitial sites within this lattice. This situation usually leads to somewhat more precise stoichiometry, since the lattice of the compound must be formed from scratch rather than being modified from an existing metallic lattice.

In surveying the occurrence and chemical reactivity of these systems, some interesting and perhaps familiar minerals were discussed: besides gem stones such as ruby, sapphire, and zircon, the more mundane silicate species involved in asbestos, mica, vermiculite, concrete, and clay were discussed. Because minerals nearly all fall in this category of partly ionic compounds, the structure of the compounds is perhaps more interesting than their reactivity (although chemical synthesis has to be applied to any mineral that is an ore, of course).

In the next chapter we shall take up covalent compounds — those in which there is very little or no electrostatic influence on the solid lattice the compound adopts. Extended polymeric structures will still be possible, some occurring naturally as in biological systems, but to a much greater degree it will be possible to examine chemical reactions in detail and apply them to synthetic problems of interest to our technological society.

Study Problems

1. One of the crystal forms of Sb_2O_3 is a linear polymer. In it each O atom is bonded to two Sb atoms; how many O's are bonded to each Sb? Draw a possible single-chain structure and a double-chain structure for Sb_2O_3, both consistent with this requirement. Use the VSEPR method from Chapter 6 to predict approximate bond angles around Sb and O. Does one of the two polymeric structures you have proposed allow these bond angles to be met more easily than the other?
2. Show that only two MoO_6 octahedra would need to be moved in the $[Mo_7O_{24}]^{6-}$ structure to give a flat sheet of linked octahedra that would be identical with a section of the $CdCl_2$ layer structure. The flat sheet is characteristic of the 6-molybdotellurate ion, $[TeMo_6O_{24}]^{6-}$.
3. Steel objects that must show great resistance to wear, such as ball-bearing parts, are sometimes heated to about 500°C in gaseous ammonia for several hours. On cooling, the surface appears no different but resists scratching or other mechanical penetration about 10 times as well as before treatment. What has happened chemically?
4. Why does zinc carbonate yield zinc oxide on heating even though the process is endothermic?
5. When $TlBr_3$ prepared at low temperatures is allowed to warm to room temperature it decomposes to a compound that has two Br's for each Tl. However, magnetic

measurements (as in Chapter 6) show that the thallium is *not* Tl^{2+}. What is the true formula for the new compound, and how do magnetic measurements rule out Tl^{2+}?

6. Ion-exchange methods can be applied to either cations or anions in a solution passing over the ion-exchange structure. An anion-exchange structure must have positively charged groups permanently bonded into it, with negative ions held electrostatically in open spaces in the structure. Describe a way such a structure could be used to separate a "hard" from a "soft" cation (e.g., Mg^{2+} from Zn^{2+}).

7. Discuss the trends apparent in the following enthalpies of solution (given in kilocalories per mole):

KCl	$CaCl_2$	$GaCl_3$	$GeCl_4$
−4.2	−17.4	−45.1	−24.1

CsCl	$BaCl_2$	$TlCl_3$	$PbCl_2$
+4.8	−2.1	−8.4	+6.8

8. Ordinary pottery is a mixture of kaolin (m.p. = 1700°C) and feldspar (m.p. = 1300°C). In firing, the temperature is raised high enough to melt the feldspar, which forms a glass cementing the solid kaolin particles together into a somewhat porous mass. What changes in composition would give translucent, nonporous chinaware? How could a transparent glaze be applied to the china?

9. Why should layer structures be rare for oxides and fluorides but quite common for sulfides and bromides?

10. Both oxides and sulfides tend in many cases to be approximately close packed in the anion, with cations in octahedral or tetrahedral sites. But while substitution of one cation for another is strictly governed by ionic size relationships in oxides, sulfides often show substitution by a wide range of cations with little size correlation. What is the origin of this difference?

Some Further Reading

The references from Chapter 13 generally apply here; in particular, the books by Cotton and Wilkinson and by Phillips and Williams are useful (see, e.g., Chapters 8 and 14 of Phillips and Williams). This is also a particularly good place to use the reference by Wells from Chapter 4, since much of these compounds' interest lies in their structures.

Since many minerals fall into this category, two references are given below that are completely outside the scope of this book but in which some students may be interested:

Krauskopf, K. B., *Introduction to Geochemistry*, New York: McGraw-Hill, 1967. The theory of the formation of minerals and geological bodies. Requires no more chemistry than this book provides, but assumes some introductory geology; pretty readable.

Johnstone, S. J. and Johnstone, M. G., *Minerals for the Chemical and Allied Industries*, London: Chapman and Hall, 1961. This is for the engineer or perhaps even for the economist; it is a list of minerals that are important industrially, with a brief discussion for each of production, uses, and so on. More like a handbook than a text.

15

Covalent Compounds

Covalent bonding is the last type of bonding remaining to us in our survey. As Section 14-1 indicated, there is a type of chemical bonding in which the simple electrostatic attraction of a positively charged atom for a negatively charged atom plays a negligibly small role in the total bond energy. If two elements having very similar differential ionization energies form a bond, the total amount of charge transfer from one atom to the other will be very small; since we can relate electronegativity to differential ionization energy in an approximate way, this corresponds to saying that the charge transfer is small if the electronegativity difference is small. Looking back to Chapter 14 again, we state rather arbitrarily that the energy effect of charge transfer is insignificant if the electronegativity difference between the bonded atoms is less than 1.0 unit. For compounds in which small electronegativity differences are displayed, then, we expect to see the structure and chemical properties governed by molecular-orbital electron distributions, with little or no Madelung lattice energy. What differences will this make?

15-1 Characteristics of Covalent Bonding

We remember first of all that the extended structure of ionic and partly ionic compounds is due primarily to the advantages of close packing (more or less) in achieving greatest stability through a favorable lattice energy. If there is no electrostatic lattice energy, there presumably will not be any tendency to form an extended lattice, whether three-dimensional, layer, or chain. In other words, if a group of atoms combine covalently, we expect to see a finite—and usually small—molecule. Of course covalently bonded molecules can form a solid lattice structure, with molecules held in place by van der Waals forces, but as we saw in Chapter 4 this leads to a relatively low-melting, soft, volatile crystal. Relying on our ideas of electron stability in MO's from Chapter 6, we expect a stable covalent molecule to be one that, generally speaking, has electrons distributed according to all of its bonding and nonbonding orbitals, and few or no electrons distributed according to its antibonding orbitals. If relatively low-energy distributions (bonding or nonbonding orbitals) are vacant, the molecule will be relatively unstable because it will be an excellent electron acceptor. If, on the other hand, the molecule has high-energy electrons, either because it has antibonding electrons or because it has nonbonding electrons equivalent in energy to AO's from an electropositive atom, it will be relatively unstable because it will readily rearrange internally or react with another molecule as a donor to increase the electrons' stability by lowering their energy. This is essentially what we said in Sections 6-10 and 6-11 in accounting for the acceptor and donor properties of BH_3 and NH_3, respectively.

In thermodynamic terms, we are saying that the course of a reaction between covalent molecules will be determined primarily by the energy factor—ΔH^0, the enthalpy change. If an ionic or partly ionic lattice can form as one of the products, its lattice energy will usually make ΔH^0 strongly negative for the reaction, which of course is favorable. If only covalent products result, ΔH still dominates the spontaneous direction of the reaction in many cases, because the entropy contribution to the free-energy change is fairly small at ordinary temperatures; for a covalent reaction system this means that bond energies dictate the products.

Kinetic Control of Mechanism

There are many cases, however, in which the sum of bond energies is very nearly the same for many possible products of a reaction. This is where another substantial difference appears between covalent reactions and partly ionic or ionic reactions: the specific product of a covalent reaction often depends on the kinetics of the reaction rather than on the thermodynamics. In other words, the mechanism of a covalent reaction is particularly important because it often allows us to distinguish between several possible products, all of which have about the same free energy of formation. If one product can be formed by a mechanism having a relatively low activation energy, but all others require a mechanism having a very high activation energy, then the former will predominate. And as we mentioned in Chapter 11, electrostatic attraction takes control of the mechanism at this point. Most covalent reactions proceed by a "partially ionic" mechanism in which a nucleophile or potential electron donor reacts with an electrophile atom or seat of at least a small positive charge. This is because the activation energy for such a mechanism (see Fig. 11-10) is lower than it would be in the absence of the attraction. In Section 15-4 we shall look at some mechanisms and the way they dictate the detailed structure of the reaction product.

Types of Covalent Compounds

Covalent compounds fall into two broad categories. The first is fairly obvious: compounds between elements having similar electronegativities. In this case the valence-orbital ionization potentials are similar and the MO's describing the electron distribution are related about equally to the AO's of each atom; there is little charge transfer and no ionic attraction. There is a second category, however, which is a little more subtle: compounds between elements having moderate electronegativity differences, but with the more electropositive atom in a highly positive formal oxidation state. Here the moderate charge separation that might be expected (as in partly ionic compounds) is smaller than the electronegativities alone suggest because of the large number of electronegative ligand atoms that are drawing electrons from the central atom. Since the positive charge accumulating on the central atom attracts electrons strongly, the net charge transfer to any given ligand atom is quite limited. It is as if the VOIP of the central (electropositive) atom AO's became larger with positive charge, making them more nearly equal to the ligand electronegative atom VOIP's and causing the bonding to be more covalent. As an example, consider antimony pentachloride, $SbCl_5$. The electronegativity difference between Sb and Cl is $2.83 - 1.82 = 1.01$ units, which is enough to suggest that $SbCl_5$ ought to be a partly ionic compound, with extended polymeric structure, high melting point, and so on. In fact, it is a liquid at room temperature. Clearly there is little charge transfer, or the ionic attraction would cause crystallization into an extended structure. We can calculate the charges on the atoms by using the differential ionization-energy procedure, as in Chapters 12, 13, and 14, remembering that if all the bonds are identical the charge on the Sb as it participates in a particular bond is $5q$ if that on the Cl is $-q$:

$DIE_{Sb}(5q) = DIE_{Cl}(-q)$ at equilibrium in the electron flow

$5.6 + 6.3(5q) = 8.3 + 10.2(-q)$ from Table 12-3

$41.7q = 2.7$

$q = 0.065$ of an electronic charge

This is an extremely small value, suggesting that ionic attraction is negligible and the bonding substantially covalent — which is, of course, compatible with the observation that thermal energy at room temperature is more than enough to melt $SbCl_5$. There are many such compounds; we shall look at some of their properties in Section 15-6.

Covalent Fluorides

This is a good place to look at the ionic/covalent properties of fluorides. Although fluorine is the most electronegative element, it forms many surprisingly covalent compounds in addition to the expected ionic ones. For example, SF_6 is a gas at room temperature even though the S–F electronegativity difference is 1.66 units, and so is BF_3 with a difference of 2.11 units. Clearly electronegativity alone is not an adequate guide to the charge distribution. Let us use differential ionization energies to calculate approximate charges on the atoms in the series of compounds CF_4 (electronegativity difference $\Delta\chi = 1.60$ units), AlF_3 ($\Delta\chi = 2.63$ units) as a molecule and in the ionic crystal, and CaF_2 ($\Delta\chi = 3.06$ units) in the ionic crystal. For CF_4 we can write, following the same procedure as for $SbCl_5$,

$$DIE_C(4q) = DIE_F(-q)$$
$$5.0 + 10.3(4q) = 10.3 + 15.8(-q)$$
$$57q = 5.3$$
$$q = 0.093$$

In this case q is the full extent of the charge on the F, since it is bonded only to one atom, C; it thus has a charge of only $-0.093e$, which corresponds to an almost completely covalent system (CF_4 is a gas, boiling at $-128\,°C$). For AlF_3 as an isolated molecule, the same sort of calculation gives a charge on F of $-0.22e$, which is appreciable, suggesting that the molecule condenses into an extended-lattice partly ionic system. If we recalculate the charge on the atoms in the crystalline state, where Al has six F neighbors and each F has two Al neighbors, we assign only half the charge on each F to a particular Al:

$$DIE_{Al}(6q) = DIE_F(-q) \qquad -2q = \text{total F charge}$$
$$2.6 + 6.2(6q) = 10.3 + 15.8(-q)$$
$$53q = 7.7$$
$$q = 0.145e$$

$$\text{total F charge} = -0.29e \qquad \text{total Al charge} = 0.87e$$

This corresponds to a rather substantial charge separation; the compound should be partly ionic, which accords with its physical properties. AlF_3 melts at 1040 °C and generally displays the characteristics of the partly ionic systems in Chapter 14. There are very few partly ionic fluorides, however, because with only a little additional encouragement the charge separation in a solid becomes so great as to correspond to a more-or-less fully ionic compound. In CaF_2, for instance, each Ca is surrounded by eight fluorine atoms and each F by four Ca atoms (as in Fig. 4-34). To calculate the charge separation we write

$$DIE_{Ca}(8q) = DIE_F(-q) \qquad -4q = \text{total F charge}$$
$$3.7 + 4.2(8q) = 10.3 + 15.8(-q)$$
$$49.4q = 6.6$$
$$q = 0.134$$
$$\text{total F charge} = -0.53e \qquad \text{total Ca charge} = 1.07e$$

This is so great a degree of electron transfer that the compound will have pretty much fully ionic properties; in fact, it melts at 1360°C and otherwise resembles a standard ionic compound.

Another factor that prevents the positive charge on the central atom in many molecular (covalent) fluorides from having a greater effect on the interactions between molecules is that if there are several fluorines around the central atom they usually fit together to produce a nearly spherical, nonpolar molecule, as in Fig. 15-1. This also applies to other molecules with similar geometries, such as $SbCl_5$. The result is that most fluorides are either almost fully covalent or, if enough additional ligand F atoms can be shared in the solid state, almost fully ionic.

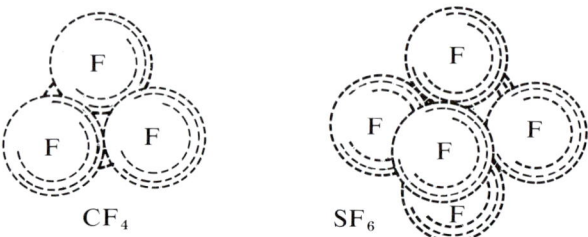

Figure 15-1 Symmetrical "ball-of-fluorine" structures of covalent fluorides.

15-2 Criteria for Catenation

There is a particularly interesting kind of covalent compound in the general category of compounds between elements having similar electronegativities. In one fashion or another, all of the elements except the rare gases can bond to themselves; this corresponds, of course, to exactly zero electronegativity difference and a completely covalent bond. Metals, in general, form delocalized "electron-sea" bonds over an entire metallic crystal, although some can also form diatomic molecules in the gas phase. There are also some interesting examples of metal–metal bonding among the transition metals, which we shall look at in Chapter 16. Many nonmetals form stable diatomic molecules such as H_2, N_2, and Br_2, although there are also molecules such as the tetrahedral P_4, an S_8 ring, and an icosahedral B_{12} unit in elemental boron. In an intermediate kind of bonding called **catenation** it is possible to form finite molecular chains of atoms bonded to other atoms of the same element. Catenated molecules are of particular importance because nearly all the molecules of biological interest — and indeed nearly all organic compounds — contain catenated carbon atoms. Since there is such an astonishing variety of these compounds, any survey of the synthesis of covalent compounds must deal with the questions of what catenated bonds are like and why carbon forms them so readily.

Since the bonds in a catenated compound are entirely covalent, with no electrostatic attraction contributing to the bond energy, we must look entirely to the MO energies for an explanation of the nature of catenated bonding. We have observed that a stable covalent molecule must have essentially all the bonding orbitals filled in order not to be a very reactive electron acceptor, but that it must not have many antibonding electrons lest it be unstable toward electron donation or rearrangement. In Chapter 6, furthermore, it developed that the number of bonding MO's that an atom can form is limited to the number of valence AO's it can use in bonding; that is, an atom with only s and p orbitals can form only four bonding MO's, no matter what system it is in. Combining these two ideas, only atoms with approximately as many valence electrons as valence orbitals can form stable catenated bonds. This follows because if two identical atoms form an electron-pair bond one electron must come from each atom, and one bonding orbital (involving one electron from a given atom) results from each valence AO. Carbon fits this requirement nicely, with four valence electrons in four valence orbitals (the $2s$ and the three $2p$), but so does silicon and the rest of the group IV elements. Since we do not observe very extensive catenation of silicon and the other elements in their compounds, there must be another condition besides the electron-to-orbital ratio.

The other condition is that the valence orbitals overlap each other well in a given catenated bond. If the overlap is poor the bond will be relatively weak, since there is no ionic contribution, and it will be quite reactive toward anything that has a different electronegativity and can thus form bonds with some electrostatic contribution toward the bond energy. That is, only unusually good overlap can give purely covalent bonds as strong as partly ionic or ionic bonds. Overlap is governed primarily by how close the two nuclei of the bonded atoms can approach each other, relative to the maximum-probability radii of the two AO's involved, as shown in Fig. 15-2. But the closeness of approach is limited by the repulsion of the inner-core electrons on each atom for those on the other

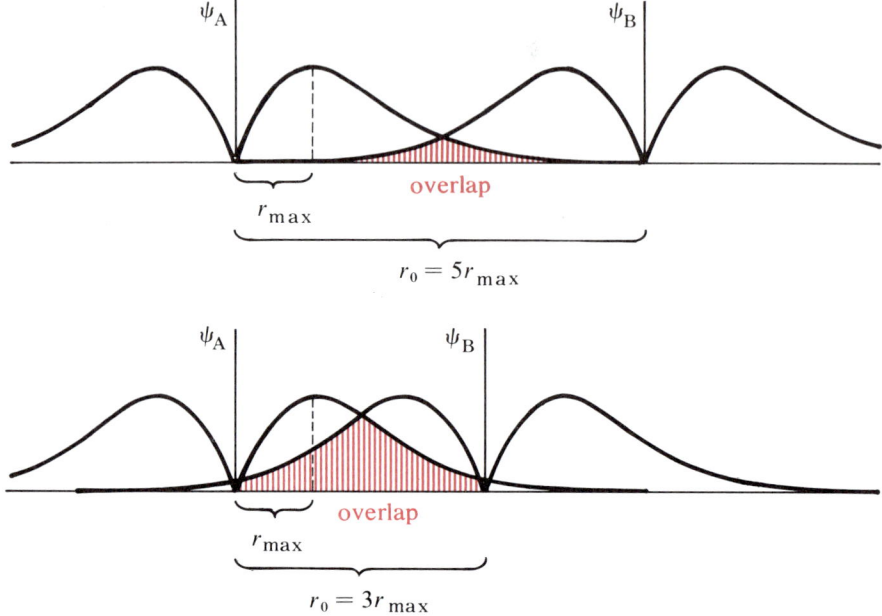

Figure 15-2   Influence of bond length on overlap.

atom, and by the repulsion of the nuclei for each other; so the smaller the inner core and the lower the nuclear charge (the atomic number) the closer the nuclei can approach each other and the better the overlap will be. Carbon is particularly fortunate in this respect, because its inner core consists of only the two $1s$ electrons, and its atomic number is only 6. It is not surprising, then, that well over 99% of all the catenated compounds we see involve carbon principally. The most noted catenating element other than carbon is sulfur, which has four valence orbitals ($3s$ and $3p$, with only modest participation by $3d$) and four p electrons. Evidence that the two s electrons do not participate strongly in bonding may be found in the bond angles in sulfur compounds, which often approximate 90°, the angle we would expect for pure p overlap. Sulfur catenation, however, is not nearly as stable or widespread as that by carbon atoms.

15-3 Molecular Orbitals for Catenated Systems

Sigma Overlap of Hybrid Orbitals

In our earlier discussion of MO's we discussed the very simple organic compound methane, CH_4 (see Section 6-11). Because H has no p orbitals, we considered only the formation of σ bonds. In Section 6-11, however, we extended the idea of σ bonding through hybrid orbitals to catenated compounds, with the hybrid orbitals on each C overlapping to form a more or less localized bond. Figure 15-3 summarizes this approach; it is the one most commonly used in simple approaches to the bonding in catenated carbon compounds, which from now on we shall call organic compounds. The simplest organic compounds are those involving only C and H (like methane), called **hydrocarbons**. In this section we shall restrict ourselves to hydrocarbon examples.

Figure 15-3 Hydrocarbon structures as chains of hybrid orbitals.

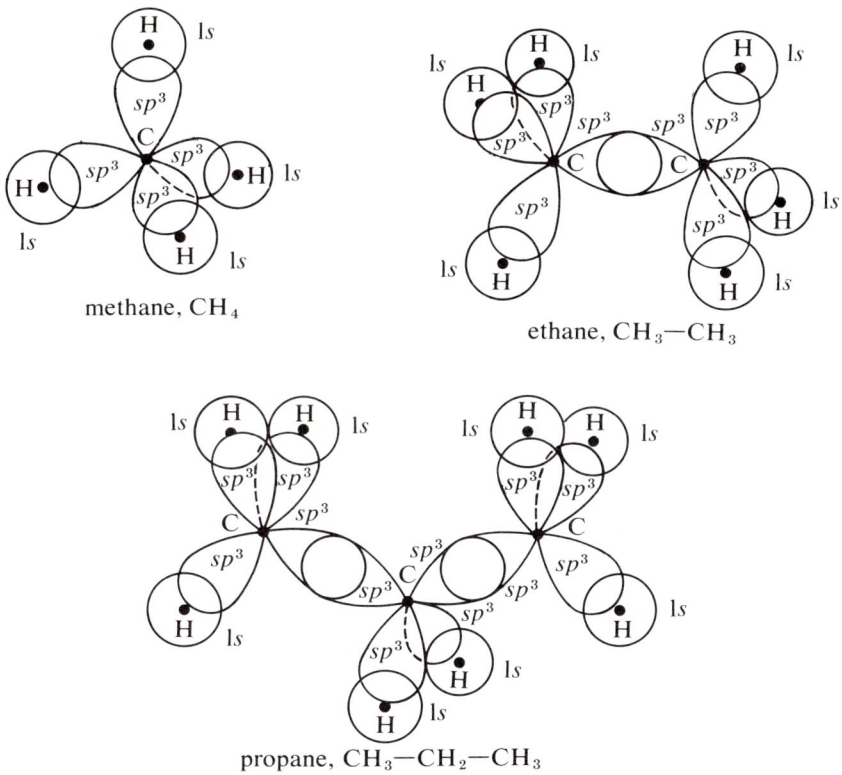

Pi Overlap and Properties of Pi Bonds

All this, however, says nothing about the possibility of π overlap of p orbitals on adjacent carbon atoms. Consider the compound ethylene, CH_2CH_2, which has the geometry indicated in Fig. 15-4. Since the bond angles are about 120°, the trigonal planar angle, we assume that they represent bonding by sp^2 hybrid orbitals on each C (see p. 325). But this leaves an untouched p valence orbital on each C—the one that is perpendicular to the plane in which the sp^2 hybrids

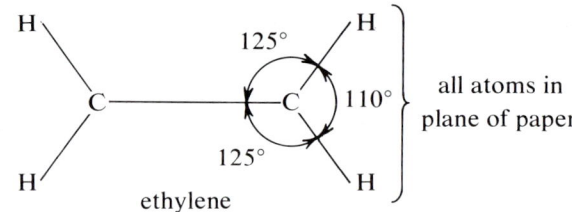

ethylene

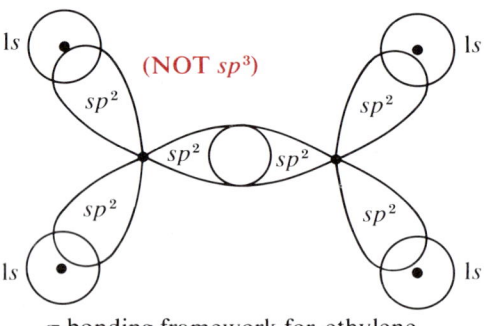

σ-bonding framework for ethylene

Figure 15-4 Hybrid-orbital overlap for σ bonding in ethylene, C_2H_4.

lie. These two p orbitals can clearly overlap in a π fashion, as in Fig. 15-5, which also shows the relative energies of the possible bonding and antibonding combinations that can be formed. As we found in Chapter 6, the energy effect of π bonding is not as great as that from σ bonding because the overlap is not as good. It is very real, however, and π bonds have a property that σ bonds do not; they resist being twisted, as seen in Fig. 15-6. We say that there is **free rotation** about a σ bond but not about a π bond. This raises an interesting structural possibility for π-bonded organic compounds: if the H atoms are partially replaced by other atoms (one replacement per C) the two new atoms can either be on the same side of the C—C bond or on opposite sides, as in Fig. 15-7. These conformations of the molecule are called **cis** and **trans**, respec-

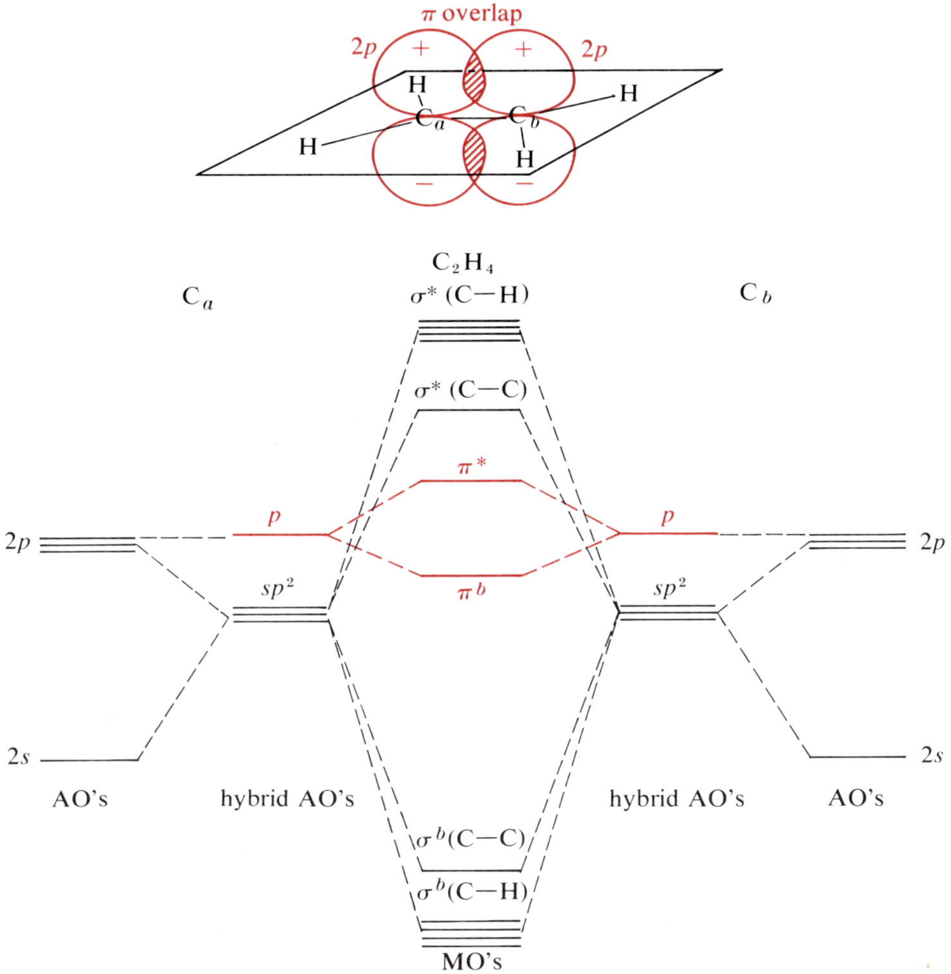

Molecular Orbitals for Catenated Systems | 771

Figure 15-5 Pi overlap and energy-level diagram for ethylene, showing energy effect of poor π overlap.

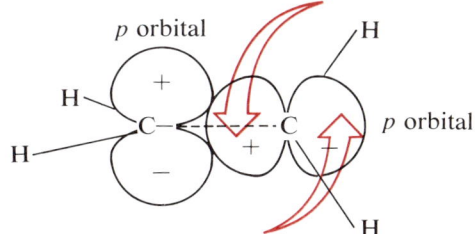

Figure 15-6 Resistance of π bonds to rotation: 90° rotation destroys π bonding overlap and 180° rotation gives π antibonding overlap, so rotation requires energy input equal to the π-bond energy.

772 | Covalent Compounds

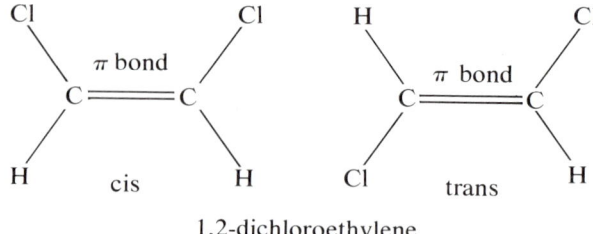

1,2-dichloroethylene

Figure 15-7 Cis and trans structural isomers.

tively. The difference may seem minor, but it can be extremely important in a reaction mechanism if both atoms or groups must react with the same incoming molecule; in this case the cis form would be capable of reacting when the trans form might not because of its unfavorable geometry. Figure 15-8 shows an example of such a reaction.

$$M^{2+} \;+\; NH_2-CH=CH-NH_2 \;\longrightarrow\; [M(NH_2-CH=CH-NH_2)]^{2+}$$

metal ion   1,2-diaminoethylene   metal complex

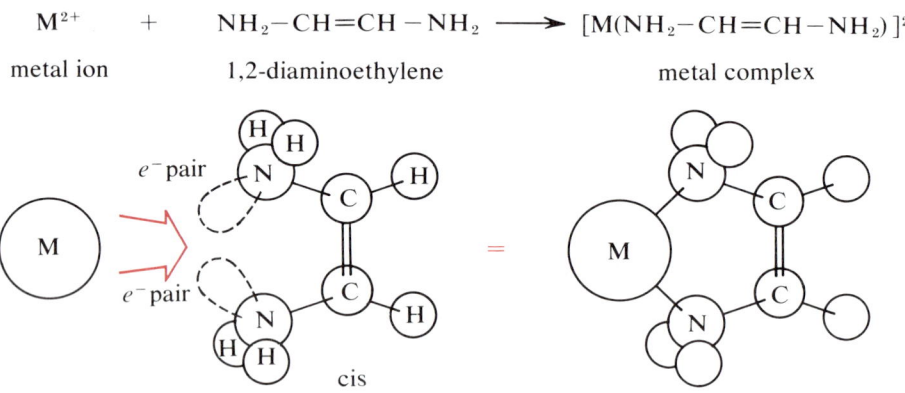

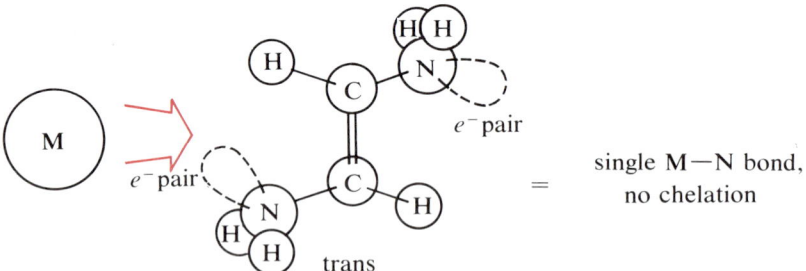

Figure 15-8 Comparison of cis- and trans-isomer reactivity.

Extended Pi Systems

What can we say about the π bonding in a string of catenated atoms? Butadiene, shown in Fig. 15-9, is an example of such a molecule. For such compounds the

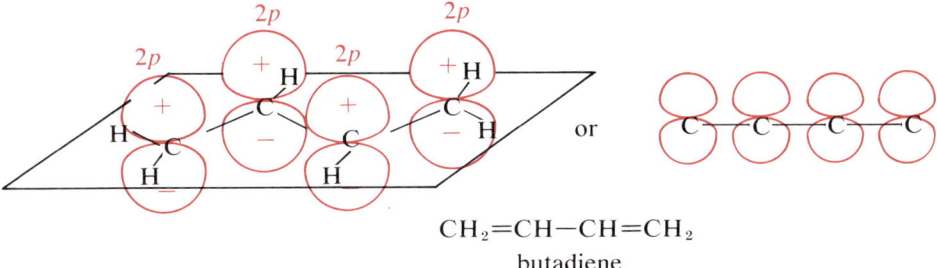

$$CH_2=CH-CH=CH_2$$
butadiene

Figure 15-9 Pi overlap in butadiene.

nature of the π MO's is particularly interesting because the highest-energy electrons and the lowest-energy vacant orbitals are both π type—so the reactions of the molecule will usually be dictated by the nature of these MO's. If the four p AO's shown in Fig. 15-9 are allowed to combine into MO's, four MO's must result, all of which are linear combinations of the four AO's. The MO's obtained from the simplest approximation, called **Hückel MO's**, have different energies that correspond to the number of nodes between atoms, in the same sense that AO's energies increase as their number of nodal surfaces increases. We expect a purely bonding MO with no nodes cutting the bond axes, and a purely antibonding MO with a node between each pair of C atoms (for a total of three nodes). Since there are four π MO's, the other two must have one and two nodes, respectively, and have intermediate energies. Figure 15-10 shows the four π MO's and their energies relative to the σ MO's. Since each C atom initially has one electron in the p orbital that combines to form these MO's, there are four electrons to be dealt with; these are best distributed in the two most strongly bonding MO's, as shown in the figure.

An even more interesting compound showing π catenation is benzene, C_6H_6, shown in Fig. 15-11. Here there are six p AO's and therefore six π MO's with different energies, depending on the number of nodes cutting the bond axes. But the cyclic nature of the compound means that a single nodal plane cuts the molecule in two places, so the strongly antibonding π MO (which we expect to have nodes between each pair of C atoms around the ring) has only three nodal

774 | Covalent Compounds

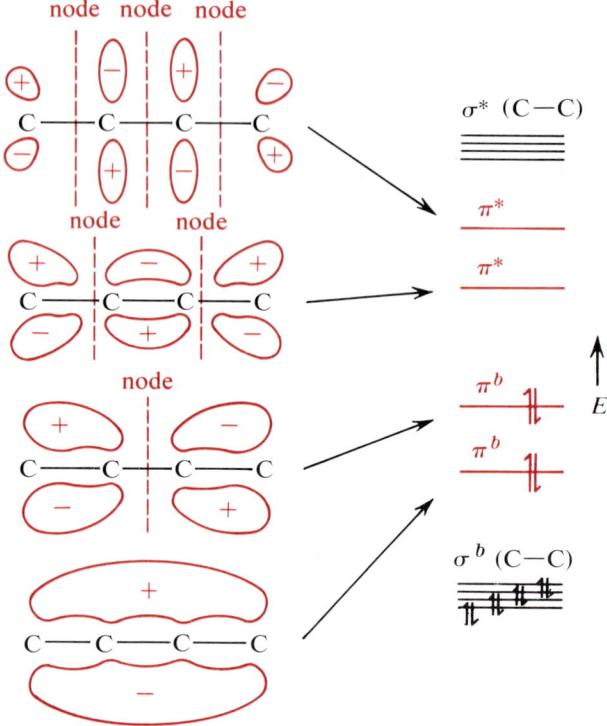

Figure 15-10 Pi MO's and energy-level diagram for butadiene.

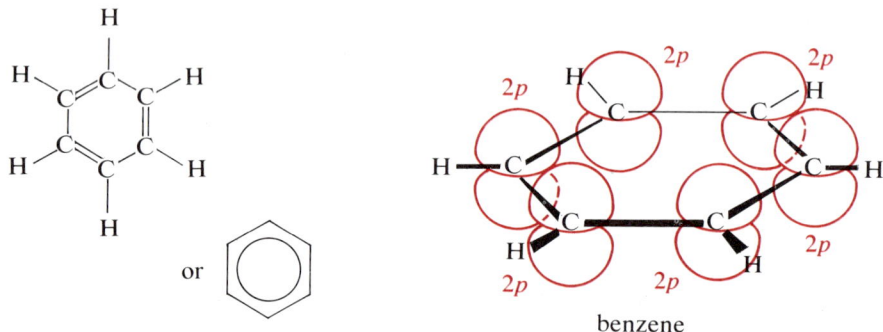

Figure 15-11 Pi overlap in benzene, C_6H_6.

surfaces, as in Fig. 15-12. If the strongly bonding MO has no nodes, then the other four must have one or two nodal surfaces. There are two ways to arrange a nodal surface cutting a benzene ring, so that it passes through nuclei or so that it passes only through bonds. Keeping this in mind, we can draw the six π MO's in Fig. 15-13; in this approximation their energies depend solely on the number of nodes, as the figure suggests, and the six electrons in the π system

Molecular Orbitals for Catenated Systems | 775

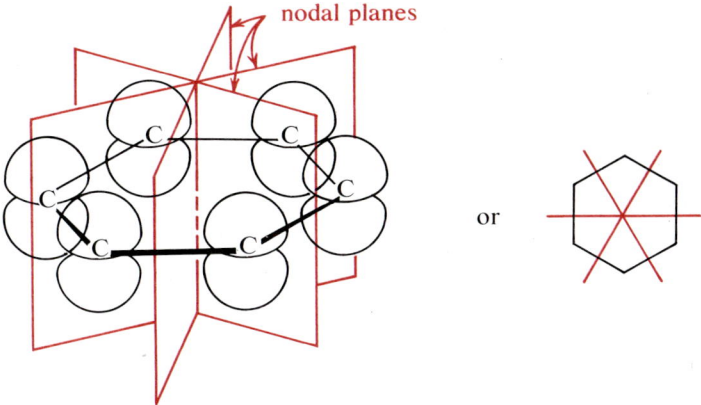

Figure 15-12 Position of nodes for the strongly antibonding π MO of benzene.

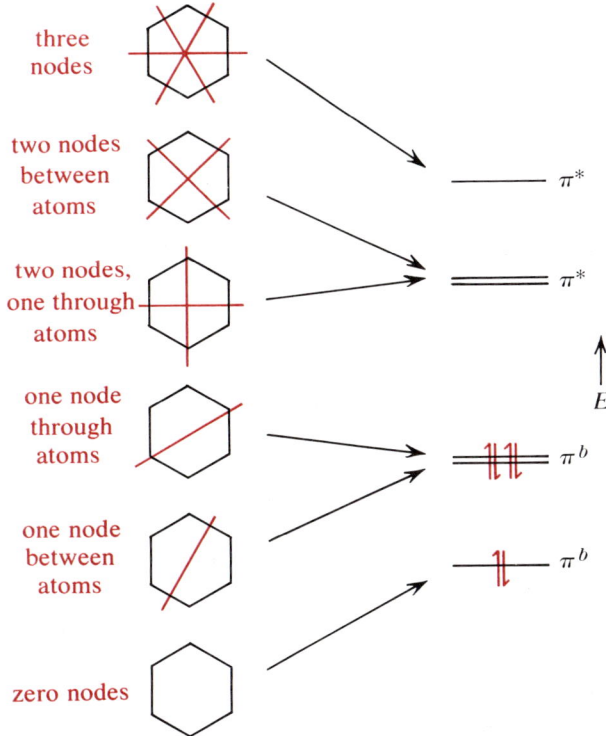

Figure 15-13 Energy-level diagram for the π MO's of benzene.

can all be accommodated in distributions corresponding to MO's that have a net bonding effect in that they have fewer nodes than nonnodes in the six C—C bond regions. Study Problem 1 suggests that cyclic π-catenated systems — like benzene — that have 2, 6, 10, 14, or in general $4n+2$ π electrons, are particularly stable in having no antibonding or nonbonding electrons. These "$4n+2$" hydrocarbon compounds form a special class called **aromatic hydrocarbons**; their reactions are very important to modern organic chemistry. Besides having larger or smaller rings, it is possible to have fused rings, as in the compounds shown in Fig. 15-14; six- and five-membered rings are the most common.

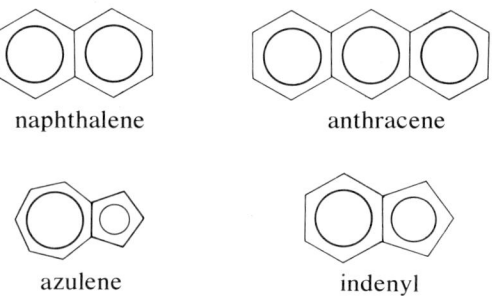

Figure 15-14 Fused-ring systems, with continuous π overlap indicated by circles.

We expect, by means of MO's such as those we have described, to be able to rationalize both the structure and the chemical reactivity of covalent compounds, whether catenated or not. In the subsequent sections of this chapter we shall try to do this for the various types of covalent compound we encounter. First let us deal specifically with the reactions of π-electron molecules.

Chemistry of Pi Bonds

The simplest type of π bond is in the molecule having only a single pair of C atoms with π overlap — a single π bond. Sometimes molecules may have two or more such pairs, separated by at least one C atom having only σ bonds; Fig. 15-15 shows some examples of both kinds. These hydrocarbons are called **alkenes** or **olefins**, in contrast to hydrocarbons having only σ bonds, which are called **alkanes** or **paraffins**. The combination of a σ bond and a π bond between two C atoms gives what is called a **double bond**; note, though, that since the π-overlap energy effect is not as great as the σ-overlap energy effect, the bond energy of a double bond is not twice as great as that of a single bond (in spite

$$CH_2=CH_2 \qquad\qquad CH_3-CH=CH_2$$
$$\text{ethylene} \qquad\qquad\qquad \text{propene}$$

$$CH_2=CH-CH_2-CH_2-CH_3 \qquad CH_2=CH-CH_2-CH=CH_2$$
$$\text{1-pentene} \qquad\qquad\qquad \text{1, 4-pentadiene}$$

Figure 15-15 Examples of olefins.

of the name). The thermochemical bond energies (see p. 461) for C—C and C=C are 82.6 kcal/mole bond and 145.8 kcal/mole bond, respectively, which suggests the general relationship. The smaller bonding effect due to the less favorable π overlap is principally responsible for the characteristic chemical reactions of alkenes; these π electrons are more available for chemical rearrangement into σ bonds than any of the other electrons in the molecule. Since these are relatively high-energy electrons (as in Fig. 15-5), the most common reactions of alkenes are those in which the alkene serves as an electron donor or Lewis base. Looked at from the other end, alkenes normally react with electrophiles, or electron-seeking species. Some examples of characteristic reactions are given in Table 15-1; most reactions involve the formation of σ bonds, usually through the attack of an electronegative atom or one with a partial positive charge.

It can be seen from Table 15-1 that most of the reactions of alkenes involve adding another atom or group of atoms to each of the two C atoms in the π bond; for this reason they are called **addition reactions**. Addition reactions destroy the π bond, but offset this unfavorable energy effect by forming two new σ bonds. If we look at catenated π systems—and particularly at aromatic systems—we find addition reactions much less often. Instead, aromatic systems usually undergo **substitution reactions**, in which one atom or group of atoms simply replaces another on the aromatic ring without destroying the π-bond system. A sampling of these reactions is given in Table 15-2. Even in the aromatic systems, the π electrons are the least stable, so the substitution reactions usually involve the attack of a positively charged species or electrophile. However, aromatic systems such as benzene have some π electrons bound more strongly than in a single π bond—see Fig. 15-13. This is true solely because the π overlap is **delocalized** over several atoms (six in the case of benzene). If an addition reaction were to occur, it would not only destroy the π bond between a given pair of C atoms in the benzene ring, but it would also render the remaining delocalized π MO's less favorable energetically. This energy cost is sufficiently great that addition reactions are rare for aromatic systems, and substitution reactions that preserve the π system occur instead.

Table 15-1
Reactions of Alkenes

Addition of H_2, X_2, or HX

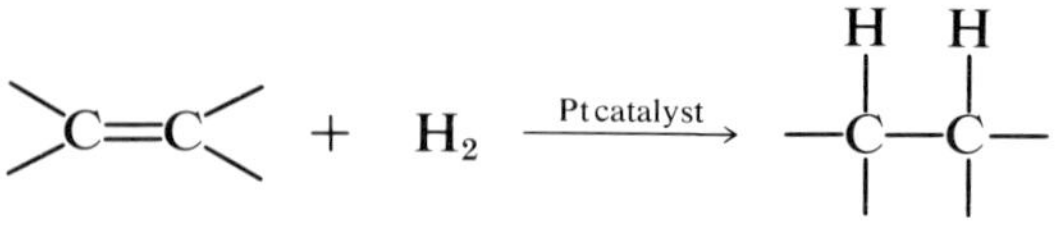

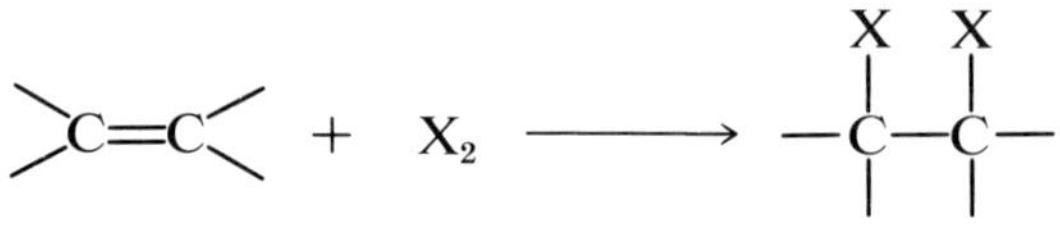

$$\ce{>C=C<} + H_2 \xrightarrow{\text{Pt catalyst}} \ce{-\underset{|}{\overset{H}{C}}-\underset{|}{\overset{H}{C}}-}$$

$$\ce{>C=C<} + X_2 \longrightarrow \ce{-\underset{|}{\overset{X}{C}}-\underset{|}{\overset{X}{C}}-} \quad X = \text{Cl or Br}$$

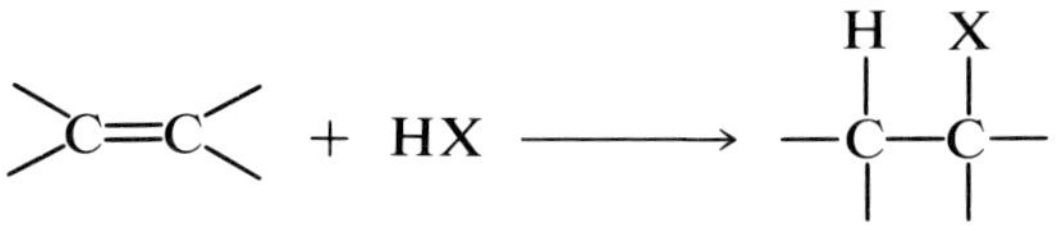

$$\ce{>C=C<} + HX \longrightarrow \ce{-\underset{|}{\overset{H}{C}}-\underset{|}{\overset{X}{C}}-} \quad X = \text{Cl, Br, or I}$$

Addition of H_2O

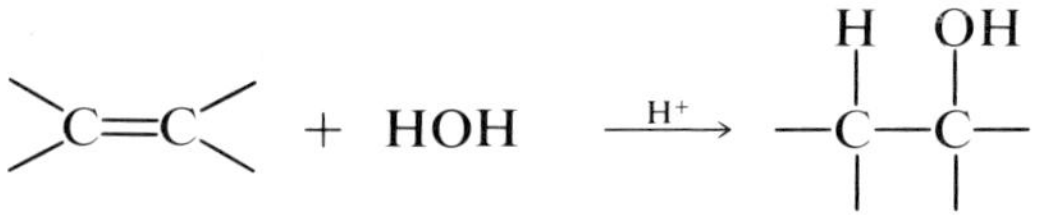

$$\ce{>C=C<} + HOH \xrightarrow{H^+} \ce{-\underset{|}{\overset{H}{C}}-\underset{|}{\overset{OH}{C}}-}$$

Formation of Halohydrins

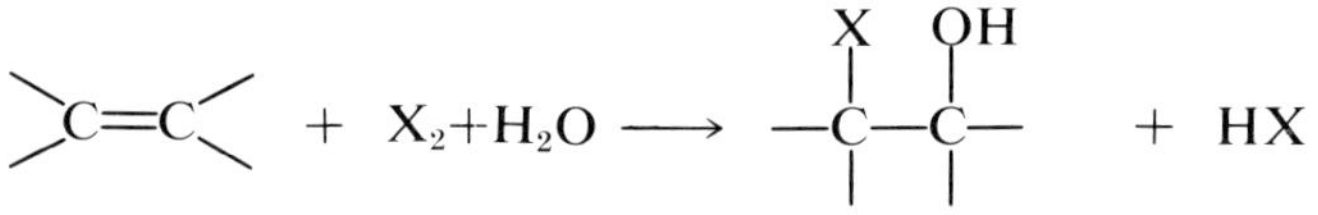

$$\ce{>C=C<} + X_2 + H_2O \longrightarrow \ce{-\underset{|}{\overset{X}{C}}-\underset{|}{\overset{OH}{C}}-} + HX \quad X = \text{Cl or Br}$$

Alkylation

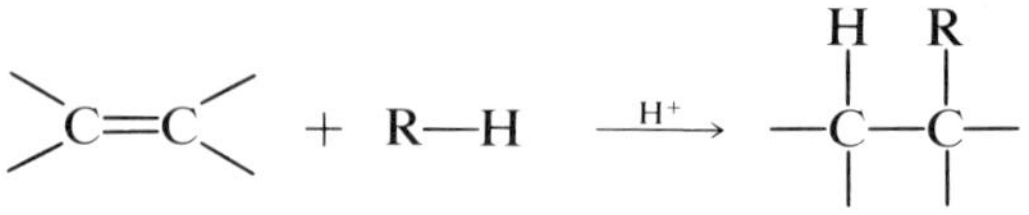

$$\ce{>C=C<} + R-H \xrightarrow{H^+} \ce{-\underset{|}{\overset{H}{C}}-\underset{|}{\overset{R}{C}}-} \quad R = \text{hydrocarbon group}$$

Pi Donation

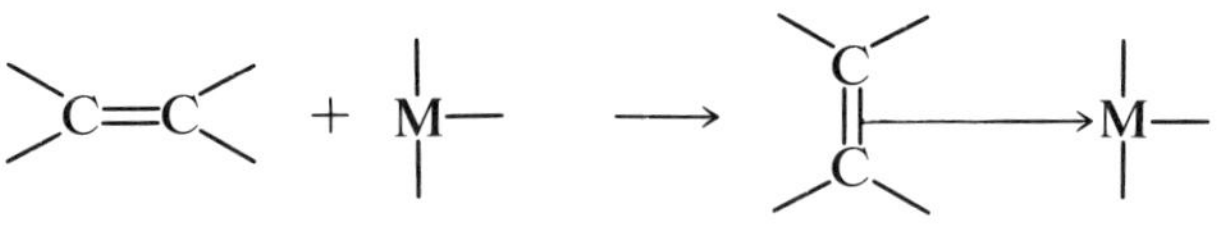

$$\ce{>C=C<} + \ce{M-} \longrightarrow \text{(π-complex)} \ce{M-} \quad M = \text{transition metal in partly ionic compound}$$

Dimerization

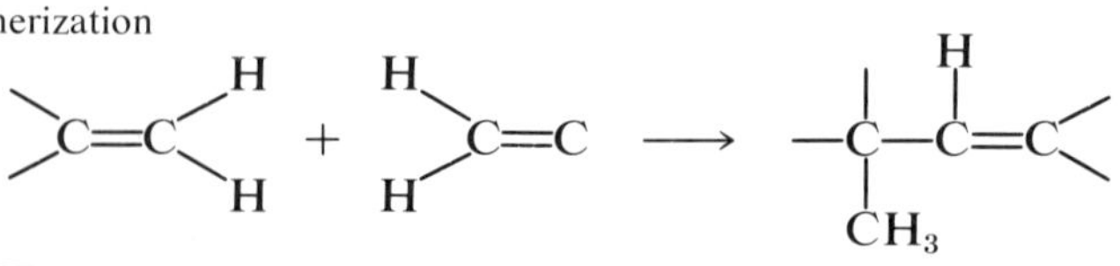

$$\ce{>C=C<^{H}_{H}} + \ce{^{H}_{H}>C=C} \longrightarrow \ce{-\underset{|}{\overset{H}{C}}-\underset{|}{\overset{}{C}}=C<} \\ \quad\quad\quad CH_3$$

Table 15-2
Reactions of Aromatic Hydrocarbons

Nitration

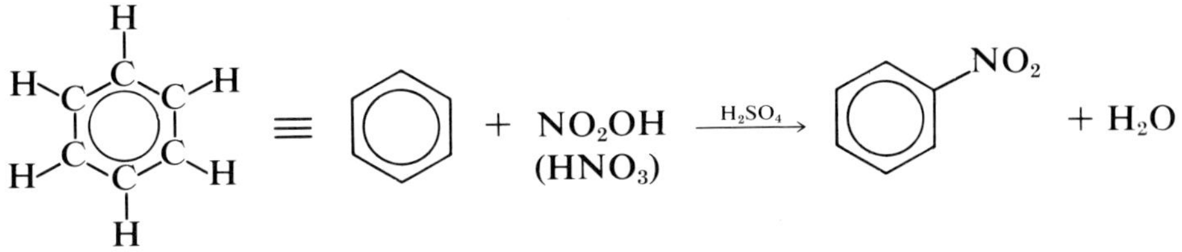

benzene + NO$_2$OH (HNO$_3$) $\xrightarrow{H_2SO_4}$ nitrobenzene (C$_6$H$_5$NO$_2$) + H$_2$O

Halogenation

benzene + X$_2$ $\xrightarrow{Fe^0}$ C$_6$H$_5$X + HX   X = Cl or Br

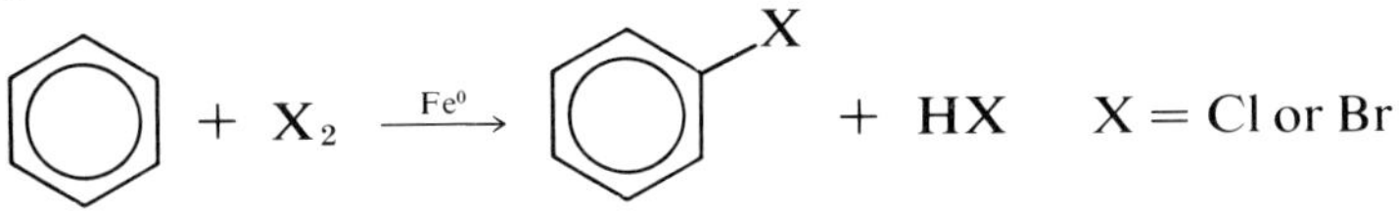

Alkylation (Friedel–Crafts)

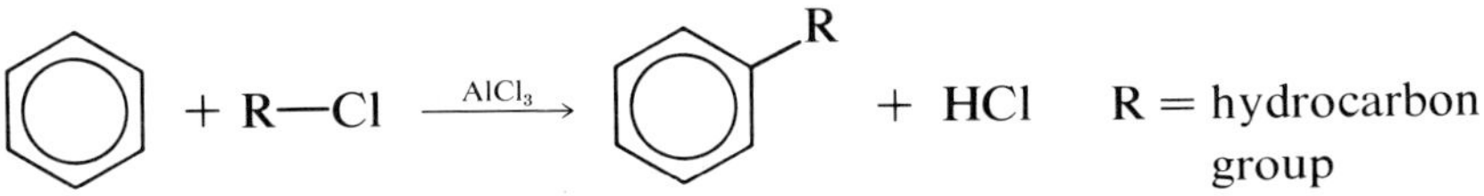

benzene + R—Cl $\xrightarrow{AlCl_3}$ C$_6$H$_5$R + HCl   R = hydrocarbon group

Mechanisms of Aromatic Reactions: Nitration

Once a single substitution has occurred the reaction can usually proceed to substitute on another aromatic C atom, but this raises an interesting question. On which C atom will the second substitution occur? All six C atoms in benzene are equivalent, so it does not matter where the first occurred — but the second can occur next to the first (the **ortho** position), one C atom away (the **meta** position), or two C atoms away (the **para** position), as in Fig. 15-16. This is one of the cases we mentioned earlier in which several products are possible and the mechanism of the reaction determines which will predominate. If we

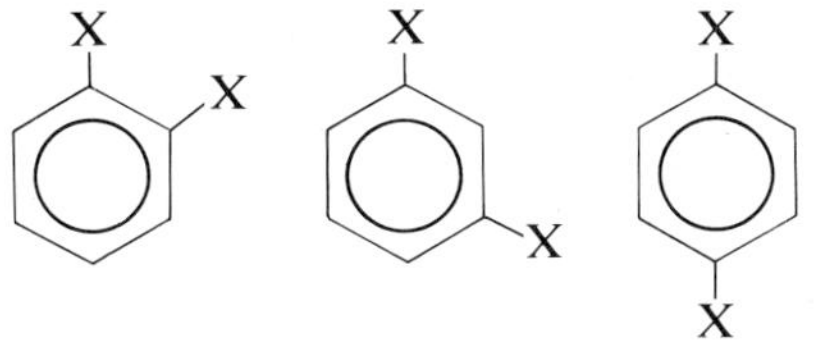

ortho (o-)   meta (m-)   para (p-)

Figure 15-16 Definition of *ortho*, *meta*, and *para* structural isomers.

consider the nitration reaction

[benzene] + HNO₃ —H₂SO₄ catalyst→ [benzene-NO₂] + H₂O —HNO₃ again→ [benzene with two NO₂ groups meta]

we find that it proceeds by attack of the electrophile NO_2^+, the nitronium ion, on the benzene π electrons. The reaction intermediate has destroyed the continuous π MO's around the ring:

[benzene] + NO_2^+ ⟶ [intermediate with H and NO₂ on same C]⁺   intermediate

The π delocalization energy is regained by transferring a proton to some neighboring species:

[intermediate]⁺ + HSO_4^- ⟶ [nitrobenzene] + H_2SO_4

The second NO_2^+ that attacks will, as an electrophile, prefer the C richest in electrons. It will not attach to the C that is already nitrated because that would destroy the continuous π system, just as in the transition state. The C atom that is ortho to the —NO₂ group will have electrons fairly strongly withdrawn from it; they will flow through the σ bonds toward the electronegative —NO₂. The meta C will be less strongly affected, and the para C still less. So on the basis of the σ-bonding electrons alone, we expect the second NO_2^+ to enter the para position. However, the π electrons are also affected. The highest-energy (most readily available) electrons in the molecule are those in the π MO having one node:

[three nitrobenzene resonance structures showing node placement] or ... or ...

node

In either of the first two possibilities, the —NO₂ is part of the π system and withdraws π electrons from one of the ortho positions, one of the meta positions, and from the para position. In the third possibility, which would withdraw electrons from the ortho and meta positions, the —NO₂ group is not part of the π system because it lies in the nodal plane of the MO, so it does not withdraw π electrons at all. So π electrons are more strongly withdrawn from the para position than from the other two positions. The net result is that the second NO_2^+ does not attack the ortho position because σ electrons are strongly withdrawn from it, giving it a partial positive charge; it does not attack the para position because π electrons are withdrawn from it. This leaves the meta position as the best of a bad lot, and indeed this is the isomer that is found experimentally, as the first equation above suggests. However, the argument also suggests that any strongly electron-withdrawing group makes further reaction more difficult, since it removes the π electrons that must be donated to the incoming NO_2^+ ion or other electrophile. By contrast, groups that donate electrons should make further reaction easier and should favor the ortho and para positions. This is entirely in agreement with experiment, and indeed the whole pattern of aromatic substitution reactions seems to be pretty well understood in terms of the MO model.

15-4 Organic Compounds and Their Reactions

In the previous sections we have seen that C is unique in its capacity to catenate, and have looked at the simpler aspects of the MO description of σ and π catenated bonds. We need now to look at some of the more common organic molecular structures and the general ways they react. In surveying them, we shall need first to develop the nomenclature, since it indicates the kinds of compounds we expect to find. Then we shall look at some of the general reaction mechanisms and products that are characteristic of organic reactions.

Hydrocarbon Nomenclature

Hydrocarbons are the simplest organic compounds; Fig. 15-17 shows the simplest of these, the "straight-chain" hydrocarbons. These have no π bonds

Figure 15-17 Nomenclature of hydrocarbons (sometimes shown as angled lines in structural formulas).

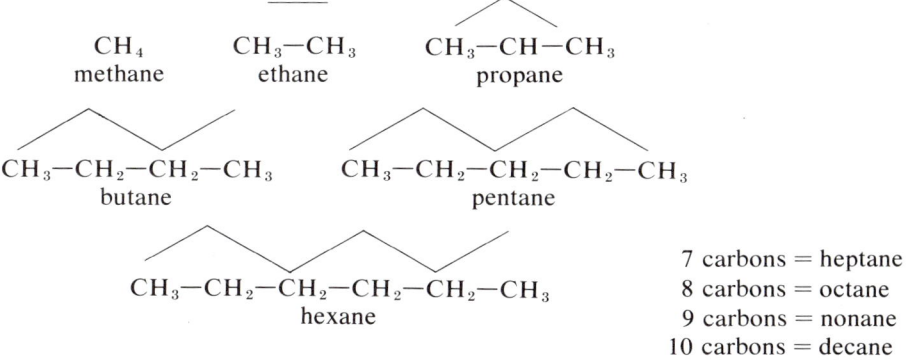

and are called **alkanes**, in keeping with our earlier discussion. Even if we restrict ourselves to alkanes, we may not find the C atoms in a straight chain; they may be branched, as in Fig. 15-18. To name these, pick out the longest possible chain of C atoms, then name the compound as that straight-chain hydrocarbon, substituted by other hydrocarbon groups (with the suffix —yl). To show where the branches are, the C atoms in the hypothetical straight-chain hydrocarbon are numbered from one end to the other and the points of attachment are given as a prefix number. Study of Fig. 15-18 should make the pattern clear. Unfor-

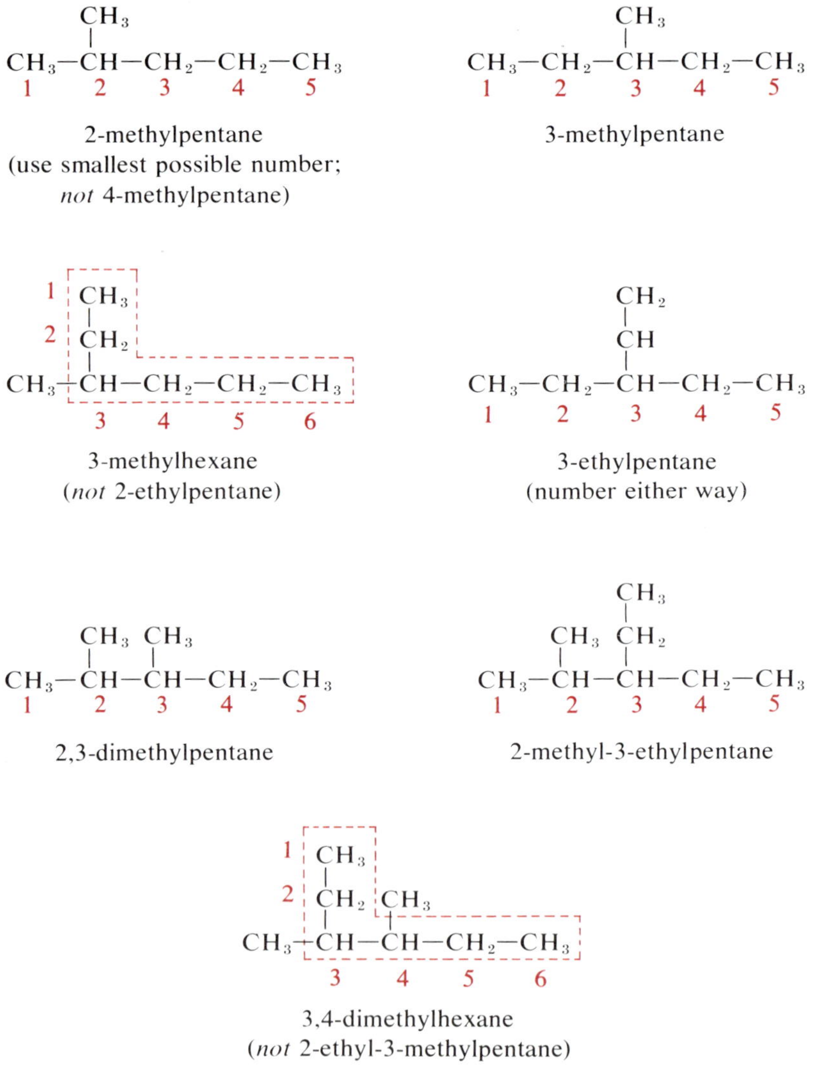

$$CH_3-\underset{\underset{1}{|}}{\overset{CH_3}{\underset{2}{C}H}}-\underset{3}{C}H_2-\underset{4}{C}H_2-\underset{5}{C}H_3$$

2-methylpentane
(use smallest possible number;
not 4-methylpentane)

$$CH_3-\underset{1}{C}H_2-\underset{\underset{3}{|}}{\overset{CH_3}{\underset{}{C}H}}-\underset{4}{C}H_2-\underset{5}{C}H_3$$

3-methylpentane

$$\underset{3}{C}H_3-\underset{\underset{}{|}}{\overset{\overset{1}{C}H_3}{\underset{}{\overset{|}{C}H_2}}}-\underset{4}{C}H-\underset{5}{C}H_2-\underset{6}{C}H_3$$

3-methylhexane
(*not* 2-ethylpentane)

$$CH_3-\underset{1}{C}H_2-\underset{\underset{3}{|}}{\overset{\overset{}{C}H_2}{\underset{}{\overset{|}{C}H}}}-\underset{4}{C}H_2-\underset{5}{C}H_3$$

3-ethylpentane
(number either way)

$$CH_3-\underset{\underset{2}{|}}{\overset{CH_3}{\underset{}{C}H}}-\underset{\underset{3}{|}}{\overset{CH_3}{\underset{}{C}H}}-\underset{4}{C}H_2-\underset{5}{C}H_3$$

2,3-dimethylpentane

$$CH_3-\underset{\underset{2}{|}}{\overset{CH_3}{\underset{}{C}H}}-\underset{\underset{3}{|}}{\overset{\overset{CH_3}{|}}{\underset{}{\overset{CH_2}{\underset{}{C}H}}}}-\underset{4}{C}H_2-\underset{5}{C}H_3$$

2-methyl-3-ethylpentane

$$\underset{3}{C}H_3-\underset{\underset{}{|}}{\overset{\overset{1}{C}H_3}{\underset{}{\overset{|}{C}H_2}}}-\underset{4}{\overset{\overset{CH_3}{|}}{\underset{}{C}H}}-\underset{5}{C}H_2-\underset{6}{C}H_3$$

3,4-dimethylhexane
(*not* 2-ethyl-3-methylpentane)

Figure 15-18 Nomenclature of branched hydrocarbons.

tunately, this systematic pattern of names is often ignored in favor of the trivial (common) names that are phonetically easier to deal with, but mean learning a different set of names. For example, a group that is a straight chain except for one methyl group on the next-to-last C is often called an **isoalkyl** group, while a straight-chain group is said to be a **normal alkyl** or *n*-alkyl group:

$$\begin{array}{ccc} & CH_3 & \\ & | & \\ CH_3-CH- & \text{or} & \begin{array}{c} H_3C \\ \diagdown \\ \diagup \\ H_3C \end{array} CH- \quad \text{isopropyl} \end{array}$$

$$\begin{array}{l} CH_3 \\ | \\ CH_3-CH-CH_2-CH_3 \quad \text{2-methylbutane or isopentane} \\ CH_3-CH_2-CH_2- \text{propyl or } n\text{-propyl} \end{array}$$

Carbon atoms bonded to only one other C are called **primary** C atoms; if they are bonded to two or three C atoms they are called **secondary** or **tertiary**, respectively:

$$\begin{array}{ccc} & & & & CH_3 \\ & & & & | \\ CH_3-CH_3 & \longrightarrow & CH_3-CH_2-CH_3 & & CH_3-CH-CH_3 \\ \uparrow \uparrow & & \uparrow & & \uparrow \\ \text{primary} & & \text{secondary} & & \text{tertiary} \end{array}$$

Substituent groups are sometimes named according to this pattern. The two most common such groups are *sec*-butyl and *tert*-butyl (or *t*-butyl):

$$\begin{array}{cc} CH_3 & CH_3 \\ | & | \\ CH_3-CH_2-CH- & CH_3-C- \\ & | \\ \text{\textit{sec}-butyl} & CH_3 \\ & t\text{-butyl} \end{array}$$

If the chain of C atoms is closed into a ring, *cyclo-* is prefixed to the name corresponding to the number of C atoms in the ring.

If the hydrocarbon contains a π bond, the same pattern of names as in Fig. 15-17 applies, except that these are now **alkenes**, not alkanes; the names end in *-ene* rather than *-ane*. In addition, the position of the π bond usually must be specified. This is done by giving the number of the C atom in the chain from which the π bond originates; the lower of the two possible numbers is used:

$$\begin{array}{cc} CH_2=CH_2 & CH_3-CH=CH_2 \\ \text{ethene} & \text{propene} \\ \text{(or ethylene)} & \text{(or propylene)} \end{array}$$

but

$$\begin{array}{cc} \underset{1}{CH_2}=\underset{2}{CH}-\underset{3}{CH_2}-\underset{4}{CH_3} & \underset{1}{CH_3}-\underset{2}{CH}=\underset{3}{CH}-\underset{4}{CH_3} \\ \text{1-butene} & \text{2-butene} \end{array}$$

$$\begin{array}{c} 1CH_2 \\ \| \\ 2CH \\ | \\ CH_3-CH_2-CH_2-\underset{3}{CH}-\underset{4}{CH_2}-\underset{5}{CH_2}-\underset{6}{CH_3} \end{array}$$

3-propyl-1-hexene (chain must include π bond)

If there is more than one π bond the compound is named as a diene, triene, or whatever is the appropriate number:

$$\underset{1 \quad 2 \quad 3 \quad 4}{CH_2{=}CH{-}CH{=}CH_2} \qquad \underset{1 \quad 2 \quad 3 \quad 4 \quad 5 \quad 6}{CH_2{=}CH{-}CH{=}CH{-}CH{=}CH_2}$$
$$\text{1,3-butadiene} \qquad\qquad \text{1,3,5-hexatriene}$$

Functional Group Nomenclature

Because the C—C and C—H bond energies are so large, the conceivable chemical reactions of alkanes (such as substitutions) are usually found to be impossible for both thermodynamic and kinetic reasons; the total free-energy change is often positive, and even when it is not, the activation energy for the reaction may be so high as to prevent any detectable reaction in a lifetime. Most reactions that are observed for organic compounds are those of **functional groups**, usually atoms (or groups of atoms) other than C and H substituted somewhere on the molecule. Because a C=C π bond is relatively high-energy and more reactive, it can also be considered a functional group. We can fairly quickly summarize the other common functional groups.

1. **Amines** contain the N atom with either H atoms or other organic groups bonded to it: —NH_2, —NHR, or —NR_2, where R stands for some organic group. All of these can be thought of as derivatives of ammonia, NH_3, in the sense that one or more H atoms have been replaced by organic groups; R—NH_2 is a **primary amine**, R_2NH is a **secondary amine**, and R_3N is a **tertiary amine**. Pursuing the ammonia analogy further, **quaternary ammonium salts**, containing the NR_4^+ ion, have already been described in Chapter 13. Amines are named either as alkyl amines or aminoalkanes:

$$CH_3-CH_2-NH_2 \qquad\qquad \begin{array}{c} H_3C \\ \diagdown \\ CH-NH-CH_3 \\ \diagup \\ H_3C \end{array}$$

ethylamine   methylisopropylamine
or aminoethane

2. **Nitro** groups contain the nitrogen atom with two oxygen atoms bonded to it: —NO_2; the collection of electronegative atoms makes it a strongly electron-withdrawing group. Its compounds are named as nitroalkanes:

$$\underset{\text{nitromethane}}{CH_3-NO_2} \qquad \underset{\text{2-nitrobutane}}{CH_3-\underset{\underset{NO_2}{|}}{CH}-CH_2-CH_3}$$

3. **Alcohols** contain the oxygen atom with a hydrogen bonded to it; —OH. They are named either as alkyl alcohols or as alkanols (using the suffix -ol):

$$\underset{\substack{\text{ethyl alcohol} \\ \text{or ethanol}}}{CH_3-CH_2-OH} \qquad \underset{\substack{\text{2,2-dimethylethanol} \\ \text{or } t\text{-butyl alcohol}}}{CH_3-\underset{\underset{CH_3}{|}}{\overset{\overset{CH_3}{|}}{C}}-OH} \qquad \underset{\substack{\text{butan-2-ol} \\ \text{or } sec\text{-butyl alcohol}}}{CH_3-CH_2-\underset{\overset{|}{OH}}{CH}-CH_3}$$

By analogy with the amines, we can imagine a "secondary alcohol" with two organic groups on the oxygen: R—OR. These are considered separately as ethers.

4. **Ethers** are named as alkyl ethers, with both alkyl groups specified if they are not the same:

$$\underset{\text{ethyl ether}}{CH_3-CH_2-O-CH_2-CH_3} \qquad \underset{\text{methyl propyl ether}}{CH_3-O-CH_2-CH_2-CH_3}$$

5. Carbonyl compounds contain an oxygen atom π-bonded to a carbon atom, $>$C=O. If the C atom in the C=O group has one organic group and one H bonded to it, the carbonyl compound is an **aldehyde**, R—CHO, named either as an alkanal (with the suffix -al) or as a derivative of an acid (see item 7):

$$\underset{\substack{\text{butanal} \\ \text{or butyraldehyde} \\ \text{(from butyric acid)}}}{CH_3-CH_2-CH_2-C{\overset{\nearrow O}{\underset{\searrow H}{}}}} \qquad \underset{\substack{\text{methanal} \\ \text{or formaldehyde} \\ \text{(from formic acid)}}}{H-C{\overset{\nearrow O}{\underset{\searrow H}{}}}}$$

6. A carbonyl compound in which the C=O carbon has two organic groups bonded to it is a **ketone**. Ketones are named either as alkyl ketones or as alkanones (with the suffix -one):

$$\underset{\substack{\text{propanone} \\ \text{or dimethyl ketone} \\ \text{(but commonly acetone)}}}{CH_3-\overset{\overset{O}{\|}}{C}-CH_3} \qquad \underset{\substack{\text{2-butanone} \\ \text{or methyl ethyl ketone}}}{CH_3-CH_2-\overset{\overset{O}{\|}}{C}-CH_3}$$

7. A carbonyl compound in which the C=O carbon also has an —OH group bonded to it withdraws the electrons bonding the OH hydrogen so strongly as to make it fairly favorable to transfer that proton to a water molecule in aqueous solution. Such carbonyl compounds are **acids**, R—COOH, named either as alkanoic acids (with the suffix *-oic acid*) or, commonly, as members of the series C_1/formic acid, C_2/acetic acid, C_3/propionic acid, C_4/butyric acid (other names exist in this series, but we can stop here):

$$CH_3-C\overset{O}{\underset{OH}{\diagup}}\qquad CH_3-\underset{\underset{}{|}}{\overset{CH_3}{C}}H-C\overset{O}{\underset{OH}{\diagup}}$$

ethanoic acid 2-methylpropanoic acid
or acetic acid or isobutyric acid

The use of the stems from these names for aldehydes, esters (item 8), and amides (item 9) should be noted.

8. **Esters** are carbonyl compounds in which the OH group that would be present in an acid is replaced by an —OR group: R—COOR. Since these are often formed by allowing an alcohol to react with an acid, they are named

$$R-C\overset{O}{\underset{OH}{\diagup}} + R'-OH \longrightarrow R-C\overset{O}{\underset{O-R'}{\diagup}} + HOH$$

as alkyl (from the alcohol) derivatives of acids, using the suffix *-ate* in place of *-ic acid*:

$$CH_3-C\overset{O}{\underset{O-CH\underset{CH_3}{\overset{CH_3}{\diagup}}}{\diagup}}\qquad CH_3-\overset{CH_3}{\underset{}{C}}H-C\overset{O}{\underset{O-CH_3}{\diagup}}$$

isopropyl ethanoate methyl 2-methylpropanoate
or isopropyl acetate or methyl isobutyrate

Esters can also be formed between alcohols and inorganic acids such as H_2SO_4 or H_3PO_4.

9. If the alcohol that would be used in forming an ester is replaced by ammonia or an amine, the product is called an **amide**, R—CONH$_2$ (or R—CONHR or R—CONR$_2$). Amides are named by replacing *-ic acid* in the acid's name with *-amide*. If the amide is made with an amine instead of ammonia, the organic groups in the amine are indicated as substituents, prefixed by *N-*:

$$CH_3-CH_2-CH_2-C\!\!\begin{array}{c}\diagup O\\ \diagdown NH_2\end{array} \qquad H-C\!\!\begin{array}{c}\diagup O\\ \diagdown N\!\!\begin{array}{c}\diagup CH_3\\ \diagdown CH_3\end{array}\end{array}$$

butanoamide   N,N-dimethylmethanoamide
or butyramide   or N,N-dimethylformamide

10. **Halides** contain a halogen atom bonded directly to carbon: R—X, since X is frequently used to represent F, Cl, Br, or I. They are named either as alkyl halides or as haloalkanes:

$$CH_3-CH_2-CH_2-CH_2-Br \qquad CH_3-\underset{\underset{Br}{|}}{CH}-CH_2-CH_3$$

1-bromobutane   2-bromobutane
or n-butyl bromide   or sec-butyl bromide

The reactions of these functional groups and of their carbon skeletons form a staggeringly large body of information, which we shall not attempt to deal with here. However, the theoretical models that have been developed for the myriad patterns of organic-compound reactivities have proven to be adequate in describing such varied reactions that chemists feel confident of their correctness. All we can do is to sketch a few ways our ideas of molecular electronic structure and reaction mechanisms can guide us in rationalizing some of the reactions observed. Perhaps a few examples will help to create a "feel" for the way we rationalize and predict reaction products.

Reaction Mechanisms: Chlorination of Hydrocarbons

One of the few reactions that alkanes will undergo is chlorination. If we mix propane and chlorine, both gases, in the dark and at room temperature they do not react; but if we heat the mixture to 250–300°C, or if we shine ultraviolet (UV) light on the mixture, we get substitution of Cl for H on the propane skeleton and an equal molar amount of HCl vapor:

$$CH_3-CH_2-CH_3 + Cl_2 \xrightarrow[\text{UV light }(h\nu)]{250\,°C \text{ or}} \begin{array}{c} CH_3-CH_2-CH_2-Cl \\ (48\%) \\ \text{and} \\ CH_3-\underset{\underset{Cl}{|}}{CH}-CH_3 \\ (52\%) \end{array} + HCl$$

With sensitive equipment it is possible to count the individual photons of UV light that a lamp emits; if we do this, we find that thousands of chloropropane molecules are formed for each photon of light absorbed by the system. And if we mix gaseous oxygen with the system, we find that the reaction is delayed for a period of time proportional to the quantity of oxygen in the system, even if its concentration is very small. How can we make sense of this pattern?

In the first place, we do not expect significant charges on any of the atoms. The electronegativities of carbon and hydrogen are so close together that we do not expect charge separation in the propane molecule, and of course there will not be any in the chlorine molecule. The reaction mechanism thus should not involve any ion–ion or ion–dipole attractions. During the reaction a C—H bond must break, but the nearly equal electronegativities and absence of charge suggest that when it breaks, one of the two σ-bonding electrons will stay with each atom—otherwise we would have to provide enough energy to pull a positive ion away from a negative ion. The energy necessary to break the C—H bond must come from somewhere, and in the absence of any electrostatic attraction we must suggest that it was provided by the simultaneous formation of another bond to either the carbon or the hydrogen. And in order for one photon of light to cause thousands of molecular reactions, the reaction step (the elementary process) that generates the product chloropropane molecule must also regenerate a reactant species—not one of the initial molecules but an activated species. The first step in the generally accepted mechanism is

$$Cl_2 \xrightarrow{h\nu \text{ or } kT} 2Cl\cdot \tag{1}$$

Since the chlorine atoms in the Cl_2 molecule are just alike, the pair of electrons in the Cl—Cl bond splits down the middle and the two chlorine atoms that result each have one unpaired electron (represented by a dot). Such species—either atoms or molecules—are called **free radicals**; this is a free-radical mechanism. The next step is

$$Cl\cdot + CH_3-CH_2-CH_3 \rightarrow CH_3-CH_2-CH_2\cdot + HCl \tag{2}$$

$$(\text{or } CH_3-\overset{\cdot}{C}H-CH_3)$$

in which the free-radical chlorine atom removes a H and one electron from a propane molecule, forming a σ H—Cl bond and leaving behind another free radical, the propyl radical. The third step involves the reaction of this propyl radical with chlorine to give a product chloropropane molecule and another free radical, $Cl\cdot$

$$CH_3-CH_2-CH_2\cdot + Cl_2 \rightarrow CH_3-CH_2-CH_2-Cl + Cl\cdot \tag{3}$$

This mechanism has the interesting property that, once step 1 has generated the free radicals, steps 2 and 3 use them to give product molecules *and more free radicals*, which can then go back and react still further. This is a **chain reaction**. Step 1 is the **chain-initiating** step, and steps 2 and 3 are **chain-propagating** steps.

The reaction does not go on without a continued supply of heat or light, which means we must continue to have a new supply of free radicals through step 1. Some radicals must be disappearing, which could happen through any of the following steps:

$$2CH_3-CH_2-CH_2\cdot \rightarrow CH_3-CH_2-CH_2-CH_2-CH_2-CH_3 \qquad (4)$$

$$2Cl\cdot \rightarrow Cl_2 \qquad (5)$$

$$CH_3-CH_2-CH_2\cdot + Cl\cdot \rightarrow CH_3-CH_2-CH_2-Cl \qquad (6)$$

These steps are quite rapid, since free radicals are very reactive, but rather infrequent, since there are few free radicals and they rarely hit each other. Steps 4, 5, and 6 are said to be **chain-terminating** steps. The effect of O_2 molecules in inhibiting the reaction can be understood by remembering that O_2 has two unpaired electrons (see p. 285), thus being in effect a stable free radical (or diradical). We can expect it to react with the propyl radicals:

$$CH_3-CH_2-CH_2\cdot + \cdot O=O\cdot \rightarrow CH_3-CH_2-CH_2-O-O\cdot$$

This free radical is more stable than the others and will not continue the chain, thus delaying the formation of any product until all the O_2 has been used up in this fashion. If the rate of production of $Cl\cdot$ radicals by elementary process 1 is more or less constant, it is easy to see that the total time during which the rate of the chlorination reaction is slowed by the presence of O_2 will be proportional to the total quantity of O_2 in the reaction mixture.

We can thus propose a mechanism for the chlorination reaction that explains the observed results and is, at the same time, compatible with the ideas of molecular electronic structure that we have already developed. However, free-radical mechanisms are relatively rare in the broad spectrum of organic reactions, because most compounds have heteroatoms with bond dipole moments, which offers the possibility for ion–dipole or at least dipole–dipole attraction as a means of lowering the activation energy of the rate-determining step in the mechanism. When this sort of attraction is possible, bond breaking is quite likely either to take both electrons away or to leave both behind, and we can expect positive or negative ions to develop. Mechanisms involving intermediate species that bear appreciable charges are called **ionic** mechanisms.

Reaction Mechanisms: Dehydration of Alcohols

To see how ionic mechanisms can apply to organic reactions, we can look at the dehydration (loss of water) of alcohols. When an alcohol is heated in the presence of a reasonably strong acid, it loses water and forms an alkene:

$$\underset{t\text{-butyl alcohol}}{CH_3-\underset{\underset{CH_3}{|}}{\overset{\overset{CH_3}{|}}{C}}-OH} \xrightarrow[90°]{H_2SO_4} \underset{\text{isobutylene}}{CH_3-C\overset{CH_2}{\underset{CH_3}{\diagdown}}} + H_2O$$

This is quite a general reaction, but the conditions under which it will occur vary depending on what the alkyl group to which the OH is bonded is like. If the C to which the OH is bonded is primary (bonded to only one other C) the reaction requires relatively high temperatures and very strong acids; if the C is secondary the conditions are milder, and if it is tertiary the conditions are very mild indeed. As a matter of fact, some tertiary alcohols dehydrate spontaneously inside a gas chromatograph, which can cause considerable confusion when a pure substance going in gives several peaks coming out. However, the presence of an acid is always necessary. We should also note the surprising result that the π bond that forms does not always involve the C atom that originally had the OH group. This is known as a **rearrangement**:

$$\underset{sec\text{-butyl alcohol}}{CH_3-CH_2-\overset{\overset{OH}{|}}{CH}-CH_3} \xrightarrow[100°]{H_2SO_4} \underset{2\text{-butene}}{CH_3-CH=CH-CH_3} \qquad \text{(principal product)}$$

but

$$\underset{n\text{-butyl alcohol}}{CH_3-CH_2-CH_2-CH_2-OH} \xrightarrow[140°]{H_2SO_4} \underset{2\text{-butene}}{CH_3-CH=CH-CH_3} \qquad \text{(principal product)}$$

Carbonium Ions

All of these features are accounted for by a mechanism involving a **carbonium ion**: a C atom in the molecular skeleton that has had both of the electrons in one of its σ bonds removed when the σ bond was broken. Since the C originally contributed one of the electrons to the bond, the removal of both electrons leaves the C atom with a net positive charge:

$$-\overset{|}{\underset{|}{C}}-X \quad \text{or} \quad -\overset{|}{\underset{|}{C}}:X \rightarrow -\overset{|}{\underset{|}{C^+}} + :X^-$$

In the dehydration reaction we assume that the first elementary process is the protonation of the OH group by the acid that is present, giving a species with a net positive charge (since a proton but no electrons were added):

$$CH_3-\underset{\underset{CH_3}{|}}{\overset{\overset{CH_3}{|}}{C}}-\ddot{O}H + \overset{\frown}{(H)}O-SO_2-OH \longrightarrow CH_3-\underset{\underset{CH_3}{|}}{\overset{\overset{CH_3}{|}}{C}}-\overset{+}{\underset{H}{\overset{H}{\ddot{O}}}}\diagdown + HSO_4^-$$

The —OH_2^+ group attracts both electrons from the C—O bond to the O, which is very electronegative anyway. Complete transfer of the pair breaks the C—O bond to give (neutral) water and leave a carbonium ion:

$$CH_3-\underset{\underset{CH_3}{|}}{\overset{\overset{CH_3}{|}}{C}}\overset{+}{\cdot\cdot}\overset{H}{\underset{H}{\overset{}{O}}} \longrightarrow CH_3-\overset{CH_3}{\underset{CH_3}{\overset{|}{C^+}}} + \overset{H}{\underset{H}{\overset{}{\ddot{O}}}}$$

The carbonium ion that results attracts electrons from one of the C—H bonds, making it favorable for that proton (H^+) to be transferred to a base such as HSO_4^- in the reaction mixture:

$$CH_3-\overset{H_2C\;\cdot\cdot\;\text{\textcircled{H}}}{\underset{CH_3}{\overset{}{C^+}}} + HSO_4^- \longrightarrow CH_3-\overset{CH_3}{\underset{CH_3}{\overset{}{C}}} + H_2SO_4$$

The result is the formation of the π-bonded alkene.

The varied conditions that are required for dehydration of primary, secondary, and tertiary alcohols simply reflect the different stabilities of the carbonium ions involved. A carbonium ion is pretty unstable, since it has a potential bonding MO vacant, which makes it an excellent electron acceptor. Any feature of the molecular electronic structure that relieves the electrostatic "strain" of the positively charged C atom in the carbonium ion (by releasing electrons toward it) will make that ion less unstable and easier to form, thereby speeding up the reaction. In effect, if the rest of the molecule can contribute electrons to reduce the positive charge on the C^+, the activation energy for the formation of the transition-state carbonium ion will be lowered. (Review Section 11-4 if these concepts and nomenclature are not clear.) A H atom is not a very good electron donor, because even in a covalent bond it has only two electrons; but a C atom is no more electronegative and has more electrons. So it is not surprising that the more C atoms are attached to the C^+, the more stable it is. This is why the tertiary alcohols react faster than the secondary, which react faster than the primary—the tertiary carbonium ion that must be formed is less unstable than the secondary carbonium ion, which is less unstable than the primary carbonium ion.

Consideration of the relative stability of the carbonium ions also gives us a guide to the rearrangements that are observed. In the rearrangement we saw earlier, n-butyl alcohol gave 2-butene:

$$CH_3-CH_2-CH_2-CH_2-OH \xrightarrow{H^+} CH_3-CH=CH-CH_3 + H_2O$$

This must have involved the formation, initially, of the very unstable primary carbonium ion, since the reactant is a primary alcohol. But if a H atom were to migrate with both bonding electrons toward the primary C⁺, the less unstable secondary carbonium ion could form:

$$CH_3-CH_2-\overset{+}{CH}-CH_2 \rightarrow CH_3-CH_2-\overset{+}{CH}-CH_3$$
$$\overset{\frown}{\underset{\textcircled{H}}{}}$$

We can fit this type of mechanism to a very wide variety of experimental results if we assume that this sort of hydride ion migration is common within the short lifetime of the carbonium ion. And it is quite generally true that the rearrangements observed in reactions that follow a carbonium ion mechanism are those that would result from the formation of a more stable carbonium ion.

Carbonium ion mechanisms have been applied to many reactions that are seemingly quite different; we have already seen one in Section 11-5. We should suspect such a mechanism any time the overall result of a reaction requires that a bond between a C atom and a more electronegative atom be broken; such a bond is going to be polar anyway, and when it breaks the electronegative atom is likely to retain the electrons. Of course, this leaves the C atom in the positively charged state that is characteristic of this mechanism.

Reaction Mechanisms: Methyllithium and Dry Ice

Whenever the electronegativity of a functional group causes even a relatively modest charge separation in an organic molecule it is likely to determine the reactions of the molecule. As we suggested in Chapter 11, this is because the electrostatic attraction between charges or polar compounds has the effect of lowering the activation energy for the particular elementary process that has the charges or dipoles arranged in the most favorable way. We have seen that this is true when the departing atom or group is more electronegative than carbon, but it is equally true when the leaving group is less electronegative than carbon. For instance, consider the compound methyllithium, CH_3Li. When it reacts with anything, we expect the mechanism of the reaction to take full advantage of the fact that lithium is much less electronegative than carbon and will thus have a charge separation

$$H-\underset{\underset{H}{|}}{\overset{\overset{H}{|}}{C^{\delta-}}}:Li^{\delta+} \qquad (\delta+ \equiv \text{partial positive charge})$$

When the Li—C σ bond is broken, we expect to see both electrons stay with the C atom:

$$H-\underset{\underset{H}{|}}{\overset{\overset{H}{|}}{C^{\delta-}}}:Li^{\delta+} \rightarrow H_3C^- + Li^+$$

Carbanions

A species containing a C atom with a net negative charge resulting from its retention of both electrons from a σ bond that has been broken is called a **carbanion**. The negatively charged C atom will react rapidly with another atom having at least a partial positive charge. If methyllithium is allowed to react with CO_2 (dry ice), we can predict the product if we think about the partial charges on the atoms in the CO_2 molecule. Since O is more electronegative than C, the C will have a partial positive charge and the O's a partial negative charge:

$$\overset{\delta-}{O}=\overset{\delta+}{C}=\overset{\delta-}{O}$$

The electrostatically favorable reaction for a methyllithium molecule bonds the CH_3^- from the methyllithium to the partially positive C atom in the CO_2 molecule:

$$Li^+ \; (CH_3^-) + \overset{\delta+}{\underset{\parallel}{\overset{\parallel}{C}}}\overset{O}{\underset{O}{}} \longrightarrow CH_3-C\overset{O}{\underset{O^-}{\diagdown}} \; Li^+$$

This product does not look very useful, but we can make it so simply by adding acid to it. Again the charges dictate the reaction:

$$CH_3-C\overset{O}{\underset{O^-}{\diagdown}} + (^+H)-O\overset{H}{\underset{H}{\diagdown}} \longrightarrow CH_3-C\overset{O}{\underset{OH}{\diagdown}} + H_2O$$

acetic acid

This is a very useful reaction for adding another C atom to an alkyl group and converting it to an acid, and we see that we can predict it exactly if we assume the formation of the CH_3^- carbanion.

Reaction Mechanisms: The Formation of Fulvene

There are some reactions that appear to involve carbanions in a more subtle way. In Chapter 11 we used the dimerization of cyclopentadiene as an example of a nonpolar reaction that proceeded at essentially the same rate in the vapor phase and in all sorts of solvents. Cyclopentadiene would thus not seem to be a very good candidate for an ionic mechanism. However, if it is allowed to react with sodium, a deep red color develops and a compound can be isolated in which the cyclopentadiene has pretty clearly lost a H atom, taken on another electron, and become a carbanion:

794 | Covalent Compounds

$$\text{cyclopentadiene} + Na\cdot \longrightarrow \text{cyclopentadienide}^- Na^+ + \tfrac{1}{2}H_2$$

This carbanion is the cyclopentadienide ion; it is quite stable, mostly because it is a cyclic π system with six π electrons, which makes it aromatic like benzene. (The extra stability of aromatic systems was discussed in Section 15-3.) The neutral compound cyclopentadiene cannot be aromatic because one of its C atoms has four σ bonds and thus cannot participate in π overlap.

The cyclopentadienide carbanion, like some others, can be produced by reaction of the neutral compound (cyclopentadiene) with a base. From Chapter 10 we remember that a base can be an electron donor—which sodium certainly is—or a proton acceptor. In other words, we ought to be able to remove a proton from cyclopentadiene with, say, OH$^-$. And we might expect, for instance, that such a carbanion would react with a carbonyl C=O group in the same way that the CH$_3^-$ ion reacted with CO$_2$. If in fact we allow cyclopentadiene and the carbonyl compound formaldehyde, H$_2$CO, to react under basic conditions, the product is fulvene:

$$\text{cyclopentadiene} + \text{formaldehyde} \xrightarrow{OH^-} \text{fulvene} + H_2O$$

We can propose a mechanism for this reaction that involves a carbanion intermediate:

There are many reactions that are base catalyzed in the same sense as this one, and a carbanion mechanism is usually proposed as we have done here.

The last few pages have shown that only a few basic mechanisms are needed to account, in general terms at least, for a very wide spectrum of organic reactions. If a compound is really nonpolar, with no significant electronegativity differences, we often find that its reactions with other nonpolar molecules proceed by some sort of free-radical mechanism. If a heteroatom functional group is present, the possibility of electrostatic attraction will usually make the functional group react by some sort of ionic mechanism—either a carbonium ion or a carbanion, depending on which way the electronegativity difference appears in the bonding of the heteroatom. If the transition state involves a carbonium ion, it will be attracted to centers of negative charge and will thus be an electrophile. A carbanion will be attracted to centers of positive charge, thus acting as a nucleophile. All of this is in the language of Section 11-5, and indeed we are only taking some specific examples and fitting them into the general concepts of that section. Since heteroatom functional groups are so common, it turns out that the great majority of organic reactions proceed through an ionic intermediate and only a few involve a free-radical intermediate, but in the next section we shall see some important applications of all three types of mechanisms.

15-5 The Formation of Polymers

So far, few examples have been presented that demonstrate the pervasive influence of organic chemistry. In fact, organic reactions are of utmost importance both industrially and biologically. Small organic molecules are quite important to us, as witness the sales of alcoholic beverages (CH_3CH_2OH) and gasoline (a hydrocarbon mixture with about eight C's per molecule). However, most of us tend to think of organic chemistry in terms of large molecules, and some illustrations of both synthetic polymers and biological macromolecules may be particularly useful.

Polyethylene and Other Free-Radical Polymers

One of the simplest synthetic polymers we can consider is **polyethylene**. We can write its formula as $(CH_2CH_2)_n$, where n indicates the number of units of the ethylene ($CH_2{=}CH_2$) **monomer** that are linked together in the polymerization reaction, usually about 1000. Of course, there has to be some group at each end of this long chain; we shall see what these are in a moment. The principal chemical reaction in preparing polyethylene has to be the reaction of one ethylene molecule with another, and in the absence of any groups on the ethylene that can donate or withdraw electrons we expect an electrically neutral free-radical mechanism. So if we supply a source of free radicals to gaseous ethylene it ought to polymerize:

$$n\text{CH}_2{=}\text{CH}_{2(g)} \xrightarrow{\text{free radicals}} {+}(\text{CH}_2\text{CH}_2{+})_{n(s)}$$

Unfortunately, at ordinary temperatures and pressures this reaction will not proceed; but taking advantage of Le Chatelier's principle, we raise the pressure, and at about 1000 atm the reaction proceeds smoothly. We propose the mechanism

$$\text{R}{-}\text{O}{-}\text{O}{-}\text{R} \rightarrow 2\text{R}{-}\text{O}\cdot \qquad \text{initiation}$$
organic peroxide

$$\left.\begin{array}{l}\text{R}{-}\text{O}\cdot + \text{CH}_2{=}\text{CH}_2 \rightarrow \text{R}{-}\text{O}{-}\text{CH}_2{-}\text{CH}_2\cdot \\ \text{R}{-}\text{O}{-}\text{CH}_2{-}\text{CH}_2\cdot + (n-1)\text{CH}_2{=}\text{CH}_2 \rightarrow \text{R}{-}\text{O}{+}(\text{CH}_2\text{CH}_2{+})_n\cdot \end{array}\right\} \text{propagation}$$

$$\left.\begin{array}{l}2\text{R}{-}\text{O}{+}(\text{CH}_2\text{CH}_2{+})_n\cdot \rightarrow \text{R}{-}\text{O}{+}(\text{CH}_2\text{CH}_2{+})_{2n}{-}\text{O}{-}\text{R} \\ 2\text{R}{-}\text{O}{+}(\text{CH}_2\text{CH}_2{+})_n\cdot \rightarrow \text{R}{-}\text{O}{+}(\text{CH}_2\text{CH}_2{+})_{n-1}\text{CH}{=}\text{CH}_2 \\ \qquad + \text{R}{-}\text{O}{+}(\text{CH}_2\text{CH}_2{+})_{n-1}\text{CH}_2{-}\text{CH}_3 \end{array}\right\} \text{termination}$$

The end groups on the polymer chain are either the organic groups from the original peroxide (free-radical source) or a π bond or a methyl group. Other commercially important polymers formed in similar reactions include Teflon (from $\text{CF}_2{=}\text{CF}_2$), Saran (from $\text{CH}_2{=}\text{CHCl}$), Orlon (from $\text{CH}_2{=}\text{CHCN}$), and Lucite (from $\text{CH}_2{=}\text{C}(\text{CH}_3)\text{COOCH}_3$).

Structural and Physical Properties of Polyethylene and Other Polymers

What are the properties of the ethylene polymer? It is a solid because kT at ordinary temperatures is not great enough to overcome the many van der Waals forces that attract an individual polymer molecule to its neighbors. With the great flexibility of a long chain of carbons, it would be surprising if the solid polymer were crystalline in the ordinary sense—there are just too many possibilities for tangling and disrupting the orderly geometry of an ideal crystal. We do find **crystallites**, however, which are small regions of the solid where a reasonably orderly array of the polymer chains exists (as in Fig. 15-19). The larger these crystallites are and the less amorphous region between them, the more crystalline the polymer is said to be. If the polymer is **amorphous** it corresponds to a very viscous fluid, and we expect it to flow (showing a plastic deformation)

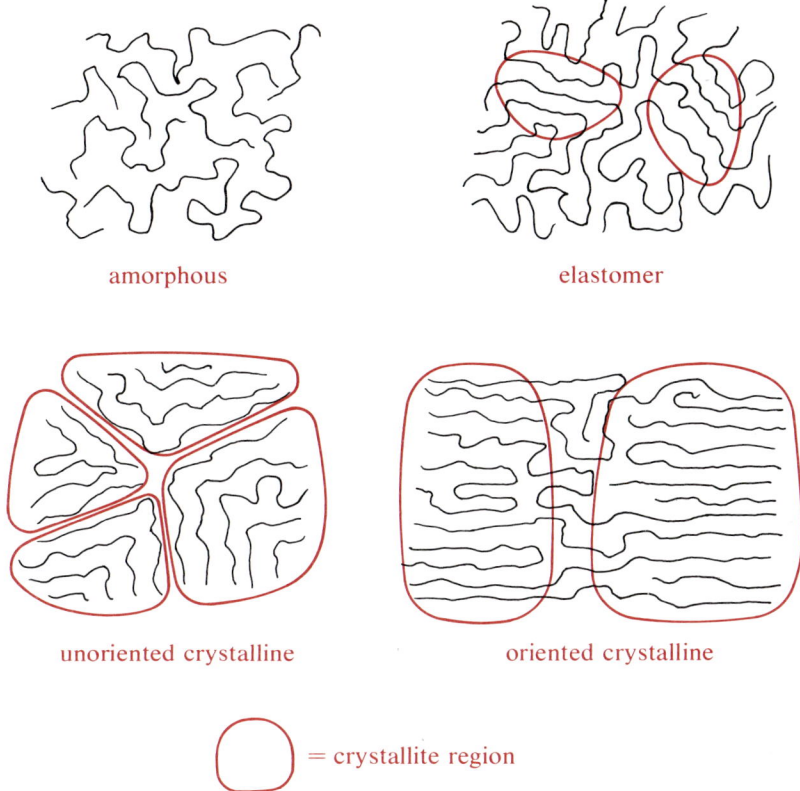

amorphous

elastomer

unoriented crystalline

oriented crystalline

◯ = crystallite region

Figure 15-19 Polymer-chain stacking and crystallites.

under an external force, like Silly Putty. If the crystallites are large but randomly oriented, the polymer will be stiff and have at least some capability for elastic deformation. It will "melt" at high temperatures, perhaps fairly sharply if it is highly crystalline, and can then be molded into desired shapes, with the crystallites reforming on cooling; this is the basic procedure for manufacturing small articles out of polyethylene, which is an **unoriented crystalline** polymer. While we are considering Fig. 15-19, let us look at **oriented crystalline** polymers, which are usually just unoriented ones that have been cold-drawn—as by extruding a fiber. The virtue of this procedure is that the tensile strength of the oriented polymer is usually higher than that of the unoriented form. Polypropylene, which is much like polyethylene but must be made by a different process, is widely used as an oriented crystalline polymer fiber for carpets as Herculon and other brand names. **Elastomers** are more fun; an elastomer is only slightly crystalline in its relaxed state, so it deforms very readily under an external force. But as the deformation (stretching, perhaps) is increased the polymer chains are forced more nearly parallel to each other—in effect, the

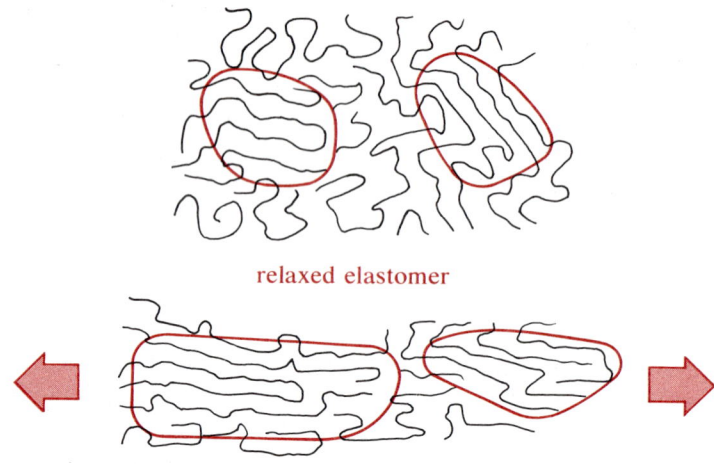

Figure 15-20   Growth of crystallites in a stretched elastomer.

crystallites grow and the polymer gets stiffer, as in Fig. 15-20. But if the right polymer is chosen, the forces between the chains will be too weak to maintain the crystallites, and when the force is released the polymer will snap back to its original form. Natural rubber, a polymer of isoprene,

$$CH_2=CH(CH_3)-CH=CH_2$$

is a familiar example of an elastomer.

What else can we expect of polyethylene? It is water repellent, since water molecules cannot hydrogen bond to it. It has the same slightly waxy feel as paraffin, which is also a long-chain hydrocarbon. If it is chilled it becomes hard and brittle; this is characteristic of all chain polymers, which are said to be glass-like at low temperatures. On a molecular scale, chilling reduces the ability of the chains to break van der Waals forces while moving past each other in response to an external force; if the force is great enough, the polymer will break instead of bending. Polyethylene (and polypropylene) cannot be dyed very easily, since dyestuffs must be held on a fiber by hydrogen bonding or more substantial electrostatic forces, none of which are possible on polyethylene. Some of these properties are desirable and some are not (plastic raincoats should be water repellent, but they should not crack in cold weather), and we shall see that other polymers can be tailored to have other properties.

Polymer-Forming Reactions

If we go to more diverse systems than polyethylene, it becomes clear very quickly that a very great variety of possibilities exists. In principle all we need to form a polymer is a single organic compound containing two functional groups that can react with each other, or two different compounds each of which contains two functional groups that can react with those on the other compound:

$$nX-R-Y \longrightarrow (R)_n + nXY$$
$$\text{or} \searrow (X-R-Y)_n$$

$$nX-R-X + nY-R'-Y \longrightarrow (R-R')_n + nXY$$
$$\text{or} \searrow (X-R-X-Y-R'-Y)_n$$

Nylon-6 and Nylon-66

A good example of both of these possibilities is **nylon**, which is made in two ways to produce slightly different polymers that still have essentially identical properties. Nylon-6 is made using the single six-carbon compound caprolactam, which is an amide within itself; it forms a polyamide:

$$n \begin{array}{c} \text{H}_2\text{C} \\ \text{H}_2\text{C} \end{array} \begin{array}{c} \text{C}-\text{NH} \\ \\ \text{C} \\ \text{H}_2 \end{array} \begin{array}{c} \text{CH}_2 \\ \text{CH}_2 \end{array} \longrightarrow (NH-\overset{O}{\overset{\|}{C}}-CH_2-CH_2-CH_2-CH_2-CH_2)_n$$

caprolactam   nylon-6

Nylon-66, on the other hand, is also a polyamide, but it is made from two compounds, one a six-carbon diacid, the other a six-carbon diamine:

$$n \underset{\text{HO}}{\overset{\text{O}}{\text{C}}}-(CH_2)_4-\underset{\text{OH}}{\overset{\text{O}}{\text{C}}} + nNH_2-(CH_2)_6-NH_2 \longrightarrow$$

adipic acid   hexamethylenediamine

$$(NH-\overset{O}{\overset{\|}{C}}(CH_2)_4\overset{O}{\overset{\|}{C}}-NH(CH_2)_6)_n + 2nH_2O$$

nylon-66

One of the nice things about nylon (either kind) is that its crystallites are held together by hydrogen bonds, as in Fig. 15-21. These provide a much greater attraction than van der Waals forces, giving nylon a higher effective melting point than polyethylene. The possibility of hydrogen bonding to the polymer molecule also reduces the water repellency of nylon, which makes it less uncomfortable than polyethylene to wear when used in clothing. For the same reason, dyeing nylon is much easier than dyeing polyethylene. Another important polymer produced in an analogous reaction is **Dacron**, which is a polyester rather than a polyamide. **Kodel** and **Mylar** are basically the same stuff.

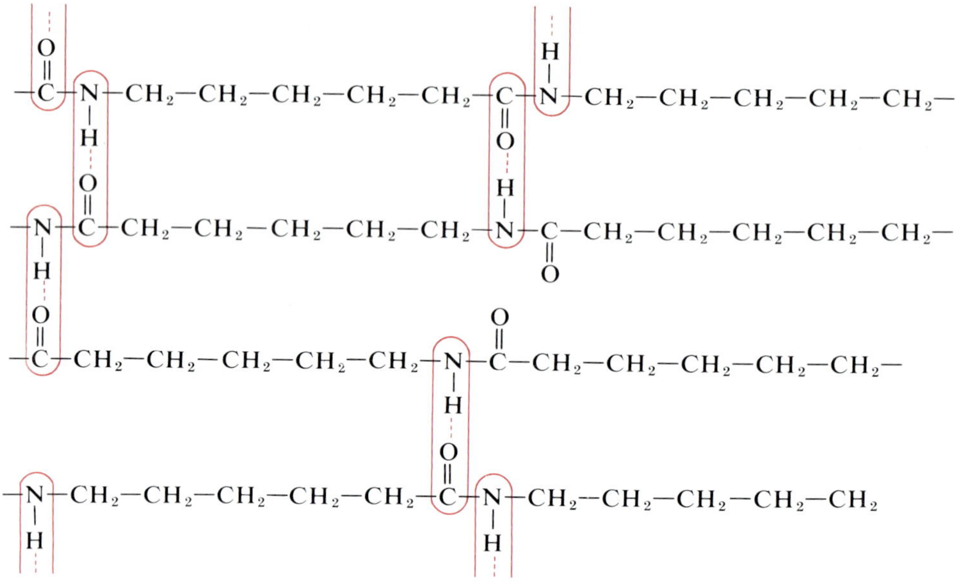

800 | Covalent Compounds

[Figure showing hydrogen-bonded nylon-6 polymer chains with repeating units of —C(=O)—N(H)—CH₂—CH₂—CH₂—CH₂—CH₂— linked by hydrogen bonds between C=O and N—H groups of adjacent chains]

Figure 15-21   Hydrogen-bond linkage of polymer chains in nylon-6.

Three-Dimensional (Cross-Linked) Polymers

The existence of hydrogen bonds between nylon molecules suggests that it ought to be possible to make polymers bonded in three dimensions if a monomer molecule or pair of molecules were used that had more than two reactive sites. The oldest such **cross-linked** polymer is **Bakelite**, which is made from phenol (hydroxybenzene) and formaldehyde. The carbon atom in formaldehyde reacts at any of three possible positions on the benzene ring in phenol (two ortho and one para position), which in effect gives the phenol three reactive sites and results in a three-dimensional bonding framework:

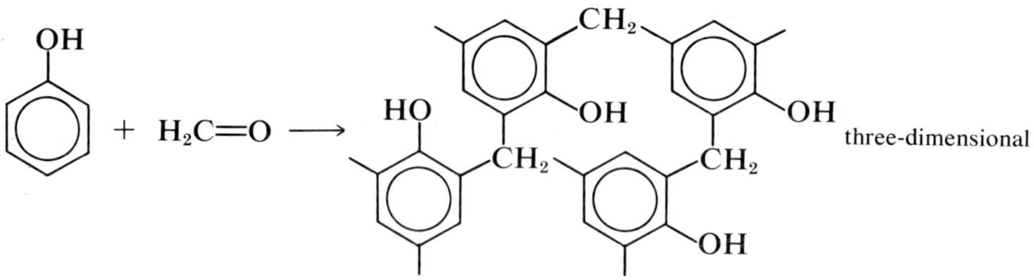

[Reaction scheme: phenol + H₂C=O → Bakelite (three-dimensional cross-linked structure of phenol rings connected by CH₂ bridges)]

phenol   formaldehyde   Bakelite

Melmac plastic dishes are produced by a related reaction between melamine and formaldehyde:

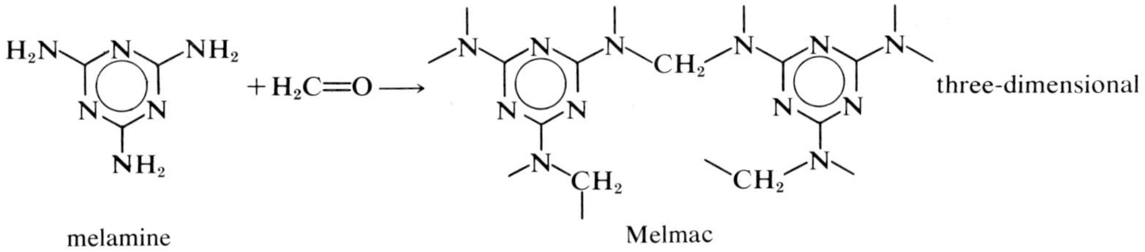

The Formation of Polymers | 801

[melamine + H₂C=O → Melmac, three-dimensional structure shown]

A particularly important property of cross-linked polymers is that if the cross linking is extensive, as in these cases, the whole sample of polymer becomes in effect one big molecule, with the physical properties we have learned to expect (in Section 4-2) for giant-molecule solids. Bakelite and Melmac are very hard and heat resistant, because the complete network of directional bonds resists deformation and also requires considerable kT energy to break up into a liquid. They are usually heated to achieve complete cross linking (hardening) and are said to be **thermosetting** polymers, while the linear polymers that melt when heated are **thermoplastic**.

Biological "Polymer" Macromolecules: Proteins

All of these latter polymers are formed by mechanisms that are either cationic or anionic, not by free radicals. As we have noted, this is a result of the charge separation due to the presence of very electronegative heteroatoms in the monomers. If we turn to biological polymers or macromolecules, we find that they must be formed in the polar aqueous medium of cell fluids. This suggests two features of these molecules (recall Section 11-5): first, they must themselves be formed by ionic mechanisms, and second, hydrogen bonding must be important in determining their structure. Effective hydrogen bonding requires the presence of oxygen and/or nitrogen atoms in the monomers. With these ideas in mind, we can look at the structure of **proteins**, a term that covers a large variety of biological macromolecules. Many proteins have been studied experimentally, and all of them can be separated into monomer units called **amino acids**:

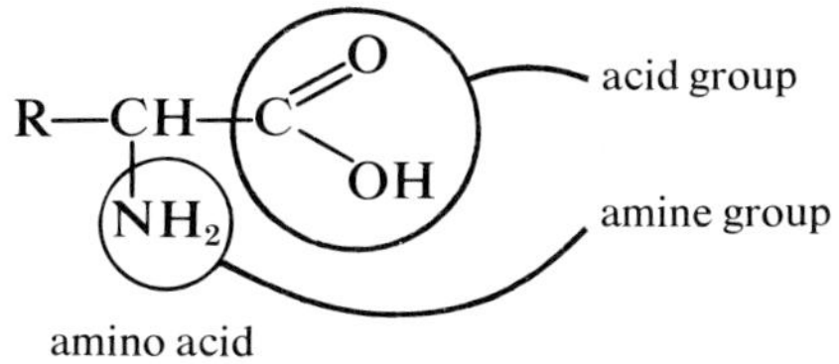

[amino acid structure: R—CH(NH₂)—C(=O)OH, with acid group and amine group labeled]

From all the proteins that have been studied, only about 21 amino acids have been isolated; that is, only about 21 R groups appear to exist in nature. However, this does not mean that only 21 proteins exist, because a protein is not, strictly speaking, a polymer. The different amino acids link together, forming amides in somewhat the same sense that nylon does, but a protein is not a sequence of a *single* repeated amino acid. Rather, many different amino acids link into a single protein, and the sequence of amino acids becomes important to any understanding of the protein's structure. From the amino acids listed in Table 15-3, we could make a "dimer" called a **dipeptide** (a sequence of two amino acids) from, say, glycine and alanine:

Table 15-3
Protein Amino Acids

glycine	Gly	 H—CHCOOH $\|$ NH_2
alanine	Ala	CH_3—CHCOOH $\|$ NH_2
valine	Val	CH_3CH—CHCOOH $\| \quad \|$ $CH_3 \; NH_2$
leucine	Leu	CH_3CHCH_2—CHCOOH $\| \qquad \|$ $CH_3 \qquad NH_2$
isoleucine	Ile	CH_3CH_2CH—CHCOOH $\| \quad \|$ $CH_3 \; NH_2$
serine	Ser	HO—CH_2—CHCOOH $\|$ NH_2
threonine	Thr	CH_3CH—CHCOOH $\| \quad \|$ $OH \; NH_2$
methionine	Met	CH_2CH_2—CHCOOH $\| \qquad \|$ $SCH_3 \quad NH_2$
cysteine	Cys	CH_2—CHCOOH $\| \quad \|$ $SH \; NH_2$
proline	Pro	$\begin{array}{c} H_2 \\ C \\ H_2C \diagup \; \diagdown CHCOOH \\ \| \qquad / \\ H_2C\text{—}NH \end{array}$

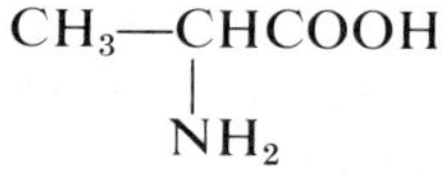

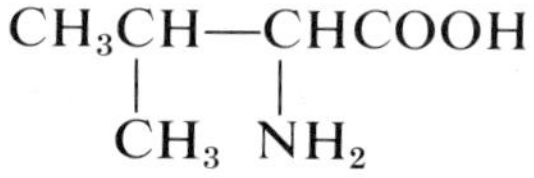

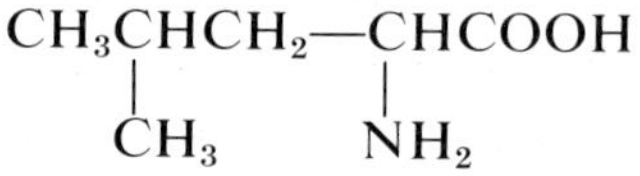

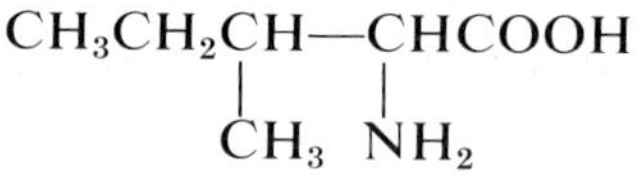

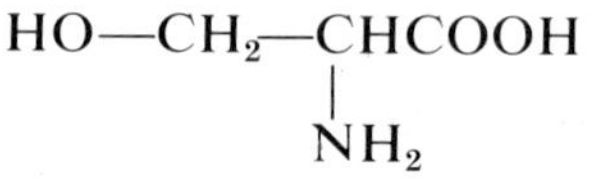

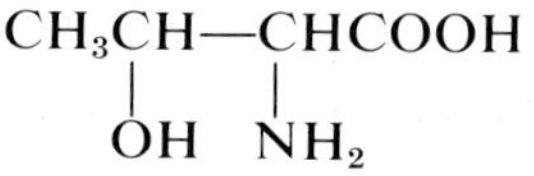

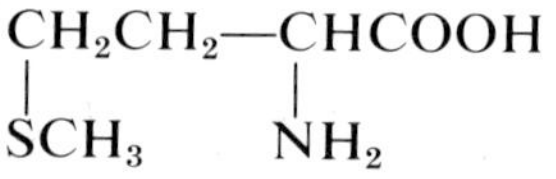

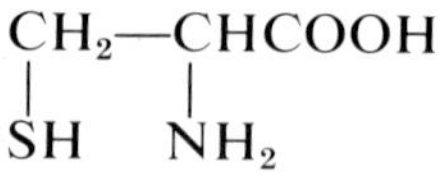

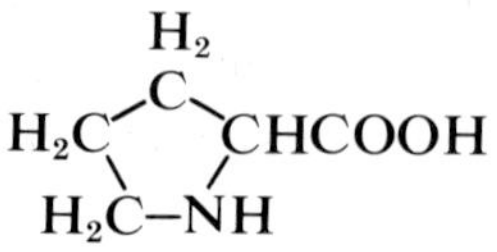

Table 15-3 (continued)

asparagine	Asn	$H_2NCOCH_2-CH(NH_2)COOH$
glutamine	Gln	$H_2NCOCH_2CH_2-CH(NH_2)COOH$
phenylalanine	Phe	$C_6H_5-CH_2-CH(NH_2)COOH$
tyrosine	Tyr	$HO-C_6H_4-CH_2-CH(NH_2)COOH$
tryptophan	Try	indolyl-$CH_2-CH(NH_2)COOH$
aspartic acid	Asp	$HOOCCH_2-CH(NH_2)COOH$
glutamic acid	Glu	$HOOCCH_2CH_2-CH(NH_2)COOH$
lysine	Lys	$H_2NCH_2CH_2CH_2CH_2-CH(NH_2)COOH$
arginine	Arg	$H_2N-C(=NH)-NH-CH_2CH_2CH_2-CH(NH_2)COOH$
histidine	His	imidazolyl-$CH_2-CH(NH_2)COOH$

$$\underset{\text{glycine}}{\text{H}-\underset{\underset{\text{NH}_2}{|}}{\text{CH}}-\text{C}\overset{\displaystyle\text{O}}{\underset{\text{OH}}{\diagdown}}} + \underset{\text{alanine}}{\text{CH}_3-\underset{\underset{\text{NH}_2}{|}}{\text{CH}}-\text{C}\overset{\displaystyle\text{O}}{\underset{\text{OH}}{\diagdown}}} \longrightarrow \underset{\text{Gly-Ala}}{\text{H}-\underset{\underset{\text{NH}_2}{|}}{\text{CH}}-\text{C}\overset{\displaystyle\text{O}}{\underset{\text{NH}}{\diagdown}}\underset{\underset{\text{C}-\text{OH}}{\overset{\displaystyle\text{O}}{\|}}}{\overset{\displaystyle\text{CH}_3}{\underset{|}{\text{CH}}}}} + \text{H}_2\text{O}$$

This would of course have different properties from another dipeptide made from, say, glycine and serine, but it would even have different properties from that formed by the same two amino acids linked in the reverse order Ala-Gly. It should be statistically clear that we could make $(21)^2 = 441$ different dipeptides, $(21)^3 = 9261$ different tripeptides, or in general $(21)^n$ different polypeptides with n amino acid groups. Since naturally occurring proteins have molecular weights of at least 10,000, with at least 50 amino acids, there are clearly a lot of possible proteins.

Besides the amino acid sequence (called the **primary structure**) of a given protein, there is an interesting secondary structure. N—H groups can hydrogen bond to C=O groups, and X-ray diffraction studies indicate that in crystalline proteins at least, the protein chain often assumes a spiral or helical structure (the **secondary structure**), with each loop of the spiral containing four amino acids so that each amino acid is hydrogen bonded to the fourth one away from it on the chain, as in Fig. 15-22 (compare with Fig. 15-21). The R— groups on the various amino acids are outside the spiral or **helix**, which leads to a third subtlety of the protein's structure. Bulky R— groups, or ones that attract or repel each other electrostatically, alter the symmetry of the helix, causing it to fold or bend at certain points. The overall structure of the protein is thus that of a linear series of amino acids bonded together as a polyamide or polypeptide, held in a helical configuration by hydrogen bonds, with the helix bent and folded into a complex and unsymmetrical three-dimensional shape (a Slinky toy with a knot tied in it?). The precise nature of the folds depends, of course, on the sequence of amino acids in the protein; since there are so many possible sequences, there are a great many possible protein configurations. Some information about their shapes can be gained from viscosity studies and light scattering, but only in a very general sense. X-ray studies can give the precise configuration in the crystalline state, but we have no assurance that the solution form has the same configuration. There is a lot still to be learned in this area, and here as in all of chemistry an understanding of chemical reactions must await an understanding of structure.

Enzyme Structure (Protein Catalysts)

One of the areas in which detailed information about the structure of proteins would be most helpful is the understanding of **enzyme** reactions. Enzymes are

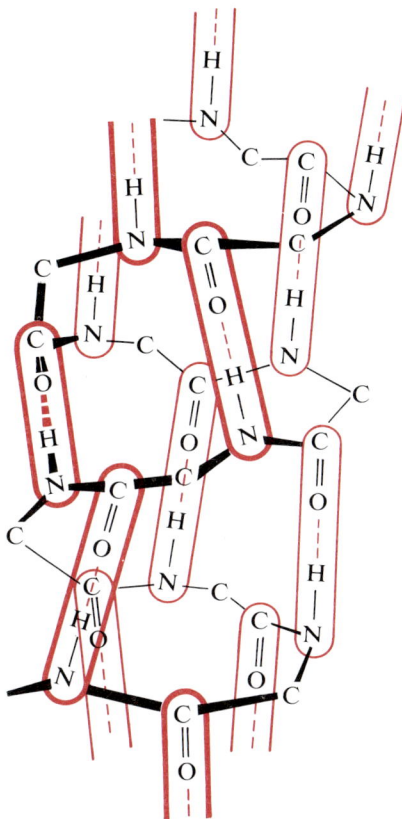

Figure 15-22 Hydrogen-bond linkage of a single polypeptide chain to itself in a helix.

proteins that, presumably by virtue of their precise geometry of folds and the pattern of polarities of R— groups in the helix, can catalyze the reaction of some substrate molecule. Nearly all biochemical reactions proceed through enzymes, and a given enzyme is usually quite specific for a particular reaction—often for a particular molecule. In Section 11-6 we dealt with the overall treatment of the kinetics of enzyme reactions, but without suggesting any mechanism. Very little detailed information is available on possible mechanisms for enzyme-catalyzed reactions. We speculate on the basis of the information we do have that the enzyme protein helix is folded so as to leave a hole or reactive site that is just the right size and shape, and has the right geometrical arrangement of polar groups around it, to accommodate a particular molecule and (by virtue of hydrogen bonding or dipole–dipole attractions) stabilize the transition state for the molecule's reaction. Some enzymes are very specific for a particular reaction and a particular molecule; presumably the hole in these cases must accommodate the whole molecule. Other enzymes are more general, accepting any of a whole class of molecules into an enzyme-substrate complex for a particular reaction, and we assume that only a particular functional group or portion of the substrate molecule has to be accommodated in the enzyme hole. Even though there are many biochemical reactions, the enormous number of possible sequences of amino acids in enzyme proteins leads to so many possible geometries for the folded helix that an enzyme can be created for, apparently, any reaction.

Nucleic Acids and Genetic Formation of Proteins

The obvious question arises of how a living organism produces the specific proteins and enzymes that are characteristic of its cellular structure; what dictates the sequences of amino acids that are constructed in a particular cell of a particular organism? In effect, we are asking how information is transferred on a molecular level to direct the synthesis of specific polypeptides. Within each cell there must be a template molecule that by virtue of its geometry and its pattern of polar groups and hydrogen-bonding groups can accept amino acids and link them in a specified order. But cells divide as an organism grows, which means that the template molecule must not only be able to reproduce protein molecules, but must also be able to reproduce itself. This is the role of **deoxyribonucleic acids** (DNA). The plural is used because different organisms, producing different proteins, must have different nucleic acids. Nevertheless, studies of different kinds of DNA have shown many common features that are worth our study. The molecular weights are very high, being over a million and sometimes over a billion; by X-ray diffraction studies the general structure has been shown to be helical, like that of a protein, but with two (rather than one) linear polymer molecules involved, spiraling around each other. Each of the two molecules is a polyester in the same sense that a protein is a polyamide; but where each amino acid contains both the functional groups necessary to form the polymer, a diprotic acid and a dialcohol are used to form a single strand of DNA. The acid is phosphoric acid, and the dialcohol is 2-deoxyribofuranose:

H₂O eliminated when ester forms

$$\underset{\text{phosphoric acid}}{HO-\underset{\underset{OH}{|}}{\overset{\overset{O}{\|}}{P}}-OH} + HO-CH_2-\underset{\underset{\underset{OH}{|}}{\overset{|}{H}}}{\overset{OH}{\underset{O\diagdown C \diagup CH_2}{\overset{|}{HC}-\overset{|}{CH}}}} + HO-\underset{\underset{OH}{|}}{\overset{\overset{O}{\|}}{P}}-OH$$

phosphoric acid   2-deoxyribofuranose

The result is thus an ester chain with the phosphates separated by three C atoms:

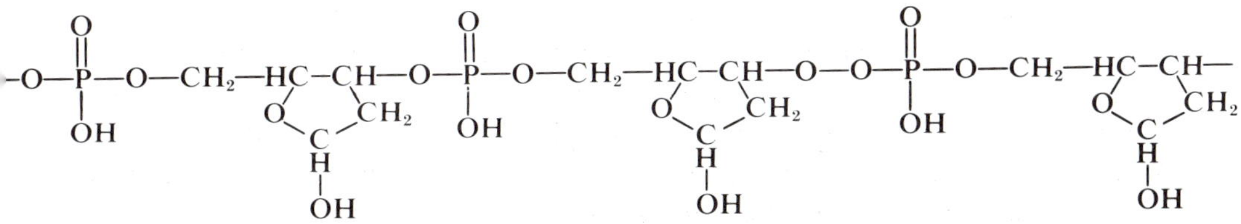

What about the other OH group on the alcohol? It does not esterify because it is not really there; the five-membered ring is the same for each dialcohol in the chain, but the third OH group has been replaced by one of four amine bases:

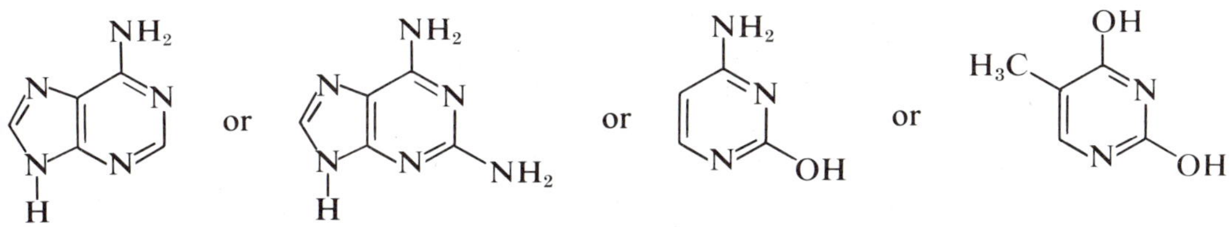

adenine   guanine   cytosine   thymine

There is thus a sequence to be specified, as for proteins—the sequence in which these bases occur along the chain. The bases hold the double helix together by hydrogen bonding. The adenine bases on one strand hydrogen bond to the thymine bases on the other strand, and the guanine on one strand to the cytosine on the other, as in Fig. 15-23.

Figure 15-23 Hydrogen-bond linkage of amine bases on one nucleic acid strand to those on the other strand in the double-chain helical structure.

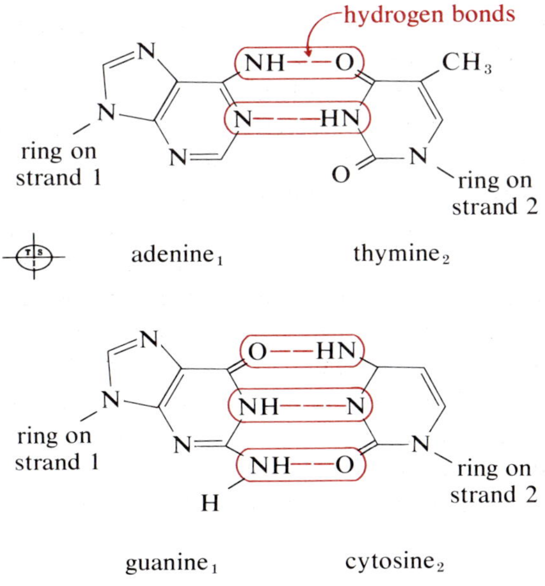

Drawn on paper, there is no reason for this to lead to a helical structure for the pair of strands. However, the geometric requirements of the hydrogen bonds that must form (given that the amine bases are rather bulky) cause the strands to twist spontaneously as they link together. The result is a double helix with a wide groove and a narrow groove running the length of the helix; the model in Fig. 15-24 may be helpful in visualizing the result.

DNA can replicate itself, which is the most basic genetic function, but it does not synthesize proteins directly. It serves as a template for RNA, **ribonucleic acid**, which is a smaller single-strand chain of phosphate polyesters very similar to DNA but having the amine base uracil instead of thymine:

$$\underset{\text{thymine}}{\text{(structure)}} \qquad \underset{\text{uracil}}{\text{(structure)}}$$

Three forms of RNA are produced, all of which are necessary to induce the specific reaction of one amino acid with another to produce a protein. One form with a large molecular weight, ribosomal-RNA, combines with existing protein to produce a site (a ribosome) where the other two smaller forms, transfer-RNA and messenger-RNA, can combine to yield certain specific sequences of hydrogen-bonded base pairs (as in DNA). There is a transfer-RNA molecule corresponding to each amino acid; the appropriate one complexes the amino acid (with the aid of an enzyme) and joins a strand of messenger-RNA at a ribosome. A single strand of messenger-RNA can be attached to several ribosomes at once, apparently, and thereby assist in building several protein molecules at once. The particular combined sequences of RNA bases serve to specify which amino acid can be accepted in the progressive formation of a polypeptide or protein. A sequence of three particular amine bases on the transfer-RNA molecule is sufficient to specify a particular amino acid; for instance, if three successive ester elements of the RNA chain have the uracil base, the protein that is being constructed will, when those bases are hydrogen bonded to the messenger-RNA strand, always have phenylalanine amino acid incorporated into it. Other sequences in the other transfer-RNA molecules are similarly specific for other amino acids. This control of reactivity is probably similar to the control an enzyme exercises, in that the RNA probably accepts amino acids in a fashion dictated by the size, shape, and polarity or hydrogen-bonding capacity of the "holes" or helical grooves in the RNA. This whole subject is one of the most interesting and rapidly growing areas of modern chemistry. For this reason it is important to discuss it at least briefly, but also for just this reason it is impossible to give a complete account—this part of the logical structure has yet to be built.

The Formation of Polymers | 809

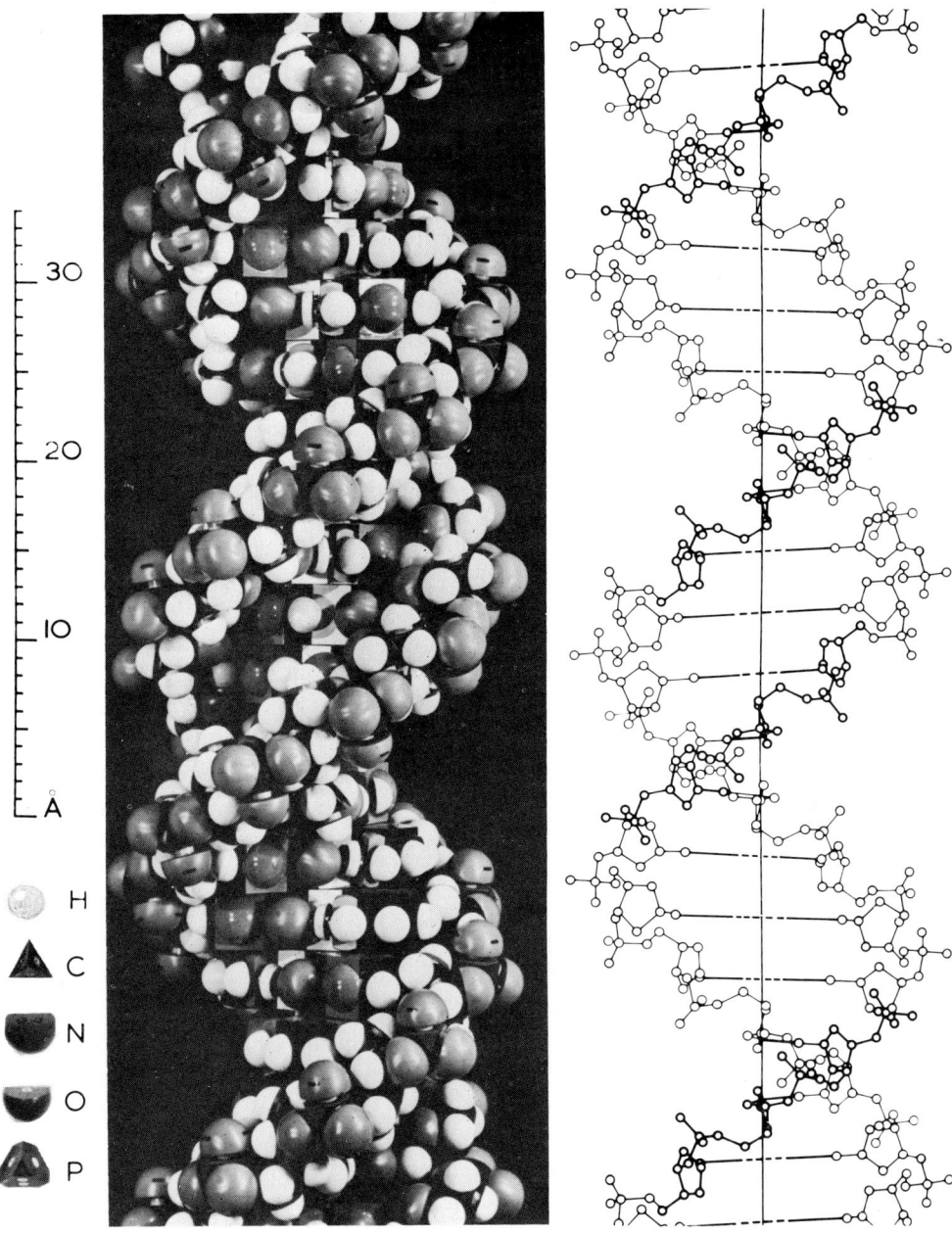

Figure 15-24 (Reprinted from Wilkins, M. H. F., *Science*, vol. 140, pp. 941–950, May 31, 1963. Copyright 1963 by the American Association for the Advancement of Science.)

15-6 Nonorganic Covalent Systems

Although carbon has unique powers of catenation, there are many covalent compounds that do not contain carbon. As we saw in Section 15-1, there are in general two categories of covalent compounds: those in which the elements bonded together have nearly the same electronegativity, and those in which the formal oxidation state of the central atom in a polyatomic species is very positive. We can look at some of these compounds, grouping them in general categories in terms of their approximate electronegativity.

Metal–Metal Compounds

Covalent compounds between atoms with low, equal electronegativities are intermetallic compounds, since all the elements with low electronegativities are metals. The simplest "catenated" case of this would be a single metallic element, whose properties we discussed in Chapters 4 and 6. If we substitute another element of similar size and electronegativity (so that the AO overlap and relative energies do not change much), the resulting solid is very much like the original metal in all its properties. Usually the preparation consists simply of melting the metals together. If the two elements are really similar, as, for instance, sodium and potassium, the substitution will occur in a random fashion and the result will be an alloy. However, if there are moderate size and electronegativity differences, it will prove energetically favorable to arrange the crystal in an orderly fashion, with more-electronegative and less-electronegative atoms alternating in some fashion. This suggests that some degree of charge transfer and ionic attraction must occur in such intermetallic compounds as Li_3Sb. This really represents a fairly extreme case of electronegativity difference in intermetallic compounds ($\Delta \chi = 0.85$); Li_3Sb probably should be considered a partly ionic giant-molecule compound. The bonding electrons are not localized in individual bonds, however, because nearly all of the intermetallic compounds show a rather high electrical conductivity in the solid state, and even the most extreme "salt-like" compounds, such as KSb, are semiconductors. The stoichiometry of the compounds is governed primarily by size relationships, in that the relative atomic diameters must allow a favorable crystal geometry, but it is also influenced by the relative number of valence electrons per orbital in the bands of solid state MO's (see Section 6-12). The compounds' chemical properties are usually closely analogous to those of the individual elements, and this is sometimes handy. Metallic sodium, for instance, is a spectacularly strong reducing agent since its $3s$ valence electrons are distributed in high-energy, unstable orbitals. This has many valuable synthetic applications (for instance, if we want to make a carbanion), but the electron transfer can be uncontrollably vigorous and dangerous. So sodium intermetallic compounds or alloys are sold by various suppliers; the presence of the other more inert metal, usually lead, dilutes the sodium and moderates the local heating that can make sodium reductions explosive.

Organometallic Compounds

Other covalent compounds involving metals are possible if the nonmetal is a fairly good electron donor (which reduces its effective electronegativity), or if there are many nonmetals to raise the formal oxidation state of the metal. In the first class there are the **organometallic** compounds. These involve a metal atom bonded directly to carbon in an organic group of some sort. We have already seen one example of these compounds in the organolithium compounds of Section 15-4. As with intermetallic compounds, the properties of these differ depending on the electronegativity difference between carbon and the metal involved. When the metal is cesium, the simpler compounds such as methylcesium are almost entirely ionic in their properties, but with metals such as tin and lead the normal organometallic compounds (such as the familiar but vanishing gasoline antiknock additive tetraethyllead) are volatile liquids and are almost entirely covalent.

Organometallic compounds can be formed in many ways, but two of the most common involve, respectively, the substitution of a metal for a halogen on an organic halide, and the donation of organic π-system electrons to a positively charged metal atom in an existing compound. For example,

$$2\,Li^0_{(s)} + CH_3\text{—}CH_2\text{—}CH_2\text{—}Cl_{(l)} \xrightarrow{\text{solvent}} CH_3\text{—}CH_2\text{—}CH_2\text{—}Li_{(\text{sol'n})} + LiCl_{(s)}$$
$$\text{\textit{n}-propyl chloride} \qquad\qquad\qquad\qquad \text{\textit{n}-propyllithium}$$

and

$$PtCl_{4(aq)}^{2-} + CH_2\!=\!CH_{2(g)} \longrightarrow \underset{CH_2}{\overset{CH_2}{\Vert}}\!\!\longrightarrow PtCl_{3(aq)}^{-} + Cl_{(aq)}^{-}$$
$$\qquad\qquad \text{ethylene}$$

The first of these is typical of nontransition metals, and the second of transition metals (to which we shall return in Chapter 16). There are some variations on the first reaction, depending on the electronegativity and number of valence electrons on the metal:

$$CH_3\text{—}I_{(\text{sol'n})} + Mg^0_{(s)} \xrightarrow{\text{ether}} CH_3\text{—}Mg\text{—}I_{(\text{sol'n})}$$
$$\text{iodomethane}$$

and

$$CH_3-CH_2-Br_{(sol'n)} + Zn^0_{(s)} \rightarrow CH_3-CH_2-ZnBr_{(sol'n)}$$
bromoethane

$$2CH_3CH_2ZnBr_{(sol'n)} \rightarrow (CH_3CH_2)_2Zn_{(sol'n)} + ZnBr_{2(s)}$$

In all of these compounds the bond between the carbon and metal has σ symmetry. These organometallic species are extremely useful to the synthetic organic chemist, as our discussion of methyllithium reactions suggested. Table 15-4 gives a few of the more interesting and useful reactions, all of which are interpreted fairly easily if we remember that the partial positive charge on the metal atom is attracted to the electronegative atom in the organic molecule. In this sense the reactions in the table are very similar to each other. Note that most organometallic compounds are very sensitive to hydrolysis because of the readiness of the partially negative organic group to attack the partially positive H in H_2O. This makes organometallic compounds rather difficult to work with, since they must be protected from the atmosphere's humidity at all times.

The second type of reaction (π donation) will be discussed more fully in the next chapter, since it involves primarily transition metals. However, the analogous reaction with trialkylaluminum is important because it leads to a type of reaction called **insertion**:

$$(R = \text{alkyl group}) \quad R_3Al + CH_2=CH_2 \longrightarrow \begin{array}{c} R \quad R \\ \diagdown \diagup \\ Al \\ \diagup \quad \diagdown CH_2 \\ R \quad \| \\ \quad CH_2 \end{array}$$

$$\begin{array}{c} R \quad R \\ \diagdown \diagup \\ Al \\ \diagup \quad \diagdown CH_2 \\ R \quad \| \\ \quad CH_2 \end{array} \longrightarrow R_2Al-\underbrace{CH_2-CH_2-R}_{\text{a new R group}}, \text{etc.}$$

A little imagination will suggest that ethylene could be polymerized this way, and indeed the **Ziegler–Natta low-pressure polyethylene process** relies on such an insertion reaction at a solid catalyst whose structural nature is still somewhat uncertain, but that is made from $TiCl_4$ and AlR_3. This process has to a considerable extent replaced the high-pressure free-radical process, since it is much cheaper to operate at lower pressures. In addition, subtle changes in the catalyst composition allow propylene and other alkenes to polymerize, even with specific cis or trans geometry if desired.

High-Oxidation-State Covalent Compounds of Metals

Metals can also form covalent compounds, even with oxygen or halogens, if they are in sufficiently high oxidation states. What is "sufficiently high" depends

Table 15-4
Reactions of Organometallic Compounds[a]

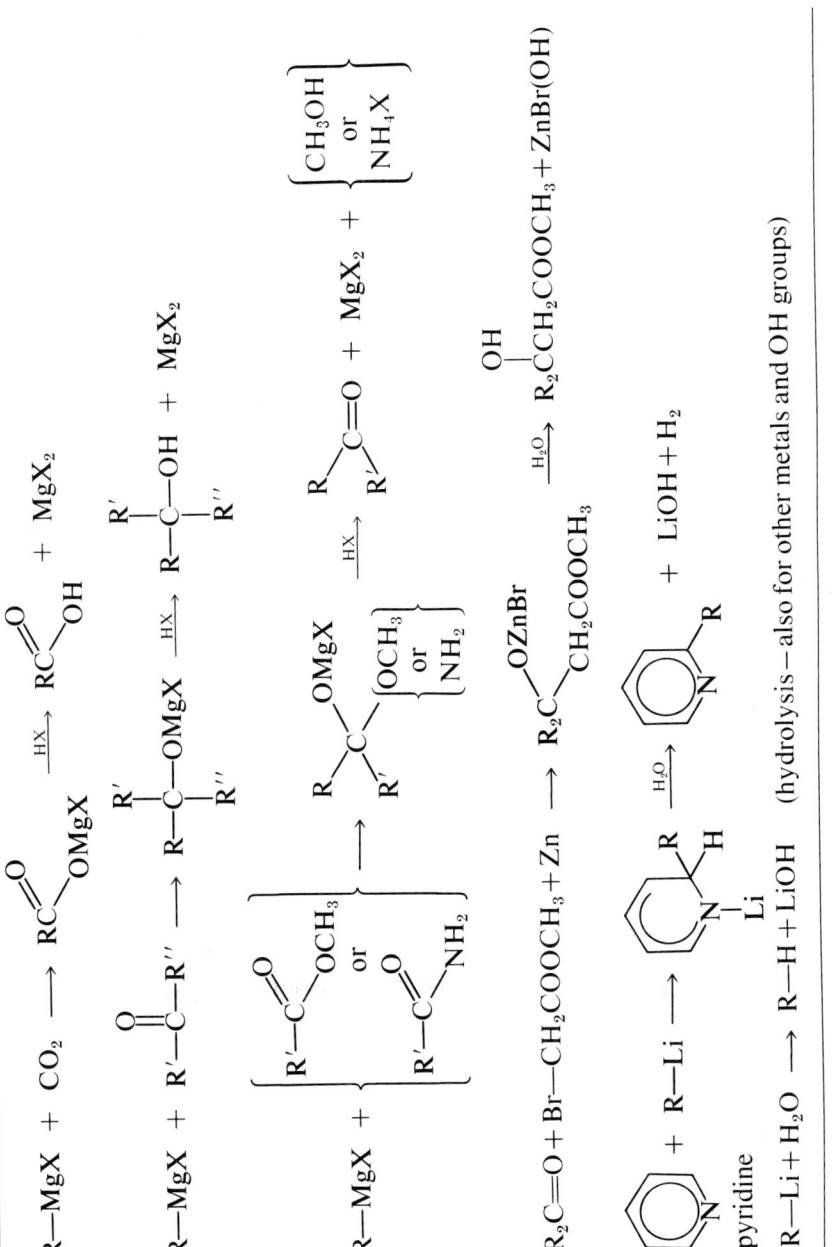

[a]R = hydrocarbon group, X = Cl or Br.

on the metal; in Section 13-3 we defined some rough limits for the halides of metals based on the idea that a system will be ionic only if the ionic-lattice energy is large enough to pay the total ionization-energy price. Generally speaking, ionization energies are relatively low for the first one or two electrons on a metal—but the third, fourth, and subsequent electrons are extremely difficult to remove. Thus there is no use looking for covalent compounds of groups I or II, although for the IIb metals the halides, for instance, will not be very ionic either. But for metals in groups III, IV, and V, covalent halides are quite possible in the higher formal oxidation states. We expect very different properties for high-oxidation-state and low-oxidation-state compounds between the same two elements. For example, there are two lead chlorides, $PbCl_2$ and $PbCl_4$, for which we can calculate an average ionization energy per chloride: the total ionization energy to the 2+ state is 518 kcal/mole, which amounts to about 260 kcal per mole of Cl. On the other hand, the total ionization energy to the 4+ state is 2330 kcal/mole, or about 580 kcal/mole Cl. Using the 300–350 kcal guideline from Section 13-3, $PbCl_2$ should be ionic (barely) while $PbCl_4$ should be completely covalent. And sure enough, $PbCl_2$ is a solid melting at 501 °C, while $PbCl_4$ is a liquid at room temperature (m.p. = −15 °C). The same pattern of properties persists in the transition metals, as we shall see.

The covalent metal halides are usually made either by direct combination of the elements or by heating an ionic halide with the oxide:

$$Sb^0_{(s)} + \text{excess } Cl^0_{2(g)} \xrightarrow{\text{heat}} SbCl_{5(l)}$$

$$6P_2O_{5(s)} + 5CaF_{2(s)} \xrightarrow{\text{heat}} 2PF_{5(g)} + 5Ca(PO_3)_{2(s)}$$

These halides resemble the organometallic compounds in hydrolyzing vigorously because of the attraction of the partly negative halogen atoms for the partly positive hydrogens in water:

$$GeCl_{4(l)} + 6H_2O_{(l)} \rightarrow GeO_{2(s)} + 4H_3O^+_{(aq)} + 4Cl^-_{(aq)}$$

They also serve as excellent electron acceptors if the coordination number in the halide is less than the effective maximum (usually 6), because an incoming electron donor can get at the strongly positively charged metal atom in the middle:

$$:NH_3 + SbCl_5 \rightarrow Cl_5Sb\text{—}NH_3$$

$$:F^- + BF_3 \rightarrow BF_4^-$$

Metalloids: Both Kinds of Covalent Compounds

Two of our recent examples have involved elements that are not really metals—boron and phosphorus. The elements running in a diagonal band down the periodic table from boron to tellurium are often called **metalloids**. These solid elements have a metallic luster, but they are rather brittle and conduct electricity poorly. Still, their behavior in the examples is analogous to what we expect for metals in high oxidation states, so we may as well include them. The important thing to notice is that there is a strong resemblance between compounds that are covalent because the central atom is in a high formal oxidation state and those that are covalent for the more obvious reason that the electronegativities of the bonded atoms are similar.

Boranes

If we turn our attention to covalent compounds of these metalloids — elements with intermediate electronegativities — we find the chemical resemblance just mentioned between their compounds and those of metals in high oxidation states, but they also have the ability to catenate into some interesting compounds (at least the lighter elements such as B, Si, and P do). A particularly interesting example is found in the **boranes**, which are catenated boron hydrides. Boranes are somewhat analogous to hydrocarbons, but because B has four valence orbitals and only three valence electrons, the electron distributions and the resulting geometries of the boranes are quite different from the straightforward tetrahedral-carbon-atom geometry of hydrocarbons. Chapter 6 dealt with the reasons why there is no BH_3, the simplest boron hydride. Figure 15-25 shows the structures of some of the boranes that are known, from which some interesting observations can be made. One observation is that there are some H atoms in every existing borane that are not σ bonded to a single B atom, but rather bridge the space between two B atoms. Figure 6-61 and the related discussion introduced the idea of three-center bonds for **diborane**, B_2H_6; this is generally the starting point for theoretical discussions of the electronic structure of boranes and their bridging hydrogens. Of course, a MO can involve a very large number of AO's, as for instance the π MO's in benzene, so the idea of multicenter bonds should not seem unusual. What is new here is that σ bonds instead of π bonds are being delocalized. Another interesting aspect of Fig. 15-25 is that the geometry of an icosahedron, for reasons we still do not fully understand, seems to be particularly favorable for the multicenter bonds of the polyboranes. All of the structures shown can be visualized as fragments of an icosahedron.

There are several reactions that yield boranes, usually diborane. The most convenient one uses lithium aluminum hydride, $LiAlH_4$:

$$3\,LiAlH_{4(s)} + 4\,BCl_{3(g)} \rightarrow 2\,B_2H_{6(g)} + 3\,AlCl_{3(s)} + 3\,LiCl_{(s)}$$

Nearly all hydrides except those of the most electronegative elements are reducing agents, because the electrons on a neutral or slightly negative hydrogen atom are not strongly bound and can be stripped away readily by a good accep-

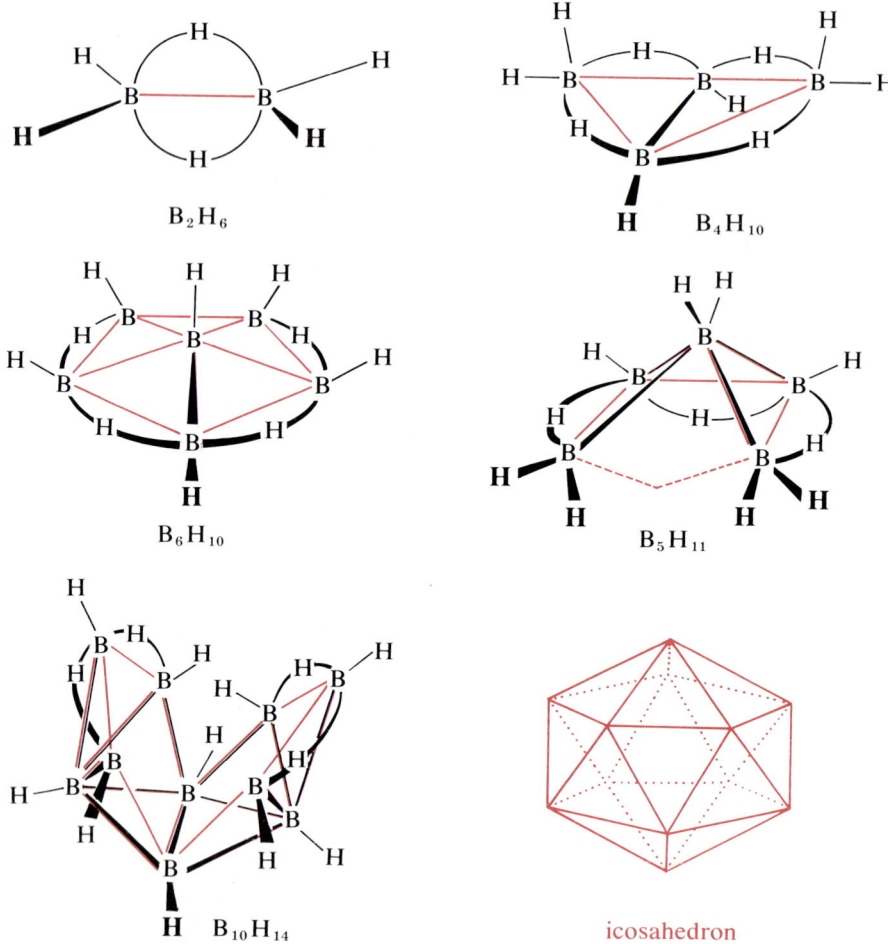

Figure 15-25 Icosahedron-fragment structures of polyboranes. Bonds are shown in black, icosahedron fragments in red.

tor. In keeping with this, the lighter boranes hydrolyze easily, with considerable evolution of heat:

$$B_2H_{6(g)} + 6H_2O_{(l)} \rightarrow 2B(OH)_{3(aq)} + 6H_{2(g)}$$

They also burn in a spectacularly exothermic manner, so much so that they have been proposed as rocket propellants:

$$B_2H_6 + 3O_2 \rightarrow B_2O_3 + 3H_2O \qquad \Delta H^0 = -483 \text{ kcal/mole reaction}$$

Pseudocatenation

Besides the type of catenation just discussed, it is possible to take advantage of the donor properties of nitrogen in ammonia and the acceptor properties of the electron-deficient boranes to make B—N compounds in which a B—N pair is equivalent to (and isoelectronic with) two C atoms:

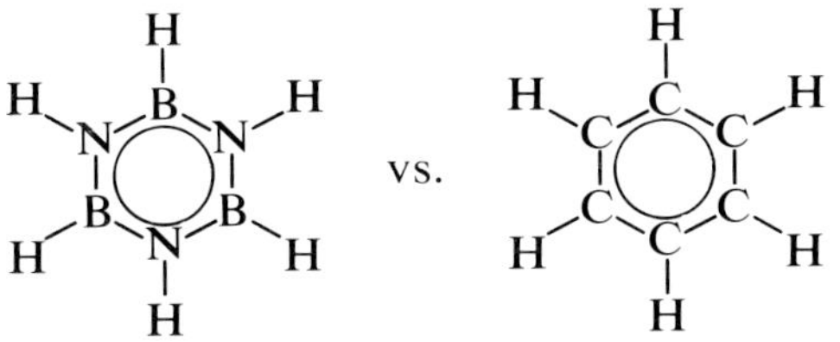

borazene vs. benzene

These compounds tend to have properties resembling those of their carbon analogs, but they are more reactive because of the more polar nature of the B—N bond.

Another example of this sort of "pseudocatenation" is seen in the **phosphonitrilic polymers**. If PCl_5 and NH_4Cl are allowed to react in boiling tetrachloroethane solvent, a mixture of species results, all with the empirical formula $PNCl_2$. The lighter products include a trimer, whose structure makes an interesting comparison with borazene and benzene:

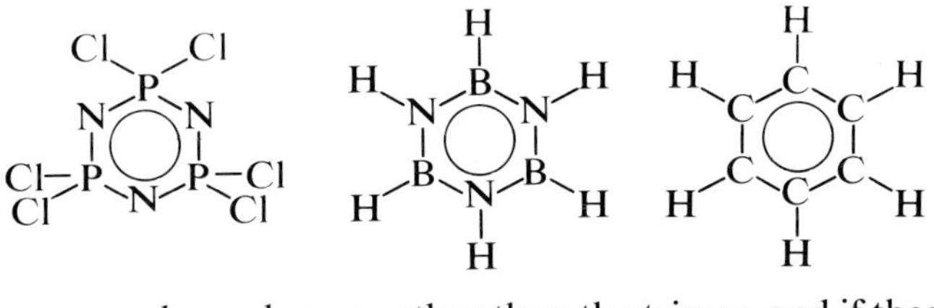

However, there are also polymers other than the trimer, and if these are heated further, true inorganic polymers with molecular weights as high as 20,000 can be formed. These apparently bear a close analogy to the carbon polymers, since the trimer shown above even seems to be aromatic—with delocalized π electrons like benzene. It is worthwhile to look at the positions of P and N in the periodic table in light of this; they lie astride the line between carbon and sulfur, the two elements that show fairly extensive true catenation. From the electronegativities, it might be assumed that phosphorus is serving as a donor and nitrogen as an acceptor in these compounds, but not all of the features of the bonding are well understood.

A similar "balancing" effect might perhaps be expected from silicon and oxygen, and it is indeed true that by far the stablest polymers involving Si are the **silicones**, which are organic-substituted groups having the general formula $(-R_2SiO-)_n$, with alternating Si and O atoms in a chain structure. Silicones are made by reacting organic chlorides with silicon (as a copper alloy), hydrolyzing the resulting product, and heating the hydrolyzed species to polymerize it:

$$2R-Cl + Si \cdot Cu \longrightarrow R_2SiCl_2 \quad (\text{and } RSiCl_3, R_3SiCl)$$

$$R_2SiCl_2 + 4H_2O \longrightarrow R_2Si(OH)_2 + 2H_3O^+ + 2Cl^-$$

$$2nR_2Si(OH)_2 \xrightarrow{\text{heat}} \left(\begin{array}{cc} R & R \\ | & | \\ Si-O-Si-O \\ | & | \\ R & R \end{array}\right)_n + 2nH_2O$$

Silicones can be made in all the ranges of physical properties that carbon polymers seem to have, and they have the additional virtue of being much less temperature sensitive in their properties than most carbon polymers. Thus they are frequently used as rubbers for high- or low-temperature gaskets, or for high-temperature electrical insulation.

Compounds Between Electronegative Elements

Finally, there are covalent compounds among the most electronegative elements; we can consider the hydrides, oxides, and halides of N, O, S, F, Cl, Br, and I. Nearly all of these compounds are gases at room temperature, which is a testimony to their covalent nature and the very weak van der Waals forces between these molecules. These are the most familiar covalent compounds, and in spite of their uniformly covalent nature they show considerable chemical diversity.

Hydrides

Taking the hydrides first, we have the familiar ammonia, water, hydrogen sulfide, and the hydrogen halides. All of these are gases except water; the extent of hydrogen bonding in these compounds has been shown in Fig. 12-19. All of them can be made by the direct combination of H_2 gas and the other element, although this is difficult for NH_3 and H_2S and explosively dangerous for HF; the direct synthesis of NH_3 by the **Haber process** is the most important source of chemically combined nitrogen for fertilizers and other uses, however. On a laboratory scale, the most convenient source of these gases is the addition of an ionic salt to a strong nonvolatile acid (or base, for the weak base NH_3):

$$NaCl_{(s)} + H_2SO_{4(l)} \xrightarrow{\text{heat}} NaHSO_{4(s)} + HCl_{(g)}$$

$$CaF_{2(s)} + H_2SO_{4(l)} \xrightarrow{\text{heat}} CaSO_{4(s)} + 2HF_{(g)}$$

$$NH_4Cl_{(s)} + OH^-_{(aq)} \xrightarrow{heat} H_2O_{(l)} + Cl^-_{(aq)} + NH_{3(g)}$$

These preparations rely on the entropy increase that accompanies the formation of a gas, and on the proton-transfer characteristics of the hydrides.

One of the most interesting chemical properties of these hydrides is their proton-transfer capability in solution, whether aqueous or not. In Chapter 13 we saw how a Born–Haber cycle can be used to analyze the differences in acidity between the hydrogen halides. To this we can immediately add a rationalization of the relative acidities of the other hydrides in this series. We expect H_2O to be a weaker acid than H_2S for exactly the same reasons that HF is weaker than its analogs (or **congeners**) – the high enthalpy of solution of the H_2O molecule due to its hydrogen-bonding capability, and the high H—O bond energy due to the better match of atomic orbital sizes. Both H_2O and H_2S should be much weaker acids than the corresponding HF and HCl, because the lower electronegativity of O and S means that the electrons in the H—O and H—S bonds will not be withdrawn as strongly from the H atoms, and the proton will not be transferred as readily to a H_2O (or other solvent) molecule. And for the same reason, NH_3 should be a much weaker acid than H_2O; since NH_3 has a nonbonding pair of electrons, the lower electronegativity will also allow it to function as a base – a proton acceptor or electron donor – in water solution. By contrast, CH_4 is a weaker acid than H_2O or even NH_3; but its lack of any nonbonding electrons prevents it from ever acting as a base.

Oxides

Most of these elements form a rather surprising variety of oxides, usually corresponding to different positive formal oxidation states of the nonmetal. This unusual circumstance is due to the great electronegativity of oxygen, which causes even these nonmetals to transfer electrons toward it. Fluorine, of course, is the exception to this, and it forms only one reasonably stable oxide, which we might better think of as oxygen fluoride. Table 15-5 shows the more common oxides in this group, with carbon included for comparison. Some can be made by direct combination of the elements, although fluorine is a sufficiently powerful oxidizing agent that it will react even with OH^-:

$$S + O_2 \xrightarrow{heat} SO_2$$

$$2SO_2 + O_2 \xrightarrow{Pt\ catalyst} SO_3$$

$$2Cl_2 + O_2 \xrightarrow{HgO\ catalyst} Cl_2O$$

$$Cl_2 + 2O_3 \xrightarrow{h\nu} Cl_2O_6$$

$$2F_2 + 2OH^- \longrightarrow OF_2 + 2F^- + H_2O$$

Table 15-5
Covalent Oxides

IV	Vb	VIb	VIIb	0
CO	N_2O	O_3	OF_2	
CO_2	NO	O_2		
	N_2O_3			
	$NO_2(N_2O_4)$			
	N_2O_5			
		SO_2	Cl_2O	
		SO_3	ClO_2	
			Cl_2O_6	
			Cl_2O_7	
			Br_2O	
			BrO_2	
			I_2O_4	XeO_3
			I_4O_9	XeO_4
			I_2O_5	
			I_2O_7	

Other preparative techniques include dehydration of the oxyacid of the nonmetal, thermal decomposition of salts of the oxyacid, and redox reactions:

$$2HNO_3 + P_2O_5 \longrightarrow 2HPO_3 + N_2O_5$$

$$2Pb(NO_3)_2 \xrightarrow{\text{heat}} 2PbO + 4NO_2 + O_2$$

$$2ClO_3^- + 2H_2C_2O_4 \longrightarrow 2ClO_2 + 2CO_2 + C_2O_4^{2-} + 2H_2O$$
$$\text{oxalic acid}$$

Nearly all of these nonmetal oxides are acidic in water solution, following the ideas developed in Chapter 12. A few are neutral or fairly inert toward attack by water, but because of the high electronegativity none are basic. Most show varied redox reactions; the oxides showing high formal oxidation states for the nonmetal (and their corresponding acids) are strong oxidizing agents, since the overall high electronegativity makes the molecule an excellent electron acceptor:

$$N_2O_5 + I_2 \rightarrow I_2O_5 + N_2$$

$$SO_3 + 2HBr \rightarrow Br_2 + SO_2 + H_2O$$

In the lower oxidation states redox reactions are still common, but the oxide can often serve either as an oxidizing or a reducing agent, depending on the nature of the other reacting molecule:

$$SO_2 + C \xrightarrow{1100°} S + CO_2 \quad \text{oxidizing agent}$$

$$SO_2 + I_2 + 6H_2O \longrightarrow SO_4^{2-} + 2I^- + 4H_3O^+ \quad \text{reducing agent}$$

$$4NO + 3Sn^0 + 10H_3O^+ + 8H_2O \longrightarrow 4H_3NOH^+ + 3Sn(OH)_2(OH_2)_4^{2+} \quad \text{oxidizing agent}$$
hydroxylammonium ion

$$2NO + Cl_2 \longrightarrow 2NOCl \quad \text{reducing agent}$$
nitrosyl chloride

Halides

All of the electronegative elements we are discussing here form covalent halides; most are gases but a few are liquids or volatile solids. Table 15-6

Table 15-6
Covalent Halides of Electronegative Elements

	N	O	S	F	Cl	Br	I	Xe
F	NF_3 N_2F_2	OF_2	SF_6 SF_4	F_2	ClF_3 ClF	BrF_5 BrF_3 BrF	IF_7 IF_5 IF_3	XeF_6 XeF_4 XeF_2
Cl	NCl_3	Cl_2O	SCl_4 SCl_2 S_2Cl_2	—		Cl_2	$BrCl$	ICl_3 ICl	none
Br	$NBr_3 \cdot NH_3$	Br_2O	S_2Br_4	—	—	Br_2	IBr	none
I	$NI_3 \cdot NH_3$	(see Table 15-5)	none	—	—	—	I_2	none

indicates the more common halides of this type, and it can be seen that they show the same sort of diversity as the oxides. Nearly all of these halides are made by direct combination of the elements; only those of nitrogen cannot be, because of the very great stability of the N_2 molecule. Just as the oxidation of ammonia provides a route to NO and NO_2, the halogens react with ammonia to give NX_3 with or without coordinated NH_3. Since fluorine must be obtained electrolytically anyway, NF_3 is usually prepared by the electrolysis of molten ammonium hydrogen fluoride, combining the reactions into a single step:

$$3I_2 + 5NH_3 \longrightarrow NI_3 \cdot NH_3 + 3NH_4^+ + 3I^-$$

$$NH_4^+ + 7HF_2^- \xrightarrow{\text{electrolysis}} NF_3 + 11HF + 6e^-$$

Many of these compounds are excellent halogenating agents, reflecting a general lack of stability:

$$3\text{SiO}_2 + 4\text{BrF}_3 \rightarrow 3\text{SiF}_4 + 2\text{Br}_2 + 3\text{O}_2$$
$$3\text{CH}_3\text{—CH}_2\text{—CH}_2\text{—OH} + \text{PBr}_3 \rightarrow 3\text{CH}_3\text{—CH}_2\text{—CH}_2\text{—Br} + \text{P(OH)}_3$$

The instability is sufficient to make NCl_3, NBr_3, and NI_3 particularly vicious explosives. Nitrogen trifluoride, however, is extremely stable toward all kinds of physical and chemical provocation; in Section 12-7 we analyzed this surprising difference, showing that the enthalpies of formation are quite different. The inertness is also partly kinetic, since NF_3 ought, thermodynamically, to hydrolyze in cold water but does not. SF_6 shares this inertness, and OF_2 and IF_7 are not as vigorous fluorinating agents as the other fluorides in this group. It should be added that all of the interhalogen ions also fall in this group, since the bonding is the same whether or not there is a net charge on the molecule, but their chemistry has already been considered in Chapter 13.

15-7 Summary

In the past three chapters the physical and chemical characteristics of ionic, partly ionic, and covalent compounds have been surveyed. Starting with a procedure for predicting charge separation in compounds, the structures of the compounds have been described and to some extent rationalized, together with their chemical properties. Some important naturally occurring materials have been mentioned, along with some interesting synthetic ones—in the process some of the outlines of chemical synthesis have begun to emerge. Here "chemical synthesis" is a term chosen to represent a twofold design process: first, the prediction of the kind of compound that might best be expected to have a desired set of physical and/or chemical properties; and second, how to choose reactants and conditions that favor the formation of this compound. As a generalization, the more ionic the compound is, the more important thermodynamic considerations will be (so that we try to make ΔG for the formation reaction as favorable as possible). For largely covalent systems, the reaction product is often determined by kinetic considerations in that the least unstable reaction intermediate usually gives the predominant product; even here the structure is important, since we can usually judge the stability of various possible intermediates from their structures.

In all this, however, we have restricted ourselves to what are sometimes called the representative elements—those having only s and p valence electrons on the free atom. Nearly a third of the elements we know, however, have d valence electrons; these are the transition metals, which we have avoided bringing into the discussion because the nature of d orbitals causes some significant differences in chemical properties. In the last chapter we shall look at these differences, and in so doing try to apply most of these same principles of chemical synthesis to the same goals of making chemistry do what we want it to, for our comfort or convenience. In some ways, transition-metal chemistry is the most varied of any of the areas that chemists are currently pursuing, and we shall find full scope for our ideas—and perhaps range for our interests.

Study Problems

1. Benzene MO's have the relative energies shown in Fig. 15-13. These correspond, because of the symmetry of the molecule, to the positions of the corners of a regular hexagon inscribed in a circle of radius 2β with one corner at the bottom, as shown in Fig. 15-26a. The same symmetry treatment can be applied to any planar cyclic π system (see Fig. 15-26b). Place one electron per carbon atom in these orbitals; show that the three- and seven-carbon systems have a π^* electron, that the four- and eight-carbon systems have nonbonding electrons, and that the five- and nine-carbon systems have a π^b vacancy. What about the three-carbon and seven-carbon positive ions? What about the five-carbon and nine-carbon negative ions? What "favorable numbers" of electrons emerge from this discussion? Show that this fits the "$4n+2$" rule.

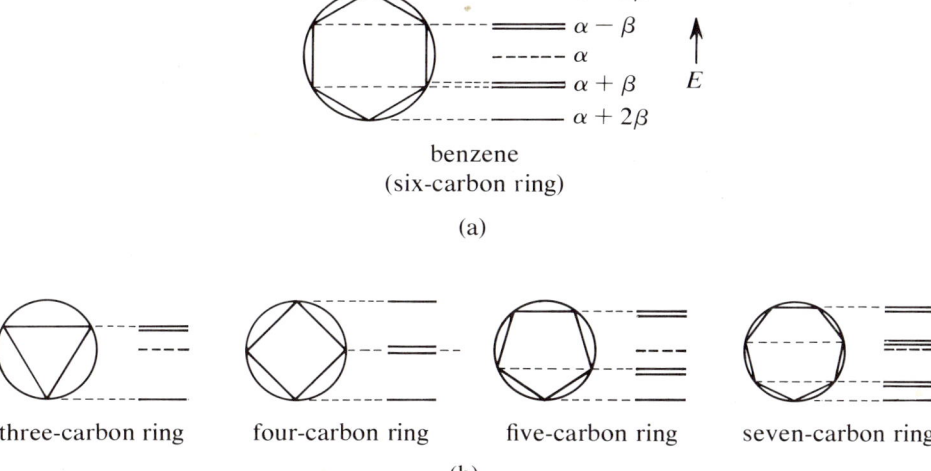

Figure 15-26 Cyclic π systems.

2. Why should methyllithium be more sensitive to hydrolysis than tetramethyllead?
3. Calculate ΔH^0 for the hydrolysis of NF_3. Assume that the reaction occurs in liquid water, producing $NO_{(g)}$, $No_{2(g)}$, and aqueous HF. Should the HF be considered

molecular or as separate ions? Thermodynamic data may be found in Tables 9-1, 13-2, and on p. 644; a cyclic treatment will be necessary for HF. Discuss the features of the molecular structure of NF$_3$ that might be responsible for the fact that hydrolysis does not occur.

4. Name the following compounds with correct spelling:

(a) $\underset{H_2C}{\overset{H_2C}{\Big>}}C=O$

(b) 1,3,5-trimethylbenzene structure with CH$_3$ groups

(c) $CH_3-CH_2-CH_2-\underset{\underset{Cl}{|}}{\overset{\overset{Cl}{|}}{C}}-\underset{\underset{Cl}{|}}{CH}-CH_2-CH_3$

(d) $CH_2=CH-\underset{\underset{CH_3}{|}}{C}=CH_2$

(e) $CH_3-CH_2-\underset{\underset{}{|}}{\overset{\overset{CH_3}{|}}{CH}}-C\underset{O-CH-CH_2-CH_3}{\overset{\nearrow O}{\diagdown}}$ with CH$_3$ on the O-CH

(f) cyclohexyl-NH-CH$_2$-CH$_3$

(g) $CH_3-\overset{\overset{O}{\|}}{C}-CH_2-\underset{\underset{}{|}}{\overset{\overset{CH_3}{|}}{CH}}-CH_3$

(h) $CH_3-CH_2-O-\overset{\overset{O}{\|}}{\underset{\underset{O}{\|}}{S}}-O-CH_2-CH_3$

(i) $CH_3-C\underset{NH-CH_2-CH_3}{\overset{\nearrow O}{\diagdown}}$

(j) $CH_3-\underset{\underset{H_3C}{|}}{\overset{\overset{CH_3}{|}}{C}}-\underset{\underset{OH}{|}}{CH}-CH_3$

(k) $CH_3-CH_2-\underset{\underset{CH_3}{|}}{CH}-CH_2-\overset{\overset{O}{\|}}{CH}$

(l) $CH_3-CH_2-CH_2-CH_2-\underset{\underset{HO}{\overset{\overset{}{|}}{C}}\overset{\nwarrow O}{}}{CH}-CH_2-CH_2-CH_3$

5. If hydride ion migrations are necessary in some ionic mechanisms for organic reactions, it is possible that other small groups could migrate. With this in mind, propose and discuss a mechanism for the following reaction:

$$CH_3-\underset{\underset{H_3C}{|}}{\overset{\overset{CH_3}{|}}{C}}-\underset{\underset{OH}{|}}{CH}-CH_3 \xrightarrow{acid} CH_3-\underset{\underset{CH_3}{|}}{\overset{\overset{CH_3}{|}}{C}}=C-CH_3 + CH_2=\underset{\underset{CH_3}{|}}{\overset{\overset{CH_3}{|}}{C}}-CH-CH_3$$

70% 30%

6. The organometallic functional group R—Mg—X, where X is a halogen (usually Br or I), is a very important tool called a *Grignard reagent* for organic synthesis, since it reacts with carbonyl groups to give alcohols:

$$\underset{R'}{\overset{R}{\Big>}}C=O + R''-Mg-X \xrightarrow{ether} \underset{R'}{\overset{R}{\Big>}}\underset{R''}{\overset{OMgX}{C\diagdown}} \xrightarrow{H_2O} R'-\underset{\underset{R''}{|}}{\overset{\overset{R}{|}}{C}}-OH + Mg(OH)X$$

Discuss the probable mechanism of this reaction. Propose a synthetic procedure for 1,1-diphenylethanol. What would happen if the initial ether solution of the carbonyl compound were wet? Could the compound

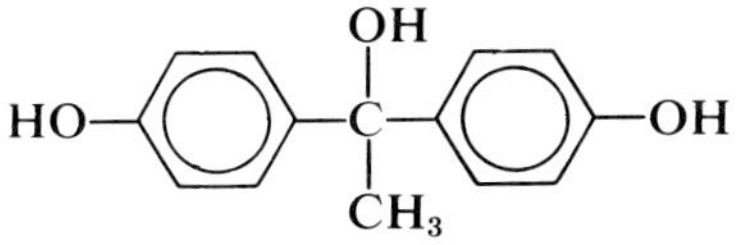

HO—⟨⟩—C(OH)(CH₃)—⟨⟩—OH

be made this way?

7. The K_a constants for the three possible nitrophenols are as follows:

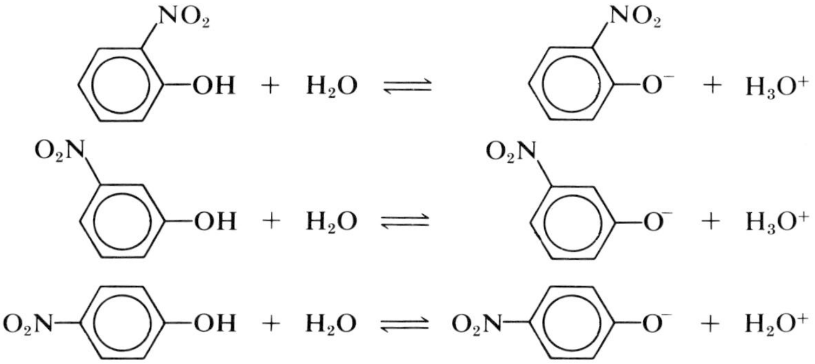

(o-nitrophenol) —OH + H_2O ⇌ (o-nitrophenoxide) —O⁻ + H_3O^+ $K_a = 7 \times 10^{-8}$

(m-nitrophenol) —OH + H_2O ⇌ (m-nitrophenoxide) —O⁻ + H_3O^+ $K_a = 5 \times 10^{-9}$

(p-nitrophenol) O_2N—⟨⟩—OH + H_2O ⇌ O_2N—⟨⟩—O⁻ + H_2O^+ $K_a = 7 \times 10^{-8}$

Why is the *meta* form a weaker acid?

8. Sketch the MO's resulting from π overlap in cyclobutadiene and discuss their energies. According to this model, should the compound be paramagnetic or diamagnetic?

9. There are some reactions of benzene compounds that apparently require attack on the ring by a nucleophile:

⟨⟩—Cl + OH⁻ ⟶ ⟨⟩—OH + Cl⁻

What effect should electron-withdrawing groups like —NO_2 have on this type of reaction? Does it matter where they are on the ring?

10. Cellulose, the polymer in cotton, is a polyether:

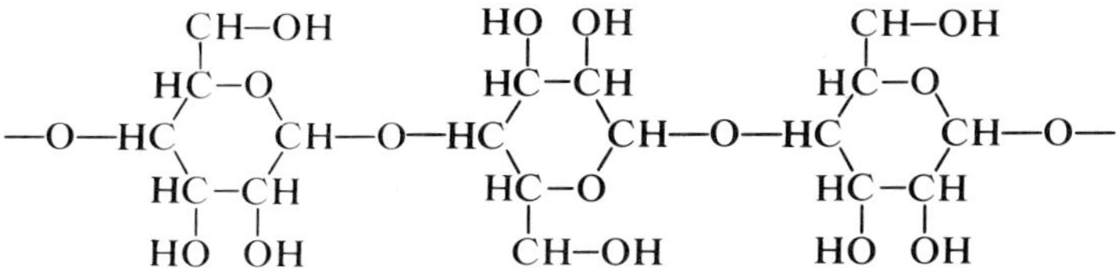

(cellulose structure with repeating glucose units connected by ether linkages)

Discuss the chemical basis for the fact that cotton sweatshirts are more comfortable than nylon ones.

11. How many protein molecules with 50 amino acid groups could be made? If the average molecular weight of the assortment were 7500 g/mole, how would the total

mass of the molecules (one of a kind) compare with the total mass of the universe, which some estimates place at 10^{55} g?

12. One of the interesting features of amino acid structure is the fact that a single amino acid contains two proton-transfer groups:

$$H_2N-CH_2-COOH + H_2O \rightleftharpoons H_2N-CH_2-COO^- + H_3O^+ \qquad K_a = 4.6 \times 10^{-3}$$

$$H_3N^+-CH_2-COO^- + H_2O \rightleftharpoons H_2N-CH_2-COO^- + H_3O^+ \qquad K_a = 2.8 \times 10^{-10}$$

Set up equilibrium-constant expressions for these two equilibria. For the first, calculate the pH at which the neutral molecule is only 1% as concentrated as the anionic form. At that pH, what fraction of the total glycine is in the $H_3N^+-CH_2-COO^-$ (*zwitterion* or *dipolar ion*) form? How basic can the solution become before 1% of the glycine is in the anionic form? In what form does glycine predominantly occur in body fluids, which have pH $\cong$ 7.2? What other proton-transfer equilibria could be written for glycine?

13. By analogy with the other metalloids, and considering the relative electronegativities of Si and H, propose a preparative reaction for silane, SiH_4. What other hydrides might your reaction prepare?

14. Sulfides of electronegative elements are most commonly made by direct reaction of the elements:

$$Cl_2 + 2S \rightarrow S_2Cl_2$$

However, because of the very high N≡N bond energy, elemental nitrogen will not react with sulfur. How can the principal nitrogen sulfide, N_4S_4, be made? Write a balanced equation.

15. One of the more familiar uses of silicones is as waterproofing materials; for instance, an aerosol spray can of silicones is sold to waterproof shoes. What features of the molecular structure of silicones make them good waterproofing agents?

Some Further Reading

The most obvious reference for this chapter is an organic chemistry text. Many students will go on to take organic, and there is no need for them to look further here. For those who are not going to go further, however, two good texts are the following:

Morrison, R. T., and Boyd, R. N., *Organic Chemistry*, Boston: Allyn and Bacon, 1966 (2nd ed.).

Hendrickson, J. B., Cram, D. J., and Hammond, G. S., *Organic Chemistry*, New York: McGraw-Hill, 1970 (3rd ed.).

Other more specific topics are covered in the following references, besides Cotton and Wilkinson and Jolly from Chapter 13:

Roberts, J. D., *Notes on Molecular Orbital Calculations*, New York: Benjamin, 1961. This is really sort of a lab manual for a theoretical topic. It tells you how to make

Hückel MO calculations to get real numbers without much trouble. It does not tell you strongly enough that this is a very crude model.

O'Driscoll, K. F., *The Nature and Chemistry of High Polymers*, New York: van Nostrand Reinhold, 1964. About the level of this chapter and easy to read. Much more information than is here, but still compact.

Light, R. J., *A Brief Introduction to Biochemistry*, New York: Benjamin, 1968. Many more topics than are presented here, but still intended for about the same level audience. There is a good deal of current interest in this.

Readings from Scientific American: Bio-Organic Chemistry, San Francisco: Freeman, 1968. This is a survey of current interests. Although the articles were not written to be read together and are of varying dates, the collection holds up well.

16

d-Electron Compounds

The compounds of the transition metals show many of the characteristics of the various kinds of compounds discussed in Chapters 13–15. Indeed, in a few cases we have used transition-metal compounds as examples in those chapters. So we do not expect to see any great surprises when we look at the detailed chemistry of transition metals. There will be some differences because of the different nature of the *d* AO's that describe the distribution of valence electrons in the transition metals, and we shall examine these differences. But first let us consider the similarities of the transition metals to the metals whose compounds we have already studied.

16-1 The Transition Metals as Metals

If we check the valence electron configurations in the periodic table inside the front cover, we find that the elements from Sc in group IIIa to Cu in group Ib have *d* valence electrons, which defines them as transition metals. These elements have differing electronegativities, ranging from 1.20 to 1.75, but generally they cluster around 1.50, which is just about the electronegativity of aluminum (1.47). So the physical properties of the elements themselves and the chemical properties of their compounds (at least those with the transition metal near a formal oxidation state of 3+, like aluminum in its compounds) should resemble those of aluminum. This is generally true, although there are substantial variations within this large group of elements.

Electrical Conductivity

For example, consider the electrical conductivity of the elements as indicated in Table 16-1. The values for the first-row transition metals are clearly metallic,

Table 16-1
Electrical Conductivity of Transition Elements (ohm-cm^{-1})

K	Ca	Sc	Ti	V	Cr	Mn
0.14×10^6	0.21×10^6	0.015×10^6	0.024×10^6	0.04×10^6	0.08×10^6	0.05×10^6

Fe	Co	Ni	Cu	
0.10×10^6	0.16×10^6	0.15×10^6	0.59×10^6	compare S, 10^{-17} Al, 0.38×10^6

suggesting that the bonding in the solid metals involves the "electron-gas" or partly filled band structure of Chapter 6. Suggesting that this delocalization of some valence electrons into partly filled bands must occur, however, does not preclude the use of other valence electrons in directional bonds. In Chapter 14 we noted that fully ionic crystals have fairly high melting points, but that the extremely high melting points of refractory oxides, for instance, reflect a considerable amount of directional bonding, which of course must be broken to allow the disorderly liquid structure to form. The melting points of the transition metals, then, give us at least a qualitative idea of the extent to which directional covalent bonding is occurring within the solid metal. Table 16-2 indicates that it must be important, since the melting points of the transition metals are considerably higher than that of aluminum and are even in the range of diamond and the refractory oxides. This, together with its low density, is why titanium is being used as a structural replacement for aluminum in very-high-speed aircraft—above about 1400 mi/hr the air friction heats aluminum dangerously near its melting point.

Table 16-2
Melting Points of Transition Elements (°C)

K	Ca	Sc	Ti	V	Cr	Mn	Fe	Co	Ni	Cu
64	838	1539	1668	1900	1875	1245	1536	1495	1453	1083

compare Al, 660°C

Malleability

Another property that correlates with directional bonding in much the same way as melting point is malleability. If a metal is malleable it will undergo severe plastic deformation without breaking. The transition metals are generally quite malleable, although some of them are very hard (so that a good deal of force is required). A few of the transition metals with very high melting points, such as tungsten, are rather brittle, presumably because of an extensive network of strong directional bonds. It follows that they are also very hard. The softest transition metals are copper, silver, and gold, all of which are also very malleable and have relatively modest melting points. The property of malleability is extremely important to our technological society, although we usually see these metals as alloys when they are used in manufacturing or as construction materials. The malleability of steel, for instance, permits it to be rolled into sheet and then bent into beer cans, automobile fenders, and tennis rackets.

Reduction Potentials

The transition metals also generally resemble aluminum in chemical properties. A good example of this is their aqueous reduction potentials. Table 16-3 indicates these values for the 2+ oxidation state that all these metals display;

Table 16-3
Aqueous Reduction Potentials for $M^{2+} \rightarrow M^0$ (V)

K	Ca	Sc	Ti	V	Cr	Mn	Fe	Co	Ni	Cu
−2.9	−2.9	−2.1	−1.6	−1.2	−0.9	−1.2	−0.4	−0.3	−0.3	+0.3
(1+)		(3+)								

compare Al, −1.66 (3+)

going across the periodic table from titanium the values become steadily more positive, even though at the beginning the value for titanium corresponds very closely to that for aluminum. The trend toward more positive potentials is explainable partly in terms of the ionization potential of these electrons, which increases going to the higher atomic numbers, but this is not the whole answer. In the next section we shall see that at least some of the existing reduction potentials can be placed in a reasonable pattern, in light of what we think we know about the energies of electrons in AO's.

Acidity of Hydrated Ions

Another example of a chemical property that resembles the corresponding property in aluminum is the acidity of the hydrated ion species, expressed in Table 16-4, as K_a for several of the 3+ transition-metal species. All of the species are weak, but not extremely weak, acids. Again this should not surprise us, since we suggested in Chapter 12 that the degree of positive charge on any hypothetical free positive ion should determine the acidity of the protons on the hydrated form of the ion. So all of these values are similar because they all refer to a 3+ ion.

Table 16-4
Acidity of Hydrate Ions (as K_a for 3+ species)
$$M(OH_2)_6^{3+} + H_2O \rightleftharpoons M(OH)(OH_2)_5^{2+} + H_3O^+$$

Species	K_a
$Al(OH_2)_6^{3+}$	1.4×10^{-5}
$Sc(OH_2)_6^{3+}$	1.2×10^{-5}
$V(OH_2)_6^{3+}$	3.0×10^{-3}
$Cr(OH_2)_6^{3+}$	1.2×10^{-4}
$Fe(OH_2)_6^{3+}$	4.5×10^{-3}
$Co(OH_2)_6^{3+}$	5.6×10^{-2}

So far we have only mentioned ways in which transition metals are similar to other metals, particularly the fairly electropositive aluminum. But there are differences, and they are often quite important to the chemistry of these elements. The reduction potentials have already shown a clear trend over a rather wide range, and all through this chapter other differences and trends will appear. Perhaps it will not require too great a leap of faith at this stage to indicate that these differences are basically due to the differences in the electronic distributions between metals with s or p electrons and those with d electrons—and to the trends in stability of the d electrons within the transition metals. Let us look into these differences by examining the nature of d orbitals.

16-2 The Nature of d Orbitals and d Electrons

We already know something about d orbitals from their algebraic representations in Table 5-1 and their graphical representations in Fig. 5-14. They differ, of course, from the s and p orbitals we have been considering in both their radial and angular dependence. These differences are what we want to look at in terms of their chemical significance. If we look at the radial dependence of the *d* orbitals first, we have the comparison shown in Fig. 16-1 for the orbitals of the manganese atom (as a typical transition metal). It is the 3*d* valence orbitals of Mn that are significant from our point of view; the maximum-probability radius is just about the same for the 3*d* as for the 3*s* and 3*p*, and much less than for the 4*s*. Since a greater distance from the nucleus means that the electron is held less tightly, we correctly infer that the most easily removed electrons on a Mn atom will be the 4*s* electrons — leading to a formal oxidation state of 2+.

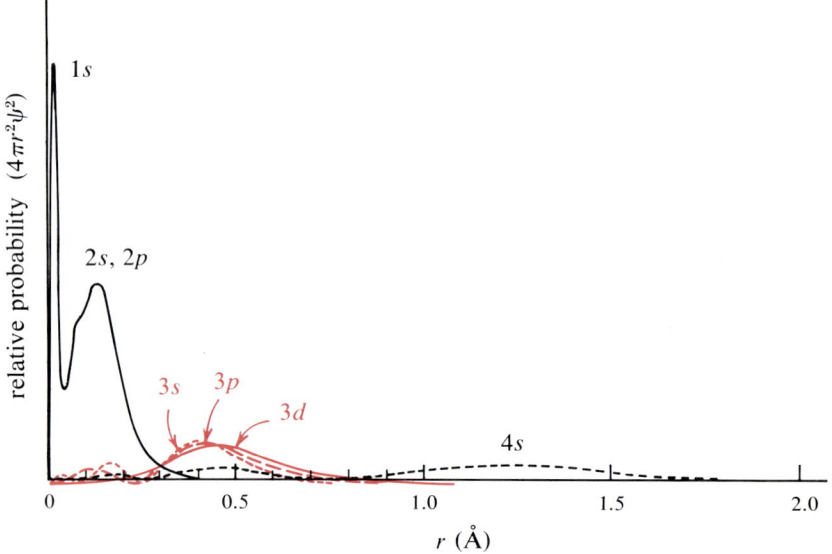

Figure 16-1 Relative size and degree of penetration for the *s*, *p*, and *d* AO's in manganese.

From Table 5-2 we can see that all of the first-row transition metals have two 4*s* electrons, and so it is not too surprising that all of them display a 2+ formal oxidation state (as Table 16-3 implied). However, it is *not* true that the 3*s*, 3*p*, and 3*d* electrons are equally easily removed. Notice that the 3*p* orbital has a small probability hump underneath the 2*p* hump and that the 3*s* has two inner probability humps, one under the 2*s* and one under the 1*s*. These will have a rather spectacular effect on the binding energy, because if an electron can really penetrate to near the nucleus (where it is not shielded very well and experiences nearly the full attraction of the nucleus with a charge of 25+), it will be very strongly bound. The 3*d* electrons show no such **penetration** effect, both because they increase from zero at the nucleus only as r^4 ($4\pi r^2 \times r^2$ in the ψ function), which is very small for small r, and because all their allowed nodes are angular nodes. Thus they are much more easily removed than the 3*s* or 3*p* electrons, and as a matter of fact (considering the penetration of the 4*s* electrons) almost as easily removed as the 4s. Figure 16-2 (which is related to Fig. 6-7) shows the magnitude of the attraction to the nucleus as a function of r; it may be helpful in judging the relative binding energies.

Figure 16-2 Effective nuclear charge attracting electrons at the maximum-probability radius of different orbitals in the Mn atom.

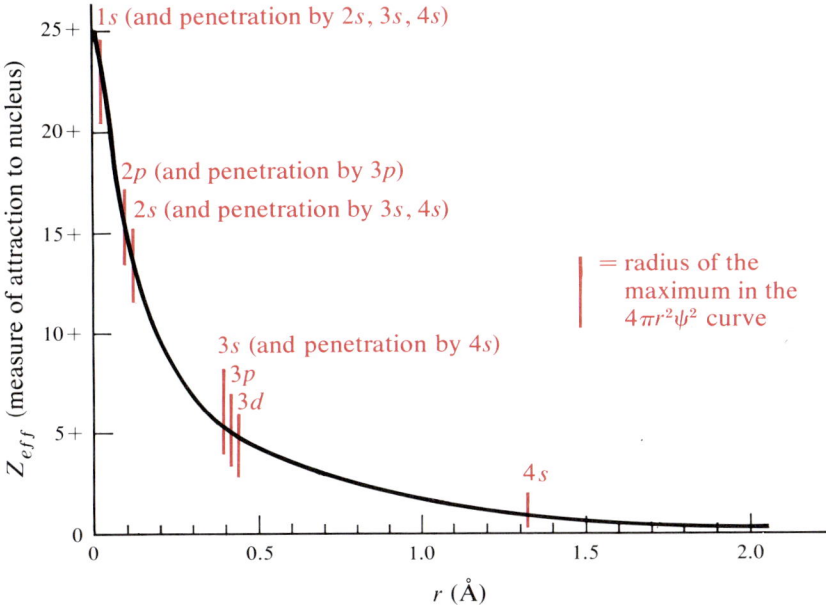

The Trend in d-Electron Stability

Since the transition metals have both d and s valence electrons, there is some importance in the relative trends in stability of these electrons as the atomic number increases going across a row of the periodic table. From Fig. 12-5 we see that the energy of the 4s orbital changes hardly at all going from Sc to Cu, while the 3d orbitals become much more stable as the atomic number increases. This is because the added d electrons (in the elements toward the right of the series) don't shield each other very well from the nuclear charge, which also increases. They can't, because they are all at about the same distance from the nucleus. On the other hand, the added d electrons are all pretty much inside the 4s electrons and shield the added nuclear charge very effectively, so that the only added stability for the 4s electrons is due to the increased nuclear charge to which they are attracted during the small fraction of the time they are found inside the average radius of the 3d electrons—that is, when they have penetrated near the nucleus. The upshot of this is that the d electrons are very readily lost to good electron acceptors for the elements Sc, Ti, and V, but are nearly impossible to remove by chemical means for the elements Co, Ni, and Cu; scandium is always Sc^{3+}, and titanium is usually Ti^{4+}, but Ni is usually Ni^{2+} and copper is almost always Cu^{2+} or even Cu^+. Another way of saying this is to note that Ti^{2+} is an extremely powerful reducing agent (electron donor), but Cu^{2+} is actually a weak oxidizing agent (electron acceptor). This trend is an important one to remember in comparing the chemistries of the different transition metals.

Symmetry of d Orbitals

The angular dependence of the d orbitals is also a point of difference and of some interest, relative to the previously discussed s and p orbitals. Figure 16-3

Figure 16-3 Symmetry of s, p, and d orbitals relative to a cube surrounding them (with octahedral ligand atoms).

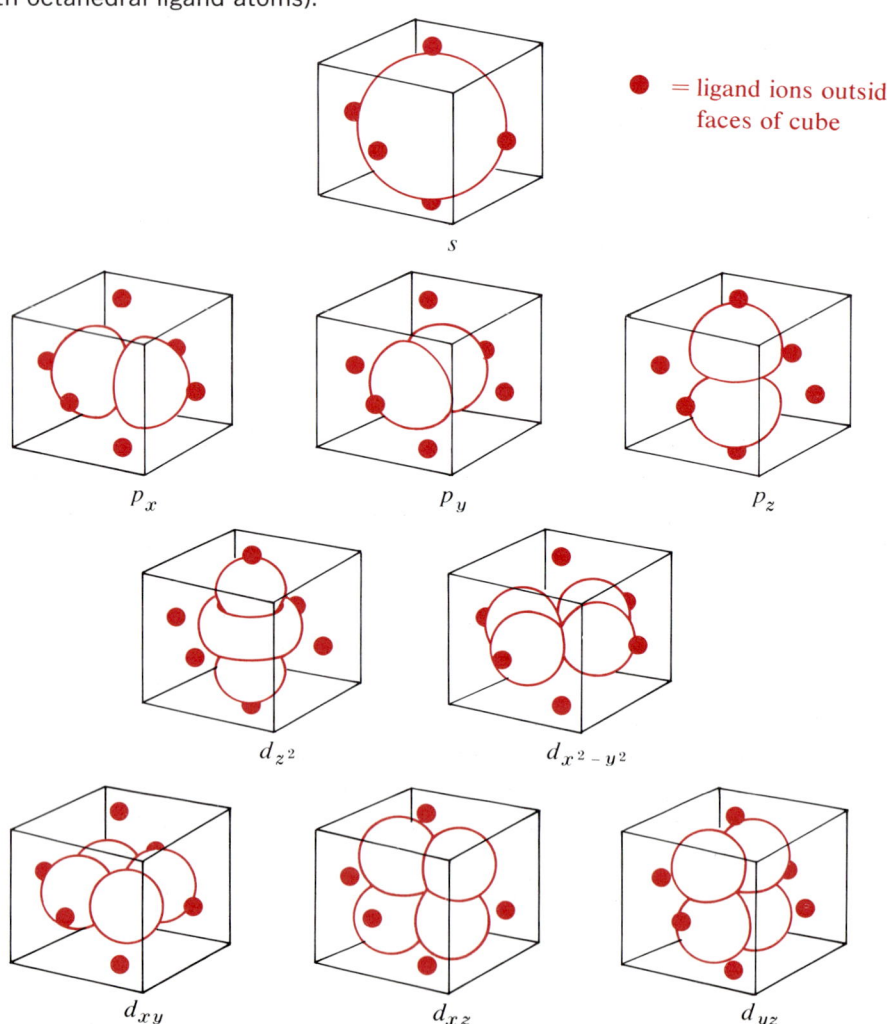

shows the geometry of these orbitals with reference to a cube surrounding the nucleus. This is a convenient reference since, for instance, octahedral ligands have cubic symmetry (centers of the faces). Notice that the s orbital, which is spherically symmetrical, meets all faces of the cube equally. So do the set of three p orbitals, considered together; one meets the north and south faces, one meets the east and west faces, and one meets the top and bottom faces. However, the d orbitals are not all equivalent in the sense that the p orbitals are. Two of them, the d_{z^2} and the $d_{x^2-y^2}$, meet the faces of the cube, but the other three (d_{xy}, d_{yz}, d_{xz}) point toward the edges of the cube and actually have a node meeting the centers of the faces of the cube. This distinction or symmetry difference has an obvious effect on the possibility of overlap of the d orbitals with other orbitals on ligand atoms in a molecule, and we shall look in the next section at the kinds of MO's that can form.

Crystal Field Theory

In a somewhat simpler approach, we can ask ourselves what the energy effect would be on a d electron if we brought up negatively charged ligands around the transition-metal atom. Since the electron is itself negatively charged it would be repelled and its energy would be raised—but how much? Figure 16-4 shows

Figure 16-4 Energy effect of bringing six octahedrally arranged ligand ions up to a transition metal.

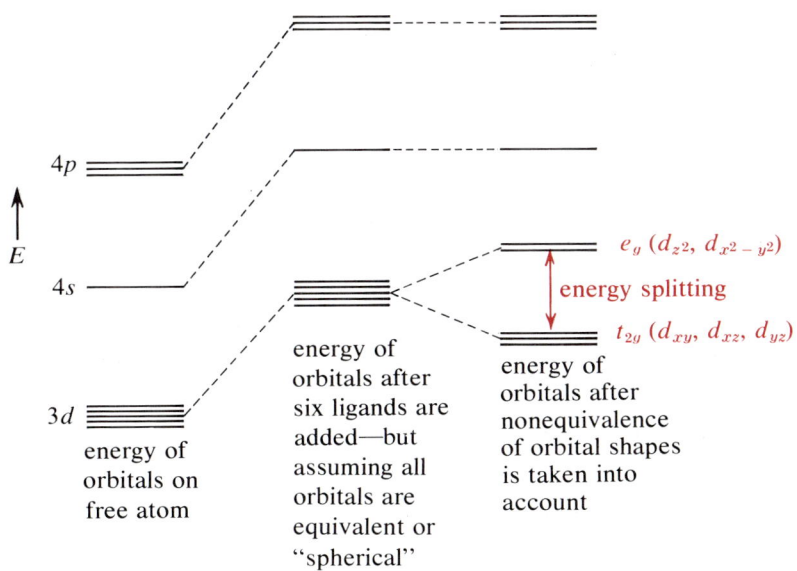

the energy effect for an electron in the presence of six octahedrally arranged ligand ions, whether the electron is distributed according to an s, a p, or a d orbital. An s electron's energy is raised by an amount equal to the overall repulsion. A p electron's energy is raised a similar amount, and it does not matter which p orbital the electron is in since all of them meet the faces of the cube similarly. But there is an energy difference involved for a d electron, depending on which orbital distribution it adopts. If it were to enter the d_{z^2} or $d_{x^2-y^2}$ orbital, either of which meets the faces of the cube and thus sticks out directly toward the ligand ions (so that the two are said to have e_g symmetry[1]), it would be very strongly repelled. The geometry of the orbital and its nodes would require the electron to stay near the negatively charged ligand ions, causing even more repulsion than a spherically distributed electron would experience. But if the electron were to enter the d_{xy}, d_{yz}, or d_{xz} orbital (which meet the edges of the cube and have nodes in the directions of the ligand ions) it would spend less time near the ligand ions than would a spherically distributed electron, and would be repelled less. These three orbitals are said to have t_{2g} symmetry. Figure 16-5 illustrates this argument as it applies to the orbitals and ligands in the xy plane around the transition-metal atom.

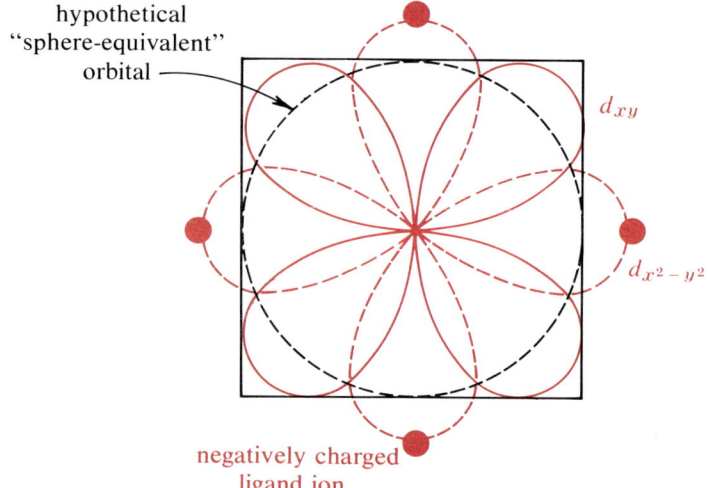

Figure 16-5 Comparison between d_{xy}, $d_{x^2-y^2}$, and a hypothetical spherical d orbital in their approach to ligand ions.

This explanation of the differences in orbital energies for d electrons in terms of electrostatic repulsions is known as **crystal field theory**. The energy differences that crystal field theory predicts are of great usefulness in correlating

---

[1] The designations e_g and t_{2g} have significance in the mathematical treatment of molecular symmetry (group theory). We introduce them here because of their wide usage, but we shall treat them only as labels.

chemical, magnetic, and electrical properties of the transition metals. For example, if we call the energy difference between the t_{2g} and e_g orbitals Δ as in Fig. 16-6, the t_{2g} orbitals will be lower in energy — more stable — than a spherical orbital would be by $\frac{2}{5}\Delta$, and the e_g orbitals will be less stable by $\frac{3}{5}\Delta$. We can thus add up a **crystal-field stabilization energy** (CFSE) for any transition-metal ion.

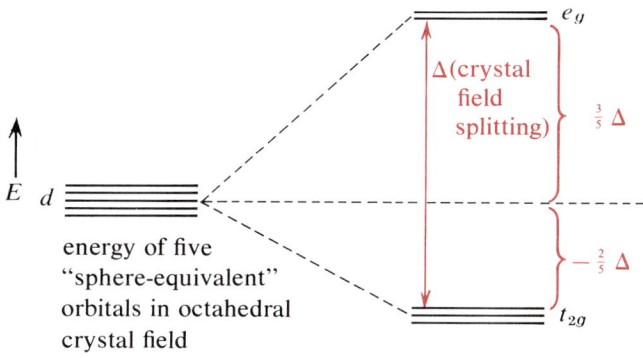

Figure 16-6 Definition of crystal-field energy splitting Δ.

Table 16-5 indicates the relative d-orbital energies for several different geometries of ligand ions, and Table 16-6 gives the number of d electrons and their possible configurations for octahedral ligand geometry, along with the net CFSE. In each case the net CFSE is obtained by adding up $\frac{2}{5}\Delta$ for each t_{2g} electron and $-\frac{3}{5}\Delta$ for each e_g electron.

Table 16-5
Energies of d Orbitals in Crystal Fields of Different Symmetries (in fractions of the octahedral Δ energy)

Coordination Number	Structure	$d_{x^2y^2}$	d_{z^2}	d_{xy}	d_{xz}	d_{yz}
1	linear	−0.314	0.514	−0.314	0.057	0.057
2	linear	−0.628	1.028	−0.628	0.114	0.114
3	trigonal planar	0.546	−0.321	0.546	−0.386	−0.386
4	tetrahedral	−0.267	−0.267	0.178	0.178	0.178
4	square planar	1.228	−0.428	0.228	−0.514	−0.514
5	trigonal bipyramidal	−0.082	0.707	−0.082	−0.272	−0.272
5	square pyramidal	0.914	0.086	−0.086	−0.457	−0.457
6	octahedral	0.600	0.600	−0.400	−0.400	−0.400
7		0.879	0.139	−0.151	−0.260	−0.608

Table 16-6
Crystal-Field Stabilization Energies (CFSE) for Octahedral d^n Configurations

	Ti²⁺ d^2	V²⁺ d^3	Cr²⁺ d^4	Mn²⁺ d^5	Fe²⁺ d^6	Co²⁺ d^7	Ni²⁺ d^8	Cu²⁺ d^9
CFSE	$\frac{4}{5}\Delta$	$\frac{6}{5}\Delta$	$\frac{3}{5}\Delta$	0	$\frac{2}{5}\Delta$	$\frac{4}{5}\Delta$	$\frac{6}{5}\Delta$	$\frac{3}{5}\Delta$

$+\frac{3}{5}\Delta$ / $-\frac{2}{5}\Delta$

or if Δ is large enough to make spin pairing preferable,

	Ti²⁺	V²⁺	Cr²⁺	Mn²⁺	Fe²⁺	Co²⁺	Ni²⁺	Cu²⁺
CFSE	$\frac{4}{5}\Delta$	$\frac{6}{5}\Delta$	$\frac{8}{5}\Delta$	$\frac{10}{5}\Delta$	$\frac{12}{5}\Delta$	$\frac{9}{5}\Delta$	$\frac{6}{5}\Delta$	$\frac{3}{5}\Delta$

	Ti³⁺ d^1	V³⁺ d^2	Cr³⁺ d^3	Mn³⁺ d^4	Fe³⁺ d^5	Co³⁺ d^6	Ni³⁺ d^7	Cu³⁺ d^8
CFSE	$\frac{2}{5}\Delta$	$\frac{4}{5}\Delta$	$\frac{6}{5}\Delta$	$\frac{3}{5}\Delta$	0	$\frac{2}{5}\Delta$	$\frac{4}{5}\Delta$	$\frac{6}{5}\Delta$

or for large Δ,

	Ti³⁺	V³⁺	Cr³⁺	Mn³⁺	Fe³⁺	Co³⁺	Ni³⁺	Cu³⁺
CFSE	$\frac{2}{5}\Delta$	$\frac{4}{5}\Delta$	$\frac{6}{5}\Delta$	$\frac{8}{5}\Delta$	$\frac{10}{5}\Delta$	$\frac{12}{5}\Delta$	$\frac{9}{5}\Delta$	$\frac{6}{5}\Delta$

The Magnetism of d Electrons

There are a couple of sets of experimental results that we can correlate immediately on the basis of Table 16-6. One is the magnetic moments of octahedral complexes. Remember that unpaired electrons give an atom a net magnetic moment, due to their spin. The magnetic moment can be determined by measuring the force attracting a sample of the compound to a magnetic field. By a derivation that is not too difficult, the magnetic moment in **Bohr magnetons** (a convenient unit for this quantity) can be shown to be approximately equal to $\sqrt{n(n+2)}$, where n is the number of unpaired electrons. So we can readily find out something about the size of Δ by measuring the magnetic moment of the compound. The FeF_6^{3-} ion [containing Fe^{III}] has a magnetic moment of 5.98 Bohr magnetons, and the $Fe(CN)_6^{3-}$ ion has a magnetic moment of 2.40 Bohr magnetons. Which has the higher Δ value? For the fluoride complex, the magnetic moment corresponds quite well to that expected for a small Δ, with five unpaired electrons:

$$\text{magnetic moment} \equiv \mu = 5.98 = \sqrt{n(n+2)}$$

$$35.76 = n(n+2)$$

$$n^2 + 2n - 35.76 = 0$$

$$n = \frac{-2 \pm \sqrt{4 + 4(35.76)}}{2} = \frac{-2 + 12.12}{2} = 5.06 \text{ electrons}$$

For the cyanide complex, the magnetic moment corresponds at least approximately to that expected for a large Δ, with only one unpaired electron:

$$\mu = 2.40 = \sqrt{n(n+2)}$$

$$5.76 = n(n+2)$$

$$n^2 + 2n - 5.76 = 0$$

$$n = \frac{-2 \pm \sqrt{4 + 4(5.76)}}{2} = \frac{-2 + 5.20}{2} = 1.60 \text{ electrons}$$

On the basis of these results, we expect Δ to be larger for the cyanide complex, since that would cause the smaller number of unpaired electrons.

Reduction Potentials of Transition-Metal Ions

Another interesting correlation involves the reduction potentials for the transition metals in the half-reaction

$$M(OH_2)_{6(aq)}^{3+} + e^- \rightarrow M(OH_2)_{6(aq)}^{2+} \qquad (\mathscr{E}^0)$$

We can break this down into individual steps by setting up a Born–Haber cycle:

$$\begin{array}{ccc}
M_{(g)}^{3+} + 6H_2O_{(g)} & \xrightarrow{-IP_3} & M_{(g)}^{2+} + 6H_2O_{(g)} \\
{\scriptstyle -\Delta H_{\text{hydr 3+}}}\Big\uparrow & & \Big\downarrow{\scriptstyle \Delta H_{\text{hydr 2+}}} \\
M(OH_2)_{6(aq)}^{3+} + e^- & \xrightarrow[{=-n\mathscr{F}\mathscr{E}^0}]{\Delta G^0} & M(OH_2)_{6(aq)}^{2+}
\end{array}$$

Since the number of free atoms or molecules does not change in the half-reaction we can assume that the entropy change is small, so that ΔG^0 is the sum of the ionization potential IP_3 and the difference between the enthalpies of hydration for the 2+ and 3+ ions:

$$\Delta G^0 = -\mathrm{IP}_3 - (\Delta H_{\mathrm{hydr}\,3+} - \Delta H_{\mathrm{hydr}\,2+})$$

The number of d electrons on each ion is changing—one electron is being added in each half-reaction. This means that the CFSE is changing by either $+\frac{2}{5}\Delta$ or $-\frac{3}{5}\Delta$ for each case. If we assume that this change in the CFSE (from Table 16-6) is the principal difference between the enthalpies of hydration of the 2+ and 3+ species for different elements, then when we add this change in CFSE to the IP_3 we should get a good match for the ΔG^0 or $-1\cdot\mathscr{F}\mathscr{E}^0$ value. Figure 16-7 shows the result of this approach; apparently in most cases we really can get a good match for $\mathscr{E}^0$. This sort of approach is useful in accounting for many patterns of chemical reactivity among the transition metals.

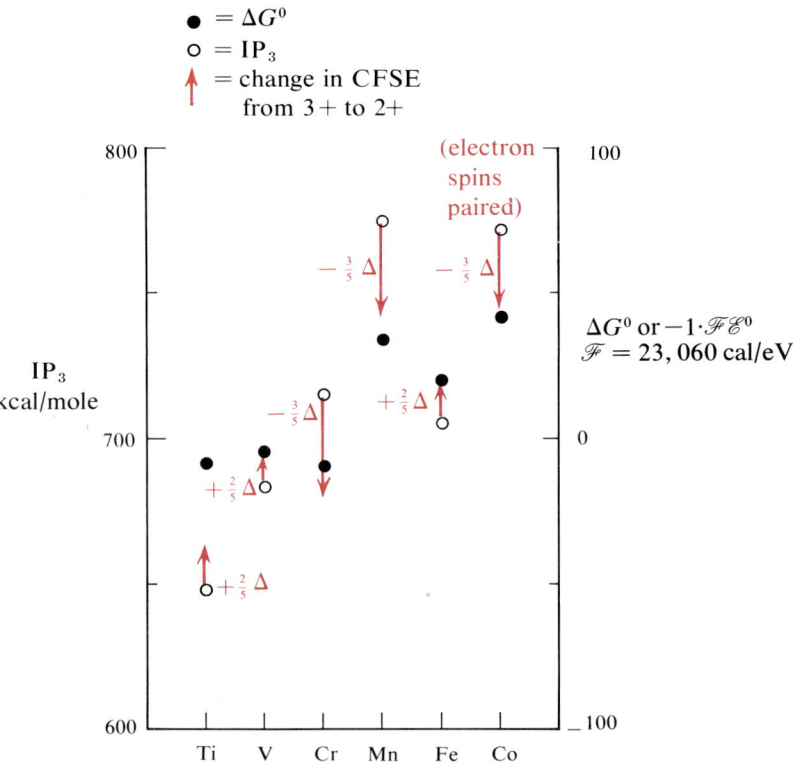

Figure 16-7 Rationalizing the difference between reduction potentials and ionization potentials through changing CFSE for each element. (The vertical scale difference allows for uniform increase of Δ for all ions from 2+ to 3+.) (Redrawn from Dunn, T. M., McClure, D. S., and Pearson, R. G., *Some Aspects of Crystal Field Theory*, New York: Harper & Row, 1965.)

16-3 Molecular Orbitals for *d*-Electron Compounds

The concept of a crystal field, as we have seen, is useful in rationalizing some of the properties of transition metals. It might even be a fairly good model for the electronic structure of the *d*-electron compounds if those compounds were really ionic, so that only straightforward electrostatic attractions and repulsions were involved in holding the molecule together. But transition metals are not electropositive enough to form completely ionic compounds with anything. Electron sharing, through orbital overlap, is involved in all of their compounds, so we need to look at the kinds of MO's to which *d* AO's can contribute. Of course, this does not preclude partly ionic bonding, because we expect some charge transfer from one AO to another if the coefficients of the two orbitals are different in the LCAO MO. The MO description of bonding gives us a general approach to the electron distribution, whether predominantly ionic or covalent.

Regardless of the formal stoichiometry of the compound they are in, we usually find transition-metal atoms surrounded by either four or six ligand atoms. In a crystal this is achieved by sharing ligands; in a solution, by solvating. For the ML_6 species, we usually find octahedral geometry; for ML_4, the geometry can be either tetrahedral or square planar, as in Fig. 16-8 (where

Figure 16-8 Coordinate axes for common geometries of transition-metal compounds.

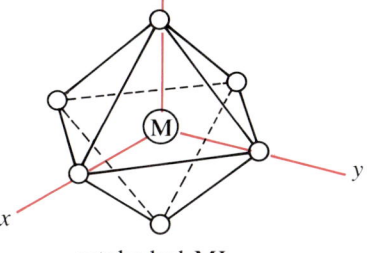

octahedral ML_6

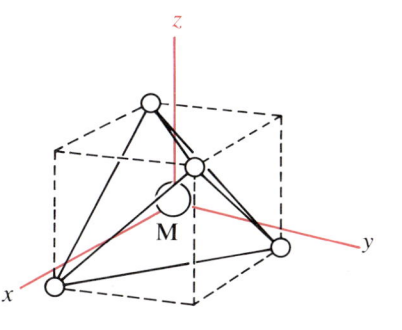

tetrahedral ML_4

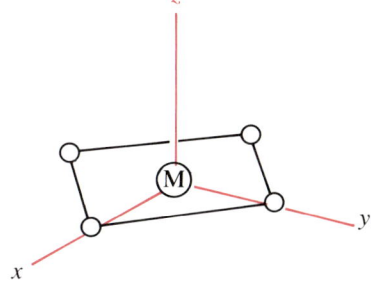

square planar ML_4

842 | *d*-Electron Compounds

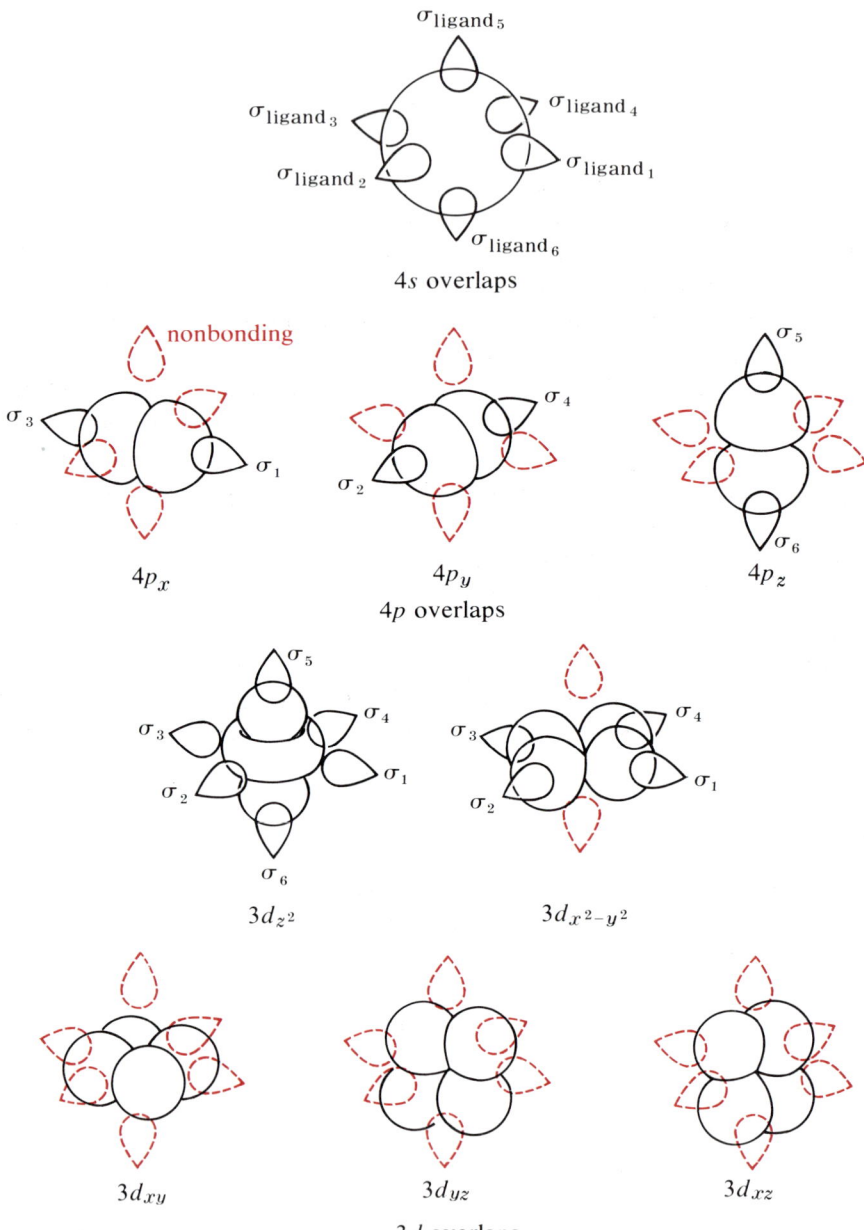

Figure 16-9 Overlaps of ligand σ orbitals with transition-metal valence orbitals for octahedral geometry.

coordinate axes are indicated to help identify the d orbitals that will overlap ligand orbitals). Let us take these one at a time and develop the MO energy levels that are appropriate for σ overlap. Pi overlap can exist, but it does not influence the energy levels very strongly in most cases.

Octahedral Molecular Orbitals

For the octahedral ML_6 system, we need to consider the $3d$, $4s$, and $4p$ orbitals on the transition-metal atom as valence orbitals. There are no $4p$ electrons, but that set of orbitals is close enough in energy to take a meaningful part in influencing the energy of the MO's. Figure 16-9 shows the overlaps of these AO's with a single σ-orbital lobe from each ligand atom. In Fig. 16-10 we construct the energy levels: (a) shows the combination of the s orbital with

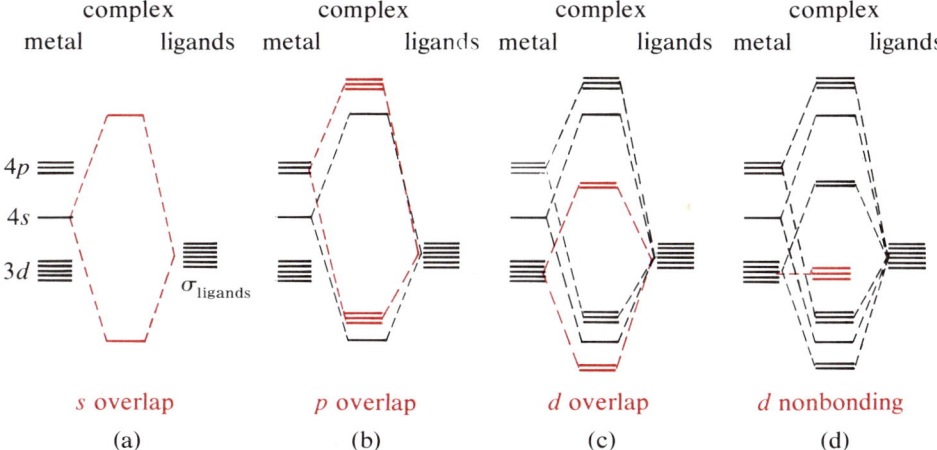

Figure 16-10 Development of the energy-level diagram for an octahedral molecule, considering only σ overlap.

ligand orbitals in a bonding and an antibonding combination; (b) shows the combination of the p orbitals with ligand orbitals into three equivalent bonding combinations and three equivalent antibonding combinations; (c) shows the combination of the $d_{x^2-y^2}$ and d_{z^2} orbitals, which stick out in just the right directions for σ overlap, into two equivalent bonding and two equivalent antibonding combinations with ligand orbitals; and (d) adds on the three remaining d orbitals, which are nonbonding because they have nodes through all the ligand σ lobes. Note, however, that these last are the ones that can show π overlap. Checking to see that we have done this properly, we started with 15 AO's (1+3+5+6) and we have created 15 MO's (6 bonding + 6 antibonding + 3 nonbonding), which is correct. Now what about the electrons in these orbitals? If each ligand serves as an electron donor or Lewis base, contributing a pair of electrons to the transition metal atom (which will certainly be an acceptor if it has a 2+ or 3+ charge), then the MO's must accommodate these six pairs of electrons plus however many valence electrons are left on the transition-metal atom in its formal oxidation state. In, say, the 2+ oxidation state, the 4s electrons are gone and all the d electrons remain; Fig. 16-11 shows that these must be distributed in the nonbonding orbitals (which we previously called t_{2g}) and/or the antibonding MO's formed from what we previously called the e_g d orbitals. So this particular section of the MO energy-level diagram is equivalent to the crystal-field splitting of Fig. 16-6.

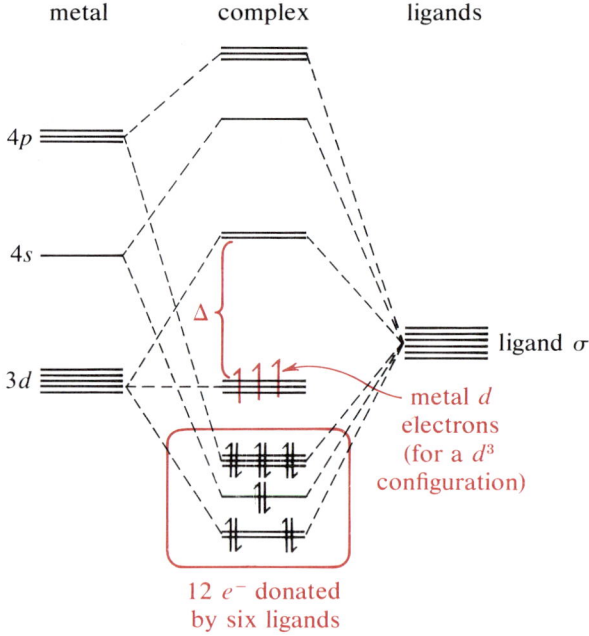

Figure 16-11   Population of octahedral energy levels by electrons for a d^3 ion such as Cr^{3+}.

Tetrahedral Molecular Orbitals

Before going on to the uses of these MO's in correlating or rationalizing experimental results, let us develop the corresponding MO energies for the 4-coordinate ML_4 systems. For the tetrahedral geometry, Fig. 16-12 shows the overlaps of the same transition-metal valence orbitals with the σ ligand orbitals. Figure 16-13 shows the stepwise construction of the energy-level diagram, and we should note that again the p orbitals form three equivalent bonding and three equivalent antibonding MO's, but the d orbitals are again not

Molecular Orbitals for *d*-Electron Compounds | 845

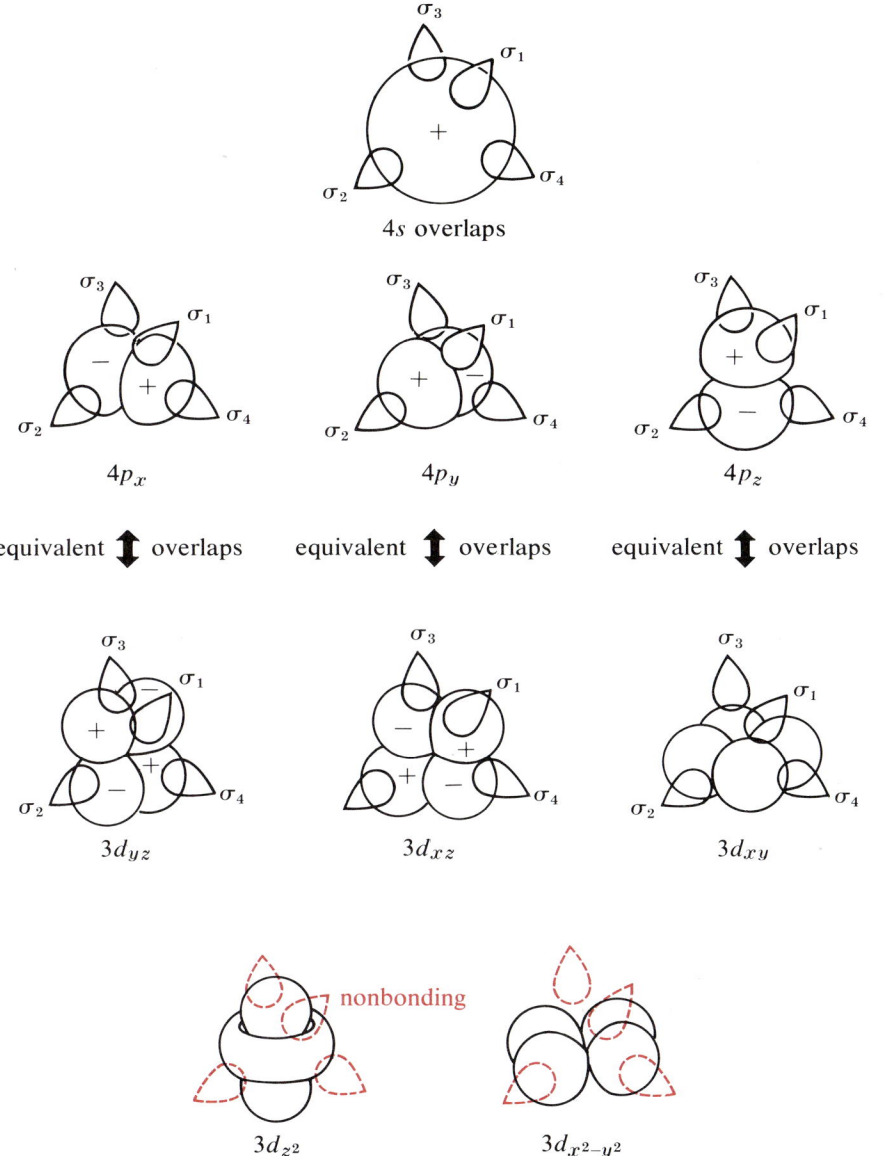

Figure 16-12 Overlaps of ligand σ orbitals with transition-metal valence orbitals for tetrahedral geometry.

equivalent. The three that were nonbonding in the octahedral case, d_{xy}, d_{yz}, and d_{xz}, have σ overlap with the tetrahedral ligands, while the other two have nodes through the ligand σ orbitals and must be nonbonding. There is a slight added complication in that the set of orbitals d_{xy}, d_{yz}, d_{xz} combine with the same com-

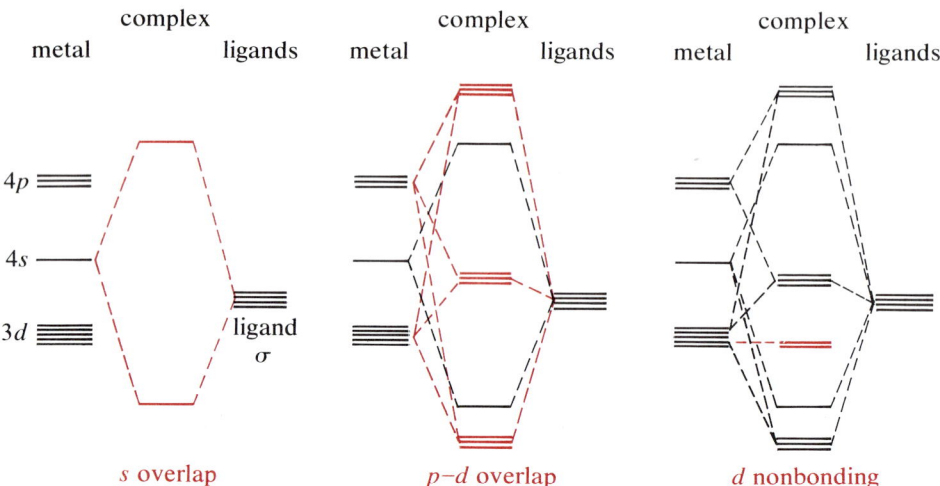

Figure 16-13 Development of the energy-level diagram for a tetrahedral molecule, considering only σ overlap.

bination of ligand-orbital signs as does the set p_z, p_x, p_y (respectively). This means that there must be three sets of three equivalent orbitals: one in which both the d and p AO's are favorably overlapping the ligand orbitals because they have the same sign, which gives a strongly bonding set of MO's; one in which both the d and p AO's have opposite signs from the ligand orbitals and yield strongly antibonding MO's; and one in which the d and p AO's do not both match but one set does and the other does not, which will be nonbonding but slightly higher in energy than the nonbonding $d_{x^2-y^2}$ and d_{z^2}. This gives the necessary 13 MO's arising from the original 13 MO's. The MO's can be populated with electrons using the same procedure as in the octahedral case.

Square-Planar Molecular Orbitals

For the square planar geometry, Fig. 16-14 shows the overlaps of the transition-metal valence orbitals with the ligand orbitals, and Fig. 16-15 builds up the energy-level diagram. If the ligands are on the x and y axes, the p_z AO will be nonbonding, but the other two will form equivalent bonding and antibonding sets of MO's. Among the d AO's, the $d_{x^2-y^2}$ will form a strong bonding and a strong antibonding combination, while the other four will be nearly nonbonding (although consideration of π overlap will split them up since they are not equivalent). If electrons donated by the four ligands fill the four bonding MO's, leaving the four nonbonding MO's for metal d electrons, it is easy to see why the square planar geometry is favored by metals with eight d electrons, such as Ni^{II}, Pt^{II}, and Au^{III}.

Molecular Orbitals for *d*-Electron Compounds | 847

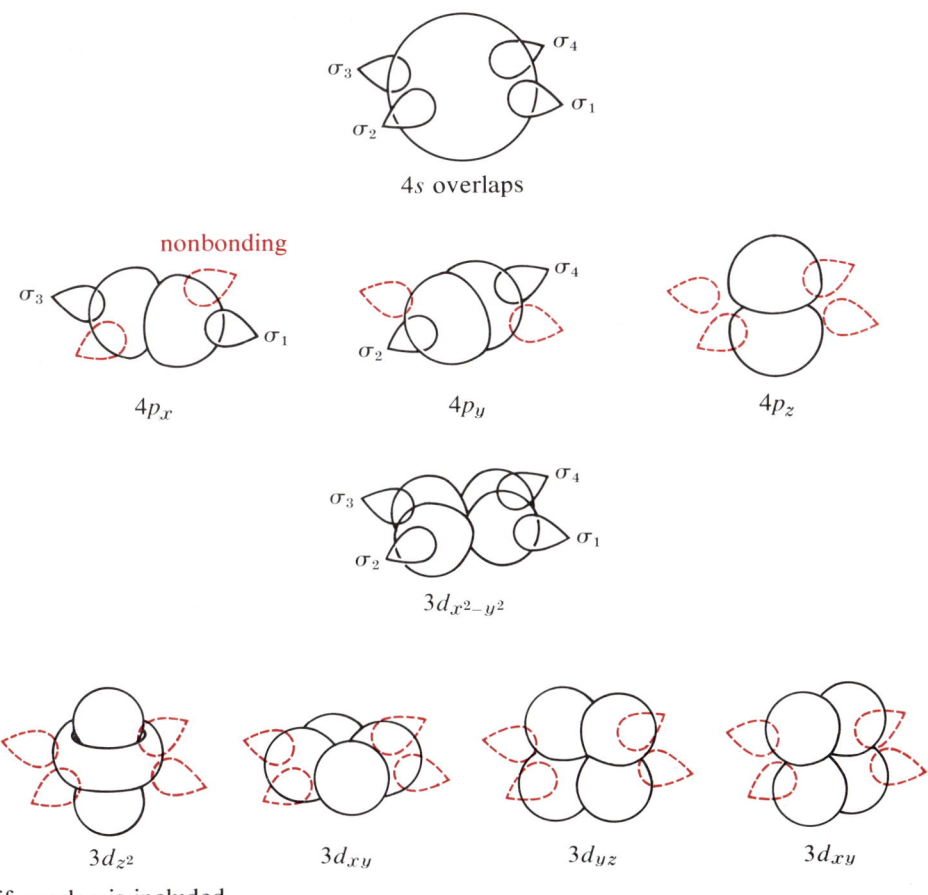

Figure 16-14 Overlaps of ligand σ orbitals with transition-metal valence orbitals for square planar geometry.

Hydration Energies

One of the ways we can conveniently tie these MO energies into our previous discussion is to consider what is actually happening when a transition-metal ion dissolves in water and forms a hydrate (usually an octahedral 6-hydrate). In Chapter 13 we produced a fairly extensive discussion of hydration energies by considering the interaction between a metal ion and water molecules as a

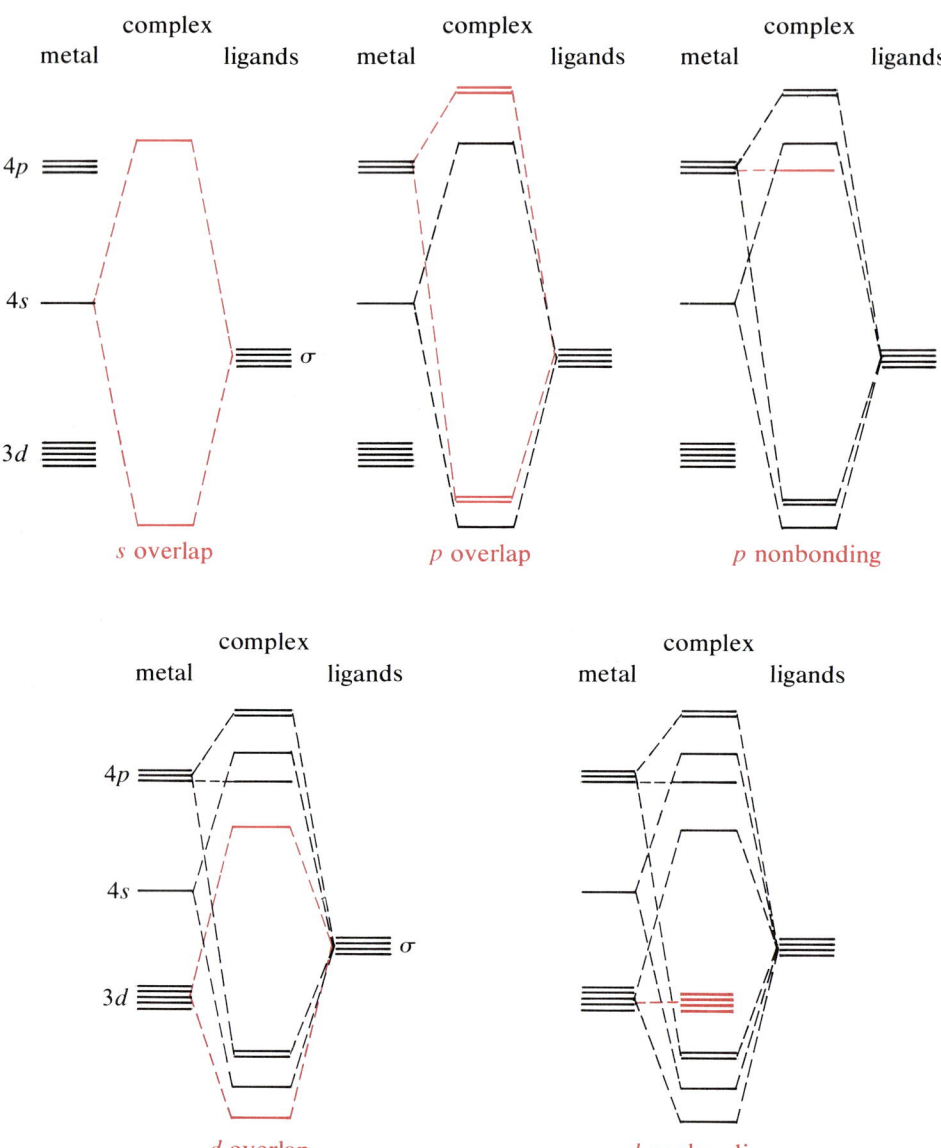

Figure 16-15 Development of the energy-level diagram for a square planar molecule, considering only σ overlap.

polarization effect. From Fig. 16-11 we can see that all six pairs of electrons that were originally in nonbonding oxygen orbitals in the water molecule have gained stability by forming the molecular hydrate system. This lowering of energy is the counterpart of the component of the hydration energy that we called U_1 in Chapter 13. The only trouble with the MO approach is that it is almost always a lot of trouble to calculate these MO energies—much more trouble than it is to go through the polarization approach. But as computer programs for approximate MO calculations become more and more widely available, accurate, and convenient, this objection is likely to disappear.

Color: Visible–Ultraviolet Spectroscopy

One of the most familiar properties of transition-metal compounds is their color. Almost all inorganic compounds that do *not* contain a transition metal are white or colorless, although some halides, particularly iodides, tend to be colored. But almost all transition-metal compounds are colored, and most of the exceptions are compounds that, like TiO_2, have no remaining d electrons. Clearly the presence of d electrons has a strong influence on the possibility of light absorption to give colors. This is a situation in which the MO energies are particularly useful. Consider the $Ti(OH_2)_6^{3+}$ ion, which is violet in water solution. Titanium has four valence electrons, but in Ti^{3+} only one d electron remains, which we can distribute in the octahedral MO energy levels as in Fig. 16-16. If we shine white light (in other words, light of all frequencies) on the Ti^{III} solution, there will be a particular frequency that, through the relationship $E = h\nu$, carries just enough energy that a d electron can absorb the light—and energy—and redistribute itself according to the next-higher-energy MO. Thus

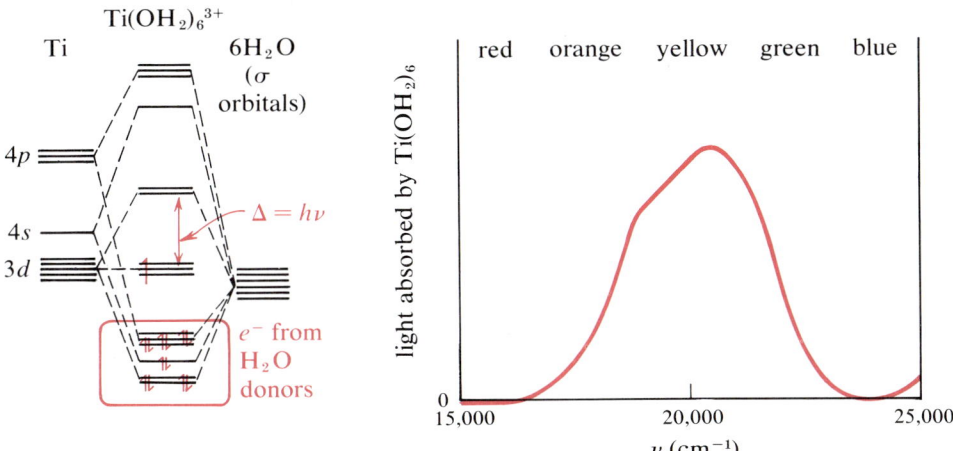

Figure 16-16 Electron distribution and the visible-range absorption spectrum for $Ti(OH_2)_6^{3+}$.

the energy-level separation Δ is a measure of the frequency of light that will be absorbed by the system. If the frequency that is absorbed falls within the visible range (about 15,000–25,000 cm^{-1}), the light that is either reflected or transmitted will not have all frequencies represented with equal intensity and so will not be white any more. The color will be the sum of the colors corresponding to those frequencies that are not absorbed. The ion $Ti(OH_2)_6^{3+}$ is violet because yellow-green light is absorbed when the d electron is excited from a t_{2g} orbital to a e_g orbital. If the MO energy separation were larger than 25,000 cm^{-1} the light absorption would occur in the ultraviolet (UV) region of the spectrum and would be invisible to the naked eye. If the separation were less than 15,000 cm^{-1}, absorption would occur in the infrared (IR) region. Generally speaking, MO energies for nontransition-metal systems are separated by larger energies than that corresponding to 25,000 cm^{-1} — so everything absorbs in the UV range even though many compounds are colorless.

The absorption peaks, as recorded on an absorption spectrophotometer, are not perfectly sharp spikes even though the energy levels are defined precisely for the MO's, as Fig. 16-17 shows. What happens is that the thermal energy kT that is available at room temperature causes the hydrate molecule to vibrate,

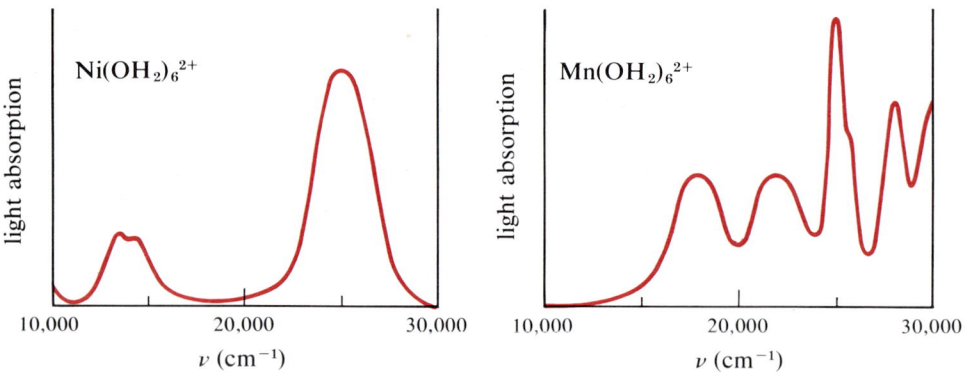

Figure 16-17 Typical absorption spectra of transition-metal complexes.

and when the internuclear distance changes between bonded atoms the overlap of their AO's changes. This changes the bonding-energy effect for the MO's, and since the MO energy separation is now changing as the molecule vibrates, the light absorption spreads out over a range of frequencies. Sometimes in order to analyze complex absorption spectra it is helpful to obtain them at very low temperatures (like that of boiling liquid helium, 4.2°K) in order to reduce kT and sharpen the absorption peaks. Quantum effects would prevent bond vibrations from stopping completely even at 0°K, however, so the peaks can never become perfectly sharp.

The electronic transition we assigned to the yellow-green absorption of $Ti(OH_2)_6^{3+}$ essentially took an electron out of a nonbonding d orbital and placed it in an antibonding MO that fairly strongly resembled another d orbital. For this reason, such transitions (which are very common in transition-metal compounds) are often called **d–d** transitions. They occur most often in the visible range, but depending on the exact details of the bonding in the MO's, they can occur in the UV or near-IR range. Since absorption can occur anywhere in the visible range, literally any color is possible. Another kind of electronic transition is possible, however, in which one of the electrons in the bonding MO's (which resemble ligand orbitals) is excited to an antibonding MO, which usually resembles a metal valence orbital. Since the electron in question is essentially being transferred from the ligands to the metal, such transitions lead to so-called **charge-transfer** absorption peaks. Charge transfer peaks usually occur in the UV, because they correspond to a greater excitation energy. Sometimes they occur in the visible range, however, and compounds for which this is true (if they are satisfactorily insoluble and weatherproof) make excellent pigments, since only a small amount of the compound is required to give the paint or other medium a brilliant hue. The great intensity of charge-transfer absorptions results from the fact that the intensity of absorption is proportional to the extent to which the electron distribution of the molecule changes between the ground state and the excited state; since an electron is being moved all the way from one atom to another, the interaction of the molecule with the incident radiation is very strong. Examples of commonly used transition-metal pigments are cobalt blue (Al_2CoO_4), chrome yellow ($PbCrO_4$), and burnt umber ($Fe_2O_3 + MnO_2$); there are many others. They bear an interesting relation to dyestuffs for fabric, which are usually organic compounds with an extended π-delocalized structure. In these dyes the strong colors are due to transitions of electrons from π bonding orbitals to π antibonding orbitals ($\pi \to \pi^*$), or sometimes from nonbonding orbitals to π antibonding orbitals ($n \to \pi^*$). Only if the π delocalization is quite extensive will the transitions fall in the visible range, so reasonably large molecules are necessary, but by adding substituent groups the dye molecule's absorption can be tailored to almost any desired frequency or color. There are two rather similar ways to account for the appearance of strong colors in our surroundings—either as electronic transitions (either d–d or charge-transfer) between d orbitals in transition-metal compounds or as the same sort of transition between π orbitals in organic compounds.

16-4 Coordination Numbers and Molecular Geometries

Thus far we have assumed without any real justification that the geometry of ligands surrounding a transition-metal atom in a molecule is, generally, octahedral or tetrahedral. What governs the number of ligands that can form a stable compound with a transition-metal atom, and what geometry will they adopt? First of all, energy considerations dictate the whole situation. If we consider ligands as electron donors, they will add to the transition-metal atom as long as there are vacant low-energy orbitals from which MO's can be formed. The more ligands there are, the more bond energy is released.

Donor-Acceptor Compounds: Carbonyls and π-Donors

Since we have used nine valence orbitals for transition metals in forming MO's, the maximum number of electron-pair donor ligands should be nine—a **coordination number** of nine. Although a coordination number of nine is very rare, there are many compounds whose stoichiometries correspond to a total of 18 valence electrons around the transition metal. For example, carbon monoxide is a two-electron donor using nonbonding electrons on the carbon, and iron (which has six d and two s electrons) forms the compound $Fe(CO)_5$, iron pentacarbonyl. The Fe atom in $Fe(CO)_5$ has 18 valence electrons around it: eight of its own and two from each of five ligands. $Ni(CO)_4$ and $Cr(CO)_6$ follow the same rule, which is sometimes called the **18-electron rule**—the transition metal will surround itself with a total of 18 electrons in filling all the possible bonding MO's it can form. There are one-electron donors, such as the halogens: the compound $Mn(CO)_5I$ follows the rule, since Mn has seven valence electrons, five CO contribute 10, and I contributes one. There are three-electron donors, of which NO is a good example; it is like CO except that it has one more electron. Compounds such as $Mn(CO)_4NO$ and $Co(NO)_2Br$ have electronic structures that follow the rule, and Table 16-7 gives some examples of these compounds—carbonyls and nitrosyls—almost all of which follow the 18-electron rule.

Table 16-7
Transition-Metal Carbonyl and Nitrosyl Compounds

V	Cr	Mn	Fe	Co	Ni
$V(CO)_6$	$Cr(CO)_6$	$Mn_2(CO)_{10}$	$Fe(CO)_5$	$Co_2(CO)_8$	$Ni(CO)_4$
			$Fe_2(CO)_9$		
$V(NO)(CO)_5$		$Mn(NO)_3CO$	$Fe(NO)_2(CO)_2$	$Co(NO)(CO)_3$	
		$Mn(NO)(CO)_4$			
$V(CO)_6^-$	$Cr(CO)_5^{2-}$	$Mn(CO)_5^-$	$Fe(CO)_4^{2-}$	$Co(CO)_4^-$	$Ni_2(CO)_6^{2-}$
			$Fe_2(CO)_8^{2-}$		
		$Mn(CO)_6^+$			
	$Cr(CO)_5I^-$	$Mn(CO)_5Br$	$Fe(CO)_4I_2$	$Co(CO)Br_2$	$Ni(CN)_3CO^{2-}$
	$Cr(CO)_5CN^-$				

There is another way to provide two-, three-, four-, five-, or even six-electron donor ligands, which we have already touched on in Chapter 15.

Organic π systems can serve as donors, with the organic system aligning itself so that one lobe of the π orbital points toward the metal, as in Fig. 16-18.

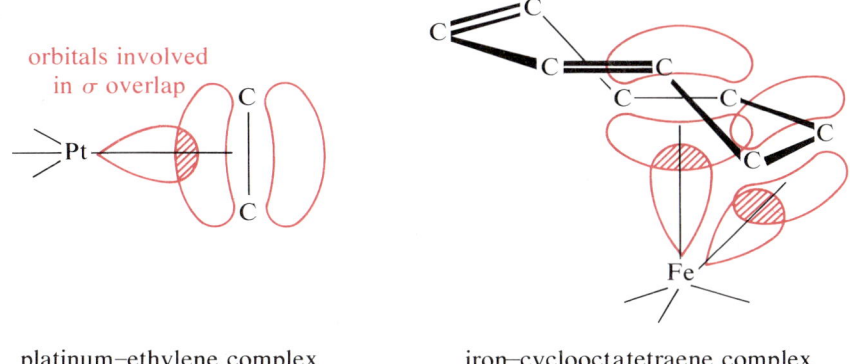

platinum–ethylene complex   iron–cyclooctatetraene complex

Figure 16-18   Orbital overlap in transition-metal compounds with π-donor molecules.

Ethylene, C_2H_4, has two π electrons and is a two-electron donor; the allyl radical, $CH_2\text{---}CH\text{---}CH_2$, serves as a three-electron donor; butadiene, C_4H_6, as a four-electron donor; cyclopentadienyl free radical, C_5H_5, as a five-electron donor; and benzene, C_6H_6, as a six-electron donor. Table 16-8 gives examples of all of these types of compounds, most of which follow the 18-electron rule.

Table 16-8
Transition-Metal Pi-Complex Compounds

Ti	V	Cr	Mn
$(C_5H_5)_2TiCl_2$	$(C_5H_5)V(CO)_4$	$Cr(C_6H_6)_2$	$(C_5H_5)Mn(CO)_3$
	$(C_7H_7)V(CO)_3$	$Cr(C_6H_6)(CO)_2(C_2H_4)$	$(C_3H_3)Mn(CO)_4$
		$(C_5H_5)Cr(CO)_2(NO)$	

Fe	Co	Ni
$Fe(C_5H_5)_2$	$(C_5H_5)Co(CO)_2$	$(C_5H_5)Ni(NO)$
$(C_4H_4)Fe(CO)_3$	$(C_3H_5)Co(CO)_3$	$(C_3H_5)_2Ni$
$(C_4H_6)Fe(CO)_3$	$(C_5H_5)_2Co^+$	$(C_3H_5)NiBr_2$
$(C_5H_5)Fe(CO)_2(C_2H_4)^+$		

One of these of particular importance is the $TiCl_3$–$AlCl_3$ catalyst for the low-pressure polymerization of ethylene. The titanium in this catalyst almost certainly forms this type of π complex with a free ethylene molecule, which then undergoes an insertion reaction.

What are the conditions under which carbonyls and π-donor molecules will form compounds with transition metals? It is necessary first of all for the metal atom to be in a formal oxidation state near zero. If it were in a very high positive oxidation state, the electrically neutral ligands CO, NO, or C_2H_4 could not spread the positive charge out enough to keep it from being very reactive toward other donors with a negative charge. But more important is the role of the d electrons, since carbonyls (and most other compounds) have only been observed to form with transition metals. Figure 16-19 shows what seems to be happening in the bonding; in effect a double bond, σ and π, is forming, but instead of one electron coming from each atom in each bond, the ligand is contributing both of the σ electrons into a vacant metal σ orbital while the metal is contributing two d electrons into a vacant π antibonding orbital on the ligand.

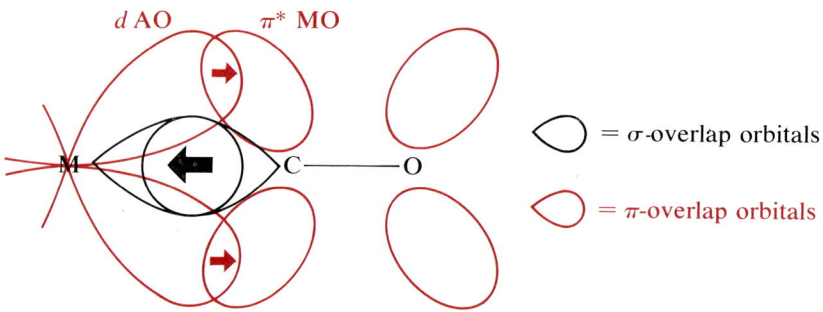

Figure 16-19   Patterns of electron donation in carbonyl bonding.

These compounds will form only if the metal has d electrons and a low charge so that it can contribute them. This is called $d\pi$–$p\pi$ bonding, and it seems to be necessary for many molecules. Another condition is that the ligands must be electrically neutral and must have vacant π orbitals. If the ligands carry a negative charge, the electrostatic repulsion of, say, ML_3^{3-} for L^- may prevent ML_4^{4-} from forming — and, of course, the vacant π orbitals are necessary to form the π bond.

Donor-Acceptor Compounds: Anion Ligands

Suppose these conditions are not met by the metal ion and the ligands; then what governs the coordination number? If the metal has, say, a substantial positive charge, or the ligands a negative charge, or the ligands have no vacant π orbitals, then the coordination number will increase to a maximum that is consistent with the repulsions between ligands. That is, more ligands will add (because of the release of the additional bond energy) until the repulsions of the existing ligands for another potential ligand that tried to shoulder its way in outweigh the possible bond energy that might be released. For most metal ions and ligands, the maximum number is six and the geometry is octahedral. There are exceptions, as for instance ZrF_7^{3-} and TaF_7^{2-}, or $Mo(CN)_8^{4-}$, but there are only a few of these and they always form with very small ligands such as F^- or CN^-. The vast majority of transition-metal structures, whether molecular complexes or simply the metal's surroundings in a crystal, show octahedral 6-coordination. In a few cases 5-coordinate species form, particularly with Cu^{2+}, which has nine valence electrons and with six two-electron-donor ligands would have 21 electrons in its valence shell—rather more than the 18 that could reasonably be accommodated in bonding orbitals. With a coordination number of five, however, the number is only 19, which is a reasonable compromise. Five-coordinate species are also important as reaction intermediates, as we shall see in Section 16-7.

For very large, bulky ligands the coordination number will be four, since the ligand–ligand repulsion or **steric hindrance** will be too great for 6-coordination. Usually these systems have tetrahedral geometry, which minimizes the repulsions. We also expect this pattern for Cu^I, Ag^I, and Au^I, which have 10 valence electrons and could accept only four two-electron donors within the 18-electron limit. Square planar compounds (with a coordination number of four) are common for d^8 ions like Ni^{II}, Pd^{II}, Pt^{II}, and Au^{III}. If we look back at those MO's (Fig. 16-15), we can see that this particular geometry is not adopted to minimize ligand–ligand repulsion, but to get the most favorable electronic energy in the necessary MO distribution. Table 16-9 gives the stoichiometries and geometries of some transition-metal complexes, for comparison with this discussion. Many of the deviations can be rationalized by considering electrostatic attractions or repulsions for additional ligands.

16-5 Multiple Oxidation States and Nonstoichiometric Compounds

Multiple Oxidation States in Molecules

In complexes such as those in Table 16-9, or in any molecular system, transition metals normally have a perfectly well-defined oxidation state. For instance, the species $Fe(CN)_6^{3-}$ has exactly one iron, exactly six cyanides, and a negative charge of exactly three electronic charges. Since the cyanide ion is 1−, we say confidently that the formal oxidation state of the iron is 3+. However, it is frequently true for transition-metal compounds that electrons can be added or removed without changing the molecular structure; in this complex, if we take the iron to be Fe^{3+} it will have five valence electrons. Adding 12 electrons from

Table 16-9
Predicted and Experimental Geometries of Transition-Metal Compounds

---

Predicted octahedral: metal with six or fewer electrons, two-electron donors

TiF_6^{2-}	0 metal electrons	octahedral
$V(OH_2)_6^{3+}$	2 metal electrons	octahedral
$Cr(NH_3)_6^{2+}$	4 metal electrons	octahedral (distorted)
$Cr(CO)_6$	6 metal electrons	octahedral
$Fe(CN)_6^{4-}$	6 metal electrons	octahedral
$FeBr_4^{2-}$	6 metal electrons	tetrahedral

Predicted 5-coordinate: metal with nine electrons, two-electron donors

$CuCl_5^{3-}$ — 9 metal electrons — trigonal bipyramidal

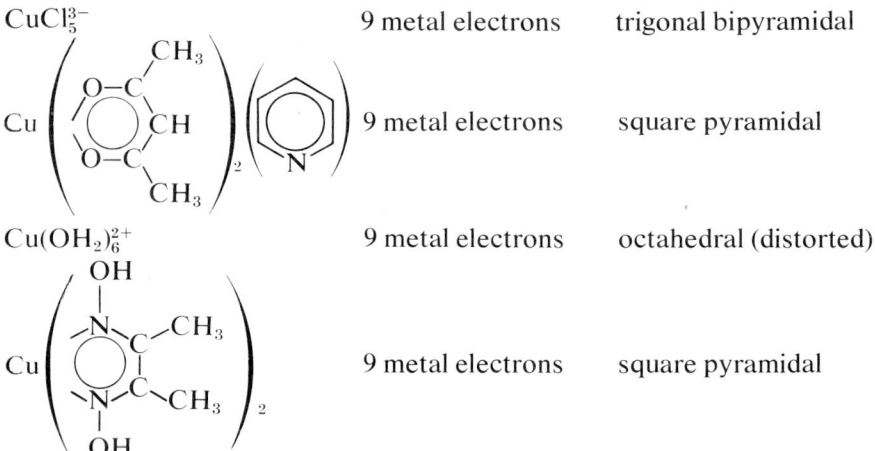

$Cu(\text{acac})_2(\text{py})$ — 9 metal electrons — square pyramidal

$Cu(OH_2)_6^{2+}$ — 9 metal electrons — octahedral (distorted)

$Cu(\text{dmg})_2$ — 9 metal electrons — square pyramidal

Predicted square planar: metal with eight electrons, two-electron donors

$AuBr_4^-$	8 metal electrons	square planar
$Ni(CN)_4^{2-}$	8 metal electrons	square planar
$Fe(CO)_5$	8 metal electrons	trigonal bipyramidal
$PtCl_4^{2-}$	8 metal electrons	square planar
$Ni(OH_2)_6^{2+}$	8 metal electrons	octahedral (distorted)

Predicted tetrahedral: metal with 10 electrons (eight or nine with bulky ligands)

$CuCl_4^{2-}$	9 metal electrons	tetrahedral (distorted)
$Ni(CO)_4$	10 metal electrons	tetrahedral
$NiCl_2(P(C_6H_5)_3)_2$	8 metal electrons	tetrahedral
$CuCl(OP(C_6H_5)_3)_2$	9 metal electrons	tetrahedral
$Cu(CN)_4^{3-}$	10 metal electrons	tetrahedral
$Cu(NH_3)_2^+$	10 metal electrons	linear
$AuCl(NH_3)_3$	10 metal electrons	tetrahedral (?)

six two-electron donors gives a total of 17, and it ought to be possible to add another electron to give the approximate limit of 18 and the species $Fe(CN)_6^{4-}$ — which does exist. But the formal oxidation state of iron in this new complex is 2+, and so we have to deal with the possibility that Fe will display more than one oxidation state. This is true for all transition metals, and the ones that, like Mn, have several d electrons but not at such low energy as to make them essentially part of the core (as they are for Ni and Cu) often display an amazing variety of oxidation states, both in molecular complexes and in extended crystals. To use Mn as an example, we have MnO_4^-, MnO_4^{2-}, $Mn(CN)_6^{2-}$, $Mn(CN)_6^{3-}$, $Mn(CN)_6^{4-}$, $Mn(CN)_6^{5-}$, $Mn_2(CO)_{10}$, and $Mn(CO)_5^-$. These species have Mn in formal oxidation states of 7+, 6+, 4+, 3+, 2+, 1+, 0, and 1−, respectively. Note that the high oxidation states, which render the metal atom an extremely "hard" acid (electron acceptor) require the O^{2-} electron donor, which is a very "hard" base. Conversely, the very low oxidation states that correspond to the metal atom as a "soft" acid form compounds only with CN^- or CO, which are quite "soft" bases. The string of cyanide complexes is unusual in having exactly the same stoichiometry with four different formal oxidation states of the central metal atom; they are, in general, not made from each other but are made by reducing a high formal oxidation state of manganese such as MnO_4^- in the presence of cyanide ion with reducing agents (electron donors) of different strengths.

Multiple Oxidation States in Crystals

In crystals with an extended polymeric structure, such as ionic or partly ionic crystals, transition metals also exhibit multiple oxidation states, sometimes in a very nonstoichiometric way. To understand the reasons for this we need to look at the energy relationships that govern the oxidation state of a metal ion in a crystal. Suppose we have a stoichiometric crystal of a metal chloride, MCl; what energy changes are involved in changing an individual metal ion in the lattice from 1+ to 2+? An easy way to classify them is to set up a Born–Haber cycle for the disproportionation reaction:

$$2M_{(g)}^+ + 2Cl_{(g)}^- \xrightarrow[+IP_2]{-IP_1} M_{(g)}^{2+} + 2Cl_{(g)}^- + M_{(g)}^0$$

$$-2U_1 \uparrow \qquad \qquad \downarrow U_2 \qquad \swarrow -\Delta H_{sub}$$

$$2MCl_{(s)} \xrightarrow{\Delta H} MCl_{2(s)} + M_{(s)}^0$$

ΔH for this reaction will probably be pretty close to ΔG at room temperature, at least, since no gas is evolved or absorbed, so it will be an approximate guide to the spontaneity of the oxidation-state change. Using the cycle, we get

$$\Delta H = (U_2 - 2U_1) + (IP_2 - IP_1) - \Delta H_{sub}$$

How does this work out numerically for a metal ion that is always 1+ (such as Na), one that is always 2+ (such as Ca), and for a transition metal that can be either, such as Cu?

For NaCl and "NaCl$_2$" we can take the lattice energies as -184 kcal/mole and -596 kcal/mole, respectively, the latter being approximately correct for MgCl$_2$. This, together with the ionization energies and heat of sublimation, gives

$$\text{Na} \quad \Delta H = (-596 + 368) + \underbrace{(1091 - 119)}_{972} - 26$$

$$= +718 \text{ kcal/mole}$$

This is very positive and thus very unfavorable thermodynamically, which is no surprise. Note that the principal positive term is the difference between ionization energies. For "CaCl" and CaCl$_2$ we can take the lattice energies as -167 kcal/mole and -522 kcal/mole, respectively, the former being a value for KCl. These give

$$\text{Ca} \quad \Delta H = (-522 + 334) + \underbrace{(274 - 141)}_{133} - 46$$

$$= -99 \text{ kcal/mole}$$

While not as dramatic in size, this value is quite negative and suggests that the electron transfer from Ca$^+$ to Ca^{2+} should be favorable and completely spontaneous; the ionization-energy difference in this case is relatively small, but the lattice energies do not differ too much from the sodium case. For the copper case, the ionic lattice energies (although the compounds cannot really be fully ionic) are -662 kcal/mole for CuCl$_2$ and -236 kcal/mole for CuCl. These and the appropriate other values give an interesting result:

$$\text{Cu} \quad \Delta H = (-662 + 472) + \underbrace{(468 - 178)}_{290} - 81$$

$$= 19 \text{ kcal/mole}$$

This very small value suggests that, because of the intermediate nature of the ionization-energy difference in this case, the lattice-energy increase for the more highly charged metal ion just about balances the energy cost of the ionization. The near-zero ΔH (and probably ΔG) suggests that changes in oxidation state could take place fairly readily *in either direction* for copper in crystals of its partly ionic compounds. This is generally true of transition metals. We can formulate a very rough rule of thumb: if the *difference* between two successive ionization potentials is in the vicinity of 300–350 kcal/mole, both of the corresponding oxidation states of that element will be fairly stable in crystals, perhaps even both in the same crystal in a nonstoichiometric fashion. If the difference is less than about 300 kcal/mole, the higher oxidation state will be much more stable than the lower one; and if it is greater than about 350 kcal/mole the lower oxidation state will be much more stable than the higher one.

The interesting thing about transition metals is that, because of the partly shielded nature of d orbitals, the differences in their successive ionization energies tend to fall in this 300–350 kcal/mole pattern. Thus Ti^{II} and Ti^{III} are separated by 335 kcal/mole difference in ionization energies, Ti^{III} and Ti^{IV} by 348 kcal/mole, Fe^{II} and Fe^{III} by 334 kcal/mole, but Sc^{II} and Sc^{III} by only 274 kcal/mole and Zn^{II} and Zn^{III} by 501 kcal/mole.

Disproportionation

One form this multiplicity of oxidation states takes is a readiness to undergo disproportionation reactions similar to the one for which we wrote the cycle. For instance, VCl_3, a partly ionic solid, disproportionates when heated to 800°,

$$2VCl_3 \xrightarrow{heat} VCl_2 + VCl_4$$

even though it is made from VCl_4 at lower temperatures:

$$2VCl_4 \rightarrow 2VCl_3 + Cl_2$$

Copper(II) oxide or halide crystals dissociate at high temperatures,

$$2CuO \xrightarrow{heat} Cu_2O + \tfrac{1}{2}O_2$$

but the Cu^I ion disproportionates at room temperature in solution:

$$2Cu^+_{(aq)} \rightarrow Cu^{2+}_{(aq)} + Cu^0_{(s)}$$

These reactions have a fairly reliable stoichiometry, but frequently partly ionic crystals of transition-metal compounds are nonstoichiometric over a rather wide range just because the balance between lattice energy and ionization energy is maintained. Co_3O_4, a mixed oxide of Co^{II} and Co^{III} that is a spinel (see p. 744), will absorb oxygen gas into its structure without changing its basic lattice over a continuous composition range all the way up to Co_2O_3; the nonstoichiometric intermediate compositions are stable because of this energy balance. This is why nonstoichiometric compounds are more common among transition metals than other metals, in analogous compounds.

16-6 Ionic and Partly Ionic Transition-Metal Compounds

At the beginning of the chapter it was suggested that as a very rough guide we could assume that the chemistry of transition metals was about like that of aluminum, which is electropositive but not extremely so. If we try to place an average transition metal with an electronegativity of 1.5 into our pattern in which an electronegativity difference of 2.0 units was required to produce an ionic compound, we see that only fluorine and possibly oxygen could form ionic transition-metal salts. As a matter of fact, the transition metals with the lowest electronegativities, such as Ti, also have rather high characteristic formal oxidation states—which, as we saw in Chapter 15, tends to reduce charge separation. So generally speaking, only the fluorides of transition metals tend to be ionic. The other halides, however, as well as the oxides and sulfides, are presumably partly ionic, which leads to chemical properties generally similar to those of ionic compounds. One difference is that partly ionic compounds can frequently be sublimed, since their framework of directional bonds (which is not present in ionic compounds) makes it almost as easy to vaporize molecules as to distort the crystal structure into the disorderly liquid phase. Most transition-metal halides can be sublimed, which is sometimes useful in purification, but the fluorides cannot, at least not the ones in relatively low formal oxidation states (2+ or 3+).

The Effect of Oxidation State on Physical and Chemical Properties

This raises the whole question of exactly what the effect of changing formal oxidation state is on the properties of otherwise similar transition-metal compounds. It is a particularly pertinent question for transition metals since they show so many different oxidation states. A good example of the effect is the set of compounds $TiCl_2$, $TiCl_3$, and $TiCl_4$, which are the only chlorides of Ti. Titanium(II) chloride is a partly ionic crystalline compound with a layer structure in which each Ti is surrounded by six Cl's (see Fig. 14-4). The Ti atoms are distributed very regularly in the layers of Cl atoms, but the layers themselves are held together only by van der Waals forces, which assures us that the compound cannot be fully ionic. Titanium(II) chloride exists only in the crystalline state; it disproportionates if heated to the temperature at which it begins to sublime:

$$2TiCl_2 \xrightarrow{475°} TiCl_4 + Ti^0$$

The readiness of the compound to undergo this reaction can be understood in terms of the ionization-energy relationships mentioned in Section 16-5. Titanium(III) chloride might be expected to be less ionic than $TiCl_2$ due to its higher oxidation state of Ti, but the difference between the compounds is rather subtle. Titanium(III) chloride also has a layer structure (the $CrCl_3$ layer structure in Fig. 14-4), but is somewhat more stable as a gaseous molecule in subliming, although it also disproportionates at a higher temperature:

$$2TiCl_3 \xrightarrow[\substack{\text{disproportionate}\\600°}]{\text{sublime }430°} TiCl_2 + TiCl_4$$

Titanium(IV) chloride on the other hand, is distinctly different. It forms a tetrahedral molecule within a molecular crystal that melts at $-30°C$ and boils without decomposition at $136°C$. The increasing number of ligand atoms in the neutral molecule has decreased the net charge on each until the attraction to another positively charged Ti is negligible or at least small compared to kT at room temperature. The resulting species must be classed as a covalent compound, since it does not have an extended polymeric structure.

Preparation of Partly Ionic Compounds

The halides, oxides, and sulfides of transition metals in reasonably low oxidation states—up to about 3+—are partly ionic; in high oxidation states they become covalent. It is often possible to make compounds involving oxyanions such as SO_4^{2-}, CO_3^{2-}, and NO_3^-, particularly if the metal in its given oxidation state is not too strong a reducing agent. These, as we noted in Chapter 13, are chemically and physically rather like the oxide, with extended polymeric structures and directional bonding. All of these partly ionic compounds have similar chemical properties and similar preparations. Usually they must be prepared at fairly high temperatures, either by direct combination of the elements or by reaction of an element and carbon or hydrogen with the oxide:

$$2Cr_{(s)} + 3Cl_{2(g)} \xrightarrow{high\ T} 2CrCl_{3(g)}$$

$$4Co + S_8 \xrightarrow{high\ T} 4CoS_2 \quad (Co^{2+}, S_2^{2-})$$

$$2Fe_2O_3 + 6Cl_2 + 3C \xrightarrow{high\ T} 4FeCl_3 + 3CO_2$$

Many of the compounds having the metal in the 2+ oxidation state are strong reducing agents, since the electrons in the d orbital distributions are not bound strongly if the positive charge is low. Preparing these compounds frequently involves the use of the HX compound to make the metal—X compound, because the hydrogen that is formed is itself a good reducing agent and prevents further oxidation or provides the reduction:

$$VO + H_2S \xrightarrow{700°} VS + H_2O$$

$$Cr + 2HF \longrightarrow CrF_2 + H_2$$

$$W + \tfrac{5}{2}Br_2 \xrightarrow{700°} WBr_5 \xrightarrow[400°]{H_2} WBr_2$$

Precautions against oxidation are necessary only with the metals toward the left of the transition-metal group in the periodic table; with the others, the d electrons are more tightly held and the metal in the 2+ state is not such a good reducing agent.

Electron-Acceptor Reactions

All of these partly ionic transition-metal compounds have a chemical characteristic that almost completely dominates their reactions: they are excellent electron acceptors. This is electrostatically favorable because of the partial positive charge on the metal atoms in these compounds, but is fundamentally possible because of the metal atom's capacity to accommodate 18 electrons in bonding MO's. We have mentioned compounds of these metals that have 18 valence electrons around the metal atom (such as the carbonyls), but these are somewhat unusual in having ligands that contribute many electrons without taking up much space around the metal atom. In compounds like $TiCl_3$, however, steric hindrance limits the number of ligands to six in the crystal structure, and since Ti^{III} has only one remaining valence electron, the six Cl ligands cannot come close to providing the metal with a total of 18 valence electrons. So the compound $TiCl_3$ is still a good electron acceptor. The reactions it undergoes with donors do not usually increase the number of electrons around the metal, because that number is essentially fixed by the steric limit on the possible number of ligands. However, the electron deficiency provides a convenient mechanism for the approach of a donor, which is a kinetic effect. The favorable thermodynamic effects are found in the lowering of energy of the donor's nonbonding electrons when they form a bond with the metal, as we saw in Section 16-3. The chlorides that are displaced (in the case of $TiCl_3$) do not gain in energy too much because they are stabilized by electrostatic attraction in a lattice or by solvation in a polar solvent. So the reactions of these partly ionic compounds frequently involve the acceptance of nonbonding electrons from a ligand atom or molecule, with the coordination number of the reaction product being limited for geometric reasons to six. Toward the right of the group of transition metals the elements have more valence electrons, and it becomes possible for six two-electron-donating ligands to meet or exceed the 18-electron limit. This is particularly true for nickel(II) and copper(II), and their donor-acceptor compounds often show coordination numbers of five or four. There does not seem to be any doubt that more ligands would fit geometrically, as Table 16-9 indicates, but there are no longer any low-energy acceptor orbitals to accommodate their electrons. In the next section we shall look at some of these reactions in considering the formation of transition-metal complexes.

16-7 Covalent Transition-Metal Compounds and Coordination Compounds

As we observed in the last chapter, compounds can be covalent for either of two reasons: compounds between two elements that have nearly the same electronegativity will be covalent, and so will compounds in which electronegativities differ but the more electropositive atom is in a highly positive formal oxidation state, with many ligand atoms withdrawing electrons from it. Transition metals are involved in both kinds of covalent compounds, and coordination compounds with electron donors form a particularly important class of the second kind, even if the metal's formal oxidation state is not very high. Let us start, however, with the more obvious group—those with similar electronegativities.

Intermetallic Covalent Compounds and Alloys

The most obvious kind of transition-metal compound in which the electronegativities are similar is an intermetallic compound. Transition metals can form compounds with each other and with other metals, but the formation of clear-cut stoichiometric compounds depends on the relative sizes of the metals (which usually must differ noticeably), the relative electronegativities, and the relative number of valence electrons per atom. Remember that this last factor governs the filling of the bands of metallic orbitals that we discussed in Section 6-12. Many compounds of this sort exist, usually with obscure stoichiometries and no particular application, but two interesting cases are worth our notice. The first is the pattern of combination that results when the sizes and electronegativities are close together, which usually results in a solid solution rather than a compound. Figure 7-52 showed the phase diagram for such a combination, the continuous series of solid solutions between Cu and Ni, and as in all solutions the composition of such a mixture is variable over wide limits. Such combinations find countless applications in our society as **alloys**.

Metals are important because of their mechanical, electrical, and chemical properties. We use them because of their strength and malleability, because they conduct electricity well, and because they are resistant to many kinds of chemical attack. In different applications, different combinations of these properties are important. The different metallic elements have different mechanical, electrical, and chemical properties, and the formation of solid solutions between them gives us a means of adjusting the combination of properties for a particular purpose — hence the importance of alloys.

A very common alloy is **brass**, which is a Cu/Zn solid solution. Brass compositions usually run from 10 to 40% Zn, the rest Cu. Varying the composition changes the appearance of the metal (since Cu is colored) and the melting point; brazing rod is about 50–50 Cu/Zn. The addition of zinc also makes the resulting alloy much easier to work using machine tools than copper itself would be. The resulting ease of fabrication, together with its attractive appearance, make brass common in a host of uses. Other commonly encountered Cu alloys include **bronze**, which is 5–15% Sn (sometimes with other elements in small amounts) and **nickel coinage**, which is 75% Cu and only 25% Ni.

The most commonly used alloy is one principally between iron and a nonmetal, carbon: **steel**. Because of the fairly substantial size and electronegativity differences between Fe and C, it is possible to form an iron carbide compound, but there is a limited range of mutual solubility from about 0–5% C, and this is the composition range of ordinary steels. Iron ores are principally oxides, which in the process of steelmaking are reduced to elemental iron using coke (elemental carbon). The resulting pig iron, by virtue of the limited solubility of carbon, is saturated with carbon; it is very hard, which is desirable for some applications, but it is also very brittle. The carbon is removed from the molten metal by bubbling O_2 through it (forming CO), and the properties change as the composition changes. Pure iron is quite malleable, but only fairly hard, so some carbon is deliberately left in it to increase hardness. Mild steel has about 0.2% C, high-carbon steel about 0.8%, and tool steel about 1.2%. In addition, the toughness or resistance to wear of the steel can be improved by adding various quantities of V, Cr, Mo, W, or Mn, usually in amounts less than 5%. If large amounts of chromium (10–20%) and nickel (5–10%) are added to the iron, the resulting alloy is moderately hard but very tough, and also resistant to corrosion by ordinary weathering processes, which gives it the name stainless steel. The corrosion resistance arises from the fact that the oxides of chromium and nickel have crystal lattices matching that of the metal at the surface well enough to be strongly bonded to it, so that only a thin film of oxide forms, which is colorless and does not change the appearance of the metal. By contrast, ordinary iron oxides do not match the lattice of iron metal and in growing, strain the metal–oxygen bonds enough eventually to fall away from the surface as flakes of rust. Recently U.S. Steel has introduced a structural steel called Cor-Ten which, by virtue of its composition, also grows a closely adhering oxide, although the oxide is a dark brown color. The effect again is to render the steel corrosion resistant, and the product is finding increasing use in contemporary architecture.

Metal–Metal Bonding in Conventional Compounds

The other particularly interesting case of intermetallic compounds that we can examine is the formation of metal–metal bonds, sometimes even clusters of such bonds, in compounds with nonmetallic elements. It appears that most transition metals can form bonds to themselves under appropriate circumstances (and, of course, they are bonded to themselves in the bulk metal). A few such compounds have been known for a long time; many others have been prepared recently, and the investigation of the circumstances under which metal–metal bonds are stable is an active area of research. One of the more or less classic examples of metal–metal bonding is a compound whose empirical formula is $MoCl_2$, made by passing gaseous Cl_2 over hot Mo metal. This in itself is a little unusual, since Mo^{II} ought to be very much like Cr^{II}, and we have already seen that Cr^{II} is such a strong reducing agent that chromium plus chlorine gives $CrCl_3$—HCl instead of Cl_2 must be used to make $CrCl_2$. The "$MoCl_2$" compound dissolves in aqueous KOH, and from the resulting solution exactly one-third of the chloride can be precipitated by adding Ag^+. The appearance of this interesting fraction for a compound whose empirical formula indicates only two chlorides suggests immediately that more than one $MoCl_2$ unit must be involved in the structure. Specifically, the formula must be Mo_3Cl_6, Mo_6Cl_{12}, or some higher multiple of three $MoCl_2$ units. The compound is insoluble in water, but from colligative-property measurements on solutions in organic solvents the formula can be established as Mo_6Cl_{12}. Since only one-third of the 12 chlorines are precipitated by Ag^+, the other eight must be bound to the Mo atoms in a particularly stable manner. This is verified by several reactions of the compound in which the $Mo_6Cl_8^{4+}$ unit is preserved and also by X-ray diffraction data, which indicate the structure shown in Fig. 16-20. Since each Mo atom is in a formal oxidation state of 2+ it has four valence electrons, so the Mo framework must accommodate 24 electrons. Several theoretical models yield 12 bonding orbitals, and the theoretical electron structure of this and similar structures is still a matter of some interest. Figure 16-20 also shows several other molecular structures in which metal–metal bonding has been proposed; the diversity of the structures offers an open field for theoretical modeling.

Organometallics

Organometallic compounds, which we discussed for nontransition metals in Chapter 15, are another type of covalent compound that transition metals can form. Some of them have what we can by this time call conventional σ bonds from the metal atom to a C atom; these are closely analogous to the nontransi-

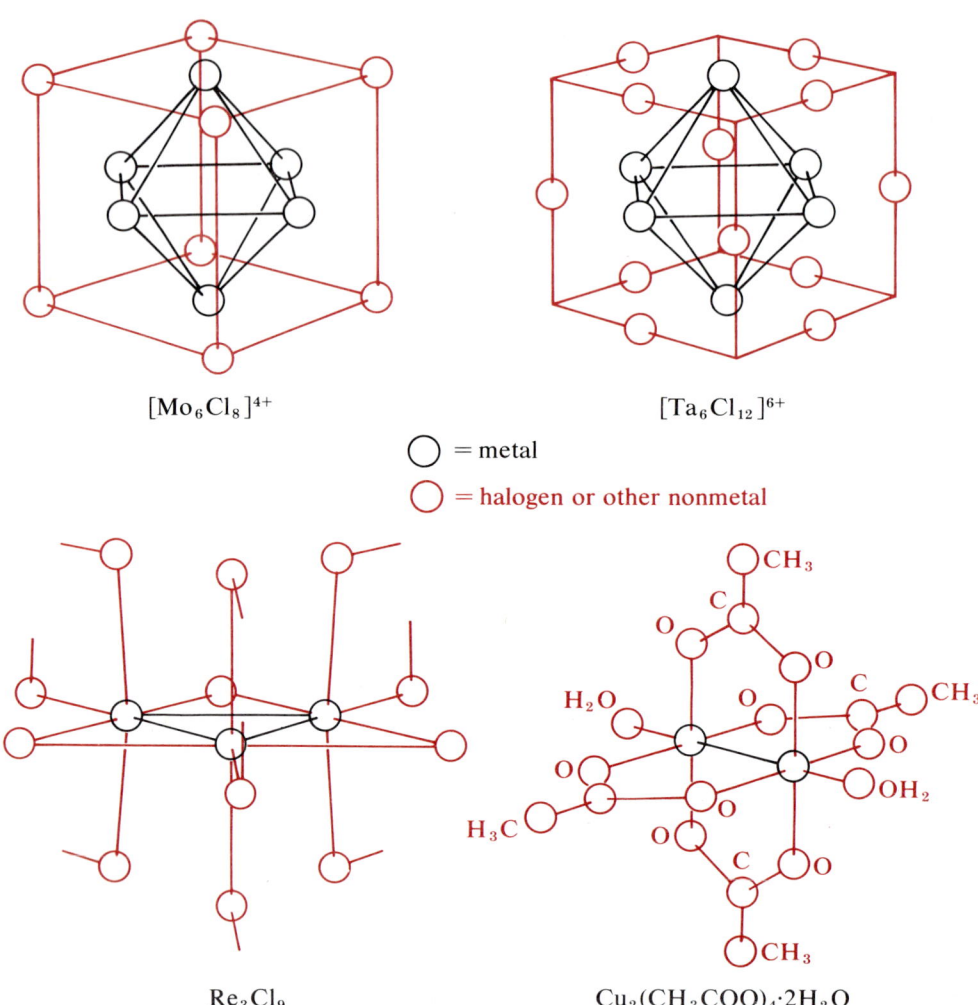

Figure 16-20 Structures of some transition-metal compounds that show metal-atom clusters.

tion metals' compounds in their physical and chemical properties. There are only a few in which no other kind of bonding is present, but many in which a σ bond is formed by a metal that is also bonded to CO groups or organic π donors. Generally they are made by using a metal halide and an organolithium or organosodium compound. The favorable free-energy change for the reaction comes from the large ionic lattice energy of the alkali metal halide that is also formed:

$$TiCl_4 + 4LiCH_3 \rightarrow Ti(CH_3)_4 + 4LiCl$$

or
$$NaMn(CO)_5 + CH_3I \rightarrow CH_3\text{—}Mn(CO)_5 + NaI$$

Another possibility for the formation of transition-metal σ organometallics is through an insertion reaction of a π-bonded hydrocarbon with a metal–hydrogen bond, rather like the catalytic reaction for polyethylene:

$$H\text{—}Co(CO)_4 + CH_2\text{=}CH_2 \rightarrow H\text{—}CH_2\text{—}CH_2\text{—}Co(CO)_4$$

Carbonyls

The frequent use of metal carbonyls in these sample reactions raises the question of how they are made. A few carbonyls can be made simply by treating the finely powdered metal with CO gas under pressure:

$$Ni^0 + 4CO \rightleftharpoons Ni(CO)_4$$
$$Fe^0 + 5CO \rightarrow Fe(CO)_5$$

More often they are made from halides, with some reducing agent such as sodium metal or copper metal added to reduce the metal and take on the halide ions:

$$VCl_3 + 6CO + 3Na \xrightarrow[\text{organic solvent}]{\text{high pressure}} V(CO)_6 + 3NaCl$$

$$2CoI_2 + 8CO + 4Cu \longrightarrow Co_2(CO)_8 + 4CuI$$

Compounds such as dicobalt octacarbonyl are said to be **binuclear**, meaning that ligands surround two transition-metal atoms in a single molecule. They undergo some interesting reactions that involve cleavage of the two metal atoms to form new bonds with one-electron donors, still satisfying the 18-electron rule:

$$Co_2(CO)_8 + 2Na^0 \rightarrow 2NaCo(CO)_4$$
$$Mn_2(CO)_{10} + 2Li^0 \rightarrow 2LiMn(CO)_5$$
$$Mn_2(CO)_{10} + H_2 \xrightarrow{\text{high } P} 2HMn(CO)_5$$
$$Mn_2(CO)_{10} + Br_2 \rightarrow 2BrMn(CO)_5$$

These compounds can in turn be allowed to react, often in ways that take advantage of the lattice energy of an alkali metal halide:

$$NaCo(CO)_4 + BrMn(CO)_5 \rightarrow (OC)_4Co\text{—}Mn(CO)_5 + NaBr$$
$$NaMn(CO)_5 + Cl\text{—}SCN \rightarrow (OC)_5Mn\text{—}SCN + NaCl$$

Pi-Donor Complexes

The chemistry of metal carbonyls is extensive and diverse, and we can only hint at its scope here. One last type of reaction is interesting, however, since it leads to the π complexes we mentioned earlier:

$$Fe(CO)_5 + \underset{\text{butadiene}}{CH_2=CH-CH=CH_2} \longrightarrow (OC)_3Fe(\eta^4\text{-}C_4H_6) + 2CO$$

$$Cr(CO)_6 + C_6H_5CH_3 \longrightarrow (OC)_3Cr(\eta^6\text{-}C_6H_5CH_3) + 3CO$$

Sometimes these can be formed from halides:

$$PtCl_4^{2-} + \underset{\text{ethylene}}{CH_2=CH_2} \longrightarrow [Cl_3Pt \leftarrow \| \begin{smallmatrix} CH_2 \\ CH_2 \end{smallmatrix}] + Cl^-$$

Occasionally an organometallic compound can show both σ and π bonding; a rather striking example of this is a tetrameric gold compound:

$$(CH_3)_3C-C\equiv C-Au \leftarrow \| \begin{smallmatrix} C \\ C \end{smallmatrix} - C(CH_3)_3$$
(structure with four Au atoms bridged by σ and π bonds to tert-butylacetylide ligands)

The student is forgiven for thinking that this represents a form of circular reasoning. As we have suggested earlier, these compounds are of considerable practical importance in the production of many plastics; the problem of rationalizing their electronic structures also makes them interesting from a theoretical point of view.

High-Oxidation-State Covalent Compounds

Transition metals also form covalent compounds by virtue of being surrounded by many ligands when they are in a high positive formal oxidation state. Some examples of such compounds are Mn_2O_7, which instead of being a refractory oxide is an oily liquid at room temperature (and a violent explosive); WCl_6, which is solid but dissolves readily in organic solvents such as CCl_4; and WF_6, which is actually a gas at room temperature. These compounds are excellent electron acceptors—Lewis acids—because of their high overall electronegativity. This means that they hydrolyze vigorously in water; Mn_2O_7 is an extremely acidic oxide. So are CrO_3, MoO_3, and WO_3, although the last two do not dissolve appreciably in water. They dissolve in base, however, to form the polymeric molybdates and tungstates that we discussed in Chapter 14. Properties that are at least roughly analogous are found for most transition-metal compounds with formal oxidation states of 5+ or more for the metal. This apparently limits the range of the discussion to elements to the left of Fe in the first row, Pd in the second, and Au in the third, because the increasing stability of the d electrons going to the right in a given row prevents the elements at the far right from adopting high formal oxidation states. However, all of the transition metals are good electron acceptors even in relatively low oxidation states. They will accept electrons from at least four and almost always six donor ligands, regardless of their formal oxidation state. In these compounds the bonding remains largely covalent, again almost regardless of the metal's oxidation state.

Bonding in Coordination Compounds

The difference between the bonding in ML_6^0 and ML_6^{4-}, say, where L is a negative ion L^-, comes down to this comparison: in ML_6^0 the metal is in quite a high oxidation state (6+), which causes the species to display largely covalent bonding; in ML_6^{4-} the additional four electrons are distributed according to MO's that are delocalized over the whole molecule so that they are still shared and the bonding is still covalent. The degree of covalent bonding, in other words, depends mostly on the number of ligands and not very strongly on the oxidation state of the metal. With this in mind, we can look at the whole array of **donor–acceptor complexes** or **coordination compounds** formed by transition metals, considering them to be properly described by MO's for which the charge separation is small.

If a transition metal can serve as an electron acceptor in forming one of these complexes, presumably any atom (whether alone or in a molecule) that has unpaired electrons can serve as a donor. There are some limits to this, because the rare gases have not been observed to serve as donors, and they certainly have nonbonding electrons; but every other nonmetal has been observed to serve as a donor to a transition metal, with varying degrees of stability. Many halide complexes have been prepared, from FeF_6^{3-} to CuI_4^{2-}, with the "softest" halide, I^-, forming its most stable complexes with the "soft" acceptor metals that have numerous d electrons and a low positive charge or formal oxidation state. These are theoretically interesting because they are relatively simple model compounds for which to calculate chemical or physical properties. Far more important, both in numbers and practical importance, are complexes in which the donor is an O or N atom. The most familiar of these are the hydrates, which we have frequently written as, for example, $Fe(OH_2)_6^{3+}$, reversing the normal order of atoms in H_2O to indicate that the O serves as the bonding atom to the metal ion. Many other oxydonors (molecules that donate electrons through an O atom) have been investigated, as Fig. 16-21 indicates. The classic molecule with a N donor atom is the analogous NH_3 molecule: $Cr(NH_3)_6^{3+}$, $Co(NH_3)_6^{3+}$, $Cu(NH_3)_4^{2+}$, $Ag(NH_3)_2^{+}$, and many others. As Fig. 16-21 shows, organic amines are also frequently used as donors, and even polyamines are frequently used because of their great stability (with many donor atoms in the molecule). A glance back at Fig. 10-1 will show that EDTA is a derivative of a polyamine, ethylenediamine, and has a total of six donor atoms. Before looking at the synthesis and reactions of these coordination compounds, let us pause to establish the system of nomenclature for these compounds.

Coordination-Compound Nomenclature

As formulas already written will suggest, the formula of a coordination compound is written with the central atom indicated first and the ligands written after it, unless there is some reason to write the formula in a manner that corresponds more closely to the structure: $Fe(CN)_6^{3-}$, but

$$(CO)_5MnNa + I-CH_3 \rightarrow (CO)_5MnCH_3 + NaI$$

On the other hand, in names the ligands should be designated first and the central atom mentioned last. If the complex is positively charged or neutral it is simply named as a derivative of the metal (or other central atom), but if it is negatively charged the suffix -*ate* is added:

$Cr(OH_2)_6^{3+}$ hexaaquochromium(III)

$[Co(NH_3)_3Cl_3]$ trichlorotriamminecobalt(III)

$CuBr_4^{2-}$ tetrabromocuprate(II)

These names suggest several other features of the nomenclature. The same set of Greek prefixes (tri-, tetra-, hexa-, in these examples) is used to indicate the number of each kind of ligand as is used in the ordinary inorganic nomenclature system. Water as a ligand is called **aquo**, and ammonia as a ligand is called **ammine** (two m's). Negatively-charged ligands are named first if there are several kinds, and are named by dropping the final e and adding o — thus sulfate

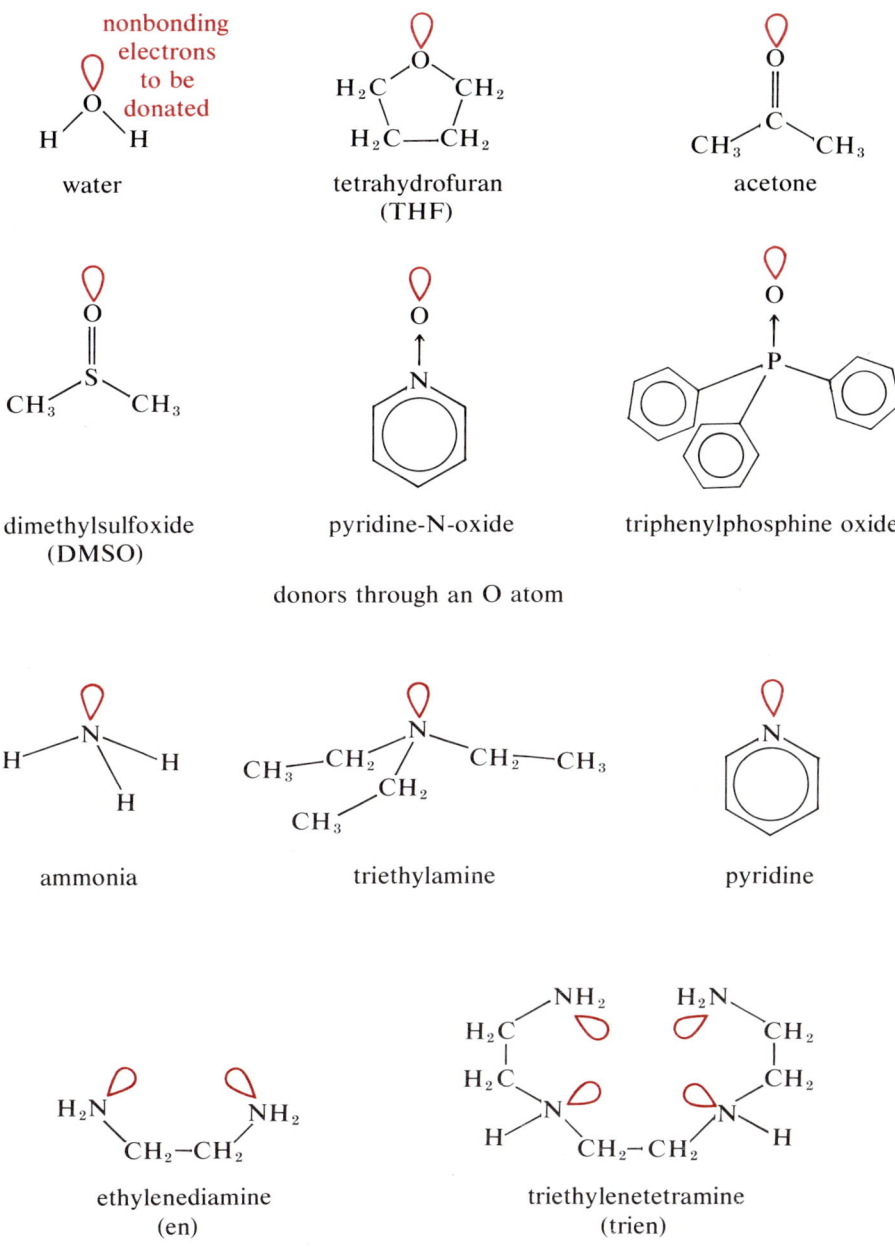

Figure 16-21 Typical donor molecules in transition-metal complexes.

as a ligand becomes sulfato, for instance. Unfortunately for the simplicity of this rule, common usage has dictated the following exceptions:

Fe^-	fluoride (as an ion)	fluoro- (as a ligand)
Cl^-	chloride	chloro-
Br^-	bromide	bromo-
I^-	iodide	iodo-
CN^-	cyanide	cyano-
OH^-	hydroxide	hydroxo-
O^{2-}	oxide	oxo-
S^{2-}	sulfide	thio-

Note that most of these just drop the -ide for -o. The use of the suffix -o applies only to negatively charged ligands, not to neutral ones. The formal oxidation state of the central atom is indicated by Roman numerals in parentheses immediately following the name of the central atom. And note that if any doubt could arise as to what species are true ligands (i.e., more or less covalently bound) the formula of the whole complex should be enclosed in square brackets. When the species CO (carbon monoxide) and NO (nitrogen oxide or nitric oxide) serve as ligands, they are considered to be electrically neutral for oxidation-state purposes and are called carbonyl and nitrosyl, respectively:

$Fe(CO)_2(C_5H_5)I$ iododicarbonylcyclopentadienyliron(I)

$Mn(CO)_4NO$ nitrosyltetracarbonylmanganese

The nomenclature must also include some means of distinguishing between different structural isomers of complexes with the same ligands that are arranged in different geometries. For example, there are two different ways to arrange the ligands of the octahedral dichlorotetramminecobalt(III) ion:

cis   or   trans

In one of these the chlorides are next to each other, while in the other they are opposite each other; because of the great symmetry of the octahedral geometry there are no other possibilities. Borrowing a term from organic nomenclature for the same sort of structural isomers, we call the first type, with neighboring chlorides, *cis*, and the other *trans*. For square planar complexes the same sort of isomerism is possible:

$$\begin{array}{ccc} Cl & NH_3 & Cl & Cl \\ & Pt & \text{or} & Pt \\ H_3N & Cl & H_3N & NH_3 \\ \text{trans} & & \text{cis} \end{array}$$

However, cis and trans isomers cannot occur for tetrahedral complexes, since each of the four tetrahedral positions is equally a neighbor to the other three. Other kinds of isomerism exist, but this discussion is adequate for our purposes.

Preparation of Coordination Compounds

When the ligand is a molecule of a normally liquid substance, a convenient means of preparing the complex is often simply to dissolve a partly ionic or ionic compound of the metal in the liquid ligand. Hydrates, ammines, and other types of complexes are often made this way:

$$FeCl_{3(s)} + 6H_2O_{(l)} \xrightarrow{\text{evaporate}} [Fe(OH_2)_6]^{3+} \; 3Cl^-_{(s)}$$

$$CrCl_{3(s)} + 6NH_{3(l)} \xrightarrow{\text{evaporate}} [Cr(NH_3)_6]^{3+} \; 3Cl^-_{(s)}$$

$$2CoCl_{2(s)} + 6CH_3-\overset{\overset{\displaystyle O}{\|}}{S}-CH_{3(l)} \xrightarrow{\text{evaporate}} [Co(CH_3SCH_3)_6]^{2+}[CoCl_4]^{2-}_{(s)}$$

If this procedure is inconvenient (if the ligand is a solid, for example) a handy variation is to dissolve both the partly ionic transition-metal compound and the ligand in a solvent whose molecules are not particularly good donors and mix the solutions. Ethanol, for instance, is polar enough to dissolve most of these species, but is a poor enough donor that most potential donor ligands will completely replace it in the **coordination sphere** of ligand molecules around the metal atom. Another variation, for halide complexes, is to use a molten halide salt as a solvent.

Sometimes it is possible to interchange ligands, if the new ligand is a better donor than the old one. This is convenient when it is possible, because the most common forms of the transition-metal compounds are the hydrates: $CoCl_2 \cdot 6H_2O$ is really $[Co(OH_2)_6]^{2+} \; 2Cl^-$. But the water molecule is a good electron-donor ligand, and only the strongest donors can replace it:

$$\text{Ni(OH}_2)_{6(aq)}^{2+} + \underset{\substack{\text{dihydrogen} \\ \text{EDTA ion}}}{\text{H}_2\text{Y}_{(aq)}^{2-}} \rightarrow \text{NiH}_2\text{Y}_{(aq)} + 6\text{H}_2\text{O}_{(l)}$$

So generally it is impossible to work in water solution when preparing these species. Partial ligand exchange is usually the way complexes with different ligands must be prepared:

$$\text{PtCl}_4^{2-} + \underset{\text{ethylenediamine (en)}}{\text{H}_2\text{N}-\text{CH}_2-\text{CH}_2-\text{NH}_2} \rightarrow [\text{PtCl}_2(\text{en})]^0$$

Reaction Mechanisms: Ligand Chemical Properties

The reactions of transition-metal complexes are interesting principally because of the different chemical properties of the complex ligands, relative to their free-molecule properties. Of course, the reactions of complexes also have synthetic value, because interesting combinations of ligands and theoretically interesting coordination geometries often result from these reactions. But some of the most surprising chemical results and useful chemical information have resulted from an understanding of the difference between a free molecule's reactions and those it displays as a ligand. A commercial example of this usefulness is the titanium catalyst for the polymerization of ethylene; without the catalyst, ethylene simply does not undergo the polymerization reaction except at very high pressures, even though it is thermodynamically favorable.

We can get a more detailed idea about the way ligand reactions are changed by reviewing some studies of the mechanism of a fairly simple reaction, the hydrolysis of halopentamminecobalt(III) complexes:

$$\text{Co(NH}_3)_5\text{X}_{(aq)}^{2+} + \text{H}_2\text{O}_{(l)} \rightarrow \text{Co(NH}_3)_5(\text{OH}_2)_{(aq)}^{3+} + \text{X}_{(aq)}^- \qquad (\text{X} = \text{Cl, Br, I})$$

Although the reaction is simple the establishment of its mechanism is not, and some very ingenious methods have been used in its elucidation. The reaction is obviously a substitution reaction, with water replacing X^- as a ligand. Furthermore, the Co atom unquestionably has at least a partial positive charge, so the mechanism of the substitution reaction involves the attack of a positive-charge seeker or nucleophile on the Co^{III} complex. Using the nomenclature from Chapter 12, we expect an S_N1 or an S_N2 reaction. S_N1, in this context, means a reaction mechanism in which the slow or rate-determining step of the reaction mechanism is first order—the basic mechanism of the elementary process is a single molecule breaking one of its bonds. S_N2 refers to a second-order rate-determining step, in which two species are coming together and forming a bond. Of course, in a substitution reaction both kinds of process eventually happen, and most mechanisms are not extreme cases of either S_N1 or S_N2, but are intermediate in that an approaching new ligand makes it more favorable energetically for the old ligand to leave. So the two names really refer to mechanisms that are *primarily* bond-breaking or bond-making, as the case may be.

In the cobalt(III) case, the rate law has the following form:

$$\text{hydrolysis rate} \equiv \frac{-d[\text{Co(NH}_3)_5\text{X}^{2+}]}{dt}$$

$$= k_a[\text{Co(NH}_3)_5\text{X}^{2+}] + k_b[\text{Co(NH}_3)_5\text{X}^{2+}][\text{OH}^-]$$

Experimentally, k_b is perhaps 10^5 times as large as k_a, but even so the OH⁻ concentration in acid solution is so small as to make the second term negligible in acid hydrolysis, while the larger k_b makes the first term negligible in basic hydrolysis. Looking first at acid hydrolysis, we note that since the reaction is being run in water, which is 55 M in water, the water concentration is effectively constant during the reaction. This means that whether water is involved in the rate-determining step or not, it will not show up in the rate law but will seem to be part of the constant. Its absence from the rate law for acid hydrolysis does not mean that we can automatically assume an S_N1 mechanism. However, several other kinds of evidence suggest that the acid mechanism does involve predominantly bond breaking (for instance, replacing the NH_3 ligands by bulkier amines ought to speed up a bond-breaking mechanism through steric hindrance, but slow down a bond-making one; this one speeds up). In basic hydrolysis it is equally true that we cannot take a mechanism for granted from the form of the rate law. The rate law suggests that the mechanism ought to be S_N2, involving both the Co^{III} complex and the OH⁻ ion in bond making. But consider the following possible mechanism, called S_N1CB (for *Conjugate Base*):

(1) $\text{Co(NH}_3)_5\text{X}^{2+} + \text{OH}^- \xrightleftharpoons{\text{fast equilibrium}} \text{Co(NH}_3)_4(\text{NH}_2)\text{X}^+ + \text{H}_2\text{O}$
(2) $\text{Co(NH}_3)_4(\text{NH}_2)\text{X}^+ \xrightarrow{\text{slow}} \text{Co(NH}_3)_4(\text{NH}_2)^{2+} + \text{X}^-$
(3) $\text{Co(NH}_3)_4(\text{NH}_2)^{2+} \xrightarrow{\text{fast}} \text{Co(NH}_3)_5\text{OH}^{2+}$

Step 2 is the rate-determining step, and it involves only bond breaking. Its rate is proportional to the concentration of $\text{Co(NH}_3)_4(\text{NH}_2)\text{X}^+$, which through the equilibrium constant for step 1 is proportional to the *product* of the concentrations of $\text{Co(NH}_3)_5\text{X}^{2+}$ and OH⁻ (like the case discussed on p. 576). Several kinds of evidence (principally involving the relative acidities of H atoms on different amine ligands—the more acidic the ligand H atom is, the faster the reaction goes by this mechanism) suggest that the S_N1CB mechanism, rather than a simple S_N2, is correct. In fact, it seems to be applicable to a number of different metals and ligands, as long as the ligand has a transferrable proton.

What is unusual about the S_N1CB mechanism? The free ammonia molecule absolutely cannot form the amide ion in water—but the ammonia ligand apparently does so readily. From Table 10-3 we see that NH_2^- is a stronger electron donor than OH^-, so the leveling effect will prevent it from ever existing in aqueous solution. But when it is already donating electrons to a transition metal atom it is no longer a stronger base than OH^-, and it can participate—as a ligand—in a chemical process that would be impossible if free NH_2^- were required.

Reactions of the Oxygen-Molecule Ligand

Now let us pursue the notion of altered ligand reactions in transition-metal complexes a little further. The most important complex of Fe^{II} to all of us is **hemoglobin**, which is responsible for oxygen transport from our lungs by the blood to all our living cells. A very great amount of effort has been expended on establishing the structure and function of hemoglobin, and thanks to some brilliant X-ray crystallography a good deal is known about it. The hemoglobin molecule consists of four protein chains, each with a molecular weight of about 16,000. Each of the chains has a helical structure that is interrupted by being bent at several points so that the chain has a fairly compact, more-or-less spherical overall geometry. Figure 16-22 shows the result in a schematic

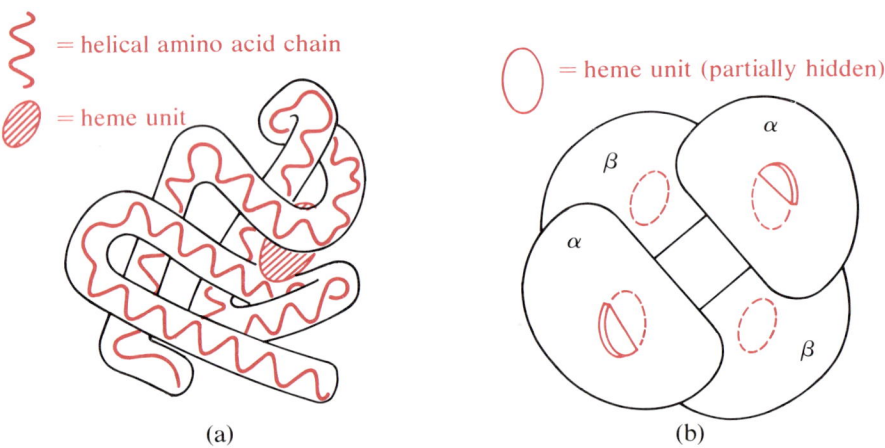

Figure 16-22 Schematic representation of hemoglobin structure. (a) The β chain of hemoglobin structure. (b) Tetrahedral assembly of four chains, two α and two β.

fashion, and also shows the approximately tetrahedral assembly of the four protein chains. Each of the chains has a **heme** unit in it, which is the site of the Fe^{II}. Heme is a square-planar chelate complex of iron with the basic structure shown in Fig. 16-23 (with some other groups on the outside of the ring). Now

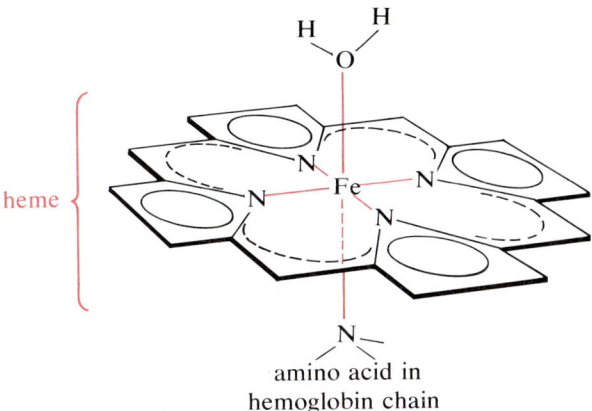

Figure 16-23 Structure of a planar heme unit, plus two other bonds around the octahedral Fe^{II}.

Fe^{II}, with six remaining valence electrons, can readily accept 12 more; that is, it can accept from six two-electron donors. Since the heme unit provides only four donor atoms, two coordination positions on the Fe are unoccupied. The heme unit is held inside the protein chain (but near the surface) by another N donor atom, which occupies a fifth coordination position on the Fe atom. The sixth position is occupied by a water molecule, as in Fig. 16-23. When the whole hemoglobin protein takes up an O_2 molecule, forming oxyhemoglobin, the tetrahedral structure of the four protein chains is distorted because one of them has flexed enough to change the electronic environment of the Fe atom and replace the coordinated water by an O_2 molecule. Apparently the protein chain is able to control the coordination of the Fe atom by bending and altering the geometry of the ligand atoms.

A very crucial feature of this O_2 coordination is that it is reversible. If hemoglobin is to serve as an oxygen-transporting species, it has to be possible to remove the oxygen in a cell at the destination. This, of course, is what we observe. But if it were not for the existence of the protein chain and its effect on the coordination geometry of the Fe^{II}, this reversibility would be impossible. Specifically, if heme (without the protein) is exposed to O_2 the Fe^{II} is oxidized to Fe^{III} irreversibly, and the product is no longer active biologically. So again we see that the reactions of a coordinated molecule, O_2 in this case, are different in a particular coordination environment than they would be if the molecule were free. In fact, the body cannot use O_2 directly, but only if it is transported by hemoglobin. If the atmosphere contains any significant amount of CO, the hemoglobin exposed to air in the lungs will form a very stable CO complex in preference to the O_2 complex. When the hemoglobin is all used up by forming this "carbonyl," the person breathing the atmosphere will suffocate even though there is still plenty of oxygen available, unless he is immediately removed from the CO-containing atmosphere. The process of oxygen transport by hemoglobin is one of the more intriguing reactions of a coordination compound.

Reactions of the Nitrogen-Molecule Ligand

Another similar area of coordination chemistry that is being actively investigated, but about which less is known, is the mechanism of biological nitrogen fixation. Prior to World War I the world depended principally on large deposits of nitrate salts in dry regions of Chile for its supply of fixed nitrogen (meaning nitrogen in compounds with other elements, as opposed to the chemically unreactive N_2 in the atmosphere). Since 1914 fixed nitrogen has been readily available for fertilizers and other uses through the Haber synthesis, involving the high-pressure catalytic combination of N_2 and H_2 to NH_3. It has remained rather frustrating, however, to realize that bacteria in some plants such as legumes can fix atmospheric nitrogen at atmospheric pressure and ordinary soil temperatures. Chemists have suspected that this surprising ability, which is necessary to provide the nitrogen in all living things, probably represented a chemical reactivity in the nitrogen molecule complexed as a ligand that it does not have as the free molecule. This feeling has been strengthened by the discovery of the immense variety of transition-metal carbonyl compounds, because the molecules CO and NN are isoelectronic and have rather similar electron distributions; CO has $4+6=10$ valence electrons, while N_2 has $5+5=10$, and the appropriate MO energy levels can be seen in Figs. 6-26 and 6-28. It ought to be possible to form N_2 complexes, then, with much the same properties as the carbonyls—but all attempts failed until 1965, when Allen and Senoff observed the following reaction:

$$2[RuCl_5(OH_2)]^{2-} + 7N_2H_5^+ + 3H_2O \rightarrow 2[Ru(NH_3)_5(N_2)]^{2+} + 5H_3O^+$$
pentachloroaquoruthenate(III)

Further studies, including X-ray diffraction and spectroscopic studies, indicated that this was a genuine complex of molecular nitrogen—but it was not made from atmospheric nitrogen. However, shortly after this original discovery it was found that when N_2 passed through a solution of $Ru(NH_3)_5Cl^{2+}$ that has been reduced [probably to $Ru(NH_3)_5(OH_2)^{2+}$] by Zn metal, the same product forms along with another remarkable product:

$$\begin{array}{c} H_3NNH_3 H_3NNH_3 \\ \backslash/ \backslash/ \\ H_3N-Ru-N-N-Ru-NH_3 \\ /\backslash /\backslash \\ H_3NNH_3 H_3NNH_3 \end{array}$$

This binuclear complex is quite stable in water solution, and indeed the coordinated —N—N group on one Ru atom seems to be a better donor than H_2O toward another Ru atom. Other N_2 complexes have been prepared more recently, including species with cobalt, rhodium, iridium, and osmium. Apparently they all involve the NN bonded end-on, like a CO molecule, and with the same sort of bonding as in carbonyls: σ-donation/π-acceptance by the ligand. The only metal missing from the above group in the periodic table is Fe. However, biochemists working with extracts of nitrogen-fixing cells had already established that an iron-containing protein, **ferredoxin**, is always associated with biological nitrogen reduction. Ferredoxin contains between four and seven Fe atoms, depending on the origin of the cell, and the Fe atoms appear to be complexed by S atoms in surrounding amino acids in such a way as to be near each other. A structure that has been proposed is

It is not yet clear how ferredoxin is involved in the molecular mechanism of nitrogen fixation, but it almost surely serves as an acceptor in much the same way that the Ru and other metal atoms do in the "model compounds" with molecular N_2. This is still a challenging area in which much remains to be understood; when we do understand it, we shall have taken an important step toward understanding how life maintains itself on our planet.

16-8 Summary

This chapter has attempted to show how d-electron compounds resemble — and how they differ from — the ionic, partly ionic, and covalent compounds we discussed in Chapters 13–15. Probably the most important difference is simply the greater number of valence orbitals that are available for combination in their compounds. The availability of nine valence orbitals means that up to 18 electrons (including those from the metal atom itself) can be accommodated in bonding or nonbonding orbitals. This fact completely changes the possible stoichiometries and geometries of the transition-metals compounds. There exists a wide variety of compounds with electrically neutral donor species, in which the concept of oxidation state seems not to bear any relation at all to the stoichiometry of the compound. We can rationalize the stoichiometry of many of these, however, in terms of the 18-electron rule. An important sense in which the transition metals resemble lighter elements such as C and N but differ from other metals is their capacity for π bonding. Heavy s- and p-electron elements generally do not π bond because the π overlap of p orbitals is so poor when the atom has a large inner core of nonvalence electrons. However, d orbitals are oriented in space about the nucleus of the transition-metal atom so as to substantially improve the possibility of π overlap, as Fig. 16-19 suggests. The resulting π bonding, with the metal contributing $d\pi$ electrons, strongly influences the stability of many transition-metal compounds.

In a simple approach to the relative stability of transition-metal compounds, we observed that the energies of d electrons differ in their various coordination geometries, depending on their particular orbital distribution. The concept of crystal-field stabilization energy provides a useful correlation of chemical properties from one metal to another, as we saw in the case of reduction potentials. The more general treatment of transition-metal compounds, involving MO's, allowed us to account for the characteristic colors or light-absorption patterns of the compounds, which is one of the principal experimental tools for probing the electronic structure of the compounds.

We developed a rationale for the common observation that transition metals often show many different formal oxidation states, and saw that this leads not only to different formulas or electric charges for covalent (molecular) compounds, but to what is frequently a substantial nonstoichiometry in partly ionic compounds. We examined some of the descriptive chemistry of the transition-metal compounds, including not only the simpler partly ionic compounds but also some of the more interesting covalent ones such as alloys and organometallic compounds. And in closing not only this chapter but the book, we attempted to show the ways in which some of the reactions of transition-metal complexes unite not only the concepts presented in this chapter but some of the ideas from many preceding chapters.

Much is known about the chemistry of ourselves and our surroundings, but much remains to be learned. In the last century, chemistry divided into several specialties; these are now in the process of growing together again as new concepts are seen to be generally applicable in many areas. Whether or not the reader chooses to become a part of this learning and growing process, perhaps at least he can now appreciate the sense in which the chemist feels as A. E. Housman did about his discipline:

> And the pleasure of discovery differs from other pleasures in this, that it is shadowed by no fear of satiety on the one hand or of frustration on the other. Other desires perish in their gratification, but the desire of knowledge never: the eye is not satisfied with seeing nor the ear filled with hearing... the sum of things to be known is inexhaustible, and however long we read we shall never come to the end of our storybook.

Study Problems

1. Propose a structure for $Mn_2(CO)_{10}$ whose bonding would satisfy the 18-electron rule.
2. How many isomers would the compound $Co(NH_3)_4Cl_2$ have if its structure were hexagonal planar? Draw each one. How many isomers would a triangular-prism structure have? Two forms of the compound are known, one green, the other violet. Discuss the way this suggests an octahedral structure.
3. Name the following compounds or ions with correct spelling:

   (a) TiF_6^{3-}
   (b) $[Co(NH_3)_4Cl_2]^+Cl^-$
   (c) $Cs_3V(C_2O_4)_3$
   (d) $[Mn(CN)_5NO]^{3-}$
   (e) $KCrF_5$
   (f) $Fe(CN)_6^{3-}$
   (g) $Ni(P(C_6H_5)_3)_2Cl_2$
   (h) $(NH_4)_2TaF_7$
   (i) $[W(CO)_5Br]^-$
   (j) $[RuNO(NH_3)_4Cl]^{2+}$

4. The phosphor that emits green light from an oscilloscope screen or TV screen is ZnS containing a very small amount of Cu as a substitutional impurity for the Zn ions. The Cu is thought to be present as Cu^+, which requires some sulfide vacancies to keep the electrical charge neutral. A model for the emission of light when the phosphor is struck by cathode rays (electrons) is shown in Fig. 16-24 in terms of the band theory of solid ZnS. Discuss the mechanism in terms of the relative energies of Cu^+, Cu^{2+}, and the ZnS lattice.

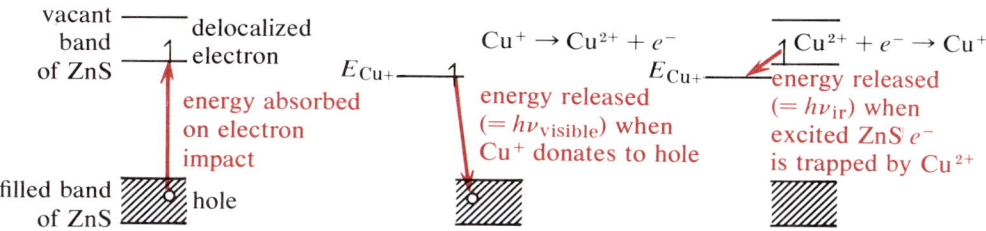

Figure 16-24

5. Show that the string of manganese complexes in different oxidation states on p. 857 does not exceed the 18-electron rule.
6. Using Tables 16-5 and 16-6, calculate the change in CFSE when each of the ions from Ti^{3+} across to Ni^{3+} in an octahedral complex loses one ligand to become 5-coordinate (in a square pyramid). Assume Δ is large enough to cause spin pairing and assume that in losing a ligand each electron stays in its same d orbital. If the rate-determining step in a ligand-substitution reaction involves this loss of one ligand, so

that this ΔCFSE might represent the activation energy for the reaction, what should be the order of reactivity of the 3+ ions? Cr^{3+} and Co^{3+} are usually observed to be most inert kinetically; is the model in accord with this?

7. The lattice energies of the transition-metal dichlorides do not increase uniformly from $CaCl_2$ across to $ZnCl_2$ as the decreasing metal-ion radius would suggest:

	Ca^{2+}	(Sc^{2+})	Ti^{2+}	V^{2+}	Cr^{2+}	Mn^{2+}	Fe^{2+}	Co^{2+}	Ni^{2+}	Cu^{2+}	Zn^{2+}
$U =$	535	—	590	610	615	600	625	640	655	660	645

kcal/mole

Plot these figures against the atomic number of the metal, draw in a smooth curve for the three ions $Ca^{2+} - Mn^{2+} - Zn^{2+}$, and comment on the relation of CFSE to the deviations of the experimental points from the smooth curve.

8. Suggest the most symmetrical structure for the tetrameric molybdenum(V) fluoride Mo_4F_{20}.

9. In the substitution reaction

$$Co(NH_3)_4Cl_2^+ + H_2O \rightarrow Co(NH_3)_4(OH_2)Cl^{2+} + Cl^-$$
trans

the rate constant is reduced by a factor of almost 100 if the four ligands are replaced by two ethylenediamine molecules:

$k = 1.9 \times 10^{-3}$ min^{-1}

$(k_{NH_3\,ligands} = 110 \times 10^{-3}$ min$^{-1})$

This is thought to be due to the reduced solvation energy of the larger ion. However, if methyl groups are added to the ethylenediamines, the rate increases:

$k = 3.7 \times 10^{-3}$ min^{-1} $k = 8.8 \times 10^{-3}$ min^{-1} $k = 250 \times 10^{-3}$ min^{-1}

Why should there be such a great difference between the second and third species' rates when only the arrangement of the groups and not their number has changed? Does this pattern imply an S_N1 or S_N2 mechanism?

10. Set up a σ-overlap MO energy-level diagram for a trigonal bipyramidal system having the axes and AO energies shown in Fig. 16-25. First, allow for overlap by the metal s orbital in a σ^b and a σ^* fashion. Second, note that the metal p_z orbital combines differently from p_x and p_y; allow for σ_z^b and σ_z^*. By sketching the AO's,

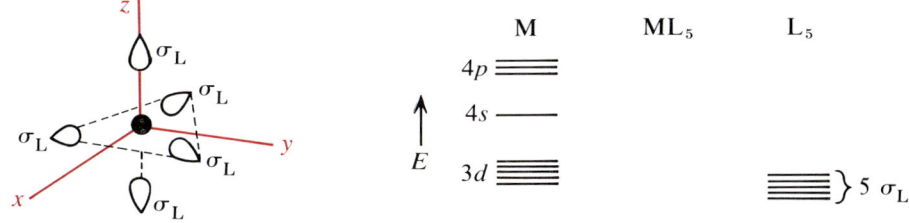

Figure 16-25

convince yourself that $d_{x^2-y^2}$ and d_{xy} combine with the same set of σ_L algebraic signs as p_x and p_y, respectively. As in the BH_3 case (Section 6-10), these p overlaps are equal; show that this leads to two σ^b, two σ, and two σ^* MO's. Now allow for d_{z^2} overlap: $\sigma_{z^2}^b$ and $\sigma_{z^2}^*$. Show that the two remaining d orbitals are nonbonding if only σ overlap is considered. If ligand electrons occupy the lowest-energy MO's, show that the d electrons must be distributed in the following energy levels:

11. The neutral ligand molecule abbreviated *qas* has four arsenic donor atoms connected by benzene rings; it forms the trigonal bipyramidal complex $Fe(qas)Cl^+$, shown in Fig. 16-26. Using the MO energy levels from Study Problem 10, explain why $Fe(qas)Cl^+$ has a magnetic moment of about 2.8 Bohr magnetons.

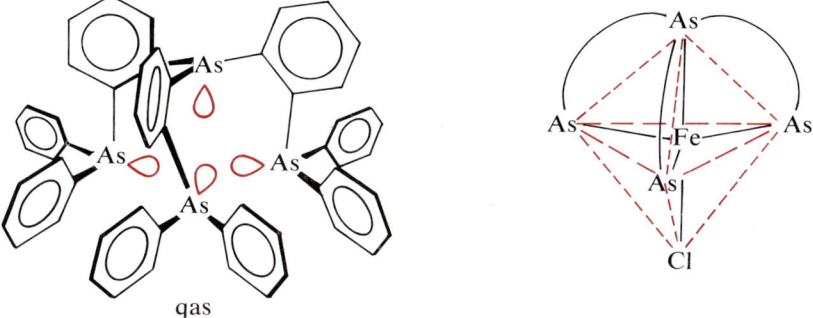

Figure 16-26

12. Using the phase diagram for a solid solution shown in Fig. 7-52 (although it is slightly oversimplified for this purpose), explain why brazing rod, which is used for welding brass objects together, is 50% Cu/50% Zn, when ordinary brass is about 70% Cu/30% Zn.

13. Predict the product when "cobaltocene" or biscyclopentadienylcobalt, $Co(C_5H_5)_2$, reacts with Br_2.
14. Which ion of the pair $VBr_6^{2-}/CuBr_4^{2-}$ can be prepared in aqueous solution? What prevents the other one from forming?
15. Nichrome alloy, which is used for electrical heating elements, has the composition 60% Ni/25% Fe/15% Cr. What chemical features of this composition might make it particularly desirable for high-temperature applications?
16. In a very important industrial reaction called the *oxo process*, olefins are converted to aldehydes by CO and H_2 over $Co_2(CO)_8$ catalyst:

$$\text{C=C} + CO + H_2 \xrightarrow{Co_2(CO)_8} H-\overset{|}{\underset{|}{C}}-\overset{|}{\underset{|}{C}}-C\overset{O}{\underset{H}{\diagdown}}$$

Assume that the first step in the process is the reaction of the catalyst with H_2, followed by the loss of a CO molecule, and propose a mechanism for the reaction. Keep in mind that the 18-electron rule effectively limits the number of ligands and that species with less than 18 electrons will be good acceptors. Since $Co_2(CO)_8$ is a catalyst, the overall mechanism must regenerate it.
17. It seems to be generally true that halide complexes are limited to tetrahedral geometry, with few exceptions (see Table 16-9). What might limit the coordination number to four when the complex is well below the 18-electron limit?

Some Further Reading

The best reference for this chapter is Cotton and Wilkinson (see Chapter 13 list). Both authors are interested in transition-metal chemistry, and the book is loaded with it. A very extensive and authoritative guide.

Larsen, E. M., *Transitional Elements*, New York: Benjamin, 1965. At a lower level than Cotton and Wilkinson; very easy to read, but still informative.

Pauson, P. L., *Organometallic Chemistry*, London: Arnold, 1967. At about the level of Cotton and Wilkinson, but still readable with the background of this chapter.

Basolo, F., and Johnson, R. C., *Coordination Chemistry*, New York: Benjamin, 1964. On the level of this chapter, but longer, more restricted, and more informative as a result.

Tables

t Values for Various Sample Sizes and Confidence Levels

Sample Size (n)	Percent Confidence Level						
	50	60	70	80	90	95	99
2	1.000	1.376	1.963	3.078	6.314	12.706	63.657
3	0.816	1.061	1.386	1.886	2.920	4.303	9.925
4	0.765	0.978	1.250	1.638	2.353	3.182	5.841
5	0.741	0.941	1.190	1.533	2.132	2.776	4.604
6	0.727	0.920	1.156	1.476	2.015	2.571	4.032
7	0.718	0.906	1.134	1.440	1.943	2.447	3.707
8	0.711	0.896	1.119	1.415	1.895	2.365	3.499
9	0.706	0.889	1.108	1.397	1.860	2.306	3.355
10	0.703	0.883	1.100	1.383	1.833	2.262	3.250
20	0.688	0.861	1.066	1.328	1.729	2.093	2.861
30	0.683	0.854	1.055	1.311	1.699	2.045	2.756
40	0.681	0.851	1.050	1.303	1.684	2.021	2.704
50	0.680	0.849	1.048	1.299	1.676	2.008	2.678
60	0.679	0.848	1.046	1.296	1.671	2.000	2.660
120	0.677	0.845	1.041	1.289	1.658	1.980	2.617
∞	0.674	0.842	1.036	1.282	1.645	1.968	2.576

Fundamental Physical Constants

Avogadro's number, N_0	6.02217×10^{23} particles/mole
Gas constant, R	8.3143×10^7 erg/mole-deg
	8.3143 joule/mole-deg
	0.08205 l-atm/mole-deg
	1.987 cal/mole-deg
Standard gas volume, V_0	22.4136 l/mole (1 atm, 273 °K)
Boltzmann's constant, k	1.38062×10^{-16} erg/deg
thermal energy, kT	1.987×10^{-3} kcal/mole-°K
	0.592 kcal/mole at 298 °K
	208 cm^{-1} at 298 °K (per molecule)
Faraday, $\mathscr{F}$	9.64867×10^4 coulomb/mole of charge
Electron charge, e	1.60219×10^{-19} coulomb
	4.80325×10^{-10} esu (electrostatic units)
Electron mass, m_e	9.10956×10^{-28} g
Proton mass, m_p	1.67261×10^{-24} g
Velocity of light (vacuum), c	2.997925×10^{10} cm/sec
Planck's constant, h	6.62620×10^{-27} erg-sec
Rydberg constant, R_∞:	1.0973731×10^5 cm^{-1}
Bohr magneton, μ_B	9.2741×10^{-21} erg/gauss per molecule
	(the gauss is a unit of magnetic field)
Bohr radius, a_0	5.29167×10^{-9} cm (for H 1s electron)
Acceleration due to gravity, g	980.665 cm/sec^2 (45° latitude)

Statistical Formulas and Tables

Standard deviation $= S = \sqrt{\dfrac{\Sigma(X_i - \bar{X})^2}{n-1}}$

Confidence interval $= \bar{X} \pm tS$

Standard deviation of mean $= \dfrac{S}{\sqrt{n}}$

Confidence interval for mean $= \bar{X} \pm t\left(\dfrac{S}{\sqrt{n}}\right)$

The following tables, which have appeared elsewhere in the text, are reprinted here for your convenience.

Absolute Molar Entropies[a] (cal/mole-°K at 298°K)

Solids

Al	6.77	Cr_2O_3	19.4	$MgCl_2$	21.4		
$AlCl_3$	40	$CoCl_2$	25.4	MgO	6.4		
Al_2O_3	12.19	CoO	10.5	$NiCl_2$	25.6		
NH_4Cl	22.6	Cu	7.96	NiO	9.22		
$BaCl_2$	30	CuCl	20.2	KCl	19.76		
BaO	16.8	Fe	6.49	NaCl	17.30		
BeO	3.37	$FeCl_2$	28.6	$SrCl_2$	28		
$CaCl_2$	27.2	Fe_2O_3	21.5	SrO	13.0		
CaO	9.5	Pb	15.51	S (rhombic)	7.62		
$CaCO_3$	22.2	$PbCl_2$	32.6	Zn	9.95		
C (diamond)	0.58	PbO	16.2	$ZnCl_2$	25.9		
$CrCl_3$	30.0			H_2O (ice, 0°C)	9.87		

Liquids

$AsCl_3$	55.8	H_2O	16.72	$SiCl_4$	57.2
BCl_3	50.0	Hg	18.5	$SnCl_4$	61.8
Br_2	36.4	HNO_3	37.2	$TiCl_4$	60.4
CH_3OH (methyl alcohol)		30.3	CCl_4 (carbon tetrachloride)		51.2
CH_3CH_2OH (ethyl alcohol)		38.4	C_6H_{12} (cyclohexane)		47.3
CH_2Cl_2 (dichloromethane)		42.7	C_6H_6 (benzene)		37.4
$CHCl_3$ (chloroform)		48.5			

Gases

NH_3	46.01	CO	47.30	F_2	48.6
Ar	36.98	Cl_2	53.29	N_2	45.77
Br_2	58.64	H_2	31.21	O_2	49.00
CO_2	51.06	HCl	44.62	NO_2	57.47
		H_2O	45.11		
CH_4 (methane)		44.50	$CH_2{=}CH_2$ (ethylene)		52.54
CH_3Cl (chloromethane)		55.97	$CH{\equiv}CH$ (acetylene)		48.00
CH_3CH_3 (ethane)		54.85	$CH_3CH_2CH_3$ (propane)		64.51

[a] NBS Circular 500.

Bond Energies (kcal/mole)[a]

	H	Li	Be	B	C	N	O	F	Cl	Br	I	S
H	104.2											
Li	58	25										
Be	53		17									
B	70			69								
C	98.7	57		89	82.6— 145.8= 199.6≡							
N	93.4			92	72.8— 147= 212.6≡	60— 100= 225≡						
O	117.5	84	124	128	85.5— 176=	48—	51— 118=					
F	134	137	92	154	107	65	45	36				
Cl	102.2	115	109	109	81	46	60	60	57			
Br	86.5	101	89	90	65	28	56	55	52	45		
I	70.5	81	69		51		56	46	50	42	36	
S	83				65— 128=			119=	68	61	52	54

[a] Principally from Cottrell, T. L., *The Strengths of Chemical Bonds*, London: Butterworths, 1958, 2nd ed.).

Equilibrium Constants (K_a) for Aqueous Proton-Donor Reactions

Reaction	K_a (25°C)
Cl_3CCOOH(trichloroacetic acid) $+ H_2O \rightleftharpoons Cl_3CCOO^- + H_3O^+$	2.0×10^{-1}
$HOOCCOOH$(oxalic acid) $+ H_2O \rightleftharpoons HOOCCOO^- + H_3O^+$	5.9×10^{-2}
$HSO_4^- + H_2O \rightleftharpoons SO_4^{2-} + H_3O^+$	1.2×10^{-2}
$H_3PO_4 + H_2O \rightleftharpoons H_2PO_4^- + H_3O^+$	7.5×10^{-3}
$Fe(OH_2)_6^{3+} + H_2O \rightleftharpoons Fe(OH_2)_5OH^{2+} + H_3O^+$	3.4×10^{-3}
$HF + H_2O \rightleftharpoons F^- + H_3O^+$	3.5×10^{-4}
$HCOOH$(formic acid) $+ H_2O \rightleftharpoons HCOO^- + H_3O^+$	1.8×10^{-4}
$Cr(OH_2)_6^{3+} + H_2O \rightleftharpoons Cr(OH_2)_5OH^{2+} + H_3O^+$	1.6×10^{-4}
C_6H_5COOH(benzoic acid) $+ H_2O \rightleftharpoons C_6H_5COO^- + H_3O^+$	6.5×10^{-5}
$HOOCCOO^-$(hydrogen oxalate ion) $+ H_2O \rightleftharpoons OOCCOO^{2-} + H_3O^+$	6.1×10^{-5}
$HN_3 + H_2O \rightleftharpoons N_3^- + H_3O^+$	1.9×10^{-5}
CH_3COOH(acetic acid) $+ H_2O \rightleftharpoons CH_3COO^- + H_3O^+$	1.8×10^{-5}
$Al(OH_2)_6^{3+} + H_2O \rightleftharpoons Al(OH_2)_5OH^{2+} + H_3O^+$	1.0×10^{-5}
$C_5H_5NH^+ + H_2O \rightleftharpoons C_5H_5N$(pyridine) $+ H_3O^+$	6.6×10^{-6}
$CO_{2(aq)} + 2H_2O \rightleftharpoons HCO_3^- + H_3O^+$ ($H_2CO_3 + H_2O \rightleftharpoons HCO_3^- + H_3O^+ : 2 \times 10^{-4}$)	4.3×10^{-7}
$H_2S + H_2O \rightleftharpoons HS^- + H_3O^+$	1.0×10^{-7}
$N_2H_5^+ + H_2O \rightarrow N_2H_4$(hydrazine) $+ H_3O^+$	3.3×10^{-9}
$NH_4^+ + H_2O \rightleftharpoons NH_3 + H_3O^+$	5.6×10^{-10}
$HCO_3^- + H_2O \rightleftharpoons CO_3^{2-} + H_3O^+$	5.6×10^{-11}
$(C_2H_5)_2NH_2^+ + H_2O \rightleftharpoons (C_2H_5)_2NH$(diethylamine) $+ H_3O^+$	8.0×10^{-12}
$HS^- + H_2O \rightleftharpoons S^{2-} + H_3O^+$	1.1×10^{-13}
$H_2O + H_2O \rightleftharpoons OH^- + H_3O^+$	1.0×10^{-14}

Standard Enthalpies and Free Energies of Formation (298 °K)[a]

Compound or Atom	ΔH_f^0 (kcal/mole)	ΔG_f^0 (kcal/mole)
$AlCl_{3(s)}$	−166.2	−152.2
$Al_2O_{3(s)}$	−399.1	−376.8
$AsCl_{3(l)}$	−80.2	−70.5
$BCl_{3(l)}$	−100.0	−90.6
$B_2O_{3(s)}$	−297.6	−280.4
$BaCl_{2(s)}$	−205.6	−193.8
$BaO_{(s)}$	−133.4	−126.3
$BeO_{(s)}$	−146.0	−139.0
$CaCO_{3(s)}$	−288.5	−269.8
$CaCl_{2(s)}$	−190.0	−179.3
$CaO_{(s)}$	−151.9	−144.4
$CoCl_{2(s)}$	−77.8	−67.5
$CoO_{(s)}$	−57.2	−51.0
$CrCl_{3(s)}$	−134.6	−118.0
$Cr(OH_2)_6Cl_{3(s)}$	−579.0	
$Cr_2O_{3(s)}$	−269.7	−250.2
$CrO_{3(s)}$	−138.4	
$CuCl_{(s)}$	−32.5	−28.2
$CuCl_{2(s)}$	−49.2	
$FeCl_{2(s)}$	−81.5	−72.2
$Fe_2O_{3(s)}$	−196.5	−177.1
$HCl_{(g)}$	−22.06	−22.77
$H_2O_{(g)}$	−57.80	−54.64
$H_2O_{(l)}$	−68.32	−56.69
$HNO_{3(l)}$	−41.4	−19.1
$KCl_{(s)}$	−104.2	−97.6
$MgCl_{2(s)}$	−153.4	−141.6
$MgO_{(s)}$	−143.8	−136.1
$NH_{3(g)}$	−11.04	−3.98
$NH_4Cl_{(s)}$	−75.4	−48.7
$NO_{(g)}$	21.60	20.72
$NO_{2(g)}$	8.09	12.39
$N_2O_{5(s)}$	−10.0	
$NaCl_{(s)}$	−98.2	−91.8
$NiCl_{2(s)}$	−75.5	−65.1
$NiO_{(s)}$	−58.4	−51.7
$PbCl_{2(s)}$	−85.9	−75.0

Compound or Atom	ΔH_f^0 (kcal/mole)	ΔG_f^0 (kcal/mole)
$PbO_{(s)}$	−52.4	−45.2
$SOCl_{2(l)}$	−49.2	
$SO_{2(g)}$	−71.0	−71.8
$SiCl_{4(l)}$	−153.0	−136.9
$SiO_{2(s)}$	−205.4	−192.4
$SnCl_{4(l)}$	−130.3	−113.3
$SrCl_{2(s)}$	−198.0	−186.7
$SrO_{(s)}$	−141.1	−133.8
$TiCl_{4(l)}$	−179.3	−161.2
$TiO_{2(s)}$	−218.0	−203.8
$ZnCl_{2(s)}$	−99.4	−88.3
$ZnO_{(s)}$	−83.2	−76.1
carbon tetrachloride, $CCl_{4(l)}$	−33.3	−16.4
chloroform, $CHCl_{3(l)}$	−31.5	−17.1
dichloromethane, $CH_2Cl_{2(l)}$	−28	−15.1
methyl alcohol, $CH_3OH_{(l)}$	−57.0	−39.7
ethyl alcohol, $CH_3CH_2OH_{(l)}$	−66.4	−41.8
acetone, $(CH_3)_2C=O_{(l)}$	−60.5	−37.2
methane, $CH_{4(g)}$	−17.9	−12.1
ethane, $CH_3CH_{3(g)}$	−20.2	−7.9
ethylene, $CH_2=CH_{2(g)}$	12.5	16.3
acetylene, $CH\equiv CH_{(g)}$	54.2	50.0
benzene, $C_6H_{6(l)}$	19.8	31.0
carbon monoxide, $CO_{(g)}$	−26.4	−32.8
carbon dioxide, $CO_{2(g)}$	−94.0	−94.3
H	52.1	48.6
Li	37.1	29.2
C	171.7	160.9
N	112.5	108.4
O	59.2	55.0
F	18.3	14.2
Cl	29.0	25.2
Br	26.7	19.7
I	25.5	16.8
S	53.2	43.6

a From NBS Circular 500.

Reduction Half-Cell Potentials[a]

Half-Reaction	$\mathscr{E}^0$ (V at 25°C)
$H_4XeO_6 + 2H_3O^+ + 2e^- \rightarrow XeO_3 + 5H_2O$	3.0
$F_2 + 2e^- \rightarrow 2F^-$	2.87
$O_3 + 2H_3O^+ + 2e^- \rightarrow O_2 + 3H_2O$	2.07
$PbO_2 + SO_4^{2-} + 4H_3O^+ + 2e^- \rightarrow 6H_2O + PbSO_4$	1.686
$MnO_4^- + 8H_3O^+ + 5e^- \rightarrow Mn(OH_2)_6^{2+} + 6H_2O$	1.491
$ClO_4^- + 8H_3O^+ + 8e^- \rightarrow Cl^- + 12H_2O$	1.36
$Cl_2 + 2e^- \rightarrow 2Cl^-$	1.358
$Cr_2O_7^{2-} + 14H_3O^+ + 6e^- \rightarrow 2Cr(OH_2)_6^{3+} + 9H_2O$	1.33
$O_2 + 4H_3O^+ + 4e^- \rightarrow 6H_2O$	1.229
$Br_2 + 2e^- \rightarrow 2Br^-$	1.066
$2Hg^{2+} + 2e^- \rightarrow Hg_2^{2+}$	0.910
$ClO^- + H_2O + 2e^- \rightarrow Cl^- + 2OH^-$	0.90
$Hg^{2+} + 2e^- \rightarrow Hg^0$	0.854
$Ag^+ + e^- \rightarrow Ag^0$	0.800
$Hg_2^{2+} + 2e^- \rightarrow 2Hg^0$	0.798
$Fe(OH_2)_6^{3+} + e^- \rightarrow Fe(OH_2)_6^{2+}$	0.770
$I_2 + 2e^- \rightarrow 2I^-$	0.535
$Cu(OH_2)_4^{2+} + 2e^- \rightarrow Cu^0 + 4H_2O$	0.346
$Hg_2Cl_2 + 2e^- \rightarrow 2Hg^0 + 2Cl^-$ (in saturated KCl solution; saturated calomel electrode)	0.242
$AgCl + e^- \rightarrow Ag^0 + Cl^-$	0.222
$Cu(OH_2)_4^{2+} + e^- \rightarrow Cu(OH_2)_4^+$	0.153
$2H_3O^+ + 2e^- \rightarrow H_2 + 2H_2O$	0 (defined)
$Pb(OH_2)_4^{2+} + 2e^- \rightarrow Pb^0 + 4H_2O$	−0.125
$Ni(OH_2)_6^{2+} + 2e^- \rightarrow Ni^0 + 6H_2O$	−0.250
$Co(OH_2)_6^{2+} + 2e^- \rightarrow Co^0 + 6H_2O$	−0.277
$PbSO_4 + 2e^- \rightarrow Pb^0 + SO_4^{2-}$	−0.356
$Cd(OH_2)_4^{2+} + 2e^- \rightarrow Cd^0 + 4H_2O$	−0.402
$Cr(OH_2)_6^{3+} + e^- \rightarrow Cr(OH_2)_6^{2+}$	−0.41
$Fe(OH_2)_6^{2+} + 2e^- \rightarrow Fe^0 + 6H_2O$	−0.440
$Cr(OH_2)_6^{3+} + 3e^- \rightarrow Cr^0 + 6H_2O$	−0.74
$Zn(OH_2)_4^{2+} + 2e^- \rightarrow Zn^0 + 4H_2O$	−0.763
$Cr(OH_2)_6^{2+} + 2e^- \rightarrow Cr^0 + 6H_2O$	−0.91
$V(OH_2)_6^{2+} + 2e^- \rightarrow V^0 + 6H_2O$	−1.18
$Mn(OH_2)_6^{2+} + 2e^- \rightarrow Mn^0 + 6H_2O$	−1.182
$Al(OH_2)_6^{3+} + 3e^- \rightarrow Al^0 + 6H_2O$	−1.66
$Al(OH_2)_2(OH)_4^- + 3e^- \rightarrow Al^0 + 4OH^- + 2H_2O$	−2.36
$Mg(OH_2)_6^{2+} + 2e^- \rightarrow Mg^0 + 6H_2O$	−2.37
$Na(OH_2)_n^+ + e^- \rightarrow Na^0 + nH_2O$	−2.71
$Ca(OH_2)_6^{2+} + 2e^- \rightarrow Ca^0 + 6H_2O$	−2.87
$K(OH_2)_n^+ + e^- \rightarrow K^0 + nH_2O$	−2.92
$Li(OH_2)_4^+ + e^- \rightarrow Li^0 + 4H_2O$	−3.05

[a] In acid solution unless OH⁻ appears in the half-reaction

Problem Answers

Chapter 1

1. 24,300 wells.
2. From the wave numbers indicated in the text, the wavelength limits for visible light are about 6.7×10^{-5} cm and 4.0×10^{-5} cm — more commonly indicated as 670–400 nm or 6700–4000 Å.
3. 31.7° Chaud.
4. 0.00416 in.
5. Nil-arbol²/pood, to maintain dimensional consistency.
6. 0.01934 cm/sec, or 4.32×10^{-4} mph.
7. $As_4 + 6Cl_2 \rightarrow 4AsCl_3$
   $5As_4O_6 + 12Cl_2 \rightarrow 8AsCl_3 + 3As_4O_{10}$
   $As_4O_6 + 12NaCl + 12H_2SO_4 \rightarrow 4AsCl_3 + 6H_2O + 12NaHSO_4$
   $2As_4O_6 + 3S_2Cl_2 + 9Cl_2 \rightarrow 8AsCl_3 + 6SO_2$.
8. New empirical formula has only 2 Cl per Mo, so some Cl has disappeared. If the empirical formula contains one Mo, its molecular weight is 356 g/mole.
9. $m = M\left(\dfrac{V_M}{V_{M+m}} - 1\right) = 0.07292$ amu for C_3H_8, CO_2; ^{16}O has a mass of 15.99484 amu by this measurement.
10. (a) $3SCl_2 + 4NaF \rightarrow SF_4 + S_2Cl_2 + 4NaCl$; (b) $KSCN + KHSO_4 \rightarrow HSCN + K_2SO_4$; (c) $2Hg_2CrO_4 \rightarrow 4Hg + 2CrO_3 + O_2$; (d) $CoCl_2 + 2HF \rightarrow CoF_2 + 2HCl$.
11. $4FeS_2 + O_2 \rightarrow 2Fe_2O_3 + 8SO_2$; 2.295 tons SO_2/ton Fe; 3.515 tons H_2SO_4/ton Fe.
12. $CsPS_2F_2$.
13. $CrCl_3P_2C_{12}H_{30}$; 2 donor $P(C_2H_5)_3$ per Cr; $Cr_2Cl_6P_4C_{24}H_{60}$.
14. $As_4S_4 + 7O_2 \rightarrow As_4O_6 + 4SO_2$; $As_4S_6 + 9O_2 \rightarrow As_4O_6 + 6SO_2$; 0.848 g As_4S_4; 0.152 g As_4S_6.
15. 13,200 miles deep!

Chapter 2

1. $\dfrac{dP}{P} = \dfrac{\Delta H_{vap}}{RT^2} dt$

   $\ln \dfrac{P_2}{P_1} = \dfrac{\Delta H_{vap}}{R}\left(\dfrac{1}{T_1} - \dfrac{1}{T_2}\right).$

2. $\dfrac{dC}{dt} = -k_1 C$

   $\dfrac{dC}{C} = -k_1 dt; \dfrac{C}{C_0} = e^{-k_1(\Delta t)}.$

3. $\dfrac{d\Psi}{dx} = A\left[\cos\left(\dfrac{2\pi}{h}\sqrt{2mE}\right)x\right]\dfrac{2\pi}{h}\sqrt{2mE}$

   $\dfrac{d^2\Psi}{dx^2} = A\left[-\sin\left(\dfrac{2\pi}{h}\sqrt{2mE}\right)x\right]\dfrac{8\pi^2 mE}{h^2} = -\dfrac{8\pi^2 mE}{h^2}\Psi$ or $\dfrac{d^2\Psi}{dx^2} + \dfrac{8\pi^2 mE}{h^2}\Psi = 0.$

4. 410°.
5. 60.6 units; 2×10^{-20} units; mighty unlikely — about one chance in a hundred million trillion!
6. 0.000496 — about one chance in 2000.
7. 4-2; there are 30 ways to achieve it against only 20 ways to achieve a 3-3 split.
8. $S = (x-a_1)^2 + (x-a_2)^2 + (x-a_3)^2 + \cdots + (x-a_n)^2$

   $\dfrac{dS}{dx} = 2(x-a_1) + 2(x-a_2) + 2(x-a_3) + \cdots + 2(x-a_n) = 0$ at most prob. x, "x_{mp}".

   Dividing by 2,
   $$x_{mp} - a_1 + x_{mp} - a_z + x_{mp} - a_3 + \cdots + x_{mp} - a_n = 0$$
   Collecting terms,
   $$nx_{mp} = a_1 + a_2 + a_3 + \cdots + a_n = \sum_i a_i$$
   $$x_{mp} = \dfrac{\sum_i a_i}{n} = \text{algebraic mean of measured values}$$

9. mean = 80.7; average deviation = 0.7; standard deviation = 0.844; 90% confidence level gives confidence interval for mean of 0.99 unit.

10. $Q = \displaystyle\int_0^\infty e^{-ax} dx = \dfrac{1}{a}.$

Chapter 3

1. $PV_0(P,T) \cdot n = H(m)T$

   $\dfrac{n}{H(m)} = \dfrac{T}{PV_0(P,T)}$

vary m keeping P and T constant:

$$\frac{n}{H(m)} = \text{const} \equiv K$$

$$H(m) = \frac{n}{K}$$

$$PV = \frac{n}{K}T$$

define $\frac{1}{K} \equiv R$:

$$PV = nRT.$$

2. mass Hg = vol × density = 76.0 cm × 1.000 cm² × 13.5955 g/cm³ = 1033.26 g mass. wt Hg = mass Hg × gravitational acceleration = 1033.26 g × 980 cm/sec² = 1.0126 × 10⁶ g-cm/sec² (or dyne)

M erg/l-atm = dyne-cm/l-atm = 1000 cm³/l × 1.0126 × 10⁶ $\frac{\text{dyne/cm}^2}{\text{atm}}$ = 1.0126 × 10⁹ erg/l-atm

R = 0.08205 l-atm/mole-deg × 1.0126 × 10⁹ erg/l-atm
= 8.314 × 10⁷ erg/mole-deg.

3. fraction = 2×10^{-293} (about 1 molecule in 10^{270} moles!) having $v = 11$ km/sec, fraction = 0.23 having $v = 500$ m/sec.
4. The gas would have to have a molecular weight of 0.06 g/mole. This is impossible, since even H atoms weigh 1.00 g/mole, but H_2 and He are light enough to allow a small fraction to escape.
5. Since Ne (AW = 20) is much lighter than Hg (AW = 200), it should be less viscous, because m appears in the numerator of expression 3-14 for η.
6. The pressure will rise initially in the xenon container, because helium atoms effuse in faster than xenon atoms effuse out. At equilibrium, however, the pressures will be equal again with the same composition in each container.
7. 2.45×10^9 molecules/ml — two and a half billion!
8. Helium is a very good heat conductor — its low mass appears in the denominator of the v_{rms} expression.
9. 740 grams; about 1½ pints of liquid water.
10. One liter at 10^{-1} torr is equivalent to 10^5 l at 10^{-6} torr; pumped at 100 l/sec, 10^5 l would last 1000 sec.
11. C_6H_{12}.
12. The total volume of the capillary is only 0.00785 cm³, which is a negligibly small part of the total volume compared to the 100 cm³ of the bulb. A one-mm length of the capillary is 7.85×10^{-7} of the total volume of the bulb plus the capillary; since at that degree of compression of the bulb's gas the pressure difference between arms of the mercury reservoir is 1 mm Hg or 1 torr, the original bulb pressure must have been 7.85×10^{-7} torr.

Chapter 4

1. Weigh pycnometer empty, then full of *m*-xylene. Subtract to get weight of *m*-xylene and calculate volume of pycnometer from the weight and known density of *m*-xylene. Weigh dry pycnometer with NaCl sample in it and subtract empty weight to get NaCl weight. Add *m*-xylene, which does not dissolve NaCl, to the partially-filled pycnometer till it is completely filled (by NaCl+*m*-xylene). Weigh and subtract previous weight to get weight of *m*-xylene used. Calculate volume of *m*-xylene present and subtract from pycnometer volume to get NaCl volume. Calculate NaCl density from its weight and volume.

2. From Bragg equation $d = 2.814$ Å. Volume of a "formula unit" is the volume of two adjacent cubes with centers 2.814 Å apart, or $2 \times 2.814^3 = 44.56$ Å^3. This corresponds to 4.456×10^{-23} cm^3 per NaCl unit. The molar volume of NaCl is 58.44 g/mole ÷ 2.165 g/cm^3 = 26.993 cm^3/mole. Comparing these numbers, $N_{Av} = [(26.993$ cm^3/mole)/$(4.456 \times 10^{-23}$ cm^3/formula unit$)] = 6.058 \times 10^{23}$ formula unit/mole.

3. Counting spheres inside the cube, there are $(6 \times \frac{1}{2}) + (8 \times \frac{1}{8}) = 4$. The volume of the four spheres is $4 \times \frac{4}{3}\pi r^3 = 16.75 r^3$. The volume of the cube is R^3, and from the drawing of a single face and the Pythagorean theorem $2R^2 = 16r^2$, or $R = \sqrt{8}r$, and $R^3 = 8\sqrt{8}r^3 = 22.61 r^3$. The spheres thus occupy 16.75/22.61 or 0.740 of the cubic volume.

4. $(2r_{met} + 2r_{cp})^2 = 2(2r_{cp})^2$
$2r_{met} + 2r_{cp} = \sqrt{2}(2r_{cp})$
$2r_{met} = (1.414 - 1)(2r_{cp})$
$2r_{met} = 0.414(2r_{cp})$
$r_{met} = 0.414 r_{cp}$

5. The larger h_1 and h_2 are, the closer the parallel lines are, and the fewer atoms are in a unit length of each line. Fewer atoms per unit area of a plane would mean less intense diffraction for planes with large h_1 and h_2.

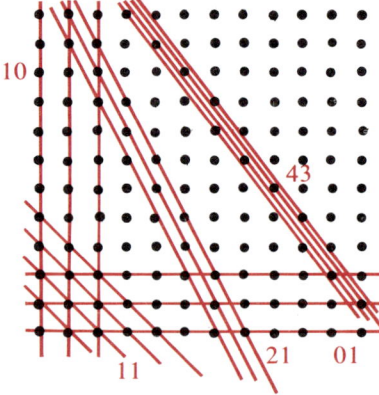

6. They should be the same, since the planes in which atoms lie define intermediate planes containing no atoms which will cleave easily.

7. Since all ions in each of these planes have the same charge, no repulsion develops when the planes move over each other. Plastic flow can occur.

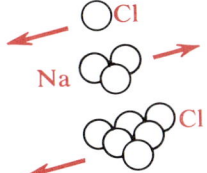

8. All wavelengths must be supplied in the Laue method since d is fixed and only a few θ angles are observable in a given crystal orientation. With all θ angles accessible in the rotating-crystal and powder methods, λ can be fixed at a single value and still be observed to diffract; if white radiation were used, the combination of all λ and all θ would expose the photographic plate (or counter) everywhere and destroy any structural information it contained.

9. To obtain the total defect energy we integrate $C_v(T)$ over T from 300° to 705°: $\int_{300}^{705} 23 e^{-10.8+0.0153T}\, dT = 1510$ cal/mole atoms or 3020 cal/mole bond. This corresponds to $3.02/69 = 0.044$ fraction of atom positions as defects, or about 1 in 20.

Chapter 5

1. Probability $= (\frac{1}{2})^{180} =$ antilog $(\log 1 - 180 \log 2) =$ antilog $(-54.0) = 10^{-54}$. Since there are 6×10^{23} α-particles in a mole, the probability of a 180° scattering occurring in the bombardment by a mole of α-particles is $6 \times 10^{23} \times 10^{-54}$, or 6×10^{-31}. This suggests that about $\frac{1}{6} \times 10^{+31}$, or 1.7×10^{30}, moles would be needed in order to expect a single 180° scattering.

2. Convert all wavelengths to wave numbers and express them as fractions of the Rydberg constant, then look for the proper set of integers n_1 and n_2:

$$\nu'_1 = \frac{1}{1215.7 \text{ Å}} = 82{,}257 \text{ cm}^{-1} = 0.7500R = \left(1 - \frac{1}{4}\right)R \qquad \therefore n_1 = 1,\ n_2 = 2$$

$$\nu'_2 = \frac{1}{1025.7 \text{ Å}} = 97{,}494 \text{ cm}^{-1} = 0.8889R = \left(1 - \frac{1}{9}\right)R \qquad n_1 = 1,\ n_2 = 3$$

$$\nu'_3 = \frac{1}{972.5 \text{ Å}} = 102{,}828 \text{ cm}^{-1} = 0.9375R = \left(1 - \frac{1}{16}\right)R \qquad n_1 = 1,\ n_2 = 4$$

$$\nu'_4 = \frac{1}{949.7 \text{ Å}} = 105{,}296 \text{ cm}^{-1} = 0.9600R = \left(1 - \frac{1}{25}\right)R \qquad n_1 = 1,\ n_2 = 5$$

3. $v = nh/2\pi mr = (nh/2\pi m)(4\pi^2 m Z e^2 / n^2 h^2) = 2\pi Z e^2 / nh$. For $1s(H)$,

$$v = \frac{2(3.14)(1)(4.80 \times 10^{-10})^2}{1(6.63 \times 10^{-27})} = 2.18 \times 10^8 \text{ cm/sec}$$

Since an orbit diameter is approximately 1 Å, $v = 2.18 \times 10^8$ cm/sec $\times 10^8$ Å/cm = 2.18×10^{16} orbit diameters/sec. A spectroscopic transition involving electron motion across an atom would take about 10^{-16} sec; by contrast, at 300°K the H atom moves at $\dfrac{\sqrt{3(8.314 \times 10^7)(300)}}{1} = 2.74 \times 10^5$ cm/sec, about a thousand times slower than the electron moves.

4. Radial nodes for $\Psi(3s)$ will occur where the factor $27 - 18\rho + 2\rho^2 = 0$, or at $\rho = 7.1$ and 1.9. By comparison, the radial node for $\Psi(2s)$ occurs at $2 - \rho = 0$ or $\rho = 2.0$, very nearly coinciding with the inner $3s$ node.

5. The $4f$ wave function has 3 total nodes ($n = 4$); an f orbital has 3 angular nodes ($l = 3$), so there can be no radial nodes. If the angular nodes are taken as planes regularly spaced, as they are in d_{xy}, d_{yz}, d_{xz}, or $d_{x^2y^2}$, the angle would be 60°.

6. Regions in which overlaps have same sign (++ or --) are symmetrically equal in area to the regions in which overlaps have opposite signs. Thus total overlap cancels to zero, which is the orthogonality condition.

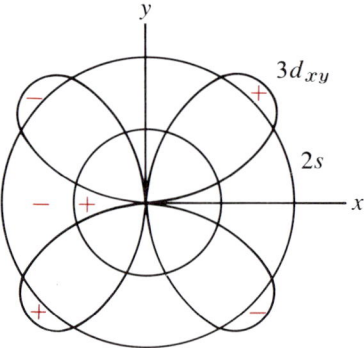
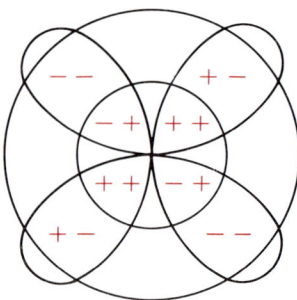

7. Since the atoms are isoelectronic in their ionized condition the total screening — the quantity σ — is constant. But going from O^{2-} to Mg^{2+} Z increases from 8 to 12, so the effective nuclear charge $(Z - \sigma)$ increases sharply. Since the Slater orbital has a factor $e^{-\frac{Z-\sigma}{Z}\rho}$, where ρ is proportional to r, the increased negative exponent for Mg^{2+} causes the electron density to drop off more sharply with distance. This means that Mg^{2+} will be considerably smaller than O^{2-}, with other species in a smooth trend.

8. Although n increases from 5 to 6, the $6s$ electrons in Ba penetrate close to a more highly charged nucleus than Sr has: $Z_{Ba} = 56$ vs. $Z_{Sr} = 38$. The $6s$ electrons are thus drawn in closer to the nucleus than they would be if they experienced the same nuclear charge as in Sr.

9. As an electron is removed the screening constant σ shrinks, which changes the average radius, and thus the energy, of the remaining electrons.

10. $IP_1(Be) > IP_1(Li)$ because a $2s$ electron is being removed in each case and Be $2s$ electrons are more stable than Li $2s$. However, $IP_1(Be) > IP_1(B)$ because boron is ionizing a less stable $2p$ electron.

Chapter 6

1. (a) tetrahedral; (b) bent, with H—O—H angle slightly smaller than 109°; (c) trigonal planar; (d) pyramidal, with F—N—F angle slightly smaller than 109°; (e) tetrahedral; (f) pyramidal, with O—Cl—O angle slightly greater than 109° and O—Cl—F angle slightly smaller; (g) square planar; (h) pyramidal, with H—As—H angle smaller than 109°; (i) linear; (j) bent, with S—S—S angle slightly smaller than 109°.

2. VSEPR predicts a bent conformation, since the 8 valence shell electrons around the S are arranged as tetrahedral pairs, two of which bond to the Cl atoms; the Cl—S—Cl angle should be somewhat smaller than 109° due to the two bulky nonbonding pairs. The MO model assigns the 20 valence electrons (2Cl×7 plus 1S×6) either to the linear-geometry MO's (Figure 6-40) or to the bent-geometry MO's (Figure 6-49). Comparison of these favors the bent conformation, since bending produces some σ-bonding character in one of the π^* orbitals obtained for linear geometry. If this MO weren't occupied, the linear geometry would be preferred; but since it is, the bent geometry is preferred. Both of these approaches are consistent with the experimental angle of 100.3°.

3.

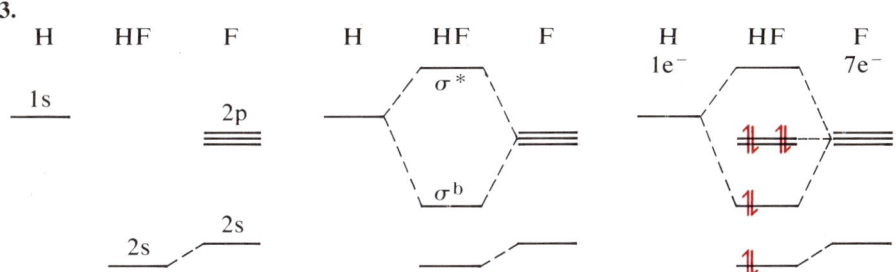

From the relative energies of MO's vs. AO's it is apparent that the fluorine atom has the predominant share of the electrons.

4. All of these species are tetrahedral and have the same number of electrons, filling the MO's of Figure 6-59 identically. But whereas the figure suggests about equal electron ownership by the C and the four H's, BH_4^- would have the B AO's at higher energy than the C AO's shown, which would tend to give the lower-energy H AO's a greater share of the bonding electrons and a partial negative charge. NH_4^+, on the other hand, has lower-energy N AO's which acquire a greater share of the bonding electrons and leave the H's with a partial positive charge. There is a trend in atomic charges from B through C to N, because of the trend in VOIP's for those atoms.

5. The normalization condition is $\int_{\text{all space}} \Psi^2 \, d\tau = 1$. If $\Psi = N^*(1s_a - 1s_b)$,

$$\int (N^*)^2 (1s_a^2 - 2 \cdot 1s_a 1s_b + 1s_b^2) \, d\tau = 1$$

$$(N^*)^2 \left[\int 1s_a^2 \, d\tau - 2 \int 1s_a 1s_b \, d\tau + \int 1s_b^2 \, d\tau \right] = 1$$

$$(N^*)^2 [\quad 1 \quad - 2S_{ab} \quad + \quad 1 \quad] = 1$$

$$(N^*)^2 (2 - 2S_{ab}) = 1$$

$$(N^*)^2 = \frac{1}{2-2_{ab}}$$

$$N^* = \sqrt{\frac{1}{2-2S_{ab}}}$$

6. The orthogonality condition is $\int_{\text{all space}} \Psi_a \Psi_b \, d\tau = 0$. For these two MO's:

$$\int \frac{1}{\sqrt{2}} (2p_{ya} + 2p_{yb}) \cdot \frac{1}{\sqrt{2}} (2s_a + 2s_b) \, d\tau \stackrel{?}{=} 0$$

$$\frac{1}{2} \int (2p_{ya} + 2p_{yb})(2s_a + 2s_b) \, d\tau \stackrel{?}{=} 0$$

$$\frac{1}{2} \left[\int 2p_{ya} 2s_a \, d\tau + \int 2p_{ya} 2s_b \, d\tau + \int 2p_{yb} 2s_a \, d\tau + \int 2p_{yb} 2s_b \, d\tau \right] \stackrel{?}{=} 0$$

Of these four integrals, the first and fourth equal zero because AO's on the same atom are always orthogonal to each other. The second and third also equal zero because they represent the situation shown in Fig. 6-23. Thus all four AO integrals equal zero, and the MO's are seen to be orthogonal.

7. $\left. \begin{array}{l} \dfrac{c_2}{c_1} = -\dfrac{\alpha_A - E}{\beta} \\[2mm] \dfrac{c_2}{c_1} = -\dfrac{\beta}{\alpha_B - E} \end{array} \right\}$ $\dfrac{\alpha_A - E}{\beta} = \dfrac{\beta}{\alpha_B - E}$ or $\alpha_A \alpha_B - (\alpha_A + \alpha_B)E + E^2 = \beta^2$

$$E^2 - (\alpha_A + \alpha_B)E + (\alpha_A \alpha_B - \beta^2) = 0$$

$$E = \frac{\alpha_A + \alpha_B \pm \sqrt{(\alpha_A + \alpha_B)^2 - 4(\alpha_A \alpha_B - \beta^2)}}{2}$$

$$= \frac{\alpha_A + \alpha_B \pm \sqrt{\alpha_A^2 - 2\alpha_A \alpha_B + \alpha_B^2 + 4\beta^2}}{2}$$

$$= \frac{\alpha_A + \alpha_B \pm (\alpha_A - \alpha_B)\sqrt{1 + \dfrac{4\beta^2}{(\alpha_A - \alpha_B)^2}}}{2}$$

But $1 \gg \dfrac{4\beta^2}{\alpha_A - \alpha_B}$, so

$$E \cong \frac{\alpha_A + \alpha_B \pm (\alpha_A - \alpha_B)\left(1 + \dfrac{2\beta^2}{(\alpha_A - \alpha_B)^2}\right)}{2}$$

$$= \begin{cases} \dfrac{\alpha_A + \alpha_B + \alpha_A - \alpha_B + \dfrac{2\beta^2}{\alpha_A - \alpha_B}}{2} = \alpha_A + \dfrac{\beta^2}{\alpha_A - \alpha_B} \\[4mm] \dfrac{\alpha_A + \alpha_B - \alpha_A + \alpha_B - \dfrac{2\beta^2}{\alpha_A - \alpha_B}}{2} = \alpha_B - \dfrac{\beta^2}{\alpha_A - \alpha_B} \end{cases}$$

The farther apart in energy orbital A and orbital B are, the less the corresponding MO will differ in energy from a given AO.

8.

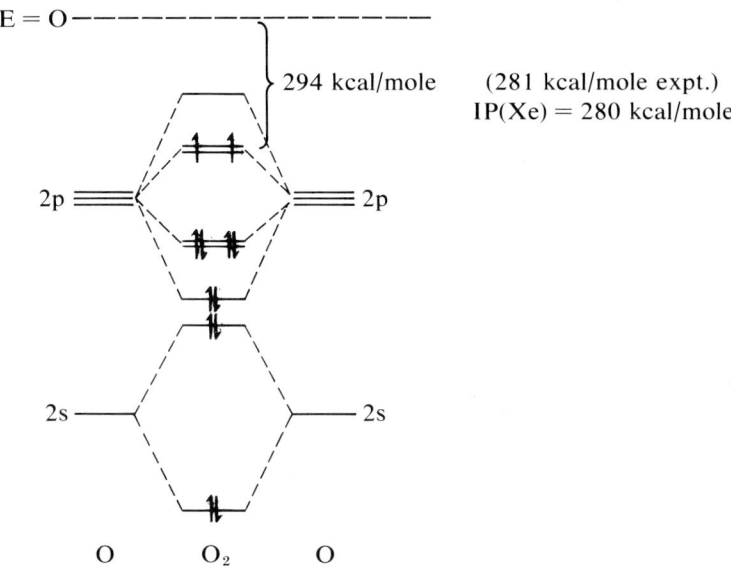

9. H$^+$ has no electrons, hence no electron energy.

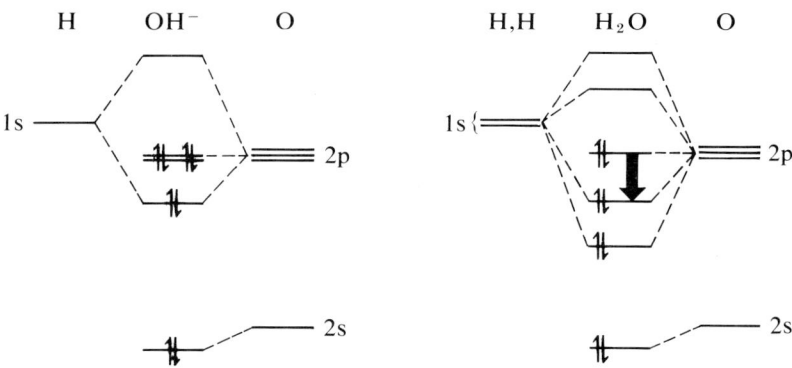

Going from OH$^-$ to H$_2$O one pair of electrons is considerably stabilized, as shown by the heavy arrow; this stabilization yields energy to the molecule's surroundings as heat.

10. NO has 11 valence electrons; filling the MO's of Fig. 6-28 with these, σ_s^b and σ_s^* cancel out and σ_s^b, π_x^b, and π_y^b contribute bonding while one electron in π_x^* (or π_y^*) "contributes antibonding." The result is a bond order of $2\frac{1}{2}$. NO$_2$ has 17 valence electrons and the bent geometry; filling the MO's of Fig. 6-48 with these, the highest-energy electron will be in the σ_{zy} MO, which has a slightly antibonding quality. Nonbonding orbitals include the two σ O(2s) AO's, σ_{yy}, and π_x, leaving four filled bond-

ing orbitals for a total bond order of just under four, or just under two per N—O bond. NO_2^- has 18 valence electrons and is still bent, but with a different (smaller) bond angle, which changes the precise MO energies. The additional electron is also in the σ_{zy} and is slightly antibonding, reducing the bond order somewhat to $3-3\frac{1}{2}$.

11.

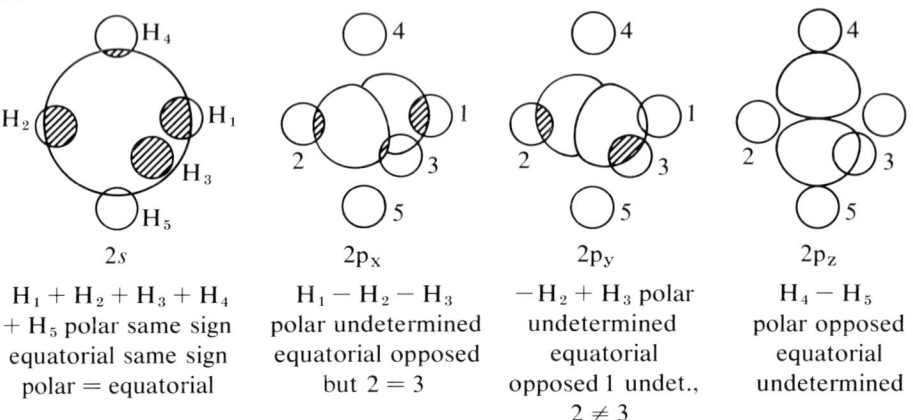

2s	$2p_x$	$2p_y$	$2p_z$
$H_1 + H_2 + H_3 + H_4$ $+ H_5$ polar same sign equatorial same sign polar = equatorial	$H_1 - H_2 - H_3$ polar undetermined equatorial opposed but $2 = 3$	$-H_2 + H_3$ polar undetermined equatorial opposed 1 undet., $2 \neq 3$	$H_4 - H_5$ polar opposed equatorial undetermined

Inspection of the above sign combinations shows that the only different combination remaining is: polar same sign, equatorial opposed, polar $\neq$ equatorial, as $H_1 + H_2 + H_3 - H_4 - H_5$. But for any of the four central-atom AO's shown, this will involve both favorable and unfavorable overlap, which indicates a net nonbonding quality. So a fifth bonding AO can't be formed.

Chapter 7

1. Because the proportion of "solid-like" molecules, which can store energy in vibrational potential energy as well as in kinetic energy, decreases with increasing temperature.
2. (a) 1.711 m, 1.667 M, 0.0299 x_{NaCl}, (b) 57.85 m (H_3PO_4 in water solvent), 14.64 M, 0.510 $x_{H_3PO_4}$, (c) 7.35 m, 4.93 M, 0.117 $x_{(CH_3)_2CO}$; since the volume of the solution is less than the combined volumes of the liquids, the two types of molecule must be attracting each other more strongly in solution than in the pure liquids. This added stability means that heat must have been released when the liquids were mixed.
3. (a) Only the center (shaded) molecule is entirely inside the two cubes. Four molecules surround it in the common face of the two cubes; each of the ten remaining cube faces has a molecule in its center, and the eight outer corners each have a molecule for a total of 23 molecules. In terms of the fractional-molecule content of the cubes, these correspond to $1 + \frac{1}{4}(4) + \frac{1}{2}(10) + \frac{1}{8}(8) = 8$ molecules. (b) If the only three neighbors are the red-shaded ones there can be no others present except the two outer faces, since the others between those faces would all touch the central molecule. There are thus $1 + \frac{1}{4}(3) = 1\frac{3}{4}$ molecules present between the outer faces. (c) $\frac{4}{13} \times 10 \cong 3$ molecules in the outer faces, for a fractional-molecule contribution of $\frac{3}{8}$. Added to $1\frac{3}{4}$, this corresponds to $2\frac{1}{8}$ molecules in the cubes. (d) If one of the three neighbors is a

facial molecule ($\times \frac{1}{2}$) instead of an edge molecule ($\times \frac{1}{4}$) the two cubes now contain an extra $\frac{1}{4}$ molecule, for a total of $2\frac{3}{8}$. If two of the three neighbors are facial the total is $2\frac{5}{8}$, and if all three neighbors are facial the total is $2\frac{7}{8}$. There aren't any other possibilities. (e) The diagonal of a face of each cube is $2d$, so the side of the cube is $\sqrt{2}d$, corresponding to a cube volume of $2^{3/2}d^3$ or $2.828d^3$ units. Two cubes thus occupy $5.656d^3$ units, which contain $2\frac{1}{2}$ molecules for an average occupied volume of $2.26d^3$ units per molecule. (f) If $d = 2^{1/6}\sigma$, then $d^3 = \sqrt{2}\sigma^3$ and $2.26d^3 = 2.26(1.414)\sigma^3 = 3.19d^3$. If the situation of three neighbors per molecule corresponds to the critical point, the critical volume is thus $3.19\sigma^3$.

4. The decrease in the solid's vapor pressure equals ΔP plus the decrease in the solution's vapor pressure:

$$\Delta T_f \times \left(\frac{dP}{dT}\right)_{sub} = \Delta P + \left[\Delta T_f \times \left(\frac{dP}{dT}\right)_{vap}\right]$$

$$\frac{d}{dT}(\ln P) = \frac{d}{dT}\left(-\frac{\Delta H_{sub}}{RT}\right) = \frac{d}{dT}\left(-\frac{\Delta H_{sub}}{R}T^{-1}\right) = +\frac{\Delta H_{sub}}{RT^2}$$

$$\frac{1}{P}\frac{dP}{dT} = \frac{\Delta H_{sub}}{RT^2} \quad \text{or} \quad \frac{dP}{dT} = \frac{P\Delta H_{sub}}{RT^2} = \frac{P\Delta H_{fus}}{RT^2} + \frac{P\Delta H_{vap}}{RT^2}$$

Returning to the first expression,

$$\frac{\Delta T_f P \Delta H_{fus}}{RT^2} + \frac{\Delta T_f P \Delta H_{vap}}{RT^2} = \Delta P + \frac{\Delta T_f P \Delta H_{vap}}{RT^2}$$

$$\Delta T_f = \frac{RT^2 \Delta P}{P \Delta H_{fus}}$$

But from Raoult's Law,

$$\frac{\Delta P}{P} = \frac{m \cdot MW_{solvent}}{1000}$$

So

$$\Delta T_f = \frac{RT^2(MW_{solvent})}{1000 \Delta H_{fus}} \cdot m = K_f \cdot m$$

5. Suppose we have exactly one gram of solute dissolved in a volume V_0 of solvent B, and that we extract it with successive portions of solvent A each having volume V_0. In any given extraction, then, if K_d refers to g/l concentrations and W^A and W^B refer to the weights of solute dissolved in solvents A and B respectively:

$$K_d = \frac{W^A/V_0}{W^B/V_0} = \frac{W^A}{W^B}$$

In the first extraction, $W^A + W^B = 1.000$ g, so

$$W^B = \frac{W^A}{K_d} = \frac{1-W^B}{K_d}$$

$$W^B(K_d+1) = 1$$

$$W^B = \frac{1}{K_d+1}$$

In the second extraction only this amount of solute ($1/(K_d+1)$) is available to be distributed: $W_2^B + W_2^A = 1/(K_d+1)$

$$W_2^B = \frac{W_2^A}{K_d} = \frac{\frac{1}{K_d+1} - W_2^B}{K_d}$$

$$W_2^B(K_d+1) = \frac{1}{K_d+1}$$

$$W_2^B = \frac{1}{(K_d+1)^2}$$

Obviously the extension of this to n extractions will give

$$W_n^B = \frac{1}{(K_d+1)^n}$$

For $K_d = 4$, if only 0.1% or 0.001 fraction is to remain,

$$0.001 = \frac{1}{5^n}$$

$$5^n = 1000$$

$$n \log 5 = \log 1000 = 3$$

$$n = \frac{3}{\log 5} = \frac{3}{0.699} = 4.29 \text{ extractions}$$

Thus 5 extractions would be required to extract better than 99.9% of the solute into solvent A if $K_d = 4$. If $K_d = 0.1$, $n \log 1.1 = 3$ or $n = 73$ extractions required for the same degree of removal of solute.

6. Honey is such a concentrated sugar solution (in water) that mold cells are killed by osmosis of the water in their cell fluid out through the cell-wall membrane.

7. Glycerine is the most viscous of the three liquids because of the many opportunities for hydrogen bonding between liquid molecules due to its three OH groups. Water also offers an opportunity for hydrogen bonding, but only to one oxygen. Ethyl ether has an oxygen atom but no H atoms attached to a strongly electronegative element, so no hydrogen bonding can occur and this liquid's viscosity is the least of the three.

8. The surface tension represents the force of attraction between neighboring molecules in the liquid; as the number of nearest neighbors per molecule decreases with increasing temperature, the surface tension decreases. Note that the critical temperature corresponds roughly to that temperature at which the surface tension vanishes (by extrapolation).

9. No change. Even though the total pressure has increased, the partial pressure of the vapor is just what it was before and the liquid-vapor equilibrium is unchanged.

10. Graph $\log P$ vs. $1/T$; the slope of the resulting line is $-\Delta H_{vap}/2.303R$. For these data, $\Delta H_{vap} = -8060$ cal/mole.
11. Boiling-point elevation gives MW = 133.0 g/mole. Vapor-pressure depression gives MW = 798 g/mole. Since the formula weight of $CuClL_3$ is $63.54 + 35.45 + 300.0 = 399.0$ g/mole, it appears that the compound forms a dimer in chloroform solution at room temperature ($2 \times 399 = 798$), but dissociates a mole of compound into three moles of particles in boiling chloroform ($399/3 = 133$). Since many CuClL coordination compounds are known, we surmise $CuClL_3 \xrightarrow{heat} CuClL + 2L$.
12. $\Delta T_b = K_b \cdot m$. The molality of the solution is

$$\frac{\text{wt rubber/MW rubber}}{\text{kg toluene}} = \frac{3.65/(9.08 \times 10^4)}{1000(0.857) - 3.65} = 4.71 \times 10^{-8} \, m.$$

This very small figure gives

$$\Delta T_b = 3.33 \times 4.71 \times 10^{-8} = 1.57 \times 10^{-7} \text{ degrees}.$$

This expected value for the boiling point elevation is far too small to measure; only the osmosis technique offers any hope of measuring high molecular weights.

Chapter 8

1. $C_2H_2 \; + \; 2\tfrac{1}{2}O_2 \; \rightarrow \; 2CO_2 \; + \; H_2O$
     $S = 48.00 \quad 2\tfrac{1}{2} \times 49.00 \qquad 2 \times 51.06 \quad 45.11$
     $\underbrace{\qquad\qquad\qquad\qquad}_{170.50 \text{ cal/deg}} \qquad \underbrace{\qquad\qquad\qquad}_{147.23 \text{ cal/deg}}$

$$\Delta S = 147.23 - 170.50 = -23.27 \text{ cal/deg}$$

The chemical combustion reaction in an oxyacetylene torch is spontaneous even though ΔS for the reaction system is negative because the surroundings are being strongly heated, so that their entropy is increasing substantially:

$$\Delta S_{\text{surroundings}} = C_v \ln \frac{T_2}{T_1}$$

2.

Compound	No. C	$S°$ (cal/mole-deg)	$S°$ difference between molecules
CH_4	1	44.50	
CH_3-CH_3	2	54.85	10.35
$CH_3-CH_2-CH_3$	3	64.51	9.66
$CH_3-CH_2-CH_2-CH_3$	4	?	?

If we hypothesize that the $S°$ *difference* from one molecule to the next is roughly constant, then we might expect $S°$ for butane to be about 9.66 cal/mole-deg greater than that for propane: $64.51 + 9.66 = 74.17$ cal/mole-deg. This matches the observed

74.10 cal/mole-deg quite well and suggests a general formula:
$$S°_{\text{straight-chain hydrocarbon}} = 54.85 \times 9.60(\text{No. C-2}) \text{ cal/mole-deg}$$

Apparently an additional —CH_2— group is worth about 9.60 cal/mole-deg.

3. $\Delta S_{\text{fusion}} = 16.72 - 9.87 = +6.85$ cal/mole-deg; $\Delta S_{\text{vaporization}} = 45.11 - 16.72 = +28.39$ cal/mole-deg. When ice melts, the orderly crystal structure breaks down into a relatively disorderly liquid, which corresponds to a positive ΔS. However, when the liquid vaporizes the large increase in volume indicates a much greater increase in disorder and thus a much larger positive ΔS.

4. Most of the elements in Table 8-1 for which both an oxide and a chloride are listed have the element in a 2+ oxidation state, which means more atoms in a mole of compound for the chloride: MCl_2 vs. MO. This additional complexity increases the total molar entropy. Chlorine is also a heavier element than oxygen, with more electrons and more closely spaced energy levels. On the other hand CO_2 has very nearly the same entropy as CCl_4, even with fewer atoms, because it is a gas while CCl_4 is a liquid. For HCl and H_2O vapor, the increased number of atoms and bonds favors the oxide.

5. (a) Spontaneous to the right because more moles of high-entropy gas are being formed than are being consumed; (b) not spontaneous; gas is being consumed; (c) not spontaneous; number of particles is decreasing sharply (7 → 2) although one mole of gas is being formed; (d) not spontaneous; gas is disappearing and the charged ions will tend to create order in the liquid water because of its very polar molecules. To be spontaneous, (b), (c), and (d) must heat their surroundings as they proceed; actually all four evolve heat.

6. $Q_1(1 \to 2) = nRT_1 \ln \dfrac{V_2}{V_1}$ (heat input)

$Q_2(2 \to 4) = 0$ (adiabatic definition)

$Q_3(4 \to 3) = nRT_2 \ln \dfrac{V_3}{V_4} = -nRT_2 \ln \dfrac{V_4}{V_3}$ (heat produced)

$Q_4(3 \to 1) = 0$ (adiabatic)

$Q_{\text{total}} = nR\left(T_1 \ln \dfrac{V_2}{V_1} - T_2 \ln \dfrac{V_4}{V_3}\right) + 0 + 0$

$= nR(T_1 - T_2) \ln \dfrac{V_2}{V_1}$

If $W_{\text{total}} = Q_{\text{total}}$, the efficiency equals the ratio of the work performed to the heat input:

$$\dfrac{W_{\text{total}}}{\text{heat input}} = \dfrac{Q_{\text{total}}}{Q_1} = \dfrac{nR(T_1 - T_2) \ln (V_2/V_1)}{nRT_1 \ln (V_2/V_1)} = \dfrac{T_1 - T_2}{T_1}$$

For a simple steam engine in which the isothermal expansion (process 1) occurs at 100°C (373°K) and the isothermal compression (3) at 25°C (298°K):

$$\text{efficiency} = \dfrac{373 - 298}{373} = \dfrac{75}{373} = 0.201 = 20.1\%$$

7. Nuclear plant:

$$\text{efficiency} = \frac{(290+273)-(40+273)}{290+273} = 0.444 = \frac{1000 \text{ megawatt-sec}}{1000 \text{ megawatt-sec} + \text{heat rejection}}$$

$$\frac{\text{heat rejection}}{\text{per second}} = \frac{10^9 \text{ joule}}{0.444} - 10^9 \text{ joule} = 1.25 \times 10^9 \text{ joule} = 2.98 \times 10^8 \text{ cal}$$

$$\frac{\text{temperature rise}}{\text{of river}} = \frac{\text{heat rejection}}{\text{total heat cap.}}$$

$$= \frac{2.98 \times 10^8 \text{ cal/sec}}{4000 \text{ ft}^3/\text{sec } (2.83 \times 10^4 \text{ cm}^3/\text{ft}^3)(1.00 \text{ cal/cm}^3\text{-deg})}$$

$$= 2.63 \text{ deg}$$

coal-fired plant:

$$\text{efficiency} = 0.498; \text{ heat rejection/sec} = 1.01 \times 10^9 \text{ J/sec};$$
$$\text{temperature rise} = 2.10 \text{ deg}$$

The thermal pollution is about 20% less for the coal-fired plant — but against this must be balanced the contribution which the coal-fired plant makes to air pollution from its stack gases even if all fly ash is removed.

8. The total energy throughput of the plant is 2.01×10^9 J/sec or 4.80×10^8 cal/sec, corresponding to $4.80 \times 10^8/6.00 \times 10^3 = 8.00 \times 10^4$ g of coal per sec. If 2%, or 1.60×10^3 g of this is sulfur which is burned according to the reaction equation

$$\text{S} \quad + \quad \text{O}_2 \rightarrow \quad \text{SO}_2$$

$$32.06 \text{ g/mole} \qquad\qquad 64.06 \text{ g/mole}$$

then $(64.06/32.06) \times 1.60$ kg or 3.20 kg/sec of SO_2 are released into the atmosphere. This corresponds to 12.7 ton/hr or 158,000 ft³/hr of pure SO_2 at 30°C. This problem has little thermodynamic significance but an increasing importance to our environment.

9. Three red balls can be placed in three boxes in only one way; if one ball is distinguishable there are three ways to place it (and the other two) in three boxes; if all three balls are different there are six ways to place them in three boxes. In our molecules, if we let the —CH_2Cl group define the "boxes":

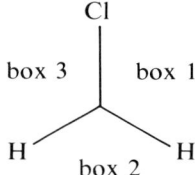

then CH_3—CH_2Cl corresponds to the three red balls, since the three H atoms on CH_3— are indistinguishable. CH_2Cl—CH_2Cl corresponds to two red and one green ball, and $CHBrCl$—CH_2Cl to a red, a blue, and a green ball. Since there are more possible configurations for $CHBrCl$—CH_2Cl, it should have a higher entropy than CH_2Cl-CH_2Cl, whose entropy in turn should be higher than CH_3-CH_2Cl. Of course, the higher molecular weight of the last two compounds will also tend to increase their absolute entropy.

Chapter 9

1. For process 1 we may calculate ΔS_{AB} as the sum of ΔS_{Ap} and ΔS_{qB}:

$$\left. \begin{array}{l} \Delta S_{Ap} = C_v \ln \dfrac{T_c}{T} \\ \\ \Delta S_{pB} = C_p \ln \dfrac{T}{T_c} \end{array} \right\} \Delta S_{AB} = (C_p - C_v) \ln \dfrac{T}{T_c}$$

Similarly, for process 2:

$$\left. \begin{array}{l} \Delta S_{Aq} = C_p \ln \dfrac{T_h}{T} \\ \\ \Delta S_{qB} = C_v \ln \dfrac{T}{T_h} \end{array} \right\} \Delta S_{AB} = (C_p - C_v) \ln \dfrac{T_h}{T}$$

Since ΔS_{AB} is constant,

$$(C_p - C_v) \ln \frac{T}{T_c} = (C_p - C_v) \ln \frac{T_h}{T}$$

$$\frac{T}{T_c} = \frac{T_h}{T}$$

Subtract T/T from the right side of this expression and 1 from the left:

$$\frac{T}{T_c} - 1 = \frac{T_h - T}{T}$$

$$T \left(\frac{T}{T_c} - \frac{T_c}{T_c} \right) = T_h - T$$

$$\frac{T}{T_c}(T - T_c) = T_h - T$$

Since $T \neq T_c$, $T - T_c \neq T_h - T$. Consequently, dQ for process 1 is *not* equal to dQ for process 2, since that equality would require $T - T_c = T_h - T$.

2. For the reaction $CO_2 \to C + 2O - \Delta H_{\text{atomization}}$
$\Delta H_{at} = +171.70 + 2(+59.16) - (-94.0) = +384.0$ cal/mole CO_2
Similarly, for $H_2O_{(g)} \to 2H + O - \Delta H_{at}$
$\Delta H_{at} = 2(+52.09) + 59.16 - (-57.80) = +221.1$ cal/mole H_2O

For the butane combustion
$C_4H_{10} + 6\tfrac{1}{2}O_2 \to 4CO_2 + 5H_2O + 682.5$ kcal/mole reaction
if 20% of each product atomizes the total $\Delta H_{\text{combustion}}$ is reduced by the atomization energy for 0.8 mole CO_2 and 1.0 mole H_2O:

$$\Delta H_{\text{combustion with atomization}} = -682.5 + 0.8(384.0) + 1.0(221.1)$$
$$+ -154.1 \text{ kcal/mole reaction}$$

$$T_{\text{flame}} = \frac{154{,}100 \text{ cal}}{85 \text{ cal/deg}} = 1815° \text{ increase}$$

$$= 1815 + 25 = 1840°C$$

3. $\Delta G = \Delta H - T\Delta S$

$\quad = \Delta E + \Delta(PV) - T\Delta S$

$\quad = 0 + \Delta(nRT) - T\Delta S$

$\quad = 0 + 0 - T\Delta S$

$\quad = -T\left(nR \ln \dfrac{V_2}{V_1}\right)$

$\quad = nRT \ln \dfrac{V_2}{V_1}$

Only if V_2 is greater than V_1 will $\ln(V_2/V_1)$ be positive and ΔG negative, which says that no spontaneous compression of a gas can occur. This is, of course, exactly what our experience tells us.

4. Zero — a liquid in contact with its vapor at 1 atm pressure is at equilibrium, which indicates $\Delta G = 0$. Above the BP,

$$\Delta G = \Delta H - T\Delta S = \Delta H_{\text{vap}} - T\left(\frac{Q_{\text{rev}}}{T_{\text{rev}}}\right)$$

$$= \Delta H_{\text{vap}} - \frac{T}{T_{\text{rev}}}(\Delta H_{\text{vap}}) = \left(\frac{T_{\text{rev}} - T}{T_{\text{rev}}}\right)\Delta H_{\text{vap}}$$

If T is greater than T_{rev} (the boiling point at the existing pressure) the quantity in parentheses, and thus ΔG, will be negative. This indicates spontaneity.

5. $\Delta H = C_p(\Delta T)$

For butane, the ΔH for heating is $0.47(2100 - 298) = +847$ cal/mole
For O_2, $\Delta H_{\text{heating}} = 0.22(2100 - 298) = +396$ cal/mole
For CO_2, $\Delta H_{\text{heating}} = 0.25(2100-298) = +450$ cal/mole
For H_2O, $\Delta H_{\text{heating}} = 0.50(2100 - 298) = +901$ cal/mole
In the reaction

$C_4H_{10} + 6\tfrac{1}{2}O_2 \rightarrow 4CO_2 + 5H_2O - \Delta H°_{298}$

$\Delta H°_{298} = 4\Delta H_f°(CO_2) + 5\Delta H_f°(H_2O) - \Delta H_f°(C_4H_{10}) - 0$

$\Delta H°_{2100} = 4(\Delta H_f° + \Delta H_{\text{heat}})_{CO_2} + 5(\Delta H_f° + \Delta H_{\text{heat}})_{H_2O} - (\Delta H_f° + \Delta H_{\text{heat}})_{C_4H_{10}}$

$\qquad\qquad\qquad\qquad\qquad\qquad\qquad\qquad\qquad\qquad - 6\tfrac{1}{2}(0 + \Delta H_{\text{heat}})_{O_2}$

$\quad = \Delta H°_{298} + [4(450) + 5(901) - 1(847) - 6\tfrac{1}{2}(396)]$

$\quad = -682{,}500 + (6305 - 3425)$

$\quad = -679{,}600$ cal or -679.6 kcal/mole reaction

This is a change of about 0.5% in ΔH^0 for the combustion, which would correspond to about 0.5% change in the flame temperature as calculated on p. 456 if 2100°K were close to the flame temperature. Actually 8300°K was calculated, which would give a change in ΔH^0 (and thus of the flame temperature) of about 2%. In Study Problem 2, however, the available ΔH^0 was reduced to -154.1 kcal/mole reaction at a flame temperature of about 2100°K. The change in ΔH^0 of $+2880$ cal is about 1.9% of this figure, which would lead to a 1.9% reduction in the flame temperature. The approximation that ΔH^0 is constant with temperature is a pretty good one.

6. $\Delta G^0 = -RT \ln K_w = -2.303 RT \log_{10} K_w$
$= -2.303 (1.987)(273)(0.134-15)$
$= 18590$ cal or $+18.59$ kcal/mole reaction at 0°C

To get ΔS^0 we must use the definition $\Delta G^0 = \Delta H^0 - T\Delta S^0$ and calculate ΔH^0:

$$\ln \frac{K_2}{K_1} = -\frac{\Delta H^0}{R}\left(\frac{1}{T_2} - \frac{1}{T_1}\right)$$

$$2.303 \log_{10} \frac{5.50 \times 10^{-13}}{1.36 \times 10^{-15}} = -\frac{\Delta H^0}{1.987}\left(\frac{1}{373} - \frac{1}{273}\right)$$

$$\Delta H^0 = +12.2 \text{ kcal/mole reaction}$$

Now $\Delta S^0 = -(\Delta G^0 - \Delta H^0)/T = -6400/273 = -23.4$ cal/deg-mole reaction. Since a negative ΔS^0 is associated with increasing order in the reaction system, it must be that the electrically-charged ions on the right of the reaction equation are ordering the polar water molecules surrounding them more than the reactant water molecules do.

7. $C_2H_4 + H_2 \rightarrow C_2H_6 - \Delta G^0$
$\Delta G^0 = \Delta G_f^0(C_2H_6) - \Delta G_f^0(C_2H_4) - \Delta G_f^0(H_2)$
$= -7.9 - 16.3 - 0 = -24.2$ kcal/mole reaction

Similarly,
$\Delta H^0 = -20.2 - 12.5 - 0 = -32.7$ kcal/mole reaction

At 298°,

$$\Delta G^\circ = -RT \ln K_{eq}$$

$$K_{eq} = 10^{-\Delta G^0/2.303RT} = 10^{+24200/2.303(1.987)(298)} = 10^{17.78}$$
$$= 6.0 \times 10^{17}$$

At 1200°C (1473°K),

$$2.303 \log_{10} \frac{K_{1473}}{K_{298}} = +\frac{32700}{1.987}\left(\frac{1}{1473} - \frac{1}{298}\right) = -44.05$$

$$K_{1473} = (6.0 \times 10^{17})(10^{-44.05/2.303}) = 4.3 \times 10^{-2}$$

The value of K_{eq} at the higher temperature is much smaller (almost 20 powers of 10!) because ΔS^0, which is multiplied by T in the ΔG^0 expression and thus becomes more important at higher temperatures, is negative due to the decreasing number of gas molecules in the reaction.

8. $K_{eq} = \frac{[\text{gauche}]}{[\text{anti}]} = \frac{20}{80} = 0.25$

Problem Answers | 911

$\Delta G^0 = -RT \ln K_{eq} = -1.987(300)(2.303)(-0.602) = +828$ cal/mole reaction
We can calculate ΔS^0 from the number of configurations:

$$\Delta S^0 = R \ln \left(\frac{\text{no. of gauche config.}}{\text{no. of anti config.}}\right) = R \ln \left(\frac{2}{1}\right) = 1.987(2.303) \log 2$$

$$= +1.379 \text{ cal/deg-mole reaction}$$

$$\Delta H^0 = \Delta G^0 + T\Delta S^0$$

$$= 828 + 300(1.379)$$

$$= 1242 \text{ cal/mole reaction}$$

The ΔH^0 value of 1242 cal/mole reaction is a better measure of the repulsion between gauche Cl atoms than the ΔG^0 value, because ΔG^0 contains the statistical weighting due to the ΔS^0 influence.

9. $2NH_3 + \tfrac{3}{2}O_2 \rightarrow N_2 + 3H_2O - \Delta H^0$
$\Delta H^0 = \Delta H_f^0(N_2) + 3\Delta H_f^0(H_2O) - 2\Delta H_f^0(NH_3) - \tfrac{3}{2}\Delta H_f^0(O_2)$
$= 0 + 3(-57.80) - 2(-11.04) - 0$
$= -151.32$ kcal/mole reaction (for 34.0 g NH_3 burned)
$\Delta H^0/\text{g } NH_3 = 151.32/34.0 = -4.45$ kcal/g NH_3
Comparing with the reaction for the combustion of ethane:
$C_2H_6 + 3\tfrac{1}{2}O_2 \rightarrow 2CO_2 + 3H_2O - \Delta H^0$
$\Delta H^0 = 2(-94.0) + 3(-57.80) - (-20.2) - 0$
$= 341.2$ kcal/mole reaction (for 30.0 g C_2H_6 burned)
$\Delta H^0/\text{g } C_2H_6 = -341.2/30.0 = -11.37$ kcal/g C_2H_6
Although the ammonia combustion is intrinsically cleaner than that of ethane (natural gas), it yields only about a third as much energy per gram—or per ton.

10. $Fe_2O_{3(s)} + \tfrac{3}{2}C_{(s)} \rightarrow 2Fe_{(l)} + \tfrac{3}{2}CO_{2(g)} - \Delta G^0$
$\Delta G^0 = 2\Delta G_f^0(Fe_{(l)}) + \tfrac{3}{2}\Delta G_f^0(CO_2) - \Delta G_f^0(Fe_2O_3) - \tfrac{3}{2}\Delta G_f^0(C)$
$= 2(+3.6) + \tfrac{3}{2}(-94.3) - (-177.1) - 0$ (taking $\Delta G_{fus} = \Delta H_{fus}$ for Fe)
$= +42.8$ kcal/mole reaction (plus sign indicates reaction unfavorable)
$\Delta H^0 = 2(+3.6) + \tfrac{3}{2}(-94.0) - (-196.5) - 0$
$= 62.7$ kcal/mole reaction
$\Delta S^0 = (\Delta H^0 - \Delta G^0)/T = (62.7 - 42.8)/298 = +0.0667$ kcal/deg
$= +66.7$ cal/deg
$\Delta G = 0 = \Delta H - T\Delta S$ at the "break-even" point for the reaction
$0 = +62700 - T(66.7)$
$$T = \frac{62700}{66.7} = 940°K \text{ or about } 670°C$$

11. $CrCl_{3(s)} + 6H_2O_{(l)} \rightarrow Cr(OH_2)_6Cl_{3(s)} - \Delta G^0$
$\Delta G^0 = \Delta G_f^0(Cr(OH_2)_6Cl_3) - \Delta G_f^0(CrCl_3) - 6\Delta G_f^0(H_2O)$
$\cong -579.0 - (-134.6) - 6(-56.7)$ (taking $\Delta G_f^0 \cong \Delta H_f^0$ for Cr species)
$\cong -104.2$ kcal/mole reaction
This large negative value indicates that the hydration reaction is very favorable thermodynamically. Since it does not proceed (in the absence of catalysts) we assume that no mechanism exists for the reaction.

Chapter 10

1. $N_2 + 3H_2 \rightleftharpoons 2NH_3$ $\hspace{2cm}$ $N_2 + O_2 \rightleftharpoons 2NO$

$$K_P = \frac{P_{NH_3}^2}{P_{N_2} \cdot P_{H_2}^3} \quad \text{(in atm)} \hspace{1cm} K_P = \frac{P_{NO}^2}{P_{N_2} \cdot P_{O_2}} \quad \text{(in atm)}$$

$$= \frac{(P_{NH_3}/760)^2}{(P_{N_2}/760)(P_{H_2}/760)^3} \text{(in torr)} \hspace{0.5cm} = \frac{(P_{NO}/760)^2}{(P_{N_2}/760)(P_{O_2}/760)} \text{(in torr)}$$

$$= \frac{760^4}{760^2} \left\{ \frac{P_{NH_3}^2}{P_{N_2} \cdot P_{H_2}^3} \text{(torr)} \right\} \hspace{0.5cm} = \frac{760^2}{760^2} \left\{ \frac{P_{NO}^2}{P_{N_2} \cdot P_{O_2}} \text{(torr)} \right\}$$

$$= 5.78 \times 10^5 \left\{ \frac{P_{NH_3}^2}{P_{N_2} \cdot P_{H_2}^3} \text{torr} \right\} \hspace{0.5cm} = \frac{P_{NO}^2}{P_{N_2} \cdot P_{O_2}} \text{(torr)}$$

The K_P value for the NH_3 equilibrium will change if the 5.78×10^5 factor due to the 760-torr standard state is absorbed into it; the K_P value for the NO equilibrium is unchanged by the change of units. The condition for this independence of units is that the same total number of gaseous molecules appear on each side of the balanced reaction equation.

2. The unstable compounds should be those involving a combination of a hard acid and soft base, or a soft acid and a hard base: AuF, $FeSe_4^{2-}$, CdO_4^{6-}, Cr_2O, $MoCl_6$, BF_3Br^-, and SI_6 would be expected to be unstable by these criteria and are; most have never been prepared.

3. Mass balance:
$$[NH_4^+] + [NH_3] = [HAc] + [Ac^-]$$

$$K_a(NH_4^+) = \frac{[NH_3][H_3O^+]}{[NH_4^+]} \hspace{1cm} K_a(HAc) = \frac{[Ac^-][H_3O^+]}{[HAc]}$$

$$[NH_3] = \frac{K_a(NH_4^+) \cdot [NH_4^+]}{[H_3O^+]} \hspace{1cm} [HAc] = \frac{[Ac^-][H_3O^+]}{K_a(HAc)}$$

Substituting these in the mass-balance expression:

$$[NH_4^+] + [NH_4^+] \frac{K_a(NH_4^+)}{[H_3O^+]} = [Ac^-] + [Ac^-] \frac{[H_3O^+]}{K_a(HAc)}$$

Multiplying through by $[H_3O^+]$:

$$[NH_4^+]\{[H_3O^+] + K_a(NH_4^+)\} = [Ac^-]\left\{[H_3O^+] + \frac{[H_3O^+]^2}{K_a(HAc)}\right\}$$

Rearranging:

$$[H_3O^+]^2 + K_a(HAc)\left\{1 - \frac{[NH_4^+]}{[Ac^-]}\right\}[H_3O^+] - \frac{[NH_4^+]}{[Ac^-]}K_a(NH_4^+)K_a(HAc) = 0$$

In the second term (the $[H_3O^+]$ coefficient), $K_a(HAc)$ is already quite a small number, and if $[NH_4^+]/[Ac^-] \cong 1$ the coefficient will be still further reduced. To a good approximation, then, we can ignore the second term:

$$[H_3O^+]^2 \cong \frac{[NH_4^+]}{[Ac^-]} K_a(NH_4^+)K_a(HAc)$$

$$\cong 1 \cdot K_a(NH_4^+)K_a(HAc)$$

$$[H_3O^+] \cong \sqrt{K_a(NH_4^+)K_a(HAc)}$$

4. $Pb^0 + PbO_2 + 4H_3O^+ + 2SO_4^{2-} \overset{2e^-}{\rightleftharpoons} 2PbSO_4 + 6H_2O - \Delta G^0$

$$\Delta G^0 = -n\mathscr{F}\mathscr{E}^0 = -2(23061)(2.0420)$$

$$= -94{,}181 \text{ cal/mole reaction}$$

$$\Delta S^0 = n\mathscr{F}\frac{d\mathscr{E}^0}{dT}$$

Graphing $\mathscr{E}^0$ vs. T and drawing in a tangent at 25°, we obtain a slope $(d\mathscr{E}^0/dT)$ of $+2.048 \times 10^{-4}$ v/deg:

$$\Delta S^0 = 2(23061)(2.048 \times 10^{-4})$$

$$= +9.446 \text{ cal/deg-mole reaction}$$

$$\Delta H^0 = \Delta G^0 + T\Delta S^0$$

$$= -94{,}181 + 298.2(9.446)$$

$$= -91{,}364 \text{ cal/mole reaction}$$

Note that the precision possible in voltage measurement makes comparably precise values of thermodynamic values available.

5. (a) $3Cu^0 + 2NO_3^- + 8H_3O^+ \to 3Cu(OH_2)_4^{2+} + 2NO$
   (b) $Cu^0 + 2NO_3^- + 4H_3O^+ \to Cu(OH_2)_4^{2+} + 2NO_2 + 2H_2O$
   (c) $5ClO_4^- + 4SnS + 12H_2O \to 5Cl^- + 4Sn(OH_2)_2(OH)_2^{2+} + 4SO_4^{2-}$
   (d) $10Cr_2O_7^{2-} + 3(CH_3)_2CHCOOH + 80H_3O^+ \to 20Cr(OH_2)_6^{3+} + 12CO_2 + 12H_2O$
   (e) $3S_2O_3^{2-} + 2NH_3 + 6OH^- \to 6SO_4^{2-} + N_2 + 6H_2O$
   (f) $4S_2O_8^{2-} + NH_4^+ + 13H_2O \to 8SO_4^{2-} + NO_3^- + 10H_3O^+$
   (g) $2Se_2Cl_2 + 7H_2O \to H_2SeO_3 + 3Se^0 + 4Cl^- + 4H_3O^+$
   (h) $P_4 + 3H_2O + 3OH^- \to PH_3 + 3H_2PO_2^-$
   (i) $2(CH_3)_2As(O)OH + H_3PO_2 + 2Cl^- + H_3O^+ \to 2(CH_3)_2AsCl + 3H_2O + H_2PO_4^-$
   (j) $2MoCl_4 + 9H_2O \to Mo(OH_2)_6^{3+} + MoOCl_3 + 5Cl^- + 2H_3O^+$
   (k) $3Cl_2 + Br^- + 6OH^- \to BrO_3^- + 6Cl^- + 3H_2O$
   (l) $2HN_3 + 5Zn + 13H_3O^+ \to 3N_2H_5^+ + 5Zn^{2+} + 13H_2O$

6. 4.83.

7. The necessary $C_2O_4^{2-}$ concentration is 6.3×10^{-7} M, which is maintained in a 0.0100 M $Na_2C_2O_4$ solution all the way down to pH = 0.01.

9. The total pressure in the reaction system will increase as the reaction proceeds because the number of moles of gas is increasing. If ΔP is the amount by which the

pressure has changed from the initial pressure of $1 + P^0$ when equilibrium is reached, then $K_P = 4(\Delta P)^3/(P^0 - 2\Delta P)$.

10. About 35 milligrams.

11. $CO_{2(aq)} + CO_{3(aq)}^{2-} + H_2O_{(l)} \rightleftharpoons 2HCO_{3(aq)}^{-}$

$$K_{eq} = \frac{[HCO_3^-]^2}{[CO_2][CO_3^{2-}]} = \frac{[HCO_3^-][H_3O^+]}{[CO_2]} \cdot \frac{[HCO_3^-]}{[CO_3^{2-}][H_3O^+]} = K_1 \cdot \frac{1}{K_2} = \frac{K_1}{K_2}$$

$$K_1 = \frac{[HCO_3^-][H_3O^+]}{[CO_2]} \qquad K_2 = \frac{[CO_3^{2-}][H_3O^+]}{[HCO_3^-]}$$

$$[CO_2] = \frac{[HCO_3^-][H_3O^+]}{K_1} \qquad [CO_3^{2-}] = \frac{K_2[HCO_3^-]}{[H_3O^+]}$$

If the primary source of CO_2 and CO_3^{2-} in the ocean is the ocean's HCO_3^- content, then $[CO_2] \cong [CO_3^{2-}]$:

$$\frac{[\cancel{HCO_3^-}][H_3O^+]}{K_1} \cong \frac{K_2[\cancel{HCO_3^-}]}{[H_3O^+]}$$

$$[H_3O^+]^2 \cong K_1 K_2$$

$$[H_3O^+] \cong \sqrt{K_1 K_2} = \sqrt{4.3 \times 10^{-7} \times 5.6 \times 10^{-11}} = 4.9 \times 10^{-9} M$$

$$pH = 9 - 0.69 = 8.31$$

This is fairly close to the experimental 8.1.

12. From the Nernst equation for the Hg^{2+}/Hg couple the concentration of Hg^{2+} is calculated as $1.7 \times 10^{-7}\ M$, which indicates a β_4 value of 2.8×10^{30}.

13. The hydrated-lead-ion concentration in $0.100\ M$ PbY is $3.5 \times 10^{-10}\ M$; since K_{sp} for $PbCrO_4$ is 3×10^{-13}, $0.100\ M$ CrO_4^{2-} is sufficiently concentrated to precipitate the lead.

14. Only about 300,000. (Calculate $[H_3O^+]$ for each pH and take the difference.)

15. Using the quadratic solution to the $K_a(HIO_3)$ expression, $[IO_3^-] = 0.142\ M$. This concentration, squared and multiplied by the indicated $[Cu^{2+}]$ of $1.57 \times 10^{-5}\ M$, does exceed the K_{sp} for $Cu(IO_3)_2$, so precipitation should occur (but it's close).

16. The concentration of $H_2PO_4^-$ must be $0.0667\ M$ to achieve the desired $[H_3O^+]$ of $4.2 \times 10^{-8}\ M$. This requires 9.20 g of $NaH_2PO_4 \cdot H_2O$.

17. $\mathscr{E}^0 = -0.037$ V.

18. Applying the Nernst equation to both the dilute and the saturated solutions, we get $\mathscr{E}(0.01\ M) = +0.287$ V and $\mathscr{E}(2.0\ M) = +0.355$ V. The measured voltage will thus be 0.068 V. The $2M$ electrode half-reaction has the more positive potential so it goes to the right, absorbing electrons from its electrode and reducing the Cu^{2+} concentration in that part of the solution. At the $0.01\ M$ electrode the half-reaction is driven to the left, metallic copper is oxidized to Cu^{2+}, and the Cu^{2+} concentration increases. This pattern of concentration change is exactly the same as the spontaneous change through diffusion from the more concentrated to the more dilute regions.

19. $K_{eq} = 2 \times 10^{55}$, which explains why Cr^{II} solutions are very unstable in air.

20. At 1 M concentrations the $\mathscr{E}^0$ values apply; iron will be reduced since its reduction

potential is higher than that of the platinum couple. Cl⁻ appears in the numerator of the concentration fraction in the Nernst equation for the platinum couple; since that fraction is subtracted from $\mathscr{E}^0(Pt)$, the Cl⁻ concentration should be made smaller to increase $\mathscr{E}(Pt)$. At $[Cl^-] = 0.626\ M$, $\mathscr{E}(Pt) = \mathscr{E}^0(Fe)$ and no reaction will occur.

21. 321 minutes are required for the deposition of nearly a pound of silver.

22.

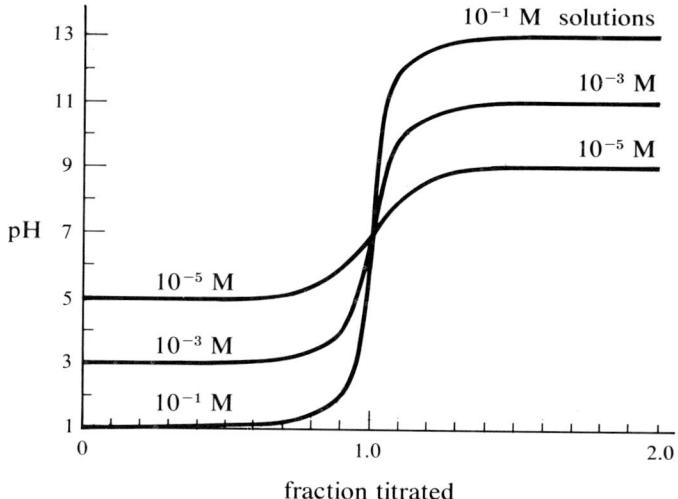

At very low concentrations the break at the equivalence point is much less pronounced. In effect, the water present (which can serve as a proton donor and acceptor) is "buffering" the solution at very low concentrations of other acids or bases.

Chapter 11

1. First-order differential rate law:

$$-\frac{d[A]}{dt} = k[A]$$

$$\frac{d[A]}{[A]} = -k\,dt$$

$$\ln [A] \Big|_{C_A}^{C_A/2} = -kt \Big|_0^{t_{1/2}}$$

$$\ln (C_A/2) - \ln C_A = -kt_{1/2} - 0$$

$$\ln (C_A/2C_A) = -kt_{1/2}$$

$$\ln 1/2 = -kt_{1/2}$$

$$\ln 2 = kt_{1/2}$$

$$t_{1/2} = (\ln 2)/k = 0.693/k$$

2. $k = Ae^{-E_a/RT}$

$k_{300} = Ae^{-20,000/(1.987 \times 300)} = Ae^{-33.55}$

$k_{310} = Ae^{-20,000/(1.987 \times 310)} = Ae^{-32.47}$

$k_{310}/k_{300} = \dfrac{Ae^{-32.47}}{Ae^{-33.55}} = e^{+1.08} = 10^{1.08/2.303} = 2.94$

3. $-\dfrac{d[A]}{dt} = k[A]^2$

$\dfrac{d[A]}{[A]^2} = -k\,dt$

$[A]^{-2}d[A] = -k\,dt$

$\dfrac{[A]^{-1}}{-1} = -kt$

$-1/[A] = -kt$ or $1/[A] = kt$

4.

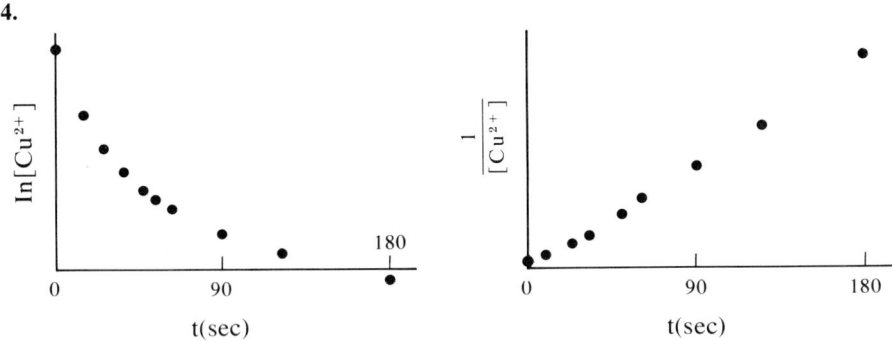

The second-order plot is clearly the one showing linear dependence. Its rate constant is given by the slope of the line, which we can determine from the coordinates of two selected points, say $t = 20$ sec and $t = 120$ sec:

$$k = \text{slope} = \dfrac{1/1.20 - 1/7.16}{120 - 20} = 1.08 \times 10^{-2}\ \text{l/mole-sec}$$

The reaction is also first-order in Sn²⁺ and inversely dependent on Cu⁺:

$$\text{rate} = k\dfrac{[Cu^{2+}]^2[Sn^{2+}]}{[Cu^+]}$$

The appearance of a fraction makes us suspect a rapid equilibrium followed by a rate-determining step, perhaps

(1) $Cu^{2+} + Sn^{2+} \underset{k_{-1}}{\overset{k_1}{\rightleftarrows}} Cu^+ + Sn^{3+}$; $K_{eq} = \dfrac{k_{-1}}{k_1} = \dfrac{[Cu^+][Sn^{3+}]}{[Cu^{2+}][Sn^{2+}]}$

(2) $Cu^{2+} + Sn^{3+} \xrightarrow{\text{slow}\ k_2} Cu^+ + Sn^{4+}$

$$\text{rate} = k_2[\text{Cu}^{2+}][\text{Sn}^{3+}]$$

Substituting for $[\text{Sn}^{3+}]$ from the K_{eq} expression:

$$\text{rate} = k_2[\text{Cu}^{2+}] \cdot K_{eq} \frac{[\text{Sn}^{2+}][\text{Cu}^{2+}]}{[\text{Cu}^+]}$$

$$= k_2 K_{eq} \frac{[\text{Cu}^{2+}]^2[\text{Sn}^{2+}]}{[\text{Cu}^+]}$$

This mechanism is thus consistent with the observed rate law; proof would require evidence of the intermediate Sn^{3+}.

5. Initially the pressure P^0 is entirely due to arsine; when a fraction f has reacted, $P_{\text{AsH}_3} = P^0(1-f)$. For two reacting molecules, three product H_2 molecules are formed, so when $P^0 f$ pressure of arsine has reacted, $\frac{3}{2}P^0 f$ pressure of hydrogen will have formed. The total pressure will thus be $P^0 - P^0 f + \frac{3}{2}P^0 f$, or $P^0(1+\frac{1}{2}f)$. At time 4.33 hr, we can calculate P_{AsH_3}:

$$P_{\text{total}} = P^0(1+\tfrac{1}{2}f)$$

$$403.0 = 392.0(1+\tfrac{1}{2}f)$$

$$f = 2\frac{403.0-392.0}{392.0} = 0.056$$

$$P_{\text{AsH}_3} = P^0(1-f) = 392.0(1-0.056) = 370 \text{ torr}$$

Calculating P_{AsH_3} at the other times in the same way and plotting $\log P_{\text{AsH}_3}$ and also $1/P_{\text{AsH}_3}$ against time, we can determine the reaction's order in AsH_3:

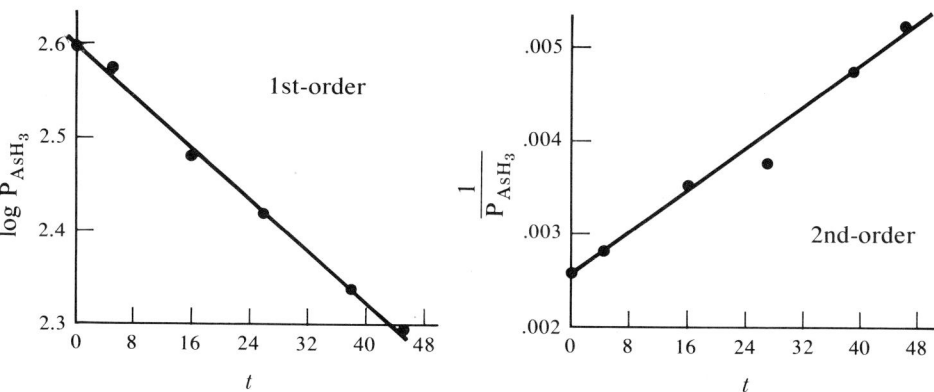

The straight-line fit is better in the first-order plot, for which the slope (k) is -6.74×10^{-3} hr^{-1}.

6.
$$\frac{\Delta(\log k)}{\Delta(1/T)} = \frac{7.096-(7.245)}{0.0032-0.0031} = -1490 = -\frac{E_a}{2.303 \times 1.987}$$

$$E_a = 1490 \times 2.303 \times 1.987 = 6820 \text{ cal/mole}$$

$$k = pN_0(2\pi)\left(\frac{RT}{3M}\right)^{1/2}\sigma^2 e^{-E_a/RT}$$

$$0.79 \times 10^7 = p\left[6.02 \times 10^{23} \times 6.28\left(\frac{8.314 \times 10^7 \times 300}{3 \times 30}\right)^{1/2}(3.5 \times 10^{-8})^2 e^{-6820/(1.987 \times 300)}\right]$$

$$p = 0.010$$

About one collision of every hundred is effective in causing reaction.

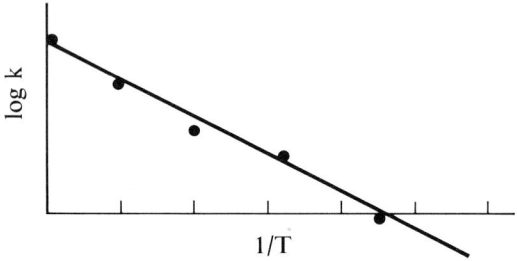

7. What we want here is a value for t, the time elapsed since the death of the tree with $[A]^0 = 15.3$ disintegrations/min-g C. We need a value for k, but that can be obtained from the half-life (Study Problem 1): $k = 0.693/t_{1/2}$.

$$\log\frac{[A]^0}{[A]} = \frac{0.693}{2.303\, t_{1/2}}t$$

$$t = \frac{2.303 \times 5720}{0.693}\log\frac{15.3}{10.0} = 3510 \text{ yr}$$

This is the time elapsed since the death of the tree (and Atlantis?) when the measurement was made in 1967. The date of destruction is thus fixed at $1967 - 3510 \cong 1543$ B.C.

8. $\dfrac{d[O]}{dt} = 0 = k_1[NO_2^-][O_2] - k_2[NO_2^-][O]$

$$[O] = \frac{k_1[NO_2^-][O_2]}{k_2[NO_2^-]} = \frac{k_1}{k_2}[O_2]$$

$$\text{rate overall} = \frac{d[NO_3^-]}{dt} = k_1[NO_2^-][O_2] + k_2[NO_2^-][O]$$

$$= k_1[NO_2^-][O_2] + k_2[NO_2^-]\left(\frac{k_1}{k_2}[O_2]\right)$$

$$= 2k_1[NO_2^-][O_2]$$

This is an alternative approach to Study Problem 4, with Sn^{3+} the intermediate and $k_2[Cu^{2+}] \ll k_{-1}[Cu^+]$.

9. S_N1:

$$(CH_3)_3S^+ \xrightarrow{\text{rate-determining}} (CH_3)_2S + CH_3^+$$

The charged product, CH_3^+, is much smaller than the charged reactant, which should make CH_3^+ more stable relative to $(CH_3)_3S^+$ in a more polar solvent due to increased solvation energy. The S_N1 mechanism would thus predict a speeding up in a more polar solvent.

S_N2:

$$(CH_3)_3S^+ + OH^- \longrightarrow HO \cdots CH_3 - S(CH_3)_2$$

Since the product here is larger and less polar than the reactants, increasing solvent polarity would actually favor the reactants (by increasing solvation energies) and slow the reaction down. The experimental rates correspond to the S_N2 mechanism.

10. The uniform trend of increasing rates with increasing bulk on the diamine ligands strongly suggests a dissociative mechanism (S_N1) for all diamines. The increasing number of —CH_3 groups crowds the Cl ligands more and more, making it more likely that the separation of Cl^- will occur.

11. A catalytic afterburner must adsorb hydrocarbons to increase their reaction rate. Since hydrocarbons are "soft" (see Table 10-2), the catalyst sites must be soft to adsorb the hydrocarbons effectively for oxidation. In tetraethyllead, the lead atom is essentially neutral, which makes it a very soft acid. The lead atoms, then, will poison the catalyst sites, particularly since the lead oxidation product (PbO or Pb_3O_4) is non-volatile even at high temperatures.

Chapter 12

1. $DIE(+q)(Li) = DIE(-q)(F)$
   $2.0 + 6.1q = 10.3 + 15.8(-q)$
   $21.9q = 8.3$
   $q = 0.264$ electron-charge units; $Li^{+0.264}F^{-0.264}$

2. $DIE(+4q)(C) = DIE(-q)(F)$
   $5.0 + 10.3(4q) = 10.3 + 15.8(-q)$
   $57.0q = 5.3$

   $q = 0.093:$ $>C^{+0.372}\!\!-\!\!F^{-0.093}$

   $DIE(+4q)(C) = DIE(-q)(Cl)$
   $5.0 + 10.3(4q) = 8.3 + 10.2(-q)$
   $51.4q = 3.3$

   $q = 0.064:$

Since carbon is less electronegative than either fluorine or chlorine, both of these atoms will withdraw electrons from it. However, since Cl is not as electronegative as F, we don't expect it to withdraw electrons as strongly. This is consistent with the lower charges calculated above.

3. $DIE(+3q_{Cl}+q_F)(C) = DIE(-q_{Cl})(Cl)$ $DIE(+3q_{Cl}+q_F)(C) = DIE(-q_F)(F)$
   $5.0 + 10.3(3q_{Cl}+q_F) = 8.3 + 10.2(-q_{Cl})$ $5.0 + 10.3(3q_{Cl}+q_F) = 10.3 + 15.8(-q_F)$
   $41.1q_{Cl} + 10.3q_F = 3.3$ $30.9q_{Cl} + 26.1q_F = 5.3$

   $-q_{Cl} = -0.0415, -q_F = -0.154, q_C = +0.278$

   The C atom has a charge intermediate between that calculated for CCl_4 and that for CF_4, which is consistent with the structure. Each Cl is less negative than in CCl_4 because all three are now having to compete with a F atom for the C electrons, instead of only with another Cl atom. The F atom is more negative than in CF_4, conversely, because it is only competing with three Cl atoms instead of with three other F atoms.

4. $E_{Zn-C} = \sqrt{82.6 \times 6} + 0.84(23.06)$
   $= 22.2 \text{ (covalent)} + 19.4 \text{ (ionic)}$
   $= 41.6 \text{ kcal/mole-bond total}$
   $E_{Zn-I} = \sqrt{6 \times 36.1} + 0.55(23.06)$
   $= 14.7 \text{ (covalent)} + 12.7 \text{ (ionic)}$
   $= 27.4 \text{ kcal/mole-bond total}$
   $E_{C-I} = \sqrt{82.6 \times 36.1} + 0.29(23.06)$
   $= 54.6 \text{ (covalent)} + 6.7 \text{ (ionic)}$
   $= 61.3 \text{ kcal/mole-bond total}$

   The large ionic contribution to the Zn—C bond energy makes it likely that $Zn(CH_3)_2$ would hydrolyze most vigorously of the three compounds given. This is experimentally verified: $Zn(CH_3)_2$ hydrolyzes vigorously, ZnI_2 dissolves readily but does not completely hydrolyze the Zn—I bonds, and CH_3I is unaffected by water.

5. The important size to consider is not the atomic radius but the bonding radius. Since AlI_3 involves a much greater electronegativity difference than BI_3 the Al—I bonds are considerably shortened while the B—I bonds are not. In effect, this "compresses" the AlI_3 molecule and reduces its molar volume.

6.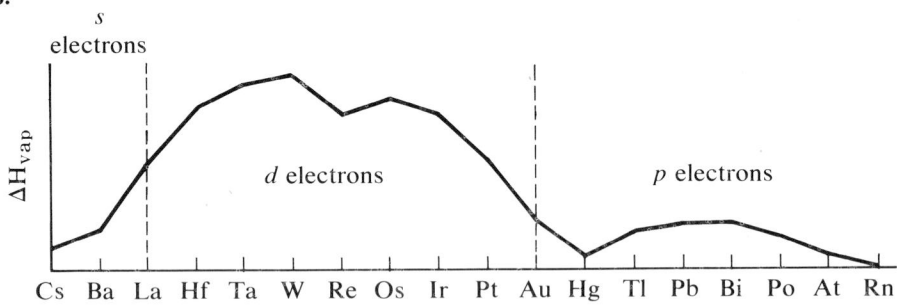

   A high ΔH_{vap} for an element means that it bonds strongly to itself, which raises the question of the type of valence orbitals and the number of valence electrons present. The curve starts low (at Cs) because there is only one valence orbital ($6s$) and one valence electron, peaks at W with six valence orbitals ($6s + 5d$) and six electrons (thus six possible bonds), drops off because adding more electrons fills more AO's and makes them unavailable for bonding, reaches a minimum at Hg with only one valence orbital (6s again, the 5d being now bound too firmly to participate in bonding) and two electrons, peaks again at Pb—Bi with four valence orbitals ($6s+6p$) and four or five electrons, and drops off to Rn with all valence orbitals filled, which is held to other Rn atoms only by van der Waals forces.

7. The discussion of "S^{6+}" centered about the fact that such a high charge would attract surrounding electrons very strongly and partially neutralize itself at the expense of surrounding atoms' electrons. In fact, a differential-ionization-energy calculation on SO_3 suggests a charge of only $+0.13$ on the S atom, which suggests almost complete neutralization of the "S^{6+}". The point is that any substantial electric charge causes redistribution of surrounding electron distributions.

8.
$$Li_3N + 8H_2O \rightarrow Li(OH_2)_4^+ + N(H\text{—}OH)_4^{3-}$$

$$N(H\text{—}OH)_4^{3-} \rightarrow NH_3 + 3OH^- + H_2O$$

Since $3OH^-$ and no H_3O^+ are produced, the solution will be strongly basic.

9. Mg^{2+} is much smaller than Na^+ due to the increased attraction for the electrons by the larger nuclear charge, while S^{2-} is only slightly larger than Cl^-. This means that the Mg—S distance in the crystal will be shorter than the corresponding Na—Cl distance, which leads to a greater density for MgS since there are more atoms per unit volume.

Chapter 13

1. Average Madelung constant $= 0.840$ per ion in formula.

$$U = -\frac{Aq_+q_-}{R}\left(1-\frac{1}{n}\right)\left(1.440 \times 10^{13} \frac{\text{kcal/mole}}{\text{erg/molecule}}\right)$$

$$\cong -\frac{0.840(n_+ + n_-)(Z_+e)(Z_-e)}{r_+ + r_-}\left(1-\frac{1}{n}\right)(1.440 \times 10^{13})$$

Here Z_+ and Z_- are the numbers of electronic charges on the positive and negative ions, n_+ and n_- are the numbers of each in the formula, and r_+ and r_- are their ionic radii. Take 8 as an average value for the Born exponent n and insert the value of the electronic charge:

$$U \cong \frac{Z_+Z_-(n_+ + n_-)}{r_+ + r_-}(0.840)(4.803 \times 10^{-10} \text{ esu})^2(\tfrac{7}{8})(1.440 \times 10^{13})(10^8 \text{ Å/cm})$$

$$\cong -244\frac{Z_+Z_-(n_+ + n_-)}{r_+ + r_- \text{ (in Å)}} \text{ (in numbers of electronic charges)}$$

For LiF, $Z_+ = Z_- = 1$, and $r_+ + r_- = 2.01$ Å:

$$U \cong -244\frac{1 \times 2}{2.01} = -243 \text{ kcal/mole (compare Born–Haber } -229.7)$$

2.

$$Sc_{(g)} + 3H_{(g)} \xrightarrow{\frac{IP_1+IP_2+IP_3}{+3EA}} Sc^{3+}_{(g)} + 3H^-_{(g)}$$

$$S\uparrow \quad \tfrac{3}{2}D\uparrow \qquad\qquad\qquad \downarrow U$$

$$Sc_{(s)} + \tfrac{3}{2}H_{2(g)} \xrightarrow{\Delta H_f} ScH_{3(s)}$$

Thus $\Delta H_f = S + \tfrac{3}{2}D + IP_1 + IP_2 + IP_3 + 3EA + U$. These quantities may be obtained from sources in this and the previous chapter: $S \cong \Delta H_{\text{vap}}$ (in endpaper periodic table) = 81 kcal/mole; D (from Table 12-4) = 104 kcal/mole; $IP_1 + IP_2 + IP_3$ (from Table 13-2) = 1019 kcal/mole; $3EA = -50$ kcal/mole. U can be calculated from the ionic radii in the endpaper periodic table and the approximate lattice-energy treatment of Problem 1: ScH_3 would have $n_+ + n_- = 4$, $Z_+ = 3$, $Z_- = 1$, $r_+ = 0.81$ Å, $r_- = 2.08$ Å, giving $U \cong -1014$ kcal/mole.

$$\Delta H_f \cong +81 + 155 + 1019 - 50 - 1014 \cong +191 \text{ kcal/mole}$$

The large amount of energy required to dissociate H_2 and triply ionize Sc can't be provided by the lattice energy, given the large size and low electron affinity of H^-.

3. A reducing agent is an electron donor: BH_4^- and BeH_4^{2-} are isoelectronic, which should make them comparable, but Be is less electronegative than B and the BeH_4^{2-} ion has a more negative charge. Both of these factors loosen the electrons involved and make BeH_4^{2-} the better reducing agent.

4. Consider the MO energy levels for diatomic and triatomic molecules shown in Figs. 6-24 and 6-39. Nitrogen has 5 valence electrons, N_2 10; either N_2^- or N_2^{2-} would have unpaired electrons in the π_x^*, π_y^* MO's (Fig. 6-24) which would leave the ion very chemically reactive. In Fig. 6-39, however, the 16 electrons of N_3^- fill all bonding and nonbonding MO's but leave the antibonding MO's vacant – a relatively stable configuration. Similarly, I_2^- would have 15 electrons, one unpaired, and I_2^{2-} would not be bonded at all; I_3^-, however, has 22 electrons filling all but one antibonding MO in Fig. 6-39 just as the 14 electrons of I_2 fill all but one antibonding MO in Fig. 6-24. In effect, an odd number of valence electrons on an electronegative atom leads to an odd number of atoms in its polyatomic anion.

5. The fact that $K_{sp} < 1$ for all five compounds suggests that the lattice energy for each exceeds the total hydration energies. The "surplus" lattice energy increases from Ba up to Be in keeping with the decreasing size of the cation. This is true in spite of the fact that the total hydration energy also shows the same trend; hydration energy trends *alone* don't determine solubility trends.

6. In the reaction of isolated molecules

$$CH_3-Na + CH_3-I \longrightarrow Na-I + CH_3-CH_3 - \Delta H^0$$

the ΔH^0 would be the energy difference between a Na—I bond plus a C—C bond and a Na—C bond plus a C—I bond:

$$\Delta H^0 = -\{[\sqrt{17\times 36} + 23(1.20)]_{\text{Na-I}} + 82.6_{\text{C-C}} - [\sqrt{17\times 82} + 23(1.49)]_{\text{Na-C}} - [\sqrt{36\times 82} + 23(0.89)_{\text{C-I}}]\}$$

$$= -52.3 - 82.6 + 71.7 + 61.2$$

$$= -2.0 \text{ kcal/mole}$$

If this is approximately equal to ΔG^0 we can calculate K_{eq}:

$$\Delta G^0 = -RT \ln K_{eq} = -2.303 RT \log K_{eq}$$
$$K_{eq} = 10^{-\Delta G^0/2.303(1.987)(300)} = 10^{+2000/1370} = 30$$

U for NaI is:

$$U = -\frac{Ae^2}{R}\left(1-\frac{1}{n}\right); A = 1.748, R = 3.11 \times 10^{-8}, n = 9.5$$

$$U = -167 \text{ kcal/mole}$$

Added to the "molecular" ΔH^0 above, this lattice energy produces a much more negative estimate for ΔG^0:

$$\Delta G^0 = -2.0 - 167 = -169 \text{ kcal/mole}$$
$$K_{eq} = 10^{-\Delta G^0/2.303RT} = 4 \times 10^{+122}!$$

Obviously the lattice energy is primarily responsible for the success of the reaction.

7. HF will be a very polar molecule because F is so electronegative; it is also quite small. These two factors make it a good solvent for ionic solids because they enable it to provide a high solvation energy for the separated ions. SF_6, on the other hand, is nonpolar as a molecule (because of its symmetry) even though the individual bonds are polar. It is also fairly large. We therefore do not expect it to be a good solvent for ionic solids.

8. Only with a very small positive ion can a M_2O lattice be formed without substantial repulsion between neighboring M^+ ions. Li^+ is such an ion, but Na^+ requires the larger O_2^{2-} ion to avoid Na^+—Na^+ contact, and the larger K^+, Rb^+, and Cs^+ require the O_2^- ion which forms a lattice with only one M^+ per oxygen ion, thereby reducing repulsion still further.

9. $S_2O_8^{2-}$ is a powerful oxidizing agent (electron acceptor) because of the peroxide group at the center of the ion. The other three redox couples shown all tend to go from right to left (as reducing agents); of the three, the least negative $\mathscr{E}^0$ corresponds to the increasing of the number of ligand atoms on the central S to four ($SO_3^{2-}/S_2O_3^{2-}$), while in the other two (SO_4^{2-}/SO_3^{2-} and $SO_3^{2-}/S_2O_4^{2-}$) the products have only three ligand atoms per S. The greater number of ligand atoms tends to stabilize the product and thus raise the potential.

10. For the six central atoms shown, the total ionization energies per halide in the indicated compounds are:

K	Ca	Ga	Ge	As	Se
4.3	9.0	19.1	25.9	37.0	47.0 eV/halide

From the argument in Section 13-3, we expect the fluorides of K and Ca to be ionic, Ga to be borderline, and the others to be covalent. Iodides are not mentioned, but they shouldn't be too different from bromides; on that assumption, the iodides of K and Ca ought to be ionic (though CaI_2 is close to the borderline) but all the others ought strictly to be covalent. Since we expect ionic compounds to be quite high-melting (600–1200°C) and covalent molecular compounds to be low-melting, the observed pattern of melting points is quite consistent.

11. The strong hydrogen-bonding capability of the F atom is causing the following reaction to proceed to the right:

$$HF + HF + H_2O \rightleftharpoons H_3O^+ + FHF^-$$

12. Potash must be what results from adding carbonic acid to "potash deprived of carbonic acid":

$$2KOH + H_2CO_3 \rightarrow K_2CO_3 + 2H_2O$$

That is, potash is potassium carbonate. K_2CO_3 is fairly soluble in water because the product of the K^+ and CO_3^{2-} ions' charges isn't too high; the solution contains hydrated K^+ and CO_3^{2-} or HCO_3^- ions (since HCO_3^- is a very weak acid). When CaO is added to the K_2CO_3 solution, it hydrates the oxide ion and dissolves:

$$CaO_{(s)} + H_2O_{(l)} \rightarrow Ca(OH)_{2(s)} \rightarrow Ca(OH_2)_{6(aq)}^{2+} + 2OH^-_{(aq)}$$

Since the formula weight of CaO is 56 g/mole and that of K_2CO_3 is 138 g/mole, using "two or three times its weight of quicklime" corresponds to equimolar amounts — just about one Ca^{2+} for each CO_3^{2-}. These two ions precipitate out because the product of their charges leads to a high lattice energy:

$$Ca(OH_2)_{6(aq)}^{2+} + CO_{3(aq)}^{2-} \rightarrow CaCO_{3(s)} + 6H_2O_{(l)}$$

This leaves in the solution only K^+, OH^-, and small concentrations of Ca^{2+} and CO_3^{2-}. Evaporation of the solution (after filtering off the $CaCO_3$) eventually exceeds the solubility of KOH and its crystals begin to form. "Close vessels" are necessary for the evaporation because in an open current of air KOH reacts with atmospheric CO_2:

$$OH^-_{(aq)} + CO_{2(aq)} \rightarrow HCO_{3(aq)}^-$$

and the HCO_3^- ion is just what we are trying to remove.

Chapter 14

1. If each O is bonded to two Sb atoms, then each Sb "owns" half that O for stoichiometric purposes. But the formula indicates $\tfrac{3}{2}O$ per Sb, so each Sb must be bonded to three O.

single chain:

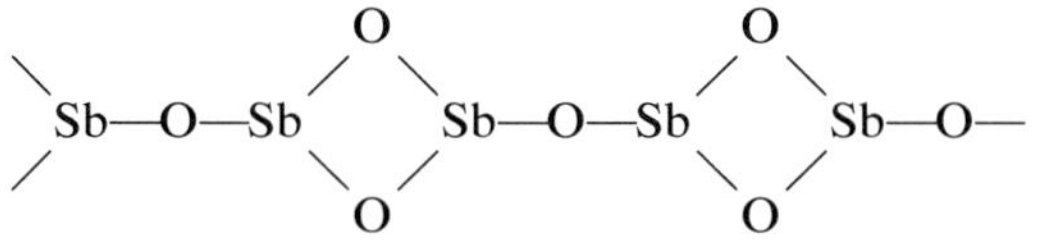

double chain:

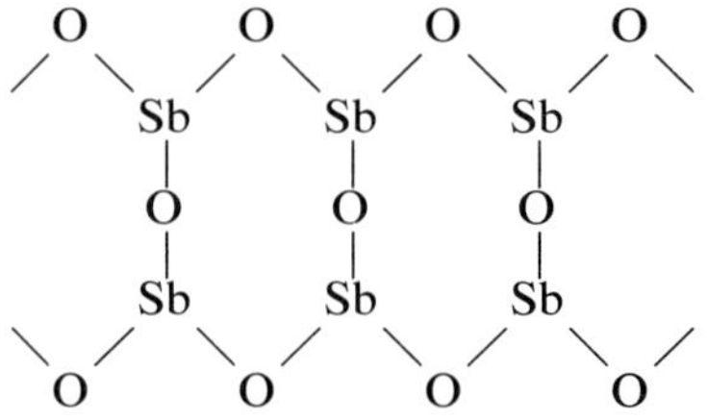

VSEPR for Sb:

$$\text{Sb: 5 valence } e^-$$
$$3(\tfrac{1}{2}\text{O}): (3 \times 2e^-)\tfrac{1}{2} = 3e^-$$
$$\underline{\text{net charge: } 0e^-}$$
$$\text{total: } 8e^- \text{ or four pairs}$$

$\left.\begin{array}{l}\text{3 bonding pairs to the three O atoms}\\ \text{1 nonbonding pair}\end{array}\right\}$ tetrahedral electron geometry (about 109°)

VSEPR for O:

$$\text{O: 6 valence } e^-$$
$$2(\tfrac{1}{3}\text{Sb}): 2 \times 1 = 2e^-$$
$$\underline{\text{net charge: } 0e^-}$$
$$\text{total: } 8e^- \text{ or four pairs}$$

$\left.\begin{array}{l}\text{2 bonding pairs to the 2Sb atoms}\\ \text{2 nonbonding pairs}\end{array}\right\}$ tetrahedral electron geometry (about 109°)

In the 4-membered rings of the single-chain structure, either all bond angles are 90° or two are larger and two are smaller. In either case, the predicted 109° angle could not be achieved for all atoms. However, since the 8-membered rings of the double-chain structure can be puckered, the 109° angle can be achieved at each atom, favoring that structure. The double chain is the observed structure.

2.

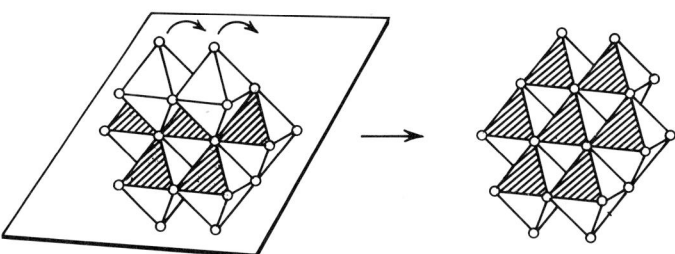

shaded faces lie in indicated plane

3. $2\text{Fe}_{(s)} + \text{NH}_{3(g)} \xrightarrow{\text{heat}} \text{Fe}_2\text{N}_{(s)} + \tfrac{3}{2}\text{H}_{2(g)}$

The nitride, as a partly-ionic compound on the surface of the object, is extremely hard and protects the softer steel interior.

4. $\text{ZnCO}_{3(s)} \xrightarrow{\text{heat}} \text{ZnO}_{(s)} + \text{CO}_{2(g)} - \Delta G^0$

$\Delta G^0 = \Delta H^0 - T\Delta S^0$. At high temperatures the favorable ΔS^0 due to gas generation makes the $-T\Delta S^0$ term sufficiently negative to overcome a positive ΔH^0.

5. Tl_2Br_4 or $\text{Tl}^+\text{TlBr}_4^-$ [mixture of Tl^{I} and Tl^{III}]. If the compound really were TlBr_2 there would be an odd number of electrons (since Tl has an odd atomic number) and thus at least one unpaired electron. In the discussion of B_2 and O_2 in Chapter 6 it was

noted that the presence of one or more unpaired electrons makes a molecule paramagnetic (attracted to a magnetic field) while with all electrons paired it is diamagnetic (repelled by a magnetic field). Since the Tl compound proved to be diamagnetic, the presence of Tl^{2+} could be ruled out.

6. The anion-exchange structure is converted to a form in which the open spaces all contain a soft anion such as Br^- (by passing a concentrated Br^- solution through the ion exchanger). If a Mg^{2+}—Zn^{2+} solution is then passed through the exchanger, the soft Zn^{2+} will be complexed by the Br^- present but the hard Mg^{2+} will pass through since it is more strongly complexed by the hard base H_2O. The Zn^{2+} can then be separately removed from the ion exchanger by passing a concentrated solution of some other anion through the exchanger.

7. The horizontal trend of increasingly negative ΔH_{soln} values suggests that the hydration energy is increasing, due to the higher charge on the central atom, faster than the lattice energy. Apparently the decreasing ionic character of the compounds partly counteracts the effect of an increasing formal oxidation state on the lattice energy. At $GeCl_4$, however, the increasingly soft nature of the central atom reduces the energy yield of the hydrolysis (ΔH_{soln} more positive). In the second row the same trends are seen, but the values are uniformly more positive than in the top row. The increasing size and softness of the central atom reduce the hydration energy in each case and increase the M—Cl bond strength, making the ΔH_{soln} less favorable.

8. A larger proportion of the lower-melting feldspar would decrease the porosity and increase the translucence of the product because the molten feldspar would be able to coat and fill the spaces between the kaolin particles more completely. After this initial firing, a glaze can be applied to the finished product by coating it with a thin paste of pure feldspar and firing again to melt the surface layer into a glass.

9. O and F are the two most electronegative elements. They thus have a greater tendency to form ionic-type close-packed structures with a maximum number of nearest-neighbor ions for a given ion. Br and S are much less electronegative, and their partly-ionic compounds have their structures determined to a much greater degree by bond-angle considerations. That is, ionic bonding in a three-dimensional array isn't as important for the less-ionic Br and S compounds.

10. In the first place, the octahedral and tetrahedral sites in a close-packed S^{2-} lattice are larger than those in a O^{2-} lattice, which makes the size requirement less stringent. Beyond this, however, the reduced ionic character of the sulfides (in which the electronegativity difference is smaller) causes the substitutions to be influenced more by covalent bond energies than strictly by sizes.

Chapter 15

1. The analogous diagrams for the eight- and nine-membered rings are:

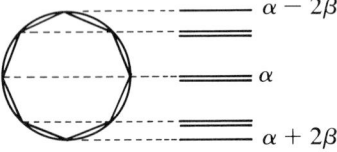 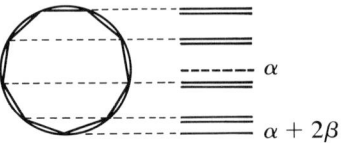

In the 3-carbon case we insert 3 electrons in the MO's beginning with the lowest-energy MO ($\alpha + 2\beta$) and observing the Exclusion Principle:

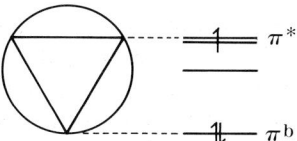
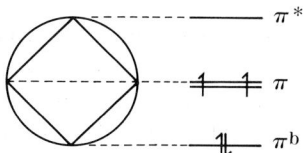

and the other electron configurations follow the same pattern. The 3-carbon positive ion, however, has one less electron than the 3-carbon neutral molecule, and the 5-carbon negative ion has one extra over the 5-carbon neutral molecule:

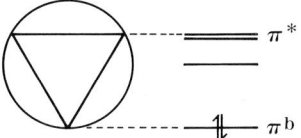
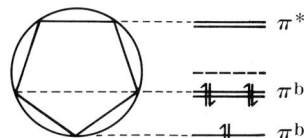

Both of these, like the 6-carbon neutral molecule benzene, have all the bonding levels filled but no nonbonding or antibonding electrons. The same is true of the 7-carbon positive ion and the 9-carbon negative ion. This is a particularly favorable sort of configuration. We infer that having 2, or 6, or 10, or in general $4n+2$ (n an integer) π electrons is energetically favorable in a planar cyclic pi-electron system.

2. Even though both of these compounds are essentially covalent, the greater electronegativity difference in $LiCH_3$ causes a greater degree of charge separation than in $Pb(CH_3)_4$ and consequently a greater attraction for the polar water molecule.

3. $2NF_{3(g)} + 3H_2O_{(l)} \rightarrow NO_{2(g)} + NO_{(g)} + 6HF_{(aq)} - \Delta H^0_{\text{hydrolysis}}$

$\Delta H^0_{\text{hyd}} = \Delta H_f^0(NO_2)_{(g)} + \Delta H_f^0(NO)_{(g)} + 6\Delta H_f^0(HF)_{(aq)} - 2\Delta H_f^0(NF_3)_{(g)} - 3\Delta H_f^0(H_2O)_{(l)}$

$= +8.09 + 21.60 + 6\Delta H_f^0(HF)_{(aq)} + 52 + 205.0$

We should assume that $HF_{(aq)}$ is still in the molecular form since it is known to be a weak acid. We can write a cycle (Hess's Law) treatment for $\Delta H_f^0(HF)_{(aq)}$:

$$
\begin{array}{ccccc}
H_{(g)} & + & F_{(g)} & \xrightarrow{\text{bond } E} & HF_{(g)} \\
\uparrow \frac{1}{2} \text{ bond } E & & \uparrow \frac{1}{2} \text{ bond } E & & \downarrow \text{Aq(HF) from Table 13-2} \\
\frac{1}{2} H_{2(g)} & + & \frac{1}{2} F_{2(g)} & \xrightarrow{\Delta H_f^0} & HF_{(aq)}
\end{array}
$$

$\Delta H_f^0(HF)_{(aq)} = \frac{1}{2} E_{H-H} + \frac{1}{2} E_{F-F} - E_{H-F} + Aq(HF)$

$= 52.1 + 18.3 - 127.8 - 5.7$

$= -63.1 \text{ kcal/mole}$

For ΔH^0_{hyd} we have:
$$\Delta H^0_{hyd} = +8.09 + 21.60 - 378.6 + 52 + 205.0$$
$$= -91.9 \text{ kcal/mole reaction}$$

This is quite favorable, as is the entropy change, since the number of gaseous molecules is unchanged and the total number of molecules is increasing—yet hydrolysis does not occur. Apparently there is no favorable mechanism for bringing a water O up to the NF_3N atom. Although the three F atoms tend to create a positive charge on the N which would attract the O, the N has a nonbonding pair of electrons which tend to prevent the approach of the O since the O atom in water has nonbonding pairs itself.

4. (a) cyclopropanone; (b) 1,3,5-trimethylbenzene; (c) 3,4,4-trichloroheptane; (d) 2-methyl-1,3-butadiene; (e) sec-butyl 2-methylbutyrate; (f) ethyl cyclohexyl amine; (g) 4-methyl-2-pentanone; (h) diethyl sulfate; (i) N-ethylacetamide; (j) 3,3-dimethyl-2-butanol; (k) 3-methylpentanal; (l) 2-propylhexanoic acid.

5.
$$CH_3-\underset{\underset{CH_3}{|}\;\underset{OH}{|}}{\overset{\overset{CH_3}{|}}{C}}-CH-CH_3 + H^+ \rightarrow CH_3-\underset{\underset{CH_3}{|}}{\overset{\overset{CH_3}{|}}{C}}-\overset{+}{CH}-CH_3 + HOH$$

Migration of a methyl group would give the more stable tertiary carbonium ion:

$$CH_3-\underset{\underset{CH_3}{|}}{\overset{\overset{CH_3}{|}}{\overset{+}{C}}}-CH-CH_3 \rightarrow \overset{1}{CH_3}-\underset{\underset{CH_3}{|}}{\overset{\overset{\overset{2}{C}H_3}{|}}{C}}-\overset{+}{CH}-CH_3$$

This ion can revert to a neutral pi-bonded molecule by losing a H^+ from either C^1, C^2, or C^3. C^1 and C^2 give the same product:

$$CH_2=\underset{\underset{CH_3}{|}}{\overset{\overset{CH_3}{|}}{C}}-CH-CH_3$$

C^3 gives the alternate product:

$$CH_3-\underset{\underset{CH_3}{|}}{\overset{\overset{CH_3}{|}}{C}}=C-CH_3$$

The 2:1 ratio of H^+ sources accounts for the 70:30 product ratio.

6. Since Mg is much more electropositive than the organic group R″, it will have a partial positive charge and will be attracted to the partially negative O on the ketone RR′CO:

$$\underset{R'}{\overset{R}{>}}\overset{\delta+}{C}=\overset{\delta-}{O} + \overset{\delta-R''}{\underset{\delta-X}{\overset{\delta+}{|}}{Mg}} \longrightarrow \underset{R'}{\overset{R}{>}}\overset{\delta+}{C}-O-\overset{\delta-R''}{\underset{}{Mg}}X^{\delta-}$$

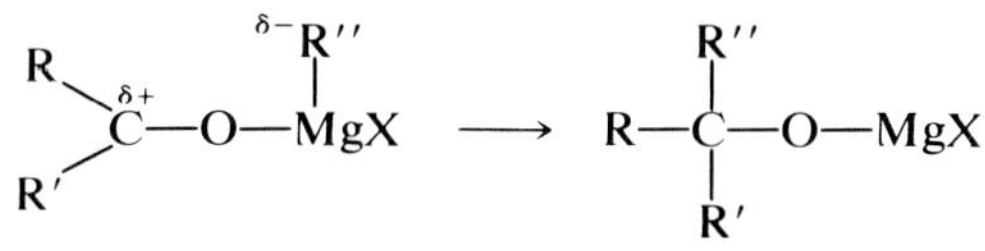

$$\underset{R'}{\overset{R}{>}}\underset{\delta+}{C}-O-\overset{\delta-R''}{\underset{|}{M}gX} \longrightarrow R-\underset{R'}{\overset{R''}{\underset{|}{C}}}-O-MgX$$

The great electronegativity difference between O and Mg causes the OMgX group to be very sensitive to hydrolysis, with the H attacking the negative O and the water O attacking the positive Mg:

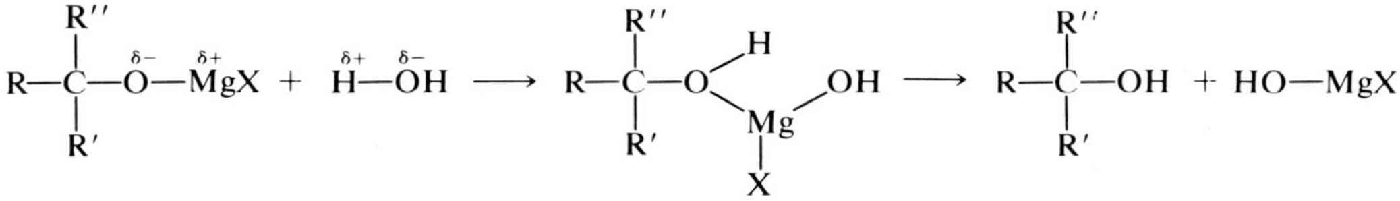

$$R-\underset{R'}{\overset{R''}{\underset{|}{C}}}-\underset{\delta-}{O}-\underset{\delta+}{M}gX + \underset{\delta+}{H}-\underset{\delta-}{O}H \longrightarrow R-\underset{R'}{\overset{R''}{\underset{|}{C}}}-O\underset{\underset{X}{\overset{|}{Mg}}}{\overset{H}{\diagdown}}OH \longrightarrow R-\underset{R'}{\overset{R''}{\underset{|}{C}}}-OH + HO-MgX$$

To prepare 1,1-diphenylethanol, we can use the diphenyl ketone (called benzophenone) and a methyl Grignard reagent:

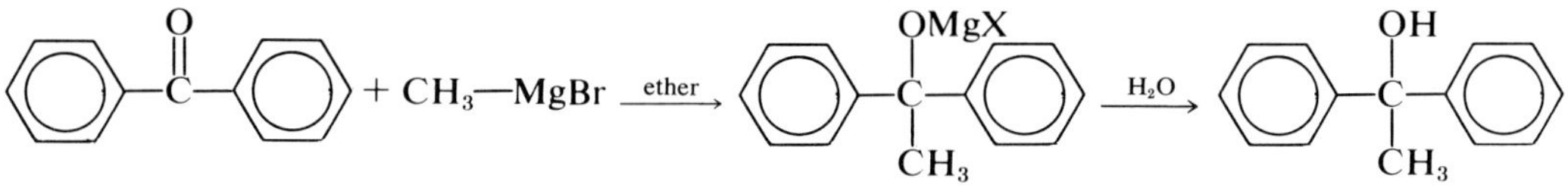

Ph—CO—Ph + CH₃—MgBr $\xrightarrow{\text{ether}}$ Ph—C(OMgX)(CH₃)—Ph $\xrightarrow{H_2O}$ Ph—C(OH)(CH₃)—Ph

If the initial ether solution were wet the water would react directly with the CH₃—MgBr:

$$CH_3-MgBr + HOH \longrightarrow CH_4 + HO-MgBr$$

preventing the desired reaction. Similarly, the OH groups on HO—C₆H₄—CO—C₆H₄—OH

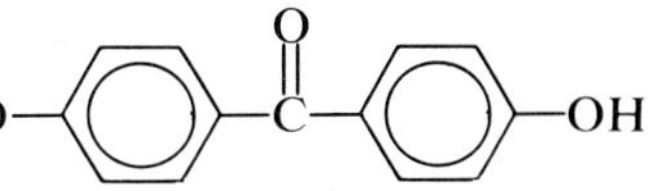

would react preferentially with the CH₃—MgBr, again preventing the desired reaction:

$$CH_3-MgBr + HO-R \longrightarrow CH_4 + RO-MgBr$$

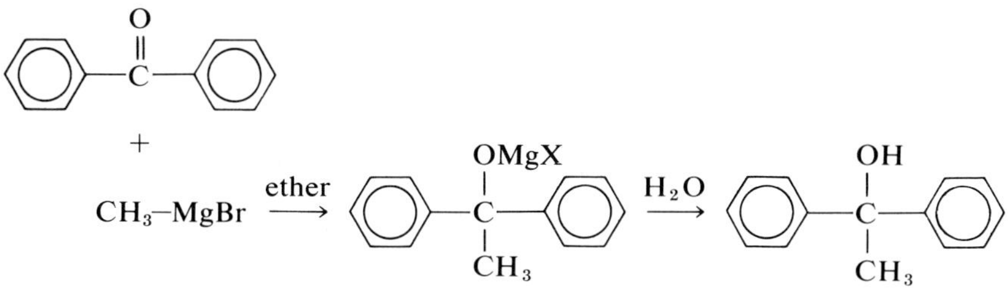

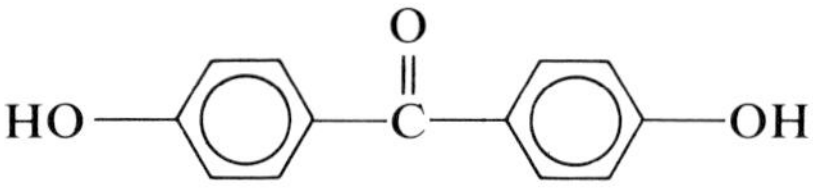

Ph—CO—Ph
+
CH₃—MgBr $\xrightarrow{\text{ether}}$ Ph—C(OMgX)(CH₃)—Ph $\xrightarrow{H_2O}$ Ph—C(OH)(CH₃)—Ph

HO—C₆H₄—CO—C₆H₄—OH

7. In the *ortho* and *para* forms the —NO₂ withdraws electrons very strongly from the carbon bearing the —OH group, which promotes the formation of a positive charge on the H and makes it a stronger acid. In the *meta* form the electron withdrawal is weaker (because neither sigma nor pi electrons are strongly affected) and the H develops less positive charge, thus is a weaker acid.

8. Cyclobutadiene:

$$\begin{array}{c} HC\!=\!CH \\ \| \quad \| \\ HC\!-\!CH \end{array}$$

This has pi energy levels given in Study Problem 1:

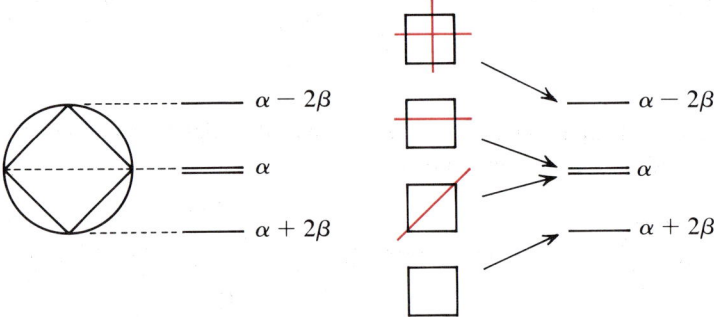

The lowest energy level ($\alpha+2\beta$) corresponds, by analogy with benzene, to a MO having no nodal planes cutting the bond axes; the middle two (α) to MO's having a single node, either through two atoms or through two bonds; the highest ($\alpha-2\beta$) to a MO having two nodes at 90° to each other and cutting every bond:

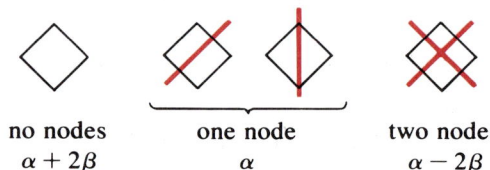

Each carbon atom contributes a single *p* electron to the pi system, which means that the energy levels will be filled as in Problem 1, with two unpaired electrons (remember Hund's rule). The existence of unpaired electrons should make the compound paramagnetic (and probably very reactive). It has been synthesized but is quite unstable.

9. The nucleophile will attack more rapidly the greater the positive charge on the ring. Therefore an electron-withdrawing group already on the ring will speed up the nucleophilic attack. Since the *ortho* and *para* positions are affected most, the rate will be enhanced most if the —NO₂ is in one of those positions relative to the leaving

group.
10. The polyamide chain in nylon has relatively few hydrogen-bonding sites compared to the large number of —OH groups on the cellulose polymer. The cellulose will thus absorb sweat much better than the nylon, since each hydrogen-bonding site can accept a water molecule.
11. Since 21 amino acids are available at each combination, the possible number of 50-acid polypeptides or proteins is

$$N = 21^{50}$$
$$\log N = 50 \log 21 = 50(1.322) = 66.1$$
$$N = 10^{66.1} = 10^{66} \text{ molecules or } 10^{43} \text{ moles}$$

If a mole weighs 7500 g, the 10^{43} moles would weigh nearly 10^{47} g, which is almost comparable to the estimated mass of the universe! Since 50 amino acids constitute quite a small protein, it is clear that only a very few of the immense number of possible proteins have ever been made.

12. $H_2N-CH_2-COOH + H_2O \rightleftharpoons H_2N-CH_2-COO^- + H_3O^+$;

$$K_a = 4.6 \times 10^{-3} = \frac{[H_2NCH_2COO^-][H_3O^+]}{[H_2HCH_2COOH]}$$

$H_3N^+-CH_2-COO^- + H_2O \rightleftharpoons H_2N-CH_2-COO^- + H_3O^+$;

$$K_a = 2.8 \times 10^{-10} = \frac{[H_2NCH_2COO^-][H_3O^+]}{[H_3N^+-CH_2COO^-]}$$

$$4.6 \times 10^{-3} = [H_3O^+]\left(\frac{100}{1}\right)$$

$$[H_3O^+] = 4.6 \times 10^{-5} \, M; \, pH = 4.34$$

Using the second equilibrium,

$$2.8 \times 10^{-10} = \frac{[H_2NCH_2COO^-][H_3O^+]}{[H_3N^+-CH_2COO^-]}$$

$$\frac{[H_2NCH_2COO^-]}{[H_3N^+-CH_2COO^-]} = \frac{2.8 \times 10^{-10}}{4.6 \times 10^{-5}} = 6.1 \times 10^{-6}$$

This very small number means that nearly all the glycine is in the dipolar-ion form (the denominator), since the neutral-molecule form is only 1% of 6.1×10^{-6} at this pH. When 1% of the total glycine is in the anionic form, we have to a good approximation

$$2.8 \times 10^{-10} = \frac{[H_2NCH_2COO^-][H_3O^+]}{[H_3N^+-CH_2COO^-]} \cong \left(\frac{1}{100}\right)[H_3O^+]$$

$$[H_3O^+] = 100 \times 2.8 \times 10^{-10} = 2.8 \times 10^{-8} \, M; \, pH = 7.55$$

At the pH of body fluids (~7.2) over 99% of the glycine is in the dipolar-ion form, judging from the above calculation. Other equilibria which might be included are:

$$H_3N^+\text{—}CH_2\text{—}COOH + H_2O \rightleftharpoons H_2N\text{—}CH_2\text{—}COOH + H_3O^+$$

$$H_3N^+\text{—}CH_2\text{—}COOH + H_2O \rightleftharpoons H_3N^+\text{—}CH_2\text{—}COO^- + H_3O^+$$

However, these are a minor influence on the observed concentrations.

13. We can use a reaction like that for the preparation of B_2H_6, since Si and B have similar electronegativities:

$$LiAlH_{4(s)} + SiCl_{4(l)} \rightarrow SiH_{4(g)} + AlCl_{3(s)} + LiCl_{(s)}$$

This reaction relies both on the entropy increase due to gas formation and the lattice energy of the LiCl and $AlCl_3$ which are formed. It should be effective in preparing other hydrides of metalloids: GeH_4, SnH_4, PH_3, AsH_3, SbH_3, BiH_3. Other methods are more convenient for some of these, however.

14. Just as ammonia can be oxidized to NO_2 and halogenated to NX_3, it will react with sulfur to give N_4S_4:

$$10S + 4NH_3 \rightarrow N_4S_4 + 6H_2S$$

Using ammonia instead of N_2 avoids having to break the $N\equiv N$ triple bond.

15. Silicones have numerous alkyl groups, which are nonpolar and will not hydrogen-bond to water molecules. The high molecular weight of the polymers and the polar Si—O bonds make the silicones very non-volatile, which enhances their permanence as a coating.

Chapter 16

1. Mn has 7 valence electrons, so 11 must be donated to it. If each Mn has 5 CO ligands 10 electrons are available, and an eleventh is available if the two Mn atoms each donate a single electron to a bond between them: $(OC)_5Mn\text{—}Mn(CO)_5$.

2. Three:

[Three square planar structures with Cr center, showing different arrangements of H_3N and Cl ligands]

Three:

[Three trigonal prismatic/octahedral structures with Cr center showing different arrangements of H_3N and Cl ligands]

Since there does not seem to be any reason for one isomer to be less stable than the other two in either of these cases, we infer from the existence of only two isomers that the octahedral structure (which can yield only two isomers) is more probable.

3. (a) hexafluorotitanate(III); (b) dichlorotetramminecobalt(III) chloride; (c) cesium tris(oxalato)vanadate(III); (d) pentacyanonitrosylmanganate(II); (e) potassium pentafluorochromate(IV); (f) hexacyanoferrate(III)—also ferricyanide; (g) dichlorobis(triphenylphosphine)nickel(II); (h) ammonium heptafluorotantalate(V); (i) bromopentacarbonyltungstate(0); (j) chlorotetramminenitrosylruthenium(III).

4. Cu^+ and Zn^{2+} are isoelectronic, but the electrons on Cu^+ are less tightly bound because of the smaller nuclear charge. This corresponds to higher-energy electrons on Cu^+. If visible (green) light is to be emitted, the energy of the Cu^+ electron must be higher than that of the vacancy created in the ZnS band by an amount equivalent to the frequency of green light. Removing the electron from Cu^+ creates Cu^{2+}, which has a vacancy in a $3d$ orbital and therefore is a good acceptor for the excited ZnS electron. Since infrared radiation is emitted in this process, the energy difference between the normally-vacant band of ZnS and the Cu^{2+} is smaller than the first difference; infrared radiation has a lower frequency than visible light.

5. MnO_4^-: Mn^{7+} has 0 valence electrons; if O^{2-} is regarded as a 4-electron donor MnO_4^- has 16 electrons about the Mn.

   MnO_4^{2-}: Mn^{6+} has 1 valence electron; MnO_4^{2-} thus has 17 electrons about the Mn.

   $Mn(CN)_6^{2-}$: Mn^{4+} has 3 valence electrons; $6CN^-$ ligands donate 12 more for a total of 15.

   $Mn(CN)_6^{3-,4-,5-}$: These have 1, 2, and 3 electrons more (respectively) than $Mn(CN)_6^{2-}$, for totals of 16, 17, and 18 electrons.

   $Mn_2(CO)_{10}$: Problem 1 indicates an 18-electron structure.

   $Mn(CO)_5^-$: Mn^{1-} would have 8 valence electrons; five CO ligands add 10 more for a total of 18.

6. CFSE:

Ti^{3+}	V^{3+}	Cr^{3+}	Mn^{3+}	Fe^{3+}	Co^{3+}	Ni^{3+}
-0.057Δ	-0.114Δ	$+0.200\Delta$	$+0.143\Delta$	$+0.086\Delta$	$+0.400\Delta$	-0.114Δ

Since a large activation energy means a slow reaction, this series suggests *decreasing* rates: $V^{3+} = Ni^{3+} > Ti^{3+} > Fe^{3+} > Mn^{3+} > Cr^{3+} > Co^{3+}$, which is consistent with the observed inertness of Cr^{3+} and Co^{3+} complexes.

7. The increase in the lattice energy over the "smooth-curve" value is just about proportional in each case to the total CFSE given in the top row of Table 16-6; since Mn^{2+} and Zn^{2+} have all d orbitals equally filled they show no net stabilization.

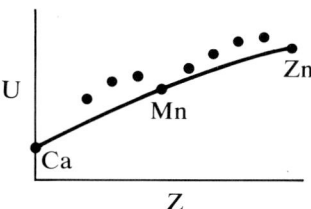

934 | Problem Answers

8. The Mo atom should be octahedrally coordinated by fluorines:

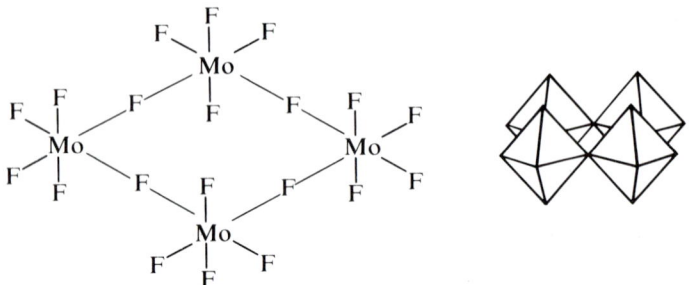

9. Having all four methyl groups on the same side of the complex as one of the Cl atoms greatly increases the steric hindrance to that atom and makes it considerably more favorable for that Cl to leave (as Cl⁻). This suggests a dissociative or S_N1 mechanism.

10.

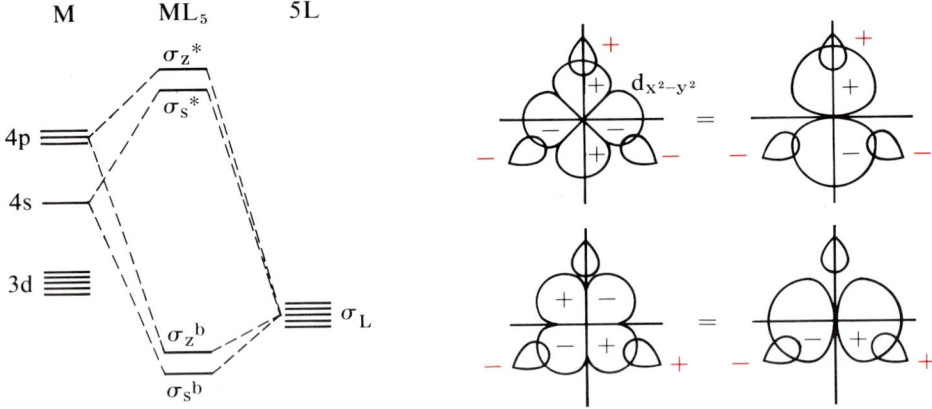

For the σ_L combination which matches $d_{x^2-y^2}$ and p_x, three MO's will result: one in which $d_{x^2-y^2}$ and p_x both match the σ_L combination in signs (σ_x^b), one in which neither match (σ_x^*), and one in which they oppose each other (σ_x). The same is true for d_{xy} and p_y, and since the overlaps are equivalent the resulting MO's are degenerate:

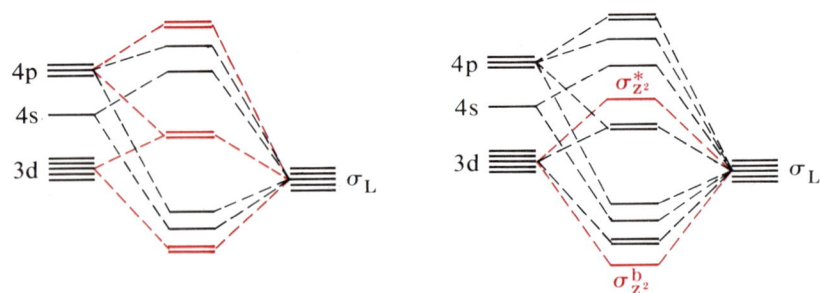

d_{xz}, d_{yz} remain as (nonbonding) orbitals:

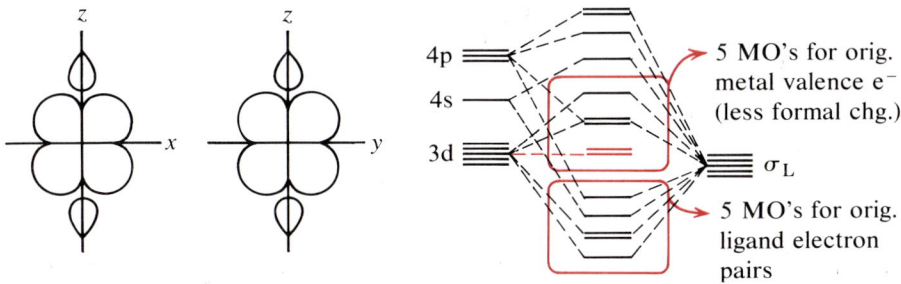

11. Since *qas* is a neutral molecule, Fe in Fe(qas)Cl$^+$ is Fe^{II} with 6 valence electrons. Four of these will go in the π levels (of Problem 10) as pairs; the other two will go *with parallel spins* (Hund's rule) in the degenerate σ nonbonding levels. With two unpaired electrons, the magnetic moment should be $\sqrt{2(2+2)} = 2.83$ B.M., very near the observed value.

12. By reference to Fig. 7-52 we see that brazing rod (50 Cu/50 Zn) melts at a lower temperature than ordinary brass (70 Cu/30 Zn). This allows welding brass objects together without melting them.

13. The cobalt atom in cobaltocene has 19 electrons around it (9 from Co^0, 5 from each C$_5$H$_5$ ring). It will therefore be a good one-electron donor in order to have no antibonding electrons. Since a Br atom is a good one-electron acceptor, we expect (and observe):

$$2\text{Co}(C_5H_5)_2 + Br_2 \rightarrow 2\text{Co}(C_5H_5)_2^+Br^-$$

14. V in VBr$_6^{2-}$ is in a 4+ formal oxidation state, while Cu in CuBr$_4^{2-}$ is only in a 2+ formal oxidation state. The high formal charge makes V^{4+} a "hard" acid — it will thus react preferentially with the hard base H$_2$O rather than the soft base Br$^-$, and VBr$_6^{2-}$ can't be formed in aqueous solution.

15. The composition resembles that of stainless steel, but with even more nickel. Nichrome, then, should be resistant to oxidation even at red heat, where ordinary iron alloys are rapidly oxidized by atmospheric O$_2$.

16. $(OC)_4Co\text{—}Co(CO)_4 + H_2 \rightarrow 2H\text{—}Co(CO)_4$
$H\text{—}Co(CO)_4 \rightarrow H\text{—}Co(CO)_3 + CO$

$$H\text{—Co(CO)}_3 + \underset{}{\overset{}{\text{C}{=}\text{C}}} \rightarrow \underset{H\text{—Co(CO)}_3}{\overset{}{\text{C}{=}\text{C}}}$$

$$\underset{H\text{—Co(CO)}_3}{\overset{}{\text{C}{=}\text{C}}} \rightarrow \underset{H\quad\text{Co(CO)}_3}{-\text{C}-\text{C}-}$$

$$\underset{H\quad\text{Co(CO)}_3}{-\text{C}-\text{C}-} + CO \rightarrow \underset{H\quad\text{Co(CO)}_4}{-\text{C}-\text{C}-}$$

```
    |  |                   |  |
  —C—C—          →       —C—C
    |  |                   |   \
    H  Co(CO)₄             H    C=O
                                /
                             Co(CO)₃


    |  |   O                     |  |   O
  —C—C—C             + CO  →   —C—C—C
    |  |   \                     |  |   \
    H      Co(CO)₃               H      Co(CO)₄


    |  |   O                         |  |   O
  —C—C—C             + HCo(CO)₄ →  —C—C—C         + Co₂(CO)₈
    |  |   \                         |  |   \
    H      Co(CO)₄                   H      H
```

Alternate elementary processes are possible at several points, of course.

17. Since halide ions are soft bases (except for F^-), stable complexes must usually have the metal in a relatively low formal oxidation state so as to be a soft acid. In this case the MX_4 species will have a net charge of 1—, 2—, or even 3—, and there will be considerable electrostatic repulsion of the negatively-charged 4-coordinate species for another X^-.

Index

Absorption spectra, 273, 849
  and crystal field theory, 849
Abstraction, 600
Acetic acid, 507
Acetylide, 694
Acid-base definitions, 502
  hard and soft, 503
Acid-base equilibria, 506
Acid-base reactions, 500, 654, 693, 741, 819
Acids
  carboxylic, 786
  catalytic character of, 601–605
Acids and bases
  Arrhenius theory of, 502
  Lewis theory of, 502
  Lowry-Bronsted theory of, 502
  strength of, 505–507, 680–681, 700, 831
    inorganic oxy acids, 700
    leveling effect, 505
Activated complex, 583, 586, 598
Activated molecule, 574
Activation energy, 579, 594, 763
Activity, 457, 493
Activity coefficient, 458
Addition reactions, 777
Adiabatic process, 416, 448–450
Adsorption, 607
  Langmuir theory of, 607–608
Age, from radioactivity, 615
Alcohols, 785
Aldehydes, 785
Alkali halides, 668, 706
  gaseous molecules, 647, 666, 676
Alkali-metal oxides, 698
Alkali metals, reduction potentials of, 682
Alkaline earth metals, 676
Alkanes, 776, 782
Alkenes, 776, 783
Alkyl radicals, 783
Alloys, 862

Alpha-particle scattering, 203
Aluminum, electrolytic reduction, 554
Amalgam, 359, 547
Amides, 697, 786, 799
Amine
  primary, 784
  secondary, 784
  tertiary, 784
Amino acids, 801, 826
Ammine, 870
Ammonia, 506, 818, 871
Ammonium, 37, 325, 507, 710
Amorphous materials, 152, 796
Angle of diffraction, 171
Angle of incidence, 171
Angular momentum, 209, 240
Anisotropic properties, 156
Anode, 542
Antibonding orbitals, 266, 273, 763, 773, 775, 843–846
Antifluorite structure, 183
Aquo complexes, 493, 870
Arithmetic mean, 81
Aromatic hydrocarbons, 776
  substitution reactions of, 777, 779–781
Arrhenius activation energy, 581
Arrhenius equation, 581, 587
Arrhenius theory of acids and bases, 502
Asbestos, 746
Atmospheric pressure, 100, 370
Atomic mass unit, 42
Atomic models, 197, 203
  Bohr model, 207–211
  Rutherford model, 204
Atomic number, 213–214, 622–632, 833
Atomic number determination, 213–214
Atomic orbitals, 222–240, 770, 831–837
  d type, 230, 831–836, 841–847
  hybrid, 325–326, 769–771

molecular orbitals from, 262–324, 769–776, 841–846
order of filling of, 241–245, 622–626
overlap of, 270–327, 660, 708, 769–773, 842–847, 882
p type, 226–227, 229, 770–771
s type, 225, 660, 708
Atomic radii, 622, 645
Atomic spectra, 211–213
Atomic volume, 649
Atomic weights, 21–25, 42
Atomization energy E, 305
Atomization process, 305
ATP, 751
Average, 81
from distribution functions, 89
Avogadro's hypothesis, 21, 105
Avogadro's number, 25–26, 173, 195
Azeotrope, 388
Azides, 695

Bakelite, 800
Balmer series, 211–212
Band theory of solids, 329
Barometer, 101
Barometric pressure, 101
Base, 502–507, 657, 794–795, 875
hard and soft, 503, 609, 857
Battery, 550
Benzene
electronic structure, 773–776
substitution reactions, 777–781
Bimolecular process, 574–575
Binuclear compounds, 867
Black-body radiation, 205–207
Body-centered cubic lattice, 179
Bohr magnetons, 838
Bohr model for atom, 207–217, 626
Boiling point, 368, 370, 375–377
Boiling-point elevation, 375
Boiling point elevation constants, derivation of, 375–376
Boltzmann energy distribution law, 117, 365, 579
Boltzmann's constant, 113, 428
Bomb calorimeter, 454
Bond angles, 253, 256, 309, 313
table of, 256
Bond dipoles, 684
Bond-dissociation energy D, 305
Bond-dissociation process, 304
Bond energies, electronegativities from, 640, 642–644
Bond energy, 261, 460–462, 642, 667
table of, 461, 642
Bond lengths, 253, 644–648
multiple bonds and, 647
table of, 647
Bond order, 282, 284, 648

Bonding
covalent, 293, 667, 762–763
ionic, 292, 667, 672
Bonding orbitals, 266–269, 769
pi, 277, 771, 774–775
sigma, 276, 769, 325–326
Boranes, 815
Born-Haber cycle, 673, 676–683, 857
Boyle's law, 102, 107–111
Bragg equation, 172
Brass, 864
Bromate ion, 37, 699
Bromic acid, 37
Bromine, 778, 779
Brönsted equation, 604
Bronze, 864
Bubble-cap column, 385
Buffer solution, 522, 523, 526
Butane, 456, 781

Cage effect, 596
Calculus, 61–78
Calomel electrode, 543
Calorie, 13
Calorimetric measurements, 454
Cannizzaro, 22
Capacity variables, 409
Capillary-rise tensiometer, 347–348
Carbanion, 793–795
Carbides, 693–694, 734–738
Carbon dating, 615
Carbon dioxide, 304–305, 507, 702, 820
Carbon monoxide, 289, 304, 820, 852–854, 867
Carbonate ion, 38, 507, 699–702, 741
Carbonic acid, 38, 507, 518, 713
Carbonium ion, 790–792
Carbonyl compounds
inorganic, 852, 867, 872
organic, 785
Carnot cycle, 439
Catalysis, 601–611
acid-base, 601–605, 795
enzyme, 609–611, 805
Catalyst, 605, 884
poisons, 609
Catenation, 704, 767
Cathode, 199, 544, 552
Cathode rays, 199
Cell, electrical, 541
Cell conventions, 542
Centrosymmetric force, 719
Cesium chloride structure, 182
CGS units, 9
Chain-initiating step, 788
Chain-propagating step, 788
Chain reaction, 788
Chain rule, 67
Chain structure, 728, 746

Chain-terminating step, 789
Charge-balance condition, 514, 545–547
Charge-transfer spectra, 851
Charles' law, 102–105
Chemical bond, origin of, 260–262
Chemical dynamics, 99, 339
Chemical equilibrium, 33, 471, 481–538
Chemical kinetics, 567–613
Chemisorption, 607
Chlorate ion, 37, 699–701
Chloric acid, 37
Chlorine, 743, 756, 778, 779
Chlorine oxides, 820
Chromate ion, 38, 700–701, 730
Chromatography, gas, 388
Chromite, 740
Chromium, 864
Cinnabar, 754
Cis conformation, 770, 873
Cleavage planes, 156, 158
Close packing, 176–185, 728
   cubic, 178
   hexagonal, 177
Close-packed lattices, 177
   octahedral sites in, 180, 727–728, 735, 744
   structures related to, 738, 744
   tetrahedral sites in, 180, 734, 744
Coefficient of viscosity, 119, 121, 344
Colligative properties, 374, 372–380
Collision number, 120, 577
Collision theory, 577–582
Collisions
   three-body, 574
   two-body, 574
Color and d electrons, 849–851
Column, 385, 389
Complex ions
   bonding in, 835–837, 841–849, 869
   definition of, 493
   equilibria among, 492–499
   nomenclature of, 870–873
   stability of, 493, 494
   stereochemistry of, 837, 855
Composition equilibrium, 412
Compositional interaction, 407, 412
Compound, definition of, 15
Compressibility, 153, 341
Compressibility factor, 132
   chart of, 139
Concentration, 355, 412, 466–468
   units of, 355, 356
Concentration cell, 565
Condensation, 730
Confidence intervals, 82
Confidence levels, 82
Configurations, statistical, 427
Congeners, 819
Conjugate acid-base pair, 506

Constant of integration, 72
Constant, proportionality, 51
Constants, tabulated physical, 885
Converging series, 670
Conversion factors, energy, 10, 885
Cooling curve, 391
Coordination compounds, 869
   electron configuration in, 838, 841–849
   kinetic properties of, 600, 601, 874–876, 881, 882
   ligand atoms in, 869–871, 874
   magnetic properties of, 838, 839
Coordination number, 179, 852
Coordination sphere, 873
Coordination theories
   crystal field, 835–838
   ligand field, 841–849
Copper, 864
Coulomb integral, 272
Coulomb Law, 7, 12, 667
Coulombic attraction, 208, 667
Couple, oxidation-reduction, 540
Covalent bonds, 293
   directional character of, 325, 725
   molecular orbital theory of, 262–333, 769–776, 841–859
   multicenter, 327–331, 769–776, 815
   multiple, 647, 648
Covalent compounds, chemistry of nonorganic, 810–822, 863–879
Covalent crystals, 162, 165, 166
Covalent radii, 645, front endpaper
Cristobalite, 745
Critical constants, 370, 371
   critical pressure, 136, 371
   critical temperature, 133, 371
   critical volume, 136, 403
Critical point, 370
Cross-linked polymer, 800
Crystal classification, 161, 162
   covalent-molecular, 162
   ionic, 162
   metallic, 162
   partly ionic, 162
Crystal cleavage, 156, 158
Crystal defects, 190–193
Crystal-field stabilization energy, 837
Crystal field theory, 835–840
   kinetics and, 881, 882
   limitations of, 841
   magnetic properties and, 838
   splitting parameter, 837
   thermodynamic properties and, 839
Crystal lattice
   body-centered cubic, 179
   cubic close packed, 178
   hexagonal close packed, 177
Crystal lattice geometry, 176–179
Crystal structures, close packed, 176

Crystal symmetry, 159–161
Crystal systems, 159, 161
Crystallinity in polymers, 796–798
Crystallites, 796
Crystallographic axes, 159, 160
Crystals
  compressibility of, 153
  covalent molecular, 162, 165, 166
  ionic, 162–164
  metallic, 162, 166, 167
  partly-ionic, 162–165
Cubic close-packed lattice, 178
Cyanamide ions, 696
Cyanate, 38, 704
Cyanide ions, 38, 696, 855
Cyclobutadiene, 825

Dacron, 799
Dalton's law of partial pressures, 130
Datum, definition of, 6
De Broglie wavelength, 215
Debye, 684
Definite proportions, law of, 17
Definitions, operational, 5
Degrees of freedom, 110, 125
Delocalized orbitals, 777
Density, 154, 340
Deoxyribonucleic acids, 806–809
Derivative, 63, 66
Derivatives, partial, 219
Deuterium, 24
Diamagnetism, 285
Diamond, 725
Diatomic molecules, 253, 262–296
  heteronuclear, 288
  molecular orbitals for, 262–296
Diborane, 324, 815
Dichromate ion, 543, 700, 730
Dielectric constant, 686
Differential calculus, 61–71
Differential ionization energy, 633–640
Differentials, 71
Differentiation, 63
Diffraction of waves, 168–172
Diffusion, 123, 345
Diffusion coefficient, 123, 346
Dilatometer, 571
Dimensional analysis, 8
Dimeric 6-tungstocobaltate, 732
Dipeptide, 802
Dipolar ion, 826
Dipole, 294, 683
Dipole moment, 361, 684–686
Disproportionation reaction, 546, 859
Dissociation constants, acid, 507
Dissociation energy, 305
Distillation, 382–387
Distillation column, 385–387
Distribution constant, 381

Distribution of solute, two phase, 381–382, 404
Disulfide ion, 704
DMP (dimethoxypropane), 758
DNA, 806–809
Donor-acceptor complexes, 869
Double bond, 776
  carbon-oxygen, 785
Dulong and Petit law, 22, 23, 185, 189
Du Nouy ring tensiometer, 347

Effective atomic number rule, *see* 18-electron rule
Effective nuclear charge, 214, 238–240, 640, 833
  Slater rules, 238–240
Effusion, 128
18-electron rule, 852
Einstein theory of heat capacities, 186–190
Elastic deformation, 155
Elasticity, 155, 798
Elastomers, 797
Electric potential, 12, 247, 539
Electrical nature of matter, 162–167, 197–199
Electrochemical equivalent, 198, 553
Electrode, 199, 541–550
  active, 547
  amalgam, 547
  gas, 549
  glass, 550
  indicating, 562
  inert, 549
  ion-selective, 549
  membrane, 549
  metal-solution, 547
  negative, 542
  positive, 542
  redox, 549
  reference, 549
  sensing, 549
Electrode potential, 540–543
  and equilibrium constants, 554–555
  sign of, 544
  standard states for, 542, 557–558
Electrolytic cell, 553
Electromagnetic waves, *see* Light
Electron
  antibonding, 266–267, 763, 767
  bonding, 266, 267, 763, 767
  charge, 199, 203, 885
  charge-to-mass ratio, 200–201
  definition, 199
  delocalized, 167, 327–331, 777, 815
  excitation, 273
  localized, 167, 769
  mass, 202, 885
  nonbonding, 255, 267, 763
  penetration, 231, 626, 832

repulsion, 237–238, 623
screening, 214, 239, 624–626, 833
spin, 240, 285, 838
wave character of, 214–215
Electron affinities, 631–633
Electron configuration
  of atoms, 241–245
  table of, 244–245
  periodic table and, 242–245, 622–628
Electron deficient compounds, 324
18-electron rule, 852
Electron volt, 9, 12, 247, 540
Electronegativity, 639–651, 664, 691, 715–717, 764–765
  scales
    Allred and Rochow, 640, front endpaper
    Mulliken, 640
    Pauling, 640, 642–644
    Sanderson, 640
  table of, front endpaper
Electroneutrality principle, 293
Electrons, paired, 268, 632
Electrophile, 589
Electroplating, 552–554
Electropositive atoms, 666
Element, 15
Elementary process, 574
Elements, characteristic, 658
Emerald, 746
Encounter, kinetic, 596
End point, 498
Endothermic reactions, 453, 472
Energy
  conservation of, 140, 444, 475
  conversion factors, 9, 12, 13, 885
  first ionization, 246, 634–635
  and first law of thermodynamics, 444, 475
  second ionization, 246, 634–635
Energy levels
  of hydrogen atom, 210, 237
  of molecular orbitals, 268, 282, 284, 303, 307, 310, 314, 318, 321, 323, 774, 775, 843, 846, 848
  of multi-electron atoms, 243
Enthalpy, 446–462
  and bond energies, 460–462
  change, 447, 451
    from electrical measurements, 557
  of combustion, 454
  of formation, 451
  of fusion, 365
  standard state for, 451
  of sublimation, 673
  temperature dependence, 478
  of vaporization, 365, 447
Entropy, 421–438
  absolute, 434–436

change, from electrical measurements, 557
  of elements and compounds, 430
  of fusion, 435, 438
  of ice, 432–434
  influence on equilibria, 472
  and molecular chaos, 429, 431
  and probability, 427–429
  units of, 430
  of vaporization, 435, 438
Enzyme catalysis, 609–611
Enzyme reactions, 804–806
Equal-probability contours, 224
Equation
  balanced, 31, 544–547
  ionic, 34
  molecular, 34
  virial, 109
Equation of state, 106, 133, 135, 136, 408
  ideal gas, 106, 408
  van der Waals, 133–137
  virial, 135
Equilibrium condition, 411–413, 464, 471
Equilibrium constant, 471, 481–538
  and cell potentials, 554
  temperature dependence of, 471–474
Equilibrium, liquid-vapor, 362, 407
Equipartition principle, 110, 206
Equivalence point, 498
Equivalent proportions, law of, 19
Error function, 84
Errors
  random, 79
  systematic, 79
Esters, 786
Ethane, 769, 781
Ethanol, 388, 785
Ethers, 785
Ethylene
  electronic structure of, 771
  geometry of, 770, 772
Ethylenediaminetetra-acetic acid, 497, 870
Eutectic, 398
Exothermic reactions, 453, 472
Expectation value, 90
Exponentials, 59, 68
Extensive variables, 409
Extraction process, 381, 382
Eyring hole theory of liquids, 352

Face-centered cubic lattice, see Cubic close-packed lattice
Faraday (unit), 198, 540, 553
Feldspar, 750
Ferredoxin, 879
First law of thermodynamics, 444, 475
First-order reaction, 572, 613, 615
Flame temperatures, adiabatic, 456, 478
Flashlight cell, 551

Fluorine, 718, 765
Fluorite structure, 182
Force
　compression, 153
　shear, 153
　tension, 153
Formal oxidation state, 659
Formaldehyde, 785
Formality, 356
Formic acid, 507, 786
Formula
　chemical, 20
　empirical, 27
　molecular, 29, 30
Free energy, 464–472, 554–558, 598
Free expansion, 424
Free-radical mechanisms, 600, 788
Free rotation of bonds, 770
Freezing point, 369, 391–401
　constant, 378
　depression of, 377, 378
Frenkel defect, 192
Frequencies, statistical, 80
Frequency of electromagnetic radiation, 11, 212, 849
Fuel cells, 552
Fugacity, 457
Function
　concept of, 41, 46
　cosine, 57
　exponential, 59
　linear, 50
　logarithmic, 60
　normalized to unity, 90, 234
　quadratic, 51
　representations of, 46–50
　sine, 57
　velocity distribution, 112
Functional groups, 784

Galena, 754
Galvanic cell, see Cell, electrical
Gas chromatography, 388, 389
Gas constant, R, 106, 885
Gas density and molecular weight, 30, 150
Gas equilibria, 482–492
Gaseous diffusion, 123
Gaussian distribution, 84
Gay-Lussac, J.
　law of combining volumes, 21
　law of volume and temperature, 102–104
Geometrical isomers, see Structural isomers
Geometry of molecules, 252–260, 313–314, 323–325, 769, 851–855
Gibbs free energy, 464
　and electrical work, 540, 553, 557
　and equilibrium constants, 471
　of formation, 468
　standard, 468, 469
Goldschmidt reaction, 743

Graham's law, 128
Gram-atomic weight, 21
Gram-molecular weight, 22
Graphite, 737
Grignard reagent, 824
Ground state, 275

ΔH, see Enthalpy
Haber process, 818
Half-cell reaction, 542
Half-life, 613, 615
Half-reactions, 544–547
Halides
　covalent, 821, 822
　hydrogen, 681, 818, 819
　ionic, 706–709
　organic, 787
　partly-ionic, 755–759
Heat, 7, 9, 13, 441–446
　of combustion, 454
　of formation, 451
　of fusion, 378
　sign of, 444
　of vaporization, 365
Heat capacity, 124
　and constant pressure, 447, 448, 478
　and constant volume, 125, 434–436, 447, 448
　Einstein theory of, 186–190
　of gases, 124–127
　　influence of temperature on, 125–127
　　at low temperature, 185–190
　of solids, 22–23, 185–190
　　temperature dependence of, 185–190
Heat, conversion to work, 438, 439, 442–444
Heat engine, 438, 439
Heisenberg uncertainty principle, 216
Helix, 804
Hematite, 740
Heme, 876
Hemoglobin, 876
Henry's law, 380
Hess' law, 452, 453, 468, 555, 673
Heterogeneity, 14
Heterogeneous reactions, 605–611
Heteronuclear ions, 696, 699, 704, 707
Heteropoly ions, 732
Histogram, 80
Hole theory of liquids, 351–353
Homogeneity, 14
Homogeneous reaction, 605
Hougen-Watson chart, 138, 139
Hückel MO's, 773
Hund's rules, 242, 285
Huygens' principle, 168
Hybrid atomic orbitals, 325, 326, 769–771
Hybridization
　sp, 325, 326
　sp^2, 325, 770, 771
　sp^3, 326, 769

Hydrated ions, 493, 847–849, 873
Hydration energy, 357, 683–690, 847–849
Hydrazine, 697
Hydrides, 692–693, 733–734, 818–819
Hydroacetylides, 694
Hydrocarbons, 769, 781–784
Hydrogen atom, 208–233, 622
  Bohr model of, 208–213
  energy states of, 236, 237
  wave equation for, 219–220
Hydrogen bond, 297, 362, 653–655, 799–800, 804–809
Hydrogen electrode, 542, 549
Hydrogen halides, 681, 818–819
Hydrogen molecule, 263
  wave functions for, 263, 265, 271
Hydrogen spectrum, 211–213, 250
Hydrogen sulfide, 819
Hydrolysis, 512
Hydronium ion, 500
Hydrous oxide, 535, 742

Ideal gas law, 106
Ideal gas temperature scale, 103
Ideal solutions, 386
Imide, 697
Indicator, 498, 524
Infinite series, 670
Infinitesimal process, 415
Insertion reactions, 812
Insoluble substances, 357
Integral
  definite, 76
  indefinite, 76
Integral calculus, 72–78
Integration, 71
Intensity variables, 409
Intensive variables, 409
Intercept, 50
Interfacial tension, 350
Interference
  constructive, 170
  destructive, 170
Interhalogen compounds, 707, 709, 821
Intermolecular forces, 140–144
Internal energy, 444
Interstitial position, 192
Interstitial structures, 734–735
Iodate ion, 37
Iodic acid, 37
Iodine pentoxide, 820
Ion exchange, 750, 761
Ion-solvent interaction, 357, 683–690
Ionic bonding, 293, 667, 680
Ionic compounds, chemistry of, 690–711, 859–862
Ionic crystals, 162–164, 668–683
Ionic equilibria, 492–538
Ionic mechanisms, 600, 789
Ionic radius, 622, 678, front endpaper

Ionic strength, 493
Ionization energy, 211, 246
  table of, 634, 635
Ionization potentials, 247
  valence orbital, 247–249, 622–630, front endpaper
Ions, 14, 36–40, 162, 668
Iron, 864
Irreversible process, 415
Isoalkyl group, 783
Isoelectronic species, 656
Isomerism, *cis-trans,* 771, 772, 872, 873, 881
Isopoly ions, 732
Isoteniscope, 363
Isothermal expansion, 417
Isothermal reversible process, 417
Isotherms, pressure-volume, 139
Isotopes, 24
Isotropic substances, 156

Joule-Thomson expansion, 450

K_a, 507
K_b, 510
K_h, 513
K_p, 483
K_{sp}, 528
K_{st}, 493
K_w, 510
Kaolin, 749
Kelvin scale of temperature, 104, 419–421
Ketones, 785
Kilocalorie, 13
Kinetic control of reaction products, 590, 763
Kinetic theory of gases, 107-127
Kinetics
  chemical, 567–617, 787–795, 874–876
  of gas-phase reactions, 574–594
  of solution reactions, 595–613
Kodel, 799

Lamellar compounds, 737
Langmuir isotherm, 607
Lattice energy, 672
  from Born-Haber cycle, 673
  from electrostatic model, 668–674
  and formation of ionic compounds, 680
  and multiple oxidation states in crystals, 857–859
  for semicovalent compounds, 721–725
  and solubility, 689
Laue method of X-ray diffraction, 174
Law
  of the conservation of mass, 32
  of definite proportions, 17
  of equivalent proportions, 19
  of multiple proportions, 18
  natural, 4
  of spontaneous processes, 414

LCAO principle, 263
Lead storage battery, 551
LeChatelier's principle, 369
  and chemical equilibria, 470
  and liquid-vapor equilibria, 369
Leclanche dry cell, see Flashlight cell
Lennard-Jones potential, 142, 185, 261, 365
Leveling effect, 505
Lewis theory of acids and bases, 502
Ligand, 254, 494, 835, 841–849, 852–855, 870–874
  altered chemical properties of, 874–879
Ligand-field theory, 841–847
Light, 10, 211, 274, 574, 787, 849–851
  and chemical reactivity, 574, 787
  and energy-level transitions, 211, 274, 849–851
Limits, 55
Linear combination of atomic orbitals, 263
Linear free-energy relationships, 604
Liquefaction of gases, 450
Liquid-liquid extraction, 381, 382
Liquid-solid equilibrium, 390–401
Liquid structure, 342, 343, 351
Liquids
  atomic distribution in, 343
  hole theory of, 350–353
  supercooling of, 394–397
  vapor pressures of, 362–390
Lithium fluoride, bonding in, 666–676
Logarithms
  common, 60
  natural, 60
Lowry-Bronsted acid-base theory, 502

Madelung constant, 670
  table of, 672
Magnetic moment, 838, 839
Magnetite, 740, 744
Malleability, 162
Manganese, 864
Manganese dioxide, 740
Manometer, 101
Many-body problem, 238
Mass-balance, 514
Mass spectrometer, 24, 247
Maxwell-Boltzmann velocity distribution function, 116
Mean free path, 120
Measurement
  distribution, 79
  reliability of, 79
Mechanical equilibrium, 412
Mechanical interaction, 407, 412
Mechanism of reaction, 574
  and molecularity, 575
  and rate law, 576
Melmac, 800
Melting point, 163, 390, 651
Meniscus, 101, 347

Meta position, 779
Metallic bond, 327–331
Metallic carbides, 759
Metallic elements
  crystal structures in, 179
  and periodic table, 327, 649, 650
Metallic hydrides, 759
Metalloids, 814
Methane, 321–323, 326, 769, 781
Methylammonium ion, 710
Mica, 749
Michaelis-Menten equations, 611
Microstates, 429
Millikan oil-drop experiment, 202, 203
Miscible substances, 357
Molal concentration, 356
Molar concentration, 356
Molar heat capacities of gases, 124–127
Molar volume of gases, 106
Mole, 26
Mole fraction, 130, 356, 373, 457, 458
Mole percent, 355
Molecular-beam collision studies, 590–594
Molecular collisions, gas, 112, 119–121, 573–594
Molecular orbitals, 262
  antibonding, 267
  for aromatic compounds, 769–775
  bonding, 266, 267
  for coordination compounds, 841–848
  energy levels of, 267–274
  for extended solid lattices, 327–331
  LCAO method, 263
  nonbonding, 267, 301
  pi, 277
  sigma, 276
Molecular sieves, 750
Molecular speed
  distribution function, 116
  most probable, 116
  root-mean-square, 111
Molecular weight determination, 373–380
Molecular weights, 26
Molecularity, 574
Molybdenum, 864
12-molybdophosphate, 732
Monomer, 724, 795
Monopole, 294
Moving phase, 388
Mulliken electronegativity, 640
Multiple bonds, 284, 776
Multiple proportions, law of, 18
Mylar, 799

Naphthalene, 358, 776
Nernst equation, 558
Nickel, 864
Nickel coinage, 864
Nitrate ion, 38, 701, 702
Nitric acid, 38, 505, 700

Nitrides, 695–698, 734–738
Nitro groups, 784
Nitrobenzene, 779–781
Nitrogen, 284, 287, 450, 821
Nitronium ion, 259, 710, 780
Nitronium perchlorate, 710
Noble-gas compounds, 543, 820, 821
Nodes, in wave functions, 222
Nomenclature
 of chemical elements, 16
 of complex ions, 870–873
 of organic compounds, 781–787
 of simple compounds, 36–40
Nonaqueous solvents, 360–362, 506, 873
Nonideal behavior of gases, 131–144
Nonstoichiometric compounds, 733–739, 855–859
Normal alkyl group, 783
Normal distribution, 80
Normality, 356
Normalization, 90, 234
Normalization constant, 234
Nuclear model of atom, 198–204
Nucleation, 394
Nucleophile, 589, 763, 795, 874
Nylon, 799

Octahedral complexes, 497, 855
Octahedral site, 180, 734, 735, 744
Octamolybdate, 731
Olefins, 776
Orbitals, 222
 antibonding, 267
 atomic, 221–233
 bonding, 266
 non-bonding, 266, 301
 Slater, 238
Order of a reaction, 572
 determination of, 570, 613–615
 of elementary processes, 574, 575
Organic compounds and reaction mechanisms, 781–795
Organometallic compounds, 811, 865
Oriented crystalline polymer, 797
Orpiment, 754
Ortho position, 779
Orthogonality condition, 235
Osmosis, 378
 reverse, 380
Osmotic pressure, 378
Overlap integral, 270
Oxalic acid, 507, 519
Oxidation, 542
Oxidation potentials, 542
Oxidation-reduction reactions, 544
Oxidation states, 657
 of characteristic elements, 658–660
 and periodic table, 659
 stability in ionic lattices, 678–680
 of transition elements, 855–859

Oxides
 acidic, 657
 basic, 657
 bonding in, 698, 738–750, 819
 enthalpy of formation, 743
 and periodic table, 657
Oxidizing agent, 559, 699, 701, 820
Oxygen, 284, 285, 287, 543, 699, 741
Ozone, 698
Ozonide, 698

Pairing energy, 632
Para position, 779
Paraffins, 776
Paramagnetism, 285, 838
Paramolybdate, 731
Partial pressures, 130
Partly-ionic compounds, chemistry of, 732–759
Pauli exclusion principle, 241
 role in chemical bonding, 268, 269
Penetration effect, 231, 626, 832
Perchlorate ion, 37, 543, 559, 701
Perchloric acid, 37, 505, 700
Periodic classification, basis of, 242–245
Periodic properties
 acidity of charged ions, 654–658
 atomic radii, 249, 645–648, front endpaper
 atomic volume, 649
 differential ionization energy, 633–639
 electron affinity, 631
 electronegativity, 640, 639–651, front endpaper
 ionization energy, 248, 622–630
 melting points, 651, 653, back endpaper
 oxidation states, 658–660
 vertical similarities, 630
Periodic table, 242, 621, endpapers
 electron configuration and, 242–245, front endpapers
Permanganate ion, 38, 543, 700–703
Perovskite, 744
Peroxide ion, 698
Peroxydisulfate ion, 705
Perxenate ion, 40, 543, 700, 701
pH, 512
 meter, 560
Phase, 14
Phase diagram, 370
Phosphate ion, 38, 700, 701
Phosphine, 743
Phosphite ion, 699, 701
Phosphonitrilic polymers, 817
Phosphoric acid, 38, 507
Phosphorus, 767
Phosphorus pentoxide, 40, 740, 820
Photochemistry, 547
Photons, 216, 574
Physical adsorption, 606

Physical constants, 885
Pi bonding, 277–279, 770–776
    in coordination compounds, 843, 853, 854
Piezoelectric, 745
Planck's constant, 186, 885
Planck's quantum hypothesis, 186, 206
Plastic deformation, 155
Points of inflection, 86
Poise (unit), 345
Poiseuille's equation, 344
Polar bond, 294
Polarizability, 294, 503
Polarization
    of ions, 294
    of solvent molecules, 686
Polarizing power, 295
Polyanion, 730
Polyatomic ions, 695, 698, 704, 706
Polyethylene, 795
Polymeric structures of partly-ionic
        compounds, 725–733, 736, 744
Polymers
    biological, 801–809
    synthetic, 795–801
Polyprotic acids, 518
Portland cement, 746
Potential
    for spontaneous composition interaction, 413
    for spontaneous mechanical interaction, 413
    for spontaneous thermal interaction, 413
    for spontaneous transition, 413
Potential energy, 140, 185, 210, 584–587
Potential surface, 584
Potentiometric endpoint, 560
Precipitate, 34
Pressure
    atmospheric, 101
    interaction with gas volume, 102
    units of, 101, 149
Pressure-volume work, 443
Primary C atoms, 783
Primary structure of proteins, 804
Probability, 86
Probability distribution, 88, 221, 234
Product of chemical reaction, 31
Proteins, 801
Proton, 202, 500
Protonated anions, 697, 703
Pycnometer, 154
Pyrites, 753
Pyrolysis, 702
Pyrophosphate, 730
Pyroxene, 746
Pyrrhotite, 753

Quadratic formula, 52, 487
Quantization, 127, 190, 209, 222, 236
Quantum mechanics, 219
Quantum number, 186, 210, 222
    principal, 222
    spin, 240
Quantum numbers for hydrogen atom, 222
Quartz, 162, 745, 750
Quaternary ammonium salts, 784
Quenched reactions, 589

R (gas constant), 106, 885
Radial distribution functions, 223–232, 832
Radian, 58
Radiation, wave theory of, 10, 11, 168–171
Radii
    covalent, 622, 645, front endpaper
    ionic, 622, 678, front endpaper
    periodic trends in, 644
    van der Waals, 143, 144
Rainbow angle, 592
Random-walk problem, 91
Raoult's law, 373, 382, 457
    deviations from, 386–388
Rate of change, 53
Rate constant, 569
Rate-determining step, 575
Rate law, 569
    differential, 613
    integral, 613
    and reaction mechanism, 573–577
Rate of reaction, 568
    in condensed phases, 595–611
    dependence on temperature, 579, 580
    and equilibrium, 580
Reactant, 31
Reaction coordinate, 585
Reaction cross section, 594
Reaction intermediate, 576
Reaction order, 572
Realgar, 754
Rearrangement, molecular, 790
Redox reactions, 544
Reduced variables
    pressure, 137
    temperature, 137
    volume, 137
Reducing agent, 693, 701, 704
Reduction, 542
Reduction potential, 542, 543
Reference electrode, 549
Refractory oxides, 724
Reliability, 79
Resonance integral, 272
Reverse osmosis, 380
Reversible process, 414
Ribonucleic acid, RNA, 808
Root-mean-square deviation, 82
Root-mean-square velocity, 111
Roots of polynomials, 52
Rotating-crystal method (X-ray diffraction), 174
Rutherford's scattering experiment, 203, 204
Rutile structure, 184, 726
Rydberg constant, 212

Saddle point, 587
Salt bridge, 541
Sapphire, 744
Schottky defect, 191
Schrödinger equation, 219
Screening constant, 214, 239
Screw dislocation, 396
Second law of thermodynamics, 414, 475
Second-order reactions, 572
Secondary C atoms, 783
Secondary structure of proteins, 804
Self-consistent field method, 240
Self-diffusion coefficient, 345
Semiconductors, 331
Semipermeable membrane, 378
Sigma orbitals, 276, 277
Significant figures, 83
Silicates, 745
Silicon dioxide, 745, 750
Silicones, 818
Siloxene, 737
Silver plating, 552
Slater orbitals, 238
Slope, 50
Sodium, 810
Sodium chloride crystal lattice, 181, 669, 672
Solid solutions, 399–401
Solids
    amorphous, 152
    crystalline, 152
Solubility, 357, 702
    of ideal solutes, 358–360
    of ionic substances, 688–690
    solvent polarity and, 361, 362
Solubility equilibrium, 527–538
Solubility-product constant, 528
Solubility of slightly soluble salts, table of, 529
Solute, 355
Solutions, 15, 355, 456, 683
    ideal, 386, 457
        boiling points of, 375
        freezing points of, 377
        vapor pressures of, 373
    nonideal, 386
Solvation energies, 357, 683–689
Solvent, 355
Solvent cage, 596
Sorption, 606
Spin pairing, 241
Spinel, 744
Spontaneity of changes, 410–426, 462–472
    criteria for, 421–423, 464–465
Spontaneous processes, law of, 414
Stability constant, 493
Standard cell potential, 554, 557
    and equilibrium constant, 554
Standard deviation, 82
Standard free energy of formation, 468
Standard half-cell potentials
    definition and measurement of, 542
    sign convention for, 542
    and standard free energy change, 540, 554
    table of, 543
Standard states, 451, 456–459
State of a system, 408
State variables, 408
Stationary phase, 388
Statistics, 79–94
Steady state approximation, 610
Steel, 864
Steric factor, 581
Steric hindrance, 855
Stirling's approximation, 93
Stoichiometry, 17
Structural isomers, 772, 872, 873
Sublimation, 368
Substitution reactions, 777, 874
Substrate, 610
Sulfate ion, 38, 529, 699–701, 705
Sulfite ion, 38, 699–701, 705
Sulfur, 704, 754, 767
Sulfur dioxide, 819–821
Sulfur hexafluoride, 713, 766, 821, 822
Sulfur trioxide, 657, 819–821
Sulfuric acid, 38, 505, 507
Sulfurous acid, 38, 700
Supercooled liquids, 394
Superoxide ion, 698
Surface-catalyzed reactions, 606–609
Surface energy, 349
Surface tension, 347
Surroundings, thermodynamic, 407
Symmetry properties, orbitals, 275–279, 773–775, 843, 844
System, thermodynamic, 407

Tait equation, 353
Talc, 748
Tangent, 64
Temperature, 102, 409, 419
    absolute, 103, 419
Temperature-composition diagrams, 397–401
Temperature scales, 102, 419
Tertiary C atoms, 783
Tetrahedral site, 180, 734
Theoretical plates, 385
    height equivalent (HETP), 385
Theory, nature of, 4
Thermal conductivity of gases, 124
Thermal equilibrium, 411
Thermal interaction, 407, 411
Thermite reaction, 743
Thermochemical bond energies, 306, 460, 461
Thermochemistry, 446–456
Thermodynamic efficiency, 439
Thermodynamic process, 408
Thermodynamic system, 407

Thermodynamic temperature scale, 419
Thermodynamic variables, 408
Thermodynamics, 406
Thermolecular process, 574
Thermoplastic polymers, 801
Thermosetting polymers, 801
Thiocyanate ion, 38, 704
Thiosulfate ion, 704
Third law of thermodynamics, 432, 475
Tin, 864
Titanium, 829
Titanium chlorides, 860
Titanium dioxide, 184, 672, 726
Titration, 498, 524, 560
Titration curves, weak acids, 524–527
Titrations
   acid-base, 524
   oxidation-reduction, 560
Torr, 9, 101
Trajectories, molecular-beam, 592, 593
Trans conformation, 770
Transition metals, 828–884
   electron configurations, 244, 245
   organometallic compounds, 852–854, 865–868
   oxidation states, 855–861
   reduction potentials, 543, 830
Transition state, 585
Translational degrees of freedom, 110
Translational kinetic energy, 110, 351
Transport properties, of gases, 123
Trigonal pyramidal geometry, 257
Trimetaphosphate, 730
Triphosphate, 730
Trithiocarbonate, 704
12-molybdophosphate, 132

Uncertainty principle, 216
Unimolecular process, 574
Unoriented crystalline polymers, 797

Valence-shell electron-pair repulsion theory, 257, 253–260
Van der Waals' constants, 134
Van der Waals equation, 133
   reduced terms, 137
Van der Waals forces, 140–144, 166
Vanadium, 864
Vapor phase, 362
Vapor pressure, 362
   binary solution, 373, 382
   dependence on temperature, 366–372
   diagrams, 367–370
   lowering, 373
   measurement of, 362
   nonideal solutions, 386
   of a solid, 368
Variable
   dependent, 47
   independent, 47

Velocity
   most probable, 116
   root-mean-square, 111, 116
Velocity gradient, 119
Vermiculite, 749
Virial equation of state, 135
Virial quantity and kinetic energy, 109
Viscometer, Ostwald, 344
Viscosity coefficients
   of gases, 119, 118–123
   of liquids, 344
Volt, definition, 12
VSEPR theory, 253–260

Water
   hydrogen-bonding in, 298, 653–655
   self-ionization of, 509–510
   solvent properties of, 357, 505, 683–689
Wave
   amplitude of, 58
   frequency of, 11
   period of, 10
Wave equation
   classical, 68, 217
   for hydrogen atoms, 218
Wave function, 219–236
   angular dependence of, 222
Wave numbers, 11, 212
Wave properties of matter, 214
Wavelength λ, 10, 57, 171
Wavelength-momentum relation, 215
Waves, electromagnetic, 10, 173
Weight percent, 17, 355
Work, 7, 9, 141, 441
Work-heat equivalence, 7, 444
Wurtzite structure, 183, 672, 717

Xenon fluorides, 821
Xenon oxides, 820
X-ray diffraction, 168–176, 342, 343
   Laue method, 174, 175
   powder method, 175
   rotating crystal method, 174, 175
X-ray spectra and atomic number, 213, 214
X-rays
   and crystal structure, 172–176
   and electron density, 173
   and liquid structure, 342–343
   monochromatic, 174

Yield, percent in stoichiometric reactions, 35

Zeolites, 750
Ziegler-Natta low-pressure polyethylene process, 812, 853
Zinc, 864
Zincblende, 754
Zincblende (ZnS) structure, 183, 672, 717, 720
Zircon, 746

PERIODIC TABLE OF BULK PROPERTIES

KEY

atomic number	1
atomic weight	1.0080 H symbol
name	hydrogen
density (g/cm³)(25°C)	8.99 × 10⁻⁵ ... −259.2 melting point (°C)
	... −252.7 boiling point (°C)
electrical conductivity (microhm⁻¹)	(gas) 0.108 enthalpy of vaporization (kcal/mole)

Ia	IIa	IIIa	IVb	Vb	VIb	VIIb		VIII		
1 1.0080 **H** hydrogen −259.2 / 8.99×10⁻⁵ / −252.7 / (gas) 0.108										
3 6.941 **Li** lithium 180 / 0.534 / 1330 / 0.108 32.5	**4** 9.01218 **Be** beryllium 1277 / 1.85 / 2770 / 0.25 73.9									
11 22.9898 **Na** sodium 98 / 0.97 / 892 / 0.218 23.4	**12** 24.305 **Mg** magnesium 650 / 1.74 / 1107 / 0.224 32.5									
19 39.102 **K** potassium 64 / 0.86 / 760 / 0.143 18.9	**20** 40.08 **Ca** calcium 838 / 1.54 / 1440 / 0.218 38.6	**21** 44.9559 **Sc** scandium 1540 / 2.992 / 2730 / 0.015 81	**22** 47.90 **Ti** titanium 1670 / 4.50 / 3260 / 0.024 106	**23** 50.9414 **V** vanadium 1890 / 5.96 / 3400 / 0.04 106	**24** 51.996 **Cr** chromium 1890 / 7.20 / 2480 / 0.078 73	**25** 54.9380 **Mn** manganese 1245 / 7.20 / 2100 / 0.054 53.7	**26** 55.847 **Fe** iron 1535 / 7.86 / 3000 / 0.10 84.6	**27** 58.9332 **Co** cobalt — / 8.9 / — / 0.16		
37 85.4678 **Rb** rubidium 39 / 1.532 / 688 / 0.080 18.1	**38** 87.62 **Sr** strontium 768 / 2.6 / 1380 / 0.043 33.8	**39** 88.9059 **Y** yttrium 1500 / 4.34 / 2930 / 0.019 93	**40** 91.22 **Zr** zirconium 1850 / 6.49 / 2580 / 0.024 120	**41** 92.9064 **Nb** niobium 2470 / 8.57 / 3300 / 0.080 ~170	**42** 95.94 **Mo** molybdenum 2610 / 10.2 / 5560 / 0.19 128	**43** 98.9062 **Tc** technetium ~2200 / 11.5 / — / — 120	**44** 101.07 **Ru** ruthenium 2250 / 12.30 / 3900 / 0.10 148	**45** 102.9055 **Rh** rhodium — / 12.4 / — / 0.22		
55 132.9055 **Cs** cesium 29 / 1.873 / 690 / 0.053 16.3	**56** 137.34 **Ba** barium 714 / 3.5 / 1640 / 0.016 35.7	**57** 138.9055 **La** lanthanum 920 / 6.18 / 3470 / 0.017 96	**72** 178.49 **Hf** hafnium 2150 / 13.29 / 5400 / 0.031 155	**73** 180.9479 **Ta** tantalum 3000 / 16.6 / 5400 / 0.081 ~170	**74** 183.85 **W** tungsten 3410 / 19.3 / 5900 / 0.181 185	**75** 186.2 **Re** rhenium 3180 / 21.02 / 5600 / 0.051 152	**76** 190.2 **Os** osmium 3000 / 22.57 / 5000 / 0.11 162	**77** 192.22 **Ir** iridium — / 22.42 / — / 0.189		
87 (223) **Fr** francium ~27 / — / — / —	**88** 226.0254 **Ra** radium 700 / 5.0 / ~1700 / — 27	**89** (227) **Ac** actinium 1050 / 10.07 / ~3200 / —								

58 140.12 **Ce** cerium 795 / 6.78 / 3470 / 0.013 95	**59** 140.0977 **Pr** praseodymium 935 / 6.78 / 3100 / 0.015 79	**60** 144.24 **Nd** neodymium 1025 / 6.80 / 3000 / 0.013 69	**61** (147) **Pm** promethium 1035 / — / ~2700 / —	**62** 150.4 **Sm** samarium — / 7.54 / — / 0.0?
90 232.0381 **Th** thorium 1700 / 11.66 / 4000 / 0.055 130	**91** 231.0359 **Pa** protactinium ~1230 / 15.37 / — / — ~130	**92** 238.029 **U** uranium 1132 / 18.95 / 3820 / 0.034 110	**93** 237.0482 **Np** neptunium — / 20? / — / —	